UNDERSTANDING VIRUSES

SECOND EDITION

UNDERSTANDING VIRUSES

TERI SHORS, PhD
Professor
Department of Biology and Microbiology
University of Wisconsin-Oshkosh

JONES & BARTLETT
LEARNING

World Headquarters
Jones & Bartlett Learning
5 Wall Street
Burlington, MA 01803
978-443-5000
info@jblearning.com
www.jblearning.com

Jones & Bartlett Learning books and products are available through most bookstores and online booksellers. To contact Jones & Bartlett Learning directly, call 800-832-0034, fax 978-443-8000, or visit our website, www.jblearning.com.

Substantial discounts on bulk quantities of Jones & Bartlett Learning publications are available to corporations, professional associations, and other qualified organizations. For details and specific discount information, contact the special sales department at Jones & Bartlett Learning via the above contact information or send an email to special-sales@jblearning.com.

Production Credits
Chief Executive Officer: Ty Field
President: James Homer
SVP, Editor-in-Chief: Michael Johnson
SVP, Chief Technology Officer: Dean Fossella
SVP, Chief Marketing Officer: Alison M. Pendergast
Publisher, Higher Education: Cathleen Sether
Senior Associate Editor: Megan R. Turner
Editorial Assistant: Rachel Isaacs
Associate Production Editor: Jill Morton
Senior Marketing Manager: Andrea DeFronzo
V.P., Manufacturing and Inventory Control: Therese Connell
Composition: Circle Graphics
Cover Design: Kristin E. Parker
Associate Photo Researcher: Lauren Miller
Cover Image: © Dennis Kunkel Microscopy, Inc./ Visuals Unlimited, Inc.
Printing and Binding: Courier Kendallville
Cover Printing: Courier Kendallville

To order this product, use ISBN: 978-1-4496-4892-3

Library of Congress Cataloging-in-Publication Data

Shors, Teri.
Understanding viruses / by Teri Shors. — 2nd ed.
p. ; cm.
Viruses
Includes bibliographical references and index.
ISBN-13: 978-0-7637-8553-6
ISBN-10: 0-7637-8553-9
1. Virus diseases. 2. Viruses. I. Title. II. Title: Viruses.
[DNLM: 1. Viruses. 2. Virus Diseases. 3. Virus Physiological Phenomena. QW 160]
RC114.5.S46 2012
616.9'1—dc23
2011027487

6048

Printed in the United States of America
15 14 13 12 11 10 9 8 7 6 5 4 3 2 1

This book is dedicated to the late

Elaine (Motschke) Gross,

my mother.

BRIEF CONTENTS

CONTENTS

PREFACE

Understanding Viruses is the product of 14 years of teaching introductory virology to undergraduate students majoring in biology, microbiology, medical technology, and to pre-med and other pre-professional students. Some of the students had never taken a microbiology course or had not taken an introductory molecular and cellular biology course prior to virology. Also, I found that some students were not adequately prepared in terms of fundamental concepts for my course and needed some form of "refresher" to aid them through the course material. I struggled to find a textbook that combined the clinical or medical aspects of viral diseases as well as the molecular biology aspects of viral replication, as I had found that students were more enthusiastic to learn the molecular aspects of viral diseases if the clinical perspective was presented with it. Unfortunately, the textbooks available did not offer this blend of educational material. The result of this experience is ***Understanding Viruses, Second Edition***, which does not simply focus on viruses as tools to study molecular biology, but instead generates connections by covering molecular biology, pathobiology (the observed nature of disease, its causes, processes, development, and consequences), epidemiology, and the historical perspectives of virology.

Virology is a dynamic discipline. Most recently, this field has gained increased importance because of the 2009 H1N1 influenza pandemic, avian influenza, gene therapy, new cancer therapies, vaccine production, development of pharmaceuticals, global change, as well as emerging viruses such as novel strains of Ebola hemorrhagic fever virus, HIV, hepatitis C, noroviruses, hantaviruses, canine and bat rabies, Nipah, Zika, Chikungunya, Wika, and West Nile encephalitis viruses. Significant outbreaks of these viral diseases have occurred worldwide and portray a future in which new and reemerging infectious viral diseases will challenge the current arsenal of prophylaxis and treatment. My intent was to create a book that provides a "big picture" or global approach to understanding these viruses, host-virus interactions, and molecular biology concepts. While my focus is on human viral diseases, I have included examples of viral diseases of other animals as well as chapters on the history of clinical trials, gene therapy, and xenotransplantation (animal to human transplants); prions and viroids; plant viruses; and bacteriophages.

Special Features

I wanted to create a virology textbook that contained all of the educational "bells and whistles" that books for first-year biology students have; for example, a full-color format, case studies, and other current pedagogy that would engage students in the learning process. Because I am a visual learner and many students also learn better visually, the book includes carefully rendered visual information. In addition, the text has a number of special features:

- Each chapter opens with a contemporary case study that presents real-world scenarios. The chapters end with follow-up questions connecting the case study to the chapter's concepts to enhance student learning. A number of chapters have additional case studies and questions at the end.
- A series of *Virus Files* are presented throughout the text, each connecting a chapter topic to current research being done or to virology techniques often used in the field.
- *Refresher Boxes* and a glossary are also available to assist students who may need to brush up on biological concepts that are typically introduced in introductory biology courses.
- Comprehensive literature and video resources are listed at the end of each chapter.
- The terms important to the understanding of virology are in **boldface** type.

This book can be used at the sophomore or junior level. Virology is a very exciting and fast-paced field with many challenging issues. I hope this textbook will inspire some students to become virologists.

Updates and Revised Material in This *Second Edition*

Understanding Viruses is a work in progress. The text is evolving to present new information as well as improving ways to explain key concepts. For this reason, Chapters 5 (Laboratory Diagnosis of Viral Diseases and Working with Viruses in the Research Laboratory) and 7 (Host Resistance to Viral Infections) of this *Second Edition* have undergone major revision. Chapter 15 (Herpesviruses) has also been modified extensively. New or updated topics in other chapters include:

- an updated model on the origin of viruses (Chapter 1)
- the 2009 H1N1 influenza pandemic (Chapters 1 and 12)
- viruses that challenge the definition of a virus (giruses and virophages; Chapters 1, 3, and 4)
- a new section on viral infections and pregnancy (Chapter 6)
- a new section on oncolytic viruses that includes the history of cancer therapy, intelligent design of oncolytic viruses, and the challenges of virotherapy (Chapter 10)
- a new section on Kaposi's sarcoma virus (Chapter 10)
- added information on the passage history used to develop the Sabin poliovirus vaccine (Chapter 11)
- a rare, newly recognized route of HIV transmission: pre-mastication (Chapter 16)
- a new section on chronic wasting disease of cervids (Chapter 19)
- new sections on HIV hotspots around the world: central Asia, Eastern Europe, and China (Chapter 16)
- viral cassava pathogens (Chapter 20)
- biofilm development and the role of bacteriophages (Chapter 21)

Besides the addition of new topics and sections, new tables and figures were created and some tables and figures were revised or updated. Additional cases studies and Virus Files were included. Topics for the new Virus Files are listed below:

- Virus File 4-2: Unraveling the Life Cycle of Mimivirus
- Virus File 4-3: Stamping Down Flu Viruses
- Virus File 5-1: The Mystery of Chronic Fatigue Syndrome
- Virus File 6-2: Is Groundwater Safe to Drink?
- Virus File 8-1: Today's Virus Hunters: C. J. Peters and W. Ian Lipkin
- Virus File 8-2: Impact of Viruses on War and Religion
- Virus File 9-1: Was Timothy Ray Brown's HIV Infection Cured?
- Virus File 11-2: Using Google Earth to Track Polioviruses Down the Congo River
- Virus File 12-2: The Perfect Storm of Cytokines That Can Kill You
- Virus File 14-2: Farmer Jesty and the Importance of Self-Promotion
- Virus File 15-1: Are Oyster Herpes Outbreaks a Symptom of Global Warming?

The primary literature, popular press, book, and video lists have been updated for every chapter.

Ancillaries Accompanying the Text

To assist you in teaching this course and supplying your students with the best in teaching aids, Jones & Bartlett Learning has prepared a complete ancillary package available to all adopters of *Understanding Viruses, Second Edition*. Additional information and review copies of any of the following items are available through your Jones & Bartlett Learning sales representative.

For the Instructor

Compatible with Windows and Macintosh platforms, the Instructor's Media CD provides adopters with the following ancillaries.

- The *PowerPoint Lecture Outline* presentation package, prepared by the author, Teri Shors, provides lecture notes, graphs, and images for each chapter of *Understanding Viruses, Second Edition*. Instructors with Microsoft PowerPoint software can customize the outlines, art, and the order of presentation. The PowerPoint files have also been prepared in HTML format for use in online course management systems.
- The *PowerPoint Image Bank* provides the illustrations, photographs, and tables (to which Jones & Bartlett Learning holds the copyright or has permission to reprint digitally) inserted into PowerPoint. With Microsoft PowerPoint, you can quickly and easily copy individual slides into your existing lecture slides. If you do not own a copy of Microsoft PowerPoint or a compatible software program, a Microsoft PowerPoint Viewer is included on the CD.
- A *Test Bank* containing over 2000 questions, prepared by James Collins of University of Arizona, is available as an Instructor download.

For the Student

The website we developed exclusively for *Understanding Viruses, Second Edition*, can be found at **go.jblearning.com/shors2**. The site contains an online study guide with chapter outlines, animated flashcards, interactive glossary, crossword puzzles, animations, research and reference links, and links to news sources.

Encounters with Microbiology, Volumes 1 and 2, edited by Jeffrey C. Pommerville of Glendale

Community College, brings together "Vital Signs" articles from *Discover* magazine in which health professionals use their knowledge of microbiology to solve their puzzling medical cases.

Guide to Infectious Diseases by Body System, Second Edition, by Jeffrey C. Pommerville, is an excellent tool for learning about microbial diseases. Each of the 15 body system units presents a brief introduction to the anatomical system and the bacterial, viral, fungal, or parasitic organisms capable of infecting the system.

20th Century Microbe Hunters, by Robert Krasner of Providence College, offers a dramatic portrayal of the achievements and lives of microbiologists such as Charles J. Nicolle (typhus epidemic), Barry Marshall and J. Robin Warren (*Helicobacter pylori*), Luc Montagnier and Robert Gallo (HIV), and Donald R. Hopkins (guinea worm).

How Pathogenic Viruses Work, by Lauren Sompayrac, is a concise summary of the basics of virology written in an understandable and entertaining manner. The book comprises nine lectures covering the essential elements of virus–host interactions with descriptive graphics, helpful mnemonic tactics for retaining the concepts, and brief lecture reviews.

Acknowledgments

Creating a book for educational purposes involves many people who labor behind the scenes. First and foremost, I want to thank the editorial staff at Jones & Bartlett Learning, publisher Cathleen Sether and senior associate editor Megan Turner, for their care in preparing the *Second Edition* of the manuscript for production. Both of them have been instrumental in getting the *Second Edition* off the ground. Their tireless work ethic and support for the *Second Edition* has been phenomenal. It is an honor knowing they have faith in the book and are dedicated to it. Without them, this revision would not have been possible.

The team at Jones & Bartlett Learning has been an energetic team to work with. Special thanks are expressed to associate production editor Jill Morton, the quiet force and glue of production. She has been the consummate professional as she multitasks between the associate editor, author, copyeditor, illustrator, photo researchers, permissions, and marketing team. Copyeditor Shellie Newell provided a fresh pair of eyes to smooth over the rough spots of the manuscript. She has saved me from communicating errors. I really appreciated her technical editing by pointing to places in the manuscript in which there could be updates that I missed. She took the extra time to direct me to resources that have truly enhanced the quality of the book. Thanks also to permissions and photo researchers Christine Myaskovsky and Lauren Miller. Christine's detailed figure logs from the *First Edition* were invaluable in preparing the *Second Edition* manuscript. Thanks also to the rest of the production team: illustrator Erica Beade, proofreader Mike Boblitt, indexer Maureen Johnson, and editorial assistant Rachel Issacs who secured numerous colleagues to peer evaluate chapters of the book and the entire marketing team. The book also benefited from scientific photographer James Gathany, who went out of his way to provide images not yet catalogued at the CDC's PHIL site. Research technician Margaret Schuelke also skillfully set up plaque assays that were used as figures in this *Second Edition*.

Peer evaluation was critical in revising and improving the accuracy of the book. Peer evaluations of chapters secured by Dr. Jeffrey Pommerville after the *First Edition* was published were critical in making revisions. I am grateful to the following peers for their evaluation of the first edition of *Understanding Viruses*: Drs. Jeffrey J. Byrd, Bret Clark, Kristin Pederson Gulrud, Don Lehman, Sandra Daise Adams, Michael Hoffman, and Royce Lee. Special thanks is extended to Dr. Erik S. Barton (Purdue University) and to Nancy Boury for her immunology technical expertise.

The book benefitted from these additional reviewers, who spent time commenting on the drafts of *Understanding Viruses, Second Edition*:

James K. Collins
University of Arizona

A. L. N. Rao
University of California, Riverside

Virologist, former undergraduate mentor, colleague, and friend John Cronn was a strong positive influence in the writing of the epidemiological and historical perspectives. John has kept me focused for over 30 years.

Other important influential mentors along the way are Bernard Moss (chief of the Laboratory of Viral Diseases at the National Institutes of Health), Bertram Jacobs, Jean M. Schmidt, Vern Winston, Gordon Schrank, and the late Denise McGuire.

Other individuals I would like to thank for various reasons are Linda Freed, Dr. Jim Armentrout, Dr. Ethan Everett, Dr. Susan McFadden, Dr. Jeremiah D. Jackson, Dr. David N. Speranza, Andy Keech, Jeremy Kroll, Amy Killpatrick, Jennifer Doris, Connie Whittaker, Rebecca Reiger, Molly Steinbach, and Steve and Sharon Osbon.

Special thanks to research students Patrick Fisher, Rhonda Mesko, and Jaime Antonio Castillo. Balancing time between research, teaching, and revising this manuscript was challenging at times. These students were patient and supportive of the book.

I thank all my former and present students who attended an 8 AM class on viruses during the winter/spring semesters in Oshkosh, Wisconsin, for the past 14 years. In addition to regular instruction, each class embraced with enthusiasm group projects on the 1918 flu, polio, and pandemic flu planning. Many students at the University of Wisconsin–Oshkosh encouraged me and appreciated my approach toward teaching. I hope that some of these students will be inspired to rise to the challenge of being virus hunters in the field. I hope those who read this text will continue to be critical and provide suggestions for improving it. Special thanks to Bim and Patti Carey for their support of this project. Bim often asked how the book was coming along, which I really appreciated.

I could not finish without acknowledging Darin Reiger for his support during revision and production of the *Second Edition*. He has witnessed the long hours I spent at the computer, dealt with the transition to progressive eyeglasses, and put up with my "zoning." All books are a product of people—*Understanding Viruses, Second Edition* is no exception.

Teri Shors
Oshkosh, Wisconsin
June 2011

ABOUT THE AUTHOR

Teri Shors received her doctoral degree from Arizona State University, Tempe, Arizona. Following her doctoral degree, she served as a postdoctoral fellow under the direction of Dr. Bernard Moss in the Laboratory of Viral Diseases, National Institute of Allergy and Infectious Diseases at the National Institutes of Health, Bethesda, Maryland. For the past 14 years, Dr. Shors has been a professor of microbiology at the University of Wisconsin–Oshkosh where she teaches introductory microbiology and virology. She has authored peer-reviewed research papers in virology. Her research specialty is poxviruses. She was the principal investigator or coprincipal investigator for a Merck/American Association for the Advancement of Science (AAAS) and several National Science Foundation (NSF) grants, including a grant to develop and write an undergraduate virology textbook to better prepare students for the rapidly changing field of viruses and viral diseases. She is a recipient of a University of Wisconsin–Oshkosh Distinguished Teaching Award and two Endowed Professorships, and is a Wisconsin Teaching Scholar. She served on the American Society for Virology's Education and Career Development Committee. Professor Shors has presented papers at scientific and educational conferences consistently throughout her career. She has led a number of cross-disciplinary projects at the University of Wisconsin–Oshkosh that have received national recognition.

VIRAL PESTILENCE TIMELINE

1500–1960

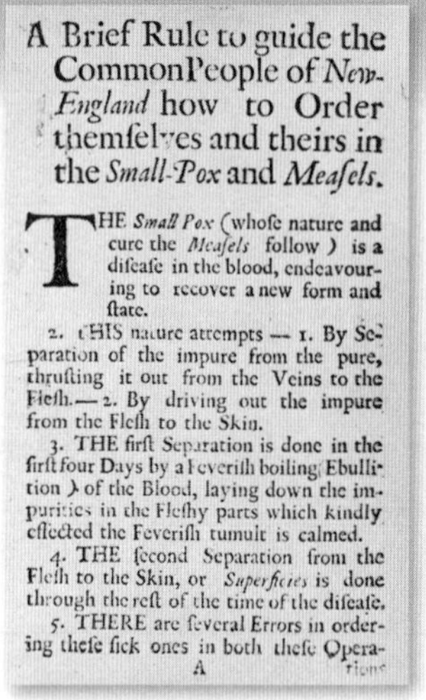

A Brief Rule to guide the CommonPeople of *New-England* how to Order themſelves and theirs in the *Small-Pox* and *Meaſels.*

THE *Small Pox* (whoſe nature and cure the *Meaſels* follow) is a diſeaſe in the blood, endeavouring to recover a new form and ſtate.

2. THIS nature attempts — 1. By Separation of the impure from the pure, thruſting it out from the Veins to the Fleſh.— 2. By driving out the impure from the Fleſh to the Skin.

3. THE firſt Separation is done in the firſt four Days by a Feveriſh boiling (Ebullition) of the Blood, laying down the impurities in the Fleſhy parts which kindly effected the Feveriſh tumult is calmed.

4. THE ſecond Separation from the Fleſh to the Skin, or *Superficies* is done through the reſt of the time of the diſeaſe.

5. THERE are ſeveral Errors in ordering theſe ſick ones in both theſe Opera-

A

1518–1520
Smallpox outbreaks, Aztec empire (Mexico)

1527–1530
Smallpox outbreaks, Inca empire (Peru)

1616
Smallpox, New England, Native American Indian population hit hard.

1649
Smallpox, Boston hit hard.

1657
Measles, Boston

1759
Measles, most places inhabited, North America

1794
1796–1797
Yellow Fever outbreaks, Philadelphia

1798
Yellow Fever outbreak, Philadelphia (one of the worst)

1818
First recorded epidemic of Dengue Fever, Peru

1847–1848
Influenza pandemic

1894
First Polio outbreak in the U.S. (Vermont)

1912–1916
Dengue Fever pandemic

1916
Large Polio epidemic, New York City

1918–1919
Spanish Influenza pandemic

1952
U.S. highest record of Polio epidemics

1500 | 1520 | 1540 | 1560 | 1580 | 1600 | 1620 | 1640 | 1660 | 1680 | 1700 | 1720 | 1740 | 1760 | 1780 | 1800 | 1820 | 1840 | 1860 | 1880 | 1900 | 1910 | 1920 | 1940 | 1950

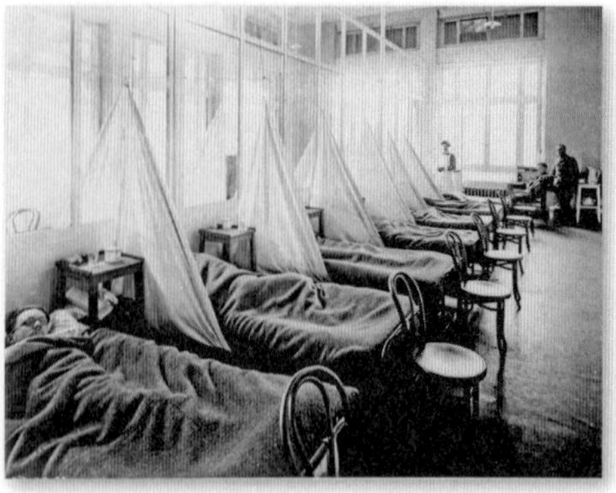

1960–present

1957
Asian influenza pandemic

1962–1965
Rubella (German Measles) pandemic

1967
First Marburg Hemorrhagic Fever outbreak. Initial cases were laboratory workers handling infected African green monkeys imported from Uganda; Marburg, Germany; and the former Yugoslavia.

1968
Hong Kong Influenza pandemic

1976
First Ebola outbreaks, Democratic Republic of Congo and Sudan

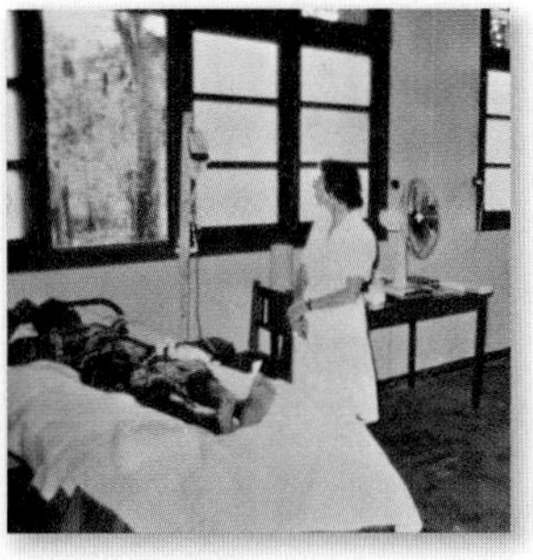

1989
First Ebola-Reston virus outbreaks at monkey quarantine facilities, Reston, Virginia (no humans became ill)

1993
Sin Nombre (hantavirus) outbreak, 4-corners disease (U.S.)

1997
Avian Influenza, Hong Kong

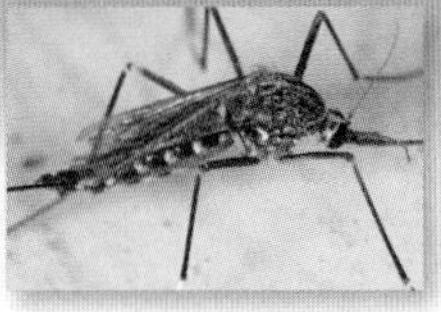

1999
Nipah Virus outbreak, Malaysia (pig and mainly human mortalities)

First recognized West Nile Virus outbreak in the U.S. (New York City)

2000–2001
Ebola outbreak, Uganda (highest mortality of all Ebola outbreaks)

2001
Foot and Mouth Disease Virus outbreak, United Kingdom (no human cases)

2003
SARS pandemic (first appeared in southern China in November 2002, recognized as a global threat in March, 2003)
Monkeypox multistate outbreak

2003–present
Avian influenza, rare but highly lethal in humans, Hong Kong, China, Indonesia, Vietnam, Thailand, Cambodia, Turkey, Iraq, Egypt, Azerbaijan, Djibouti

2005
Largest and deadliest outbreak of Marburg Hemorrhagic Fever, Angola

First outbreak of Polio in the U.S. in 26 years unvaccinated children in an Amish community, Minnesota

2006
Mumps outbreaks, midwestern U.S.

Chikungunya Virus outbreak, La Reunion Island (France)

2006–2007
Porcine Reproductive and Respiratory Syndrome Virus (also called blue ear pig disease), China

2007
Foot and Mouth Disease Virus outbreak, Southern England
Viral Hemorrhagic Septicema spreads to the Great Lakes region, U.S.

2009
H1N1 Influenza pandemic, Epicenter: Mexico

2011
Many imported Measles outbreaks, often spread by air travel; U.S.

1960 1970 1980 1990 2000 2010 2020

CHAPTER 1

Introduction to Viruses

On April 30th, 2009, newspaper headlines in Mexico City announced that the number of H1N1 influenza deaths had risen to 236 in Mexico. Mexico was the epicenter of the 2009 H1N1 (Swine flu) pandemic.

"Nothing brings us so close to the riddle of Life—and its solution—as viruses."

Wolfhard Weidel, virologist

OUTLINE

CASE STUDY: HEMORRHAGIC FEVER

Recently you were inspired by an instructor in a virology course to subscribe to ProMED mail. While searching the ProMED mail archives for outbreaks of viral hemorrhagic fever, you found a ProMED mail post entitled "Undiagnosed Hemorrhagic Fever—Angola: Request for Information." It was dated March 15, 2005. The post describes the death of two nurses working at the same Huige Province hospital. At the time of the post, 56 people with similar symptoms had died in Angola. Hospital workers told journalists that the illness began with a 2-day fever. Then the patient would cough up blood, lapse into a coma, and die within 4 days. The World Health Organization was planning on investigating the outbreak.

Coincidentally, you have learned that a college friend has joined the Peace Corps. At the end of the semester, he will graduate and then be deployed to Angola to work on HIV/AIDS awareness and youth outreach. You have asked him if he is aware of the viral hemorrhagic fevers such as Ebola, Lassa, or any new hemorrhagic fever viruses in Angola (**FIGURE CS 1-1**). He said he is very naïve about hemorrhagic fever viruses and would appreciate it if you would do some more research on this subject on his behalf.

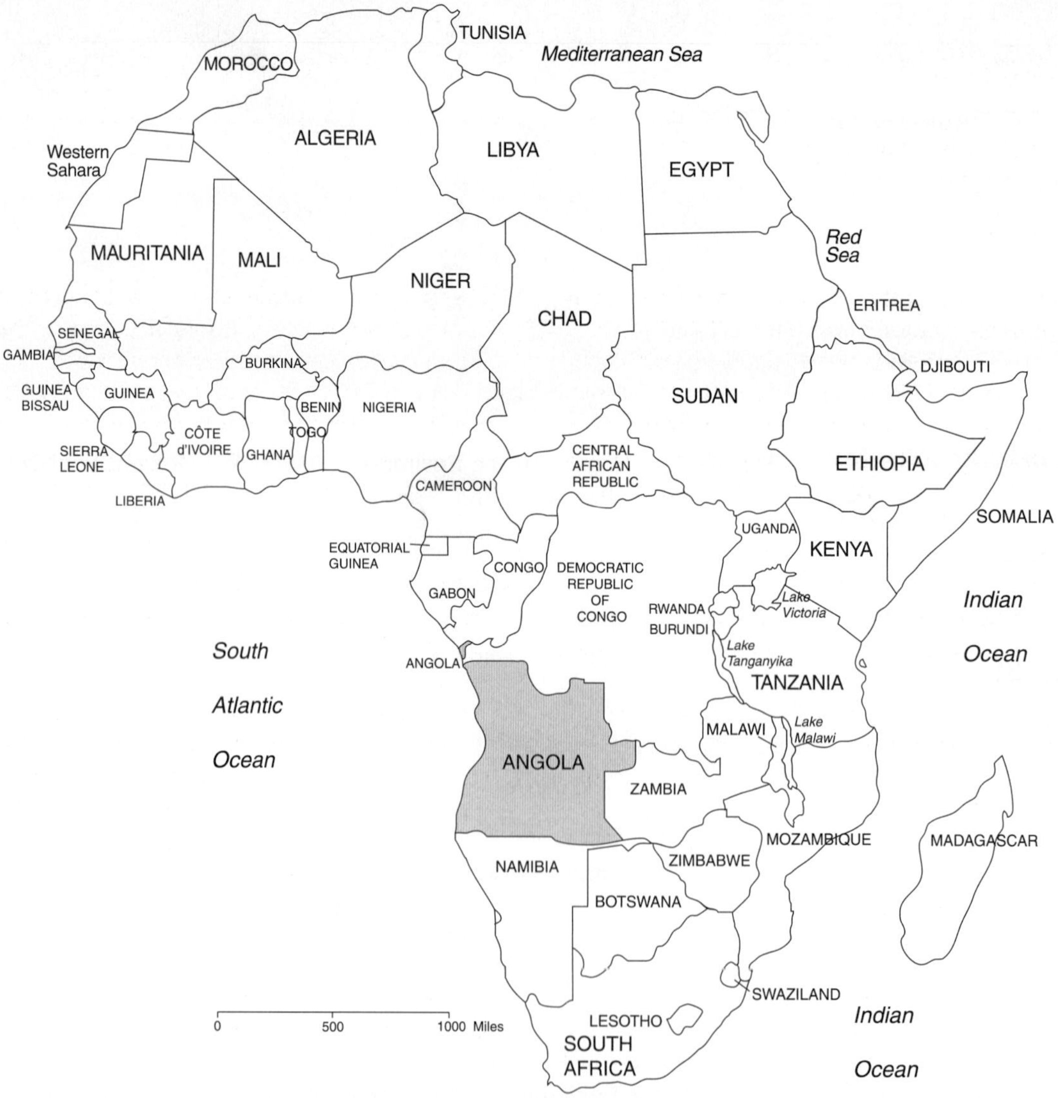

FIGURE CS 1-1 Angola is located on the Atlantic coast of southern Africa. It is bordered by Namibia to the south and Zambia and Zaire to the east and north. Angola's 994 mile-long coastline and its four major ports make it a natural trans-shipment point for the entire region. Angola has one of the world's worst child mortality rates, with one in four failing to live beyond their fifth birthday. Routine immunizations rarely happen. Yellow fever, Dengue fever, and Crimean Congo hemorrhagic fever outbreaks are **endemic** to Angola.

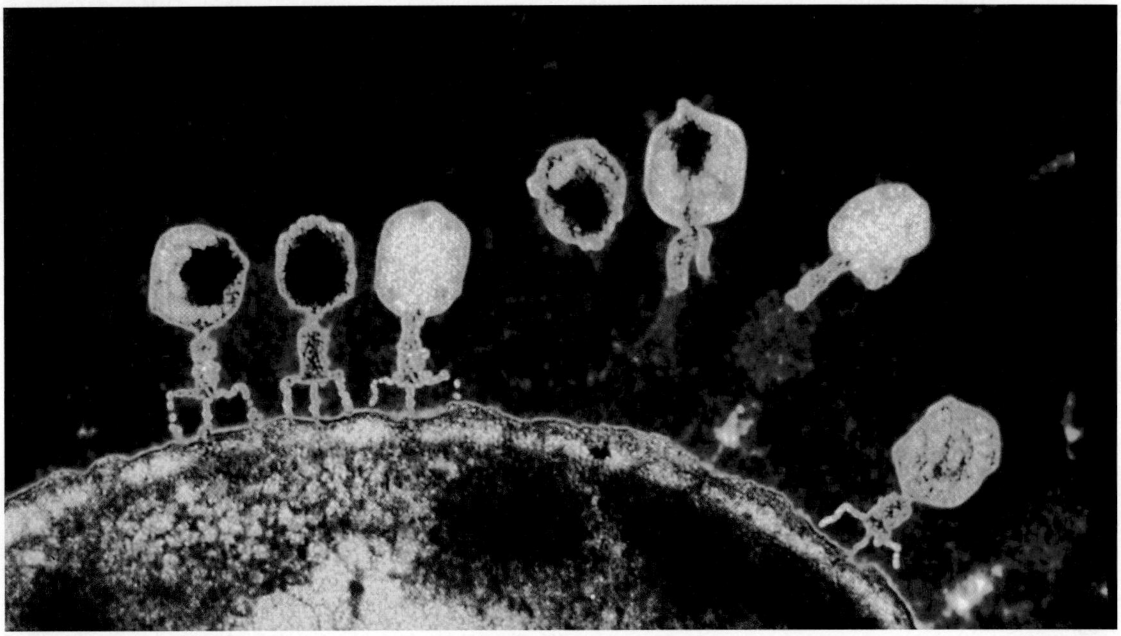

FIGURE 1-1 Colored electron micrograph of viruses (referred to as bacteriophages) attacking the bacterium *Escherichia coli*. Many bacteriophages have a "lunar lander" or head-and-tail structure. The bacteriophage infects the bacterium by attaching to the cell wall. Subsequently, its genetic material enters the cell.

When you think of the word *virus*, what comes to mind? The 2009 H1N1 influenza virus? An invisible entity responsible for the common cold? The cause of sexually transmitted diseases? A mysterious and insidious killer, such as the one that causes Ebola hemorrhagic fever? Biological agents that look like "lunar lander" spaceships (**FIGURE 1-1**)? Have you ever been sickened by images of animals or humans suffering from viral diseases?

This chapter begins your journey through the invisible world of viruses. The word *virus* has been used in the medical world for almost 200 years. In the early 1790s, it simply meant *poison* or *poisonous slime*. A short definition of viruses describes them as submicroscopic agents that are capable of growth in living cells. These biological entities are both beneficial and harmful to humankind. This overview of viruses provides a basis for understanding their impact on all living organisms.

1.1 Viral Impact on the Environment, Research, and Disease

Viruses and Aquatic Ecosystems

Viruses are usually associated with negative effects. Without viruses, though, the earth would be a very different place—maybe even a planet without life! For example, did you know that viruses are the most abundant biological entity in both freshwater and seawater? More than one million viruses are present in a teaspoon of water. Most of these viruses are known as **bacteriophages** (viruses that infect bacteria). For each bacterium in the water there are 15 to 25 virus particles. Bacteria and microscopic plankton are food for nearly all aquatic animals. The discovery of viral abundance in natural water has led to the conclusion that viruses are significant biological agents involved in both the mortality (death) of aquatic microorganisms and in the building of aquatic communities. How can this be?

Dramatic experiments led by Gunnar Bratbak in 1990 were performed in which bacteriophages were selectively removed from seawater and the growth rates of the remaining bacterial and planktonic organisms measured. It was expected that the bacterial and planktonic populations would dramatically increase because they *were freed from viral infection*. Surprisingly, the bacterial populations stopped growing completely, because they depended upon nutrients released as the bacteria were killed by viruses (**FIGURE 1-2**). Without the death of these microbes via viruses, there was no "fuel" to keep the aquatic community running. These viruses are essential for regulating both saltwater and freshwater ecosystems.

During 2009, scientists studying the ecosystems of the pristine freshwater lakes of Antarctica

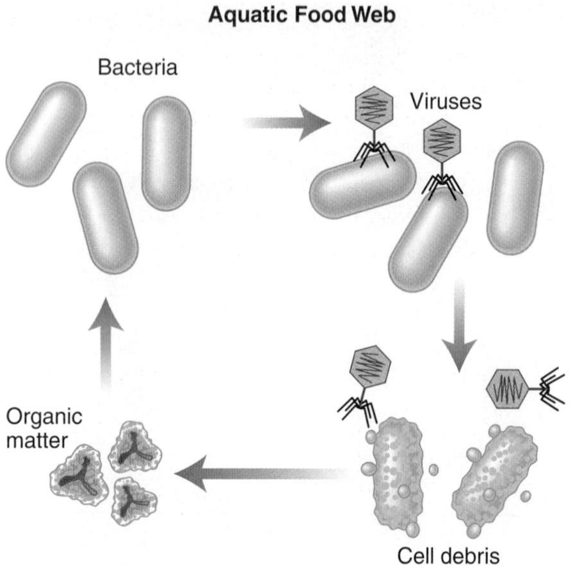

FIGURE 1-2 Viruses free up organic matter so that new life can be generated.

found incredibly diverse types and high numbers of previously unidentified viruses present in the icy lakes. These lakes contained very little animal life (**FIGURE 1-3**). They were void of penguins or seals. The vast majority of life present in the lakes was microbes: bacteria, algae, protozoa, and zooplankton. The microscopic communities adapted to very extreme conditions. In the winter, there was nearly complete darkness and the lakes remained frozen. During the summer, the ultraviolet radiation was intense and the ice melted, resulting in open water in parts of the lake. In contrast to what scientists found in other aquatic systems, the viruses most abundant in the Antarctic lakes during the summer ranged from 50–150 nanometers in diameter. These viruses were mainly tailed bacteriophages and algal viruses. During the spring, there was a reduction of the larger viruses and an increase in viruses that were less than 30 nanometers in diameter. The viruses likely play a role in controlling the microbial populations during the seasonal transition of the ice-covered lakes in the spring to open-water lakes in the summer.

The Hershey-Chase Blender Experiment

Since their discovery, bacteriophages have taught us many things about the molecular biology of cells. Viruses of bacteria were termed bacteriophages (the suffix "phage" comes from the Greek for "eating") because of their ability to eat, or **lyse**, bacteria. One of the first bacteriophages used in the laboratory, T2, infects the host bacterium *Escherichia coli*. The T2 bacteriophage consists almost entirely of a tightly condensed piece of DNA that is surrounded or packaged by a protein coat. The

FIGURE 1-3 Researchers' tents are dwarfed by Lake Bonney and the Taylor Glacier, one of Antarctica's dry valleys. Scientists are trying to find out what viruses can survive such extreme conditions.

bacteriophage infects *E. coli* and utilizes the cell to reproduce more T2 bacteriophages (**FIGURE 1-4**).

In 1952, geneticists Alfred Hershey and Martha Chase provided evidence that DNA was the hereditary material. They set up an experiment in which they asked the question, What component of a T2 bacteriophage (protein or DNA) enters the host bacterium?

In their experiment, Hershey and Chase grew two cultures of *E. coli* in the laboratory. One flask of *E. coli* was infected with T2 bacteriophages, in which the protein coat of the bacteriophage was labeled with radioactive sulphur [^{35}S]. The second *E. coli* flask was infected with T2 bacteriophages in which the genetic material (DNA) of the virus had been labeled with radioactive phosphorous [^{32}P].

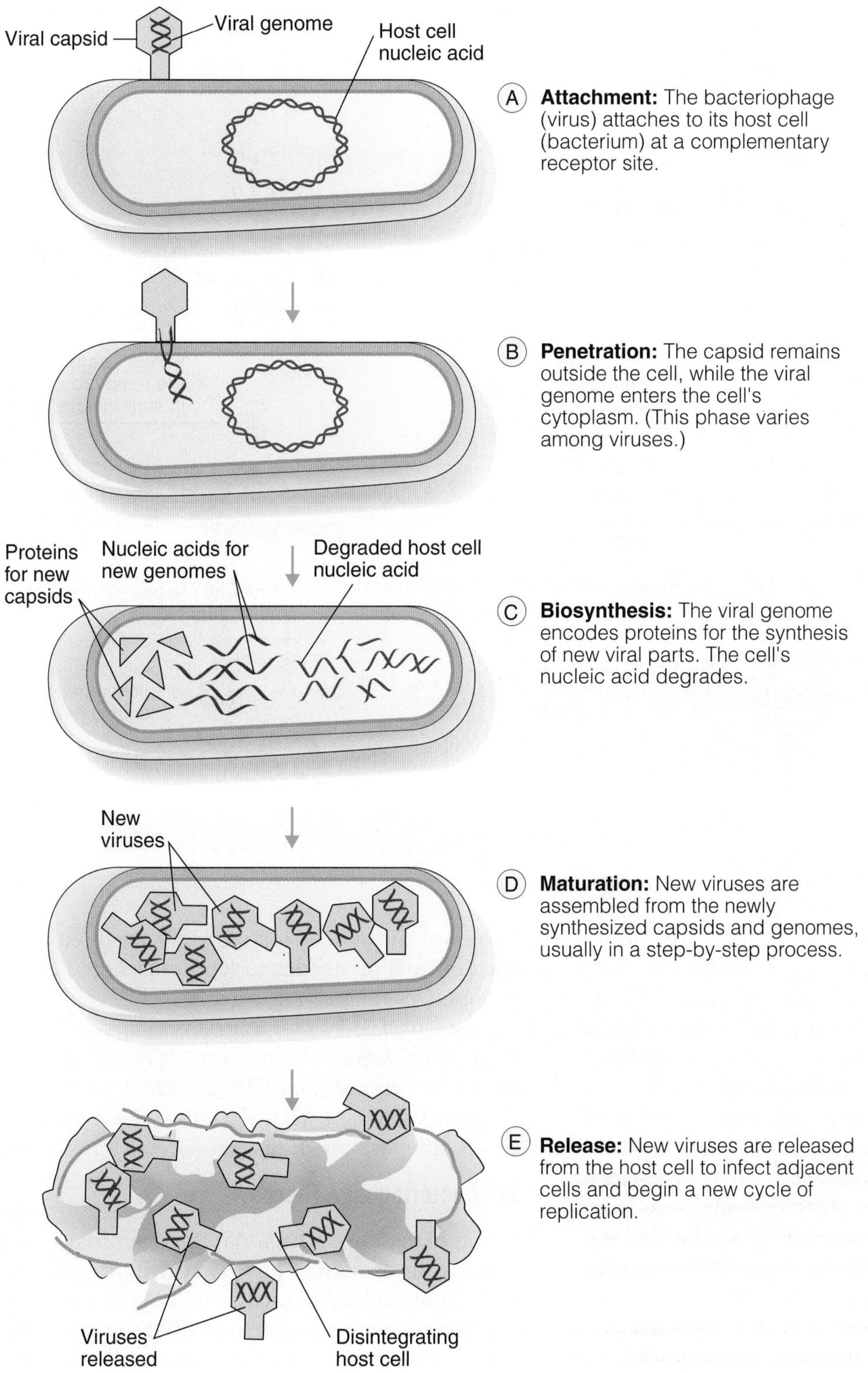
Viral capsid
Viral genome
Host cell nucleic acid
(A) **Attachment:** The bacteriophage (virus) attaches to its host cell (bacterium) at a complementary receptor site.
(B) **Penetration:** The capsid remains outside the cell, while the viral genome enters the cell's cytoplasm. (This phase varies among viruses.)
Proteins for new capsids
Nucleic acids for new genomes
Degraded host cell nucleic acid
(C) **Biosynthesis:** The viral genome encodes proteins for the synthesis of new viral parts. The cell's nucleic acid degrades.
New viruses
(D) **Maturation:** New viruses are assembled from the newly synthesized capsids and genomes, usually in a step-by-step process.
(E) **Release:** New viruses are released from the host cell to infect adjacent cells and begin a new cycle of replication.
Viruses released
Disintegrating host cell

FIGURE 1-4 Diagram of a lytic bacteriophage infection.

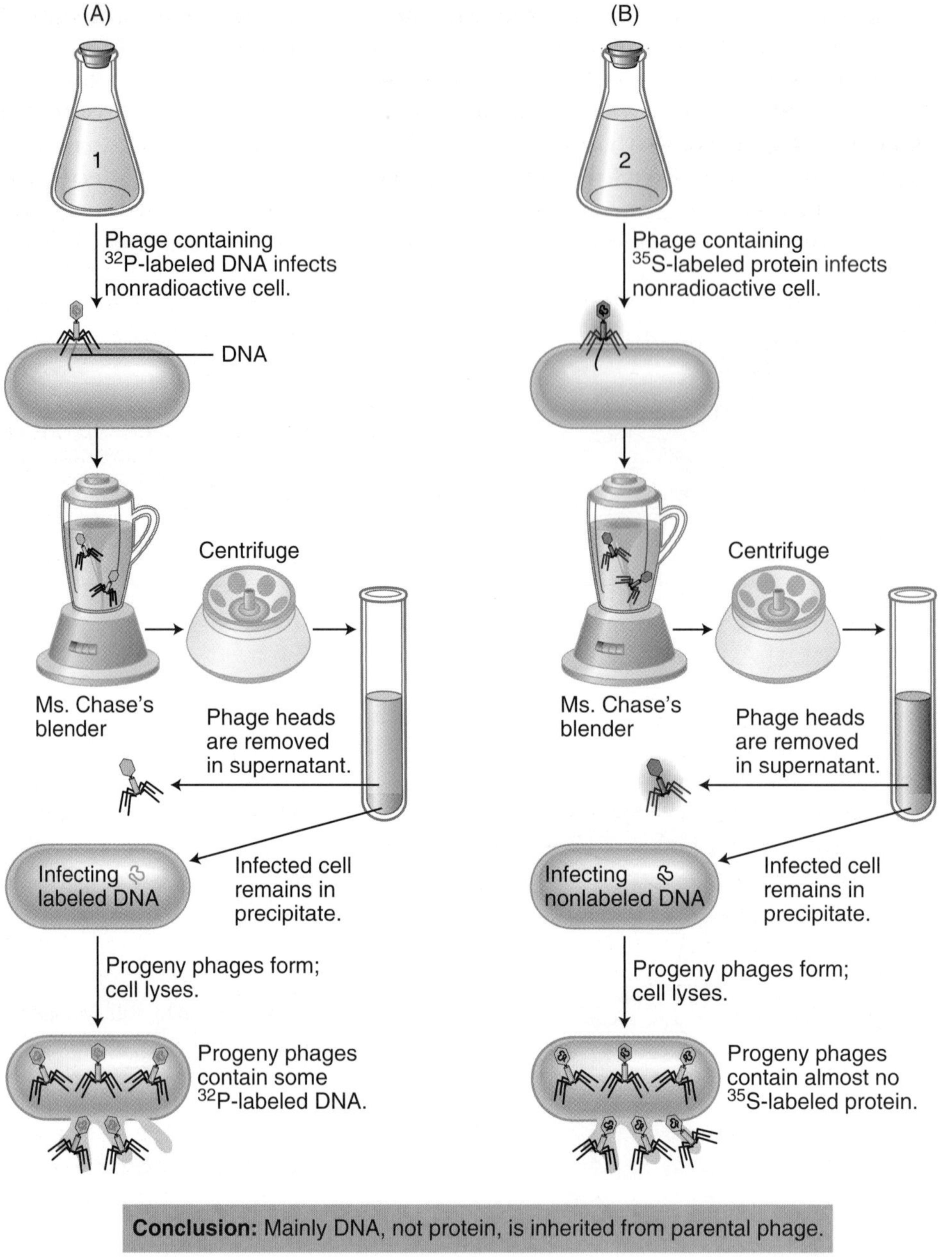
(A)
(B)
1
2
Phage containing
^{32}P-labeled DNA infects
nonradioactive cell.
Phage containing
^{35}S-labeled protein infects
nonradioactive cell.
DNA
Centrifuge
Centrifuge
Ms. Chase's
blender
Ms. Chase's
blender
Phage heads
are removed
in supernatant.
Phage heads
are removed
in supernatant.
Infecting
labeled DNA
Infected cell
remains in
precipitate.
Infecting
nonlabeled DNA
Infected cell
remains in
precipitate.
Progeny phages form;
cell lyses.
Progeny phages form;
cell lyses.
Progeny phages
contain some
^{32}P-labeled DNA.
Progeny phages
contain almost no
^{35}S-labeled protein.
Conclusion: Mainly DNA, not protein, is inherited from parental phage.

FIGURE 1-5 Diagram depicting the design of Alfred Hershey and Martha Chase's experiment. It provided evidence that DNA was associated with the genetic material of the bacteriophage.

The radioactive T2 bacteriophages were allowed to attach and infect the *E. coli* (**FIGURE 1-5**). After infection, the *E. coli* radioactively labeled T2 bacteriophage mixtures were poured into separate blenders. The blender dislodged the bacteriophage particles from the bacterial host cells in each mixture. The mixtures were then centrifuged (concentrated). This separated the bacteriophage particles in the supernatant liquid from the bacterial cells in the precipitate. Hershey and Chase found that the supernatant contained [^{35}S] (the viral protein) and the precipitate (cell portion) contained the [^{32}P] (the viral DNA). This experiment concluded that the DNA labeled with [^{32}P] transmitted the infective component of the bacteriophage. The DNA was the genetic material that specified all of the information needed to synthesize new T2 bacteriophages (Figure 1-5).

Bacteriophages and other types of viruses continue to be used as molecular biology tools to study host-cell interactions and the molecular biology of cells. Viruses have taught us much of what we know about the processes of replication, transcription, and translation.

Bacteriophage Therapy

If you experience a bacterial infection that is not healing on its own, a physician will prescribe an antibiotic such as penicillin or erythromycin to treat the infection. Before penicillin's discovery in the 1940s, though, the medical community was focusing its attention on bacteria-killing research (i.e., the use of bacteriophages to kill bacteria pathogenic to humans). Felix Twort (England) and Felix d'Herelle (Canada) first described bacteriophages ("phages" for short) in 1915 and 1917, respectively. Before the use of penicillin, it was determined that each type of bacterium can only be infected with a specific type of phage. In other words, phages exhibit a very narrow host range: a phage that infects the bacterium *E. coli* will not infect the *Streptococcus pneumoniae* bacterium.

The idea of phage therapy formed the basis for applied medical research as well as for the 1924 Pulitzer Prize-winning novel *Arrowsmith* by Sinclair Lewis. Researchers in Europe began to use bacteriophages to treat infections. In some cases, a liquid containing bacteriophages was poured onto an open wound; in others, the bacteriophages were given orally, via aerosol, or injected. The results were varied, and when antibiotics came into the mainstream, phage therapy largely faded in the West. Countries in the East such as Poland and Russia, however, continued to keep phage therapy research alive (see Chapter 21).

Phage therapy is now being reconsidered as a weapon against antibiotic-resistant "superbugs." These superbugs are lurking in hospitals and cause deadly infections that cannot be treated with the strongest antibiotics currently available. Research and development of alternative therapies are urgently needed to combat this superbug crisis.

Gene Therapy

Each of us carries a few defective genes. We remain blissfully unaware of this fact unless one of our close relatives or friends suffers from a genetic disease. Most of us do not suffer any harmful effects from defective genes because we carry two copies of nearly all genes (one inherited from each parent). This redundancy exists so that if one of our genes is defective, the second (functional) gene compensates for the defect.

In 1990, W. French Anderson, R. Michael Blaese, and Kenneth Culver—American pioneer researchers at the National Institutes of Health (NIH)—announced results of the first clinical gene therapy trial to genetically correct the adenosine deaminase (ADA) gene belonging to a four-year-old girl. ADA deficiency is a rare but very serious defect that causes immune deficiency, resulting in the lack of normal protection against bacterial and viral infections.

How was a good copy of ADA delivered to the girl? The hero was a retrovirus. The retrovirus was genetically engineered to carry a working ADA gene so that the body could produce functioning adenosine deaminase and consequently eliminate the root cause of the disease. To date, the girl is alive and well.

At least 2000 laboratories are engaged in gene therapy research worldwide. In addition to retroviruses, adenoviruses are being used in clinical gene therapy trials to deliver corrected genes. The history of gene therapy has been a roller-coaster. Gene therapy patient Jesse Gelsinger died in 1999 at the University of Pennsylvania while undergoing gene therapy. New regulations were established to patrol experiments, and new protocols were created. Even though the public is slow to hear about gene therapy developments, steady research progress continues. Gene therapy is discussed in greater detail in Chapter 9.

Vaccine Development

Vaccine development is one of the greatest advances in the history of medicine. The word vaccine is from *vacca*, the Latin word for cow. This is because the material of cowpox (a disease affecting the udders of cows) was injected into people to protect them against smallpox during the 1800s.

Attempts to deliberately protect humans against disease have a long history. Edward Jenner's work with cowpox vaccination holds the title as the first scientific attempt to control an infectious disease by means of intentional inoculation. In the late 18th century, Jenner, a rural physician, observed that milkmaids who had caught cowpox, a mild disease, didn't get smallpox. They also had beautiful complexions in contrast to others who had pitted faces after contracting smallpox. Jenner deduced that cowpox induced immunity against smallpox. To test his theory, on May 14, 1796, he removed matter from a cowpox pustule on the hand of milkmaid Sarah Nelmes and inserted it into a cut on the arm of eight-year-old James Phipps. James became mildly ill and developed a fever, but recovered after nine days.

On July 1, 1796, Jenner injected smallpox matter into James and repeated it again several months later. James did not get smallpox as a result of either of these inoculations. He was revaccinated 20 times and later died of tuberculosis at the age of 20.

Today there is a mandatory list of vaccines for schoolchildren in the United States. The list includes polio, diphtheria, measles, mumps, rubella (German measles), tetanus, pertussis, hepatitis B, and a few others. Most of these vaccines were the result of biomedical research after WWII. Read more about vaccines in Chapter 7.

Most people have experienced a cold, influenza, a cold sore on the lips, or plantar warts, many of which were caused by viruses. During the golden age of microbiology (1857–1914), rapid advances in microbiology, mainly spearheaded by scientists

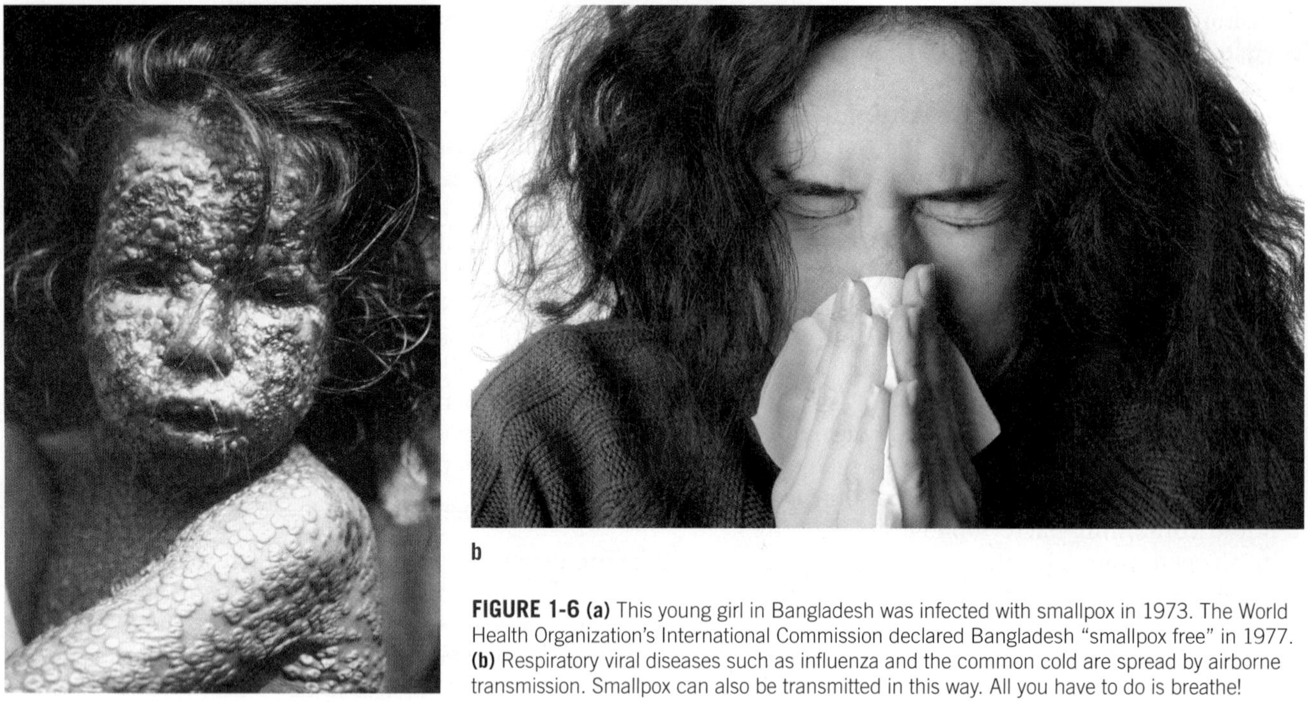

a

b

FIGURE 1-6 (a) This young girl in Bangladesh was infected with smallpox in 1973. The World Health Organization's International Commission declared Bangladesh "smallpox free" in 1977. **(b)** Respiratory viral diseases such as influenza and the common cold are spread by airborne transmission. Smallpox can also be transmitted in this way. All you have to do is breathe!

Robert Koch (Germany) and Louis Pasteur (France), determined that microorganisms cause **infectious diseases**. Additionally, Koch developed a set of experimental steps for determining which specific bacterium causes a particular disease. These steps are referred to as Koch's Postulates.

Bacteria or protozoa were identified to be the cause of certain diseases. They did not, however, cause other diseases such as measles, yellow fever, polio, and smallpox. Today we know that viruses can infect virtually every organism on earth (**FIGURE 1-6**).

Viruses and Cancer

Viral infections are linked to approximately 15% of all cancers. About 80% of these are cancers of the cervix (caused by papillomaviruses) and the liver (caused by hepatitis viruses B and C). These viruses are thought to be one factor that acts at an early stage in the process that leads to cancer.

Cancer has afflicted humans throughout recorded history. Over one hundred years ago, amateur epidemiologists reported a curious phenomenon: cervical cancer was common among prostitutes, but extremely rare in nuns. Some noted that cervical cancer was very high in women who were married to men whose first wives had died of cervical cancer. From these observations and reports, scientists speculated that a sexually transmitted agent caused cervical cancer (refer to Chapter 10 for more information on this phenomenon). In 1986, a paper entitled "Detection of Papillomavirus DNA in Human Semen" was published in *Science*. Today it is known that human papillomaviruses are responsible for 90% of all cervical cancers. A safe and effective vaccine against the most common high-risk types of human papillomaviruses is available.

1.2 Early Virus Studies

Before the invention of the electron microscope in 1931, viruses could not be seen or grown in the laboratory. *Most importantly, viruses were defined as agents so small they could pass through filters that trapped most known bacteria.* Scientists had to develop ways to observe and grow these invisible agents in the laboratory. The initial observations and methods developed to study viruses involved bacterial and plant systems. The first virus discovered was actually a plant virus known as tobacco mosaic virus (TMV), a disease that destroys tobacco crops. In 1892, the Russian botanist Dimitri Iwanowski demonstrated that extracts from diseased tobacco plants could transmit disease to healthy tobacco plants after passage through ceramic filters known to trap most bacteria. Iwanowski didn't understand the full significance of this result. In 1898 Martinus Beijerinick extended Iwanowski's experiments and was the first to develop the idea of the virus, which he called a *contagium vivum fluidum* (contagious living fluid).

Bacteriophages were first isolated from natural sources (such as human sewage) and studied by scientists in the early 20th century. Bacteriophages grow by inserting their DNA into a host bacterium, directing the host's biosynthetic machinery to make copies of the bacteriophage DNA and protein coat, and then destroying the host bacterium. The destruction of the host bacterium releases new bacteriophages, which then go on to infect fresh hosts.

Bacteriophage plaque assays are used to quantify the number of infectious bacteriophages in a given phage-containing sample. Briefly, bacteriophages are allowed to adsorb to host bacteria in a test tube. The mixture is then poured onto a solid agar plate of medium and the bacteria are allowed to grow. At this point, the bacteriophages lyse bacteria that are present on the surface of agar. The holes (called **plaques**) in the bacterial lawn are areas where the bacteria have been killed by bacteriophages.

Major advances in animal virology did not occur until 1952, when Renato Dulbecco modified the bacteriophage assay to work on animal cell cultures. **FIGURE 1-7** shows a 6-well dish containing crystal violet-stained monolayers of monkey kidney cells that were infected with vaccinia virus. Similar to the bacteriophage assays, plaques or clearings in the cell monolayer are visualized where infected cells have been destroyed by viral infection. More viruses were used to infect the cell monolayers in the wells on the left-hand side of the dish than the wells on the right side of the dish. Dulbecco was awarded a Nobel Prize for the development of animal plaque assays in 1975.

Characteristics of Viruses

Today we know that viruses share a number of common features. First of all, they are small. As previously mentioned, viruses are able to pass through filters that retain or trap most known bacteria. Hence, viruses are smaller than bacteria. As a rule, most bacteria are 100 times larger than viruses. Typically, bacteria range from 1 to 10 micrometers (μm) in length. A virus would fall in the range of 0.03–0.1 μm in length (or 30–100 nm). Of course there are always some exceptions: Some viruses, such as poxviruses, can be 200 to 400 nanometers (nm) in length, and filoviruses (such as Ebola) can be up to 1000 nm in length. **FIGURE 1-8** provides size comparisons of biological molecules, viruses, bacteria, cellular organelles, and eukaryotic cells.

A second feature used to define a virus is its complete dependence upon the host cell to reproduce itself. Viruses do not have functional organelles or ribosomes. Viruses are too small to carry enough genetic material to code for all of the gene products necessary to rebuild a virus. As a result, a virus must use its host cellular protein synthesis machinery to synthesize viral proteins. The genome or genetic material of a virus consists of one species of nucleic acid, DNA, or RNA. The DNA or RNA genome of a virus can be single- or double-stranded.

The outside of the virus particle contains a receptor-binding protein or viral-attachment protein that will allow the virus to adhere to **receptors** present on the surface of cells. For example, the common cold virus (also known as rhinovirus)

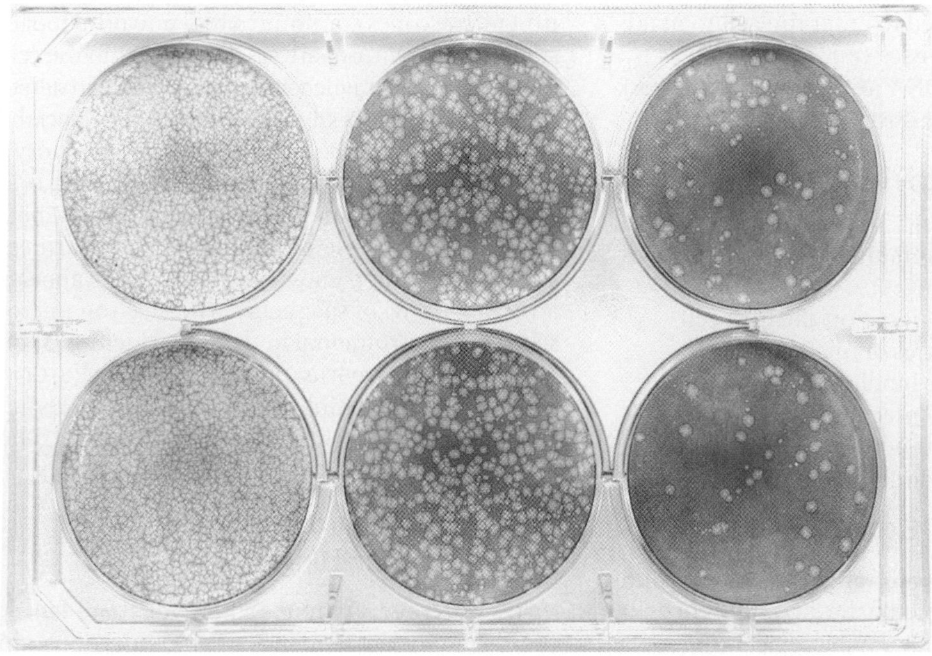

FIGURE 1-7 Plaque assays can be used to study animal or human viruses. The 6-well dish contains crystal violet-stained monolayers of monkey kidney cells that were infected with vaccinia virus.

FIGURE 1-8 Viruses are smaller than their host cells. This illustration demonstrates size comparisons of a eukaryotic cell, a bacterium, and a virus.

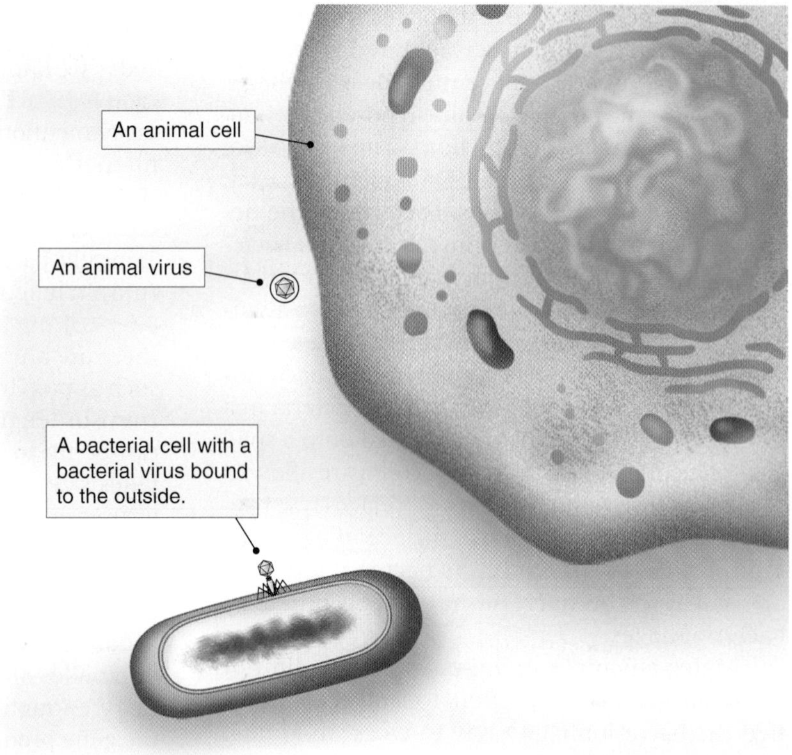

binds to a receptor on the outside of cells called the **I**nter**c**ellular **A**dhesion **M**olecule-**1** (ICAM-1).

Unlike other unicellular microorganisms, some human or animal viruses can reproduce themselves even if nothing but the viral genome is introduced into the cell. In other words, the viral genetic material is infectious. **FIGURE 1-9a** demonstrates that the RNA genome of poliovirus can be introduced into cultured cells (by a method called **transfection**) and viruses will be produced from that RNA within the cell. In this case, the polioviruses are not required to enter the cell to make more copies. Retroviruses can persist in cells by integrating their own DNA (or a copy of their RNA) into the genome of the host cell (Figure 1-9b).

Review of Virus Characteristics

- Small (nm in size)
- Pass through filters that trap most known bacteria
- Completely dependent on the host cell
- Contain one species of nucleic acid
- Contain receptor-binding protein
- Genome may be infectious
- Some viruses can persist by integrating genome into the cellular chromosome

Visualizing Viruses: Electron Microscopy

Electron microscopes were originally invented in the early 1930s to overcome the limitations of light microscopes to visualize nonbiological materials such as metals and small electronic parts. Light microscopes at that time could magnify specimens as high as 1000 times.

Instead of light rays, electron microscopes use a beam of electrons focused by magnets to resolve minute structures. With electron microscopy, it is possible to magnify structures 100,000 times and resolve them at 0.5 nm. Virologists were quick to take advantage of this new, powerful tool. Kausche, Pfankuch, and Ruska published the first electron micrograph of TMV in 1939. Today electron microscopes continue to be a powerful tool in studying how viruses are assembled within the cell, the structure of fragile viruses, and the rapid detection and diagnosis of viral infections (especially viruses that cannot be cultivated in the laboratory). The electron micrograph image in **FIGURE 1-10a** represents the first isolation and visualization of Ebola virus in 1976. Some of the filamentous particles are fused together, end-to-end, giving the appearance of a "bowl of spaghetti." The electron microscope was instrumental in the initial identification of the new coronavirus, now known as SARS-CoV. Biologist Cynthia Goldsmith is observing a viral isolate via the electron microscope from the 2003 SARS outbreak in Figure 1-10b.

Are Viruses from Outer Space? Theories of Viral Origin

Viruses are everywhere. Wherever there is life, there are viruses! Evidence of viral infections can

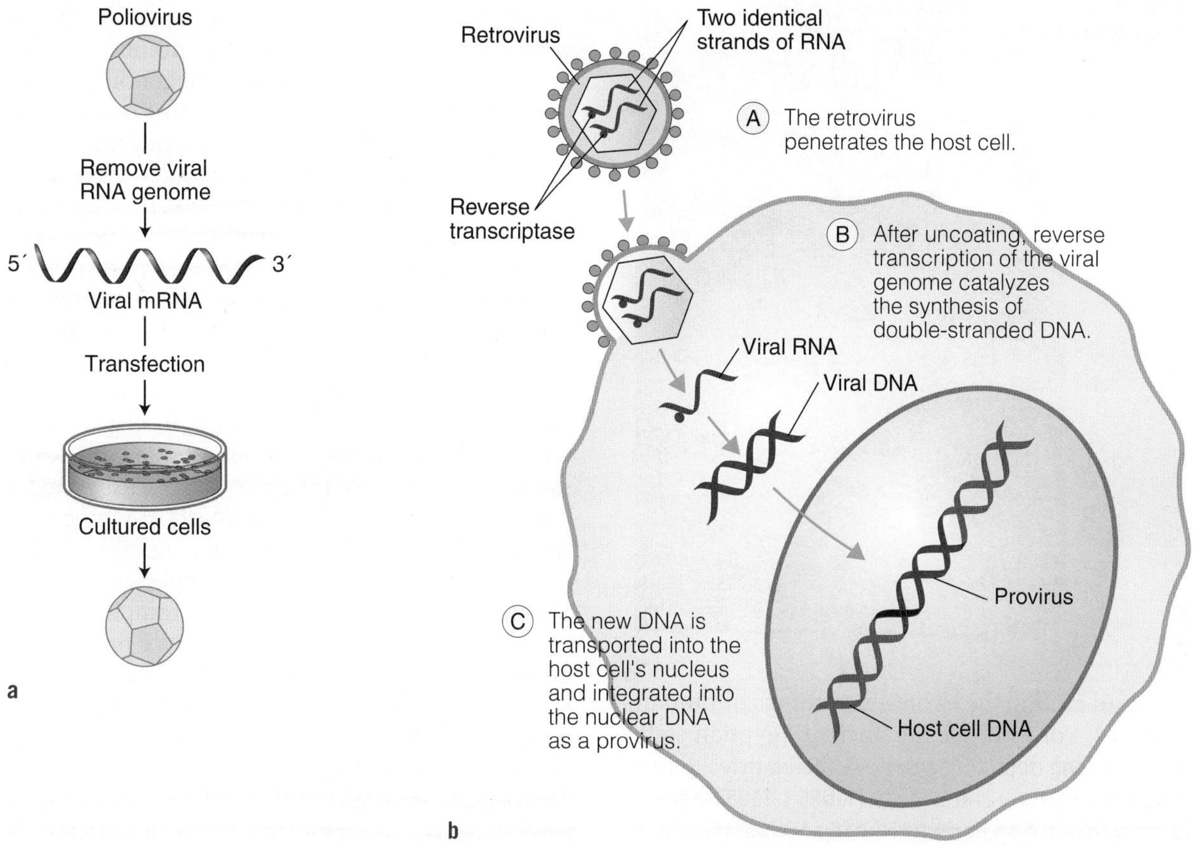

FIGURE 1-9 (a) Poliovirus RNA is infectious! Infectious particles of poliovirus can be produced even if only the genetic material (RNA) of the virus is introduced into cells growing in a culture dish. **(b)** Retroviruses can persist in cells by integrating their own DNA (or a copy of their RNA) into the genome of the host cell. The integrated viral genome is called a **provirus**. Adapted from Flint, S. J., et al. *Principles of Virology: Molecular Biology, Pathogenesis, and Control of Animal Viruses,* Second Edition. ASM Press, 2003.

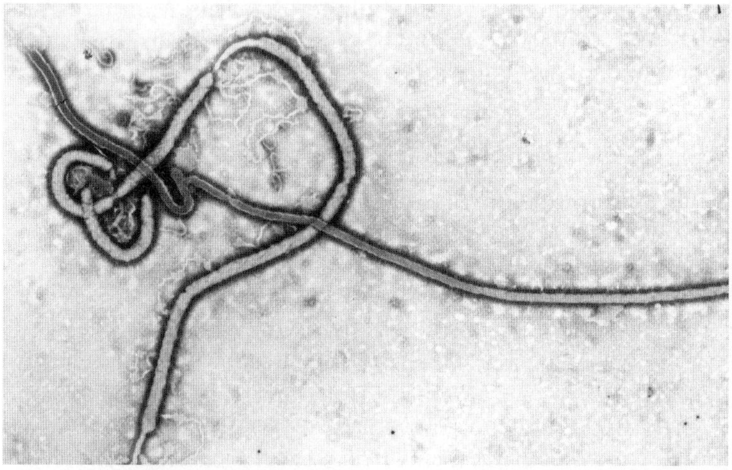

a

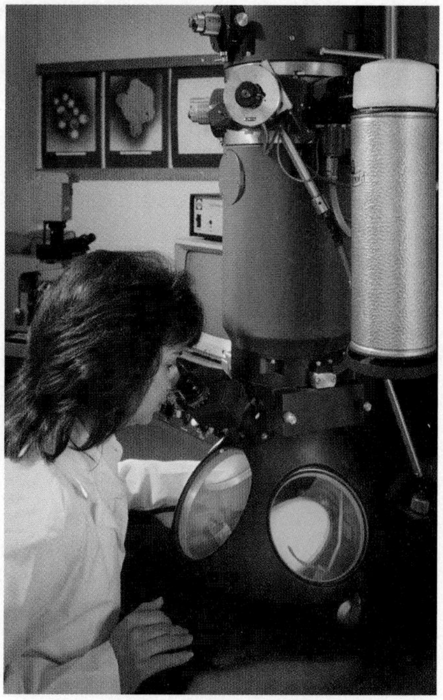

b

FIGURE 1-10 (a) Transmission electron micrograph of Ebola virus particles that were isolated from a human diagnostic specimen and then cultured in Vero cells. Magnification 40,0003. **(b)** Visualization of SARS-CoV via the electron microscope.

FIGURE 1-11 Egyptian priest from the 18th dynasty (14th century B.C.) with foot-drop deformity.

be found among the earliest recordings of human activities. For example, an ancient Egyptian stele tomb carving depicting a polio-afflicted priest (circa 14th century B.C.) is shown in **FIGURE 1-11**. The foot-drop deformity is characteristic of residual paralysis due to poliomyelitis.

Where did these viruses come from? Could a cough or sneeze be a sign of a close encounter with a tiny visitor from outer space? The late Sir Fred Hoyle (1915–2001)—a world-renowned astronomer known for being controversial—and his former student Nalin C. Wickramasinghe proposed the **panspermia hypothesis**. This hypothesis asserts that viruses or other microorganisms are raining down upon earth and contaminating it. Hoyle and Wickramasinghe proposed that these outer-space microbes were responsible for originating life on earth and cause massive contagion flowing in from space. They speculated that influenza pandemics occurred in our history when solar winds during sunspot peaks caused the viruses to be swept down through the earth's atmosphere. Hoyle speculated that diseases tend to strike during the winter season because cooler weather generates stronger downdrafts. Almost all members of the scientific community have dismissed the panspermia theory. Most scientists believe that cosmic radiation would almost certainly destroy germs in space.

Theories about the origins of viruses developed within the last couple of decades had two trains of thought based on results from research published on molecular virology studies of the 1980s. The first take was that viruses were precursors of the earliest cells. The other take was that viruses originated from cells that underwent degeneration as a result of viral parasitism. The viruses were gene robbers that "broke away" as genetic elements from cellular genomes.

As technology improved, viral genomes were sequenced. These sequences did not resolve the debate. Instead, they threw a considerable new light on viruses. It became evident that the genomes of viruses are so diverse that it is unlikely that all viruses evolved from a common single-celled ancestor termed as the **last universal common ancestor (LUCA)** that lived perhaps 3 or 4 billion years ago. As more viral sequences became available, especially the larger genomes of Mimivirus, Mamavirus, and Marseillevirus, a quiet revolution was brewing among evolutionary biologists. This is because many virus groups don't share any common genes, ruling out the idea that viruses have a common origin. Sequence analysis of viral genes reveals at least five classes of viral genes (**TABLE 1-1**).

The classes of genes the viruses possess strongly depend on viral genome size. For example, viruses with small RNA genomes often have only a few genes. The majority of these viral genes belong to the hallmark class. Viruses with larger DNA genomes such as the poxviruses possess all five classes of genes. More than 80% of the genes found in the moderate and large genomes of bacteriophages and archaeal viruses are **ORFans**. ORFans are open reading frames that have no known homologs and no known function. ORFan genes have been found in prokaryotes such as *E. coli*. Daubin and Ochman theorize the ORFan genes were acquired from bacteriophages.

The discovery of the giant Mimivirus called to question the definition of a virus. Mimivirus was first thought to be a new legionella-like bacterium

TABLE 1-1 Classes of Viral Genes

Class	Gene Description
1	Virus genes that have closely related **homologs** in cellular organisms (especially, the host of a given virus) present in a narrow group of viruses.
2	Virus genes that are conserved within a major group of viruses that have distantly related cellular homologs.
3	Virus-specific genes that have no detectable cellular homologs. These genes are referred to as ORFans.
4	Virus-specific genes that are conserved in a broad group of viruses but have no detectable homologs in cellular life forms.
5	Genes shared by many diverse groups of viruses with only distantly related homologs in cellular organisms. These are referred to as *viral hallmark genes*.

A homolog is a gene sequence that is similar to a gene sequence in the cellular (host) genome.

Source: Reprinted from Koonin, E. V., et al. 2006. "The ancient Virus World and evolution of cells." *Biology Direct* 1:29; doi10.1186/1745-6150-1-29.

isolated from a cooling tower in Bradford, England. It was discovered during an investigation to find the source causing a pneumonia outbreak in 1992. About ten years later, it was determined that it was not a bacterium, but instead a giant virus that was able to grow inside of an amoeba. Its entire DNA genome was analyzed. The structure and genome of Mimivirus were not only large in size, but it contained genes that were not found in any other known viruses. These novel genes were homologs of genes involved in protein synthesis, a process that occurs only in cellular organisms! In 2008, it was discovered that a virus 50 nm in size infected and replicated inside of Mimivirus. It was classified as a **virophage** and named Sputnik. This was the first time virologists had determined that a virus could infect a virus! These findings put another spin on the origin of large DNA viruses. Raoult and Forterre suggested that biological entities be divided into two groups: ribosome-encoding organisms and capsid-encoding organisms that include viruses.

A very crude model of the evolution of life and current viral genomics is shown in **FIGURE 1-12**. This figure represents an emerging concept. It begins with the **hydrothermal origin hypothesis**, also called the iron-sulfur (FeS) world theory that postulates the first organic chemical structures were formed at warm alkaline thermal vents or fissures (long, narrow openings) found in the ocean seafloor. Hydrothermal vents leaked hot sulfuric acid into the surrounding environment. Supporters of this theory claim that a gradient formed between the hydrothermal vent water and extremely ice cold water that surrounds the vent at the bottom of the ocean. The temperature at the cooler temperatures would be suitable for organic chemical synthesis to occur. The vents contained FeS and iron nickel sulfide (Fe-Ni-S) that acted as catalysts fueled by chemical energy (H_2 from the hydrothermal environment and CO_2 from the marine environment) resulting in the formation of the organic precursor molecules for the building blocks of life (e.g., amino acids, sugars) in the cooler surrounding environment (about 100°C). As the biological melting pot cooled off, different classes of viruses emerged from different genetic elements at different stages. These early viruses must have self-replicated even if very poorly. The RNA viruses evolved first, followed

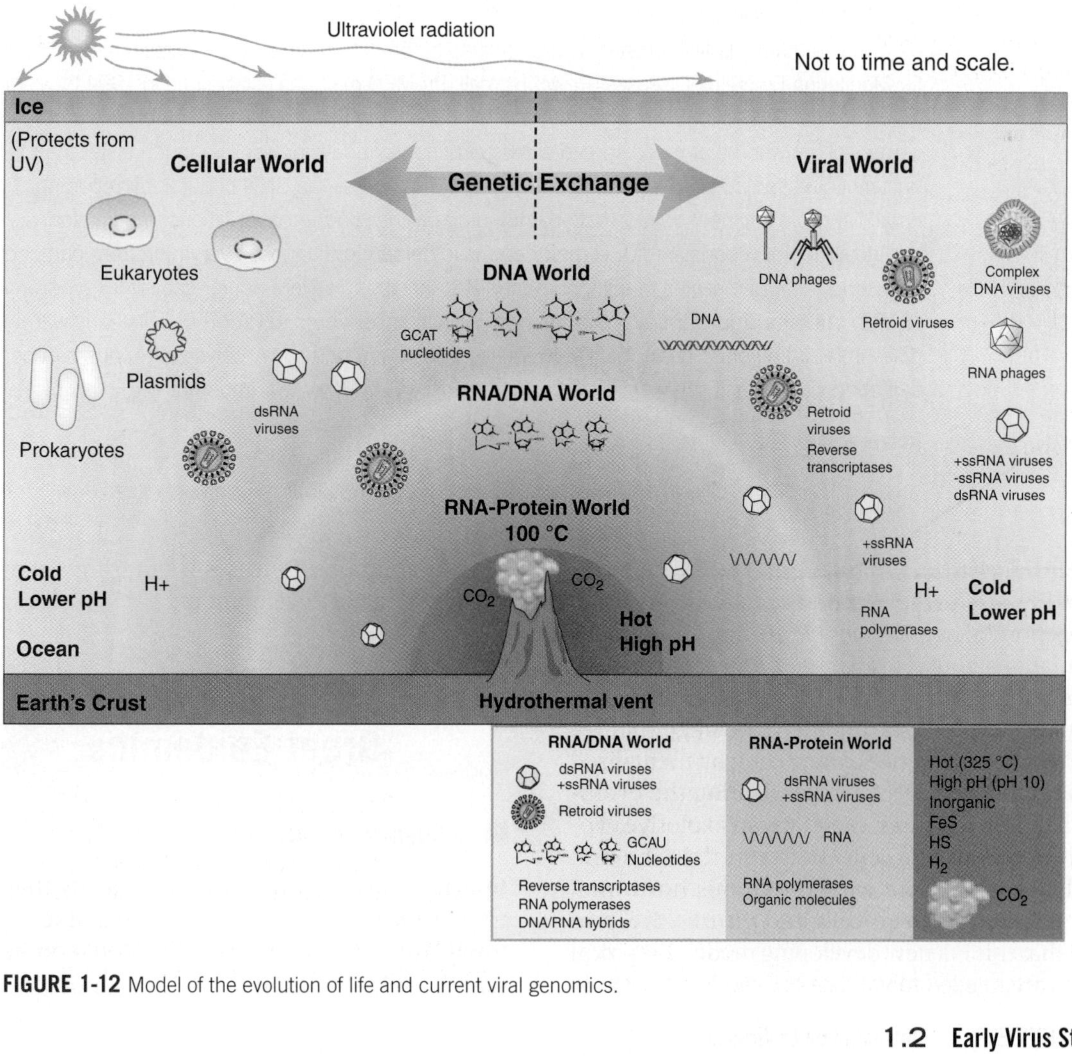

FIGURE 1-12 Model of the evolution of life and current viral genomics.

VIRUS FILE 1-1 Use of PubMed, CDC Publications, and ProMED to Research Specific Viruses or Viral Outbreak Information

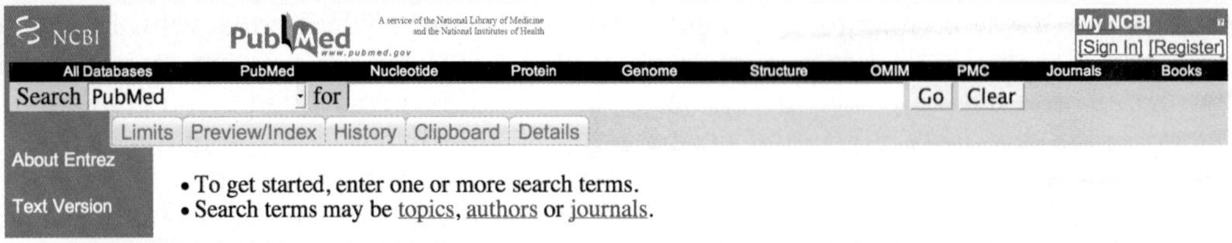

FIGURE VF 1-1 PubMed interface. PubMed is the most frequently used search engine for medical research journal articles.

Each chapter of *Understanding Viruses* will contain one or more Virus Files. These files or synopses are intended to connect students to research being conducted in the field of virology. References to the original research articles will be provided at the end of each file. Students and instructors interested in exploring topics further may opt to search for more information via PubMed, which is a service of the National Library of Medicine. It includes over 20 million citations for biomedical articles back to the 1950s. These citations are from MEDLINE and additional life science journals. PubMed searches can be done by topic, author, journal title, and other parameters. To begin a PubMed query, start at their website (which can be found easily through a Web search engine).

Many Virus Files may cite reports published by the Centers for Disease Control and Prevention (CDC), such as *Morbidity and Mortality Weekly Report* (*MMWR*) and *Emerging and Infectious Diseases* (*EID*). The archives of these publications can be quickly searched from the CDC's website (http://www.cdc.gov/).

If you are particularly interested in virus-related outbreaks, another valuable resource is Program for Monitoring Emerging Diseases (ProMED) mail. ProMED mail was established in 1994 by a small group of scientists. ProMED mail is intended to address the threat of disease outbreaks in remote corners of the world that could spread across continents in days or weeks—much faster than the doctors could spread word about the disease. ProMED mail is the CNN of outbreak reporting. This email list posts a variety of information, including on-the-ground observations, media stories, and government reports. ProMED members have helped to diagnose everything from camel pox in Saudi Arabia to measles in military men in Kazakhstan to severe acute respiratory syndrome (SARS) in China. Today there are more than 30,000 subscribers to ProMED. This early warning disease reporting often beats local authorities in disease reporting. Archives of reports can be searched via the mail site within the ProMed website (http://www.promedmail.org).

References

Check, E. 2004. "Dispatches from the front line." *Nature* 432:544–545.

Woodall, J., and Calisher, C. H. 2001. "ProMED-mail: Background and purpose." *EID* Supplement 7;(3):563.

by **retroid viruses** and subsequently DNA viruses. Retroid viruses use reverse transcriptase to replicate their genomes (see Chapters 3, 16, and 17 for more information about retroid viruses and agents). Over the course of 10–100 million years, complex gene assemblies evolved during the DNA stage that resulted in the emergence of new compartmentalized cells and the large DNA viruses. During the evolution of the DNA viruses, there was an explosive evolution of eukaryotic cells. Between the viral and cellular worlds, there was continuous horizontal gene transfer between cells and viruses. Keep in mind that this is a new developing model. Empirical studies are needed to validate the model. The study of the evolution of viruses is entering a new and exciting stage of development.

1.3 Viruses in History: Great Epidemics

Influenza

Imagine walking down a street in a city that used to be full of activity but now looks like a ghost town. You see quarantine signs posted on homes; flags or wreaths hanging on the doors indicate

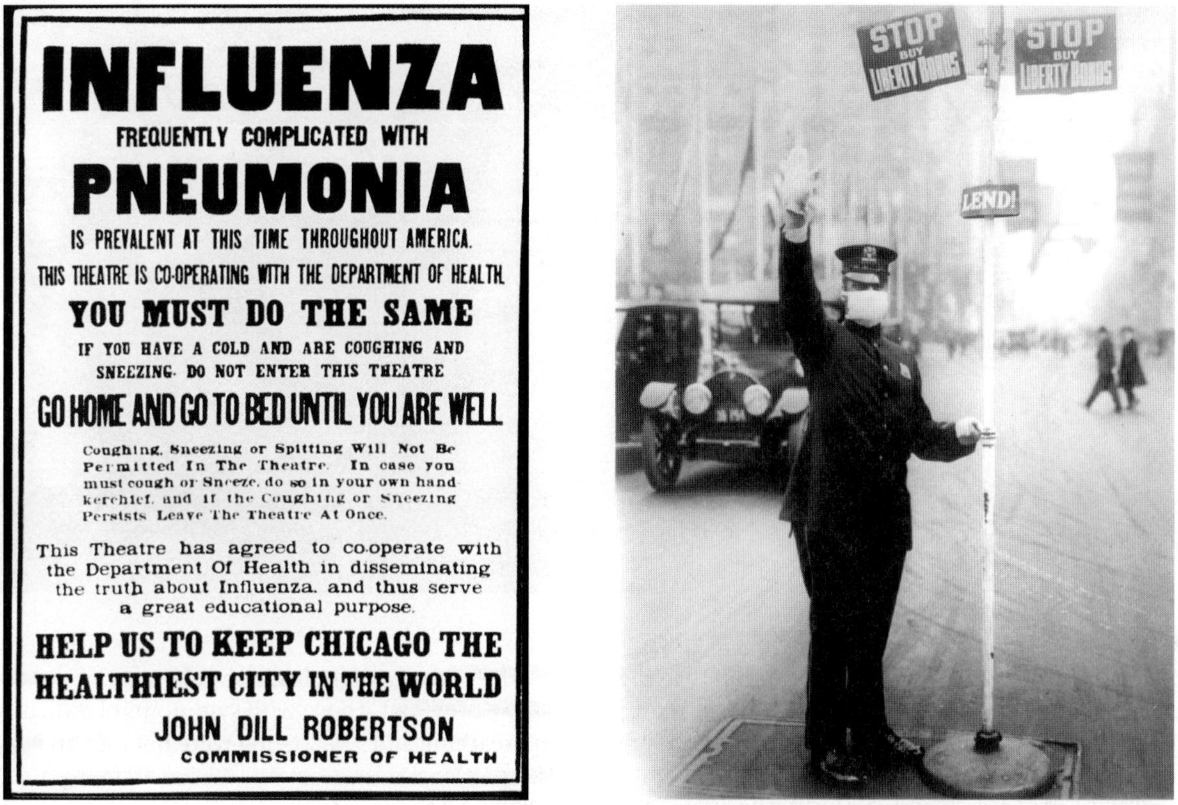

a b

FIGURE 1-13 (a) Poster used to educate Chicago residents during an influenza epidemic that occurred in the fall of 1918. **(b)** A New York City policeman wearing a cloth mask to protect him from the Spanish flu in 1918.

whether it was a parent, child, or grandparent who had recently died from a contagious disease. Schools, churches, theaters, libraries, and most restaurants and stores are closed (**FIGURE 1-13a**). There are no public gatherings, not even funerals. The few people you see have a cloth mask covering their face (Figure 1-13b). There is a shortage of coffins and the use of mass graves is common. This really did happen in 1918. The visitor to this town was the virus named *influenza*.

In 1918 the Spanish flu (also termed *la grippe*) pandemic charged across America in 7 days and across the world in 3 months. It claimed more American lives than all the major wars of the 20th century combined. Estimates of deaths range from 20 to 50 million—most in the brief period from October through the end of December. The Spanish flu was associated with high rates of morbidity, mortality, social disruption, and high economic costs, and was to be the most destructive pandemic ever known.

The Spanish flu incubation period and the onset of symptoms were so short that apparently healthy people in the prime of their lives (ages 20–40) were suddenly overcome, and within an hour could become helpless with fever, delirium, and chills. Severe headache, pains in muscles and joints, hair loss, and acute congestion accompanying temperatures of 101°F to 105°F occurred. The most unusual pathologic finding was massive pulmonary edema and/or hemorrhage. This was a unique viral pneumonia—a patient could be convalescing one day and dead the next. Those who did not die of Spanish flu often died of secondary bacterial pneumonia.

Surprisingly, before the recent concerns of a potential bird flu pandemic, not very many people were familiar with the 1918 Spanish flu. It may be that few survivors wanted to talk to historians about the terrifying and ghastly days when so many were dying that there were bodies in the streets.

Poliomyelitis

Evidence of the viral disease poliomyelitis dates to the dynasties of Ancient Egypt. In the current era, polio epidemics peaked in the United States in 1916 and the mid-1940s to 1950s. This disease was quite frightening because of its mysterious seasonal incidence (July–October). Many children were not allowed to play outside because of the fear of polio. Newspapers included word games and puzzles to occupy children while they stayed indoors. Some families fled to remote summer vacation homes.

Polio struck an industrialized nation free from the poor sanitation conditions that typically play

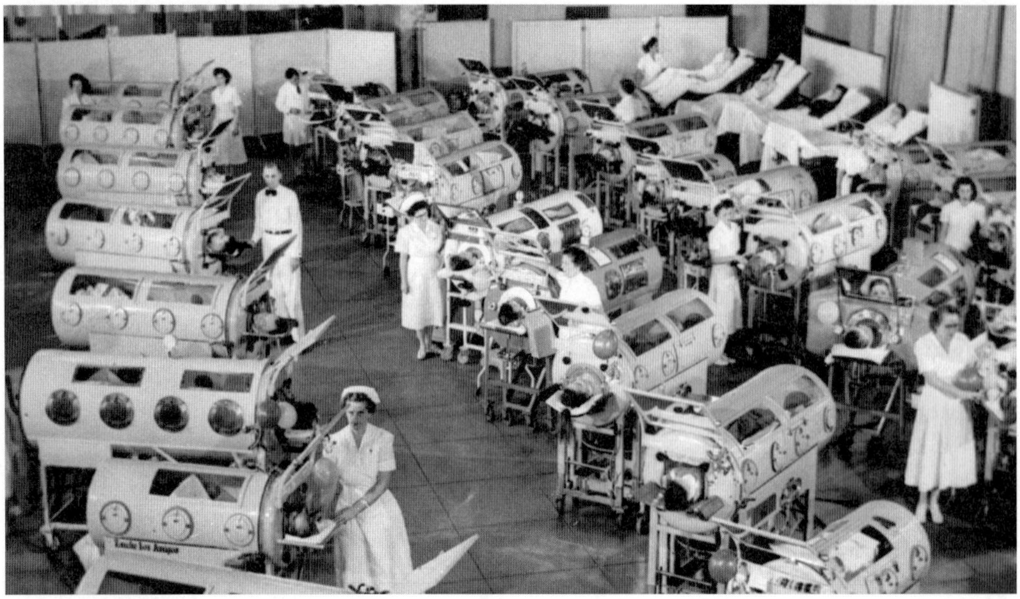

a role in epidemics. It occurred in a nation thriving with new technologies that would lead to man's control over disease. Civil engineers were creating a network of aqueducts and water purification plants to provide clean, safe drinking water for much of the nation.

Polio's transmission was a mystery. Cities responded with methods that had met with success in ridding epidemics of typhus, cholera, and diphtheria. New York City officials had the streets doused with 4 million gallons of water a day to flush the streets of their germs intentionally killing 72,000 stray cats that were thought to be virus carriers. Many cities were fogged with insecticides. None of these measures worked, though: polio was spread through human feces. At the time, officials did not know that.

Polio broke America's heart. It crippled its victims. Children were lined up in wheelchairs. Paralysis of the muscles used for respiration and

swallowing (called bulbar poliomyelitis) was sometimes fatal. Those with symptoms of difficulty in breathing and swallowing were put into an **iron lung**, also known as the "drinker respirator" (**FIGURE 1-14**).

Franklin Delano Roosevelt was likely the most famous adult who suffered from the effects of poliomyelitis. Roosevelt contracted polio in 1921, at the age of 39. He spent over half of his personal fortune to purchase Warm Springs, Georgia, a resort with warm natural springs for swimming and rehabilitation, to provide a place for the "polios." Frightened by reminders of this terrible disease, guests not suffering from polio abandoned the resort.

Roosevelt did not recover from the initial paralytic effects of polio. When he left Warm Springs to run for governor of New York in 1928 and then for the U.S. presidency in 1932, he chose to hide the effects from the public (**FIGURE 1-15a**). He was not photographed being carried or wheeled about.

FIGURE 1-15 **(a)** FDR at Warm Springs, Georgia, in 1924. **(b)** The press portrayed Roosevelt as a robust, physically strong leader.

a

b

He would lean on a cane or a companion's arm. It appeared to many that he could walk. It was a brilliantly staged deception. In Roosevelt's opinion, and in those of his advisors, a robust appearance was necessary to portray a physically strong leader, and the press cooperated (Figure 1-15b).

The March of Dimes

The March of Dimes began in 1938 as an effort to raise money for polio treatment and research. The organization selected Dr. Jonas Salk to lead research on polioviruses, and in 1941 provided the first iron lung to assist polio patients. The March of Dimes ran field trials of the Salk vaccine (an inactivated preparation of poliovirus) with 1,830,000 schoolchildren participating in 1954. In 1955, the vaccine was declared safe, effective, and potent and was licensed for general use. Later, in 1960, the Sabin vaccine was licensed for use in the United States. (Read more about polio vaccination in Chapter 11.)

Today, polio is rare due to the worldwide efforts of the World Health Organization (WHO), CDC, Rotary International, and United Nation's Children Fund (UNICEF) to eradicate it. In 2010, endemic areas of polio were reduced to four countries: India, Pakistan, Nigeria, and Afghanistan. Some countries in Africa have a "re-infected" status today.

Acquired Immunodeficiency Syndrome (AIDS)

In June of 1981, a group of physicians in Los Angeles, California, reported five unusual cases of *Pneumocystis* pneumonia in the CDC's *Morbidity and Mortality Weekly Report.* All patients were young men, and all were sexually active homosexuals. Two of the patients died from this pneumonia and all patients had experienced other rare infections such as candidiasis, a fungal infection of the throat, mouth, or in women, the vagina. *Pneumocystis* pneumonia is rare in the United States and almost exclusively found in severely immunosuppressed individuals. This first report recognized a new growing epidemic in the United States that was later termed AIDS (see Chapter 8, Virus File 8-1).

As cases were reported to the CDC, a pattern characteristic of an epidemic emerged, but the culprit was a mystery until Robert Gallo (United States) and Luc Montagnier (France) discovered the human immunodeficiency virus (HIV). Today AIDS is a worldwide epidemic, with the population in Africa being the most severely affected. There are several antiviral drugs available to prolong the lives of those suffering from AIDS, but there is no cure or vaccine. Education promoting AIDS prevention has reduced cases in some countries. (Read more on AIDS in Chapters 8 and 16.)

1.4 Recent Viral Outbreaks

The First Pandemic of the 21st Century: H1N1 Influenza A, 2009

At the end of the typical 2009 influenza season, health authorities in Mexico City recognized an unusual pattern of influenza-like illness. Dozens of individuals were suffering from an atypical flu during the middle of March. The number of cases increased dramatically within the month. On April 6, a U.S. data mining biosurveillance company posted an alert on its Website warning about a possible flu outbreak spreading in La Gloria, Mexico. The data mining company tracked thousands of website searches daily for early signs of medical problems or civil unrest anywhere in the world. The company advised the CDC that the Mexican outbreak reports had potential of a public health emergency and international concern. On April 18, Mexican health officials sent the CDC 14 mucus samples from patients suffering from severe flu or pneumonia-like infections for testing. The CDC held a press conference on April 24 to announce that 7 of the 14 Mexican samples contained the same viral strain that was also causing flulike illness in individuals located in adjacent counties in California and Texas. They indicated that containment of this outbreak was "not very likely." Preliminary laboratory tests at the CDC suggested that the infections were caused by a swine Influenza A (H1N1) based on genetic testing (see Chapter 12, Influenza Viruses, for an explanation of the "H1N1" designation). The viral isolate characterized had never been detected in humans or pigs.

That same day, the President of Mexico, Felipe Calderon, advised citizens to wear face masks when using public transportation, stay indoors and avoid crowded places, exercise frequent handwashing, cover mouths when coughing, cough or sneeze into the crook of the arm or a tissue, and avoid sharing food. Schools were closed in Mexico City. People were urged not to go to work and to seek immediate medical attention if they experienced flulike symptoms. These recommendations were repeated daily through public media announcements including newspapers. The Mexican army distributed 6 million masks, handing many out to citizens at subway stations and Metrobus lines.

Mexican President Calderon invoked emergency powers, giving the government the power to enforce quarantine and conduct home inspections. Public events were canceled on April 25. WHO Director General Dr. Margaret Chan declared the outbreak "a public health emergency of international concern." Two days later, schools and universities were closed country-wide in Mexico. Members of the public were alarmed by the dramatic changes imposed by the government in response to this epidemic. On April 26, the government ordered all gyms, cinemas, art galleries, restaurants (except for take-out orders), bars, and cantinas to close (**FIGURE 1-16**). By April 29, the Mexican government required that drivers in public transportation wear masks and gloves. The fine for not complying was around $150 U.S. dollars (40 times the daily minimum wage in Mexico). Instead of imposing the fines, the Mexican police enforced regulation by taking bribes from drivers who failed to comply. They also threatened to seize taxis or buses for 5 days for noncompliance. On this same day, the Mexican Ministry of Health reported its April monthly total of 2155 patients with severe pneumonia and 100 deaths. These government orders were not very different from the measures applied during the 1918 influenza pandemic.

Meanwhile the United States declared a public health emergency on April 26. Also, the first U.S. fatality was reported: a 23-month-old child visiting Mexico. The child died at a Houston, Texas, hospital. On April 27, Spain declared the first confirmed case of H1N1 in Europe. Hong Kong, Thailand, Singapore, Malaysia, Vietnam, and Indonesia issued travel advisories against travel to Mexico. Countries in Southeast Asia, Russia, India, and North and South America initiated airport screenings. The CDC recommended that U.S. citizens avoid all nonessential travel to Mexico. The European Union Health Commissioner Androulla Vassiliou recommended that individuals postpone nonessential travel to affected parts of the United States and Mexico. Argentina, Ecuador, Peru, and Cuba closed their boarders for travel to and from Mexico. The Philippines, China, and Indonesia banned the importation of pork from Mexico and certain states within the United States (e.g., border states like Texas and California but also Kansas). By the next day, there were confirmed or suspected cases of H1N1 in the United States, Canada, New Zealand, the United Kingdom, and Spain. By the end of April, local officials closed schools in New York City and Forth Worth, Texas, school districts due to suspected H1N1 cases. Egyptian leaders ordered the slaughter of more than 300,000 pigs farmed by the Coptic Christian minority in Egypt. Violence erupted in Cairo as Christian pig farmers clashed with the police.

The first wave of influenza continued in 2009. On June 11, WHO Director Dr. Margaret Chan declared the world situation an H1N1 pandemic that was "unstoppable." She also stated that this influenza strain was a very different virus than what we had been used to from season to season. The 2009 H1N1 influenza virus spread rapidly from person to person. It also targeted unusual risk groups: young people (aged 6 months to 19 years), children with neuromuscular diseases, pregnant women, and the obese. The H1N1 virus spread rapidly through frequent international travel. Acting Director of the CDC's Emergency Response Team Dr. Richard Besser served as the public face of the

FIGURE 1-16 All restaurants were closed except for take-out orders during the H1N1 outbreak in Mexico City, Mexico (April 29, 2009).

CDC's response to H1N1. As of June 12, 2009, the virus had spread to 74 countries around the world. Over 29,000 cases were reported, including 145 deaths. By July 1, H1N1 cases were reported in all 50 states. It was estimated that more than 1 million people had been infected.

Within a month of this outbreak, sequence data for the new virus became available in public-access databases. A report was published in *ScienceExpress* on May 22, 2009. Researchers discovered the H1N1 influenza virus contained a unique combination of gene segments from both classic Northern American and Eurasian pig influenza strains as well as gene segments from human and avian influenza strains. It was a "mutt" of a virus. The most critical gene segments that pandemic strains contain are novel H and N gene segments (refer to Chapter 12 for further explanation of influenza virus genes and pandemic strains). The 2009 H1N1 virus contained an H gene from classic American pig influenza viruses and the N gene from a Eurasian pig influenza virus. Three gene segments of the 2009 virus shared common sequences from the 1918 pandemic influenza strain. *It contained the genetic characteristics of a pandemic strain and could spread easily from person to person.*

The main challenge for scientists and health officials was to assess how severe the pandemic might be. Researchers were concerned that a second wave of influenza during the fall flu season could be of the same severity experienced in 1918. Like the 1918 influenza virus, young people were dying from influenza or complications of it, making it different from typical seasonal influenza that causes complications in infants, the immune compromised, and the elderly resulting in death. In 1918, a milder influenza appeared in the spring of the year that returned in the fall with a vengeance, making it likely the deadliest influenza strain in history. Having had two "near misses," the emergence of severe acute respiratory syndrome caused by a new coronavirus (2002–2003) and the spread of H5N1 avian influenza in the Middle East, nations began working diligently on pandemic action plans in 2005. These plans provided guidance, resources, and checklists intended to help every sector of society to reduce the impact of a pandemic on businesses, hospitals, schools, and the community. Global vaccine manufacturing and stockpiling of influenza antivirals such as Tamiflu became a priority. Regulatory authorities licensed pandemic influenza vaccines for a number of countries, including the United States and Canada. Healthcare workers were given first priority to early vaccination. Other high-risk groups were next on the list for vaccination: pregnant women and children (refer to Chapter 12, Influenza Viruses, about the 2009 seasonal and H1N1 vaccines). Healthcare facilities stockpiled medications, masks, gowns, and other supplies. Some hospitals suspended the use of student volunteers to limit patients' possible exposure to the H1N1 virus since the influenza virus spreads quickly in academic and university settings.

The 2009 H1N1 virus started a second wave in the fall. On October 24, President Obama declared H1N1 influenza a state of national emergency. This waived certain regulatory requirements for healthcare facilities in response to emergencies and access to the experimental drug peramivir to treat severe cases of H1N1. Responding to a pandemic in the 21st century had some technological advantages. The media responded with public service announcements. It updated the international situation; educated the public, especially on cough etiquette, frequent handwashing, school closings, staying home when sick, and the availability of seasonal and H1N1 vaccine; and corrected disinformation (e.g., it was safe to eat pork). Internet sites contained accurate up-to-date information for citizens and healthcare professionals (**TABLE 1-2**). Some Website H1N1 health maps and cell phone

TABLE 1-2 Pandemic Influenza Resource and Surveillance Information

Internet Resources	URL
Comprehensive U.S. Government Information	http://www.flu.gov/
CDC Seasonal Flu Information	http://www.cdc.gov/flu/
WHO Global Alert and Response: Influenza	http://www.who.int/csr/disease/influenza/en/
European Commission: Influenza Information	http://ec.europa.eu/health/ph_threats/com/Influenza/h1n1_en.htm
Pandemic Influenza: Canada	http://www.pandemicflu.ca/
Public Health Agency of Canada: Influenza	http://www.phac-aspc.gc.ca/influenza/index-eng.php
Google Flu Trends Around the World	http://www.google.org/flutrends/
Tracking the Progress of H1N1 Swine Flu	http://flutracker.rhizalabs.com/

applications tracking influenza cases were created. Users could track outbreaks reported in their region and set alerts notifying a user on their device by email when new cases were reported in their region (Table 1-2).

Like in 1918, different snake oil or quack remedies were advertised. The fear of an H1N1 pandemic resulted in Internet advertisements for a plethora of fraudulent products that cure, treat, or prevent H1N1 infection. Products included shampoos, air purifiers, herbal supplements, inhalers, and even body washes. The Food and Drug Administration developed a comprehensive list of websites listing the companies selling various gels, kits, supplements, sprays, and other unauthorized products that made unsubstantiated claims about H1N1 protection or treatment. More on the epidemiology of the H1N1 pandemic is found in Chapter 12.

Severe Acute Respiratory Syndrome (SARS), 2002–2003

The SARS coronavirus (SARS-CoV) emerged from the Guangdong Province of southern China during November and December of 2002. The first infected individuals handled, butchered, or sold food animals, or prepared and served food. They experienced influenza-like symptoms during the first week of illness: fever greater than 100.4°F (38°C) for more than 24 hours, headache, and body aches. During the second week of illness, sick individuals initially developed a dry cough and many developed diarrhea. Most sick individuals rapidly deteriorated to an **atypical pneumonia**. The fever followed by the rapid progressive respiratory compromise were the key signs and symptoms from which the syndrome derived its name as severe acute respiratory syndrome (SARS).

On February 16, 2003, a nephrologist (physician specializing in kidney diseases) working at a Chinese hospital began experiencing the early signs and symptoms of SARS. Five days later, the physician traveled from Guangzhou to Hong Kong, Special Administrative Region of China, and stayed at a hotel there. The physician felt well enough to sightsee and shop with his brother-in-law for 10 hours during the day of his arrival, but the next day sought medical attention. He was directly admitted to the intensive care unit (ICU) of a hospital with respiratory failure. He later died.

The next person to get sick was his brother-in-law. Subsequently, a nurse in the accident and emergency department at the same hospital became ill. The nurse was present in the same resuscitation room as the nephrologist who died but had no direct contact with him and she was wearing a surgical mask at the time. The fourth person to get sick was a Chinese-Canadian businessman returning to a family reunion in Hong Kong. His stay at the hotel overlapped with that of the nephrologist for one day. There was no direct contact between the businessman and the nephrologist in the common areas of the hotel. The businessman was later admitted to a different hospital than the nephrologist. The next three individuals to contract SARS were nurses who had close encounters with the Chinese-Canadian businessman. They had cleaned him after an episode of diarrhea. These nurses did not wear masks or gowns during their routine care of any patients in the hospital ward. *The SARS-CoV was efficiently transmitted in the healthcare setting among healthcare workers, patients, and hospital visitors.*

Well-documented outbreaks of SARS transmission occurred in hospitals located in Canada, China, Hong Kong, Singapore, Taiwan, and Vietnam. The concept of "**super spreading**" was proposed to explain incidents where a SARS patient infected many more persons than would normally be expected. What made SARS notorious—in contrast to other infectious diseases like influenza—was its *propensity to cause hospital outbreaks*. Healthcare workers accounted for 21% of all SARS cases during this 2002–2003 outbreak. SARS transmission studies led to a new approach to manage patients that was termed **"respiratory hygiene/cough etiquette."** SARS-CoV also infected a mobile population of people who were able to travel for several days before the onset of severe symptoms. Other emerging viral infections such as Ebola, hantavirus pulmonary syndrome, and Nipah do not spread by travelers as rapidly as SARS because individuals experience severe symptoms quickly, making them less able and likely to travel.

On March 15, 2003, the WHO issued a travel advisory that included emergency guidelines for travelers and airlines. The outbreak spread to 28 regions around the world, resulting in 8096 cases and 774 deaths (**TABLE 1-3**). The SARS outbreak is a good example of how modern technology enabled laboratory investigators to determine that SARS was caused by a new coronavirus and not a novel influenza virus or an agent of bacterial origin such as *Mycoplasma pneumoniae*, *Chlamydia pneumoniae*, or *Legionella pneumophila*. Refer to Chapter 3, Virus File 3-1 for "The Race to Characterize SARS-CoV," Chapter 5, Virus File 5-2 for the "Rapid Diagnosis of SARS," and Chapter 12, section 12.15 regarding "Lessons Learned from the Severe Acute Respiratory Syndrome (SARS) Outbreak."

TABLE 1-3 Summary of SARS Cases (November 1, 2002–July 31, 2003)*

Location	Female	Male	Total Cases	Number of Deaths	Average Age	Date(s) of First and Last Probable Cases
China	2674	2607	5327**	349	N/A	Nov. 16, 2002–June 3, 2003
China, Hong Kong Special Administrative Region	977	778	1755	299	40 (0–100)	Feb. 15–May 31, 2003
Vietnam	39	24	63	5	43 (20–76)	Feb. 23–April 14, 2003
Canada	151	100	251	43	49 (1–98)	Feb. 23–June 12, 2003
United States	13	14	27	0	36 (0–83)	Feb. 24–July 13, 2003
China, Taiwan	218	128	346***	37	42 (0–93	Feb. 25–June 15, 2003
Singapore	161	77	238	33	35 (1–90)	Feb. 25–May 5, 2003
Philippines	8	6	14	2	41 (29–73)	Feb. 25–May 5, 2003
Australia	4	2	6	0	15 (1–45)	Feb. 26–April 1, 2003
Republic of Ireland	0	1	1	0	56	Feb. 27, 2003
United Kingdom	2	2	4	0	59 (28–74)	March 1–April 1, 2003
Switzerland	0	1	1	0	35	March 9, 2003
Germany	4	5	9	0	44 (4–73)	March 9–May 6, 2003
Thailand	5	4	9	2	42 (2–79)	March 11–May 27, 2003
Italy	1	3	4	0	30.5 (25–54)	March 12–April 20, 2003
Malaysia	1	4	5	2	30 (26–84)	March 14–April 22, 2003
Romania	0	1	1	0	52	March 19, 2003
France	1	6	7	1	49 (26–61)	March 21–May 3, 2003
Spain	0	1	1	0	33	March 26, 2003
Sweden	3	2	5	0	43 (33–55)	March 28–April 23, 2003
Mongolia	8	1	9	0	32 (17–63)	March 31–May 6, 2003
South Africa	0	1	1	1	62	April 3, 2003
Indonesia	0	2	2	0	56 (47–65)	April 6–17, 2003
Kuwait	0	1	1	0	50	April, 9, 2003
New Zealand	1	0	1	0	67	April 20, 2003
India	0	3	3	0	25 (25–30)	April 25–May 6, 2003
Republic of Korea	0	3	3	0	40 (20–80)	April 25–May 10, 2003
China, Macao Special Adminstrative Region	0	1	1	0	28	May 5, 2003
TOTALS			**8096**	**774**		

*WHO statistics

**Case classification by sex is unknown for 46 cases.

N/A (not available)

***Since July, 11, 2003, 325 cases have been discarded in Taiwan, China. Laboratory information was insufficient or incomplete for 135 discarded cases, of which 101 died.

Source: Table adapted from: http://www.who.int/csr/sars/country/table2004_04_21/en/index.html

The mystery as to the origin of the SARS-CoV is still not fully resolved. Early investigations suggested the virus originated from animals because the first persons infected were traders or animal food handlers at live markets, restaurant workers, and butchers of **exotic animals** for culinary purposes. These investigations discovered that blood specimens of wild animal traders at wholesale markets in Guangzhou had significant levels of **antibodies** against the SARS-CoV compared to vegetable traders. The presence of antibodies indicated the traders had been infected by the SARS-CoV probably through capturing and marketing wild animals. For this reason, scientists concentrated their investigations toward animals being sold at markets as a likely source of the virus.

Traders who only engaged in civet cat trading were much more likely to have been infected with SARS CoV than traders who only engaged in snake or fowl marketing. All evidence pointed toward masked palm civet cats as playing a role in the transmission of SARS CoV. It led to a temporary ban on the hunting, sale, transportation, and export of all wild animals in the Guangdong Province. Over 10,000 civets, badgers, and raccoon dogs were **culled** to prevent the spread of SARS-CoV.

In the wild, masked palm civets are arboreal, taking shelter in hollow trees in the mountain and hill forests of China. They are solitary and nocturnal predators; the female can bear young (litter of 1–4) twice per year. Civets eat mainly fruits but will also eat rodents, birds, insects, and roots. The farming of masked palm civets started in the late 1950s. Breeding the civets for use as exotic food became popular in the late 1980s. In 2003, there were about 40,000 masked palm civets, raised in 660 farms all over China. The farms started from either the capture and breeding of local wild civets or from breeding civets brought in from other farms. Scientists studying farmed civet cats reported few diseases among them. Several research groups suggested that SARS-CoV entered the human population through transmission by infected civet cats but that they were not the natural **reservoir** of SARS-CoV. In 2005, two independent groups of researchers isolated and identified SARS-CoV from Chinese horseshoe bats in China or Hong Kong. They trapped hundreds of bats in their natural habitats for **zoonotic pathogens**. Their logic for testing bats was that these mammals are often persistently infected with many viruses (e.g., Nipah, Hendra, Ebola viruses) but never exhibit symptoms. In other words, they are healthy carriers of certain viruses. Many people eat bats or use bat feces in traditional medicine for asthma, kidney ailments, and general malaise in Southeast Asia. Sanitation in the live markets is often lacking. Bats, civets, and other wild animals were often in the same cages at food or traditional medicine markets, contributing to the conditions that led to the SARS outbreak.

Foot and Mouth Disease, 2001

Anyone who was in Great Britain in 2001 will remember one of the world's worst foot and mouth disease (FMD) epidemics. Tourists were prompted to follow these rules:

Do not go near cattle, sheep, pigs, or deer and never feed farm animals.

Do not go on farmland that has been or is being used by livestock.

Do not attempt to walk on footpaths that are closed.

Five days before returning home from FMD-infected areas, people were asked to avoid farms, zoos, sale barns, stockyards, animal laboratories, meatpacking plants, fairs, and other animal facilities. It was suggested to bathe and launder or dry-clean all clothing, including outerwear, and to remove any dirt or soil from shoes, followed by wiping them with a cloth that was dampened with a bleach solution.

The media stories showed photographs of individuals bleaching their shoes before entering the airports. All luggage and personal items (including watches, cameras, laptop computers, CD players, and cell phones) were supposed to be cloth-dampened with a bleach solution. Once inside the airport, travelers were questioned about where they had been during their stay and if they were carrying any food products of pigs or ruminants.

Epidemiologists determined how FMD spread to farms and the market. The outbreak began at Burnside farm, Heddon-on-the-Wall, Northumberland, England. The FMD outbreak spread to nearby farms, markets, and meat-packing facilities via vehicles transporting infected animals (**FIGURE 1-17**). Road sites were set up with automatic vehicle-spray devices to contain the spread of FMD (**FIGURE 1-18a**).

Foot and mouth disease is highly contagious among animals and causes loss of production that can have grave economic implications for the meat and dairy industries (Figure 1-18b). FMD has occurred in most parts of the world, causing extensive epidemics in cloven-hoofed animals (that is, cattle, pigs, sheep, goat, and deer). The last outbreak of FMD in the United States occurred in 1929. Slaughtering infected animals or animals in contact with infected animals, vaccination, and public cleansing and disinfection centers can control the disease. Agricultural officials use serology testing to monitor healthy animals for FMD antibodies.

Cases of FMD in humans are rare, and the symptoms are few and/or mild. Human cases have resulted from close contact with animals suffering from FMD, with the virus entering through broken skin, being ingested in unpasteurized milk, or being inhaled. FMD has not been beaten. The Ministry of Agriculture, Fisheries, and Food in England prefers to slaughter livestock during an outbreak instead of vaccinating, and as a result the scourge of FMD continues. Refusal to vaccinate is driven by the market. Animals that test positive for FMD antibodies cannot be sold to other countries. Vaccination would provide immunity to these animals, but because they would then test positive for FMD antibodies it depreciates their market value. This is a controversial issue affecting

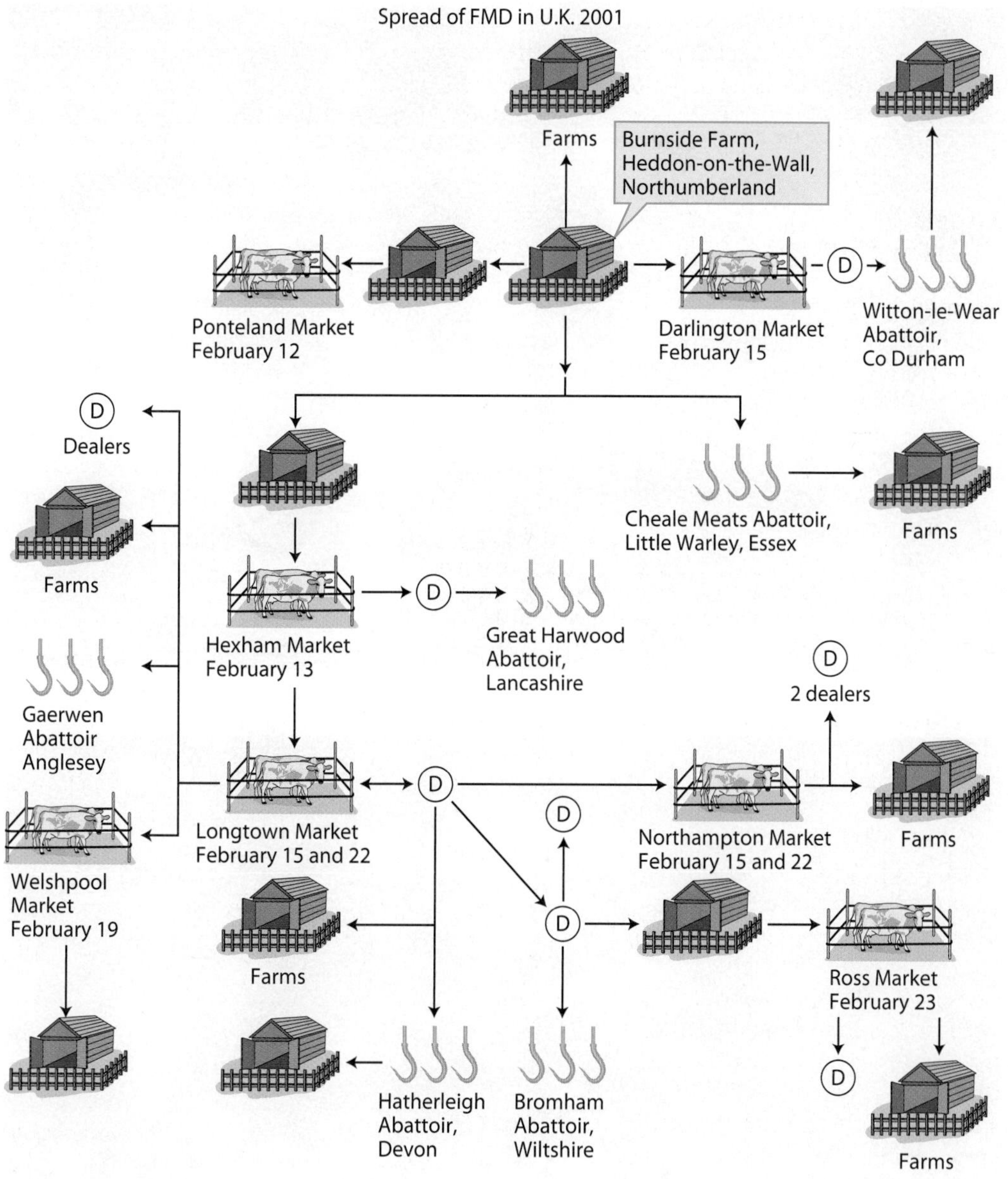

FIGURE 1-17 Epidemiologists determine how FMD spread to farms and the market during the 2001 outbreak in England.

farmers, politicians, scientists, veterinarians, and the consumer.

Hantavirus: Four Corners Disease, 1993

On May 14, 1993, a Native American marathon runner in rural New Mexico known to be in excellent health collapsed and died of respiratory failure at an Indian Health Service Hospital emergency room. Days before his collapse he visited a physician twice with flulike symptoms, but his chest x-ray was normal. He was treated with antibiotics and acetaminophen. Two days before he fell ill, his fiancé died of the same mysterious respiratory illness. Both victims died from fluid buildup in their lungs. Normally with each breath, the air sacs (aveoli) of the lungs take in oxygen and release carbon dioxide. In these circumstances, the lung aveoli were filled with fluid instead of air, preventing oxygen from being absorbed into the bloodstream, resulting in death. On May 17, the Indian Health Service reported five similar deaths. All of the individuals were previously very healthy. By June 11, 24 cases of respiratory failure following flulike symptoms were reported in the Four Corners region of Colorado, Utah, Arizona, and New Mexico.

This virulent new disease baffled medical examiners and the CDC. The CDC finally identified the virus as a strain of the hantavirus, an "old world" virus carried by rodents known to only cause hemorrhagic fever and kidney failure. *Old world* refers to those parts of the world known before the voyages of Christopher Columbus. It includes Europe, Asia, and Africa and its surrounding islands. Hantavirus pulmonary syndrome is a rare disease that is sometimes caused by the hantavirus called sin nombre virus (no-name virus in

FIGURE 1-18 (a) This site was located on the edge of the restricted area in Northumberland, England, in 2001. **(b)** The British Ministry's strategy for containment of FMD is to slaughter and burn the infected animals. This photograph shows smoke rising from pyres of burning cows slaughtered to prevent the spread of FMD.

a

b

English). Several different types of wild mice, such as deer mice and rats, can be infected with hantavirus and pass it in their droppings, urine, or saliva. Once humans come in contact with fomites such as contaminated blankets or food storage areas, the virus then enters directly into the respiratory system through breathing contaminated air particles. It kills 50% of its victims.

Why did sin nombre virus strike the Four Corners region in 1993? It is likely that the virus had long been present in rodent populations. Officials believe that the unusually mild winter and spring of 1993, the rainfall, and the abundance of pinyon nuts on which the rodents feed led to an increased rodent population in the summer and greater opportunities for people to come in contact with infected rodents. More than 20% of deer mice subsequently captured in southwestern areas of the United States, and as far north as southwestern Montana, tested positive for sin nombre virus (**FIGURE 1-19**). The detective work of the CDC led to a course of action, treatment, and prevention, thus thwarting a nationwide epidemic.

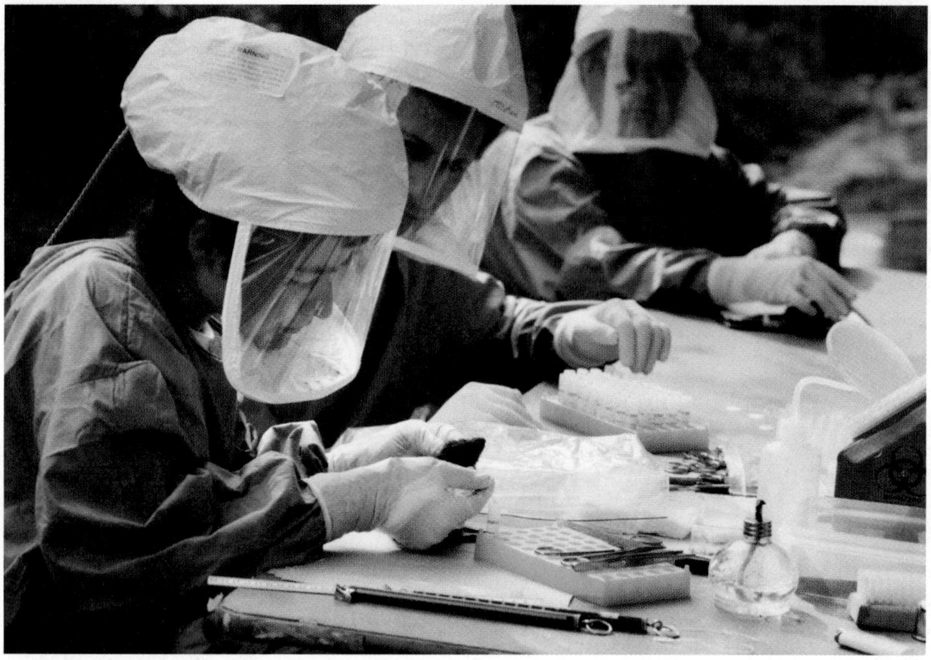

FIGURE 1-19 Rodents that carry the sin nombre virus can be deceptively cute, but suspected carriers must be handled with extreme care. This photograph shows scientists wearing protective gear while collecting and analyzing deer mice.

West Nile Virus, 1999

The next time you are outdoors where there are mosquitoes, you may change your behavior after reading this section. You might wear clothing with long sleeves and apply a bug repellent that includes the ingredient DEET (meta-N, N-diethyl toluamide) over any exposed skin. You may find yourself removing any standing water in your neighborhood (for example, water that has accumulated in old tires, buckets, and wheelbarrows). You may start reporting any dead birds you find.

During the summer of 1999 elderly people in Queens, New York, were getting sick with headaches, fever, weakness, and confusion. Routine tests from the medical laboratory came back negative for bacterial or fungal pathogens. Given the symptoms of the patients, it was suspected that they were suffering from **encephalitis** (inflammation of the brain) of viral origin. Samples of blood, spinal fluid, and tissues from those who had died after falling ill were shipped to the Vector-Borne Diseases Laboratory at the CDC for analysis. Based on a routine antibody test, the disease was first diagnosed as encephalitis caused by the St. Louis Encephalitis Virus (SLE). The positive test result, however, was a very weak reaction. Nevertheless, helicopters began hovering over major highways and residential areas, spraying misty clouds of insecticides such as malathion and pyrethroids. The City Health Commissioner advised individuals to avoid contact with mosquitoes and exposure to insecticides.

Simultaneously, people started finding more than the usual number of dead crows in the area, and several exotic species of birds at the Bronx Zoo died. Birds don't die from SLE. They can, however, die from West Nile Virus (WNV). The CDC did not test the samples for WNV because that virus had never appeared in the United States before. No one had made the connection between the human epidemic and the wildlife epidemic.

As more cases appeared, the CDC conducted additional tests on the brain tissue of deceased individuals, the dead crows, and exotic birds from the zoo. Officials at the CDC attempted to isolate and study the virus taken directly from these tissues. At the same time, the laboratory at the New York State Department of Health in Albany began sequencing the RNA genome of the virus they had isolated from infected brain tissues. These results were shared among the scientific community via a special Internet site called ProMED (Program for Monitoring Emerging Diseases). ProMED shares results about epidemics occurring anywhere in the world, and scientists and physicians who specialize in infectious diseases visit the site on a daily basis. The ProMED post of their results matched with outbreaks of encephalitis in Romania, Egypt, Israel, Italy, and South Africa. It matched exactly with a strain of WNV in Israel from an epidemic in 1998. This turned out to be the final piece of the puzzle. The disease is spread via mosquitoes that have fed upon sick birds that harbor WNV. The mosquitoes then acquire WNV from the birds and transmit the disease via biting humans or animals.

The final toll from the 1999 WNV epidemic in the New York City metropolitan was 62 cases of encephalitis and 7 deaths. The virus was believed to have infected as many as 1900 unknowing Queens residents who did not develop encephalitis. Exotic zoo birds, American crows, at least 20 other North American wild bird species, and horses were infected, and quite a few died. Why did the Israeli strain of WNV enter the United States? Scientists have suggested several reasons why WNV enters new regions. The most supported hypothesis is that migratory birds play a role in transporting the virus. Other hypotheses are:

- During migration, birds can be displaced by storms.
- Legal and illegal importation of birds (bird and pet trade).
- Birds are vagrants on ships or other transportation.

Most mammals become infected by mosquito bites. The virus is transmitted when the mosquitoes are feeding (taking a bloodmeal). Most non-bird species, including humans and horses, are dead-end hosts and do not transmit WNV to mosquitoes. Mosquitoes will feed on alligators resulting in an infection. However, outbreaks on alligator farms have also been linked to alligators being fed WNV-infected horse meat and ground beef contaminated with feces from WNV-infected wild birds. Reports have shown that birds shed WNV through their **cloaca**. The cloaca is a cavity at the end of the digestive tract into which the intestinal, genital, and urinary tracts open. It is found in birds, reptiles, most fish, and amphibians. Direct transmission has been reported between alligators in close contact, possibly by fecal contamination. Alligators shed WYN in their feces. Chipmunks and squirrels also secrete WNV in feces and urine, and might be able to spread WNV horizontally (**FIGURE 1-20**). Humans do not shed WNV in their secretions or excretions, but the virus can be transmitted by blood transfusions and in organ transplants. **TABLE 1-4** lists species found positive for WNV through surveillance efforts.

Norovirus Outbreaks, 1972 and 2002

Today, noroviruses (formerly called Norwalk-like viruses) may be the most commonly identified cause of infectious intestinal diseases. Much publicity has been given to outbreaks on cruise ships,

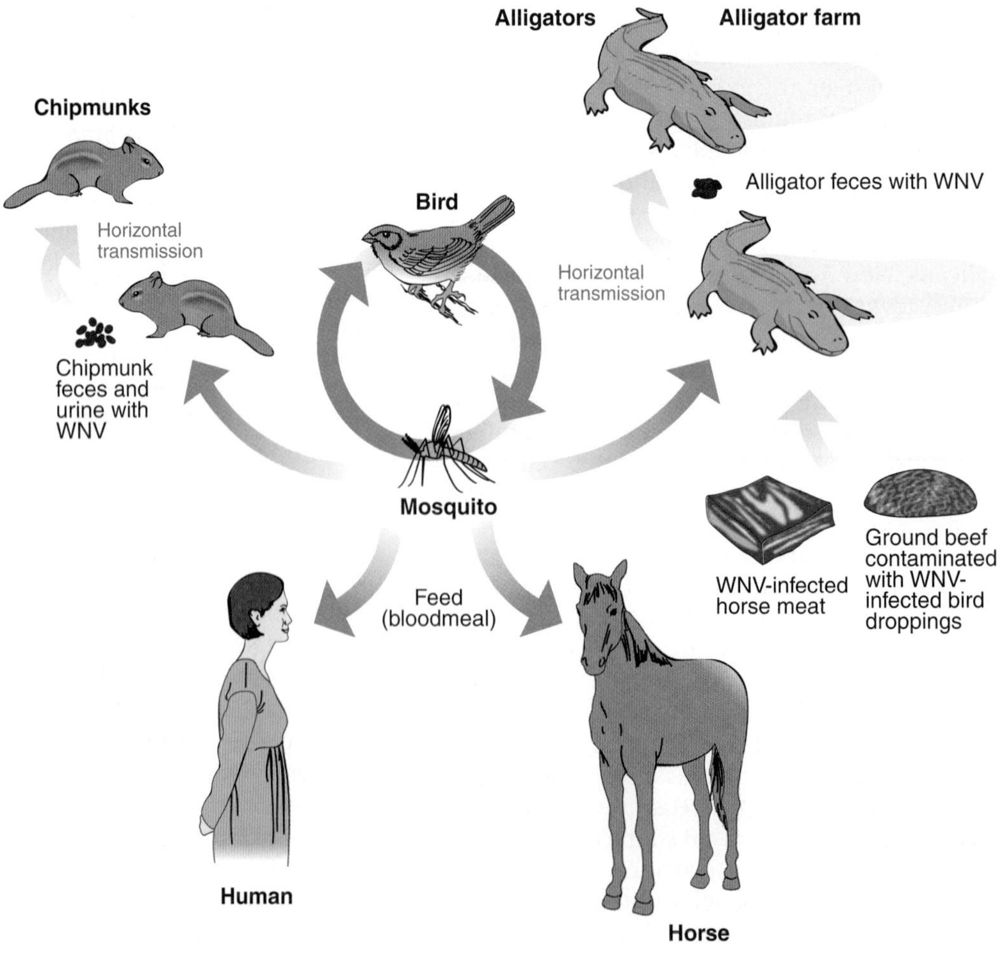

FIGURE 1-20 Diagram showing how WNV spreads from infected birds to mosquitoes to humans or other hosts such as horses, alligators, and chipmunks. Horizontal transmission may occur in chipmunks and alligators through fecal shedding.

TABLE 1-4 Species Found Positive for West Nile Virus in Surveillance Efforts

Birds	White-crowned pigeon	Demoiselle crane (c)(a)
Wood duck	Rock dove (feral pigeon)	West African crowned (a)
Eurasian wigeon (c)	Mauritius pink pigeon (c)(a)	Wattled crane (c)(a)
Mallard	Common ground-dove	Whooping crane (c)(a)
Bronze-winged duck (spectacled duck) (c)	Eurasian collared-dove	Mississippi sandhill crane (c)
Domestic goose (c)(a)	White-winged dove	Red-crowned crane (c)(a)
Canvasback	Mourning dove	Siberian crane (c)(a)
Canada goose	Luzon pigeon (bleeding heart pigeon) (c)(a)	Hooded crane (c)(a)
Barnacle goose (c)(a)	Inca dove	White-naped crane (c)(a)
Emperor goose (c)	Belted kingfisher	Black-necked crane (c)(a)
Greater Magellan goose (Andean goose) (c)(a)	Yellow-billed cuckoo	Virginia rail
Abyssinian blue-winged goose (c)(a)	Cooper's hawk	Lady Ross' turaco (lantain-eater) (c)(a)
Tundra swan (c)	Northern goshawk	Cedar waxwing
Trumpeter swan (c)(a)	Sharp-shinned hawk	Northern cardinal
Mute swan	Golden eagle	Blue grosbeak (a)
Rosy-billed duck (c)(a)	Red-tailed hawk	Rose-breasted grosbeak
Ruddy duck	Rough-legged hawk (c)	Dickcissel
Chimney swift	Red-shouldered hawk	Western scrub
Ruby-throated hummingbird	Broad-winged hawk	American crow
Common	Swainson's hawk	Common raven
Emu (c)	Northern harrier	Fish crow
Ruddy turnstone	Swallow-tailed kite	Blue jay
Killdeer	Bald eagle	Steller's jay
Piping plover	Mississippi kite	Black-billed magpie (c)
Herring gull	Osprey	Song sparrow
Laughing gull	Harris' hawk (c)	Savannah sparrow
Ring-billed gull	Merlin	Fox sparrow
Great black-backed gull	Prairie falcon	Eastern towhee field sparrow
Black skimmer	Peregrine falcon	Zebra finch (c)
Grey gull (c)(a)	American kestrel	American goldfinch
Inca tern (c)(a)	Crested guineafowl	House finch
Yellow-crowned night-heron (c)	Northern bobwhite	Purple finch
Black-crowned night-heron (c)	Chukar (c)(a)	Evening grosbeak
Great blue heron	Ruffed grouse	European goldfinch (c)
Green heron	Domestic chicken (red junglefowl) (c)	Barn swallow
Least bittern	Green junglefowl (c)(a)	Purple martin
Turkey vulture	Impeyan (Himalayan) pheasant (c)	Tree swallow
Black vulture	Bulwer's wattled pheasant (c)(a)	Red-winged blackbird
King vulture (c)(a)	Turkey (domestic and wild)	Rusty blackbird
Saddle-billed stork (c)(a)	Ring-necked pheasant	Brewer's blackbird
Marabou stork (c)(a)	Mount peacock (c)(a)	Baltimore oriole
Lesser Adjutant stork (c)(a)	Crested partridge (c)(a)	Brown-headed cowbird
Chilean flamingo (c)	Blyth's tragopan (c)	Boat-tailed grackle
Greater flamingo (American) (c)	Argus pheasant (c)(a)	Great-tailed grackle
Scarlet ibis (c)	Greater sage grouse	Common grackle
Waldrapp (c)(a)	Common loon	Loggerhead shrike

(continued)

TABLE 1-4 Species Found Positive for West Nile Virus in Surveillance Efforts (continued)

Gray catbird	Eastern kingbird	**Mammals**
Northern mockingbird	Black-whiskered vireo	Domestic cattle (c)
Brown thrasher	Warbling vireo	Mountain goat (c)
Tufted titmouse (c)	Red-eyed vireo	Domestic (Suffolk) sheep (c)
Black-capped chickadee	American white pelican	Llama (c)
Carolina chickadee	Brown pelican (c)(a)	Alpaca (suri) (c)
Black-throated blue warbler	Double-crested cormorant	White-tailed deer
Yellow-rumped warbler	Guanay cormorant (c)	Reindeer (c)
Yellow warbler	Red-headed woodpecker	Mule deer
Blackpoll warbler	Downy woodpecker	Babirusa (c)(a)
Common yellowthroat	Yellow-bellied sapsucker	Domestic dog (c)
Kentucky warbler	Pied-billed grebe	Timber wolf (c)
Northern parula	Cockatoo (unspecified) (c)	Domestic cat (feral)
Ovenbird	Cockatiel (c)	Striped skunk
Northern waterthrush	Red-crowned parrot (c)	Harbor seal (c)
Nashville warbler	Macaw (unspecified) (c)	Red panda (c)(a)
Canada warbler	Budgerigar (c)	Black bear (a)
Hooded warbler	Lorikeet (c)	Big brown bat
House sparrow	Black-footed penguin (c)	Little brown bat
White-crested laughingthrush (c)(a)	Magellan penguin (c)(a)	Domestic rabbit (c)
White-breasted nuthatch	Northern saw	Domestic horse (c)
European starling	Boreal owl (c)	Donkey (c)
Palm tanager (c)	Short-eared owl	Mule (c)
Carolina wren	Verreaux's eagle owl (milky eagle owl) (c)(a)	Great Indian rhinoceros (c)(a)
Winter wren	Great horned owl	Barbary macaque (c)
Veery	Snowy owl (c)	Ring-tailed lemur (c)
Hermit thrush	Eastern screech owl	Indian (Asian) elephant (c)(a)
Gray-cheeked thrush	Tawny owl (c)	Gray squirrel
Swainson's thrush	Great gray owl (c)	Fox squirrel
Wood thrush	Spotted owl (c)	Eastern chipmunk
Eastern bluebird	Barred owl	Human
American robin	Northern hawk owl (c)	**Reptiles**
Traill's flycatcher	Barn owl	American alligator (c)
Eastern phoebe	Ostrich (c)(a)	Crocodile monitor (c)(a)
Scissor-tailed flycatcher		

Note: Species included in this list are wild animals unless followed by a "c," which denotes either a captive or farmed animal(s). Virus or viral RNA was detected in animal tissue unless followed by an "a," which denotes detectable antibodies only have been reported.

Source: Adapted from National Wildlife Health Center. Madison, Wisconsin. Updated 9/2004.

but these viruses also pose problems in hospitals, nursing homes, schools, and homes. Norwalk-like viruses were first characterized and described in 1972 during an epidemic of diarrhea and vomiting involving students at an elementary school in the town of Norwalk, Ohio. These viruses thrive in places where humans are in high concentrations. Cruise ships happen to provide the perfect breeding environment for viruses: they provide an enclosed area with up to several thousand people in close contact with each other.

The virus is spread person to person via feces and sometimes by eating contaminated food (especially oysters) or drinking contaminated water. The intestinal illness is usually short lived (2–3 days). Norovirus outbreaks reached an all-time high in the

United States in 2002. The most important thing that anyone can do to keep from getting sick is to wash his or her hands. By frequently washing your hands, you wash away germs that you have picked up from other people, or from contaminated surfaces, or from animals and animal waste.

1.5 Smallpox: Ancient Agent of Bioterrorism

The use of biological agents in terrorism is not new. It dates back as early as the 6th century, when the Assyrians poisoned the wells of their enemies with rye ergot. Warring tribes would catapult diseased carcasses over castle walls to infect their enemies. On several occasions, smallpox was used as a biological weapon. Pizarro presented South American natives with clothing and gifts laden with the smallpox virus in the 15th century and Sir Jeffrey Amherst provided Native Americans (who were loyal to the French) with blankets tainted with smallpox during the French and Indian War (1754–1763). The epidemic killed most of the tribe, resulting in a successful British attack on Fort Carillon.

The term *smallpox* was introduced during the 16th century to distinguish it from the "great pox," or syphilis. Smallpox has been one of humankind's greatest scourges. It is believed that this virus appeared in agricultural settlements in Africa around 10,000 B.C. The disease left many disfigured with pox scars; some were blinded and many died. It nearly wiped out the Native American population in the United States, which had been free of infectious diseases before the Europeans came. Native Americans were not resistant to smallpox and other European diseases because they had never been exposed to those viruses. Yet few doctors today have seen a case of smallpox.

In 1967, the WHO launched an aggressive program to eradicate smallpox. It was very cost effective. In 1977, the last naturally occurring case was reported in Somalia. In 1980, the WHO declared the world free of smallpox and recommended to cease vaccination. Today we are once again vulnerable to this disease. It is speculated that immunity is not lifelong and many individuals have not been vaccinated. (Read more on smallpox in Chapter 14.)

Other Bioterrorism Concerns

Bioterrorism is still a clear and present danger. A series of anthrax cases, some of which resulted in death, occurred during the weeks following the terrorist attacks on the United States of September 11, 2001. As a result there has been increased security and concern over the possibility of additional attacks. The scientific community is working as a team to be prepared for any future attacks. This preparedness involves the production of vaccines, drugs, and diagnostic tests that will play a major role in reducing the threat of bioterrorism.

Summary

Viruses impact all forms of life. They play an important role in ecosystems by affecting population growth in both positive and negative ways. The origin of viruses remains a debatable topic. Early pioneers of virology studied bacteriophages, plant viruses such as TMV, and smallpox virus that caused large numbers of outbreaks and mortalities throughout recorded human history.

Gene therapy, bacteriophage therapy, and vaccine development utilize viruses in applications toward health problems. Influenza, poliomyelitis, AIDS, foot and mouth disease, West Nile encephalitis, hantavirus pulmonary syndrome, smallpox, and norovirus-related gastrointestinal distress are examples of viral diseases that have had historical and recent relevance. Biological weapons have been used for centuries. Disease-causing viruses can be effective weapons used to infect and potentially kill large populations. Understanding the biology of viruses will help the scientific community prepare for a biological attack. The production of vaccines, drugs, and diagnostic tests will play a major role in reducing the threat of bioterrorism.

CASE STUDY QUESTIONS

These questions relate to the Case Study presented at the beginning of the chapter.

1. To find general information about viral hemorrhagic fevers via the World Wide Web, what search engines and sites would you use? What keywords would you use in your searches?
2. How would you find research papers pertaining to viral hemorrhagic fever? List your resources and the titles of the journal articles. Where was the research conducted?

Resources

Primary Literature and Reviews

Asnis, D., et al. 1999. "Outbreak of West-Nile-like viral encephalitis—New York." *MMWR* 48(38):845–849.

Ball, K. 2009. "The enigma of the H1N1 Flu: Are you ready?" *AORN J* 90:6;852–866.

Berger, A., et al. 2004. "Severe acute respiratory syndrome (SARS)—Paradigm of an emerging viral infection." *J Clin Virol* 29:13–22.

Boyer, M., et al. 2009. "Giant Marseillevirus highlights the role of amoebae as a melting pot in emergence of chimeric organisms." *PNAS* doi/10.1073/pnas.0911354106.

Bratbak, G., et al. 1990. "Viruses as partners in spring bloom microbial tropho-dynamics." *App Environ Microbiol* 56(5):1400–1405.

Brehman, J. G., and Henderson, D. A. 1998. "Poxvirus dilemmas—Monkeypox, smallpox, and biologic terrorism." *N Engl J Med* 339(8):556–559.

Carlton, R. M. 1999. "Phage therapy: Past history and future prospects." *Archives Immunologiae et Therapiae Experimentalis* 47:267–274.

Chang, L. Y., et al. 2009. "Novel swine-origin Influenza virus A (H1N1): The first pandemic of the 21st century." *J Formos Med Assoc* 108:7;526–532.

Claverie, J. M., et al. 2009. "Mimivirus and Mimiviridae: giant viruses with an increasing number of potential hosts, including corals and sponges." *J Invertebr Pathol* 101:172–180.

Condon, B. J., and Sinha, T., 2009. "Who is that masked person: the use of face masks on Mexico City public transportation during the Influenza A (H1N1) outbreak." *Health Policy* published online (doi:10.1016/jhealthpol.2009.11.009).

Cramer, E. H., et al. 2002. "Outbreaks of gastroenteritis associated with Noroviruses on cruise ships—United States." *MMWR* 51(49):1112–1115.

Daubin, V., and Ochman, H. 2004. "Start-up entities in the origin of new genes." *Curr Opin Genet Dev* 14:616–619.

Echevarria-Zuno, S., et al. 2009. "Infection and death from Influenza A H1N1 in Mexico: A retrospective analysis." *Lancet* 374:2072–2079.

Forterre, P., and Prangishvili, D. 2009. "The origin of viruses." *Res Microbiol* 160:466–472.

Furman, J. A. 1999. "Marine viruses and their biogeochemical and ecological effects." *Nature* 399:541–548.

Garten, R. J., et al. 2009. "Antigenic and genetic characteristics of swine-origin 2009 A (H1N1) Influenza viruses circulating in humans." *Science* 325:5937;197–201.

Gottlieb, M. S., Schanker, H. M., Fan, P. T., Saxon, A., and Weisman, D. O. 1981. "*Pneumocystis* pneumonia—Los Angeles." *MMWR* 5(30):250–252.

Guan, Y., et al. 2003. "Isolation and characterization of viruses related to the SARS coronavirus from animals in southern China." *Science* 302:276–278.

Hope-Simpson, R. E. 1978. "Sunspots and flu: A correlation." *Nature* 275:86.

Hoyle, F., and Wickramasinghe, N. C. 1987. "Influenza viruses and comets." *Nature* 327:664.

Hoyle, F., and Wickramasinghe, N. C. 1990. "Sunspots and influenza." *Nature* 343:304.

Khan, K., et al. 2009. "Spread of a novel Influenza A (H1N1) virus via global airline transportation." *N Engl J Med* 36:2;212–214.

Koonin, E.V., and Martin, W. 2005. "On the origin of genomes and cells within inorganic compartments." *Trends Genet* 12:12;647–654.

Koonin, E. V., et al. 2006. "The ancient Virus World and evolution of cells." *Biology Direct* 1:29 doi10.1186/1745-6150-1-29.

Koonin, E. V., and Dolja, V. V. 2006. "Evolution of complexity in the viral world: the dawn of a new vision." *Virus Res* 117:1–4.

Kuenzi, A. J., Douglass, R. J., and Bond, C. W. 2000. "Sin nombre virus in deer mice captured inside homes, southwestern Montana." *EID* 6(4):386–389.

Lakshminarayan, M. I., et al. 2006. "Evolutionary genomics of nucleo-cytoplasmic large DNA viruses." *Virus Res* 117:156–184.

Lane, C. L., La Montagne, J., and Fauci, A. 2001. "Bioterrorism: A clear and present danger." *Nat Med* 7(12):1271–1273.

La Scola, B., et al. 2008. "The virophage as a unique parasite of the giant mimivirus." *Nature* 455:100–104.

Lau, S. K. P., et al. 2005. "Severe acute respiratory syndrome coronavirus-like virus in Chinese horseshoe bats." *PNAS* 102:39;14040–14045.

Li, W., et al. 2005. "Bats are natural reservoirs of SARS-like coronaviruses." *Science* 310:676–679.

Lopez-Bueno, A., et al. 2009. "High diversity of the viral community from an Antarctic lake." *Science* 326:858–861.

Lopman, B. A., et al. 2003. "Two epidemiological patterns of Norovirus outbreaks: Surveillance in England and Wales 1992–2000." *EID* 9(1):71–77.

Ostrow, R. S., et al. 1986. "Detection of papillomavirus DNA in human semen." *Science* 23:731–733.

Pearson, H. 2008. "'Virophage' suggests viruses are alive." *Nature* 454:7;677.

Raoult, D., and Forterre, P. 2008. "Redefining viruses: Lessons from Mimivirus." *Nat Rev Microbiol* 6:315–319.

Rappole, J. H., Derrikson, S. R., and Hubalek, A. 2000. "Migratory birds and spread of West Nile Virus in the western hemisphere." *EID* 6(4):319–328.

Rohwer, F., and Thurber, R. V. 2009. "Viruses manipulate the marine environment." *Nature* 459:207–212.

Shi, Z., Hu, Z. 2008. "A review on animal reservoirs of the SARS coronavirus." *Virus Res* 133:74–87.

Stone, R. 2002. "Stalin's forgotten cure." *Science* 298: 728–731.

Trifonov, V. Khiabanian, H., and Rabadan, R. 2009. "Geographic dependence, surveillance, and origins of the 2009 Influenza A (H1N1) virus." *N Engl J Med* 36:2; 115–119.

Tsang, K. W., et al. 2003. "A cluster of cases of severe acute respiratory syndrome in Hong Kong." *N Engl J Med* 348:20;1977–1985.

Wang, T. T., and Palese, P. 2009. "Unraveling the mystery of swine influenza virus." *Cell* 137:983–985.

Xu, H. F., et al. 2004. "An epidemiologic investigation on infection with severe acute respiratory syndrome coronavirus in wild animals traders in Guangzhou." [Article in Chinese] *Mar* 38:2;81–83.

Zhong, N. S., et al. 2003. "Epidemiology and cause of severe acute respiratory syndrome (SARS) in Guangdong, People's Republic of China, in February, 2003." *Lancet* 362:1353–1358.

Popular Press

Iezzoni, Lynette, and McCullough, David. 1997. *Influenza, 1918: The Worst Epidemic in American History.* New York: TV Books.

Kehret, Peg. 1996. *Small Steps: The Year I Got Polio.* Morton Grove, IL: Albert Whitman.

Lewis, Sinclair, with a new afterword by Doctorow, E. L. 1998. *Arrowsmith,* Revised Edition. New York: Signet Classics.

McCormick, Joseph B., and Fisher-Hoch, Susan, with Horvitz, Leslie Alan. *Level 4: Virus Hunters of the CDC.* 1996. Atlanta: Turner Publishing.

McKenna, Maryn. 2004. *Beating Back the Devil: On the Front Lines with the Disease Detectives of the Epidemic Intelligence Service.* New York: Free Press.

Peters, C. J., and Olshaker, Mark. 1998. *Virus Hunter: Thirty Years of Battling Hot Viruses Around the World.* New York: Anchor Books.

Preston, Richard. 1995. *The Hot Zone.* New York: Random House.

Preston, Richard. 2002. *The Demon in the Freezer.* New York: Random House.

Radetsky, Peter. 1994. *The Invisible Invaders.* Boston: Little, Brown.

Regis, Ed. 1996. *Virus Ground Zero: Stalking Killer Viruses with the Centers for Disease Control.* New York: Pocket Books.

Seavey, Nina Gilden, Smith, Jane S., and Wagner, Paul. 1998. *A Paralyzing Fear: Triumph over Polio in America.* New York: TV Books.

Video Productions

Zoonotic Viruses. 2009. Films for the Humanities.

The Final Inch (Polio in India). 2009. HBO Documentary.

H1N1. 60 Minutes. October 18, 2009. CBS.

H1N1. 60 Minutes. November 1, 2009. CBS.

The Silent Killer: SARS. 2008. Films for the Humanities.

The Invisible Enemy: Weaponized Smallpox. 2008. Films for the Humanities.

Mega Disasters: Alien Infection. 2007. The History Channel.

American Experience—Influenza 1918 (2006 DVD). PBS.

The Age of AIDS. Frontline. 2006. PBS.

A World Without Polio. 2005. Films for the Humanities.

The Age of Viruses. 2006. Films for the Humanities.

Warm Springs. 2005. HBO Production.

Bioterror, The Invisible Enemy. 2004. Films for the Humanities.

Understanding Viruses. 2001. Films for the Humanities.

Ebola—The Plague Fighters. 1996. NOVA.

Conquering an Invisible World. 1999. Films for the Humanities.

The Emerging Viruses. 1998. BBC Production, Films for the Humanities.

eLearning

go.jblearning.com/shors2

The site features eLearning, an online review area that provides quizzes and other tools to help you study for your class. You can also follow useful links for in-depth information, or just find out the latest virology and microbiology news.

CHAPTER 2

Eukaryotic Molecular Biology and Host Cell Constraints

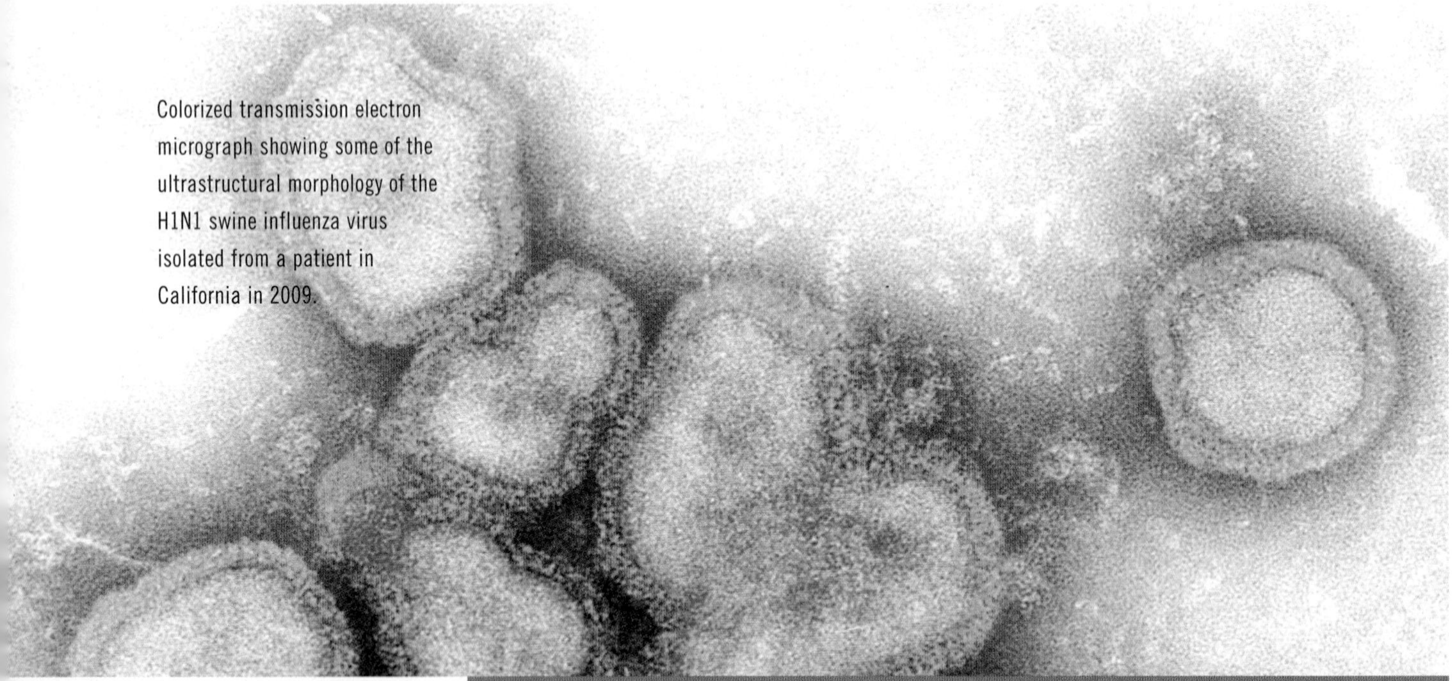

Colorized transmission electron micrograph showing some of the ultrastructural morphology of the H1N1 swine influenza virus isolated from a patient in California in 2009.

"Every path has its puddle."

English proverb

OUTLINE

CASE STUDY: MYSTERIOUS RASH

The headline in the February 6, 2002, edition of *The Seattle Times* read, "Mysterious Rash Hits School Kids Here, Across U.S." From October 2001 to June 2002, a total of 27 state health departments reported multiple groups of schoolchildren who developed mysterious rashes. The children complained of itchy, sunburn-like rashes on their cheeks and arms that lasted from a few hours to two weeks. Other symptoms included a burning sensation on the skin and hives that moved from one part of the body to another. A few children experienced other signs and symptoms such as fever, vomiting, sore throat, or headache. The outbreaks affected from 10 to 600 people at a time. A few teachers and school staff were affected, but rarely parents or siblings. Some schools were temporarily closed to clean air filters and check ventilation systems. Authorities at all of the schools that were closed reported an "exceptionally high" level of dust and dandruff in the air.

The level of parental concern and media hype prompted an investigation by the Centers for Disease Control and Prevention (CDC). They found no common cause for the outbreaks, but there were a few reports of cases associated with parvovirus B19 infection (Fifth disease). The majority of cases remain unexplained. The CDC is continuing to monitor reports of groups of children with rashes and is providing technical assistance to state and local health departments investigating the outbreaks.

Two reports on this topic published by the CDC follow:

Cartter, M., et al. 2002. "Rashes among schoolchildren—14 states, October 4, 2001–February 27, 2002." *MMWR* 51:161–164.

Kacica, M. A., et al. 2002. "Update: Rashes among schoolchildren—27 states, October 4, 2001–June 3, 2002." *MMWR* 51:524–527.

The first chapter introduced some common viral diseases present in our daily lives and reflected on viral epidemics and pandemics. Typical viruses were defined as having the following characteristics:

- small size (nm range)
- can pass through filters that trap most known bacteria
- completely dependent upon the host cell
- usually contain one type of nucleic acid (RNA or DNA)
- contain receptor-binding proteins
- genome may be infectious
- some viruses can persist by integrating their genome into the host cell's chromosome

Viruses are inert outside of their host cells. They can enter (infect) all types of cells, including animal, plant, and bacterial cells. This textbook mainly focuses on human viruses, but addresses some animal, plant, and bacterial viruses as well. Before we can study these individual viruses in depth, though, we must first analyze the basic structure of a human (or in some cases an animal) virus.

Virus particles consist of a nucleic acid genome protected by a protein shell that gives the virus particle its strong structure. The shell, referred to as the **capsid**, can also contain other proteins, such as receptor or viral attachment proteins, that allow the virus to adhere to the outside of a host cell. Human or animal virus anatomy is different from bacteriophages (read more about viral structure in Chapter 3).

The capsid protects the nucleic acid genome of the virus from a harsh environment that is laden with nucleases. The capsid and associated genome constitute the **nucleocapsid**. Many viruses also code for and carry their own polymerases. DNA polymerase, RNA polymerase, or **reverse transcriptases** catalyze the formation of polynucleotides of DNA or RNA using an existing strand of DNA or RNA as a template. This is necessary because the host may not contain a polymerase that will replicate the viral genome (read about constraints of host cells later in this chapter).

Human and animal viruses may also have an additional lipid bilayer membrane wrapped around the capsid of the virus particle. These viruses are called **enveloped viruses** (**FIGURE 2-1a**). The lipid bilayer membrane is stolen from the host cell as new viruses bud and exit from the host cell following infection and replication. This lipid bilayer may come from the host cell's plasma (outer) membrane, the nuclear membrane, or its trans Golgi network. Viruses that do not contain an envelope are called **naked viruses** (Figure 2-1b).

Enveloped viruses contain matrix proteins that are located inside the envelope. These proteins add rigidity to the virus particle and are believed to act as a bridge between the nucleocapsid and viral proteins embedded in the envelope. In doing so they secure the internal nucleocapsid to the envelope, which may be critical for budding and release

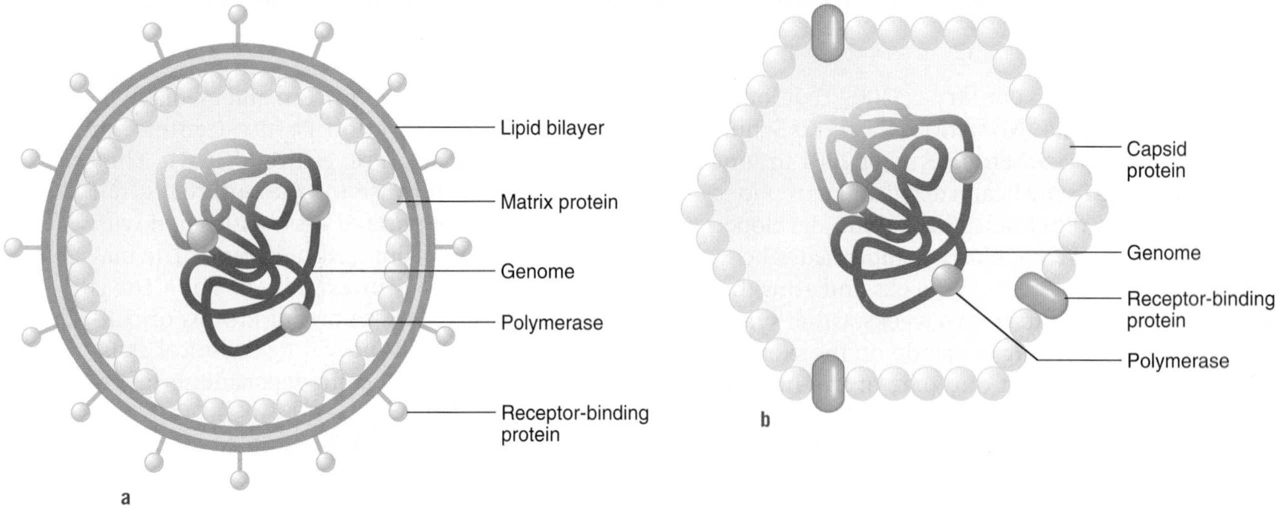

FIGURE 2-1 **(a)** Structure of a typical animal or human enveloped virus. **(b)** Diagram of a naked (nonenveloped) animal or human virus.

of viruses from host cells. The entire structure of the virus—the genome, the capsid, and (where present) the envelope—make up the **virion**, or virus particle.

2.1 How Many Genes Are Required to Build a Simple Virus Particle?

Viral genomes are small and diverse compared to their hosts. Viruses have evolved ways to replicate that are economical; for example, a simple virus may require only one type of capsid protein. This protein is used over and over again to build the capsid. Hence, the virus would need one gene encoding the capsid protein to accomplish this task.

What other genes would a simple virus need? Besides a gene encoding the capsid protein, additional genes would be needed to encode a receptor-binding protein and usually a gene for a viral polymerase. This means a viral genome could contain as few as three genes.

The genome sizes of viruses vary greatly. The size of RNA viral genomes varies from 1.7 to 27 kilobase (kbp) in length. These viral genomes generally encode fewer proteins (usually less than a dozen), whereas the genomes of DNA viruses most often are larger, ranging in length from 3.2 to over 200 kbp. These genomes roughly code for 4 to 200 gene products. An exception is the recently discovered Mimivirus, which is enormous. Its genome is believed to code for more than 1000 gene products.

Why Are Viruses Dependent Upon the Host for Replication?

Viruses can package/carry as few as 5 or 6 genes or as many as 200. What cellular processes do viruses need that would take more than this number of genes to accomplish the task of generating new viruses? To answer this question, a review of eukaryotic replication, transcription, and translation is necessary. How many gene products are necessary to do these processes?

2.2 Molecular Biology Review

To replicate, viruses must enter a host cell. Some viruses use the host's replication and transcription machinery, and others bring their own genes or proteins for viral replication and transcription. All viruses are dependent upon the host for the translation machinery and energy supplies. Let's briefly review the processes of eukaryotic DNA replication, RNA transcription, and translation.

The Central Dogma of Eukaryotic Molecular Biology

The central dogma of molecular biology is that DNA is transcribed into RNA that is translated into protein. Protein is never back-translated to RNA or DNA, and DNA is never created from RNA. DNA is never directly translated to protein. This dogma is never broken by eukaryotes. It is, however, broken by retroviruses (see Chapter 10).

It is also important to remember that the processes of replication, transcription, and translation

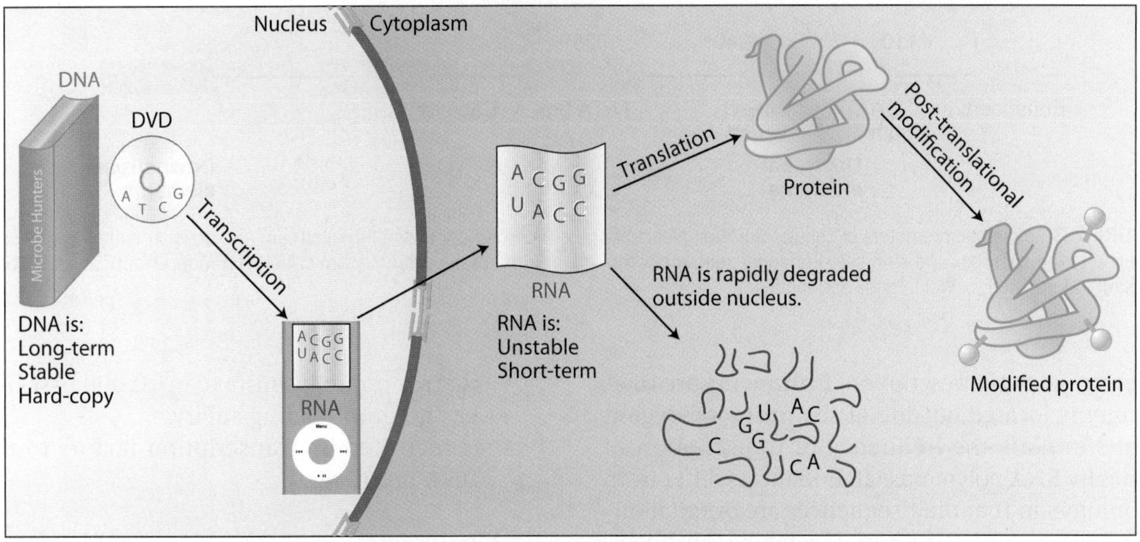

FIGURE 2-2 Replication, transcription, and translation are localized processes in host cells. Adapted from Huges, A. "The Central Dogma and Basic Transcription," *Connexions*, July 27, 2003, http://cnx.org/content/m11415/latest/.

are localized in eukaryotic host cells (**FIGURE 2-2**). DNA replication and RNA transcription occur in the **nucleus** of the host cell. Processed mRNAs exit the nucleus and are translated in the **cytoplasm** of cells. A virus in need of the cellular machinery that performs these processes must therefore find its way to the correct parts of the cell to replicate its genome, transcribe its RNA, and get its RNA translated into viral proteins.

Eukaryotic DNA Replication

Host genomes are always composed of double-stranded DNA (dsDNA), in contrast to viral genomes that may be dsDNA, single-stranded DNA (ssDNA), double-stranded RNA (dsRNA), or single-stranded RNA (ssRNA). Eukaryotic cells have more than a dozen DNA polymerases that synthesize the dsDNA. Two of these (DNA polymerase α and δ) are important for the replication of eukaryotic chromosomes. The chromosome dsDNA is copied/read $3' \rightarrow 5'$ and synthesized in a $5' \rightarrow 3'$ direction by DNA polymerases in the nucleus of the cell. DNA polymerases are sometimes referred to as DNA-dependent DNA polymerases because they synthesize DNA from a DNA template. Additional properties of cellular DNA polymerases are that they

- cannot "initiate de novo DNA synthesis" (it requires an RNA primer to do so).
- have high fidelity of copying (generating one error in every 10^9 base pair replications).
- may possess proofreading ability ($3' \rightarrow 5'$ exonuclease/editing activity to remove incorrect nucleotides).
- may possess helicase (unwinding) and primase activities.
- are localized and active in the nucleus of cells.

The process of DNA replication in eukaryotes requires additional cellular proteins such as primases (to synthesize RNA primers), ssDNA-binding proteins (to protect the dsDNA from nucleases during replication), and ligases (to join together **Okazaki fragments**/lagging strands).

Eukaryotic RNA Transcription

Eukaryotes carry out the conversion of information in DNA into RNA by the process of transcription. Enzymes called DNA-dependent RNA polymerases catalyze the RNA synthesis. Eukaryotic cells contain three different types of DNA-dependent RNA polymerases: RNA polymerases I, II, and III. **RNA polymerase I** synthesizes ribosomal RNA (rRNA). **RNA polymerase II** synthesizes pre-messenger RNA (pre-mRNA) and some small nuclear RNAs (snRNA). **RNA polymerase III** synthesizes transfer RNA (tRNA), 5S rRNA, and other small RNAs. We will focus on RNA polymerase II because it is responsible for synthesizing cellular mRNA that is translated by ribosomes. Viral mRNAs must be structurally similar to cellular mRNAs so *that the cell can recognize viral mRNAs and translate them using the host protein synthesis machinery (e.g., ribosomes).*

The cell synthesizes its own mRNA in the nucleus by transcription of its DNA, followed by posttranscriptional processing of the mRNA transcript (pre-mRNA). Eukaryotic DNA contains sequences (elements) that dictate the start of RNA transcription. A typical eukaryotic promoter region with its control elements for making cellular mRNA via RNA polymerase II is shown in **FIGURE 2-3**. The AT-rich region of the promoter binds cellular proteins (**transcription factors**) that facilitate the binding of RNA polymerase II and determines the

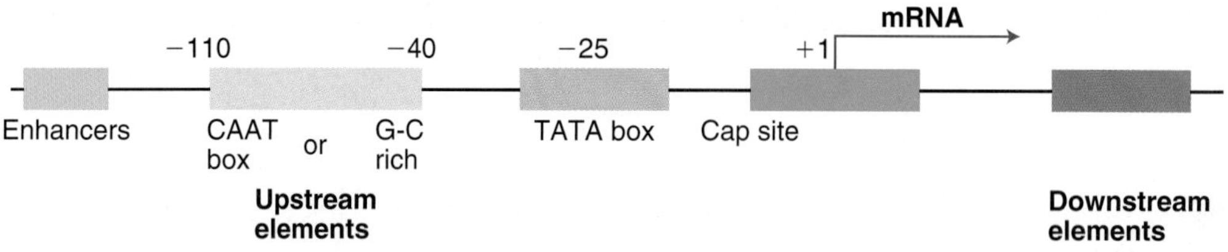

FIGURE 2-3 Eukaryotic promoters of cellular mRNAs. Notice that the promoter contains both upstream and downstream elements in addition to the TATA box and start site. Enhancer sequences that are not part of the promoter are orientation-independent and may be present before, after, or within gene sequences.

starting point of transcription. **Enhancers** are DNA elements located outside of the promoter region that stimulate the frequency of transcription of genes by RNA polymerase II. Enhancers differ from promoters in that their sequences are orientation-independent and position-independent (upstream, downstream, or within genes).

The characteristics of RNA polymerase II are that it

- binds the transcription complex in the promoter region of DNA.
- can **initiate** the de novo synthesis of RNA (it does not require a primer).
- copies/reads 3′→5′ and synthesizes RNA 5′→3′.
- synthesizes and processes RNA in the nucleus of the cell.
- is error prone (1 mistake in 10,000 bases).
- has no proofreading ability.
- is recruited by transcription factors to the DNA promoter.

RNA polymerase II synthesizes a large primary pre-mRNA transcript (**FIGURE 2-4**). As the RNA is being transcribed (after ~30 nucleotides), the 5′ end of the mRNA is "capped" by the addition of a methylated guanine nucleotide (5′m^{7}G). This "cap" is recognized by cellular ribosomes and thus plays an important role in the initiation of translation. The cap helps protect the mRNA from degradation. The mRNA transcript is cleaved about 20 to 30 nucleotides past the polyadenylation sequence, and then a poly(A) polymerase adds 100 to 200 residues of adenylic acid to the 3′ end of the mRNA. The

FIGURE 2-4 Eukaryotic transcription occurs in the nucleus. After the mRNAs are processed, the mature (capped, polyadenylated) mRNA exits the nuclear pores and enters the cytoplasm of the cell.

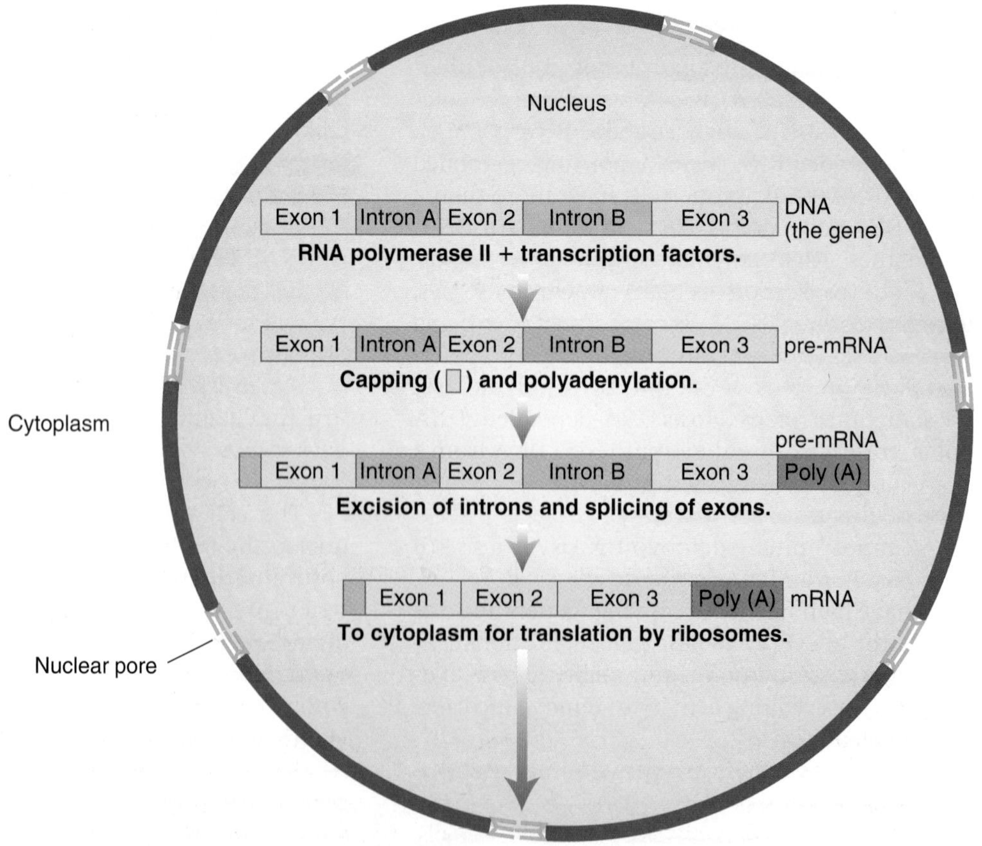

poly(A) tail may also help protect the mRNA from degradation.

The poly(A) tail affects the stability of some mRNAs. It usually becomes shorter as an mRNA ages in the cytoplasm, and when it reaches a minimal length the mRNA will be degraded by cellular nucleases. Intron removal and exon splicing of the pre-mRNAs require snRNAs along with proteins and adenosine triphosphate (ATP). After this event, the final processed mRNAs are exported from the nucleus into the cytoplasm of the cell where they will be translated. Only about 5% of the originally transcribed RNA exits the nucleus (Figure 2-4).

Review of Open Reading Frames (ORFs)

Regions of DNA are transcribed into mRNA and then translated into protein by ribosomes. For every gene, there up to 6 possible reading frames (3 in each direction). If it is an "open" reading frame then it has a start and stop codon. The reading frame that is used in translating a gene (in eukaryotes) is often the longest ORF. An ORF starts with an *atg* (the coding for the amino acid methionine) in most species and ends with a stop codon (*taa, tag,* or *tga*).

The DNA sequence shown in **FIGURE 2-5** can be read in 6 reading frames. The 3 forward reading frames are shown below the sequence, with the translated amino acids listed below each DNA sequence. Frame 1 starts with the "**a**," Frame 2 with the "**t**," and Frame 3 with the "**g**." Stop codons are indicated by an "*" in the protein sequence. The only ORF is Frame 1.

Cap-Dependent Initiation of Translation

Translation is the decoding of mRNA into protein. It involves all three types of RNA (tRNA, rRNA, and mRNA) and is a three-step process:

- **initiation** (formation of the initiation complex),
- **elongation** (synthesis of the polypeptide/protein), and
- **termination** (the mRNA contains an in-frame stop codon, and the polypeptide/protein is released).

The focus of this review will be on **initiation of translation** because *all* viruses are dependent upon the host cell translation machinery to produce viral protein products using cellular ribosomes. Initiation is the key event in translation because viruses need to hijack the host translational machinery to produce their own proteins. Viruses have developed many different strategies to get their mRNA preferentially translated over the myriad of cellular mRNAs.

Ribosomal Scanning Model

This model postulates that a ternary complex is formed (met-tRNAi + eIF-2 + GTP) and that it associates with the small ribosomal subunit (40S). This complex binds/enters the 5′ cap of the eukaryotic mRNA and then migrates and scans linearly down the mRNA, usually stopping at the first AUG it reaches. Marilyn Kozak has hypothesized that the 40S ribosome will scan along the mRNA until it encounters an AUG in the best consensus sequence (usually the first AUG, but not always). The Kozak consensus sequence is GCC A/G CC**AUG**(G). After this event, additional protein synthesis initiation factors, the large ribosomal subunit, and ATP are involved in the initiation of translation (**FIGURE 2-6**). The initiation of protein synthesis involves many cellular factors, which are listed in **TABLE 2-1**. After initiation, the steps of elongation and termination are carried out (**FIGURE 2-7**).

Translation and Open Reading Frames

Viruses use ingenious strategies to generate many protein products encoded by small genomes that contain a limited number of genes. One of these strategies is **ribosomal frameshifting**, which occurs when the ribosome shifts into another reading frame and then continues translating the mRNA into protein in that new frame. Hence, certain viruses compress their genetic information by encoding different proteins in overlapping reading frames. Ribosomal frameshifting is very rare in the eukaryotic translation of genes, but is a hallmark of the translation of many retrovirus, coronavirus, paramyxovirus, astrovirus, torovirus, and arterivirus genes.

Another strategy that viruses may use to control gene expression is **translational readthrough**

Translation and open reading frames (ORFs)

```
        5′ atgcccaagctgaatagcgtagaggggttttcatcatttgaggacgatgtataa3′   DNA
ORF
 1   atg ccc aag ctg aat agc gta gag ggg ttt tca tca ttt gag gac gat gta taa
      M   P   K   L   N   S   V   E   G   F   S   S   F   E   D   D   V   *

 2   tgc cca agc tga ata gcg tag agg ggt ttt cat cat ttg agg acg atg tat
      C   P   S   *   I   A   *   R   G   F   H   H   L   R   T   M   Y

 3   gcc caa gct gaa tag cgt aga ggg gtt ttc atc att tga gga cga tgt ata
      A   Q   A   E   *   R   R   G   V   F   I   I   *   G   R   C   I
```

FIGURE 2-5 DNA sequence with the three forward reading frames shown with the translated amino acids below each DNA sequence. Stop codons are indicated with an asterisk. Modified from an illustration by Scott Cooper, UW–La Crosse.

FIGURE 2-6 Ribosomal scanning model: initiation of translation. The ribosome scans for best context of the Kozak's consensus sequence. Adapted from Promega Corporation. *Promega Protein Guide: Tips and Techniques.* Promega Corporation, 1993.

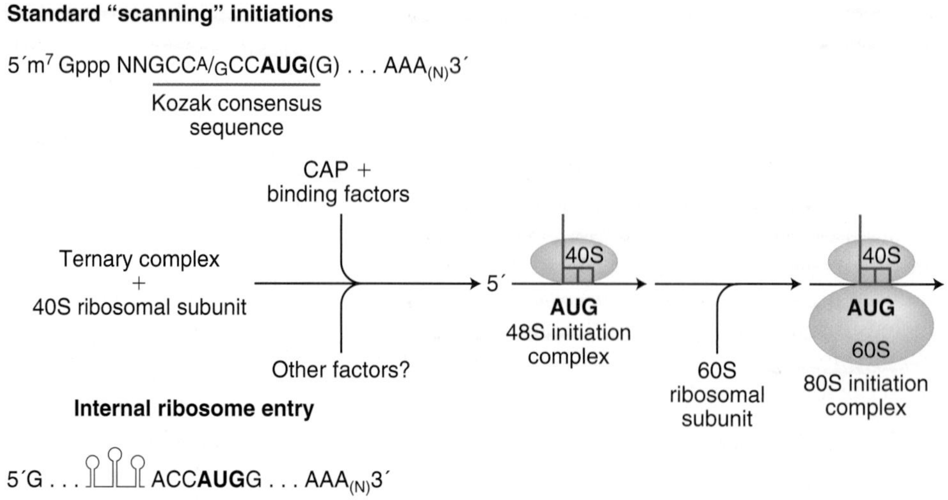

(**FIGURE 2-8**), which is a variation of the frameshifting theme. In translational readthrough, a stop translation signal may be ignored.

Why Do All Viruses Use the Host's Protein Synthesis Machinery?

Viruses do not have the genetic capacity to encode the proteins necessary for eukaryotic protein synthesis. In addition to ribosomes and rRNA, there are at least a dozen translation initiation factors (IFs) involved in the initiation of translation in eukaryotes (Table 2-1). These eukaryotic initiation factors catalyze individual steps in the cap-dependent pathway. Even though many reactions remain poorly understood, a proposed model for cap-dependent initiation has been developed.

The term **cap-dependent translation** initiation refers to the fact that with some rare exceptions, all eukaryotic initiation of translation requires the 5′ cap of the cellular mRNA to be present. The cap is considered to be the signal recognized by the ribosome to identify the 5′ end of

TABLE 2-1 Eukaryotic Translation Initiation Factors

Factor	Function(s)
eIF1A (eIF4C)	Stimulation of Met-tRNAi and mRNA binding to 40S ribosomes
eIF2	Met-tRNAi binding to 40S ribosomes
eIF2B (GEF)	GDP:GTP exchange on eIF2
eIF2C (Co-eIF2A)	Stabilization of ternary complex
eIF3	Ribosome dissociation, stabilization of ternary complex, stimulation of mRNA binding
eIF3A (eIF-6)	Ribosome dissociation
Ded1	mRNA binding, RNA helicase
eIF4A	mRNA binding, RNA helicase
eIF4B	mRNA binding, RNA helicase
eIF4E	mRNA binding, cap recognition
eIF4F (CBPII)	mRNA binding, cap recognition, RNA helicase
eIF4G	mRNA binding, anchor protein
eIF4H	mRNA binding
eIF5	Ribosomal subunit joining
eIF5D	Ribosomal subunit joining

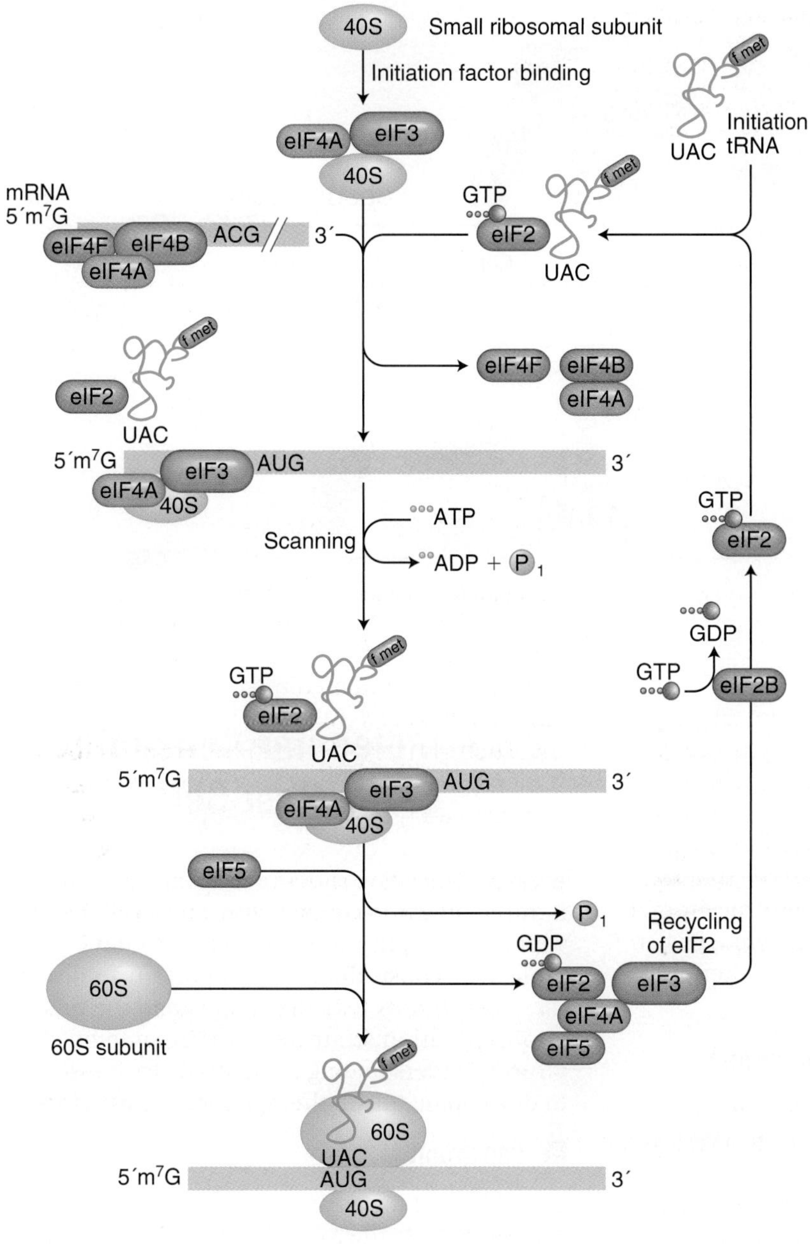

FIGURE 2-7 Detailed model of eukaryotic initiation of translation. Note the number of eukaryotic initiation factors (eIFs) involved. Their functions are listed in Table 2-1. Modified from Gilbert, S. F. *Developmental Biology*, Fifth Edition, Sinauer Associates, 1997.

mRNAs because it distinguishes mRNAs from other cellular RNAs (e.g., rRNA, tRNAs, and snRNAs). *Some viruses bypass the cap recognition requirement.* This phenomenon is called **cap-independent translation**. Viruses do this by generating viral mRNAs that contain **internal ribosomal entry sites** (IRES) that allow the ribosome to enter the 5′ end of the viral RNA independently of a cap structure at that end (**FIGURE 2-9**).

Leaky Scanning

Translation initiation can occur at one or more AUG sites near the 5′ end on a given viral RNA. The first AUG may not be in a context favorable to

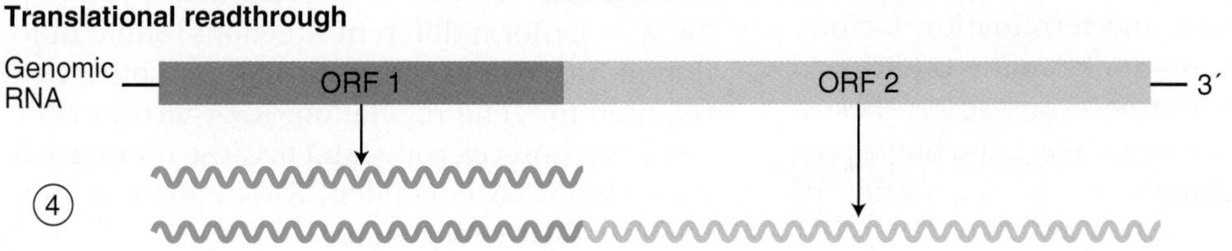

FIGURE 2-8 Translational readthrough generates two overlapping proteins from a single RNA. Adapted from Levy, J. A., Fraenkel-Conrat, H., and Owens, R. A. *Virology*, Third Edition, Prentice Hall, 1994.

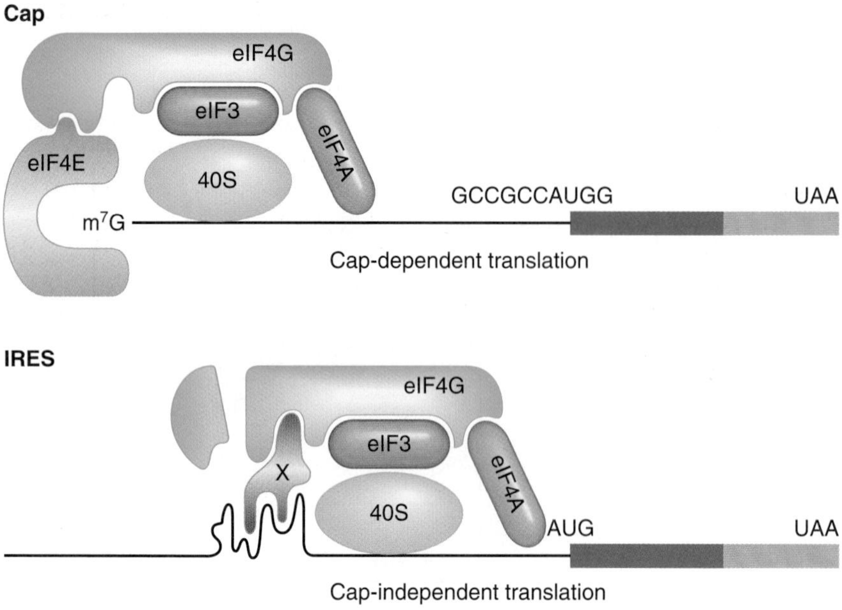

FIGURE 2-9 Models of cap-dependent and cap-independent translation. Adapted from Weaver, R. F. *Molecular Biology*, Third Edition, McGraw-Hill Higher Education, 2001.

Kozak's consensus. In such cases, the ribosome may inefficiently initiate translation at the first AUG, but more often the 40S ribosome will bypass that first AUG and initiate translation farther downstream at an AUG in better context. This is called **leaky scanning** by the ribosome. It allows multiple viral proteins to be synthesized from a single mRNA. A different protein is generated when the ribosome initiates translation at an alternative AUG (Figure 2-8).

Posttranslational Processing of Proteins

Most eukaryotic proteins undergo some form of modification following translation. **Posttranslational modifications** (such as glycosylation, phosphorylation, and proteolytic cleavage) serve many functions. Phosphorylation of viral proteins often is needed for nucleic acid binding, whereas proteolytic cleavage often is vital to the maturation and assembly of many viruses. Several viruses have exploited the presence of cell-surface carbohydrates usually associated with glycoproteins, using them as portals of entry into the cell.

Revisit the Question: Why Do All Viruses Use the Host's Protein Synthesis Machinery?

Eukaryotic translation involves more than a dozen initiation, elongation, and termination factors. Most viral genomes do not have the capacity to contain all of the genes necessary to synthesize proteins. As a result, viruses hijack the host's protein synthesis machinery for the translation of viral mRNAs.

2.3 Molecular Constraints of the Host Cell

A virus is not able to infect every cell it encounters. It must come into contact with a host cell that can support its replication. Host cells do have some molecular constraints. It is important to discuss these constraints to better understand the unique viral replication strategies that are employed by viruses. Understanding these strategies is essential to developing future therapeutics against viruses.

Constraint 1

The ability of a virus to replicate in a host cell is constrained by the availability in that cell of the specific proteins that the virus needs to infect and replicate within the cell. The host cell may be lacking an outside cell receptor for attachment of the virus, or it may be missing an internal cellular protein such as a **transcription** or **replication factor**.

Many viruses contain RNA genomes, requiring the need for an RNA-dependent RNA polymerase for genomic synthesis. *Even though a wide range of host cells contain RNA-dependent RNA polymerases, these cellular polymerases are incapable of replicating viral genomes.* Cellular RNA-dependent RNA polymerases perform different functions. Some function in host defense mechanisms, and others are required for gene regulation. RNA viruses overcome this host cell constraint because they encode their own RNA-dependent RNA polymerases to replicate their viral genome.

Constraint 2

Eukaryotic host protein synthesis machinery is equipped only to translate monocistronic RNAs (monocistronic RNAs code for one protein), and it usually does not recognize internal initiation sites within RNA.

It has been determined that viruses overcome these cellular constraints in at least two different ways: they generate separate mRNAs for each gene (functionally monocistronic mRNAs) or generate an mRNA encompassing several genes that is translated into a large precursor "polyprotein," which is then cleaved into individual proteins via viral or cellular proteases (**FIGURE 2-10**).

Constraint 3

In an infected cell, the expression of the viral genomes is in direct competition with that of the numerous cellular genes. To overcome this constraint, viruses have evolved ways to produce abundant amounts of their own proteins. Some use strategies that confer a competitive advantage toward viral mRNAs, whereas others preferentially degrade host cell mRNAs.

As an example, the yeast L-A virus produces a viral nuclease (Gag) that decapitates cellular mRNAs. The capless cellular mRNAs are then susceptible to mRNA degradation by a cellular exoribonuclease called Xrn1p. Interestingly, the

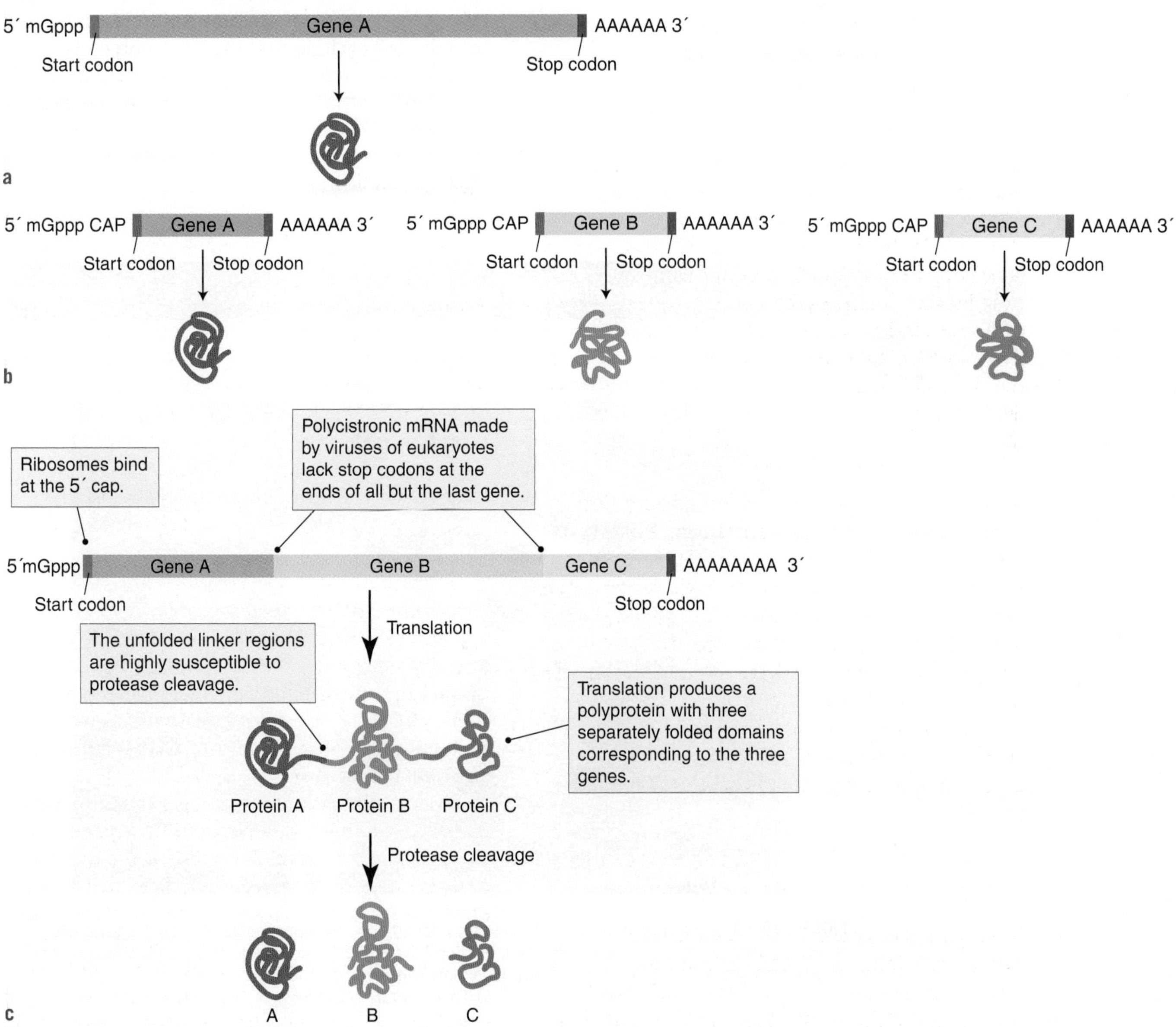

FIGURE 2-10 (a) Eukaryotic cells contain monocistronic mRNAs that encode one gene product. **(b)** Some viruses contain separate RNAs for each gene. **(c)** Some viral genomes are transcribed into one precursor RNA that is translated into a large polyprotein. The polyprotein is later cleaved into individual proteins by viral and/or cellular proteases.

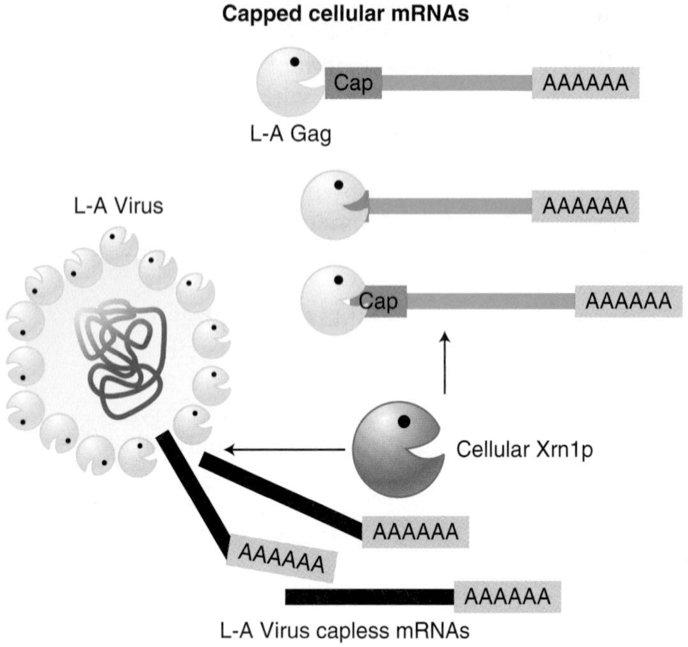

FIGURE 2-11 Model of decoy-decapitation mechanism used by yeast L-A virus.

L-A virus mRNAs are capless. This decapitation event is a "decoying event." The Xrn1p exonuclease degrades capless cellular mRNAs, partially freeing the capless viral mRNAs from Xrn1p attack (**FIGURE 2-11**).

Other viruses, such as herpes simplex virus (HSV) and poxviruses, inhibit cellular translation and use strategies to degrade host mRNAs (and sometimes viral mRNAs as a consequence) following infection. Influenza viruses cleave host cell mRNA caps and use the host cell as a primer for viral transcription.

Constraint 4

If a DNA virus requires the use of a host cell DNA polymerase to replicate its genome, it must either infect an undifferentiated dividing cell or it must have some means of pushing a differentiated cell into the cell cycle so that the DNA polymerase enzymes that it needs for replication will be available. Differentiated cells are not usually cycling through the cell cycle, which means that they will not be producing the DNA polymerases necessary for DNA replication.

VIRUS FILE 2-1 Coronaviruses Are Associated with Kawasaki Disease

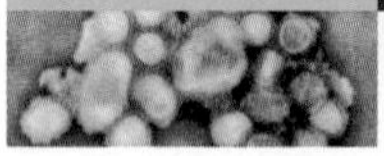

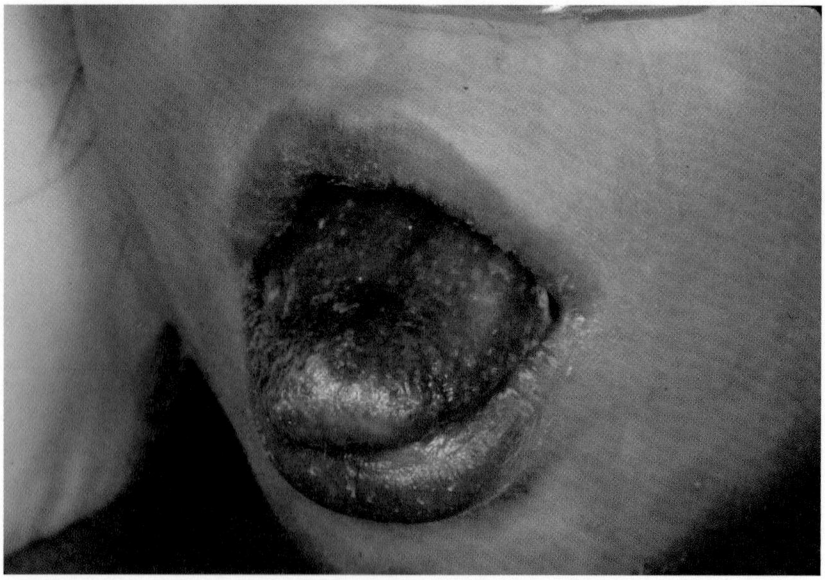

FIGURE VF 2-1 One symptom of Kawasaki disease in children is the appearance of a red "candy-coated strawberry tongue," shown here in a five-year-old boy. Kawasaki disease is the leading cause of acquired heart disease among children in the United States and other developed countries.

Tomisaku Kawasaki first reported what would later be called Kawasaki disease (KD) in 1967. He described 50 Japanese children who between the years 1961 and 1967 suffered from an

VIRUS FILE 2-1 Coronaviruses Are Associated with Kawasaki Disease (continued)

illness that was characterized by prolonged fever, rash, inflammation of the lips and oral cavity, "strawberry tongue," and redness and swelling of the hands and feet. Initially Kawasaki thought the disease was a benign childhood illness. Later it was determined that KD had in fact caused a number of deaths in children who had appeared to be improving or who had recovered. Upon autopsy, it was discovered that the children had developed coronary artery abnormalities. The cause of their deaths was heart failure or myocardial infarction (destruction of heart tissue resulting from obstruction of the blood supply to the heart muscle).

Kawasaki disease has been diagnosed in the United States since 1971, with epidemics reported to the CDC during the 1980s. The illness affects primarily preschool-aged children and is more common in boys. These epidemics have been observed in winter and spring. The disease is most commonly observed in children from the middle and upper classes. Approximately 3000 children with symptoms of KD are hospitalized in the United States annually.

The cause of KD remains unknown. There are many theories, including an infectious **etiology**, an immune system abnormality, and even a possible link to carpet shampoo. No tests exist for this disease; diagnosis is based on clinical symptoms. A history of preceding respiratory illness has been reported.

Human coronaviruses have attracted renewed interest because of the emergence of a novel strain associated with severe acute respiratory syndrome (SARS-CoV). Studies of human coronaviruses have been hampered because of the inability to cultivate them **in vitro**. The 2003 SARS outbreak suddenly brought coronaviruses back into the limelight. Molecular tools were available to quickly identify and classify the particular novel coronavirus. This led to the development of detection methods of SARS-CoV and the accumulation of coronaviral genomic sequence data that allowed researchers to search for new human coronaviruses in children with respiratory diseases.

Frank Esper and his colleagues at Yale University School of Medicine sought to determine whether novel human coronaviruses were circulating in New Haven, Connecticut, where Yale is located. They did indeed identify a novel human coronavirus present in 79 of the 895 (8.8%) children they tested. They named it SARS-CoV-NH. A six-month-old infant with classic KD tested positive for the new SARS-CoV-NH. Further investigation was conducted on control children who did not have KD and children with KD. Respiratory specimens from 8 of 11 (73%) children with KD, and 1 of 22 (4.5%) control children without KD, tested positive for human SARS-CoV-NH infection. The results suggest that there is an association between KD and SARS-CoV-NH. Further studies are needed to determine the precise role played by the novel coronavirus in the pathogenesis of KD. Advances in molecular biology are leading to many new discoveries in the field of virology.

References

CDC. 1983. "Kawasaki syndrome—United States." *MMWR* 32(7):98–100.

Esper, F., et al. 2005. "Evidence of a novel human coronavirus that is associated with respiratory tract disease in infants and young children." *J Infect Dis* 191:492–498.

Esper, F., et al. 2005. "Association between a novel human coronavirus and Kawasaki disease." *J Infect Dis* 191:499–502.

Kawasaki, T. 1967. "Acute febrile mucocutaneous syndrome with lymph node involvement with septic desquamation of the fingers and toes in children." [in Japanese] *Jpn J Allerg* 16:178–222.

Pappas, P., et al. 1985. "Multiple outbreaks of Kawasaki syndrome—United States." *MMWR* 34(3):33–35.

ProMED-Mail. Kawasaki Disease. 2001. China (Hong Kong) ProMED mail;28 May: *South China Morning Post,* by Mary Ann Benitez.

Summary

Viruses all rely, to varying degrees, on the metabolic processes of their hosts to reproduce themselves. Their hosts challenge them in several ways. The ability of the virus to replicate inside of a host cell depends on whether or not the host cell contains the appropriate receptors for entry and the proteins needed to carry out viral replication, transcription, and translation.

This chapter reviewed eukaryotic molecular biology and discussed the constraints that viruses must face. These constraints include:

1. Cellular RNA-dependent RNA polymerases cannot replicate *viral* RNA genomes.
2. Eukaryotic machinery only translates monocistronic mRNAs.
3. Viral mRNAs are in direct competition with the myriad cellular mRNAs for the translation machinery.
4. Differentiated cells may not contain the DNA polymerase enzymes required for replication of the genomes of DNA viruses.

It is intended that as a consequence of your coursework, you will begin to understand the behavior of viruses from a molecular biology perspective. In other words, you have been provided with insights to the molecular needs of viruses and with a few examples of strategies that viruses use to overcome cellular limitations.

This textbook does not emphasize the use of viruses as tools. It should be noted, though, that much of what we know today about genes, replication, DNA repair, transcription and control, splicing and processing of RNA, translation of mRNA, and protein modifications have relied extensively on viruses as critical research tools. *Viruses have been our "eyes" into cells. The intimate association between a virus and its host has allowed us to understand how cells function.* Chapter 4 will expand on replication strategies of viruses. These strategies are dictated by the nature of the viral genetic material.

CASE STUDY QUESTIONS

These questions relate to the Case Study presented at the beginning of the chapter.

1. Rashes can be explained by a variety of causes. List a few possibilities.
2. Could additional cases be caused by a new or yet-to-be-identified virus? Why or why not?
3. The CDC points to several challenges that impede investigations of reported rashes among schoolchildren and the identification of their causes. List these challenges.

Resources

Primary Literature and Reviews

Everly, D. N., Jr., Feng, P., Mian, E. S., and Read, G. S. 2002. "mRNA degradation by the virion host shutoff (Vhs) protein of herpes simplex virus: Genetic and biochemical evidence that Vhs is a nuclease." *J Virol* 76(17): 8560–8571.

Knipe, D. M., Samuel, C. E., and Palese, P., eds. 2001. "Virus–host cell interactions." In *Fundamental Virology*. Philadelphia: Lippincott Williams & Wilkins.

Kozak, M. 1999. "Initiation of translation in prokaryotes and eukaryotes." *Gene* 234:187–208.

Kozak, M. 2002. "Pushing the limits of the scanning mechanism for initiation of translation." *Gene* 299:1–34.

Maison, D. C., et al. 1995. "Decoying the cap-mRNA degradation system by a double-stranded RNA virus and poly(A)-mRNA surveillance by a yeast antiviral system." *Mol Cell Biol* 15(5):2763–2771.

Mitchell, P., and Tollervey, D. 2000. "mRNA stability in eukaryotes." *Curr Opin Genet Dev* 10:193–198.

Wassenegger, M., Krczal, G. 2006. "Nomenclature and Functions of RNA-Directed RNA Polymerases." *Trends Plant Sci* 11(3).

Zhong, J., et al. 2009. "Evolution of the RNA-dependent RNA polymerase (RdRP) genes: Duplications and possible losses before and after divergence of major eukaryotic groups." *Gene* 447:29–39.

Popular Press

Haseltine, W. A. 2001. "Beyond chicken soup." *Scientific American*, Nov. 2001, Vol. 285, Issue 5, p. 56.

Video Productions

Understanding Viruses. 2001. The Discovery Channel.

A Journey through the Cell. 1997. Cambridge Educational.

eLearning

go.jblearning.com/shors2

The site features eLearning, an online review area that provides quizzes and other tools to help you study for your class. You can also follow useful links for in-depth information, or just find out the latest virology and microbiology news.

CHAPTER 3

Virus Architecture and Nomenclature

This is a molecular model of a Simian Virus 40 particle. It is a naked virus that was first identified as a contaminant of rhesus monkey kidney cells being used to produce a vaccine against poliovirus during the 1960s.

OUTLINE

"Nature does nothing uselessly."

Aristotle

CASE STUDY: A NEW VIRAL PNEUMONIA

A 20-year-old female college student was admitted to a hospital in Alberta, Canada, in February. Five days earlier, she had experienced coldlike symptoms that were followed by a high fever and a productive cough of bloody sputum. When she arrived in the emergency room, she complained of chest pain and shortness of breath. The attending physician, Dr. Rudolph, ordered a chest x-ray. A sputum sample was collected and sent to the hospital medical technology laboratory. Assuming it was pneumonia, the patient was treated with antibiotics.

The patient did not respond to the antibiotics, even after several days of treatment. Laboratory tests were negative for bacteria that typically cause pneumonia, such as *Streptococcus pneumoniae, Haemophilus influenzae*, and *Legionella pneumoniae*. The physician suspected she had contracted a viral pneumonia. Shortly thereafter, two more college students and a professor were admitted to the same hospital. All of them had pneumonia-like symptoms.

Dr. Rudolph began researching medical journals for new causative agents of pneumonia. He read a study by Bernard La Scola and colleagues in which they reported Mimivirus in Canadian patients who suffered from hospital- and community-acquired pneumonia. During the course of his reading, he learned that Mimivirus was first discovered after a 1992 outbreak of pneumonia in Bradford, England. The virus was isolated from amoebae growing in the water of a cooling tower. The "virus" was thought to be a bacterium because of its large size: its particles were over 400 nm in diameter. The electron micrograph shown in **FIGURE CS 3-1** shows a Mimivirus particle infecting an amoeba.

Scientists determined that this new organism was a virus because all attempts to amplify the 16S rDNA from the genome of the new organism failed. The 16S rDNA gene is a universal bacterial gene. The entire genome of Mimivirus was published in *Science* in 2004. Mimivirus has the largest known viral genome. Its genome is 1,181,404 nucleotide base pairs in length. It contains 1262 possible open reading frames (ORFs) or genes. Ten percent of its genes have known functions. They include genes for an amino-acyl RNA synthetase and peptide release factor, six tRNAs, a translation elongation factor EF-TU (which prokaryotes also produce), and translation initiation factor 1, which is also a prokaryotic translation factor. These genes are relevant to the key steps of translation: tRNA charging, initiation, elongation, and termination. The virus does not contain ribosome components.

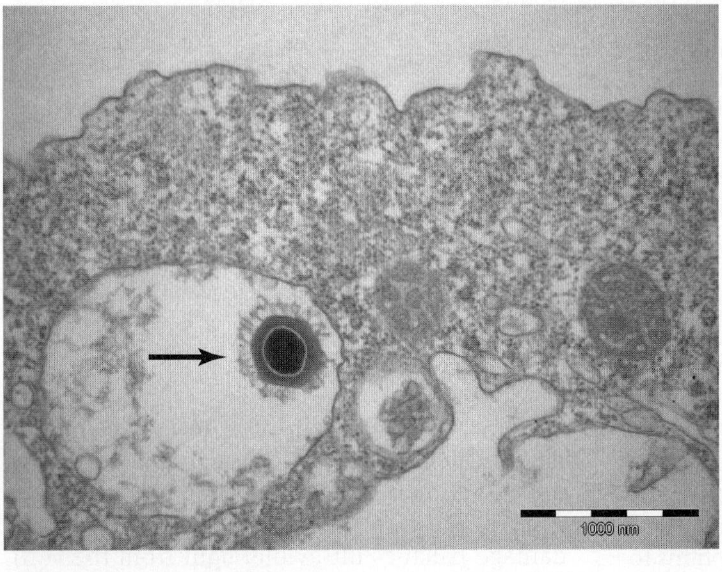

FIGURE CS 3-1 The arrow points to a Mimivirus particle infecting an amoeba.

References

Koonin, E. V. 2005. "Virology: Gulliver among the Lilliputians." *Curr Biol* 15(5):167–169.

La Scola, B., et al. 2003. "A giant virus in amoebae." *Science* 299:2033.

La Scola, B., et al. 2005. "Mimivirus in pneumonia patients." *EID* 11:449–452.

Ogata, H., Raoult, D., and Claverie, J. M. 2005. "A new example of viral intein in Mimivirus." *Virology* 2(1):8.

Pennisi, E. 2004. "The birth of the nucleus." *Science* 305:766–768.

Raoult, D. 2005. "The journey from Rickettsia to Mimivirus." *ASM News* 71(6):278–284.

Raoult, D., et al. 2004. "The 1.2-megabase genome sequence of Mimivirus." *Science* 306:1344–1350.

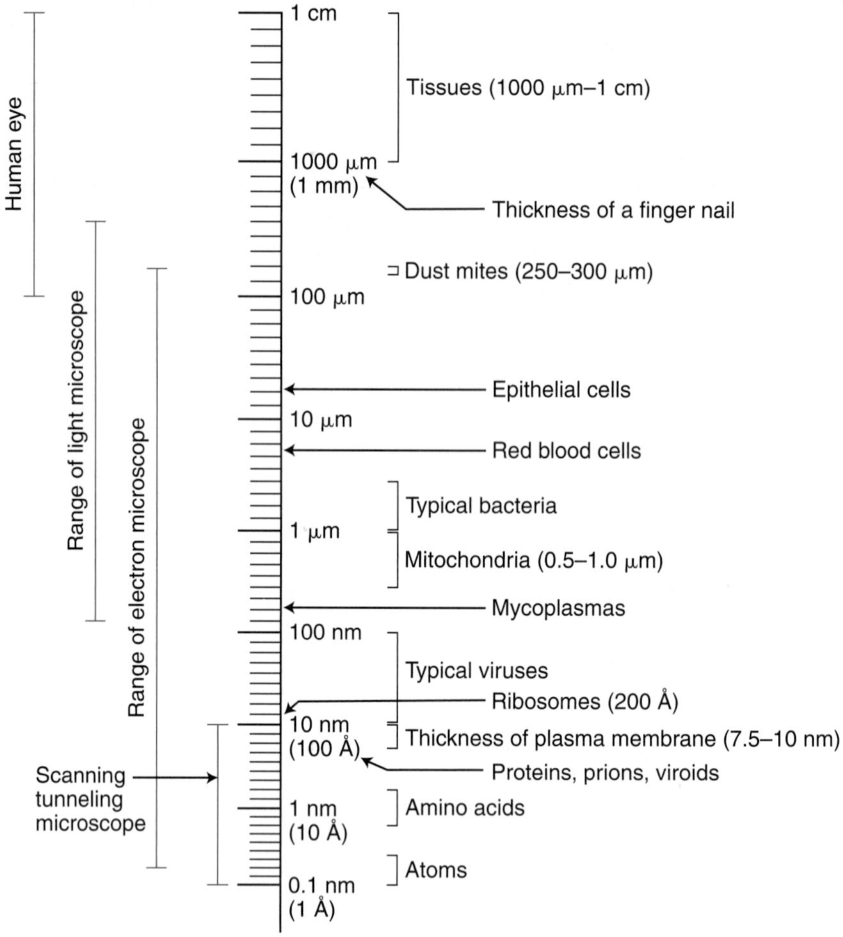

FIGURE 3-1 The limits of microscopic and human eye resolution. Each major division represents a 10-fold change in scale. To the right of the scale are approximate sizes of molecules, cells, organelles, and tissues. Adapted from Prescott, L. M., Harley, J. P., and Klein, D. A. *Microbiology,* Sixth Edition, McGraw-Hill Higher Education, 2005.

This chapter will dissect the structure and function of a virus particle and address how viruses are classified and named. As more viruses are isolated and studied worldwide, a universal nomenclature system is being implemented to tie viral characteristics to virus names. The first definition of a virus was based on early experiments in which viruses were defined as entities that passed through filters that trap known bacteria. In other words, viruses were smaller than bacteria. The size of a typical virus is in the nanometer (nm) range, as opposed to a bacterium in the micrometer (μm) range (**FIGURE 3-1**). More recently, virologists are unraveling the tremendous diversity among viruses, including their capsid architecture and size, genome size, mechanisms of replication, and interaction with host cells. There are now exceptions to this original definition of a virus that was based on size criteria. Giant viruses that grow within amoebas can be visualized with a light microscope. Mimivirus and Mamavirus contain genetic material that codes for more genes than numerous bacteria and archaea. In 2008, French researchers discovered a virus that can infect a virus! The new virus was named Sputnik. It is 50 nm in size and found inside of Mamavirus.

3.1 Virus Properties

Why are virus *particles* formed? The nucleic acid genome of a virus is fragile. After leaving the host cell, a virus enters a hostile environment as it travels between cells. The following virus properties are conducive to the survival of viruses in nature.

Protection

The environment outside of its host would quickly degrade an unprotected viral genome. The unprotected genome is sensitive to physical damage (such as shearing by mechanical forces), chemical damage (such as ultraviolet light from the sun), and enzymatic damage. The natural environment is laden with nucleases either derived from leaky or dead cells, or nucleases deliberately secreted by mammals as a defense mechanism against infection. The breaking of a single **phosphodiester bond** or chemical modification of one nucleotide may inactivate a virus containing a single-stranded genome. This event would render replication of the genome impossible, thereby inactivating the virus particle. Hence, the genetic material of the virus particle requires protection.

Recognition

The outer surface of the virus particle is involved in the recognition and interaction or attachment with the host cell during the first step of the virus life cycle. For naked viruses this interaction is mediated via the protein **capsid** of the virus particle, and for enveloped viruses a virally encoded envelope protein mediates this.

Self-Assembly

Virus particles are essentially "molecular packages." The simplest type of particle consists of a protein shell that can form around the viral genome spontaneously. This formation process is called **self-assembly**. The process requires minimal energy and is reversible. It results in a very stable structure that can assemble and disassemble itself. In addition to virus particles, biological membranes can also self-assemble (**FIGURE 3-2**). The ability to self-assemble is a characteristic of viruses with small genomes. For example, the structural proteins of polioviruses can assemble themselves into capsids (**FIGURE 3-3**).

Fidelity

DNA replication, RNA transcription, and protein synthesis are vulnerable to an occasional error. The size of the genome in a small virus particle cannot code for more than a few types of proteins of limited size. These genes code for small proteins, minimizing the chance for an occasional error. As a result, the particle composition is of high fidelity in terms of precision and accuracy.

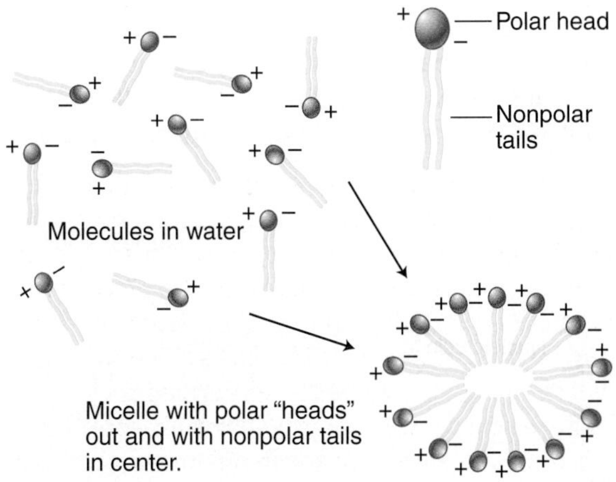

FIGURE 3-2 When put in water, these membrane molecules automatically organize themselves into more complex and biologically useful structures. This process is termed self-assembly. Modified from an illustration by Paul Decelles, Ph.D., Johnson County Community College.

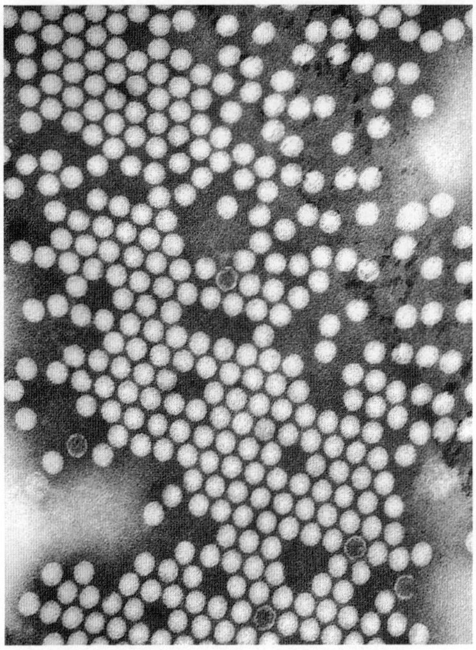

FIGURE 3-3 Electron micrograph of purified poliovirus particles. The particles self-assemble into uniform 29 nanometer (nm) virions. Magnification ~200,000×.

Economy

An icosahedron is the regular geometric figure that approaches a sphere. A sphere has the most volume for the least amount of surface area. This is economical. In addition, the same type of protein is used over and over again in creating the virus particle. An orderly, functional arrangement of protein subunits creates a structure that is frugal in expenditure of genome coding capacity. If one protein is synthesized or folded inaccurately, it can be discarded. In addition, in many viruses a single protein can have many different functions.

3.2 Virus Structure and Morphology

Stability

Why are virus particles composed of subunits? The capsid of the virus is composed of many viral structural proteins that come together to form identical subunits called **capsomers**. The capsomers are arranged symmetrically around the viral genome (**FIGURE 3-4**). The final structure is very stable. All viruses in the same family have capsids with the same number of capsomers.

FIGURE 3-4 (a) Electron micrograph and molecular model of an adenovirus. Adenoviruses consist of 252 capsomers (repeating protein subunits). Arrows indicate the penton fibers that extend outside of the particle and are used in attachment of the particle to a host cell. The fibers can easily become detached during preparation for electron microscopy. **(b)** Molecular model of adenovirus.

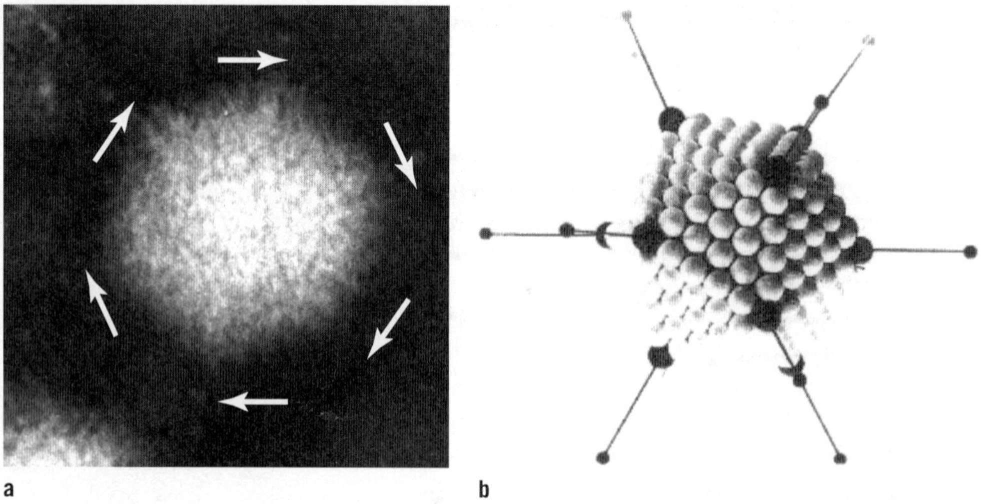

Subunit construction of the virus capsid is necessary for two reasons. First, the genome size limits the number of viral protein products that is encoded by the virus genome. Thus, the solution is to use the same protein product over and over again to generate identical subunits to trap the genome inside of the virus particle. Second, the nucleic acid genome can only code for a protein that is 15% of its weight. This is based on the premise that the average molecular weight of 3 nucleotides (a triplet codon that encodes one amino acid) is 1000 daltons. The average molecular weight of an amino acid is 150 daltons. Thus, multiple proteins (capsomers) are needed to contain the viral genome inside of the virus particle.

The term **structural protein** refers to those proteins that make up the virus particle (both inside and outside). Each type of protein functions in one or more of the following ways: by protecting the viral nucleic acid, attaching to receptors on the surfaces of host cells, penetrating the cell membrane, being involved in replication of the viral genome (for example, RNA-dependent RNA polymerases of RNA viruses), and facilitating replication and modifying the host cell.

Naked or Enveloped

Viruses that do not contain an envelope are called **naked viruses**. Some human and animal viruses have an additional lipid bilayer membrane wrapped around the capsid of the virus particle. These viruses are called **enveloped viruses**. The lipid bilayer membrane is stolen from the host cell as new viruses bud and exit from the host cell following infection and replication. This lipid bilayer may come from the host cell's plasma (outer) membrane, the nuclear membrane, or its trans Golgi network. (Refer to the introduction of Chapter 2.)

Two Types of Viral Shapes

Icosahedral (**FIGURE 3-5a**) and **helical** (Figure 3-5b) are the dominant capsid shapes of viruses. These shapes allow the virus particle to consist of a tightly constructed, stable, molecular package. The more subunits that form the structure of the particle, the more stable the virus particle becomes. A paper pattern of a three-dimensional icosahedral model of human rhinovirus type 14 is provided in **FIGURE 3-6**. Why not photocopy this illustration and construct a virus? Also, an interesting aside:

FIGURE 3-5 (a) Schematic drawing of a typical icosahedral-shaped virus particle. Adapted from G. E. Kaiser. "Viruses." *Microbiology Lecture Guide.* The Community College of Baltimore County, Catonsville Campus, 2006. http://student .ccbcmd.edu/courses/bio141/lecguide/index.html. **(b)** Schematic drawing of a typical helical-shaped virus particle. Adapted from Kaiser, G. E. "Viruses." *Microbiology Lecture Guide.* The Community College of Baltimore County, Catonsville Campus, 2006. http://student.ccbcmd.edu/courses/bio141/lecguide/index.html.

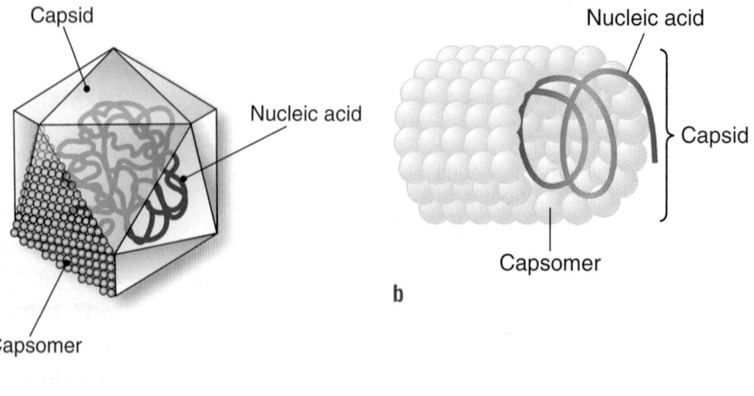

Three-dimensional model of human rhinovirus type 14

We have studied with great interest a structural projection of the human rhinovirus type 14 capsid, which recently appeared in an article by Rossmann *et al.* (*TIBS* Vol. 12, No. 8, pp. 313–318). Based upon their projection we have designed a pattern for constructing a three-dimensional model of the capsid. It is our belief that assembling and examining such a model can aid a person in understanding the spatial associations between those proteins which comprise the capsid structure. We would like to share our model with fellow readers of *TIBS* so that they may join us in having a pleasant learning experience.

The large triangular areas in the structure represent the 20 faces of the capsid which together form the shape of a regular icosahedron. Each of the 180 small quadrangles superimposed over the faces of the model represents the position of a single molecule of viral protein (VP) 1, 2, or 3. The other protein constituent, VP4, is not shown on this model for sake of simplicity.

In order to prepare a model from this illustration, we suggest that you first copy it onto another piece of paper (this will avoid damage to this issue of *TIBS*). You should then cut out your copied illustration along the outside edges, leaving all numbered tabs attached. Next, fold down all of the faces and tabs along the lines of the capsid edges (lighter lines in the model). Your model can then be assembled by matching together, one at a time, those pairs of tabs with the same identifying number.

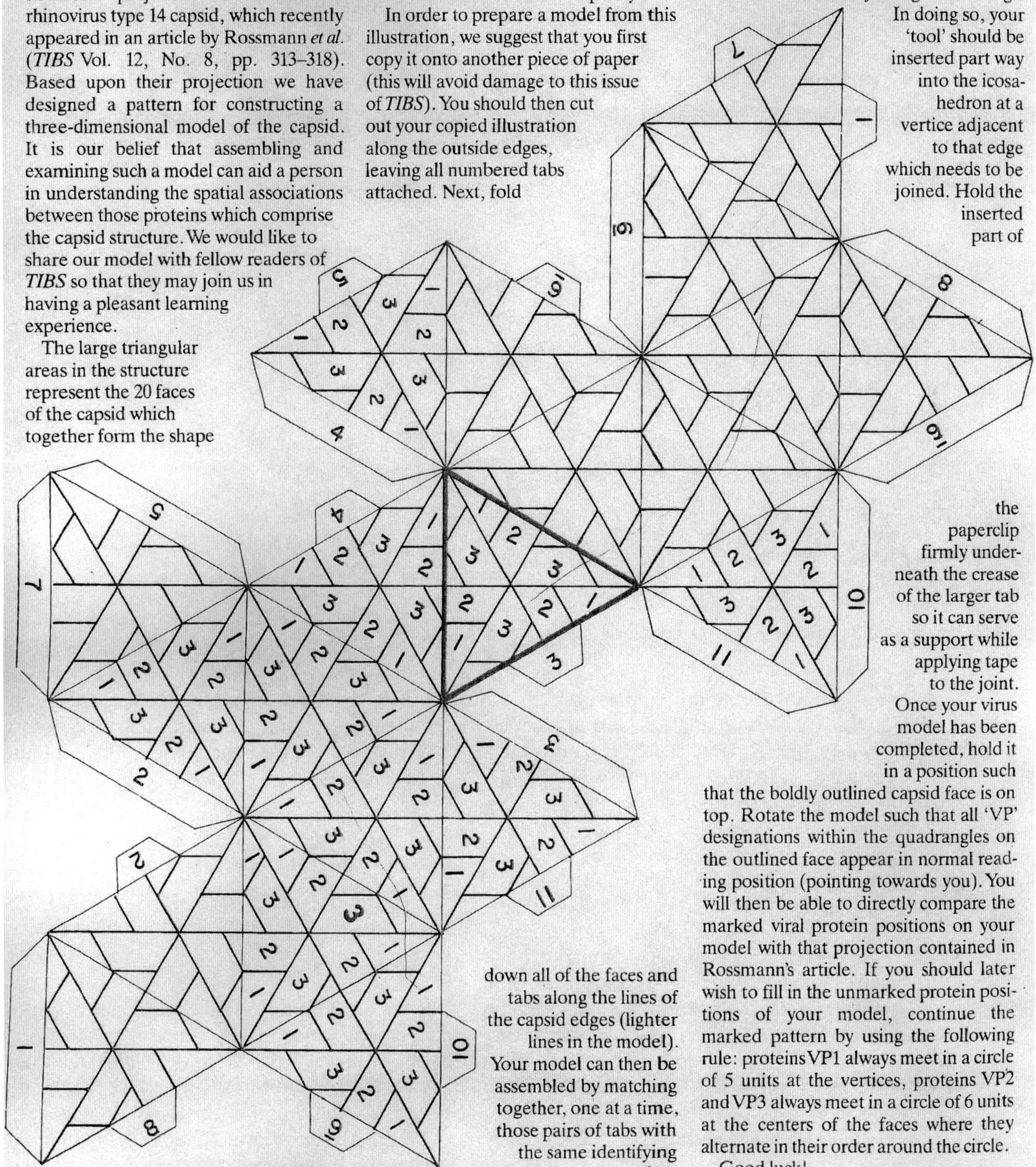

As you match together each pair of identically numbered tabs, they should be tucked to the inside of the icosahedral structure. Those edges of faces which are matched can then be joined by placing pieces of transparent tape over the outside of the model. The order in which these tabs are matched is not of importance. A paperclip which has been straightened and then bent at a right angle will serve as a handy tool when joining the last edge. In doing so, your 'tool' should be inserted part way into the icosahedron at a vertice adjacent to that edge which needs to be joined. Hold the inserted part of the paperclip firmly underneath the crease of the larger tab so it can serve as a support while applying tape to the joint.

Once your virus model has been completed, hold it in a position such that the boldly outlined capsid face is on top. Rotate the model such that all 'VP' designations within the quadrangles on the outlined face appear in normal reading position (pointing towards you). You will then be able to directly compare the marked viral protein positions on your model with that projection contained in Rossmann's article. If you should later wish to fill in the unmarked protein positions of your model, continue the marked pattern by using the following rule: proteins VP1 always meet in a circle of 5 units at the vertices, proteins VP2 and VP3 always meet in a circle of 6 units at the centers of the faces where they alternate in their order around the circle.

Good luck!

CHRISTON J. HURST
WILLIAM H. BENTON
JULIA M. ENNEKING
Health Effects Research Laboratory, U.S. Environmental Protection Agency, Cincinnati, OH 45268, USA.

FIGURE 3-6 Three-dimensional model of human rhinovirus type 14. A reprint of Hurst, C. J., Benton, W. H., and Enneking, J. M. 1987. "Three-dimensional model of human rhinovirus type 14." *TIBS* 12:460.

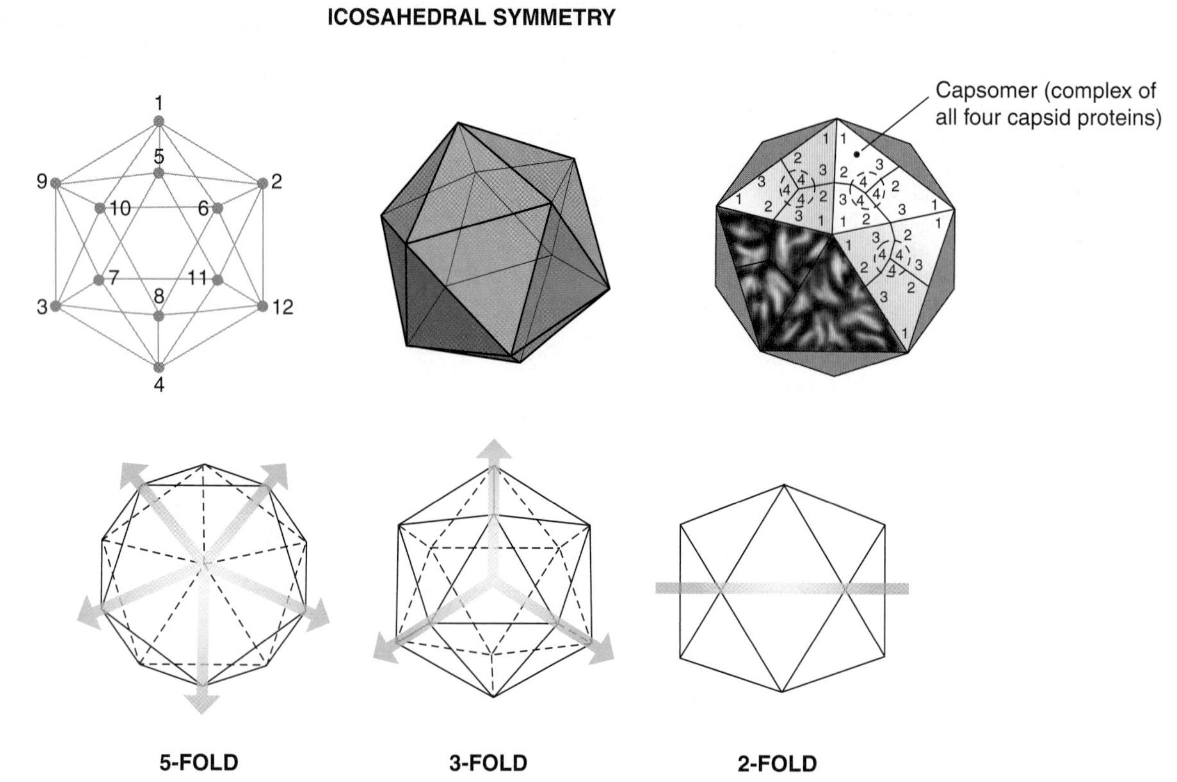

FIGURE 3-7 Schematic drawing showing the 12 vertices of an icosahedron, including a three-dimensional view of an icosahedron and its 5-3-2 symmetry. In addition, it depicts the structure of poliovirus that is composed of 60 capsomers. Each capsomer consists of 4 capsid proteins.

quilters have become fascinated with the icosahedral pattern.

An icosahedron is a polyhedron that has 20 identical equilateral triangular faces or planes (sides) and 12 vertices (**FIGURE 3-7**). If you rotate the icosahedron, it has 2-3-5 symmetry. Six 5-fold axes of symmetry pass through the vertices. Ten 3-fold axes extend through each face. Fifteen 2-fold axes pass through the edges of the icosahedron. A virus with 2-3-5 symmetry requires a multiple of 60 units to cover the surface completely. The capsid of poliovirus is made up of 60 capsomers. Each capsomer contains 1 copy of each viral structural protein.

Larger virus particles can hold a bigger and more complex genome. There are some viruses that don't fit the typical icosahedral or helical design. These viruses are called **complex** viruses. Examples of viruses with these rare shapes are members of the Poxviridae and Asfarviridae families (**FIGURE 3-8**). Members of the Poxviridae and Asfarviridae families are large dsDNA viruses (~175–270 nm in diameter) that contain large genomes (~170–190 kbp in length) and have complicated architectures.

Enveloped viruses are represented by icosahedral (**FIGURE 3-9a**), helical (Figure 3-9b), or complex shapes. Electron microscopic and x-ray crystallographic techniques allowed virologists to determine the overall morphology of a particular virus. Examples of electron micrographic images representing the three basic shapes of viruses are found in **FIGURES 3-10a–c**.

3.3 Viruses That Challenge the Definition of a Virus

Giruses

Viruses are often thought of as "bags of genes" that package enough information to interact with their host and to hijack the host's cellular machinery for multiplying tiny virus particles. The recent discovery of Mimivirus has shed a new light upon the definition of a virus. The "giant" DNA-containing viruses such as Mimivirus are called **giruses** to emphasize their exceptional size and unique properties. Giruses are not filterable through the typical "sterilizing" filters with a 0.2 to 0.3μm pore size. Mimivirus was the first virus of the same dimensions and genomic complexity as typical parasitic organisms. It was isolated from a cooling

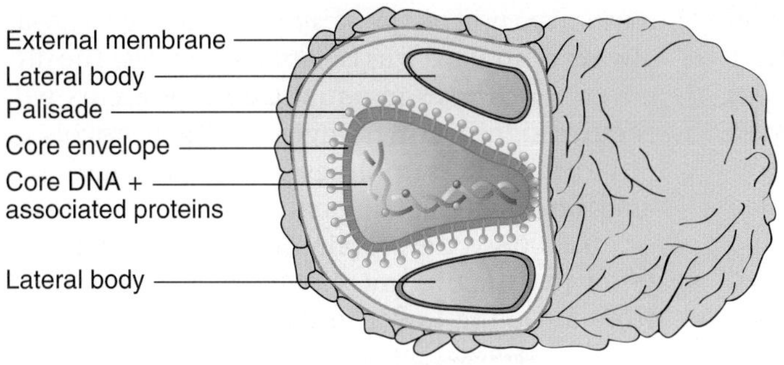

FIGURE 3-8 Schematic drawing of a complex-shaped poxvirus particle. Adapted from Decoster, A., et al. "Virology." Microbiology. Découvrir le Groupe Hospitalier de l'Institut Catholique de Lille, 2007.

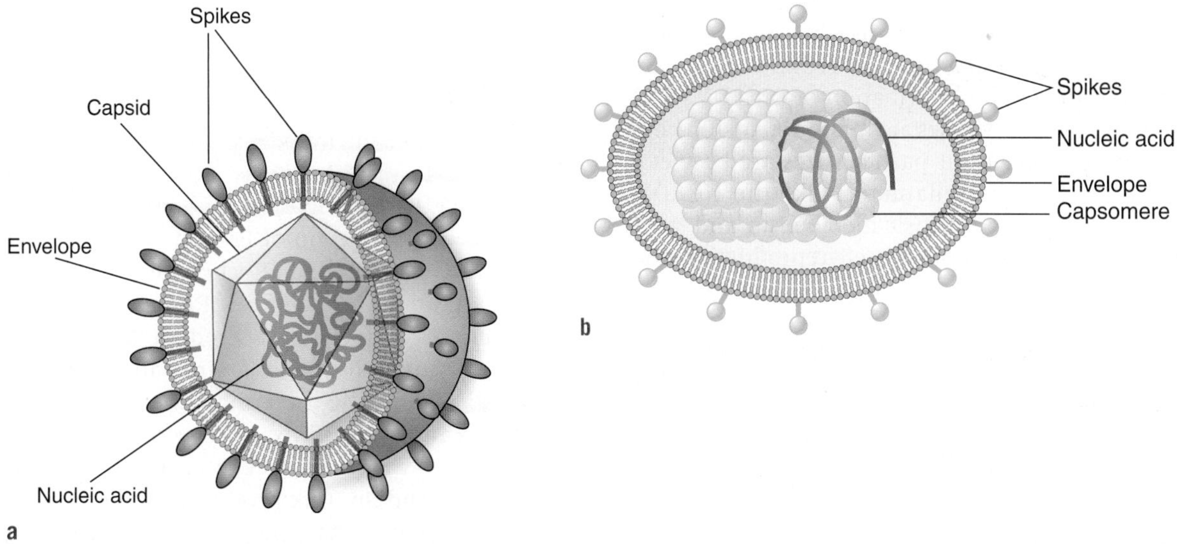

FIGURE 3-9 (a) Schematic drawing of a typical enveloped icosahedral-shaped virus particle. Adapted from Kaiser, G. E. "Viruses." *Microbiology Lecture Guide.* The Community College of Baltimore County, Catonsville Campus, 2006. http://student .ccbcmd.edu/courses/bio141/lecguide/index.html. **(b)** Schematic drawing of a typical enveloped helical-shaped virus particle. Adapted from Kaiser, G. E. "Viruses." *Microbiology Lecture Guide.* The Community College of Baltimore County, Catonsville Campus, 2006. http://student.ccbcmd .edu/courses/bio141/lecguide/index.html.

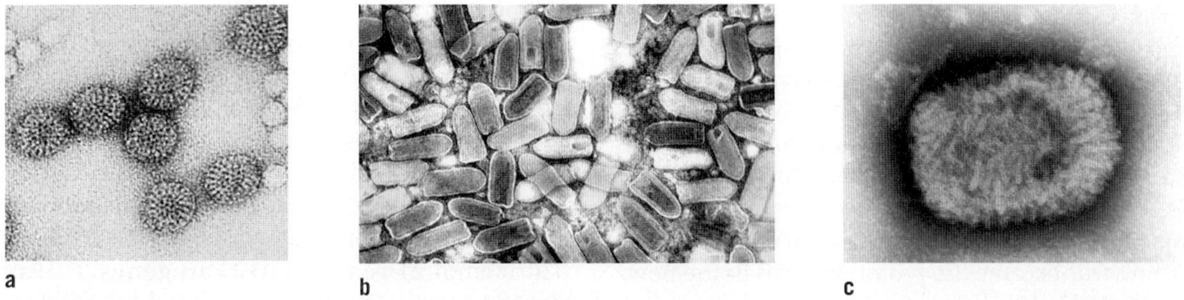

FIGURE 3-10 (a) Electron micrograph of icosahedral-shaped rotavirus particles. Rotaviruses have a wheel-like appearance. **(b)** Electron micrograph (160,000× magnification) of helical-shaped vesicular stomatitis virus (VSV). VSV is an important pathogenic virus of cattle, causing fever and vesicles in the mouth and on the feet. Magnification ~ 40,000×. **(c)** Electron micrograph of a single brick-shaped or complex-shaped smallpox particle. This virion is from a human skin lesion, which is from a diagnostic specimen that was sent to the CDC in 1966 as part of the WHO global smallpox eradication program. Magnification 150,000×.

tower in Bradford, England, during a pneumonia investigation. (Refer to the discussion in Chapter 1 on the origins of viruses.) The name Mimivirus was derived from "microbe mimicking virus" because it was initially identified and described as a Gram-positive coccus-shaped bacterium that resembled a *Legionella*-like bacterium. It was first named *Bradfordcoccus* until it was further characterized and determined to be a virus that morphologically resembles members of a group of viruses that infect eukaryotic hosts referred to as Nucleo-Cytoplasmic Large DNA Viruses (NCLDV). Iridoviruses (infect insects and cold-blooded vertebrates), asfarviruses (infect mammals), and phycodnaviruses (infect green algae) are examples of NCLDV. (Refer to Chapter 4 for more information on NCLDV.) The genomes of these viruses are in the 100 to 400 kilobase pair range. Their particle sizes are between 160 and 200 nm in diameter. Like the NCLDV, Mimivirus has no intrinsic metabolism and lacks the machinery required for self-reproduction and protein synthesis.

Mimivirus has a linear dsDNA 1200 kilobase pair (1.2 megabase pair) complex genome that codes for about 911 gene products (three times more genes than previously known large viruses and some bacteria). It is the largest known virus—a monster virus! The Mimivirus particle is a fiber-covered icosahedral protein capsid that has a diameter of 750 nm, or 0.75 μm (**TABLE 3-1**, **FIGURE 3-11**). It is twice the size of a small bacterium such as *Mycoplasma*. Mimivirus was discovered within the free-living amoeba host *Acanthamoeba polyphaga*. *A. polyphaga* is a pathogenic amoeba that is ubiquitous in air, soil, and aquatic environments. In the laboratory, Mimivirus is propagated in *Acanthamoeba castellanii*. A feeding phagocytosis pathway normally used by the amoeba to feed on bacteria takes up the virus. Mimivirus particles are encased in a 125-nm-thick polysaccharide layer, making it even more palatable for its amoebic host. The Gram-staining of the virus is likely due to its lipopolysaccharide (LPS)-like layer. Bacteria contain LPS layers. The initial step of phagocytosis by the amoeba is triggered by particle size in the μm range. Mimivirus became the prototype virus in the new Mimiviridae family.

The properties of Mimivirus that led to its misinterpretation as an amoeba-infecting bacterium might actually be important to the lifestyle of many other Giruses infecting their cellular hosts via a phagocytic route. If this is true, protozoan feeding on bacteria should be analyzed more carefully in search of new Giruses. These giant viruses are not an oddity. *Mimivirus is likely the first representative of the giant aquatic viruses, Giruses, infecting a wide variety of phagocytic protists.* Particle size was always central to virus isolation protocols. The discovery of Giruses in aquatic environments will require significant changes to the protocols of environmental samplings for studies aiming at assessing microbial diversity.

In 2008, a new strain of Mimivirus was isolated by inoculating *A. polyphaga* with water from a cooling tower in Paris, France. The new Mimivirus strain was called **Mamavirus** because it appeared to be even larger than Mimivirus when observed by transmission electron microscopy. Mamavirus shares similar particle morphology to Mimivirus.

Virophages

In 2008, an icosahedral virus, 50 nm in size, was found **inside** of Mamavirus (a close relative of Mimivirus). It is the first example of a virus that infects a virus! A group of French researchers at the Centre National de la Recherche Scientifique (National Center for Scientific Research), a basic research organization located in Marseilles, France, discovered this new virus and named it **Sputnik**, after the first man-made satellite. Sputnik cannot multiply in *A. castellanii* in the laboratory. *A. polyphaga* cells co-infected with Mamavirus and Sputnik produced fewer and deformed Mamavirus particles, however, suggesting that Sputnik hinders Mamavirus. The team suggested that Sputnik is a **virophage** because it behaves like bacteriophages that infect and debilitates bacteria. Sputnik infects and sickens the amoebic Mamavirus. It multiplies rapidly after an **eclipse** phase, in the Girus **factory** found in an amoeba co-infected with Mamavirus. (Refer to Chapter 4 about life cycles of a virus, e.g., eclipse phase.) Mamavirus acts as a "**helper virus**." A helper virus is a virus whose replication aids the development of a defective virus or **satellite virus** into fully infectious particles.

Sputnik is a 50-nm icosahedral particle that contains an 18,343 base pair circular dsDNA genome that codes for 21 gene products (**FIGURE 3-12**). Thirteen of 21 genes are **ORFan genes**. ORFan genes are genes that have no detectable homologs in the current sequence database. (Refer to the discussion in Chapter 1 on the origins of viruses.) The other 8 non-ORFan genes have viral, plasmid, bacterial, or eukaryotic homologs. Three of the Sputnik gene products are related to Mimivirus/Mamavirus gene products, suggesting horizontal gene transfer between the two viruses in the same way that bacteriophages ferry genes between bacteria.

TABLE 3-1 Virion Sizes and Shapes of Selected Families of Viruses

Family	Representative Viruses	Shape	Approximate Diameter of Virion	Enveloped or Naked	Chapter(s)
Adenoviridae	Adenovirus-36	Icosahedral	75 nm	Naked	10
Baculoviridae	*Autographa californica* nuclear polyhedrosis virus	Helical	30–60 nm diameter × 250–300 nm length	Enveloped	18
Bunyaviridae	Sin Nombre Hantavirus	Icosahedral	112 nm in diameter	Enveloped	1, 19
Caliciviridae	Norovirus	Icosahedral	27–32 nm in diameter	Naked	1, 19
Closteroviridae	Citrus Tristeza Virus	Flexus, Filamentous, Helical	12 nm in diameter × 1250–2200 nm in length	Naked	21
Coronaviridae	Severe Acute Respiratory Syndrome Associated Coronavirus	Irregular shaped	100 nm in diameter	Enveloped	1, 19
Filoviridae	Ebola Marburg	Helical	80 nm in diameter × 9000–14,000 nm in length	Enveloped	8, 19
Flaviviridae	West Nile Virus Yellow Fever Virus Dengue Fever Virus Hepatitis C Virus	Icosahedral	45–50 nm in diameter	Enveloped	1, 18, 19
Hepadnaviridae	Hepatitis B	Icosahedral	42 nm in diameter (Dane particle)	Enveloped	10, 17
Herpesviridae	Herpes Simplex 1 (causes cold sores) Herpes Simplex II (causes genital herpes)	Icosahedral	105 nm in diameter	Enveloped	15
Mimiviridae	Mimivirus	Icosahedral	750 nm in diameter	Enveloped	4
Orthomyxoviridae	Influenza A	Icosahedral, Pleomorphic	80–120 nm in diameter	Enveloped	12
Papillomaviridae	Papillomaviruses type 16 and 18	Icosahedral	52–55 nm in diameter	Naked	10
Parvoviridae	Parvovirus B19	Icosahedral	20 nm in diameter	Naked	4
Picornaviridae	Poliovirus Human Enteroviruses Hepatitis A	Icosahedral	30 nm in diameter	Naked	11
Poxviridae	Smallpox Vaccinia Molluscum contagiosum Monkeypox	Complex	270–350 nm in diameter	Enveloped	14
Rhabdoviridae	Rabies	Helical	75 nm in diameter × 180 nm in length	Enveloped	13
Reoviridae	Rotavirus	Icosahedral	77 nm in diameter	Naked	7
Retroviridae	Human Immunodeficiency Virus	Icosahedral	100 nm in diameter	Enveloped	
Siphoviridae	λ phage	Binary (head and tail structure)	63 nm in diameter, 200 nm in length	Naked icosohedral shell, flexible tail	22
Togaviridae	Chikungunya Virus	Icosahedral	65 nm in diameter	Enveloped	19
Virgaviridae	Tobacco Mosaic Virus	Helical	18 nm in diameter × 300 nm in length	Naked	1, 21

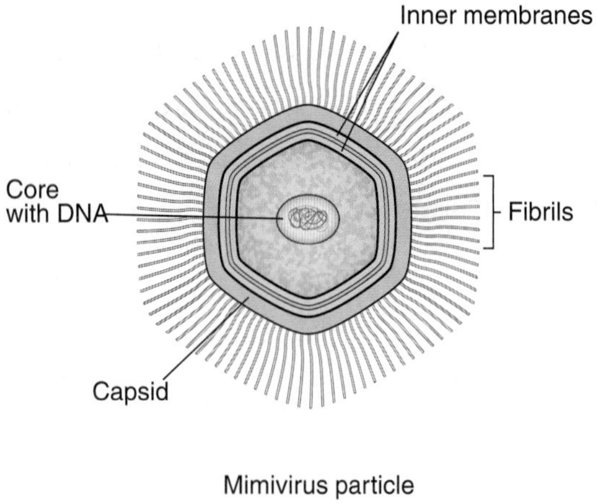

FIGURE 3-11 Structure of the Mimivirus particle.

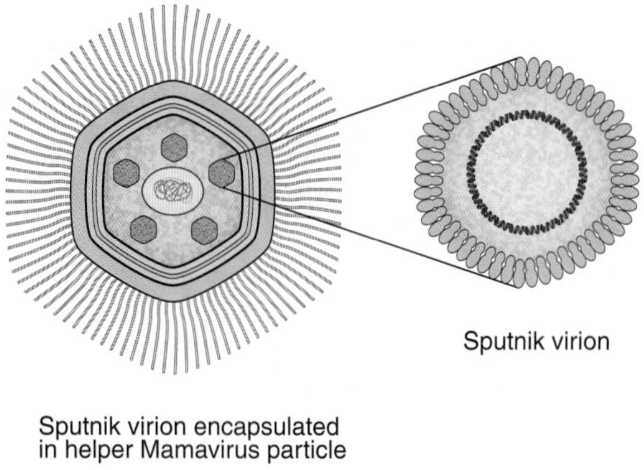

FIGURE 3-12 Structure of the virophage Sputnik found inside of Mamavirus.

3.4 Baltimore Classification

Viruses can be classified or grouped by such criteria as size, capsid symmetry, whether or not they are naked or enveloped, mode of transmission, the host they infect, the type of nucleic acid they contain, or the diseases they cause. Nobel-prize laureate David Baltimore proposed a system to classify viruses based on the type of nucleic acid the virus contains and its mechanism of replication. His scheme is based on the theme that all viruses must generate +ssmRNAs from their genomes in order to produce proteins and replicate themselves. The precise mechanism by which this is achieved is different for each virus family. Baltimore divided the viral genomes into seven groups or classes. The replication strategies for them are discussed in Chapter 4. The seven classes are (**TABLES 3-2** and **3-3**):

I. dsDNA viruses
II. ssDNA viruses
III. dsRNA viruses
IV. +ssRNA viruses
V. –ssRNA viruses
VI. RNA reverse transcribing viruses
VII. DNA reverse transcribing viruses

3.5 Taxonomy: What's in a Name?

In 1898, Friedrich Loeffler and Paul Frosch identified the first animal virus, the one that causes foot and mouth disease (FMD). Their experiments demonstrated that a filterable agent causes the disease. The virus was not isolated until the 1930s, when the first actual studies of viruses began. As described in Chapter 1, FMD virus causes one of the most infectious animal diseases known. Symptoms include the formation of fluid-filled vesicles on the mucosa, face, feet, and hairless areas that rupture to form ulcers. Hence the name **foot and mouth** disease.

In 1901, while on duty in Cuba, U.S. Army medical scientist Walter Reed and his colleagues conducted an exhaustive series of experiments demonstrating that the **yellow fever virus** was transmitted between humans by mosquitoes, and that the disease could be experimentally transferred via filtered blood. In yellow fever infections, the patient produces black-colored blood-filled vomit, and there is a deterioration of the liver, kidneys, and heart. Yellow bile pigment from the damaged liver colors the skin and gives rise to the name of the disease and its causative agent.

These early experiments in virology that led to the discovery of viruses were based upon the *size* of the infectious agent. In other words, any infectious entity that was filterable was termed a ***virus***. Porcelain Chamberland "candle-style" filters (**FIGURE 3-13**) were used by Beijerinck, Ivanovsky, Loeffler, and Frosch to isolate the first plant and animal viruses. These studies were facilitated by the advent of the transmission electron microscope. The need to classify and maintain a consistent naming system became a practical necessity. It was particularly important for the identification of those viruses causing disease in humans, domesticated animals, and plants.

The earliest attempts to *classify* viruses were based only on size. Viruses were named by their pathogenic properties, common organ tropisms, and common ecological and transmission characteristics; for example, if we used this system today, all of the viruses that cause hepatitis would belong in a "hepatitis virus" classification. Hepatitis A, B, and C viruses, yellow fever virus, and Rift Valley

TABLE 3-2 Genome Properties of Selected DNA Virus Families

Family	Representative Viruses	Genome Type	Average Genome Size	Average Number of Genes or ORFs*	Chapter(s)
Adenoviridae	Adenovirus-36	dsDNA	31,000 bp	27	10
Baculoviridae	Autographa californica nuclear polyhedrosis virus	dsDNA	130,000 bp	125	18
Hepadnaviridae	Hepatitis B	dsDNA	3,200 bp	7	10, 17
Herpesviridae	Herpes Simplex 1 (causes cold sores) Herpes Simplex II (causes genital herpes)	dsDNA	150,000 bp	150	15
Mimiviridae	Mimivirus	dsDNA	1,200,000 bp	911	4
Papillomaviridae	Papillomaviruses type 16 and 18	dsDNA	8,000 bp	7	10
Parvoviridae	Parvovirus B19	ssDNA	5,600	3	4
Polyomaviridae	Simian Virus 40	dsDNA	5,300 bp	7	10
Poxviridae	Smallpox (Variola major or minor) Vaccinia Molluscum contagiosum Monkeypox	dsDNA	200,000 bp	200	14, 19
Siphoviridae	λ phage	dsDNA	48,500 bp	68	22

*ORFs are open reading frames

TABLE 3-3 Genome Properties of Selected RNA Virus Families

Family	Representative Viruses	Genome Type	Average Genome Size	Average Number of Genes or ORFs	Chapter(s)
Bunyaviridae	Sin Nombre Hantavirus	–ssRNA 3 segments	12,500 b	4	1, 19
Caliciviridae	Norovirus	+ssRNA	7,400 b	3	1, 19
Closteroviridae	Citrus Tristeza Virus	+ssRNA	19,200 b	12	21
Coronaviridae	Severe Acute Respiratory Syndrome Associated Coronavirus	+ssRNA	29,700 b		1, 19
Filoviridae	Ebola Marburg	–ssRNA	19,000 b	7	8, 19
Flaviviridae	West Nile Virus Yellow Fever Virus Dengue Fever Virus Hepatitis C Virus	+ssRNA	11,000 b	10	1, 18, 19
Orthomyxoviridae	Influenza A	–ssRNA 8 segments	12,000 b	10	12
Picornaviridae	Poliovirus Human Enteroviruses Hepatitis A	+ssRNA	7,400 b	9	11
Rhabdoviridae	Rabies	–ssRNA	15,000 b	5	13
Reoviridae	Rotavirus	dsRNA 11 segments	18,500 bp	22	7
Retroviridae	Human Immunodeficiency Virus	+ssRNA (2 copies are present inside the genome)	9,750 b	10	16
Togaviridae	Chikungunya	+ssRNA	11,700 b	8	18, 19
Virgaviridae	Tobacco Mosaic Virus	+ssRNA	6,400 b	4	1, 21

FIGURE 3-13 **(a)** Candle Chamberland filters. **(b)** Modern Chamberland filter apparatus. If a mixture of viruses and bacteria is filtered through a bacterial-proof filter, the viruses will pass through into the filtrate in the flask.

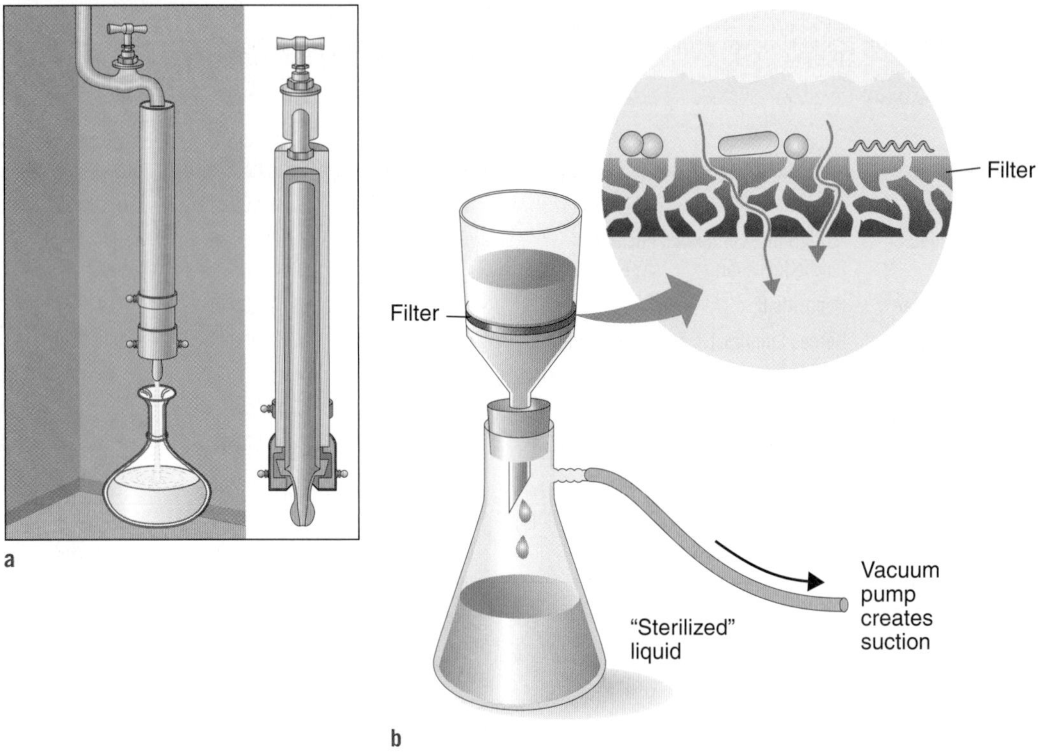

fever virus would be grouped together. Today, virologists know that genetically these viruses are very different and thus have placed them in different families based on morphology, physical and chemical characteristics (protein, lipid, and carbohydrate content), antigenic (serologic relationships) properties, and biological properties.

In the 1950s and 1960s there was a tremendous surge in the discovery of new viruses. The influx of new knowledge prompted the need for a classification scheme for these newly discovered viruses. For some time, the development of a universal taxonomic system was the subject of competing and conflicting opinions among virus researchers. Historically, viruses have been named after the diseases they caused, morphological characteristics (**FIGURE 3-14**), places the viruses were first isolated, their discoverers, and acronyms. See **TABLE 3-4** for examples.

FIGURE 3-14 Scanning electron micrograph of SARS associated coronaviruses (SARS-CoV). The viral envelope has petal-shaped spikes, giving the coronavirus an appearance of a crown (Latin *corona*). The inset is an electron micrograph of a negatively stained SARS-CoV. Note the short and stubby spikes visible on the surface of the virus.

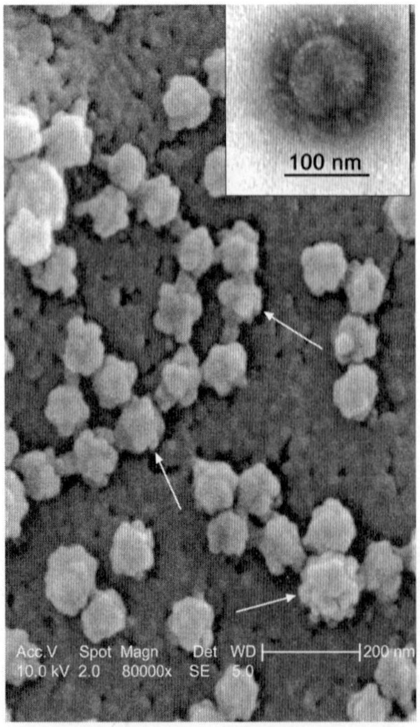

3.6 The International Committee on Taxonomy of Viruses

In 1966, the International Committee on Taxonomy of Viruses (ICTV) was formed. It was charged with developing a universal taxonomic scheme to address the nomenclature and classification of viruses. By that time there was a rapidly growing mass of data on individual viruses, and the universal taxonomic system has continued to evolve.

The system developed by the ICTV is based on the following hierarchical levels. The names are designated with suffixes:

Order (-virales)
Family (-viridae)
Subfamily (-virinae)
Genus (-virus)
Common names

TABLE 3-4 Examples of the Origins of Virus Names

Virus/Type of Virus	Origin of Name
Adenovirus	First isolated from adenoids and tonsils.
Coronavirus	Virion morphology shown in electron micrographs of negatively stained coronaviruses gave them the appearance of a *crown* (Latin *corona*; see Figure 3-14).
Coxsackie	First isolated from children with a poliolike illness in Coxsackie, New York, in 1948.
Ebola	Named after the Ebola River located in the northern Democratic Republic of the Congo (formerly Zaire), where the virus first appeared in 1976.
Epstein-Barr	Named after M. A. Epstein and Y. M. Barr, who, along with B. G. Achong, discovered the virus.
Hendra	Hendra is the suburb of Brisbane, Queensland, Australia, where the virus first appeared in 1994.
Hepatitis A, B, C	Associated with hepatitis (liver malfunction, yellowing of skin).
Herpesvirus	Term comes from the Greek word meaning "to creep." It was observed that the sores seemed to creep over the surface of the skin.
HIV	Acronym for human immunodeficiency virus.
La Crosse encephalitis	In 1963, this virus was first isolated from the brain of a four-year-old boy who died of encephalitis in *La Crosse*, Wisconsin.
Mamavirus	Relative of Mimivirus but has a larger genome than Mimivirus.
Marburg	1967 Marburg, Germany, outbreak.
Mimivirus	Microbe-mimicking virus.
Nipah	In 1999, Nipah was the first village the virus struck near Kuala Lumpur in Malaysia.
Norwalk	Was first identified in 1972 after an outbreak of gastrointestinal illness in Norwalk, Ohio.
Papillomavirus	Induces warts or papillomas.
Papovavirus	*Pa*pilloma *po*lyoma *va*cuolating agent
Picornavirus	*Pico:* small, *rna*: ribonucleic acid
SARS-CoV	Acronym for severe acute respiratory syndrome-*co*ronavirus.
Smallpox (Variola major or minor)	"Small pox" was a term used to distinguish this disease from syphilis, which was known as the "great pox."
West Nile	Virus first isolated from a woman located in the West Nile District of Uganda in 1937.

To date, only three orders have been approved by the ICTV: Mononegavirales, Caudovirales, and Nidovirales; for example, the order Mononegavirales consists of four virus families: Bornaviridae, Paramyxoviridae, Rhabdoviridae, and Filoviridae. The ICTV's most recent report, which was published in 2009, lists 6 orders, 87 families (65 virus families are not assigned to an order), 19 subfamilies, 348 genera, and more than 5450 viruses belonging to more than 2285 virus species. Each year, an ICTV Master Species list is posted to the ICTV website (http://talk.ictvonline.org). Descriptions of **viroids** and **prions** are included in the ICTV's Eighth Report. **TABLE 3-5** lists examples of full formal taxonomic terminology.

A taxonomic key used for the placement of viruses can be found in *Virus Taxonomy, Seventh Report of the ICTV*, which was published in 2000 (**FIGURE 3-15**). Many molecular characteristics and measurements have been used to create the taxonomic key. Some of the properties used in taxonomy are virion properties such as size, shape, the presence or absence of an envelope, and capsid symmetry and structure. Physical and chemical properties are also considered. These include:

- Virion molecular weight
- pH, temperature, cation, detergent, and irradiation stability

TABLE 3-5 Examples of Common and Scientific Names of Viruses

Common Name of Virus	Scientific Name
Vaccinia	Family Poxviridae, subfamily Chordopoxvirinae, genus Orthopoxvirus, *vaccinia* virus
Poliovirus 1	Family Picornaviridae, genus Enterovirus, poliovirus 1
Rabies	Order Mononegavirales, family Rhabdoviridae, genus Lyssavirus, rabies virus
Herpes simplex virus 2	Family Herpesviridae, subfamily Alphaherpesvirinae, genus simplex virus, human herpes virus 2 (herpes simplex virus 2)

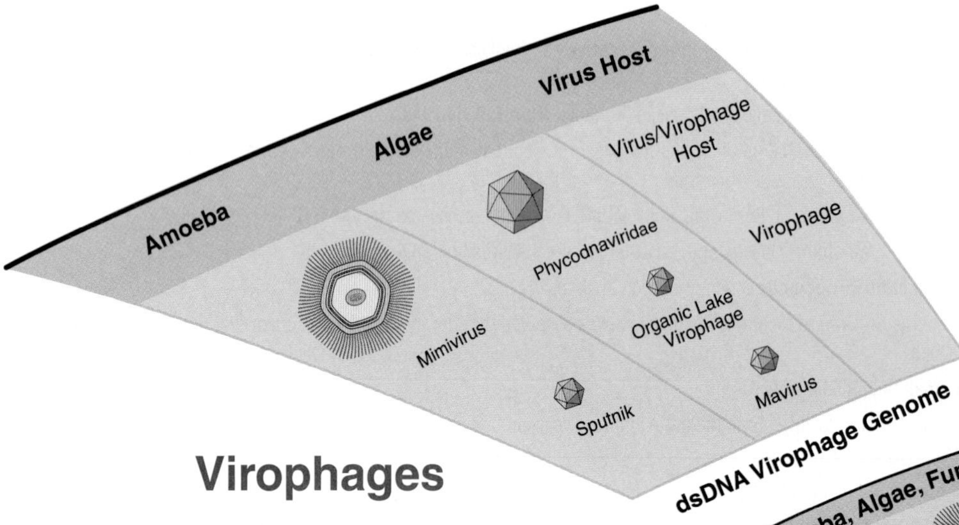

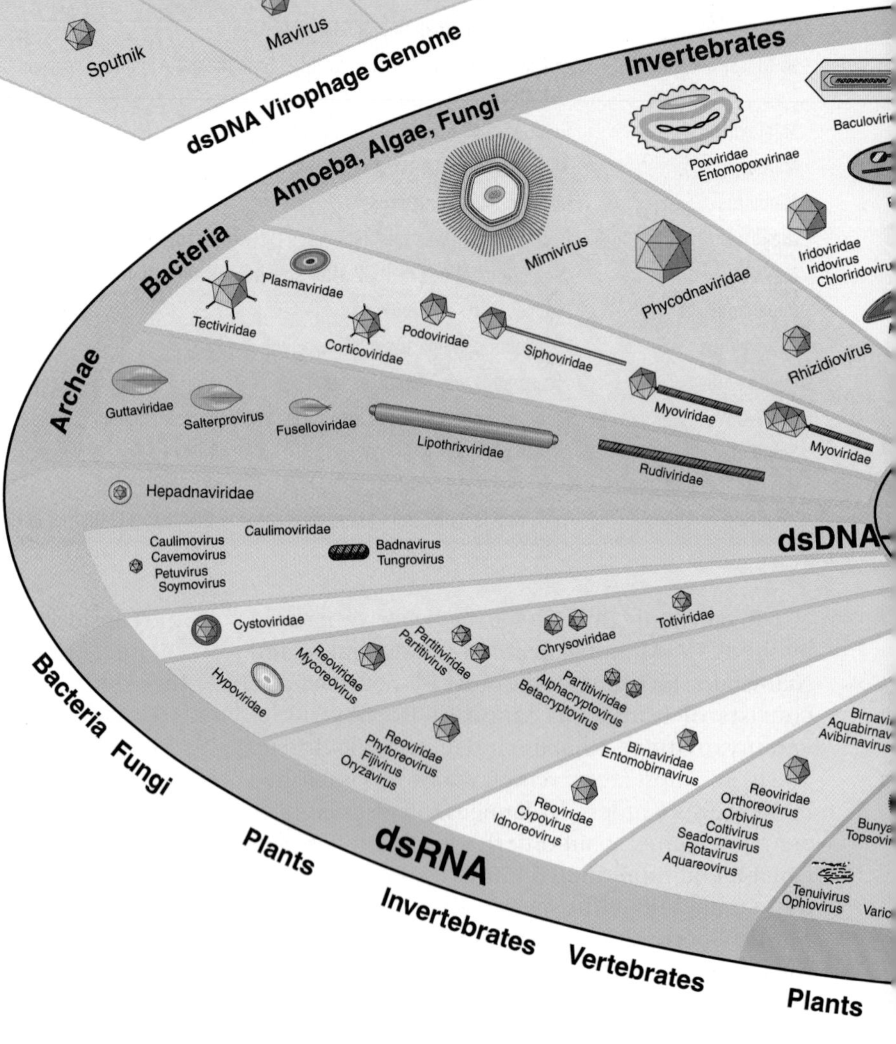

FIGURE 3-15 Compilation of viral diversity. Courtesy of Dr. Claude M. Fauquet, Director of the ILTAB.

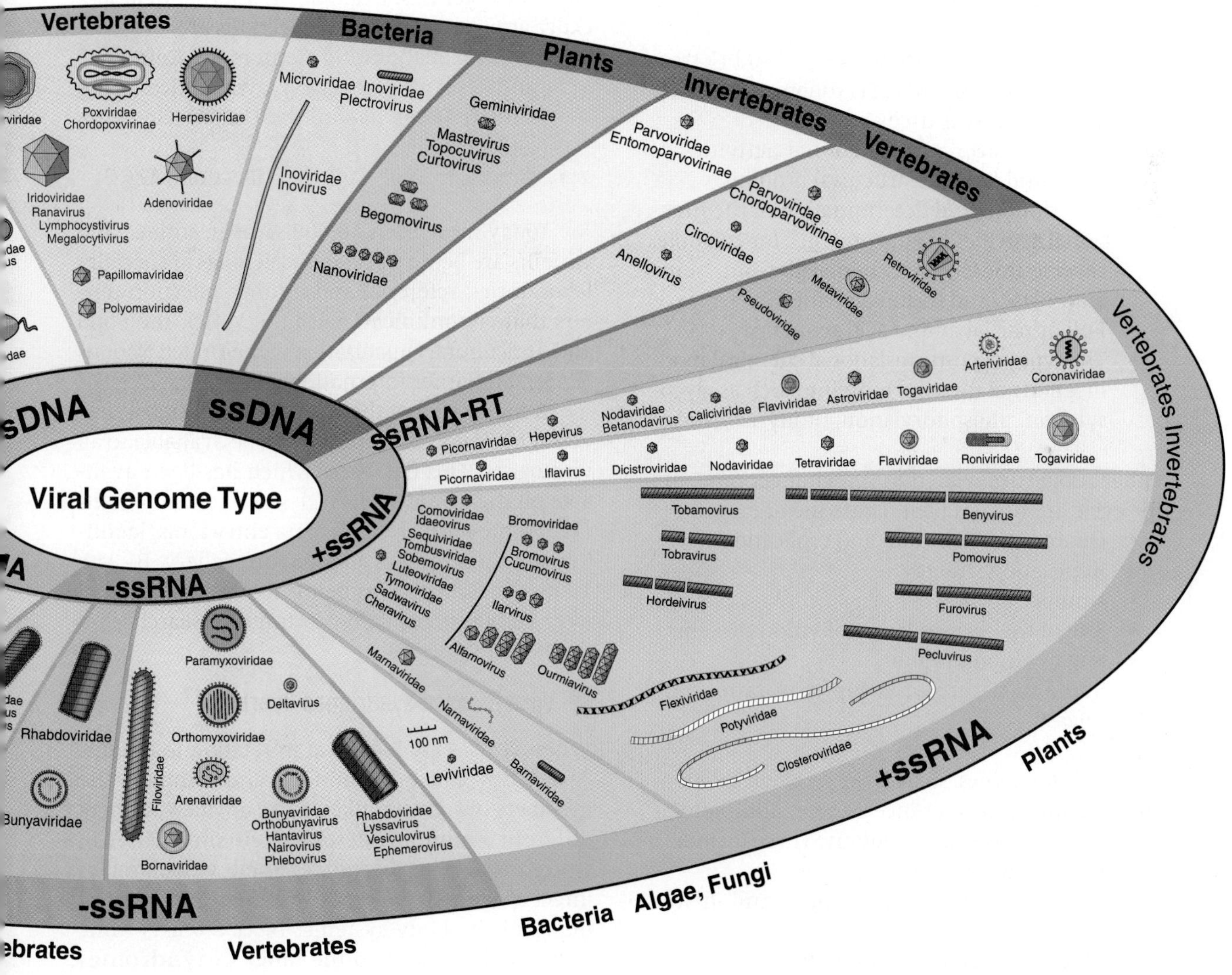

Vertebrates
Bacteria
Plants
Invertebrates
Vertebrates
Poxviridae
Chordopoxvirinae
Herpesviridae
Iridoviridae
Ranavirus
Lymphocystivirus
Megalocytivirus
Adenoviridae
Papillomaviridae
Polyomaviridae
Microviridae
Inoviridae
Plectrovirus
Inoviridae
Inovirus
Geminiviridae
Mastrevirus
Topocuvirus
Curtovirus
Begomovirus
Nanoviridae
Parvoviridae
Entomoparvovirinae
Parvoviridae
Chordoparvovirinae
Circoviridae
Anellovirus
Retroviridae
Metaviridae
Pseudoviridae
Vertebrates Invertebrates
Arteriviridae
Coronaviridae
Picornaviridae
Hepevirus
Nodaviridae
Betanodavirus
Caliciviridae
Flaviviridae
Astroviridae
Togaviridae
Picornaviridae
Iflavirus
Dicistroviridae
Nodaviridae
Tetraviridae
Flaviviridae
Roniviridae
Togaviridae
ssDNA
ssRNA-RT
Viral Genome Type
+ssRNA
-ssRNA
Comoviridae
Idaeovirus
Sequiviridae
Tombusviridae
Sobemovirus
Luteoviridae
Tymoviridae
Sadwavirus
Cheravirus
Bromoviridae
Bromovirus
Cucumovirus
Ilarvirus
Alfamovirus
Ourmiavirus
Tobamovirus
Tobravirus
Hordeivirus
Benyvirus
Pomovirus
Furovirus
Pecluvirus
Flexiviridae
Potyviridae
Closteroviridae
Marnaviridae
Narnaviridae
100 nm
Leviviridae
Barnaviridae
+ssRNA
Plants
Paramyxoviridae
Deltavirus
Rhabdoviridae
Orthomyxoviridae
Filoviridae
Arenaviridae
Bunyaviridae
Orthobunyavirus
Hantavirus
Nairovirus
Phlebovirus
Rhabdoviridae
Lyssavirus
Vesiculovirus
Ephemerovirus
Bornaviridae
-ssRNA
Vertebrates
Bacteria
Algae, Fungi

Viruses

- Genome properties: type of nucleic acid (DNA or RNA), strandedness (ss or ds), linear or circular, and polarity (positive-sense, negative-sense, ambisense)
- Complete or partial nucleotide sequence of the viral genome
- Presence of repetitive sequence elements within the genome
- Guanine + cytosine (G + C) ratio of the genome
- Presence or absence and type of 5′ end terminal cap
- Presence or absence of 5′ end terminal covalently linked protein
- Presence or absence of 3′ terminal poly(A) tail

Lipid, carbohydrate, protein content, and characteristics are also considered in taxonomy. The virion protein characteristics used are:

- Number, size, and functional activities of structural and nonstructural proteins
- Descriptions and determination of the functional activities of viral proteins: for example, reverse transcriptase, hemagglutinin, neuraminidase, and fusion activities
- Full or partial amino acid sequence
- The types of posttranslational chemical modifications of the viral proteins (such as glycosylation, phosphorylation, or myristylation)

Other properties related to the viral genome and life cycle are used as well. These include:

- Genome organization and replication
- Replication strategy
- Number of ORFs
- Site of the accumulation of virion proteins
- Site of virion assembly
- Site and nature of virion maturation and release

Antigenic properties, such as the serologic relationships obtained at reference centers and biologic properties, are used in taxonomy. These include:

- Natural host range
- Tissue tropisms, pathology, and histopathology
- Mode of transmission in nature
- Geographic distribution

International specialty groups and culture collection centers establish lower hierarchical levels. These levels and their definitions are:

- **Species**—(1991 ICTV's accepted definition) a **polythetic class** of viruses that share the same replicating lineage and occupy a particular niche. (A polythetic class is defined as a class whose members share several properties but no individual members possess all properties.)
- **Type-species**—different serotypes of the same species.
- **Strain**—different isolates of the same virus that come from different geographical locations.
- **Variant**—a virus whose phenotype differs from the original wild-type strain but for which the genetic basis is not known.

These specialty groups may be influenced or guided by the World Health Organization (WHO) and other agencies. Examples of international specialty groups and culture collections are:

- WHO Collaborating Centers for Reference and Research on Influenza, with offices in the United States, Japan, the United Kingdom, and Australia.
- American Type Culture Collection (ATCC)

Today more than 30,000 viruses, strains, and variants are being tracked by scientists at specialty laboratories, reference and culture collection centers that communicate with the WHO, the Food and Agricultural Organization of the United Nations (FAO), and other international entities.

A major goal of the ICTV was to design a universal ICTV database that would be available to all virologists. This database, which has been available on the World Wide Web since 1993, contains taxonomic software tools, data entry tools, identification tools, and even a picture gallery. By late 1998, the ICTVdB was being accessed at a rate of over 10,000 times daily via Internet searches on the World Wide Web.

Viral Disease Syndromes Overlap

When viruses are looked at in relation to the diseases they cause in humans and animals, the members of some families cause a wide variety of different symptoms despite their similarities in structure, size, and genome. Some virus families infect specific tissue types, whereas others infect many of the body systems. **TABLE 3-6** lists some families of viruses and the range of **syndromes** they cause. Syndromes are defined as sets of symptoms and physical signs occurring together. Chapter 5 will discuss laboratory diagnosis of viral diseases. One may initially think that diagnosis of viral diseases may not be important because there are few antivirals available for treatment. However, diagnosis is indeed important despite the aforementioned fact.

TABLE 3-6 Virus Families and the Range of Syndromes They Cause

Syndrome	Picornaviridae	Togaviridae	Herpesviridae	Hepadnaviridae	Rhabdoviridae
Skin rashes	+	+	+		
Diarrhea	+				
Paralysis	+		+		
Hepatitis	+	+		+	
Rabies					+
Arthritis		+			
Respiratory illness	+		+		
Conjunctivitis	+		+		
Encephalitis		+	+		+
Meningitis	+	+	+		

Source: Modified and adapted from Collier, L., and Oxford, J. 2000. *Human Virology*, 2nd ed. Oxford, UK: Oxford University Press, p. 15.

VIRUS FILE 3-1 The Race to Characterize SARS-CoV

On February 28, 2003, Dr. Carlo Urbani (**FIGURE VF 3-1**), an infectious disease specialist based at the 60-bed French Hospital in Hanoi, Vietnam, alerted the WHO about a case of atypical pneumonia. He had originally suspected an avian influenza virus, but very quickly began to suspect a different, possibly new virus.

Of the first 60 SARS patients, more than half were healthcare workers. Dr. Urbani recognized the severity of the public health threat. His appeal for assistance from the WHO, CDC, and Medecins sans Frontieres (MSF, or Doctors without Borders) resulted in the successes of infection control and rapid identification of SARS.

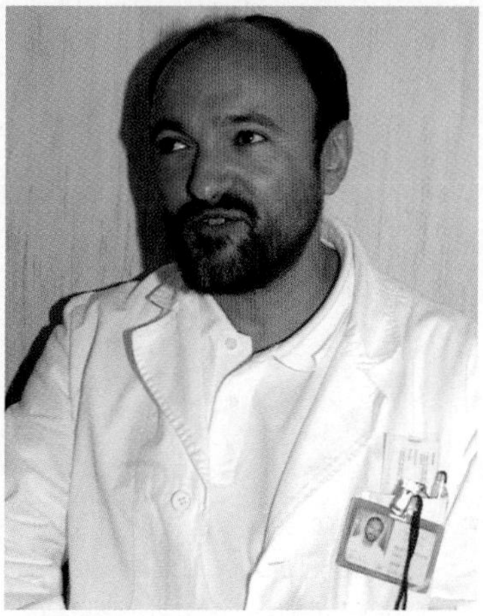

FIGURE VF 3-1 Dr. Carlo Urbani.

In just a few weeks, SARS spread via people through air travel to at least three continents. Researchers from no fewer than 10 countries collaborated to identify the virus, sequence its genome, and take steps toward rapid diagnosis. Dr. Urbani cared for SARS patients for 18 days in a hospital isolation ward in Hanoi. He died on March 29, 2003, while attending a medical meeting in Bangkok, Thailand. By dealing with the SARS outbreak intuitively and openly, the severity of the outbreak was greatly curtailed. On May 1, 2003, the genome sequence of the SARS-CoV was published via *Science Express*. These gene sequences enable researchers to more quickly develop diagnostic tests and vaccines to combat the rapidly spreading, highly contagious respiratory disease.

References

Marco, A. M., et al. 2003. "The genome sequence of the SARS-associated coronavirus." *Science* 300(5624):1399–1404.

Rota, P. A., et al. 2003. "Characterization of a novel coronavirus associated with severe acute respiratory syndrome." *Science* 300(5624):1394–1399.

Summary

Virus particles are stable molecular packages containing a protein coat wrapped around the nucleic acid genome of the virus. This protein coat provides protection to the genome and is involved in the recognition of host cells to initiate infection. There are two viral shapes that are most common in nature: icosahedral and helical. Electron microscopic and x-ray crystallographic techniques are important tools used to determine the overall structure and morphology of viruses.

The main criteria used by the ICTV to classify and name viruses were outlined in this chapter. The universal taxonomy system developed by the ICTV is based on the following hierarchical levels: order, family, subfamily, genus, and common names. Specialty laboratories currently track over 30,000 virus strains and variants. One or several different virus families may cause viral disease syndromes. Laboratory diagnosis of viral diseases will be addressed in Chapter 5.

CASE STUDY QUESTIONS

These questions relate to the Case Study presented at the beginning of the chapter.

1. Mimivirus has been proposed as the sole member of a new virus family Mimiviridae whose members are classified as large nucleo-cytoplasmic dsDNA viruses. This virus significantly challenges the evolution and definition of a virus. Discuss this.
2. Based on ribosomal RNA analysis, where would you tentatively propose to place Mimivirus in the three-domain "Tree of Life" hypothesized by Carl Woese? Dr. Woese recognized a third form of life—the archaea—whose genetic makeup is distinct from but related to both eubacteria (procarya) and eucarya.
3. Many have debated the issue of whether or not viruses are alive. Besides the inclusion of a partial translational machinery apparatus, Mimivirus carries other genes that have not been identified in any other known virus. Research these genes and their functions. (References to get you started follow.)

   Koonin, E. V. 2005. Virology: Gulliver among the Lilliputians. *Curr Biol* 15(5):167–169.

   La Scola, B., et al. 2003. A giant virus in amoebae. *Science* 299:2033.

   La Scola, B., et al. 2005. Mimivirus in pneumonia patients. *EID* 11:449–452.

   Ogata, H., Raoult, D., and Claverie, J. M. 2005. A new example of viral intein in Mimivirus. *Virology* 2(1):8.

   Pennisi, E. 2004. The birth of the nucleus. *Science* 305:766–768.

   Raoult, D. 2005. The journey from *Rickettsia* to Mimivirus. *ASM News* 71(6):278–284.

   Raoult, D., et al. 2004. The 1.2-megabase genome sequence of Mimivirus. *Science* 306:1344–1350.

   Use this research to support your hypothesis regarding whether or not Mimiviruses are alive.
4. Viruses and the nuclei of cells share some structural similarities. List these similarities.
5. From an evolutionary standpoint, could a virus have provided the first nucleus? Or was a nucleus something an early bacterial cell evolved on its own or in partnership with an archaeum?
6. Build a virus using the paper model provided in Figure 3-6.

Resources

Primary Literature and Reviews

Ackermann, H. -W., Berthiaume, L., and Tremblay, M. 1988. *Virus Diagrams*. Boca Raton, FL: CRC Press.

Adams, M. J., Antoniw, J. F., and Kreuze, J. 2009. "Virgaviridae: A new family of rod-shaped plant viruses." *Arch Virol* 154:1967–1972.

Boyer, M., et al. 2009. "Giant Marseillevirus highlights the role of amoebae as a melting pot in emergence of chimeric microorganisms." *PNAS* doi/10.1073/pnas.0911354106.

Claverie, J.-M., et al. 2006. "Mimivirus and the emerging concept of 'giant' virus." *Virus Res* 117:133–144.

Claverie, J.-M., et al. 2009. "Mimivirus and Mimiviridae: Giant viruses with an increasing number of potential hosts, including corals and sponges." *J Invert Pathol* 101:172–180.

Claverie, J.-M., and Abergel, C. 2009. "Mimivirus and its Virophage." *Annu Rev Genet* 43:49–66.

Cohen, J., and Powderly, W. G., eds. 2004. *Infectious Diseases*, 2nd ed. Philadelphia: Mosby.

Eberhard, M. 2004. "Virus taxonomy: One step forward, two steps back." *EID* 10:153–154.

Fauquet, C. M., et al., eds. 2005. *Virus Taxonomy Eighth Report of the International Committee on Taxonomy of Viruses*. London: Elsevier Academic Press.

Ghedin, E., and Claverie, J-M. 2005. "Mimivirus relatives in the Sargasso sea." *Virol J* 2 (Aug 16):62 doi:10.1186/1743-422X-2-62.

Hurst, C. J., Benton, W. H., and Enneking, J. M. 1987. "Three-dimensional model of human rhinovirus type 14." *TIBS* 12:460.

Knipe, D. M., Samuel, C. E., and Palese, P., eds. 2001. "Virus assembly." In *Fundamental Virology*. Philadelphia: Lippincott Williams & Wilkins.

La Scola, B., et al., 2008. "The virophage as a unique parasite of the giant mimivirus." *Nature* 455:100–104.

Marco, A. M., et al. 2003. "The genome sequence of the SARS-associated coronavirus." *Science* 300(5624): 1399–1404.

Matthews, R. E. F. 1985. "Viral taxonomy for the non-virologist." *Ann Rev Microbiol* 39:451–474.

Pearson, H. 2008. "'Virophage' suggests viruses are alive." *Nature* 454:7: 677.

Raoult, D., and Forterre, P. 2008. "Redefining viruses: Lessons from Mimivirus." *Nat Rev Microbiol* 6;315–319.

Rossman, M. G., et al. 1987. "Common cold viruses." *TIBS* 12:313–318.

Rota, P. A., et al. 2003. "Characterization of a novel coronavirus associated with severe acute respiratory syndrome." *Science* 300(5624):1394–1399.

Susan-Monti, M., et al. 2007. "Ultrastructural characterization of the giant volcano-like virus factory of *Acanthamoeba polyphaga* Mimivirus." *PLoS ONE* 2(3):e328. Doi:10.1371/journal.pone.0000328.

Tidona, C. A., and Darai, G., eds. 2001. *The Springer Index of Viruses*. New York: Springer.

van Regenmortel, M. H. V., and Mahy, B. W. 2004. "Emerging issues in virus taxonomy." *EID* 10(1):8–13.

van Regenmortel, M. H. V., et al., eds. 2000. *Virus Taxonomy Classification and Nomenclature of Viruses: Seventh Report of the International Committee on Taxonomy of Viruses*. San Diego: Academic Press.

Zandi, R., et al. 2004. "Origin of icosahedral symmetry in viruses." *PNAS* 101(44):15556–15560.

Popular Press

de Kruif, P., and Gonzalez-Crussi, F. 2002. "Walter Reed: in the interest of science—and for humanity!" In *Microbe Hunters*. San Diego: Harvest Books, an imprint of Harcourt Trade Publishers.

Radetsky, P. 1994. *The Invisible Invaders: Viruses and the Scientists Who Pursue Them*. Boston: Back Bay Books.

Turbak, S. 1991. "Turnip crinkle virus." *Quilting International*. July, pp. 18–22.

Villarreal, L. P. 2004. "Are viruses alive?" *Scientific American*. December 2004, Vol. 291, Issue 6, 100–105.

Video Productions

Killer Virus: Hunt for the Next Plague. 2009. Discovery Communications, Inc.

60 Minutes H1N1 Airdate 10/18/09. CBS Broadcasting Inc.

60 Minutes H1N1 Airdate 11/01/09. CBS Broadcasting Inc.

AIDS: Evolution of an Epidemic. 2008. Howard Hughes Medical Institute.

The Age of Viruses. 2006. Films for the Humanities.

AIDS and Other Epidemics. 2004. Films for the Humanities.

SARS and the New Plagues. 2003. The History Channel.

War Against Deadly Microbes and Lethal Viruses. 2003. Films for the Humanities.

Conquering an Invisible World. 1999. Films for the Humanities.

Another War: Disease and Political Strife. 1997. Films for the Humanities.

eLearning

go.jblearning.com/shors2

The site features eLearning, an online review area that provides quizzes and other tools to help you study for your class. You can also follow useful links for in-depth information, or just find out the latest virology and microbiology news.

CHAPTER 4

Virus Replication Cycles

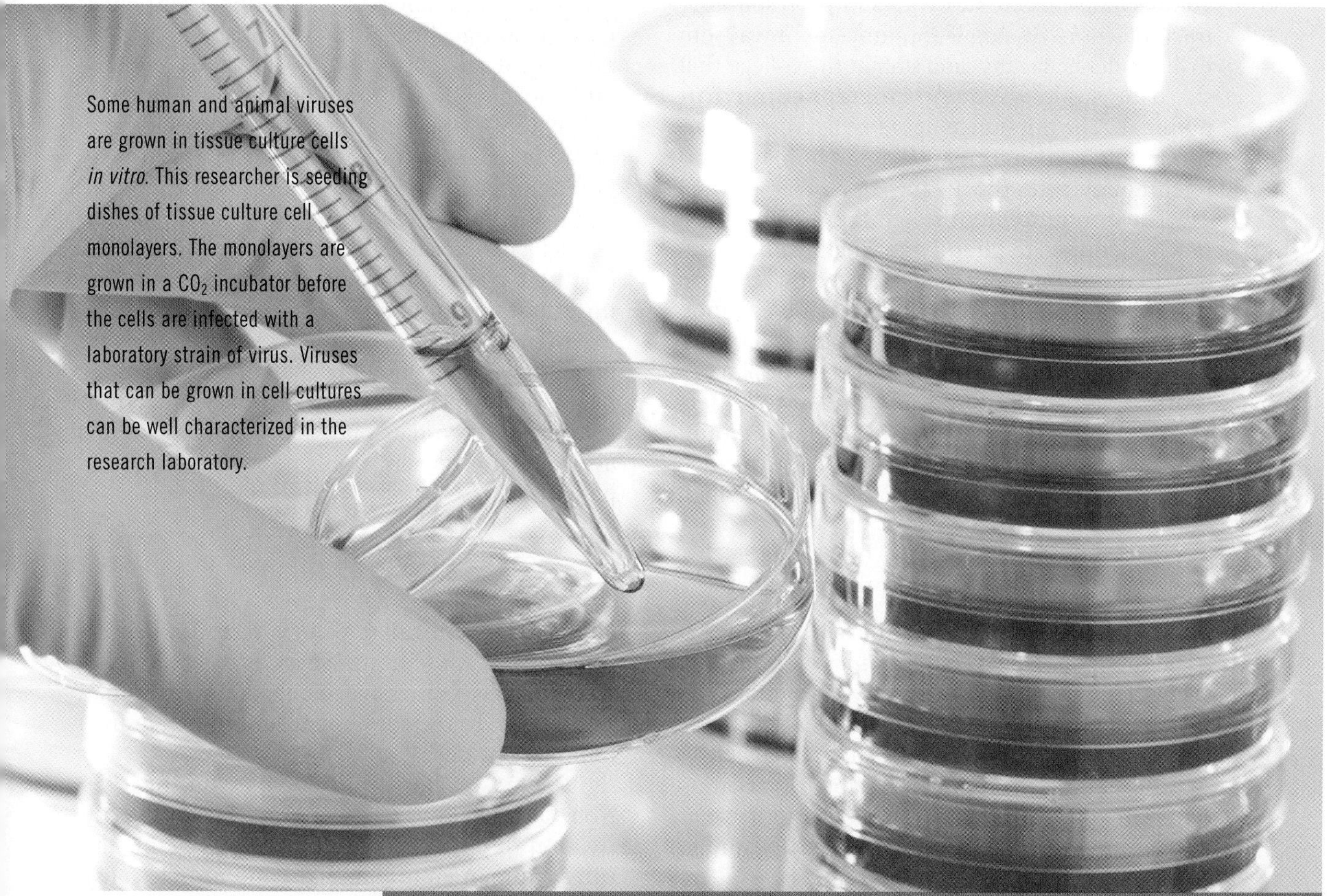

Some human and animal viruses are grown in tissue culture cells *in vitro*. This researcher is seeding dishes of tissue culture cell monolayers. The monolayers are grown in a CO_2 incubator before the cells are infected with a laboratory strain of virus. Viruses that can be grown in cell cultures can be well characterized in the research laboratory.

OUTLINE

"In the struggle for survival, the fittest win out at the expense of their rivals because they succeed in adapting themselves best to their environment."

Charles Darwin

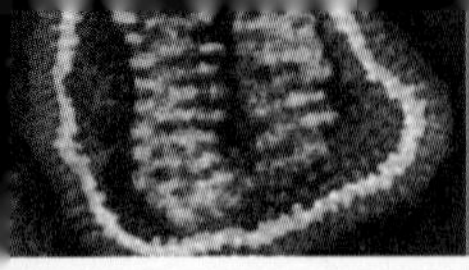

CASE STUDY: HUMAN METAPNEUMOVIRUS

The campus day care was recently closed during the peak of the winter flu season because many of the young children were sick with a lower respiratory tract infection. An email announcement was sent to all students, faculty, and staff at the college that stated the closure was due to a **metapneumovirus** outbreak. The announcement briefed the campus community with information about human metapneumonoviruses (hMPVs).

The announcement stated that hMPV was a newly identified respiratory tract pathogen discovered in the Netherlands in 2001. New tests confirm that it is one of the most significant and common viral infections in humans. It is clinically indistinguishable from a viral relative known as respiratory syncytial virus (RSV). Both RSV and hMPV infections occur during the winter. hMPV may account for 2% to 12% or more of previously unexplained pediatric lower respiratory infections for which samples are sent to diagnostic laboratories, and a lesser percentage in adults.

Both hMPV and RSV cause upper and lower respiratory tract infections associated with serious illness in the young, immunosuppressed, elderly, and chronically ill. Common symptoms include cough, fever, wheezing or exacerbation of asthma, and rhinorrhea (runny nose). It can produce severe enough symptoms to cause intensive care admission and ventilator support. Healthy adults can get a mild form of the disease that is characterized by a cough, hoarseness, congestion, runny nose, and sore throat.

Members of the campus day care staff were doing their best to limit the epidemic spread of the hMPV outbreak at the center. Primary care physicians need up-to-date knowledge and heightened awareness to recognize this new viral disease in patients.

In Chapter 2, you learned that viruses are dependent upon host cells for their reproduction, yet to do so viruses must overcome certain cellular constraints. Only those viruses that have been able to adapt to their hosts have been able to exist in nature. The previous chapter focused on taxonomy and structure of viruses and in summary stated that several different virus families cause similar viral disease syndromes.

This chapter focuses on experiments such as **one-step growth curves**, which are used to study virus–host interactions. These studies have provided information about the events that occur at each step of the infection cycle (**attachment**, **penetration**, **uncoating**, **replication**, **assembly**, **maturation**, and **release**), including the intricate details of the strategies that animal and human viruses use to express and replicate their diverse genomes.

4.1 One-Step Growth Curves

Virologists could not study animal and human viruses well in the laboratory before tissue culture methods were developed by John F. Enders, Thomas H. Weller, and Frederic C. Robbins in the late 1940s. Before their work, viruses were injected into animals and tissues were analyzed for the pathological signs of viral infection. Experimental animals were difficult to work with and expensive to maintain. Another drawback was that animals were not very permissive to infection with human viruses due to the species barrier and the animal immune response.

Enders, Weller, and Robbins won the 1954 Nobel Prize in Physiology or Medicine for the cultivation of poliovirus in nonnervous tissue cultures (human embryonic skin and muscle cells). Their observations and procedures used to grow viruses *in vitro* contributed to the refinement of tissue culture techniques and played a monumental role in the development of vaccines against poliovirus in the 1950s (Salk vaccine) and 1960s (Sabin vaccine).

One-step or **single-step growth curve experiments** are used to study a single replication cycle of viruses. Max Delbruck developed the one-step growth curve experiment while using an *Escherichia coli*-T4 bacterial system, which gave faster experimental results than did traditional methods. With the advent of cell culture systems, these experiments were carried out with viruses that infect tissue culture cells. These experiments are performed in special cell culture facilities.

Briefly, **monolayers** (or cell suspensions in liquid medium) of tissue culture cells such as monkey kidney cells are allowed to adhere and form monolayers on the bottom of plastic dishes. The monolayers of cells are subsequently infected with the virus of choice. They are infected at a high **multiplicity of infection** (**MOI**) to ensure that

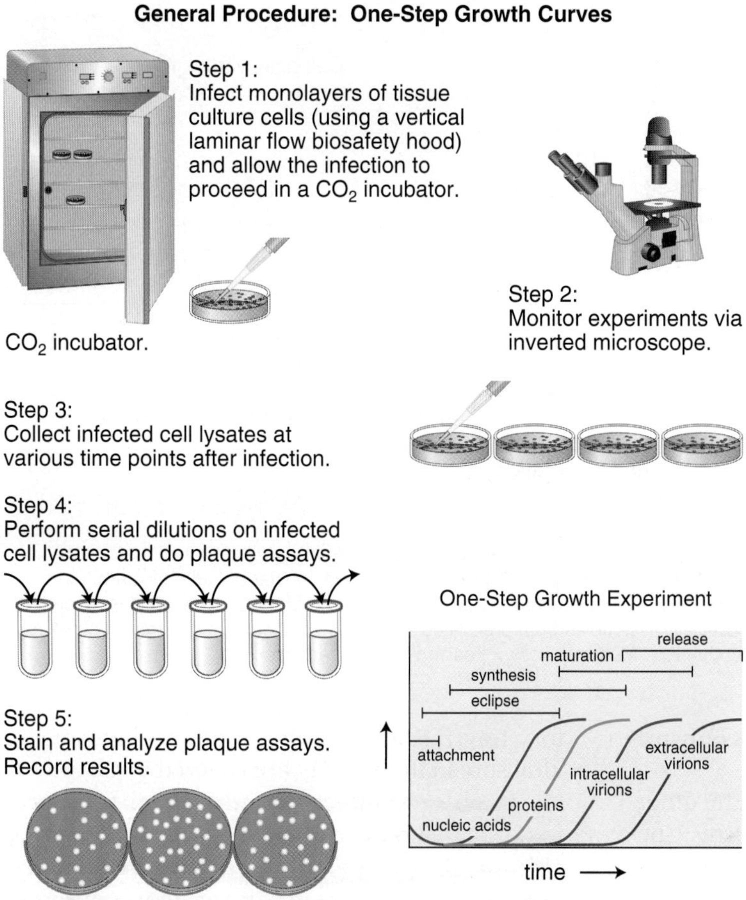
General Procedure: One-Step Growth Curves
Step 1:
Infect monolayers of tissue culture cells (using a vertical laminar flow biosafety hood) and allow the infection to proceed in a CO_2 incubator.
CO_2 incubator.
Step 2:
Monitor experiments via inverted microscope.
Step 3:
Collect infected cell lysates at various time points after infection.
Step 4:
Perform serial dilutions on infected cell lysates and do plaque assays.
Step 5:
Stain and analyze plaque assays. Record results.
One-Step Growth Experiment
release
maturation
synthesis
eclipse
attachment
extracellular virions
intracellular virions
proteins
nucleic acids
time

FIGURE 4-1 The diagram briefly outlines the steps involved in performing one-step growth experiments. Step 5 includes a photograph of viral **plaques** (clearings where the virus destroyed the cell monolayer). The plaque assay is a quantitative assay used to determine the number of viruses present in a given sample. The results of these assays can be used to generate a one-step growth curve for a particular virus. For more details about virological methods see Chapter 5, Laboratory Diagnosis of Viral Diseases and Working with Viruses in the Research Laboratory.

every cell of the monolayer is infected simultaneously. The MOI is the average number of viruses/cell. Hence, classic one-step growth experiments usually use an MOI of 10 (10 viruses/cell). The infected cells are maintained in CO_2 incubators and monitored throughout the course of infection. At various times during the infection, infected cells and/or tissue culture fluid are harvested and plaque assays are performed. **Plaque assays** are used to quantitate the number of intracellular or extracellular virus particles present during that point of infection. All viruses should be going through the same step in the viral replication cycle at the same time (**FIGURE 4-1**).

From these experiments, virologists have determined that there is a general pattern observed during the life cycle of a virus that distinguishes it from the life cycle of a bacterium. Shortly after the infection, the input or inoculated virus disappears. No virus particles are detected at this time. This is termed the **eclipse** period (**FIGURE 4-2b**). This continues until progeny viruses are detectable (anywhere from one to several hours or even days depending on the virus), which is termed the productive stage. Figure 4-2b illustrates that there is a lag phase in which few bacteria are detected but there is never a disappearance of bacteria observed during the life cycle of a typical bacterium. The viral attachment, eclipse, and productive (maturation and release) stages will be discussed in detail as these key steps of viral replication are dissected in the following section.

4.2 Key Steps of the Viral Replication Cycle

1. Attachment (Adsorption)

The first step in the life cycle of a virus is attachment. The virus must be able to attach to its host and enter the "correct" or "target" cell. The attachment event is electrostatic and does not require any cellular energy. This step is a critical step in the viral replication cycle and a great target for antiviral therapies developed to prevent viral infections. If virus attachment is blocked, the infection is prevented. A virus is said to exhibit a tropism for a particular cell type when it targets and infects that cell type. In many cases these cell types are a specific population of cells within organs. **TABLE 4-1** lists examples of viruses and their cellular tropism. Sometimes viruses also display species

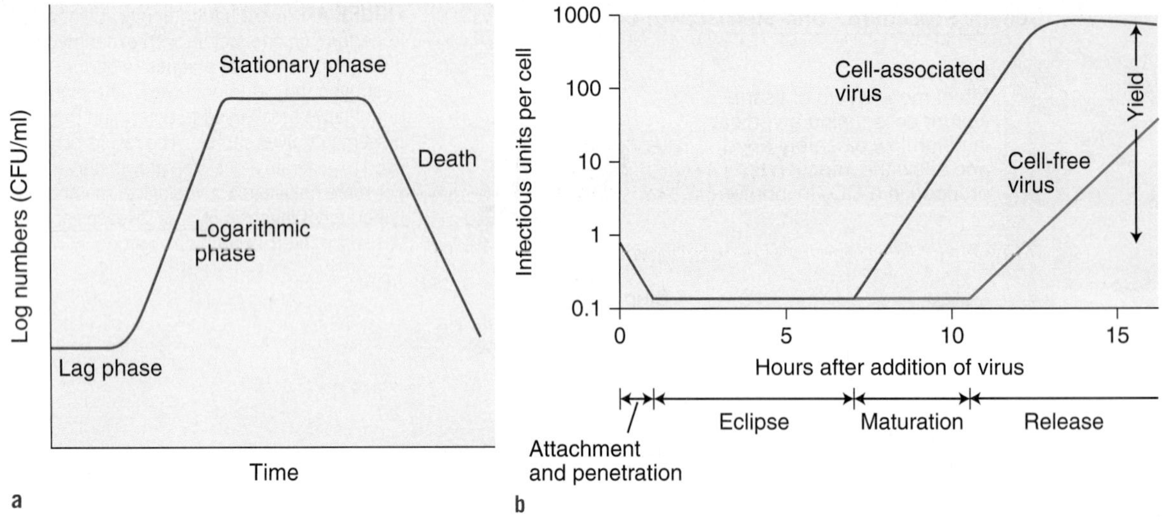

FIGURE 4-2 Typical bacterial growth versus a one-step growth curve of a naked virus. **(a)** Bacterial growth generally proceeds in a series of phases: lag, log (exponential growth in which the rate of multiplication is most rapid and constant), stationary, and death. Viruses require host cells for growth and reproduction. CFU/ml = colony forming units per milliliter. Modified from an illustration by H. Douglas Goff, Ph.D., University of Guelph. **(b)** Viruses are assembled from preformed "parts" when enough of the preformed parts have been made. Adapted from White, D. E., and Fenner, F. J. *Medical Virology,* Fourth Edition. Academic Press, 1994.

tropism. For example, poliovirus infects only primate cells.

Host range is a term that refers to the different types of tissue culture cells or organisms (species) that the virus can infect. The host range may be broad (infecting many different animals or cell lines of different species) or narrow. An example of a broad-range virus is rabies, for which all mammals have varying susceptibility. Human immunodeficiency virus (HIV), which infects humans and monkeys but causes disease only in humans, falls into the narrow range. Human viruses and animal viruses also have preferred routes of entry (e.g., influenza and rhinoviruses enter via the respiratory tract). Routes of entry and mechanisms for viral spread in the body are covered in Chapter 6.

In order to infect cells, the attachment proteins located on the outside of the virus must be able to bind to **cellular surface receptors**. Cellular receptors are usually proteins, glycoproteins, carbohydrates, or lipids. **TABLE 4-2** provides examples of viruses and their cellular receptor(s). *Viruses have evolved to use these receptors for attachment and entry to their hosts.* You may ask, "Why haven't cells evolved to keep up with the evolution of viruses?" The answer is that viruses have evolved to use essential components of the cells as receptors. Without these essential components, the cell can't exist. Cell surface receptors play important

TABLE 4-1 Viral Cell Tropism

Virus(es)	Cell Type
HIV	CD4+ T lymphocytes, macrophages
Rabies	Muscle, neurons
Human papilloma	Differentiating keratinocytes
Hepatitis A, B, C	Liver (hepatocytes)
Human herpes simplex 1 and 2	Mucoepithelium
Influenza A	Respiratory epithelium
Rotavirus	Intestinal epithelium
Norovirus	Intestinal epithelium
Cytomegalovirus	Epithelium, monocytes, lymphocytes
Rhinovirus	Nasal epithelium
Poliovirus	Intestinal epithelium
Epstein-Barr	B cell

TABLE 4-2 Cell Surface Receptors Used by Viruses to Attach and Enter Cells

Virus	Cell Surface Receptor
Influenza A	Sialic acid
HIV-1	CD4 and chemokine co-receptors (CXCR5, CCR4)
Hepatitis C	Low-density lipoprotein receptor
Rabies	Acetylcholine receptor, neural cell adhesion molecule, nerve growth factor, gangliosides, phospholipids
Rhinovirus	Intracellular adhesion molecule 1 (ICAM-1)
Hepatitis B	IgA receptor
Adenovirus Type 2	Integrins $\alpha_v\beta_3$ and $\alpha_v\beta_5$
Poliovirus	Immunoglobulin superfamily protein (CD155)

VIRUS FILE 4-1 How Are Cellular Receptors Used for Viral Attachment Discovered?

Scientists have developed several techniques to identify cell surface receptors and co-receptors to which viruses attach in order to initiate infection. These approaches may be viral receptor-interference studies or genetic techniques. Parts (a)–(c) of **FIGURE VF 4-1** illustrate the general scheme of the various methods employed.

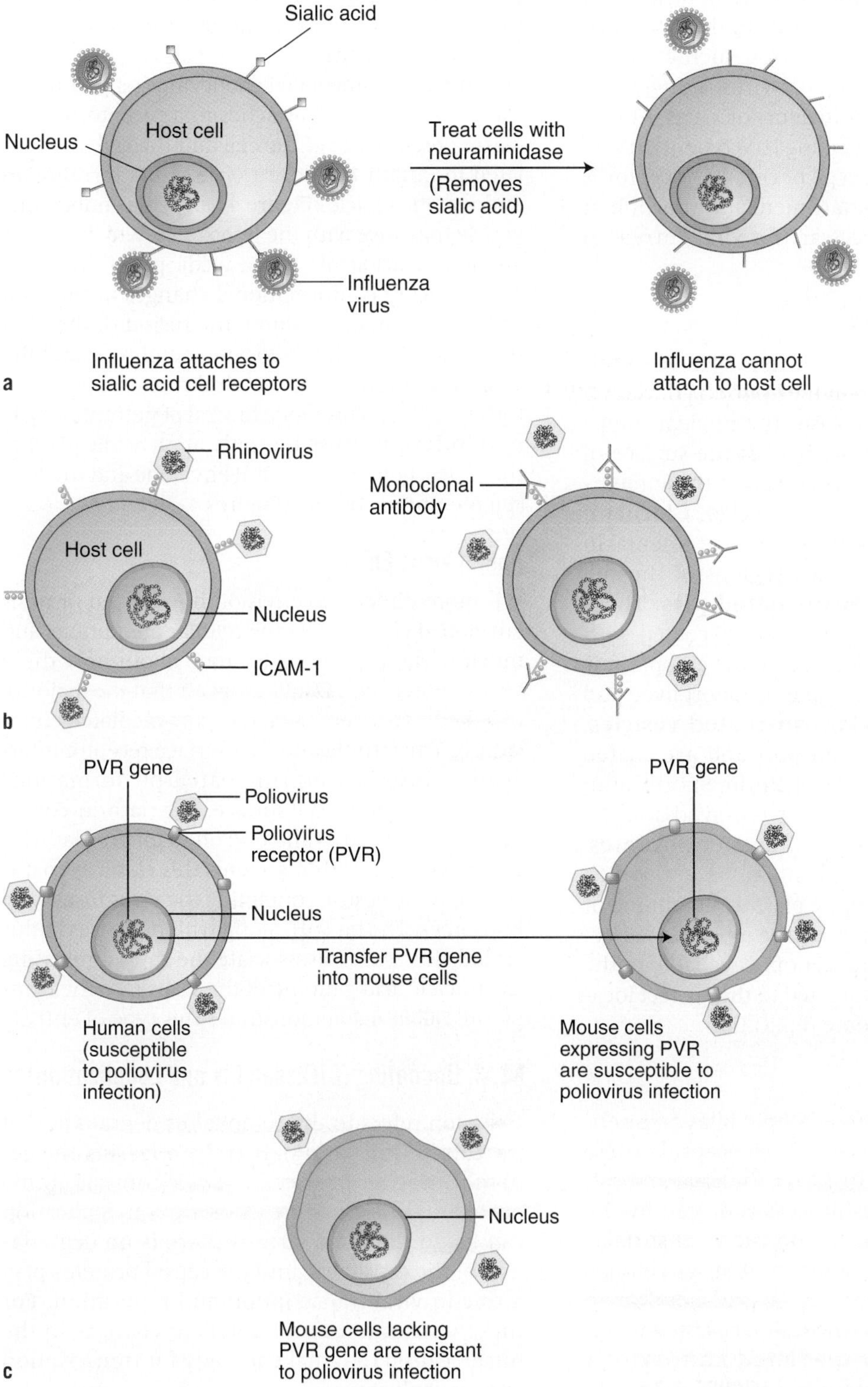

FIGURE VF 4-1 Identification of host cell receptors. **(a)** Removal of cell surface receptors. Adapted from Paulson, J. C., and Rogers, G. N. *Methods Enzymol* (1987):162–168. **(b)** Monoclonal antibodies block cell surface receptors. Adapted from Staunton, D. E., et al. *Cell* 56 (1989):849–853. **(c)** Gene-transfer experiments. Adapted from Mendelsohn, C., et al. *PNAS USA* 20 (1986):7845–7849.

roles in normal cellular activities. We do not know all of the cellular receptors for every virus; however, research continues in this area of host–virus interactions.

There are several factors that may influence the efficiency of viral attachment, such as the density of receptors present on the host cell surface, the density of the **ligands** on the viral surface, and the concentrations of virus and host cells. Temperature, pH, and the presence or absence of specific ions may also play a role in the efficiency of attachment. For some viruses, such as poliovirus, rhinovirus, and influenza virus, a single type of cellular receptor is sufficient for virus attachment. In other cases, including HIV Type 1 (HIV-1) and adenoviruses, one type of cellular receptor is required for the initial attachment and attachment to a **co-receptor** is necessary for viral entry into the cell.

2. Penetration (Entry)

After the animal or human virus attaches to a cellular receptor, it must cross the lipid bilayer plasma membrane (or in some cases the nuclear membrane) of the host cell. Activity at the surface of cellular membranes is dynamic and these membranes are constantly being recycled. **Clathrin**, which is a large, fibrous protein, is instrumental in the formation of specialized regions of the cell membrane called **clathrin-coated pits**. These pits appear as invaginations that are coated with dark material, and are located on the cytoplasmic side of the membrane. The pits are short lived and soon bud off to form **clathrin-coated vesicles**. These vesicles are for transport and are coated with a latticelike network of clathrin. Shortly after formation, the clathrin coat is removed and the resultant vesicles are referred to as **endosomes**. Sometimes these vesicles contain viruses, which can penetrate directly at the plasma membrane or via endosomes. The virus particle must then disassemble to make the viral genome available in the cytoplasm, where it is targeted to the correct location in the cell for genome replication.

Enveloped Virus Entry

Enveloped viruses contain a lipid bilayer, or envelope, that surrounds the nucleocapsid. These viruses enter cells via fusion of the viral and cellular membranes. This process is driven by the viral glycoproteins located on the viral surface. The two basic modes of entry of an enveloped human/animal virus are by **ligand-mediated fusion** of the virus and the cellular plasma membrane or by **receptor-mediated endocytotic entry** of an enveloped virus (**FIGURE 4-3a**). (In ligand-mediated fusion it is the viral ligand rather than the host receptor that mediates the fusion event.)

In ligand-receptor-mediated fusion, the virus attaches to the plasma membrane of the cell and fusion takes place between the viral and cellular membranes. The **nucleocapsid** of the virus is released inside of the cell. The remaining viral envelope remains as a "patch" on the cellular plasma membrane (Figure 4-3a). The fusion at the plasma membrane mode of penetration is **pH independent**.

In receptor-mediated endocytotis (engulfment), the enveloped virus attaches to a receptor on the plasma membrane of the cell and the cell is stimulated to engulf the entire virus, thus forming an endocytotic vesicle (Figure 4-3b). This endocytotic vesicle may fuse with the lysosomes, which possess an internal acidic pH. In the acidic pH of the endocytic vesicles, conformational changes in the viral envelope proteins facilitate the fusion of the viral membrane with the endocytic membrane and the subsequent release of the viral nucleocapsid into the cytoplasm. This mode of viral penetration is **pH dependent** because it is only at an acidic pH that the fusion between the viral envelope and the host cell membrane occurs (Figures 4-3b and 4-3c).

Naked Virus Entry

It is more difficult to envision how naked or nonenveloped viruses cross the cellular membrane, and much remains to be understood about how these viruses enter cells. Studies suggest that the majority of naked viruses enter via receptor-mediated endocytosis. The virus ligand–cell surface receptor interaction causes a clathrin-coated pit formation/invagination at the cell surface. The clathrin-coated pits encase the virus and bud off to form a clathrin-coated vesicle. Within seconds this clathrin coat is shed and the vesicle containing the virus fuses with lysosomes. The low pH, along with proteases in this endocytic vesicle, disassociate the capsid, releasing the nucleic acid genome of the virus into the cytoplasm. **FIGURE 4-4** demonstrates this type of entry.

3. Uncoating (Disassembly and Localization)

This step refers to the removal or degradation of the capsid (uncoating), thereby releasing the genome into the host cell. The genome is transported to the site where transcription/replication can begin. In some viruses there is no degradation of the capsid because the capsid proteins play a role in viral transcription and replication. For these viruses uncoating refers to changes in the nucleocapsid that make it ready for transcription and/or replication.

Receptor-mediated fusion of an enveloped virus with the plasma membrane

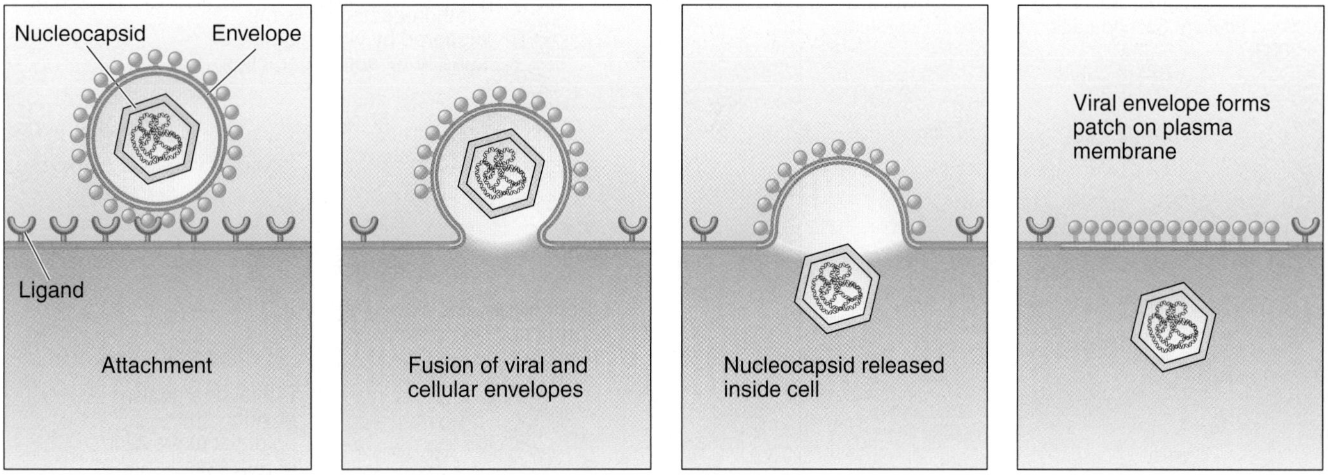

a

Receptor-mediated endocytotic entry of an enveloped virus

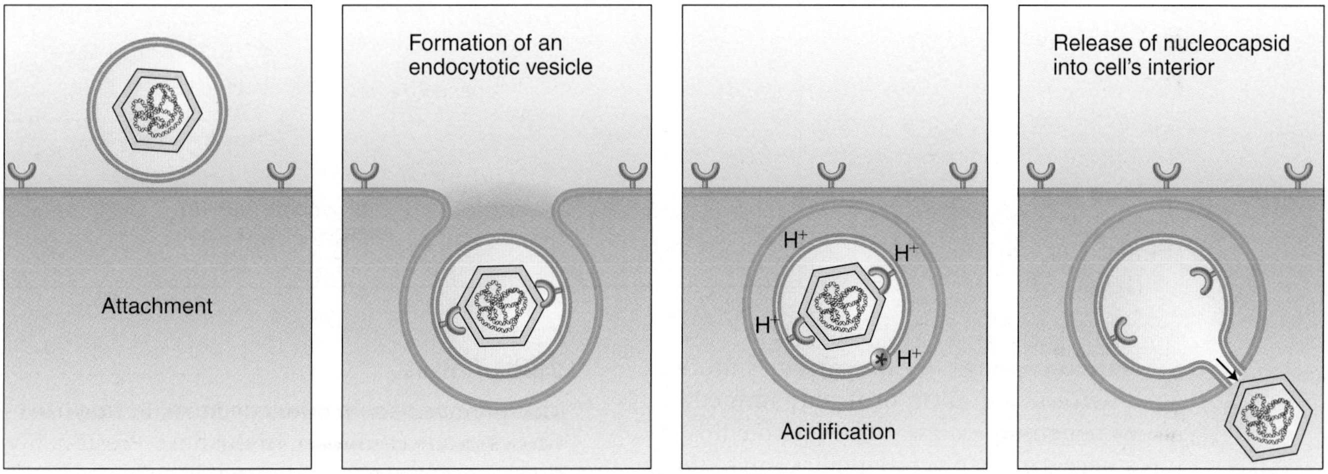

b

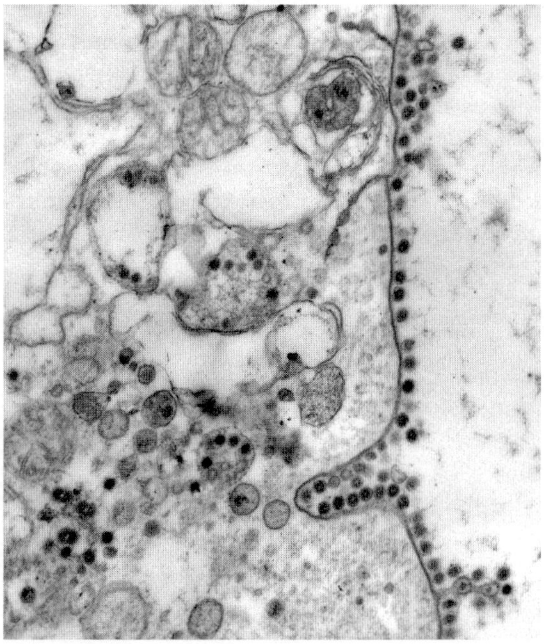

c

FIGURE 4-3 **(a)** Viral entry steps in receptor-mediated fusion. Adapted from Wagner, E. K., and Hewlett, M. J. *Basic Virology,* Second Edition. Blackwell Publishing, 2003. **(b)** Viral entry steps in a receptor-mediated endocytotic entry of an enveloped virus. Adapted from Wagner, E. K., and Hewlett, M. J. *Basic Virology,* Second Edition. Blackwell Publishing, 2003. **(c)** Electron micrograph of mouse hepatitis viruses (family Coronaviridae) are absorbed into mouse intestinal cells via receptor-mediated endocytosis. The plasma membrane is invaginated and will release the viruses inside the cell. Magnification 40,000×.

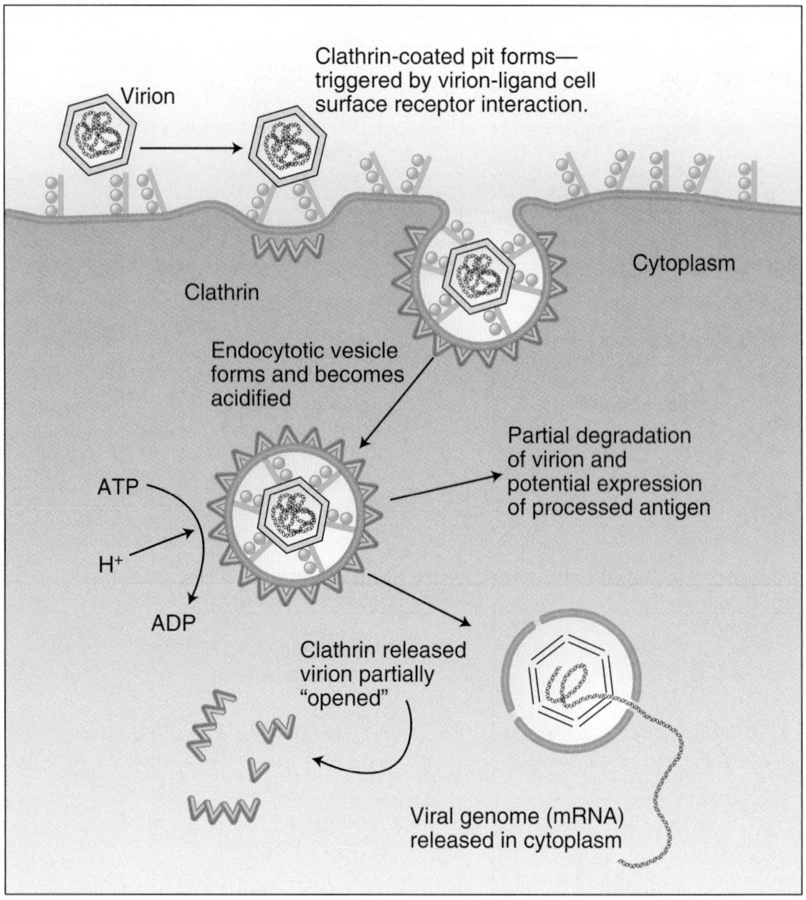

FIGURE 4-4 Steps that naked viruses use to enter cells. Adapted from Wagner, E. K., and Hewlett, M. J. *Basic Virology*, Second Edition. Blackwell Publishing, 2003.

The uncoating step may occur simultaneously with penetration or it may immediately follow penetration of the virus into the host cell. It is a necessary step before replication of the genome can occur. When the nucleic acid genome is uncoated, infectious particles are no longer detected in one-step growth experiments. This is the start of the eclipse phase, which continues until new infectious virus particles are made (see Figure 4-2b).

4. Types of Viral Genomes and Their Replication

When viruses infect cells, two important and separate events must occur:

- *the production of virus structural proteins and enzymes, and*
- *replication of the viral genome.*

The genome of a virus may consist of DNA or RNA, which may be single stranded (ss) or double stranded (ds) and linear or circular (**FIGURE 4-5**). The entire genome may occupy either one nucleic acid molecule (**monopartite** or linear genome) or several nucleic acid molecules (**multipartite** or segmented genome). The different types of genome necessitate different replication strategies.

dsDNA Viruses

The genome replication of most RNA viruses occurs in the cytoplasm of the host. Presumably, this is because their replication is associated with RNA-dependent RNA polymerases that the host cell nucleus cannot provide. In contrast, most DNA viruses replicate their genomes in the nucleus and utilize the host's DNA and RNA synthesizing machinery, along with the host's RNA processing machinery. This means the viral genome must traverse the nuclear membrane to utilize the aforementioned cellular machinery (**FIGURE 4-6**).

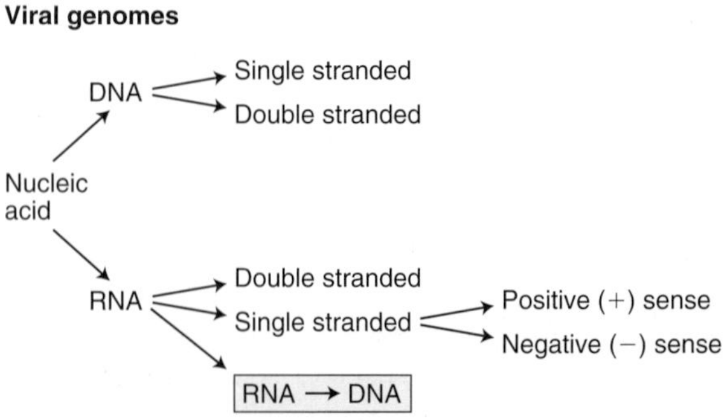

FIGURE 4-5 Types of viral nucleic acid genomes.

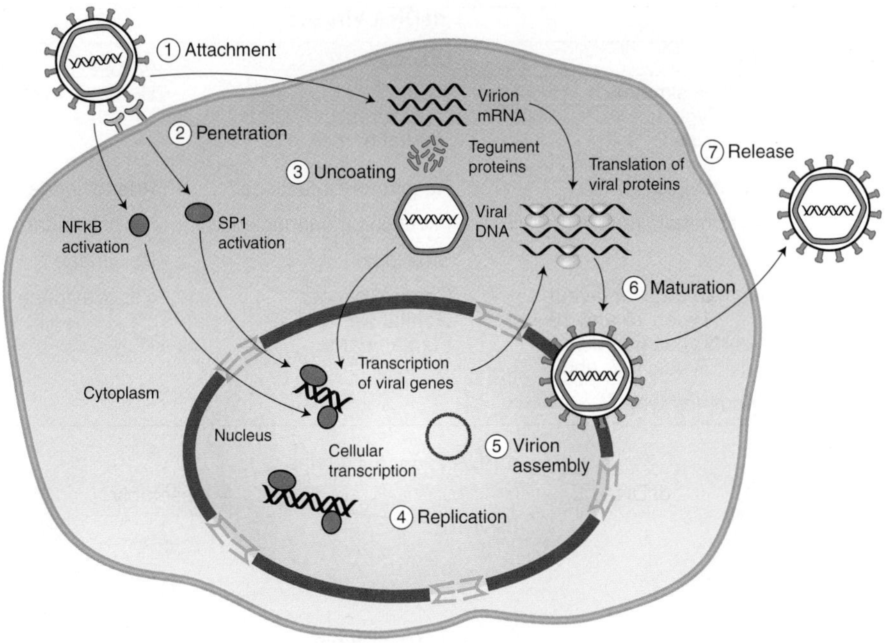

FIGURE 4-6 Example of a life cycle of a dsDNA virus (cytomegalovirus). Cytomegalovirus is a member of the Herpesviridae family. The gH and gB glycoproteins present on the outside of the cytomegalovirus particle bind to the cellular receptors. The attachment event triggers cellular transcription factors SP1 and NF-κB to migrate to the host cell nucleus. After the penetration and uncoating step, the viral cytomegalovirus dsDNA is released and enters the host cell nucleus where the DNA is replicated and transcribed with the help of host cell transcription factors SP1 and NF-κB. Viral mRNAs are exported into the cytoplasm where they are translated by the cellular machinery. Viral dsDNA, and viral and cellular proteins are packaged into the virion. The virion buds from the cell, gaining an envelope from the plasma membrane as the particle is released.

Replication of many DNA viruses involves strategies that are familiar in cell biology: DNA replication and mRNA transcription from dsDNA. Viral proteins are translated from the **monocistronic** mRNAs generated via transcription of viral mRNAs, as shown in **FIGURE 4-7a**, which also lists examples of dsDNA viruses, the diseases they cause, and the families to which they belong. Many DNA viruses have evolved ways to evade host defenses and can cause tumors in animals.

Papovaviruses and herpesviruses are dsDNA viruses and have the most straightforward replication strategy. These viruses utilize the cellular DNA-dependent RNA polymerase II located in the host's nucleus to transcribe the viral mRNAs from the dsDNA viral genome. The host cell must be cycling through the cell cycle for its DNA polymerase to be available for use by these DNA viruses. The viral RNA transcripts are spliced and cleaved via cellular machinery to produce monocistronic mRNAs that are exported into the cytoplasm and translated accordingly by the cell's translation machinery. The viral dsDNA is packaged, along with the necessary structural proteins and enzymes, resulting in the generation of the newly assembled progeny viruses.

Poxviruses differ from the other dsDNA virus families listed in Figure 4-7a in that they replicate solely in the cytoplasm. These viruses carry their own DNA-dependent DNA polymerase (to replicate the viral dsDNA genome) within the virus particle. The genomes of poxviruses are large (ranging from 130–230 kbp, or roughly 100–200 genes), allowing these viruses to be fully equipped with the genes to make them independent of the host's nuclear enzymes and machinery. The monocistronic mRNAs are transcribed directly from the viral dsDNA (Figure 4-7a). Refer to Chapter 14 for more details about poxvirus replication.

Six families of viruses comprise a group of viruses collectively designated Nucleo-Cytoplasmic Large DNA Viruses (NCLDV). These families of viruses include major pathogens of humans and other mammals as well as viruses that infect cold-blooded vertebrates, insects, and algal hosts. All of the viruses contain dsDNA genomes and replicate either exclusively in the cytoplasm of the host cells or possess both cytoplasmic and nuclear stages in their life cycle (**TABLE 4-3**).

Phycodnaviruses are members of the Phycodnaviridae family. The family name is derived from two distinguishing features: "phyco" from their algal hosts and "dna" because all of these viruses have dsDNA genomes. The phycodnaviruses are recognized as important ecological **virioplankton** in aquatic ecosystems. Virioplankton now represent the most abundant viruses in natural waters, surpassing the number of bacteria by an order of magnitude. The phycodnaviruses, along with other viruses, including bacteriophages, play roles in nutrient recycling (refer to Chapter 1, Viruses and Aquatic Ecosystems) and algal blooms. There are currently extensive **metagenomics** studies underway to determine aquatic **viromes**. Metagenomics is the genomic analysis of all DNA applied to entire communities of microbes and/or viruses, bypassing the need to isolate and culture individual microbes or viruses in the laboratory. A virome is the genomes of all viruses that inhabit a particular environment.

dsDNA Viruses

Virus	Disease	Family
Herpes simplex		Herpesviridae
Type 1	Cold sores	
Type 2	Genital herpes	
Adenovirus	Respiratory infections	Adenoviridae
Cytomegalovirus	Infectious mononucleosis	Herpesviridae
Variola	Smallpox	Poxviridae
Human Papillomavirus	Cervical cancer	Papovaviridae
*Types 16 and 18	Genital warts	
*Types 6 and 11	Plantar warts	
*Types 1, 2, and 4		
*common types		

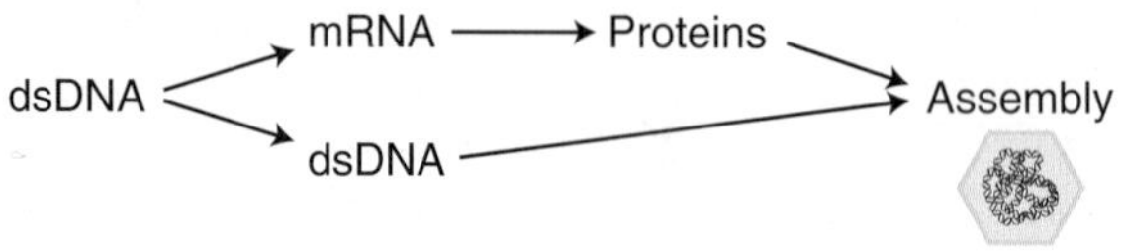

a

ssDNA Viruses

Virus	Disease	Family
Human parvovirus B19	Fifth disease (slapped-cheek syndrome)	Parvoviridae
Transfusion transmitted virus (TTV)	Hepatitis?	Circoviridae

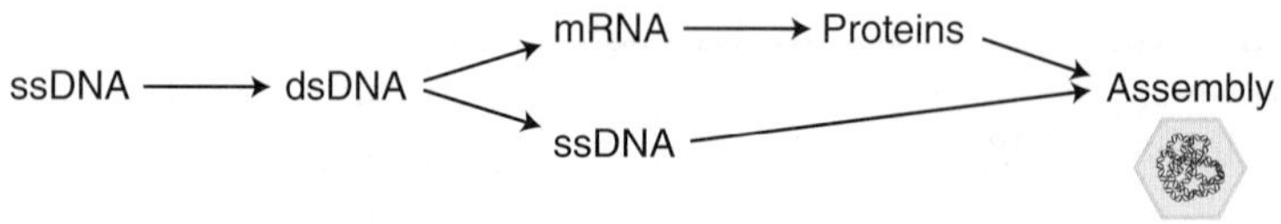

b

FIGURE 4-7 (a) List of dsDNA viruses and their replication strategy. Adapted from Harper, D. R. *Molecular Virology,* Second Edition. BIOS Scientific Publishers, 1999. **(b)** List of ssDNA viruses. Adapted from Harper, D. R. *Molecular Viorology,* Second Edition. BIOS Scientific Publishers, 1999.

ssDNA Viruses

Parvoviruses are the smallest of the human viruses (only 20–25 nm in diameter). In contrast to the dsDNA virus genomes, these ssDNA viruses contain very small linear genomes (the genome of the human parvovirus B19 is 5 kb). Parvoviruses do not carry any enzymes in the virus particle. These viruses infect cells that are in the cell cycle because they are dependent upon the host's DNA polymerase to synthesize the viral ssDNA and the cell's DNA-dependent RNA polymerase II to transcribe the viral dsDNA into viral mRNA in the nucleus. The cellular splicing machinery is also used in the production of the viral mRNAs. The general outline of their replication strategy is illustrated in Figure 4-7b.

ss/dsDNA Viruses (Using an RNA Intermediate)

Hepadnaviruses replicate via a very unique and somewhat complicated mechanism. This textbook focuses on hepatitis B virus (HBV) because it specifically infects humans. Other members of the Hepadnaviridae family infect woodchucks, ground

TABLE 4-3 Host Range and Replication Sites of Selected Families of the Eukaryotic Nucleo-Cytoplasmic Large DNA Viruses

Virus Family	Replication Site	Host Range
Ascorviridae	Nucleus and Cytoplasm	Insects, mainly noctuids (night-flying moths)
Asfarviridae	Cytoplasm	Mammals
Iridoviridae	Nucleus and Cytoplasm	Insects, cold-blooded vertebrates (reptiles, amphibians, and fish)
Mimiviridae	Cytoplasm	Acanthamoeba
Phycodnaviridae	Nucleus and Cytoplasm	Chlorella-like green algae, algal symbionts of paramecia and hydras
Poxviridae	Cytoplasm	Insects, mammals, birds, reptiles

squirrels, chipmunks, ducks, geese, chimps, gibbons, and orangutans. Interestingly, HBV-infected cells produce different forms of virus-related particles. Electron microscopy of partially purified virus particle preparations reveal three types of particles: a 42- to 47-nm mature spherical virus particle (known as **Dane particles**, named after their discoverer); 22-nm spherical particles, which are found in 10,000- to 100,000-fold excess over the Dane particle; and filamentous particles that are 22 nm in diameter and of varying lengths. All three forms contain the same surface protein, called the hepatitis B surface antigen (HbsAg). The Dane particle is the only infectious particle of HBV. The 22-nm spheres and filaments do not contain nucleic acid (**FIGURE 4-8**).

The genome of the Dane particles consists of a 3.2-kb linear DNA that is arranged in a relaxed circle. Some parts of the genome are dsDNA, whereas others consist of ssDNA regions or gaps. This partially duplexed DNA consists of a full-length (–) sense ssDNA and a shorter length (+) sense ssDNA. As a result, the gapped regions contain only (–) sense ssDNA. After the HBV has entered its host cell and the virus is partially uncoated, the partial dsDNA genome of the Dane particle migrates to the nucleus, where it is completed or repaired by a viral reverse transcriptase. The dsDNA enters the nucleus and the ends are ligated by cellular enzymes, forming a circular **episome.** (The term episome applies to a viral genome that is maintained in cells by autonomous replication.) Next, the repaired viral dsDNA associates with cellular histones and is transcribed into separate viral mRNA transcripts and a full-length ssRNA pre-genome (**FIGURE 4-9**).

The viral mRNAs are translated to yield the hepatitis B core antigens and the viral reverse transcriptase. The RNA pre-genome associates with the viral reverse transcriptase and is packaged with the core proteins to form an **immature** virus particle in the cytoplasm of the cell. The viral **reverse transcriptase** synthesizes the (–) sense ssDNA strand using the ssRNA intermediate as a template (see *Refresher:* Molecular Biology, which

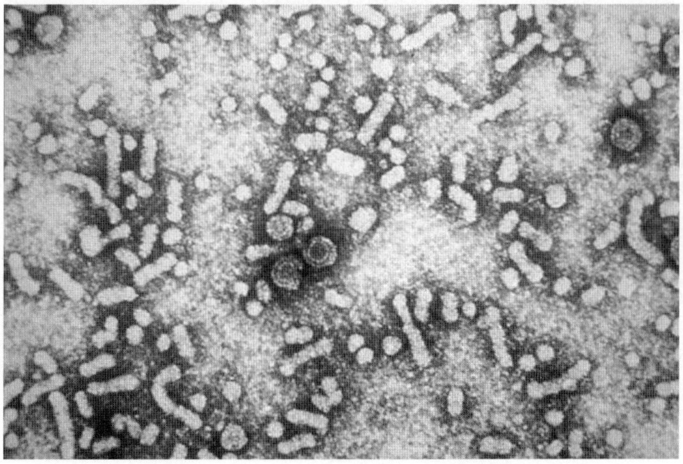

a

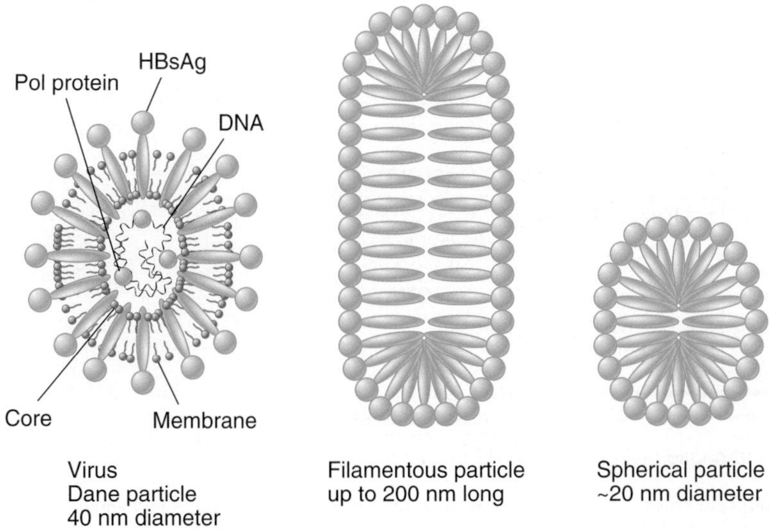

b

FIGURE 4-8 (a) HBV infection results in the formation of three different types of virus particles: 42- to 47-nm intact infectious Dane particles, 22-nm spheres, and 22-nm filaments of varying lengths. **(b)** Illustration depicting the different forms of HBV particles. Mature hepatitis B Dane particles contain dsDNA with associated protein, but their mode of replication is different from the other dsDNA viruses and their replication strategy. Adapted from University of South Carolina, School of Medicine. "Virology: Hepatitis Viruses." *Microbiology and Immunology.* University of South Carolina, 2008. http://pathmicro.med.sc.edu/virol/hepatitis-virus.htm.

FIGURE 4-9 ssDNA/dsDNA virus (that uses ssRNA as an intermediate) and its replication strategy. Adapted from Harper, D. R. *Molecular Virology,* Second Edition. BIOS Scientific Publishers, 1999.

Virus	Disease	Family
Hepatitis B	Hepatitis associated with liver cancer	Hepadnaviridae

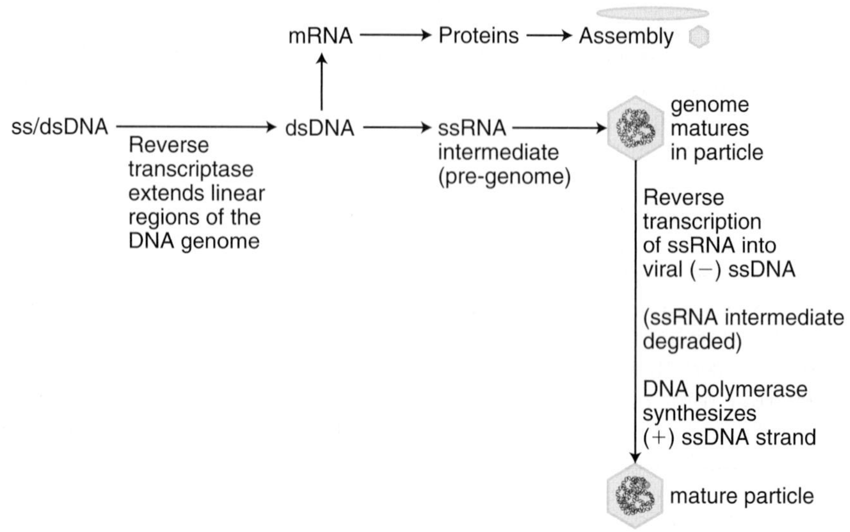

Refresher Molecular Biology

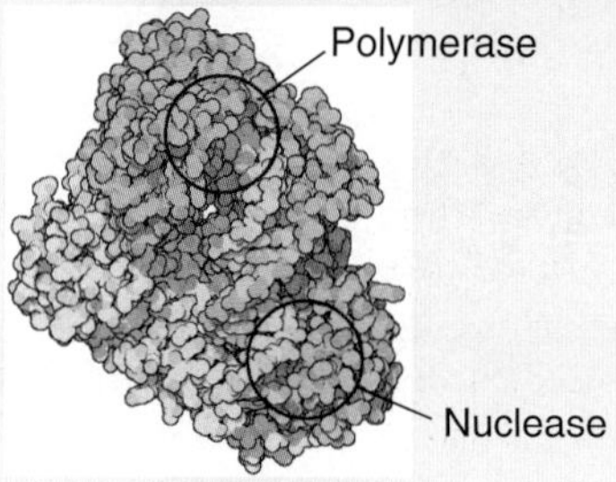

What is reverse transcriptase?
Reverse transcriptase (RT) has three distinct enzymatic activities:

1. RNA-dependent DNA polymerase
2. RNase H activity (cleaves/degrades RNA from RNA/DNA hybrids)
3. DNA-dependent DNA polymerase

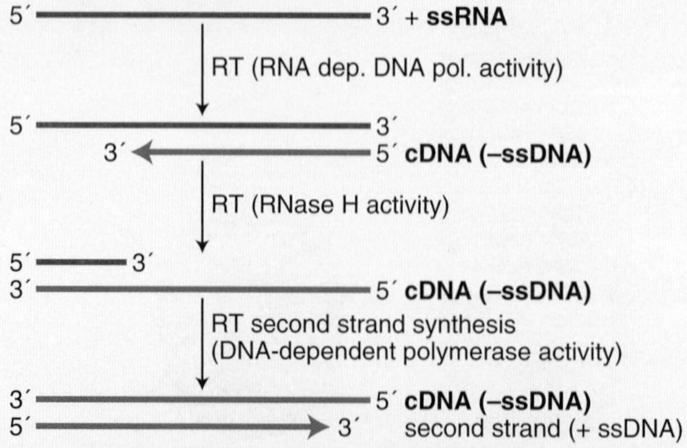

Retroviruses and hepadnaviruses utilize RT in their life cycles.

FIGURE RB 4-1 Reverse Transcriptase Courtesy of David S. Goodsell, Scripps Research Institute. Protein Data Bank: 2hmi. Reproduced from J. Ding, et al., *J. Mol. Biol.* 284 (1998): 1095–1111.

is about reverse transcriptase functions). The pregenome is degraded by the RNase H activity of the reverse transcriptase enzyme, but it leaves a short sequence of RNA at its 5′ end that acts as a primer for DNA polymerase to synthesize a complementary (+) DNA strand in the **mature** particle.

Hepatitis B is one of a few known nonretroviral viruses that uses reverse transcription as part of its replication process. Other viruses that utilize reverse transcriptase are retroviruses such as human T-cell leukemia virus (HTLV) and HIV, which possess an RNA genome. For these retroviruses reverse transcription is one of the first steps in viral replication, whereas for hepatitis B reverse transcription occurs during maturation (the latter steps) in making new virus particles. In addition, in contrast to retroviruses, HBV does not have **integrase** activity. The DNA of hepatitis B is usually not integrated into cellular DNA; it is found as an independent episome. Integrated parts of the hepatitis B genome, however, are found in the chromosomes of hepatocellular tumors from cancer patients. Retroviruses have integrase activity (see Chapter 10).

RNA Viruses

RNA viruses are unique because their genetic information is encoded in RNA. The genomes of RNA viruses are diverse [ss or ds, (+) or (–) sense, linear or segmented]. The type of RNA genome determines if the first step after uncoating will be translation, transcription, or RNA replication.

Viruses that contain +ssRNA genomes have genomes that can be directly translated using the host cell machinery because the +ssRNA acts like an mRNA (**FIGURE 4-10**). These +ssRNA viruses, however, do need to carry the gene that encodes the replicase that produces the viral genomic RNA.

All other types of RNA viruses (–ssRNA, dsRNA, linear, segmented) must be transcribed into mRNA before translation can occur. Eukaryotic host cells do not contain RNA-dependent RNA polymerases (see Section 2.3, Molecular Constraints of the Host Cell), and as a result these viruses must carry an RNA-dependent RNA polymerase that will synthesize the viral +ssRNA, mRNAs, and –ssRNA viral genomes into the host cell with them.

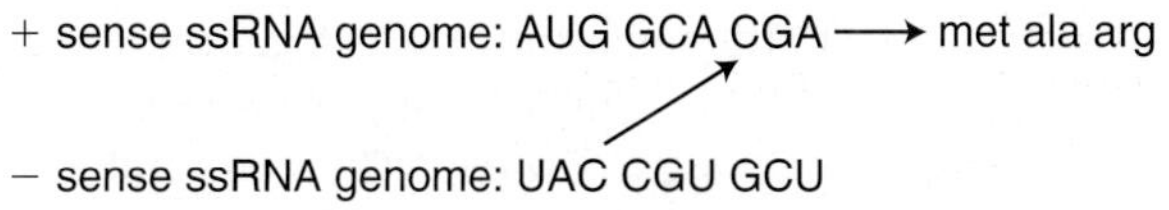

FIGURE 4-10 Differences between positive (+) and negative (–) sense ssRNA viral genomes.

dsRNA Viruses

Rotaviruses have emerged as the main agent of acute gastroenteritis in infants and children worldwide. These viruses have dsRNA **segmented** genomes. The rotavirus particle contains 11 segments or pieces of the viral dsRNA. The host does not produce RNA-dependent RNA polymerases; thus, the virus carries its own RNA-dependent RNA polymerase and the replication cycle occurs solely in the cytoplasm.

A rotavirus particle is nonenveloped and icosahedral with a double-capsid. One of its two layers is removed but the other is not; the transcription takes place inside of this single capsid and the mRNAs are released in the cytoplasm for translation. After attachment, entry, and uncoating, the virus synthesizes a +ssRNA from each of the 11 dsRNA segments (using the –ssRNA strands of the dsRNA genome as a template) via a viral RNA-dependent RNA polymerase. These viral ssRNAs are also capped via a viral capping enzyme. The RNAs are not polyadenylated. Half of the newly synthesized capped +ssRNAs strands (mRNAs) are translated by the cellular machinery in the cytoplasm. The remaining strands are packaged into a viral capsid during assembly (**FIGURE 4-11**).

At this stage of the life cycle, the RNAs inside the particles are sensitive to RNase treatment. During maturation of the virus particle, the complementary –ssRNA strands are synthesized using the capped +ssRNAs as a template within the virus

Virus	Disease	Family
Rotavirus	Gastroenteritis	Reoviridae
Reovirus	Mild respiratory and gastrointestinal symptoms	Reoviridae

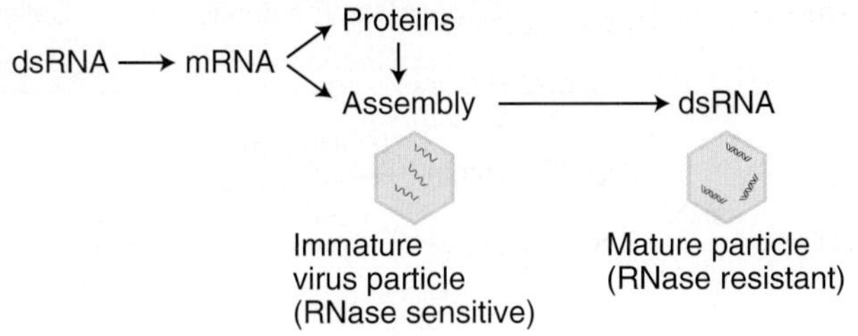

FIGURE 4-11 List of dsRNA viruses and their replication strategies. Adapted from Harper, D. R. *Molecular Virology*, Second Edition. BIOS Scientific Publishers, 1999.

particle to form the remaining dsRNA genomic segments (Figure 4-11). These final dsRNA segments are resistant to RNase treatment. There are still several remaining questions about the replication cycle of rotaviruses and other viruses of the Reoviridae family; for example, how does the virus particle manage to contain only one of copy of each of the 11 mRNAs?

+ssRNA Viruses

The +ssRNA viruses include several families of viruses. Members of the Picornaviridae, Flaviviridae, and Caliciviridae families, in particular, are ubiquitous in nature and cause a wide range of diseases. Their success and widespread distribution suggest that their replication strategy is very effective. The RNA in the virus particle itself functions as mRNA. This genomic RNA is a polycistronic mRNA that is recognized by cellular machinery and translated as one open reading frame into a single polyprotein precursor that is subsequently cleaved into individual viral proteins by viral and cellular proteases (**FIGURE 4-12**). One of the viral encoded proteins is an RNA-dependent RNA polymerase that replicates the viral genome. It transcribes the viral +ssRNA into a –ssRNA replicative intermediate, which in turn serves as a template for the genomic +ssRNA (Figure 4-12). Note that there are exceptions to this replication strategy. Not all ssRNA viruses produce a single polyprotein that is cleaved by proteases into individual proteins. Some produce more than one mRNA, allowing greater control of the production of individual proteins; for example, early replication proteins and later structural proteins are produced at different times during the viral replication cycle (**FIGURE 4-13**).

–ssRNA Viruses

Viruses in the Paramyxoviridae, Rhabdoviridae, and Filoviridae families contain –ssRNA nonsegmented genomes. All of these viruses encode their own RNA-dependent RNA polymerases that transcribe the –ssRNA genome into several different viral monocistronic +ssmRNAs that can be recognized by the host cell machinery. The different +ssRNAs are made by a complicated start–stop type of mechanism. In other words, a range of viral mRNAs are each translated to make different viral proteins rather than a polyprotein. All of the proteins are not produced to the same level, and a number of control mechanisms are used. The second function of the viral RNA-dependent RNA

FIGURE 4-12 List of +ssRNA viruses and their replication strategies. Adapted from Harper, D. R. *Molecular Virology*, Second Edition. BIOS Scientific Publishers, 1999.

Virus	Disease	Family
Poliovirus	Poliomyelitis Postpolio syndrome	Picornaviridae
Rhinovirus (many types)	Common cold	Picornaviridae
Hepatitis A	Hepatitis	Picornaviridae
Cocksackie		Picornaviridae
Group A Types 21, 24	Common cold	
Group A Types 4, 5, 9, 10, 16	Hand, foot, and mouth disease	
Group B Types 1–5	Myocarditis	
Group B Types 2, 5	Hand, foot, and mouth disease	
Echoviruses		Picornaviridae
Various Types	Diarrhea	
Types 1–7, 9, 11, 13–23, 25, 27	Aseptic meningitis	
Rubivirus	Rubella	Togaviridae
Yellow fever	Hemorrhagic fever	Flaviviridae
Hepatitis C	Hepatitis liver cancer	Flaviviridae
Dengue	Dengue fever	Flaviviridae
West Nile	Fever, rash, myalgia encephalitis	Flaviviridae
Norovirus	Gastroenteritis	Caliciviridae
Sapovirus	Gastroenteritis	Caliciviridae

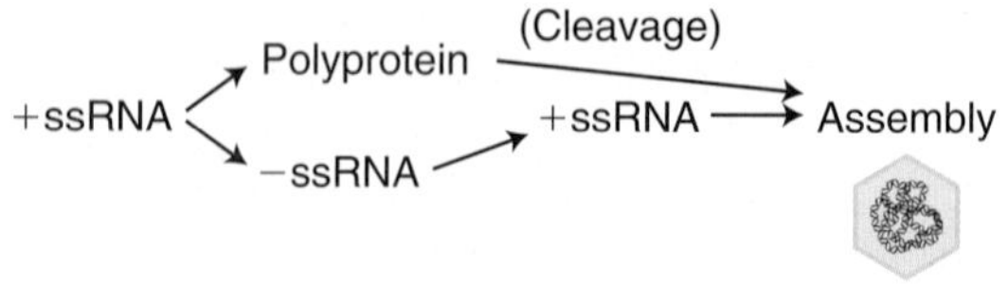

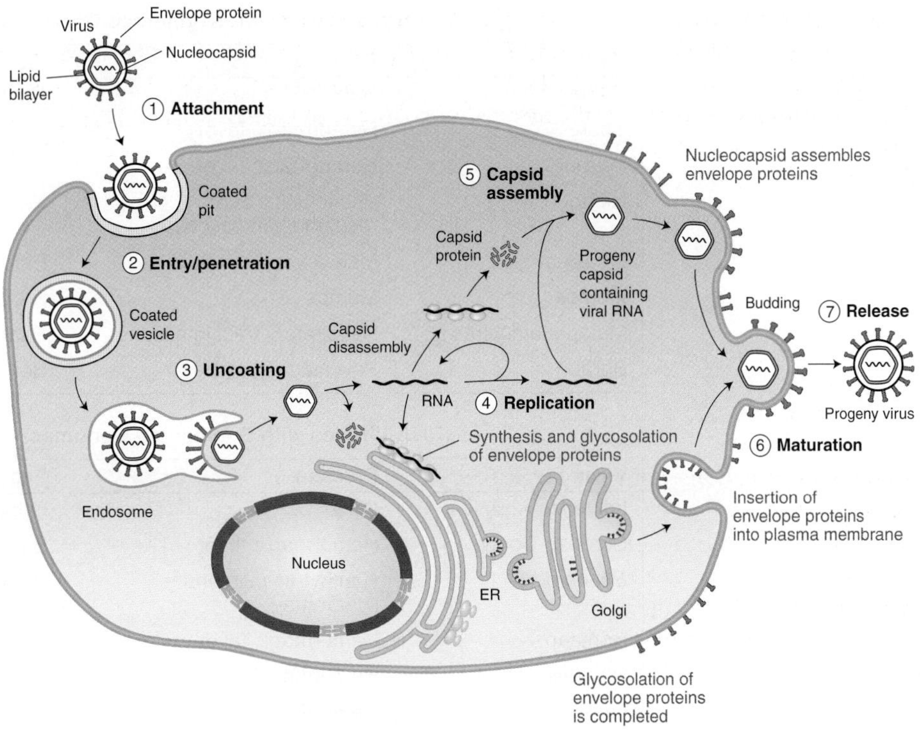

FIGURE 4-13 Example of a life cycle of a +ssRNA enveloped virus. All steps of the life cycle take place in the cytoplasm. Like a typical eukaryotic mRNA, the viral +ssRNA genome is directly translated by the cellular ribosomal machinery in the cytoplasm. Viral glycoproteins are synthesized on and inserted into the rough endoplasmic reticulum (ER) membrane, where they are subsequently transported to the trans Golgi network and then inserted into the plasma membrane. A viral polymerase synthesizes the viral genome. Newly synthesized capsid proteins bind to the replicated genome, forming a nucleocapsid that buds out of the plasma membrane to form the final enveloped virion.

polymerase is to synthesize the viral/progeny genome using the +ssRNA as a template. Hence, the RNA-dependent RNA polymerase is sometimes referred to as having a **transcriptase** and **replicase** function (**FIGURE 4-14a**).

The –ssRNA viruses containing segmented genomes also encode their own RNA-dependent RNA polymerase that functions as a transcriptase and replicase. Each segment produces a monocistronic mRNA or an RNA that is differentially spliced to make monocistronic mRNAs. The genomes of the viruses in the Arenaviridae and Bunyaviridae families are more complicated in that at least one of the viral ssRNA genomic segments are **ambisense** [the ssRNA is both (+) and (–) sense on the same ssRNA segment]. The process of replication (Figure 4-14b) and translations of these RNAs is not completely understood.

Viruses with ssRNA Genomes That Use a dsDNA Intermediate to Replicate

The Retroviridae family contains viruses that have been identified in virtually all organisms including invertebrates. This suggests that these viruses have an evolutionarily successful design. Their biology is quite unique.

The main focus on retroviruses has been on the avian (chicken) or human retroviruses: Rous sarcoma virus (RSV, discovered in 1911, see Chapter 10), HIV (discovered in 1983), and HTLVs (discovered in 1981). Retrovirus infections cause a wide spectrum of diseases including cancer, immune deficiencies, and neurological disorders. Most retroviral infections, however, occur without having any detectable, deleterious damage to the host.

The replication cycle of retroviruses includes the integration of the viral complementary DNA (cDNA) into the chromosomal DNA of the host cell. The result of this integration event is that the retroviral DNA is inherited from parent to offspring of the infected host if germline cells (sperm and egg) contain the integrated viral genome. These are termed **endogenous retroviruses** or **proviruses**, and their biologic properties and functions are still under investigation. Approximately 8% to 12% of the human genome consists of sequences of **human endogenous retroviruses (HERVs)**. Retroviruses that are not integrated in germline cells of their hosts are called exogenous retroviruses (or external viruses).

The genome of retroviruses contains two copies of a +ssRNA molecule that is reverse transcribed into

FIGURE 4-14 (a) List of –ssRNA viruses (nonsegmented and segmented) and their replication strategies. Adapted from Harper, D. R. *Molecular Virology*, Second Edition. BIOS Scientific Publishers, 1999. **(b)** Ambisense RNA viruses: Strategies for replication and mRNA synthesis of RNA genome. Adapted from Harper, D. R. *Molecular Virology*, Second Edition. BIOS Scientific Publishers, 1999.

–ssRNA Viruses with Non-segmented Genomes:

Virus	Disease	Family
Rabies	Rabies	Rhabdoviridae
Ebola	Hemorrhagic fever	Filoviridae
Marburg	Hemorrhagic fever	Filoviridae
Nipah	Encephalitis and respiratory infections	Paramyxoviridae
Measles	Measles	Paramyxoviridae
Mumps	Mumps	Paramyxoviridae
Metapneumovirus	Respiratory tract infections	Paramyxoviridae
Borna	Psychiatric disorders?	Bornaviridae

–ssRNA Viruses with Segmented Genomes:

Virus	Disease	Family
Influenza A, B, C	Influenza	Orthomyxoviridae
Crimean-Congo	Hemorrhagic fever	Bunyaviridae
Sin nombre	Hantavirus pulmonary syndrome	Bunyaviridae
Hantaan	Hemorrhagic fever	Bunyaviridae
Rift Valley fever	Hemorrhagic fever	Bunyaviridae
Lassa	Hemorrhagic fever	Arenaviridae

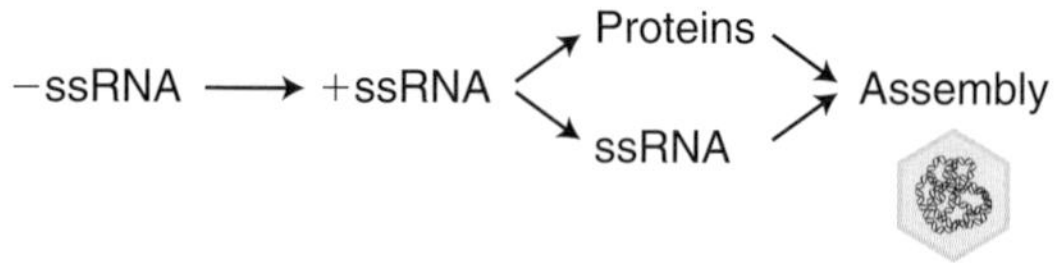

a

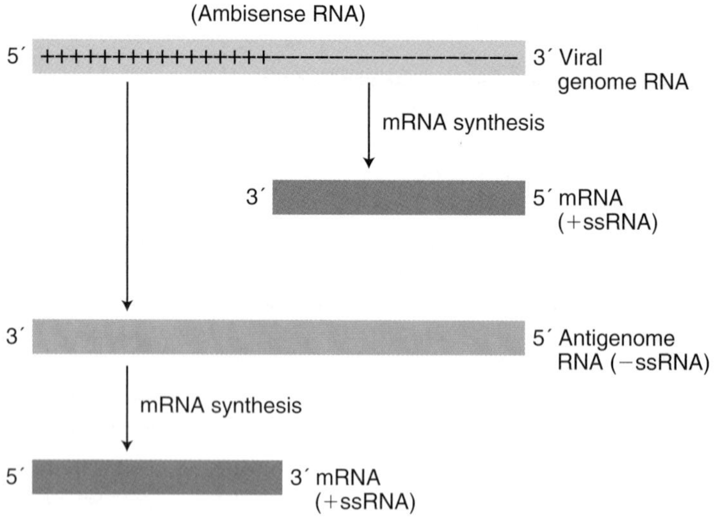

b

dsDNA by a viral RNA-dependent DNA polymerase (reverse transcriptase) to produce an RNA:DNA hybrid, which in turn is converted to dsDNA. The viral dsDNA is inserted into the host chromosomal dsDNA (**FIGURE 4-15**). The integrated DNA (provirus) is subsequently transcribed by the host's DNA-dependent RNA polymerase II. The mRNA transcripts are then spliced and exported into the cytoplasm of the cell, where they will be translated by the cellular protein synthesis machinery. Some full-length +ssRNA transcripts will be packaged into the new retrovirus particles (**FIGURE 4-16**). Refer to Chapter 10 (Sections 10.3 and 10.4) for more information.

5. Assembly

It is not always possible to identify the assembly, maturation, and release of virus particles as distinct

Virus	Disease	Family
HIV-1 and 2	AIDS	Retroviridae
HTLV I	T-Lymphocyte Leukemia	Retroviridae
HTLV II	?	Retroviridae

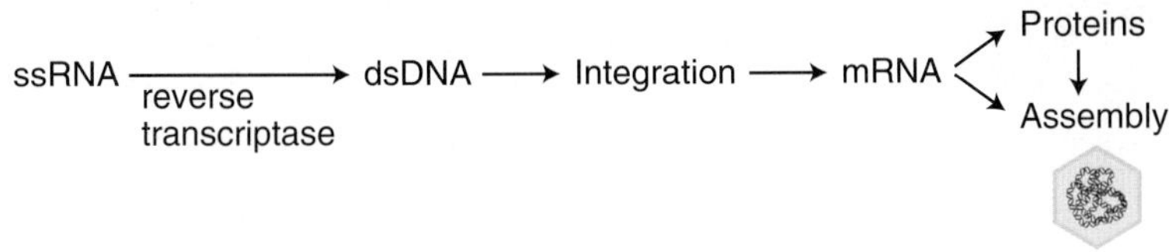

FIGURE 4-15 List of ssRNA viruses (that use a DNA intermediate) and their replication strategies. Adapted from Harper, D. R. *Molecular Virology,* Second Edition. BIOS Scientific Publishers, 1999.

and separate stages of the viral life cycle. Virus assembly is a key step in the replication cycles of viruses. It involves the process in which the immature virus particle is formed. Despite the structural diversity of virus particles, the repertoire of assembly mechanisms is limited. All of the components of the virus must be assembled to create a stable structure. At the same time, the newly assembled virus must accomplish disassembly to start a new infectious life cycle.

The assembly event occurs when an appropriate concentration of virus proteins and genomic nucleic acids is reached and localized at specific sites within the infected cell. The genomic nucleic acids are packaged into preexisting shells that form via **self-assembly** (spontaneous assembly, also refer to Chapter 3) of viral capsid proteins, or are coated with capsid proteins, or are co-assembled with capsid proteins. Assembly sites (for example, the cytoplasm, nucleus, on the inner surface of the plasma membrane of cells) differ according to the virus and have some influence on how the virus particle is released.

Historically, research directed toward virus assembly mechanisms has received less attention because of more interest in the mechanisms of viral

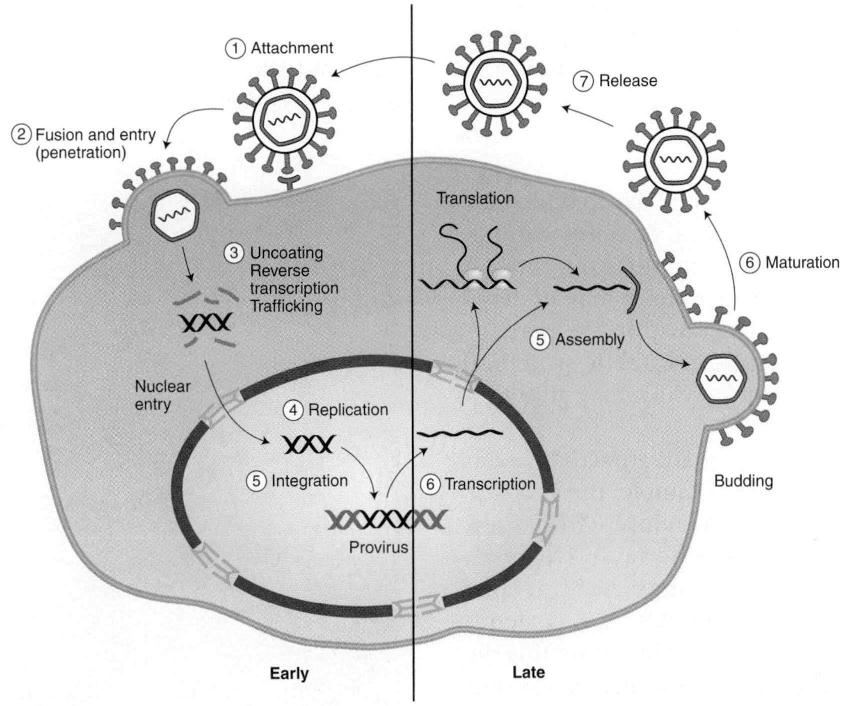

FIGURE 4-16 Basic life cycle of a typical retrovirus. Retroviruses undergo either a latent (early) phase in which the virus begins the steps of a typical virus life cycle: virus attaches to a host cell receptor(s) followed by uncoating of the virus particle in which the genome is translocated (also referred to as trafficking) to the nucleus where the genome is reverse transcribed into dsDNA that is integrated into the host's chromosome. At this point, the retroviral DNA is replicated along with the host chromosome. This early stage is latent in that no infectious particles are made while the retroviral DNA has integrated into the chromosomal DNA. After an "activation" event, the integrated retroviral DNA is transcribed. Viral mRNAs are exported to the cytoplasm where the cellular translational machinery produces the viral proteins. The capsid proteins and retroviral RNA is assembled/packaged into a nucleocapsid that buds from the plasma membrane, forming the final particle that undergoes a maturation step before the virus is fully infectious. Modified from an illustration by Kate Bishop, MRC National Institute for Medical Research, UK.

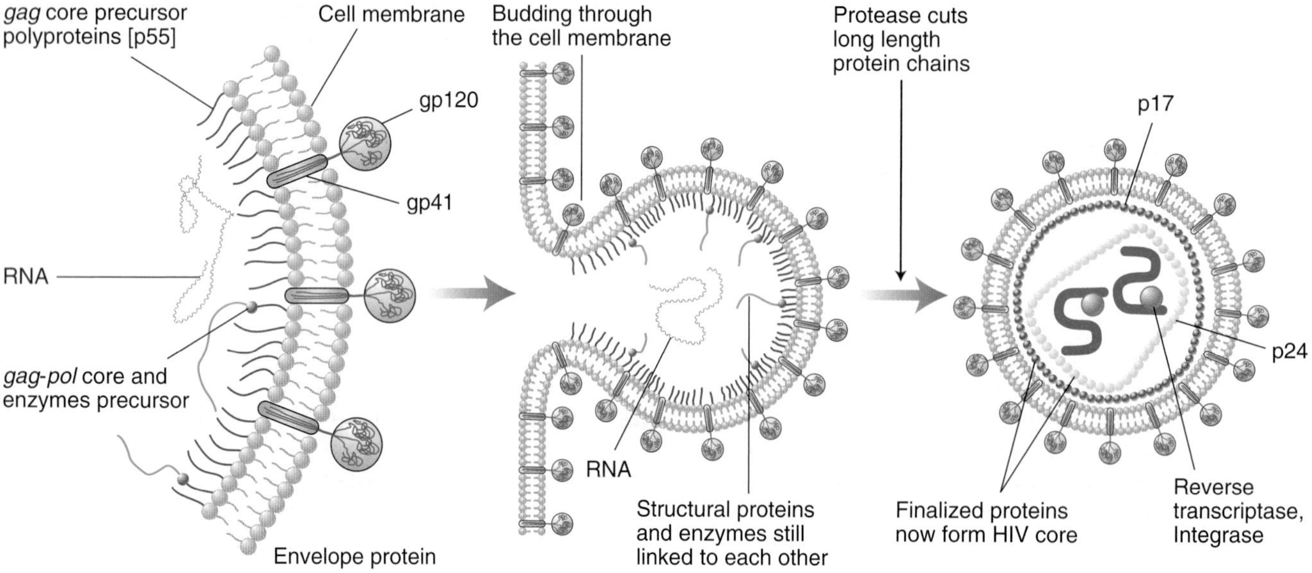

FIGURE 4-17 The structural proteins within the immature HIV virus particle must be cleaved by a viral protease inside of the particle in order for the virus to be infectious. Adapted from Vella, S., et al., *AIDS Soc. 4* (1996): 15–18.

gene expression and replication. There is now renewed interest in understanding virus assembly because of the development of new molecular technologies and the success of therapeutic agents designed to inhibit virus-specific reactions involved in the production of infectious virus particles. These advances in understanding continue at an accelerated pace.

6. Maturation

This is the stage of the virus life cycle in which the virus becomes infectious. Viral or cellular proteases are often involved in maturation. One or more **capsid** or **envelope** proteins may undergo specific proteolytic cleavage within the particle. The cleavage event results in a subtle structural change of the virus particle, which may give it increased stability.

Virus-encoded proteases are attractive targets for antiviral therapies; for example, the protease inhibitors Saquinavir mesylate (Invirase), Saquinavir (Fortovase), Ritonavir (Norvir), Indinavir (Crixivan), Nelfinavir (Viracept), Amprenavir (Agenerase), and ABT-378 (Kaletra) target the HIV-encoded protease by preventing the maturation of virions capable of infecting other cells (**FIGURE 4-17**).

7. Release

Newly formed viruses are either released to the outside environment upon lysis, escaping the cell as it disintegrates (**lytic viruses**), or are released by budding (**FIGURE 4-18**) through the plasma membrane of the cell (as is the case with retroviruses, togaviruses, orthomyxoviruses, paramyxoviruses, bunyaviruses, coronaviruses, rhabdoviruses, and hepadnaviruses). Viruses that are released via budding may damage the cell (as is the case with paramyxoviruses, rhabdoviruses, and togaviruses) or they may not (as is the case with retroviruses). Some viruses bud from other membranes and are released from the cell via a secretory-like mechanism.

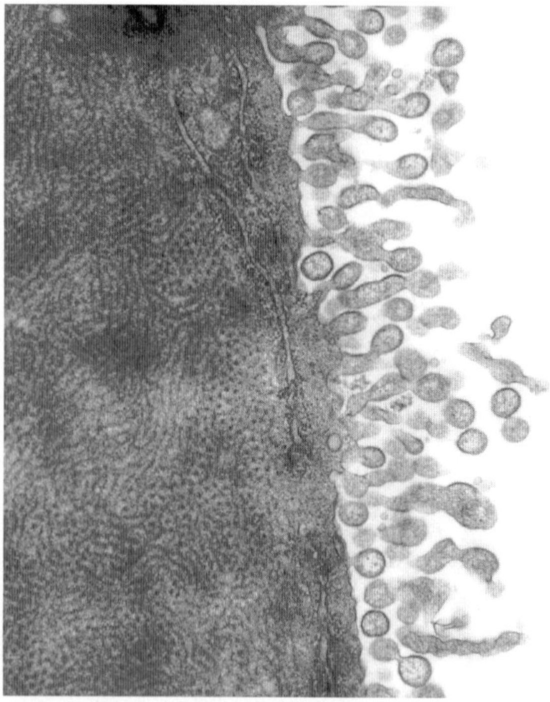

FIGURE 4-18 Transmission electron micrograph of measles virus released by budding.

Lytic Viruses

Most naked viruses are released when infected cells break open (cell lysis/destruction) due to the activity of viral enzymes, rather than distention. Poliovirus is an example of a lytic virus. A lytic life cycle is one that kills the host cell. Many enveloped viruses also do this.

Latent Eukaryotic Viruses

Retroviruses such as HIV-1 undergo a **latent** (**nonlytic**) **cycle** in which the viral DNA (provirus) becomes inserted into the host's DNA. In certain cell types, the proviral DNA replicates "silently" along with the cellular DNA, and the virus is undetected for many years. The provirus, however, can be activated at any time, allowing a productive infection or lytic cycle (one that produces infectious exogenous particles) to occur. Note that the viral DNA remains integrated and does not excise itself from the chromosome as is done in bacteriophages (see Chapter 21). **Latency** is considered to be one of the biggest reasons why drug therapy fails to eradicate HIV from patients.

Herpesviruses enter and migrate down neurons, where they become latent in the body of neurons. They do not integrate their DNA; it remains as an episome. Subsequent activation of the latently infected neurons by a variety of factors (such as extreme temperatures, physical trauma, emotional stress, and immune suppression) enables the viruses to migrate back up the nerve cell and replicate again in the epithelial cells. Two common herpesviruses are herpes simplex virus type 1 (HSV-1), which usually causes fever blisters or oral herpes, and herpes simplex virus type 2 (HSV-2), which usually causes genital herpes (see Chapter 15).

VIRUS FILE 4-2 Unraveling the Life Cycle of Mimivirus

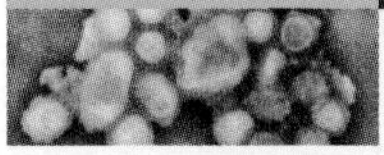

The discovery of Mimivirus in 2003 sent a shockwave through the community of virologists and evolutionists. The virus was 750 nm in size—gargantuan compared to typical viruses—and was termed a **Girus** (see Chapter 3, Viruses That Challenge the Definition of a Virus). It contained the largest known viral genome at a whopping 1,181,404 base pairs of dsDNA, coding for 911 gene products. It contained many genes not found before in other viruses such as **aminoacyl-tRNA synthetases** that are central to components of cellular translation machinery, genes that are components of cellular replication and transcriptional machinery, and genes associated with metabolic pathways, and four of its genes contained **introns**. *But Mimivirus still appears to be absolutely dependent on its host cells for synthesis of its proteins.*

Following the deciphering of the genome sequence of Mimivirus, efforts began to elucidate the life cycle of the virus. Research teams led by Didier Raoult or Abraham Minsky resulted in extensive ultrastructural studies to determine the life cycle of Mimivirus. In their studies, Mimiviruses were "fed" and allowed to infect *Acanthamoeba polyphaga* at a cell–virus ratio of 1:10. At various times post-infection within a 24-hour period, the infected cells were collected and prepared for transmission electron microscopy. Electron microscopy demonstrated that within 30 minutes, the Mimiviruses appeared to enter the amoeba cells by phagocytosis. Upon engulfment, the Mimiviruses underwent uncoating steps by fusing with the host phagosomes. During the fusion event, the viral capsids morphed into 5-fold star-shaped structures that acted like a "stargate," or portal, through which viral DNA was released into the cytoplasm of the host cell. Subsequently, the viral DNA was imported into the host nucleus where its first round of replication began. Like poxviruses, Mimiviruses possess their own transcriptional apparatus (refer to Chapter 14).

By 3 hours post-infection, Mimivirus DNA exited the host nucleus to form cytoplasmic viral replication factories surrounded by mitochondria. By 5 to 8 hours post-infection, these viral factories increased 50% in size and viral proteins were detected. In addition to electron microscopy, direct fluorescent staining was performed on the infected cells in order to study the Mimivirus factories. The viral factories appeared to have three zones: an electron-dense inner replication center, an intermediate assembly zone, and the peripheral zone

(continued)

where particles matured and acquired fibrils. At 8 hours after the infection, the viral factories contained empty, fiberless capsids that were partially assembled as well as some icosahedral capsids undergoing DNA packaging through a transient "stargate" or aperture. By 10 to 12 hours post-infection, mature fibril-coated particles budded from the viral factories and were released through cell lysis (**FIGURE VF 4-2**). Overall, the takeover of the cellular machinery by Mimivirus was rapid and efficient. Many of the viral events in the life cycle took place in a giant volcano-like viral factory.

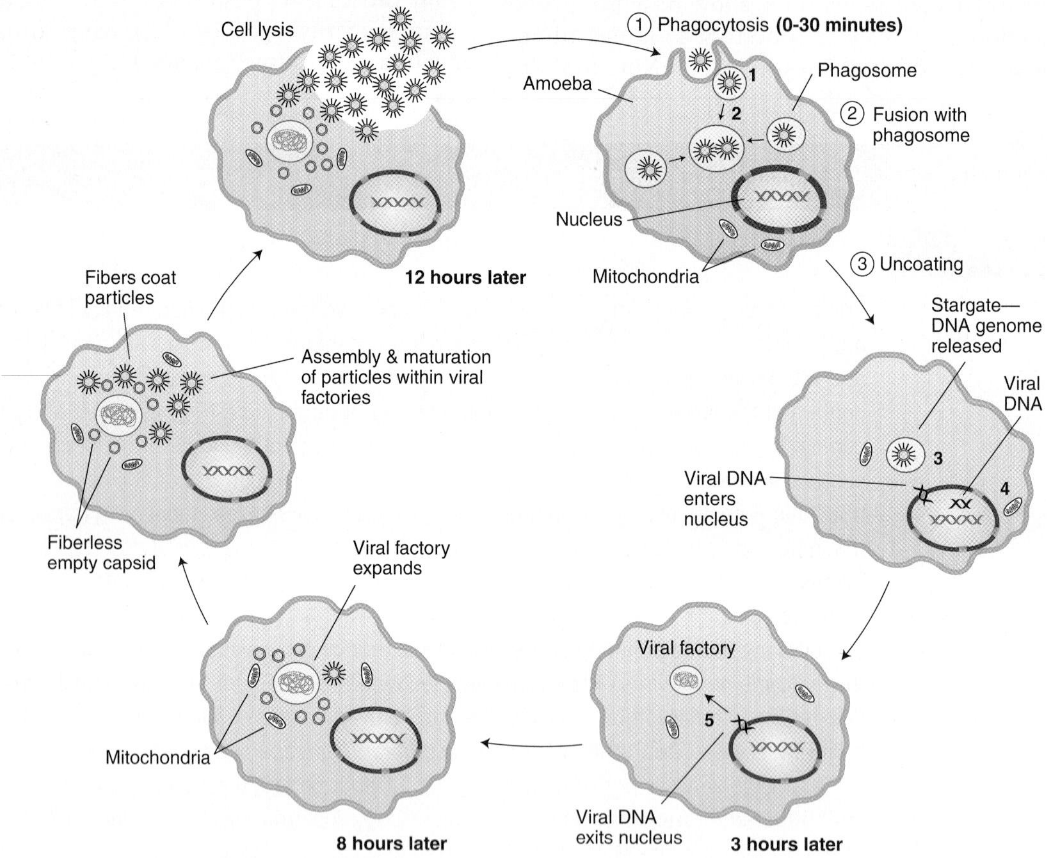

FIGURE VF 4-2 The Mimivirus life cycle based on observations of infected amoeba cells at various times after infection. Adapted from Suzan-Monti, M., et al. 2007. "Ultrastructural characterization of the giant volcano-like virus factory of *Acanthamoeba polyphaga* Mimivirus." *PLoS ONE* 3:e328. Figure 8.

References

Claverie, J. M., et al. 2009. "Mimivirus and Mimiviridae: Giant viruses with an increasing number of potential hosts, including corals and sponges." *J Invertebr Pathol* 101:172–180.

Claverie, J. M., and Abergel, C., 2009. "Mimivirus and its virophage." *Ann Rev Gen* 43:49–66.

Suzan-Monti, M., et al. 2007. "Ultrastructural characterization of the giant volcano-like virus factory of *Acanthamoeba polyphaga* Mimivirus." *PLoS ONE* 3:e328.

Zauberman, N., et al. 2008. "Distinct DNA exit and packaging portals in the virus *Acanthamoeba polyphaga* Mimivirus." *PLoS Biol* 6:1104–1114.

Why Don't the Viruses Get Stuck on the Cellular Receptors as They Are Released?

The release of virus particles poses an interesting dilemma. Viruses are designed to enter rather than leave cells. Some viruses, such as influenza A, produce a protein during their life cycle to destroy the cellular receptors as they exit the cell. The protein, **neuraminidase**, cleaves the sialic-acid receptors on the outside of cells as the infectious particles are released (see Chapter 12). As a result, the viruses do not aggregate at the cell surface.

4.3 The Error-Prone RNA Polymerases: Genetic Diversity

Viruses replicate rapidly. During the process of replication, an error, or point mutation, may occur. The mutation rate of DNA viruses is usually similar to those of their cellular hosts because most of these polymerases used to replicate DNA genomes possess proofreading ability; for example, the mutation rate of herpesviruses (with **proofreading ability**) is one error in every 10^8 to 10^{11} bases. The genome sizes of herpesviruses range from 1.3×10^5 to 2.0×10^5 base pairs in length. As a result, herpesviruses potentially evolve very slowly because few mutations will be made, if any, during the infection cycle of the virus.

RNA viruses possess mutation rates as high as one error in 10^3 to 10^4 bases. The RNA-dependent RNA polymerases and RNA-dependent DNA polymerases (for example, reverse transcriptases) used by the RNA viruses for genome replication do not possess proofreading ability. Hence, the presence of mutants in a virus population during each replication cycle of the virus occurs much more rapidly than in DNA viruses or cellular organisms. HIV and coronaviruses (such as severe acute respiratory syndrome-associated coronavirus, or SARS-CoV) are excellent examples of RNA viruses with high mutation rates. These viruses misincorporate a nucleotide into their genome once in 10^3 to 10^4 bases. For coronaviruses, this means that there are three mutations that occur during the replication of one viral genome. For HIV, this means one to two mutations in every genome copied. The high mutation rate probably limits the size of most RNA virus genomes to approximately 10^4 nucleotides.

Many mutations are lethal because the mutated virus is unable to replicate. Nonlethal mutations may give the mutated virus a selective advantage. Mutations have been associated with the development of antiviral drug resistance (such as the drug-resistant strains of HIV), changes in virulence (for example, the 1997 Hong Kong avian influenza A was highly virulent), changes that allow the virus to evade the host's immune system (such as the influenza viruses and antigenic drift, which will be discussed in Chapter 12), and changes in host range.

FIGURE 4-19 A Vietnamese woman weighs a chicken at a market in Hanoi, Vietnam (January 2004). At the time, the avian influenza virus infected several humans who had direct contact with chickens. The virus infected millions of chickens, raising fears that it might mutate to a strain that could readily pass from human to human, leading to the next human influenza pandemic.

A good example of a change in host range adaptation occurred in 1978. Canine parvovirus suddenly began killing large numbers of dogs globally in 1978. The parvovirus originally had infected only cats, foxes, raccoons, and minks. A small number of changes in the capsid genes adapted the virus for efficient spread among dogs (**FIGURE 4-19**).

4.4 Targets for Antiviral Therapies

Modern technology allows scientists to deliberately design drugs. To do this, they need to understand "the enemy." Knowledge of the life cycle of viruses and the mapping of them using computer-aided design (CAD) is applied toward antiviral development by scientists at pharmaceutical companies. Any of the seven stages of the viral life cycle can be targeted for antiviral intervention. The stages are:

1. Attachment specificity
2. Penetration
3. Uncoating
4. Replication
5. Assembly
6. Maturation
7. Release

TABLE 4-4 lists the current antiviral therapies available along with a list of which viruses they target and their mechanism of action. Key to antiviral drug development is that the drug must *target a process essential for viral replication and it must be active against the virus without being "toxic" to the host organism*. It has been difficult to develop antivirals that have no toxic side effects for the host because viruses use some of the host cellular processes for replication. Hence, drugs cannot target those cellular processes because of toxicity problems for the host.

TABLE 4-4 Prevention and Treatment of Human Viral Diseases: Antiviral Drugs

Drug	Virus/Disease	Target
Idoxuridine Trifluridine	Herpes simplex keratoconjunctivitis	Viral and cellular DNA synthesis
Vidarabine	HSV-1, HSV-2	Viral DNA polymerase
Acyclovir	HSV-1, HSV-2, VZV, EBV, CMV	Virus DNA polymerase
Tromantadine	HSV	Viral DNA polymerase
Famciclovir	HSV-1, HSV-2, VZV, some activity against EBV, CMV, and HBV	Viral thymidine kinase and DNA polymerase
Penciclovir	HSV-1, HSV-2	Viral DNA polymerase
Valacyclovir	HSV-1, HSV-2, VZV, modestly active against EBV and CMV	Viral DNA polymerase
Gancyclovir	CMV retinitis in HIV patients	Virus polymerase
Foscarnet	Acyclovir-resistant HSV/VZV strains Ganciclovir-resistant CMV HSV-1, HSV-2, HHV-6, EBV, VZV, parainfluenza virus CMV retinitis in HIV disease	Viral DNA polymerase and reverse transcriptase
Docosanol (10% topical cream)	HSV-labialis episodes	Fusion inhibitor
Brivudine (approved for use in Germany and other European countries)	HSV-1 and VZV	Viral thymidine kinase and DNA polymerase
Entecavir, Adefovir	Hepatitis B	Hepatitis B reverse transcriptase
Abacavir	HIV-1 and HIV-2	HIV-1 reverse transcriptase
Didanosine (ddI)		
Emtricitabine (FTC)		
Lamivudine (3TC)		
Lamivudine + zidovudine (Combivir)		
Stavudine (d4T)		
Tenofovir + emtricitabine (Truvada)		
Tenofovir DF		
Zalcitabine (ddC)		
Zidovudine (AZT)		
Nevirapine	HIV-1	HIV-1 reverse transcriptase
Delavirdine	HIV-1	HIV-1 reverse transcriptase
Efavirenz	HIV-1	HIV-1 reverse transcriptase
Amprenavir	HIV-1	HIV-1 protease
Atazanavir		
Fosamprenavir		
Saquinavir		
Lopinavir + ritonavir		
Indinavir		

(continued)

TABLE 4-4 Prevention and Treatment of Human Viral Diseases: Antiviral Drugs (continued)

Drug	Virus/Disease	Target
Darunavir	HIV-1	HIV-1 protease
Nelfinavir		
Ritonavir		
Tipranavir		
Enfuvirtide	HIV-1	Binds to HIV-1 gp41 surface protein, inhibiting viral entry
Selzentry	HIV-1	Binds to CCR5 co-receptor
Raltegravir	HIV-1	Integrase strand transfer inhibitor
Atripola: cocktail of efavirenz, tenofovir, and emtricitabine	HIV-1	A fusion inhibitor and two reverse transcriptase inhibitors
Ribavirin	Broad spectrum (inhibits DNA and RNA viruses): RSV, Influenza A and B, HCV, HSV-1 and HSV-2, measles, mumps, Lassa fever	mRNA mutagen
Amantadine	Influenza A	Inhibits penetration and uncoating of virus
Rimantidine		
Relenza and Tamiflu	Influenza A	Neuraminidase inhibitor
Peramivir (not FDA approved)	Influenza A (H1N1)	Neuraminidase inhibitor
Arbidol (not FDA approved)	Influenza A and B, hepatitis C	Virus-mediated membrane fusion, immune modulator?
Cidofovir	Broad spectrum (inhibits DNA viruses): HSV-1, HSV-2, VZV, CMV, EBV, adenovirus, HPV, CMV retinitis in HIV patients, experimentally used to treat poxvirus infections	DNA polymerase
Interferons	Hepatitis B & C Hairy cell leukemia, HPV, respiratory viruses	Cell defense proteins activated
Fomivirsen	CMV	Inhibits viral replication and translation (antisense molecule)
Podophyllotoxin	HPV genital warts	Binds E2 and inhibits E2/E3 interaction

Abbreviations of viruses: Herpes simplex type 1 (HSV-1), herpes simplex type 2 (HSV-2), varicella zoster virus (VZV), Epstein-Barr virus (EBV), cytomegalovirus (CMV), hepatitis B virus (HBV), hepatitis C virus (HCV), human herpes Type 6 virus (HHV-6), human immunodeficiency virus Type 1 (HIV-1), respiratory syncytial virus (RSV), hepatitis C virus (HCV), human papillomavirus (HPV). Abbreviations of antivirals: zidovudine (AZT), didanosine (ddI), zalcitabine (ddC), stavudine (d4T), lamivudine (3TC). Unless otherwise specified, drugs are approved by the FDA.

VIRUS FILE 4-3 Stamping Down Flu Viruses

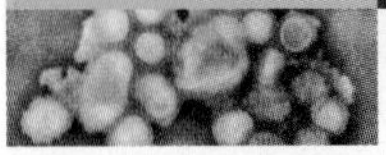

On a global scale, there are about 3.5 million severe cases of influenza illness and 300,000–500,000 deaths annually. Most at risk are the elderly, young children, and immunocompromised patients. Influenza virus can be controlled in two ways: vaccination and use of antivirals. It takes at least 6 months to produce sufficient vaccine to vaccinate a large proportion of the population upon the emergence of a new strain. For this reason, antivirals are necessary to mitigate an influenza pandemic.

Currently, two types of anti-influenza drugs are available: M2 ion channel blockers (amantidine and rimantadine) and neuraminidase inhibitors: tamiflu (oseltamivir phosphate) and relenza (zanamivir). M2 ion channel blockers prevent the uncoating step of the life cycle of influenza viruses. After influenza A viruses bind to sialic acid receptors, the virus is internalized by receptor-mediated endocytosis. The low pH of the endosome causes the viral and endosomal membranes to fuse and an influx of H+ ions enter the M2 ion channels present on the influenza virus particle, allowing the viral RNA segments to be released into the cytoplasm of the host cell (uncoating). Antivirals that block the M2 ion channel prevent this uncoating step (**FIGURE VF 4-3**).

(continued)

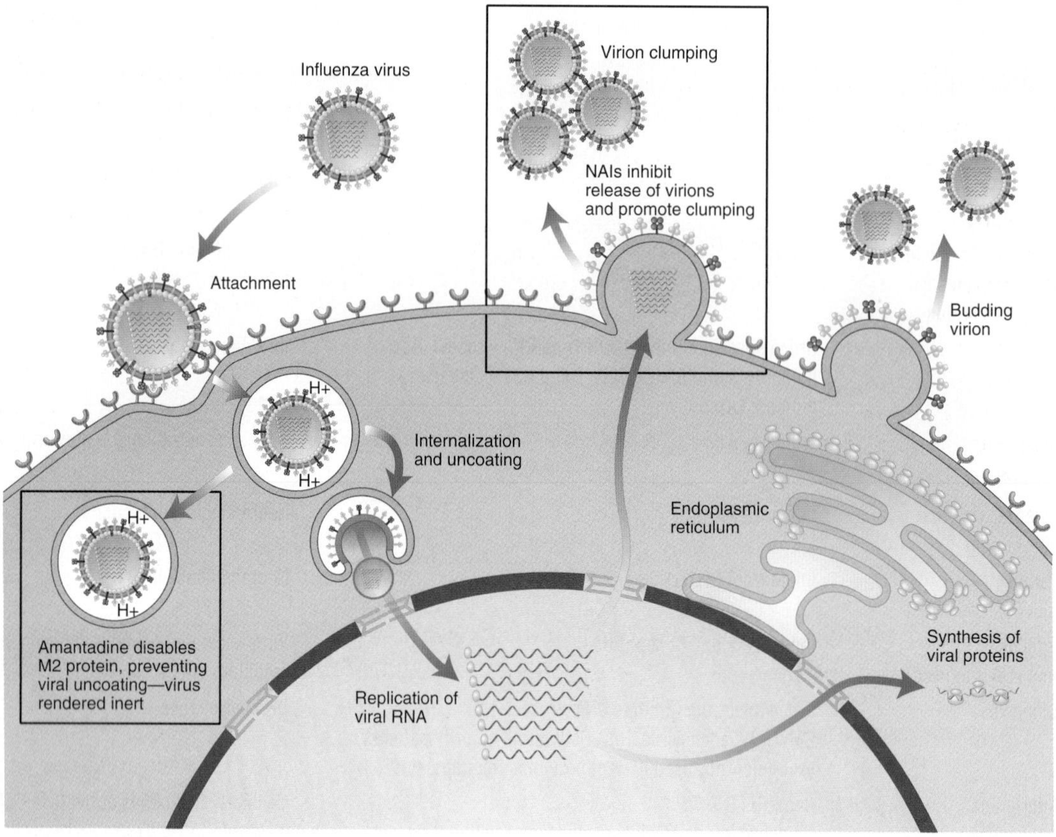

FIGURE VF 4-3 Influenza antivirals block the M2 ion channel or neuraminidase activity of influenza virus.

Antiviral drugs that block neuraminidase cause influenza viruses to clump at the surface of host cells, preventing its spread to other host cells at the end of the viral life cycle. Neuraminidase is an influenza protein present on the outside of the virus particle. As newly assembled viruses bud from host cells and are released, neuraminidase cleaves or removes the host cell sialic acid receptors present on the outside of host cells. When neuraminidase is blocked, influenza viruses stick to the sialic acid receptors present on the plasma membrane of the infected host cells, causing viruses to clump and not spread to other cells of the respiratory tract (Figure VF4-3).

Unfortunately, resistant strains of M2 inhibitors rapidly emerge. Resistant strains of oseltamivir (Tamiflu) are also emerging. It is believed that indiscriminate use of these drugs has led to the appearance of drug-resistant viruses. In response to the need for new anti-influenza drugs, scientists are developing and testing promising new antiviral candidates. Structure-based design, targeting the conserved active site amino acids of neuraminidase, is being used to synthesize new high-affinity neuraminidase inhibitors.

Two such candidates are CS-8958 and peramivir. Both of these are neuraminidase inhibitors. During the 2009 H1N1 pandemic, emergency authorization was given to use the FDA-unapproved peramivir to treat severe cases of H1N1 influenza. Patients receiving peramivir were hospitalized with life-threatening influenza including such complications as kidney failure. Peramivir could only be used in an injectable form.

References

Bai, G. R., et al. 2009. "Amantadine- and oseltamivir-resistant variants of influenza A viruses in Thailand." *Biochem Biophys Res Comm* 390:897–901.

Kiso, M., et al. 2010. "Efficacy of the new neuraminidase inhibitor CS-8958 against H5N1 influenza viruses." *PLoS Pathog* 6:2e1000786.

RNA Virus Mutagens: A New Class of Antiviral Drugs?

The mechanism of action of the antiviral drug ribavirin was a mystery for over 30 years. Drs. Joseph T. Witkowski and Roland K. Robins synthesized ribavirin in the laboratory in 1970. In 1972, it was reported that ribavirin inhibited a wide spectrum of viruses in vitro but its mechanism was unknown. Today it is used to treat hepatitis C virus infections in combination with interferon-α, SARS-CoV infections, human rabies (as in the case of the Wisconsin patient in 2004), Lassa fever, and some herpesvirus infections. It is used experimentally to treat HBV, Hantaan, Dengue, and parainfluenza virus infections.

In 2001 the mechanism of action was proposed and demonstrated by Shane Crotty, Craig E. Cameron, and Raul Andino. Their poliovirus experiments suggested that ribavirin's antiviral activity is exerted through lethal mutagenesis. In other words, the drug overwhelms the virus with a high mutation rate that in turn drives the virus into a genetic meltdown. Interestingly, this high mutation rate does not allow the virus to escape the inhibitory effects of the drug. This discovery provides pharmaceutical companies with an entirely new drug strategy: RNA virus mutagens.

Summary

With the development of animal cell culture techniques, scientists have been able to study virus replication via one-step growth experiments. The knowledge today is very detailed and continues to progress so rapidly that it is impossible to cover the viral replication strategies used by every family of viruses in a single chapter. This chapter presents an overview of viral replication that includes the seven key steps in the life cycle of a virus. They are:

1. Attachment specificity
2. Penetration
3. Uncoating
4. Replication
5. Assembly
6. Maturation
7. Release

Virus attachment occurs when a virus particle attaches to a target cell surface receptor and is the first event in the viral life cycle. Viral receptors are discovered via experiments that involve:

- *the removal of surface receptors,*
- *the use of monoclonal antibodies to block cell surface receptors, and*
- *gene-transfer experiments.*

Viruses enter their target cells via fusion or receptor-mediated endocytosis. ADE has been shown to be an alternative mechanism of viral entry into host cells. This phenomenon has been observed with Dengue, West Nile, HIV-1, HIV-2, and Influenza A viruses.

Replication strategies are carried out based on the nature of the viral genome (DNA or RNA, ss or ds, segmented or nonsegmented). It is often difficult to distinguish between the steps of assembly, maturation, and release of virus particles.

Some viruses destroy their host cells (lytic viruses), whereas others do not. Some viruses have developed ways to overcome the conundrum of entry versus exiting cell surfaces that are coated with viral receptors.

Viruses are masters of mutation. Mutations can be lethal or nonlethal. Nonlethal mutations may give rise to mutants that may increase their survivorship; for example, mutations may increase their infectivity, result in viral drug resistance, antagonize the host immune system, or broaden the host range of the virus.

Scientists are applying the knowledge obtained through molecular biology toward the design of antiviral drugs. These antivirals must target the virus specifically to be effective in eliminating the virus without causing harm to its host.

CASE STUDY QUESTIONS

These questions relate to the Case Study presented at the beginning of the chapter.

1. hMPV and RSV belong in the Paramyxoviridae family. What type of nucleic acid genome do they have? Do they encode their own viral polymerase? (see Figure 4-14a).
2. Draw a flowchart of the viral life cycle of hMPV.
3. How many genes does hMPV have in its genome?
4. The first report of a fatal encephalitis case associated with hMPV was published in the CDC's *Emerging and Infectious Diseases* publication in 2005. The authors recommended screening for patients, especially children with encephalitis symptoms of unknown origin. What is encephalitis?
5. Have neurological symptoms been associated with other viruses in the Paramyxoviridae family? If so, which virus(es)?

Note: You may need to do further research to find the answers to these questions. The following references will provide some helpful information.

Alto, W. A. 2004. "Human metapneumovirus: A newly described respiratory tract pathogen." *JABFP* 17(6):466–469.

Schildgen, O., et al. 2005. "Human metapneumovirus RNA in encephalitis patient." *EID* 11(3): 467–470.

Van Den Hoogen, B. G., et al. 2001. "A newly discovered human pneumovirus isolated from young children with respiratory tract disease." *Nat Med* 7(6):719–724.

CASE STUDY 1: RABIES

In February 2009, a 17-year-old Texas teenager arrived at a hospital emergency room presenting with symptoms of a severe headache, neck pain, dizziness, **photophobia**, nausea and vomiting, and tingling of her face and forearms. A **lumbar puncture** was performed, which revealed a very high white blood cell count suggesting the teen might be fighting a bacterial infection. She was treated intravenously with the broad-spectrum antibiotic ceftriaxone. No bacteria were cultured from her cerebral spinal fluid so the antibiotic was discontinued. After a 3-day hospital stay, she was released because her symptoms had resolved. About a week later, she returned to the hospital with the same symptoms. She also had a rash on her arms and back. **Magnetic resonance imaging** (MRI) was performed on her head. The MRI showed enlarged lateral ventricles inside her brain. The size was abnormal for her age and she was given the diagnosis of **encephalitis**. She was hospitalized and treated with ceftriaxone and antibiotics used to treat tuberculosis.

Four days later, she continued to weaken. She became agitated and "combative." At this time, clinicians did an extensive workup, collecting the patient's history, and created a list of possible etiologies to explain the encephalitis/aseptic meningitis. A breakthrough came when the teenager mentioned she had been hiking and spelunking during December (2008) and had brushed up against some bats but didn't recall being bitten. Serum, saliva, and cerebral spinal fluid were drawn or collected from the patient, a **nuchal skin biopsy** was performed, and the clinical specimens were sent to the CDC to test for rabies antibodies, rabies antigens, or the RNA genome of rabies virus.

The next day, no rabies RNA or virus antigens were detected. Four serum samples tested positive for anti-rabies antibodies by direct fluorescent antibody tests, however. Anti-rabies antibodies were also found in the cerebral spinal fluid. Four days later, the teenager was given 1 dose of rabies vaccine and 1,500 international units (IU) of human rabies immunoglobulin.

Eight days later, her symptoms resolved and she was released from the hospital. Another week passed and she returned to the emergency room of a hospital with recurring headache but left before a lumbar puncture was performed. A few more days passed. Again, she had a recurring headache. A lumbar puncture was performed. Her headache resolved and she has not been rehospitalized since. Only her boyfriend met the criteria of requiring a series of rabies vaccinations.

1. Create a hypothesis as to why this teenager survived a rabies virus infection.
2. What do the CDC researchers speculate as to why this patient survived rabies? Refer to the CDC report related to this case published February 26, 2010 (*MMWR* 59[7]; *MMWR* can be found at: http://www.cdc.gov/mmwr/).
3. Assuming the teenager was not bitten, what could be another plausible explanation for how the teenager contracted this infection?
4. Rabies virus is a member of the Rhabdoviridae family. What are the molecular characteristics of this family of viruses (e.g., particle characteristics, life cycle)?
5. How many genes do rabies viruses carry and what are their function(s)?
6. What types of cells do rabies viruses replicate in (hint: Table 4-1)?
7. Rabies bind to what host receptor(s)?
8. Who should receive a rabies vaccination? (hint: http://www.cdc.gov/rabies/)

For more information see Chapter 13 (Rabies).

Resources

Primary Literature and Reviews

Ackermann, H-W., Berthiaume, L., and Tremblay, M. 1998. "The replication cycle." In *Virus Life in Diagrams*. Boca Raton, FL: CRC Press.

Audelo-del-Valle, J., et al. 2003. "Infection of cultured human and monkey cell lines with extract of penaeid shrimp infected with Taura syndrome virus." *EID* 9(2):265–266.

Avril, R. M. C-C., Dixon, D. W., Vzorov, A. N., Marzilli, L. G., and Compans, R. W. 2003. "Prevention of poxvirus infection by tetrapyrroles." *BMC Inf Dis* 3:9. http://www.biomedcentral.com/1471-2334/3/9.

Baranowski, E., Ruiz-Jarabo, C.M., and Domingo, E. 2001. "Evolution of cell recognition by viruses." *Science* 292:1102–1105.

Boriskin, Y. S., et al. 2006. "Arbidol: A broad-spectrum antiviral that inhibits acute and chronic HCV infection." *Virol J* 3:56.

Coffin, J. M. 1995. "HIV population dynamics in vivo: Implications for genetic variation, pathogenesis, and therapy." *Science* 267:483–489.

Cohen, J., and Powderly, W. G., eds. 2004. *Infectious Diseases*, 2nd ed. Philadelphia: Mosby.

Crotty, S., Cameron, C. E., and Andino, R. 2001. "RNA virus error catastrophe: Direct molecular test by using ribavirin." *PNAS USA* 98(12):6895–6900.

De Clercq, E. 2002. "Cidofovir in the treatment of poxvirus infections." *Antiviral Res* 55(1):1–13.

Dimiter, S. D. 2000. "Cell biology of virus entry." *Cell* 101:697–702.

Drake, J. W., and Holland, J. J. 1999. "Mutation rates of RNA viruses." *PNAS USA* 96(24):13910–13913.

Enders, J. F., Robbins, F. C., and Weller, T. H. Nobel Lecture, December 11, 1954. "The cultivation of the poliomyelitis viruses in tissue culture."

Flint, S. J., Enquist, L. W., Krug, R. M., Racaniella, V. R., and Skalka, A. M., eds. 2000. "Genome replication and mRNA production by RNA viruses." In *Principles of Virology: Molecular Biology, Pathogenesis, and Control*. Washington, DC: ASM Press.

Fu, Y-X. 2001. "Estimating mutation rate and generation time from longitudinal samples of DNA sequences." *Mol Biol Evol* 18(4):620–626.

Junge, R. E., Duncan, M. C., et al. 1999. "Clinical presentation and antiviral therapy for poxvirus infection in Pudu (Pudu Pudu)." *J Zoo Wildlife Med* 31(3):412–418.

Knipe, D. M., and Howley, P. M., eds. 2001. "Virus assembly." In *Fields Virology*, 4th ed. Philadelphia: Lippincott Williams & Wilkins.

Knipe, D. M., Samuel, C. E., and Palese, P., eds. 2001. "Principles in virology and virus entry and uncoating." In *Fundamental Virology*. Philadelphia: Lippincott Williams & Wilkins.

Martina, B. E. E., Haagmans, B. L., Kuiken, T., et al. 2003. "SARS virus infection in cats and ferrets." *Nature* 425:915.

Mendelsohn, C. L., et al. 1986. "Transformation of a human poliovirus receptor gene into mouse cells." *PNAS USA* 83(20):7845–7849.

Mendelsohn, C. L., Wimmer, E., and Racaniello, V. R. 1989. "Cellular receptor for poliovirus: Molecular cloning,

nucleotide sequence, and expression of a new member of the immunoglogulin superfamily." *Cell* 56:855–865.
Mothes, W., et al. 2000. "Retroviral entry mediated by receptor priming and low pH triggering of an envelope glycoprotein." *Cell* 103:679–689.
Paulson, J. C., and Rogers, G. N. 1987. "Resialylated erythrocytes for assessment of the specificity of sialyloligosaccharide binding proteins." *Methods Enzymol* 138: 162–168.
Pluong, C. X. T., et al. 2004. "Clinical diagnosis and assessment of severity of confirmed dengue infections in Vietnamese children: Is the World Health Organization classification system helpful?" *Am J Trop Med Hyg* 70(2): 172–179.
Schaechter, M., ed. "Antiviral agents." In *Encyclopedia of Microbiology*, 3rd Edition. Amsterdam: Elsevier Ltd. 2009.
Sieczkarski, S. B., and Whittaker, G. R. 2005. "Viral Entry." *Curr Top Microbiol Immunol* 285:1–23.
Smith, A. E., and Helenius, A. 2004. "How viruses enter cells." *Science* 304:237–242.
Staunton, D. E., Merluzzi, V. J., Rothlein, R., et al. 1989. "A cell adhesion molecule, ICAM-1, is the major surface receptor for rhinoviruses." *Cell* 56:849–853.
Stine, G. J. 2002. "Anti-HIV therapy." In *AIDS Update 2002*. Upper Saddle River, NJ: Prentice-Hall.
Tyler, K. L. 2004. "Isolation and molecular characterization of a novel type 3 reovirus from a child with meningitis." *J Infect Dis* 189(9):1664–1675.
Wilkin, T. J., et al. "HIV Type 1 chemokine coreceptor use among antiretroviral-experienced patients screened for a clinical trial of a CCR5 inhibitor: AIDS clinical trial group A5211." *Clin Infect Dis* 44:591–595.

Popular Press

Haseltine, W. A. 2001. "Beyond chicken soup." *Scientific American*, November 2001, Vol. 285, Issue 5, 56–63.
Tiollais, P., and Buendia, M-A. 1991. "Hepatitis B virus." *Scientific American*, April 1991, Vol. 264, Issue 4, 116–123.
Villarreal, L. P. 2004. "Are viruses alive? Although viruses challenge our concept of what 'living' means, they are vital members of the web of life." *Scientific American*, December 2004, Vol. 291, Issue 6, 100–105.
Wong, K. 2002. "Oral drug halts smallpoxlike virus in mice." *Scientific American, News Online*, March 2002.
Wright, L. "To vanquish a virus." 2003. *Scientific American in Focus Online*, July 21, 2003.

Video Productions

Understanding Viruses. 1999. The Discovery Channel.
The Next Plague: The Nipah Virus. 1998. The Discovery Channel.
Viruses: The Deadly Enemy. 1996. Available at www.insight-media.com. Human Relations Media.
The Emerging Viruses. 1991. British Broadcasting Corporation.

eLearning

go.jblearning.com/shors2

The site features eLearning, an online review area that provides quizzes and other tools to help you study for your class. You can also follow useful links for in-depth information, or just find out the latest virology and microbiology news.

CHAPTER 5

Laboratory Diagnosis of Viral Diseases and Working with Viruses in the Research Laboratory

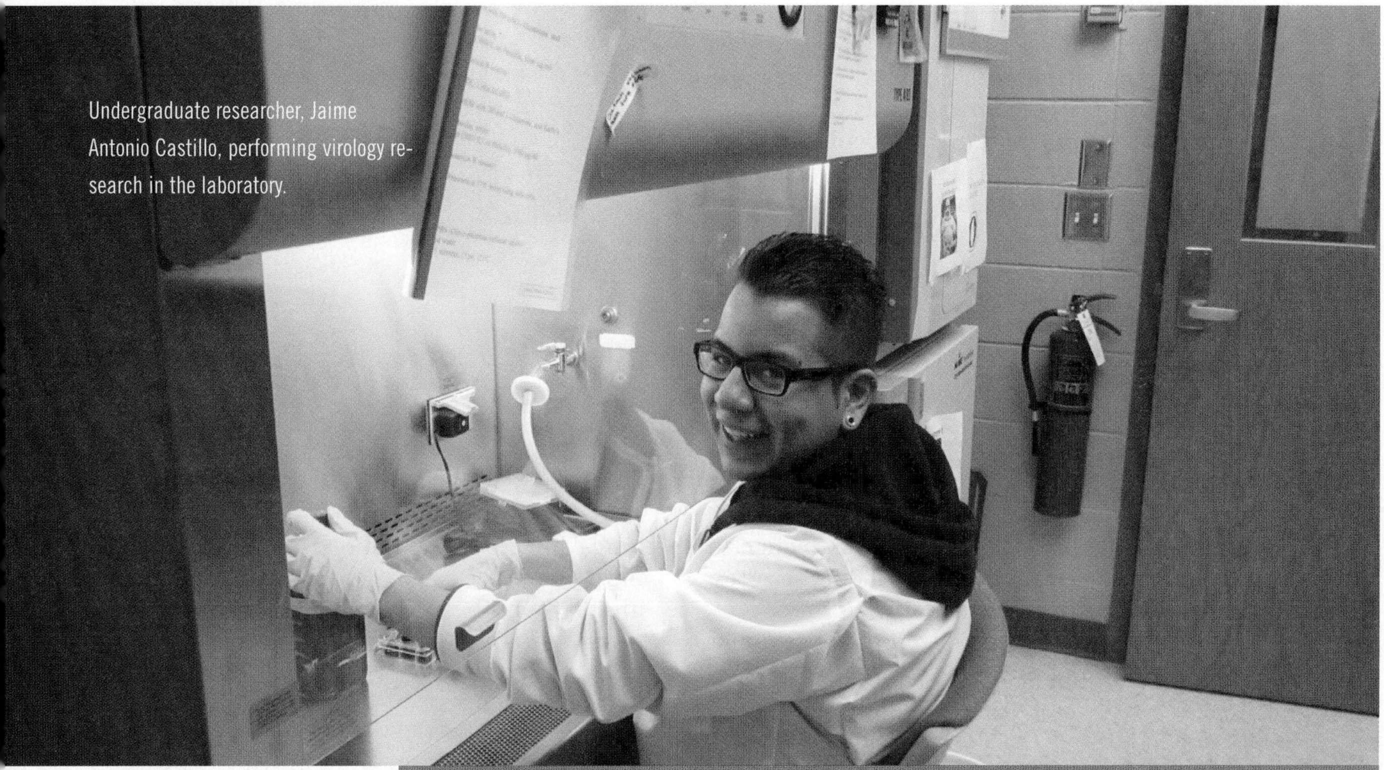

Undergraduate researcher, Jaime Antonio Castillo, performing virology research in the laboratory.

OUTLINE

“Where observation is concerned, chance favors only the prepared mind.”

Louis Pasteur

CASE STUDY: CLUSTER OF VIRAL MENINGITIS AND ENCEPHALITIS CASES

Dr. Cronn is an inpatient infectious disease consultant for an academic medical center located in the Midwest. He has returned to work after a three-week vacation. During his absence, there had been three cases of **aseptic meningitis** and/or **encephalitis** within a two-week time span. Infection of the meninges, the membranes surrounding the brain and spinal cord, is called **meningitis**. The inflammation of the brain itself is called **encephalitis**. The term "**aseptic**" is used when medical laboratory technicians have been unable to culture an organism present in the cerebrospinal fluid (CFS) of patients suffering from a central nervous system syndrome. Early symptoms of meningitis/encephalitis include headache, high fever, and stiff neck. Other symptoms can include confusion, irritability, memory loss, nausea, vomiting, sensitivity to looking at bright lights, and sleepiness. Determining whether the cause is viral or bacterial is important because it will determine the course of treatment. Without treatment, aseptic central nervous system infections may cause a variety of outcomes from benign to severe neurological problems, including seizures, brain damage, paralysis, and death.

A **spinal tap**, also referred to as a lumbar puncture, was performed to collect CFS from patients for diagnostics. A needle is inserted into the lumbar portion of the back to withdraw CSF (**FIGURE CS 5-1**). In all of these patient cases, the CSF was used for preliminary tests to determine red blood cell, neutrophil and lymphocyte counts, and protein and glucose levels. Findings are shown in **TABLE CS 5-1**. Dr. Cronn reviewed the results of the CSF analysis. An abnormal level of cell counts is a valuable diagnostic window to the central nervous system. For example, unusually high levels of white blood cells (e.g., more than 1000) occur in the majority of patients who have bacterial meningitis. Having white blood cell counts in the 100 range or less is more common in patients with viral meningitis. High protein levels may be a sign of injury, nerve inflammation, a tumor, bleeding, or infection. Low or high levels of glucose may indicate the presence of inflammation, bacterial or fungal infection, or a tumor. Dr. Cronn noted that the preliminary tests in Table CS 5-1 were supporting the hypothesis that the patients may suffer from a viral infection or some other condition (e.g., inflammation). Additional tests included magnetic resonance imaging (MRI) and computerized tomography (CT) scan images of the brain and a chest X-ray (looking for tubercles in the lungs caused by the bacterium *Mycobacterium tuberculosis,* which may cause tuberculosis meningitis). The charts referred to any results available as "unremarkable." In other words, there were no unusual findings (**TABLE CS 5-2**). Gram stains of the CSF to determine if bacteria were present in the CSF were negative for all three patients. All cultures for bacteria, viruses, fungi from CSF, blood, and urine samples were negative (Table CS 5-2).

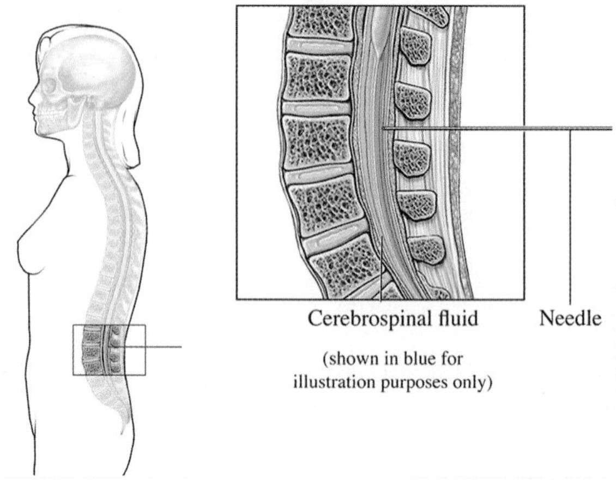

FIGURE CS 5-1 Cerebrospinal fluid is collected by performing a "spinal tap," also known as a lumbar puncture. © Nucleus Medical Art, Inc./Alamy

As Dr. Cronn looked up from the charts and paperwork, a young intern, Marvi Valentine stepped into his office. They exchanged information about these cases. Marvi remembered these

TABLE CS 5-1 Cerebral Spinal Fluid Preliminary Analysis in Case Patients

	Age/Sex	Red Blood Cells/mm^3	White Blood Cells/mm^3	% Neutrophils	% Lymphocytes	Glucose mg/dl	Protein mg/dl
Normal Values		*0*	*0–5*	*0*	*60–80*	*50–70*	*15–45*
Patient #1	56, male	34	154	77	12	63	82
Patient #2	72, male	0	120	4	64	53	106
Patient #3	86, female	24	3	4	20	87	40

Source: Adapted from Big, C., et al., 2009. "Viral infections of the central nervous system: A case-based review." *Clin Med Res* 7(4):142–146.

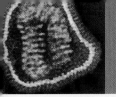

TABLE CS 5-2 Additional Laboratory Tests

Patient	Gram Stain of CSF	Cultures for Bacteria, Viruses, Fungi from CSF, Blood and Urine Samples	Head CT Scan	MRI	Chest Radiograph
#1	Negative	Negative	Mild abnormalities	Not performed	Not performed
#2	Negative	Negative	Unremarkable	Unremarkable	Unremarkable
#3	Negative	Negative	Unremarkable	Not performed	Unremarkable

TABLE CS 5-3 Patient History, Diagnostic, and Treatment Summary

Patient	History	Initial Treatment	Diagnostics/Method	Final Treatment
#1	Recurrent aseptic meningitis 2 years prior to this episode, chronic headaches	Intravenous vancomycin, meropenum, and acyclovir	PCR on CSF negative for herpes simplex viruses 1 and 2, positive for enterovirus	Antibiotics and antiviral discontinued, continued supportive treatment
#2	History of polymyalgia rheumatica and giant cell arteritis (inflammatory diseases)	Intravenous ceftriaxone, vancomycin, ampicillin, and acyclovir	PCR on CSF for varicella zoster was positive	Continued on acyclovir while hospitalized, went home with 3-week valacyclovir treatment
#3	Prior urinary tract infection and gastroesophageal reflux	Intravenous ceftriaxone, vancomycin, and acyclovir	PCR on CSF positive for herpes simplex virus type 1	Continued on a 21-day course of acyclovir and anti-epileptic drugs

patients very well. Dr. Cronn asked Marvi to describe early signs and symptoms experienced by these patients. She carefully described each patient. The first was a 56-year-old male who suffered from a fever of 103°F (39.4°C) during the initial examination, chills, headache, and a stiff neck. He was uncomfortable but did not have any neurological problems or a rash at the time of his examination. The second patient, a 72-year-old man, also had a fever and headache, no rash, and some neurological problems. He was only able to answer simple questions during examination. The third patient was an 86-year-old woman who, five days earlier, was diagnosed with a urinary tract infection. She was treated and released. During her second visit to a local emergency room, she had a 103°F (39.4°C) fever and was confused. All three of the patients had clinical symptoms of aseptic meningitis or encephalitis with varying degrees of severity. More details about the patients' histories are listed in **TABLE CS 5-3**. Initial treatment of all patients involved the administration of intravenous antibiotics such as vancomycin, ceftriaxone, ampicillin, or meropenem, which are used to treat bacterial infections. All of these patients were also treated with acyclovir, an antiviral drug used to treat herpesvirus infections. Table CS 5-3 also lists the molecular diagnostic results. **Polymerase chain reaction** (PCR; refer to **Refresher: PCR**) was performed on the CSF specimens to screen for different viral pathogens. Each patient suffered from a different viral infection: enterovirus, varicella zoster, or herpes simplex virus type 1.

Dr. Cronn and Marvi continued conversing about these cases. He told her that when he started practicing medicine over 30 years ago, PCR testing was not available and acyclovir became available to physicians in 1982. Central nervous system infections have a wide spectrum of causes. The use of molecular diagnostics today to determine the cause of these infections was very important for management and prognosis of these patients. He was impressed with the speed of the laboratory tests and predicted that molecular diagnostics will remain important in guiding clinicians as well as identifying potential pathogens.

References

Big, C., et al. 2009. "Viral infections of the central nervous system: A case-based review." *Clin Med Res* 7(4):142–146.

Glaser, C. A., et al. 2006. "Beyond viruses: Clinical profiles and etiologies associated with encephalitis." *Clin Infect Dis* 43:1565–1577.

Refresher PCR

PCR

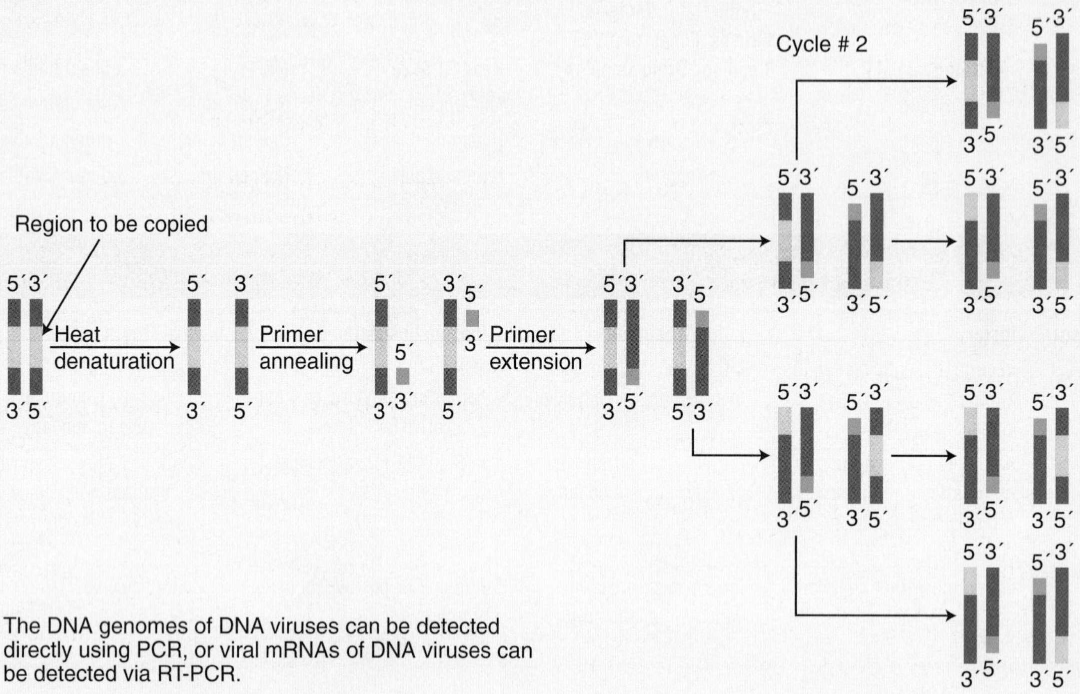

The DNA genomes of DNA viruses can be detected directly using PCR, or viral mRNAs of DNA viruses can be detected via RT-PCR.

RT-PCR 1st strand synthesis

First strand synthesis

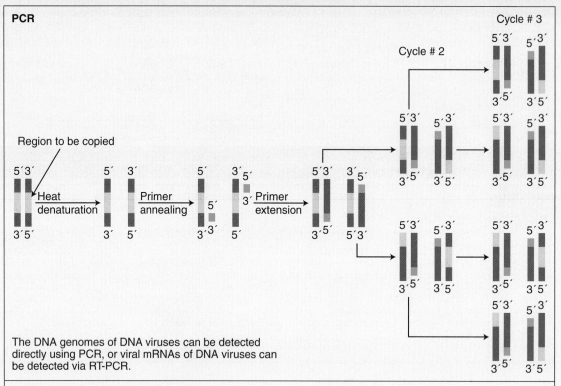

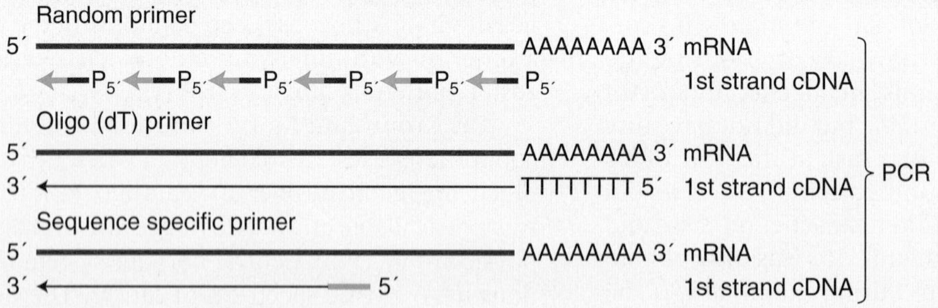

RNA viruses must be detected via RT-PCR. The primers used for cDNA synthesis can be either random primers, oligo-dT primers, or primers that target specific viral RNA sequences. This method involves the amplification of RNA into cDNA via reverse transcriptase. After the new strands of DNA are synthesized, the DNA can be directly amplified via PCR.

Real time PCR vs. conventional PCR

Real time: As the PCR product is being amplified, it is being detected by a PCR monitor (FRET technology) during each cycle (= real time).

Conventional: Involves a post-processing method to quantitate PCR products (e.g., agarose gel electrophoresis).

FIGURE RB 5-1 Review of PCR, RT-PCR, and Real Time PCR vs. conventional PCR.

Chapter 4 presented an overview of viral replication including one-step growth experiments. How, then, do virologists determine which viruses are causing a particular syndrome? If some viral diseases are clinically obvious and there are so few antivirals available to combat them, why have definitive laboratory diagnostic tests been developed? There are several reasons:

- Patient management is determined by diagnosis; for example, a delivery via cesarean section may be prudent if a pregnant woman has HIV, genital herpes, or genital warts at the time of delivery.
- Some antiviral therapies are now available. In particular, several chemotherapeutic agents are available to treat herpesvirus, influenza, and HIV infections (see Table 4-4.).
- Rapid advancements in the development of antiviral drugs are expanding the range of viral diseases for which a precise diagnosis will become necessary; for example, it was announced in *The Lancet* (December 12, 2003) that scientists at the U.S. Army Medical Research Institute of Infectious Diseases (USAMRIID) were able to treat rhesus macaques with an experimental drug called recombinant nematode anticoagulant protein c2 (rNAPc2) after the macaques were given a high-dose lethal injection of Ebola virus. To date there is no cure for the disease, but this experimental drug could become an accepted treatment for Ebola virus infection (**FIGURE 5-1**).

Viral infections may demand public health measures to prevent spread to others for several reasons, including:

- Blood donors may be asymptomatic carriers of viruses. This is why screening blood banks for HIV, hepatitis B and C, and West Nile virus reduces the spread of these diseases by blood transfusion.
- Documenting novel strains of influenza allows officials to publicly announce health measures that can be implemented (e.g., vaccination of vulnerable members of the community, including children and the elderly).
- The positive identification of viral diseases (such as West Nile or Yellow fever) linked to encephalitis and mosquito (vector-borne) transmission enables authorities to initiate antimosquito control measures. The introduction of exotic or newly emerging diseases, such as the monkeypox outbreak in the United States in 2003, demanded containment, surveillance, and other control measures.
- Continuous surveillance of viral infections in the community enables officials to establish means of controlling, monitoring, and evaluating immunization programs. It provides the evidence for new epidemics and new virus–disease associations. It is interesting to note that 90% of all viral diseases were completely unknown at the end of World War II (1945).

This chapter focuses on laboratory diagnosis of viral infections and working with viruses in a research setting. Technology has changed the clinical diagnostic laboratory since the 1990s. Implementing molecular technology is now routine in the form of commercialized kits that give rapid and reliable results.

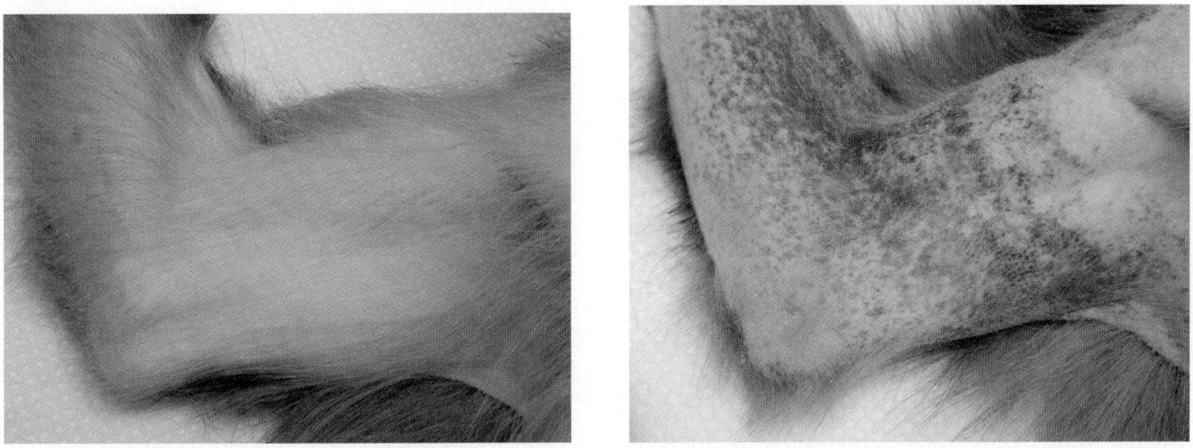

a **b**

FIGURE 5-1 **(a)** Right arm of rhesus monkey treated with rNAPc2 10 days after Ebola infection. The monkey survived. **(b)** Right arm of untreated rhesus monkey 13 days after infection with Ebola virus. The monkey did not survive.

5.1 Proving Causation of Viral Diseases

Robert Koch, a German physician investigating the cause of anthrax and tuberculosis in cattle during the 1880s came up with a series of requirements or postulates to prove that both of these diseases were caused by different bacterial infections. The four criteria to prove that a specific microbe causes a disease became known as **Koch's Postulates**. The criteria that Koch used are:

1. The microbe must always be associated with every case of the disease but not in healthy hosts.
2. The microbe must be isolated from the diseased host and grown in pure culture.
3. Upon inoculation of the cultured microbe into a healthy host, the same disease is reproduced.
4. The microbe is reisolated from the intentionally infected host.

Koch's Postulates were guidelines that applied well to bacterial diseases that caused obvious clinical signs and symptoms and could be reproduced in animals experimentally. The requirements reflected the techniques of the 19th century at a time when bacteriology was in its infancy. Even during Koch's time, it was realized that fulfillment of these postulates couldn't be used to demonstrate causality in all cases. There are many exceptions. For example, there can be asymptomatic carriers of typhoid fever and cholera. Not all individuals (or animals experimentally) exposed to the pathogenic bacterium that causes tuberculosis will be infected. The discovery of other pathogens, especially viruses, made fulfilling Koch's Postulates more difficult.

Many viral infections are correlated with an asymptomatic carrier state or subclinical infections such as those caused by HIV, hepatitis C, herpes simplex viruses, polioviruses, and West Nile viruses. It is known that a very small percentage of individuals who are infected with polioviruses will be paralyzed, yet the polio vaccine that prevents poliovirus infection demonstrates that poliovirus is the causative agent of poliomyelitis. Therefore, Koch's Postulates could not be applied rigidly to every disease.

In the 1930s, Thomas Rivers, an animal virologist, pointed out the limitations of Koch's Postulates for viruses and modified them to prove viruses caused nonbacterial meningitis in two patients at the Hospital of the Rockefeller Institute for Medical Research. He published papers describing his identification of the **etiological agent**.

Rivers isolated viruses from the spinal fluid of the patients. His experimentation included the inability to culture a bacterial agent from the patients, filterability of the virus-like agent, and the presence of protective antibodies in the serum from mice and guinea pigs inoculated with the filterable virus-like agent. His six criteria to be met to prove that a virus causes a disease were:

1. Isolation of virus from diseased hosts.
2. Cultivation of virus in host cells.
3. Proof of filterability (to exclude larger pathogens).
4. Production of a comparable disease when the cultivated virus is used to infect experimental animals from the same species as the original host or in related ones.
5. Reisolation of the same virus from the infected experimental host.
6. Detection of a specific immune response to the virus.

The revolution in biotechnology is helping to make leaps in medicine. It has been especially useful in discovering pathogens that cannot be propagated in pure culture. During the 1990s, Fredricks and Relman modified Koch and Rivers's criteria to reflect the application of new technologies. They proposed criteria for nucleic acid-based evidence of a pathogen using polymerase chain reaction (PCR; see Refresher Box: PCR) that must be fulfilled to prove causation. They offer the following guidelines to establish the relationship between disease and a causative agent:

1. A nucleic acid sequence of the pathogen should be present in all cases of infectious disease. The microbial nucleic acids should be detected in diseased organs and not in unaffected organs.
2. No pathogen-associated nucleic acid sequences should be present in healthy hosts or tissues.
3. After resolution of disease, nucleic acid sequences of the pathogen should no longer be detected. However, if there is a disease relapse, the opposite should occur.
4. The nucleic acid sequence copy number that correlates with the severity of disease is more likely to be the cause of the disease.
5. The clinical features and pathologies observed are consistent with the biological properties of the suspected pathogen.
6. The pathogen or its antigens can be detected in diseased tissue that contains the nucleic acid sequences of the suspected pathogen.
7. The sequence-based evidence of the pathogen should be reproducible.

Changing technology makes it possible to update Koch's Postulates for defining a causal relationship between a microbe and a disease. The ability to detect nucleic acid sequences in microbes is a powerful tool for identifying previously undiscovered microbial pathogens.

5.2 Viral Diagnostics in the Clinical Laboratory

Over 60% of all infectious disease cases seen by physicians are due to viral infections. Accurate and rapid detection and diagnosis are essential for successful antiviral treatment. In addition to a patient's travel history, symptoms, and the season of the year, the clinical microbiology laboratory plays an important role in the diagnosis and control of viral diseases.

The quality of the specimen collected from the patient and its transportation from the patient to the laboratory limit the ability of the laboratory personnel to perform diagnostic tests. Some diagnostic tests require the virus to be isolated and grown in cell cultures; therefore, transport conditions must ensure that the virus will be viable. It is critical that the collected specimen is representative of the site of infection. The shorter the interval between the collection of the specimen and its delivery to the laboratory, the greater the potential for isolating a virus.

Storage and Collection of Biological Specimens for Viral Testing

Collecting specimens for viral diagnosis varies. During a medical examination, the physician makes an educated guess as to the likely virus responsible for the patient's symptoms. The physician collaborates with a clinical microbiologist who determines the appropriate specimens that will be collected for laboratory testing (**FIGURE 5-2**). Timing of specimen collection is vital to the correct test result. For example, for many viral infections such as influenza, rhinovirus, and West Nile virus, viral shedding begins shortly before symptoms appear and then rapidly decreases. In contrast, with chronic viral infections, virus shedding is prolonged even when the patient appears asymptomatic (e.g., cytomegalovirus, hepatitis C, and HIV). Specimens collected with swabs are placed into viral transport medium whereas liquid specimens such as CSF, blood, or urine are not diluted in viral transport medium. If specimens can be processed immediately, they are stored in the refrigerator (4°–8°C). If a specimen must be kept for retrieval at a later date, it is best to store the samples potentially containing viruses at 70°C or lower. Storing human and animal viruses at 15°C to –20°C in frost-free environments results in rapid loss of infectivity. A guide for specimen collection for viral diagnosis of human infections is provided in **TABLE 5-1**.

Nearly 50% of all specimens collected are from the respiratory tract, and about 30% of viruses diagnosed in the clinical laboratory are related to respiratory infections. Herpes simplex viruses (cause of oral and genital herpes; 70%) and varicella zoster virus (cause of chickenpox and shingles; 30%) can be cultured on a routine basis from dermal lesions. Detection of viruses from blood specimens represent disseminated, invasive infections that may result in systemic disease. Cytomegalovirus (CMV), HIV, hepatitis C and B viruses, parvovirus B19, and adenoviruses are examples of viruses that can be detected from blood specimens. The new gold standard for diagnosis of central nervous system disease is molecular diagnostics performed on CSF specimens (refer to chapter opener case study). Feces are used for cases of gastroenteritis in identifying viruses that cannot be grown in cell cultures. Mumps, CMV, and adenoviruses are commonly cultured from urine. Conjunctival swabs or corneal scrapings are used to identify corneal infections. Herpes simplex viruses and adenoviruses are the most common ocular viral pathogens. Occasionally lung, liver, spleen, kidney, and brain tissues may be used for molecular testing to identify viral pathogens. The fresh tissue is digested with proteases in the presence of a detergent and the nucleic acid is extracted for PCR testing. PCR technology in the clinical laboratory will continue to impact processing of specimens in the coming years.

The Five General Approaches for Laboratory Diagnosis of Viral Infections

Similar to detection and identification methods for bacteria, five approaches for laboratory diagnosis of viral infections are used. They are:

1. Microscopy
2. Detection of viral antigens
3. Culture
4. Nucleic acid detection
5. Antibody detection

TABLE 5-2 lists detection approaches used to routinely detect human viruses in the clinical virology laboratory.

Microscopy

Microscopic detection of viruses has taken two directions: light microscopy to observe intracellular **inclusions** or clumps of viruses within cells, or electron microscopy used to observe individual virus particles. Light microscopic examination of

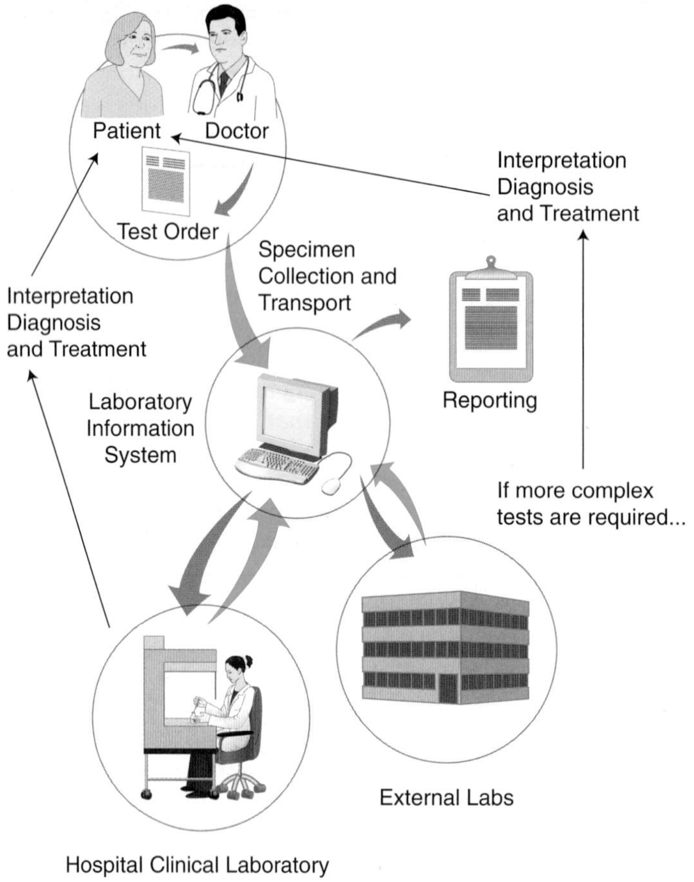

FIGURE 5-2 Overview of viral diagnosis.

TABLE 5-1 Specimen Information for Diagnostics of Human Viral Diseases

General Disease	Potential Virus	Specimen(s)
Respiratory (e.g., sore throat, bronchitis, pneumonia, croup)	Influenza, rhinovirus, respiratory syncytial virus (RSV), adenovirus, SARS-CoV, hantavirus, metapneumonia virus	Nasalpharyngeal aspirates, throat swab, sputum, bronchoalveolar lavage for lower respiratory tract infections
Skin rashes (maculopapular* or vesicular**)	Herpes simplex viruses, measles virus, rubella, varicella zoster virus, human herpesvirus 6, human herpesvirus 8, monkeypox	Swab or aspirate fluid in vesicle and scrape cells at the base of the lesion
Central nervous system (e.g., aseptic meningitis and encephalitis)	Herpes simplex virus, cytomegalovirus (CMV), coxsackievirus A, enterovirus, rabies, varicella zoster virus, arboviruses, arenaviruses	Cerebrospinal fluid, brain tissue; blood for arboviruses and arenaviruses
Hepatitis	Hepatitis A, B, C	Serum
Congenital infections	CMV, herpes simplex virus	Serum, urine (for CMV), amniotic fluid
Genital infections	Papillomaviruses, herpes simplex viruses	Genital or vesicle swab, vesicle scrapings, endocervical swab and biopsy tissue (for papillomaviruses)
Infectious mononucleosis	Epstein-Barr virus, CMV	Blood, plasma, peripheral blood lymphocytes
Eye infections	Herpes simplex virus, adenovirus	Corneal scraping, dermal swab, throat swab, eye swab
Immunodeficiency	HIV-1, HIV-2	Blood, plasma

*A maculopapular rash is usually a large area that is red and has small, confluent bumps.

**A vesicular rash contains blister-like lesions that can be filled with clear, cloudy, or bloody fluid.

TABLE 5-2	Detection and Identification of Human Viruses in the Clinical Laboratory[a]					
Virus	**Microscopy**	**Antigen Detection**	**Nucleic Acid Detection**	**Culture**	**Antibody Detection**	**Comments**
Adenoviruses	2	1	1	1	3	IFA and culture are often used for respiratory specimens. IHC for tissue specimens. NAATs used to monitor viral load in compromised hosts.
Hantaviruses	3	1	2	3	1	BSL-4 lab needed for culture. Serology and NAATs in specialized labs useful for diagnosis. IHC used in fatal cases.
Influenza viruses	3	1	1	1	2	Rapid antigen tests widely used but not optimal in sensitivity and specificity. NAAT is most sensitive but not widely used. Serology is used for epidemiological or retrospective studies.
Coronaviruses	3	1	3	3	2	NAATs and antibody tests for SARS-CoV must be confirmed by reference laboratory.
Parainfluenza viruses	1	2	1	1	3	IFA is the most widely accepted detection method.
Rhinoviruses	3	2	3	1	3	Most common testing in the clinical laboratory is culturing. NAAT is used in epidemiologic studies and research.
Respiratory syncytial virus (RSV)	2	1	1	1	3	Rapid antigen tests, especially IFA most widely used.
Human metapneumonovirus	3	1	3	2	3	NAAT is the main diagnostic method. Conventional culture is difficult
Noroviruses	2	2	3	3	3	EM used in equipped laboratory. NAAT is challenging due to strain variability.
Enteroviruses	3	1	1	1	3	Enterovirus RNA detection used for central nervous system infections.
Rotaviruses	2	2	1	3	3	Direct antigen detection is the top choice for diagnosis. EM useful if available.
Epstein-Barr virus	3	1	2	3	1	Serology is used routinely. NAATs used for viral-related tumors. IHC or ISH used on tumor biopsy specimens.
Papillomaviruses	3	1	3	3	3	NAAT is used for detection and genotype differentiation.
Human immuno-deficiency virus	3	2	1	3	1	Serology is the primary diagnostic method. Proviral DNA and plasma RNA levels are used to monitor viral load to guide therapy.
Hepatitis C and G viruses	3	1	3	3	1	Serology is used for diagnosis. NAATs are used to monitor viral load in response to therapy.
Human T-cell lymphotropic virus	3	2	3	3	1	Serology is the test for diagnosis. NAAT is useful for virus identification in HTLV Western blot-positive but untypeable specimens.
Hepatitis A virus	3	3	3	3	1	Serology is the standard diagnostic test.
Hepatitis B virus	3	1	1	3	1	Viral antigens and antibodies as well as NAAT is used for monitoring the course of infection and therapy.
Hepatitis D virus	3	2	2	3	3	Testing only done by reference laboratories. Diagnosis is only used if patient also has hepatitis B infection. IHC of biopsy tissue is useful for diagnosis.
Hepatitis E virus	3	2	3	3	1	Serology is the standard diagnostic test.

(continued)

TABLE 5-2 Detection and Identification of Human Viruses in the Clinical Laboratory[a] (continued)

Virus	Microscopy	Antigen Detection	Nucleic Acid Detection	Culture	Antibody Detection	Comments
Herpes simplex virus	2	1	1	1	2	Shell vial is used for rapid determination of viral replication. IFA and IHC used for rapid detection in skin or mucous membrane lesions. NAAT is used for central nervous system infections.
Varicella zoster virus	1	1	1	1	2	IFA and NAATs are commonly used as rapid tests.
Cytomegaloviruses	2	1	1	1	2	IHC is used on tissue specimens. Serology primarily used to determine prior infection. NAATs used to assess risk of disease and response to therapy.
Herpesviruses 6 and 7	3	1	1	3	2	NAAT is choice for diagnostics.
Herpesvirus 8	3	1	1	3	2	Serology is used to identify infected persons. IHC used to for Kaposi's sarcomas.
Filoviruses and arenaviruses	1	1	1	1	2	BSL-4 lab needed for culturing. Testing confined to specialized laboratories. Lymphocytic choriomeningitis virus (LCM) diagnosed by serology.
Measles virus	2	3	2	2	1	Serology is most useful for diagnosis and determination of immunity.
Mumps virus	2	3	3	1	1	
Rubella virus	3	2	3	1	1	
Parvoviruses	3	1	3	3	1	
Arboviruses	3	2	3	3	1	
Hendra and Nipah viruses	3	1	2	2	1	
Rabies virus	2	2	1	2	2	
Poxviruses	1	1	3	1	1	

[a]Key: 1 = approach is useful for diagnostic purposes; 2 = approach is useful under certain circumstances or for the diagnostics of specific infections; 3 = a test that is seldom useful for diagnostic purposes.

Abbreviations: IHC, immunohistochemistry; IFA, immunofluorescence assay; NAAT, nucleic acid amplification assay; ISH, in situ hybridization; EM, electron microscopy; BSL, biosafety level.

Source: Adapted from Table 17.18. pp. 259–260. *Mandell, Douglas, and Bennett's Principles and Practice of Infectious Diseases*, Seventh Edition, 2010, Gerald L. Mandell, John E. Bennett and Raphael Dolin (eds.), Churchill Livingstone Elsevier.

stained histological tissues or lesions for viral inclusions is a rapid test used for such viruses as measles, herpes simplex viruses, and varicella zoster viruses. For more than 60 years, **immunofluorescence** has been one of the primary technologies used by diagnostic virology laboratories. Specimens are usually fixed, frozen tissues prepared on slides; tissue cultures also can be grown on glass slides. Virus-specific antibodies containing a fluorescent tag bound to the Fc region of the antibody are allowed to react with the specimen. **Antibodies** are proteins produced by the host in response to a particular pathogen (see **Refresher: Immunology Terms**). These **fluorescent antibodies** are used as a stain to detect specific viral **antigens** present in the clinical specimen or virus-infected cell cultures. Any unbound antibody is washed away and the specimen is observed with a fluorescent microscope. Fluorescent microscopes have the same resolving power as a light microscope. Immunofluorescence relies on differential cytopathic effects for virus detection (refer to section on Working with Viruses in the Research Laboratory). **FIGURE 5-3** shows the detection of Epstein-Barr viral antigens using indirect immunofluorescence. An application of this method that has more limited use is **immunohistochemistry** (IHC). The advantage of IHC is that viral antigens can be detected in the absence of well-defined viral inclusions. It too uses antibodies to localize viral proteins in fixed tissue sections or cell cultures.

Individual virus particles cannot be seen with the light microscope and are considered invisible or ultramicroscopic because of their small size (30–300 nm range). The development of electron microscopy, which resolves structures at 0.5 nm, became a rapid way to visualize virus particles in the clinical laboratory (see Chapter 1: Visualizing

Refresher — Immunology Terms

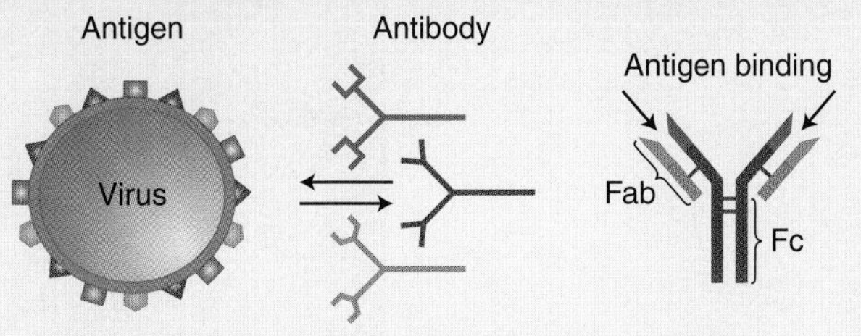

Antigen: A foreign substance that stimulates an immune response in the form of antibody formation or a cell-mediated response. Bacteria and viruses are common sources of antigens.

Antibody: A three-lobed globulin molecule found in the blood or other body fluids that can be produced by the presence of an antigen (invading microorganism). It has a destructive influence on the antigen.

FIGURE RB 5-2 Immunology definitions: antibody, epitope, B cell, antigen, and immunogen.

Viruses: Electron Microscopy); however, *few diagnostic laboratories currently use this technique.* This method directly visualizes virus particles contained in a clinical specimen. The morphology (shape) of the particle enables an examiner to assign the virus to a particular family of viruses. It is a rapid method (involving a simple and fast negative-staining procedure) that is especially useful for examining viruses that cannot be grown in cell cultures. Application of the electron microscope during the late 1940s involved distinguishing poxviruses (which include the following human viral pathogens: smallpox, monkeypox, and molluscum contagiosum) from varicella zoster virus (a herpesvirus that causes chickenpox). In the 1970s, electron microscopy led to unexpected findings, especially the detection of viruses present in stool specimens from young children suffering from gastroenteritis.

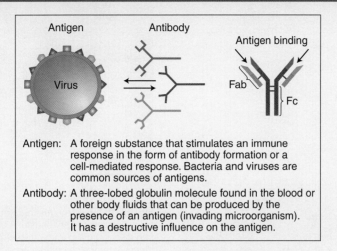

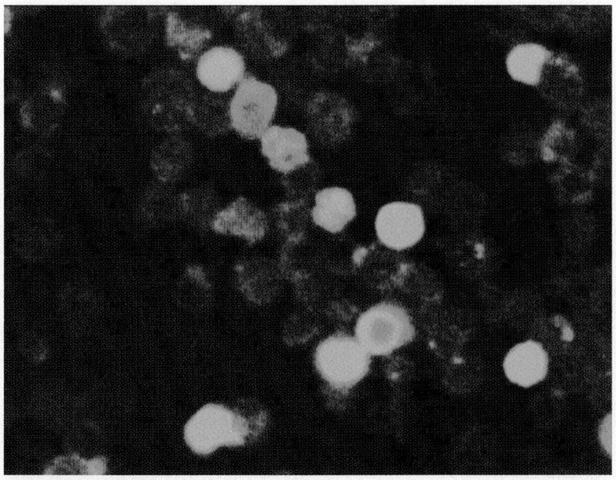

FIGURE 5-3 Rapid diagnosis of Epstein-Barr virus (EBV) infection by indirect immunofluorescent staining of infected cells. Patient serum was collected to determine whether the individual had a primary diagnosis of infection (infectious mononucleosis, IgM EBV antibodies) or reactivation of a latent virus (IgG antibodies).

Electron microscopy does have its limitations, though. For one, it is not a sensitive method. At least 1,000,000 to 10,000,000 viruses must be present in 1 milliliter (ml) of the sample to visualize the virus by electron microscopy. It also requires a skilled microscopist and an expensive electron microscope for examination of samples. Certain clinical specimens (feces and mucous samples) do contain high enough concentrations of virus particles that can be visualized via electron microscopy.

Immunoelectron microscopy is used to concentrate the number of virus particles in a given sample. Specific antibodies to a suspected viral pathogen are added to the specimens. The virus–antibody complexes are concentrated by centrifugation directly onto a specimen grid. Subsequently, the sample is negatively stained and viewed via the electron microscope. **FIGURE 5-4** is an electron micrograph of specimens prepared in this manner. This technique can be slightly modified to increase the sensitivity of the assay. Labeling antibody molecules with electron-dense markers such as colloidal gold may be used in various protocols to enhance visualization of the virus/antibody reaction. Rotaviruses, noroviruses, coronaviruses, torovirus-like particles, astroviruses, Hepatitis A, and adenoviruses have been identified this way in clinical specimens (**FIGURE 5-5**).

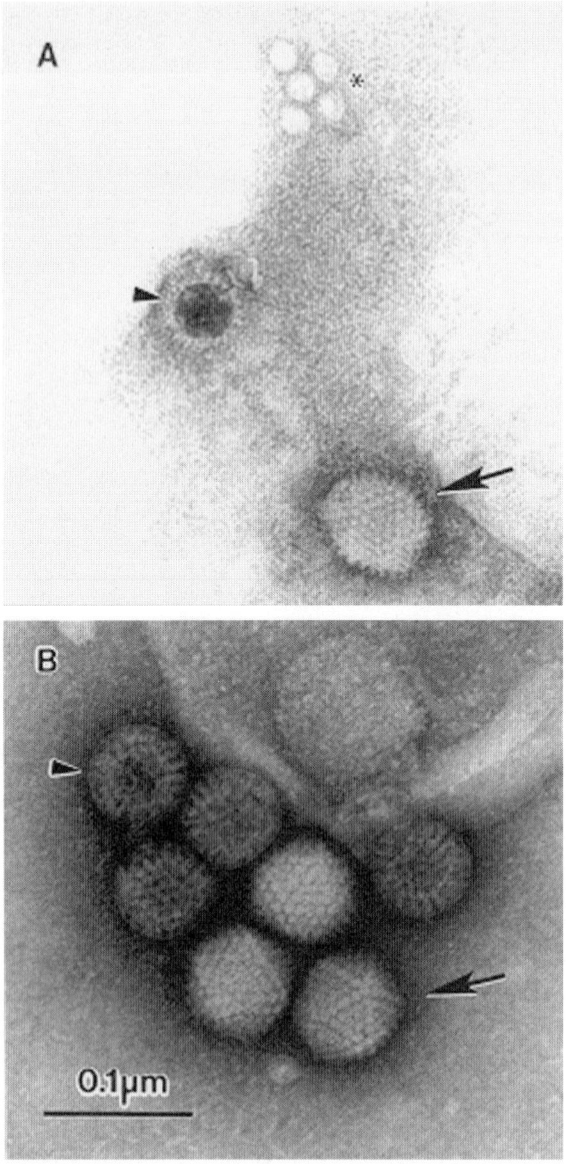

FIGURE 5-4 A fecal sample was collected from a one-year-old patient with diarrhea. The images show that multiple agents were observed in the sample. An arrowhead points to adenovirus particles, and an arrow points to incomplete rotavirus particles.

Detection of Viral Antigens

A large number of commercially available test kits are available for viral antigen detection and routinely used in clinical laboratories. The tests are inexpensive, technically easy to perform, and permit a rapid turnaround time. Examples of antigen detection are **enzyme-linked immunosorbent assays** (ELISAs) and hemagglutination assays (refer to section on Working with Viruses in the Research Laboratory). Examples of viruses detected for rapid detection within an hour of the receipt of a given specimen are tests for respiratory syncytial virus (RSV) and influenza A and B viruses. These tests are very useful during peak months of transmission.

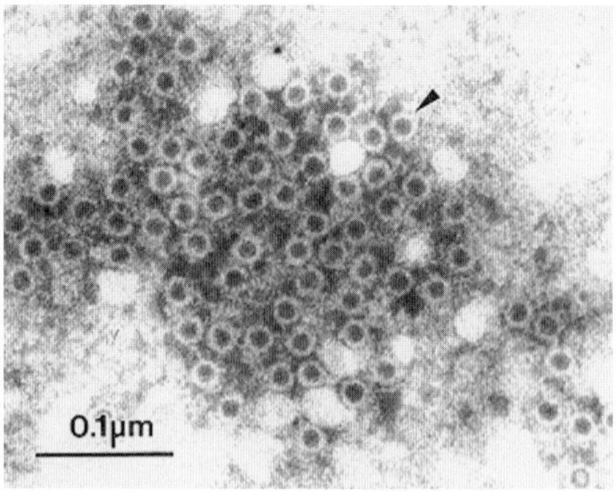

FIGURE 5-5 Electron micrograph of parvovirus B19 (the causative agent of erythema infectiosum, also known as fifth disease) using immunoelectron microscopy procedure. Virions/immune complexes were centrifuged directly onto a specimen grid. The arrowhead points to a genome-defective virus particle.

ELISAs are based on antibodies binding to their antigens and the detection of this reaction using a commercial antibody conjugated with an active enzyme. This enzyme reacts with its substrate to produce a color change. Observing and measuring color determine the test result. **FIGURE 5-6** illustrates the detection of viral antigens or serum antibodies in clinical samples. **FIGURE 5-7** is a photograph of an ELISA test used to determine if HIV antibodies were present in a person's serum. The ELISA is a very sensitive method and will result in a reaction even if only one or two antibodies are present in the serum sample. If the ELISA test is reactive, or positive, it is repeated. If a second ELISA test is also positive, the HIV-positive status must be confirmed by Western immunoblotting (see section on Antibody Detection). Using the two tests together ensures extreme accuracy. Thus, the ELISAs are used to screen for viruses and antibodies present in patient serum samples.

Culture

Viruses require a living host cell for replication. Cell cultures are derived from cells taken from original tissue using an enzymatic, mechanical, or chemical process. The cells are grown in culture medium as monolayers and used for routine virus isolation and identification. **Cell cultures** were used in research laboratories to study viruses by the early 1960s but were not used in diagnostic laboratories until the early 1970s (see section on

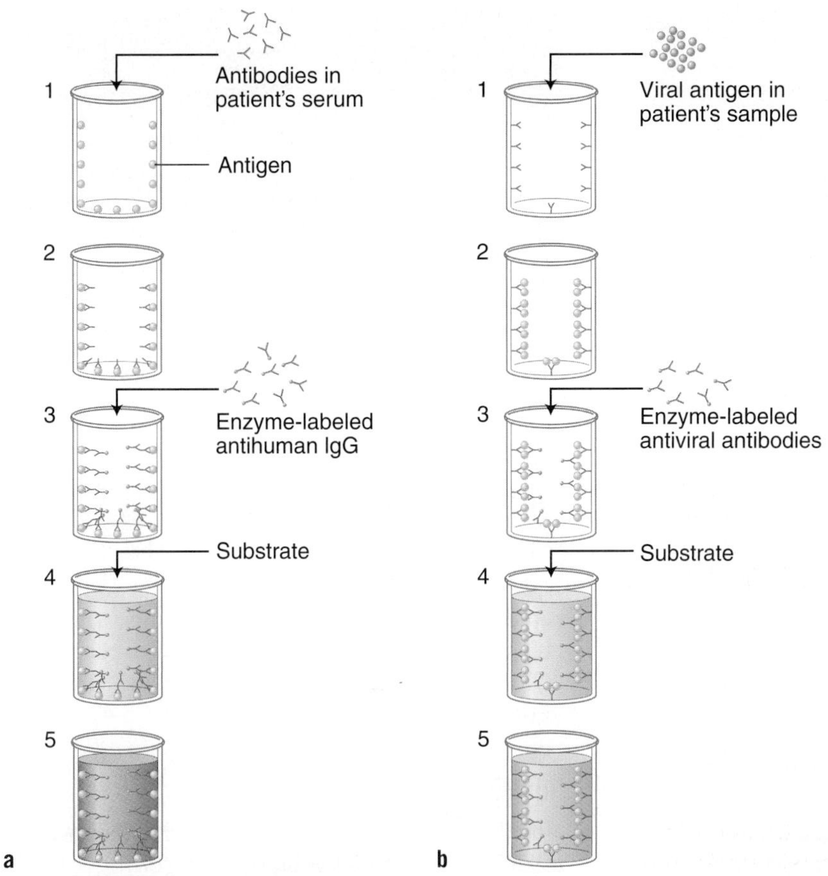

FIGURE 5-6 **(a)** Schematic illustrating the ELISA procedure used to detect antibodies in patient serum. Briefly, patient serum is added to wells (typically 96-well plastic plates) precoated with known viral antigens. If virus-specific antibodies are present, they will bind to the viral antigens. Next, antihuman IgG enzyme-labeled antibodies are added, which react with human antibodies. The enzyme substrate solution is then added. If a color develops, it indicates the reaction is positive. Note: Rigorous washing follows each incubation step. Adapted from Specter, S. C., Hodinka, R. L., and Young, S. A., *Clinical Virology Manual*, Third Edition. ASM Press, 2000. **(b)** Schematic illustrating the ELISA procedure used to detect viral antigens present in a clinical sample (for example, feces, throat, or nasal wash). The clinical sample is added to wells precoated with virus-specific antibodies. If viral antigens are present in the clinical sample, they will bind to the virus-specific antibodies. Next, enzyme-labeled antiviral antibodies are added, followed by substrate solution. If a color develops, it indicates the reaction is positive. Note: Rigorous washing follows each incubation step. Adapted from Specter, S. C., Hodinka, R. L., and Young, S. A. *Clinical Virology Manual,* Third Edition. ASM Press, 2000.

Working with Viruses in the Research Laboratory). No particular cell culture line can support the replication of all viruses, so diagnostic laboratories use multiple cell lines. Cell cultures (**TABLE 5-3**) permissive to infection by suspected viruses in the clinical specimen are inoculated with the clinical specimen. The inoculated cultures are monitored daily for **cytopathic effects** (CPEs). CPEs are visual changes in the infected cell culture that are caused by viral replication. The inoculated cultures are monitored for CPEs and compared to uninoculated control cell cultures. After 50% of the monolayer has CPEs, these virus-induced changes can be visualized with an inverted light microscope under low power or the infected cells will be analyzed by other techniques such as PCR and immunofluorescence (see section on Working with Viruses in the Research Laboratory).

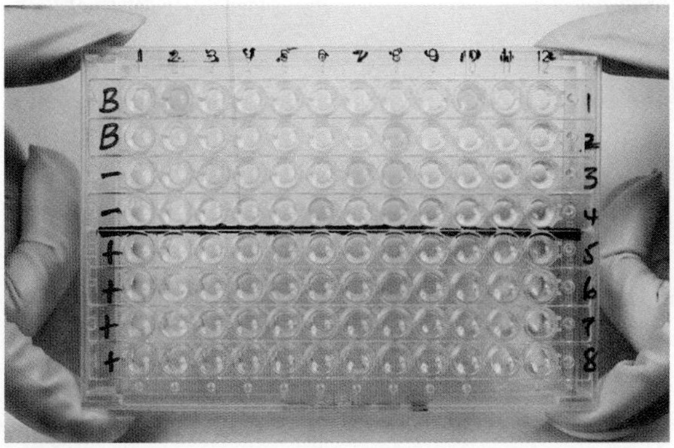

FIGURE 5-7 HIV ELISA test. This 96-well plate is read via a spectrophotometer. The yellow color indicates a positive antigen–antibody reaction. The ELISA test involves adding a patient's serum to wells coated with lysed T-lymphocytes infected with HIV. Any anti-HIV antibodies in serum will bind to the viral antigens from the T-cell. After a washing step, enzyme-labeled antihuman IgG is added, followed by a chromagenic substrate for the enzyme that when acted upon by the enzyme produces a colored product if anti-HIV antibodies are present.

It can take days to weeks for CPEs to occur. Rapid diagnosis is increasingly important in patient management. The **centrifugation culture (shell vial technique)** is used in clinical labs for rapid diagnosis of viral infections. This method allows the laboratory technician to detect viral antigens before CPEs are present in a given clinical specimen. The basic protocol entails adding culture cells grown in suspension to shell vials containing a cover slip on the bottom of the vial, allowing cells to form monolayers at the bottom of the vial, inoculation with patient specimen (virus), a slow speed centrifugation step (1000 rpm for about 45 minutes) to increase infectivity of the virus, an incubation step (about 36 hours), and then the monolayers are fixed with acetone. Cover

TABLE 5-3 Cell Lines Commonly Used in Diagnostic Virology Laboratories

Virus	Replication in Cell Cultures PMK	HDF	HEp-2	A549
Adenovirus	+	++	+++	+++
Cytomegalovirus	–	+++	–	–
Enterovirus	+++	++	+/–	+/–
Herpes simplex virus	+	++	++	+++
Influenza virus	+++	+	–	–
Parainfluenza virus	+++	+	+/–	+/–
Rhinovirus	+	+++	+	–
Varicella zoster virus	+	+++	–	+++

Abbreviations: PMK, primary monkey kidney cell line; HDF, human diploid fibroblasts; HEp-2, human epidermoid larynx carcinoma cell line; A549, human heteroploid lung carcinoma cell line

Key: +++ Replicates very well, ++ Moderate replication, + Replicates poorly, +/– Variable replication, – No replication

Source: Adapted from *Clinical Virology Manual*, Fourth Edition. S. Specter, R. L. Hodinka, S. A. Young, D. L. Wiedbrauk, Editors, Chapter 3: Primary Isolation of Viruses, p. 37, Table 2.

TABLE 5-4 Detection of Virus: Traditional CPE Method Versus Shell Vial Technique

Virus	Days to Detect CPE Conventional Cell Culture Method [avg (range)]	Days to Detect CPE Shell Vial Centrifugation Method (range)
RSV	6(2–14)	1–2
Influenza A	2(1–7)	1–2
Influenza B	2(1–7)	1–2
PIV 1–4	6(1–14)	1–2
Adenovirus	6(1–14)	2–5
HSV	2(1–7)	1–2
VZV	6(3–14)	2–5
CMV	8(1–28)	1–2

slips from the shell vials are stained with fluorescent monoclonal antibodies and examined under a fluorescence microscope (**FIGURE 5-8**). Some laboratories use a mixture of cell lines such as mink lung cells (strain Mv1Lu) and A549 cells in the same shell vial, which is especially suited for the rapid detection of early viral antigens. Shell vial cultures can be stopped and negative results reported at 2 days rather than 7–14 days using conventional cell culture methods (**TABLE 5-4**).

Nucleic Acid Detection

Nucleic acid detection methods used to be confined to university research laboratory settings. Today these methods are a necessary and important part of many clinical laboratories. **Nucleic acid-**

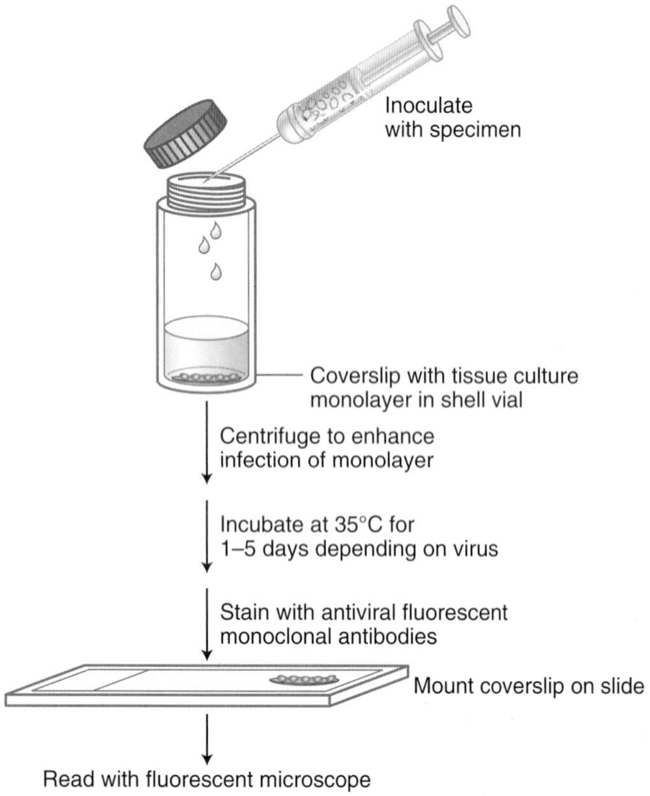
Inoculate
with specimen
Coverslip with tissue culture
monolayer in shell vial
Centrifuge to enhance
infection of monolayer
Incubate at 35°C for
1–5 days depending on virus
Stain with antiviral fluorescent
monoclonal antibodies
Mount coverslip on slide
Read with fluorescent microscope

a

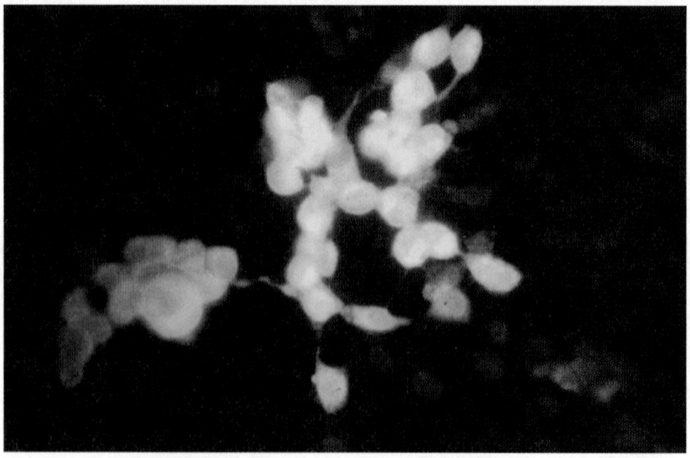

b

FIGURE 5-8 (a) Tissue culture cells are grown on coverslips on the bottom of shell vials. Clinical specimens are allowed to adsorb for a specified time period. The shell vials are then centrifuged at a slow speed and overlaid with culture medium. The cells typically are fixed to the coverslips for two to three days and then virus antigens (infected cells) are detected using immunofluorescence microscopy. Modified from J. H. Shelhamer, et. al., *Ann. Intern. Med.* 124 (1996): 585–599. **(b)** Detection of herpes simplex virus type 1 using the shell vial technique and immunofluorescence. The brightly fluorescing cells are infected with HSV-1.

amplification tests (NAATs) detect viral nucleic acids in order to diagnose and manage patients suffering from viral diseases. The majority of NAATS utilize some form of PCR. Kary Mullis developed PCR technology while working at Cetus Corporation (Emoryville, California) in 1983. This method is used to replicate DNA using a thermostable DNA polymerase in a test tube (*in vitro*). PCR to amplify viral genomes of DNA viruses and reverse transcriptase-PCR (RT-PCR) technology to amplify genomes of RNA viruses have become routine in the clinical virology laboratory (**FIGURE 5-9**). Nucleic acid based amplification (NASBA) and transcription-mediated amplification (TMA) are non-PCR methods used to amplify viral sequences, preferably viral RNAs. Both methods are similar to PCR in that they utilize primers that are complementary to the sequences of the viral RNAs of interest. However, at least one of the primers in the amplification reaction contains a promoter sequence for T7 RNA polymerase. NASBA uses three separate enzymes (T7 RNA polymerase, reverse transcriptase, RNase H), while TMA uses two (T7 RNA polymerase and a native reverse transcriptase that also has RNase H activity). **TABLE 5-5** lists the commercial types of NAATs available for detection of common viruses in clinical specimens.

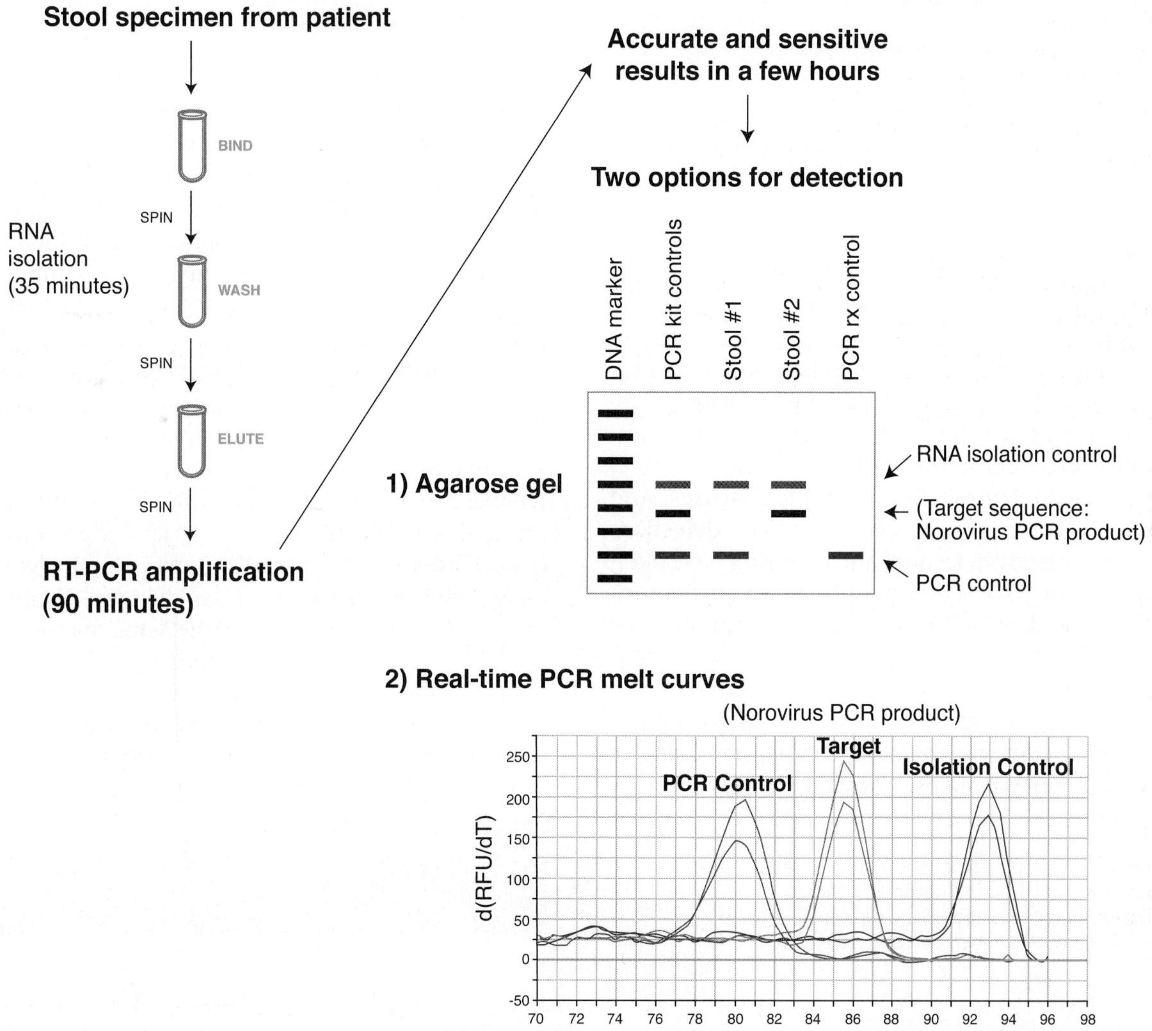

FIGURE 5-9 Schematic of the procedure used to isolate and detect noroviruses from stool samples using an RT-PCR kit in a molecular diagnostics laboratory. Patient stool samples are collected. Total RNA is purified from stool using a rapid isolation kit. RT-PCR is performed on the stool samples and control RNAs provided in the diagnostic kit. Kit RT-PCR controls include an RNA isolation control and a PCR control. These controls are necessary in order to confirm the integrity of the RNA isolated is adequate, that the PCR reaction is working properly and a norovirus PCR positive control. The RT-PCR products are electrophoresed through an agarose gel, stained and visualized or detected using real-time PCR melt curves. Norovirus is very difficult to culture *in vitro*, making identification via standard microbiological assays challenging. Molecular diagnostic kits circumvent the need to culture and isolate noroviruses *in vitro*. In this illustration, stool sample 2 tests positive for norovirus. Stool sample 1 does not contain any noroviruses. Adapted from Norgen Biotek Corp. "Norovirus RT-PCR Detection Kit" and "Total RNA Purification Kit". Ontario, Canada: Norgen Biotek Corp; June 23, 2011. http://www.norgenbiotek.com/display-product.php?ID=68; http://norgenbiotek.com/display-product.php?ID=449.

TABLE 5-5 NAATs Used in Clinical Virology Laboratories

Virus	PCR	TMA	NASBA
	Type of NAAT		
Cytomegalovirus	+		+
Enterovirus	+		+
Hepatitis B virus	+		
Hepatitis C virus	+	+	
Herpes simplex virus 1, 2			+
Human immunodeficiency virus	+	+	+
Human papillomavirus	+		+
Influenza A, B			+
Metapneumovirus			+
Respiratory syncytial virus			+

Abbreviations: PCR, polymerase chain reaction; TMA, transcription-mediated amplification; NASBA, nucleic acid-based amplification. All of these methods amplify viral nucleic acids present in a clinical specimen.

Source: Adapted from Table 17.9. pp. 246. *Mandell, Douglas, and Bennett's Principles and Practice of Infectious Diseases*, Seventh Edition, 2010, Gerald L. Mandell, John E. Bennett, and Raphael Dolin (eds.), Churchill Livingstone Elsevier.

NAATs have replaced the conventional definitive diagnosis of central nervous system (CNS) infections (see chapter opener case study). In the past the definitive diagnosis relied on a brain biopsy. Now the detection of viral nucleic acids in CSF by PCR is as sensitive as histologic stains of brain biopsy specimens. Traditional methods to detect cytomegaloviruses, herpes simplex viruses, and varicella zoster viruses such as antigen detection, serum antibody detection, and culturing have been replaced by NAATs because of their exquisite sensitivity compared to the other methods. Commercial NAATs are also available for HIV-1, HIV-2, CMV, and hepatitis C to monitor viral loads that are an important indicator of disease progression and response to antiviral therapy (see Chapter 16, 16.12: Managing HIV Patients: Antiviral Therapy).

Antibody Detection

Serology was the most used diagnostic test in clinical laboratories before the development of viral antigen tests and NAATs. Serological methods are especially useful in identifying viruses that are difficult to grow in cell culture and viruses that cause diseases with slow courses. Patient serum (which contains antibodies) can be used to identify virus strains or serotypes, evaluate the course of infection, and determine whether the infection is recent or chronic. Detecting antibodies against viruses is an **indirect** measure of viral infection because it indicates a host antiviral response rather than viral particles. When an individual encounters a virus for the first time, his or her body responds by producing a class of virus-specific antibodies termed IgM. (Antibody classes are discussed in more detail in Chapter 7.)

IgM antibodies are proteins found in serum (the liquid portion of blood after removal of clotting factors from plasma) during the first two to three weeks of a primary infection. The presence of specific IgM antibodies indicates a recent infection. Next, IgG antibodies appear. IgG antibodies are the most common antibody. Approximately 70% to 75% of antibodies are of IgG class. If a person is reinfected with the same virus, the IgM antibody response is the same as for the primary response, but for the IgG response, memory cells are activated, producing IgG-specific antibodies. Within a day, very high levels of IgG are detected (**FIGURE 5-10**). By examining antibody profiles, a recent or chronic infection can be identified. Slow-course diseases—such as infectious mononucleosis caused by Epstein-Barr virus and hepatitis B and C viruses—can be diagnosed in the same manner.

There is a variety of methods/tests developed that use IgM or IgG determination in diagnostic virology. Methods covered in this chapter include IFA and ELISA (see sections on microscopy and antigen detection for these applications) and

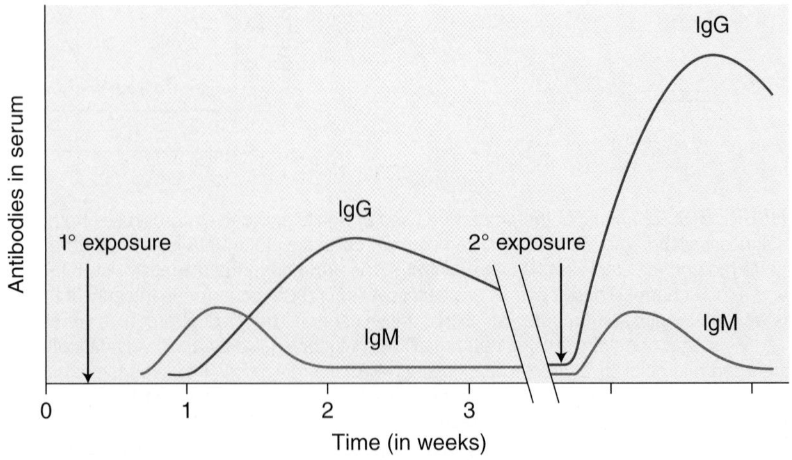

FIGURE 5-10 Primary (1°) and secondary (2°) antibody responses toward a viral pathogen. The most common serological techniques used in the clinical laboratory are enzyme-linked immunosorbent assays (ELISAs).

Western blotting. All of these tests are based on antibodies binding their antigens and the detection of this reaction using a commercial antibody conjugated with an active enzyme. This enzyme reacts with its substrate to produce a color change. Observing and measuring color determine the test result. Serology plays an important role in the detection of HIV infection. All reactive rapid HIV tests must be confirmed by a Western immunoblot or immunofluorescent assay. Negative results by these tests should be repeated 4 weeks after the initial reactive test result. The confirmatory test must be done by trained personnel at laboratories equipped to run more complex tests.

The basic approach to Western blotting involves the separation of viral proteins (from purified virions available commercially) through an acrylamide gel. This protein separation is referred to as **SDS polyacrylamide gel electrophoresis**, or SDS-PAGE. SDS stands for sodium dodecyl sulfate. It is a strong detergent that denatures proteins. As the proteins migrate through the polyacrylamide gel, they are separated by size. Charge plays a very small role in this process, too, because the negatively charged SDS coats the proteins, thus giving every protein the same mass-to-charge ratio. Characteristically, smaller proteins migrate through the gel faster than larger ones. After the proteins are separated, they are transferred to nitrocellulose in the same pattern that they were on the gel. The proteins bind to the nitrocellulose, which then is cut into strips and probed with patient serum, or positive or negative serum as controls (**FIGURE 5-11**).

Any viral antibodies present in the serum sample will react with the viral proteins present on the nitrocellulose membrane. A second antibody, such as antihuman Ig coupled with an enzyme, is then incubated with the nitrocellulose. Stringent washing of the nitrocellulose occurs between each step in the procedure. Finally, the blot is incubated with

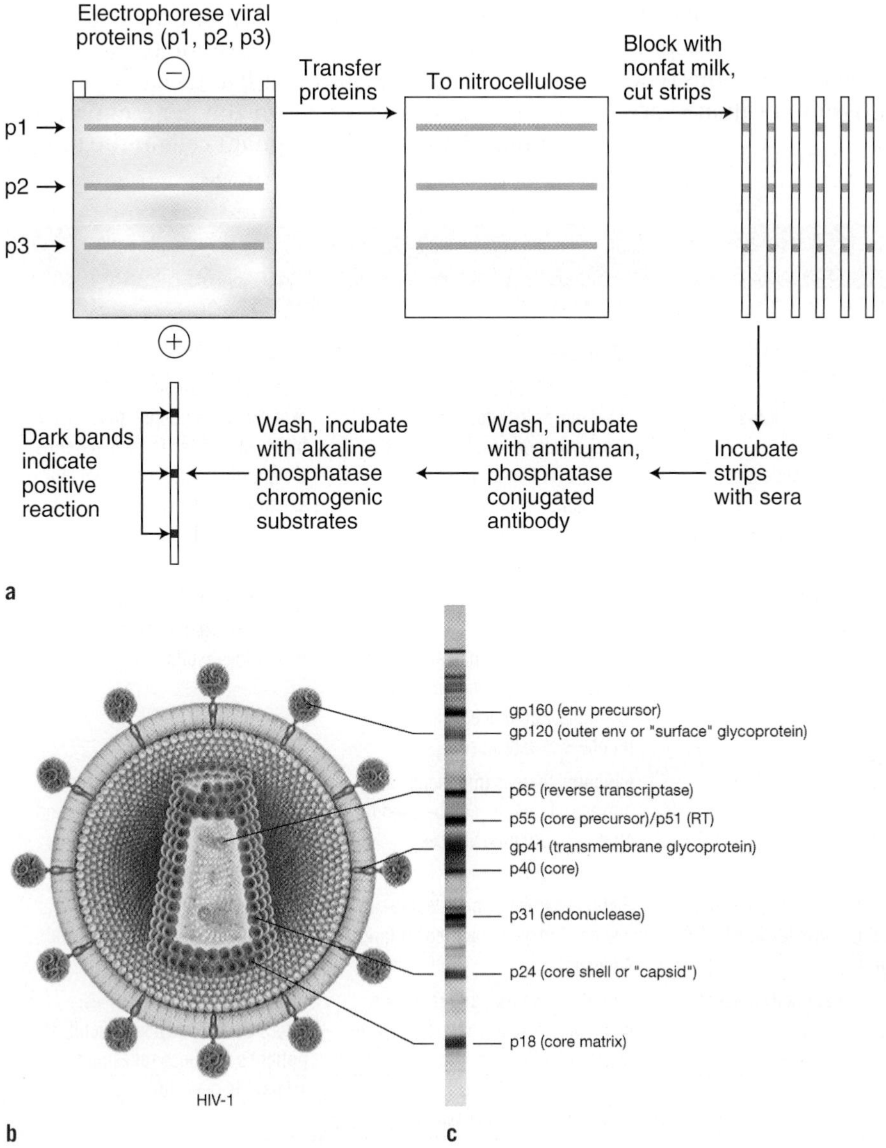

FIGURE 5-11 **(a)** Diagram showing the basic principles behind the Western blotting procedure. Modified from Specter, S. C., Hodinka, R. L., and Young, S. A. *Clinical Virology Manual*, Third Edition. ASM Press, 2000. **(b)** Schematic of the structure of HIV-1. SDS disrupts HIV-1 virions and the viral proteins are separated in a polyacrylamide gel as part of the Western blot procedure. Courtesy of Bio-Rad Laboratories. **(c)** These are the typical results of a Western blot testing patient serum for HIV-1 antibodies. Lane 1 is a positive control and lane 3 is a negative control. Lane 2 represents blots probed with patient sera. Image courtesy of Bio-Rad Laboratories.

the appropriate enzyme substrate. Protein bands reacting with antiviral antibodies from the serum sample, which are recognized by the secondary antihuman-enzyme antibodies, will change color upon the addition of substrate. Figure 5-11c is an example of the type of Western blot profile seen with HIV-1 testing.

The overall sensitivity of serological tests is determined by the timing of specimen collection. If a blood sample is collected before antibodies are detectable, a test will have poor sensitivity. If a viral infection is defined by documenting a rise in antibodies in a patient, the acute and convalescent serum must be sufficiently timed to document the increase. Also, this assumes that a patient can mount an antibody response to infection for serological testing to be useful.

Besides diagnostics, serology plays an important role in measuring the immunity to specific viruses such as measles, mumps, rubella, and varicella zoster viruses. For example, immunized patients have persistent IgG antibodies. Therefore, detectable levels of antibodies indicate that the patient is immune to infection. Data from results of this type of serological testing can be used to evaluate vaccination programs.

Advantages and Disadvantages of Methods

Accurate and rapid diagnosis is essential for successful treatment. **TABLE 5-6** compares the advantages and disadvantages of diagnostic methods. Within the past 10 to 15 years, additional antivirals have become available. Recall that Table 4-4 lists antivirals approved by the FDA.

Microarrays and Protein Arrays: New Options in Viral Diagnostics

A new revolutionary tool applied to cancer biology and drug and therapy development is the DNA microarray or DNA chip. Microarrays were developed by a team of researchers at Stanford University during the early 1990s and first introduced commercially in 1996. DNA chips are used to assess and classify tumors of cancer patients; for example, certain tumor cells are more aggressive than others. The DNA chip has allowed researchers to determine which patients will need the most intensive treatment based on information provided by the chip. The chip is used to determine the molecular abnormalities that distinguish long- and short-term cancer survivors based on the gene expression pattern exhibited in malignant cells at the time of

TABLE 5-6 Comparison of Diagnostic Methods

Diagnostic Method/Approach	Basic Concept of the Test	Advantages	Disadvantages
Microscopy	Electron microscopy: Uses high resolution to visualize virus particles directly in specimen. Immunofluorescence: Uses fluorochrome-tagged antisera to detect viral antigens in cells from specimen.	Electron microscopy: rapid (15 minutes to 3 hours), works for viruses that cannot be cultured, suitable for "unknowns." Immunofluorescence: Quick (1–3 hours), semiquantitative, good sensitivity.	Electron microscopy: Requires expensive electron microscope and highly skilled technician; requires a separate microscope room; not sensitive and specific.
Antigen Detection	Similar to immunofluorescence but uses an enzyme as a tag and usually done in microtiter plates.	Rapid (1–3 hours). Can be read by a machine. Very sensitive. No special skills needed, many commercial kits available, can process large numbers of samples.	Requires specific antisera, problems over false positives and borderline results.
Culture	Growth of live virus in cultured cells *in vitro*.	Indicates there is infectious virus present in sample; can be used to study and quantify viruses.	Slow (results in days to weeks). Labor intensive; requires skill and experience (both to prepare cells and to read CPEs).
Nucleic Acid Detection	Amplification of viral nucleic acid to assessable levels (PCR, RT-PCR, NASBA, etc.).	Highly sensitive, can be "same-day" results. Can be automated for large numbers.	May be oversensitive, expensive, requires good technique to avoid cross-contamination.
Serology	Detection of an antibody response to a virus.	Can be automated and read by a machine; presence of IgM indicates recent infection.	Not applicable to all viruses; interpretation may be difficult; the patient's response takes up to 10 days to develop.

Source: Adapted from Madely, C. R. 2008. "Is it the Cause?—Robert Koch and Viruses in the 21st Century." *J Clin Virol* 43:9–12.

diagnosis. The risk for developing certain cancers can be assessed and a prognosis made based on microarray test results.

Currently, the power of the microarray in virology can best be seen in research studies; for example, microarrays can be used to study viral gene expression and promoter activities. There are many potential applications for arrays in the clinical virology laboratory. These applications include:

- diagnostic microarrays, including novel virus identification,
- detecting agents of bioterror (viral, bacterial, and eukaryotic pathogens),
- detecting the presence or absence of viral pathogenicity genes,
- patient management (monitoring antiviral drug-resistant mutations, i.e., probing HIV protease and reverse transcriptase genes),
- vaccine quality control (ensuring vaccines are free of contaminating microbes by detecting their genetic sequences in the vaccine preparation), and
- studying genes involved in host response to viral infection.

To date, microarrays have been designed to detect genomes of enteroviruses, rhinoviruses, adenoviruses, herpesviruses, orthomyxoviruses, paramyxoviruses, nidoviruses (arteriviruses and coronaviruses, i.e., SARS-CoV), retroviruses, hepadnaviruses, and papillomaviruses.

How do microarrays work? A DNA chip or microarray consists of a small nylon membrane, silicon chip, or glass microscope slide that contains samples of immobilized DNA of many genes arranged in a specific pattern. The microarray works by exploiting the ability of a given mRNA molecule to specifically hybridize to the DNA template from which it originated.

The following general procedure represents how microarrays might fit routinely into a diagnostic laboratory in the near future:

1. Nucleic acid is extracted from a biological sample (such as sputum, blood, or CSF) by an automated instrument.
2. The RNA and/or DNA is amplified via RT-PCR or PCR that will incorporate fluorescently labeled nucleotides into the genetic material of any virus present in the sample.
3. The amplified, labeled nucleic acid would be applied to a DNA chip and allowed to hybridize to complementary sequence probe(s) representing known sequences of pathogenic viruses; for example, there may be specific microarray chips developed for respiratory pathogens, gastrointestinal pathogens, or blood-borne viruses.
4. A detection instrument (microarray reader) and specialized software would be used to interpret the fluorescent hybridization signals on the chip and provide a readout that identifies the virus present in the original biological specimen (**FIGURE 5-12**).

Prototypes of DNA chips developed for the diagnosis of similar infectious diseases are used in some laboratories. Before microarrays can be

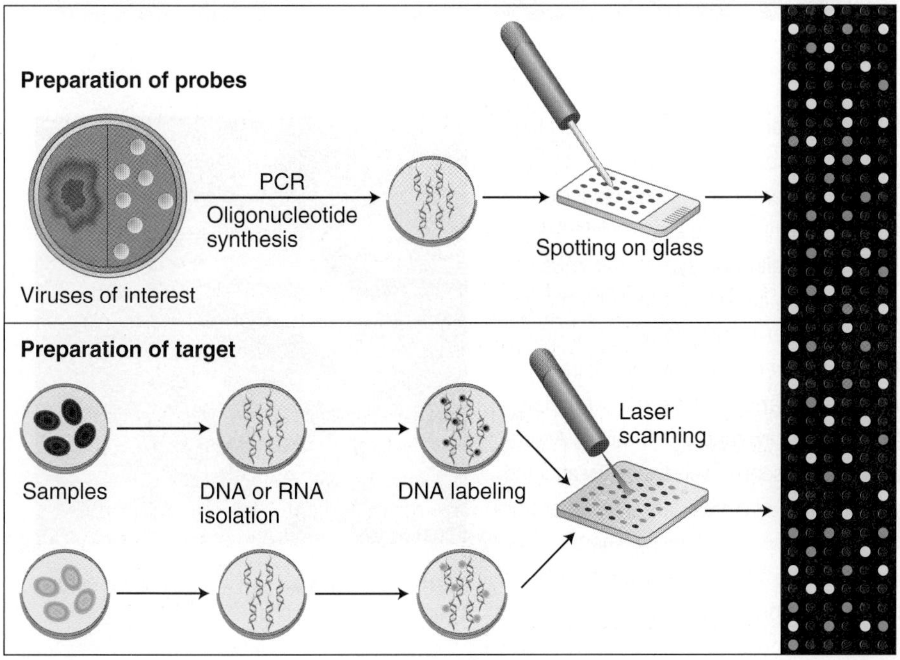

FIGURE 5-12 The process involved in creating a microarray. In the upper panel, probes of the genome sequences of the viruses of interest (such as viruses that cause respiratory tract infections) are generated and spotted onto a "chip." In the lower panel, the DNA or RNA of biological samples (such as sputum or throat swab) is extracted and amplified with either fluorescently labeled Cy3-dCTP (green) or Cy5-dCTP (red) dye. The labeled nucleic acid from the specimens is spotted onto the microarray chip that contains the immobilized probes. Complementary nucleotides of the genetic material will anneal. Green spots represent hybridization of the probe with Cy3-dCTP, red spots represent spots hybridized with Cy5-dCTP, and yellow spots represent hybridization with both Cy3 and Cy5-dCTP target sequences. Adapted from Clewley, J. P., 2003. "A role for arrays in clinical virology: Fact or fiction?" *J Clin Virol* 29:2–12.

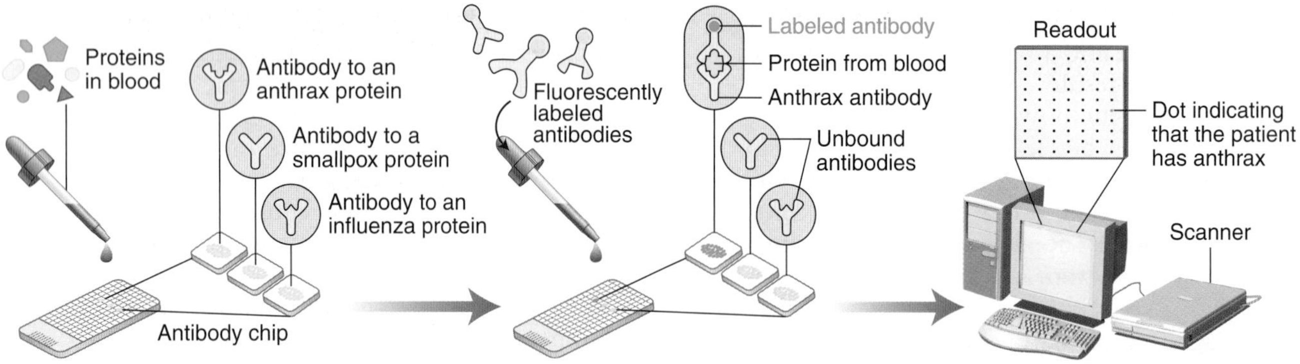

FIGURE 5-13 Protein arrays contain bound protein probes (proteins of infectious disease agents or proteins expressed during diseases states, such as cancers and genetic diseases). Next, a blood sample from a patient is incubated with the protein microarray. Subsequently, fluorescently labeled antibodies are incubated with the chip. A protein array reader will detect the presence of antibody "sandwiches." Adapted from Friend, S. H., and Stoughton, R. B. 2002. "The magic of microarrays." *Scientific American,* February, 286(2):44–53.

routinely used in the virology clinical laboratory, though, the tools must contain probe sequences that hybridize with high sensitivity and specificity. This will allow for the precise detection of their intended targets. Results must be reproducible, and quality control and assurance systems must be established.

Protein arrays are a spin-off of the DNA chip. Arrays of antibodies instead of DNA probes are immobilized on the chip. The chips are probed with a blood sample from a patient, which will contain proteins from the disease-causing microorganism or cellular proteins induced during the disease state. Subsequently, specific fluorescently labeled antibodies are added to the chip. A "sandwich" of antibodies will form wherever a protein from the patient's blood is bound to the microarray chip. The chip is then scanned and interpreted by a microarray reader instrument (**FIGURE 5-13**). Before protein arrays become commonplace in a doctor's office, though, technology needs to be developed that enables the simultaneous mass production of different antibodies that retain their ability to bind tightly and specifically to target proteins that may be present in minute quantities in a patient's biological sample. Protein arrays will be limited by which proteins the chips can detect. More knowledge needs to be gained regarding new and emerging viruses and what proteins represent disease markers present in tissues when screening for noninfectious diseases before the technology will be ready for the commercial market.

VIRUS FILE 5-1 The Mystery of Chronic Fatigue Syndrome

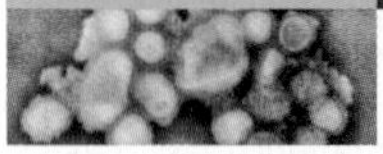

Chronic fatigue syndrome (CFS) is defined as fatigue that is not improved by bed rest and a condition that worsens with mental or physical activities. Besides fatigue, an individual also suffers from four or more of the following symptoms: muscle pain, sore throat, impaired short-term memory or concentration, tender lymph nodes, multi-joint pain without swelling or redness, headaches, unrefreshing sleep, and post-exertional malaise (**FIGURE VF 5-1**). An average of 2.5 million Americans and 17 million people worldwide are affected by CFS. At least 25% of Americans who suffer from CFS are unemployed or on disability. There is no known specific cause and no specific diagnostic test for it. The condition can persist for years.

FIGURE VF 5-1 A woman with symptoms of CFS (e.g., fatigue, headache).

VIRUS FILE 5-1 The Mystery of Chronic Fatigue Syndrome (continued)

A number of pathogens have been linked to CFS, such as herpesviruses, Epstein-Barr virus (cause of mononucleosis), parvovirus B19, enteroviruses, or bacteria such as *Coxiella burnetti* or *Mycoplasma* species. However, none of these associations have been confirmed. Many patients claim that their symptoms appeared after a viral infection. For years, skeptics questioned whether CFS was a "real" disease. A viral trigger could justify and explain their symptoms.

Well-conducted studies are needed to determine the etiology of CFS. In 2009, a study linking CFS to a virus was reported by Lombardi et al. in *Science.* The authors detected DNA sequences of a retrovirus named **xenotropic murine leukemia virus-related virus** (XMRV) in the **peripheral blood mononuclear cells** in 68 of 101 patients (67%) with CSF. XMRV was recently discovered in prostate cancer cells from patients who had an inherited mutation in the *RNaseL* gene (see Chapter 7, The Interferon Response). Lombardi's study also noted that 3.7% of healthy controls test positive for this virus. These results were an incredible mood-lifter for individuals desperate for a biological cause and cure for CFS. Meanwhile, Paprotka et al. and Singh et al. published studies that showed XMRV replication was inhibited *in vitro* by the following HIV antiviral drugs: reverse transcriptase inhibitors AZT, tenofovir, and the integrase inhibitor raltegravir. Some of these licensed drugs for treating AIDS can be immediately tested for their efficacy against CFS.

Unfortunately, more than a handful of other research groups could not repeat Lombardi's connection of XMRV to CFS. Lombardi and colleagues' study has now been thoroughly disproven that XMRV is involved due to the unfortunate generation of this virus through the early passage of prostrate cancer cells through mice. Find some of the articles and comments listed below. Does it offer false hopes? Why or why not?

References

Erlwein, O., et al. 2010. "Failure to detect the novel retrovirus XMRV in chronic fatigue syndrome." *PLoS ONE* 5(1):e8519.

Enserink, M. 2010. "Conflicting papers on hold as XMRV frenzy reaches new heights." *Science* 329:18–19.

Groom, H. C. T., et al. 2010. "Absence of xenotrophic murine leukaemia virus-related virus in UK patients with chronic fatigue syndrome." *Retrovirology* 7(10).

Kean, S. 2010. "An indefatigable debate over chronic fatigue syndrome." *Science* 327:254–255.

Lombardi, V. C., et al. 2009. "Detection of an infectious retrovirus, XMRV, in blood cells of patients with chronic fatigue syndrome." *Science* 326:585–589.

Lloyd, A., et al. 2011. Comment on "Detection of an infectious retrovirus, XMRV, in blood cells of patients with chronic fatigue syndrome." *Science* 328:825-b.

Paprotka, R. et al. 2011. "Recombinant Origin of the Retrovirus XMRV." Science Express/10.1126/science.1205292.

Paprotka, T., et al. 2010. "Inhibition of xenotropic murine leukemia virus-related virus by APOBEC3 proteins and antiviral drugs." *J Virol* 84(11):5719–5729.

Schlaberg, R., et al. 2009. "XMRV is present in malignant prostatic epithelium and is associated with prostate cancer, especially high-grade tumors." *PNAS* 106(38):16351–16356.

Silverman, R. H., et al. 2010. "The human retrovirus XMRV in prostate cancer and chronic fatigue syndrome." *Nat Rev Urol* 7(7):392–402.

Singh, I. R., et al. 2010. "Raltegravir is a potent inhibitor of XMRV, a virus implicated in prostate cancer and chronic fatigue syndrome." *PLoS ONE* 5(4):e9948.

Sudlow, C., et al. 2010. Comment on "Detection of an infectious retrovirus, XMRV, in blood cells of patients with chronic fatigue syndrome." *Science* 328:825-a.

Urisman, A., et al. 2006. "Identification of a novel gammaretrovirus in prostate tumors of patients homozygous for R462Q RNASEL variant." *PLoS Pathog* 2(3):e25. PMID:16609730.

Van der Meer, Jos W. M. et al. 2011. Comment on "Detection of an Infectious Retrovirus, XMRV, in Blood Cells of Patients with Chronic Fatigue Syndrome." *Science* 328:825.

Van der Meer, J. W. M., et al. 2011. Comment on "Detection of an infectious retrovirus, XMRV, in blood cells of patients with chronic fatigue syndrome." *Science* 328:825-c.

5.3 Viral Load Testing and Drug Susceptibility Testing

Testing methods used as *in vitro* susceptibility assays measure the inhibitory effects of a particular antiviral agent on the entire virus population in a patient or in cell culture systems used to develop new drugs. Methods to do this often involve PCR-based strategies and sometimes plaque reduction assays.

PCR-Based Strategies

PCR or RT-PCR is used to determine the relative amounts of viral nucleic acid present in cell cultures maintained in different drug concentrations following *in vitro* infection. This is referred to as drug susceptibility testing. Reduced levels of viral nucleic acid indicate inhibition of the virus. This method can be used to discover or synthesize new antiviral drugs.

PCR or RT-PCR is also used to monitor **viral loads**. Determining viral loads is especially important to assess the effects of antiviral therapy used to manage patients infected with HIV (see Chapter 16) or hepatitis C virus (see Chapter 17). HIV and hepatitis C viral load refers to the number of virus particles floating in the blood. The quantity of RNA genomes of the virus is an indirect measure of the number of viral particles. Monitoring viral loads becomes a guide for treatment choices. For example, a viral load test is taken before an HIV or hepatitis C patient begins treatment and is called a baseline measurement. Another viral load measurement is taken 4 to 8 weeks after treatment is started to determine the patients' response to the medications. If the treatment is working, the amount of viral RNA in the blood will go down. If it is not working, the viral RNA will stay the same or even increase.

Plaque Reduction Assays

Plaque reduction assays are the gold standard for directly measuring the extent to which an antiviral drug inhibits the effects of viral infection in tissue culture. Cell monolayers in 6- or 12-well dishes are infected with virus [50–100 plaque-forming units (PFU) per well] and incubated in the presence of the antiviral compound. Depending upon which virus is being tested, some assays will need to be overlaid with medium containing the antiviral compound and agarose (a solidifying agent) so that discrete plaques are formed. This helps prevent secondary spread of the virus to the remainder of the cell monolayer. Following two to four days of incubation, the cells are fixed and stained with crystal violet dye and the plaques are counted (**FIGURE 5-14**). The lowest concentration of the drug causing 50% inhibition in plaque formation represents the 50% inhibitory dose (ID^{50}). A plaque reduction assay is accurate, reliable, and relatively simple to perform. A disadvantage of this method is that the enumeration of plaques is tedious, time-consuming, and subjective. These assays are best suited for small numbers of specimens as they are difficult to automate. The infectivity titer of the virus must be determined before the plaque reduction assay can be performed.

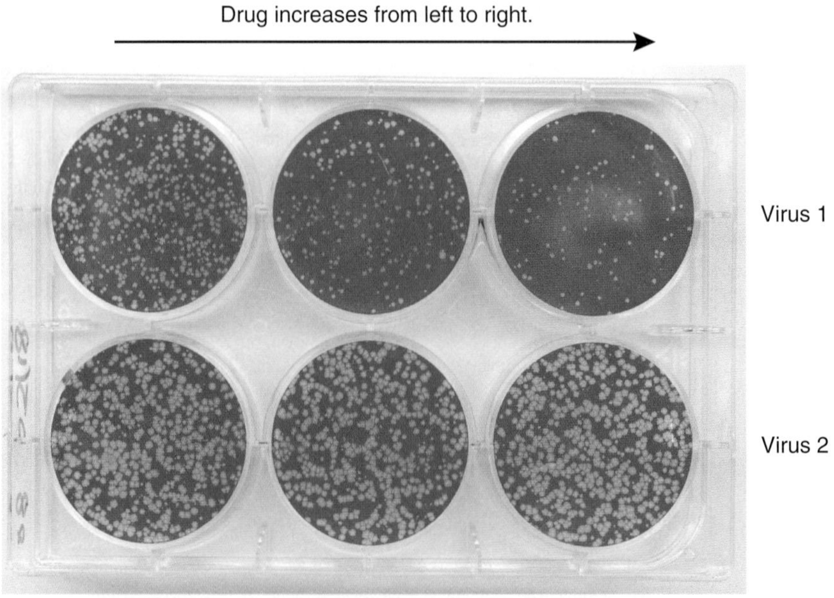

FIGURE 5-14 Plaque reduction assay. Two different strains of vaccinia virus were tested in this assay. Virus 1 is sensitive to the effects of the compound, whereas virus 2 is resistant to increasing concentrations of the compound being tested.

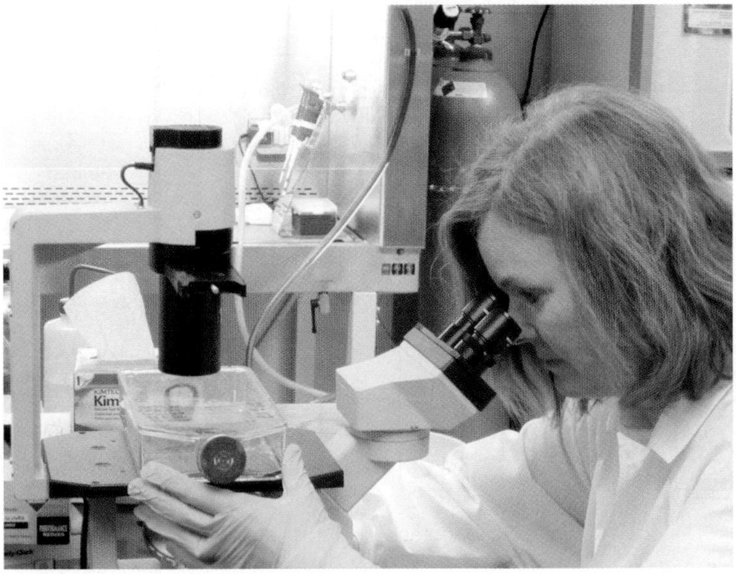

a

b

FIGURE 5-15 **(a)** Research technician Margaret Schuelke looking into an inverted microscope to visualize cell culture cells grown as monolayers in flasks. **(b)** Research technician Paul Kennedy holding suspension in spinner flasks of cells prior to infecting the cells with a virus for research experiments.

Plaque reduction assays can be completed within two to four days for poxviruses, one to two weeks for herpes simplex viruses, and as much as six weeks for slow-growing viruses such as CMV and VZV.

5.4 Working With Viruses in the Research Laboratory

Viruses must be grown in a "host" system. They can be grown in animals, embryonated eggs, or tissue (cell) cultures. Some viruses will only grow in animals. Mice, cotton rats, rabbits, and guinea pigs are most often used to cultivate viruses. Sometimes, the only animal models available to grow human viruses are monkeys. It is very expensive to grow viruses in animals. Therefore, the preferred method, when possible, is cell culture.

Optimal growth conditions for viruses differ tremendously. Cell cultures are grown as monolayers attached to the bottom of a flask or suspension medium. Cell cultures grown in suspension medium are grown in liquid (similar to growing bacteria in a broth) that contains a spinning magnet that keeps the cells suspended in the liquid (**FIGURE 5-15**) instead of falling to the bottom of a cell culture flask. Cells grown in culture vary in the number of times they can be "subpassaged." For example, when primary cell cultures are used to seed a flask, these cells typically will live only one generation. Diploid cells can be subpassaged, and heteroploid cells can be subpassaged indefinitely. These cells are immortal (**TABLE 5-7**; see Chapter 10: Viruses and Cancer for characteristics of immortalized cell lines).

Cultured cells provide the most versatile method of growing and assaying viruses. A detailed description of the theory and practice of cell culture techniques is provided in *Culture of Animal Cells* by R. Ian Freshney. A typical cell culture facility is shown in **FIGURE 5-16**.

TABLE 5-7 Cell Cultures Commonly Used in the Virology Laboratory

Cell Culture	Examples	No. of Subpassages
Primary	Kidney tissues from experimental mammals, such as monkeys and rabbits, and embryos from experimental birds, such as chickens	1 or 2
Diploid (limited passage)	Human embryonic lung or human newborn foreskin	20–50
Heteroploid	Human epidermoid carcinoma of larynx (HEp-2) or of lung (A549)	Indefinite

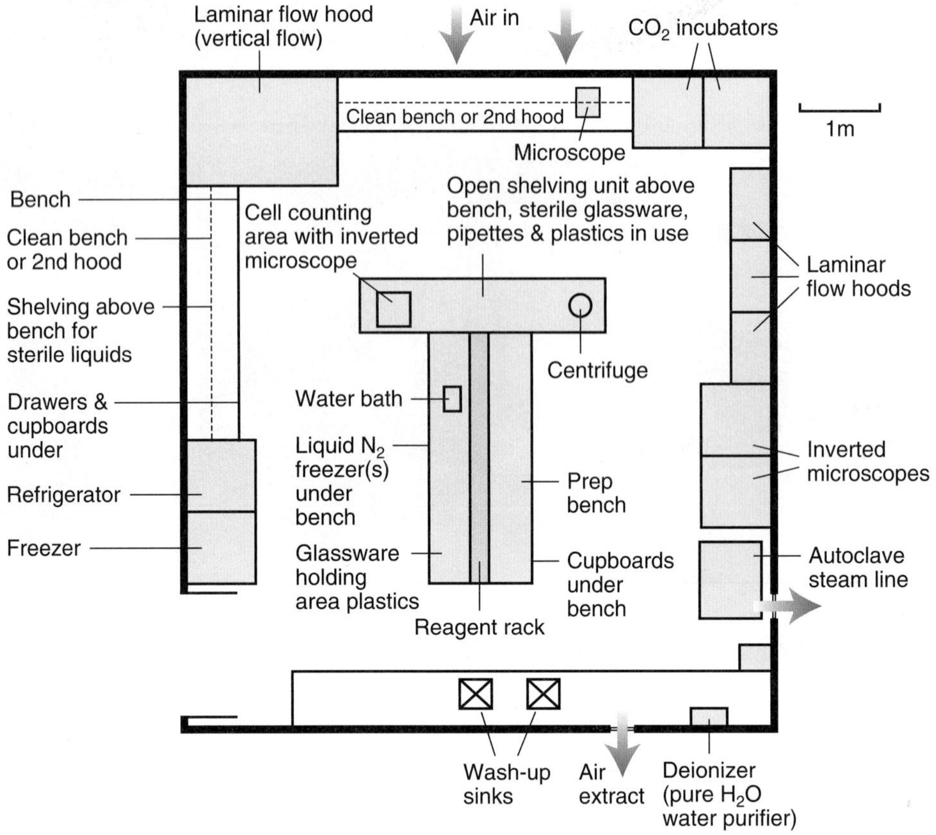

FIGURE 5-16 Diagram of a typical cell culture facility staffed by two to three investigators. Adapted from Freshney, I. R. *Freshney's Culture of Animal Cells*, Third Edition. John Wiley & Sons, 1998.

Laminar Flow Hoods

The pillar of a virologist's facility is a laminar vertical flow biosafety hood (**FIGURE 5-17**). These facilities must be kept very clean and free from drafts that carry dust, spores, and bacteria, which could contaminate cultures. Clean air (sterile handling) is critical in the facility. Biosafety hoods provide clean air to the working area and provide a constant flow of air out of the work area to prevent room air from entering it. The air flowing out from the hood suspends and removes contaminants introduced into the work area by personnel. The most important part of the laminar flow hood is a high-efficiency bacteria-retentive filter. Once air is passed through a prefilter that removes gross contaminants (i.e., lint and dust), it is then channeled through a high-efficiency particulate air filter (HEPA) filter that removes 99.97% of particles of 0.3 μM or higher from the air.

Cytopathic Effects

In general, viruses cause several common morphological changes in the cells that they infect. Cytopathic effects (CPEs) are visual changes in the host that are due to viral infection. CPEs, including inclusion body formation and hemadsorption assays, are the most common observations or methods used to recognize cells in culture infected by viruses.

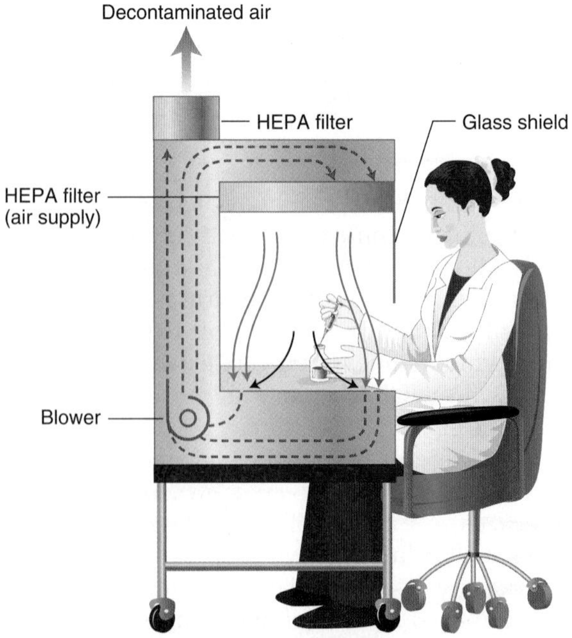

FIGURE 5-17 Vertical flow laminar hood. Air circulates within the hood; clean air is passed through HEPA filters, downward onto the working area. All air entering the room is also passed through a HEPA filter.

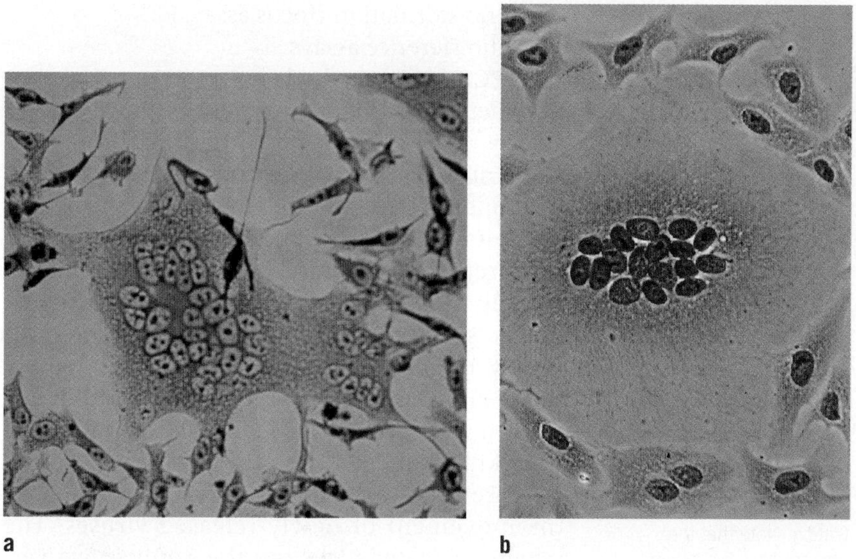

FIGURE 5-18 (a) BS-C-1 (monkey kidney) cells infected with a modified vaccinia virus that expresses measles virus genes, resulting in syncytia and fusion of mouse L cells. (b) Photograph shows a giant or "multinuclear" cell (also referred to as a syncytia), a mechanism of viral spread. Several viruses, including HIV, can induce syncytia formation. Syncytia-inducing variants of HIV have been correlated with rapid disease progression in HIV-infected individuals.

Virus-induced CPEs can be observed with an inverted light microscope under low power. The observations include rounding of the cells, shrinkage, increased refractility, fusion/syncytia formation, aggregation, loss of adherence, and cell lysis/death (**FIGURES 5-18** and **5-19**).

CPEs occur as a result of:

- entry into the host (crossing the cellular plasma membrane),
- inhibition of cellular transcription or stimulation of cellular RNA polymerase activity,
- virus interactions with RNA processing pathways, and
- virus interactions with the translational apparatus.
- host responses to viral infection (see The Interferon Response in Chapter 7).

CPEs are the easiest and most widely used criteria for recognizing viral infection. Not all viruses cause CPEs, though. For this reason, other methods must be used to detect viral infections.

Inclusion bodies are subtle intracellular abnormalities that only occur in infected cells. They are intracellular granules that are a visible site of viral replication or assembly. The presence of a specific type of inclusion body may be indicative or diagnostic of a specific viral infection. To visualize inclusion bodies, infected cells are fixed to the culture dish, stained by a dye such as eosin, and observed by light microscopy (**FIGURE 5-20**). Poxviruses, paramyxoviruses, reoviruses, rabies virus, herpesviruses, adenoviruses, and parvoviruses induce striking inclusion bodies in cells.

Hemadsorption is the adherence of red blood cells (erythrocytes) to other cells, including cells infected by viruses. Influenza viruses, parainfluenza viruses, or togaviruses contain an envelope protein called hemagglutinin on their surface that is able to agglutinate, or bind, to human or animal red blood cells. In the laboratory, a suspension of red blood cells is added to an infected cell monolayer. The mixture is incubated and then observed microscopically for adsorption of red blood cells

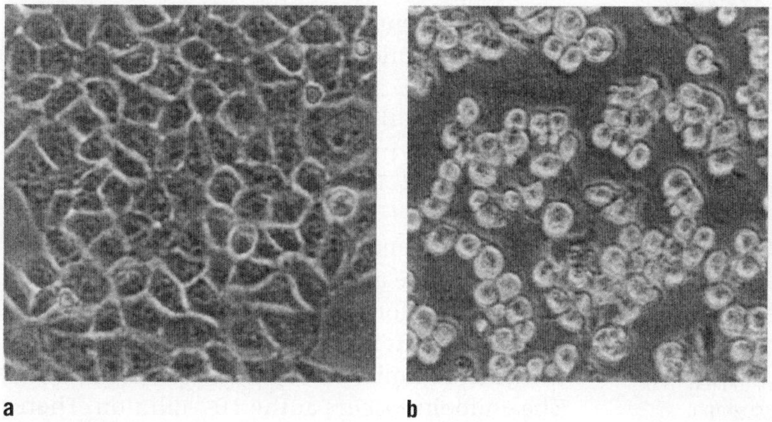

FIGURE 5-19 (a) Uninfected monkey kidney cells (BS-C-1) viewed by low-power light microscopy. The confluent monolayer of cells remains attached and intact to its substrate. (b) BS-C-1 cells infected with vaccina virus. CPEs include rounding and detachment of the cells from the tissue culture dish.

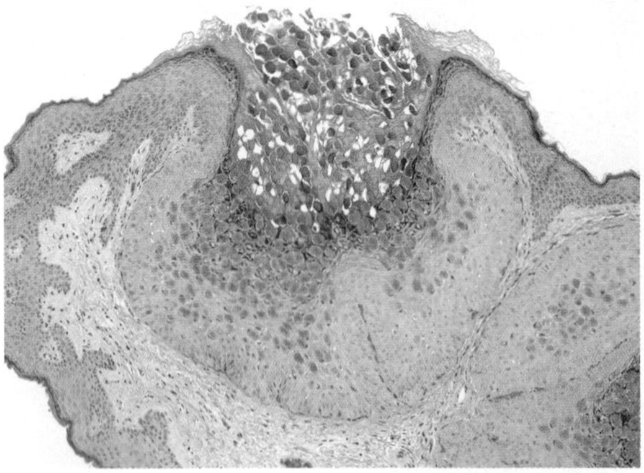

FIGURE 5-20 Epidermis with lesions containing inclusion bodies stained by eosin and hematoxylin. The lesions were caused by molluscum contagiosum virus (MCV) infection. Note that the inclusion bodies become darker toward the crater of the lesion.

to the cells in the monolayer (**FIGURE 5-21**). Even though the replication of some of these viruses does not cause CPEs, the presence of the virus can be detected by hemadsorption.

Common Methods Used to Study Viruses in the Research Laboratory

The following methods are most often used to study viruses in research laboratories:

- plaque assays
- the endpoint dilution method, or tissue culture infectious dose $(TCID)_{50}$
- neutralization, hemagglutination, and hemagglutination inhibition assays
- transformation (focus assays)
- interference assays
- PCR-based methods
- detection of viral enzymes

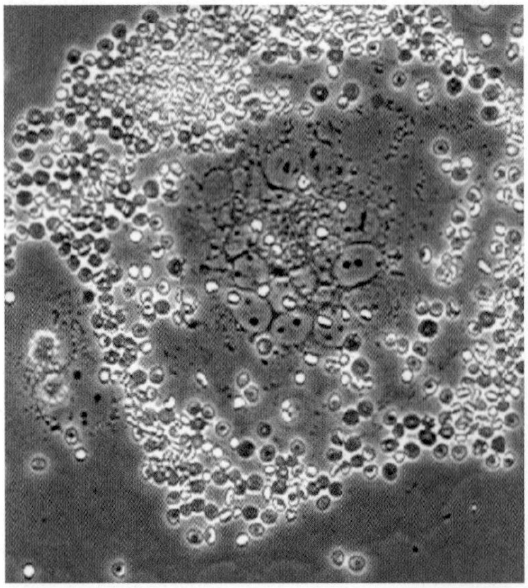

FIGURE 5-21 Hemadsorption of monkey red blood cells to cells infected by measles virus. Note the attachment of the red blood cells to the infected cells in the middle of the photograph.

A plaque assay is a quantitative assay used to determine the number of viruses in a given sample. It is used to quantitate the number of viruses in a prepared virus stock or viruses present at a specific time during single-step growth experiments (see Chapter 4, Virus Replication Cycles). Briefly, cell monolayers are infected with tenfold serial dilutions of virus. A plaque is produced when a virus particle infects a cell, replicates, and then kills that cell. The monolayers of infected cells are usually overlaid with agarose to restrict the movement of newly released viruses. The newly replicated virus can then only infect surrounding cells, and they, too, are killed. After several rounds of replication, a visible area, or plaque, of killed cells forms.

Plaques are visualized by staining cells with dyes such as neutral red or crystal violet. Theoretically, each virus in the original inoculation gives rise to a clearing (plaque) or plaque-forming unit (PFU), as described in **FIGURE 5-22**. In other words, a plaque assay measures infectivity. That being said, though, it is possible that viral particles are present in a given sample but are not infectious. More specifically, a plaque assay measures only the number of infectious particles in a sample, not the total number of particles.

The endpoint dilution assay $(TCID)_{50}$ was used before the development of the plaque assay and is still used for viruses that do not form plaques but do produce CPEs in cell cultures. It is also used in animal experiments to determine virulence. Cell culture assays are performed by infecting cell monolayers in replicates with serial dilutions of virus. After incubation, the infected cells are observed for CPE and are scored + (positive or infected) or – (negative or uninfected). An example of endpoint dilution assay results is shown in **TABLE 5-8**.

Fifty percent of the cell cultures displayed CPE at the 10^{-4} dilution. At high dilutions, none of the cell cultures become infected because the dilution is so high that there are no infectious particles in those dilution tubes. At low dilutions, every culture is infected. Therefore, the virus stock contains 10^4 $TCID_{50}$ units.

When an endpoint dilution assay is performed in animals, the animals are inoculated with virus and observed for disease and death. **TABLE 5-9** provides the results from a typical animal experiment, in which ten mice were inoculated per dilution. The endpoint occurs at the 10^{-4} dilution. There-

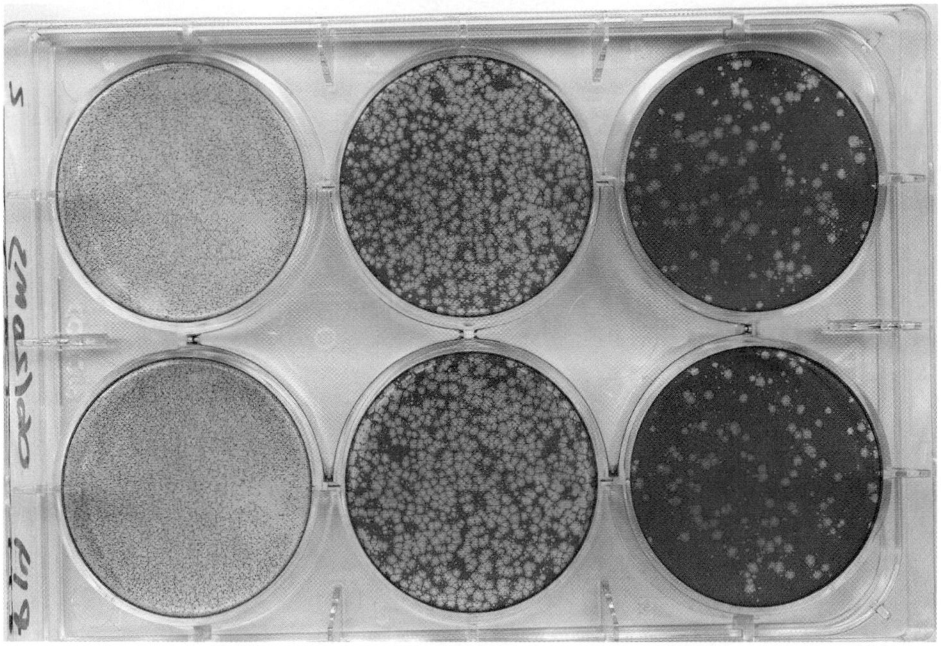

FIGURE 5-22 Plaque assay used to quantitate a vaccina virus stock. After incubation, the liquid medium of the infected cells was removed and the uninfected cell monolayer was stained with a crystal violet/ethanol solution. Clear areas (dead, infected cells) represent plaques. Note that tenfold serial dilutions of virus were plated on confluent monolayers of cells.

fore, the virus stock contains 10^4 $TCID_{50}$ units. Quite often, results will fall between dilutions. When this occurs, statistical methods developed by Reed and Munch are used to calculate the lethality endpoint.

As stated previously in the discussion on hemadsorption assays, some viruses have the ability to agglutinate or form a red blood cell lattice around the virus when virus particles are in high enough concentration. Agglutinated red blood cells can be easily distinguished from cells that are not agglutinated: Red blood cells that are not agglutinated are free to roll to the bottom of the well, forming a dense, easily recognized button at the bottom of the well. Those red blood cells that are agglutinated are not free to roll to the bottom of the well and form an even coat on the bottom of the surface of the well. This assay is simple and can be used to quantitate hemagglutinating virus. The virus required to cause the agglutination may be infectious or noninfectious. An example of a hemagglutination assay is shown in **FIGURE 5-23**.

TABLE 5-8 Example of Endpoint Dilution Data

Virus Dilution	CPE Results									
10^{-2}	+	+	+	+	+	+	+	+	+	+
10^{-3}	+	+	+	+	–	+	+	+	+	+
10^{-4}	–	+	–	+	+	–	–	+	–	+
10^{-5}	–	–	–	–	+	–	–	–	–	–
10^{-6}	–	–	–	–	–	–	–	–	–	–
10^{-7}	–	–	–	–	–	–	–	–	–	–

Neutralization and hemagglutination inhibition assays are methods used for the detection or quantitation of virus and strain-specific neutralizing antibodies. Neutralizing antibodies block the virus from binding to host cells, resulting in no CPE. These assays are done by incubating dilutions of serum (that contains antibodies) with virus. Subsequently the virus/antibody mixture is added to cell cultures permissive to infection or erythrocytes (**FIGURE 5-24**). If neutralizing antibodies are present, no CPE occurs (no infection) and hemagglutination is inhibited, allowing the erythrocytes to pellet. Without neutralizing antibodies, CPE occurs and so does hemagglutination (**FIGURE 5-25**).

Focus assays are used to determine if a virus is capable of "transforming" or immortalizing cells rather than killing them. This phenomenon most notably occurs with tumor viruses (e.g., certain retroviruses). Characteristics of transformed cells include immortalization of cells, loss of contact inhibition anchorage independence (cells no longer need to attach to a solid surface to grow), and tumorigenicity (the ability to form tumors if injected into an animal model). Transformed cells no longer grow as a monolayer, but instead have lost their contact inhibition and pile on top of each other to form densely packed areas of cells called **foci** (Figure 5-25; refer to Chapter 10, Viruses and Cancer, for more information regarding characteristics of cancer cells).

Interference assays are used to detect viruses that do not cause visual CPEs in cell cultures, but

TABLE 5-9	Lethality of Poliovirus in Mice					
Virus Dilution	**Alive**	**Dead**	**Total Alive**	**Total Dead**	**Mortality Ratio**	**% Mortality**
10^{-2}	0	10	0	25	0/25	100
10^{-3}	0	10	0	15	0/15	100
10^{-4}	5	5	5	5	5/10	50
10^{-5}	10	0	10	0	10/10	0
10^{-6}	10	0	10	0	10/10	0

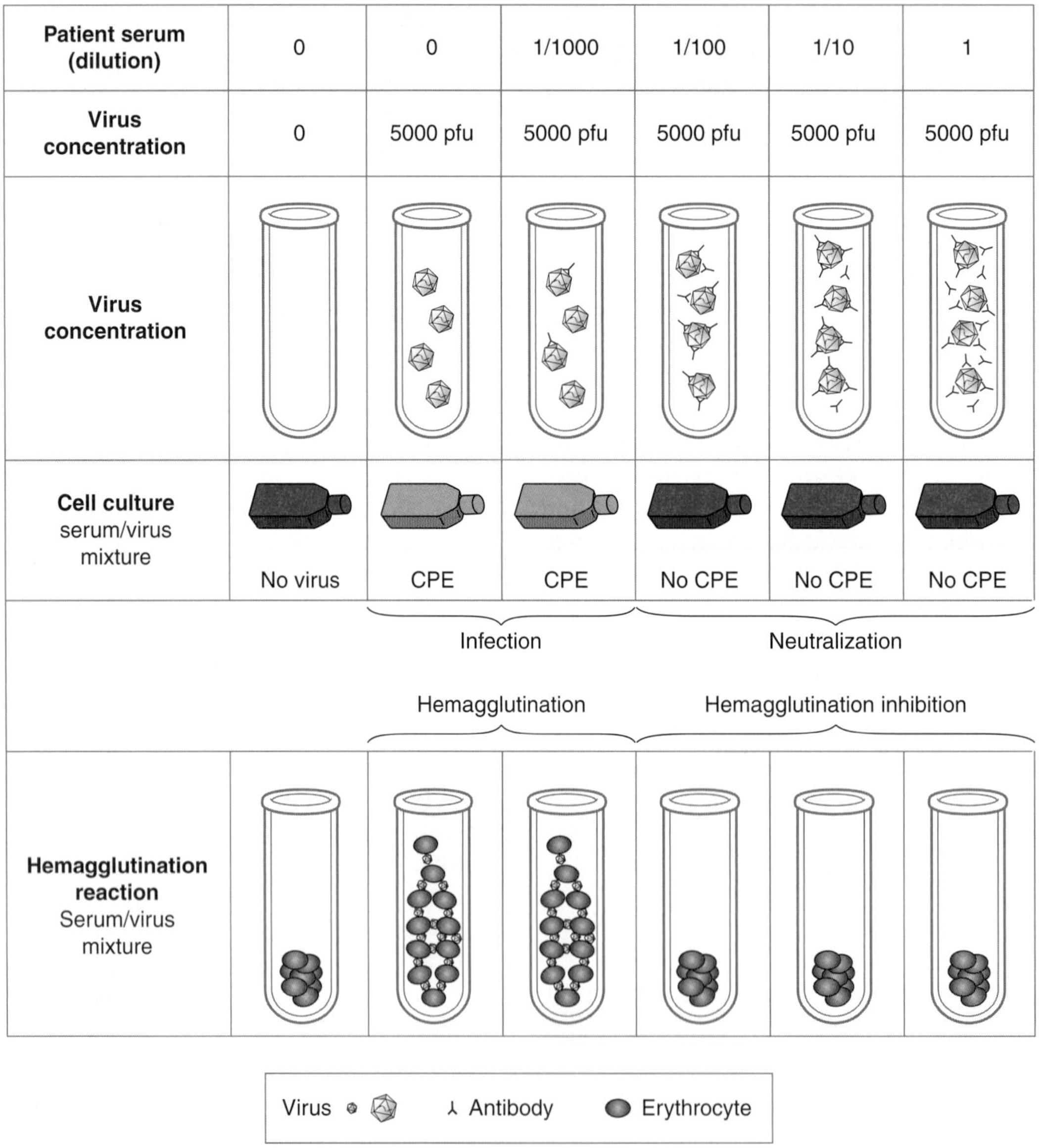

FIGURE 5-23 Neutralization, hemagglutination, and hemagglutination inhibition assays. Dilutions of serum are mixed with a known concentration of virus and then added to cell cultures or erythrocytes. Cultures are observed for CPE (infection). Viruses not neutralized will bind to erythrocytes, causing hemagglutination. Samples containing a high enough titer of neutralizing antibodies will bind to the virus causing hemagglutination to be inhibited. The titer of the antibody is 100. (PFU stands for plaque forming units.) Adapted from Murphy, P. R., Rosenthal, K. S., and Pfaller, M. A. *Medical Microbiology*, Sixth Edition. Elsevier, 2009.

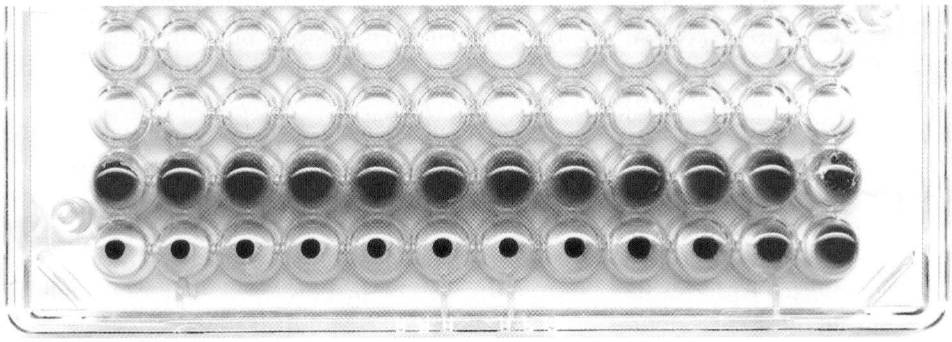

FIGURE 5-24 Different samples of influenza viruses were serially diluted and incubated with chicken red blood cells. Wells in the first bottom 9 wells contain no virus. The buttons/dots are nonagglutinated chicken red blood cells in the bottom 9 wells. The virus titer would be the last dilution that shows hemagglutination activity in an assay.

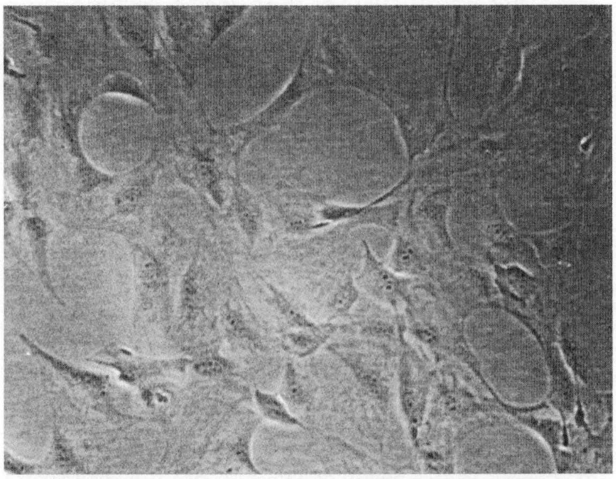

a

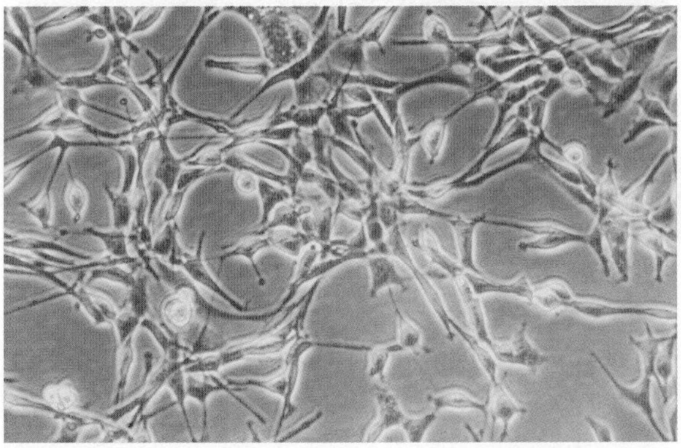

b

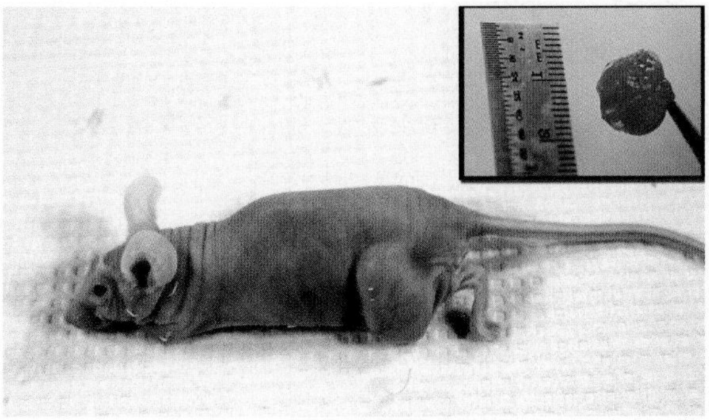

c

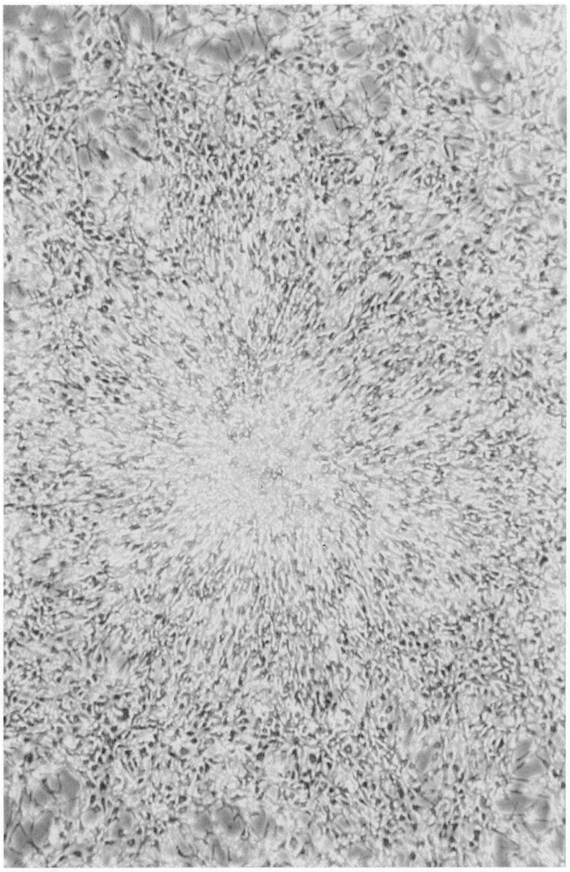

d

FIGURE 5-25 (a) Mouse cells (NIH 3T3). These cells are not transformed. Note their flat, dull appearance. **(b)** Transformed mouse NIH 3T3 cells. The cells appear refractile. These transformed cells were injected into nude mice. **(c)** Nude mice developed tumors that were excised and allowed to grow in a culture dish. **(d)** Transformed NIH 3T3 mouse cells removed from the nude mouse tumor growing in culture. Note the focus formed in the center of the photograph.

may be detected by their ability to interfere with the replication of a second virus added to the same cell culture. This is referred to as an **interference phenomenon**. Molecular mechanisms for the viral interference are discussed in Chapter 7. An interference assay is frequently used to detect rubella virus. When rubella virus infects African green monkey kidney cells, CPEs do not occur. The cells are incubated for 10 days and then reinfected or "challenged" with a second virus, such as echovirus 11. Controls that are not infected with rubella virus are run side by side during this experiment. After two days, no CPE is observed in co-infected cells, but there is CPE in cells infected with echovirus 11. Thus, the rubella virus interfered with or inhibited the replication of echovirus 11.

PCR is used extensively in virology research labs to discover new viruses, to generate viral nucleic acid sequence information, and to study virus replication *in vitro*. Research into viral ecology and evolution is driving the need to develop large datasets of complete viral genomes for virus surveillance to predict viral evolution and pandemics. Viral nucleic acid is isolated from natural reservoirs of viruses, cell cultures, or purified virions and subsequently used in PCR or RT-PCR reactions. In some instances, PCR products are electrophoresed in an agarose gel and the DNA fragments are analyzed. Sometimes the DNA is also digested with restriction enzymes and the digested fragments are analyzed for their electrophoretic pattern using restriction fragment length polymorphism (RFLP) analysis (see **Refresher: Restriction Enzymes**).

In the early 1990s researchers showed that the 5′ nuclease activity of Taq DNA polymerase could be exploited as a method to indirectly determine the level of DNA amplification with the use of specific fluorescent probes that are sensitive to 5′ nuclease degradation. They combined this method with the real-time monitoring of the DNA amplification in the PCR reaction tube in which amplified dsDNA would bind dyes. This advanced technology bypasses the need for gel electrophoresis because fluorescence from the dyes in the PCR reaction tube of amplified DNA can be measured instead of visualizing DNA fragments in a gel. The combined processes are called fluorescence detection of **real-time PCR**.

A detection system that uses a combination of real-time PCR and **fluorescence resonance energy transfer** (**FRET**) is used in biomedical research that monitors molecular activities inside of infected cells. In this method, two fluorescently labeled FRET probes are used, one with a fluor at the 5′ end (such as Cy3) and the other with a fluor at the 5′ end (such as Cy5). When the PCR product is produced, the labeled probes hybridize and bring the two fluors next to each other (**FIGURE 5-26**).

Even though PCR technology is sensitive, rapid, and can be adjusted to suit many applications, it does have some disadvantages. It is an expensive method and trace amounts of DNA contaminants could serve as DNA templates, resulting in the amplification of the wrong template nucleic acid (false positives). Laboratory benches, equipment, and pipetting devices can be contaminated by previous DNA preparations. Sterile technique

Refresher: Restriction Enzymes

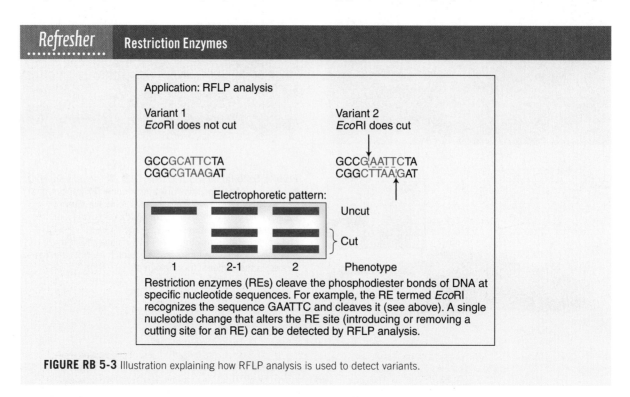

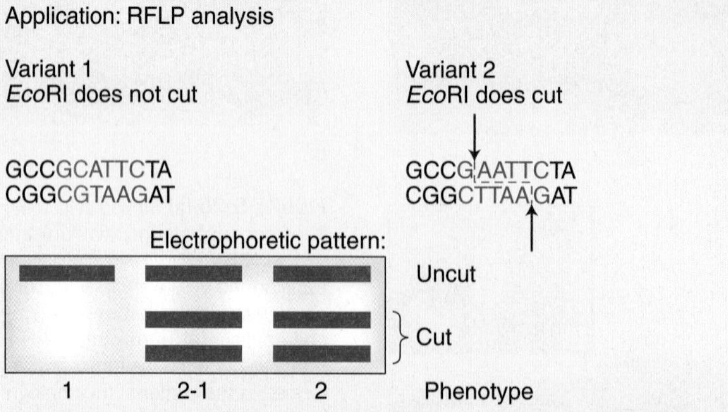

Restriction enzymes (REs) cleave the phosphodiester bonds of DNA at specific nucleotide sequences. For example, the RE termed *Eco*RI recognizes the sequence GAATTC and cleaves it (see above). A single nucleotide change that alters the RE site (introducing or removing a cutting site for an RE) can be detected by RFLP analysis.

FIGURE RB 5-3 Illustration explaining how RFLP analysis is used to detect variants.

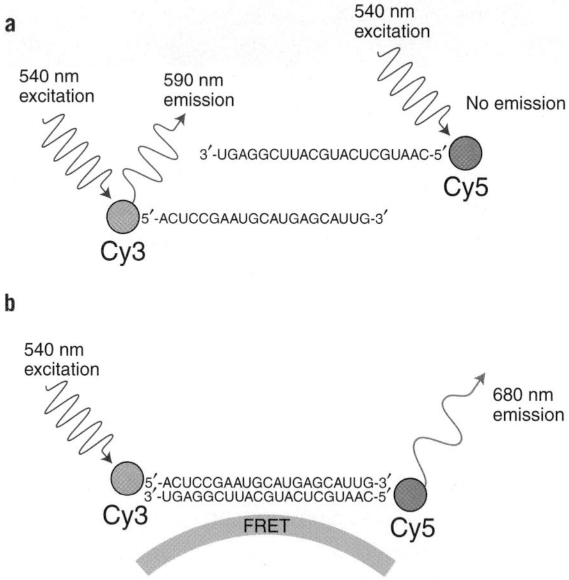

FIGURE 5-26 Diagram of FRET occurring between Cy3 and Cy5 fluorescent moieties when labeled primers are annealed. Adapted from Held, P. An Introduction to Fluorescence Resonance Energy Transfer (FRET) Technology and Its Application in Bioscience. BioTek Instruments, Inc. 2005.

and the use of new or sterile plasticware to prepare PCR reagents and template DNA are essential. Control reactions must be included with all reactions; for example, a negative "no DNA" control that contains all of the reaction components except the template DNA and a positive control that was used successfully in previous PCR reactions ensures reproducibility and reliability of the test.

Viral proteins, including enzymes produced during viral replication, can be detected by molecular biological, biochemical, and immunological methods. Detection of enzyme activities characteristic of specific viruses can be used to quantitate and identify these viruses; for example, reverse transcriptase activity in serum or cell cultures is indicative of the presence of a retrovirus such as HIV or HTLV.

5.5 Laboratory Safety

Precautions are necessary to prevent laboratory hazards. Workers in clinical virology laboratories test human serum and other body fluids that may contain HIV-1 or hepatitis C or B virus. This requires caution and safety procedures at all times. The safest procedure is to treat all clinical specimens as potentially infectious. Proper laboratory procedure involves using rigorous aseptic technique, and gloves and laboratory coats are worn at all times. There are different classes of biosafety cabinets that provide increasing levels of containment. If available, workers should be vaccinated against the potential infectious agents they may be working with. There are four designated Biosafety Level (BSL) laboratories. From minimum to maximum containment requirements, they are BSL-1, BSL-2, BSL-3,

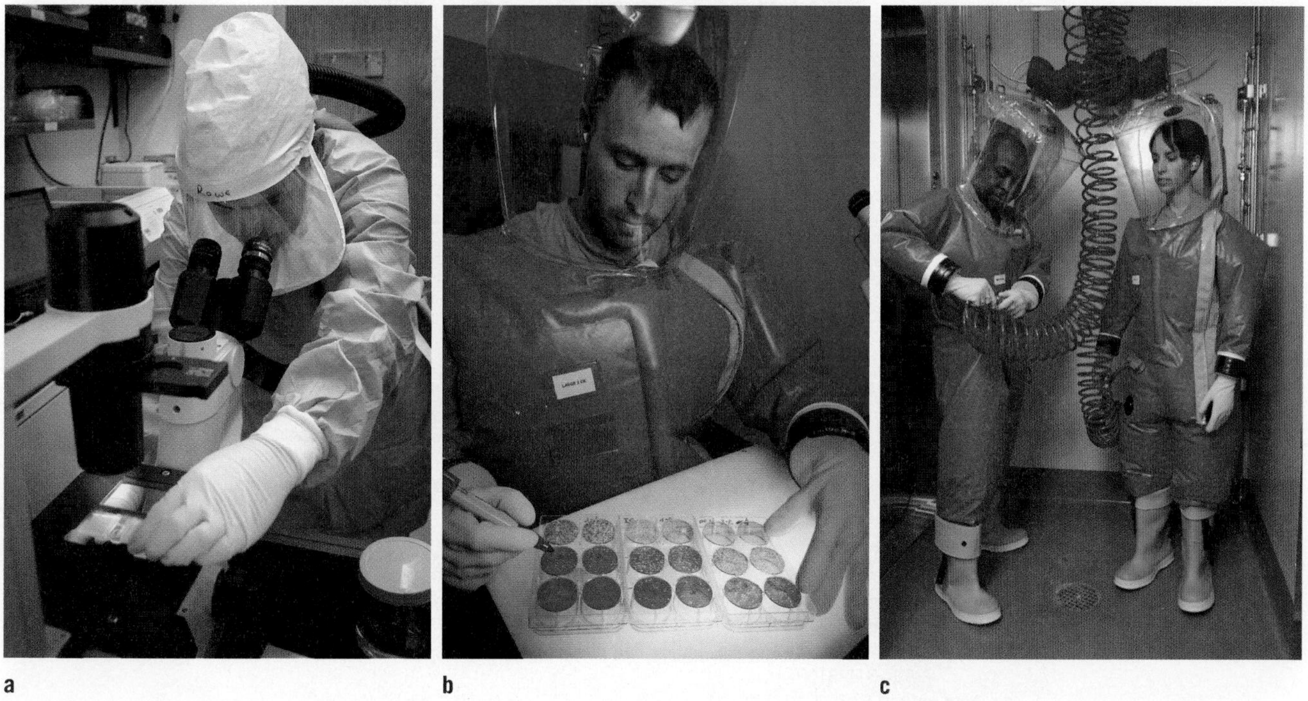

FIGURE 5-27 (a) A CDC researcher working in a BSL-3 lab on highly lethal influenza strains. **(b)** A CDC researcher counting viral plaques present in a fixed monolayer of cells in dishes that has been set up on a light box inside of a BSL-4 laboratory. **(c)** CDC scientist attaching his supportive air hose, which will provide a supply of filtered breathable air as well as maintain positive pressure inside of his air-tight suit while working in a BSL-4 laboratory.

VIRUS FILE 5-2 Rapid Diagnosis of SARS

The largest outbreak of SARS occurred in Beijing, China, in 2003. It began March 5 and by July 4 there were 8439 probable SARS cases and 812 deaths. On March 15 the World Health Organization (WHO) recommended postponing nonessential travel to the affected areas and screening of airline passengers. At the time, SARS was a newly emerging infectious disease caused by a novel coronavirus. There was an immediate need for a rapid and reliable diagnostic test. Dr. Christian Drosten, a physician and research scientist, and his research team at the Bernhard Nocht Institute for Tropical Medicine in Hamburg, Germany, were the first to develop a diagnostic test for SARS.

On March 15—the same day as the WHO's global alert—a doctor and his wife, both infected with SARS, landed at Frankfurt airport in Germany. Initial tests on sputum samples from the doctor were conducted at a clinical laboratory in Frankfurt. The tests included various PCR tests for known viruses and bacteria and involved looking at the sputum sample with an electron microscope (**FIGURE VF 5-2**). Although the virus appeared to be a typical paramyxovirus, the lab was unable to identify the virus. A second sample taken on March 17 was sent to Drosten's laboratory to determine if the illness was caused by a tropical virus. Drosten quickly ruled out paramyxoviruses and sent samples to Europe's leading paramyxovirus expert, Dr. Albert Osterhaus at Erasmus University in Rotterdam, the Netherlands. Dr. Osterhaus confirmed the negative result. On March 22, Drosten received a sample of the SARS virus that had been grown from a clinical specimen in cell cultures in the Frankfurt laboratory. Immediately, Drosten's team began a new series of PCR tests and by March 25 some of the PCR products matched the sequences from the coronavirus family. Just as this new information was received, the CDC announced that they had identified SARS as the infectious agent. Subsequently, Drosten's team developed an RT-PCR assay. Since that time, additional RT-PCR, serologic (ELISAs), and tissue culture isolation assays have been developed for the SARS-associated coronavirus.

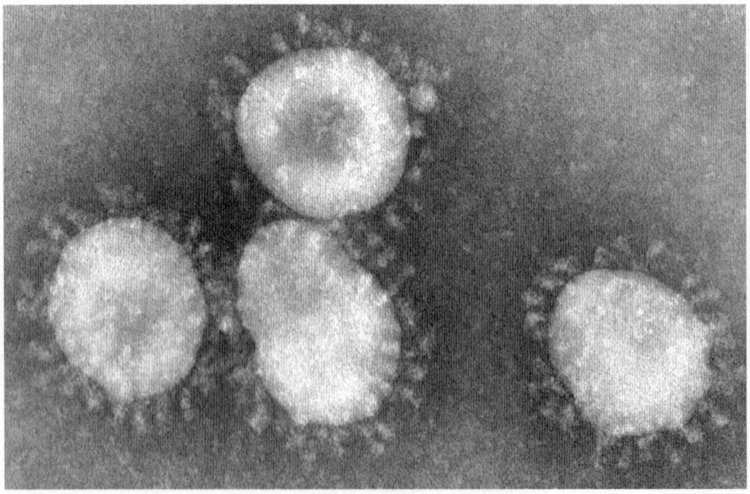

FIGURE VF 5-2 Electron micrograph of SARS-CoV, the etiological agent recognized as the cause of the SARS outbreak in 2003.

References

Chan, K. H., et al. 2004. "Detection of SARS coronavirus in patients with suspected SARS." *EID* 10:294–299.

Drosten, C., et al. 2003. "Identification of a novel coronavirus in patients with severe acute respiratory syndrome." *N Engl J Med* 348:1967–1976.

Emery, S. L., et al. 2004. "Real-time reverse transcription-polymerase chain reaction assay for SARS-associated coronavirus." *EID* 10:311–316.

Martina, B. E. E., et al. 2003. "SARS virus infection of cats and ferrets." *Nature* 425:915.

Wang, L-F, et al. 2006. "Review of bats and SARS." *EID* 12(12):1834–1840.

Zheng, S., and Zhihong, H. 2008. "A review of studies on animal reservoirs of the SARS coronavirus." *Virus Res* 133:74–87.

and BSL-4. Criteria for the different levels are available at http://www.cdc.gov/od/ohs/.

Exposure to infectious aerosols poses the most risk when working with clinical specimens. Crucial to safe working conditions are the various types of specialized equipment available as barriers between the virus and the laboratory worker. Equipment ranges from protective clothing and gloves to complex biosafety cabinets and containment devices. Workers in BSL-4 labs, which require maximum containment, work with infectious agents that have no vaccine or cure (**FIGURE 5-27**).

Summary

Viral diagnostics are critical for patient management and for surveillance of disease outbreaks. They enable infectious disease experts to establish control measures. The rapidly expanding knowledge about SARS is a prime example of how clinical, public health, and research communities worldwide mounted an aggressive response to a new disease. In this particular case, scientific journals played a major role in this endeavor by expediting online publication of peer-reviewed data.

This chapter defined a typical clinical laboratory setting, including the equipment required to cultivate viruses. The most common way to propagate viruses is in tissue culture cells. These cells usually undergo rapid morphological changes, termed CPEs, when infected by viruses. CPEs include the formation of inclusion bodies, syncytia and fusion, cell shrinkage, refractility changes, rounding, aggregation, and lysis (cell death).

Common methods used to study viruses in the research laboratory are single-step growth experiments, plaque assays, hemagglutination assays, and transformation assays.

The quality of clinical specimens collected from the patient and its transportation from the patient to the laboratory limit the ability of the laboratory personnel to perform diagnostic tests. Hence, there are important rules about the storing and collecting of specimens and the type of specimens collected. Rapid diagnosis is critical for patient management. There are five general approaches to laboratory diagnosis: microscopy, detection of viral antigens, culture, nucleic acid detection, and antibody detection. The chapter discussed the electron microscopic, molecular, and serological methods used in the clinical virology laboratory. Many diagnostic laboratories now perform drug susceptibility testing to monitor AIDS patients because of viral resistance to the various drug regimes patients must follow.

Lastly, safety in the clinical laboratory is of the utmost importance. Precautions are necessary to prevent laboratory hazards. Chapter 6 addresses the mechanisms of infection and how diseases are spread through the body and to others.

CASE STUDY QUESTIONS

These questions relate to the Case Study presented at the beginning of the chapter.

1. Early recognition of meningitis and/or encephalitis requires rapid diagnosis. Why?
2. Besides the viruses presented in the case study, what other viruses can cause meningitis or encephalitis in humans?
3. Besides nucleic acid amplification, what other methods can be used to detect viruses that cause central nervous system infections and how long do these test results take?

Resources

Primary Literature and Reviews

Athmanathan, S., Bandlapally, S. R., and Gullapalli, N. R. 2002. "Comparison of the sensitivity of a 24 h-shell vial assay, and conventional tube culture, in the isolation of herpes simplex virus-1 from corneal scrapings." *BMC Clin Path* 2(1):1. Epub.

Barenfanger, J., et al. 2001. "R-Mix cells are faster, at least as sensitive and marginally more costly than conventional cell lines for the detection of respiratory viruses." *J Clin Virol* 22:101–110.

Caceda, E., and Kochel, T. J. 2007. "Application of modified shell vial culture procedure for arbovirus detection." *PLoS ONE* 2(10):e1034.

Chan, K. H., et al. 2004. "Detection of SARS Coronavirus in patients with suspected SARS." *EID* 10:294–299.

Cherkasova, E., et al. 2003. "Microarray analysis of evolution of RNA viruses: Evidence of circulation of virulent highly divergent vaccine-derived polioviruses." *PNAS USA* 100(16):9398–9403.

Clewley, J. P. 2003. "A role for arrays in clinical virology: Fact or fiction?" *J Clin Virol* 29:2–12.

Crist, G. A., et al. 2004. "Evaluation of the ELVIS plate method for the detection and typing of herpes simplex virus in clinical specimens." *Virology* 49:173–177.

Djikeng, A., et al., 2008. "Viral genome sequencing by random priming methods." *BMC Genomics* 9:5. doi:10.1186/1471-2164-9-5.

Drosten, C., et al. 2003. "Identification of a novel coronavirus in patients with severe acute respiratory syndrome." *N Engl J Med* 348:1967–1976.

Emery, S. L., et al. 2004. "Real-time reverse transcription-polymerase chain reaction assay for SARS-associated coronavirus." *EID* 10:311–316.

Enders, J. F., et al. 1949. "Cultivation of the Lansing strain of poliomyelitis virus in cultures of various human embryonic tissues." *Science* 109:85–87.

Fouchier, R. A., et al. 2003. "Koch's Postulates fulfilled for SARS virus." *Nature* 423:240.

Freshney, R. I. 1994. *Culture of Animal Cells: A Manual of Basic Technique*, 3rd ed. New York: Wiley.

Fredricks, D. N., and Relman, D. A. 1996. "Sequence-based identification of microbial pathogens: A reconsideration of Koch's Postulates." *Clin Microbiol Rev* 9(1):18–33.

Geisbert, T. W., et al. 2003. "Treatment of Ebola virus infection with a recombinant inhibitor of factor VIIa/tissue factor: A study in rhesus monkeys." *Lancet* 362:1953–1958.

Hazelton, P. R., and Gelderblom, H. R. 2003. "Electron microscopy for rapid diagnosis of infectious agents in emergent situations." *EID* 9:294–303.

Kato-Maeda, M., Gao, Q., and Small, P. M. 2001. "Microarray analysis of pathogens and their interaction with hosts." *Cell Microbiol* 3(11):713–719.

Leland, D. S., and Ginocchio, C. C. 2007. "Role of cell culture for virus detection in the age of technology." *Clin Microbiol Rev* 20(1):49–78.

Lin, B., et al. 2004. "Use of oligonucleotide microarrays for rapid detection and serotyping of acute respiratory disease-associated adenoviruses." *J Clin Microbiol* 42(7):3232–3239.

Long, W-H., et al. 2004. "A universal microarray for detection of SARS coronavirus." *J Virol Methods* 121:57–63.

Madeley, C. R. 2008. "Is it the cause?—Robert Koch and viruses in the 21st century." *J Clin Virol* 43:9–12.

Malmsten, A., et al. 2003. "HIV-1 viral load determination based on reverse transcriptase activity recovered from human plasma." *J Med Virol* 71:347–359.

Martina, B. E. E., et al. 2003. "SARS virus infection of cats and ferrets." *Nature* 425:915.

McNair Scott, T. F., and Rivers, T. M. 1936. "Meningitis in man caused by a filterable virus. I. Two cases and the method of obtaining virus from their spinal fluids." *J Exp Med* 63(3):397–414.

Murray, P. R., et al., eds. 1995. *Manual of Clinical Microbiology*, 6th ed. Washington, DC: ASM Press.

Petricoin, E. F. III, et al. 2002. "Medical applications of microarray technologies: A regulatory science perspective." *Nat Genet Suppl* 32:474–479.

Reed L. J., and Muench, H. 1938. "A simple method of estimation of 50% endpoints." *Am J Hygiene* 27(3):493–497.

Richman, D. D., Whitley, R. J., and Hayden, F. G., eds. 2002. *Clinical Virology*, 2nd ed. Washington, DC: ASM Press.

Rivers, T. M. 1937. "Viruses and Koch's Postulates." *J Bacteriol* 33:1–12.

Rivers, T. M., and McNair Scott, T. F. 1936. "Meningitis in man caused by a filterable virus. II. Identification of the etiological agent." *J Exp Med* 63(3):415–432.

Rodriguez Roche, R., et al. 2000. "Comparison of rapid centrifugation assay with conventional tissue culture method for isolation of Dengue 2 virus in C6/36-HT cells." *J Clin Microbiol* 38(9):3508–3510.

Szollosi, J., Damjanovich, S., and Matyus, L. 1998. "Application of fluorescence resonance energy trans-

fer in the clinical laboratory: Routine and research." *Cytometry* 34:159–179.
Specter, S., Hodinka, R. L., and Young, S. A., eds. 2000. *Clinical Virology Manual.* Washington, DC: ASM Press.
Van Doornum, G. J. J., and De Jong, J. C. 1998. "Rapid shell vial culture technique for detection of enteroviruses and adenoviruses in fecal specimens: Comparison with conventional virus isolation method." *J Clin Microbiol* 36:2865–2868.
Walker, N. J. 2002. "A technique whose time has come." *Science* 296:557–559.
Wang, D., et al. 2002. "Microarray-based detection and genotyping of viral pathogens." *PNAS USA* 99(24): 15687–15692.
Wang, Y., and Wang, N. 2009. "FRET and mechanobiology." *Integr Biol (Camb)* 1(10):565–573.
Weinberg, A., Brewster, L., Clark, J., and Simoes, E. ARIVAC Consortium. 2004. "Evaluation of R-Mix shell vials for the diagnosis of viral respiratory tract infections." *J Clin Virol* 30:100–105.
Wilson, W. J., et al. 2002. "Sequence-specific identification of 18 pathogenic microorganisms using microarray technology." *Mol Cell Probes* 16:119–127.

Popular Press

Barry, J. M. 2004. *The Great Influenza: The Epic Story of the Deadliest Plague in History.* New York: Viking, Penguin Group.
Bear, G. 2000. *Darwin's Radio.* New York: Ballantine.
Duncan, K. 2006. *Hunting the 1918 Flu: One Scientist's Search for a Killer Virus.* Toronto, Canada: University of Toronto Press.
Friend, S. H., and Stoughton, R. B. 2002. "The magic of microarrays." *Sci Am,* February, 286(2):44.
Garrett, L. 1994. *The Coming Plague.* New York: Penguin.
Harper, D. R., and Meyer, A. S. 1999. *Of Mice, Men, and Microbes: Hantavirus.* San Diego: Academic Press.
McCormick, J. B., and Fisher-Hoch, S., with Horvitz, L. A. 1996. Level 4: *Virus Hunters of the CDC.* Atlanta: Turner Publishing.
O'Brien, T. J. 2000. *West Nile Outbreak in New York.* Philadelphia: Knopf.
Peters, C. J., and Olshaker, M. 1998. *Virus Hunter: Thirty Years of Battling Hot Viruses Around the World.* New York: Knopf.
Preston, R. 1995. *The Hot Zone.* New York: Anchor.
Regis, E. 1996. *Virus Ground Zero: Stalking the Killer Viruses with the Centers for Disease Control.* New York: Pocket Books.
Regush, N. 2000. *The Virus Within: A Coming Epidemic.* New York: Penguin Putnam.
Sloot, R. 2010. *The Immortal Life of Henrietta Lacks.* New York: Crown Publishers.
Zimmerman, B. E., and Zimmerman, D. J. 1996. *Killer Germs: Microbes and Diseases That Threaten Humanity.* Chicago: Contemporary Books.

Video Productions

Bad Blood: A Film by Marilyn Ness. 2010. http://badblood documentary.com/
Killer Virus: Hunt for the Next Killer Flu. 2009. Discovery Channel.
H1N1. 2009, November 1. CBS *60 Minutes.*
H1N1. 2009, October 18. CBS *60 Minutes.*
The Great Fever. 2007. PBS; Bosch and Company, Inc., Film for American Experience.
The Age of AIDS. Frontline. 2006. PBS Home Video.
SARS and the New Plagues. 2003. History Channel.
SARS: The True Story. 2003. Films for the Humanities.
Mosquito Nightmare: West Nile Virus. 2000. Films for the Humanities.
The Next Plague: The Nipah Virus. 1998. Films for the Humanities.
Ebola: The Plague Fighters. 1996. Public Broadcasting Service.
Invisible Enemies: The Hanta Virus. 1996. Films for the Humanities.
On the Trail of a Killer Virus. 1996. A&E Production.

eLearning

go.jblearning.com/shors2

The site features eLearning, an online review area that provides quizzes and other tools to help you study for your class. You can also follow useful links for in-depth information, or just find out the latest virology and microbiology news.

CHAPTER 6

Mechanisms of Viral Entry and Spread of Infection in the Body

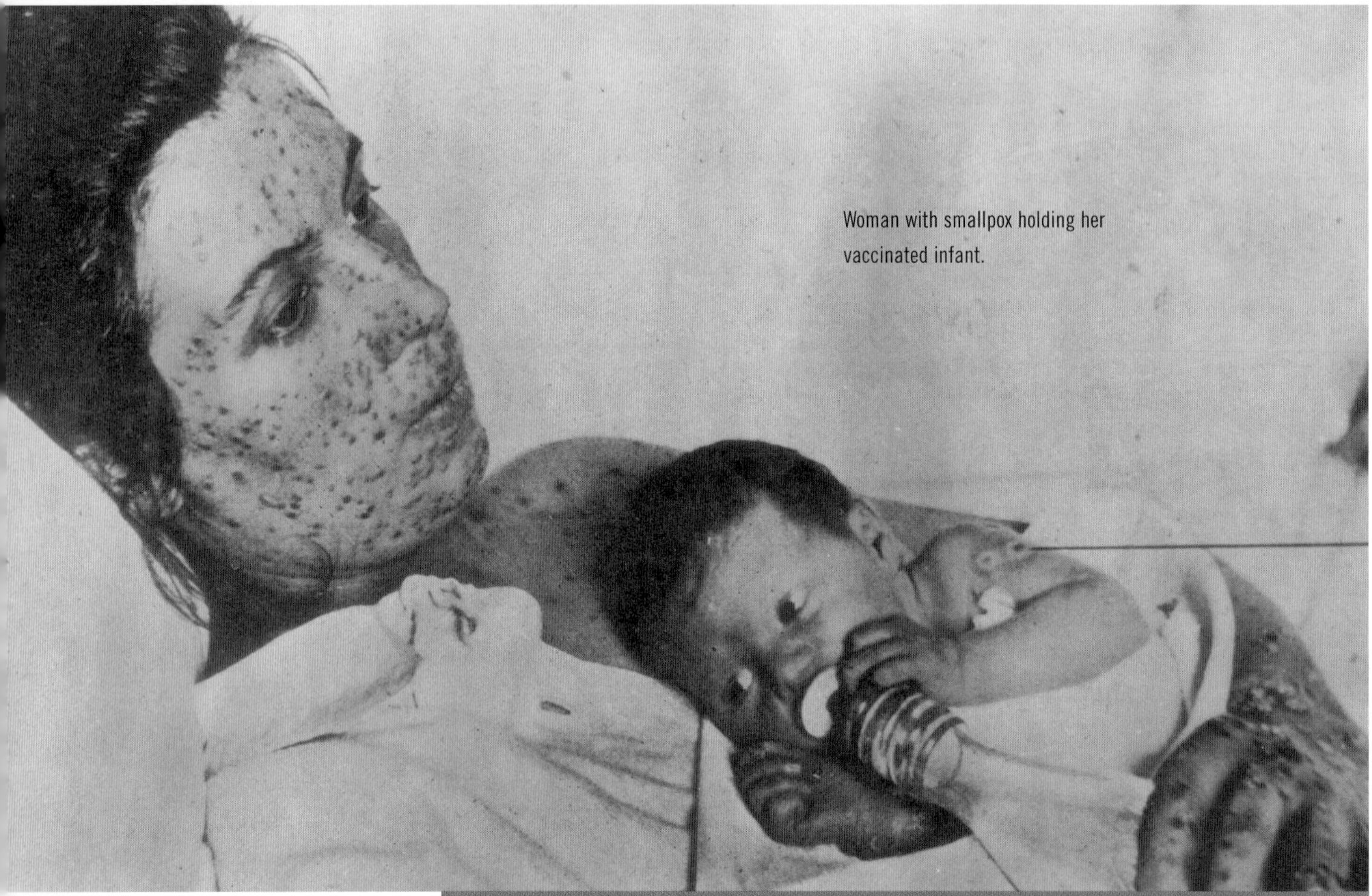

Woman with smallpox holding her vaccinated infant.

OUTLINE

"... no man dared to count his children as his own until they had had the disease."

Comte de la Condamine, 18th-century mathematician and scientist referring to smallpox

Sam Bugert, a computer consultant, was traveling by air from Hong Kong to Japan during the height of the severe acute respiratory syndrome (SARS) outbreak in 2003. While on the plane to Japan, Sam planned a business trip to Toronto, Canada. Just the thought of flying to Canada made Sam feel tired. He asked the flight attendant for several cups of coffee, which he drank hastily as he planned his trip to Canada.

After Sam got off the plane at Tokyo International Airport, he passed through a security checkpoint, which included walking in front of a thermal imaging scanner (**FIGURE CS 6-1**). He was briefed on the plane about the scanners that were being used to detect passengers with elevated temperatures because a presenting symptom of SARS was fever. The use of the infrared thermal imaging scanners provided a noninvasive, fast, and cost-effective means for mass screening of potential SARS-infected persons.

After Sam walked through the scanner, an airport staff member pulled him aside, checked his temperature orally with a thermometer, and asked him to fill out a standardized questionnaire for SARS. He didn't feel ill and urgently wanted to get to his next flight.

The staff member was very patient. She explained that the imaging cameras detect the temperature of an object by measuring the amount of infrared radiation emitted by that object. It provides a digital image showing temperature patterns. Scanners have primarily been used in the diagnosis of breast cancer (detecting inflammation) and nerve dysfunctions. The scanners measure the skin and not the core body temperature. The accuracy of the scanner is +/− 2°C (+/− 3.6°F). The accuracy of the screening can be affected by such factors as physical activity, stress, caffeine, nicotine, alcohol consumption, and circulatory problems. Other factors affecting their accuracy include equipment variables, the operator's training and experience, and environmental factors such as room temperature and air currents.

Sam told the staff member that he had been consuming large volumes of coffee during the flight to Japan and this was probably causing his elevated scanner results. The staff member noted that he was asymptomatic but followed through with airport procedures and sent Mr. Bugert to the hospital for further tests.

References

Centre for Emergency Preparedness and Response (CCDR), Population and Public Health Branch, Health Canada. 2004. "Thermal image scanners to detect fever in airline passengers, Vancouver and Toronto, 2003." *CCDR* 30:165–167.

Ng, E. Y. 2005. "Is thermal scanner losing its bite in mass screening of fever due to SARS?" *Med Phys* 32:93–97.

FIGURE CS 6-1 A nurse in Taiwan's Chiang Kai-shek International Airport gestures a passenger to move on to immigration after being cleared of high body temperature by a thermal scanner.

Chapter 5 focused on the techniques used to grow, study, and identify viruses from clinical specimens. This chapter examines an earlier stage, when the virus enters the host, multiplies and spreads within the host's body, and then exits (or is "shed") into the environment.

6.1 Preferred Routes of Entry

All viruses target a specific population of cells. Before viruses can do that, though, they must somehow enter the body of the human or animal host (**FIGURE 6-1**). Viruses gain access to their hosts through the skin or the following mucous membranes:

- respiratory tract
- gastrointestinal tract (also referred to as the alimentary canal)
- genital tract
- conjunctiva (eyes)

In addition, a fetus can be infected when certain blood-borne viruses cross the placenta and reach the fetal circulation, causing congenital (present at birth) malformations. A baby also may be infected as it passes through the birth canal. Viral infections and pregnancy are addressed at the end of Section 6.2.

Respiratory Tract Entry

The average person inside a ventilated building inhales at least eight microorganisms per minute, or about 11,520 per day. Although most of the inhaled microorganisms are nonpathogenic bacteria or molds, it is not surprising that inhalation is the most common route of virus infection. To combat the huge numbers of microorganisms inhaled each day, the human body employs within its respiratory tract a very effective cleansing mechanism designed to remove and dispose of inhaled particles: a blanket of mucus and ciliated cells that line the nasal cavity and most of the lower respiratory tract (**FIGURE 6-2**). Particles inhaled into the lower respiratory tract (the lungs) are often trapped

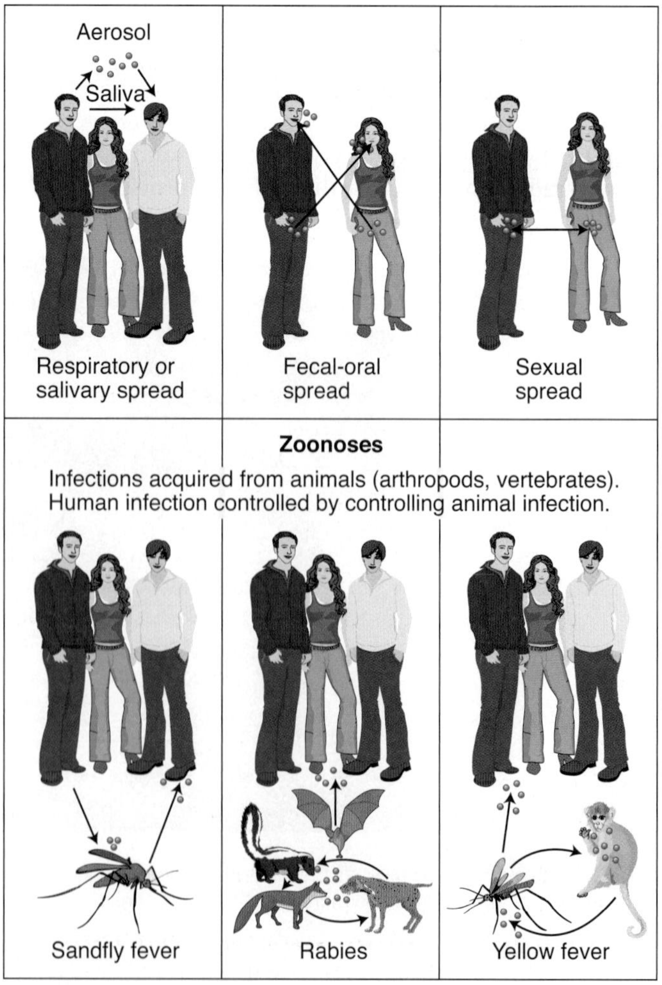

FIGURE 6-1 Viruses may be spread via respiratory or salivary routes of transmission, which are not readily controllable. Oral-fecal route transmission is controllable by public health measures. Venereal diseases are difficult to control because of social factors. Controlling vectors or animal infections can control human infections related to zoonoses.

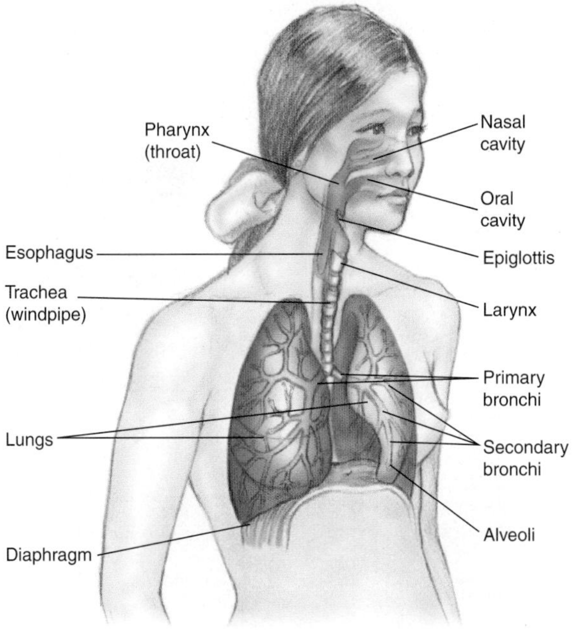

FIGURE 6-2 Schematic of the upper and lower respiratory tracts.

TABLE 6-1 Viruses That Initiate Infection via the Respiratory Tract That Cause Respiratory Symptoms

Human Virus	Symptoms
Rhinovirus	Respiratory (common cold)
Coronavirus (SARS)	Respiratory
Influenzas A and B	Respiratory
Parainfluenza virus	Respiratory
Respiratory syncytial virus	Respiratory
Adenovirus	Respiratory
Echovirus (some types)	Respiratory
Cocksackie viruses A type 21 and B types 4 and 5	Respiratory
Sin nombre virus (hantavirus)	Hantavirus pulmonary syndrome (HPS)

in the mucus, moved upward to the back of the throat via **ciliary action**, and then coughed out. Particles entering the nasal cavity are trapped in mucus and carried to the back of the throat by ciliary action and then swallowed. The average person produces 10 to 100 ml (0.04–0.40 cups) of mucus from the nasal cavity each day and a similar amount from the lungs.

Particles such as viruses and bacteria that are smaller than 5 μM are likely to enter the lungs and reach the **alveolar sacs**. The alveolar sacs of the lungs do not contain mucus or ciliated cells. Instead, they are lined with immune cells called macrophages. These macrophages move along the alveolar surface of the lung, engulfing airborne particles that reach the alveolar surface. Antibodies (IgG and IgA) are present in the upper and lower respiratory tracts. They provide specific defenses against invading microbes. Normally the lungs are nearly sterile because microbes inhaled on a continual basis are also being destroyed by macrophages or removed via mucociliary action. (Chapter 7, Host Resistance to Viral Infections, provides more information about immune defense.)

When the mucociliary defenses are defective or compromised, a highly successful virus may initiate an infection by avoiding entrapment in the mucus, being carried to the back of the throat, and then being swallowed. If it reaches the alveolar sacs, the virus must then resist being engulfed by macrophages or survive being digested and killed by macrophages. Viruses such as influenza and rhinoviruses that can infect their hosts via the respiratory tract do so by attaching to receptors located on the ciliated epithelial cells of the respiratory tract. **TABLES 6-1** and **6-2** list the viruses that can initiate human infections via the respiratory tract.

Gastrointestinal Tract Entry

This mode of infection is known as the **oral-fecal route**. This means that viruses shed in feces have gotten into someone's mouth and are ingested with swallowed materials and then carried to the intestinal tract. In the stomach, acid and proteolytic enzymes may inactivate viruses. Successful viruses entering the gastrointestinal tract (**FIGURE 6-3a**) are usually resistant to acid and bile (bile is secreted by the liver).

TABLE 6-2 Viruses That Initiate Infection via the Respiratory Tract and Cause Generalized Symptoms, Usually without Respiratory Symptoms

Human Virus	Examples of Symptoms
Mumps virus	Fever, swelling of salivary gland(s)
Measles	Cough, fever, rash
Varicella zoster	Chickenpox, shingles lesions
Hantaan virus	Hemorrhagic symptoms, kidney failure
Variola virus	Fever, malaise, headache, vomiting, smallpox (deep-seated rash)
Coxsackie A16	Hand, foot/mouth disease
Enterovirus 71	

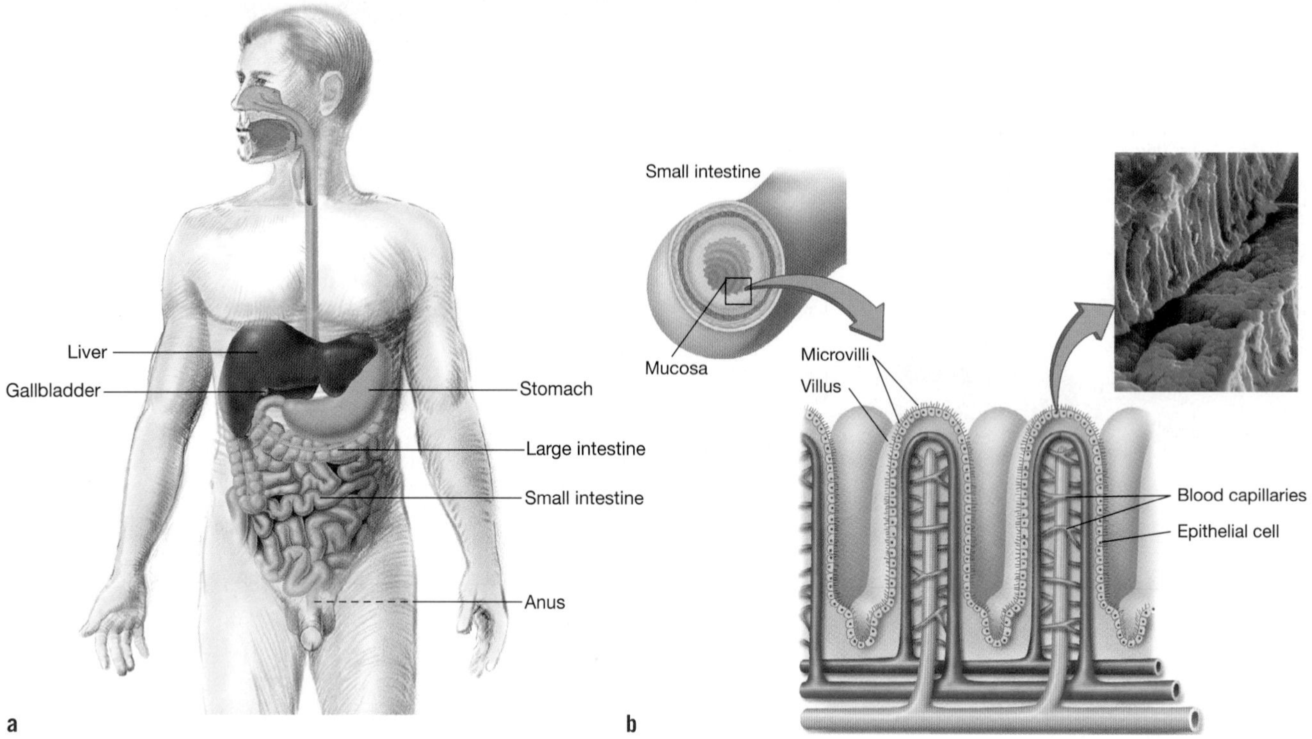

FIGURE 6-3 (a) Schematic of the gastrointestinal tract. **(b)** Panels depicting the surface area expansion of the small intestine. Note the villi and the microvilli of the epithelial cells.

As food is liquefied in the stomach, it is slowly released into the small intestine for further processing. The intestinal tract is protected by mucus that contains secretory IgA antibodies. The constant movement of liquefied food through the intestinal tract provides viruses in the liquid the exposure necessary for attachment to cellular receptors on the intestinal epithelial cells. (The absorptive surface area of the small intestine is roughly 250 square meters—the size of a tennis court!) This is possible because the inner surface of the small intestine is not flat. It contains villi, which are projections that are covered by epithelial cells. The epithelial cells in turn contain densely packed microvilli (Figure 6-3b). The villi contract and expand to mix the material in the intestine, which provides opportunities for virus particles to encounter living intestinal epithelial cells. The enteric viruses bind to receptors located on the epithelial cells. Examples of human viruses that infect via the gastrointestinal route are listed in **TABLE 6-3**.

TABLE 6-3 Viruses That Initiate Infection via the Gastrointestinal Tract

Virus	Symptoms
Rotavirus	*Gastroenteritis, dehydrating diarrhea, nausea, vomiting
Sapovirus	*Gastroenteritis, rotavirus-like illness in children
Norovirus	*Gastroenteritis, diarrhea, nausea, acute vomiting
Astrovirus	*Gastroenteritis, watery diarrhea (usually short in duration)
Enteric adenoviruses	*Gastroenteritis, prolonged diarrhea, nausea, vomiting
Poliovirus	Severity of symptoms vary: Mild symptoms include fever, nausea, vomiting and abdominal pain; severe symptoms include paralysis, difficulty swallowing, and muscle wasting
Hepatitis A	Short, mild flu-like illness, jaundice
Other adenoviruses	Usually symptomless
Some enteroviruses	Usually symptomless

*Inflammation of the lining of the membrane of the stomach and intestines.

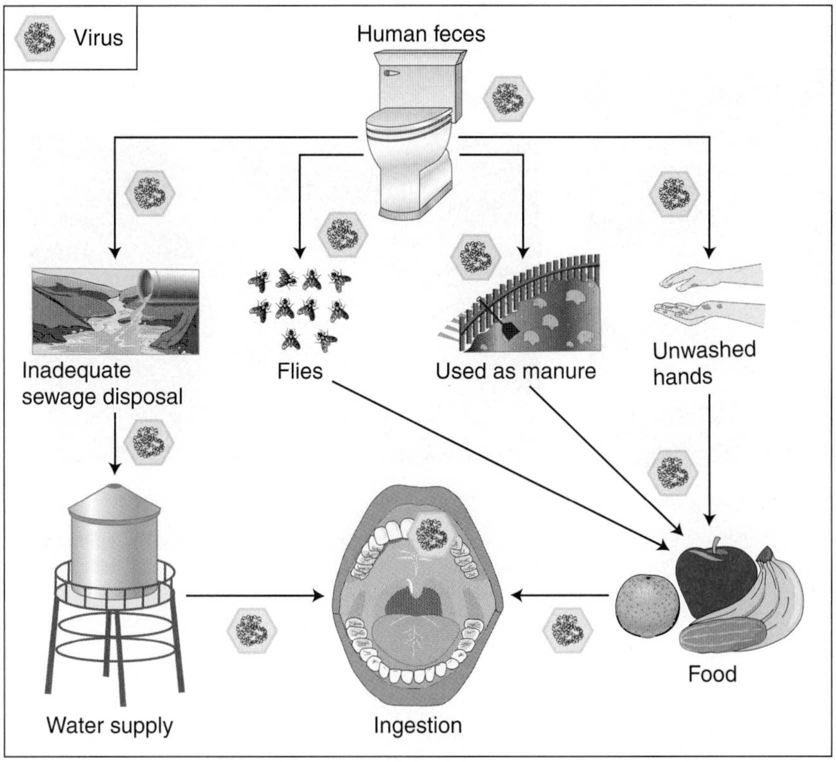

FIGURE 6-4 Schematic of oral-fecal transmission of enteric viruses.

Acute gastroenteritis is the major cause of morbidity and mortality worldwide. Infants and children less than five years of age average three to ten episodes per year. In the developing world, this means more than 1.5 billion cases and 1.5–2 million deaths per year. These deaths are mainly caused by dehydration, but malnutrition also plays a very important role. As economics and sanitation improve in the developing world, the morbidity and mortality caused by diarrheal diseases decrease. Approximately 38 million pediatric cases still occur in the United States annually, resulting in over 1.5 million doctor visits, 200,000 hospitalizations, and 300 deaths per year. **FIGURE 6-4** shows different ways in which the oral-fecal route of transmission by viruses can occur.

Genital Tract Entry

Sexually transmitted diseases (STDs) are infections that can be transferred from one person to another through sexual contact. STDs have large social and economic consequences, especially in the developing parts of the world. In the United States, adolescents and young adults (ages 15–24) are at greatest risk for acquiring an STD. **TABLE 6-4** lists viruses that initiate infection via the genital tract.

Conjunctiva Entry

The **conjunctiva** is the thin, transparent tissue that covers the outer surface of the eye. It is constantly cleansed by tear secretions and is wiped by the eyelids. It is a rare route of entry. When we get something in our eye, the conjunctiva suffers minor injuries. These injuries are susceptible to infection by viruses or other microbes through contaminated towels, fingers, or even ophthalmologist instruments during eye exams (**iatrogenic diseases**) rather than by airborne organisms. Eye injuries from metal chips and rust are a prevalent injury among factory workers and those in similar industries. Foreign bodies in the eye can result from workers blowing into their safety glasses, goggles, welding hoods, or respirators to remove debris.

Foreign bodies in the eye are associated with a condition called epidemic keratoconjunctivitis (EKC), which is known colloquially as "shipyard eye" because of its prevalence among shipbuilders. Symptoms include the inflammation of the cornea

TABLE 6-4 Examples of STDs Caused by Viruses

Virus	Symptoms/Disease
HIV-1 and HIV-2	AIDS
Hepatitis B	Liver damage, possibly cancer
Hepatitis C	Liver damage, possibly cancer
Herpes simplex-2	Herpetic lesions of cervix and urethra
Papillomavirus	Genital warts, possibly cancer

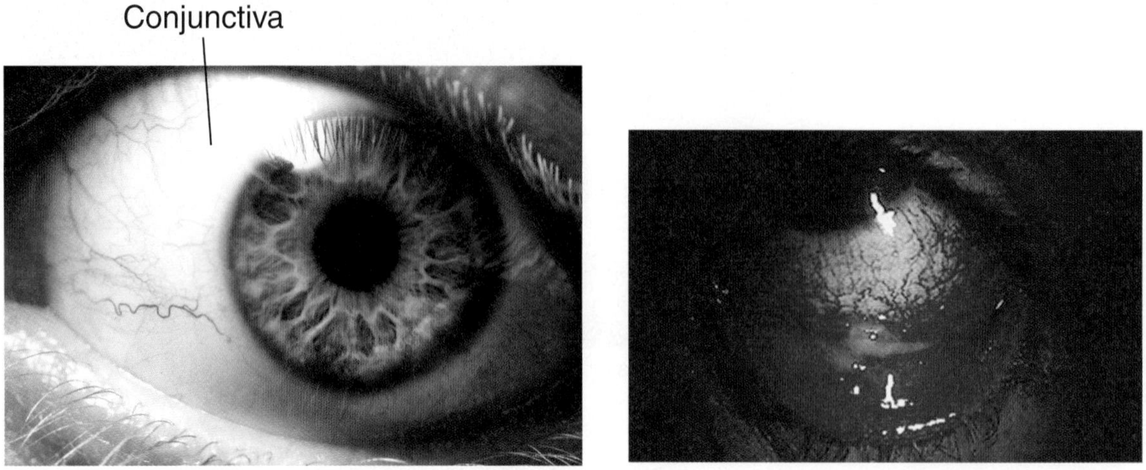

a b

FIGURE 6-5 **(a)** Photograph depicting the conjunctiva of the eye. Two common self-limiting forms of viral conjunctivitis are **epidemic keratoconjunctivitis** (EKC) and **pharyngoconjunctival fever** (PCF). **(b)** This image shows the pseudomembranes that can develop with EKC. These are additional membranes made up of fibrin and leukocytes caused by the body's immune response to viral antigens. An ophthalmologist can remove the pseudomembranes.

and conjunctiva caused by an adenovirus 8 infection. It is epidemic due to the speed at which the infection spreads. Conjunctivitis, or "pink eye," is sometimes caused by enterovirus Type 70. Photographs in **FIGURE 6-5** show the physical signs of viral eye infections.

Other Routes of Entry

Other routes of virus entry include:

- skin
- transplants
- blood transfusions and blood products
- iatrogenic induction

Skin Entry

The skin is a natural barrier to microorganisms. The outside of the skin is a rather hostile environment due to its dryness, acidity, and bacteria that reside on its surface. The outermost layer of skin consists of dead skin cells (**FIGURE 6-6**). Viruses require living cells to replicate themselves, which means that successful viruses must enter through breaks in the skin. These occur as a result of insect, animal, or human bites; needlestick injuries by doctors, tattooists, drug addicts, acupuncturists, or as a result of body piercing; puncture wounds and abrasions; and damage by a vaccinator's needle (as in cases of introduction of live vaccinia virus to vaccinate against smallpox and monkeypox). The poxvirus vaccines are administered with a bifurcated needle that is poked several times into the skin to abrase the skin during the administration of the virus (see Chapter 14). Other vaccines are administered with standard needles. In rare cases, a person can contract the disease for which he or she has received a vaccination. Biting arthropods such as mosqui-

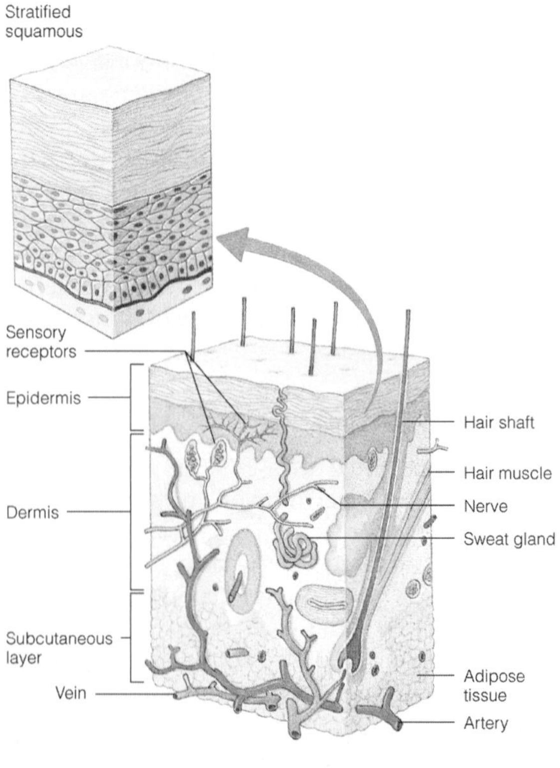

FIGURE 6-6 Diagram showing the layers and structures of the skin.

TABLE 6-5 Examples of Viruses That Enter Through the Skin

Virus	Route	Disease
Papillomavirus	Minor trauma	Warts, possibly cancer
Cowpox	Minor trauma	Vesicular or nodular lesions on milkers' fingers
ORF virus	Minor trauma	Lesions on hands, arms, or face of farmers, shepherds, veterinarians
Molluscum contagiosum	Minor trauma	Firm, donut-shaped flesh-colored bumps
Herpes simplex virus	Minor trauma	Herpetic lesions on face, genitals, fingers
Hepatitises B and C	Injection	Hepatitis, possibly cancer
HIV-1 and HIV-2	Injection	AIDS
Rabies	Animal bite	Acute encephalitis
LaCrosse encephalitis virus	Mosquito bite	Encephalitis
Dengue fever	Mosquito bite	Dengue fever and Dengue hemorrhagic fever
West Nile virus	Mosquito bite	Encephalitis
Colorado tick fever virus	Tick bite	Fever, occasional encephalitis, myocarditis, or tendency to bleed
Sand fly fever virus	Sand fly bite	Fever, nausea, and vomiting
Ebola virus	Injection	Hemorrhagic fever

toes, ticks, and sandflies penetrate the skin during feeding and can, therefore, introduce viruses into the body. Sometimes the mouthparts of the arthropod are contaminated with infectious virus and there is no multiplication of the virus in the arthropod. In other cases, the virus multiplies in the arthropod and is present in the saliva or in the feces of the arthropod. During feeding, the saliva is transferred to a susceptible human or animal host, or the animal defecates and the virus in the feces gains entry via the wound made by the feeding arthropod. The virus also may enter a blood capillary and thus be introduced into the vascular system.

The classical infectious disease transmitted by a biting animal is rabies. Virus is shed in the saliva of infected animals such as dogs, bats, raccoons, or skunks and is introduced into bite wounds. **TABLE 6-5** lists examples of viruses that enter through the skin.

Viral Entry via Transplants

An increasingly common way of acquiring viral infections is from solid organ (especially of the kidneys) or bone marrow transplants. Members of the general population are often persistent asymptomatic seronegative carriers of viruses such as Epstein-Barr and cytomegalovirus (CMV). These viruses create notorious problems for the immune-suppressed transplant recipient. Examples of transplant-related viral infections are listed in **TABLE 6-6**.

Transfusions and Blood Products

Before 1995, when the American Red Cross began screening donated blood, HIV, hepatitises B and C, and viral infections acquired via blood transfusions were more prevalent. Most of these infections were related to blood transfusions required during surgical procedures or contaminated blood products containing Factor VIII, a clotting factor needed by hemophiliacs. Advances in donor screening and blood testing have dramatically improved blood safety (**FIGURE 6-7**).

All blood donated at the American Red Cross Centers in the United States—approximately 50% of the nation's blood supply—is tested by National

TABLE 6-6 Examples of Viruses Associated with Transplants

Virus	Disease
Cytomegalovirus	Mononucleosis, disseminated infection
Epstein-Barr virus	Mononucleosis, Burkitt's lymphoma
HIV-1 and HIV-2	AIDS
Hepatitis B and C	Hepatitis, possibly cancer
HTLV-1	Adult T-cell leukemia-lymphoma
West Nile virus	Encephalitis
Lymphocytic choriomeningitis virus infection (LCMV)	Asymptomatic or mild disease (e.g., headache, fever, chills, muscle aches; rarely meningitis)
Rabies virus	Rabies (hydrophobia)

VIRUS FILE 6-1 Rabies Transmission: Human Rabies Caused by Tiny Bat Bites

a b

FIGURE VF 6-1 **(a)** Silver-haired bat and **(b)** Mexican free-tailed bat. The majority of human rabies cases in the United States are due to bat bites, most of which are generated by these two species of bats.

Rabies comes from the Latin *rabere,* which means "to rage or to rave." The Greeks used the term *lyssa* ("madness") for rabies.

Human rabies cases in the Unites States are very rare. Rabies virus is transmitted via a bite from an infected animal or human. The virus replicates in muscle cells and migrates to the neurons of the nerves near the bite. It can take the virus 12 to 180 days to travel from the peripheral to the central nervous system (CNS). Once the virus reaches the CNS, the person exhibits the classic signs and symptoms of rabies—difficulty swallowing, hydrophobia, and tremors. With the exception of the 2004 Wisconsin patient mentioned in Chapter 13, the disease at this point is fatal, and human victims usually die within 6 to 14 days.

In the past ten years, most of the fatal rabies cases in the United States have been caused by bat bites, and of these bites, most are caused by the silver-haired bat and the Mexican free-tailed bat (**FIGURE VF 6-1**). Bat bites feel like sharp needle jabs, but because a bat's teeth are small, its bite does not always leave a detectable mark on the skin. During 1995–2009, there were 23 cases of human rabies in the United States correlated with possible bat exposure. In the following *Morbidity and Mortality Weekly Report* case reports, bat bites were considered the probable cause: a 49-year-old man in California in 2000, a 69-year-old man in Wisconsin in 2000, a 71-year-old man in Texas in 1997, and a 13-year-old girl in Connecticut in 1995. None of the individuals reported being bitten by a bat; however, the victims were generally too ill to provide enough accurate information about what kind of contact they had with a bat or any animal, or were not diagnosed until an autopsy was performed. In some cases, family members thought the victims may have had contact with bats in their homes or, in the case of the Texas man, at a hotel one to two months prior to hospitalization.

In cases where a victim may not have been aware of being bitten by a bat, authorities must rely on the scant information that can be obtained from others. Any bats found in a room with an unattended child or a sleeping, intoxicated, or otherwise mentally impaired person should be caught and tested for rabies virus infection. The rabies incubation period is fairly long, so administering the vaccine immediately after a bite from a rabid animal can save the individual. In situations in which an individual cannot reasonably exclude the possibility of a bite exposure,

VIRUS FILE 6-1 Rabies Transmission: Human Rabies Caused by Tiny Bat Bites (continued)

postexposure prophylaxis should be given unless capture and testing of the animal in question yields a negative result.

References

Cases of Rabies in Human Beings in the U.S., by Circumstances of Exposure and Rabies Virus Variant, 1995–2009 http://www.cdc.gov/rabies/location/surveillance/human_rabies.html

Centers for Disease Control and Prevention. 1995. "Human rabies—Connecticut." *MMWR* 45(10):207–209.

Centers for Disease Control and Prevention. 1996. "Human rabies—Florida." 1996. *MMWR* 45(33):719–728.

Centers for Disease Control and Prevention. 1997. "Human rabies—Montana and Washington." *MMWR* 46(33):770–774.

Centers for Disease Control and Prevention. 1997. "Human rabies—Texas and New Jersey." *MMWR* 47(1):1–5.

Centers for Disease Control and Prevention. 1998. "Human rabies—Virginia." *MMWR* 48(5):95–97.

Centers for Disease Control and Prevention. 2000. "Human rabies—California, Georgia, Minnesota, New York, and Wisconsin." *MMWR* 49(49):1111–1115.

Centers for Disease Control and Prevention. 2002. "Human rabies—Tennessee." *MMWR* 51(37):828–829.

Fedler, H. M., Nelson, R., and Reiher, H. W. 1997. "Bat bite?" *Lancet* 350:1300.

Testing Laboratories (NTLs) located throughout the country. These laboratories are designed to adapt to rapidly changing technology and new scientific and medical advancements (**TABLE 6-7**). Continual vigilance is critical to protecting the blood supply from known pathogens and the emergence of new infectious agents. CMV testing is performed on blood designated for patients who require CMV-negative blood, including organ recipients and immune-suppressed patients.

Iatrogenic Induction

Infections resulting from iatrogenic induction (in other words, infections generated by a physician) are rare. There have been reports of rabies deaths after patients received corneal transplants and other organs from donors infected with rabies. There have also been a small number of iatrogenic cases of Creutzfeldt-Jakob Disease (CJD) in patients who received contaminated pituitary growth hormone, corneas or dura mater from human cadavers, or who were operated on with

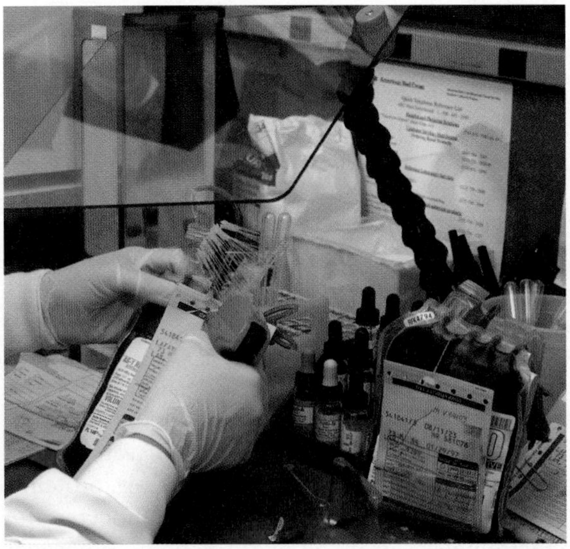

FIGURE 6-7 Donated blood is taken to a transfusion center where it must undergo testing for blood-borne pathogens.

TABLE 6-7 American Red Cross Viral Disease Testing

Virus	Test	Implemented	Discontinued
HIV-1	HIV-1 antibody test	1985	
HIV-1 and HIV-2	HIV-1/2 antibody test	1992	
HIV-1	HIV-1 p24 antigen test	1996	2003
HBV	Hepatitis B surface antigen	1971	
HBV	Hepatitis B core antigen	1987	
HCV	Anti-HCV	1990	
HTLV-1	HTLV-1 antibody test	1989	
HTLV-1 and HTLV-2	HTLV-1/2 antibody test	1998	
HCV and HIV	*NAT	1999	
West Nile virus	*NAT	2003	

*Nucleic acid testing (NAT)

contaminated neurosurgical instruments. CJD is a transmissible spongiform encephalopathy that is related to diseases caused by an unconventional agent called a **prion** (see Chapter 19, What About Prions and Viroids?).

6.2 Mechanisms of Viral Spread or Pathogenesis

The pathogenesis of many viral infections is not well understood because of our complex physiology. Humans have many specialized organs and tissues in which viruses proliferate, and in many cases viruses overcome natural barriers within the body, such as those that separate organs and tissues. Our knowledge is limited for several reasons. In some instances there is not a suitable animal model to study the pattern of viral spread. Some viruses, such as those that cause hemorrhagic fevers, must be studied in BSL-4 laboratories. Experimentation on dangerous pathogens is slow and deliberate, making extensive investigations difficult. Lastly, some viral pathogens cannot be grown in the laboratory, which severely limits our ability to study them. Success stories include our understanding of the pattern of viral spread for a number of childhood diseases—including measles, mumps, rubella, and poliomyelitis—through the teachings and research of pioneers of viral pathogenesis such as Frank Fenner, David Bodian, and Cedric Mims.

Replication and Infection Within the Host

After gaining entry into a potential host, virus replication may remain **localized** at or near the site of entry or spread to regional lymph nodes. This causes a **primary viremia**—the initial presence of the virus in the host's bloodstream—and results in the spread to other susceptible organs such as the liver and spleen. This latter type of spread to other organs is called a **systemic infection**. **TABLES 6-8** and **6-9** list examples of viruses that cause localized or systemic infections.

TABLE 6-8 Viruses That Cause Localized Infections

Virus	Primary Replication Site
Rhinovirus	Upper respiratory tract
Rotavirus	Intestinal epithelium
Papillomavirus (warts)	Epidermis

TABLE 6-9 Viruses That Cause Systemic Infections

Virus	1° Replication	2° Replication
Enterovirus	Intestinal epithelium	Lymphoid tissues, CNS
HSV-2	Urogenital tract	Lymphoid cells, CNS
HSV-1	Oropharynx (the throat, including the tonsils)	Lymphoid cells, CNS

Localized Viral Infections

A good example of a localized viral infection is the one caused by a papillomavirus. Papillomaviruses that cause warts enter the skin through an abrasion and infect cells in the basal cell layer of the epidermis. As the viral genes are expressed, the virus causes the host cell to become keratinized or thick. These cells move upward, toward the skin surface, resulting in a papilloma. Direct contact or abrasion of the papilloma can shed the virus, resulting in transmission and spread to others. In this case, papillomaviruses cause a localized infection and virus spreading is based on cell-to-cell contact.

Systemic Viral Infections

In addition to cell-to-cell contact, there are two main mechanisms for spread of viruses throughout the host: the bloodstream and the peripheral nervous system (PNS) (**FIGURE 6-8a**). The era of viral pathogenesis began with Frank Fenner's investigation of mousepox (also called infectious ectromelia), which causes a smallpox-like infection in mice.

From his experiments in 1948, Fenner was able to describe the sequential course of mousepox infection in mice. He injected mousepox into the epidermis (viral entry) of the footpads of mice and tracked the course of infection as a viremia spread to the liver and spleen. Within a few hours of mousepox injection, virus was detected in the regional lymph nodes of the mice. The entire lymphatic system flows toward the bloodstream. Its fine vessel network beneath the skin is present in all parts of the body except the central nervous system (CNS). The lymphatic system's function is to return fluids from body tissues to the blood. The lymph vessels drain **lymph**—a watery fluid that consists of water, salts, glucose, urea, proteins, and some immune cells such as lymphocytes and macrophages—from all over the body. Wherever there are lymph vessels, there are blood vessels (Figure 6-8b). The two systems work together.

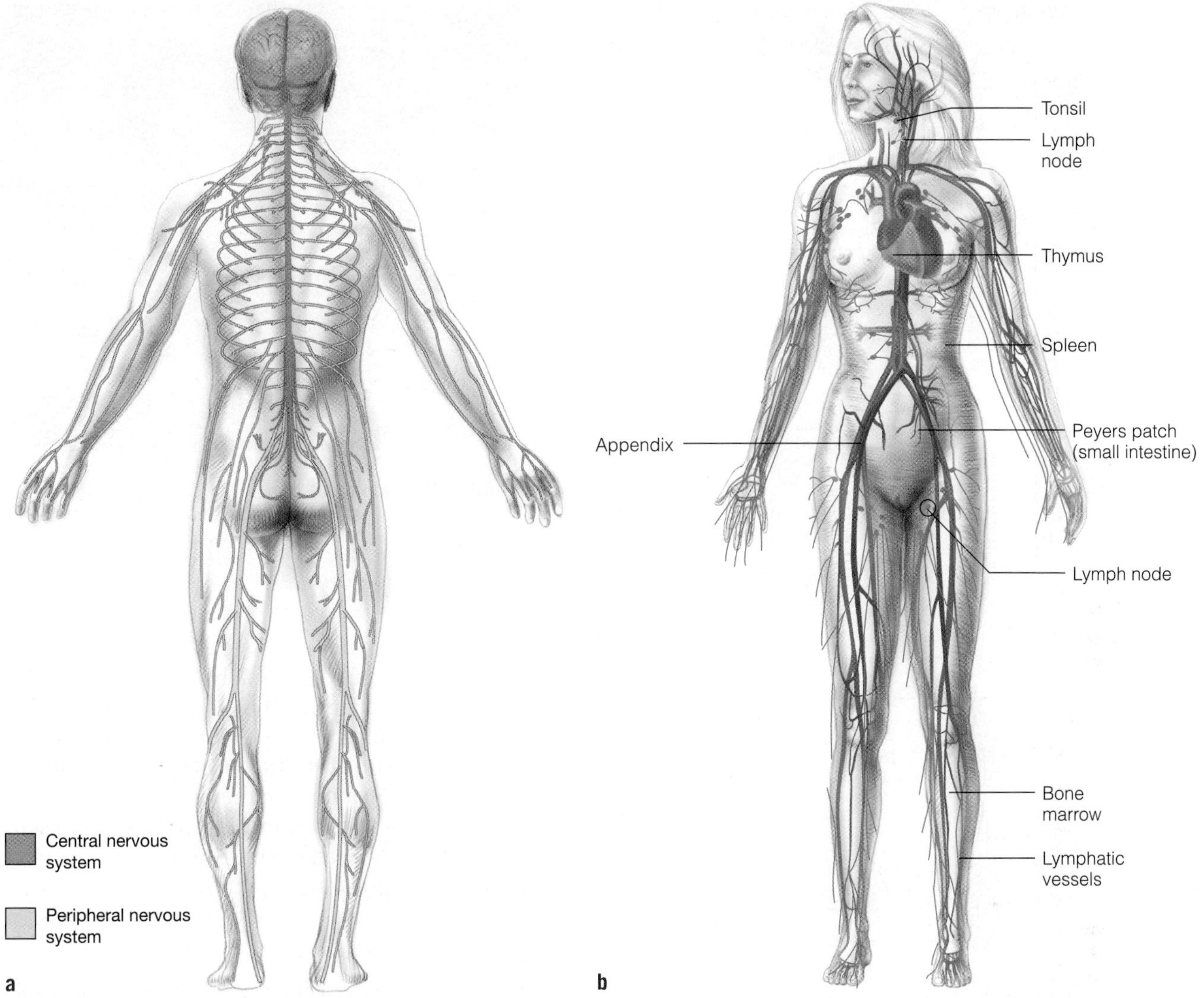

FIGURE 6-8 **(a)** Schematic distinguishing the peripheral nervous system from the central nervous system. Adapted from Morphonix, LLC, "Brain Specimens," *Journey into the Brain*, http://www.morphonix.com/software/education/science/brain/game/specimens/nervous_system.html. **(b)** The lymphatic system.

The mousepox viruses in Fenner's study multiplied in the lymphoid cells of the nodes and entered the bloodstream to give a primary viremia, in which the spleen and liver were infected with virus. Prolonged viremia resulted in more spillover of virus into the bloodstream, leading to the infection of other internal organs (such as the intestines, pancreas, and salivary glands) and to the focal infection of the skin, thus causing a rash. Mousepox pathogenesis is an example of a systemic viral infection (**FIGURE 6-9**).

Not all infections result in a systemic infection. Within the lymph nodes present near the site of viral entry, viruses may be engulfed by macrophages (a type of immune cell) and inactivated, or the viruses may enter the bloodstream (primary viremia). Typically an infected individual develops symptoms of **malaise**—a vague feeling of bodily discomfort at the beginning of an illness—and fever during primary viremia.

If viruses enter the blood via the lymphatic system, biting arthropod vectors, blood transfusions, or intravenous drug abuse, they may travel free in the plasma or in association with red blood cells, platelets, lymphocytes, or monocytes. *The blood is the most effective and rapid vehicle for the spread of virus through the body.* From the blood, the virus gains access to the liver, spleen, and bone marrow, where the virus replicates again. High titers of virus are produced and many reenter the bloodstream (secondary viremia). Some viruses, such as hepatitis B, circulate in the blood. More often, the virus replicates inside of lymphocytes or macrophages before it reaches its target organ (the liver)—an **incubation period** that averages two weeks for these types of viral infections. (An incubation

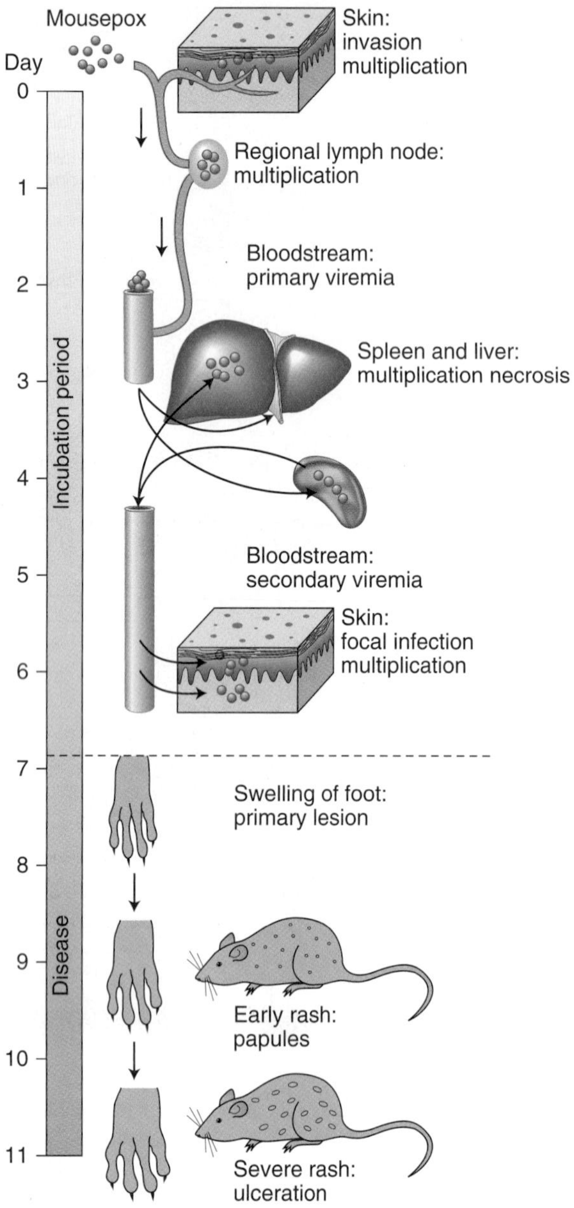

FIGURE 6-9 The dynamics of mousepox pathogenesis. Reprinted from *The Lancet*, 262, Fenner, F., *The pathogenesis of the acute exanthems. An interpretation based on experimental investigations with mousepox*, 915–920. Copyright 1948, with permission from Elsevier.

period is defined as the period between infection and the appearance of symptoms of the disease.)

Important Target Organs

Skin

The skin is involved in many viral systemic infections. Rashes (exanthems) are produced after viruses have invaded the skin via blood vessels. Rashes have been described as macules, papules, vesicles, or pustules. A macule is a patch of skin that is discolored but not usually elevated (**FIGURE 6-10**). Papules are small, solid, usually inflammatory (red) elevations of the skin that do not contain pus. Vesicles are blisters on the skin containing fluid, ranging up to one centimeter in size (**FIGURE 6-11**). Pustules are elevations on the skin containing pus. **TABLE 6-10** lists some examples associated with skin rashes. A rash that affects mucous membranes may accompany some viral infections.

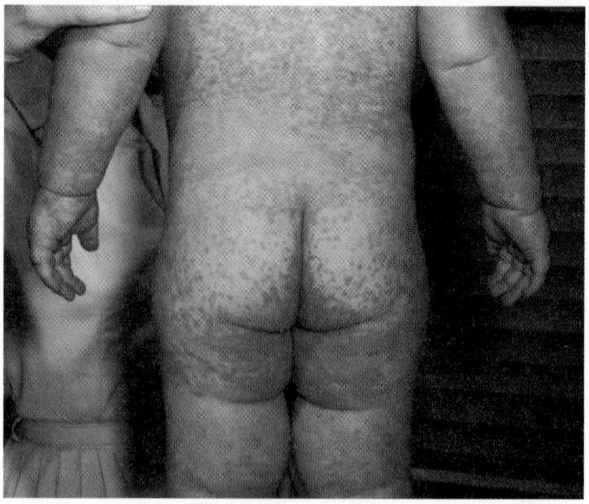

FIGURE 6-10 Rash caused by rubella virus (German measles). This child has the characteristic red blotchy rash on his buttocks and back during the third day of the rash.

Lungs and Liver

Most respiratory infections are caused by the localized spread of respiratory viruses such as rhinoviruses (the viruses that cause the common cold). Sometimes a virus that can cause a systemic infection may infect the lungs; for example, both measles and varicella zoster viruses (the latter of

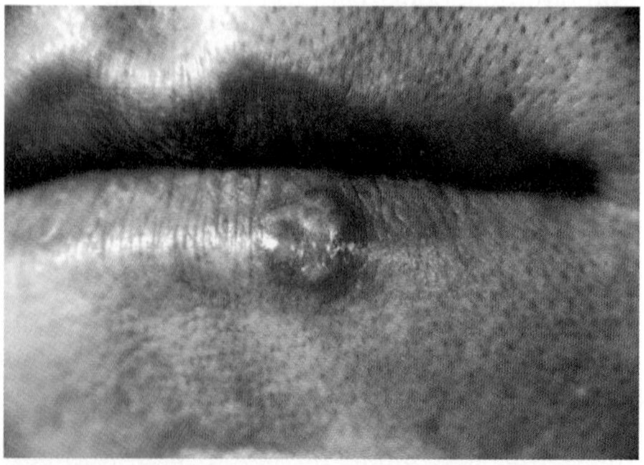

FIGURE 6-11 Cold sore (vesicular lesion) caused by herpes simplex virus type 1.

TABLE 6-10 Examples of Viruses Associated with Skin Rashes (Exanthems)

Virus	Type of Rash
Rubella (German measles)	Macular (Figure 6-10)
Dengue fever virus	Macular
Measles	Maculopapular
Herpes simplex virus	Vesicular (Figure 6-11)
Herpes zoster (causes chickenpox and shingles)	Vesicular
Coxsackie virus	Vesicular
Enterovirus	Vesicular
Variola (causes smallpox)	Vesicular

which causes chickenpox and shingles) may cause pneumonia in patients whose immunity is compromised. Almost any organ may be infected via the bloodstream with some kind of virus. Clearly, viruses that infect the lungs, brain, heart, and liver (causing pneumonia, encephalitis, myocarditis, and hepatitis) are the most life-threatening viral infections.

Viruses that infect the liver, such as the hepatitis viruses, can cause severe hepatitis. This could potentially result in a life-threatening situation. Hepatitis B and C infections can lead to chronic carrier-state infections that may eventually result in hepatocellular (liver) cancer. Besides the hepatitis viruses, other viruses that produce systemic infections—such as yellow fever, rubella, mumps, CMV, Epstein-Barr, and herpes simplex viruses—may damage the liver.

Only a few viruses infect the kidneys or heart. CMV and hantaviruses can infect the kidneys. Coxsackie B viruses may infect the heart and are also implicated as a cause of juvenile-onset insulin-dependent diabetes mellitus type-1 (IDDM1), which results from the destruction of pancreatic beta cells. The development of diabetes in twins is only about 40%, suggesting an environmental factor plays a role in this disease. To date, 14 different viruses have been reported to be associated with the development of IDDM1 in humans and animals.

Neurotropic Viruses

Neurotropic viruses are viruses that infect the nervous system. In reality, though, most neurotropic viruses infect other types of cells as well; for example, herpes simplex virus types 1 and 2 can replicate in epidermal cells as well as macrophages of the dermis. The first neurons to be infected are located in the peripheral nervous system (PNS). Poliovirus, herpes simplex virus types 1 and 2, varicella zoster, and rabies viruses are examples of neurotropic viruses. (In fact, rabies virus spreads only through a neural route.) Once in the PNS, the viruses can spread to other neurons to the spinal cord and to the brain with devastating results. **FIGURE 6-12** illustrates an infection by varicella zoster. Herpes zoster virus replicates in a variety of cell types as well as in neurons.

CNS Entry: Crossing the Blood-Brain Barrier

In 1885, bacteriologist Paul Ehrlich used aniline dyes to stain and thus visualize fine structures of cells. He discovered that when he injected aniline dye into an animal, all of the internal organ systems

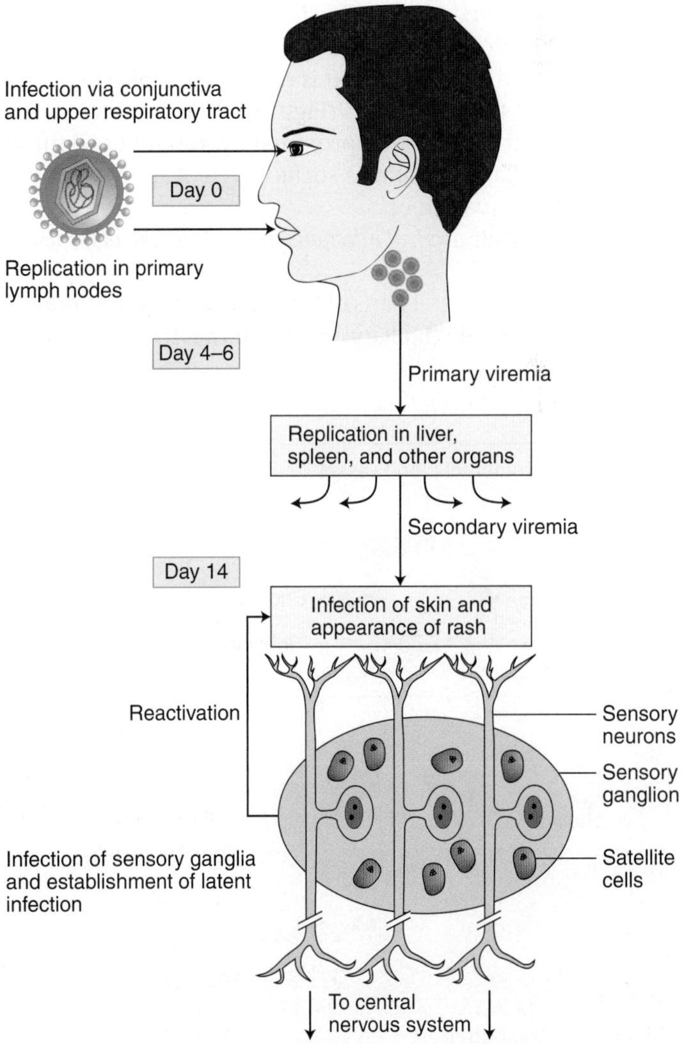

FIGURE 6-12 Pathogenesis of varicella zoster virus. Adapted from Flint, S. J., et al. *Principles of Virology: Molecular Biology, Pathogenesis, and Control of Animal Viruses*, Second Edition. ASM Press, 2003.

except the brain would stain. Later, in 1913, his graduate student Edwin Goldwin injected trypan blue dye directly into the cerebrospinal fluid of rabbits and dogs and discovered that only the brain became stained. The combination of these two experiments demonstrated the existence of a "barrier" between the brain and the other internal organs—the **blood-brain barrier.**

The blood-brain barrier exists at the interface between the bloodstream and CNS tissues. It is semi-permeable. Most large molecules, including toxins and viruses are prevented from crossing it. Thus, the blood-brain barrier protects the brain from potentially toxic substances or microbes. It also makes it difficult for therapeutics used to treat chemical imbalances to cross this barrier. The blood-brain barrier does have transport systems that move the substances the brain needs across the barrier and into brain tissue (**FIGURE 6-13**).

The anatomical basis of the blood-brain barrier was deciphered using electron microscopy in the 1960s. It revealed a vast capillary network surrounding the brain. It is made up of epithelial cells with tight junctions (Figure 6-13a). In the human brain, there are approximately 400 miles of these capillaries totaling a surface area of approximately 100 square feet.

How do viral infections spread from the blood to the brain?

In several parts of the brain such as the choroid plexus, the capillary endothelium is fenestrated (has window-like openings) and the basement membrane is sparse (Figure 6-13b). During viremia, mumps and some togaviruses pass through the capillary epithelium into ™the choroid plexus and then enter the cerebrospinal fluid that is produced by the choroid plexus. Other viruses (e.g., poliovirus and some togaviruses) may directly infect or be transported across the capillary endothelium. Some viruses such as HIV and measles virus cross the endothelium within infected monocytes or lymphocytes.

Neural Entry

Viruses gain access to the CNS via three different routes: via the nerves connecting the PNS of the body to the CNS, via the olfactory nerves, or via the bloodstream during viremia (discussed earlier). Some viruses can enter via one or more of these routes. There is limited knowledge available as to how viruses move in and among the cells of the CNS.

The structure of neurons is much different than that of other cells. They are polarized, possessing two structurally distinct ends: the axon and the dendrites (**FIGURE 6-14**). Neurons transmit electrochemical signals over long distances up to several feet and pass messages to each other. Protein synthesis does not occur in the axon of the neuron; hence, the virus needs to travel long distances to reach its site of replication. For example,

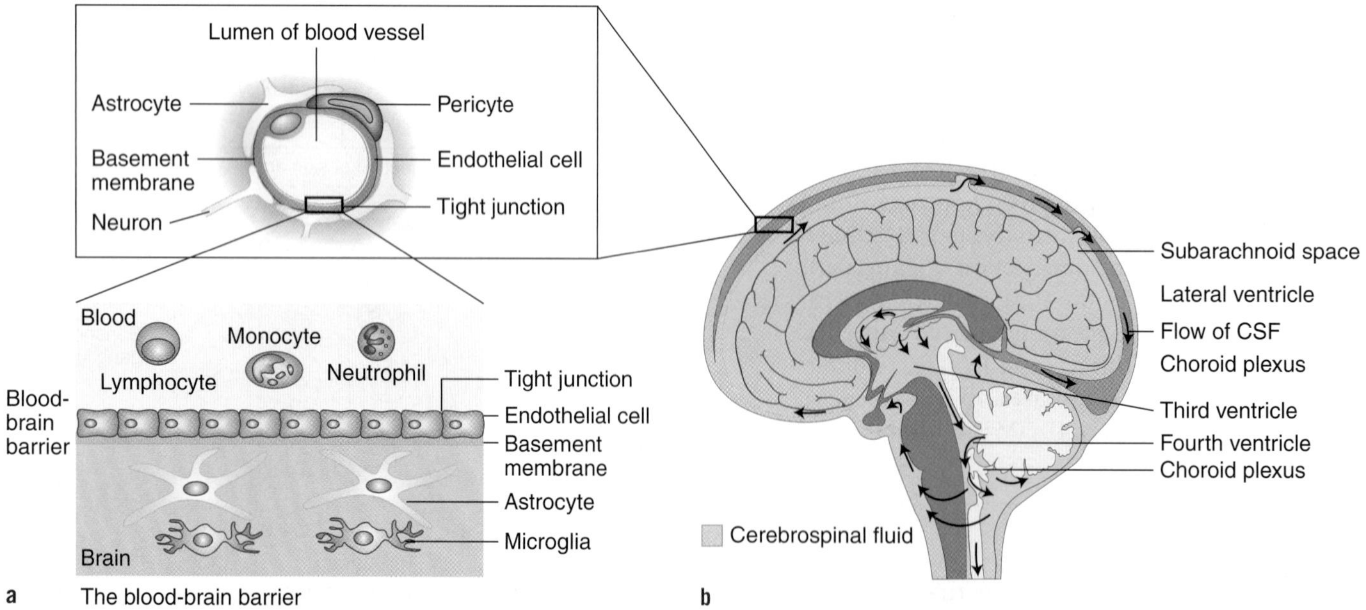

FIGURE 6-13 Cross-section through the blood-brain barrier. **(a)** Epithelial cells tightly line the capillaries of the brain, forming the blood-brain barrier between the astrocytes and microglia of the brain. Adapted from K. Francis, et al., *Expert Rev. Mol. Med.* 5 (2003): 1–19. **(b)** Vertical cross-section of the brain showing the choroid plexus, which produces the cerebrospinal fluid (CSF) that is found within the ventricles of the brain and in the subarachnoid space around the brain and spinal cord.

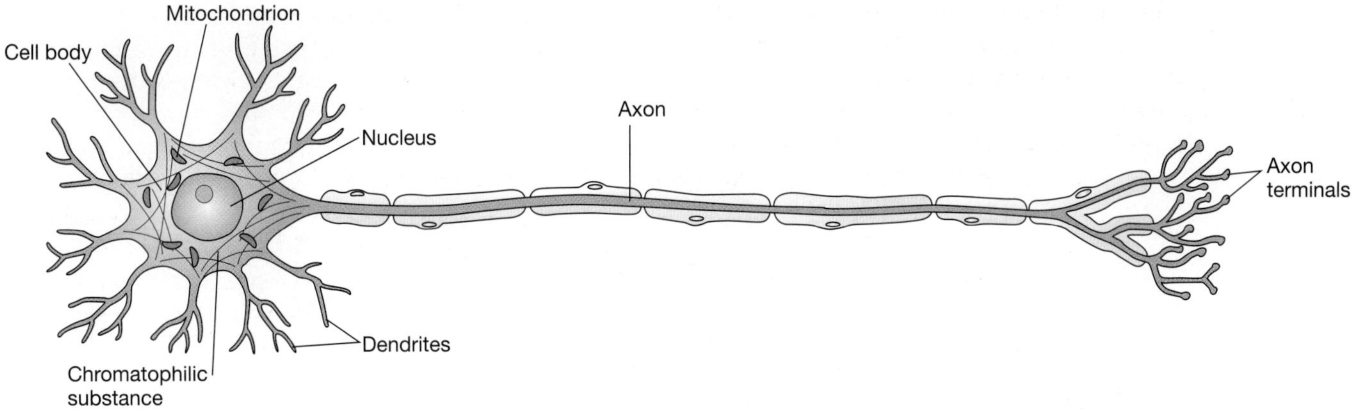

FIGURE 6-14 Structure of neurons. Neurons have three basic parts: the cell body, the axon, and the dendrites or nerve endings. The cell body contains the major portion of the cell, which includes the dendrites and nucleus. Axons transmit the signals. Viruses can only replicate in the cell body because that is where the cellular protein synthesis machinery is located.

in the adult human, axons that control the stomach muscles can be as far as 20 inches away from the cell bodies and dendrites in the brain stem.

Viral Infections and Pregnancy

Viral infections during pregnancy can develop by three different mechanisms: **transplacental**, **perinatal**, or **postnatal** transmission. Transplacental transmission occurs when viruses may cross the placenta, infecting the fetus. These viruses are passed *in utero*. Perinatal transmission occurs during labor or vaginal delivery. Viruses present in vaginal secretions or blood infect the newborn baby during natural childbirth. Postnatal transmission occurs when the newborn is infected through breast milk, through transfused blood, by hands or instruments, or by the respiratory (airborne transmission) route from infected contacts.

The **placenta** is an organ that serves as a protective interface between the mother and the developing fetus. The fetus is connected by the umbilical cord to the placenta. It is involved in the exchange of nutrients and waste products and gas exchange between the blood supply (circulatory system) of the mother and fetus through the blood vessels of the umbilical cord (**FIGURE 6-15**). The placenta keeps the fetus separate from the mother's blood supply and provides a link between the two that allows the placenta to carry out functions that the fetus cannot perform by itself. A thin membrane called an amniotic sac containing amniotic fluid surrounds the fetus (**FIGURE 6-16**).

Women may experience infections during pregnancy caused by a variety of pathogens. Often the infections pass unnoticed. Surprisingly, only a few viruses infect the developing embryo and fetus resulting in severe damage. Severe infections of the fetus can result in spontaneous abortion, stillbirth, and neonatal death. These consequences were a common result of smallpox infections. Less lethal viruses, such as rubella viruses and CMV, cause severe congenital malformations (**teratogenic** effects).

The majority of severe transplacental infections are caused by CMVs, parvovirus B19, and rubella viruses (**TABLE 6-11**). Less frequently, effects on a developing fetus are caused by measles, herpes simplex viruses, lymphocytic choriomeningitis viruses, varicella zoster, enteroviruses, mumps, adenoviruses, hepatitis viruses, and HIV (Table 6-11). The availability of vaccines against some of these viruses has reduced the adverse effects on pregnancy.

The placenta is a good barrier against the majority of viruses. Why is it that only a few viruses cross the placenta? Very little is known about how viruses cross the placenta. Size is not a factor because CMV is a large virus (230 nanometers in diameter) and parvovirus B19 is a smaller virus (22–24 nanometers in diameter). Both efficiently cross the placenta. Most viruses disseminate through the bloodstream. The majority of transplacental infections of the fetus occur when the mother is **viremic**. However, some viruses like Epstein-Barr virus (EBV) are commonly found in maternal blood, yet EBV rarely crosses the placenta. There must be one or more virus-specific factors or host changes or factors during pregnancy that allow certain viruses to cross the placenta.

The placenta is created from the embryo. Therefore, the fetus and placenta are *foreign tissues* within the mother's body. There are many unanswered questions about maternal immunity during pregnancy. At one time, it was thought that the placenta was immunologically inert. In other words,

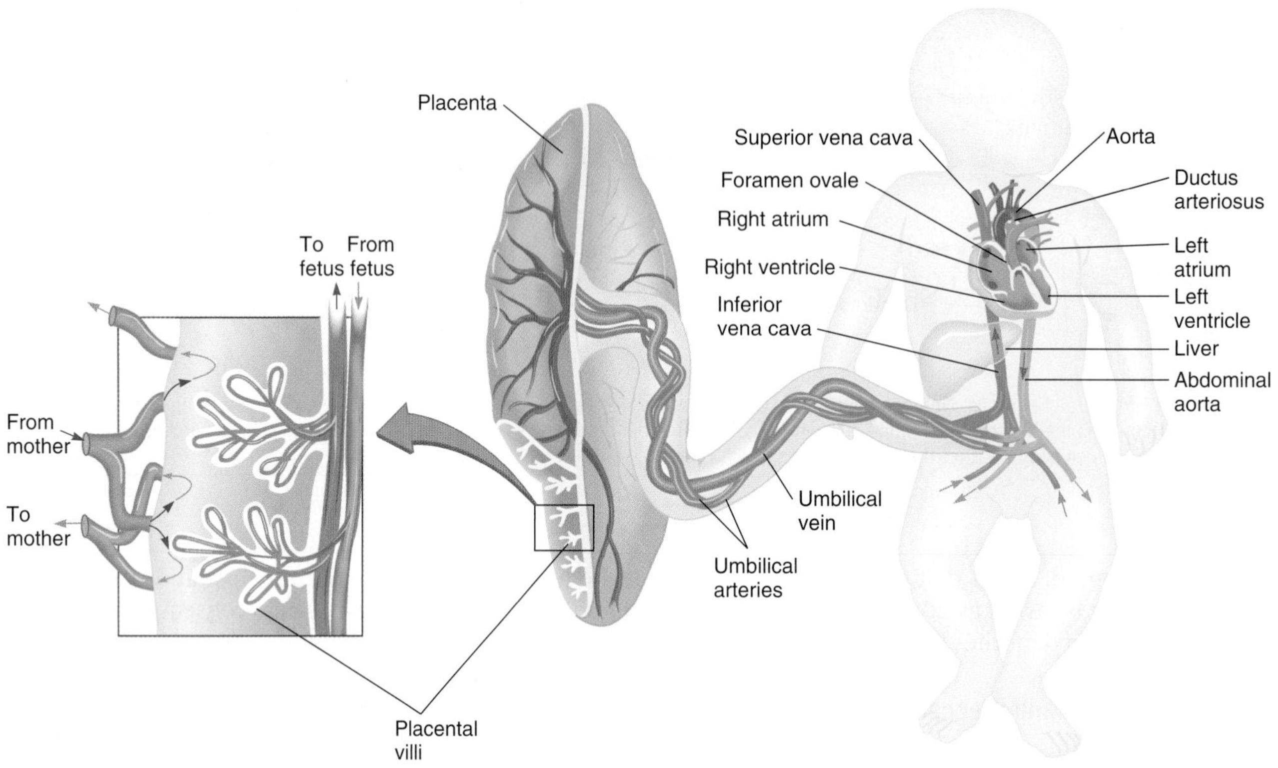

FIGURE 6-15 Detailed anatomy of the placenta showing gas exchange within the placenta.

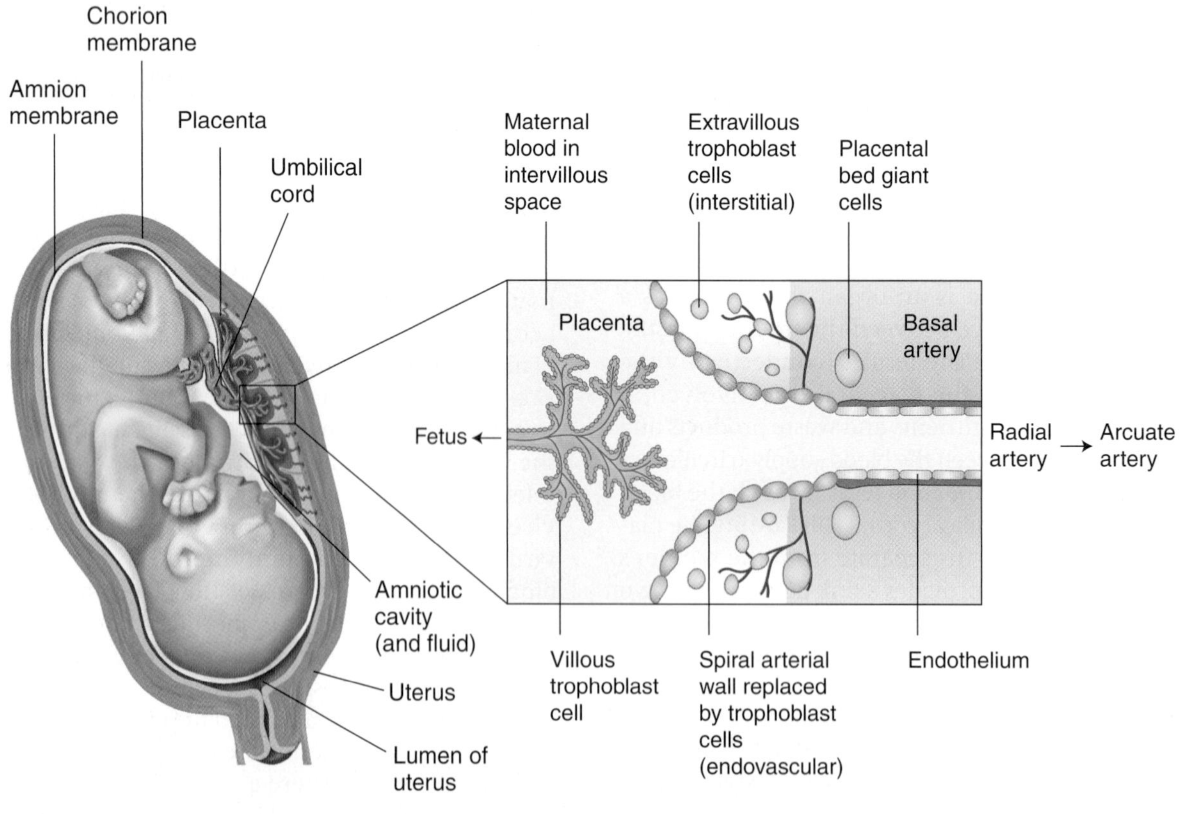

FIGURE 6-16 Normal pregnancy showing the maternal and fetal sides of the placenta.

TABLE 6-11	Selected Viral Infections and Pregnancy	
Virus	**Mode of Transmission**	**Effects on Fetus/Comments**
Variola virus (causes smallpox)	Transplacental Postnatal	Retrospective studies estimate 34% of pregnant women suffering from smallpox infections died and 40% of women miscarried or had premature births. Vaccination before pregnancy reduced the risk for death.*
Cytomegalovirus (CMV)	Transplacental Perinatal Postnatal	Most common congenital viral infection. Overall risk of infection is greatest during third trimester (transplacental transmission). 30% of infants with severe infection die; among survivors, more than 50% develop neurological problems—e.g., microcephaly, hearing loss, and **intellectual disability.** CMV is the leading congenital viral infection in the U.S. with an incidence of 1%–3% of live births, resulting in about 8,000 infants with congenital effects each year.
Herpes simplex virus-1 and -2 (HSV-1, HSV-2)	Transplacental (10% of cases) Perinatal (90% of cases; transmitted in birth canal) Postnatal**	HSV-2 causes 70% of neonatal cases.
Rubella	Transplacental Postnatal	Congenital infection causes severe problems in developing fetus if pregnant mother is infected before 20 weeks of gestation. It is vaccine-preventable. The most common congenital defects are cataracts, heart abnormalities, deafness, and intellectual disability.
Parvovirus B19	Transplacental Postnatal	30%–40% of pregnant women are seronegative for this virus and are therefore susceptible to infection. Effects on fetus include miscarriage, **hydrops fetalis** (serious abnormal fluid build up), fetal anemia, **myocarditis**, intrauterine fetal death.
Varicella zoster	Transplacental Postnatal	About 90% of women in the U.S. are immune before they become pregnant. If a pregnant woman is infected before the first 20 weeks of pregnancy, about 1%–2% of cases result in congenital varicella syndrome. The prognosis is severe if an infant is infected. Acyclovir is safe to use during pregnancy.
Enteroviruses (e.g., echoviruses and coxsackie viruses)	Transplacental? Perinatal Postnatal	These viruses may be transmitted to the fetus during delivery in 30%–50% of mothers. The rate of transplacental transmission is unknown. Mothers infected with coxsackie B viruses during pregnancy may increase the rate of heart anomalies in fetus. Coxsackie virus infections during pregnancy may result in an increase in the rate of insulin-dependent diabetes mellitus (IDDM1) in the fetus/child. A few rare reports describe fetal damage and death. There are insufficient data to correlate echoviruses with deleterious effects during pregnancy.
Measles	Transplacental Postnatal	Measles infection during pregnancy can be severe and has been associated with spontaneous abortion, premature labor and low birth rate. No congenital syndrome has been described. Neonates born to mothers with an active infection are at risk of developing neonatal measles.
Lymphocytic choriomeningitis virus (LCMV)	Transplacental Perinatal Postnatal	If a woman has an LCMV infection during pregnancy, the unborn baby can also become infected. LCMV infection can cause severe birth defects (e.g., hydrocephalus, visual impairment, or intellectual disability, or loss of pregnancy [miscarriage]).
Adenovirus	Transplacental Perinatal Postnatal	Adenoviruses have been isolated from amniotic fluid and are associated with severe neonatal infections, chorioamnionitis (inflammation of fetal membranes), and preterm birth. Detection of adenoviruses in the placenta is correlated with seasonal infections by adenoviruses.
Papillomavirus (PV)	Transplacental Perinatal (via birth canal) Postnatal	A recent study demonstrated placental infection in 23% of cases and transplacental transmission in 12% of cases.*** The following HPV types were detected in maternal, newborn, and placental samples: 6/11, 16, 42, 54, 31, 58, 18, 51 59, 34, 35, 70, 26, 33, 73, 68, 45.

(continued)

TABLE 6-11 Selected Viral Infections and Pregnancy (continued)

Virus	Mode of Transmission	Effects on Fetus/Comments
Influenza A	Transplacental (rare) Perinatal (rare) Postnatal (rare)	Mortality rates among pregnant women in the pandemics of 1918 and 1957 pandemic strains appeared to be abnormally high. In 2009, pregnant women suffering from the H1N1 virus had a greater risk of severe illness. There is an increased risk for schizophrenia neurodevelopmental disorders in offspring of mothers exposed to influenza during pregnancy. Pregnant women are given priority to receive influenza vaccination. Pregnant women infected with influenza A may receive antiviral treatment.
West Nile virus (WNV)	?	A rare case is reported in the literature in which a mother was infected during the 27th week of pregnancy. The child was delivered at term but had brain lesions and chorioretinal scarring. It is not clear if WNV caused these abnormalities.
Hepatitis B virus (HBV)	Transplacental Perinatal	Common in countries where there is a high percentage of people infected with HBV. HBV can cause fetal hepatitis. There is no risk of congenital anomalies in children of women infected with HBV during pregnancy. Transmission of HBV to the fetus can be reduced by immunization or immunoglobulin administration immediately after birth. Vaccination and injection of mother with immunoglobulin seems to be safe to the fetus.
Hepatitis C virus (HCV)	Transplacental (rare) Perinatal (rare)	About 4% of transmission from mother to child during pregnancy, childbirth, or breastfeeding. The rate of transmission is directly proportional to the level of viremia in the mothers. Mothers who did not transmit the HCV had low viral yield. No specific damage to neonates has been described.
Hepatitis E virus (HEV)	Transplacental Perinatal	Transmission from mother to fetus is high (over 50%; virus is present in cord blood). There are limited data on its complications of pregnancy. Pregnant women with HEV often develop fulminant hepatitis and sometimes liver failure.
Human immunodeficiency virus (HIV-1)	Transplacental Perinatal Postnatal	About 100–200 infants in the U.S. are infected with HIV-1 annually. Many of these infections involve women who were not tested early enough in pregnancy or who did not receive prevention services. Antivirals reduce the risk of mother-to-child transmission by lowering the plasma viral load in pregnant women or through postexposure prophylaxis in their newborns. In rich countries, highly active antiviral therapy (HAART) has reduced transmission from mother to fetus from 25% to around 1%–2%. HAART is not available to low- and middle-income countries.

*Nishiura, H. 2006. "Smallpox during pregnancy and maternal outcomes." *Emerg Infect Dis* 12(7):1119–1121.

**Infection occurs within 28 days of birth.

***Rombaldi, R. L., et al., 2008. "Transplacental transmission of human papillomavirus." *Virol J* 5:106–119.

the mother's immune system was compromised so that it would not mount an immune response to reject the *foreign* placenta or fetus. The placenta villi are covered by **villous trophoblasts** that lacks both the typical **Major Histocompatibility Complex (MHC) class I and II molecules** (Figure 6-16). This layer of trophoblasts acts as a boundary separating the mother's uterine epithelium from the developing embryo and fetus.

MHC I are molecules present on all nucleated cells, and MHC II are present on **professional phagocytes**. Professional phagocytes contain receptors on their surface that detect antigens normally not found in the body (e.g., pathogens, transplanted or foreign tissues). MHC molecules are used to identify an individual as unique, distinguishing self from nonself by presenting foreign antigenic peptides to cytotoxic T lymphocytes. The cytotoxic T lymphocytes recognize foreign protein antigens in association with MHC I molecules and then kill foreign or infected cells. This mechanism is a key component of the host defenses as well as immune histocompatibility of transplanted tissue (see Chapter 7: Host Resistance to Viral Infections). During a normal pregnancy, placental trophoblast cells are not recognized by cytotoxic T lymphocytes as being foreign.

The placental **extravillous trophoblasts** *that invade the uterus do contain an unusual combination of MHC 1-like molecules* (Figure 6-16). There is a symbiotic relationship between the mother and fetus in which a compromise between placental extravillous trophoblasts invading into the maternal uterus and maternal immune resistance allows the temporary coexistence of the two individuals. It appears that the maternal **cell-mediated** arm of immune defense changes (e.g., less type 1 cytokines are produced) while retaining normal **humoral immunity** mediated by **antibodies** (see Chapter 7). During pregnancy, the mother maintains preexisting antibodies to protect her and the fetus from infections. IgG antibodies can pass through the human placenta and provide protection to the fetus *in utero* (refer to Chapter 7). So even though pregnant women are not immunosuppressed in the classic sense, immunity changes during pregnancy, making the mother more susceptible to infection by certain pathogens, including viruses.

Viruses entering the placenta will encounter macrophages called **Hofbauer cells** believed to prevent viral transmission of pathogens from mother to fetus. How is it that some viruses can overcome these defenses to cross the placenta? If we reexamine the placenta in Figure 6-16, note the blood flow to the uterus is redirected through the placenta to support fetal development. The placenta anchors the fetus to the uterus. Maternal blood enters the **intervillous space**. It is possible that a virus may enter the intervillous space and infect the layer of actively dividing villous trophoblasts that line the intervillous space, allowing the virus to invade the blood supply of the fetus. This means that viruses must attach to a receptor protein present on the outside of the villous trophoblasts. Human CMV is the most common intrauterine viral infection. Studies have shown that CMV infects the uterine wall and the adjacent placenta. The exact cellular receptors that CMV utilizes for entry is unknown. It has been suggested that more than one receptor may be involved as trophoblasts differentiate during pregnancy. Potential receptors include epidermal growth factor receptors and **integrins**.

There is still much to be learned about immune status changes during pregnancy, the key events of development, structure, and function of the human placenta and its relationship to virus transmission and replication. There are challenges to this type of research. For example, villous trophoblast developmental biology lacks an animal model system, and conventional cell culture is not possible. Research involves the analysis of biopsies from human placental tissues (availability of placental tissue from pregnant women is limited) or villous explants (placental tissues grown on **collagen** or **matrigel** surface; these trophoblasts grow for about 5 days, in contrast to conventional cell culture in which cell lines can be grown indefinitely). Human trophoblasts can be obtained from embryonic stem cells to study early stages of placental development. This type of research has required private funding and the use of discarded embryos from fertility clinics. Many outstanding comparative **placentologists** have not continued their investigations because of a lack of funding for this kind of research. Modern-day placentologists engaged in research are now taking advantage of confocal and electron microscopy to elucidate what molecules pass between the maternal and fetal tissues of the human placenta and that of other species at the ultrastructural level.

6.3 Patterns of Diseases

There are four basic patterns of viral infections; the specific variations among them will be discussed in detail in later chapters. The four patterns are:

- acute, nonpersistent infection
- acute infection, followed by persistent latent infection
- chronic infection with continuous shedding of virus
- slow infection

Most **acute viral infections** are self-limiting. These diseases have symptoms of sudden onset or last a short time as with rhinoviruses, which cause the common cold, as opposed to a chronic infection in which virus is continually shed and the disease symptoms may or may not be present. **Persistent infections with acute onset** are characteristic of DNA viruses. The chronic state of infection occurs due to a phenomenon termed **latency**. A latent infection is dormant or "concealed," and infectious virus particles are not detected unless the virus is reactivated.

Latency is the hallmark of herpesviruses. Virtually 100% of adults have antibodies in their serum against herpes simplex virus type 1, the causative agent of cold sores. Most individuals become infected in the first few years of life and the virus establishes lifelong latency and may be reactivated at any time. This persistent latent state occurs because the viral DNA remains present in the form of an **episome** (a circular molecule of DNA, independent of the host chromosome) in the nucleus of neurons. (Read more about herpesviruses in Chapter 15, Herpesviruses.) Reactivation of latent

TABLE 6-12 Examples of Latent Infections Caused by Viruses

Virus	Primary Disease	Reactivated Disease	Site of Latency
Herpes simplex virus-1	Oral herpes/cold sores	Recurrent herpes simplex	Neurons in the sensory ganglia
Herpes simplex virus-2	Genital herpes	Recurrent genitalis	Neurons in the sensory ganglia
Varicella zoster	Chickenpox	Shingles	Neurons in the sensory ganglia
Epstein-Barr virus	Mononucleosis	Burkitt's lymphoma, nasopharyngeal carcinoma	B lymphocytes
Cytomegalovirus	Subclinical (except in fetus and immunocompromised)	Subclinical (except in fetus and immunocompromised)	B lymphocytes
Hepatitis B	Hepatitis	Liver cancer	Hepatocytes
Papillomavirus	Warts	Warts and carcinomas	Epithelium (skin)
HIV-1 and HIV-2	Seroconversion illness	AIDS and AIDS dementia	T lymphocytes, macrophages, brain cells

herpesviruses can be lethal in immunocompromised individuals (for example, in AIDS or organ-transplant patients). Examples of latent viral infections are listed in **TABLE 6-12**.

A handful of virus pathogens are **chronic persistent infections**, which are characterized by continuous virus production that lasts at least six months. Hepatitises B and C are the most important chronic viral infections of humans. The virus particles of hepatitises B and C are found in plasma, genital secretions, and saliva. Approximately 20% to 40% of individuals chronically infected with hepatitis B will develop chronic liver inflammation. Over a period of years this chronic inflammation leads to cirrhosis (scarring) of the liver and liver failure, or to hepatocellular carcinoma (liver cancer). An effective vaccine is now available to prevent hepatitis B virus infection. Unfortunately, vaccines cannot help those who are already chronically infected (**FIGURE 6-17**).

Hepatitis C infection is the most common cause of chronic hepatitis, and there is no vaccine for it. Approximately 85% of acute infections become chronic. As with hepatitis B, infections with hepatitis C can result in liver inflammation, cirrhosis, and hepatocellular carcinoma. In the United States, hepatitis C is the leading specific cause of liver failure requiring liver transplantation (Figure 6-17).

Slow infections are those that have long incubation periods (measured in years). These types of infections lead to slow, progressive diseases. HIV and another group of unclassified infectious agents called *prions* fall into this category. Prions cause transmissible spongiform encephalopathies (TSEs) such as bovine spongiform encephalopathy (BSE), a fatal brain disease in cattle commonly referred to as mad cow disease. There is a similar disease to BSE in humans called new variant Creutzfeldt-Jakob Disease (vCJD).

There is strong laboratory and epidemiological evidence that vCJD is associated with BSE. It is believed that people become afflicted with the vCJD form by eating contaminated beef products from BSE-infected cows. This disease was first diagnosed in the United Kingdom, where the majority of cases occur, but cases exist worldwide. Many of the cases in other countries, though, are in individuals who resided in the United Kingdom during its BSE outbreak between 1980 and 1996. Those at increased risk for vCJD infection are individuals who lived in the United Kingdom during this time. **TABLE 6-13** lists definite and probable CJD cases in the United Kingdom through August 2010.

The United States reported its first case of vCJD in 2002. The patient was born in the United Kingdom and moved to Florida in 1992. It is believed that she consumed BSE-contaminated food before moving to the United States.

6.4 Virus Exit: Shedding

How do viruses get from one host to the next? How are virus infections maintained in populations? Viruses are in general shed from their routes of entry, usually body openings or surfaces. These are the primary infection sites. With the exception of the CNS, viruses can be shed effectively from all other main parts of the body; for example, viruses that invade the respiratory tract are shed in mucus and saliva during coughing, sneezing, and talking. Viruses that infect the gut are shed in the feces. Sexually transmitted viruses are shed via semen, cervical secretions, breast milk, and saliva.

Viruses that enter the skin through abrasions and direct contact of lesions are shed via direct

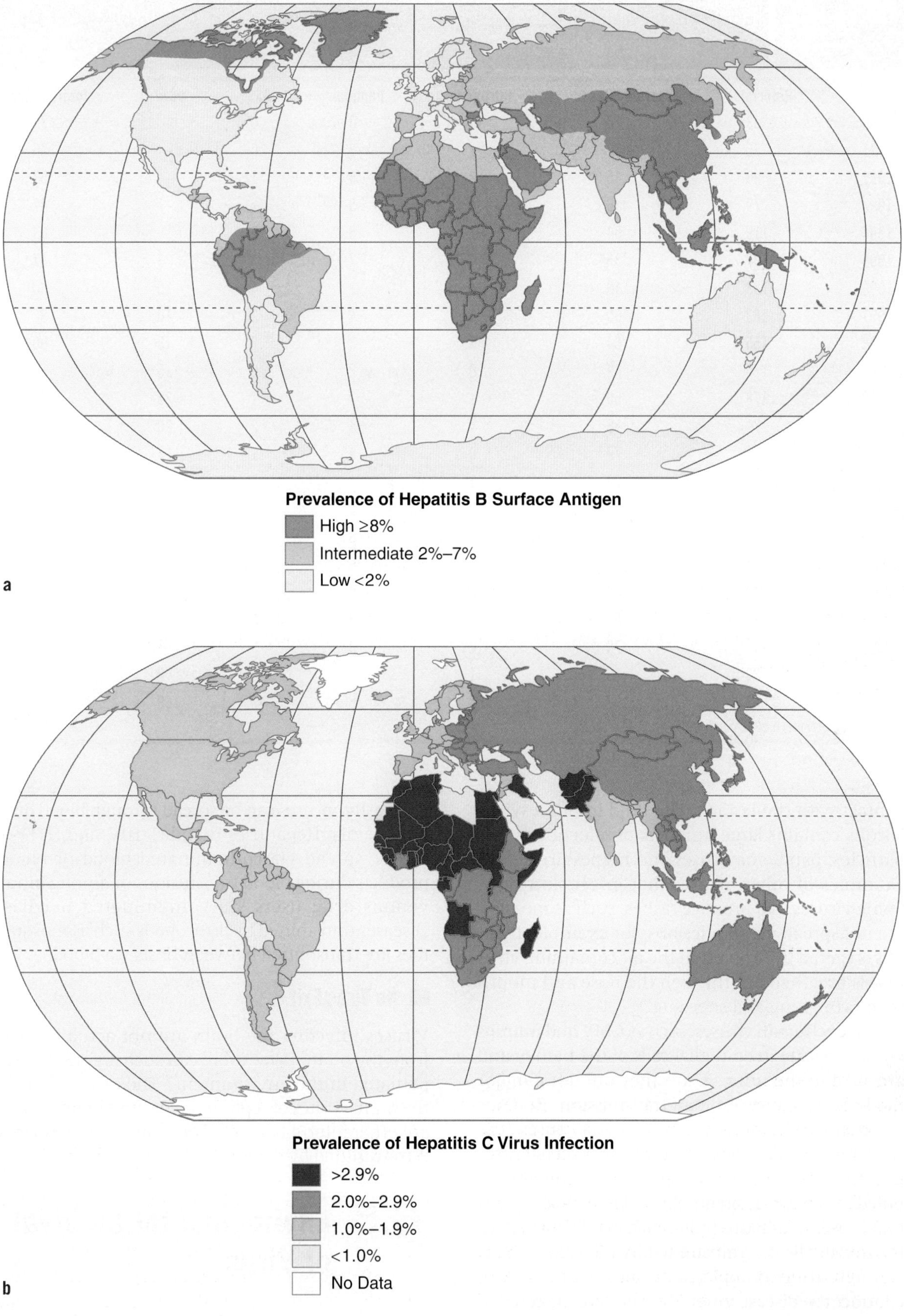

FIGURE 6-17 (a) Global distribution of chronic hepatitis B infection in millions. The estimated total chronic HBV infections worldwide is more than 350 million. **(b)** Global distribution of chronic hepatitis C infection in millions. The estimated total chronic HCV infections worldwide is 170 million. Data from World Health Organization hepatitis C prevalence, 2000, and United Nations global population.

TABLE 6-13 Definite and Probable CJD Cases in the United Kingdom: Referrals of Suspect CJD and Deaths of Definite and Probable CJD

Year	Referrals	Sporadic Cases	Iatrogenic	Familial	GSS	vCJD	Total Deaths
1990	53	28	5	0	0	–	33
1991	75	32	1	3	0	–	36
1992	96	45	2	5	1	–	53
1993	79	36	4	5	2	–	47
1994	119	54	1	5	3	–	63
1995	87	35	4	2	3	3	47
1996	133	40	4	2	4	10	60
1997	163	60	6	4	2	10	82
1998	155	64	3	3	2	18	90
1999	170	62	6	2	0	15	85
2000	178	50	1	2	1	28	82
2001	179	58	4	4	2	20	88
2002	163	72	0	4	1	17	94
2003	162	79	5	4	2	18	108
2004	114	50	2	4	2	9	67
2005	124	66	4	8	5	5	88
2006	111	69	1	6	3	5	84
2007	115	63	2	8	1	5	79
2008	147	87	5	2	3	1	98
2009	147	78	2	3	5	3	91
2010*	83	29	1	0	0	2	32
Total	2653	1157	63	76	42	169	1507

*Data are as of August 9, 2010.

contact with the lesions. The fluid in herpesvirus lesions contains large numbers of infectious virus particles; papillomaviruses and herpesviruses are examples of other viruses that infect skin cells. Some viruses that produce rashes, such as measles, are not spread via skin lesions; for example, measles is spread by droplets in the air containing virus that is breathed out through the nose and mouth when coughing and sneezing.

A few human viruses, such as CMV and mumps virus, replicate in epithelial cells of the kidney and are shed in the urine. Thus, they are not a major mode of human-to-human transmission. Rodents infected with hantaviruses and arenaviruses (the causative agent of hemorrhagic fevers) shed viruses in urine, thus contaminating the environment. As noted in Chapter 1, sin nombre virus, the hantavirus that causes hantavirus pulmonary syndrome (HPS), is a deadly disease transmitted by infected rodents through urine, droppings, or saliva. Humans can contract the disease when they breathe in aerosolized virus, so rodent control in and around the home remains the primary strategy for preventing hantavirus infection.

Lastly, viruses can be spread through blood during viremia. Hepatitises B and C, HIV, and HTLV-1 can be spread via contaminated blood or blood products during blood transfusions or among intravenous drug users via contaminated needles. Diseases transmitted by arthropods such as mosquitoes are transmitted between hosts via blood.

No Virus Exit

Viruses infecting the brain are not actually shed from the brain. How, then, can such a virus be perpetuated in the environment? The answer is that the primary sites of replication for most brain viruses are other internal organs. For example, rabies virus is transmitted by contact with infected saliva.

6.5 Environmental Survival of Virus

Virus transmission is dependent upon how well the virus survives the harsh environment outside of a host. Factors that affect virus survival in the

VIRUS FILE 6-2 Is Groundwater Safe to Drink?

If you dig deep into the ground, you'll likely find **groundwater**. It fills the spaces within bedrock and sediment. Groundwater exists almost everywhere, flowing beneath your feet, under cities, homes, farms, and even deserts. Groundwater is the source of water for many wells. Wells are created by drilling a straw-shaped hole from the land surface into the groundwater reservoir. After drilling, a pipe and pump are installed. The water is pumped from the well for use.

It has long been assumed that groundwater was free from contamination and pathogenic microbes, including viruses. From 1971 to 2002, the Centers for Disease Control and Prevention (CDC) reported 716 outbreaks associated with an infectious agent present in drinking water. Sixty percent of the outbreaks were attributable to groundwater! The infectious agent in more than half of the outbreaks was unknown and assumed to be of viral origin. One of the most notable viral outbreaks occurred in 2007.

Noroviruses were present in the well of an Egg Harbor restaurant located in Door County, Wisconsin. At least 229 people (patrons and restaurant employees) became ill, and 6 were hospitalized. Noroviruses cause about 50% of all gastroenteritis outbreaks worldwide. County health officials determined that the private well of the restaurant was contaminated with fecal material leaking from a new septic system into the groundwater below the restaurant. The restaurant well was 85 meters (278 feet) deep. Noroviruses are very stable in the environment, allowing them to survive for long periods in groundwater. Their findings were surprising because the restaurant septic system was new. It was installed by a licensed plumber and met the Wisconsin Administrative Code.

Mark Borchardt led an investigation outlining a combination of epidemiological, microbiological, and hydrogeological methods used to confirm the source of the 2007 norovirus restaurant outbreak in northeastern Wisconsin. The epidemiological investigation involved the identification of cases. Ill patrons called the Door County Public Health Department (DCPHD) and the local hospital. DCPHD nurses interviewed callers using a standardized outbreak questionnaire that was modified to include specific restaurant menu items that may have been the source of the outbreak as well (e.g., house salad, steak, ice cubes). Individuals who experienced diarrhea or vomiting within 2 to 5 days after consuming food or beverages at the restaurant were asked to submit a stool sample to the Wisconsin State Laboratory of Hygiene for pathogen analysis.

Tap water samples from the restaurant and nearby homes were collected during the microbiological investigation. Quantitative reverse transcriptase polymerase chain reaction (RT-PCR) measured 50 copies of norovirus genomes per liter of well water. Nucleotide sequencing of the viral genome determined that noroviruses present in restaurant well water were identical to the viruses present in patrons' stool samples and in sewage in the septic tank.

A hydrological investigation was done by injecting fluorescent dyes into the restaurant toilet (eosin dye) and a dosing chamber (fluorescein dye) on the septic system line that moves sewage effluent from the septic system to an infiltration (leaching) field (**FIGURE VF 6-2**). This tracer testing showed that the effluent was traveling from the septic tanks (through a leaky fitting) and infiltration field to the well in 6 to 15 days.

The analyzed data collected from the epidemiology, microbiology, and hydrogeology methods pointed toward the restaurant's own septic system as the source of noroviruses. Without this interdisciplinary approach, it is unlikely that the source of the outbreak would have been resolved.

To date, approximately 71 municipal water systems in Wisconsin use no form of water disinfection. Borchardt led another investigation that evaluated a sandstone aquifer located in south-central Wisconsin for viruses. This aquifer is approximately 75 meters (246 feet) deep and serves as a source of groundwater for wells in the Madison, Wisconsin, area. Enteroviruses were

(continued)

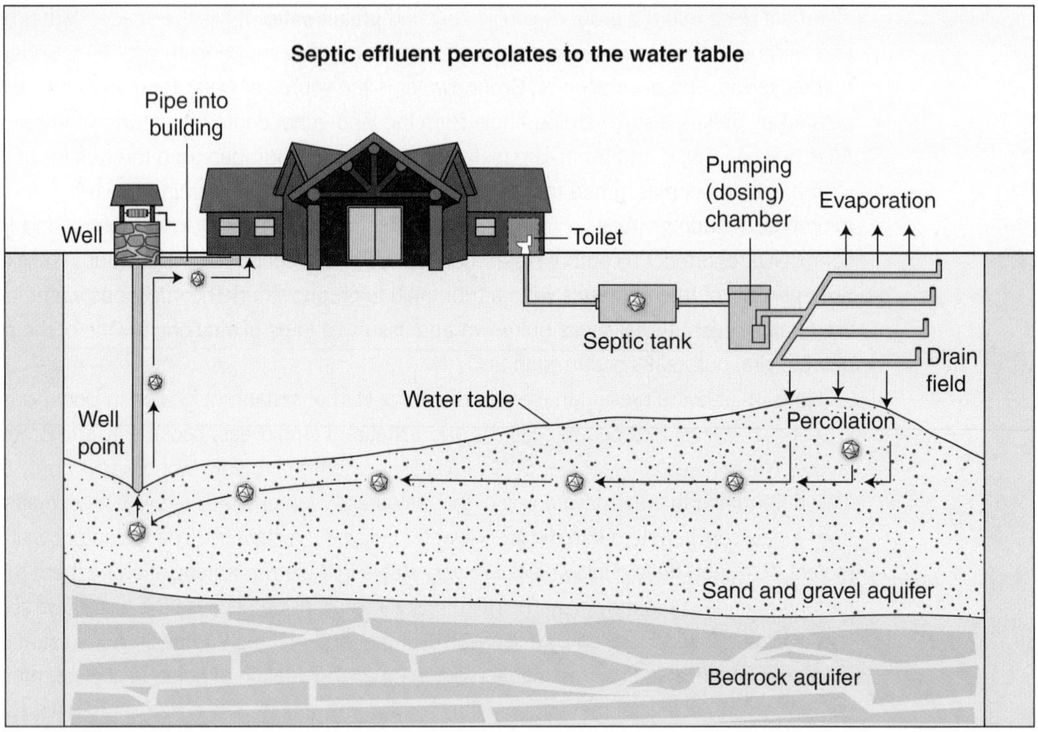

FIGURE VF 6-2 Septic effluent percolates to the water table. Tracer dyes uncovered a leak in the new septic system of the restaurant.

found in every well tested using RT-PCR. The presence of enteroviruses was correlated to rainfall and snowmelt events during early 2008. Detected enteroviruses included echovirus 3, echovirus 6, echovirus 11, Coxsackie A16 and B4, Adenoviruses 2, 6, 7, and 41 as well as noroviruses and rotaviruses. Leaky sewers were the likely virus source. They determined that viruses can travel hundreds of meters through porous bedrock and remain viable.

References

Blackborn, B. G., et al. 2004. "Surveillance for waterborne-disease outbreaks associated with drinking water–United States, 2001–2002." *MMWR* 53(SS08);23–45.

Borchardt, M. A., et al. 2007. "Human enteric viruses in groundwater from a confined bedrock aquifer." *Environ Sci Technol* 41(8):6606–6612.

Borchardt, M. A., et al. 2010. "Norovirus outbreak caused by a new septic system in a dolomite aquifer." *Ground Water* 49(1):85–97.

Patel, M. M., et al. 2009. "Noroviruses: A comprehensive review." *J Clin Virol* 44:1–8.

environment are the composition of the virus (enveloped viruses are more sensitive to degradation than naked viruses), whether the viruses are present in human and animal wastes (organic matter protects the virus), temperature, humidity, and pH. **TABLES 6-14** and **6-15** list different kinds of viruses, their environmental conditions in an experimental setting, and their survival times as recorded from numerous pilot scale or field studies.

6.6 Human Viruses in a Water Environment

Environmental (water) virology is a separate scientific discipline and began with efforts to detect poliovirus in water in the 1950s. Human enteric (intestinal) viruses enter the environment via

TABLE 6-14 Nonenveloped Virus Survival

Virus	Type of Sample	Log10 Reduction	Time	*Temperature	pH
Rotavirus	Tap water	2	64 days	20°C	8
Rotavirus	Distilled water	1	<1 hour	NA	8
Rotavirus	River water	2–3	64 days	20°C	7.4
Rotavirus	Feces	3	13 days	20°C	NA
Rotavirus	Feces	2	33 days	4°C	NA
Hepatitis A	Mixed human and pig wastes	NA	5 weeks	NA	NA
Hepatitis A	Mineral water	6	330 days	24°C	7
Norovirus	Surface water	NA	4 months	0–4°C	NA
Poliovirus type 1	Septic tank effluent	1.5	28 days	20°C	NA
Poliovirus type 1	Raw sewage	2	45 minutes	20°C	NA
Poliovirus type 1	Freshwater	0.7	4 days	24°C	NA
Poliovirus type 1	Seawater	1.5–4	4 days	24°C	NA
Foot and mouth disease virus	Feces	NA	21–103 days	NA	NA
Rhinovirus type 2	In cell culture medium on skin surface	0.1	3 hours	NA	NA
Rhinovirus type 2	In cell culture medium on plastic surface	<0.5	24 hours	NA	NA
Rhinovirus type 2	In cell culture medium on stainless steel surface	1	24 hours	NA	NA

*20°C = 68°F (~ room temperature). NA: Not available.

Source: Adapted from *Virus Survival in the Environment with Special Attention to Survival in Sewage Droplets and Other Environmental Media of Fecal or Respiratory Origin 2003* by Mark D. Sobsey and John Scott Meschke, World Health Organization Meeting 23–25 September 2003, Rome, Italy.

TABLE 6-15 Enveloped Virus Survival

Virus	Type of Sample	Log10 Reduction	Time	*Temperature	pH
Herpes simplex virus type 1	Saliva on chrome tap handle	3	2 hours	23°C	7
Herpes simplex virus	Tap water	3	4 hours	37–40°C	6.8
Herpes simplex type 1	Artificial tears opthalmic solution	3	5 days	NA	NA
Herpes simplex type 1	In saliva on penny	6	2 hours	22°C	NA
Herpes simplex type 1	In saliva on glass	3	2 hours	22°C	NA
HIV (LAV strain)	Saliva (maintained wet or dried)	Stable	4 days	20°C	NA
HIV (LAV strain)	50% human plasma	NA	>15 days	20°C	NA
Yellow fever virus	Plasma	NA	90 minutes	27°C	NA
Vaccinia virus	Cell culture suspension on cotton surface	4–5	6–10 weeks	25°C	NA
Hepatitis B	Blood on stainless steel and cotton swabs	0.2	14 days	25°C	NA
Respiratory syncytial virus	In sterile water on countertops	4	45 minutes	NA	NA
SARS-CoV	Feces	NA	>2 days	NA	NA
SARS-CoV	Urine	NA	>24 hours	20°C	NA
SARS-CoV	Virus in 75% ethanol	NA	Less than 5 minutes	20°C	NA

*20°C = 68°F (~ room temperature). NA: Not available.

Source: Adapted from *Virus Survival in the Environment with Special Attention to Survival in Sewage Droplets and Other Environmental Media of Fecal or Respiratory Origin 2003* by Mark D. Sobsey and John Scott Meschke, World Health Organization Meeting 23–25 September 2003, Rome, Italy.

sewage-contaminated water. Humans may be exposed to these viruses through consumption of undercooked shellfish grown in contaminated waters, food crops grown using wastewater for irrigation or human sewage as fertilizer, and sewage-polluted recreational and drinking water. Shellfish are filter feeders and tend to concentrate viruses and bacteria in their edible tissues. A method used by Bergh and his colleagues at the University of Bergen (Norway) in 1989 to quantitate virus particles in natural unpolluted waters found up to 2.5×10^8 virus particles per milliliter of water. Most of the particles in the samples were bacteriophages, but viruses were also present.

Feces of infected individuals contain extremely high numbers of virus particles. One gram of stool from an infected individual contains 10^5 to 10^{11} virus particles. Over 100 virus species have been found in sewage-contaminated waters. These viruses can cause hepatitis, gastroenteritis, meningitis, fever, rash, conjunctivitis, and perhaps diabetes.

As a general rule, viruses persist longer than enteric bacteria in the water. Studies have shown that waterborne outbreaks related to potable water actually met bacteriological standards. Thus, it is unsafe to rely on bacteriological standards to assess the virological quality of any kind of water. In water in a natural setting, the conditions are always changing because water is always changing. Environmental virologists are faced with the challenge of applying techniques that work on clinical isolates in natural settings. Surveying viral contamination in surface and ground water has bridged the environmental and clinical virology laboratories to improve public health.

Summary

Viruses infect animals and humans by preferred routes of entry. These routes of entry occur through

- the respiratory tract
- the gastrointestinal tract
- the genital tract
- the conjunctiva
- the skin (through breaks, abrasions, and bites)
- transplants
- blood and blood product transfusions
- iatrogenic induction

Upon entry, viruses remain localized or cause systemic infections by spreading throughout the body via

- the bloodstream
- the PNS
- the lymphatic system
- organs such as the liver
- crossing the blood-brain barrier
- the CNS
- crossing the placenta to reach the fetus

The pathogenesis of infection is unique to each viral pathogen. There are four basic patterns of viral infections: acute, persistent latent infections, chronic infections, and slow infections. Viruses exit (shed) into the environment through mucus, saliva, feces, semen, cervical secretions, breast milk, skin lesions, urine, and blood. Viruses do not exit from the brain but instead are shed from a primary replication site and from internal organs.

Virus stability in the environment varies depending upon the chemical composition of the virus and whether or not it is protected in organic matter such as feces, or by temperature, humidity, and pH. Today, more attention has been placed on environmental virology. Many viruses can be found in water, leading to waterborne disease outbreaks. Molecular methods are being used to monitor aquatic environments to improve public health.

Chapter 7 will address host resistance to viral infection. This will cover physiological factors affecting resistance to viral infection, the interferon response, and immune responses toward viral infection. Vaccines and artificially induced immunity also will be discussed.

CASE STUDY QUESTIONS

These questions relate to the Case Study presented at the beginning of the chapter.

1. Elevated temperature is a common presenting symptom for a number of illnesses, including infectious diseases. List viral infections other than SARS in which thermal scanners may be useful tools for mass screening of body temperature, to delay or avert other public health crises in which widespread transmission of infection is possible.
2. Thermal image scanners were operational in Canada at Vancouver International Airport and Toronto's Pearson International Airport during part of the SARS outbreak in 2003, and 4,569,759 passengers were actually scanned. An elevated temperature was detected in 1,435 individuals (0.031%). None of them were assessed as having SARS. Approximately 36.3 million people were assessed via scanners in Hong Kong in 2003. Of these, only 1,921 had a fever. Forty of these passengers were admitted to a hospital, but none was diagnosed with SARS. From these data, does it appear that mass screening for fever is the most effective means for detecting disease in travelers during an outbreak? Should other "low-tech" approaches be used? Should a combination of both thermal scanners and other approaches be used? Defend your detailed answer.

CASE STUDY 1: HIV

A New York City man in his 40s was tested for HIV five times between September 2000 and May 2003. The results were negative each time and his T-cell counts were normal. In mid-December 2004, the man was prompted to visit his private physician because of the following symptoms: sore throat, fever, malaise, weakness, significant weight loss, and chronic fatigue. When tested for HIV this time, both the ELISA and Western immunoblot assays yielded positive results. His CD4 T-cell count was 80/µl and his plasma contained 280,000 HIV-1 RNA copies/ml. Individuals with T-cell counts below 200 and an RNA viral load of 100,000 HIV-1 RNA copies/ml of plasma are considered to be in the advanced stage of HIV infection—full-blown AIDS. During this stage, patients experience many opportunistic infections because they have too few T-cells to help their body fight the virus or any other invading infection. The infections are called **opportunistic** because they take advantage of the immune system when the virus compromises it.

These results were alarming. In a short time, the man's T-cell counts and viral RNA load matched those that occur during full-blown AIDS. The man provided a detailed history of his sexual activity. He admitted to having unprotected anal sex with hundreds of male partners, at times while under the influence of the drug crystal methamphetamine (Meth). He was referred to a team of researchers at the New York School of Medicine and Cabrini Medical Center for viral genomic studies to determine whether the HIV-1 virus in his body was susceptible to antiviral drug therapy. The researchers isolated HIV-1 from his peripheral blood cells and began drug susceptibility testing. Meanwhile, the man's health was rapidly declining because he was continuing to lose weight rapidly and was suffering from additional opportunistic infections.

The results obtained by the team of researchers showed that the man was harboring an HIV strain that was multidrug resistant. It was resistant to three of the four classes of antiviral drug: nucleoside reverse transcriptase inhibitors, nonnucleoside reverse transcriptase inhibitors, and protease inhibitors. This meant that the man would be very difficult to treat with a standard antiretroviral regimen. Fortunately, this strain was susceptible to a new class of anti-HIV drugs known as fusion inhibitors. Fusion inhibitors block the virus particles from fusing to the plasma membrane of a cell, thereby blocking their entry into cells. The HIV in the man's body was shown to be sensitive to efurvitide (T-20) and efavirenz (T-1249) fusion in hibitors, and this treatment was initiated.

1. The New York City Department of Health and Mental Hygiene (DOHMH) issued a health advisory alert in February 2005 to begin tracing the man's sexual contacts. Why?

(continued)

CASE STUDY 1: HIV (continued)

HIV enters CD4+ T-cells by binding to CD4+ using a viral protein called gp120. The CD4+ cellular protein receptor alone is not enough to allow viral entry into cells. The virus must interact with a second host protein or **co-receptor**. This co-receptor may be different for different types of cells. One is called CCR5 and another is called CXCR4 or fusin. CCR5 is present on T-cells and macrophages. Fusin is primarily found on CD4+ cells. **FIGURE CS 6-2** shows HIV binding to host cells.

CCR5 appears to be important for viral entry during the early stages of HIV infection. In contrast, cells expressing CXCR4 appear to be important for the entry of HIV strains that are prevalent in individuals with a more aggressive disease during the later stages of infection. Individuals who inherit a defective CCR5 gene are resistant to HIV infection. The multidrug-resistant HIV variant from this patient was able to infect cells with both CCR5 and CXCR4 co-receptors. In fact, the viral population from the patient consisted of CCR5- and CXCR4-tropic viruses of approximately equal proportions.

2. If this patient inherited a defective CCR5 co-receptor, would that help to explain his rapid decline in health?
3. The patient did not inherit a defective CCR5 co-receptor. Speculate as to why the viral populations in his body were so aggressive.

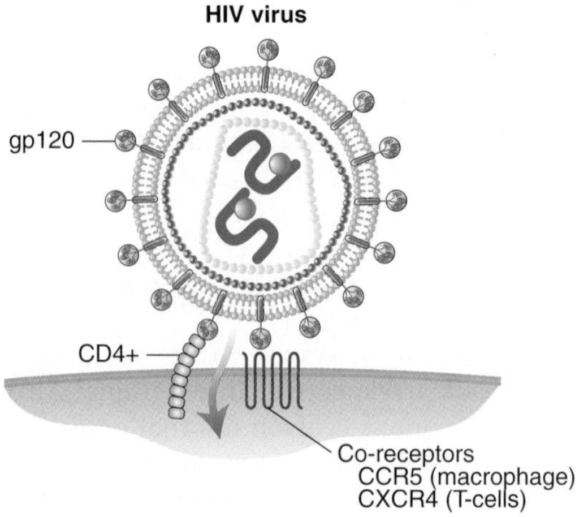

FIGURE CS 6-2 Illustration of an HIV virus particle binding to the CD4+ receptor and a chemokine receptor CCR5 or CXCR4. Adapted from Project Inform, "Co-Receptors: CCR5-Understanding HIV," *The Body: The Complete HIV/AIDS Resource*, http://www.thebody.com/content/art4978.html#ccr5.

References

Madison, P. B. 2005. "Aggressive HIV strain sets off dubious public health measure." *Nat Med* 11:360.

Markowitz, M., et al. 2005. "Infection with multidrug resistant, dual-tropic HIV-1 and rapid progression to AIDS: A case report." *Lancet* 365:1031–1038.

CASE STUDY 2: ADENOVIRUS SEROTYPE 14

Eighteen-year-old Joseph Spencer, a high school varsity swimmer at a school in Oregon, suddenly became ill with a fever, chills, and vomiting. His mother took him to the local hospital emergency room, where he was placed in an intensive care unit (ICU), sedated, and put on a ventilator. Joseph spent 18 days in the hospital and then was able to return home. Even after weeks of bed rest and physical therapy, he suffered from breathing problems.

Joseph was 1 of 140 individuals who were confirmed as suffering from an adenovirus serotype 14 (Ad14) infection during March to June of 2007. The CDC, U.S. Air Force, and state and local public health departments identified clusters of patients infected with Ad14 in Oregon, Washington, and Texas. Of these patients, 53 (38%) were hospitalized, 24 (17%) were placed in ICU, and 9 (5%) died of pneumonia. Ad14 was first isolated and identified in Holland in 1955. It caused sporadic outbreaks in Europe and Asia but not the Western hemisphere. To date, 51 adenovirus serotypes have been identified. These adenoviruses typically cause benign ailments such as colds, pinkeye, bronchitis, and stomach flu. The cases described here are unusual and were life-threatening. Doctors do not routinely test for adenoviruses. Recent surveys, including those on military bases, indicate that the virus suddenly appeared widely across the United States.

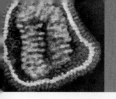

CASE STUDY 2: ADENOVIRUS SEROTYPE 14 (continued)

1. How would you determine if some people are genetically prone to Ad14 infection or have weaker immune systems to fight this infection versus being infected with a virus that is inherently more dangerous? What type of analysis would determine if Ad14 had mutated from the original 1955 strain?
2. Adenoviruses are shed in respiratory secretions and feces. They can persist for weeks on environmental surfaces. Would the stability of these viruses in the environment pose a challenge in controlling adenoviruses outbreaks? Why or why not?

References

Metgar, D., et al. 2007. "Abrupt emergence of diverse species B adenoviruses at US Military Recruit Training Centers." *JID* 196:1465–1473.

Oregon Department of Human Services, et al. 2007. "Acute respiratory disease associated with Adenovirus Serotype 14—four states, 2006–2007." *MMWR* 56(45): 1181–1184.

CASE STUDY 3: RABIES

David McRae was a wildlife artist who earned a reputation sketching mammals. While working for Scottish Natural Heritage for 15 years as a licensed bat conservationist, performing studies of bats around his home and in remote parts of Scotland, he was affectionately called "Batman." Despite the nature of his work, he was not vaccinated with a rabies vaccine that offers marginal protection against European bat lyssavirus (EBL). EBL is endemic in continental Europe but rare in the United Kingdom. It is a relative of the classic rabies virus (see Chapter 13) and is a member of the genus Lyssavirus. From 1902 to 2001, there were no human EBL cases in the United Kingdom.

In 2002, McRae was prescribed anti-inflammatory drugs by a general practitioner for pain in his left shoulder. Some time later, he began vomiting blood and was admitted to a hospital. Physicians thought the anti-inflammatory drugs caused his condition. His temperature was 101.3°F (38.5°C). McRae complained of left-arm pain, difficulty swallowing, and tingling sensations in both of his upper arms. He was lucid but had difficulty articulating words and had involuntary movement of his eyeballs. He was treated with high doses of flucloxacillin and ciprofloxacin to treat a possible bacterial infection, and high doses of acyclovir (used to treat herpesvirus infections). Computed tomography (CT) and magnetic resonance imaging (MRI) scans of McRae's head were performed on the first day of admission. Results were normal. During the second day of his hospital stay, cerebrospinal fluid (CSF) was drawn from McRae. His CSF had mildly raised protein levels but otherwise was normal. He was treated with intravenous immunoglobulin for 5 days to treat presumptive Guillain-Barré syndrome.

On day 5 of his hospital stay, McRae became confused, agitated, and aggressive. He had to be sedated. Another head CT scan was normal. The next day, his right lung collapsed. He began to hypersalivate. McRae was transferred to an intensive care unit and placed on a ventilator. His mental state deteriorated quickly and his arms and legs became flaccid (weak and soft or lacking muscle tone). An electroencephalogram indicated a nonspecific encephalitic pattern.

More patient history was gathered. At this time, it was noted that McRae had been bitten by a Daubenton's bat in Angus, Scotland, about 19 weeks earlier. He was not wearing gloves at the time and was bitten on the ring finger of his left hand. Very few doctors in the UK had seen a patient with rabies, which made a diagnosis extremely difficult. In light of the new medical history information, saliva, blood, and skin biopsy samples were taken from McRae during his seventh day of illness at the hospital. Results were available on his ninth hospital

(continued)

CASE STUDY 3: RABIES (continued)

day stay, showing low levels of lyssavirus RNA (EBL) in his saliva by reverse transcriptase polymerase reaction (RT-PCR). No antibodies were detected in his blood. The "Batman" died on day 14 of his hospitalization from a rare form of rabies. It was a bleak day for conservationists in Scotland.

1. How could this tragic outcome have been prevented? (Hint: if needed, read about rabies viruses in Chapter 13.)
2. McRae had a combination of furious and paralytic rabies symptoms. List his symptoms. Next to each symptom, include the form it belongs to (paralytic or furious; see Chapter 13 for help).
3. How was McRae infected with EBL?
4. Between 30 and 40 volunteers in Scotland count bats in caves, derelict buildings, tunnels, hollow trees, and households. Could lyssaviruses be spread by airborne transmission in these environments? (Hint: you will find helpful information in a reference paper listed below.)
5. Could McRae's viral infection spread to clinical staff? What would be the route(s) of transmission?
6. What precautions did the clinical staff take while treating and caring for McRae?
7. List other members of the Lyssavirus genus. What mammals does each member infect and where is it endemic? (Hint: use the Rupprecht reference listed below.)
8. Will classic rabies virus vaccine protect vaccinated individuals from all lyssaviruses? (Hint: you will find the answer in Rupprecht paper listed below.)
9. Bats are the reservoir for other viruses that infect humans (e.g., SARS, CoV). Create a list of viruses that cause infections in humans.

References

Daszak, P. 2010. "Bats, in black and white." *Science* 329:634–635.

Johnson, N., Phillpotts, R., and Fooks, A. R. 2006. "Airborne transmission of lyssaviruses." *J Med Microbiol* 55:785–790.

Nathwani, D., et al. 2003. "Fatal human rabies caused by European bat lyssavirus type-2a infection in Scotland." *CID* 37:598–601.

Rupprecht, C. E., et al. 1995. "The ascension of wildlife rabies: a cause for public health concern or intervention." *Emerg Infect Dis* 1(4):107–114.

Streicker, D., et al. 2010. "Host phylogeny constrains cross-species emergence and establishment of rabies virus in bats." *Science* 329:676–679.

CASE STUDY 4: SMALLPOX

During the 16th century, a tribe of tundra hunters known as Yukagirs settled along the Kolyma River in a location above the Arctic Circle and south of the East Siberian Sea. The settlement was named Pohodsk. The main activities of this tribal community were hunting and fishing. They made garments of traditional skin and fur. This region of Russia has some of the most extreme environmental conditions on Earth. Temperatures in the summer may reach 108°F (42°C) but can drop to −94°F (−70°C) in winter. Its tundra recently has been affected by a global weather change. The permafrost is now undergoing repeated cycles of freezing and thawing.

The numbers of Pohodsk villagers were decimated during the 17th and 19th centuries by smallpox strains of extraordinary lethality. Smallpox is caused by Variola virus. Early signs and symptoms include high fever, headache, fatigue, and vomiting. A macular rash follows. The rash progresses to vesicular, pus-filled blisters that crust, scab, and fall off after about three weeks, leaving a pitted scar known as a pock mark.

Today only a few dozen Yukagirs remain in the village. A mass grave near Pohodsk contains many smallpox victims. The grave contains deep pits cut into the ice that were used to store provisions. These "icehouses" kept bodies frozen, including remains of smallpox victims wrapped in deerskin shrouds. During 1991, a team of Russian scientists unearthed the frozen bodies buried over 160 years ago. Scientists told the Yukagirs that their mission was to disinfect the cemetery because the frozen bodies may contain viable smallpox virus. The scientists discovered the frozen bodies were well preserved. Their skin was blackened and pocked. Researchers carefully removed biopsy material, and then doused the tomb with disinfectant. No live virus was isolated from the biopsy material of bone and skin fragments taken from the frozen corpses.

1. Bodies of smallpox victims are buried in frozen tundra across Siberia and possibly Canada and Alaska. As the planet warms, some of these areas are thawing. Does this cause concern for anyone who may accidentally come in contact with a smallpox-infested body? Could the disease be reintroduced into the human population? Explain your answer.
2. What precautions would scientists unearthing frozen smallpox victims take to avoid contracting the disease?
3. How is smallpox spread? (For more help, see Chapter 14 on Poxviruses.)
4. Is it possible that live Variola virus exists in the permafrost? Explain your answer.
5. List occupations of individuals who may come in contact with corpses from individuals who lived long ago and those who come in contact with recently deceased people, exposing them to the hazards of an infection. Discuss appropriate precautions that should be taken when handling cadavers that are infectious hazards. (Hint: see reference below.)
6. Create a list of other infectious diseases besides smallpox that caused human deaths over a hundred years ago.

Reference

Healing, T. D., Hoffman, P. N., and Young, S. E. J. 1995. "The infection hazards of human cadavers." *CDR Rev* 5(5):R61-R68. Retrieved February 14, 2011, from http://www.hpa.org.uk/web/HPAwebFile/HPAweb_C/1200660055286

Resources

Primary Literature and Reviews

Adler, S. P., et al. 2007. "Recent advances in the prevention and treatment of congenital cytomegalovirus infections." *Semin Perinatol* 31:10–18.

Aplin, J. D. 2010. "Developmental cell biology of human villous trophoblast: Current research problems." *Int J Dev Biol* 54:323–329.

Bergh, O., et al. 1989. "High abundance of viruses found in aquatic environments." *Nature* 340:467–468.

Blackborn, B. G., et al. 2004. "Surveillance for waterborne-disease outbreaks associated with drinking water–United States, 2001–2002." *MMWR* 53(SS08);23–45.

Borchardt, M. A., et al. 2007. "Human enteric viruses in groundwater from a confined bedrock aquifer." *Environ Sci Technol* 41(8):6606–6612.

Borchardt, M. A., et al., 2010. "Norovirus outbreak caused by a new septic system in a dolomite aquifer." *Ground Water* 49(1):85–97.

Bosch, A. 1998. "Human enteric viruses in the water environment: A minireview." *Int Microbiol* 1:191–196.

Centers for Disease Control and Prevention. 2002. "Probable variant Creutzfeldt-Jakob disease in U.S. resident—Florida." *MMWR* 51(41):927–929.

Centers for Disease Control and Prevention. 2005. "Infection in organ transplant recipients—Massachusetts, Rhode Island, 2005." *MMWR* 54:1–2.

Chamberland, M. E., Alter, H. J., Busch, M. P., Nemo, G., and Ricketts, M. 2001. "Emerging infectious disease issues in blood safety." *EID* 7(3, Suppl):552–553.

Chandy, S., et al. 2010. "Congenital rubella syndrome and rubella in Vellore, South India." *Epidemiol Infect* (Jul 20):1–5.

Chin, J., ed. 2000. *Control of Communicable Diseases Manual: An Official Report of the American Public Health Association,* 17th ed. Washington, DC: American Public Health Association.

Cohen, J., and Powderly, W. G., eds. 2004. *Infectious Diseases,* 2nd ed. Philadelphia: Mosby.

Coovadia, H. 2004. "Antiretroviral agents—how best to protect infants from HIV and save their mothers from AIDS." *N Engl J Med* 351(3):289–292.

Daszak, P. 2010. "Bats, in black and white." *Science* 329: 634–635.

Dosiou, C., and Giudice, L. C. 2005. "Natural killer cells in pregnancy and recurrent pregnancy loss: Endocrine and immunologic perspectives." *Endocr Rev* 26(1):44–62.

Francis, K., et al. 2003. "Innate immunity and brain inflammation: The key role of complement." *Exp Rev Mol Med* 5:1–19.

Heazell, A. E. P., and Crocker, I. P. 2008. "Live and let die—Regulation of villous trophoblast apoptosis in normal and abnormal pregnancies." *Placenta* 29:772–783.

Healing, T. D., Hoffman, P. N., and Young, S. E. J. 1995. "The infection hazards of human cadavers." *Commun Dis Rep CDR Rev* 5(5):R61–R68.

Jamieson, D. J., Kourtis, A. P., and Rasmussen, S. A. 2006 "Lymphocytic choriomeningitis virus: an emerging obstetric pathogen?" *Am J Obstet Gynecol* 194(6): 1532–1536.

Jamieson, D. J., et al., 2006. "Emerging infections and pregnancy." *Emerg Infect Dis* 12(11):1638–1643.

Johnson, N., Phillpotts, R., and Fooks, A. R. 2006. "Airborne transmission of lyssaviruses." *J Med Microbiol* 55: 785–790.

Jun, H. S., and Yoon, J. W. 2003. "A new look at viruses in type 1 diabetes." *Diabetes Metab Res Rev* 19(1):8–31.
Longo, L. D., and Reynolds, L. 2010. "Some historical aspects of understanding placental development, structure and function." *Int J Dev Biol* 54:237–255.
Maidji, E., et al. 2007. "Developmental regulation of human cytomegalovirus receptors in cytotrophoblasts correlates with distinct replication sites in the placenta." *J Virol* 81(9):4701–4712.
Mims, C., Nash, A., and Stephen, J., eds. 2001. *Mims' Pathogenesis of Infectious Disease,* 5th ed. San Diego: Elsevier Academic Press.
Moffett, A., and Loke, C. 2006. "Immunology of placentation in eutherian mammals." *Nat Immunol Rev* 6:584–594.
Moffet-King, A. 2002. "Natural killer cells and pregnancy." *Nat Immunol Rev* 2:656–663.
Nathanson, N., ed. 2002. *Viral Pathogenesis and Immunity.* Philadelphia: Lippincott Williams & Wilkins.
Nathwani, D., et al. 2003. "Fatal human rabies caused by European bat lyssavirus type-2a infection in Scotland." *CID* 37:598–601.
Nelson, R., and Reiher, H. W. 1997. "Bat bite?" *Lancet* 350:1300.
Nishiura, H. 2006. "Smallpox during pregnancy and maternal outcomes." *Emerg Infect Dis* 12(7):1119–1121.
Nuray, A., Ustunel, I., and Demir, R. 2009. "Uterine natural killer (uNK) cells and their missions during pregnancy: A review." *Acta Histochem* 113(2):82–91.
Ornoy, A., and Tenenbaum, A. 2006. "Pregnancy outcome following infections by coxsackie, echo, measles, mumps, hepatitis, polio and encephalitis viruses." *Reprod Toxicol* 21:446–457.
Panda, B., et al. 2010. "Selected viral infections in pregnancy." *Obstet Clin N Am* 37:321–331.
Patel, M. M., et al. 2009. "Noroviruses: A comprehensive review." *J Clin Virol* 44:1–8.
Pereira, L., et al. 2005. "Insights into viral transmission at the uterine-placental interface." *Trends Microbiol* 13(4): 164–165.
Regan, N., et al. 2008. "Emerging and zoonotic infections in women." *Infect Dis Clin North Am* 22(4):755–772.
Repprecht, C. E., et al. 1995. "The ascension of wildlife rabies: A cause for public health concern or intervention." *Emerg Infect Dis* 1(4):107–114.
Reuter, J. D., et al. 2004. "Systemic immune deficiency necessary for cytomegalovirus invasion of the mature brain." *Virology* 78(3):1473–1487.
Rhode Island Hospital, et al. 2005. "Lymphocytic choriomeningitis virus infection in organ transplant recipients—Massachusetts, Rhode Island, 2005." *MMWR* 54:1–2.
Rombaldi, R. L., et al. 2008. "Transplacental transmission of human papillomavirus." *Virol J* 5:106–119.
Shi, L., Nora, T., and Patterson, P. H. 2005. "Maternal influenza infection is likely to alter fetal brain development indirectly: The virus is not detected in the fetus." *Int J Dev Neurosci* 23:299–305.
Short, S. J., et al. 2010. "Maternal influenza infection during pregnancy impacts postnatal brain development in the rhesus monkey." *Biol Psychiatry* 67(10):965–973.
Sinha, S., and Kumar, M. 2010. "Pregnancy and chronic hepatitis B infection." *Hepatol Res* (40):31–48.
Streicker, D., et al. 2010. "Host phylogeny constrains cross-species emergence and establishment of rabies virus in bats." *Science* 329:676–679.
Thornburg, K. L., and Hunt, J. S. 2010. "Contemporary comparative placenta research an interview with Allen Enders." *Int J Dev Biol* 54:231–236.
Tsekoura, E. A., et al. 2010. "Adenovirus genome in the placenta: Association with histological chorioamnionitis and preterm birth." *J Med Virol* 82(8):1379–1783.
Townsend, C. L., et al. 2010. "Antiretroviral therapy in pregnancy: Balancing the risk of preterm delivery with prevention of mother-to-child HIV transmission." *Antivir Ther* 15(5):775–783.
Veenstra, A. L., et al. 2003. "The immunology of successful pregnancy." *Hum Reprod Update* 9(4):347–357.
Volmink, J., et al. 2007. "Antiretrovirals for reducing the risk of mother-to-child transmission of HIV infection." *Cochrane Database Syst Rev* 24(1):CD003510.
Wong, K. 2002. "Oral drugs halt smallpoxlike viruses in mice." *Sci Am News,* March 2002.

Popular Press

Barry, J. M. 2004. *The Great Influenza: The Epic Story of the Deadliest Plague in Human History.* New York: Viking Penguin Group.
Duncan, K. 2006. *Hunting the 1918 Flu: One Scientist's Search for a Killer Virus.* Toronto: University of Toronto Press.
Ewald, P. W. 2002. *Plague Time: The New Germ Theory of Disease.* New York: Anchor.
Finley, D. 1998. *Mad Dogs: The New Rabies Plague.* College Station, TX: Texas A&M University Press.
Gould, T. 1995. *A Summer Plague: Polio and Its Survivors.* New Haven, Connecticut: Yale University Press.
Henderson, D. A. 2009. *Smallpox—the Death of a Disease: The Inside Story of Eradicating a Worldwide Killer.* Amherst, New York: Prometheus Books.
Kimball, A. M. 2006. *Risky Trade: Infectious Disease in the Era of Global Trade.* Burlington, Vermont: Ashgate Publishing Company.
Mims, C. 2000. *When We Die.* New York: St. Martin's Griffin.
Oldstone, M. B. A. 2009. *Viruses, Plagues, and History: Past, Present and Future.* Oxford, UK: Oxford University Press.
Pendergrast, M. 2010. *Inside the Outbreaks: The Elite Medical Detectives of the Epidemic Intelligence Service.* Geneva, Illinois: Houghton Mifflin Harcourt.
Schwartz, M. 2003. *How the Cows Turned Mad.* Berkley: University of California Press.
Sherman, I. W. 2007. *Twelve Diseases That Changed the World.* Washington, D.C.: ASM Press.

Viruses in Fiction Books

Case, J. 1998. *The First Horseman.* New York: Ballantine Books.
Chrichton, M. 1969. *The Andromeda Strain.* Harper: New York.
Christofferson, A. 2001. *Clinical Trial.* New York: Forge Books.
Cook, R. 1988. *Outbreak.* New York: Berkley Books.
DeMille, N. 1997, 2002. *Plum Island.* New York: Grand Central Publishing.
Fuller, J. G. 1974. *Fever! The Hunt for a New Killer Virus.* New York: Ballantine Books.
Gipson, F. 1965. *Old Yeller.* New York: Scholastic Book Services.
Kall, D. 2005. *Pandemic.* New York: Forge Books.
King, Stephen 1991. *The Stand.* New York: Signet Classics.
Marr, J. S. 1999. *The Eleventh Plague.* New York: Avon Books.
Matheson, R. 2007. *I am Legend.* New York: Tor Books.

McCusker, P., and Larimore, W. M. D. 2009. *The Gabon Virus (Time Scene Investigators).* New York: Howard Books.

McCusker, P., and Larimore, W. M. D. 2010. *The Influenza Bomb (Time Scene Investigators).* New York: Howard Books.

Niles, S. 2007. *28 Days Later: The Aftermath.* Harper: New York.

Perry, S. D. 1998–2004. *Resident Evil Series.* New York: Pocket Books.

Preston, D. 1997. *Mount Dragon.* New York: Bantam Books.

Preston, R. 1998. *The Cobra Event.* New York: Ballantine Books.

Video Productions

Bad Blood: A Cautionary Tale. A Film by Marilyn Ness. 2010. Necessary Films, available at http://badblooddocumentary.com/.

H1N1. 60 Minutes. November 1, 2009. CBS.

Killer Virus: Hunt for the Next Killer Flu. 2009. Discovery Channel.

H1N1. 60 Minutes. October 18, 2009. CBS.

The Age of AIDS. Frontline. 2006. PBS Home Video.

Childhood Exanthems: Measles, Chickenpox & Scarlet Fever. 2004. Medivision Film Series.

Dog Bites and Rabies Prevention. 2004. Medivision Film Series.

New Blood. 2002. Films for the Humanities.

Tainted Blood. 2002. Films for the Humanities.

Hepatitis, Cytomegalovirus, and Epstein-Barr Virus. 1999. Films for the Humanities.

eLearning

go.jblearning.com/shors2

The site features eLearning, an online review area that provides quizzes and other tools to help you study for your class. You can also follow useful links for in-depth information, or to just find out the latest virology and microbiology news.

CHAPTER 7

Host Resistance to Viral Infections

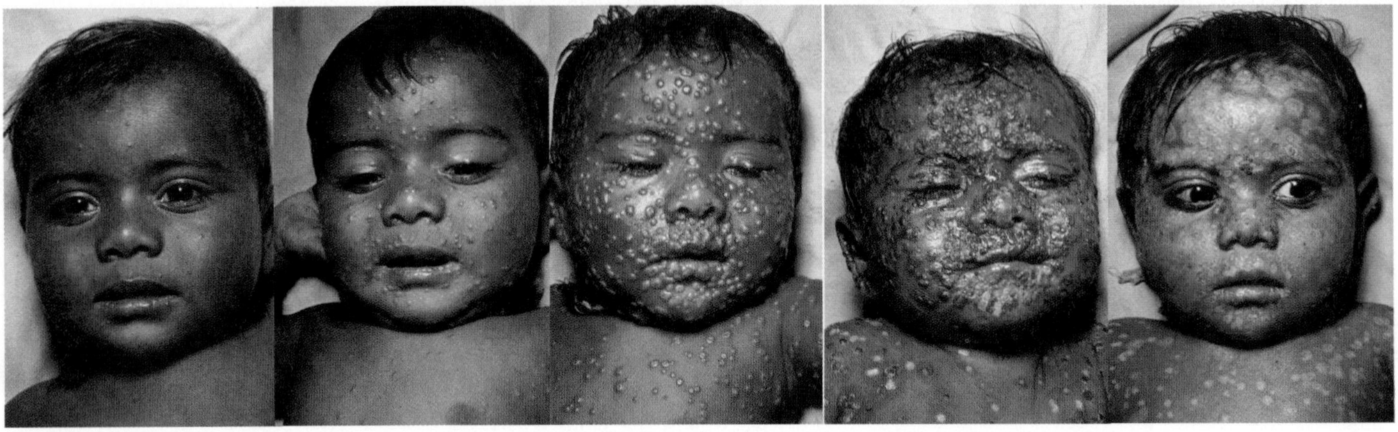

Development and healing of smallpox rash skin lesions in a 9-month-old unvaccinated Pakistani child. From left to right, photographs of the child were taken on days 2, 4, 7, 13, and 20 of the disease. This boy has survived a smallpox infection. He has now developed a natural immunity toward the smallpox virus.

"That which does not kill you makes you stronger."

Viktor Frankl

OUTLINE

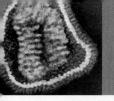

CASE STUDY: LYMPHOCYTIC CHORIOMENINGITIS VIRUS

One day Sally Cronn's college roommate purchased a hamster at the local pet store. Sally was fond of pets but felt slightly uneasy about the purchase because she recalled hearing about a hamster virus that killed three organ transplant recipients in 2005 (**FIGURE CS 7-1**). Sally asked her roommate if she knew anything about hamster viruses. The roommate responded, "I find it hard to believe that such a cute, furry, little pet hamster could cause the spread of a deadly virus." This increased Sally's suspicion.

Sally, a microbiology major with an interest in viruses, decided to research the hamster virus that infected the organ transplant recipients. Within the June 3, 2005, issue of the *Morbidity and Mortality Weekly Report* published by the Centers for Disease Control and Prevention (CDC), she found an article that described the epidemiological investigation regarding the deaths of four of the transplant recipients.

The organ donor had a history of high blood pressure and died suddenly of a stroke. Family members consented to donating her organs, and her liver, lungs, kidneys, and corneas were removed. Within three weeks of transplantation, the four individuals who received the kidneys, liver, and lungs experienced various symptoms, such as abnormal liver functioning, kidney failure, fever, rash, diarrhea, and hypoxia (an absence of oxygen in tissues). Three of the four transplant recipients died approximately three weeks after receiving a donor organ. Interestingly, the two donors who received corneas were asymptomatic.

After extensive diagnostic testing, public health officials suspected that the solid organ recipient deaths were caused by a transplant-transmitted infection. Tissue and blood samples from the donor and recipients were sent to the CDC. Subsequently, a lymphocytic choriomeningitis virus (LCMV) infection was identified as the cause of illness of all four solid organ transplant recipients. About 5% of common house mice, hamsters, and other rodents carry LCMV, and 2% to 5% of the general population has antibodies against the virus. Hospital staff, organ bank staff, and family members of the donor and recipients were interviewed to determine the source of LCMV. Family members of the donor revealed that a pet hamster had been acquired recently and that there were limited opportunities for the donor to be exposed to wild rodents.

FIGURE CS 7-1 Wild and pet rodents can carry viruses that may infect humans. Garbage storage areas with open garbage cans attract rodents and insects. This creates a situation in which infectious organisms may be passed on to humans.

References

Barton, L. L., Peters, C. J., and Ksiazek, T. G. 1995. "Lymphocytic choriomeningitis virus: An unrecognized teratogenic pathogen." *EID* 1(4):153–153.

Childs, J. E., et al. 1991. "Human-rodent contact and infection with lymphocytic choriomeningitis and Seoul viruses in an inner-city population." *Am J Trop Med Hyg* 44(2):117–121.

Childs, J. E., et al. 1992. "Lymphocytic choriomeningitis virus infection and house mouse (*Mus musculus*) distribution in urban Baltimore." *Am J Trop Med Hyg* 47(1):27–34.

Gregg, M. B. 1975. "Recent outbreaks of lymphocytic choriomeningitis in the United States of America." *Bull World Health Organ* 52:549–553.

Rhode Island Hospital, et al. 2005. "Lymphocytic choriomeningitis virus infection in organ transplant recipients—Massachusetts, Rhode Island." *MMWR* 54(21):537–539.

Rousseau, M. C., et al. 1997. "Lymphocytic choriomeningitis virus in Southern France: Four case reports and a review of the literature." *Eur J Epidemiol* 13:817–823.

Centers for Disease Control and Prevention (CDC). LCMV FAQS Lymphocytic Choriomeningitis. Retrieved February 16, 2011, from http://www.cdc.gov/ncidod/dvrd/spb/mnpages/dispages/lcmv.htm

Each pathogen has its own infection strategy, resulting in the development of a disease with distinct symptoms. Upon viral entry, the infected host elicits a number of responses toward infection. This chapter discusses factors that affect host resistance and host defenses. It includes nonspecific (innate) host defenses such as the interferon response and specific (adaptive) immune responses (antibody production and cell-mediated immunity). After you finish studying this chapter, you should be able to answer the following questions:

1. How are viruses initially detected by the innate immune response?
2. How are the specific identity of the virus and the location of the infection communicated to the adaptive immune response?
3. How do the molecules identified by the innate system, and the adaptive system, differ?
4. How do the viral structures identified by B lymphocytes, T helpers (T_H), and cytotoxic T lymphocytes (T_C) differ?
5. Once these viral structures are identified, how do the humoral and cell-mediated arms of the adaptive immune system response act to clear infection? How do these actions actually function to clear viral infection?
6. What is the nature of "memory," which all vaccines seek to produce?

7.1 Physiological Factors and Barriers Affecting Resistance

The **epidemiological triad model** of disease addresses the interactions among the host, agent, and environment that produce disease (**FIGURE 7-1**). The triad model dictates that the condition of the host impacts its susceptibility to viral infection. Prevention efforts can be directed toward any or all of the three different elements depending upon the disease in question and what type of intervention is available and appropriate. Several physiological factors and barriers affect resis-

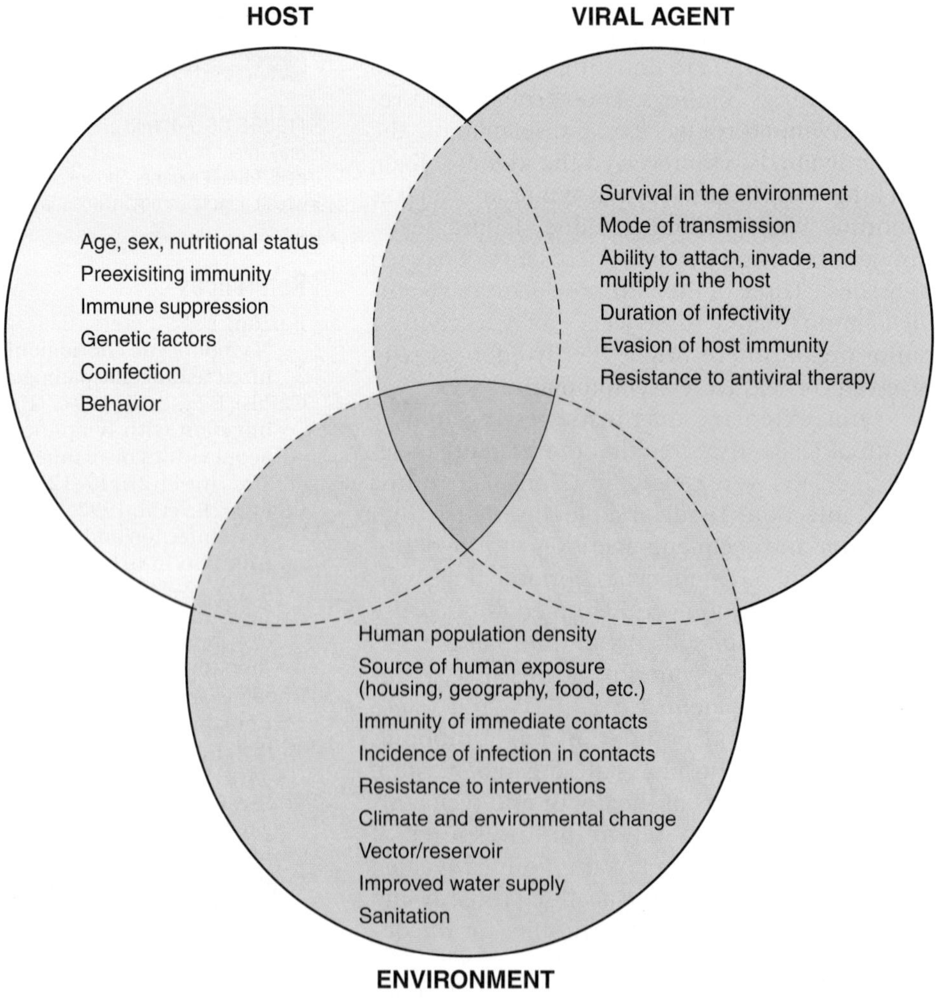

FIGURE 7-1 Triad model of disease causation.

tance to viral infection. These factors include but are not limited to:

- age
- nutrition
- hormones
- fever
- genetic factors
- dual infections
- species resistance

Age

Newborns and the elderly are typically the most vulnerable to viral infections, and they tend to experience the severest symptoms while infected. Newborns are protected from common viral infections immediately after birth through antibodies passed to them from their mothers. These antibodies decline rapidly in the first few months of life, though, as the newborn begins to develop its own immune system. During this period, viral infections are frequently seen in infants. Elderly individuals become vulnerable to infection as their immune systems decline with age. There are some exceptions to this generality. Herpesvirus infections are always more severe in adults than in the young. Another exception of vulnerability to viral infection occurred during the 1918 Spanish influenza pandemic. During that time, influenza A viruses caused severe symptoms in individuals during the prime of their lives (18–40 year olds) with developed and fully functioning immune systems (see Chapter 12).

Shingles is often a disease of the elderly. It is caused by varicella zoster, the same virus that causes chickenpox. After someone is infected with chickenpox, the virus DNA remains in the nuclei of sensory (skin) neurons. The virus becomes reactivated when the immune system is weakened due to aging, severe illness, or long-term use of corticosteroids. When the immune system cannot suppress the dormant virus any longer, the virus infection becomes productive again, causing an infection along the nerve (**FIGURE 7-2**).

Shingles begins with symptoms including fatigue, chills, and fever. On the third or fourth day, the individual begins to experience a bruised feeling on one side of his or her face or body, and the infected skin area becomes excessively sensitive. Pain, tingling, itching, and prickling sensations begin to affect the skin, and an inflamed rash appears. The rash consists of small blisters along the path of a nerve, providing a visual symptom that is a critical part of the diagnosis (Figure 7-2).

The rash is usually limited to one side of the body and rarely affects the lower part of the body or the face. The blisters usually crust and heal after five days. Half of shingles patients, however, experience pain—often quite severe—that persists for months and sometimes years. This is called **postherpetic neuralgia**. Overall health and nutrition often determine the severity of the illness and length of recovery.

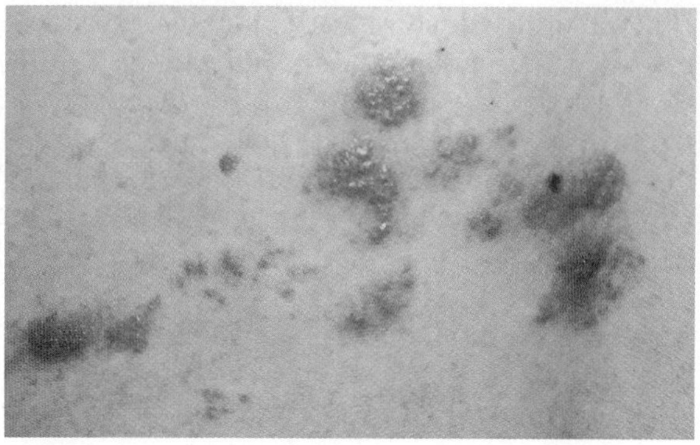

FIGURE 7-2 Shingles rash. Note: It follows a nerve, usually on one side of the body. Reactivation of Herpes zoster virus following treatment for Hodgkin's disease.

Nutrition

Nutritional deficiencies and malnutrition interfere with the integrity of skin and mucous membranes. This can result in an increase of infectious disease incidence in a population and in the severity of disease in an infected individual. Blood, mucus, milk, and other body fluids contain a wide variety of natural inhibitors, some of which can coat viruses and impede their attachment to cells. Hence, any reduced activity of these natural inhibitors may increase the success of an invading virus. Besides interfering with the integrity of physical barriers and fluids that protect the host against pathogens, it has been documented that nutritional deficiencies interfere with innate immunity (e.g., functioning of complement system and phagocytes) and adaptive immunity (e.g., antibody production and cell-mediated responses). These immune responses are discussed in detail later in this chapter.

In addition to severe malnutrition, a deficiency in essential vitamins such as vitamin A may have substantial effects on immunity. Between 100 and 140 million children in developing countries (especially those in Africa and Southeast Asia) ingest inadequate amounts of vitamin A. These children are at a 23% higher risk of dying from common infectious diseases. During a famine, the mortality rate can be as high as 50%.

Lack of vitamin A has shown a striking correlation with increased risk of common childhood diseases and even death. This is especially true for children afflicted with measles and diarrheal diseases. This recognition has resulted in the World Health Organization's (WHO) goal to eliminate vitamin A deficiency (VAD) through interventions that include a combination of breastfeeding and vitamin A supplementation, coupled with education and promotion of vitamin A-rich diets and food fortification. In 1998, the WHO and its major partners began to deliver vitamin A supplements through immunization programs. In 2001, vitamin A was given during Polio National Immunization days in over 60 countries, predicted to prevent over 250,000 childhood deaths. Vitamin A has been shown to act as an **adjuvant** to vaccines and, under certain circumstances, supplementation has been shown to enhance both cellular and humoral responses in animals and humans. An adjuvant is an agent added to enhance its effects (e.g., adjuvant enhances the immune response stimulated by an antigen when injected with an antigen or immunogen; see Section 7.7, Vaccine Additives).

Hormones

Patients undergoing treatment with hormones such as **glucocorticoids** (a group of corticosteroids) are at increased risk for the development of infections; for example, organ recipients undergo chronic steroid therapy. The therapy is intended to prevent rejection of the donor organ by the immune system. Common complications of this therapy include viral infections, especially cytomegalovirus and virus-related malignancies such as human herpes 8-associated Kaposi's sarcoma.

A recent study led by Medzhitov found that influenza virus infection in mice triggered an increase in glucocorticoid levels, which is part of a generalized stress response. In their study, the induction of glucocorticoids caused suppression of the immune response toward bacterial infection. The mice infected with influenza were challenged with a model bacterial pathogen (*Listeria*) and then studied for changes in the immune system. Influenza virus infection is known to increase one's susceptibility to secondary bacterial infections, particularly those opportunistic bacterial pathogens of the respiratory tract. Medzhitov's research suggests that the mechanism of influenza virus-mediated immune suppression was due to the increase in glucocorticoid levels. It showed that surgical removal of the adrenal glands from the mice, which are the main sources of glucocorticoids, eliminated most of the immunosuppressive effects of influenza infection.

In addition to patients undergoing hormone therapies to suppress the immune system, individuals under stress may produce hormones that decrease immunity. A 2001 study published by Stowe on immune responses and short-term spaceflight concluded that astronauts experienced reactivated latent herpesvirus (Epstein-Barr virus and cytomegalovirus) infections. This reactivation was correlated with the increase of neuroendocrine hormones produced by astronauts during flight. Perhaps neuroendocrine hormones play a role in cellular immunity. Further studies are needed to determine if immune suppression and herpesvirus reactivation will have important health consequences during long-term spaceflights.

Fever

Almost all viral infections are accompanied by a fever or febrile response. A fever is one of the body's defense mechanisms against some invading viruses and other microbes. A normal, healthy individual maintains a constant body temperature of 98.6°F (37°C) via the hypothalamus region of the brain, which acts as the body's thermostat.

Fever induced by viral infections is elicited by substances called **pyrogens**. Immune cells such as phagocytes produce pyrogens in response to invading organisms. Pyrogens include interleukin 1 (IL-1), interleukin 6 (IL-6), and tumor necrosis factor-α (TNF-α). **Interleukins** are cytokines that act to regulate the immune response, especially T lymphocyte proliferation. **Cytokines** are regulatory proteins made by cells that act on other cells to stimulate or inhibit their function. These pyrogens stimulate the hypothalamus of the brain to produce prostaglandins that in turn raise the body's temperature, resulting in fever (**FIGURE 7-3**). Influenza viruses, myxoviruses, adenoviruses, herpesviruses, paramyxoviruses, coxackieviruses, and Western equine encephalitis viruses induce pyrogenic effects. Other viruses, such as vacinia virus and polioviruses, do not. Several vaccinia virus strains prevent fever by producing IL-1 receptors. These receptors sequester cellular IL-1, thereby moderating the fever response.

Genetic Factors

Certain populations of humans are more susceptible to viral infections than others; for example, individuals with two defective chemokine receptor 5 genes (*CCR5*) are more resistant to being infected with HIV compared to those with a functional *CCR5* gene. Those infected with HIV that possess an altered chemokine receptor 2 (*CCR2*) gene may progress to full-blown AIDS at a much slower rate than those who do not.

Researchers have identified host genetic factors that influence the outcome of viral infections. The MHC/HLA (major histocompatibility complex/human leukocyte antigen) is located within a large gene region that serves as a resistance locus for

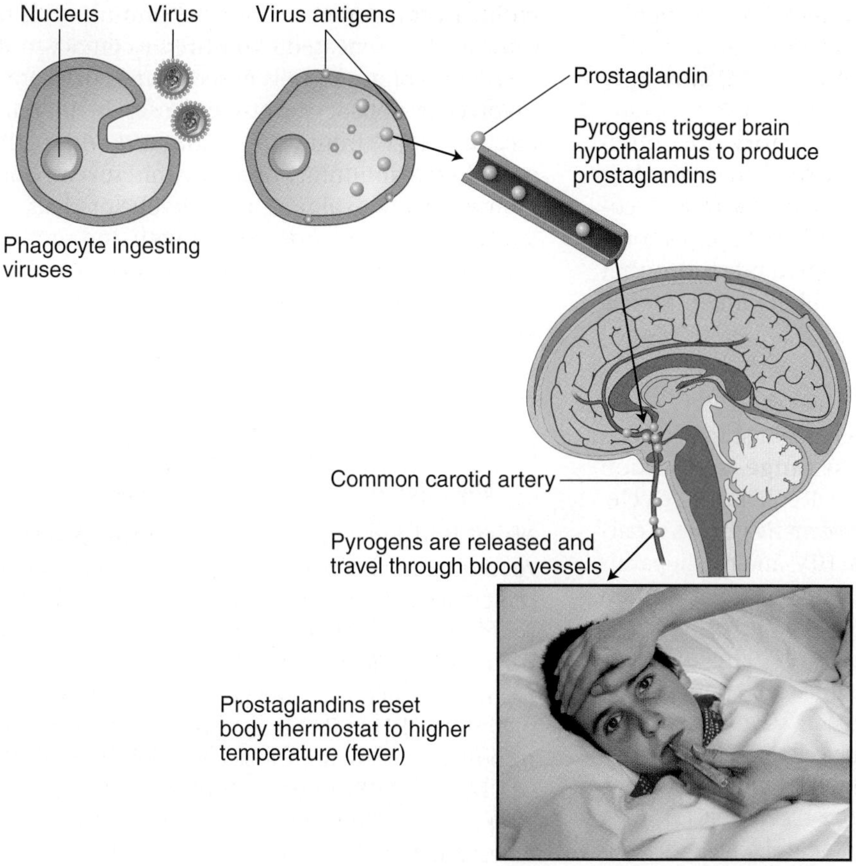
Nucleus
Virus
Virus antigens
Prostaglandin
Pyrogens trigger brain hypothalamus to produce prostaglandins
Phagocyte ingesting viruses
Common carotid artery
Pyrogens are released and travel through blood vessels
Prostaglandins reset body thermostat to higher temperature (fever)

FIGURE 7-3 Fever response.

nearly every human infectious disease or autoimmune disease that has been studied. This super locus is located on chromosome 6. It contains many genes related to immune system function in humans.

Several viral resistance genes have been mapped in laboratory (inbred) mice. These include the *mx* (myxovirus) and *Flv* (flavivirus) genes. The *mx* gene product protects the mouse host during influenza A or B virus infection by interacting with the viral polymerase, thereby blocking synthesis of viral messenger RNAs (mRNAs). Later, it was discovered that the *mx* gene in mice also confers resistance to other RNA viruses such as members of the Orthomyxoviridae, Paramyxoviridae, Rhabdoviridae, Bunyaviridae, and Togaviridae families. Over the years, human **homologs** of the *mx* genes have been discovered. Some of these mx proteins possess antiviral activities and some have unknown functions within the interferon system (see The Interferon Response in Section 7.2).

The inbred mouse *Flv* gene plays a role in resistance to viruses within the genus Flavivirus. This genus includes the human pathogens West Nile virus, Dengue fever virus, and Yellow fever virus. The *Flv* resistance gene encodes for a full-length protein within a family of proteins termed Oas1b, or 2′5′-oligoadenylate synthetase 1B. Susceptible mice encode a truncated form of the protein that lacks 30% of the C-terminal due to a premature stop codon encoded by the *Flv* gene. The known function of Oas1b is to produce 2′5′ oligo A polymers that in turn activate a latent ribonuclease, RNase L. RNase L degrades viral and cellular single-stranded RNAs (ssRNAs). This is part of the host's interferon response, which will be discussed in detail later in the chapter (see The Interferon Response in Section 7.2). The Oas1b proteins from both susceptible and resistant mice differ by one unique change within the protein: In the P-loop motif of the protein, a region involved in RNA recognition and binding, four amino acids are missing. It is believed that the Oas1b protein may specifically recognize and bind RNA structures unique to flavivirus RNAs.

Dual Infections

Probably the first virus that comes to mind when thinking of altering host resistance is human immunodeficiency virus (HIV). Opportunistic infections are the hallmark of HIV infections. Patients suffering from full-blown acquired immunodeficiency syndrome (AIDS) usually die as a result of a number of secondary infections that are difficult to control. These infections most often are caused by *Candida albicans*, *Pneumocystis carinii* (pneumonia), *Toxoplasma gondii*, *Cryptosporidium coccidi*, *Myco-*

bacterium avium, *Mycobacterium tuberculosis* (tuberculosis), and hepatitis viruses. These opportunistic infections occur because of the decreased immunity in HIV-infected individuals.

Viral infections of the respiratory tract often lower host resistance to secondary bacterial infections (e.g., hormonal changes: induction of glucocorticoids levels that cause immune suppression), resulting in secondary pneumonia infections that can be fatal. Notable examples of this are influenza and measles virus infections.

Species Resistance

The range of cells that can act as a host to a virus is referred to as the virus's **host range** (see Section 4.2, Key Steps of the Viral Replication Cycle: Attachment). The host range of many viruses is narrow; for example, poliovirus, HIV, and the hepatitis viruses have a very limited host range (e.g., human or other primate cells). This can be the case because host cells that are resistant to specific viral infections do not contain conserved cellular receptors located on the outside of cells used for viral entry, and, therefore, penetration into the cell to cause an infection cannot be achieved. Poliovirus only infects primates. Poliovirus binds to the receptor CD155. A recent study by Ida-Hosonuma and colleagues determined that poliovirus is restricted to primates because the gene sequence of the CD155 poliovirus receptor remains conserved in primates but is highly variable in other mammalian species such as rabbits and ring-tailed lemurs, which are not susceptible to poliovirus infection. Their work has suggested that rapid changes in the *CD155* gene during mammalian evolution determined the host range restriction of polioviruses.

Besides virus-receptor interactions, viruses may be dependent upon intracellular host factors for uncoating, nuclear import, and viral RNA transcription/translation. Host range is almost always determined by intracellular factors for certain DNA viruses and retroviral families. Intracellular host factors also play a critical role in determining the host range of hepatitis C virus.

7.2 Host Defenses Against Viral Invaders: Nonspecific Host Defenses (Innate Immunity)

Nonspecific immunity protects us against any pathogen, regardless of the species or type of microbe. Host resistance (or innate immunity) is not improved by repeated exposure or contact to the pathogen. Immune cells and enzymes or proteins involved in nonspecific host defenses do not retain any "memory" from prior encounters with pathogens. Innate immunity occurs within several hours after exposure to almost any microbe or virus. It is the immunity that one is born with and serves as the first response to eliminate pathogens and prevent infection. This section will answer the questions: How are viruses initially detected by the innate immune response? What kinds of viral molecules are recognized by the innate system?

Nonspecific Defense: Mechanical Immunity and Phagocytosis

Most body defense cells contain **pattern-recognition receptors** (PRRs) that detect **nonself** molecules or "patterns" shared by viruses that are referred to as **pathogen-associated molecular patterns** (PAMPs). Viral PAMPs are essential for replication such as a viral genome that may consist of DNA or RNA. PAMPs trigger innate immunity responses. Following the recognition of viral RNA or DNA, the PRRs undergo conformational changes or specific modifications that drive signaling pathways that directly inhibit viral infection (e.g., induction of interferon, discussed later in this chapter).

The first line of defense against invading viruses is probably the trapping of viruses by mucus and phagocytes in the mucosal tract. This is often referred to as **mechanical immunity**. Phagocytosis is the engulfment and ingestion of foreign material (i.e., viruses and bacteria) by **phagocytes**. Examples of phagocytic cells are **macrophages**, **neutrophils**, and **blood monocytes**. These cells are analogous to the video game character Pacman that runs through a maze while eating all of the characters in its path. Phagocytic cells patrol the tissues or blood, eating and digesting foreign invaders they encounter.

Monocytes are precursors to macrophages. They are large white blood cells containing a nucleus and granulated cytoplasm. They circulate in the bloodstream for about eight hours, during which time they enlarge, migrate to tissues, and differentiate into macrophages. Macrophages are 5 to 10 times larger than monocytes and contain more lysosomes (organelles that contain many hydrolytic enzymes) than the monocytes. Some macrophages become fixed in tissues, whereas others are wandering macrophages that travel in tissues by amoeboid movement. Macrophages can ingest viruses and other foreign invaders, dead cells, cell debris, and other cellular matter. Macrophages are attracted and move toward substances via **chemotactic** behavior.

Refresher Immunology

The primary function of major histocompatibility complex (MHC) is to present a sampling of all peptides being produced in a nucleated cell of the body (for MHC I) or that were engulfed by a phagocyte (for MHC II) to T cells. Healthy cells will be ignored while cells containing foreign proteins (e.g., cells infected by viruses) will be attacked by the immune system.

It is these peptides that are recognized by T helper cells (MHCII) and T killer (cytotoxic) cells (MHCI).

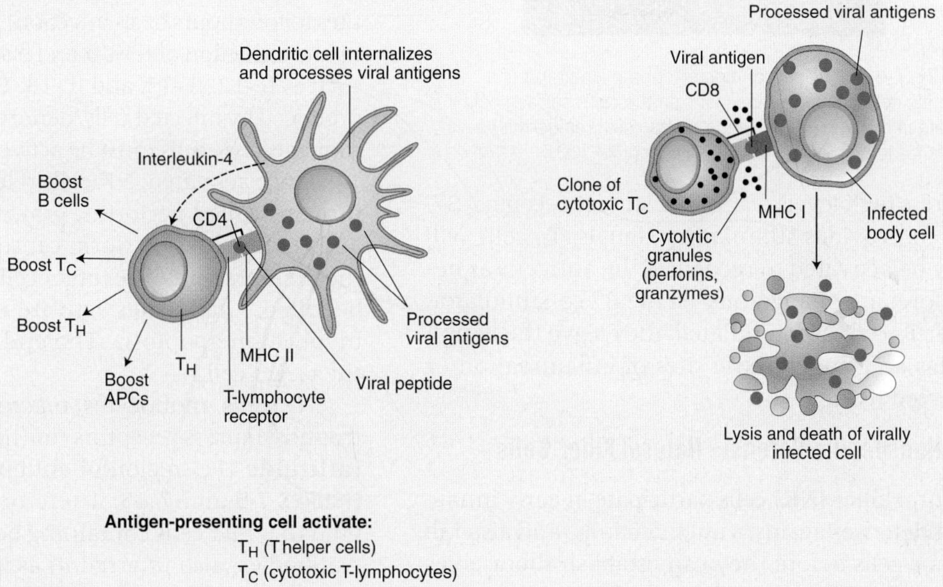

Macrophages patrol tissues for foreign invaders, whereas neutrophils and blood monocytes patrol circulating blood. Neutrophils are white blood cells that begin their two-week lifespan in the bone marrow, which contains high numbers of neutrophils. This enables the body to respond to viral or other microbial challenges with a massive out-pouring of neutrophils.

Some of the neutrophils migrate to the peripheral blood vessels and circulate in the blood for 7 to 10 hours before they migrate into tissues. Chemotactic factors released from inflamed areas attract neutrophils to the inflammation site in the body. Neutrophils contain specialized granules that possess peroxidases, lysozyme, hydrolytic enzymes, collagenase, and lactoferrins that aid in their phagocytic activity.

Nonspecific Defense: Dendritic Cells

Dendritic cells are covered with long spiky arms that resemble the dendrites of nerve cells. Dendritic immune cells are present in tissues that have contact with the environment, such as the skin (where they are called Langerhan's cells), mucous membranes, and the lining of the nose, lungs, and gastrointestinal tract. Langerhan's cells were first described by medical graduate student Paul Langerhan, who was working under the mentorship of Professor Rudolf Virchow at the Institute of Pathology of the University of Berlin in 1868. In his 1868 landmark paper, "Uber die Nerven der menslichen Haut," Langerhan described the cells as nerve endings of the skin. In 1882 he corrected his interpretation, stating: "I am now convinced that my cells are in no way essential for nerve endings."

Dendritic cells express MHC I and MHC II molecules on their surfaces (see **Refresher: Immunology**). They act as scouts that are very efficient at identifying foreign invaders such as viruses and bacteria, even when they are present in minute numbers (**FIGURE 7-4**). Dendritic cells internalize the pathogen, digest it, and display or present the foreign peptides on their surface via the MHC II molecules to T helper (T_H) cells. (T_H cells are described later in this chapter as part of the discussion on specific immune defenses.) Activated dendritic cells also

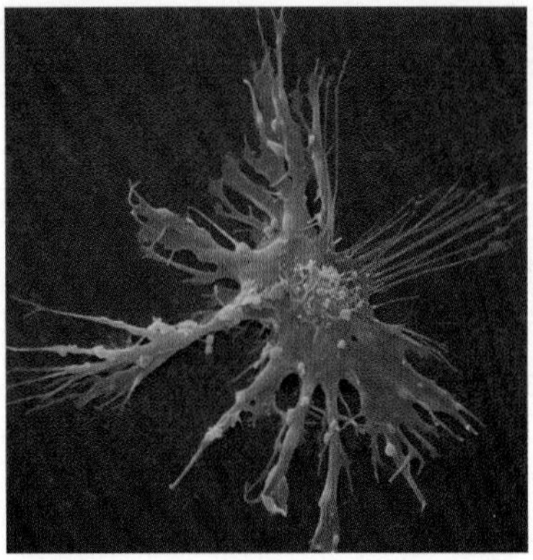

FIGURE 7-4 Colorized scanning electron micrograph of a dendritic cell. The long projections on the cell's surface help it migrate in the upper layer of the human skin (epidermis) as it scouts for foreign antigens. Magnification: x3,500.

express high levels of another surface receptor, B7, which provides stimulatory signals. T_H cells will only be activated to recognize the antigens as being foreign or dangerous via the B7 co-stimulator. After T_H cells are stimulated, they leave the lymph nodes and travel to the sites of inflammation or infected tissues.

Nonspecific Defense: Natural Killer Cells

Natural killer (NK) cells participate in early innate host defenses against viral infections, and also kill tumor cells before they can establish themselves as cancers. They recognize cells that undergo a declined expression of MHC molecules or cells that contain surface antigens displayed by some tumor cells or virally infected cells. NK cell deficiency in humans is associated with recurrent herpesvirus and cytomegalovirus infections, the rapid progression of cancers (especially lymphomas), hepatitis, AIDS, chronic fatigue syndrome, and various immunodeficiency and autoimmune diseases.

In healthy people, NK cells represent 5% to 15% of the total lymphocyte population. They respond immediately to viral infections (within minutes to four hours of an infection). They are activated by the induction of IFNs α and β and other cytokines such as IL-12, IL-15, and IL-18. These cytokines are produced by infected cells, dendritic cells, or by macrophages. NK cells must be active to do their job.

Once activated, NK cells release pore-forming proteins called perforins, granzymes (proteases), and chemokines. Perforin causes pores to form in the membranes of the target cells that are going to be killed. Granzymes activate enzyme pathways involved in apoptosis. The end result is death of the target cell.

NK cells, monocytes, macrophages, and neutrophils express receptors for the **fragment crystallizable** (Fc) region of antibodies, as shown in **FIGURES 7-5** and 7-15. Therefore, these cells can bind to target cells containing bound **antibodies**. Antibodies, also referred to as immunoglobulins, are proteins typically found in blood or other

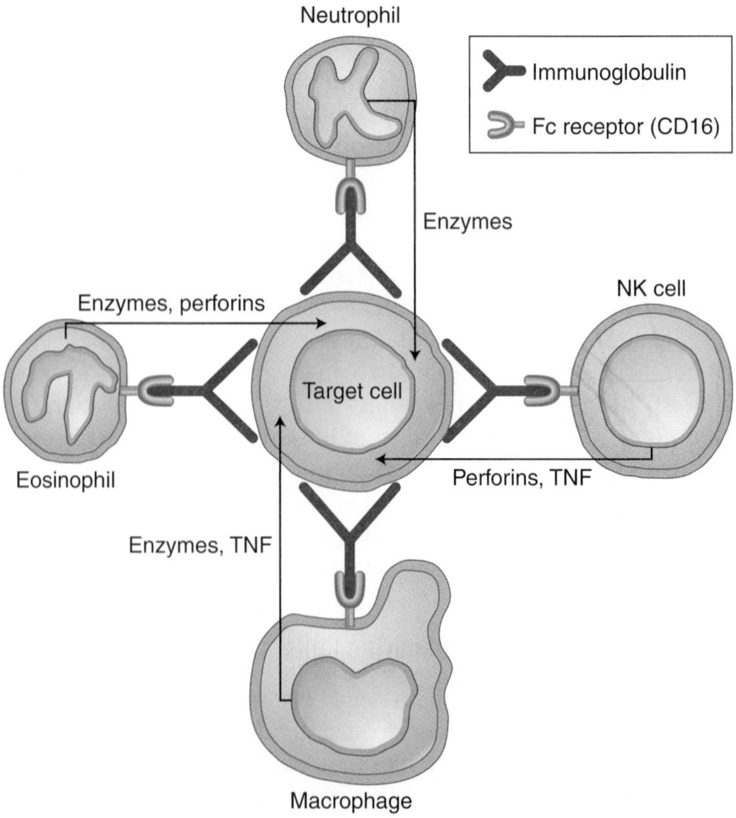

FIGURE 7-5 Natural killer cells, eosinophils, macrophages, and neutrophils can bind to target (virally infected) cells containing antibodies attached to their surfaces via their Fc receptors. Adapted from A. J. Cann, "Humoral Immunity," *Microbiologybytes: Infection & Immunity*, http://www.microbiologybytes.com/iandi/3b.html.

bodily fluids that are part of the adaptive immune system. Antibodies identify and neutralize viruses or other foreign entities. If NK or any of the aforementioned cells interact with these target cells, they will kill them. This type of cytotoxicity is referred to as antibody-dependent cell-mediated cytotoxicity (ADCC).

The Interferon Response

Before specific immune system responses mount, the body begins both subtle and dramatic nonspecific immune responses that are induced by proteins called **interferons** (IFNs) at the early stage of viral infection. The ultimate outcome of this response is cellular suicide of infected cells. This is a clever way to localize infections and prevent a virus from spreading rapidly throughout the body before the immune system can finish the job. It is sometimes referred to as an infection "firebreak."

History of IFN: The Magic Bullet

In 1957, researchers Alick Isaacs and Jean Lindenmann of the National Institute for Medical Research (NIMR) in London first described interferon in an article entitled "Virus Interference I. The Interferon," published in the *Proceedings of the Royal Society of London*. Their pioneering work spurred clinical trials because interferon was speculated to be the "magic bullet" for treating viral infections, similar to what penicillin did for bacterial infections.

The Isaacs and Lindenmann experiment demonstrated that a substance "secreted by" influenza-infected cells interfered with subsequent influenza virus infections in neighboring cells (**FIGURE 7-6**). It was later discovered that IFNs are naturally occurring proteins that possess antiviral activity. Host cells secrete IFNs in response to viral invasion. The secreted IFNs bind to receptors on neighboring cells. This receptor-binding event acts as a signal to other cells. The end result is the creation of a cellular antiviral state that will be discussed in greater detail later in this chapter.

There Are a Multitude of IFNs

IFNs are **cytokines**—protein factors made by cells to act on other cells. In other words, they play a role in cell-to-cell communication. The number of proteins in the cytokine family continues to grow as newly discovered proteins that act on immune and nonimmune cells are added to it. Most cytokines are biochemically and structurally similar to hormones and growth factors, and they exit host cells and bind to their cell-specific cytokine receptors located on the outside of cell surfaces. This binding event activates intracellular signaling pathways, ultimately resulting in altered gene expression of target cells.

Cytokines differ from hormones and growth factors in that they are produced at very low levels within cells and are induced in response to host challenges, especially microbial infections. This text focuses on the antiviral activities of IFNs; however, IFNs affect a number of other processes, including the regulation of cell growth, cell differentiation, cell survival and death, and roles in infectious and inflammatory diseases, autoimmunity, and cancer.

IFN research has been complicated by the fact that there are a multitude of human IFNs, which have been classified into three types: I, II, and III. Their division is based on differences in amino acid sequence composition and the cellular receptor to which the IFN molecule binds.

Type I IFNs share amino acid sequence similarities. All Type I IFNs bind to a receptor complex termed the human IFN-$\alpha\beta$ receptor (IFN-$\alpha\beta$R). There are at least eight different subtypes of IFN: alpha (α), beta (β), delta (δ), epsilon (ε), zeta (ζ), omega (ϖ), kappa (κ), and tau (τ). Type I IFNs exhibit antiviral and some antitumor activities. IFN-α has been approved for use as therapy to treat viral infections (e.g., hepatitis C and B infections) and many types of hematological cancers and solid tumors. IFN-β has been used in the treatment of multiple sclerosis. Homologs of Type I IFNs are found in many species, including rats, mice, birds, reptiles, and fish; for example, IFN tau (τ) is a unique subclass of Ω IFNs that are transiently produced in ruminants during early pregnancy.

There is only one human Type II IFN: IFN gamma (γ), which binds to the IFN-γ receptor (IFN-γR). Interferon gamma is known as immune interferon, and plays a different role, that of up-regulating the immune response, compared to all other interferons. The antiviral activities of IFN-γ are very potent for some DNA viruses, including herpesviruses. However, the antitumor activities of IFN-γ are weak and have not been very useful in treating cancer.

Discovery of members of a new family of IFNs, denoted by lambda (λ), were reported in 2003. They bind to a novel receptor complex known as the class II chemokine receptor (IFN-λR). IFN-λ's possess antiviral activity and enhance MHC class I antigen expression. (MHC will be discussed later in this chapter.)

The Type I IFN Pathway

This chapter focuses on Type I IFNs α and β because of their important roles in the induction of the antiviral state. Almost any virus-infected cell will produce IFN-α and IFN-β.

The presence of viral dsRNA, a crucial player of the IFN pathway, triggers many of the cellular responses to virus infection. Viral double-stranded

Isaacs and Lindenmann 1957 experiment

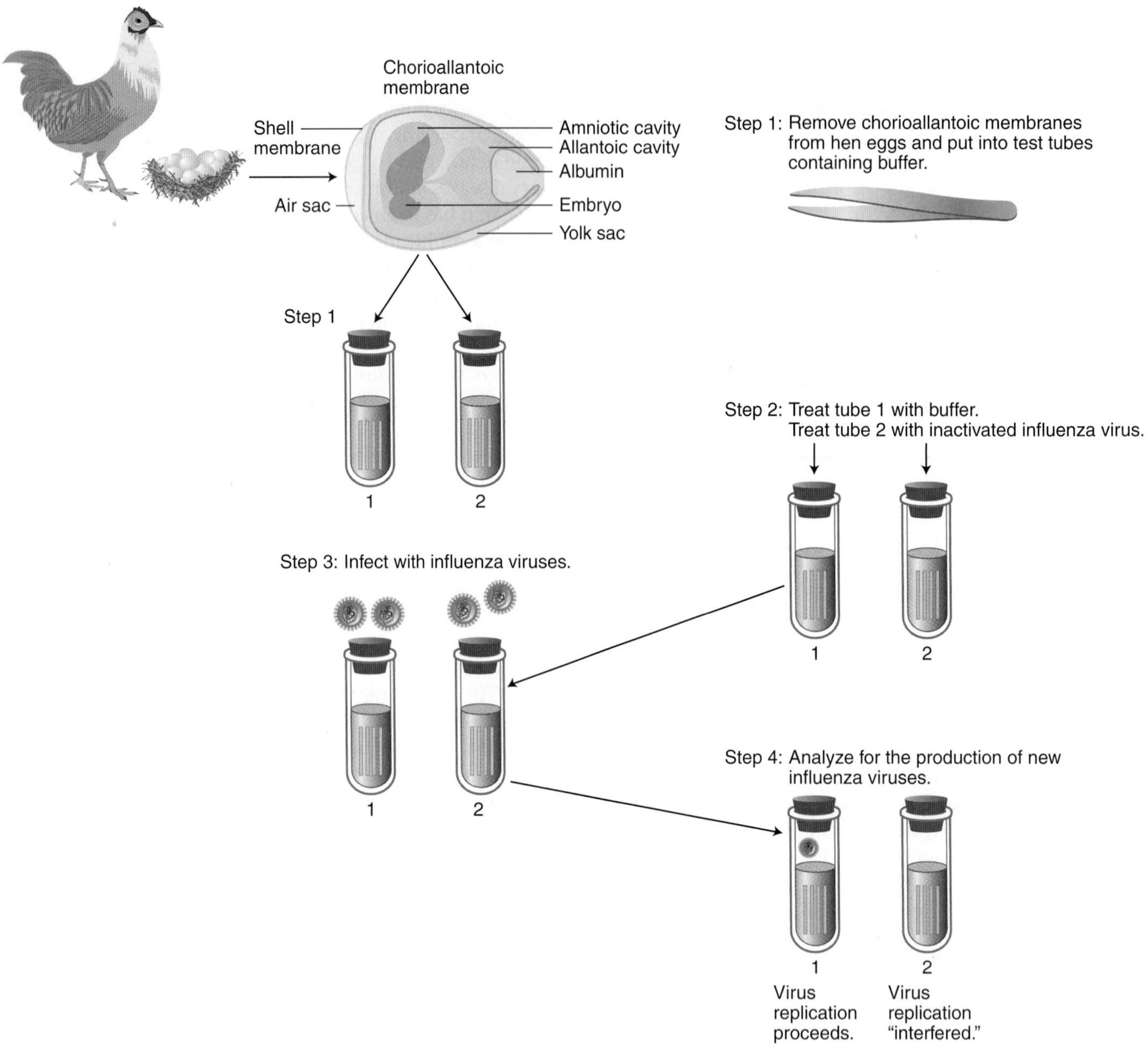

Chorioallantoic membrane
Shell membrane
Air sac
Amniotic cavity
Allantoic cavity
Albumin
Embryo
Yolk sac
Step 1: Remove chorioallantoic membranes from hen eggs and put into test tubes containing buffer.
Step 1
1
2
Step 2: Treat tube 1 with buffer.
Treat tube 2 with inactivated influenza virus.
1
2
Step 3: Infect with influenza viruses.
1
2
Step 4: Analyze for the production of new influenza viruses.
1
2
Virus replication proceeds.
Virus replication "interfered."

FIGURE 7-6 A graphic representation of the Isaacs and Lindenmann experiment, which demonstrated that host cells produce substances that interfere with the replication of viruses.

RNA (dsRNA) is a replication intermediate produced within cells during viral genome replication and also during transcription in viruses that have overlapping open reading frames (ORFs). The dsRNA must be at least 30 nucleotides in length to activate the IFN response. These long dsRNA molecules are not produced in uninfected cells. Hence, long dsRNA molecules act as a "viral infection signal" that tells the host cell to set up its antiviral state.

The long viral dsRNA molecules should not be confused with short interfering RNAs (siRNAs). siRNAs play a role in the RNA interference (RNAi) pathway, a recently discovered gene-silencing mechanism observed in fungi, plants, invertebrates, and some vertebrates. The siRNAs are less than 25 nucleotides in length, thus escaping the IFN pathway.

dsRNA "sets up" the antiviral state by stimulating virally infected cells to produce and secrete Type I IFNs, which then bind to neighboring IFN-αβRs present on the same cell or neighboring cells. The autocrine effects are quite significant during early infection.

IFNs do two key things. They directly interfere with viral replication in infected cells and they upregulate molecules required to alert the adaptive immune system to viral infection (e.g., chemokines, pyrogens, MHC-1 molecules). The IFN receptor-binding event initiates signaling pathways that result in increased transcription of many genes. Their gene products possess diverse functions such as antiviral activity, chemokine activity, transcription regulation, and antigen presentation (discussed later in this chapter). In addition to being involved in setting up the antiviral state, IFNs halt cell proliferation of uninfected cells and alter transcription of MHC genes.

The list of **interferon-stimulated genes** (ISGs) with direct antiviral activity is now over 20. More than 300 ISGs exist but most with no known function. The ISGs list is being compiled through the use of extensive **microarray analysis** (refer to Chapter 5 for this method) of RNA samples collected from experiments on human and mice cell lines treated with IFN α, IFN β, or IFN γ. An INTERFEROME open access database contains more than 28 publicly available microarray datasets (http://www.interferome.org).

Here we will discuss three well-characterized IFN-induced *antiviral proteins*: the dsRNA-activated protein kinase (PKR), Oas1b, and RNase L. These proteins are normally produced in an inactive form at low levels in the cell. After IFN-induction, the cell begins to make high levels of these proteins.

PKR is a serine/threonine kinase. When active, it functions as a kinase to phosphorylate serine or threonine residues on itself (termed autophosphorylation) or on amino acid residues of other protein targets. PKR becomes active when it binds to dsRNA (a by-product of viral replication). Once activated, PKR autophosphorylates and subsequently phosphorylates eukaryotic initiation factor-2 α (eIF-2α). eIF-2α is involved in the initiation of protein synthesis. Phosphorylation of eIF-2α blocks translation both at the level of host and viral protein synthesis. Protein synthesis is vital to the survival of cells and therefore affects viral production. PKR's activity shuts down host cell activities, leading to cellular and viral incapacitation (**FIGURE 7-7**). 2′5′-oligo(A) synthetase 1B and RNase L are IFN-induced antiviral proteins that act in concert. The enzyme Oas1b is activated by viral dsRNA, converting adenine triphosphate (ATP) into 2′5′-oligomers of adenylic acid. The 2′5′-A oligomers activate a latent ribonuclease, RNase L. RNase L catalyzes the degradation of both viral and host cellular RNAs resulting in a cellular altruistic event (Figure 7-7).

IFN-Resistant Viruses: The Battle Against PKR

Different viruses have evolved mechanisms to overcome the antiviral effects of PKR. Strategies that some viruses use involve the synthesis of viral products, RNA, or proteins, which bind and sequester dsRNA or bind directly to PKR, disabling inhibition of the cellular protein synthesis machinery. Details pertaining to specific IFN-resistant viruses are discussed in Chapters 12 (Influenza Viruses), 14 (Poxviruses), and 15 (Herpesviruses), which focus on individual viruses.

Marketing IFN

The IFN story represents the complexities involved in the development of commercial drug therapies. Immediately after their discovery, IFNs were viewed as a potential antiviral panacea. Researchers who tried to repeat Isaacs and Lindenmann's experiments, though, were unsuccessful. Skeptics nicknamed interferon "misinterpreton." In spite of this, there remained a small group of believers and the press coverage about interferon did not go unnoticed. By May 1958, patent applications were filed in the United States, Canada, and Germany.

Public interest in the United Kingdom was slow to develop, but foreign interest prompted the United Kingdom's Medical Research Council (MRC) and NIMR to consider collaborating with foreign and British companies to develop IFN as an antiviral agent. When penicillin was developed earlier in the United Kingdom, their scientists were not able to control patents and many believed that penicillin was lost to American drug companies. The "missed opportunity" to develop penicillin and control its associated patents in the United Kingdom was fresh in the memory of British drug companies and scientists. Thus, in July 1959 a collaborative agreement was formed between the NIMR, MRC, the NRDC (National Research Development Corporation), and the British pharmaceutical companies Glaxo Laboratories, Imperial Chemical Industries-Pharmaceuticals (ICI), and Burroughs Wellcome. One result of this collaboration between government and industry was the formation of the Scientific Committee on IFN, which consisted of representatives from each party. By September of that year, the committee's meetings, which were chaired by Isaacs, concentrated on four major IFN research problems:

- The species-species effect of IFNs. IFN isolated from rabbits had the best antiviral effect when tested in rabbit tissues, whereas IFN isolated from chick cells had the best antiviral activity in chicken tissues. This phenomenon remains true today.
- Standardization of biological IFN activity assays. Each collaborating laboratory needed to use the same assays to produce comparable measurements of IFN activity.

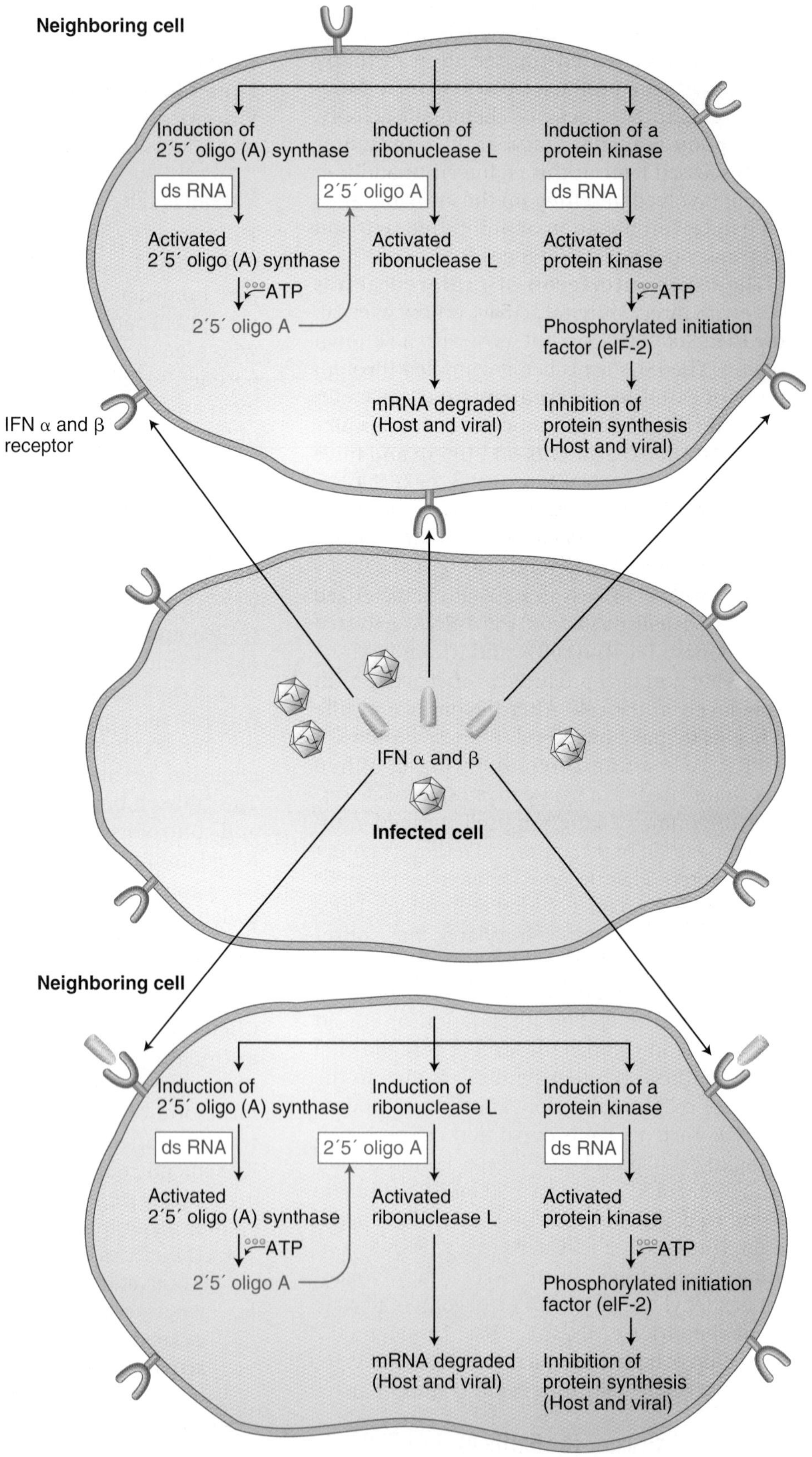

FIGURE 7-7 The type 1 interferon pathway: "cellular altruism." Infected cells secrete IFNs that bind to receptors on neighboring cells. This binding event causes the cell to express several genes, including three that are involved in setting up an antiviral state: PKR, ribonuclease L, and 2′5′-oligo(A) synthetase. The viral replication intermediate, dsRNA, activates PKR and RNase L. The end result is the destruction of viral and host mRNAs and the inhibition of protein synthesis.

- Large-scale production of IFN. Only minute quantities of IFN prepared from monkey kidney cells could be produced at that time.
- The application of IFN from the test-tube to animal testing. At the time, scientists were trying to demonstrate IFN's antiviral effects in mice. Determining the doses needed, however, proved difficult, and the results of the tests were not reproducible.

In June 1961, volunteers of the IFN scientific committee initiated the first human experiments. David Tyrell and another committee member at the MRC's common cold unit in Salisbury had inoculated themselves intranasally with a small amount of monkey IFN that they had prepared themselves. A couple of days later, they inoculated themselves with a strain of Coxsackie virus that was known to cause the common cold in humans 90% of the time. Months later, the Coxsackie virus could be isolated from nasal washings of both of them. Tyrell had experienced a cold but the other committee member did not. The IFN had not caused any side effects and the committee members agreed that small-scale human trials were doable.

At this time, there were no mandatory safety standards in the United Kingdom for testing drugs, nor any toxicity testing or clinical trial regulations for experiments on humans. The committee agreed to apply safety and toxicity standards used for vaccines (i.e., polio vaccine) toward IFN. This required that IFN be injected into mice and guinea pigs and if there were no adverse effects, it would be injected into the skin of committee members. This turned out to be problematic because guinea pigs died 3 to 4 days after injection with IFN. The scientists argued this was due to minute amounts of penicillin that had contaminated the IFN preparation.

IFN was isolated from monkey cells that were grown in medium containing penicillin to prevent bacterial contamination of the monkey cell cultures. It was known that guinea pigs were highly susceptible to small amounts of penicillin. After some debate, small-scale human trials continued in 1962. The committee chose to investigate IFN protection against vaccinia virus and common cold viruses. Unfortunately, IFN protection could not be demonstrated and interest in IFNs plummeted. The collaboration continued, but resources declined because of the expense involved in producing monkey IFN and the lack of understanding of IFN's mechanism of action.

During the late 1960s and into the early 1970s, the second and third waves of human trials were held. About this time, Matilda Krim, a tumor biologist at New York's Memorial Sloan-Kettering Cancer Center, began searching the scientific literature for promising new developments in the area of cancer research, and IFN came up in the database as showing an inhibitory effect on tumor viruses and tumors in mice and rats. Another study suggested that IFN had an inhibitory effect on tumors unrelated to virus infection.

News relating to IFN's potential as a new anticancer drug resulted in worldwide media hype. This was during a time in which conventional cancer therapies consisted of radiation, radical surgery, and harsh chemotherapy with severe side effects. The National Cancer Institute (NCI) was under public attack because of the low success rate associated with "battling cancer," and there was a growing demand for effective and less toxic remedies. IFN looked promising because it was a "natural" inhibitor that reportedly did not have toxic side effects.

Krim began applying for funding toward IFN research from the NCI and the National Institute of Allergy and Infectious Diseases (NIAID). Requests for funding were denied in part because experimental data were weak. IFN preparations were not "pure," and reviewers raised concern over funding such a high-risk project. Documentation that IFN inducers (e.g., dsRNA) were highly toxic also raised concerns by the grant review panel. Meanwhile, the United Kingdom's Scientific Committee on IFN dissolved due to IFN's disappointing results.

The early preparations of IFN used in scientific experiments and clinical studies consisted of a crude protein fraction that contained less than 10% interferon by weight. In addition to blocking antibody synthesis and other activities, these crude preparations not only contained antiviral activity, but also antiprotozoal, antibacterial, and antiproliferative activities. Unanticipated side effects occurred, ranging from flu-like symptoms to gastrointestinal effects that included anorexia, vomiting, diarrhea, and abdominal pain. There were central nervous system effects such as depression, dizziness, and altered mental states (confusion), and cutaneous adverse reactions such as rashes, dry skin, and alopecia (hair loss). Occasionally, some individuals experienced cardiac difficulties. It was not known if these were caused by IFN or contaminants in the preparation.

In 1974, Hans Strander, a physician at the Karolinksa Hospital in Stockholm, Sweden, admitted to Krim that since 1971 he had been treating osteosarcoma patients with purified IFN supplied by Finnish physician Kari Cantell (**FIGURE 7-8**). One of Strander's patients had gone into remission and the growth of tumors in others seemed to show signs of halting. Strander was optimistic about the use of IFN and stated that he did not observe toxic side effects in patients.

FIGURE 7-8 Hans Arthur Strander worked in Kari Cantell's research laboratory in the 1960s.

Later, Strander's clinical study was controversial because he did not use a control group that was not treated with IFN concurrently during his investigation; however, his results and enthusiasm were enough to motivate Krim to persist in her work on IFN. Results of further clinical trials trickled in, and IFN began to look like "the miracle drug looking for a disease." It turned out that the key was using purified IFN, but unfortunately doing so turned out to be no easy task, and several years passed before the process was perfected. Success lay in the recombinant DNA technology of the early 1980s, which allowed scientists to clone several different species of the human IFN gene(s) and produce large amounts of it in *Escherichia coli.* After 20 years at a snail's pace, IFN research finally had the opportunity to quickly move forward. In addition, a number of companies were involved in altering the IFN gene(s) to make it more effective and with fewer side effects. IFN research helped genetic engineering move forward in the discipline of biotechnology.

The IFN story is one of persistence. Development and randomized controlled clinical trials required commitments between drug companies, governments, regulators, physicians, laboratory researchers, and patients and their families. Today IFN is used to treat a diverse range of viral diseases, immune disorders, nonmalignant tumors, and cancers. **TABLE 7-1** lists the therapeutic uses of IFN.

TABLE 7-1 IFN Clinical Trials

Type of IFN	Disease Treated
IFN α	Chronic hepatitis B (HBV) and C (HCV)
IFN α and β	AIDS-related Kaposi's sarcoma and HIV infection
IFN α and γ	Genital papillomavirus infection, laryngeal papillomatosis
IFN α and β	Myeloproliferative disorders (cancer spreads to the blood; e.g., childhood leukemia)
IFN α	Multiple myeloma (cancer involving plasma cells)
IFN α	Thrombocythemia (blood disease characterized by the production of too many platelets, resulting in bleeding because the blood cannot clot)
IFN α	Non-Hodgkin's lymphoma and Hodgkin's disease
IFN α	Gastrointestinal tumors
IFN α	Colorectal cancer
IFN α, β, and γ	Melanoma and basal cell carcinoma
IFN α and γ	Ovarian cancer
IFN α and β	Bladder cancer
IFN α, β, and γ	Kidney cancer
IFN α and γ	Mesothelioma (rare form of lung cancer, often associated with asbestos exposure)
IFN α	Hemangiomas (reddish-colored benign tumor with many dilated blood vessels connected to it)
IFN α and γ	Nonmalignant scleroderma (autoimmune disease of connective tissue that results in the buildup of scar tissue usually of the skin but can occur in organs)
IFN α, β, and γ	Multiple sclerosis
IFN γ	Chronic granulomatous disease (rare inherited disease of the immune system resulting in defective phagocytes), leishmaniasis (parasitic infection), mycobacterial infections (bacterial infection), osteopetrosis (rare congenital disorder in which bones are too dense), Omenn's syndrome (inherited immune disorder), hyperimmunoglobulinemia E (disease characterized by recurrent skin and lung infections), rheumatoid arthritis, atopic dermatitis, and psoriasis

Apoptosis and Viral Infection: A Double-Edged Sword

The term apoptosis has a long history in medicine, dating back to Greek physicians over 2000 years ago. It was used in the context of "the falling off of bones" by the Greek physician Hippocrates of Cos, and "the dropping of the scabs" by Galen, one of the first experimental Greek physiologists. In the 1960s, John Foxton Ross Kerr described the morphological changes of dying rat liver cells. He hypothesized that this type of cellular death was attributed to a deliberate sequence of events and coined the term apoptosis in a landmark paper coauthored with A. H. Wyllie and A. R. Currie that was published in the *British Journal of Cancer* in 1972. In the paper, the authors suggested that cell death might occur by a controlled mechanism that they called apoptosis. Kerr and his colleagues continued apoptosis research; however, the topic of apoptosis was largely ignored until researchers documented a correlation between apoptosis and immune system function and regulation. The immunology connection spearheaded an explosion of research. Today, over 199,100 papers have been published in the field of apoptosis, including research in premiere journals such as *Science* and *Nature*.

Apoptosis, or programmed cell death (PCD), can serve as a cellular defense mechanism to combat viral infection. It is either a way to remove cells that are no longer needed (e.g., the resorption of the tadpole tail at the time of its metamorphosis, or the sloughing off of the inner lining of the uterus—the endometrium—at the start of menstruation) or are harmful to the body (e.g., virally infected cells or cancer cells).

Apoptosis is a common cellular response to a virus infection. Virally infected cells usually undergo PCD by "dying from within," whereas the removal of cancer cells is induced from outside of the cell, such as when a T lymphocyte attacks a cancer cell (**FIGURE 7-9**). It is possible that a virally infected cell could cause a neighboring uninfected cell to undergo PCD if it were to shut off the production of growth factors to its neighboring cell, thereby causing cell death. A virally infected cell may also secrete factors to induce a neighboring cell to activate a PCD pathway (i.e., a Type I IFN pathway).

Apoptosis is triggered by events related to the disruption of the cell cycle or virus infection. The apoptotic response involves a cascade of cellular signaling pathways. The morphological and nuclear changes associated with apoptosis are usually caused by the activation of a family of proteases called **caspases**. Caspases are typically dormant in healthy cells but in response to cell-death stimuli, are converted into active enzymes that cause morphological changes associated with protease and nuclease activity (**FIGURE 7-10**). For example, caspase 3 causes the breakdown of several cytoskeletal proteins, cleavage of poly ADP-ribose polymerase and degradation of inhibitor of caspase-activated DNase (ICAD), releasing CAD, which cleaves cellular DNA. Apoptosis changes include cell shrinkage from its neighboring cells, chromatin condensation to the periphery of the nucleus, nucleolus disintegration, nuclear fragmentation, membrane blebbing, and the formation of apoptotic bodies containing cytoplasm, organelles, and nuclear DNA fragments (which resolve into a characteristic DNA ladder pattern when separated via agarose gel electrophoresis). Apoptotic bodies are engulfed by macrophages.

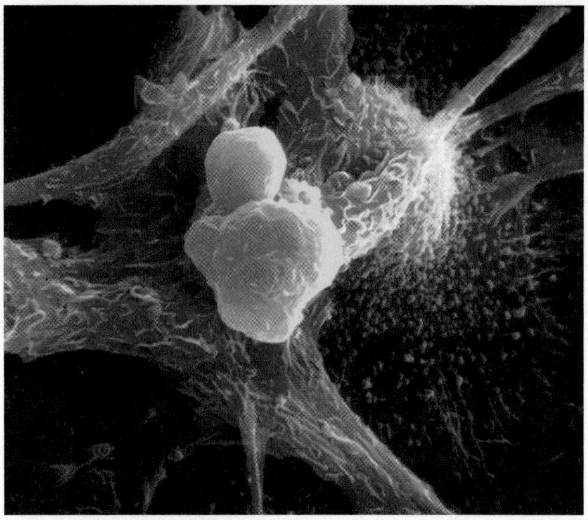

FIGURE 7-9 A colored scanning electron micrograph that shows a T_C lymphocyte (orange cell) inducing a cancer cell (pink) to undergo apoptosis. Note the pink apoptotic bodies emerging from the cancer cell. T_C lymphocytes are part of the body's immune response system. Their job is to survey, attach, and release chemicals to kill cancer cells in the body. T_C lymphocytes recognize surface markers on other cells in the body that label those cells for destruction. In this way, they help to keep virus-infected or malignant cells in check.

A number of viruses can trigger apoptosis in a wide range of infections including those of the central nervous system and heart (**TABLE 7-2**). Apoptosis can be detected in the brains of patients suffering from viral infections leading to central nervous system diseases, including patients with herpes simplex virus encephalitis, HIV-1 associated dementia, and cytomegalovirus encephalitis. Also, apoptotic cardiomyocytes have been observed in diseased human hearts in patients with active viral mycocarditis. Research using animal models reveals a positive correlation between apoptosis and disease severity, suggesting that apoptosis is a pathogenic mechanism in virus-induced disease. Apoptosis and mortality rates are high in mice infected with Sindbus virus or paralysis

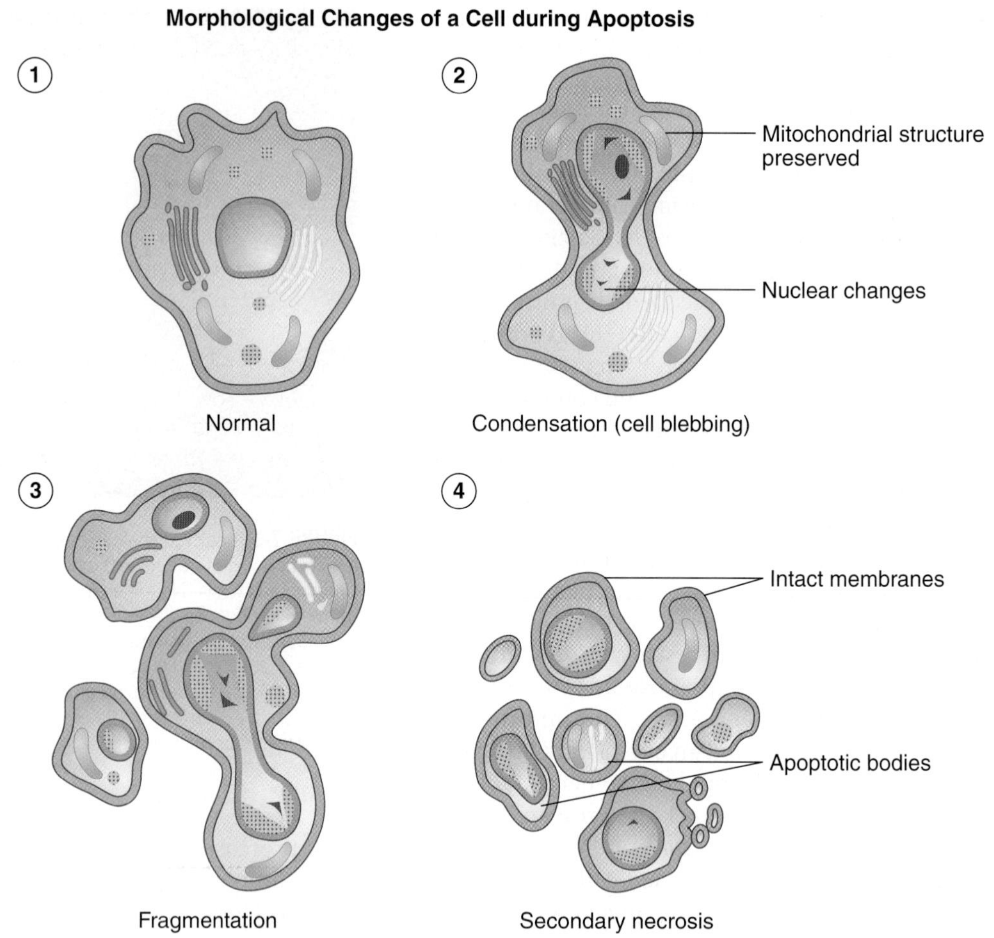

FIGURE 7-10 Cellular apoptosis occurs as a result of signals that activate a cascade of proteases and nucleases that contribute to the death of the cell. Many morphological changes are associated with apoptosis, such as membrane blebbing, DNA condensation, and fragmentation, and the formation of apoptotic bodies. Viruses can trigger and counteract apoptosis by mechanisms mediated by a number of different viral genes. Adapted from an illustration by Bender MedSystems.

with mice infected with West Nile virus. There are exceptions in which apoptosis may not be required for pathogenesis. For example, it has been shown that the brains of mice experimentally infected with a street rabies virus isolated from silver-haired bats showed little or no apoptosis even though the silver-haired bat rabies virus has been associated with most human cases of rabies in the United States.

It seems logical that viruses would be negatively affected by apoptosis. Viruses need metabolically active cells to ensure their reproduction and survival. In response to a virus infection, the host cell naturally responds by producing cytokines, including IFNs, nucleases, and proteases, all of which promote an apoptotic or cellular "suicide" event. This event localizes and prevents the spread of vi-

TABLE 7-2 Apoptosis Detected in Animal Models of Human Disease

System/Organ	Virus	Animal Model	Apoptosis in Acute Disease Detected in Animal Model(s)	Human Disease
Central nervous system (CNS; mainly brain and spinal cord)	West Nile virus	Infects neurons of mice or golden hamsters, causing **encephalitis** and **myelitis**	Apoptosis detected in same areas as viral antigens	Fever; may progress to encephalitis, meningitis, or myelitis
CNS (mainly brain and spinal cord)	Sindbis virus	Infects neurons of mice, causing acute CNS disease	Apoptosis detected in neurons of brain	Causes fever, rash, meningitis, arthritis, and encephalomyelitis
Heart	Coxsackievirus B3	Causes acute myocarditis associated with a low level of infected myocytes (1%–3%) in some strains of mice	Apoptosis detected in cardiomyocytes	Myocarditis

Source: Adapted from Clarke, P., Tyler, K. L. 2009. "Apoptosis in Animal Models of Virus-Induced Disease." *Nature Microbiology Reviews* 7: 144–155. (Table 1 on page 148).

ruses to other cells. The host also uses immune cells such as macrophages to engulf apoptotic cells or other immune cells like T lymphocytes to induce apoptosis. It is not surprising, then, that a virus would produce an antagonist of apoptosis, but what would be the advantage of promoting apoptosis during viral infection?

When apoptotic bodies are formed during PCD (as shown in step 4 of Figure 7-10), they pinch off of the dying cell and are consumed or engulfed by neighboring immune cells. The engulfment of apoptotic bodies that contain virus particles provides a mechanism for virus dissemination or release into extracellular fluids. Experiments by Mi and colleagues, and Teodoro and Brandon, show that if PCD is induced during viral genomic replication, virus production is compromised. If PCD is induced after virion assembly, though, there is enhanced viral release from infected cells and dissemination. It therefore appears that if viruses are able to manipulate apoptotic pathways, they can ensure their survival.

Complement System

A living organism's ability to recognize "nonself" or foreign invaders that may cause damage is believed to have evolved over 700 million years ago. The complement system consists of more than 30 different serum- and membrane-bound glycoproteins that act in sequence (i.e., one protein activates another, which in turn activates another, and so forth). The system has three distinct pathways: the classical, lectin, and alternative. Virus infection can activate all three pathways of the complement cascade, but only enveloped viruses are adversely affected by complement (a lipid bilayer is required). Complement activation ends in the formation of a membrane attack complex (MAC), which inserts into lipid membranes of bacteria, eukaryotic cells, or viruses and causes osmotic lysis.

The complement proteins present in blood serum are inactive. They are activated by

- the binding of complement protein C1q to viral antigen-antibody complexes (the classical pathway),
- the interaction of viral surface carbohydrates or virally infected cells with mannan-binding lectin (the lectin pathway), and
- other viral antigens or infected cells that cause the breakdown of an important complement protein termed C3, leading to release of lysosomal enzymes (the alternative pathway) and resulting in lysis of enveloped viruses.

Both the lectin and alternative pathways are "antibody-independent" pathways. The end result of all of these pathways is the final destruction of the invading viruses and infected cells (**FIGURE 7-11**).

A number of viruses have developed different strategies to avoid complement-mediated destruction. Herpesviruses and coronaviruses interfere with the classical complement activation pathway. Some poxviruses and herpesviruses produce gene products that mimic cellular complement proteins, preventing the binding and activation of cellular complement proteins and thereby preventing the cascade from continuing its pathway.

Hypocomplementemia is a rare condition in which components of the complement system are lacking or reduced in concentration. Patients with this malady have been shown to be more susceptible to infection and disease caused by certain viruses; for example, patients with severe and frequent intraoral (inside the mouth) herpes simplex 1 cold sores have been linked to C4 complement protein deficiency.

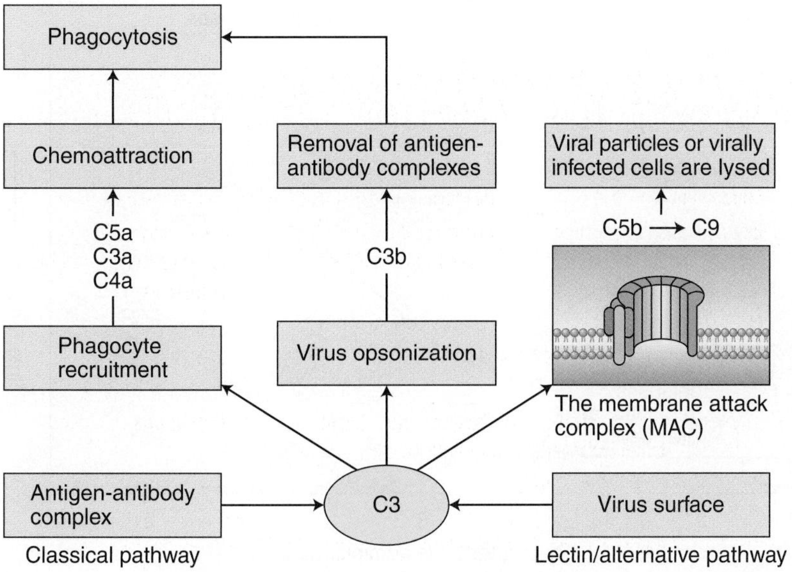

FIGURE 7-11 Complement cascade pathways. Most proteins in the complement system are designated with a "C" followed by a number. The numbers are not indicative of the order in which they act within the cascade, but rather the order in which they were discovered. When a complement protein is cleaved with a protease, two fragments are formed and referred to as "a" and "b." Opsonization is the process by which viruses or other pathogens are altered in such a manner that phagocytes more readily and efficiently engulf them. Adapted from S. Smith. "Immunologic Aspects of Organ Transplantation." *Organ Transplant.* Medscape, 2002.

7.3 Immunity Takes Time: Specific Immune System Responses (Adaptive Immunity)

Nonspecific defenses play an important role shortly after viral infection. Specific immune responses require days to weeks before they are induced or effective in clearing viruses. **FIGURE 7-12** illustrates the time frame in which specific and nonspecific defenses are induced after influenza virus infection. Due to the extremely complex nature of the immune system, this chapter focuses only on antibody (also known as humoral) and cell-mediated responses (T-cell responses) used to neutralize, control, and eliminate viruses from the host. This section attempts to address which viral molecules are identified by the adaptive system. It discusses how the viral structures identified by B lymphocytes, T helpers (T_H) and cytotoxic T lymphocytes (T_C) differ and, once these viral structures are identified, how the humoral and cell-mediated arms of the adaptive immune system response act to clear infection. It covers how these actions actually function to clear viral infection. It also briefly addresses how the specific identity of a virus and the location of infection are communicated to the adaptive immune response. Eliminating cancer cells, bacteria, and parasitic pathogens often involves even more immune function complexity.

Key Players of the Immune System: The Lymphocytes

Lymphoid organs, soft tissues, and a network of lymphatic vessels are positioned throughout the body. (The prefix "lymph" comes from the Latin word *lympha,* meaning a clear, watery, fluid.) Lymph contains white blood cells that circulate throughout the lymphatic system. It acts to remove foreign invaders and certain proteins from the tissues and supplies **lymphocytes** to the bloodstream. Lymphocytes congregate in the lymph nodes and exit through outgoing lymph vessels. Lymphocytes are white blood cells that are formed in the lymphoid organs and tissues and make up 25% to 33% of all white blood cells in adult peripheral blood. Lymphocytes exhibit exquisite recognition in distinguishing pathogens from host cells.

The main types of lymphocytes are **B** and **T cells**, which are the key players of the immune system. B lymphocytes are born and mature in the bone marrow, whereas T lymphocytes mature in the thymus gland. They are the second line of defense, tailoring their activities toward individual threats created by invading pathogens.

Humoral Response: Antibody Production

The word humoral means arising from bodily "humors," or fluids—especially serum. When blood is drawn and allowed to clot, the remaining fluid is called serum (**FIGURE 7-13**). Serum contains many proteins, including albumin, complement proteins, hormones, growth factors, transferrin, and

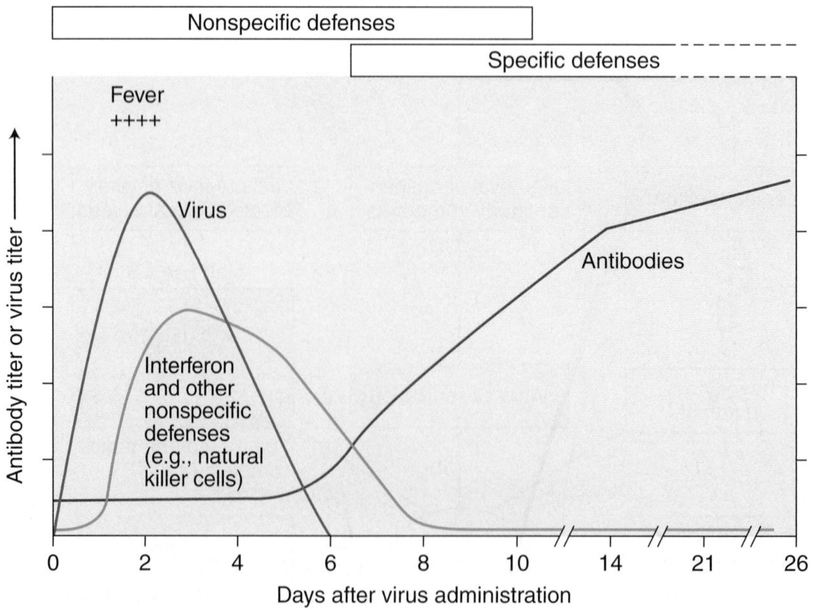

FIGURE 7-12 Nonspecific host defenses occur immediately after infection, whereas specific immune responses take days to weeks to mount and respond to a viral infection. The graph relates to the primary immune response of a typical influenza infection. Adapted from S. Baron. *Medical Microbiology,* Fourth Edition. Microbiology & Immunology, 1996.

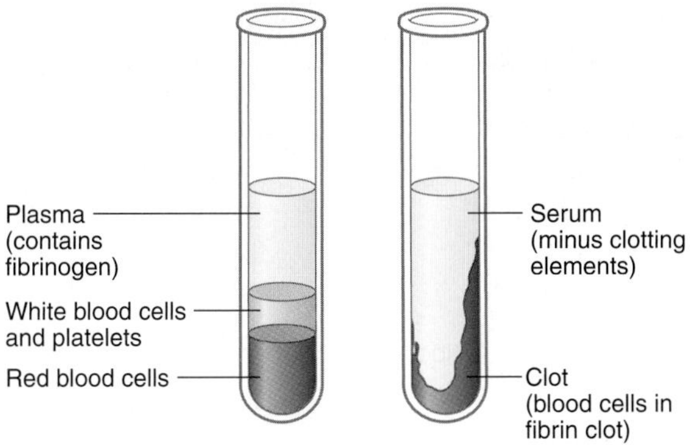

FIGURE 7-13 Blood specimens with anticoagulant (on the left) and without anticoagulant. Adapted from V. Apanius, *Field Collection and Processing of Blood*, http://www2.fiu.edu/~ animals/field_pro.html.

antibodies. In the laboratory, serum often is used to cultivate tissue cultures. Antibodies are Y-shaped glycoproteins produced by plasma cells and belong to a family of proteins called immunoglobulins. (Occasionally the terms antibody and immunoglobulin are interchanged.)

B lymphocytes can directly recognize **antigens** of foreign microbial invaders. An antigen is any molecule that is recognized by antibodies. It is usually a protein but can be a polysaccharide or other smaller molecules. A viral antigen can act as an **immunogen**. An immunogen is any substance that provokes an immunity when introduced into the body. Each mature B lymphocyte displays about 150,000 membrane-bound antibodies that contain identical antigen-binding sites. In addition to antibodies on the surface of mature B lymphocytes, there are other surface proteins that allow B lymphocytes to "bind" to other immune cells such as T_H cells.

In particular, class II MHC molecules are present on the surface of B cells. These molecules permit the B cell to function as an antigen-presenting cell (APC). APCs are defined as immune cells that express MHC II on their surface and associate with T_H cells (see **Refresher: Immunology**). T_H cells, which are infected and destroyed by the HIV virus, are T lymphocytes that contain CD4 receptor molecules on their surface.

Antibody Production: Humoral Response

Mature B cells patrol the lymphatic system. When the B cells encounter invading viruses, any B cell that expresses antibody that can specifically bind to **epitopes** of the virus invader is activated to differentiate into high populations of clones of **plasma cells** and a population of **memory cells** that express the correct antibody. Epitopes represent a region on the surface of an antigen (e.g., amino acid residues of a protein molecule) recognized by an antibody binding site. Each epitope on a molecule such as protein elicits the synthesis of a different antibody. Plasma cells are large B cells that are specialized in rapidly secreting large numbers of antibodies that respond to the same antigen that matched the B cell receptor. Memory cells are formed from activated B cells that are specific to the antigen encountered during a primary response. These cells are able to live for a long time, and can respond quickly following a second exposure to the same virus invader/immunogen.

The activated B cells continue to divide and differentiate for 4 to 5 days. Secreted antibodies bind to the viral antigens, facilitating their clearance from the host. Memory cells will retain the correct antibody on their cell surfaces, whereas most of the plasma cells will secrete the correct antibodies (**FIGURE 7-14a**). Most plasma cells usually live approximately 1 to 2 weeks. A subset of plasma cells are "long lived" and can live from months to years in the bone marrow and play a role in humoral memory of the immune system. Memory cells are longer-lived. When memory cells encounter the same antigen, they generate a secondary antibody response to it. Research suggests that memory B cells require periodic exposure to antigen for their maintenance. Antibody-producing B cells work in concert with T_{H2} cells (Figure 7-14b; explained in Section 7.4 on T cell-mediated immunity).

Antibodies are made up of four peptide chains: two identical light (L) chains and two larger, identical heavy (H) chains. Each L chain is bound to a heavy chain via a disulfide bridge. The exact number of disulfide bridges varies among the different antibody classes. The amino termini of the antibody molecule vary greatly in amino acid composition among antibodies. This is what gives antibodies their "specificity" to bind many different epitopes of viral antigens or antigens of other origins (**TABLE 7-3**). Hence, these regions are labeled as variable or Fab (fragment antigen binding). The variable region is 110 to 130 amino acids in length. The carboxyl termini of antibodies consist of constant regions that are the same sequence for every antibody. This is the Fc receptor of the antibody mentioned earlier (Figure 7-5). The H chain carboxyl termini are labeled Fc. The term was derived from the observation that this part of the antibody molecule would crystallize during cold storage. Experiments involving digestion with papain and pepsin proteases, along with mercaptoethanol reduction and alkylation of disulfide bonds, have made it possible to determine the multichain structure of antibodies (**FIGURE 7-15**).

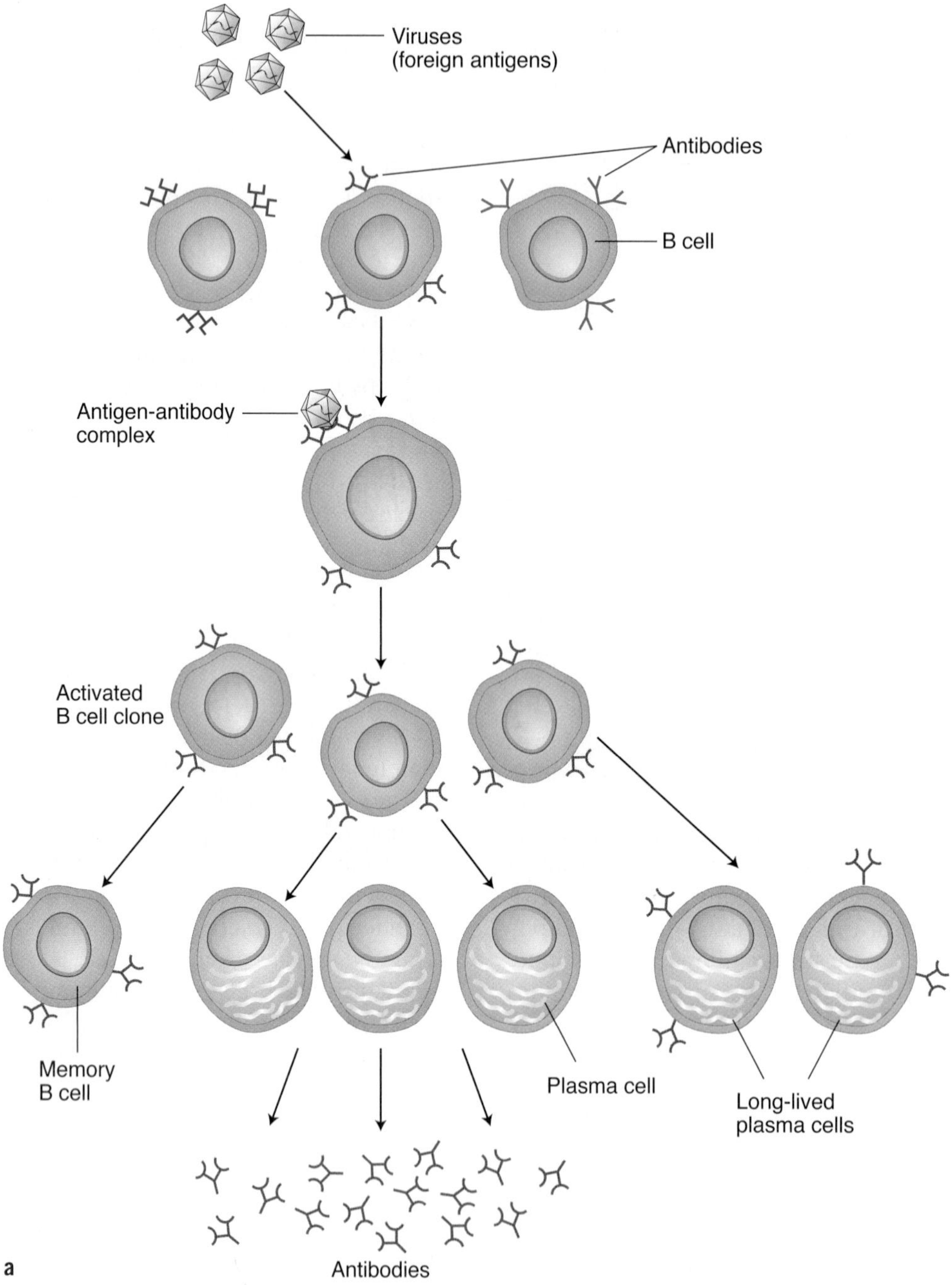

FIGURE 7-14 (a) The humoral response is characterized by antibodies produced by B lymphocytes. B lymphocytes differentiate into memory B cells and plasma cells. The latter has one task: to produce and release antibodies in large quantities. These antibodies bind specifically to the invading virus antigens. **(b)** Antibody-producing B cells work in concert with T_{H2} cells to clear viral antigens of a primary viral infection or viral antigens that have escaped or bypassed initial antibody defenses. This occurs through the presentation of viral antigens on MHC II receptors located on B cells and APCs. In this case, viral antigens displayed by B cells via the MHC II receptor-viral antigen complex stimulate activated T_{H2} cells to bind to them and stimulate the B cells to differentiate into plasma and memory B cells. The plasma B cells produce large quantities of specific antibodies that circulate and patrol the body to eliminate viral antigens.

(continued)

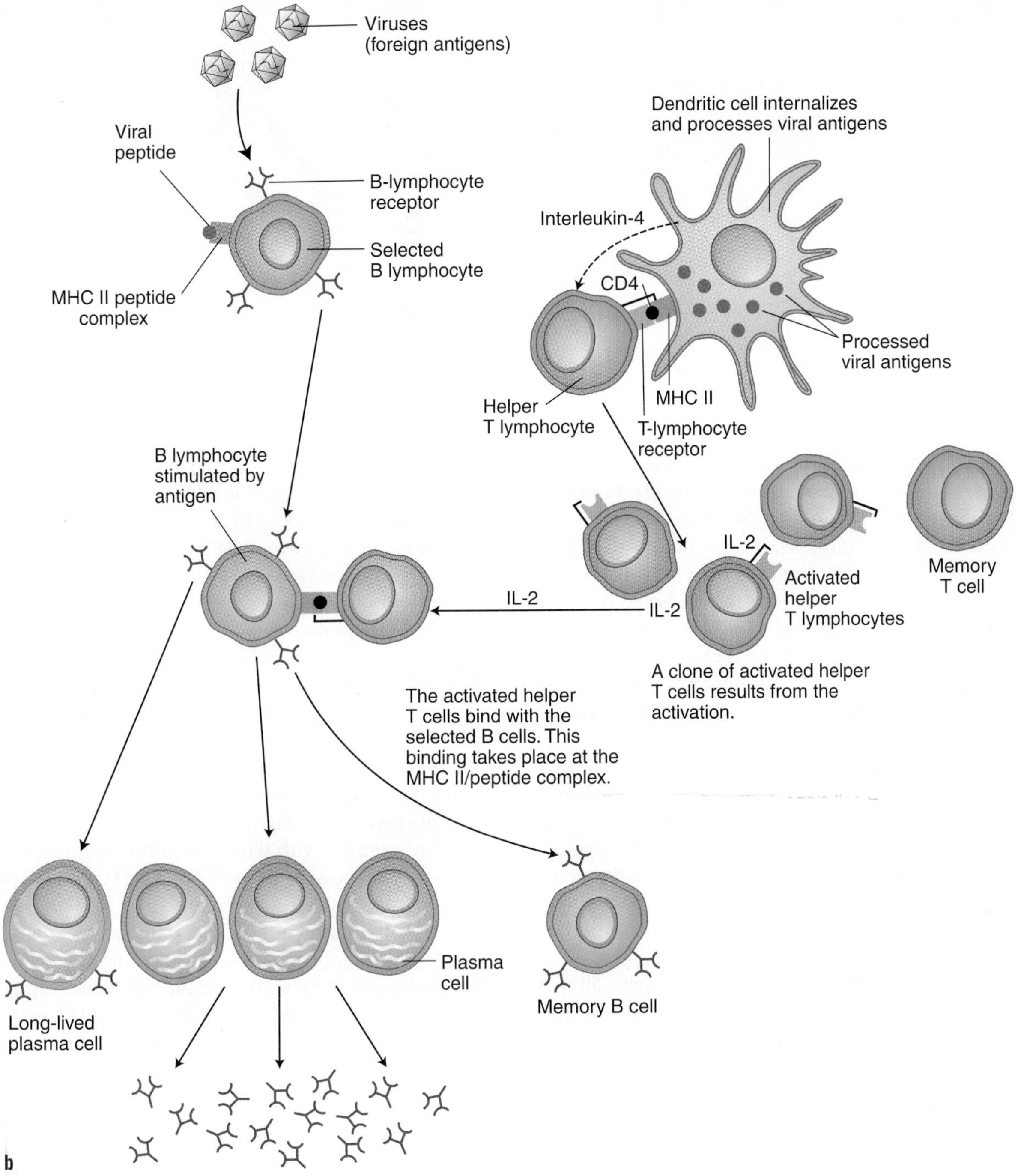

FIGURE 7-14 (Continued)

TABLE 7-3 Characteristics of Human Classes of Immunoglobulins

Property/Biological Activity	IgA	IgG	IgM
Molecular weight	150,000–600,000	150,000	900,000
Crosses placenta	–	+	–
Shape/structure	Dimer	Monomer	Pentamer
Binds to Fc receptors of phagocytes	–	+	?
In vivo serum half-life	6 days	23 days	5 days
Percent of total serum immunoglobulin	10%–15%	80%	5%–10%
First antibodies to appear after immunization or exposure to a pathogen	–	–	+
Much higher concentration in secretions	+	–	–

FIGURE 7-15 Structure of IgG antibody. Note the following features of these Y-shaped antibodies: light and heavy chains, disulfide bonds, and variable and constant regions.

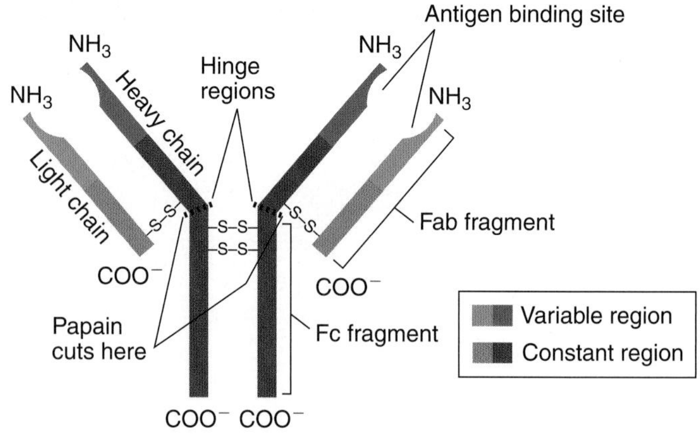

The constant region of the antibody molecule determines what mechanism will be used to **neutralize** its attached antigen. Neutralizing antibodies react with viral antigens, destroying or inhibiting the infectivity and virulence of the virus. Antibodies are divided into five major isotypes based on their constant regions: IgA, IgD, IgE, IgG, and IgM. IgE plays a role in hypersensitive allergic reactions, and the function of IgD is unknown. The focus of this discussion is on the functions of IgA, IgG, and IgM during viral infections.

The structures of the antibodies IgG, IgA, and IgM are shown in **FIGURE 7-16**. Notice that IgA and IgM are composed of more than one immunoglobulin molecule linked by disulfide bonds, resulting in larger molecules and the inability to cross the human placenta. The Fc receptor of IgG is responsible for the transfer of IgG from mother to fetus across the placental membrane.

IgM has a pentamerous structure, with all five of the Fc receptors radiating in the center of the pentamer. The individual monomer units are held together by disulfide bonds and a polypeptide called the J (joining) chain. IgM is the first immunoglobulin to respond to a viral or other microbial antigen (**FIGURE 7-17**). Whenever IgM is found in serum samples in the clinical laboratory setting, it indicates that the patient has had a recent or has a current infection. The IgM pentamer has ten binding sites, allowing it to neutralize viral infectivity using less IgM than IgG antibodies to do the job. IgM antibodies last between only a few weeks or a few months, as opposed to IgG, which provides years of protection against subsequent infections caused by the same virus. Eventually, the production of IgM is shut off, and IgG takes over (Figure 7-17).

The strength with which an antibody molecule binds an epitope is called its **affinity**. IgM antibodies are generally low affinity, but all other isotypes are high affinity because B cells mutate their variable domains during maturation in the lymph nodes, allowing for production and selection of higher affinity mutated immunoglobulins. This is called somatic hypermutation and affinity hypermutation.

FIGURE 7-16 Structure comparison of IgG, IgA, and IgM antibodies. Note the size differences. IgA and IgM contain J chains in addition to disulfide bonds that link their structures together.

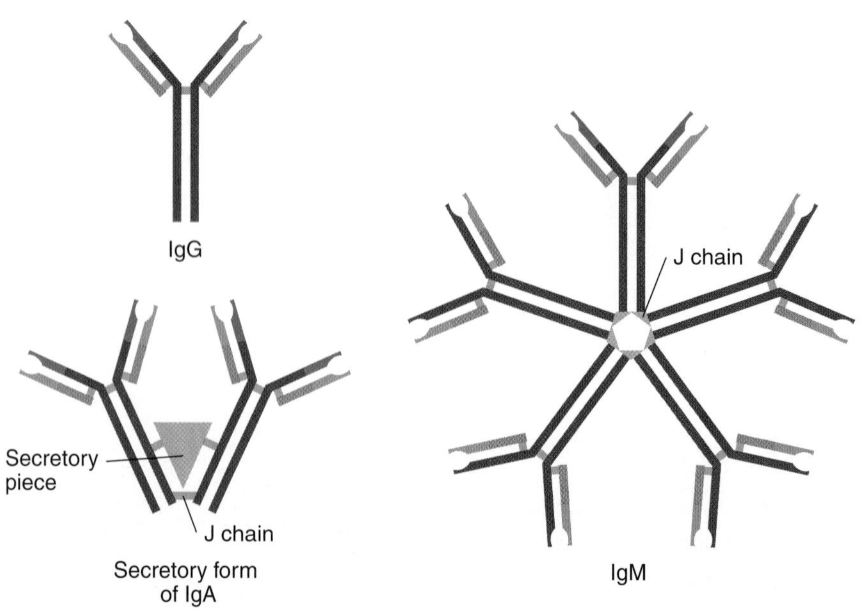

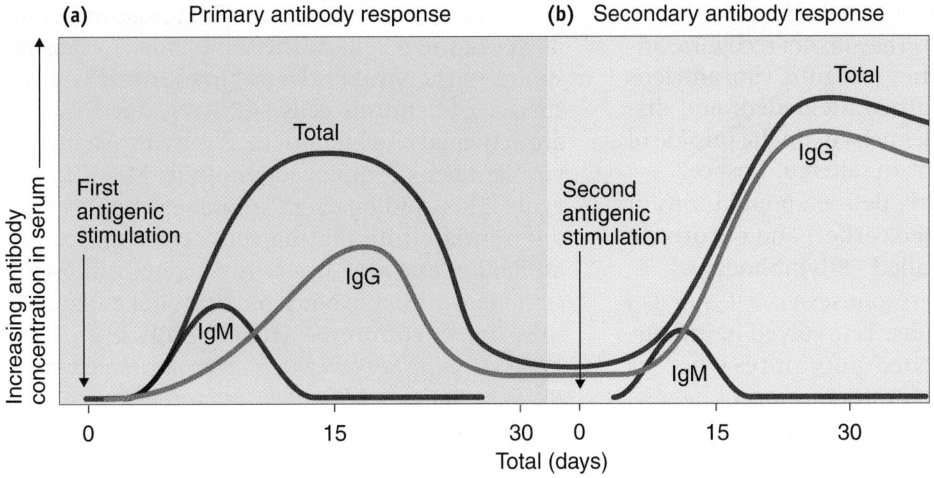

FIGURE 7-17 Graph **(a)** represents a typical immune response to a bloodborne viral infection encountered for the first time. Note: IgG or IgA levels never decline to zero. There is memory in the form of these circulating immunoglobulins for life. Graph **(b)** represents an immune response after a subsequent exposure to the virus encountered in the first graph. Note the more rapid IgG response in this panel.

IgG antibodies are the most abundant class of antibody present in serum (Table 7-3). As mentioned previously, IgG is produced for very long periods of time, offering long-term protection against encounters with prior pathogens. In epidemiological surveys that test for antibodies in serum, IgGs indicate past infections. IgG antibodies can cross the placenta, thus protecting newborns for the first few months of life.

IgA antibodies are found in secretions of the body, especially the mucous membranes of the respiratory tract, the gastrointestinal tract, and other mucous surfaces. Hence, they play an important role in viruses that enter by these routes. IgA is present in colostrum, the fluid produced by a mother's mammary glands during lactation before true milk appears. These maternal antibodies help to protect the child from infections early in life. IgA is also present in breast milk but in lower concentrations than in colostrum. Hence, breastfeeding plays an important role in maintaining the health of newborns.

IgA is also present in saliva, tears, and mucus. IgA usually exists as a monomer; however, secretory IgA, which is present in external secretions, is of dimer or tetramer form and is linked by disulfide bonds and a J chain(s). Each day, an individual secretes 5 to 15 grams of secretory IgA in the form of mucus.

Antibodies play an important role in preventing viral infections, whereas the cell-mediated response is an important defense in eliminating virally infected cells from the body. (Cell-mediated responses will be discussed in greater detail later in the chapter.)

Not all antibodies that bind to viral antigens will neutralize or interfere with the infectivity of viruses. Typically, the antibodies that bind to viral surface antigens are able to block attachment and entry, preventing cell-to-cell spread of infection. A representative of the concept behind the viral neutralization assay is shown in **FIGURE 7-18**.

Antibodies may promote the aggregation of virus particles, thereby reducing their infectivity. Antibodies cannot enter cells; hence they must neutralize viruses before the viruses are replicating inside of the host. Nonneutralizing antibodies bind to viral antigens, but do not interfere with infectivity. In some cases, they enhance infectivity.

T Cell-Mediated Immunity

T cells are born in the bone marrow and migrate to the thymus, where they mature. T cells differ from B cells in that they recognize cells that contain for-

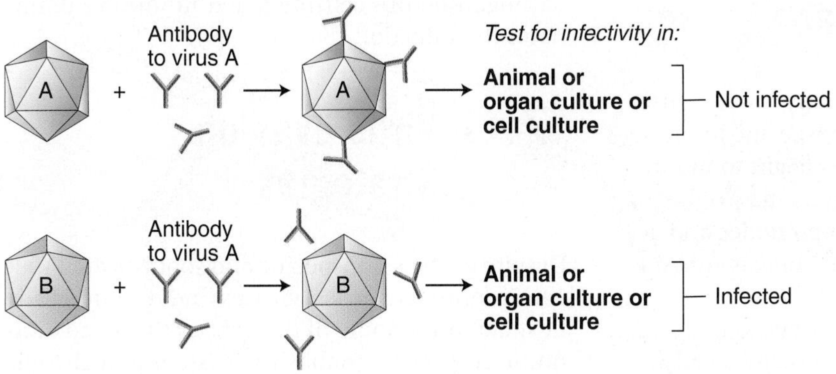

FIGURE 7-18 The viral neutralization assay is a sensitive and specific assay used to identify virus-specific antibodies in animals and humans. The assay is performed in two steps. In the first step the virus is incubated with antibodies, and in the second step the mixture in step one is inoculated into a host system. The absence of infectivity constitutes a positive neutralization assay.

eign antigens (i.e., virally infected cells, tumor cells, and foreign tissue grafts). They do not recognize antigens directly. Instead, they recognize viral antigens that have been internalized, degraded, and displayed on MHC molecules present on the outside of APCs or on the surface of an "altered" host cell.

There are two clearly defined populations of T cells: T_H cells (described earlier) and **cytotoxic T cells** (T_c cells, also called T8 lymphocytes). T_H cells promote immune responses via a T_{H1} or T_{H2} response. The T_{H1} response is involved in inflammation and cell-mediated immunity, whereas T_{H2} response activates immune responses dependent on antibodies. The T_{H2} response antagonizes the T_{H1} response and vice versa. Tc cells kill in an antigen-specific manner.

Like B cells, which express unique antibodies on their surfaces, T cells express their own unique type of receptors on their surfaces. T_H cells express CD4 in association with the T cell receptor (TCR) on their surface and recognize only antigen bound to MHC II molecules. T_C cells express CD8 in association with TCR on their surfaces and recognize only antigens associated with MHC I molecules. As a general rule, T_H cells act as helper cells and T_C cells act as cytotoxic cells.

The T cell-mediated immune response begins when T_H cells recognize and bind to an antigen-MHC II complex present on APCs (i.e., macrophages, dendritic cells, or B cells displaying viral antigens). This binding event activates the T_H cell, causing it to secrete cytokines that in turn activate various cells that participate in the immune response (i.e., B cells, macrophages, and T_C cells). Some cytokines promote the growth of more T cells (for example, interleukin 2 stimulates the production of T_H, memory T cells, and T_C cells). **FIGURE 7-19** illustrates the process of T cell-mediated immunity. This form of adaptive cellular immunity is specific and involves memory and self–nonself recognition.

7.4 B and T Cells Work Together

When a viral infection occurs, many host defenses are mobilized. In addition to nonspecific immune defenses, cell-mediated defenses begin to mount. B cells, T cells, and APCs such as macrophages usually travel to the nearest lymph nodes and act in concert. The intended final outcome is containment of the viral infection.

When a B cell internalizes a virus, the viral antigens are digested by cellular proteases and displayed on surface MHC II antigen presenting complexes of the B cell. At the same time, T_H cells respond to the viral antigens "**presented**" on the surface of dendritic cells (APCs). Cloned T_H cells are activated and bind to the B cells displaying the viral antigens complexed within its MHC II molecule. This binding event stimulates the B cells to differentiate into plasma cells secreting specific antibodies and memory cells. Secreted antibodies circulate within the body, binding viral antigens as they are encountered (**FIGURE 7-20**). *B cells and T cells recognize fundamentally different viral structures. B cells recognize whole protein antigens where as T cells recognize processed short viral peptides.* Probably the most significant effector for elimination of virus-infected cells is the action of T_C cells. T_C cells contain receptors that examine each cell it encounters for short viral peptides found on cellular surfaces. If a cell is infected with a virus, it is swiftly killed.

7.5 Some Final Comments on Cell-Mediated Immunity

Adaptive cell-mediated immunity appeared about 450 million years ago. Contemporary research is unraveling this very sophisticated host defense system that protects us against specific viral and microbial pathogens. Animal studies in mice have shown that this same system is used to clear tumor cells and other abnormal cells and antigens from the animal. The major players are B and T cells. Cytokines play an important role in activating or communicating between B and T cells. Repeated exposure of the same pathogens results in the maintenance of memory B and memory T cells. Figure 7-20 illustrates a comprehensive view of how cell-mediated immunity is used to eliminate virally infected cells from the body. It is important to remember that it does take some time for this response to mount; hence, nonspecific immunity provides immediate action against any invading pathogen before cell-mediated immunity eliminates the infection (Figure 7-20).

7.6 Virus Evasion Strategies

Despite a host's nonspecific and sophisticated specific immune defenses, viruses find ways to evade host defenses. Some of the viruses that evade immune responses establish persistent and chronic

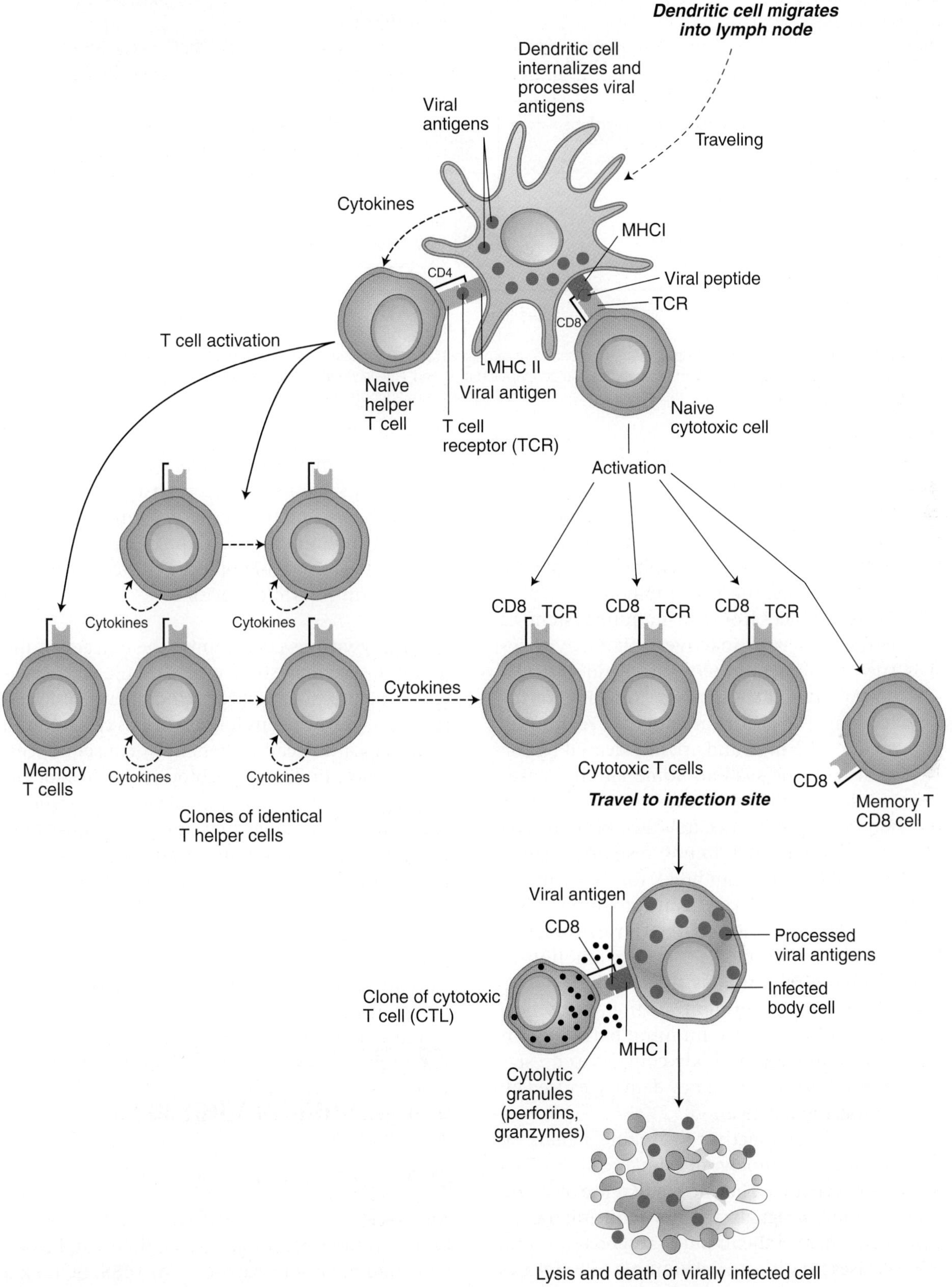

FIGURE 7-19 T_{H1} response system. T_{H1} cells recognize and bind to APCs containing viral antigens bound to MHC II molecules on the APC cell surface. This results in the activation of cytokines such as IL-12. IL-12 stimulates the T_{H1} cells to differentiate into more T cells, including T_C cells. The T_C cells migrate to the site of infection where they destroy the virally infected cells.

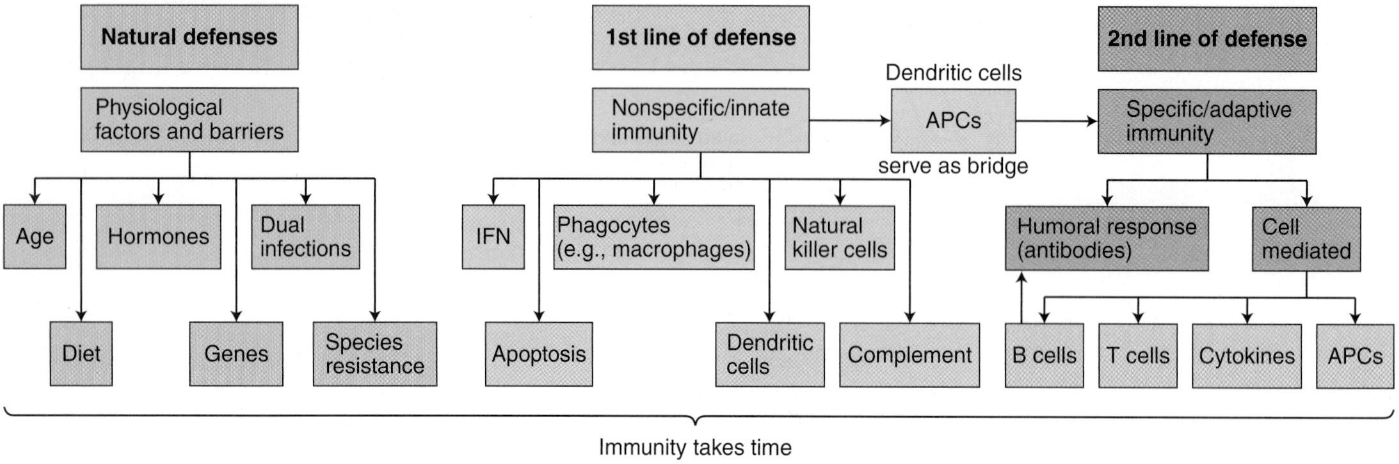

FIGURE 7-20 Humans and animals have more than one form of defense system against viral pathogens. Simpler defense systems immediately mount upon viral infection, followed by the cooperation of B and T cells within days or at most a few weeks. B and T cells learn from each viral attack, retaining memories to respond when subsequent encounters occur.

infections. There are several basic mechanisms by which viruses evade elimination by the host:

- The most significant form of immune evasion strategy has to do with viruses that mutate rapidly, thereby escaping neutralizing antibodies and the T_C cell responses.
- The virus hides from the immune system, as is the case with inactive or latent viruses such as herpes simplex virus and varicella zoster virus. Latent viruses cause lifelong infections. The viruses are not fully replicating most of the time and are not cleared by the host immune system. Some latent viruses contain viral peptides that are very close or identical in sequence to a host protein. The result is that the immune response against the viral protein can break tolerance to the self protein and result in an autoimmune disease. This is termed **molecular mimicry**.
- A few may synthesize excessive amounts of soluble viral antigens that bind all of the host's circulating neutralizing antibodies. This can also cause immune pathology in which antigen-antibody complexes do significant tissue and kidney damage caused by complement activation.
- Viruses may inactivate cytokine signals by producing homologs of cellular cytokines or their receptors. The end result is the blocking of immune signals, thereby reducing the effectiveness of the immune response.
- Viruses may inactivate immune cells; for example, viruses may infect immune cells, blocking their activity and causing immune suppression. HIV does this by infecting T_H cells, resulting in their depletion during full-blown AIDS.
- Viruses may block cellular pathways that normally kill infected cells such as the apoptosis, complement, or interferon pathways. Viruses may produce protein or nucleic acid products that act as anti-apoptotic, anti-complement, or anti-interferon factors.

Many viruses compromise the efficacy of the immune system in some way. Chronic and persistent viral infections yield large numbers of infected cells and viral antigens for years. This can result in pathological instead of beneficial consequences. Exciting new findings have brought the field of immune evasion to the forefront of viral pathogenesis, vaccinology, and molecular immunology. As new mechanisms of host immunity are discovered, new strategies by which viruses evade these mechanisms are discovered. The balance between host responses and immune evasion determines the outcome of any viral infection. Chapters 11 to 18 will cover specific viral evasion strategies used by viruses.

7.7 History of Immunotherapy

Passive Antibody Therapy

After receiving his medical degree in 1878, German Emil von Behring worked as a military and troop physician in various garrisons. In 1888, he took a position at the Hygiene Institute of Berlin as an assistant to Robert Koch (a German pioneer in bacteriology who proposed Koch's postulates, the criteria used to determine if a given bacterium was causing a particular disease). Along with Japanese

VIRUS FILE 7-1 Ebola Virus Immune System Evasion Tricks

Ebola virus outbreaks have been documented since 1976. These outbreaks have occurred in small communities located in very remote areas of Africa such as southern Sudan, the Democratic Republic of Congo (formerly Zaire), Uganda, and Gabon. Ebola infections can cause high mortality rates (as much as 90% in some outbreaks). The most striking symptom is uncontrolled bleeding from any orifice. Progress in understanding the pathogenesis of Ebola virus infections has been slow, primarily because research of these viruses requires Bio-Safety Level 4 (BSL-4) containment. Public awareness of bioterrorism—heightened by an Ebola scare in Canada in February 2001, events relating to the September 11, 2001, terrorist attacks in the United States, and concomitant knowledge that the former Soviet Union had been evaluating Ebola virus as a bioweapon during the Cold War—has changed our perspective regarding this pathogen and reinforced the need for an effective vaccine.

During infection, the Ebola virus accumulates in high numbers in the liver, spleen, lymph nodes, and lungs, severely damaging these organs during the course of the disease. Individuals who succumb to the infection die with a very high **viremia** (presence of virus in the bloodstream). There is an absence of phagocytes infiltrating to the sites of infection and little evidence of a specific antibody response. Cellular immunity is poor. Recent research points to several possible mechanism(s) for this immune suppression.

The Ebola virus genome consists of seven genes. The glycoprotein (GP) viral gene product is found in two forms: a secreted form (sGP), which is found in large amounts in the blood of infected individuals early in the course of infection, and a nonsecreted form (GP), which is incorporated into the membrane wrapped around the virus particle (**FIGURE VF 7-1**). Some evidence suggests that the sGP form binds to and inhibits the activation of neutrophils. Ebola virus infects endothelial cells and immune cells such as macrophages and dendritic cells, resulting in immune suppression. Research by Harcourt and colleagues suggests that Ebola virus also selectively inhibits the host's interferon response. Thus, it appears that Ebola virus has evolved several mechanisms to avoid its elimination

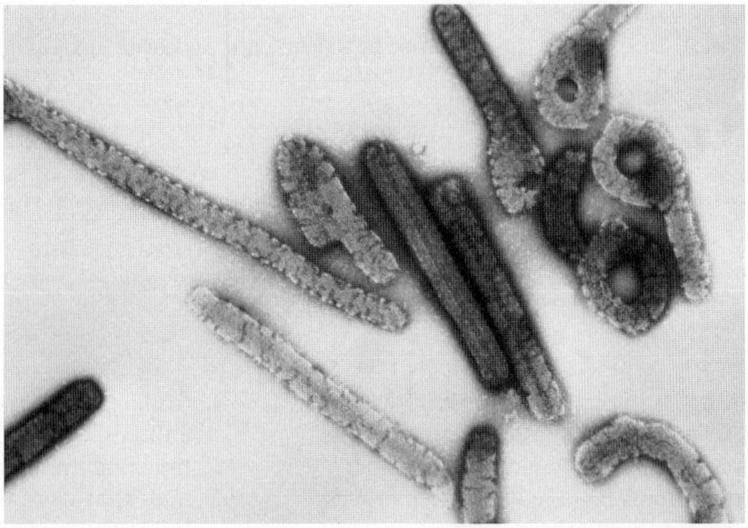

a

FIGURE VF 7-1 (a) TEM of Ebola virus particles.

(continued)

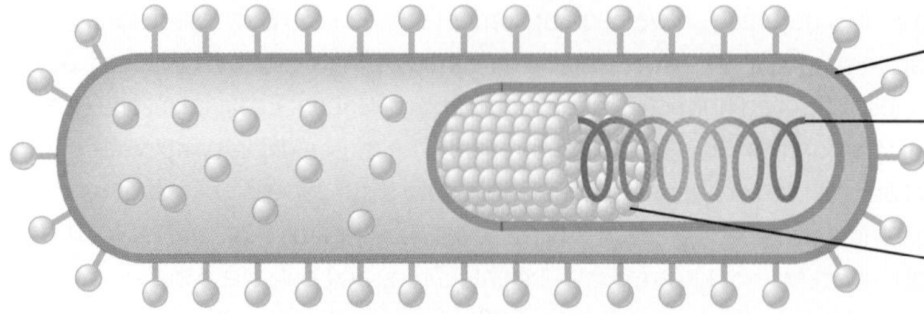

b

c

FIGURE VF 7-1 (Continued) **(b)** Structure of Ebola virus. Modified from the CDC. **(c)** Ebola testing during an outbreak.

by the host. To date, several different Ebola vaccines are being tested in humans and monkeys. The Vaccine Research Center at the NIH has prepared a vaccine for human testing via a program set on an accelerated track, boosted by an infusion of biodefense funds from the CDC.

References

Geisbert, T. W., and Jahrling, P. 2003. "Towards a vaccine against Ebola virus." *Expert Rev Vaccine* 2:777–789.

Harcourt, B. H., et al. 1999. "Ebola virus selectively inhibits responses to interferons, but not to interleukin-1b, in endothelial cells." *J Virol* 73:3491–3496.

Mahanty, S., et al. 2003. "Cutting edge: Impairment of dendritic cells and adaptive immunity by Ebola and Lassa viruses." *J Immunol* 170:2797–2801.

Vastag, B. 2004. "Ebola vaccines tested in humans, monkeys." *JAMA* 291:549–550.

Yang, Z., et al. 1998. "Distinct cellular interactions of secreted and *trans*-membrane Ebola virus glycoproteins." *Science* 279:1034–1037.

physician and bacteriologist Shibasaburo Kitasato, he began experiments leading toward the development of therapeutic serum in Koch's laboratory.

Their experiments involved a six-step process (**FIGURE 7-21**):

1. Growing the bacteria that cause diphtheria or tetanus in pure culture.
2. Inactivating the bacteria with a disinfectant.
3. Injecting the attenuated/killed bacteria into rats, guinea pigs, or rabbits.
4. Collecting the serum produced by these animals.
5. Injecting the serum into nonimmunized animals that were previously infected with the fully virulent bacteria.
6. The mice are injected with immune serum or control serum from nonimmunized mice.

In 1890, they published their discoveries, in which they noted that their sterilized broth cultures

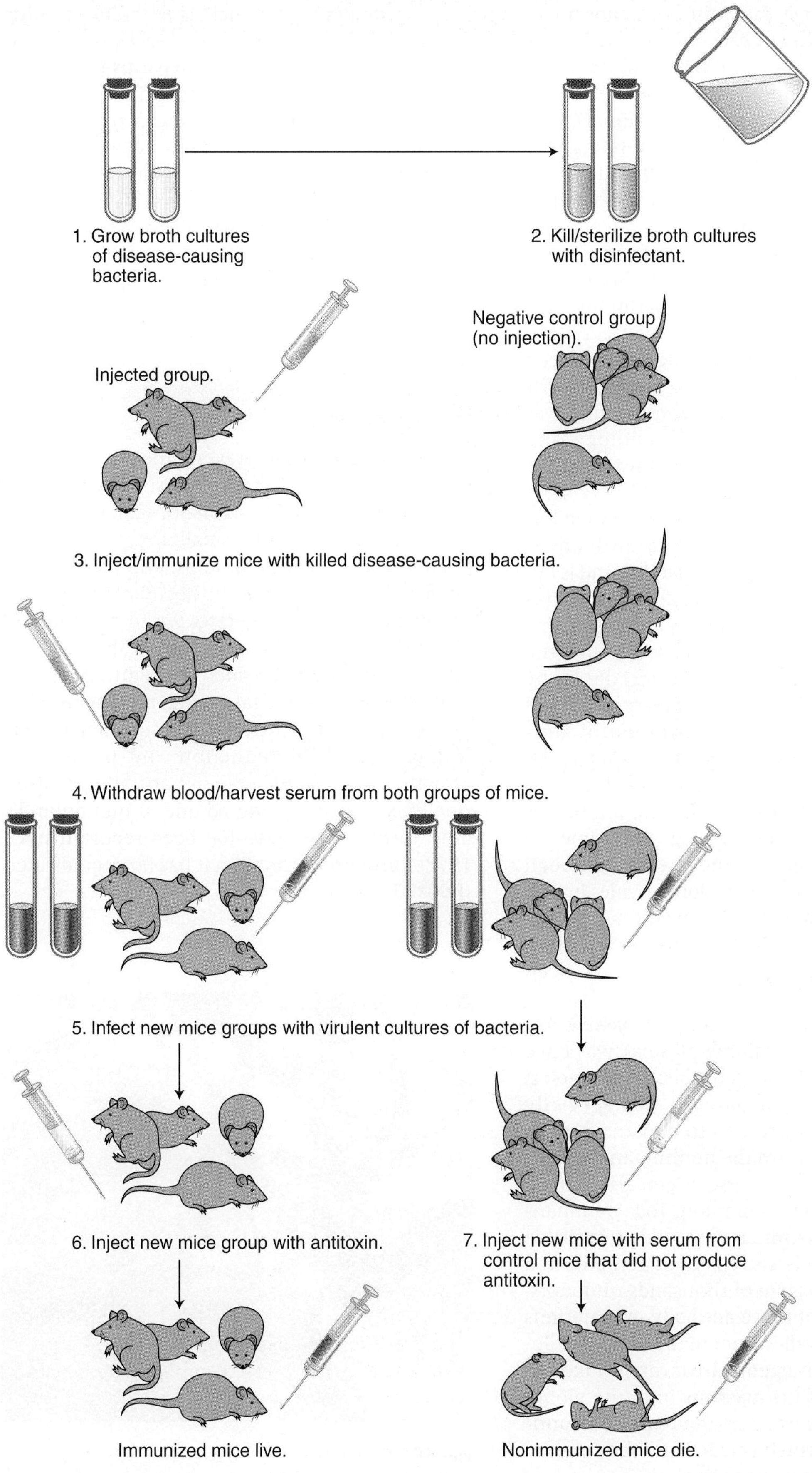
1. Grow broth cultures of disease-causing bacteria.
2. Kill/sterilize broth cultures with disinfectant.
Negative control group (no injection).
Injected group.
3. Inject/immunize mice with killed disease-causing bacteria.
4. Withdraw blood/harvest serum from both groups of mice.
5. Infect new mice groups with virulent cultures of bacteria.
6. Inject new mice group with antitoxin.
7. Inject new mice with serum from control mice that did not produce antitoxin.
Immunized mice live.
Nonimmunized mice die.

FIGURE 7-21 Preliminary experiments by Emil von Behring and Shibasaburo Kitasato paved the way for passive immunization. Passive immunization through intravenous immunoglobulin is used as therapy in many acute viral infections.

of bacteria caused animals to produce a neutralizing substance against the poisonous substances (toxins) that the virulent bacteria produced. They termed these substances antitoxins and were able to show that these antitoxins could be used to immunize animals and prevent these animals from showing symptoms of diphtheria or tetanus after being challenged or exposed to infectious bacteria. Specifically, the researchers had shown that antitoxins neutralized the toxins.

In late 1891 von Behring injected an experimental serum containing diptherial antitoxins into a little girl dying of diphtheria in a Berlin hospital. The girl experienced a rapid recovery. Within three years, 20,000 children in Berlin had been immunized against the diphtheria bacterium. For this remarkable achievement, von Behring was awarded the Nobel Prize in Physiology and Medicine in 1901, and today he is considered the founder of immunotherapy.

Serum therapy is now known as passive immunity and antitoxins are called antibodies. Passive immunization is still used today and is defined as an impermanent form of acquired immunity in which antibodies against a disease-causing organism are acquired naturally (as through the placenta to an unborn child) or artificially (as by injection of antiserum). The most common use of antiserum today is the administration of antiserum following a poisonous snake or spider bite.

The antiserum is derived from actively immunized human beings or animals. The antibodies last for two to three weeks, during which time the person is protected against the disease. Although short-lived, passive immunization provides immediate protection; active immunization can take weeks to develop. Consequently, passive immunization can be lifesaving when a person has been infected with a deadly organism.

The use of immunoglobulin for passive immunity (antibody-containing plasma) was once derived solely from pooled plasma of horses. Although this animal material was specially treated before administration to humans, serious allergic reactions toward the nonhuman antigens present in antiserum or horse origin, also termed **serum sickness**, were common. Today, human-derived immune globulin is more widely available and the risk of side effects is reduced. It is prepared from the pooled plasma of thousands of donors. Hence, recipients of these antibody preparations are receiving antibodies from many people in defense of diverse pathogens. Great care is taken in the preparation of intravenous immune globin (IVIG). The process involves the removal of antibody aggregates (which can activate the complement pathway), treatment with solvents such as ethanol, and the use of detergents to inactivate bloodborne pathogens such as HIV and hepatitis C. Passive immunization is used as protection against such viral diseases as hepatitis (caused by the hepatitis A virus), respiratory disease (caused by RSV), measles, rabies, and Ebola and Lassa fever viruses. In 2003, the largest hepatitis A outbreak in United States history was found to be associated with green onions in salsa from a restaurant in Monica, Pennsylvania. As a result of the investigation, immune globin was administered to approximately 9000 individuals who ate food from the restaurant during the time of the outbreak or had exposure to ill persons involved in the outbreak.

Vaccination

Deliberate attempts to prevent infectious diseases have a long history. During the last 200 years, vaccination has controlled six major viral diseases in most parts of the world: smallpox, yellow fever, poliomyelitis, measles, mumps, and rubella. The impact of vaccination on the health of the world's people is hard to exaggerate. Susan and Stanley Plotkin, coauthors of the chapter "A Short History of Vaccination" in *Vaccines* (3rd edition), state, "with the exception of safe water, no other modality, not even antibiotics, has had such a major effect on mortality reduction and population growth." Not a single case of naturally acquired smallpox—once considered one of humankind's most terrible scourges—has been reported since 1978. Through vaccination it has been eradicated (**FIGURE 7-22**).

FIGURE 7-22 This Nigerian child is receiving a smallpox vaccination during the WHO Smallpox Eradication Project, 1968.

Attempts to vaccinate date to 1100 A.D. inoculation practices against smallpox in China. Five practices have been described:

1. The nose was plugged with powdered smallpox scabs laid on cotton wool.
2. Powdered smallpox scabs were blown into the nose.
3. A healthy child wore the undergarments of an infected child for several days.
4. A piece of cotton containing contents of a smallpox vesicle was stuffed into the nose.
5. White cow fleas were ground into powder and made into pills. This may have been the first attempt at an oral vaccine.

Greater details regarding the history of smallpox vaccination is covered in Chapter 14.

Today, vaccination of large populations is routine. Progress in understanding the immune system, though, has been slow. The early discovery of antibodies held the interest of scientists as being the main mechanism of disease resistance. We now know that antibody production is only part of the immune response against pathogens. Russian scientist Ilya Metchnikoff discovered phagocytic immunity in 1883, but it was not until the 1940s that cell-mediated immunity revived interest and discovery. At present, cellular immunology is central in the interest of most immunologists.

Traditional vaccines sought to produce the nature of immune system "memory," especially the dependence upon the elicitation of neutralizing antibodies to prevent viral infections. At high concentrations, the neutralizing antibodies either prevented the infection or limited it to a transient, subclinical episode. The current generation of vaccines incorporates important alternative immunizing strategies that revolve around T_C cell responses. This is being done because viruses that cause chronic diseases such as AIDS and hepatitis C produce proteins that are hypervariable, making them poor targets for neutralizing antibodies, for T_C and T_H responses and even antiviral drugs. T_C cells can detect viral proteins made in infected cells and are not limited to detecting only those viral proteins accessible on cell surfaces. In addition, T_C cells have the ability to lyse infected cells.

Active Immunization Elicits Long-Term Protection

Passive immunity provides transient protection against a specific infectious agent, whereas **active immunity** provides protection that is **long-term** or has **immunologic memory**. Active immunity means that the immune system is active. It may be induced artificially via vaccination or through natural infection. Active immunity developed by childhood vaccination programs dramatically reduces the number of deaths from infectious diseases.

Traditional Vaccines: Killed or Inactivated Viruses

Whole-virus particle vaccines are commonly used today (**TABLE 7-4**). In some cases, vaccine production involves the inactivation of the virus by heat. This rarely works because heat denaturation often destroys viral antigens, thereby reducing their immunogenicity. Other inactivation methods are ultraviolet radiation and chemicals such as formalin (the most common method), β-propiolactone, and phenol. These preparations are killed or inactivated vaccines (**FIGURE 7-24**). The Salk vaccine against poliovirus is an example of a vaccine produced by formaldehyde inactivation (Chapter 11). Inactivated vaccines usually require a booster (another dose), and more vaccine agent must be injected as compared to live vaccines.

Traditional Vaccines: Live, Attenuated Viruses

Live or attenuated vaccines involve the use of avirulent mutant viruses. The live virus multiplies inside of the recipient host and elicits a long-lasting immune response while causing little or no disease. Viruses may be attenuated in virulence through the repeated culturing of the virus in nonhuman cultured cells or animals (**FIGURE 7-23**;

TABLE 7-4 Examples of Traditional Vaccines Still Used in Vaccination Programs Against Viruses Today

Virus	Type of Vaccine
Vaccinia (smallpox)	Live attenuated
Measles	Live attenuated
Mumps	Live attenuated
Poliovirus (Sabin)	Live attenuated
Varicella zoster (chickenpox)	Live attenuated
Yellow fever	Live attenuated
Hepatitis A	Inactivated
Rubella	Live attenuated
Influenza	Inactivated or Live attenuated (LAIV, FluMist, intranasal)
Polio (Salk)	Inactivated
Human papillomavirus	Inactivated
Japanese encephalitis	Inactivated
Varicella zoster (shingles)	Live attenuated
Rabies	Inactivated
Rotavirus	Live attenuated (Rotarix) Live attenuated (RotaTeq, reassortant of 5 rotaviruses)

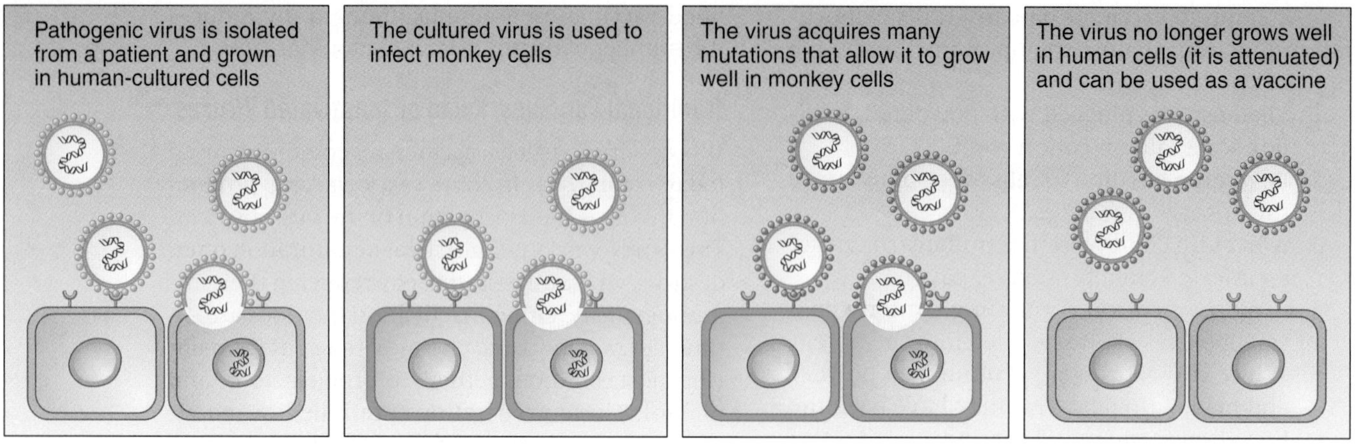
Pathogenic virus is isolated from a patient and grown in human-cultured cells
The cultured virus is used to infect monkey cells
The virus acquires many mutations that allow it to grow well in monkey cells
The virus no longer grows well in human cells (it is attenuated) and can be used as a vaccine

FIGURE 7-23 Production of attenuated viral vaccines. The pathogenic viruses are grown in human cell cultures and then passaged many times in other cell lines such as monkey cells. The serial passage of the virus in monkey cells causes the virus to accumulate adaptive mutations suited for replication in monkey cells and not human cells. The attenuated virus will not cause disease in the human host, but the host will produce immunity against it. Adapted from C. Janeway, et al. *Immunobiology: The Immune System in Health and Disease,* Second Edition. Garland Publishing, 1996.

also refer to Sabin vaccine, Chapter 11) or at different temperatures. Temperature-sensitive mutants are unable to replicate satisfactorily at temperatures much higher than normal (above 104°F, or 40°C). These live vaccines tend to revert toward virulence. Cold-adapted mutants replicate well at lower temperatures (91.4°F, or 33°C) but not at normal human temperature (98.6°F, or 37°C). Most of these mutations do not revert.

Live attenuated vaccines produce strong antibody and T cellular responses that are similar to natural infection. In general, only one or two doses are required. Severe reactions to live attenuated vaccines are possible in immunosuppressed individuals. Table 7-4 lists examples of inactivated and live vaccines currently in use.

Advantages and Disadvantages of Traditional Vaccines

Even though traditional vaccines have been very successful, there are limitations to these vaccines:

- Not all infectious agents can be grown in animals or cell cultures. There are a number of viral diseases for which no vaccines have been developed.
- Batches of vaccine may be insufficiently killed or attenuated during the manufacturing process, thereby introducing virulent viruses into the vaccine and inadvertently spreading the disease.
- Attenuated strains of virus may revert. Continual testing is necessary to ensure that virulence has not occurred.
- Not all viral diseases are preventable by traditional vaccines (e.g., AIDS and hepatitis C).
- The yield and rate of production of human and animal viruses in cell cultures are often low, making vaccine production costly.
- The vaccines have a limited shelf life and often require refrigeration. This creates storage problems in countries with large, rural areas without electricity.

FIGURE 7-24 Production of inactivated viral vaccines. Inactivated viruses cannot replicate inside the host; however, the viral antigens may be "preserved" by chemicals, causing an immunogenic response to the host. Adapted from C. Janeway, et al. *Immunobiology: The Immune System in Health and Disease,* Second Edition. Garland Publishing, 1996.

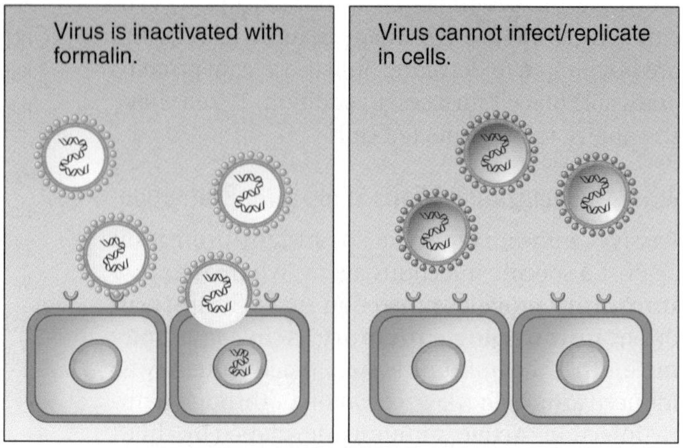
Virus is inactivated with formalin.
Virus cannot infect/replicate in cells.

- Circulating viruses may mutate, requiring frequent new vaccine development (e.g., seasonal influenza vaccines).

The New Generation of Vaccines

Recombinant DNA technology has provided a means for creating vaccines that overcome some of the drawbacks of traditional vaccines. It has allowed researchers to manipulate pathogenic viruses or to use specific viral proteins (parts of the virus) as vaccines, making the vaccines much safer because any risk of infection is eliminated. Our new understanding of the two lines of immune defenses (antibody and cell mediated) has resulted in a renewed interest in creating more vaccines.

There are several types of technology being applied toward potential vaccines. The current approaches to these vaccines are based on the following:

- recombinant subunits
- peptides
- live vectors
- reassortant viruses
- naked DNA
- edible transgenic plants for viral proteins

Recombinant Subunit Vaccines

Hepatitis epidemics date to 2000 B.C. Before World War II, physicians did not know that hepatitis B virus (HBV) was the primary cause of hepatitis. Frequent yellow jaundice outbreaks swept through the United States armed forces. Physicians ultimately linked the outbreaks to the human serum used in the vaccine that had been administered to prevent yellow fever.

A series of scientific observations from the 1950s through the 1970s contributed toward the development of an HBV vaccine. These observations pertained to the transmission of the disease via blood, especially blood transfusions. Subsequently, it was discovered that HBV carriers contained a specific viral surface antigen in their blood, termed HbsAg. Once it was discovered that HbsAg was part of the virus that causes hepatitis B, blood screening was revolutionized. In the meantime, HBV infection emerged as a significant global health threat, with between 10 and 30 million individuals becoming infected each year.

The challenge, of course, was the creation of a vaccine. Medical research revealed that HBV infection may establish chronic hepatitis and cirrhosis. Approximately 80% of all hepatocellular carcinomas (liver cancers) were related to HBV infection. Early studies showed that passive immunization using HBV immune globulin was successful in preventing or modifying the course of HBV infection. The production of a HBV vaccine was hampered until the advent of recombinant DNA technology, because it was not possible to grow HBV in cell culture and no animals with the exception of higher primates are susceptible to HBV infection. A plasma-derived vaccine prepared by harvesting the HbsAg from plasma was licensed in 1981, but many chose not to be vaccinated because of fears regarding the safety of plasma-derived products and the high cost of the vaccine. As a result, several vaccine manufacturers began to apply recombinant DNA technology toward the development of a new HBV vaccine. This technology had the potential to create an unlimited supply of a safe and effective vaccine that could eventually be produced at much lower cost than the plasma-derived vaccines.

It took nine years of research, including failure to produce the recombinant viral antigens in *E. coli*, before a successful vaccine was produced using genetically engineered *Saccharomyces cerevisiae* (a yeast; **FIGURE 7-25**). When creating the vaccine, the HbsAg or HBV viral coat protein is glycosylated. Bacteria cannot glycosylate proteins, therefore a yeast (eukaryotic) expression system was used to produce a glycosylated form of HbsAg that could act as an effective immunogen. In 1986 the Food and Drug Administration (FDA) licensed and approved the first recombinant DNA vaccine, an improved version of the HBV vaccine, for universal infant vaccination, and in 1996 approved it for adolescent vaccination. The vaccine is 80% to 100% effective and is administered in a series of three doses. (Additional doses, sometimes referred to as boosters, are not recommended.) The duration of immunity is expected to last more than 15 years.

Peptide Vaccines

Whereas subunit vaccines use an entire viral protein that contains several antigenic determinants to immunize an individual, peptide vaccines use a specific domain of a protein that contains a single epitope or antigenic determinant. The use of chemically synthesized viral peptides to induce a protective immune response has great potential because virtually any peptide sequence can be produced. The disadvantage is that peptides in general are not very immunogenic. The amount of peptide required to elicit an immune response may be 1000× more than vaccinating with an inactivated virus. To overcome this, the peptides are linked to a carrier protein to improve their immunogenicity. Unfortunately, peptide vaccines are the least effective and most expensive to produce. None are currently in use.

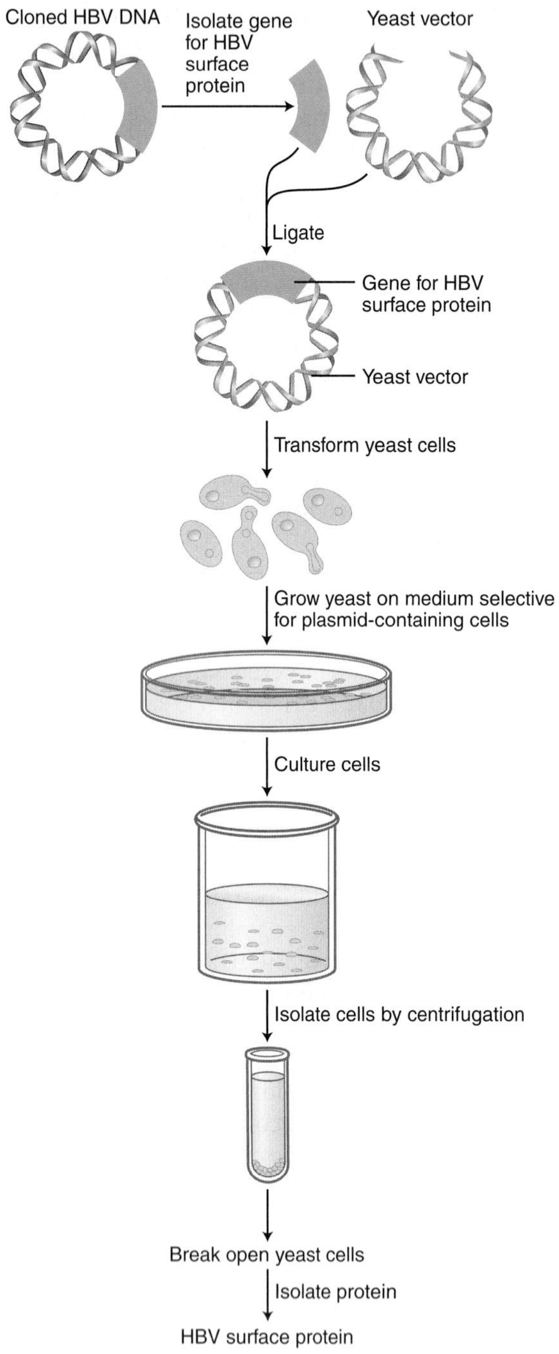

FIGURE 7-25 General procedure used to create recombinant HBV vaccine. Adapted from Marks, D. B., Marks, A. D., and Smith, C. M. *Basic Medical Biochemistry: A Clinical Approach.* Lippincott, Williams & Wilkins, 1996.

Recombinant (Live) Vector Vaccines

Molecular biology techniques are being used to develop new live vaccines. Genes coding for surface proteins of pathogenic viruses can be inserted into safe viruses (such as vaccinia and canarypox). Upon infection the new hybrid virus expresses and initiates an immune response toward itself and the inserted antigens (**FIGURE 7-26**).

In 1980, the WHO declared the worldwide eradication of smallpox and recommended discontinuing vaccination with vaccinia virus. Ironically, that same year, recombinant DNA technology was being used to manipulate vaccinia virus as an expression vector and candidate vector vaccine against unrelated infectious diseases such as rabies virus.

The first vaccinia vector recombinant vaccine used in the field was a rabies vaccine (vaccinia rabies glycoprotein, or VRG). Wildlife populations harbor the rabies virus, which presents a threat to humans when they have contact with infected animals either directly or indirectly, through their pets. In the year 2000, the U.S. Department of Agriculture—Animal and Plant Health Inspection Service reported 7369 cases of rabies in animals, of which 6.1% were foxes, 37.7% were raccoons, 30.2% were skunks, and 6.9% were domesticated dogs. The remaining 19% were rabies in bats and other animals not specified in the report. (No human cases were reported to the CDC that year.)

The vaccinia vector recombinant vaccine was used in large-scale field conditions to vaccinate foxes in several European countries beginning in 1987. By 2001, Belgium was officially declared rabies-free.

There was an explosive outbreak of raccoon rabies in Pinellas County, Florida, in 1995, when rabid raccoons spread the disease to otters, cats, and a horse. A total of 145 people were treated for exposure to rabid animals. In response to this, Pinellas County Animal Services implemented a five-year vaccination plan for distributing VRG oral rabies vaccine, in baits, in areas of high raccoon concentrations. The baits consisted of capsules of VRG inserted into blocks of fish meal, which then were distributed via air by helicopters and by ground staff. The baits were designed to be ruptured when bitten, releasing the immunizing dose of VRG into the animal's mouth (**FIGURE 7-27**). Within four years, the number of reported rabies cases in Pinellas County decreased to one.

The complications of smallpox vaccination via vaccinia virus, although rare (572 complications in 14.2 million vaccines in the United States in 1968), did raise some important safety issues about the use of recombinant vector vaccines because of the possibilities of accidental laboratory infections and side effects of vaccination. Two major steps were taken to enhance the safety of vaccinia virus, both of which have led to safe vaccinia virus strains:

- deletion or disruption of genes involved in nucleic acid metabolism, host interactions, and extracellular virus formation via recombinant DNA technology, and
- attenuation of the virus through serial passage in cell culture or an unnatural host.

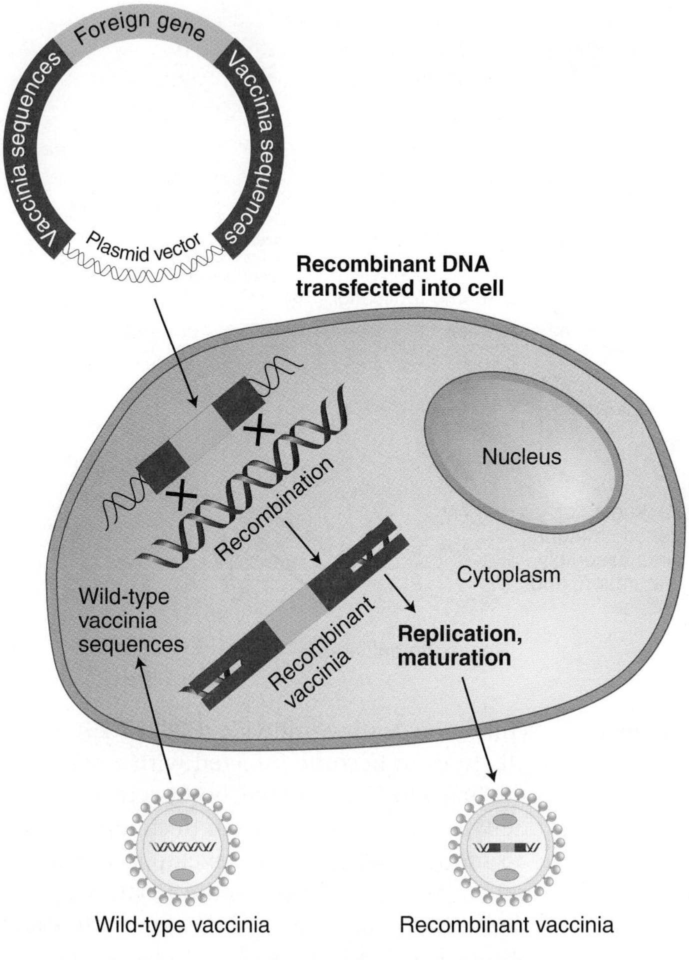

FIGURE 7-26 A recombinant vaccinia virus expressing the rabies glycoprotein was an early example of a successful poxvirus vector used in immunization. The vector was created by the insertion of a cDNA for the rabies virus glycoprotein gene into the thymidine kinase (*TK*) locus of vaccinia virus. Disruption of the *TK* gene allowed a biochemical selection of the recombinant as well as attenuation of the vector. Adapted from an illustration by the College of Veterinary Medicine, University of Florida.

Avipoxviruses only replicate in avian species. They have been isolated from a number of bird virus species and include fowlpox, canarypox, pigeonpox, penguinpox, and quailpox. Attenuated virus strains more effectively stimulate mucosal and cell-mediated immunity, and attenuated strains of avipoxviruses have a long and safe history of testing. Attenuated strains of canarypox have been tested extensively and their safety has been demonstrated in a variety of species and immune-compromised animals and human volunteers. The attenuated canarypox-based vectors can infect human cells but cannot replicate or grow in them. Hence, they produce a self-limited infection that does not cause any harm or symptoms in humans. The European Medicines Evaluation Agency (EMEA) has accepted at least two vaccines for cats derived from this technology, which are marketed in the European Union but are not yet available in the United States.

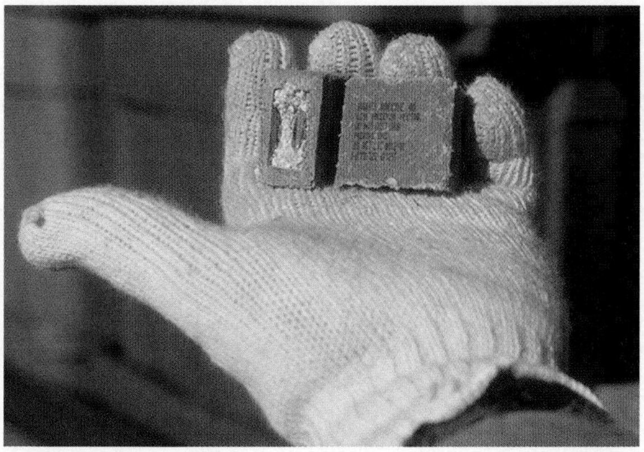

FIGURE 7-27 Rabies vaccine bait used in the field to vaccinate wildlife.

Reassortant Vaccines

When two viruses containing segmented genomes infect the same cell, reassortants may be created. The reassortant progeny viruses contain segments from both parent viruses (**FIGURE 7-28**). Reassortant plays an essential role in nature by generating novel progeny viruses. It has also been exploited in the laboratory in assigning functions to different segments of the genome; for example, in a reassorted virus, if one segment comes from virus A and the rest from virus B, one can determine which proper-

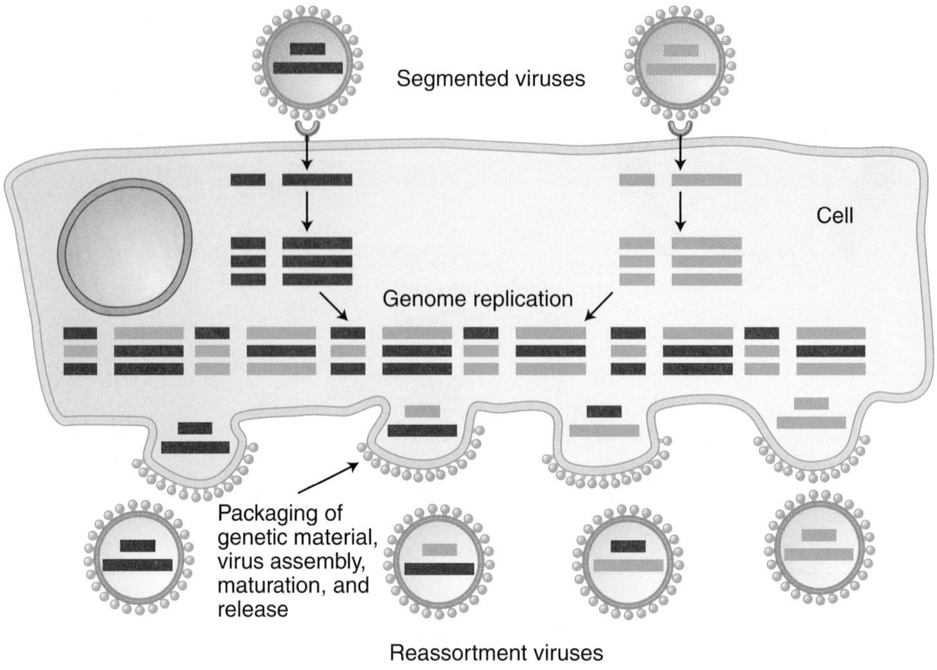

FIGURE 7-28 Simplified diagram showing the natural generation of reassortant viruses. Adapted from University of South Carolina, School of Medicine. "Virology: Viral Genetics." *Microbiology and Immunology.* University of South Carolina, http://pathmicro.med.sc.edu/mhunt/genet.htm.

ties resemble virus A and which resemble virus B thereby mapping a phenotype to a viral gene. The identification of the gene functions has made possible techniques to intentionally reassort rotavirus strains or influenza strains and prepare candidate vaccines strains of each type of virus from the desired phenotypic traits of different parent viruses.

Rotaviruses are the leading cause of severe dehydrating diarrhea in infants and young children in the United States, and are a major cause of childhood death worldwide (**FIGURE 7-29**). Virtually all children become infected with a rotavirus in the first 3 to 5 years of life, but severe diarrhea and dehydration primarily occur in children aged 3 to 35 months. Research to develop a safe, effective rotavirus vaccine began in the 1970s when it was discovered that previous infection with rotavirus strains isolated from wild animals protected laboratory animals from experimental infection with human rotaviruses.

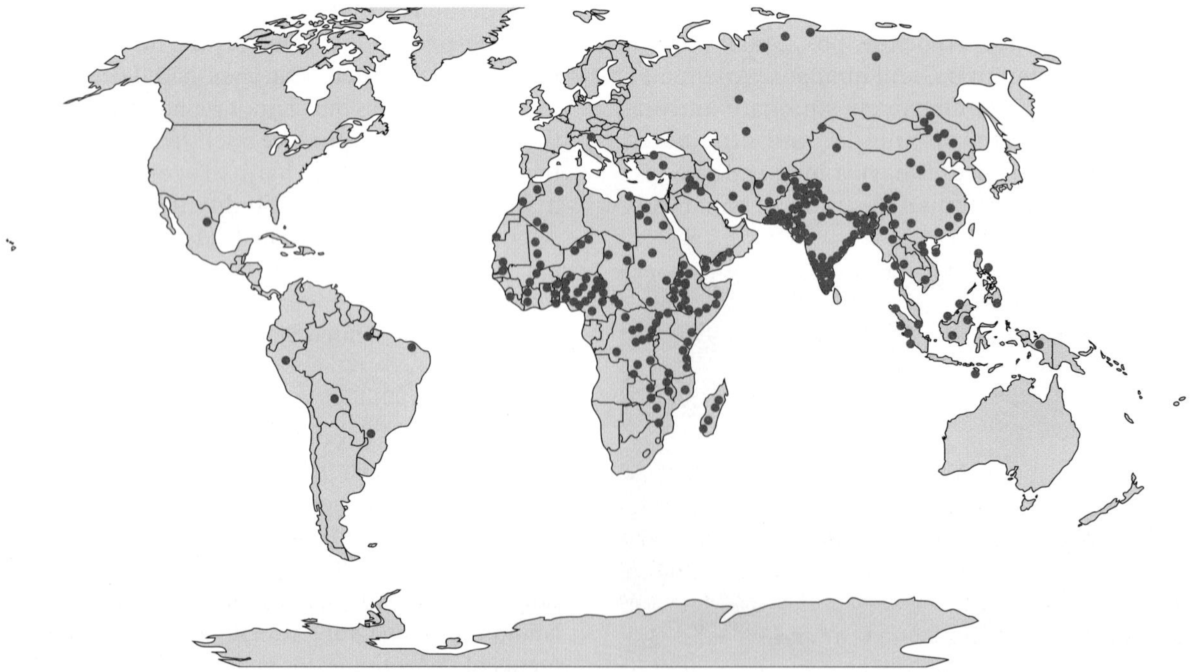

FIGURE 7-29 Number of worldwide deaths related to rotavirus infection. One dot represents 1000 deaths. It is estimated that there is a global distribution of 440,000 annual deaths. Adapted from Parashar, U., et al. *Emerg Infect Dis* 9 (2003): 565–572.

Rotashield, a live oral human–rhesus reassortment-based rotavirus vaccine produced by Wyeth-Lederle Vaccines and Pediatrics, was licensed for use in the United States in August 1998 after sailing through all the necessary clinical trials and approval processes. Three months later, the CDC and the American Academy of Pediatrics recommended universal vaccination of all American infants. By July 1999, more than 600,000 infants in the United States had received at least one immunization. In October 1999, though, the use of the vaccine was suspended because 15 children had developed intussusception after administration of the vaccine. Intussusception occurs when one portion of the bowel slides into the next, much like the pieces of a telescope. When this occurs, it creates an obstruction in the bowel, with the walls of the intestines pressing against one another. This in turn leads to swelling, inflammation, and decreased blood flow to the intestines. If diagnosed, the condition is readily treatable with a barium or air enema or by surgery.

Approximately 0.05% of vaccinated children were potentially affected. This was also very close to the rate of intussusception in unvaccinated children. The consensus was that this vaccine was safe and was withdrawn due to public perceptions, not hard data. The cost of withdrawing it from the market was approximately $80 million, and antivaccine groups viewed this event as another example of how vaccines can damage the health of children. Other manufacturers developing similar rotavirus vaccines worried that their vaccines might also contain the risk of intussusception. At present there are two licensed rotavirus vaccines available for use in the United States (RotaTeq manufactured by Merck and rotaRix manufactured by GlaxoSmithKline). The estimated risk of intussusception is 1 in 100,000 infants. RotaTeq is an attenuated human-bovine reassortant virus licensed in 2006, and rotaRix is an attenuated human reassortant licensed in 2008. Both vaccines were suspended for a brief period in 2010 after it was discovered that they were contaminated with pig circovirus DNA.

Naked DNA Vaccines

Naked DNA vaccines consist of plasmid DNAs that have been manipulated to contain a gene encoding a viral antigen of interest. In addition, promoter/terminator DNA sequences are introduced into the plasmid to facilitate the gene expression of the viral gene of interest in human or mammalian cells. The objective of a DNA vaccine is not to raise an immune response to the DNA itself, but rather to the expressed antigen carried on the plasmid DNA.

Plasmid DNA vaccines have several distinct advantages over other vaccines. Plasmids can easily be manufactured, are very stable, and resist temperature extremes, making storage and transport practical and inexpensive. The DNA sequence of the plasmid can easily be changed in the laboratory (**FIGURE 7-30**), and mixtures of plasmids can also be used that encode many viral protein fragments, creating a broad vaccine. Thus, vaccinologists can respond to changes in the viral pathogen of interest. Administering a DNA vaccine causes the recipient's cells to produce and process the viral antigen(s) of interest in the same way as those of the proteins of the virus against which protection is to be produced. This makes a much better antigen over purifying that protein and using it as an immunogen. As a result of the way the antigen is presented, there is also a T_C response (**FIGURE 7-31**). Clinical trials of candidate plasmid DNA HIV-1 are in progress. DNA vaccines have been tested in a variety of animal models for different viral infectious viruses such as influenza, Ebola, West Nile, and SARS-CoV.

Edible Vaccines

One of the major reasons that vaccines are not available worldwide is that many developing nations simply cannot afford them. As a result, 30 million children born each year are not adequately immunized by modern standards. Plant-made vaccines offer an alternative system for stable, low-cost, rapid scalability vaccines that are free of human pathogens. Vaccine proteins will have similar

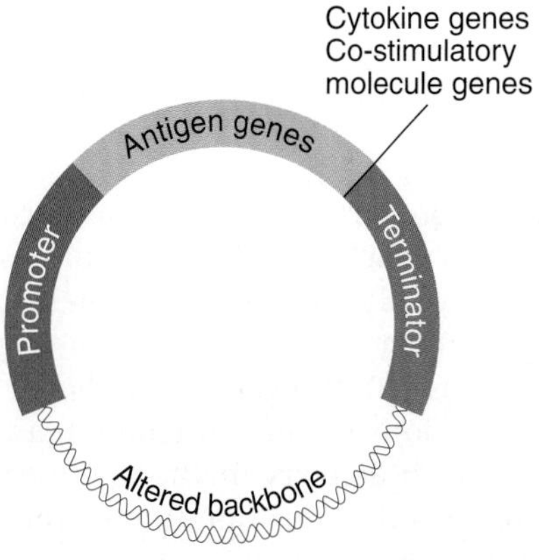

FIGURE 7-30 Schematic of a typical DNA vaccine plasmid. A gene encoding the viral antigen gene of interest is flanked between a strong viral promoter and transcriptional terminator sequences. Additional sequences may be inserted into the vector, which stimulate cytokines or other immune stimulatory molecules, and also may be inserted into the plasmid as well as an antibiotic resistance gene to facilitate plasmid production in bacteria. Adapted from de Quadros, C. A., ed. *Vaccines: Preventing Disease and Protecting Health.* Pan American Health Organization, 2004.

FIGURE 7-31 Plasmid DNAs are injected into vaccinees. Adapted from Campbell, M. K., and Farrell, S. O. *Biochemistry*, Fifth Edition. Brooks/Cole, 2006.

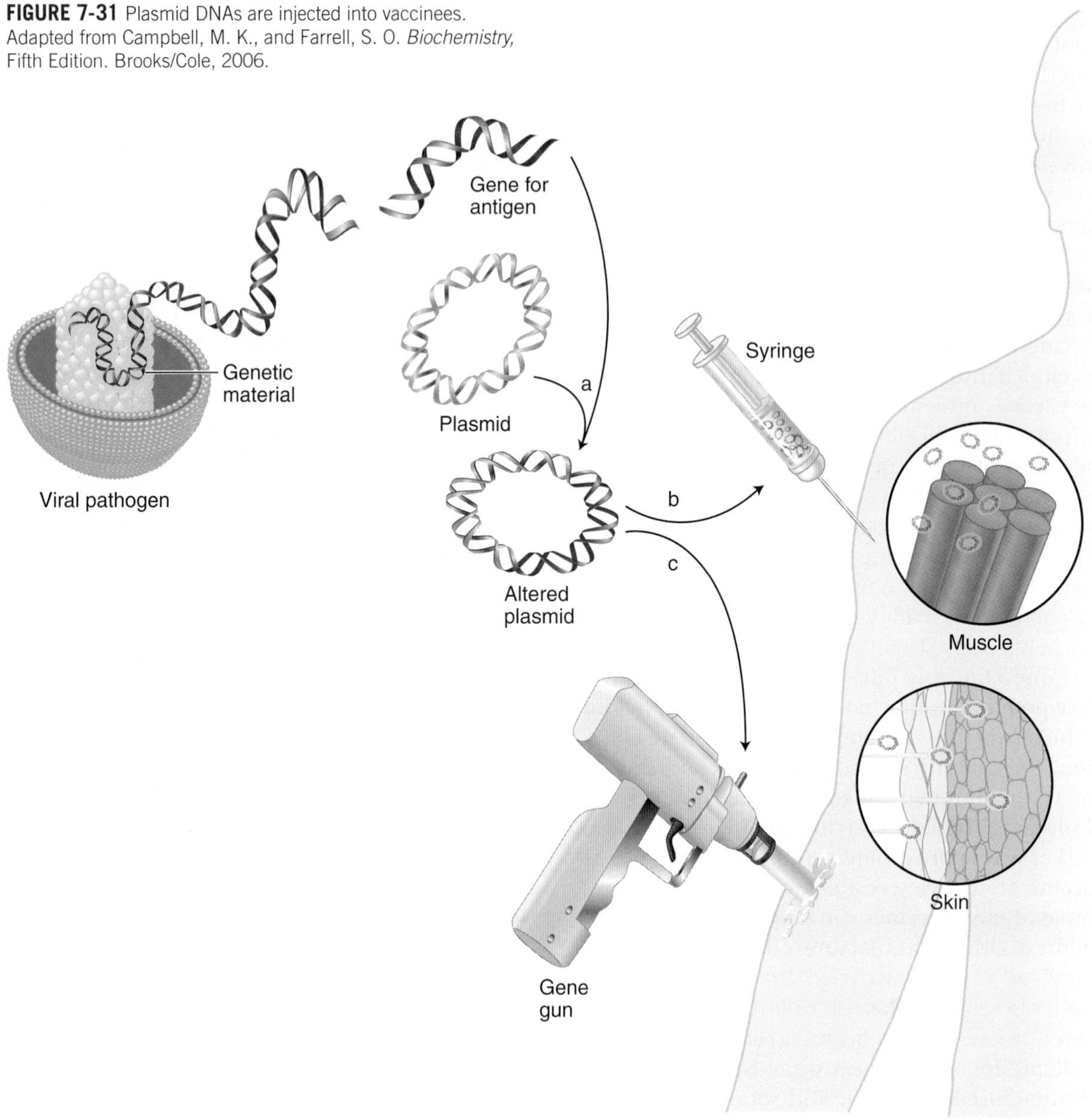

biochemical properties to pathogens because the plant cellular machinery can fold and assemble proteins correctly. Hence, these plant-derived proteins should be effective in stimulating an immune response. To the general public, of course, the most obvious benefit of edible plant-based vaccines is their delivery method: you eat them. The problem is that our immune system almost always tolerates what we ingest—so called **oral tolerance**.

Plant-made pharmaceuticals were possible only after the rapid advances in molecular biology and plant biotechnology, and the first research publications relating to the production of plant-made pharmaceuticals were not published until 1990. Today over 40 worldwide laboratory research teams have explored plant-based antigens as vaccines, with the idea of delivering oral vaccines to recipients in developing countries.

Early studies of plant-derived subunit vaccines focused on transformation systems for tobacco, but the toxicity of tobacco precluded its use for oral delivery. Other plant systems under consideration are bananas, tomatoes, carrots, lettuce, corn, and potatoes. To date, the most attractive plant systems are the tomato and potato because of their wide use in global diets, ease of genetic transformation, and the relatively short time needed to obtain fruit or tubers for bioassays (**FIGURE 7-32**).

One of the major issues that needs to be addressed with plant-based vaccines is the ability to provide a consistent dose. Scientists are approach-

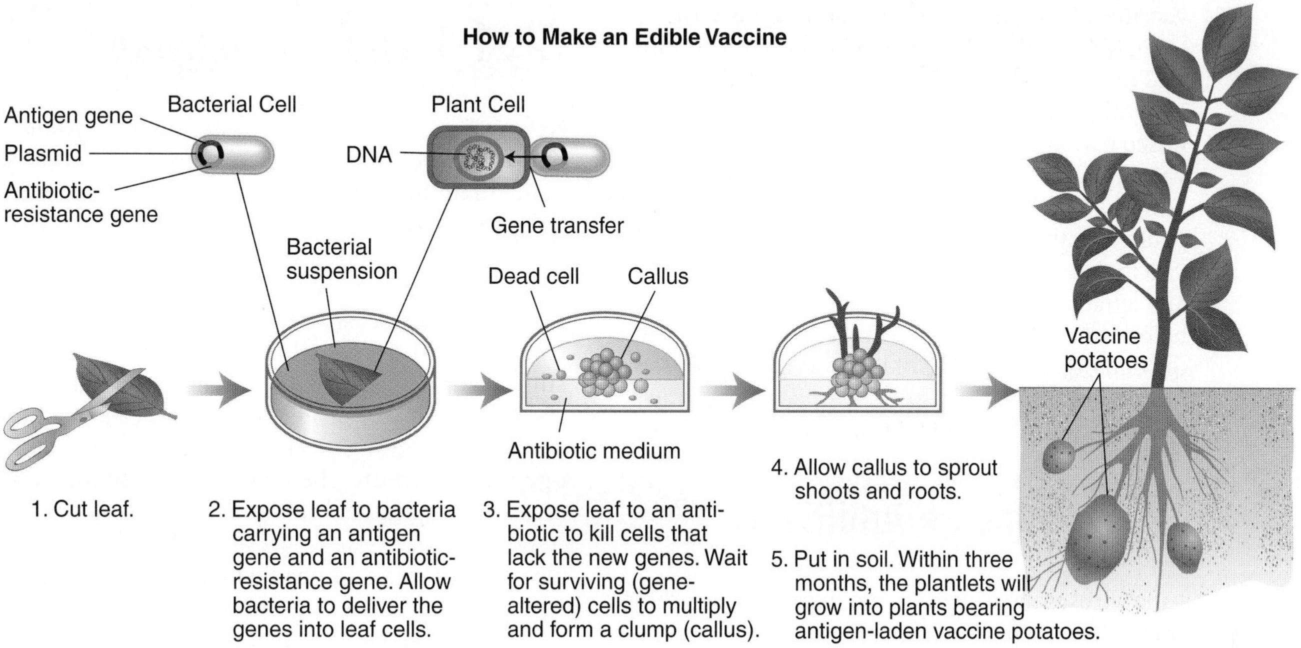
How to Make an Edible Vaccine
Antigen gene
Bacterial Cell
Plasmid
Antibiotic-resistance gene
Plant Cell
DNA
Gene transfer
Bacterial suspension
Dead cell
Callus
Antibiotic medium
Vaccine potatoes
1. Cut leaf.
2. Expose leaf to bacteria carrying an antigen gene and an antibiotic-resistance gene. Allow bacteria to deliver the genes into leaf cells.
3. Expose leaf to an antibiotic to kill cells that lack the new genes. Wait for surviving (gene-altered) cells to multiply and form a clump (callus).
4. Allow callus to sprout shoots and roots.
5. Put in soil. Within three months, the plantlets will grow into plants bearing antigen-laden vaccine potatoes.

FIGURE 7-32 Transgenic plants are an alternative for the expression of viral antigens for use as vaccines as well as antibodies and other therapeutics. Adapted from Langridge, W. H. R., 2000. Edible vaccines. *Sci Amer.* 283(3): 48–53.

ing this subject by applying food processing techniques. Plants expressing a viral gene of interest are dried and made into a powder material that is then put into gelatin capsules for oral intake.

The key advantage of plant-based systems is that they can synthesize complex proteins that can then be mass-produced with minimal cost and with enhanced vaccine stability. Even though the concept of plant-based vaccines is attractive for the aforementioned reasons, no major pharmaceutical company has aggressively pursued a development effort because there are doubts about sufficient return on investment. It is anticipated that the development of a single product would cost between $110 and $800 million and take 12 to 15 years for licensure. Other concerns of pharmaceutical companies include environmental safety issues, such as transgene spread and the rigorous scientific evaluation required of the product, and the difficulty of finding personnel with a background in plant biology research and development expertise. These are major challenges for a commercial venture. If these hurdles can be overcome, though, it might be possible to make plant-derived vaccines available to everyone who needs them and at an affordable price.

In July 2004, the Pharma-Planta Project, a consortium of 39 academic laboratories and industrial partners representing 31 organizations from 11 European countries and South Africa, was formally introduced to the public. The project was funded by a European Union contribution of 12 million euros, and its ongoing mission is to optimize techniques for the production of antibodies and vaccines that can be used to prevent important human infectious diseases. A priority product under development is the production of an antibody produced in maize that can be used to block HIV transmission.

7.8 Immunizing the Compromised Host

Vaccines are one of the most powerful tools for preventing viral infections. Infectious diseases remain the leading cause of death worldwide, and the number of immunocompromised patients has increased dramatically since the mid-1970s. Patients may be immunocompromised after organ transplantation, infections with HIV, and intensive therapy for cancer(s).

Infections have been the major obstacle to successful transplantations. Transplant patients remain on immunosuppressive drugs to prevent organ rejection, and HIV-infected patients become progressively more immunosuppressed over time. Thus, immunizing the compromised individual is important from both vulnerability and public health standpoints.

In general, it appears that there is no harm in vaccinating HIV-positive patients with inactivated vaccines. The safety of immunization with live attenuated vaccines is a matter of concern regarding HIV-infected patients, though, and live vaccines are generally not recommended for immunocompromised patients. Few studies are available regarding vaccination of cancer patients undergoing radiation or intensive chemotherapy. Immunizations of patients with certain vaccines such as the influenza vaccine are often recommended before lymphoma and leukemia patients begin cancer treatments and before transplant recipients receive bone marrow or solid organs. The use of live vaccines is not recommended for patients actively undergoing cancer chemotherapy or after transplantation.

7.9 Vaccine Additives

After viruses have been grown in cell culture, they undergo a purification process that involves ultrafiltration (to remove cellular debris and concentrate the virus), centrifugation, chromatography, and other extraction procedures. There may be some impurities remaining in the vaccine, such as antibiotics (e.g., neomycin or kanamycin), which are used to prevent the growth of bacteria in cell cultures or egg proteins from viruses cultured in embryonated chicken eggs. An adjuvant such as aluminum hydroxide is often added to stimulate immunity. Vaccines may also contain other additives such as sulfites or thimerosal (mercury) as preservatives. Manufacturers are moving away from thimerosal as a preservative. Many epidemiological studies have found no adverse long-term consequences for childhood vaccines, even those containing thimerosal.

7.10 Side Effects

The majority of side effects to vaccines are limited to localized reactions such as swelling, redness, pain, and itching at the site of injection. Allergic responses (i.e., as anaphylaxis) may occur because of overreactive immune responses to vaccine components such as antibiotics, egg proteins, animal protein, preservatives, and stabilizers. Severe side effects are rare. The cofounders of the U.S. National Vaccine Information Center (NVIC, formed in 1982) worked with federal legislators to help create and pass the National Childhood Vaccine Injury Act of 1986. The act requires physicians and other healthcare providers who administer vaccines to maintain permanent immunization records and to report occurrences of any untoward effects to the U.S. Department of Health and Human Services.

7.11 Vaccine Delivery

New ways to deliver vaccines are under consideration. Needle-and-syringe systems are woefully inadequate for worldwide vaccination programs. Safety is an issue in developing nations where HIV and hepatitis B and C are epidemic, and reusing needles is often practiced out of habit. In the United States at least 14 needles are needed to vaccinate a child from birth to adolescence. As a result, needleless systems are being designed with the idea of injury protection. Another consideration is the investigation of new delivery technology that will more effectively stimulate mucosal and cell-mediated immune responses. Lastly, newer delivery systems are attractive to individuals who are afraid of needle injections, a fear that can compromise vaccine compliance.

Jet injectors have been used in several mass vaccination programs (**FIGURE 7-33**). In 1986, the CDC's *Morbidity and Mortality Weekly Report* linked a Med-E-Jet device to an outbreak of HBV infection. In this case, the device was administering injections of human chorionic gonadotropin hormone (HCG)—not a vaccine—to attendees at a weight loss clinic in California. Injectors were improved to contain disposable nozzles, but because of the risk of transmitting bloodborne pathogens, the WHO stopped recommending multiple-use jet injectors for immunization programs.

Several companies are currently working on new vaccine and drug-delivery technologies. These technologies include nasal sprays, skin patches (the vaccine is swabbed onto the skin and temporarily covered with an adhesive patch, or

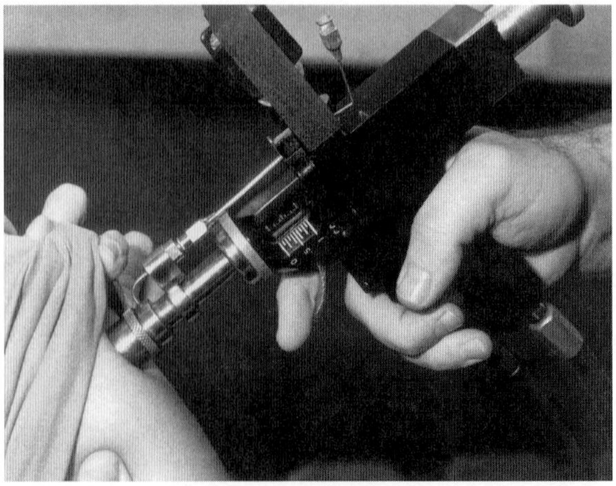

FIGURE 7-33 Photograph showing a clinician using a jet injector gun during adult immunization. The device uses gaseous pressure to propel drugs or vaccines through the outer layers of the skin and into the underlying subcutaneous tissues without the use of needles.

the vaccine may be incorporated into the patch), and time-release pills. To date, none of these technologies has been applied to licensed vaccines.

7.12 Manufacturing and Quality Assurance of Vaccines

All vaccines available in the United States are subjected to a well-defined regulatory process for approval. Each vaccine is tested for safety and effectiveness through preclinical (animal) and clinical (human trial) studies. After a vaccine is developed and preliminary preclinical research has been conducted, a report is prepared and submitted as part of an Investigational New Drug Application (IND) for approval. The IND application asks for the FDA's permission to proceed with testing the vaccine on human subjects. The FDA reviews the application and decides whether or not the vaccine is safe for human trials. The FDA provides guidelines pertaining to the necessary studies needed to satisfy the efficacy provisions of the law and informal guidance on specific methods used in vaccine manufacture and how to conduct the clinical studies. These regulations are referred to as good clinical practices (GCPs).

Clinical trials proceed in at least three phases. As the phases proceed, more human subjects are used and the patient history of the subjects will change. By phase III, full-scale production of the vaccine occurs. Trials are conducted using consistent lots of the vaccine, and the vaccines must be manufactured according to current good manufacturing practices (CGMPs).

At the end of phase III human trials, two forms of FDA-approved licenses are applied for: establishment and product. Establishment licenses must be granted to each location or facility where the vaccine will be manufactured. All locations are inspected and the product is examined during all phases of manufacturing. A product license is granted after an application to the Center for Biologics Evaluation and Research (CBER) has been approved. Approval means that the manufactured vaccine product meets the prescribed standards of safety, purity, and potency (**TABLE 7-5**). **TABLE 7-6** is a list of companies and sponsors that produced licensed viral vaccines for immunization and distribution in the United States in 2004.

After FDA establishment and product licenses have been approved, the so-called "marketing" phase of the vaccine occurs. CBER and the CDC jointly manage the Vaccine Adverse Event Reporting System (VAERS), a cooperative program for vaccine safety. The process of research, testing, review, and approval helps to ensure that drugs marketed in the United States are safe and effective. Similar mechanisms around the world provide the same assurances.

The Economics of Vaccines

There is disparity between the vaccines that are developed and produced and the ones that are actually needed. The market must meet the demands of the customer. In this case, the largest customers of vaccines are children, and the majority of children in need are located in impoverished parts of the world. Children should receive 8 to 12 different vaccines in the first years of life. Historically, vaccine development and commercialization were supported by private investments, but there are very few private investors interested in supporting

TABLE 7-5 Vaccine Development Overview

Preclinical		Clinical				
Stage I		Stage II		Stage III		Stage IV
Animal studies	FDA IND	Phase I	Phase II	Phase III	FDA LICENSE	Phase IV
		Use healthy volunteers. Small studies (20–80 subjects). Test low and high dose ranges for safety.	Limited number of human trials (100–200 subjects). Determine optimal dose/immunogenicity. Controlled investigations. Use of placebo group.	Full-scale manufacture. Numerous clinical studies at different sites. Large number of subjects (100s–1000s). Include subjects with preexisting conditions CBER application.		Vaccines available to physicians and patients. Large-scale safety and clinical trials. Trials on subjects not tested in Phase III. Monitor adverse effects.
		2 to 10 years.				

IND = Investigational new drug application.

TABLE 7-6 Virus Vaccines Licensed for Immunization and Distribution in the United States

Virus Vaccine	Sponsor
Vaccinia (smallpox)	Acambis
Influenzas A and B, poliovirus, rabies	Aventis Pasteur, Inc.
Rabies	BioPort Corp.
Rabies, influenza	Chiron Behring GmbH & Co.
Influenza	CSL
Hepatitis A and B, rotavirus, human papillomavirus, influenza	GlaxoSmithKline
Influenza (intranasal form)	MedImmune Vaccines, Inc.
Hepatitis A and B, measles, mumps, rubella, rotavirus, varicella zoster (shingles), varicella zoster (chickenpox), human papillomavirus	Merck & Co., Inc.
Influenza, Japanese encephalitis virus	Novartis
Influenza, Japanese encephalitis virus, poliovirus, rabies, yellow fever virus	sanofi pasteur
Smallpox (Variola)	Acambis

Source: Courtesy of the Centers for Disease Control and Prevention (CDC), March 2010.

the development and manufacture of vaccines that would benefit developing countries. The investments of the Bill and Melinda Gates Foundation are an exception.

The vaccine industry is dominated by only a handful of companies. Ongoing bioterrorism concerns and new biotechnology advances are generating new interest in the vaccine market, though.

The major consumer of vaccines is the UNICEF. Vaccine procurement has become complex in recent years as the market has changed. Vaccine manufacturers adjust their prices to ensure that developing countries can afford them while charging higher prices in industrialized nations. Stage III of clinical research and development is the most expensive stage in the process. Hence, differential pricing of licensed vaccines allows the manufacturer to recoup their research and development expenses.

The GAVI Alliance (http://www.gavialliance.org/) formerly called the Global Alliance for Vaccines and Immunization (GAVI) was created in 1999 in response to disparities in vaccine immunization programs and access to vaccines in low-income countries. It forms close partnerships between the private and public sectors with the goal of saving children's lives and saving the "people's health through the widespread use of vaccines." GAVI's partners include UNICEF, WHO, the Bill and Melinda Gates Foundation, the Developing Country Vaccine Manufacturers Network, and others. GAVI's objectives are to expand the reach of immunization services, introduce priority vaccines for developing countries, and establish tools and systems to promote sustainable financing in developing countries.

The Bill and Melinda Gates Foundation has committed more than $1.5 billion to help support GAVI immunization goals. Since then, the governments of Australia, Canada, Denmark, Ireland, Italy, Germany, France, Spain, Luxembourg, Norway, the Netherlands, Sweden, the United States, the Russian Federation, Republic of Korea, European Commission, and the United Kingdom have also pledged their support. GAVI's support for programs provide new and underused vaccines to more of the world's children. By understanding the current vaccine market, manufacturers, and consumers, and the investment decisions during the research and development pathway of new candidate vaccines, one can understand how economics has shaped the vaccine world. Mechanisms like GAVI may change the economics of high-priority vaccines for the developing world. Partnerships and economic savvy are needed to meet the biggest viral challenges facing the world: vaccines that prevent the transmission of HIV and hepatitis C.

Summary

The immune system is amazingly complex. It is responsible for defending the host from viral and other microbial invasions as well as tissue rejection and growth of cancerous cells. This host defense system contains two separate arming mechanisms: nonspecific responses (innate) and specific immune responses (adaptive). Interactions between the host, viral pathogen, and environmental factors determine the extent of a particular viral disease. This is referred to as the triad model of disease causation; for example, the host factors that impact susceptibility to viral infection include age, nutritional state, hormonal balance, fever responses, genetic factors, presence of other microbial infections, and species resistance.

Immediately after an infection, nonspecific host defenses provide protection against any pathogen they encounter, but they do not retain any "memories" from prior encounters. Examples of nonspecific responses are the IFN response, apoptosis, phagocytosis, the complement system, and the action(s) of dendritic and natural killer cells.

Specific immunity requires days to weeks before it is induced or effective in clearing viruses and other pathogens or cancerous cells. Specific defenses result in memory cells that patrol the host and facilitate clearance of pathogens after subsequent encounters with the same pathogen. The key players of the specific immune defense system are B and T lymphocytes, which are involved in antibody and cell-mediated responses to neutralize, control, and eliminate foreign invaders.

There are five classes of antibody molecules. Three of them participate in fighting pathogenic infections: IgA, IgG, and IgM. Whenever IgM is found in patient serum samples analyzed in the clinical laboratory, it indicates that the patient has had a recent or current infection. IgA is found in secretions of the body. IgG is the most abundant class of antibody present in serum. It is produced for long periods of time, resulting in long-term protection against invasion by prior pathogens.

Cell-mediated immunity is directed by T lymphocytes that recognize viral invaders via MHC I and II antigen-presenting complexes located on the surfaces of APCs. Cytokines communicate between the B and T cells, promoting the growth of more T and B cells. Repeated exposure of the same pathogens results in the maintenance of B and T memory cells.

Despite these host defensive mechanisms, many viruses have evolved ways to evade them. Many viruses can establish persistent and chronic infections that compromise the immune system in some way. The balance between host responses and immune evasion determines the outcome of any viral infection.

The knowledge gained from scientific experiments regarding host defenses has allowed researchers to develop immune therapies such as serum therapy or passive immunity and vaccination. Traditional (live or killed viruses) or newer generation vaccines that use genetic engineering techniques to manipulate viruses for vaccine development are used to prevent viral infections today. Vaccines are now available for the following viruses: vaccinia (smallpox), measles, mumps, rubella, varicella zoster (chickenpox), hepatitis A, poliovirus, and certain types of papillomaviruses, rotavirus, and influenza. The United States has formed the NVIC to monitor reports of any untoward effects related to vaccination.

New ways to deliver vaccines are under consideration because needle-and-syringe type systems are woefully inadequate for worldwide vaccination programs. New technologies being considered are nasal sprays, skin patches, time-release pills, and edible (plant-derived) vaccines.

Vaccines are costly to produce and manufactured by a limited number of companies. All vaccines must be subjected to a well-defined regulatory process for approval. Approval stages occur both at the preclinical and clinical levels. Clinical approval can take between two and ten years before licensure occurs. There is a disparity between what vaccines are developed and produced and which vaccines are needed. The largest consumer group of vaccines is children and the majority of them in need are located in developing countries. Very few private investors support the development and manufacture of vaccines that would benefit impoverished parts of the world. Organizations like GAVI and the Bill and Melinda Gates Foundation have responded to this disparity by forming partnerships between private and public sectors to develop new priority vaccines for developing countries. Bioterrorism concerns and new technologies are rejuvenating the vaccine market as well.

CASE STUDY QUESTIONS

1. LCMV infection is either asymptomatic or causes a mild, self-limiting illness in healthy persons. Discuss reasons why four of the organ recipients were at higher risk for LCMV. Refer to the Epidemiological Triad of Disease in your answer.
2. Who else is at risk for asymptomatic LCMV infection? Why?
3. Family members of the donor tested positive for LCMV antibodies. LCMV was isolated from the pet hamster. The family member who cared for the pet hamster had both IgM- and IgG-specific LCMV antibodies in her blood. What does IgM indicate?
4. How will scientists determine whether the LCMV isolated from the hamster is the same LCMV that infected the organ recipients?
5. How is LCMV transmitted to rodents?
6. If you own a pet hamster, what precautions should you take after handling it and cleaning the cage?
7. How would you clean a rodent-infested area such as a dusty garden shed?

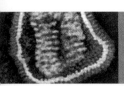

CASE STUDY 1: MEASLES

John Kostman, a sophomore sociology major at a midwestern university, woke up on a Tuesday morning with a fever. He had been studying hard and not sleeping as much as he should. He assumed he had some kind of flu bug. Three days later his fever had not gone down and he developed a runny nose and cough. He also noticed his eyes were very red (conjunctivitis). He thought he had pinkeye and decided to go to the campus health clinic to get something for his eyes. He seemed to recall that pink eye was very contagious.

At the clinic, a physician, Dr. McDermott, carefully examined him. He noticed a rash developing at John's hairline. Dr. McDermott asked John a lot of questions, which annoyed John because he was feeling worse by the minute. Dr. McDermott asked John for his vaccination records and if he had been in contact with anyone with similar symptoms. "I'll have to get my shot records from my mother over the weekend," said John. He did recall that he had volunteered to babysit his sister's 18-month-old son Jamie over the winter holiday break two weeks earlier. At the time, Jamie had had a runny nose and fever. John hadn't followed up with his sister to find out how Jamie was feeling because "kids get sick often." Dr. McDermott ordered blood tests and told John he should take some Tylenol and to rest. He said he would call him with the test results as soon as they came in.

Dr. McDermott suspected that John might have measles. Measles is very rare and children are vaccinated for measles, though, and Dr. McDermott had never seen a case of measles himself. He thought it was likely that John had been vaccinated but needed his vaccination records to confirm it.

Over the weekend, John received his vaccination records from his mother and faxed a copy to Dr. McDermott. His rash had progressed down his entire body. Dr. McDermott reviewed the records and noticed that John did not have a measles vaccine booster as a child. The blood tests confirmed his suspicion of measles. He began asking John more questions about his contacts, given that measles is highly contagious and that there had been a few reports of school outbreaks of measles in the United States and Canada—countries that practice routine measles vaccination. Dr. McDermott asked John if he had traveled during his vacation break or if he had any contact with someone from a foreign country. John said that his only contact was his new nephew, who had been adopted from an orphanage in China two weeks earlier. Shortly after this conversation, John's sister called and mentioned that his nephew had just gotten over the measles. She called the adoption agency and found out that other new adoptees from China were also suffering from measles. Efforts were being made to contact airline passengers who had traveled from China to the United States on the flights that the adoptees and their families had been on.

1. What would John's blood be tested for (e.g., antibodies, virus)?
2. Most persons in the United States are immune to measles; therefore, airline passengers traveling on planes with infected individuals are at low risk. Should Dr. McDermott make a decision to offer measles-containing vaccine to all staff and students who attended the same university as John? Why or why not?
3. Could John have contracted measles from someone else? Explain your answer.
4. What are some of the complications of measles infection? What age group(s) are more at risk for complications? (Hint: Perform an Internet search via the CDC's Web site or other credible Web sites.)

References

Alaska Department of Health and Social Services, et al. 2004. "Brief report: Update: Measles among adoptees from China–April 14, 2004." *MMWR* 53(14):309.

Alaska Department of Health and Social Services, et al. 2004. "Multistate investigation of measles among adoptees from China–April 9, 2004." *MMWR* 53:14;309–310.

Bell, A., King, A., Pielak, K., and Fyfe, M. 1997. "Epidemiology of measles outbreak in British Columbia–February 1997." *CCDR* 1997. Apr 1, 23(7):49–51.

Lurie, P., et al. 2003. "Measles outbreak in a boarding school—Pennsylvania, 2003." *MMWR* 53(14): 306–309.

Reynolds, A., et al. 2002. "Measles outbreak among internationally adopted children arriving in the United States, February—March 2001." *MMWR* 51(49): 1115–1116.

Resources

Primary Literature and Reviews

Agol, V. I., and Gmyl, A. P. 2010. "Viral security proteins: Counteracting host defenses." *Nat Rev* 8:867–878.

Atencia, R. 2007. "Differential expression of viral PAMP receptors mRNA in peripheral blood of patients with chronic hepatitis C infection." *BMC Infect Dis* 7:136.

Berzofsky, J. A. 2004. "New strategies for designing and optimizing vaccines." *ASM News* 70(5):219–223.

Biron, C. A., Byron, K. S., and Sullivan, J. L. 1989. "Severe herpesvirus infections in an adolescent without natural killer cells." *N Engl J Med* 320(26):1731–1735.

Blue, C. E., Spiller, O. B., and Blackgourn, D. J. 2004. "The relevance of complement to virus biology." *Virology* 319:176–184.

Brinton, M. A., and Perelygin, A. A. 2003. "Genetic resistance to flaviviruses." *Adv Virus Res* 60:43–85.

Brown, M. G., et al. 2001. "Vital involvement of a natural killer cell activation receptor in resistance to viral infection." *Science* 292:934–937.

Centers for Disease Control and Prevention (CDC). 1999. "Prevention of hepatitis A through active or passive immunization: Recommendations of the Advisory Committee on Immunization Practices (ACIP)." *MMWR* 48(RR12):1–37.

Chen, R. T., et al. 2004. "Suspension of rotavirus vaccine after reports of intussusception—United States, 1999." *MMWR* 53(34):786–789.

Clarke, P., and Tyler, K. L. 2009. "Apoptosis in animal models of virus-induced disease." *Nat Rev Microbiol* 7:144–155.

Coates, P. T. H., Clovin, B. L., Hackstein, H., and Thomson, A. W. 2002. "Manipulation of dendritic cells as an approach to improved outcomes in transplantation." *Exp Rev Mol Med* 18:1–21.

Cohen, J., and Powderly, W. G., eds. 2004. *Infectious Diseases*, 2nd ed. Philadelphia: Mosby.

Das, A., Maini, M. K., 2010. "Innate and adaptive immune responses in hepatitis B virus infection." *Dig Dis* 28:126–132.

Dato, V., et al. 2003. "Hepatitis A outbreak associated with green onions at a restaurant—Monaca, Pennsylvania, 2003." *MMWR* 52(47):1155–1157.

de Blecourt, W., and Usborne, C., eds. 2004. *Cultural Approaches to the History of Medicine: Mediating Medicine in Early Modern and Modern Europe.* New York: Palgrave McMillan.

de Quadros, C. A., ed. 2004. *Vaccines: Preventing Disease & Protecting Health.* Washington, DC: Pan American Health Organization.

De Veer, M. M., et al. 2001. "Functional Classification of Interferon-Stimulated Genes Identified Using Microarrays." *J Leukoc Biol* 69:912–920.

Diamond, M. S. 2003. "Evasion of innate and adaptive immunity by flaviviruses." *Immunol Cell Biol* 81:196–206.

Favoreel, H. W., Van de Walle, G. R., Nauwynck, H. J., and Pensaert, M. B. 2003. "Virus complement evasion strategies." *J Gen Virol* 84(Pt1):1–15.

Garcia-Sastre, A., and Biron, C. A. 2006. "Type 1 interferons and the virus-host relationship: A lesson in détente." *Science* 312:879–882.

Geisbert, T. W., and Jahrling, P. B. 2003. "Towards a vaccine against Ebola virus." *Exp Rev Vaccines* 2(6):777–789.

Glass, R. I., et al. 2004. "The future of Rotavirus vaccines: A major setback leads to new opportunities." *Lancet* 363:1547–1550.

Goldsby, R. A., Kindt, T. J., Osborne, B. A., and Kuby, J. 2003. *Immunology*, 5th ed. New York: W. H. Freeman.

Haller, O., Frese, M., and Kochs, G. 1998. "Mx proteins: Mediators of innate resistance to RNA viruses." *Rev Sci Tech Off Int Epiz* 17(1):220–230.

Itoh, K., et al. 1994. "Hypocomplementemia associated with hepatitis C viremia in sera from voluntary blood donors." *Am J Gastroenterol* 89(11):2019–2024.

Jacobs, B. L., and Langland, J. O., 1996. "When two strands are better than one: The mediators and modulators of the cellular responses to double-stranded RNA." *Virology* 219:339–349.

Jamieson, A. M., et al. 2010. "Influenza virus-induced glucocorticoids compromises innate host defense against a secondary bacterial infection." *Cell Host Microbe* 7(2):103–114.

Kawai, T., and Akira, S. 2006. "Innate immune recognition of viral infection." *Nat Immun* 7(2):131–137.

Kawai, T., and Akira, S. 2010. "The role of pattern-recognition receptors in innate immunity: Update on toll-like receptors." *Nat Immunol* 11(5):373–384.

Knipe, D. M., and Howley, P. M., eds. 2001. *Fundamental Virology*, 4th ed. Philadelphia: Lippincott Williams & Wilkins.

Kohlmeier, J. E., Woodland, D. L. 2009. "Immunity to respiratory viruses." *Annu Rev Immunol* 27:61–82.

Koyama, S., et al. 2008. "Innate immune response to viral infection." *Cytokine* 43(3):336–341.

Langland, J. O., et al. 2006. "Inhibition of PKR by RNA and DNA viruses." *Virus Res* 119:100–110.

Leclerc, C. 2003. "New approaches in vaccine development." *Comp Immunol Microbiol Infect Dis* 26:329–341.

Ma, J. K. C. 2003. "The production of recombinant pharmaceutical proteins in plants." *Nat Gen* 4:794–805.

Mahanty, S., et al. 2003. "Cutting edge: Impairment of dendritic cells and adaptive immunity by Ebola and Lassa viruses." *J Immunol* 170:2797–2801.

McFadden, G., and Barry, M. 1998. "How poxviruses oppose apoptosis." *Sem Virol* 8:429–442.

McFadden, G., et al. 2009. "Cytokine determinants of viral tropism." *Nat Rev Immun* 9:645–655.

Munz, C., et al. 2009. "Antiviral immune responses: Triggers of or triggered by autoimmunity?" *Nature* 9:246–258.

Netea, M. G., Kullberg, B. J., and Van der Meer, J. W. M. 2000. "Circulating cytokines as mediators of fever." *Clin Infect Dis* 31(Suppl 5):S178–S184.

Orange, L. O., 2003. "Disseminated Varicella infection due to the vaccine strain of Varicella zoster virus in a patient with a novel deficiency in natural killer T cells." *J Infect Dis* 18(7):948–953.

Paolazzi, C. C., Perez, O., and De Flippo J. 1999. "Rabies vaccine. Developments employing molecular biology methods." *Mol Biotechnol* 11(2):137–147.

Parashar, U. D., et al. 2003. "Global illness and deaths caused by rotavirus disease in children." *Emerg Infect Dis* 9(5):565–572.

Pastoret, P. P., and Vanderplasschen, A. 2003. "Poxviruses as vaccine vectors." *Comp Immunol Microbiol Infect Dis* 26:343–355.

Pestka, S. 2001. "The human interferon: A species and receptors." *Biopolymers* 55:254–287.

Pestka, S., et al. 2004. "Interleukin-10 and related cytokines." *Ann Rev Immunol* 22:929–979.

Pichlmair, A., et al. 2007. "Innate recognition of viruses." *Immunity* 27:370–383.

Pieters, T. 1993. "Interferon and its first clinical trial: Looking behind the scenes." *Med Hist* 37:270–295.

Plotkin, S. A., Orenstein, W. A., and Offit, P. A., eds. 2003. *Vaccines*, 4th ed. Philadelphia: W. B. Saunders.

Powledge, T. M. 1984. "Interferon on Trial." *Biotechnology* 2:214–228.

Pyzik, M., et al. 2010. "Self or nonself? That is the question: Sensing of cytomegalovirus infection by innate human receptors." *Mamm Genome* 22(1–2):6–18.

Rizzi, M., et al. 2010. "Long-lived plasma and memory B cells produce pathogenic anti-GAD65 autoantibodies in stiff person syndrome." *PLoS ONE* 5:e10838. doi:10.1371/journal.pone.0010838.

Ross, A. C. 1992. "Vitamin A status: Relationship to immunity and the antibody response." *Proc Soc Exp Biol Med* 200:303–320.

Rouse, B. T., and Sehrawat, S. 2010. "Immunity and immunopathology to viruses: What decides the outcome?" *Nat Rev Immun* 10:514–526.

Samarajiwa, S. A., et al. 2009. "INTERFEROME: The database of interferon regulated genes." *Nucl Acids Res* 37:D852–D857.

Samuel, C. E. 2001. "Antiviral actions of interferons." *Clin Microbiol Rev* 14(4):778–809.

Seet, B. T., et al. 2003. "Poxviruses and immune evasion." *Ann Rev Immunol* 21:377–423.

Singer, P. A., et al. 2007. "A tough transition: What is holding back biotechnology in the developing world?" *Nature* 449(13):160–163.

Singhera, B. K., et al. 2006. "Apoptosis of viral-infected epithelial cells limit viral production and is altered by corticosteroid exposure." *Respir Res* 7:78 doi10.1186/1465-9921-7-78.

Spear, G. T., et al. 2001. "The role of the complement system in virus infections." *Curr Top Microbiol Immunol* 260:229–245.

Stephensen, C. B. 2001. "Vitamin A, infection, and immune function." *Annu Rev Nutr* 21:167–192.

Stoitzner, P., Pfaller, K., Stossel, H., and Romani, N. 2002. "A close-up view of migrating Langerhan's cells in the skin." *J Invest Dermatol* 118(1):117–125.

Stowe, R. P., et al. 2001. "Immune responses and latent herpesvirus reactivation in spaceflight." *Aviat Space Environ Med* 72(10):884–891.

Stratov, I., DeRose, R., Purcell, D. F., and Kent, S. J. 2004. "Vaccines and vaccine strategies against HIV." *Curr Drug Targets* 5(2):71–88.

Stuart-Harris, R., and Penny, R., eds. 1997. *Clinical Applications of the Interferons*. New York: Chapman & Hall Medical.

Takeuchi, O., Akira, S., 2009. "Innate immunity to viral infection." *Immunol Rev* 227(1):75–86.

Turner, P. C., and Moyer, R. W. 1998. "Control of apoptosis by poxviruses." *Sem Virol* 8:453–469.

Vastag, B. 2004. "Ebola vaccines tested in humans, monkeys." *JAMA* 291(5):549–550.

Voelker, R. 1999. "Eradication efforts need needle-free delivery." *JAMA* 281:20 1879–1881.

Wang, J. P., et al. 2007. "Innate immunity to respiratory viruses." *Cell Microbiol* 9(7):1641–1646.

Warrell, M. J., and Warrell, D. A. 2004. "Rabies and other lyssavirus diseases." *Lancet* 363:959–969.

Welsh, R. M., et al. 2010. "Heterologous immunity between viruses." *Immunol Rev* 235(1):244–266.

Wilkins, C., and Gale, M. Jr. 2010. "Recognition of viruses by cytoplasmic sensors." *Curr Opin Immunol* 22(1):41–47.

Wolstenholme, G. E. W., and O'Connor, M., eds. 1967. *Interferon*. Boston: Little, Brown.

Yu, J. Y., DeRuiter, S. L., and Turner, D. L. 2002. "RNA interference by expression of short-interfering RNAs and hairpin RNAs in mammalian cells." *PNAS USA* 99(9):6047–6052.

Zhi-yong, Y., et al. 2004. "A DNA vaccine induces SARS coronavirus neutralization and protective immunity in mice." *Nature* 428:561–564.

Zinkernagel, R. M., and Hengartner, H. 2001. "Regulation of the immune response by antigen." *Science* 293:251–256.

Zuniga, E. I., et al. 2005. "Immunosuppressive viruses and dendritic cells: A multifront war." *ASM News* 71(6): 285–290.

Popular Press

Allen, A. 2007. *Vaccine: The Controversial Story of Medicine's Greatest Lifesaver.* New York: W. W. Norton.

Banchereau, J. 2002. "The long arm of the immune system." *Sci Am* November, 287(5):52–59.

Cantell, K. 1998. *The Story of Interferon: The Ups and Downs in the Life of a Scientist.* Hackensack, NJ: World Scientific Publishing.

Ewald, P. W. 2002. *Plague Time: The New Germ Theory of Disease.* New York: Anchor Books.

Finley, D. 1998. *Mad Dogs: The New Rabies Plague.* College Station, Texas: Texas A & M University Press.

Hall, S. S. 1997. *A Commotion in the Blood: Life, Death, and the Immune System.* New York: Owl Books.

Johnson, H. M., Bazer, F. W., Szente, B. E., and Jarpe, M. A. 1994. "How interferons fight disease." *Sci Am* May, 270:68–75.

Langridge, W. H. R. 2000. "Edible vaccines." *Sci Am* 283(3): 66–71.

Oldstone, M. B. A. 2009. *Viruses, Plagues & History: Past, Present, and Future.* New York: Oxford University Press.

Schrof, J. M. 1998. "Miracle vaccines." *U.S. News & World Report,* November 23rd, 1998.

Video Productions

Bad Blood. 2010. Necessary Films.

American Experience: The Polio Crusade. 2009. PBS.

The Final Inch: In the Fight to Eliminate Polio the Next Step is the Biggest. 2009. HBO Documentary Films.

Unseen Enemies, A Close-Up Look at the World's Deadliest Diseases: Rotavirus. 2006. Films for the Humanities.

Influenza. 2006. Films for the Humanities.

In Search of the Polio Vaccine (History Channel). 2005. A & E Production.

Dog Bites and Rabies Prevention. 2004. Medivision Film Series.

The Immune System at Work. 2003. Films for the Humanities.

Tainted Blood. 2002. Films for the Humanities.

New Blood. 2002. Films for the Humanities.

Your Immune System. 2001. Films for the Humanities.

eLearning

go.jblearning.com/shors2

The site features eLearning, an online review area that provides quizzes and other tools to help you study for your class. You can also follow useful links for in-depth information, or just find out the latest virology and microbiology news.

CHAPTER 8

Epidemiology

This man is sneezing into a handkerchief while riding public transportation. This sneeze could potentially be the start of a viral epidemic.

OUTLINE

"I hate definitions."

Benjamin Disraeli (1804–1881)

CASE STUDY: VIRAL GASTROENTERITIS

You have returned to your parents' home in Wisconsin for spring break. Once home, your four-year-old sister, Tina, has a swimming lesson at the local swimming club and you decide to take her to her lesson. The pool was very busy. There was a birthday party of five ten-year-old children, preschool swimming classes, mother–infant swimming classes, and a few other private groups in attendance. You noticed that the pool water appeared cloudy, but assumed that the group instructor would not allow children to swim if there was a problem with the pool maintenance. Other than swallowing a fair amount of water, Tina enjoyed her lesson.

Thirty-six hours later, Tina experienced an onset of symptoms: vomiting, diarrhea, nausea, and chills. You took her temperature and noted that she had a fever of 102.2°F (56.8°C). Your mother was concerned that Tina could become dehydrated and decided that a physician should examine her. An epidemiologist intern was present during Tina's examination and asked her mother to fill out a questionnaire about Tina's recent gastrointestinal illness. Specimens were collected from Tina and sent to a clinical laboratory. Tina was hospitalized and given intravenous normal saline. Two days later, Tina was rehydrated and able to tolerate oral feedings before being discharged. All cultures for bacterial pathogens gave negative results, but a rapid diagnostic test was positive for a viral pathogen.

8.1 What Is Epidemiology?

Prior chapters provided an introduction to virus life cycles at the molecular level as well as background on the naming of viruses, laboratory methods used to cultivate viruses, diagnosis of viral diseases, mechanisms of viral spread, and the immune system. This chapter expands on the concept of the **triad model of disease causation** described in Chapter 7 and introduces the scientific field of **epidemiology**, which is linked to virology (**FIGURE 8-1**).

Why does an infectious disease occur in one population rather than another? Why are certain age groups more at risk for a particular viral disease? Why aren't viral diseases randomly distributed within a specific geographic area? These are questions studied by epidemiologists.

What is the definition of epidemiology? If we break the word apart into its Greek root definitions, the prefix **epi-** means "on or upon, befall" and **demos** means "people, population." If the term is literally defined by this etymology, epidemiology means, "the study of that which befalls man."

Specifically, epidemiology is the study of how diseases affect whole communities. Its objectives are to determine the distribution, cause, control, and prevention of diseases in populations. In this era of bioterrorism, epidemiology has become more important than ever. The detection of covert and terrorist biological attacks will most likely occur at a local level, where disease-tracking systems and experts need to be in place. Epidemiologists must be able to respond to reports of rare, unusual, or unexplained illnesses so that critical decisions can be made rapidly.

Epidemiology Definitions

The field of epidemiology contains its own language and terminology; thus, it is important to define some of these terms before applications of epidemiology can be discussed.

Endemic. The prefix **en-** means "in or within." The term endemic applies to diseases that persist at a moderate and steady level within a given geographic area; for example, many individuals suffer from cold sores caused by herpes simplex 1. This would be considered an endemic illness (**FIGURE 8-2**).

Sporadic. Disease outbreaks that have no pattern of occurrence in time or location.

Epidemic. An unusually high number of cases in excess of normal expectation of a similar illness in a population, community, or region (**Figure 8-2**).

Pandemic. The prefix **pan-** means "all or across." Pandemic refers to a worldwide epidemic; for example, acquired immunodeficiency syndrome (AIDS) is a pandemic disease.

Morbidity. The illness or disease state.

Mortality. Numbers of deaths correlated with a particular disease. Mortality rates are expressed in quantitative terms; for example, the annual death rate or mortality rate is calculated as follows:

{[total number of deaths from all causes in 1 year]/[number of persons in the population at midyear]} × 1000.

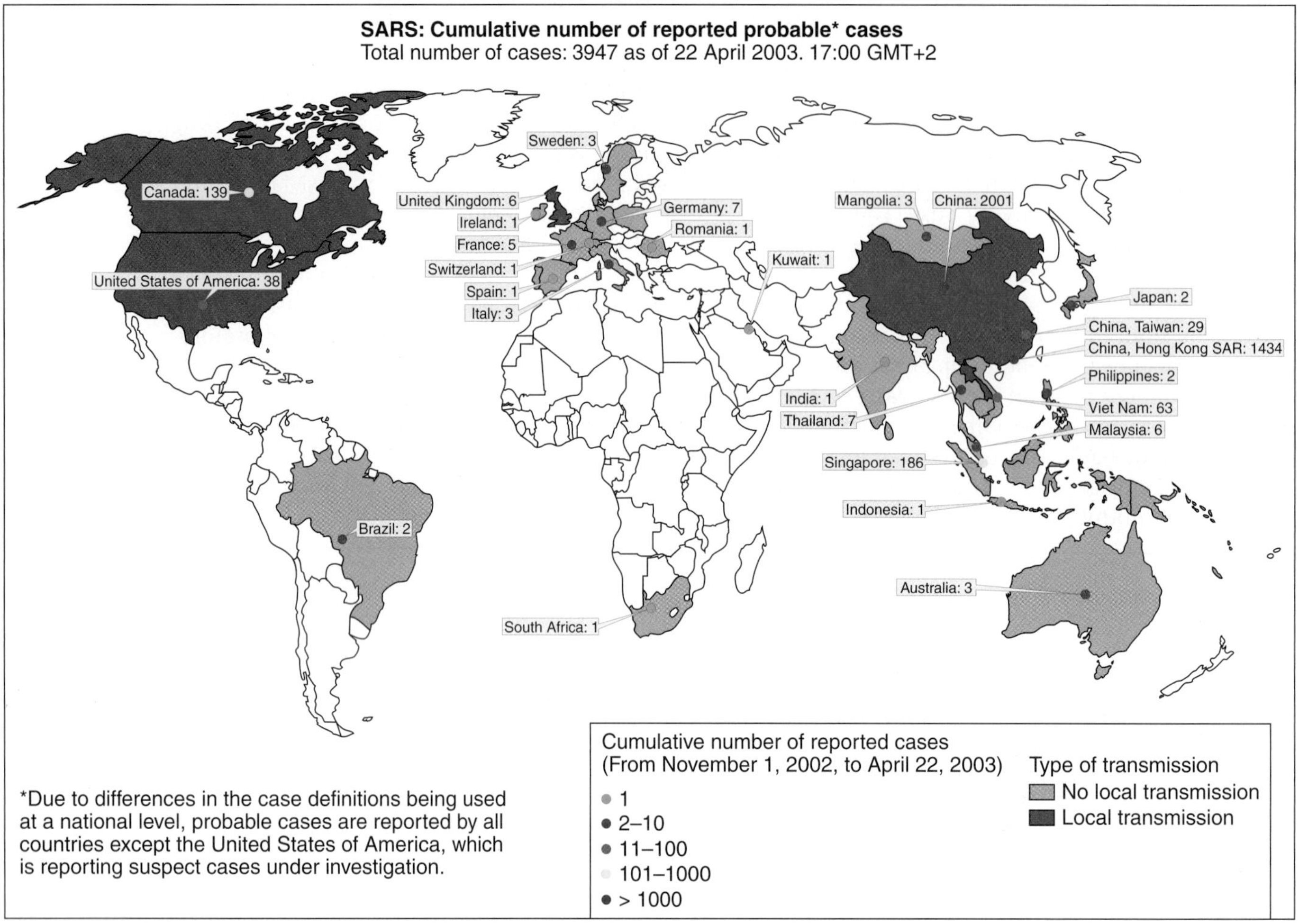

FIGURE 8-1 Epidemiology is the branch of science that deals with disease diagnostics, trends, and control measures. Epidemiologists map epidemics and pandemics and would be on the frontlines of a terrorist-caused biological attack. Epidemiologists on the Centers for Disease Control and Prevention's (CDC's) Public Health Mapping Team created this severe acute respiratory syndrome (SARS) pandemic map. Increased global travel contributed to the spread of SARS. Reproduced from World Health Organization, "Severe Acute Respiratory Syndrome," Epidemic and Pandemic Alert and Response, Courtesy of the World Health Organization.

Incidence. A measurement of morbidity. Incidence is the number of new cases of a disease that occur in a specified period of time in a susceptible population. The incidence rate is expressed per 1000 persons:

Incidence per 1000 = {[number of ***new*** cases]/[number of persons at **risk**]} × 1000.

Prevalence. A measurement of morbidity that refers to the number of cases existing in a population at a specified time. It does not take into account the duration of the disease. Prevalence rates are also recorded as prevalence per 1000.

Carrier. An individual who is a carrier harbors the virus but is not infected as measured by serologic methods (i.e., there is no evidence of an antibody response) or by evidence of clinical illness. This person can still infect others. A carrier status may be of limited duration or chronic (i.e., lasting for months or years).

Incubation period. The time between infection with a virus and the onset of symptoms (**FIGURE 8-3**). These times can vary; for example, the incubation period for influenza is 1 to 2 days, whereas the incubation period for chickenpox is 14 to 16 days (**TABLE 8-1**).

Prodromal period. The first appearance of mild or nonspecific signs and symptoms of an illness (Figure 8-3).

Period of illness. The time span of when a patient experiences defined symptoms and signs of illness.

Mode of transmission. This defines how an infectious disease is spread or passed on. Modes can be **direct** (e.g., through touching, kissing, or sexual contact, or by droplet and oral–fecal transmission) or **indirect**

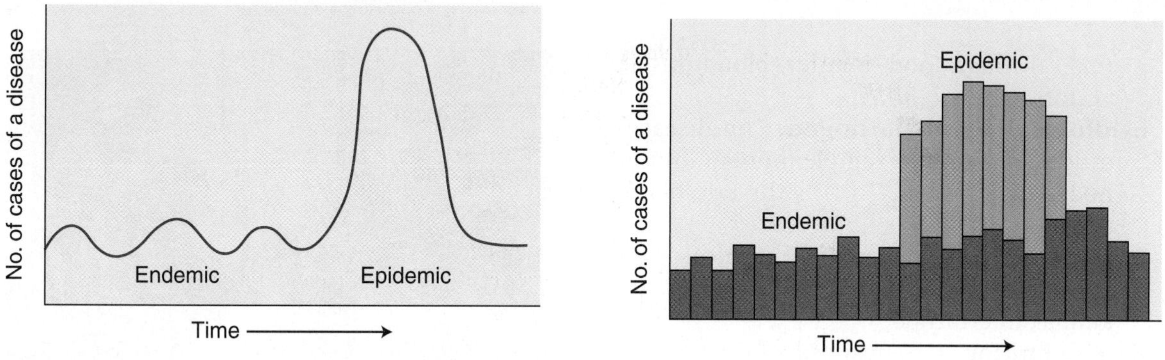

FIGURE 8-2 Comparison of endemic versus epidemic diseases. Adapted from Gordis, L. *Epidemiology*, Second Edition. W. B. Saunders, 2000.

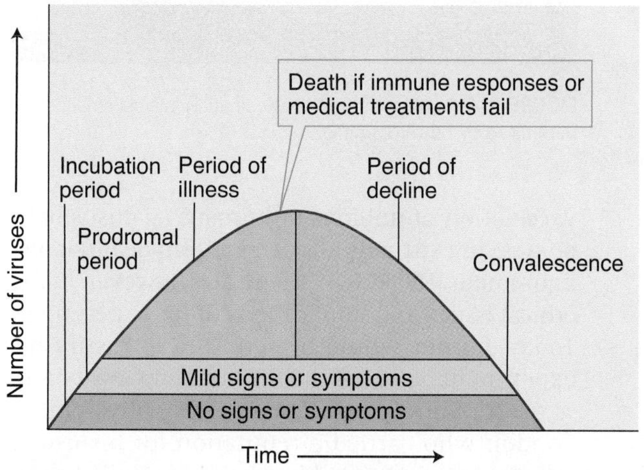

FIGURE 8-3 Course of infectious disease. Tortora, G. T., Berdell, R. F., and Case, C. L. Fig. 14.5, p. 417 from *Microbiology*, Eighth Edition, Copyright © 2004 by Pearson Education, Inc. Reprinted with permission.

TABLE 8-1 Incubation Periods for Communicable Viral Diseases (Partial List)

Viral Disease	Incubation Period	Period of Communicability
Influenza	1–3 days	3–5 days from clinical onset in adults; up to 7 days in children
SARS	1–10 days	Not completely understood. Early studies suggest no transmission before symptoms appear; transmission more likely during the 2nd week of illness. Healthcare workers at greatest risk, especially during procedures like intubation or nebulization.
Chickenpox	2–3 weeks (commonly 14–16 days)	As long as 5 days but usually 1–2 days before onset of rash, continuing until lesions are crusted (5 days)
Common cold	12–72 hours (usually 24 hours)	From 1 day before to 5 days after onset
Measles	10–14 days until rash appears	From 1 day before the beginning of the prodromal period (usually about 4 days before rash onset) to 4 days after the onset of the rash
Smallpox	7–19 days, commonly 10–14 days to onset of illness; and 2–4 days more to onset of rash	From the time of development of the earliest lesions to disappearance of all scabs (about 3 weeks)
Mumps	12–26 days (usually 18)	From 6 days before to 9 days after symptoms appear
Rubella	14–17 days with a range of 14–21 days	1 week before and to 4 days after onset of rash
Rabies	2–8 weeks or longer	From animals, 3–5 days before symptoms and during the course of the disease
Ebola	2–21 days	As long as blood and secretions contain virus. Can be as long as 61 days after onset of illness from care of corpses.
West Nile encephalitis	3–12 days	Not directly transmitted from person to person
Hepatitis B	45–180 days (average 60–90 days)	From many weeks before the onset of the first symptoms and throughout the clinical course of the disease

(e.g., through food, fomites, blood, insect vectors, or air).

Etiological agent/pathogen. The disease-causing agent; for example, human immunodeficiency virus (HIV) is the etiological agent of AIDS.

Reservoir. The site where the etiological agent lives, grows, and multiplies (e.g., human, animal, or arthropod).

Case definition. A standard set of criteria used to identify who has the disease being studied.

Communicable period. The time span of when an infected individual or animal is contagious and can directly or indirectly infect another person, animal, or arthropod (Table 8-1).

Convalescence. The recovery period after an illness. Though the individual is feeling much better, he/she may still be infectious and able to transmit the disease during this stage of infection (Figure 8-3).

Zoonosis. Any infection or infectious disease transmissible from animals to humans; for example, rabies virus can be transmitted to humans from a rabid animal.

8.2 History of Epidemiology: From Observational Data to Preventative Action

Three pioneers of epidemiology—Edward Jenner, John Snow, and Florence Nightingale—were not knowledgeable about the pathogenesis of disease. Their important observations and actions, however, led to the prevention of smallpox (1796), and cholera (1854), and a reduction in mortality rates of wounded men in the British Army during the Crimean War (1855), respectively.

As described in Chapter 1, Jenner was a British general practitioner and surgeon in the late 18th century who became very interested in smallpox. At the time he was practicing, smallpox was a worldwide scourge, killing thousands each year in England alone. Survivors were badly scarred and disfigured, and many became blind as a result of corneal infections.

Jenner had observed that milkmaids appeared to be immune to smallpox, and deduced that the pustules or lesions they occasionally got on their hands were caused by a relative of smallpox called cowpox and hypothesized that exposure to the cowpox virus would protect individuals from contracting smallpox. The results of Jenner's experiment on an eight-year-old "volunteer" led to the vaccination of millions of humans against smallpox, saving virtually all of them from death or disfigurement (**FIGURE 8-4**). There are, however, many ethical issues and implications of his experiment. Today Jenner would be required to justify his experiment before an institutional review board.

FIGURE 8-4 Painting of the first vaccination against smallpox performed by Edward Jenner.

John Snow was a 19th-century physician in London who earned a reputation for his use of anesthetics, most notably when he supplied chloroform to Queen Victoria during the birth of her children. Snow was later recognized as an epidemiological hero. He possessed excellent critical thinking skills, recognizing that a disease prevention strategy could be conceived if one could locate its source. Snow believed in the **germ theory of disease** during a time period when the majority of his colleagues believed in the **miasmatic theory**, which held that a disease was caused by *miasma* or polluted gases that rose from swamps and decaying or putrid matter. Many of the concepts and epidemiological methods Snow discovered are still useful today.

In London, during the first ten days of September 1854, approximately 500 individuals died from cholera. At the time of the epidemic, Snow's rival, the Registrar General William Farr, collected data to support the hypothesis that the people were contracting cholera from a *miasma.* He believed there were more deaths in the parts of London located at the lowest elevations. Residents began to panic and spread rumors that cholera was being transmitted by *miasmata* emanating from the ground where Black Death (plague) victims were buried during a pandemic in the previous century, a site upon which homes had since been built.

Snow had experience in attending and observing cholera patients as a practicing physician prior to this outbreak. He observed young children who shared hospital beds in which they caught cholera from each other. This reinforced his affirmation that cholera was transmitted by "direct contagion." He became a cholera expert by combining his clinical observations with microscopic analysis of different water sources. Suspicious that cholera was transmitted by contaminated water, Snow compared water sources used by those who got cholera and those who didn't. His scientific observations allowed him to develop a classical experimental design to study the 1854 London cholera epidemic.

The Registrar General cooperated with Snow by providing him with addresses of those individuals who died from cholera, and Snow plotted the deaths on a map of the area. From this he could immediately determine that cholera deaths were not localized to the area where plague victims had been buried. He also knew that each household signed up randomly for water sources provided by water companies, so he walked door to door in order to determine who supplied water to the homes of cholera victims. He developed a test using silver chloride to distinguish among the three water sources available, and became so proficient that he could distinguish water sources on the basis of appearance and odor alone.

Two water companies were located in a water intake area where the water was frequently contaminated by sewage outlets. One of the companies moved upstream and away from London's polluted and contaminated part of the Thames River; Snow hypothesized that there would be fewer cholera deaths and cases occurring in households that used water from this company—an accurate hypothesis. Households that used water drawn from the two other companies, which were located downstream of the sewage outlets, had very high mortality rates from cholera, and the largest cluster of deaths and cases at the beginning of the epidemic occurred at a public water pump located on Broad Street (**FIGURE 8-5**).

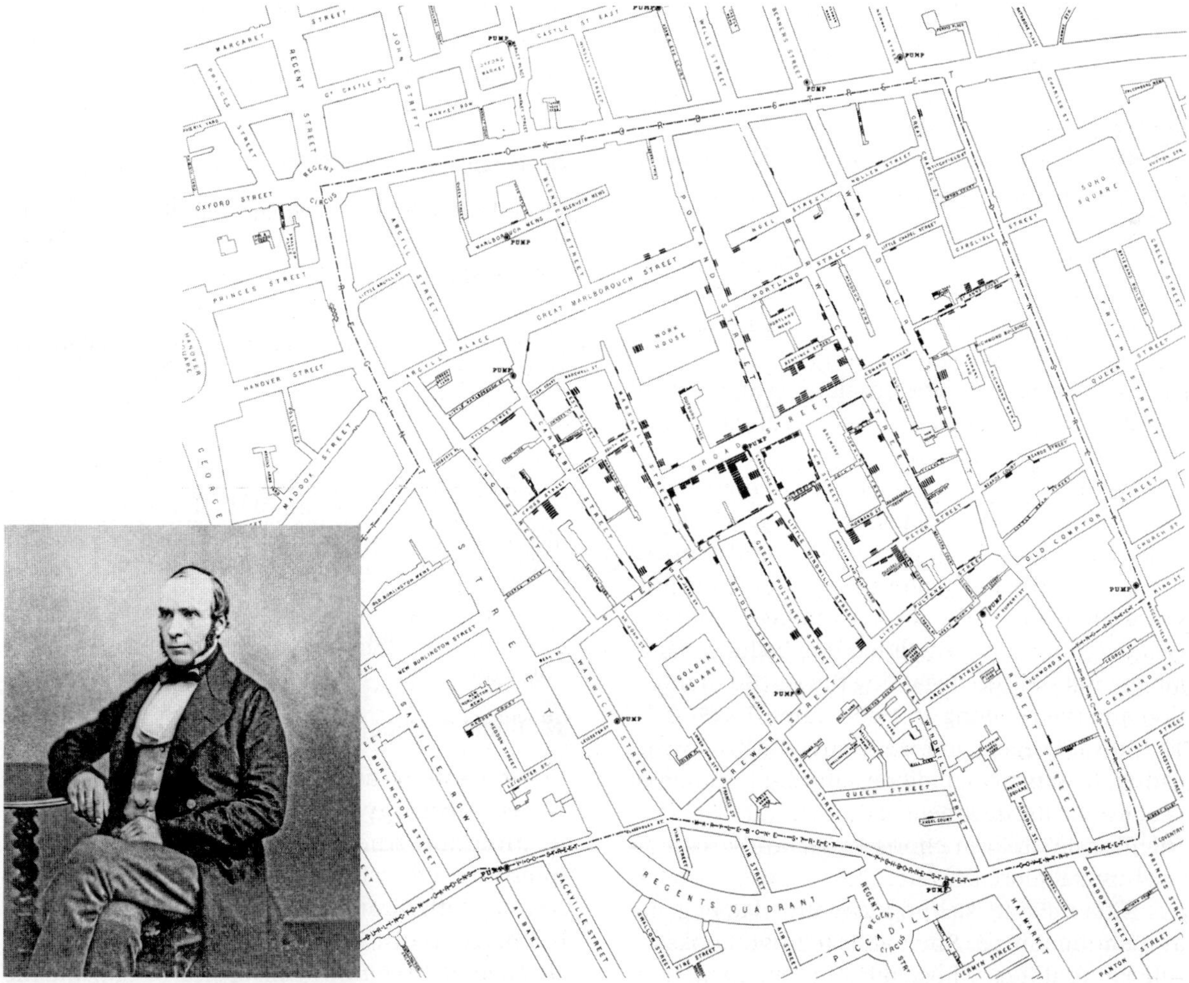

FIGURE 8-5 At the beginning of the 1854 London cholera epidemic, clusters of deaths occurred along the Broad Street water pump. This is a portion of the original map created by Dr. John Snow (inset) and published in 1855. Deaths are plotted as lines parallel to the front of the building in which people died. Courtesy of Frerichs, R. R. John Snow website: http://www.ph.ucla.edu/epi/snow.html, 2006.

Snow compiled mortality rates and locations of deaths, compared these to the location of water sources, and then submitted a report to Parliament. On September 7, 1854, he spoke publicly before the Board of Guardians, a political group charged with the safety and welfare of the area. They documented his concerns and authorized the removal of the pump handle on the Broad Street Water Pump the next day. Removal of the pump handle led to an immediate, measurable reduction in the number of cholera deaths. Thus, Snow was able to prove his hypothesis: that a waterborne disease was spread via contaminated water that traveled both on the surface (the Broad Street Pump) and via groundwater supplies (water piped from the two companies located downstream of London's sewage outlets). What he had dubbed his Grand Experiment had laid the groundwork for descriptive epidemiological investigation.

Florence Nightingale is consistently credited for her establishment of modern nursing practices. In addition, she was also a brilliant organizer, statistician, and one of the most influential women in the 19th century. She was born into a wealthy family in 1820, a time when women did not attend universities or pursue careers. Nightingale was driven to lead a professional rather than domestic life and refused marriage proposals from several prominent suitors. She instead pursued a career in nursing, a choice to which her parents were opposed.

Nightingale's mark on epidemiology was linked to her collection of statistics and graphical representation of mortality rates of the British Army soldiers during the Crimean War.

She had observed the unsanitary conditions of the army hospital, where infectious diseases such as typhus, cholera, and dysentery were rampant. Many of the wounded soldiers died of these infections instead of battle wounds. Nightingale believed these infections were preventable and worked hard to improve the quality of sanitation. She provided members of the Royal Commission on the Health of the Army with briefs that included facts, tables, and statistics regarding the spread of infection and the resulting mortality rates (**FIGURE 8-6**). The Royal Commission was charged with investigating the sanitary conditions, the organization of its staff, and the treatment of the sick and wounded. Nightingale's reform efforts dramatically reduced the hospital mortality rate.

After the war, Nightingale returned to England as a national hero. She published two books on reforming military hospitals in order to address the hygienic conditions that the wounded men in the British Army were exposed to during the Crimean War.

8.3 The Complexities of Disease Transmission

The epidemiological triangle represents one of the fundamental concepts behind disease causality. Before a viral disease actually occurs, the interaction of the viral agent, host, and environment may alter the host's susceptibility. For example, 80% of the population suffers from cold sores, which are caused by herpes simplex virus 1. The virus usually remains dormant; however, it may be reactivated during times of physical or emotional stress, menstruation, pregnancy, immune deficiency, or physical trauma. This change in the host increases his or her susceptibility to herpes simplex 1 infection.

Factors Associated with Increased Risk

Epidemiologists gather information to identify populations at risk for disease and what factors contribute to human susceptibility. **TABLE 8-2** lists some of the environmental factors, host and pathogen characteristics that influence the pattern of viral infections in humans.

Modes of Transmission

Viral diseases can be transmitted directly or indirectly. In direct transmission, the virus is passed from **person to person** through direct physical contact such as touching with contaminated hands, by sexual contact, kissing, or skin-to-skin contact. Viruses may be shed via saliva, other secretions (such as sperm), and skin.

Indirect transmission occurs when the virus is transferred or carried by an intermediate vehicle to a host by one or more of the following means: airborne, vectorborne, foodborne, waterborne, or vehicleborne. Examples of viral diseases and their indirect modes of transmission are listed in **TABLE 8-3**. (See Chapter 6 for a review of portals of viral entry and exit.)

Chain of Infection

There is a strong association between the **triangle of epidemiology** and a concept termed the **chain of infection**. The term chain of infection is frequently used in hospitals and other healthcare settings with regard to the control and prevention of infectious diseases. A viral pathogen leaves its source or reservoir through a portal of exit, and the virus then spreads by one or more modes of transmission. The virus enters the body (susceptible host) through its preferred portal of entry. The chain of

a

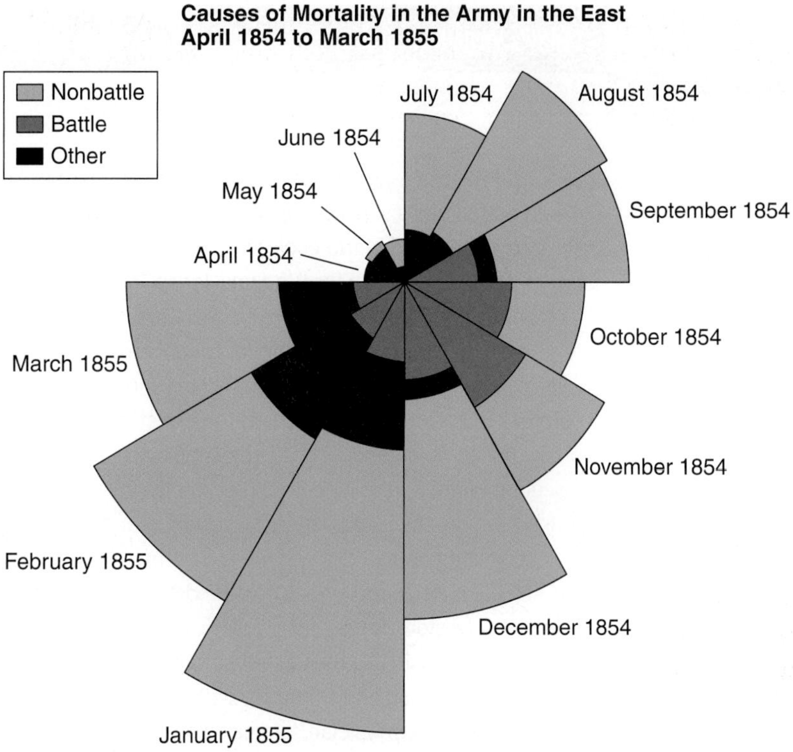

b

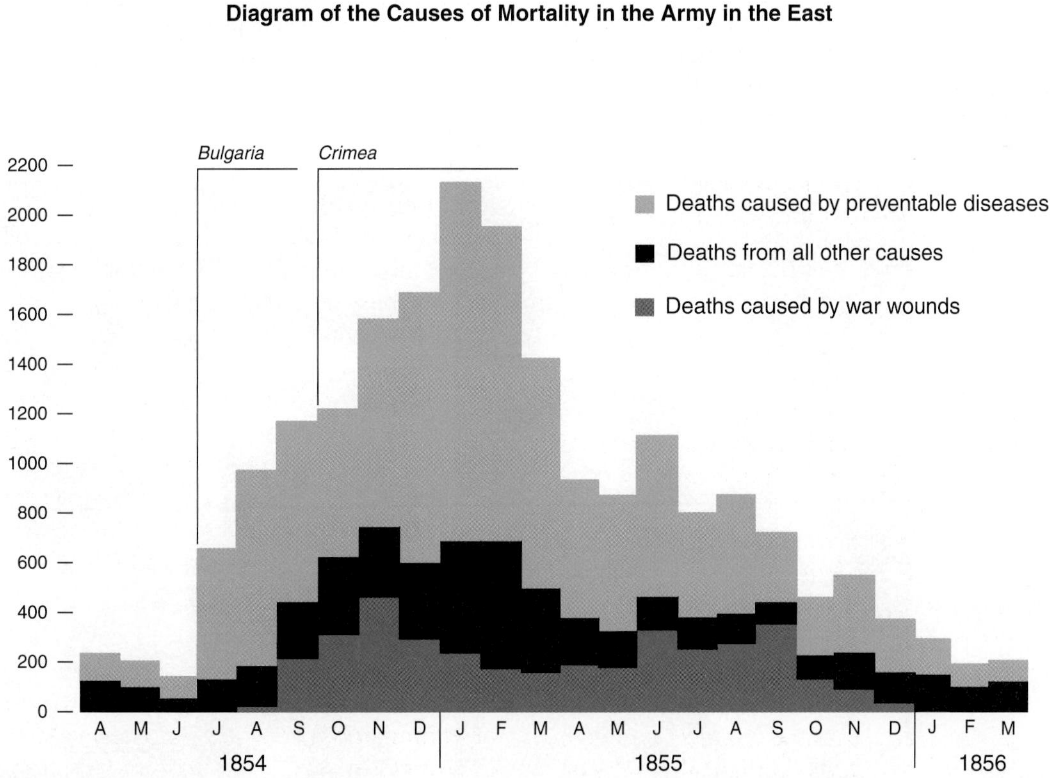

c

FIGURE 8-6 (a) Florence Nightingale (unknown date). **(b)** Re-creation of a polar area (wedge) diagram invented by Florence Nightingale to show the different causes of death of British soldiers during the Crimean War. The blue wedges measured from the center of the circle represent the deaths from war wounds. The black wedges represent deaths from all other causes. The green wedges represent the deaths from preventable (infectious) diseases. Note: The data are not proportional to the area. Reproduced from F. Nightingale. Notes on Matters Affecting the Health, Efficiency and Hospital Administration of the British Army. Harrison and Sons, 1858. **(c)** A contemporary version of the wedge diagram is depicted with the same color scheme: blue represents deaths caused by war wounds; green represents deaths caused by preventable (infectious) diseases; and black represents deaths from all other causes.

Dr. C. J. Peters is a physician, field virologist, and former U.S. Army colonel (**VF 8-1a**). He is an expert on the basic virology, pathogenesis, and epidemiology of hemorrhagic fever viruses. His life story is depicted in the book *Virus Hunter: Thirty Years of Battling Hot Viruses Around the World*. Considered to be the "Indiana Jones" of virus hunters, Dr. Peters has been involved in biodefense and hazardous virus research, including the investigation of a Junin virus outbreak in Argentina, Machupo virus in Bolivia, Nipah virus in Malaysia, and Rift Valley Fever virus in Egypt. After that he led a CDC team during a New Mexico hantavirus outbreak (1993) and Ebola virus outbreak in Kikwit, Zaire (1995). In these instances, he was part of a team responding to a major public health emergency. These epidemiological investigations involved listening to those at the epicenter of the outbreak, collecting samples for analysis, and containing the viral outbreaks. Author of more than 300 papers on research and the control of viral diseases, Dr. Peters serves as a consultant to the CDC and the U.S. Army Medical Research Institute of Infectious Diseases (USAMRIID). He is also Director for Biodefense and Professor of the University of Texas Medical Branch, Galveston, Texas.

Dr. W. Ian Lipkin is a master virus hunter in the laboratory (**VF 8-1b**). He is the Director of the Centers for Infection and Immunity of the Mailman School of Public Health at Columbia University. Dr. Lipkin is a physician/scientist who is known for the creation of cutting edge molecular techniques used to quickly identify and study new and emerging viral pathogens. Dr. Lipkin and his colleagues developed **MassTagPCR**. MassTagPCR is a sensitive polymerase chain reaction (PCR) that allows for the nucleic acid detection of 20 to 30 different pathogens simultaneously in clinical samples. Specific nucleic acid sequences are amplified using Mass Tag degenerate

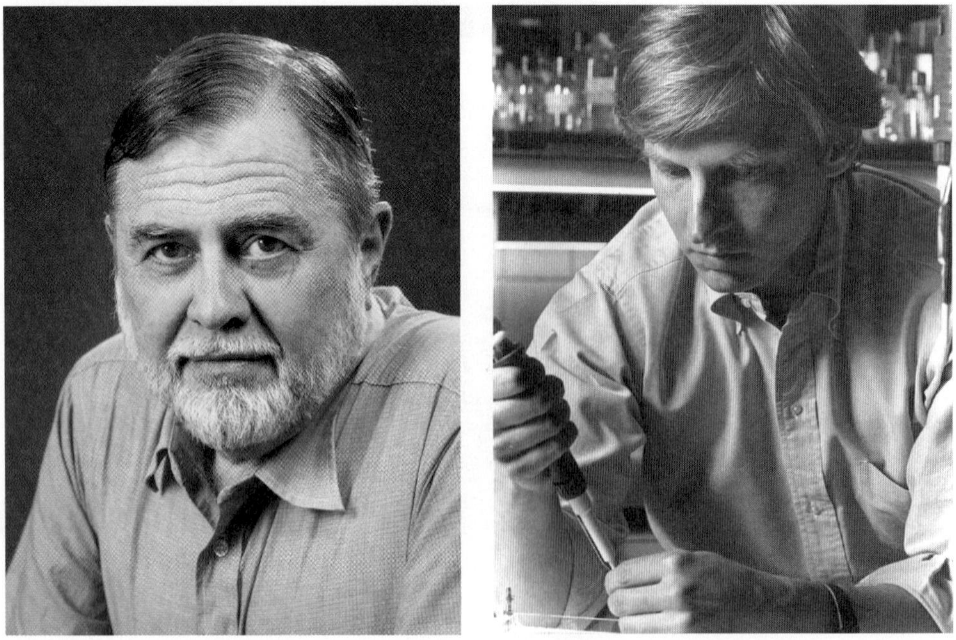

a

b

FIGURE VF 8-1 (a) C. J. Peters. **(b)** W. Ian Lipkin.

primers. Subsequently, the Mass tags are cleaved and the PCR products are analyzed by mass spectrometry.

Using these new techniques, Dr. Lipkin has worked with a team of scientists, discovering at least 400 new viruses since 2002. They have also expanded the use of this technique to answer other epidemiological questions such as whether there is a link between autistic children and intestinal disorders related to high levels of measles viruses in their intestines. At the time of this writing, Lipkin was asked to organize a large-scale investigation involving three laboratory teams to determine if the retrovirus named **x**enotropic Moloney **m**urine leukemia virus-**r**elated **v**irus (XMRV) is linked to chronic fatigue syndrome (CFS). XMRV nucleotide sequences have been found in patients with prostate cancer and CFS. Four independent studies conducted by research teams from the United States, Japan, and the United Kingdom were published in the December 2010 issue of *Retrovirology* that suggest mouse DNA contamination was present with the human tissue tested for XMRV. The American Red Cross, the largest supplier of blood in the United States, issued a statement on December 3, 2010, that blood donations are suspended indefinitely from any donors who have been diagnosed with CSF to ensure patient safety. At the time of this statement, data were insufficient to determine the frequency of XMRV in the donor population, whether XMRV can be transmitted via a transfusion, and if XMRV causes CFS, certain cancers, or other maladies.

References

Briese, T., et al. 2005. "Diagnostic system for rapid and sensitive differential detection of pathogens." *EID* 11(2):310–313.

Enria, D., and Peters, C. J. 2006. "Overview of viral hemorrhagic fevers." In *Tropical Infectious Diseases,* Second Edition, pp. 726–733.

Hornig, M., et al. 2008. "Lack of association between measles virus vaccine and autism with enteropathy: A case-control study." *PLoS ONE* 3(9):e3140.

Lipkin, W. I. 2008. "Pathogen discovery." *PLoS Pathogens* 4(4):e1000002.

Lipkin, W. I. 2010. "Microbe hunting." *Microbiol Mol Biol Rev* 74(3):363–377.

Our Best Defense: Fighting Emerging Diseases. Austin, TX: University of Texas Foundation. Accessed February 23, 2010, from http://www.stateoftomorrow.com/stories/biosafety/peters.htm

Palacios, G., et al., 2006. "Mass tag polymerase chain reaction for differential diagnosis of viral hemorrhagic fevers." *EID* 12(4):692–695.

Peters, C. J., Olshaker, M. 1998. *Virus Hunter: Thirty Years of Battling Hot Viruses Around the World.* New York, Anchor Books/ Doubleday.

Smith, R. A. 2010. "Contamination of clinical specimens with MLV-encoding nucleic acids: Implications for XMRV and other candidate human retroviruses." *Retrovirology* 7:112.

Stoltenberg, C., et al. 2010. "The Autism Birth Cohort: A paradigm for gene-environment-timing research." *Mol Psychiatry* 15:676–680.

TABLE 8-2 Factors Associated with Viral Disease Causation

Host Characteristics	Pathogen Characteristics	Environmental Factors
Age	Stability in the environment	Population density (crowding results in increased host exposure to the pathogen)
Sex	Virulence factors (enhanced ability to attach, invade, or multiply in the host)	Sanitary conditions
Race and genetic factors	Presence of immune evasion genes	Climate changes
Immune status (decreased)	Resistance to antiviral therapy	Environmental changes
Nutritional status (lack)	Enhanced mode of transmission	Blood products (contaminated)
Behavior (e.g., occupation, lifestyle, religion, customs)		Geographical locations
Previous infections or current coinfections		Existence of zoonotic infections/vectors

TABLE 8-3	Human Viral Diseases and Their Most Common Modes of Transmission
Viral Disease	**Predominate Mode of Transmission**
Influenza	Airborne
Common cold	Direct contact or airborne
Poliomyelitis	Waterborne (oral–fecal route)
West Nile encephalitis	Vector (mosquitoes)
Rabies	Direct (bite laden with saliva containing rabies virus)
Oral cold sores	Contact with saliva laden with HSV-1 is the most common mode
AIDS	Sex with HIV-positive individual, vehicle-transmission (HIV-contaminated needles/blood transfusions)
Chickenpox	Direct contact or airborne (droplet transmission)
Measles	Direct contact or airborne (droplet transmission)
Hepatitis A	Foodborne (oral–fecal route)

infection can be broken at each step in the chain through any of the following practices:

- Rapid identification of the viral pathogen
- Proper sanitation
- Disinfection and sterilization of fomites
- Use of barrier techniques
- Handwashing
- Proper trash and waste disposal
- Proper food handling
- Aseptic technique
- Recognition of high-risk individuals

The Concept of Herd Immunity

The concept of **herd immunity** is based on the premise that if the majority of the population is mostly protected from a disease through immunizations or genetic resistance, the chance of a major epidemic is highly unlikely. Jonas Salk, developer of the polio vaccine during the 1950s, suggested that if herd immunity was at 85%, a polio epidemic would not occur. Herd immunity provides a barrier to direct transmission of infectious diseases through a population (**FIGURES 8-7** and **8-8**).

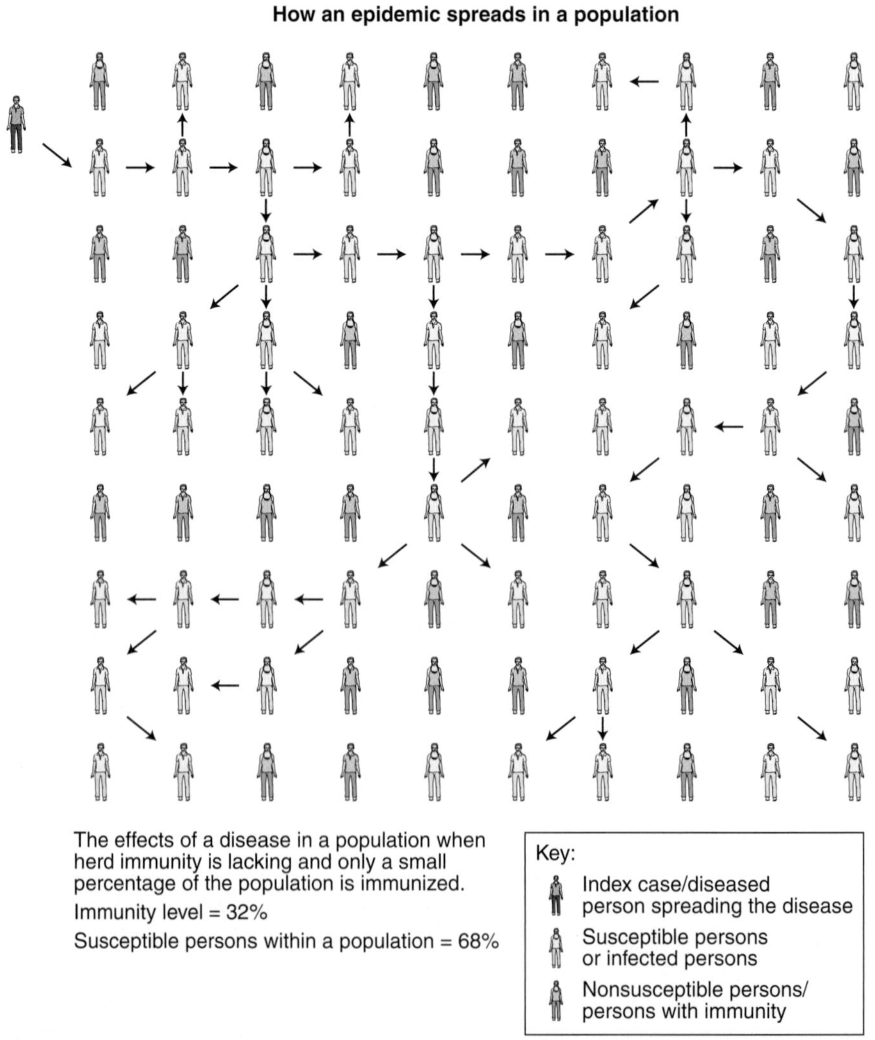

How an epidemic spreads in a population
The effects of a disease in a population when herd immunity is lacking and only a small percentage of the population is immunized.
Immunity level = 32%
Susceptible persons within a population = 68%
Key:
Index case/diseased person spreading the disease
Susceptible persons or infected persons
Nonsusceptible persons/ persons with immunity

FIGURE 8-7 Diagram showing a population with a low immunization level, resulting in a low level of protection of the population.

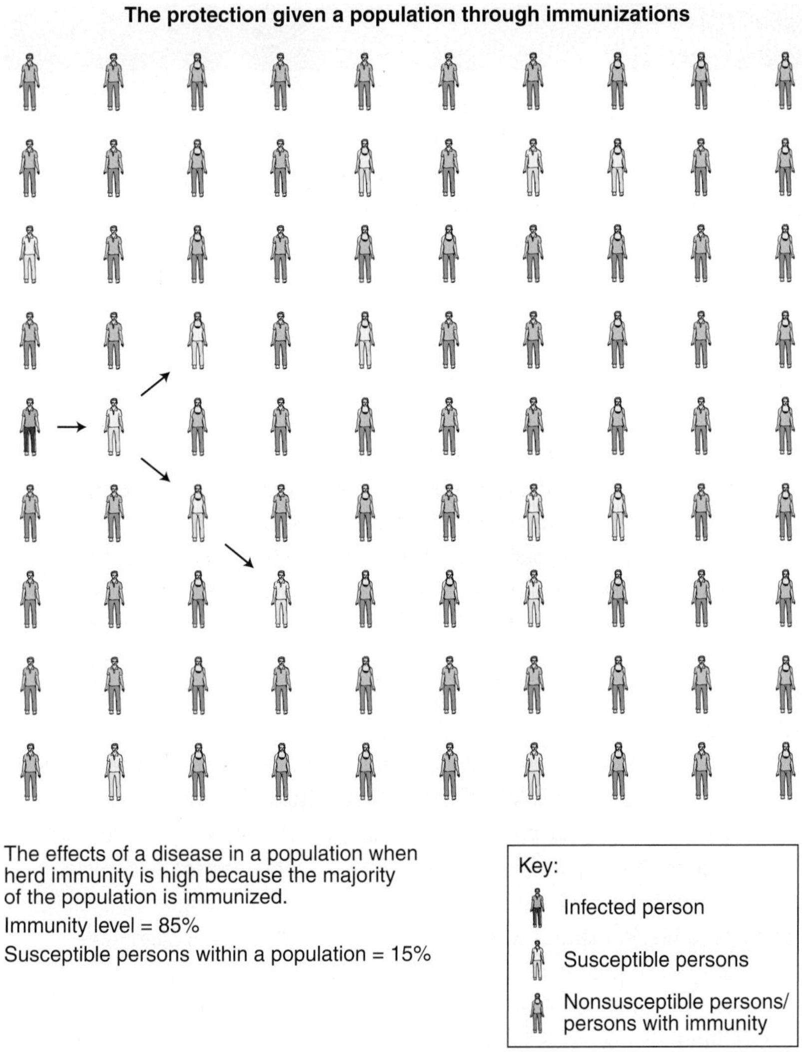
The protection given a population through immunizations
The effects of a disease in a population when herd immunity is high because the majority of the population is immunized.
Immunity level = 85%
Susceptible persons within a population = 15%
Key:
Infected person
Susceptible persons
Nonsusceptible persons/ persons with immunity

FIGURE 8-8 Diagram showing a population with a high immunization level, resulting in good protection of the population.

During the 1950s and 1960s, it was not uncommon in the United States for parents to encourage their healthy children to interact with children with several infectious diseases, such as mumps and the chickenpox. They wanted their children to get a case of the disease and become immune before they reached adulthood because it was known that contracting such illnesses in adulthood resulted in a severe illness with complications. Today, the goal of any public health immunization program is to reach 100% immunity in a population.

8.4 Epidemiology Today

The Goals of Epidemiology: What Does an Epidemiologist Want to Know?

During a real epidemic, an epidemiologist wants to know the following:

- Case definition (what)
- Person (who)
- Place (where)
- Time (when)
- Risk factors (how and why)

Gathering Data

Investigation starts with the gathering of information and confirmation of the epidemic. Diagnostic tests are conducted and a case definition is set in place. Next, **descriptive studies** are performed after the epidemic occurs. Descriptive studies are usually inexpensive and of short duration. These studies are used to generate a testable hypothesis or enough evidence to indicate sufficient cause for conducting lengthier and more costly analytical studies.

Dr. John Snow's search for the cause of the cholera epidemic in London in 1854 is an example of descriptive epidemiology. Descriptive studies are observations of when and where (time and place) the disease occurs and who is affected (person). The studies take into account the onset and duration of the illness. Descriptions frequently

VIRUS FILE 8-2 Impact of Viruses on War and Religion

Viruses and other infectious agents have shaped world history. Entire tribes of native American Indians during the 1700-1800s were killed by smallpox (variola virus). Documentary evidence of influenza, measles, smallpox, yellow fever, and norovirus epidemics were recorded as military losses in wars dating from 212 B.C. to 2002 A.D. It is now widely accepted that battle casualties were fewer than those caused by pathogens. Doyle and Lee reported that some military engagements in wars from 430 B.C. to the 1940s were critical to the outcome of the war. The French were particularly prone to epidemics. For example, in 1779, King Louis XVI had an army of 40,000 positioned on the coast of England, but the troops never disembarked because a smallpox epidemic killed at least 8,000 of them. Later, in 1802, Napoleon sent 33,000 troops to suppress resistance in Santo Domingo, Haiti. Almost all of their forces died of yellow fever. In 1812, Napoleon sold the Louisiana Purchase to the United States. Doyle and Lee suggest that Napoleon might have been fearful of more epidemics in the New World. When Napoleon's armies marched into Russia, the army retreated into a countryside with no food. Many died from malnutrition, starvation, and disease. Realizing that unspoiled food was central to maintaining an armada, the French government offered 12,000 francs to anyone who could develop a method to preserve food. Canning is partly attributable to Napoleon and relevant to microbiology today.

Many plagues decimated populations throughout the Middle Ages. During the first centuries of Christianity, plagues were considered a "test" to develop faith. Epidemics influenced the religion of most cultures. The impact of smallpox on human cultures is reflected by the genesis of the patron saint of smallpox sufferers, St. Nicaise. Jews were often blamed for epidemics, forcing their migration. Throughout history, whenever humans traveled as immigrants, soldiers, traders, explorers, or adventurers, epidemics were a camp follower. The annual pilgrimage known as the Hajj to Mecca, Saudi Arabia, resulted in many infectious disease outbreaks. The Hajj continues to be an obligation for all able-bodied Muslims who can afford to do so at least once in a lifetime (**FIGURE VF 8-2**). This journey has been ongoing for centuries and was an especially hazardous trek for early pilgrims. Thirst and disease were common. In 1926,

FIGURE VF 8-2 Pilgrims visiting the Kaaba, in Mecca, Saudi Arabia.

VIRUS FILE 8-2 Impact of Viruses on War and Religion (continued)

about 12,000 Syrian pilgrims died during their journey. Today, the Saudi Arabian government estimates that 2 to 3 million pilgrims from all over the globe travel to Mecca each year. As the pilgrims migrate en masse, problems such as overcrowding, safety, and poor sanitation occur. Cramped quarters contribute to the spread of infectious diseases. Poor sanitation can lead to gastrointestinal illness. Participants are from different countries, bringing with them infectious agents endemic of their native lands. Today, The Saudi Arabian government requires that pilgrims have up-to-date routine vaccinations. Participants from Africa and South America must show proof of yellow fever vaccination. In 2010, the Saudi government also required individuals to be vaccinated against H1N1 influenza. Epidemics caused by viral respiratory pathogens such as influenza, respiratory syncytial virus, parainfluenza, and adenoviruses are common with the Hajj.

Before the mid-1800s, routine vaccination for childhood diseases and antibiotics did not exist. Proper sanitation was challenging. Infectious diseases caused life-threatening infections among the public at large. The U.S. average life expectancy was 49 years in 1900 and rose to 78.3 years in 2010 (an increase of 29.3 years). This life expectancy increase is attributed to improved sanitation, vaccination, and eventually antibiotics.

References

Doyle, R. J., and Lee, N. C. 1985. "Microbes, warfare, religion, and human institutions." *Can J Microbiol* 32:193–200.

Piqueras, M. 2007. "Microbiology: A dangerous profession?" *Int Microbiol* 10:217–226.

Smallman-Raynor M. R., and Cliff, A. D. 2004. "Impact of infectious diseases on war." *Infect Dis Clinics North Am* 18:341–368.

include the age, sex, ethnicity, socioeconomic status, and overall health status (diet and immune status) of the affected individuals. Sometimes other information, such as environmental exposures and personal lifestyle characteristics, may be recognized as risk factors.

Descriptive studies may be published as case reports (anecdotal or clinical observations that may indicate a causal relationship) or surveillance studies. Surveillance studies may rely on data generated by death certificates or other traditional channels. After descriptive studies, control and prevention measures may be implemented, and the results of these studies and measures are reported. Examples of graphs generated from descriptive studies are shown in **FIGURE 8-9**. **Analytical epidemiology** may be needed to determine the causes and effects (the why and how) of the epidemic.

Analytical Epidemiology

Analytical studies are more informative than descriptive studies. These investigations are hypothesis-driven studies that use descriptive research to determine the why and how pertaining to the variations in incidence and prevalence of diseases in different populations, and requires good research design. In true analytical epidemiology, individuals in a population suffering from a disease are compared to a **control group** that does not have the disease and has not been exposed to the disease. The control group may be matched by age, sex, location, and socioeconomic status. The two groups are studied, compared, and analyzed in what is called a **case-control** method (**TABLE 8-4**).

The **cohort method** used by analytical epidemiologists studies two similar populations or **cohorts**. Cohort members are grouped by common characteristics, especially birth year. One of the cohorts has been exposed to the disease-causing agent and the other has not. The cohorts may be studied as a group, backward in time (**retrospectively**), at a single point in time (**cross-sectionally**), or forward in time (**prospectively**) (**FIGURE 8-10**). Most cohort studies require large numbers of subjects that are followed over long periods of time (years).

During a college football game between teams from Florida and North Carolina in 1999, the members of the North Carolina team suffered from gastrointestinal distress during the game.

FIGURE 8-9 (a) A graph accompanying a descriptive study to characterize the epidemiology of measles in the United States from 1985-2003. This report summarized that no endemic measles was circulating in the United States at that time. Imported measles cases do, however, continue to occur and can result in limited indigenous transmission. Adapted from Dayan, G., et al. *MMWR* 53 (2004):713–716. **(b)** These are the results of an investigation of a hepatitis A outbreak that occurred at a restaurant in Monaca, Pennsylvania, in 2003. The outbreak was associated with contaminated green onions that were in large batches of salsa. The source was determined to be one or more farms in Mexico. At least 9000 people who were at the restaurant or who were in contact with ill individuals were provided immunoglobulin as a preventative measure. The restaurant was closed. Adapted from Dato, V., et al. *MMWR* 52 (2003):1155–1157. **(c)** Description study showing the distribution of West Nile virus infection by area of residence, province of Quebec, Canada, 2002–2003. Adapted from Public Health Agency of Canada, *Canada Communicable Disease Report* (*CCDR*), June 1, 2004. 30(11):97–104.

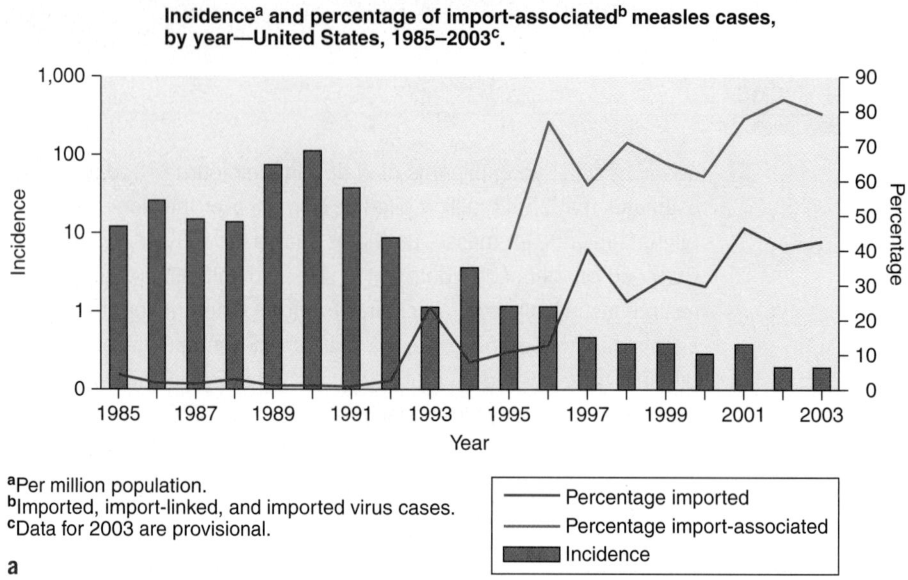

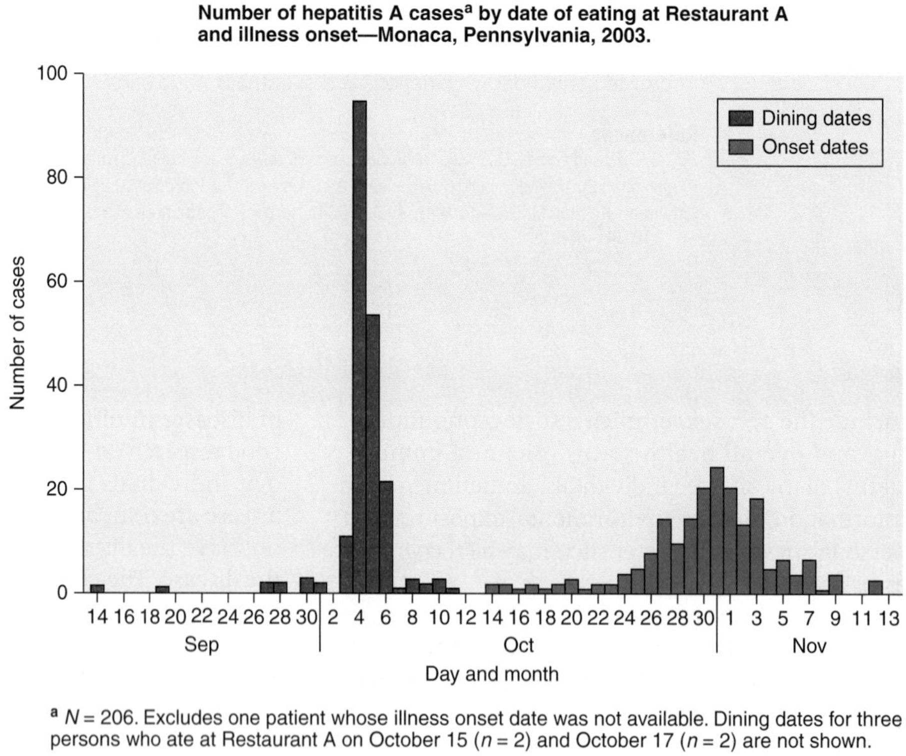

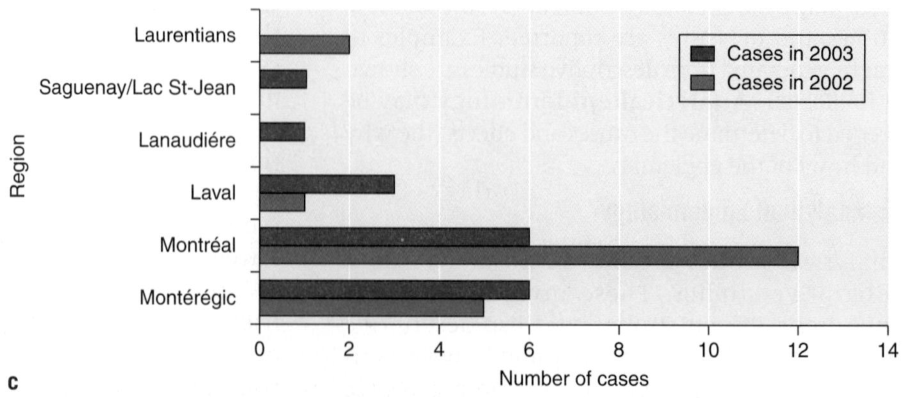

TABLE 8-4 Case-Control Study Investigating the Risk Factors for Avian Influenza A (H5N1 strain) Disease in Hong Kong, 1997

The study compared 15 human cases of human Avian Influenza to a control group that did not have antibodies against Avian Influenza A and was matched by age, sex, and neighborhood. Results of the investigation suggested that exposure to live poultry in the markets during the week prior to the illness was significantly associated with Avian Influenza disease (64% of the cases and 29% of the control group). Travel, eating or preparing poultry products, and recent exposures to individuals with respiratory illnesses including Influenza A were not associated with Influenza A illness.

	No. (%) with Reported Exposure			
Activity and Exposure	**Case Patients (*n* = 15)**	**Controls (*n* = 41)**	**OR (95% CI)**	***P*[a]**
Exposure to poultry				
Exposed to live poultry in market[b]	9/14 (64)	11/38 (29)	4.5 (1.2–21.7)	.045
Consumed poultry in restaurant	6/12 (50)	9/31 (29)	2.9 (0.6–14.9)	.375
Consumed poultry organs or poultry	1/14 (7)	4/38 (11)	0.6 (0.0–7.5)	.999
Consumed undercooked poultry products	5/11 (45)	10/29 (34)	1.9 (0.4–11.2)	.707
Household member cooked poultry products	2/13 (15)	11/33 (33)	0.2 (0.0–1.3)	.193
Household member in poultry industry	0/14	0/38	Undefined	NA
Exposure to human illness				
Anyone in flat had influenza-like illness[c]	5/15 (33)	15/40 (38)	0.8 (0.2–2.8)	.999
Contact with known H5N1 case	1/13 (8)	0/35	+Inf (0.1 to +Inf)[d]	.500
Other exposure				
Travel outside Hong Kong	0/14	3/37 (8)	0.0 (0.0–4.9)	.800
Selected outdoor activities[e]	2/14 (14)	8/37 (22)	0.0 (0.0–2.5)	.225
Live birds in home	6/12 (50)	14/31 (45)	1.4 (0.3–6.4)	.901
Other animals in home	2/13 (15)	1/33 (3)	2/5 (0.1–97.4)	.999
Played in indoor playground	0/15	13/41 (32)	0.0 (0.0–0.5)	.013
Household uses soap to clean[f]	5/14 (36)	25/38 (66)	0.3 (0.1–1.0)	.058

Note: OR, odds ratio; CI, confidence interval; NA, not available.

[a]Exact method.

[b]Includes visiting poultry stall, retail or wholesale market selling live poultry, or poultry farm in week before illness.

[c]Influenza-like illness includes anyone with fever and cough or sore throat.

[d]+Inf, an unknown upper bound (positive infinity) for the CI.

[e]Includes activities such as visiting aviary, feeding wild birds in park, and/or having picnic in park.

[f]Uses soap and water to clean knife after preparation of poultry.

Source: Adapted from Mounts, et al. 1999. *J Infect Dis* 180:505–508.

The next day, members of the Florida team developed similar symptoms. This retrospective cohort study concluded that the source of the disease-causing agent, Norwalk virus, was a turkey sandwich in a box lunch (this accounted for 95% of the primary cases). The researchers hypothesized that the virus was transmitted across the scrimmage line by person-to-person contact during the football game. The North Carolina team physician and coaches reported that their players were vomiting on the sidelines and were playing in uniforms soiled with vomit and feces. Given

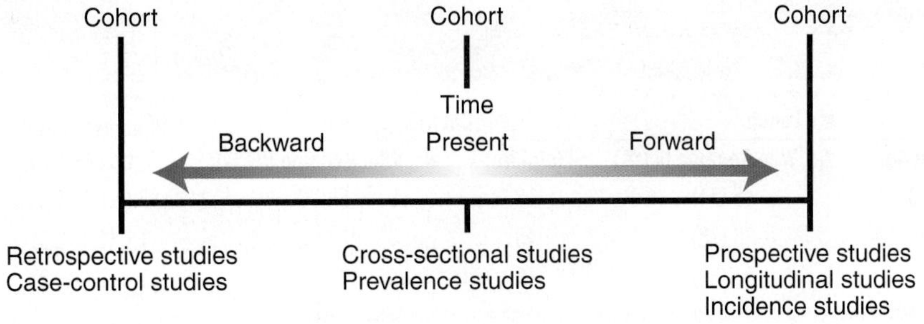

FIGURE 8-10 The relationship of cohort studies to other types of studies.

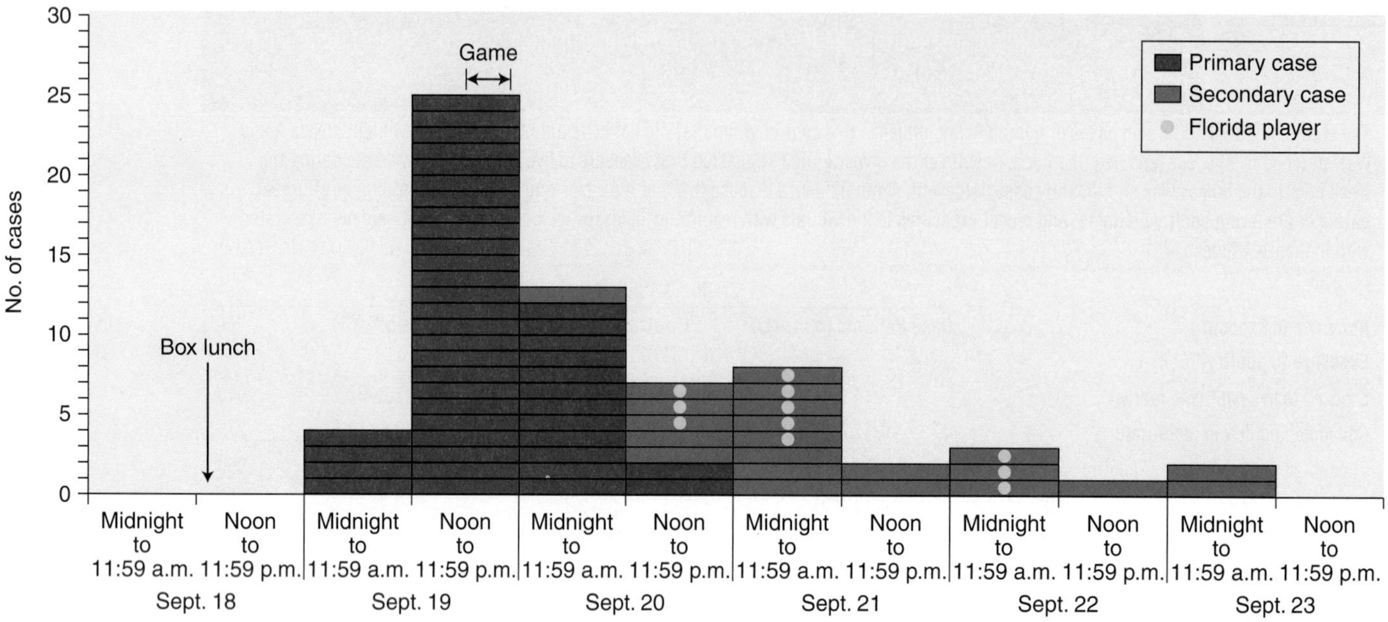

FIGURE 8-11 Person-to-person transmission of the Norwalk virus during a college football game. Adapted with permission from Becker, Moe, Southwick, and MacCormack. *The New Journal of Medicine* 343 (2000):1223–1227. © 2005 Massachusetts Medical Society.

the intense physical contact inherent in the game, transmission likely occurred via oral–fecal transmission and aerosol transmission of the vomitus. Based on their findings, researchers recommended that players with acute gastroenteritis be excluded from competition to avoid transmitting the disease to other players (**FIGURE 8-11** and **TABLES 8-5** and **8-6**.).

Surveillance and Serological Epidemiology

Surveillance programs and reporting systems targeted at infectious diseases play a fundamental

TABLE 8-5 Rate and Risk of Illness Among Players of the North Carolina Football Team Based on Meals Eaten

		Ate Meal		Did Not Eat Meal		
Meal	**Date**	**Total No.**	**No. Who Became Ill (%)**	**Total No.**	**No. Who Became Ill (%)**	**Relative Risk (95% CI)[a]**
Lunch	9/18	81	50 (62)	27	4 (15)	4.1 (1.6–10.0)
Dinner	9/18	87	45 (52)	21	9 (43)	1.2 (0.7–2.2)
Late dinner	9/18	63	34 (54)	45	20 (44)	1.2 (0.8–1.8)
Breakfast	9/19	85	42 (49)	23	12 (52)	0.9 (0.6–1.5)
Lunch	9/19	76	39 (51)	32	15 (47)	1.1 (0.7–1.7)

[a]CI denotes confidence interval.

Source: Reproduced with permission from Becker, Moe, Southwick, and MacCormack. *The New England Journal of Medicine* 343 (2000):1223–1227. © 2005 Massachusetts Medical Society.

TABLE 8-6 Food-Specific Rates and Risks of Gastrointestinal Illness Among Individuals Who Ate Components of the Box Lunch

	Ate Lunch		Did Not Eat Lunch			
Meal	**Total No.**	**No. Who Became Ill (%)**	**Total No.**	**No. Who Became Ill (%)**	**Unadjusted Odds Ratio (95% CI)**	**Adjusted Odds Ratio (95% CI)[a]**
Sandwich	63	45 (71)	18	5 (28)	2.6 (1.2–5.5)	4.9 (1.3–18.9)
Apple	36	28 (78)	45	22 (49)	1.6 (1.1–2.3)	2.4 (0.6–9.3)
Candy bar	63	43 (68)	18	7 (39)	1.8 (1.0–3.2)	1.6 (0.5–5.0)

[a]The adjusted odds ratios were adjusted for each of the other foods in the box lunch. CI denotes confidence interval.

Source: Reproduced with permission from Becker, Moe, Southwick, and MacCormack. *The New England Journal of Medicine* 343 (2000):1223–1227. © 2005 Massachusetts Medical Society.

role in public health. Surveillance can take several forms:

- Monitoring available data from mandated reports on reportable diseases (morbidity and mortality statistics).
- Active field surveillance by epidemiologists at healthcare facilities (such as interviewing physicians and patients or reviewing medical records).
- Serological screening of populations (e.g., monitoring the success of vaccination programs). Antibodies represent the "footprints" of disease, exposure to disease, and protection against disease.

Communicable Disease Surveillance Organizations and Their Publications or Reporting Mechanisms

Official organizations from the United States, Canada, the European Union, and other member states of the WHO share the same mission: that all people attain the highest possible level of health. A short list of some of the major organizations,

VIRUS FILE 8-3 Descriptive Epidemiology and AIDS

Observations made by physicians, technicians, and epidemiologists were essential in recognizing a new viral epidemic in the 1980s. The sad history of AIDS is one of the finest examples of epidemiological investigations. The following timeline depicts some of the key events in HIV/AIDS history.

March 1981 Eight cases of a rare form of aggressive Kaposi's sarcoma (KS) in young gay men in New York were reported. (Prior to the AIDS epidemic, the annual incidence of KS was 0.2 to 0.6 cases per million population and it was usually found in older men of Hebrew or Italian ancestry. By the end of 1999, the CDC recorded 46,684 HIV cases in the United States that had a confirmed diagnosis of KS.)

April 1981 CDC drug technician Sandra Ford observes an increase in prescriptions for pentamidine, a drug used to treat *Pneumocystis carinii* pneumonia (PCP), a rare pneumonia recognized in the elderly. In this instance, however, a young 20-year-old gay man was requesting refills to treat PCP. (A 1967 survey revealed that only 107 cases of PCP had been recorded in the medical literature in the United States, and that all of those patients suffered from immune suppression. By the end of 1999, of the 800,000 to 900,000 people in the United States living with HIV infection, the CDC recorded 166,368 had a confirmed diagnosis of PCP.)

June 1981 The CDC published a report about five young, active homosexual men in Los Angeles, California, who were later treated for PCP and had experienced cytomegalovirus infection and candidal mucosal infection. Two of the men had died (**FIGURE VF 8-3**).

July 1981 The CDC published a report about 26 young, active homosexual men in California and New York who were then suffering from PCP.

December 1981 A new population was being affected by the rare PCP: intravenous drug users.

The documentation of these critical observations unraveled the mystery of a new disease referred to as Gay-Related Immune Deficiency (GRID). By 1983, this new disease was striking adult heterosexuals and children. Subsequently, the name of the disease was changed to Acquired Immune Deficiency Syndrome (AIDS). In 1983 the etiological agent, human immunodeficiency virus-1 (HIV-1), was isolated. Between 1982 and 1983, there were 2042 AIDS cases reported in the United States, and 63% of these individuals had died of severe immunosuppression characterized by a loss of T lymphocytes.

References

Centers for Disease Control and Prevention (CDC). 1981. "Kaposi's sarcoma and *Pneumocystis* pneumonia among homosexual men—New York City and California." *MMWR* 4(30):305–309.

Gottlieb, M. S., et al. 1981. "*Pneumocystis* pneumonia—Los Angeles." *MMWR* June 5 30:250–252.

Hymes, K. B., et al. 1981. "Kaposi's sarcoma in homosexual men: A report of eight cases." *Lancet* 2:598–600.

Masur, H., et al. 1981. "An outbreak of community acquired *Pneumocystis carinii* pneumonia: Initial manifestation of cellular immune dysfunction." *N Engl J Med* 305:1431–1438.

(continued)

CENTERS FOR DISEASE CONTROL

June 5, 1981 / Vol. 30 / No. 21

MMWR

MORBIDITY AND MORTALITY WEEKLY REPORT

Pneumocystis Pneumonia — Los Angeles

In the period October 1980-May 1981, 5 young men, all active homosexuals, were treated for biopsy-confirmed *Pneumocystis carinii* pneumonia at 3 different hospitals in Los Angeles, California. Two of the patients died. All 5 patients had laboratory-confirmed previous or current cytomegalovirus (CMV) infection and candidal mucosal infection. Case reports of these patients follow.

Patient 1: A previously healthy 33-year-old man developed *P. carinii* pneumonia and oral mucosal candidiasis in March 1981 after a 2-month history of fever associated with elevated liver enzymes, leukopenia, and CMV viruria. The serum complement-fixation CMV titer in October 1980 was 256; in May 1981 it was 32.* The patient's condition deteriorated despite courses of treatment with trimethoprim-sulfamethoxazole (TMP/SMX), pentamidine, and acyclovir. He died May 3, and postmortem examination showed residual *P. carinii* and CMV pneumonia, but no evidence of neoplasia.

Patient 2: A previously healthy 30-year-old man developed *P. carinii* pneumonia in April 1981 after a 5-month history of fever each day and of elevated liver-function tests, CMV viruria, and documented seroconversion to CMV, i.e., an acute-phase titer of 16 and a convalescent-phase titer of 28* in anticomplement immunofluorescence tests. Other features of his illness included leukopenia and mucosal candidiasis. His pneumonia responded to a course of intravenous TMP/SMX, but, as of the latest reports, he continues to have a fever each day.

Patient 3: A 30-year-old man was well until January 1981 when he developed esophageal and oral candidiasis that responded to Amphotericin B treatment. He was hospitalized in February 1981 for *P. carinii* pneumonia that responded to oral TMP/SMX. His esophageal candidiasis recurred after the pneumonia was diagnosed, and he was again given Amphotericin B. The CMV complement-fixation titer in March 1981 was 8. Material from an esophageal biopsy was positive for CMV.

Patient 4: A 29-year-old man developed *P. carinii* pneumonia in February 1981. He had had Hodgkins disease 3 years earlier, but had been successfully treated with radiation therapy alone. He did not improve after being given intravenous TMP/SMX and corticosteroids and died in March. Postmortem examination showed no evidence of Hodgkins disease, but *P. carinii* and CMV were found in lung tissue.

Patient 5: A previously healthy 36-year-old man with a clinically diagnosed CMV infection in September 1980 was seen in April 1981 because of a 4-month history of fever, dyspnea, and cough. On admission he was found to have *P. carinii* pneumonia, oral candidiasis, and CMV retinitis. A complement-fixation CMV titer in April 1981 was 128. The patient has been treated with 2 short courses of TMP/SMX that have been limited because of a sulfa-induced neutropenia. He is being treated for candidiasis with topical nystatin.

---

*Paired specimens not run in parallel.

FIGURE VF 8-3 The CDC published a summary of the five homosexual men who had been treated for microbial infections that were documented to only occur in individuals with suppressed immune systems. Courtesy of Gottlieb, M. S., et al. 1981. "*Pneumocystis* pneumonia–Los Angeles." *MMWR* June 5 30:250–252.

their headquarter locations, and epidemiological publications relating to infectious diseases is shown in **TABLE 8-7**. These organizations perform ongoing surveillance of many viral diseases, respond to disease outbreaks, and develop and publish infectious disease prevention and control recommendations.

In 1951, following the start of the Korean War and concurrent with its threat of biological warfare, the CDC formed the Epidemic Intelligence Service (EIS). EIS is a training program composed of physicians, researchers, and scientists that plays a pivotal role in combating the causes of major epidemics. EIS investigation sites are located worldwide (http://www.cdc.gov/eis/index.html).

A private organization, **Doctors Without Borders** (in French, Médecins Sans Frontières, or MSF; http://www.doctorswithoutborders.org/), was founded in 1971 by a group of French physicians who believed that everyone has a right to medical care, including those in remote areas and locations where there is political unrest. Every year MSF unites more than 2500 volunteer doctors, nurses, medical professionals, logistic experts, water sanitation experts, and administrators from all over the world in response to epidemic emergencies in unstable areas of the world. There are also organizations that focus on the health of wildlife, such as the National Wildlife Health Center (http://www.nwhc.usgs.gov/). Part of its mission is to support the development of advanced diagnostics to detect zoonotic pathogens and the control of wildlife disease dynamics.

ProMED-mail: An Online Program Used to Monitor Disease Outbreaks

On August 19, 1994, with modest support from the Federation of American Scientists and SatelLife, Dr. Jack Woodall founded Pro-MED-mail as an Internet reporting tool to monitor emerging infectious disease outbreaks (http://www.promedmail.org/). Its aim was to assist in disseminating outbreak information (such as location or confirmed disease) as rapidly as possible. On that day, the first ProMED-mail message was posted via email to 40 subscribers in seven countries. The first posts on ProMED consisted of queries relating to media/newspaper reports and local observations.

TABLE 8-7 Government Organizations Involved in Surveillance and Publication of Epidemiological Information and Other Important Epidemiology Links

Organization	Headquarters	Publications Available Online
Centers for Disease Control and Prevention (CDC)	Atlanta, Georgia, United States	*Morbidity and Mortality Weekly* (*MMWR*) http://www.cdc.gov/mmwr/ *Emerging and Infectious Diseases Journal* (*EID*) http://www.cdc.gov/ncidod/EID/index.htm
Health Canada	Ottawa, Canada	*Canada Communicable Disease Report* (*CCDR*) http://www.hc-sc.gc.ca/pphb-dgspsp/publicat/ccdr-rmtc/
World Health Organization (WHO)	Geneva, Switzerland	*Weekly Epidemiological Report* (*WER*) http://www.who.int/wer/en/
European Commission (Europa)	Brussels, Belgium	*EuroSurveillance Weekly and Monthly* http://www.eurosurveillance.org/index-02.asp
Health Protection Agency (HPA)	London, United Kingdom	*Communicable Diseases Report Weekly* (*CDR*) http://www.hpa.org.uk/cdr/ *Communicable Disease and Public Health Journal* http://www.hpa.org.uk/cdph/
Pan American Health Organization (PAHO)	Washington D.C., United States	*Pan American Journal of Public Health* http://new.paho.org/journal/
Doctors without Borders/ Médecins Sans Frontières (MSF)	U.S. Headquarters: New York, NY International Office: Brussels, Belgium	*Alert Quarterly* http://www.doctorswithoutborders.org/publications/alert/
Caribbean Epidemiology Centre	Port of Spain Trinidad and Tobago	*CAREC Surveillance Report* http://www.carec.org/
CDC Wonder	CDC, Atlanta, Georgia, United States	*Online Data for Epidemiological Research* http://wonder.cdc.gov/
WWW Virtual Epidemiology site	Dept. of Epidemiology and Biostatistics, University of California, San Francisco	*The WWW Virtual Library: Medicine and Health: Epidemiology* http://www.epibiostat.ucsf.edu/epidem/epidem.html

VIRUS FILE 8-4 Sentinel Chicken Surveillance Programs

In 1999, the first outbreak of West Nile virus was recognized in the United States. It began in New York City and neighboring counties in New York state in late August and September. Initially, the outbreak was attributed to St. Louis encephalitis virus (SLE), based on positive enzyme-linked immunosorbent assay tests on cerebrospinal fluid of patients. (SLE and West Nile viruses are antigenically very similar and cross-reactivity was being observed in most serologic tests.) Later, the isolation of virus and genomic sequences from dead crows, other species of birds from zoos, mosquito pools, and human brain tissues correctly identified the cause of the outbreak.

In May 2000, Canadian Health authorities stationed cages of sentinel chickens along 2500 km (1550 miles) of the border with the United States in an effort to identify the presence of West Nile virus in susceptible animals before the disease was detected in humans in Canada. Ultimately, the sentinel chickens were key in detecting a new viral epidemic (**FIGURE VF 8-4**).

FIGURE VF 8-4 Photograph showing a caged sentinel chicken flock used to detect the presence of West Nile virus.

Compared to other birds with the disease, chickens with West Nile virus develop a very short viremia without symptoms. Chickens can spread the virus from the cloaca and the virus may be found in feces. Animal handlers used precautions while handling the sentinel chickens during the weekly testing for West Nile Virus antibodies.

Today West Nile virus sentinel-chicken surveillance programs are not only being used in Canada but many locations throughout the United States. In addition, mosquito populations are being monitored for density, species, and presence of West Nile virus. Surveillance systems provide information that is used to determine where and how mosquito abatement and control programs should commence.

References

Asuis, D., et al. 1999. "Outbreak of West Nile-like viral-encephalitis—New York, 1999." *MMWR* October 1 48(38):845–849.
Centers for Disease Control and Prevention (CDC). West Nile Virus website: http://www.cdc.gov/ncidod/dvbid/westnile/index.htm.
Fine, A., et al. 1999. "Update: West Nile-like viral encephalitis—New York, 1999." *MMWR* October 8 48(39):890–892.
Health Canada. West Nile Surveillance Program website: http://www.hc-sc.gc.ca/pphb-dgspsp/wnv-vwn/index.html.

Current subscribers participate in dialogues addressing rumors, actual disease outbreaks, and questions pertaining to emerging infectious diseases. Since 1999, the International Society for Infectious Diseases has furnished new servers and software to handle the large volume of subscribers. Today Pro-MED email posts are read by tens of thousands of viewers in over 160 countries, and thousands more refer to its website. This forum has been successful in alerting local, national, and international organizations about new outbreaks; for example, before official reports were published, three individuals—Stephen Cunnion (M.D., Ph.D., M.P.H., International Consultants in Health, Inc.), Jesse Huang (State Health Dept. in Nashville Tennessee), and Dan Silver (Intellibridge Corporation)—posted a request for information (RFI) on ProMED about SARS, a new viral disease epidemic in China.

8.5 A Word About Prevention and Containment of Contagious Diseases: Quarantine

The practice of **quarantine** is still used today to prevent person-to-person transmission of infectious diseases, such as during the SARS outbreak of 2003. Quarantine is the segregation of healthy persons who are not ill but have been exposed to an individual who suffers from a communicable disease from the general population. This is necessary because the exposed healthy individual is at risk of becoming infected. In May 2007, a man infected with the bacterium *Mycobacterium tuberculosis* was ordered into isolation by the U.S. government. The man was a U.S. citizen who suffered from a multidrug-resistant form of tuberculosis. This was the first time since 1963, when an individual infected with smallpox was placed in isolation, that someone was quarantined in the United States.

The word quarantine comes from the Latin word *quaresma,* meaning forty. This definition dates to the 14th century when ships arriving in Venice, Italy, sailing from plague-infected ports were required to sit at anchor for 40 days before landing. Before the Federal Quarantine Legislation was passed in 1878 to prevent the transmission of yellow fever, little was done to prevent the importation of infectious diseases in the United States. More laws were passed regarding quarantine requirements when cholera arrived in the United States in 1892. By 1921, quarantine stations were set up under federal government authority. In 1967, quarantine practices were authorized by the National Communicable Disease Center (now known as the CDC). In the late 1960s, over 500 staff oversaw 55 quarantine stations located at every port and major border crossing. Today the number of stations and staff has been reduced and the CDC has changed its focus. Instead, emphasis is placed on routine inspections and enhanced surveillance systems to meet the changing needs of international traffic. Quarantine operations involve the cooperation of the EIS, state and local health departments, the United States Citizenship and Immigration Service (USCIS), Customs and Border Protection, the United States Department of Agriculture (USDA), and the United States Fish and Wildlife Service (USFWS).

Isolation is still commonly used in modern public health. Isolation refers to the separation of ill/infected individuals from healthy individuals. Infected individuals can be isolated in a hospital, at home, or in designated community-based facilities. During the early 1900s and through the 1950s, as soon as the Health Officer gained knowledge pertaining to cases of scarlet fever, yellow fever, smallpox, diphtheria, measles, whooping cough, chickenpox, cholera, bubonic plague, membranous croup, and cerebral meningitis, **placards** or notices from the state's board of health were posted on entrances of the homes of infected individuals (**FIGURE 8-12**). It was unlawful for anyone other than a physician or trained nurse to enter or leave the premises in which sick individuals were being isolated. The placard was typically displayed for at least 21 days after scarlet fever or smallpox was

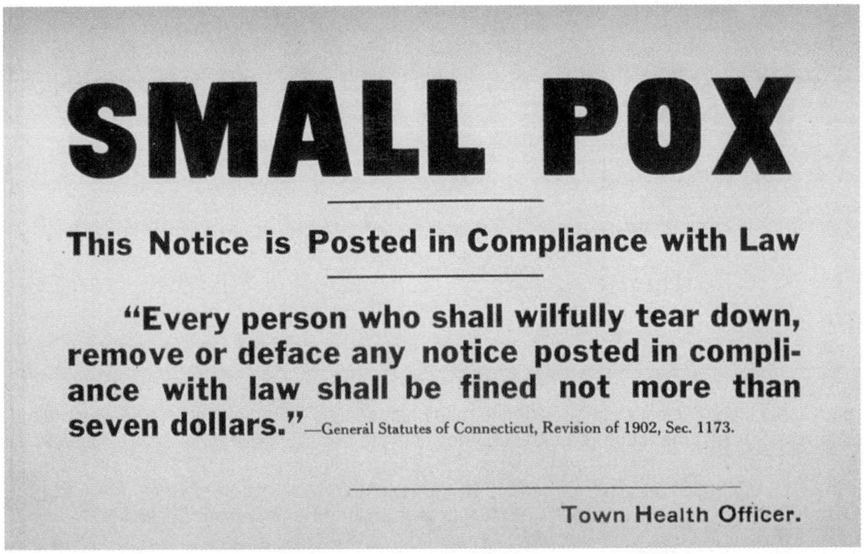
SMALL POX
This Notice is Posted in Compliance with Law
"Every person who shall wilfully tear down, remove or deface any notice posted in compliance with law shall be fined not more than seven dollars."—General Statutes of Connecticut, Revision of 1902, Sec. 1173.
Town Health Officer.

FIGURE 8-12 In the United States, local health officers posted placards such as this on the entrances of homes where sick individuals had been reported. Mutilation or removal of the card was unlawful without the authority of the health officer or board of health.

reported and 14 days after the report of other infectious diseases. There were also specific guidelines for the disinfection of clothing, removal of infected articles, and disinfection of homes or facilities upon death or convalescence.

8.6 Travel Medicine

We live in a global world in which many people are traveling without adequate travelers' health preparation. If you were to plan a trip to Tanzania (central Africa) or Uzbekistan (central Asia), how would you determine what infectious diseases are endemic in those countries and the vaccines that you need as a result (**FIGURE 8-13**)? What other information should you know? For example, should you bring insect repellent? A mesh netbed or headnets? Bottled water? Are there special requirements if you are traveling with children, the elderly, or someone who is pregnant? Will you be camping? Are you traveling via an international adoption program? How long will you be there? Will you need any blood products? Antivirals or antibiotics? Are there environmental health risks? What will be your mode of transportation (ship, air)?

Today, many individuals travel globally on business. A 1991 study published in the *Annals of Emergency Medicine* pertaining to the morbidity and mortality of U.S. World Bank staff and consultants traveling in developing countries described how approximately 100,000 World Bank staff and consultants filed medical insurance claims related to health problems associated with exposure to infectious diseases. Overall, medical claims are increasing as the frequency of travel increases.

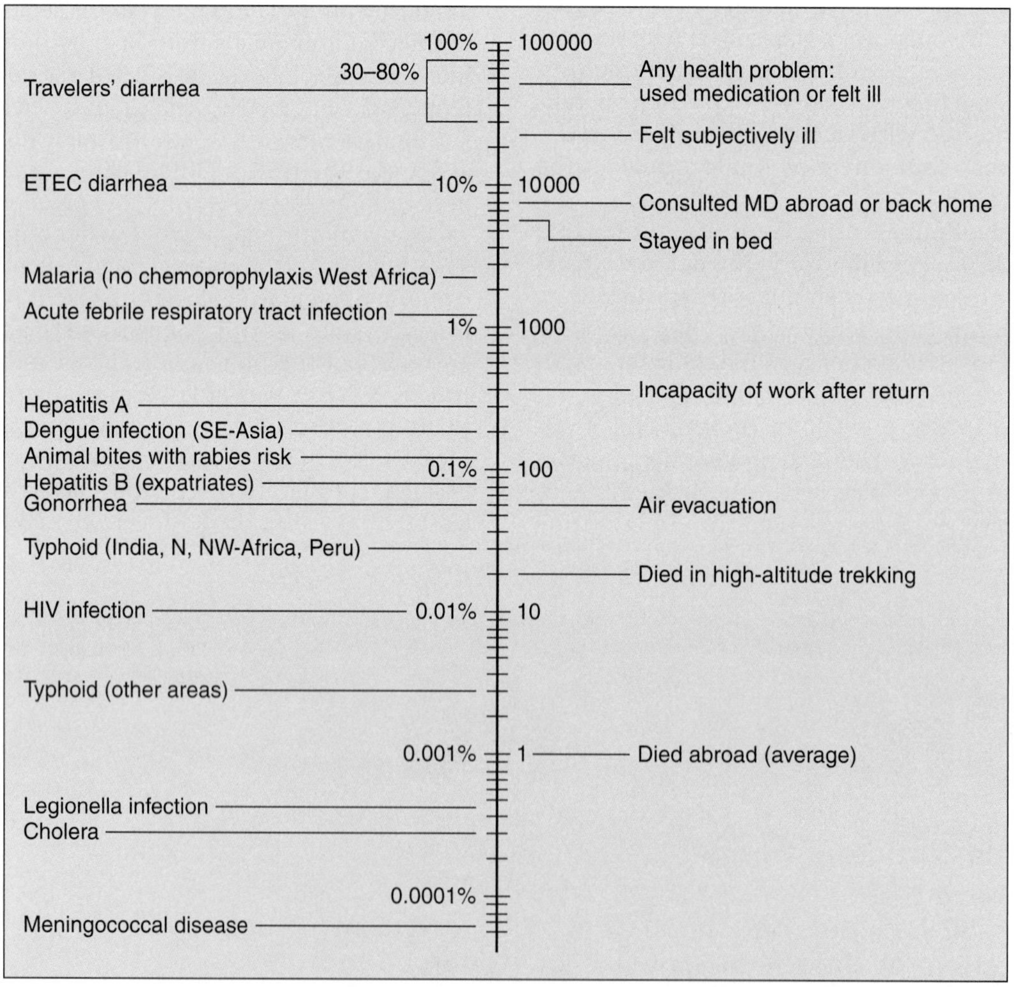

FIGURE 8-13 World Bank staff and consultants filed medical claims relating to many bacterial and viral infectious diseases while traveling in developing countries. The figure represents the monthly incidence rates of these health problems. Note: ETEC refers to diarrhea caused by Enterotoxigenic *E. coli* (ETEC). The word febrile means fever. According to this survey, 400 per 100,000 individuals were unable to work upon their return from developing countries. Reprinted from Hargarten, W. et al., *Ann Emerg Med*, vol. 20, "Overseas Fatalities of United States Citizen Travelers," pp. 622–626, © 1991, with permission from American College of Emergency Physicians.

FIGURE 8-14 This traveler is avoiding disease-ridden water by filtering water from a mountain stream.

The International Society of Travel Medicine (ISTM) and the American Society of Tropical Medicine and Hygiene (ASTMH) provide directories of private travel clinics throughout the United States and other countries. The CDC also maintains an updated Web page on Travelers' Health. It contains information on specific diseases that affect travelers, vaccinations, special needs, traveling with children, how to avoid illness associated with contaminated food or drinking water, travel via air or cruise ships, and updated information on current outbreaks (**FIGURE 8-14**). Important resources are listed in **TABLE 8-8**. The immune status of the traveler should always be considered before traveling to areas where exposure to infectious diseases is high.

TABLE 8-8 Travelers' Health Information

Internet Resources for Travelers	URL
CDC's National Center for Infectious Diseases Travelers' Health	http://www.cdc.gov/travel/
The International Society of Travel Medicine	http://www.istm.org/
American Society of Tropical Medicine and Hygiene (ASTMH)	http://www.astmh.org/
CDC's National Immunization Program Vaccine Publications	http://www.cdc.gov/vaccines/pubs/vis/default.htm
WHO's International Travel and Health	http://www.who.int/ith/

Arthropod Bites

Adventure travel packages have become increasingly popular. Dive tours, wilderness safaris, and sailing to remote areas often expose tourists to blood-feeding arthropods that can carry viruses or parasites (e.g., *Plasmodium sp.*, which causes malaria) and sting injuries by marine life (**TABLE 8-9**). Tourists should educate themselves about such risks and take adequate measures to avoid injuries and infections; for example, the use of DEET, wearing

TABLE 8-9 Examples of Viral Diseases in Humans Caused by Arthropod-Borne Viruses*

Viral Disease	Endemic Geographic Location	Vector	Disease Symptoms
West Nile encephalitis	North America, Africa, Indian subcontinent, Middle East, former Soviet Union, Europe	Mosquito	Fever, rash, encephalitis
Yellow fever	South and Central America, Africa	Mosquito	Hemorrhagic fever, hepatitis
Dengue fever (Dengue virus 1–4)	Throughout the tropics	Mosquito	Fever, hemorrhage, rash
St. Louis encephalitis	The Americas	Mosquito	Encephalitis, hepatitis
La Crosse encephalitis	North America	Mosquito	Encephalitis
Japanese encephalitis	Asia, Pacific Islands, northern Australia	Mosquito	Encephalitis, fever
Rift Valley fever	Africa, Arabia	Mosquito	Fever, hemorrhage, encephalitis, retinitis
Epidemic polyarthritis (Ross River virus)	Australia, South Pacific	Mosquito	Fever, arthralgia (severe joint pain), rash
Eastern and Western equine encephalitis	The Americas	Mosquito	Encephalitis
Chikungunya fever	Philippines, SE Asia, Africa	Mosquito	Fever, arthralgia (severe joint pain), rash
Kyasanur Forest disease	India	Tick	Hemorrhage, fever, meningoencephalitis
Powassan encephalitis	Canada, Russian Federation, North America	Tick	Encephalitis
Colorado tick fever	Canada, USA	Tick	Fever
Crimean–Congo hemorrhagic fever	Central Asia, Africa, Middle East, Europe	Tick	Hemorrhagic fever
Vesicular stomatitis	The Americas	Sand fly	Fever, encephalitis

*Note that this list is not complete. Many of the viral diseases caused by arthropods cause encephalitis, an inflammation of the brain. Initial symptoms of encephalitis consist of a headache, nausea, vomiting, fever, and lethargy. Meningoencephalitis refers to inflammation of the brain and its membranes; it is sometimes referred to as encephalomeningitis.

protective clothing, and use of bed nets are very effective methods for preventing arthropod-transmitted diseases.

Contaminated Food

Foodborne illnesses in remote areas are most often associated with the following:

- Ciguatera fish poisoning (common in the Pacific and Indian oceans and the Caribbean sea)
- Paralytic shellfish poisoning (often following a dinoflagellate bloom)
- Outbreaks in tropical and subtropical waters, which have become more frequent in places including Mexico, Malaysia, Guatemala, El Salvador, Papua New Guinea, India, and the Solomon Islands
- Poisoning related to toxic and unfamiliar mushrooms (nontoxic mushrooms in Europe and North America closely resemble highly toxic mushrooms in other parts of the world and vice versa)

Hepatitis A outbreaks associated with food consumption (oral–fecal route of transmission) are highest in developing countries that have poor sanitation and food hygiene. Individuals traveling to Eastern Europe, Greece, Turkey, and the Commonwealth of Independent States (including Russia) are at immediate risk for contracting hepatitis A. The United States, Canada, Scandinavian countries, Australia, Japan, New Zealand, and developed parts of Europe are considered low risk. Serological studies have indicated that persons from industrialized countries born after World War II have very low titers of anti-hepatitis A antibodies, increasing their risk of contracting hepatitis A upon exposure. The case fatality rate of hepatitis A infection is estimated to be 0.15% but increases to 2.0% or higher in those over 40 years of age. There is a recommended and effective vaccine available for those persons from industrialized countries who are traveling to countries where hepatitis A is endemic.

Outbreaks of norovirus have occurred aboard cruise ships and in hotels. These viruses, which are transmitted via oral–fecal and aerosol routes, are the most common causes of nonbacterial diarrhea in travelers. Cruise ships bring together large numbers of people from different parts of the world, and the confined quarters of the cruise ship is a factor in the spread of gastrointestinal disease. The incubation period of norovirus infection is also short (one to two days). Frequent handwashing helps to prevent the spread of infectious diseases on cruise ships. The CDC investigates outbreaks of gastrointestinal nature if 3% or more of the passengers or crewmembers of a passenger ship become ill.

Influenza is common among travelers. The risk of exposure to influenza depends upon the time of year and the destination of the traveler. In the tropics, influenza occurs throughout the year. In the southern hemisphere (e.g., parts of Africa, South America, and Australia), most influenza activity occurs from April through September. The influenza season in the northern hemisphere is November through April. Influenza ranks next to hepatitis A as one of the most common vaccine-preventable diseases of travelers.

Contaminated Water

Examples of viral waterborne pathogens are hepatitis A, hepatitis E, poliovirus, and norovirus. It is not wise to assume that tap water is potable in many areas. Travelers should learn basic field techniques—such as heat, filtration, and chemical disinfection—that can be used to improve water quality, as well as the limitations of each method.

Medical Facilities in Remote Areas

Employees of petroleum, mining, and construction companies, members of the armed services and intelligence communities, remote area public health personnel, missionaries, individuals on pilgrimages, adventure and health spa travelers, and scientists travel to remote areas of developed and developing countries. Travel in rural areas may place individuals outside of protective law enforcement and out of reach of personal security and healthcare facilities in the event of medical emergencies. Besides pretravel immunizations and education, travelers should be counseled by a physician about carrying a specialized medical kit. Sometimes self-treatment may be necessary to combat diarrheal diseases, acute respiratory diseases, and other physical ailments. Some companies will require pretravel medical screening because medical facilities in foreign countries often lack the resources (i.e., medications, vaccines, antibiotics, antivirals, sterile needles, and a safe blood supply) and experience in treating chronic illnesses.

Systematic screening of blood donations is not yet feasible in all developing countries, and thus the safety of the blood supply ultimately depends on the quality of blood transfusion services in the host country. Blood should be transfused only when absolutely necessary, especially in those countries where screening of blood for transmissible diseases is not yet widely performed. Travelers should take active steps to minimize the risk of injury and establish a plan for dealing with medical emergencies. The risk of bloodborne transmission of hepatitis B can be minimized with hepatitis B vaccination prior to travel.

8.7 Tracking Diseases from Outer Space: Early Warning Systems

Environmental factors often play a pivotal role in outbreaks of viral diseases like Dengue fever, hantavirus pulmonary syndrome, West Nile encephalitis, and Rift Valley fever; for example, heavy rainfall, temperature changes, and vegetation changes may set the stage for a population surge of disease-carrying pests such as mosquitoes, ticks, or rodents, and the diseases they carry spread rapidly.

NASA's Global Hydrology and Climate Center in Huntsville, Alabama, is using high-tech satellites to monitor environmental changes in entire regions, countries, and even continents. Scientists from the NASA center visit and observe sites with potential disease outbreaks. They combine their field observations with the satellite images to set up early warning systems for regions. Sophisticated mathematical algorithms are computed and correlated with the observed field conditions and satellite data to simulate the landscape and its potential for vectorborne disease outbreaks. These "red flags" allow health officials to focus on mosquito eradication, vaccination programs, and other efforts to prevent an outbreak before it happens. Perfect predictions will never be possible, but reasonably accurate risk estimates can be made by combining field work with the newest satellite technologies (**FIGURE 8-15**).

Summary

Epidemiology is the branch of science that deals with how diseases affect whole communities or populations. It is concerned with disease diagnostics, trends, prevention, and control measures. The field has evolved its own terminology that can be quite confusing for those not familiar with it. Common terms include endemic, sporadic, epidemic, pandemic, morbidity, mortality, incidence, prevalence, carrier, incubation period, prodromal period, period of illness, mode of transmission, etiological agent, reservoir, case definition, communicable period convalescence, and zoonosis, which were defined at the beginning of the chapter.

Two of the early pioneers of epidemiology, Edward Jenner and John Snow, focused on the prevention of smallpox and cholera, respectively, during the 18th and 19th centuries. Snow's search for the cause of the cholera outbreak in London in 1854 laid the groundwork for descriptive epidemiological investigation.

The epidemiological triangle or triad model is used to explain the complexities of infectious disease causation. Epidemics occur when

- there is an increase in a particular pathogen.
- the pathogen becomes more virulent.
- the pathogen enters the host via a new portal of entry.
- the pathogen is introduced to a new environmental setting.
- the host has increased risks or susceptibility.
- there is increased exposure between the host and pathogen (Table 8-2).

Viral diseases are transmitted directly (person to person) or indirectly (via airborne, foodborne, waterborne, vectorborne, or vehicleborne

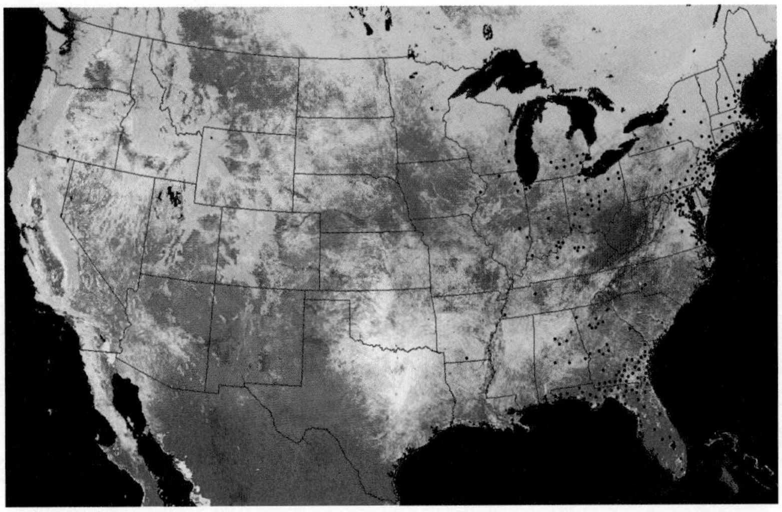

FIGURE 8-15 This satellite image helped monitor and predict the spread of West Nile virus in the United States. The land surface temperatures are represented by shades of red. Blue indicates the range of highest and lowest temperatures the region experiences during the year, with deep blue representing the greatest range. The black dots on this map represent infected crows reported in 2001. The colors on this map indicate the relative levels of risk for West Nile Virus in 2001. The image was produced by INTREPID from data taken by the National Oceanic and Atmospheric Administration's (NOAA) Advanced Very High Resolution Radiometer (AVHRR) instrument.

transmission). Viral pathogens have preferred portals of entry into their hosts and exits from them. Hospital and healthcare settings apply various practices to break the chain of infection. Practices include frequent handwashing and the use of protective barriers (i.e., gloves, masks, and coats), maintaining a clean environment, and disposing of medical wastes and used medical supplies properly.

Herd immunity acts as a "firebreak" by slowing or preventing further spread of infectious diseases. Vaccination is an effective way to create herd immunity. As the number of vaccinated people increases, the likelihood that the disease will be transmitted to unvaccinated individuals is reduced.

There are two types of epidemiological investigations: descriptive and analytical studies. Descriptive studies provide information relating to the what, when, where, and who pertain to a specific outbreak in a community. These studies are used to generate a testable hypothesis that is used in further analytical epidemiological investigations. Analytical epidemiology provides the why and how of an outbreak (testing hypotheses for causation). Epidemiologists use case-control or cohort methods.

Surveillance programs and reporting systems targeted at infectious diseases play a fundamental role in public health. Government and private organizations actively participate in surveillance and respond to disease outbreaks. Internet reporting has become a successful forum in alerting local, national, and international organizations about new disease outbreaks. Quarantine and isolation techniques may also be enforced by these organizations.

To meet the needs of the growing numbers of individuals traveling globally through business or recreation, travelers' clinics have been set up. These clinics provide vaccinations and information on specific diseases that affect travelers. The immune status of the traveler should always be considered before traveling to areas where exposure to infectious diseases is high. Medical facilities in remote areas often lack proper medical supplies, blood products, vaccines, and experienced staff in treating infectious diseases.

Changes in the environment may set the stage for new diseases; for example, increased rainfall may contribute to a surge of disease-carrying mosquitoes. The diseases they carry spread rapidly. High-tech satellites are being used to monitor environmental changes in entire regions, countries, and even continents in an effort to develop early warning systems for disease outbreaks.

CASE STUDY QUESTIONS

These questions relate to the Case Study presented at the beginning of the chapter.

1. What viral agent could be causing the infection?
2. What type of specimens would be collected from Tina to determine the cause of her GI upset?
3. What types of laboratory tests might have been performed to determine the pathogen that was causing her illness?
4. Would you take any special infection control precautions in caring for your sister or for a patient with gastroenteritis?
5. What treatment is effective?
6. Create a list of questions that were likely on the epidemiologist's questionnaire.
7. What could be the source(s) of Tina's infection?

For Questions 8–10, consider this additional information about the Case Study: During Tina's examination, several other children were being brought into the clinic for gastroenteritis symptoms. You recognized that at least one of the children was in Tina's swimming class. Another child at the clinic had been at the party held at the pool on the day of Tina's lesson. The intern epidemiologist begins a more thorough investigation.

8. What would be the epidemiologist's initial steps in her investigation?
9. How might she look for additional cases?
10. Once information about the cases was collected, how would she characterize the outbreak (descriptive epidemiology)?

For Questions 11 and 12, consider this additional information about this Case Study: The epidemiologist visited the swimming club. She interviewed the staff and swimmers and indicated that the pool water was visibly cloudy. None of the pool staff had formal training in pool disinfection.

11. Could this illness have been prevented?
12. Create an outline for a general protocol that includes monitoring and proper response protocols used to prevent failures in the environmental health system at the swimming club.

Important Epidemiology Links

The WWW Virtual Library: Medicine and Health: Epidemiology: http://www.epibiostat.ucsf.edu/epidem/epidem.html

CDC Wonder: Online Data for Epidemiological Research: http://wonder.cdc.gov/

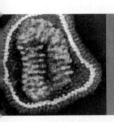

CASE STUDY 1: MUSICIANS AND VIRAL INFECTIONS

Two college students (Tammy Meyer and Larry Hoff) in a Music Appreciation course were working on a paper assignment together about the interplay between music and infectious diseases. Tammy hypothesized that early musicians during the baroque (1600–1750) and classical (1750–1830) periods were more likely to die of an infection at a higher rate than their peers. She believed that musicians and composers in this period were always poor and couldn't afford good food or living quarters. Their lifestyles contributed to infections that were fatal. Larry, on the other hand, argued that musicians and composers died at the same rate as the public at large because there were few vaccinations and no antibiotics during this time period. Tammy and Larry also disagreed in their predictions of the risk of today's rock 'n' roll musicians contracting lethal pathogens. Tammy thought that all of today's artists lead high-risk lifestyles (e.g., promiscuity, unprotected sex, alcohol abuse, and intravenous drug abuse). Larry argued that because more is known about

(continued)

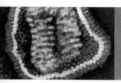

CASE STUDY 1: MUSICIANS AND VIRAL INFECTIONS (continued)

sexually transmitted diseases, along with HIV and hepatitis C education, the majority of musicians were living comfortable lives and not involved in high-risk sexual behaviors or abusing drugs and alcohol. Both found the following medical history paper authored by Jeffrey S. Sartin when researching this topic in the library: "Contagious Rhythm: Infectious Diseases of 20th Century Musicians."

1. Whom would you agree or disagree with and why?
2. Research the average age life expectancy of humans. What was the average life expectancy during the baroque period? The classical period? Today?
3. What is Jeffrey Sartin's assessment of innovative musical artists? Are they more at risk for contracting and dying from an infection? Why or why not?
4. Perform additional literature research. Create a list of at least ten famous musicians and their cause of death.
5. Compare deceased musicians from different genres of music (e.g., rap, hip-hop, jazz, rock 'n' roll, country, blues classical and baroque, modern orchestral, and opera). Is there any correlation between the type of music generated by the artist and his/her risk for contracting a lethal infectious disease? Explain your answer.

References

Sartin, J. S. 2009. "Contagious rhythm: Infectious diseases of 20th century musicians." *Clin Med Res* 8(2): 106–113.

CASE STUDY 2: YELLOW FEVER VIRUS

Three days upon his return from a holiday break visit to Brazil, 20-year-old college student Jaime Castillo began feeling ill. He called his friend to tell him he wouldn't be able to meet with him to study for a physics exam. Jaime's muscles ached. He had a headache and his symptoms quickly worsened to a high fever, chills, nausea, vomiting, and extreme back pain. His roommates took him to the emergency room of a local hospital. Given Jaime's travel history, physicians ordered a thorough laboratory workup to evaluate Jaime and admitted him to the hospital. His symptoms were treated accordingly. The laboratory workup included tests for infectious agents that cause meningitis, rickettsial infections, malaria, mononucleosis, typhoid fever, and hemorrhagic fevers. Jaime's condition appeared to be improving but then he relapsed. His kidneys and liver began to fail. He became **jaundiced**. The skin and conjunctiva (whites) of his eyes turned yellow. Abnormally high levels of bilirubin cause jaundice, which is indicative of liver or gallbladder disease. As soon as Jaime became jaundiced, doctors were able to focus more closely on a diagnosis of his condition. There are a finite number of viruses and conditions that cause jaundice. Knowing that the yellow fever vaccine is not required to enter Brazil, there was a strong possibility that Jaime could be suffering from a yellow fever virus infection. Jaime admitted that he had not been vaccinated against yellow fever virus. Test results came back positive for yellow fever virus.

1. How do you suspect Jaime contracted yellow fever virus?
2. Is the yellow fever virus enveloped? Naked? What type of nucleic acid genome does it contain? List some viral relatives of yellow fever virus.
3. How many genes does yellow fever virus contain and what does each gene code for?
4. In what countries is yellow fever endemic?
5. What could Jaime have done to prevent becoming infected with yellow fever virus?
6. What is the mortality rate of yellow fever?
7. What specimens are collected from a patient in order to confirm yellow fever as a diagnosis?
8. Besides yellow fever virus, create a list of other viruses that cause jaundice. What are

the genomic and structural characteristics of these viruses (e.g., naked vs. enveloped)?

9. When and where did the last epidemic of yellow fever occur in the United States? How was yellow fever eradicated from the United States?
10. There are fascinating accounts of how scientists determined that yellow fever was caused by a virus. What were the contributions of the following individuals to yellow fever research? When did their discoveries/work take place?
    a. Walter Reed
    b. Adrian Stokes
    c. Elihu H. Smith
    d. Jesse W. Lazear
    e. James Carroll
    f. Hideyo Noguchi
    g. William A. Young
    h. Carlos Finlay
    i. Max Theiler
11. Discuss the impact of yellow fever epidemics on the construction of the Panama Canal.
12. Despite a highly effective vaccine developed over 60 years ago, yellow fever is a reemerging infection. Speculate as to why this is the case.

References

CDC Yellow Fever Home Page http://www.cdc.gov/ncidod/dvbid/yellowfever/

Gardner, C. L., Ryman, K. D. 2010. "Yellow fever: A reemerging disease." *Clin Lab Med* 30(1):237–260.

Piqueras, M. 2007. "Microbiology: A dangerous profession?" *Int Microbiol*10:217–226.

Resources

Primary Literature and Reviews

Asuis, D., et al. 1999. "Outbreak of West Nile-like viral encephalitis—New York, 1999." *MMWR* 48(38): 845–849.

Becker, K. M., Moe, C. L., Southwick, K. L., and Newton, M. J. 2000. "Transmission of Norwalk virus during a football game." *N Engl J Med* 343(17):1223–1227.

Centers for Disease Control and Prevention (CDC). "Kaposi's sarcoma and *Pneumocystis* pneumonia among homosexual men—New York City and California." *MMWR* 4(30):305–309.

CDC. 1996. "Update: Mortality attributable to HIV infection among persons aged 25–44 Years—United States, 1994." *MMWR* 45(6):121–125.

CDC. 2003. "Hepatitis A outbreak associated with green onions at a restaurant—Monaca, Pennsylvania, 2003." *MMWR* 52(47):1155–1157.

CDC. 2004. "An outbreak of norovirus gastroenteritis at the Swimming Club—Vermont, 2004." *MMWR* 53(34): 793–795.

CDC. 2004. "150th Anniversary of John Snow and the pump handle September 3, 2004." *MMWR* 53(34):783.

Davidkin, I., Peltola, H., and Leinikki, P. 2004. "Epidemiology of rubella in Finland." *EuroSurv Mon* 9(4):11–12.

Dayan, G., et al. 2004. "Epidemiology of measles—United States, 2001–2003." *MMWR* 53(31):713–716.

Dossey, B. M. 1999. *Florence Nightingale: Mystic Visionary and Healer.* Springhouse, PA: Springhouse.

Doyle, R. J., and Lee, N. C. 1985. "Microbes, warfare, religion, and human institutions." *Can J Microbiol* 32: 193–200.

Ferichs, R. R. 2002. "History, maps, and the Internet: UCLA's John Snow Web site." *SoC Bull* 34(2):3–7. Retrieved February 25, 2010 from http://www.ph.ucla.edu/epi/snow.html.

Fine, A., et al. 1999. "Update: West Nile-like viral encephalitis—New York, 1999." *MMWR* 48(39):890–892.

Gardner, C. L., and Ryman, K. D. 2010. "Yellow fever: A reemerging disease." *Clin Lab Med* 30(1):237–260.

Gaulin, C., et al. 2004. "Assessment of surveillance of human West Nile virus infection in Quebec, 2003." *Can Comm Dis Rep* 30(11):97–104.

Gordis, L. 2000. *Epidemiology*, 2nd ed. Philadelphia: Saunders.

Gottlieb, M. S., et al. 1981. "*Pneumocystis* pneumonia—Los Angeles." *MMWR* 30:250–252.

Grimes, D. A. 2002. "Descriptive studies: What they can and cannot do." *Lancet* 359:145–149.

Hargarten, S. W., Baker, T. D., and Guptill K. 1991. "Overseas fatalities of United States citizen travelers: An analysis of deaths related to international travel." *Ann Emerg Med* 20:622–626.

Heymann, D. L., ed. 2004. *Control of Communicable Diseases Manual*, 18th ed. Washington DC: American Public Health Association.

Hoff, B., and Smith III, C. 2000. *Mapping Epidemics: A Historical Atlas of Disease.* Danbury, CT: Franklin Watts.

Keystone, J. S., Kozarsky, P. E., Freedman, D. O., Nothdurft, H. D., and Connor, B. A., eds. 2004. *Travel Medicine.* Philadelphia: Mosby.

Masur, H., et al. 1981. "An outbreak of community acquired *Pneumocystis carinii* pneumonia: Initial manifestation of cellular immune dysfunction." *N Engl J Med* 305: 1431–1438.

Meltzer, M. I. 2004. "Multiple contact dates and SARS incubation periods." *EID* 10(2):207–209.

Mounts, A. W., et al. 1999. "Case-control study of risk factors for avian influenza A (H5N1) disease, Hong Kong, 1997." *J Infect Dis* 180:505–508.

Nies, M. A., and McEwen, M. 2001. *Community Health Nursing: Promoting the Health of Populations,* 3rd ed. Philadelphia: W. B. Saunders.

Piqueras, M. 2007. "Microbiology: A dangerous profession?" *Int Microbiol* 10:217–226.

Sartin, J. S. 2009. "Contagious rhythm: Infectious diseases of 20th century musicians." *Clin Med Res* 8(2):106–113.

Smallman-Raynor M. R., and Cliff, A. D. 2004. "Impact of infectious diseases on war." *Infect Dis Clin North Am* 18:341–368.

Snodgrass, M. E. 2003. *World Epidemics: A Cultural Chronology of Disease from PreHistory to the Era of SARS.* Jefferson, NC: McFarland.

Timmreck, T. C. 2002. *An Introduction to Epidemiology,* 3rd ed. Sudbury, MA: Jones and Bartlett.

Vandenbroucke, J. P. 2001. "Changing images of John Snow in the history of epidemiology." *Soz Praventivmed* 46(5): 288–293.

Woodall, J., and Calisher, C. H. 2001. "ProMED-mail: Background and purpose." *EID* 7(3):563.

Popular Press

Barry, J. M. 2004. *The Great Influenza: The Epic Story of the Deadliest Plague in History.* New York: Viking.

Carrell, J. L. 2004. *The Speckled Monster: A Historical Tale of Battling Smallpox.* New York: Plume.

Carrol, M. C. 2004. *Lab 257: The Disturbing Story of the Government's Secret Plum Island Germ Laboratory.* New York: William Morrow.

Crawford, D. H. 2009. *Deadly Companions: How Microbes Shaped Our History.* New York: Oxford University Press.

DeSalle, R., ed. 1999. *Epidemic! The World of Infectious Disease.* New York: The New Press and The American Museum of Natural History.

Edlow, J. A. 2009. *The Deadly Dinner Party: And Other Medical Detective Stories.* New Haven, Connecticut: Yale University Press.

English, M. P. 1990. *Victorian Values: The Life and Times of Dr. Edwin Lankester.* Bristol, United Kingdom: Biopress LTD.

Harper, D. R., and Meyer, A. S. 1999. *Of Mice, Men, and Microbes: Hantavirus.* San Diego: Academic Press.

Iezzoni, L., and McCullough, D. 1997. *Influenza, 1918: The Worst Epidemic in American History.* New York: TV Books.

Johnson, S. 2006. *The Ghost Map: The Story of London's Most Terrifying Epidemic—and How It Changed Science, Cities, and the Modern World.* New York: Riverhead Books.

Levy, E., and Fischetti, M. 2003. *The New Killer Diseases: How the Alarming Revolution of Mutant Germs Threatens Us All.* New York: Crown Publishers.

Loh, C., and Exchange, C, eds. 2004. *At the Epicentre: Hong Kong and the SARS Outbreak.* Hong Kong: Hong Kong University Press.

McCormick, J. B., and Fisher-Hoch, S., with Horvitz, L. A. 1996. *Level 4: Virus Hunters of the CDC.* Atlanta: Turner Publishing.

McKenna, M. 2004. *Beating Back the Devil: On the Front Lines with the Disease Detectives of the Epidemic Intelligence Service.* New York: Free Press.

Merkel, E. 2002. *Final Epidemic.* New York: Signet Book.

Murphy, J. 2003. *An American Plague: The True and Terrifying Story of the Yellow Fever Epidemic of 1793.* New York: Clarion Books.

Oldstone, M. B. A. 2009. *Virus, Plagues and History: Past, Present and Future.* New York: Oxford University Press.

Peters, C. J., and Olshaker, M. 1997. *Virus Hunter: Thirty Years Battling Hot Viruses Around the World.* New York: Anchor Books.

Preston, R. 1995. *The Hot Zone.* New York: Random House.

Preston, R. 2009. *Panic in Level 4: Cannibalism, Killer Viruses, and Other Journeys to the Edge of Science.* New York: Random House.

Regis, E. 1996. *Virus Ground Zero: Stalking Killer Viruses with the Centers for Disease Control.* New York: Pocket Books.

Rhodes, R. 1997. *Deadly Feasts: Tracking the Secrets of a Terrifying New Plague.* New York: Simon and Schuster.

Seavey, N. G., Smith, J. S., and Wagner, P. 1998. *A Paralyzing Fear: Triumph Over Polio in America.* New York: TV Books.

Sherman, I. W. 2006. *The Power of Plagues.* Washington, DC: ASM Press.

Shilts, R. 1988. *And the Band Played On: Politics, People, and the AIDS Epidemic.* New York: St. Martins Press.

Vinten-Johnson, P., Brody, H., Paneth, N., Rachman, S., and Rip, M. 2003. *Cholera, Chloroform, and the Science of Medicine: A Life of John Snow.* Oxford: Oxford University Press.

Watts, S. 1997. *Epidemics and History: Disease, Power, and Imperialism.* New Haven, CT: Yale University Press.

Wills, C. 1996. *Yellow Fever, Black Goddess: The Coevolution of People and Plagues.* Reading, MA: Helix Books/Addison-Wesley.

Video Productions

Viral Outbreak: The Science of Emerging Disease. 2011. Howard Hughes Medical Institute (HHMI) Holiday Lecture.

Bad Blood. 2010. Necessary Films.

Killer Virus: Hunt for the Next Plague. 2009. Discovery Communications.

The Final Inch: In the Fight to Eliminate Polio, the Last Step Is the Biggest. 2009. HBO Documentary.

Why Do Viruses Kill? 2009. Films for the Humanities.

H1H1. October 18, 2009. 60 Minutes, CBS Broadcasting.

The Polio Crusade. 2009. PBS Home Video.

The Virus Empire: Silent Killers and Invisible Enemies. 2008. Films for the Humanities.

The Invisible Enemy: Weaponized Smallpox. 2008. Films for the Humanities.

The Great Fever. 2007. A Bosch and Company, Inc. Film for the American Experience.

Fighting AIDS. January 1, 2006. 60 Minutes, CBS Broadcasting.

Influenza. 2006. Films for the Humanities.

The Age of Viruses. 2006. Films for the Humanities.

The Age of AIDS. 2006. Frontline, PBS Home Video.

Fighting AIDS. 2006. 60 Minutes, CBS Broadcasting.

Unseen Enemies: A Close-Up Look at the World's Deadliest Diseases, Rotaviruses. 2006. Films for the Humanities.

SARS and the New Plagues. 2005. The History Channel.

The Great Fever. 2005. The American Experience, PBS.

Modern Marvels: City Water Systems. 2004. History Channel.

Aids and Other Epidemies. 2004. Films for the Humanities.

Emerging Diseases. A Cambridge Educational Production. 2004. Films for the Humanities.

Bioterror. 2003. NOVA, PBS.

The SARS Outbreak. May 7, 2003. Nightline, ABC News Production.

The Danger of SARS. April 21, 2003. Nightline, ABC News Production.

Mad Cow Disease. December 23, 2003. Nightline, ABC News Production.
Outbreak: Stopping SARS. 2003. Films for the Humanities.
SARS. 2003. Canada Health.
SARS: The True Story. 2003. A BBCW Production.
War Against Deadly Microbes and Lethal Viruses. 2003. Films for the Humanities.
A Never Ending War (Smallpox). September 23, 2002. Nightline, ABC News Production.
West Nile Virus. October 6, 2002. Nightline, ABC News Production.
New Blood. 2002. Anime New Network.
Foot and Mouth Disease. March 26, 2001. Nightline, ABC News Production.
The Ebola Virus. March 12, 2001. Nightline, ABC News Production.
History Uncover: Clouds of Death. The Scourge of Biochemical Warfare. 2001. An A & E Production.
The Truth Will Out: Is vCJD Caused by BSE? 2001. Films for the Humanities.
Histories Mysteries: Smallpox Deadly Again? 2000. The History Channel.
Mosquito Nightmare: West Nile Virus. 2000. A Discovery Channel Production.
Conquering and Invisible World. 1999. Films for the Humanities.
With Every Breath: The Hanta Virus. 1999. Films for the Humanities.
Brain Snatchers: Mad Cow Disease. 1998. A Discovery Channel Production.
The Emerging Viruses. 1998. BBC Production. Films for the Humanities.
Plague Fighters. 1996. NOVA, PBS.
Ebola: Chasing the Virus. 1997. Films for the Humanities.
Great Minds of Modern Medicine: Confront the Alarming Dangers of Infectious Diseases with Karl Johnsons. 1997.
Around the World: Global Immunization. 1997. Films for the Humanities
Another War: Disease and Political Strife. 1997. Films for the Humanities.
Ebola: Chasing the Virus. 1997. Films for the Humanities.
Great Minds of Modern Medicine: Confront the Alarming Dangers of Infectious Diseases with Karl Johnsons. 1997.
The Next Plague: The Nipah Virus. 1997. Films for the Humanities.
20th Century with Mike Wallace: Outbreak! The New Plagues. 1997. The History Channel.
Ebola: Diary of a Killer. 1996. Films Media Group.
The New Explorers: On the Trail of a Killer Virus. 1994. An A & E Production.
Preventing Hantavirus Disease. 1994. CDC Educational Video.
And the Band Played On. 1993. HBO Video.

eLearning

go.jblearning.com/shors2

The site features eLearning, an online review area that provides quizzes and other tools to help you study for your class. You can also follow useful links for in-depth information, or just find out the latest virology and microbiology news.

CHAPTER 9

The History of Medicine, Clinical Trials, Gene Therapy, and Xenotransplantation

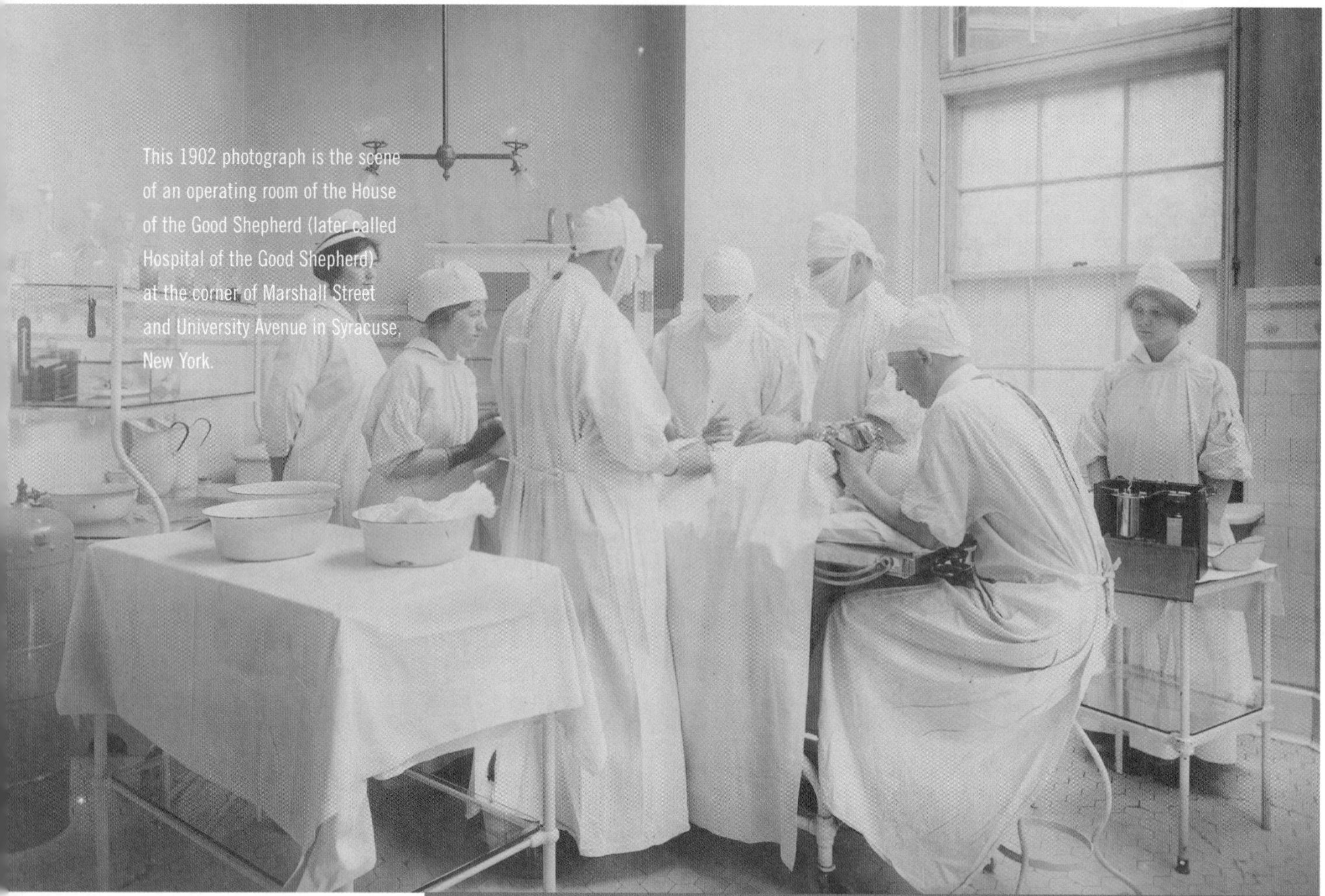

This 1902 photograph is the scene of an operating room of the House of the Good Shepherd (later called Hospital of the Good Shepherd) at the corner of Marshall Street and University Avenue in Syracuse, New York.

"The science of today is the technology of tomorrow."

Edward Teller

OUTLINE

CASE STUDY: XENOZOONOSIS

Your best friend in college, Ted, is suffering from severe kidney disease. His nephrologist has informed him that he should register for the kidney waiting list. Ted was fully aware that many people die waiting for an organ. One day, you and Ted have a conversation pertaining to xenotransplantation, a topic covered in your virology textbook. Ted asks you to answer the following questions:

1. What are the two major obstacles to confronting xenotransplantation?
2. Can the safety of human recipients be guaranteed?
3. What ethical arguments can be made against the ethical use of animals in this way?
4. Is xenotransplantation the only way in which to tackle the donor deficit problem?
5. Given the history of viruses crossing the species barrier (such as rabies, severe acute respiratory syndrome [SARS], and influenza), is xenotransplantation too risky?

How will you respond to his questions?

9.1 Why Is the History of Medicine Important?

The rationale for discussing the history of medicine is to demonstrate that our quest for improving health through the use of medicines and medical procedures has occurred since ancient times. These advancements have dramatically improved the life span and health of humans.

In 1984, Baby Fae was born with a fatal heart disease. Physicians at Loma Linda University Medical Center replaced her heart with that of a baboon, but Fae died of organ failure 21 days later. Doctors were surprised, though, to discover that Fae's body had not rejected the heart, as many had expected. This inspired doctors to continue medical neonatal heart transplant procedures.

Ashanti De Silva was born with mutations in both of her **adenosine deaminase** (ADA) genes, which are located on chromosome 20. ADA is required for metabolic function of a variety of cells, especially T-lymphocytes. ADA deficiency is one cause of severe combined immunodeficiency syndrome (SCID), which has also been referred to as the "boy in the bubble" disease. Children with SCID suffer from overwhelming infections and rarely survive childhood. At the age of 4, Ashanti was injected with genetically altered white blood cells. Her immune system has strengthened, allowing her to attend school. She continues low-dose intravenous PEG-ADA therapy.

Jesse Gelsinger was born with a rare liver disease called **ornithine transcarbamylase** (OTC) deficiency. Extreme OTC deficiency results in death, but Jesse led a relatively normal life by taking daily medications. When offered the opportunity to participate in a clinical trial at the University of Pennsylvania's Institute for Human Gene Therapy (IHGT) testing the safety of gene therapy for OTC deficiency, he agreed to participate. He died three days after being injected with an adenovirus that contained functional genes to compensate for his illness.

If you were faced with a life-threatening illness, would you opt for an experimental therapy? By participating in a new treatment for your illness, the media attention generated might increase the public awareness about the need for new therapies and better medical procedures relating to your illness. It could inspire continued research directed toward a cure for your illness. Would the increased public awareness inspire you to justify your decision? Or would costs, risks, and potential complications influence your decision? What about religious reasons?

Some of the biggest medical breakthroughs have not come without furor over prior unanticipated complications. Today, open-heart surgery is one of the most commonly performed operations in the United States, with a high overall survival rate. The mortality rate at most U.S. hospitals ranges from 1% to 3%. Open-heart surgery refers to any surgery in which the chest is opened and surgery is performed on the heart muscle, valves, arteries, or other structures.

In 1952, Toronto surgeon Wilfred Bigelow performed an open-heart operation. He applied the concept of hypothermia to perform the procedure. The animal patient was placed in a bed of ice, lowering the total body temperature so that the tissues used very little oxygen. In doing so, blood flow was interrupted for up to 15 minutes while the surgeon corrected heart defects. **FIGURE 9-1** shows this procedure being performed on a patient at

FIGURE 9-1 A medical team at the National Institutes of Health Clinical Center prepares a patient for open-heart surgery in 1955. The patient is lying on a bed of ice to lower the body temperature.

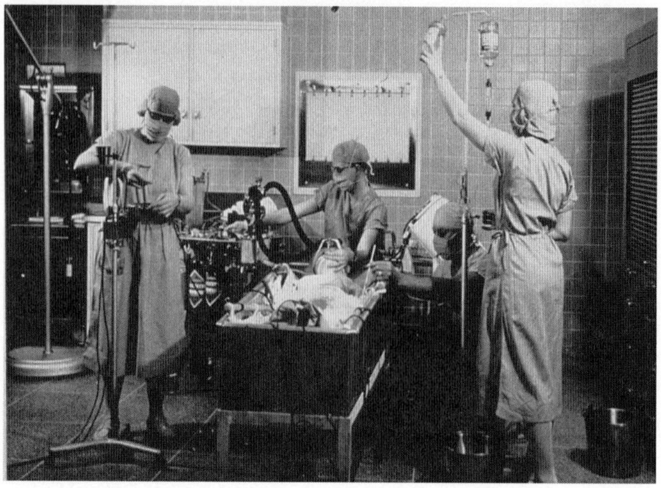

the National Institutes of Health Clinical Center in 1955. Drs. Walton Lillehei and John Lewis, at the University of Minnesota on September 2, 1952, performed the first open-heart surgery on a human, a five-year-old girl who was born with a hole in her heart. Today heart surgery is routine, and instead of a bed of ice or a cooling blanket, a heart–lung machine pumps the blood during the procedure. Former President Bill Clinton left the hospital four days after undergoing heart bypass surgery in September 2004.

History of Medicine and Clinical Research

Clinical research is evident as far back as 600 B.C. In the biblical book of Daniel, Daniel describes how he and three companions were captured in their native land of Judah and transported to Persia where they were forced to serve under the king of Babylon. Daniel wrote that he objected to the King's food and described a comparative protocol for diet and health. He showed that a diet of legumes and water made for healthier individuals than the king's diet of meat, fish, eggs, other rich foods, and wine. Two thousand years before Roger Bacon discovered the scientific method, Daniel described his experiment:

> "Please test your servants for ten days. Let us be given vegetables to eat and water to drink. You can then compare our appearance with the appearance of the young men who eat the royal rations, and deal with your servants according to what you observe." So he agreed to this proposal and tested them for ten days. At the end of ten days it was observed that they appeared better and fatter than all the young men who had been eating the royal rations. So the guard continued to withdraw their royal rations and the wine they were to drink, and gave them vegetables. (Daniel 1:12–16)

Ancient Chinese herbalists such as Shen Nung (ca. 2700 B.C.) also described clinical studies. Nung tasted and tested plants for their medicinal properties, thus gaining practical experience that allowed him to categorize medicines and record the toxicity and lethal dosage of herbal medicines.

There is evidence of surgical techniques being performed in ancient India as early as 800 B.C. Sushruta studied anatomy and described plastic surgery techniques that he developed. He specialized in cataract extraction and rhinoplasty (restoration of a mutilated nose). The details of his surgeries were recorded in *Sushruta Samahita* ("Sushruta's Compendium"). The steps Sushruta followed are remarkably similar to those used during advanced surgery today.

The practice of medicine in Europe dates to the 5th century B.C., around the time of the birth of the Greek physician Hippocrates in 460 B.C. Hippocrates believed that there was a rational explanation for all physical illnesses, rejecting the then-popular belief that the cause of disease was attributed to the disfavor of the gods or evil spirits. He founded a medical school and traveled throughout Greece practicing medicine.

Hippocrates accurately described disease symptoms of malaria, respiratory infections, diarrhea, and epilepsy. He was the first physician to use a careful, systematic approach while examining a patient's condition. His observations included facial appearance, pulse, respiration, temperature, localized pains, appearance of bodily fluids (sputum, feces, and urine), and movements of the body. Hippocrates also described the importance of cleanliness in managing patient wounds. The Hippocratic Oath, taken by all physicians today before they begin medical practice, is based on a modern version of a medical ethics doctrine authored by Hippocrates.

Galen (131–201 A.D.), who is considered the father of experimental physiology, challenged

Hippocrates' teachings. Galen was a Greek physician who traveled extensively throughout his life while practicing medicine, teaching, and studying. He was a prolific writer, authoring more than 70 books on anatomy, physiology, pharmacy, pathology, and temperaments, and was the first physician to perform animal experiments with cats, monkeys, pigs, and oxen. Through his animal experiments, Galen demonstrated that an excised heart would beat outside of the body, showing that heartbeat is not dependent upon the nervous system.

Medicine continued to improve throughout the Middle Ages (476–1453) and Renaissance (1453–1600). Hospitals were built in England, Scotland, France, and Germany starting around the 1100s, and the first medical textbooks were published in the 1470s. Andreas Vesalius (1514–1564), a Belgian physician who may have been inspired by the anatomical drawings of Leonardo da Vinci (**FIGURE 9-2**), changed the history of the study of anatomy by dissecting the human body.

Vesalius and his students resorted to grave-robbing and other activities to secure cadavers for their studies. In 1543, Vesalius published two anatomy manuals that corrected the works of Galen. His human anatomical research provided a strong foundation for all future medical research (**FIGURE 9-3**). During the 17th century, emphasis was placed on studying a patient's blood and vital statistics. Extraordinary advances in medical research and microbiology occurred in the 18th century. These included the discoveries of pioneers in microbiology and medicine such as Antoni von Leeuwenhoek (1632–1723), who invented the light microscope; James Lind, who in 1747 conducted clinical trials involving the treatment of scurvy with citrus fruits; and Edward Jenner's 1796 vaccination trials using cowpox to prevent smallpox.

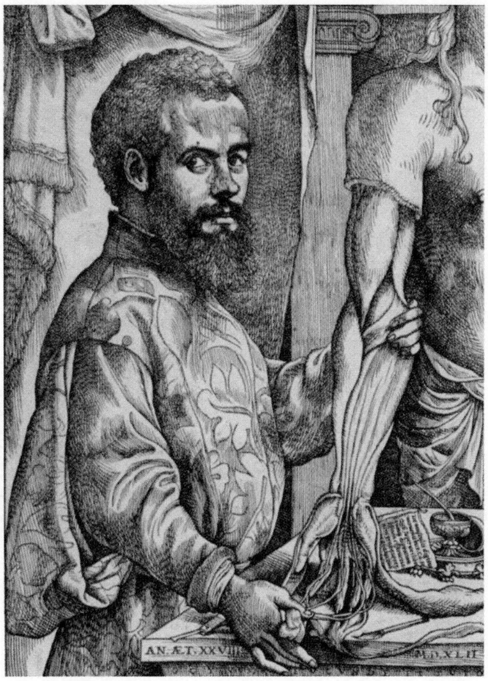

FIGURE 9-3 Andreas Vesalius (1514–1564), whose human anatomical research provided a strong foundation for all future medical research.

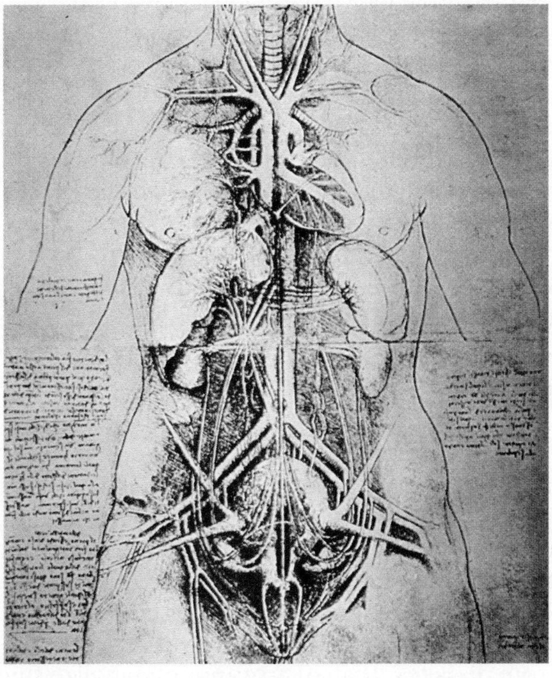
FIGURE 9-2 Anatomical drawing by Leonardo da Vinci.

Ignaz Semmelweis led the most sophisticated preventative clinical trial of the 19th century by requiring physicians to wash their hands with chloride of lime to prevent the spread of "childbed" or puerperal fever in the maternity ward of a Vienna General Hospital in 1846. Mortality rates of women in childbirth dying from puerperal fever were as high as 50% from 1841–1846. Semmelweis studied the cadavers of fever victims including a fellow physician, Jakob Kolletschka, who died of a similar infection sustained from a small cut on his finger during an autopsy of a deceased mother. Semmelweis reasoned that the disease was caused by an infectious living organism and insisted on handwashing practices. As a result of these practices, the mortality rate of the Vienna General Hospital dropped to 1.27% in 1848.

Medicine was a male-dominated field, with the exception of women's significant role in assisting childbirth. The first woman to graduate from an American medical school was Elizabeth Blackwell (1821–1910; **FIGURE 9-4a**). In her first career, Blackwell was a teacher, which was considered a suitable career for a woman. After a close friend

a b

FIGURE 9-4 Women pioneers of medicine. **(a)** Dr. Elizabeth Blackwell, the first woman in America to earn a medical degree. **(b)** Dr. Emily Stowe, the first woman in Canada to earn a medical degree.

died, her interests shifted toward caring for the sick. After applying to and being rejected by more than 20 medical schools, Blackwell gained admittance (despite the reluctance of most students and faculty) at Geneva Medical College in upstate New York. In fact, the admissions committee accepted her application as a "joke." Two years later, in 1849, she earned her degree.

After graduation, Blackwell worked in clinics in London and Paris for two years. Blackwell lost her sight in one eye after contracting a severe, purulent eye infection from a patient while working and studying midwifery at La Maternite in Paris, which ended her dreams of becoming a surgeon. She returned to New York in 1851 but was refused work as a physician. In 1853, with the help of her friends, she opened her own dispensary in a single rented room. Her sister also became a doctor, and together they opened the New York Infirmary for Women and Children in 1857, where they provided training and experience for women doctors and medical care for the poor. The hospital was staffed by women.

Another pioneer in medicine was Canadian Emily Jennings Stowe (1831–1903), the first woman to practice medicine in Canada (**FIGURE 9-4b**). Like Blackwell, Jennings became a teacher at the age of 15. She married John Stowe in 1856, and for the first several years of their marriage she stayed at home while raising their children. When her husband became ill with tuberculosis and was committed to a sanatorium, Stowe returned to teaching for financial reasons. Her husband's illness had inspired her interest in medicine, though, and she decided to pursue a medical career.

Stowe faced the same obstacles as Elizabeth Blackwell, and she was denied entrance into medical schools in Canada. She persisted and was later accepted at the New York Medical College for Women, from which she graduated with a specialty in diseases of women and children. Stowe returned to Canada in 1867 and established a private practice in Toronto, but although she saw patients on a regular basis, the College of Physicians and Surgeons in Ontario initially denied Stowe a medical license because it did not admit women. Eventually Stowe was admitted and received her license in 1880.

Other medical pioneers in the 19th century were Louis Pasteur (see Chapter 13), Robert Koch (see Chapter 5), Emil von Behring (see Chapter 7), and Elie Metchnikoff (see Chapter 7).

Insulin and Penicillin: Great Medical Breakthroughs in the 20th Century

The discovery of penicillin and insulin early in the 20th century was monumental in driving clinical research forward. The collaborative efforts of a team of Canadian doctors at the University of Toronto led to the discovery in 1921 of insulin, a lifesaving treatment for individuals suffering from diabetes. Diabetics cannot metabolize sugars due to the lack of insulin production, which results in a high concentration of sugar in the blood and urine. Before the discovery of insulin, diabetics were put on special restrictive diets. Patients lost weight and many died of poor nutrition.

The insulin project began when Fredrick Banting hypothesized that the pancreas secretes a hormone responsible for the metabolism of sugar. In 1921, he discussed his hypothesis with the researcher John James Rickard Macleod, an expert on carbohydrate metabolism. Macleod cautiously agreed to provide Banting with laboratory space and funds to begin testing his hypothesis. In addition, he was offered the assistance of graduate student Charles Best during the summer of 1921, and together they perfected a surgical procedure that would create diabetic dogs and invented a technique to measure the dogs' blood sugar levels. They were able to keep the diabetic dogs alive with a crude extract from the pancreas of dogs.

James Bertram Collip, a biochemist with protein purification expertise, joined Banting's team in the fall of 1921. His contribution to the project was a purified extract of the pancreas that contained "antidiabetic" properties. In January 1922, Collip's extract—insulin—was tested on 14-year-old Leonard Thompson, who had been a diabetic for three years. At the time, Thompson weighed a mere 65 pounds and was near coma and death. After receiving the extract, his blood sugar returned to normal and he regained his health. In 1923, Banting and Macleod shared the Nobel Prize in Physiology or Medicine with Best and Collip.

Penicillin may have saved the lives of your parents and grandparents (**FIGURE 9-5**). In contrast to the discovery of insulin by hypothesis-driven medical research, penicillin was an accidental discovery. Alexander Fleming earned a medical degree from St. Mary's Medical School in London in 1906. After doing research and serving in World War I as a captain of the Army Medical Corps, Fleming returned to lecturing and research at St. Mary's. His research focus was antiseptics produced by human tissues and secretions. In 1921, he discovered an enzyme secreted in tears called lysozyme. Lysozyme breaks down the cell walls of certain bacteria and can act as a mild antiseptic.

Like most scientists, Fleming was not in the habit of discarding old experimental bacterial culture plates. Many scientists keep their experimental reagents as long as possible in case they need to return to the experiment. One day in 1928, Fleming glanced at a stack of old petri plates containing staphylococci colonies. He examined a greenish mold contamination on the plate. None of the staphylococci bacteria were growing near the mold, which appeared to be producing a substance that caused the staphylococci to lyse or rupture.

FIGURE 9-5 World War II poster promoting the use of penicillin to treat infected wounds.

Fleming began experiments to test the laboratory mold, which he determined to be *Penicillium*, and other molds for their antibacterial properties. In 1929, he published a paper in the *British Journal of Experimental Pathology* entitled, "On the Antibacterial Action of Cultures of a *Penicillium*, with Special Reference to Their Use in the Isolation of *B. Influenzae*." Fleming demonstrated that the mold inhibited *Streptococcus, Staphylococcus,* and *Corynebacterium,* but that it was not toxic to many gram-negative bacteria. He named the active substance penicillin and suggested it could be used to isolate bacteria in the laboratory and as an antiseptic to remove penicillin-sensitive bacteria on patient dressings.

Fleming did not follow up on the application of penicillin in treating bacterial infections. Instead, a decade later, Howard Florey and Ernst Chain followed through on the medical applications of penicillin. Chain extracted penicillin from the *Penicillium* mold and with the aid of Florey began testing it on mice that had been infected with *Streptococcus*. All mice infected with *Streptococcus* survived after penicillin injections, whereas mice that were not given penicillin injections did not survive. These experiments were expanded to include human trials, which required the mass production of penicillin. Fleming, Florey, and Chain shared the Nobel Prize in Physiology or Medicine in 1945 "for their discovery of penicillin and its curative effect in various infectious diseases."

Before the 20th century, the infant mortality rate in the United States was as high as 20%, and many children died before the age of one. The rate is now 6.7%, and the reduction has been attributed to vaccine-preventable diseases, the introduction of antibiotics, handwashing, and other methods used to prevent and control infections. In the 1950s the average life expectancy in the United States was 62 years; today it is 78.

The past 50 years have seen tremendous advances in biomedical research. Therapies being developed today will gradually become commonplace in treating future health problems. Each new application will require all of us—researchers, policymakers, and the public—to consider the benefits, risks, and implications of new treatments.

9.2 Clinical Trials Today

Today is a golden age of medical research. Chapter 7 introduced the necessary steps to obtain manufacturing approval and quality assurance of vaccines. The process includes preclinical and clinical trials that proceed in at least three phases. From the start of clinical trials to the end of Phase III trials, successful drug/vaccine therapies have been tested on 400 to 30,000 volunteers and cost, on average, $83 million. Information regarding active clinical trials in the United States, Canada, and other countries is available on the Internet (key word search on the Internet for "clinical trials").

Gene therapy and xenotransplantation are on the frontier of modern medicinal research. The products and procedures used in these experimental treatments must adhere to additional regulations set forth by the Food and Drug Administration (FDA) as well as National Institutes of Health (NIH) guidelines and policies relevant to gene transfer research, and public disclosure of serious side effects is mandatory. These strict measurements are in place to safeguard the volunteers participating in these types of clinical trials.

Gene Therapy

Gene therapy is an experimental treatment that involves the introduction of genes into a person's cells to replace or compensate for defective genes in a person's body that are responsible for a disease or medical problem. But what does gene therapy have to do with viruses?

The "good" genes must be delivered or find their way to the location in the body in which its product is normally active. The "good" genes can be carried and delivered to cells by customized vectors such as viruses. These engineered viruses are manipulated using genetic engineering techniques to escape the immune surveillance of the body, ensuring there is no harm to the patient. The safety of gene therapy was intensely re-evaluated after Jesse Gelsinger died of complications during the 1999 clinical trial on OTC described briefly in Section 9.1.

Gelsinger was one of two individuals in the study who were administered 300 times the normal vector dose as the initial patients. The vector was a modified adenovirus. Gelsinger died as a result of the accumulation of a **cytokine** termed interleukin-6 (IL-6). A high level of IL-6 causes adult respiratory distress syndrome (ARDS). Normally, another cytokine called interleukin-10 (IL-10) suppresses IL-6 production as does another cytokine, tumor necrosis factor alpha (TNF-α). Unfortunately, Gelsinger's IL-10 and TNF-α levels did not effectively reduce the increased IL-6 levels, resulting in an immunological complication and ultimately his death.

After Gelsinger's death, gene therapy trials at the University of Pennsylvania were shut down and an FDA investigation was performed. The FDA determined that the IHGT violated 18 specific FDA rules and regulations during the OTC study, and the field of gene therapy in general was scrutinized and

VIRUS FILE 9-1 Was Timothy Ray Brown's HIV Infection Cured?

Timothy Ray Brown, a U.S. citizen living in Germany, was infected with HIV during the 1990s. From about 2003 to 2007, he underwent highly active antiretroviral therapy (HAART). His regimen consisted of 600 mg of efavirenz (HIV reverse transcriptase non-nucleoside inhibitor), 200 mg of emtricitabine (HIV nucleoside reverse transcriptase inhibitor), and 300 mg of tenofovir (HIV nucleoside reverse transcriptase inhibitor) each day. During this time, Mr. Brown did not experience any illnesses associated with acquired immunodeficiency syndrome (AIDS). The standard HIV drug regimen kept his infection under control.

In 2007, he was diagnosed with acute myeloid leukemia (AML). AML is a type of cancer in which the bone marrow makes abnormal myeloblasts (a type of white blood cell), red blood cells, or platelets. At the time of his AML diagnosis, his T-cell count was 415 and plasma HIV-1 RNA loads were not detectable. AML is the most common type of acute leukemia in adults and worsens quickly if not treated. The standard treatment for AML is chemotherapy to destroy most of the patient's blood cells and then infusing the patient with **stem cells** from the blood or bone marrow of a HLA matching donor **(FIGURE VF9-1)**. Stem cells are cells that have the potential to develop into many different types of cells of the body. They act as a repair system of the body. There are two main types of stem cells: adult or embryonic stem cells.

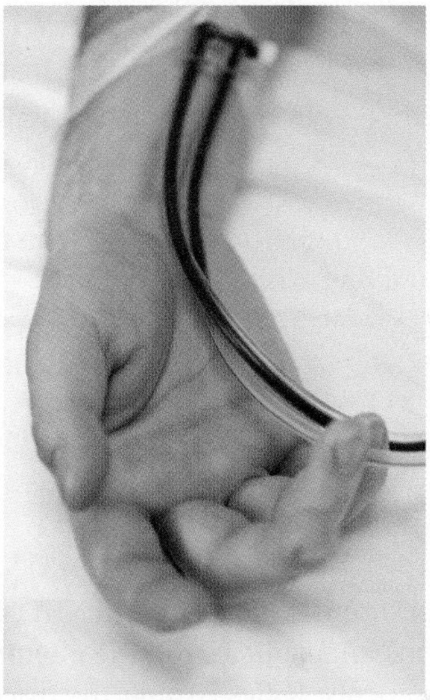

FIGURE VF 9-1 This individual is donating stem cells from blood. The blood is separated by a process called apheresis, which separates out some donor components and returns the remaining components to the donor.

HLA stands for **h**uman **l**eukocyte **a**ntigen. These antigens are present on all nucleated cells of the body, allowing the body's immune system to recognize self or foreign cells. HLA typing, also called tissue-typing, plays an important role in the compatibility of tissue, graft, stem cell, and organ transplants. The new stem cells will repopulate the immune system and kill any remaining leukemia cells that survived chemotherapy.

Mr. Brown was treated with two courses of chemotherapy followed by a final course of chemotherapy. The first course of chemotherapy treatment caused harsh side effects. He experienced liver toxicity and renal failure. Mr. Brown's viral load shot up to 6.9×10^6 copies of HIV-RNA per milliliter of plasma. Once his viral load stabilized, chemotherapy was resumed. Three months after chemotherapy treatments, his HIV-1 RNA load was not detectable.

Unfortunately, 7 months later, his AML relapsed. Mr. Brown's physician, Dr. Gero Hutter, a hematologist at the Charite University Hospital in Berlin, Germany, came up with the idea of finding a matching HLA type donor who had stem cells resistant to HIV-1 infection. Dr. Hutter knew researchers determined that persons possessing two gene copies (making them homozygous) of a defective CCR5 gene, called *CCR5-Δ32*, were highly resistant to infection by macrophage-tropic strains of HIV-1. CCR5 is a chemokine receptor that serves as a co-receptor for HIV-1 entry into target cells. The *CCR5-Δ32* gene codes for a truncated or shorter gene product (a protein that is missing 32 amino acids). HIV-1 cannot gain entry into cells because it is unable to utilize the truncated CCR5 co-receptor. About 1% of Caucasians carry the homozygous *CCR5-Δ32* gene. It is not present in Africans. Matching HLA-type donors were screened for homozygosity for the *CCR5-Δ32* allele. It was not easy to find a donor match with the double mutation.

(continued)

VIRUS FILE 9-1 Was Timothy Ray Brown's HIV Infection Cured? (continued)

Following through with this idea and the consent of Mr. Brown and a clinical Internal Review Board, Mr. Brown underwent stem cell transplantation from an HLA-identical donor who was homozygous for the CCR5-Δ32 allele in an attempt to re-populate his immune system. The procedure was repeated. Mr. Brown's AML went into remission. HIV meds were discontinued. Doctors needed to prove that Mr. Brown did not harbor any viral reservoirs in body tissues. HIV-1 reservoirs would allow the virus to re-seed the body after HAART was discontinued. The course of Mr. Brown's treatment was not easy and his recovery was long. At the time of this writing, Mr. Brown's viral RNA levels were undetectable for over 20 months and no HIV-1 reservoirs were detected during biopsy procedures of tissues. His anti-HIV antibody levels also declined, suggesting that there were no HIV-1 particles to stimulate antibody production. He remains mysteriously HIV-free! This is why Dr. Hutter posits that Timothy Ray Brown is cured.

Would this approach be practical to cure most HIV patients? Desirable? Could it lead to other approaches to curing HIV infection? Speculate as to why Mr. Brown is HIV-free.

References

Allers, K., et al. 2010. "Evidence for the cure of HIV infection by *CCR5*Δ32/Δ32 stem cell." *Blood.* DOI 10.1182/blood-2010-09-309591.

DeNoon, D. J. 2010. WebMD "A Cure for HIV?" http://www.webmd.com/hiv-aids/news/20101215/hiv-aids-cure-faq

Hutter, G., et al. 2009. "Long-term control of HIV by *CCR5* Δ32/Δ32 stem-cell transplantation." *N Engl J Med* 360(7):692–698.

questioned. The FDA and NIH created new regulations and policies to police experiments and new protocols. Rules issuing the disclosure of information regarding gene therapy trials are available to the general public via Web-based databases.

In January 2003, gene therapy suffered another major setback. The FDA temporarily halted 27 gene therapy trials in which retrovirus vectors were being used. The halt was enforced after the FDA learned that a second child had developed leukemia after being treated for X-linked combined immunodeficiency disease (X-SCID). The syndrome only affects boys. Boys with X-SCID have a defective copy of a gene that encodes the cytokine interleukin-2 (IL-2), which modulates T-helper cell production, located on their X chromosome (see Chapter 7 for a review). Without gene therapy treatment, X-SCID children usually die before their first birthday.

The second child who developed leukemia was participating in a trial led by Alain Fischer at Necker Hospital in Paris, France. A retroviral vector was used to genetically modify the boys' own marrow hematopoietic cells. Unfortunately, the vector had integrated its genome into a gene involved in normal cellular growth. This disrupted the gene's function, which resulted in uncontrolled white blood cell division (leukemia). The theoretical risk of retroviral integration was known; however, this was the second time the vector had integrated into the same gene in both patients who had gotten leukemia while being treated. This risk appeared to be much greater than researchers had thought it would be.

The role of gene therapy in the death of Jolee Mohr was determined to be unrelated to gene therapy. Mohr died on July 24, 2007, at the University of Chicago Medical Center after her second gene therapy treatment. The 36-year-old woman received two injections into her right knee to treat rheumatoid arthritis. The injections contained an adeno-associated virus containing a gene that encodes the receptor for the inflammatory protein known as TNF-α, which plays a role in rheumatoid arthritis. The leading cause of her death was a massive *Histoplasma capsulatum* fungal infection. Mohr was taking immune suppressive drugs to treat her arthritis. The drugs may have played a role in weakening her body's ability to fend off the fungal infection. Initial tests showed that only a trace amount of the genetically engineered virus was found in tissues outside of her knee joint.

Despite highly publicized rare events, gene therapy trials move forward. Gene therapy holds promise for treating genetic disorders caused by single-gene defects such as Huntington's chorea, Duchenne muscular dystrophy, polycystic kidney disease, familial hypercholesterolemia, sickle-cell

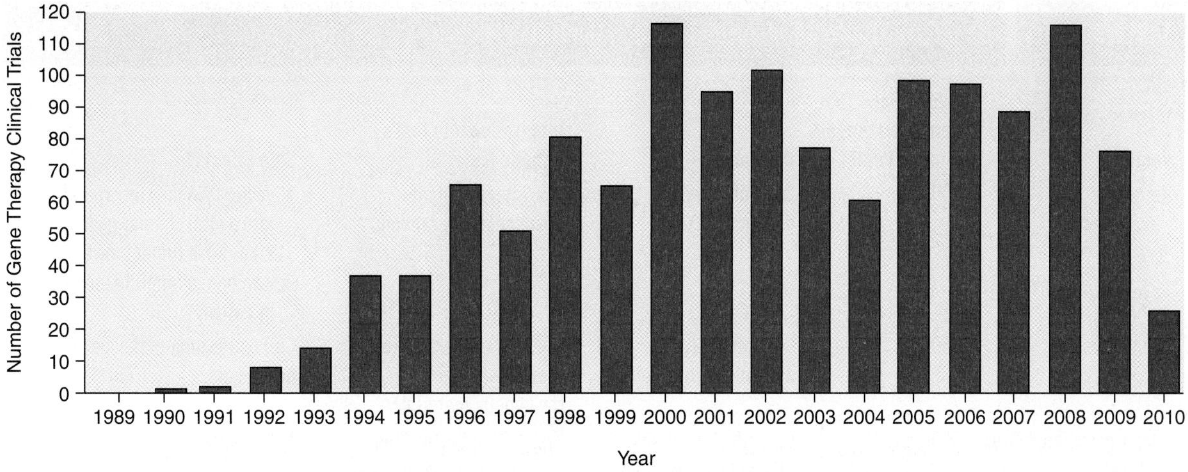

FIGURE 9-6 The number of gene therapy trials approved worldwide, January 1989–June 2010. Adapted from Edelstein, M. L., Abedi, M. R., and Wixon, J. *J Gene Med.* 9 (2007): 833–842; and Gene Therapy Clinical Trials Worldwide Home Page: http://www.abedia.com/wiley/index.html

anemia, hemophilia A and B, phenylketonuria, and cystic fibrosis. It is also being used to treat other health problems such as cancer, diabetes, high blood pressure, and heart disease. The hope is that gene therapy will provide cures for diseases in which the good gene must be working throughout one's life, such as hemophilia. In other instances, gene therapy may be used to restore eyesight (including color blindness) or hearing, repair a wound, or grow new blood vessels. In later cases, the treatment would be a temporary solution.

At the time of this writing, the majority (93%) of gene therapy clinical trials occur in the United States and Europe (**FIGURE 9-6**). Of these, 3.5% of the trials are in Phase III of clinical trials. As guidelines have become more stringent, fewer trials are being approved.

Gene Delivery by Viruses: The Key to Gene Therapy

One of the biggest challenges of gene therapy is delivering genes to the correct cells or tissues; for example, if the good gene is needed in the liver, genes must be targeted to liver cells. How do scientists ensure that the gene is targeted to the liver and not the big toe?

Viruses can act as gene delivery "vehicles" or vectors. They make good vectors because viruses can be engineered to target and enter specific types of cells. To safeguard the patient, viruses are engineered so that they cannot replicate and destroy or harm the cells of the patient. There remain some drawbacks regarding viral vectors, though. For example, they are limited in the size of gene that they can carry. Some genes may be too large for certain viral vectors. Sometimes the patient may get sick (as in the case of Jesse Gelsinger) and the patient may mount an immune response against the viral vector, preventing further treatments from working. The hallmarks of a good gene delivery system are its ability to:

- target the appropriate cells
- integrate the correct gene into the cell's nuclear DNA
- transcribe and translate the correct gene so that its gene product can function properly
- cause no toxic or harmful effects related to the gene delivery vehicle.

There are two ways to deliver the viral vectors to the patient's cells:

1. *In vivo* therapy, in which the patient's body is directly injected with the modified gene therapy viral vector.
2. *Ex vivo* therapy, in which the patient's cells are removed, grown in culture dishes in the laboratory, incubated with the viral vector to introduce the genes, and then transplanted back into the patient.

The most popular viral vectors used in gene therapy trials are:

- retroviruses
- adenoviruses
- adeno-associated virus
- herpes simplex virus
- vaccinia or modified vaccinia Ankara (MVA)

There are both advantages and limitations to each type of vector (**TABLE 9-1**).

Gene Delivery Without Viruses

Plasmid DNA or "naked" DNA is also used to deliver genes into patients' cells. Plasmids have

TABLE 9-1 Characteristics of Viral Vectors Used in Gene Therapy

Vector	Size of Gene That Can Be Packaged into the Vector	Cell Target	Integration into Cell's Genetic Material?	Side Effects
Retrovirus	8000 bp	Can only infect dividing cells.	Yes, integration into host genome is random.	1. Vector DNA may incorporate into a vital cellular gene (e.g., may cause tumor growth). 2. Can cause immune responses in patients.
Adenovirus	7500 bp	Can infect dividing and nondividing cells.	No. After a week or two, the cell will discard the genetic material.	Can cause immune responses in patients.
Adeno-associated virus	5000 bp	Can infect dividing and nondividing cells. Requires a "helper" virus to replicate inside of cells.	Yes, 95% of the time the integration will be very specific. It will integrate into a specific region on chromosome 19.	1. Less integration-specific side effects. 2. Rarely causes immune responses in patients.
Herpes simplex virus	20,000 bp	Can infect cells of the nervous system.	No, but it does stay in the nucleus for a long time as a separate DNA that replicates when the cell divides.	Can cause immune responses in patients.
Vaccinia or MVA-(poxvirus)	25,000 bp	Can infect dividing cells. Can selectively infect tumor cells.	No, DNA replicates in the cytoplasm of cell. Eventually the DNA is lost.	1. Can cause immune response in patients. 2. Limited repeated treatment.

fewer limitations regarding the size of the DNA of the corrected gene that can be inserted into the plasmid, and they don't generate an immune response. Hence, they are a safer alternative. Plasmid DNA delivery is less efficient than viral delivery, however. Research in the past five years has revolutionized the efficiency of nonviral gene transfer. To aid plasmid DNA cell entry, the DNA is either complexed with liposomes or other chemical polymers or physical energy is applied. Physical energy methods include electroporation, pressure-mediated delivery, ultrasound, laser, magnetic fields, and ballistic delivery.

Triple helix-forming nucleotides, antisense technology, ribozymes, and RNA interference systems (RNAi) are being used as a gene therapy approach when inserting a good copy of a gene will not correct the cell's problem; for example, a defective gene may code for a gene product that prevents the normal gene product from functioning correctly in the cell. In this case, a good copy or normal gene will not help. Instead, the approach to repairing the problem is to remove or inactivate the defective gene. The majority of the types of genes engineered into vectors used in today's clinical trials function as:

- specific antigens (involved in immune modulation)
- cytokines
- tumor suppressors
- receptors
- DNA replication inhibitors
- cell protection/drug resistance
- deficiency (compensation for defective genes)
- growth factors
- suicide (induce apoptosis)

A smaller percentage of clinical trials involve transferring other types of genes that function as hormones, adhesion molecules, porins, ion channels, transporters, ribozymes, gene silencers (siRNA), and transcription factors.

9.3 Xenotransplantation

The prefix **xeno-** means stranger; foreign, or different. The term **xenotransplantation** refers to any procedure that involves the use of live cells, tissues, and organs from a nonhuman animal source, transplanted or implanted into humans or used for clinical ex vivo perfusion. Xenotransplantation does not include nonliving animal products (i.e., pig insulin or pig heart valves).

History of Organ Transplants

At the turn of the 20th century, pioneer French surgeon Alexis Carrel (1873–1944) perfected the technique of vascular anastomosis, or blood-vessel suturing. Carrel became interested in repairing blood vessels after the president of the French Republic, Sadi Carnot, bled to death after being fatally stabbed by an assassin in Lyons, France, in 1894. This was a time when a surgeon's skills were limited to sewing muscles, skin, and other tissues together, but not blood vessels. Carrel's success repairing blood vessels was attributed to a number of factors:

- his unusual manual dexterity;
- his use of special materials, including tiny Vaseline-lubricated needles and fine thread;
- his insistence on avoiding trauma to the fragile vessels and surrounding tissues; and
- strict aseptic techniques and rules in the operating room, which were far more rigid than those of his colleagues.

After he achieved success in arterial and venous repairs in 1902, the idea of replacing failed organs with healthy ones could be considered. In 1904 Carrel left Lyons for Montreal, but he ultimately decided to take a position in Chicago at the University of Illinois. From 1904–1906 Carrel worked with physiologist Charles Guthrie, with whom he performed transplantation surgeries on animals, especially dogs. They published 28 papers together regarding the retransplanting or transplantation of arteries, veins, kidneys, ovaries, thyroid glands, and a thigh.

VIRUS FILE 9-2 Rabies Transmission from Solid-Organ Transplants

During the first week of May 2004, a man from Arkansas was admitted to a hospital in Texas. He had a low-grade fever and suffered from mental confusion. His neurological symptoms were not indicative of rabies. A computed axial tomography (CAT) scan of the man's brain showed bleeding consistent with a stroke. He died 48 hours after his admission into the hospital. He had not been given any blood products during his hospital stay.

The deceased individual underwent routine donor eligibility screening and testing, which did not include testing for rabies, and his family consented to organ donation. The man's liver, kidneys, and lungs were removed and transplanted into three recipients at a transplant center in Texas shortly thereafter. All three patients were discharged 5–12 days after the transplant operations but were re-admitted into hospitals 21–27 days after their initial transplant dates. All of them deteriorated rapidly, suffering from tremors, lethargy, and seizures, and subsequently died.

The Centers for Disease Control and Prevention (CDC) investigated the incident. They examined brain tissues of the diseased recipients and confirmed rabies transmission from the three solid-organ transplants and traced it to the organ donor, whose tissues also tested positive for rabies antibodies. The CDC is currently working with federal and organ donor agencies to review screening practices. The majority of rabies cases are caused by bites from rabid mammals. Rabies has also been caused by nonbite exposures such as contamination of an open wound, scratches, or direct contact with infectious material such as saliva from a rabid animal. These nonbite exposures are very rare.

Human-to-human transmission of rabies has occurred among eight recipients of transplanted corneas. The eight cases occurred in five countries: two cases each in India, Thailand, and Iran, and one case each in France and the United States. All of the donors died of rabies or a rabies-like illness.

The CDC investigated the United States' incident, which occurred in 1979 and was the first incident of rabies infections in solid organ donor and transplant recipients. As a result of the investigation, stringent medical standards were implemented among the Eyebank Association of America (EBAA) and the eye banking community. Since the implementation of these measures there has not been a rabies case associated with contaminated corneas in the more than 43,000 sight-restoring transplants using ocular tissues that are performed each year in the United States.

References

University of Alabama at Birmingham Hospital, et al. 2004. "Update: Investigation of rabies infections in organ donor and transplant recipients—Alabama, Arkansas, Oklahoma, and Texas, 2004." *MMWR* 53(26):586–589.

In 1906, Guthrie took a position at Washington University in St. Louis and Carrel went to work at the Rockefeller Institute for Medical Research (now known as Rockefeller University) in New York. Carrel documented the problems with rejecting foreign tissues as early as 1907. For his pioneering efforts in vascular suture and the transplantation of blood vessels and organs, Carrel was awarded the Nobel Prize in Physiology or Medicine in 1912.

The earliest historical account of a human-to-human organ transplant occurred in a November 14, 1911, *New York Times* article entitled, "Dr. Hammond Gives Patient New Kidney." The reporter remarked that this was the first attempt in the United States to transplant a kidney from a man who was killed in an automobile accident to a man dying from renal tuberculosis. Dr. L. J. Hammond at Philadelphia Methodist Episcopal Hospital performed the operation. The donor kidney had been maintained in cold storage one day prior to the operation. There was no follow-up story, but it is likely that the patient's immune system rejected the organ.

Most medical historians consider the first human-to-human kidney transplant to be the one performed in Russia in 1933 by Dr. S. Voronoff. The kidney donor came from a cadaver, and the transplant failed due to organ rejection. The first successful human-to-human kidney transplant occurred in 1954 between kidney donors who were identical twins. The donor and recipient were of course perfectly matched, circumventing organ rejection. This first successful operation, which was performed by Dr. Joseph Murphy at Brigham and Women's Hospital in Boston, and a second successful operation that was performed in 1960 by Sir Michael Woodruff in Edinburgh, Scotland, are milestones in the history of transplantation.

Sporadic attempts at cross-species transplants occurred during the early 1960s:

- In 1963 and 1964, surgeon Keith Reemtsma at Tulane University in New Orleans transplanted 13 chimpanzee kidneys into humans. Twelve of the patients survived between 9 and 60 days. One patient survived for 9 months with no signs of rejection while on primitive immunosuppression drugs.
- In 1964, Thomas Starzl of the University of Colorado transplanted 6 baboon kidneys into humans. Patients survived 19–98 days, and most died of infections related to immune suppression.
- In 1964, James Hardy of the University of Mississippi transplanted a chimpanzee heart into a 68-year-old comatose man. The chimpanzee heart was too small to support the patient's circulatory system and functioned for only two hours.

Sir Peter Medawar (1915–1987) and Sir Frank Burnet (1899–1985) were jointly awarded the 1960 Nobel Prize in Physiology or Medicine for their research on immune tolerance as evidenced by skin graft acceptance and rejection in both animals and humans. Their research spearheaded studies involving tissue typing and the development of immune suppressive treatments to prevent transplant rejection.

Early efforts to minimize this rejection included radiation therapy and hormone/steroid treatments to reduce the immune system function, but that would often lead to a much greater risk of infection for transplant recipients. The survival rates for transplant procedures were very low, and research efforts included a search for therapies to overcome or prevent organ rejection. In 1983, the FDA approved cyclosporin, the first antirejection drug.

9.4 Organ Need, Supply, and Demand

The development of xenotransplantation has been driven in part by the shortage of donors. Continued advancements in the development of better drugs and regimens for suppressing cross-species immune-mediated rejection has resulted in the increasing use of animals to provide insulin, skin grafts, and heart valves.

More than 25% of patients on waiting lists die before receiving an organ. **FIGURE 9-7** demonstrates this by comparing the number of registered organ recipients versus the number of donors available in the United States and Canada between 1993 and 2009. The waiting list grows by approximately 300 people each month. As of 2010, approximately 101,000 individuals were awaiting organ transplants in the United States. Each day, approximately 77 people receive and organ transplant but19 people on average died while waiting for an organ, and over $40 billion is spent each year on patients with end-stage organ failure. The average wait time for a new kidney in 2010 was 10 years.

A bill before the Illinois legislature that would allow HIV-infected persons to donate their organs to others with the virus was under consideration. The Illinois House passed it 95-22 on March 2, 2004, and the state Senate approved it 55-2 on May 2, 2004. Illinois House Bill 3857 was signed by the governor that year.

The illicit trade in human organs is a problem throughout China, India, Turkey, Israel, Egypt, Brazil, Iraq, Africa, and Russia. Organ trafficking is illegal in all of these countries. According to the

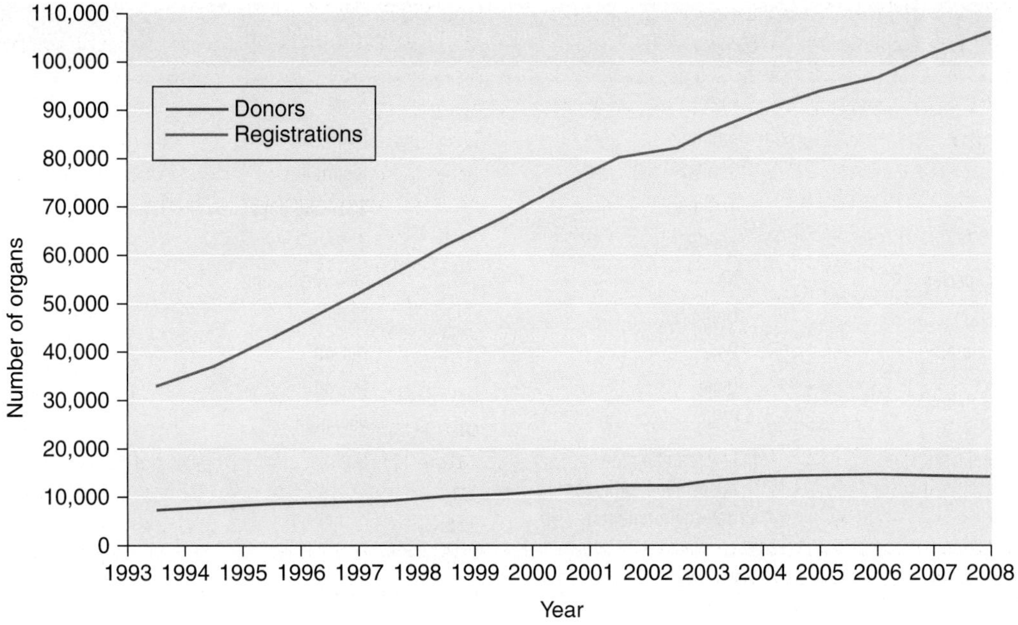

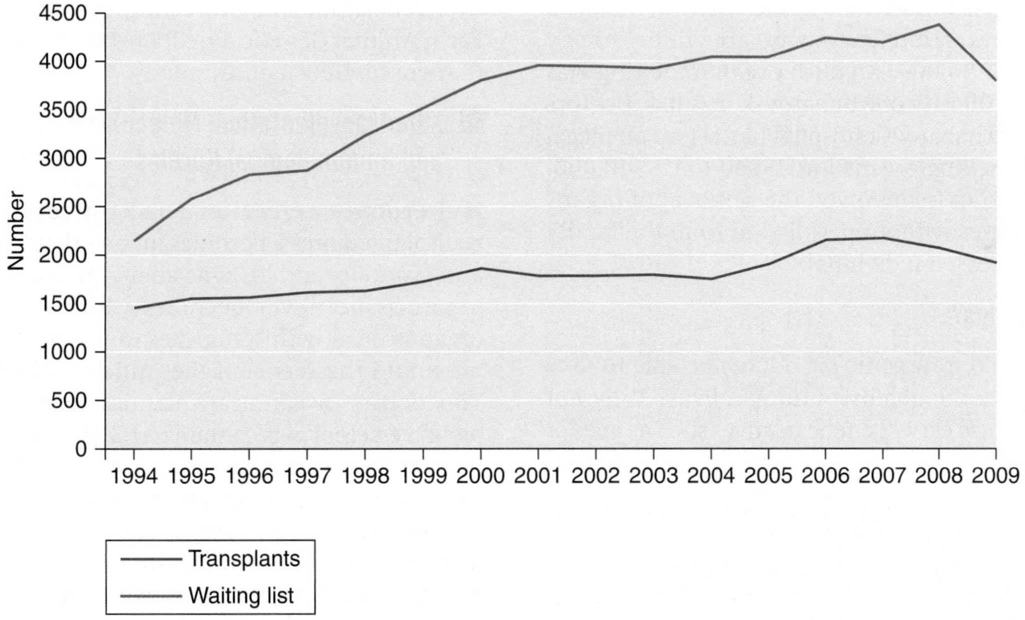

FIGURE 9-7 (a) United States organ registration and donor lists (1993–2008). Adapted from Scientific Registry of Transplant Recipients, 2003 *OPTN/SRTR Annual Reports,* http://www.ustransplant.org/annual_Reports/default.aspx. **(b)** Transplants and waiting list at year-end in Canada, 1994–2009. Adapted from Canadian Institute for Health Information, *Canada's Organ Donation Rate Still Too Low to Meet the Need,* http://secure.cihi.ca/cihiweb/dispPage.jsp?cw_page=media_14apr2004_e.

WHO, the search for organs intensified because of the increase in kidney disease and the lack of kidneys available to meet the demands. Only 10% of the need was met in 2005, spurring the illegal kidney trade. Poverty is one of the main reasons why sellers give up their organs. Selling a kidney can be a quick way out of debt or to keep from getting deeper into debt. In organ trade hot spots like Iraq, where unemployment has been as high as 18%, people sell their organs to survive. International organ trafficking is flourishing. The majority of illegal trading involves kidneys but there is also some trading of half-livers, eyes, blood, and skin.

In April 2004, two doctors and two individuals working in the city's transplantation service at a Moscow hospital were arrested after removing

TABLE 9-2 Comparison of Potential Species for Donor Xenotransplantation

	Pigs	Apes	Monkeys
Physiology	Similar	Nearly identical	Similar
Organ size	Similar	Identical	Too small
Gestation	100 days	251–289 days	170–193 days
Progeny	10–18	1–2	1–2
Protection of species	No	Yes, very strong	Yes
Availability	Unlimited	None	Low
Transgenic animals	Done	Not yet	Done
Cloning	Done	Not yet	Done
Costs	Low	Very high	High
Microbiological risks	Low (can be grown in specified pathogen-free environments)	High	High

Source: Adapted from Denner, J. 2003. "Porcine Endogenous Retroviruses (PERVs) and Xenotransplantation: Screening for Transmission in Several Clinical Trials and in Experimental Models Using Non-Human Primates." *Ann Transplant* 8(3): 39–48.

organs from patients who were technically alive. Young, impoverished men in such states as Moldova sell kidneys to foreigners who are willing to pay up to US$250,000. An illicit organ trade ring was busted in the city of Khabarovsk in 2004. Doctors from the Khabarovsk hospital had set up an illegal practice, selling organs for US$60 to US$40,000. At least 15 cases involved the removal of organs from corpses without permission from the family of the deceased individual.

Why Pigs?

Why would transgenic pig donors be able to save or improve the quality of human lives? Why not choose to use closer genetic relatives such as apes or monkeys? There are many advantages of using pigs as donor organs for human transplants (**TABLE 9-2**). Pig organs are similar in size to human organs, and pig physiology is similar to that of humans. A pig kidney will act similar to a human kidney, and the circulatory system is similar; the pig supplies the best size-matched heart that would meet cardiac output of a human.

Another advantage is that pigs are readily available and can be bred quickly. They have a short gestation period and can carry up to 18 piglets in a litter. They can be raised in special pathogen-free environments following cesarean protocols to exclude serious human pathogens with minimal costs. In addition, pigs have been genetically modified to decrease immunological rejection and were successfully cloned in 2000. Overall, these genetic modifications result in a more "humanized" pig.

Organizations working on small-scale pig cloning projects include the University of Missouri–Columbia,, PPL Therapeutics PLC of Scotland (the company that helped clone Dolly, the sheep) and the Department of Biotechnology, Institute of Farm Animal Genetic, Friedrich-Loeffler-Institute, Mariensee, Neustadt, Germany.

Xenotransplantation: Molecular Roadblocks and Immunological Hurdles

Hyperacute rejection (HAR) occurs when a transplanted organ becomes inflamed, turns dusky and cyanotic, and loses its ability to contract. The organ tissue never becomes vascularized. The organ is dead within minutes to hours of a transplant and the recipient dies quickly (**FIGURE 9-8**). This occurs in xenotransplantation because the blood vessels of pigs contain **α-1,3-galactose**, a carbohydrate epitope unique to pigs. The human immune system recognizes the α-1,3-galactose as foreign and rejects these molecules by quickly destroying the newly transplanted organ. Naturally occurring antibodies and complement activation is directed against the transplanted organ tissues.

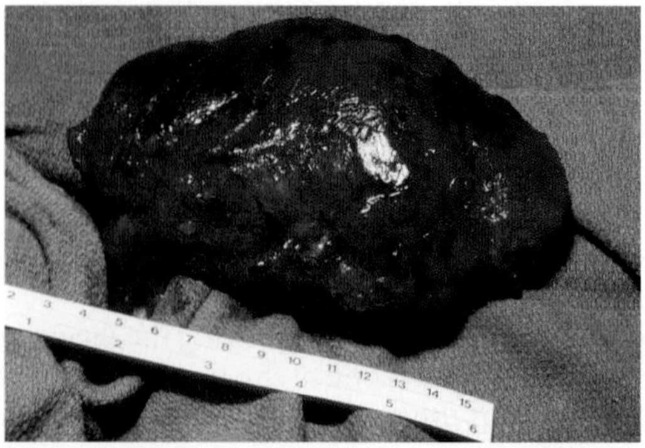

FIGURE 9-8 Transplant rejection of a human donor kidney.

The second type of immune rejection is **acute vascular rejection** (AVR). This term applies to rapid vascular injury that occurs within a few days to several weeks of the transplant. It, too, is triggered by complement and antibody response toward the organ tissues. There is also an influx of macrophages, neutrophils, and natural killer cells.

A third type of rejection, chronic, occurs within months to years following transplantation. It is believed that chronic rejection is really the result of continued and prolonged multiple acute rejections. Immune suppressive drug treatment is used to treat chronic rejection.

One might imagine that there would be a wide array of antigens on pig organs that natural human antibodies would recognize. Interestingly, it has been shown that 80% of human antibodies recognize a single α-1,3-galactose carbohydrate that is present on the surfaces of pig cells. The α-1,3-galactosyltransferase enzyme is responsible for an α-1,3 linked galactose to N-acetyl lactosaminide. This enzyme is found in New World monkeys and lower-order mammals such as pigs. In contrast, humans and higher-order primates such as chimpanzees and baboons have lost the α-1,3-galactosyltransferase gene and they produce antibodies against α-1,3-galactose.

In order to prevent HAR and AVR, genetically modified pigs have been produced using gene knockout strategies to eliminate the α-1,3-galactosyltransferase gene. Without this gene's function, the α-1,3-galactose residues are no longer produced on the surface of pig cells, thereby reducing the recipient's production of anti-α-1, 3-galactose antibodies. An alternative approach involves treating the transplant recipient with a continuous intravenous infusion of synthetic or natural α-1,3-galactose to sequester any circulating anti-α-1,3-galactose antibodies.

Another approach used to prevent rejection of pig tissue involves the insertion of the human α-1,2-fucosyltransferase [H transferase (*HT*)] gene and human complement inhibitor CD59 gene (*hCD59*) into the pig chromosome. The *HT* gene product synthesizes α-1,2-fucoselinkages on the cell surfaces of human cells. Hence, donor pig tissues expressing the *HT* gene on their surfaces will appear like human tissues to the transplant recipient. The α-1,2-fucose linkages are also tolerated by the pig donor.

The *hCD59* gene product (sometimes referred to as protectin) is expressed in many types of human cells. It binds to complement factors, preventing cell lysis after complement activation. Pig kidneys, hearts, and lungs expressing *hCD59* have been shown to function longer than control organs that did not express *hCD59*, but the approach did not completely protect organs from antibody-mediated rejection. Scientists have produced transgenic pigs that have been engineered to express both *hCD59* and *HT*. These transgenic pigs have been shown to have strong protection against the antibody and complement activation response of xenotransplant rejection. Transgenic pigs have also been developed to express two additional complement regulatory proteins: human decay-accelerating factor (hDAF) and membrane cofactor protein (MCP). Overall, the progress in overcoming the immunologic hurdles in the past decade has been remarkable. **TABLE 9-3** summarizes the genes that are being manipulated in the development of transgenic and cloned pigs.

9.5 Xenozoonosis

The word **zoonosis** refers to a disease being transmitted from an animal to a human; the term **xenozoonosis** is defined as a disease transmitted from an animal to a human recipient after transplantation of an animal organ. Viruses and other microbes lurk within the donor animal's organs and the infectious disease risks of xenotransplantation pose a problem for organ recipients and the public at large. Transplant patients must receive immunosuppressive treatment to prevent organ rejection, and as a result even nonpathogenic

TABLE 9-3 Genes Manipulated in Pig Donors to Prevent Immune-Rejection

Gene Product	Function of Gene Product	Insertion or Deletion in Pig Genome
α-1,3-galactosyl transferase	Adds α galactose to N-acetyl lactosamine	Deletion
α1,2-fucosyltransferase	Synthesizes α1,2 fucose linkages on human cell surfaces	Insertion
hCD59	Prevents human complement activation	Insertion
Human decay-accelerating factor	Prevents human complement activation	Insertion
Membrane cofactor protein	Prevents human complement activation	Insertion

microbes in pigs might adapt to the human recipient and cause disease. Viruses are of major concern because, unlike infections caused by other microorganisms, they are not usually treatable by antibiotics. A major concern is that animal viruses may jump to humans and cause a pandemic.

At one time scientists felt that viral risks could be overcome by breeding pigs in disease-free environments. The blow to this theory came in 2000 when two independent research labs discovered that pigs harbor porcine endogenous retrovirus sequences (PERVs) in their genomes. These PERVs were shown to be activated and able to produce infectious virus in human 293 cells in vitro (**FIGURE 9-9**). As a result, breeding virus-free pigs would be difficult, if not impossible. This became a turning point in xenotransplantation regulation by the FDA in the United States.

To date, sophisticated methods have been developed to detect PERV transmission. So far, there is no evidence for PERV transmission in the more than 200 patients who have received pig xenotransplants or in butchers who were frequently exposed to pig tissues. Rats, guinea pigs, minks, macaques, and baboons inoculated with high doses of PERVs and given strong daily immunosuppressive treatment failed to exhibit any signs of infection. To further prevent PERV transmission, a team of researchers at the Robert Koch Institute in Berlin, Germany, developed an RNAi strategy to increase the safety of xenotransplantation.

Other viruses have been discovered in pigs, with herpes virus reactivation being one of the most common infectious complications of immune-suppressed patients. Three novel porcine herpes viruses—porcine lymphotrophic herpes viruses I, II, and III (PLHV-I, PLHV-II, and PHLV-III)—have been detected in pigs.

Another herpes virus, porcine cytomegalovirus (PCMV), has been found in pigs. PCMV is very similar to human cytomegalovirus. Pigs acquire PCMV early in life and often show no clinical symptoms of the infection. In commercial pig herds and breeding colonies for xenotransplantation research, prevalence for PCMV antibodies is over 95%. Researchers have tried to wean piglets within the first two weeks of birth and isolate them from other animals to prevent PCMV exposure. Raising the pigs in qualified pathogen-free environments has been successful in eliminating PCMV in a proportion of the pigs when compared to pigs bred with conventional methods.

Strategies to control or exclude PCMV infection from the donor have been developed via in vitro studies. The experiments were performed using herpesvirus antiviral drugs such as ganciclovir, acyclovir, leftunomide, cidofovir, and foscarnet, but only cidofovir and foscarnet proved effective against

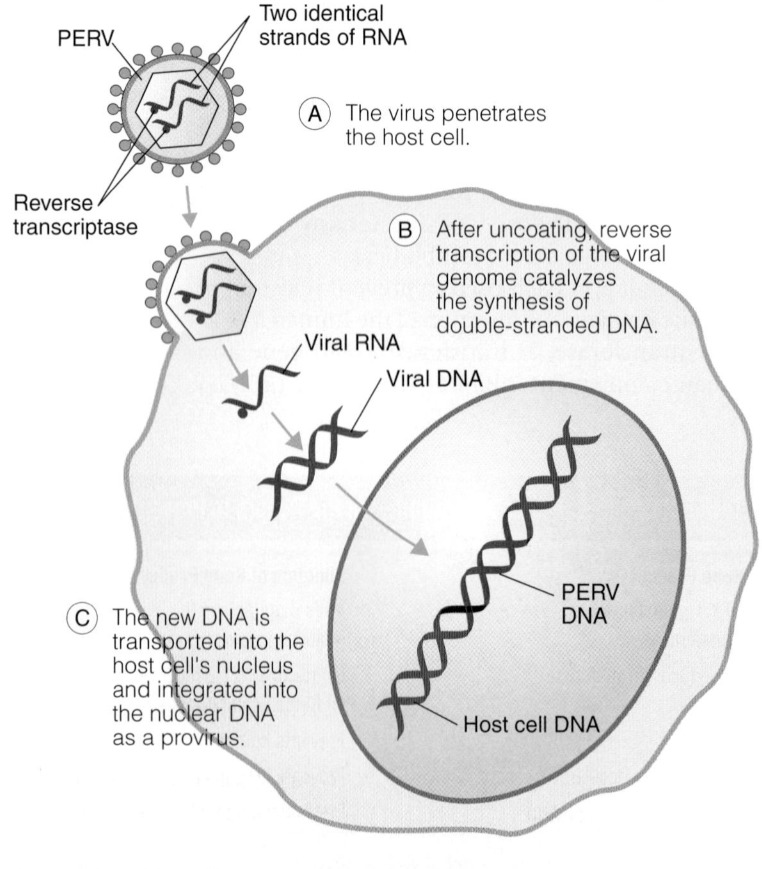
PERV
Two identical strands of RNA
A The virus penetrates the host cell.
Reverse transcriptase
B After uncoating, reverse transcription of the viral genome catalyzes the synthesis of double-stranded DNA.
Viral RNA
Viral DNA
PERV DNA
C The new DNA is transported into the host cell's nucleus and integrated into the nuclear DNA as a provirus.
Host cell DNA

FIGURE 9-9 When a pig retrovirus infects a pig cell, the RNA of the retrovirus is transcribed into DNA by reverse transcriptase, then is inserted into the host genome by an integrase. The pig genome contains many PERV sequences. The proviral DNA is not active while integrated into the host genome. Instead, it is passively replicated along with the host genome and passed onto the original cell's offspring. Eventually, in response to changes in the host's environmental condition or health (e.g., after organ transplantation), the provirus will excise itself from the genome and resume activity as a virus.

PCMV. Both of these antivirals have significant toxic effects in transplant recipients, though, so pharmacological control of PCMV might be achieved but at the cost of significant toxic side effects.

Another virus that has been shown to cross species barriers is a hepatitis E-like swine virus, which is genetically similar to human hepatitis E (HEV) and can infect both pigs and nonhuman primates. HEV causes acute hepatitis and jaundice in 30 to 80 percent of humans infected. It is rare in the United States and is usually found in countries where sanitation conditions are poor. HEV and HEV-like swine virus are transmitted via an oral–fecal route and may occur through direct contact with food or water contaminated with pig feces containing HEV-like swine virus. Recipients of pig organs, pig farmers, veterinarians, meat handlers, and others in close contact with pigs are at risk for HEV-like swine virus or other closely related viral infections.

Lastly, circoviruses have been discovered in pigs. Porcine circovirus Type 1 is ubiquitous in pig populations, but to date has not been known to cause any disease in pigs or humans. Porcine circovirus Type 2 is the likely suspect for a postweaning multisystemic wasting syndrome (PMWS) in pigs. It generally affects 25% of the pigs in a pen one to two weeks after weaning. Infected pigs develop a fever and waste rapidly, and may develop a hairy coat and runty appearance. Some pigs develop a cough and diarrhea and many pigs die after several weeks. The potential risk of porcine circovirus Type 2 transmission via xenotransplantation is unknown.

Xenotransplantation is at a crossroads. Tremendous progress has been made regarding the genetic engineering and cloning technology developed for pigs, which now can be bred with low PERVs and α-1,3-galactosyl-transferase deficiency. Clinical trials using pig cell xenotransplantation are ongoing. The most up-to-date information regarding FDA applications and scientific literature regarding pig cell transplantation being considered for neurodegenerative disorders such as Parkinson's and Huntington's disease, dates to 2002.

TABLE 9-4 Health Information Resources

What Are You Searching For?	Resource	Internet URL
Journal articles and reviews	PubMED ScienceDirect	http://www.ncbi.nlm.nih.gov/pubmed http://www.sciencedirect.com/
Gene therapy trials	Gene Therapy Clinical Trials Worldwide	http://www.abedia.com/wiley/index.html
Disease and vaccine/health information	Centers for Disease Control and Prevention (CDC) Health Canada World Health Organization	http://www.cdc.gov http://www.hc-sc.gc.ca/index-eng.php http://www.who.int/en/
Medical research	National Institutes of Health (NIH) Wellcome Trust Institut Pasteur	http://www.nih.gov/ http://www.wellcome.ac.uk/index.htm http://www.pasteur.fr
Health information for patients and medical professionals	Mayo Clinic	http://www.mayoclinic.com/
Transplant community facts	U.S. Transplant.org Canadian Institute for Health Information UK Registration and Transplants	http://www.ustransplant.org/ http://www.cihi.ca/CIHI-ext-portal/internet/EN/Home/home/cihi000001 http://www.uktransplant.org.uk/ukt/
Disease outbreaks	ProMED mail	http://www.promedmail.org/
Regulatory information on food, drugs, medical devices, cosmetics, radiation-emitting products	U.S. Food and Drug Administration (FDA)	http://www.fda.gov/
Clinical trials registry in the United States and worldwide	Clinical Trials.gov Search for Cancer Clinical Trials	http://clinicaltrials.gov/ http://www.cancer.gov/clinicaltrials/search/
Cancer	Oncolink National Cancer Institute (NCI)	http://www.oncolink.org/ http://www.cancer.gov/
Cross-cultural health	EthnoMed	http://ethnomed.org/
Xenotransplantation	Xenotransplantation (Official Journal of the International Xenotransplantation Association)	http://onlinelibrary.wiley.com/journal/10.1111/(ISSN)1399-3089

Putting these cell therapies into practice is not of course a straightforward task. It must be done at the pace of laboratory-based research rather than corporate share price. Today, there is less commercial support for xenotransplantation than there was in 2002. At the time of this writing, large companies such as Novartis and Baxter have withdrawn from the field, and smaller companies such as Alexion, Geron, Circe Biomedical, HepatAssist, Diacrin, Immerge, and Nextran have greatly reduced their interest. Funding for xenotransplantation via other sources such as government and foundations has proved difficult, perhaps in anticipation of the scientific obstacles and the high costs associated with *in vivo* studies. The list of resources in **TABLE 9-4** is aimed to provide students with a good starting list of Internet resources available to research questions related to medicine, clinical trials, gene therapy, and xenotransplantation.

Summary

Humans have strived to improve their health for centuries. There is documented evidence of surgical techniques and the development of herbal medicines that occurred more than 2000 years ago. Shen Nung, Sushruta Samahita, Hippocrates, and Galen are among the earliest medical pioneers. In the 16th century Andreas Vesalius changed the history of human anatomy by studying dissected humans, which laid the groundwork for further investigation into blood, medical research, and microbiology.

Women played a part in the history of medicine and health care as caretakers and nurses. It was not until the middle of the 19th century, though, that women were permitted to join men in what were considered the elite, literate, and scholarly ranks of medical doctors. Elizabeth Blackwell and Emily Jennings Stowe broke down barriers for women physicians during the 19th century when they became the first licensed physicians in the United States and Canada, respectively, and as adamant supporters of women's rights. Both specialized in midwifery and diseases of women and children.

The discovery of penicillin and insulin early in the 20th century was monumental in driving clinical research forward. Today's routine therapies and breakthroughs have come after the culmination of discoveries and practices of many medical trailblazers.

This is the golden age of medical research. Many preclinical and clinical trials are active throughout the world. Gene therapy and xenotransplantation are on the frontier of modern medicine.

Gene therapy is an experimental treatment that involves the introduction of genes into a person's cells to replace or compensate for defective genes in a person's body that are responsible for a disease or medical problem. One of the biggest challenges of gene therapy is the delivering of genes to the correct cells or tissues, and viruses act as gene-delivery vehicles. Retroviruses, adenoviruses, adeno-associated viruses, herpes simplex virus, and poxviruses such as vaccinia or MVA are the most popular viral vectors used in gene therapy trials. Each type of viral vector has both advantages and limitations.

Side effects related to gene therapy include immune responses toward the viral vector and the interruption of vital cellular genes, which can result in cancer. The field of gene therapy continues to evolve, and regulations and policies to police experiments and protocols are constantly being designed and updated to ensure the safety of these treatments.

Despite rare, highly publicized events, gene therapy trials continue to hold promise for treating genetic disorders caused by single-gene defects such as Huntington's chorea, Duchenne muscular dystrophy, polycystic kidney disease, familial hypercholesterolemia, sickle-cell anemia, hemophilia A and B, phenylketonuria, and cystic fibrosis. Gene therapy is also being used to treat other health problems such as cancer, diabetes, high blood pressure, and heart disease. The hope is that ultimately gene therapy will provide a cure for diseases in which the good gene must be working throughout one's life, such as with hemophilia. In other instances, gene therapy may be used to restore eyesight (including color blindness) or hearing, or needed to repair a wound or grow new blood vessels, in which case the later treatment would be a temporary solution.

Xenotransplantation—the procedure that involves the use of live cells, tissues, and organs from a nonhuman animal source—is a field being driven forward due to the shortage of organ donors and the continued advances toward the development of immunosuppressive drugs used to prevent cross-species immune-mediated rejection. This technology would not exist without the early work of organ transplant pioneers like Alexis Carrel, Sir Peter Medwar, Sir Frank Burnet, Keith Reemtsma, Thomas Starzl, and James Hardy.

The use of pigs organs rather than nonhuman primate organs for human transplants is attractive for several reasons:

- Pig organ size and physiology are similar to humans.
- There have been many advances in pig genetics and breeding, and genetic modifica-

tions of pigs have resulted in a more "humanized" pig.
- Pigs can be raised in special pathogen-free environments at minimal cost.

Hyperacute rejection (HAR) and acute vascular rejection (AVR) present major immunological hurdles for xenotransplantation. The genomes of pigs have been modified to prevent these types of rejection. Despite this, chronic rejection still threatens xenotransplantation and normal human-to-human organ transplantation. This type of rejection requires continued advancements in the development of more immune suppressive drugs. In addition to immune system rejection, xenozoonosis poses a problem for organ recipients.

Diseases transmitted via xenotransplantation may put the human community at risk. Viruses are a major concern because infections caused by other microorganisms are usually treatable by antibiotics, whereas those caused by viruses are not. Pathogens known to cause disease in pigs need to be eliminated from donor herds.

Pigs harbor PERVs in their genomes. PERVs may be reactivated during immunosuppressed conditions. Herpes virus infections are one of the most common complications of immune-suppressed patients. Several herpes viruses have been isolated from pigs. These include PLHV-I, II, and III, and PCMV.

Other porcine viruses of concern are HEV-like swine viruses and circoviruses. Xenotransplantation of animal organs to humans could allow microbes present in the donor organs to cross the species barrier. In doing so, nonpathogenic microbes in a natural host (pigs) may cause disease in a human recipient. Efforts are being made to breed pathogen-free or pathogen-reduced pigs. Methods are being developed to screen and monitor animal sources and recipients for known viruses. New and better diagnostic tests are being developed as new viruses are discovered. Guidelines and policies concerning xenotransplantation safety with specific attention to individual and societal risks and benefits have been prepared and are updated continuously. Safety will dictate the future directions of xenotransplantation.

CASE STUDY 1: GENE THERAPY

Sally Goodwin is a premed student volunteering in a cancer ward at a local hospital. She's met many patients and was particularly inspired by 16-year-old Sharon Smith, who suffered from a rare cancer. Sharon confided to Sally that she was researching her cancer via the Internet and had stumbled across a Web page pertaining to gene therapy clinical trials. Sally was very interested in what Sharon had found because she had been learning about gene therapy in her virology course. Soon, Sally and Sharon began discussing this topic regularly during Sally's visits. Sharon approached Sally with the following questions. How might you answer them if you were Sally?

1. Is gene therapy a permanent cure for any condition?
2. Can gene therapy be used for disorders that arise from mutations in more than one gene?
3. What types of viruses are used in gene therapy? What are their advantages and limitations?
4. Will gene therapy patients mount an immune response against the virus vector used in treatments?
5. Are there toxic effects related to gene therapy treatments?
6. Are there examples of patients who had major complications related to gene therapy and how was the problem resolved?
7. What are some of the ethical considerations for using gene therapy?
8. Gene therapy is expensive. Who has access to these therapies and who pays for their use?

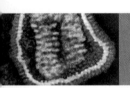

CASE STUDY 2: ONCOLYTIC VIRUSES

At the age of 76, U.S. Senator Ted Kennedy was diagnosed with a cancerous brain tumor in 2008. After experiencing seizures, Kennedy was hospitalized and put through a battery of tests, including a biopsy that revealed a **malignant glioma**. The cancer comes from glial cells. Glial cells function like glue, providing support and protection for neurons. Gliomas respond poorly to chemotherapy and radiation therapy. Chemotherapeutic drugs do not efficiently cross the blood–brain barrier to destroy the cancerous glial cells. Patients with glioma have a short survival rate. Brian gliomas usually start in the brain and do not metastasize (spread to other organs in the body). Kennedy underwent treatments at the Washington Brain and Spine Institute. He survived 15 months. Alternative therapies are needed because brain gliomas are virtually untreatable.

A study analyzing 790 of the deadliest type of brain gliomas, known as glioblastomas for genetic aberrations, was published in the *New England Journal of Medicine* (December, 2010). They found that the NF-κB inhibitor-α (*NFKBIA*) gene is often deleted and the epidermal growth factor receptor (*EGFR*) gene is amplified in the chromosomes of glioblastomas. *NFKBIA* is a tumor-suppressor gene and the *EGFR* gene codes for a receptor that binds epidermal growth factor (EGF). EGF drives many cellular responses, including changes in gene expression, anti-apoptosis, and cellular growth and proliferation. The *NFKIA* gene product suppresses the cellular signaling activities of NF-κB and EGFR pathways (refer to Chapter 10, Section 10.2 about the involvement of cellular genes and cancer).

Reference

Bredel, M., et al., 2011. "*NFKBIA* deletion in glioblastomas." *N Engl J Med* 364:627–637.

1. Could the deletion of *NFKIA* and amplification of the *EGFR* genes explain why glioblastomas are difficult to treat? Explain your answer.
2. Theoretically, could gene therapy be used to target these molecular defects? Why or why not? What types of gene "delivery" vehicles could be used to replace abnormal genes involved in cell cycle function?

Oncolytic virus therapy is being explored as treatment for glioblastomas. Oncolytic viruses are able to infect and lyse cancer cells.

3. Perform a literature search using PubMED or ScienceDirect and create a list of oncolytic viruses and what type of cells they infect.
4. Why are Herpes Simplex 1 viruses offering promise in treating gliomas? (For example, it is a neurotropic virus?) List some other features of this virus that make it an attractive candidate for virotherapy.
5. What methods are used to deliver an oncolytic virus to a glioblastoma?

Additional references to help you get started:

Friedman, G. K., et al. 2009. "Herpes simplex virus oncolytic therapy for pediatric malignancies." *Mol Ther* 17(7):1125–1135.

Kyritsis, A. P., et al. 2009. "Viruses, gene therapy and stem cells for the treatment of human glioma." *Cancer Gene Ther* 16:741–752.

Ramirez, M., et al. 2010. "Oncolytic virotherapy for neuroblastoma." *Discovery Med* 10(54): 387–393.

Resources

Primary Literature and Reviews

Allers, K., et al. 2010. "Evidence for the cure of HIV infection by CCR5Δ32/ Δ32 stem cell." *Blood.* DOI 10.1182/blood-2010-09-309591.

Andreansky, S. S., et al. 1996. "The application of genetically engineered herpes simplex virus to the treatment of experimental tumors." *PNAS USA* 93:11313–11318.

Bredel, M., et al. 2011. "NFKBIA deletion in glioblastomas." *N Engl J Med* 364:627–637.

Buning, H., Braun-Falco, M., and Hallek, M. 2004. "Progress in the use of adeno—associated viral vectors for gene therapy." *Cells Tissues Organs* 177(3):139–150.

Caplan, N. J. 2004. "Gene therapy and prospects. Down-regulating gene expression: The impact of RNA interference." *Gene Ther* 11:1241–1248.

Chain, E., et al. 1940. "Penicillin as a chemotherapeutic agent." *Lancet* ii:226–228.

Comroe, J. H. Jr. 1998. "Retrospective redux: Who was Alexis who?" *Respir Care* 43(2):131–40.

Costa, C., Zhao, L., and Burton, W. V. 2002. "Transgenic pigs designed to express human CD59 and H-transferase to avoid humoral xenograft rejection xenotransplantation." *Xenotransplantation* 9:45–57.

Denner, J. 2003. "Porcine endogenous retroviruses (PERVs) and xenotransplantation: Screening for transmission in several clinical trials and in experimental models using non-human primates." *Ann Transplant* 8(3):39–48.

Di Domenico, M., et al. 2010. "Towards gene therapy for deafness." *J Cell Physiol* Dec. 29 (Epub ahead of print).

Fleming, A. 1929. "On the antibacterial action of cultures of a *Penicillium,* with special reference to their use in the isolation of *B. influenzae.*" *Br J Exper Pathol* 10:226–236.

Friedman, G. K., et al. 2009. "Herpes simplex virus oncolytic therapy for pediatric malignancies." *Mol Ther* 17(7): 1125–1135.

Gallin, J. I., ed. 2002. *Principles and Practice of Clinical Research,* New York: Academic Press.

Gridley, D. S., and Slater, J. M. 2004. "Combining gene therapy and radiation against cancer." *Curr Gene Ther* 4(3):231–148.

Harrington, R. K., and Pandha, H. 2004. "Recent developments and current status of gene therapy using viral vectors in the United Kingdom." *BMJ* 329(7470):839–842.

Hutter, G., et al. 2009. "Long-term control of HIV by CCR5 Δ32/Δ32 stem-cell transplantation." *N Engl J Med* 360(7):692–698.

Karlas, A., Kurth, R., and Denner, J. 2004. "Inhibition of porcine endogenous retroviruses by RNA interference: Increasing the safety of xenotransplantation." *Virology* 325:18–23.

Kyritsis, A. P., et al. 2009. "Viruses, gene therapy and stem cells for the treatment of human glioma." *Cancer Gene Ther* 16:741–752.

Le Tissier, P., Stoye, J. P., Takeuchi, Y., and Weiss, R. A. 1997. "Two sets of human-tropic pig retrovirus." *Nature* 389:681–682.

Meng, X. J., et al. 1998. "Genetic and experimental evidence for cross-species infection by the swine hepatitis E virus." *J Virol* 72:9714–9721.

Mudur, G. 2004. "Kidney trade arrest exposes loopholes in India's transplant laws." *BMJ* 328:246.

Mueller, N. J., and Fishman, J. A. 2004. "Herpesvirus infections in xenotransplantation: Pathogenesis and approaches." *Xenotransplantation* 11:486–490.

Ogata, K., and Platt, J. L. 2004. "Cardiac Xenotransplantation: Future and Limitations." *Cardiology* 101:144–155.

Pang, J. J., et al. 2010. "Achromatopsia as a potential candidate for gene therapy." *Adv Exp Med Biol* 664:639–646.

Phelps, C. J., et al. 2003. "Production of α1,3-galactosyltransferase-deficient pigs." *Science* 299:411–414.

Platt, J. L., ed. 2001. *Xenotransplantation,* Washington DC: ASM Press.

Ramirez, M., et al., 2010. "Oncolytic virotherapy for neuroblastoma." *Discovery Med* 10(54):387–393.

Roberts, J. P. 2004. "Gene therapy's fall and rise (again)." *Scientist* 18:22–24.

Ruediger, H., and Madsen, J. C. 2004. "Feasibility of xenotransplantation." *Surg Clin North Am* 84:289–307.

Schneider, M. K. J., and Seebach, J. D. 2010. "Xenotransplantation literature update: June–October 2010." *Xenotransplantation* 17:481–488.

Sournia, J.-C. 1992. *The Illustrated History of Medicine.* London: Harold Starke.

Stieger, K., and Lorenz, B. 2010. "Gene therapy for vision loss—Recent developments." *Discovery Med* 10(54): 425–433.

Teranishi, K., et al. 2002. "Depletion of anti-Gal antibodies in baboons by intravenous therapy with bovine serum albumin conjugated to Gal oligosaccharides." *Transplantation* 73(1):129–139.

Tischer, I., et al. 1986. "Studies on epidemiological and pathogenicity of porcine circovirus." *Arch Virol* 91:271–276.

Warren, J., ed. 2004. "Xenotransplant News." *Xenotransplantation* 11:387–389.

Wells, D. J. 2004. "Gene therapy progress and prospects: Electroporation and other physical methods." *Gene Ther* 11:1363–1369.

Wolfe, N. D., et al. 2004. "Naturally acquired simian retrovirus infections in central African hunters." *Lancet* 363:932–937.

Yong-Guang, Y., and Sykes, M. 2007. "Xenotransplantation: Current status and a perspective on the future." *Nat Rev Immunol* 7:519–531.

Yu, J.-Y., DeRuiter, S. L., and Turner, D. L. 2002. "RNA interference by expression of short-interfering RNAs and hairpin RNAs in mammalian cells." *PNAS USA* 99(9):6047–6052.

Popular Press

Anand, G. 2006. *The Cure: How a Father Raised $100 Million and Bucked the Medical Establishment in a Quest to Save His Children*, New York: Harper.

Cherry, M. J. 2005. *Kidney for Sale by Owner: Human Organs, Transplantation, and the Market*, Washington D.C.: Georgetown University Press.

Fenster, J. 2003. *Mavericks, Miracles, and Medicine: The Pioneers Who Risked Their Lives to Bring Medicine into the Modern Age*, New York: Carrol and Graf.

"Gene Therapy. Special Report." 1997, *Sci Amer*, 276(6): 95–123.

Kennedy, M. 2004. *A Brief History of Disease, Science, and Medicine, Mission Viejo*, CA: Asklepiad Press.

Kimball, A. M. 2006. *Risky Trade: Infectious Disease in the Era of Global Trade*, United Kingdom: Ashgate Publishing.

Lanza, R. P., Cooper, D. K. C., and Chick, W. L. July 1997. "Xenotransplantation" *Sci Amer* 276(7):54–59.

Newton, J. 1987. *Uncommon Friends: Life with Thomas Edison, Henry Ford, Harvey Firestone, Alexis Carrel & Charles Lindbergh*. New York: Harcourt.

Skloot, R. 2010. *The Immortal Life of Henrietta Lacks*. New York: Crown.

Wells, H. G. 1994. *The Island of Dr. Moreau*, Reprint Edition. New York: Bantam Classics.

Williamson, J. 1951. *Dragon's Island*. New York: Simon and Schuster. (Fiction)

Zivin, J. Z. 2000. "Understanding clinical trials." *Sci Amer* 282(4):69–75.

Video Productions

Bad Blood. 2010. Necessary Films.

Extraordinary Measures. 2010. Sony Pictures.

I am Legend. 2007. Directed by Francis Lawrence.

History's Mysteries—Body Snatchers. 2006. A&E Home Video.

The American Experience: Partners of the Heart. 2003. PBS.

Dangerous Prescription. 2003. Frontline. PBS.

Mavericks, Miracles and Medicine. 2003. History Channel.

The Land of Moreau. 2002. Directed by Richard Stanley and John Frankenheimer.

Mutter Museum: Strange but True Medical Mysteries. 2002. Discovery Channel.

Organ Farm. 2001. PBS.

Conquering an Invisible World. 1999. Films for Humanities.

History of Gene Therapy. 1999. Films for the Humanities.

Another War: Disease and Political Strife. 1997. Films for Humanitites.

Human Gene Therapy and the Future of Modern Medicine. 1993. Films for the Humanities.

eLearning

go.jblearning.com/shors2

The site features eLearning, an online review area that provides quizzes and other tools to help you study for your class. You can also follow useful links for in-depth information, or just find out the latest virology and microbiology news.

CHAPTER 10

Viruses and Cancer

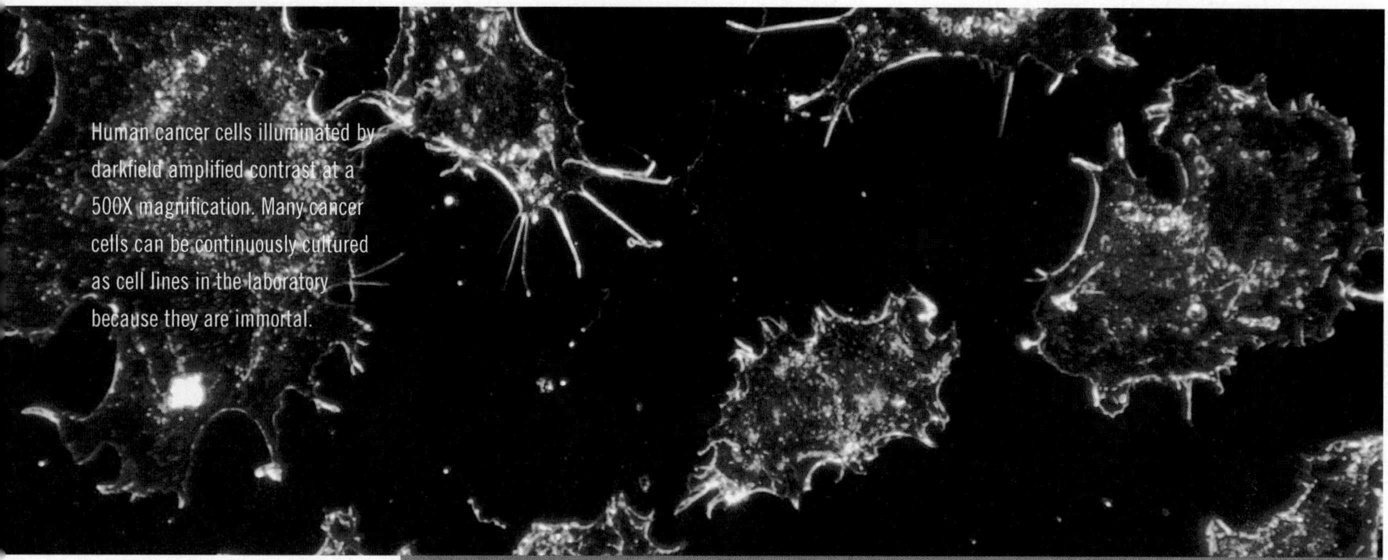

Human cancer cells illuminated by darkfield amplified contrast at a 500X magnification. Many cancer cells can be continuously cultured as cell lines in the laboratory because they are immortal.

"Growth for the sake of growth is the ideology of the cancer cell."

Edward Abbey (1927–1989), radical environmentalist

OUTLINE

CASE STUDY: MYSTERIOUS HEPATITIS SYMPTOMS

Tom Kent, a college junior studying biology, traveled home for a family reunion during spring break. On the day of the reunion, he noticed that his grandfather, John, looked very tired. Tom remembered that John had a heart bypass operation in 1990. Tom coaxed his grandfather to schedule a complete physical examination with his primary care physician as soon as possible.

At his appointment the following week John told his doctor that he didn't have much of an appetite, had some abdominal pain, and occasionally felt nauseous. He was worried that it could be his heart again and discussed the operation he underwent back in 1990. He also mentioned that he had been feeling like this for quite some time but blamed it on having a bad "ticker."

Upon examination, the doctor found that John's heart appeared normal. To be thorough, though, he decided to set up a treadmill test and carefully examined John's medical records. He noticed that John required a blood transfusion during his 1990 heart operation. He asked Tom if he ever followed up with any blood tests after the heart operation, to which John replied no. The doctor examined his abdomen and found that it was tender.

A week later John's doctor called with some test results. He told John that his liver enzymes were elevated and given his discomfort, a liver biopsy should be done.

One in three people will develop cancer at some time in their lifetimes. There are over 200 different types of cancer that affect all of the different body tissues. **Cancer** is the result of the uncontrolled growth of a single cell that eventually forms a clone of tumor cells that have the potential to **metastasize** to other sites of the body. Each of us is composed of approximately 10^{14} cells, all of which have the potential to become cancerous, but only one cell in one-third of the population turns cancerous. The chance of an individual cell becoming cancerous, therefore, is very small—3×10^{-14}. We now know that for the change of a healthy cell into a cancerous cell to occur, a series of changes slowly releases the cell from the multiple checks and balances that control its normal growth.

Over the years, researchers have identified many factors, including viruses, that may increase the risk of developing certain types of cancers. Viruses can cause genetic changes in cells that make them more likely to become cancerous. The link between viruses and cancer was one of the crucial discoveries in cancer research.

10.1 History of Cancer Viruses and Tumors

Cancer has afflicted humans throughout history. The earliest evidence of cancer is found in fossilized bone tumors of human mummies in ancient Egypt. The origin of the word cancer was credited to the Greek physician Hippocrates (460–370 B.C.), who used the terms *carcinos* and *carcinoma* to describe nonulcer-forming and ulcer-forming tumors.

When asked what causes cancer, the average individual responds with answers such as **carcinogens** such as tobacco, asbestos, or radiation. This is indeed true for about 80% of all cancers; however, **about 20% of human cancers are associated with viruses**. In fact, during the 17th and 18th centuries many believed cancer was a contagious disease. The first cancer hospital in Reims, France, was moved from the city because of the fear of the spread of cancer throughout the city. Then, in 1775, British surgeon Percival Pott (1714–1788) reported his observation that chimney sweeps often developed scrotal cancer. Chimney sweeps were children and men who climbed inside the flues of chimneys and brushed them clean. Children were often placed in this occupation because of their smaller size, and in the case of boys their scrotal skin was in prolonged contact with chimney soot, which consisted of carcinogens such as coal, tar, pitch, creosote, and other oils. Many developed scrotal and testicle cancer (**FIGURE 10-1**), and the correlation between occupation, environment, and disease was made.

One of the most important discoveries in cancer biology was the demonstration that injecting "filterable agents" or viruses into chickens could produce tumors. In 1908, the Danish veterinarians Wilhelm Ellerman and Olaf Bang demonstrated that a filterable agent caused a leukemia common in chickens. At the time, though, leukemia was not recognized as a type of cancer and their discovery was largely ignored. In 1911, Dr. Peyton

FIGURE 10-1 Chimney sweeps were common between the 18th and 19th centuries. Surgeon Percival Pott was the first to describe scrotal cancer in chimney sweeps in England who were constantly exposed to soot. The photograph shows chimney sweeps working in Sembach, Germany, in 1965.

(Francis) Rous, a pathologist at the Rockefeller Institute in New York who later became one of the most influential cancer researchers, reported that a filterable agent caused a slow-growing solid tumor or **sarcoma** in healthy Plymouth Rock hens. He injected healthy chickens with a cell-free, bacteria-free liquid filtrate of sarcoma tissue and the chickens developed sarcomas at the site of injection. This demonstrated that a transmissible agent could cause tumors. The agent was a **retrovirus**, later named **Rous sarcoma virus** (**RSV**).

Rous's discovery came at a time when the idea that cancer was a contagious disease was waning. The connections between certain work environments and the onset of cancer had been clearly drawn. In 1966, 55 years after his initial findings, Rous was awarded a Nobel Prize in Physiology or Medicine for his discovery of tumor-inducing viruses.

Rous sarcoma virus has a simple viral genome that encodes only three genes (***gag***, ***pol***, and ***env***) that are needed for replication and formation of virus particles. Evidence from several laboratories in the 1970s demonstrated that RSV had a fourth gene, *src,* that was not required for viral production but was required to transform cells, a process termed oncogenic transformation. Viral genes that could cause cells to become transformed or cancerous became known as viral tumor-causing genes or **oncogenes**.

During the mid-1970s, Drs. J. Michael Bishop and Harold E. Varmus investigated whether the RSV *src* cancer-causing gene could be found in the DNA of normal cells. They conducted a set of experiments that today would be considered daunting because gene manipulation technology had not yet been fully developed. Bishop, Varmus, and their colleagues at the University of California–San Francisco prepared a ^{32}P radioactively labeled single-stranded complementary DNA (cDNA) probe of the *src* gene. The *src* probe was allowed to hybridize with denatured genomic DNA from chickens and other bird species (see Figure 10-3a).

They found *src* sequences in the genomes of normal chicken cells and the normal cells of other birds such as turkeys, ducks, and quail, and published their results in a 1976 *Nature* article. Their work continued and the *src* gene was also found in normal cells of mammals (including humans) and in fish. Their work demonstrated that oncogenes were cellular genes that were hijacked by viruses from cells. Later it was determined that the functions of these cellular genes were found to be cytokines, cytokine receptors, protein kinases, G-proteins, transcription factors, and other nuclear proteins that regulate cell growth. Both Bishop and Varmus were awarded a Nobel Prize in Physiology or Medicine in 1989.

10.2 Cancer Today

Prostate cancer is the leading cancer among men in the United States, followed by lung and colon cancer, regardless of the man's race or ethnicity. Breast cancer is the leading cancer among women in the United States, followed by lung and colon cancer, regardless of the woman's race or ethnicity. **TABLE 10-1** lists data from the American Cancer Society Surveillance Research Team regarding the estimated new cases of cancers and estimated deaths by gender in the United States in 2010. At least six viruses are thought to contribute to 20% of cancers. They are:

- Hepatitis B virus (HBV)
- Hepatitis C virus (HCV)
- Human papillomavirus (HPV)
- Epstein-Barr virus (EBV)

TABLE 10-1 Estimates of the Leading Body Sites of New U.S. Cancer Cases and Deaths in 2010

	Estimated New Cases*		
Male		**Female**	
Prostate	240,890 (29%)	Breast	230,480 (30%)
Lung and bronchus	115,060 (14%)	Lung and bronchus	106,070 (14%)
Colon and rectum	71,850 (9%)	Colon and rectum	69,360 (9%)
Urinary bladder	52,020 (6%)	Uterine corpus	46,470 (6%)
Melanoma of the skin	40,010 (5%)	Thyroid	36,550 (5%)
Non-Hodgkin's lymphoma	36,060 (4%)	Non-Hodgkin's lymphoma	30,300 (4%)
Kidney and renal pelvis	37,120 (5%)	Melanoma of the skin	30,220 (4%)
Oral cavity and pharynx	27,710 (3%)	Kidney and renal pelvis	23,800 (3%)
Leukemia	25,320 (3%)	Ovary	21,990 (3%)
Pancreas	22,050 (3%)	Pancreas	21,980 (3%)
All sites	822,300 (100%)	All Sites	774,370 (100%)
*Excludes basal and squamous cell skin cancers and *in situ* carcinoma (except urinary bladder).			
	Estimated Deaths		
Male		**Female**	
Lung and bronchus	85,600 (28%)	Lung and bronchus	71,340 (26%)
Prostate	33,720 (11%)	Breast	39,520 (15%)
Colon and rectum	25,250 (8%)	Colon and rectum	24,130 (9%)
Pancreas	19,360 (6%)	Pancreas	18,300 (7%)
Liver and intrahepatic bile duct	13,260 (4%)	Ovary	15,460 (6%)
Leukemia	12,740 (4%)	Non-Hodgkin's lymphoma	9,570 (4%)
Esophagus	11,910 (4%)	Leukemia	9,040 (3%)
Non-Hodgkin's lymphoma	9,750 (3%)	Uterine corpus	8,120 (3%)
Urinary bladder	10,670 (4%)	Liver and intrahepatic bile duct	6,330 (2%)
Kidney and renal pelvis	8,270 (3%)	Brain and other nervous system	5,670 (2%)
All sites	300,430 (100%)	All sites	271,520 (100%)

Source: Courtesy of the American Cancer Society. *Cancer Facts and Figures 2011*. Atlanta: American Cancer Society, Inc.

- Kaposi's sarcoma virus
- Human T-lymphotropic virus types 1 and 2 (HTLV-1 and -2)

Eighty percent of viral-associated cancers are **cervical cancer** (caused by HPV) and **liver cancer** (caused by HBV and HCV).

Important Cancer Biology Definitions

The following terms are used frequently in the field of cancer biology.

Oncogene—a gene that has the potential to convert a normal cell to a cancerous or transformed cell.

Viral oncogene (*v-onc*)—a viral gene responsible for the oncogenicity of the virus. The *v-onc* genes present in RNA tumor-inducing retroviruses are altered cellular genes that are acquired via recombination with the host genome. The *v-onc* genes from DNA tumor viruses are not cellular counterparts and the gene product usually functions in viral replication.

Proto-oncogene—cellular genes that promote normal growth and division of cells. Their expression is tightly regulated.

Cellular oncogene (*c-onc*)—mutated form of proto-oncogenes that cause tumors.

Tumor suppressor genes—genes that suppress or inhibit the conversion of a normal cell into a cancer cell. These genes cause cancer when they are turned off in a cell.

Cell transformation—the change in the morphological, biochemical, or growth properties of a cell.

Cancer—a term for diseases in which abnormal cells divide without control. Cancer cells can invade nearby tissues and can spread through the bloodstream and lymphatic system to other parts of the body.

Metastasis—when a cell or clump of cells separates from a tumor and spreads to another location. Metastasis is one of the main reasons why some tumors are difficult to control.

Characteristics of Cancer Cells

Much of what we now know about the genes involved in the development of cancer is attributed to research on RNA and DNA tumor viruses. Three assays have been developed to determine whether or not viruses or chemicals can transform cells in culture. They are:

- **Focus-forming assays:** Transformed cells lose contact inhibition and form densely packed cells that pile on top of each other (called foci) instead of forming a monolayer of cells that do not grow on top of each other. Normal cells form a monolayer of cells that do not grow on top of each other and do not form foci (see Chapter 5).
- **Soft agarose assays:** Transformed cells will be able to divide and form free colonies when suspended in a methyl cellulose or agarose medium, a semisolid agar gel. Normal cells do not proliferate in soft agarose.
- **Reduced serum requirement:** Certain transformed cells may grow in medium containing reduced serum or growth factors; for example, normal NIH 3T3 cells will not grow in medium containing less than 5% serum. If Simian virus-40 (SV-40) is used to transform NIH 3T3 cells, the abnormal cells grow in medium containing 0.5% serum.

The properties of cancer or **transformed** cells in culture are different from normal cells. Some transformed cells actually form tumors when injected into animals. Some of the characteristics seen in transformed cells in culture may also be seen in tumor cells growing *in vivo* in experimental animals or patients. Some of these phenotypic changes—characteristic of transformed cells in culture (*in vitro*)—are:

- Cells undergo **genetic changes** such as an increase in the number and size of nuclei, resulting in instability. Transformed cells may become **polyploidy** and have **elevated levels of telomerase** that maintain telomere length.
- Cells become **immortalized** (cells can be passaged indefinitely) or have an unlimited life expectancy in culture.
- Cells undergo **metabolic changes** such as **dividing and growing rapidly**.
- Cells display a **lack of contact inhibition**. Normal cells stop replicating when they come in contact with another cell, whereas abnormal cells continue to divide and pile up on top of each other into a focus. The foci originate from the same cell.
- Cells display **anchorage-independent growth/loss of need for adhesion**. When freshly isolated normal cells are suspended in a liquid medium and they come in contact with a suitable solid surface, such as the bottom of a culture dish, they will attach, spread, and proliferate as a monolayer on the surface. In contrast, cells derived from tumors or transformed cells grow without the need to attach to a surface.
- Transformed cells can grow independently without the addition of serum or growth factors such as cytokines.

Refresher Cell Cycle and Cancer

The cell cycle contains checkpoints in the cycle of events in a normal eukaryotic cell from one cell division to the next. It consists of interphase, mitosis, and cell division. Checkpoints can block the progression through the phases of the life cycle if certain conditions are not met. Cyclins and cyclin-dependent kinases regulate the cell cycle. Cancer cells have mutations in the genes that control the cell cycle, causing uncontrolled growth and cell division.

Telomeres are DNA sequences found at the ends of eukaryotic chromosomes. During the normal aging process, telomeres become shorter with each cycle of cell division. The telomerase enzyme synthesizes telomeric DNA on chromosome ends. A short telomere signals the cell to stop dividing. Cancer cells contain increased levels of telomerase activity. They maintain telomere length so that the cell never receives a signal to stop dividing (**FIGURE RB 10-1**).

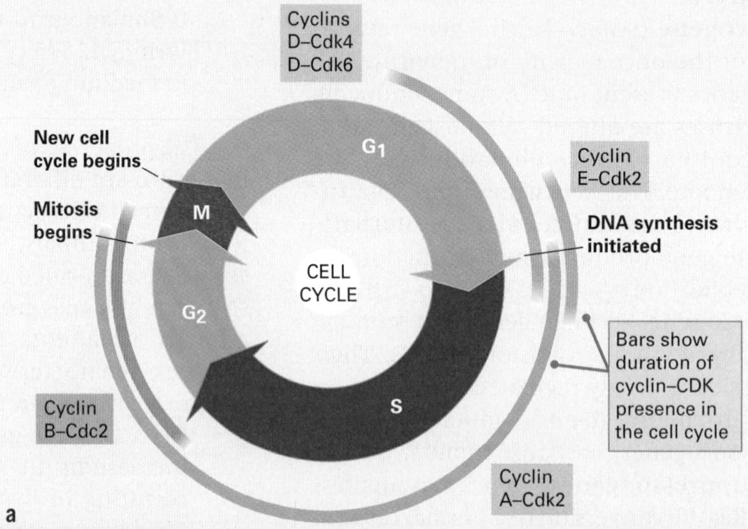

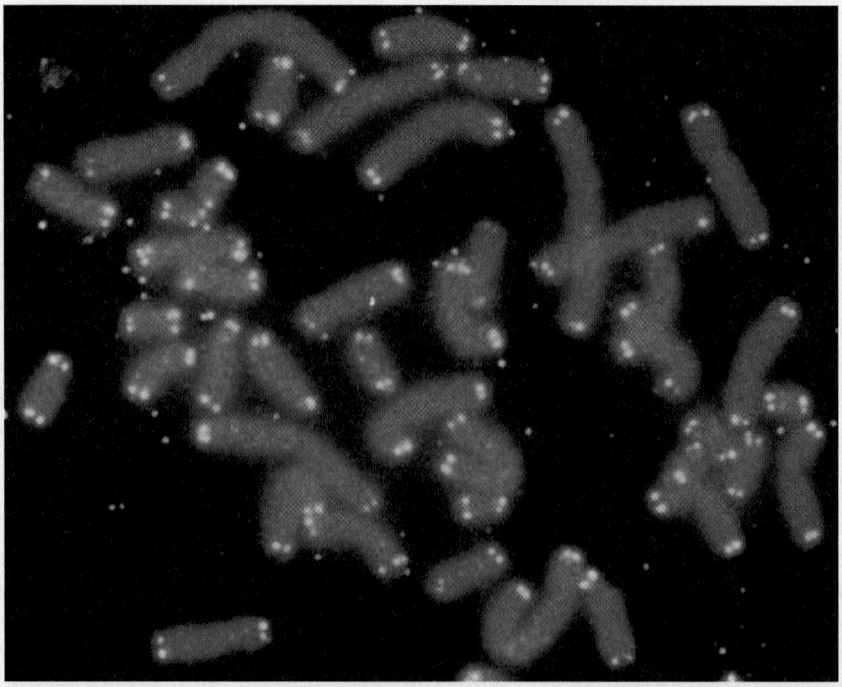

FIGURE RB 10-1 (a) A schematic of the eukaryotic cell cycle. **(b)** The red chromosomes hybridized with fluorescent yellow probes, which in turn hybridized to the telomere DNA.

- Cells undergo a **loss of cell cycle control**. Transformed cells fail to stop at cell cycle checkpoints in the cell cycle. The growth of normal cells is restricted by these checkpoints.
- Cells may have altered cell surfaces. There will be **changes in membrane structure and function**; for example, transformed cells display **tumor-associated carbohydrate antigens**.
- When transformed cells are injected into experimental immunosuppressed animals (i.e., nude mice), they may produce tumors. Tumor formation defines malignant transformation (**FIGURE 10-2**).

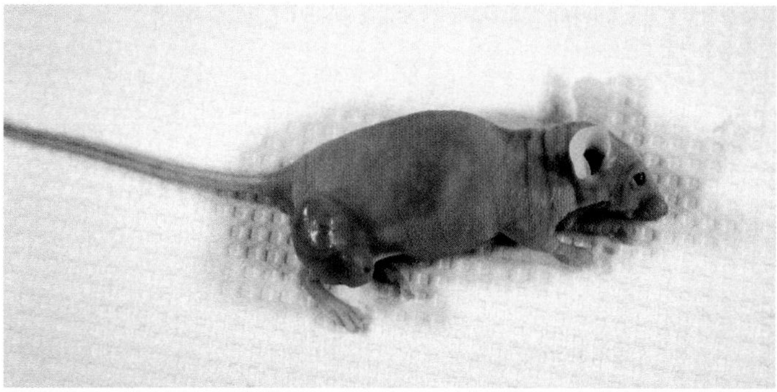

FIGURE 10-2 This nude mouse was injected with transformed mouse NIH 3T3 cells that contained a viral oncogene, and a tumor formed at the site of injection. Nude mice are born without a thymus and therefore cannot generate mature T lymphocytes. This immunodeficiency makes nude mice useful for investigating cancers. The mouse does not reject the cancer cells and actual tumors will form quickly in the mice.

The following alterations are observed *in vivo* in cancer cells:

- There is an increase in oncogene protein expression because the cellular oncogenes have undergone chromosomal translocations, amplifications, or mutations.
- The cells lose tumor suppressor gene function due to a deletion or mutation in the tumor suppressor gene.
- DNA methylation patterns are altered.
- Cells produce an increased or unregulated level of growth factors.
- Cells divide uncontrollably.
- Cells have increased levels of enzymes involved in nucleic acid synthesis and lytic enzymes (i.e., proteases, collagenases, and glycosidases).
- Telomerase activity is reactivated.
- Cells can avoid the host immunosurveillance response.

Cancer Is a Multistep Process

Tumors arise from a series of events that lead to greater loss of regulation of cell division. For cancer to occur, the following must happen:

- The cell must bypass apoptosis (programmed cell death).
- The cell circumvents the need for growth signals from other cells.
- The cell escapes immunosurveillance.
- The cell commands its own blood supply.
- The cell may possibly metastasize.
- Mutations in cellular tumor suppressor genes may also be required for full malignancy to occur.

The association of viruses and human cancers is not causal and is often correlative. Viruses are thought to be cocarcinogens in the development of human tumors.

10.3 Molecular Mechanisms of Virally Induced Tumor Formation by RNA Tumor Viruses (Retroviruses)

A retrovirus may cause cancer in three ways: it may carry an oncogene (*v-onc*) into a cell, it may activate a cellular proto-oncogene, or it may inactivate a tumor suppressor gene. Analysis of the retrovirus genome is central to understanding the molecular mechanism of virally induced tumor formation.

The Retrovirus Genome

Retroviruses differ from DNA tumor viruses in that their genome consists of RNA in contrast to DNA. The RNA must be copied into DNA prior to integration into the host chromosome. The retrovirus *pol* gene that encodes reverse transcriptase performs the conversion of RNA into DNA. David Baltimore and Howard Temin discovered reverse transcriptase independently. Temin was working on his Ph.D. under Renato Dulbecco at the California Institute of Technology. In 1963, he showed that RSV could not infect or replicate in cells in the presence of drugs that inhibited RNA transcription, such as actinomycin D and α amanitin. This was not unexpected because it was known that the genome of RSV consisted of RNA.

To his surprise, though, Temin discovered that DNA replication inhibitors such as 5-bromodeoxyuridine and cytosine arabinoside inhibited the replication of RSV. For this reason, Temin proposed

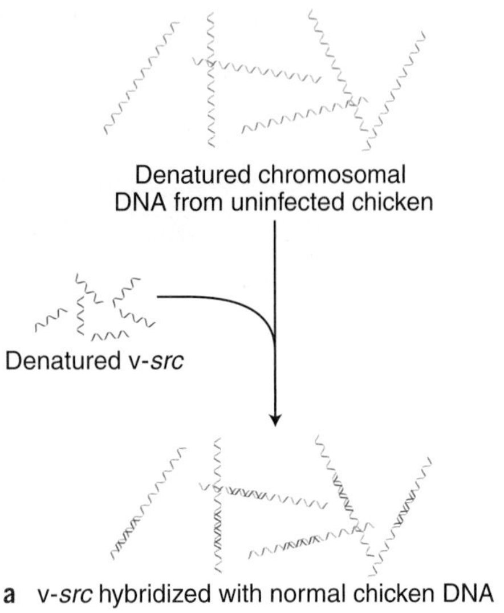

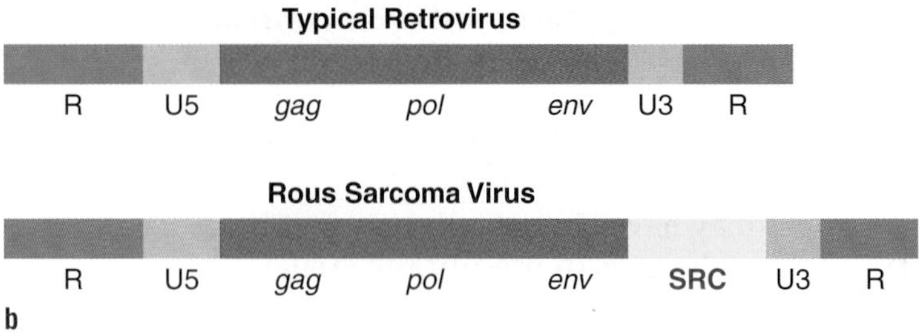

FIGURE 10-3 **(a)** Bishop-Varmus experiment: oncogenes hybridize with normal cellular genes. **(b)** The typical structure of a retrovirus genome. Retroviruses contain at least three genes—*gag, pol,* and *env*—in addition to repetitive and unique sequences at the ends of the genome. Retroviruses that can cause tumor formation may contain an additional *v-onc* such as *v-src.* C is a packaging signal. The packaging signal is needed so that newly synthesized RNA molecules can be incorporated into new virus particles. Adapted from University of South Carolina, School of Medicine. "Virology: Oncogenic Viruses." *Microbiology and Immunology.* University of South Carolina, 2008. http://pathmicro.med.sc.edu/lecture/retro.htm.

that the retrovirus reverse-transcribed its RNA into DNA, an idea that contradicted the contemporary dogma that DNA is transcribed into RNA. Once the RNA of the retrovirus is copied into DNA, the retroviral DNA is integrated into the host cell chromosome. In 1970, Temin and Baltimore isolated and described reverse transcriptase, an accomplishment for which they shared the Nobel Prize in 1975.

Retrovirus particles contain two copies of the RNA genome. The genomes of most retroviruses consist of three or four genes that are located between unique and repetitive sequences located within its RNA genome. The genome lengths of replication-competent retroviruses (a retrovirus with a functional *pol* gene) range in size from 7 to 12 kb in length. The main genes are labeled *gag, pol,* and *env* (**FIGURE 10-3**). *Gag* is an acronym for group antigens. The *gag* gene encodes matrix and core proteins of the virus, and these proteins protect the viral RNA genome. The *pol* gene encodes a protein that has reverse transcriptase, RNase H, helicase, and integrase functions. This enzyme carries out the reverse transcription process of converting the viral RNA template into DNA. The *env* gene encodes a protein that is embedded within a lipid bilayer that surrounds the nucleoprotein core particle of the virus. These proteins are the "spikes" that bind to host cell receptors. **FIGURE 10-4** shows a typical retrovirus.

The retroviral genome is flanked at each end by repetitive sequences (designated **R**) and unique sequences (designated **U**). These sequences will enable the DNA copy of the genome to be inserted into the DNA of the host. The genome can act as an enhancer as well, causing the host nucleus to transcribe the DNA copies of the retroviral genome at a rapid rate.

Each viral genome contains a unique (U5) sequence located at the 5′ end of the genomic RNA. The opposite end of the viral RNA (3′ end) contains a U3 region. The U5 region contains a primer binding sequence. The U3 region contains promoter-enhancer sequences that control viral RNA transcription from the 5′ long terminal repeat (LTR).

Some retroviruses contain an additional gene, *v-onc.* It is not an essential viral gene because it is unrelated to the strategy of viral replication. The *v-onc* gene was hijacked from the genome of its host during the course of evolution. It encodes a protein that is capable of inducing cellular transformation (cancer). Replication of the retroviral genome will be further explained in Chapter 16 (HIV).

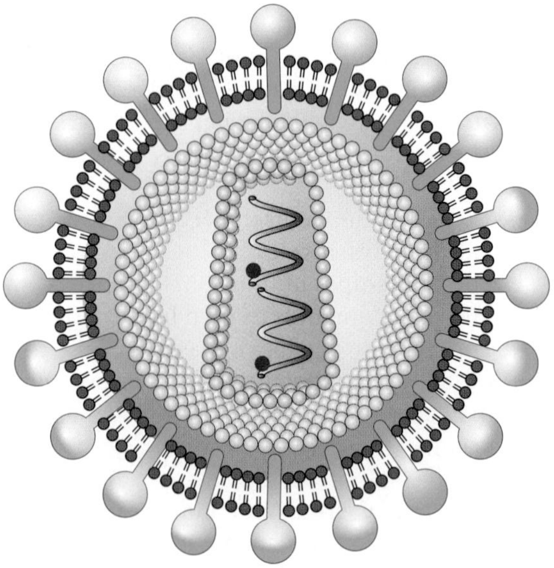

FIGURE 10-4 The structure of a retrovirus particle.

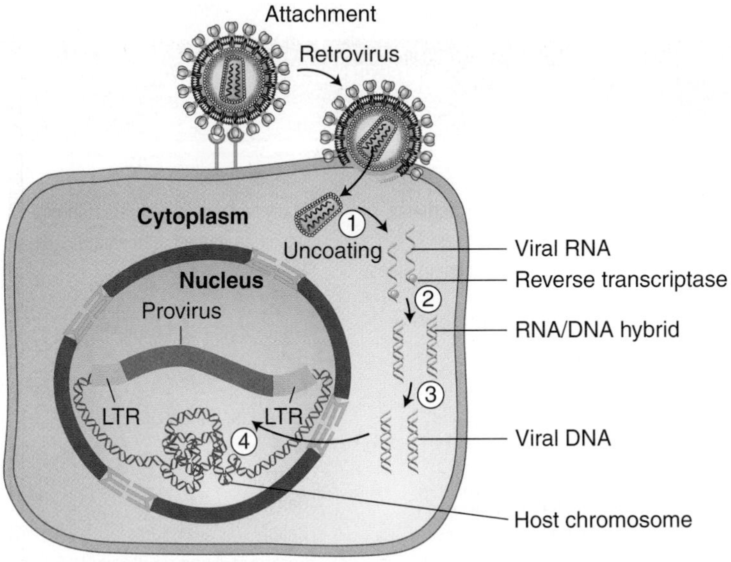

FIGURE 10-5 Integration of proviral DNA into the host. The integrase function of the *pol* gene product is involved in this event. Adapted from an illustration by the Institut National de Recherche Pédagogique, "The Replication of HIV in the Target Cell," *Immunologie—AIDS,* http://www.inrp.fr/Acces/biotic/immuno/html/repvih.htm.

Molecular Mechanisms of How Retroviruses Can Cause Cancer: Proviral Integration

There are several mechanisms by which retroviruses may play a role in converting normal host cells into tumor cells. When a provirus integrates into the host's genome (**FIGURE 10-5**), it is random with respect to host cell DNA. The provirus may integrate within or near a *c-onc,* though, thereby altering the expression of it. The gene products of *c-onc*'s normally function in signal transduction and cell cycle regulation (**TABLES 10-2** and **10-3** and **FIGURES 10-6** and **10-7**). Therefore, a provirus integration event may disrupt the normal restraints on normal growth and division of cells, causing tumors to form.

The integration of proviral DNA into the host genome can result in the following:

- Insertional **activation** of the expression of a *c-onc* when viral promoters or enhancer elements cause abnormal expression of an unaltered *c-onc,* resulting in uncontrolled growth of normal cells. The *c-onc* may undergo hyperactive expression or it may be expressed at inappropriate times during the cell cycle, causing growth at inappropriate times.

TABLE 10-2 Examples of Cellular Genes Involved in Cancer (Genes Involved in the Stimulation of Cell Division)

Class	Gene	Function	Cancer Type (result of one mutant allele of the gene)
Growth factors or receptors for growth factors	*PDGF*	Platelet-derived growth factor	Brain and breast
	RET (*re*arranged during *t*ransfection)	Growth factor receptor	Thyroid, brain, and breast
	erb-B	Receptor for epidermal growth factor	Glioblastoma (a brain cancer) and breast
	erb-B2	Receptor for growth factor	Breast, ovarian, and salivary
Cytoplasmic signaling proteins	*c-src*	Tyrosine kinase	Breast
	ras (*r*at *s*arcoma)	GTPase	Breast, colon, lung, and pancreatic
	bcl-1	Stimulates cell cycle	Breast, head, and neck
Nuclear DNA binding proteins	*c-jun*	Transcription factor	Breast
	c-fos	Transcription factor	Breast
	c-myc	Transcription factor	Burkitt's lymphoma, leukemia, stomach, lung, breast
	c-rel	Transcription factor	Lymphoma
	c-ets-1	Transcription factor	Lymphoma
	c-hox11	Transcription factor	Acute T cell leukemia
	c-lyl-1	Transcription factor	Acute T cell leukemia
	c-lyt-10	Transcription factor (also called NFκB2)	B-cell lymphoma

TABLE 10-3 Examples of Cellular Genes Involved in Cancer: Tumor Suppressor Genes (Genes Involved in the Suppression of Cell Division)

Class	Gene	Function	Cancer Types
Nuclear proteins	*P53*	Halts cell cycle in G1, induces apoptosis	Many cancers (e.g., brain tumors, leukemia, breast, sarcomas)
	p16	Inhibits cyclin D-dependent kinase activity	Melanoma, pancreatic
	WT1	Transcription factor	Wilm's tumor of the kidney
	BRCA1	DNA repair	Breast and ovarian
	BRCA2	DNA repair	Breast
	RB1	Master brake on cell cycle	Retinoblastoma, bone, bladder, lung, breast
	MTS1	Brake on cell cycle	Many cancers
	MSH2	DNA mismatch repair	Colon
	MLH1	DNA mismatch repair	Colon
	MEN1	Intrastrand DNA crosslink repair	Parathyroid and pituitaryadenomas, islet cell tumors, carcinoid
Cytoplasmic proteins	*APC* (adenomatous—polyposis coli)	Interact with cell adhesion proteins	Colon and stomach
	DPC4	Regulation of TGF-β/BMP signal transduction	Colon and pancreatic
	NF-1 (neurofibromatosis type 1)	Inhibits cell growth and division in nerve cells, inactivates RAS	Brain, nerve, leukemia
	NF-2 (neurofibromatosis type 2)	Links cell membrane to actin skeleton	Brain and nerve
Transmembrane proteins	*DCC* (deleted in colon carcinoma)	Receptor? Mediates netrin-1 activity	Colon
	PTCH	Receptor for sonic hedgehog involved in early development	Basal cell skin carcinoma
Location unknown	*VHL*	Downregulation of cyclin D1, component of ubiquitin ligase complex	Kidney

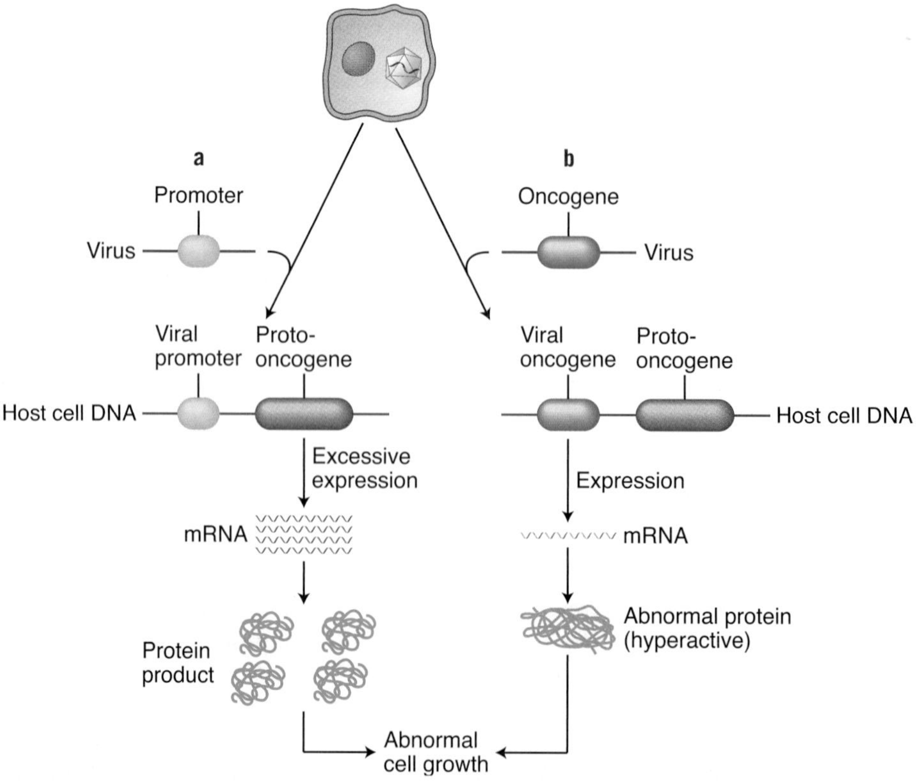

FIGURE 10-6 Host cells may be transformed when a viral promoter or oncogene is inserted into the chromosome of its host. **(a)** Insertion of a viral promoter upstream of a cellular proto-oncogene causes inappropriate expression of the normal proto-oncogene. **(b)** Insertion of a viral oncogene may promote the growth and division of cells, causing tumors to form. The *v-onc* may be a mutated form of a *c-onc* that could interfere with cellular signaling involved in growth, resulting in abnormal growth. Adapted from Marks, D. B., Marks, A. D., and Smith, C. M. *Basic Medical Biochemistry: A Clinical Approach.* Lippincott Williams & Wilkins, 1996.

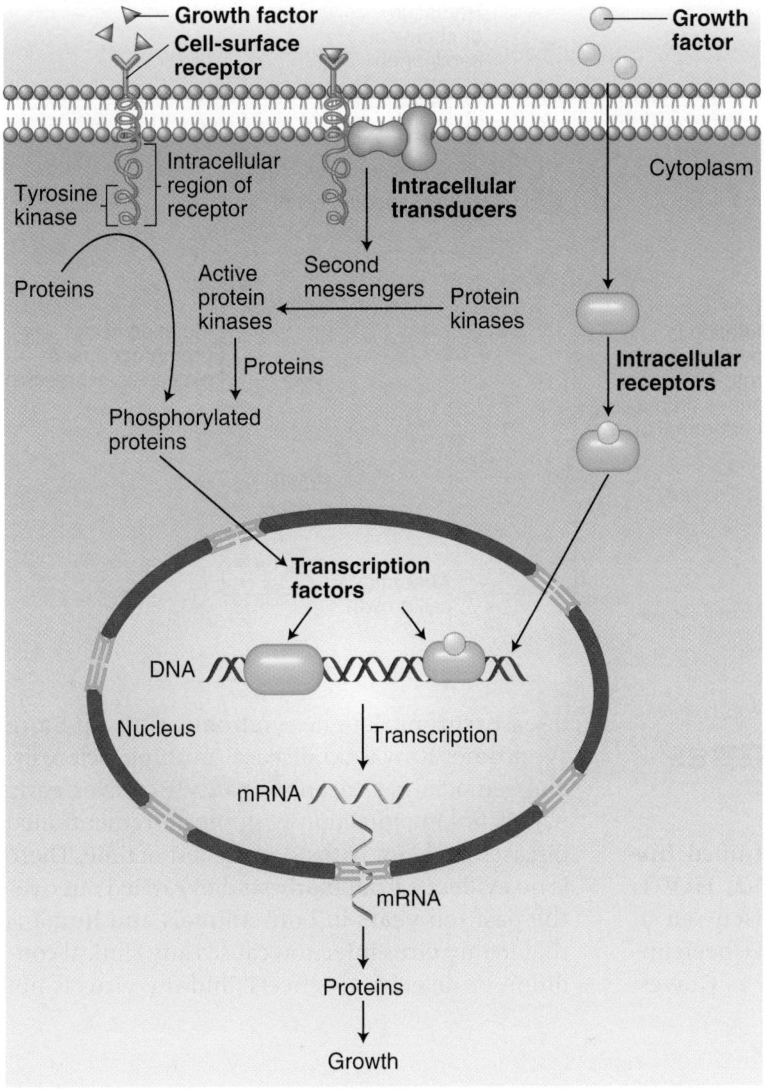

FIGURE 10-7 Cellular oncogenes (*c-onc*) promote the normal growth and division of cells via various functions (e.g., as growth factors, growth receptors, and transcription factors). Their expression is tightly regulated. Mutated forms of these genes can cause disruption of normal cellular growth and division. Adapted from Marks, D. B., Marks, A. D., and Smith, C. M. *Basic Medical Biochemistry: A Clinical Approach.* Lippincott Williams & Wilkins, 1996.

- Insertional **inactivation of a cellular tumor suppressor gene**, resulting in uncontrolled cell division.
- The integration of proviral DNA that carries a *v-onc* into the host genome. The *v-onc* may become activated and transcribed, coding for the oncogene product (e.g., oncogenic activation of the normal cellular platelet-derived growth factor [PDGF] growth receptor). The *v-onc* may promote the normal growth and division of cells, causing tumors to form (**FIGURES 10-8** and **10-9**). The *v-onc* may be a mutated form of a *c-onc* (e.g., a truncated form) that could interfere with cellular signaling involved in growth. Examples of retroviral oncogenes are listed in Table 10-3.

More About the Retrovirus Life Cycle

The integrated provirus often remains quiescent as a silent, persistent infection. The DNA may be transcribed by the cell's RNA polymerase II into viral messenger RNA (mRNA) that is translated by host cell ribosomes. The viral proteins are packaged along with the viral RNA and the progeny viruses exit by budding from the host cell surface.

Human Endogenous Retroviruses (HERVs)

Approximately 8% of the human genome contains sequences with similarity to infectious retrovruses. These are sequences that contain the easily recognized *gag, pol, env,* and LTRs. These sequences have acquired many mutations over the course of evolutionary time so that, with few exceptions, they are now defective and incapable of producing any protein. HERVs have been proposed as etiological cofactors in chronic diseases such as cancer, autoimmune diseases, and neurological diseases (**VIRUS FILE 10-1**). Despite intense efforts by many groups, there has been little direct evidence discovered to support these claims.

FIGURE 10-8 If *c-onc*'s are damaged by carcinogens, mutations in the promoter or gene may lead to transformation of the cell. Adapted from Marks, D. B., Marks, A. D., and Smith, C. M. *Basic Medical Biochemistry: A Clinical Approach.* Lippincott Williams & Wilkins, 1996.

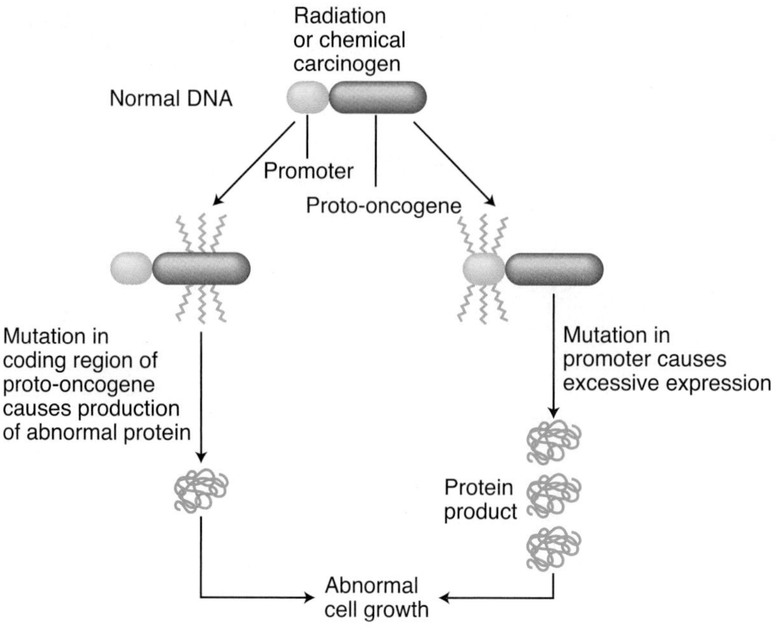

10.4 Human Retroviruses

Five human retroviruses have been identified: **human foamy virus, HTLV-1, HTLV-2, HIV-1,** and **HIV-2** (HIV-1 and -2 will be discussed in Chapter 16). Human foamy viruses have been implicated in a variety of illnesses, such as Graves' disease, chronic fatigue syndrome, Guillain Barré syndrome, Kawasaki disease, multiple sclerosis, and hemodialysis encephalopathy. However, early papers linking infection with the aforementioned diseases have not withstood the test of time. There is no evidence from many studies carried out over the past ten years in both animals and humans that foamy virus infection causes any clinical condition or deleterious effects, and the virus is not

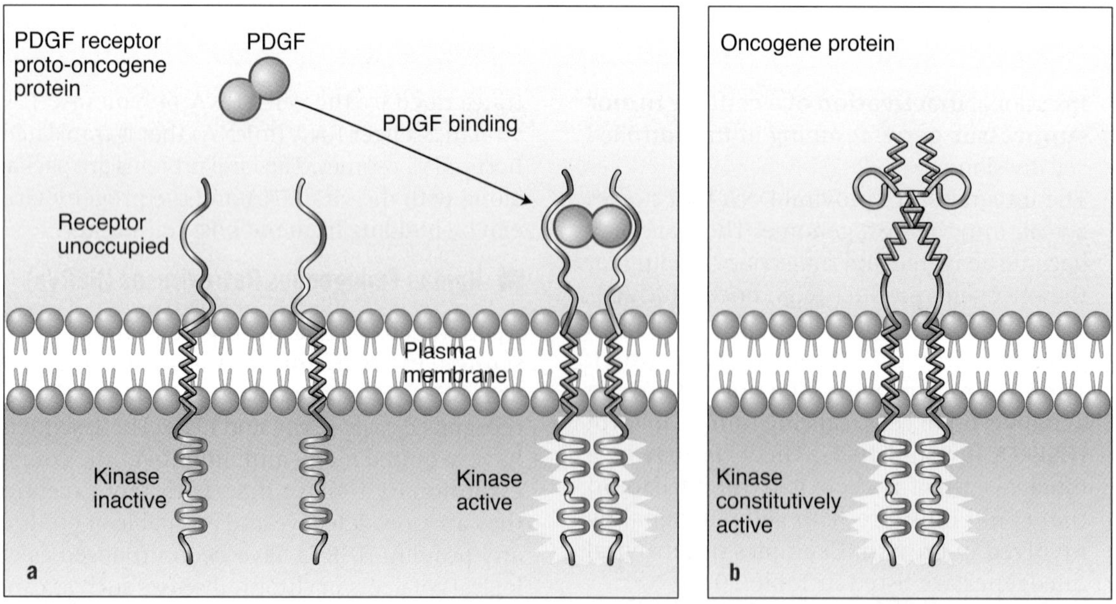

FIGURE 10-9 Mechanism for oncogenic activation of the normal PDGF growth receptor. **(a)** Normal PDGF growth receptor is activated in response to growth stimulation by PDGF. **(b)** A mutated form of the PDGF growth receptor is constitutively active, driving abnormal growth forward. Adapted from Cooper, G. M., and R. E. Hausman. *The Cell: A Molecular Approach,* Third Edition. 2003, Sinauer Associates.

VIRUS FILE 10-1 Endogenous Retroviruses Might Contribute to Some Cases of Schizophrenia

Schizophrenia is a chronic, relapsing psychotic disorder that primarily affects thought and behavior. The average age of onset is 20 to 25 years. The disease can be well controlled with antipsychotic drugs, which are sometimes called **neuroleptics**. Schizophrenia affects approximately 1% of the population. Diagnosis of the disease includes auditory hallucinations (usually voices that converse with or about the patient) and paranoid delusions (the belief that outside forces are conspiring against the patient). Other overt symptoms include the inability to pay attention, low motivation, bizarre thoughts, the loss of the sense of pleasure, disorganization, gradual depletion of thoughts and speech, and social withdrawal.

Schizophrenia has been accepted as a psychotic disorder caused by a combination of genetic predisposition and triggering factors from the environment. Twin and adoption studies indicate that when one member of a twin pair has the illness, the risk of schizophrenia in the other twin is 17% for fraternal twins and nearly 50% for identical twins. The risk of schizophrenia in the relatives of affected individuals is 2% in third-degree relatives, 2% to 6% in second-degree relatives, and 6% to 17% in first-degree relatives. No specific gene or combination of genes has been identified that confers the risk for schizophrenia.

A number of environmental factors seem to increase the risk for schizophrenia and in general they occur early in life; for example, severe malnutrition during the first trimester or maternal influenza during the second trimester of pregnancy double the relative risk of schizophrenia. Other studies show that the disease is more likely to strike individuals born in cities than on farms and to strike people born in winter (when influenza and other diseases are common) than other times of the year. Even without decisive evidence, some biologists, such as Paul Ewald, argue that microbial pathogens cause most common chronic diseases such as schizophrenia.

It has been hypothesized that retroviruses may be one of the infectious agents involved in the pathogenesis of schizophrenia. It is known that retroviruses such as HTLV and HIV can replicate within the central nervous system and cause neurological and psychiatric symptoms in some individuals.

A 2001 study by Hakan Karlsson and colleagues identified retroviral RNA, presumably of endogenous origin in the cerebrospinal fluids (CSFs), in 10 of 35 patients who suffered from recent-onset schizophrenia (mean age of 25 years). The majority of the patients (7 of 10) harbored sequences related to the HERV-W family of human endogenous retroviruses. Retroviral sequences were also detected in the CSFs of 1 of 20 individuals suffering from chronic schizophrenia. No retroviral sequences were detected in CSF samples collected from 22 patients suffering from noninflammatory neurological diseases nor from 30 individuals who did not suffer from any psychiatric disease.

In schizophrenia the frontal cortex, which is the most highly developed part of the brain, is slow and affects emotion and cognitive functioning. The frontal cortex develops last and is not complete until the mid-20s, when schizophrenia typically develops. During Karlsson's collaborative investigation, postmortem samples of brain tissue of the frontal cortex of five individuals who suffered from schizophrenia and six individuals who did not suffer from any psychotic disease were analyzed for the presence of retroviral RNA transcripts. Results showed that a low level of retroviral expression occurred in normal brains but an unexpectedly high level of expression occurred in schizophrenic brains.

Endogenous retroviruses are a part of the human genome, having inserted themselves possibly millions of years ago, but the part they play in human disease is just beginning to be understood. Replication of Karlsson's findings will motivate subsequent critical studies. Robert Yolken, a collaborator with Karlsson, believes that herpes or other viral infections during pregnancy

(continued)

may activate the dormant HERVs, disrupting brain development and increasing the risk for schizophrenia later in life. Herpes infections are common among people with schizophrenia. If his theory is correct, screening pregnant women for herpes and treating the ones at risk may prevent schizophrenia in some individuals.

References

Fox, D. "The insanity virus." *Discover*, June, 2010 issue.

Karlsson, H., et al. 2001. "Retroviral RNA identified in the cerebrospinal fluids and brains of individuals with schizophrenia." *PNAS USA* 98(8):4634–4639.

Nakamura, A., et al. 2003. "Human endogenous retroviruses with transcriptional potential in the brain." *J Hum Genet* 48(11):575–581.

Schretlen, D. J. et al. 2010. "Neuroanatomic and cognitive abnormalities related to herpes simplex virus type 1 in schizophrenia." *Schizophr Res* 118(1):224–231.

Zimmer, C. 2001. "Do chronic diseases have an infectious root?" *Science* 293:1974–1977.

transmitted to others. Humans infected by accidental or occupational exposure remain well. For this reason, human foamy viruses are being developed as potential gene therapy vectors.

The majority of the individuals infected with HTLV-1 are asymptomatic carriers. HTLV-1 is endemic in parts of Japan, South America, Africa, and the Caribbean. HTLV-1 causes two fatal diseases: Adult T-cell leukemia (ATL) and HTLV-1-associated myelopathy (HAM).

The lifetime chance of an infected individual developing ATL is about 2% to 5%. In the early 1980s HTLV-1 was isolated from patients suffering from T-cell leukemia and T-cell lymphoma by Bernard Poiesz and Robert Gallo's research laboratory in the United States and Y. Hinuma's research laboratory in Japan. Only a small number of HTLV-1 infected individuals develop leukemia, and generally it occurs decades after initial infection. The development of T-cell malignancy remains poorly understood. ATL therapy uses a combination of the antivirals zidovudine and interferon-α. The T lymphocytes of ATL patients express high levels of IL-2 receptors. Novel strategies involve treating patients with antibodies against the IL-2 receptor that are armed with toxins or radionuclides.

HTLV-1 infects primarily CD4+ T-lymphocyte. It is transmitted from mother to child either via the placenta during birth or through breastfeeding. It is transmitted from male to female during sexual intercourse, but there is no evidence of female-to-male transmission. Blood transfusions are an additional and efficient route of transmission. HTLV-2 infections have been associated with T-cell malignancies, but the epidemiology of HTLV-2 has been less well studied. In the United States donated blood is screened for the presence of HTLV-1 and -2.

10.5 Human DNA Tumor Viruses

Richard E. Shope discovered the first DNA tumor viruses during the early 1930s (rabbit fibroma in 1932 and rabbit papilloma isolated from cottontail rabbits in 1933). He minced tumor material from infected animals and inoculated it into domestic rabbits. It was shown that the growths were attributed to a filterable agent because the infectious agent passed through a Berkefeld "V" filter. The growths that formed often regressed.

DNA tumor viruses are a diverse group of viruses with different structures, genome organization, and replication strategies. Some of these viruses, such as papillomaviruses, Epstein-Barr virus, Kaposi's sarcoma-associated herpesvirus, and hepatitis B, induce tumors in the natural host. Others, such as adenovirus, polyomaviruses, and SV-40, induce tumors in experimental animals.

DNA tumor viruses differ from RNA tumor viruses in that the oncogenes of DNA tumor viruses are essential viral genes used in replication (mostly virally encoded nuclear proteins). The oncogenes of small DNA tumor viruses (such as adenoviruses and SV-40) do not have cellular counterparts or homologs. DNA tumor viruses target the Rb and p53 tumor suppressor gene products of the host. In other words, the viral-encoded gene products of DNA tumor viruses interact with proteins that have a negative regulatory role in cell proliferation

(**TABLE 10-4**). This causes an alteration of cell cycle progression. The larger DNA tumor viruses also encode cellular homologs to activate signal transduction pathways that enhance cell growth and division.

The oncogenic potential of DNA tumor viruses is low; for example, the adenoviruses, polyomaviruses and SV-40 frequency of transformation is less than 1 in 100,000 infected cells. Transformation only occurs in cells that undergo an **aborted life cycle**. During an aborted life cycle, only the early viral genes are expressed. Infectious particles are not produced and the cells are not killed during an aborted cycle. The integration of the viral genome into its host genome may enable the abortive life cycle and cellular transformation.

Epstein-Barr Virus (EBV)

EBV was the first human virus to be directly implicated in human tumors. EBV is a herpesvirus that is sometimes referred to as human herpesvirus 4 (HHV-4). All herpesviruses have two modes of infection: a lytic replication cycle, which generates infectious virus particles and destroys the host cell, and a latent replication cycle, in which the viral genome persists in its host cell but with dramatically restricted gene expression and without cell destruction (herpesviruses are discussed in Chapter 15).

EBV is named after Anthony Epstein and Yvonne Barr, the two scientists who first isolated the virus in 1964 from lymphoma samples collected by Dennis Burkitt. Dr. Burkitt, a British surgeon, was working in equatorial Africa in 1956. While there, he observed and described an unusual lymphoma that was very common in children in that region. It became known as Burkitt's lymphoma, a disease that is very similar to leukemia. It is an aggressive, malignant cancer and is characterized by a solid tumor composed predominately of aberrant B-cells (**FIGURE 10-10**). B-cells are integral to a normal healthy immune system (see Chapter 7 for a review of the immune system).

Ninety-five percent of the population of the United States between the ages of 35 and 40 are persistently infected with EBV. The infection, which is usually asymptomatic, causes a number of conditions. It usually hits teens and young adults by causing infectious mononucleosis. EBV causes 85% of mononucleosis cases. The virus is passed via the saliva of an infected person; it is often referred to as the "kissing disease." The virus can also be transmitted while sharing a glass, eating utensils, or a straw with an infected individual. Mononucleosis is not usually a life-threatening illness. Sometimes the spleen will be enlarged and could rupture. The incubation period is four to seven weeks. The initial signs of infection are:

- sore throat
- swollen glands in the neck
- fatigue
- lack of appetite
- headache
- white patches in the back of the throat

TABLE 10-4 DNA Tumor Viruses Target Cellular Tumor Suppressor Genes

Virus	Gene Product	Cellular Target
EBV	*EBNA-2*	*c-myc*
	EBNA-3A and *EBNA-3B*	Interferes with Notch signaling pathway
	EBNA-3C	Rb
	LMP1	CD40, bcl-2, A20
KSHV	*v-IL6*	Interferes with IL-6 pathway
	v-Bcl-2	c-Bcl-2 and p53
	v-cyclinD2	c-cyclin D, Rb
HBV	*HbX*	*c-src*, p53
HCV	*NS5*	c-bcl-6, p53, Ig heavy chain gene(V_H), β-catenin
Papillomavirus	*E7*	Rb
	E6	p53
	E5	Platelet-derived growth factor
Adenovirus	*E1A*	Rb
	E1B	p53
SV-40	Large T antigen	Rb and p53
Polyomavirus	Large T antigen	Rb
	Middle T antigen	*c-src* and P13K

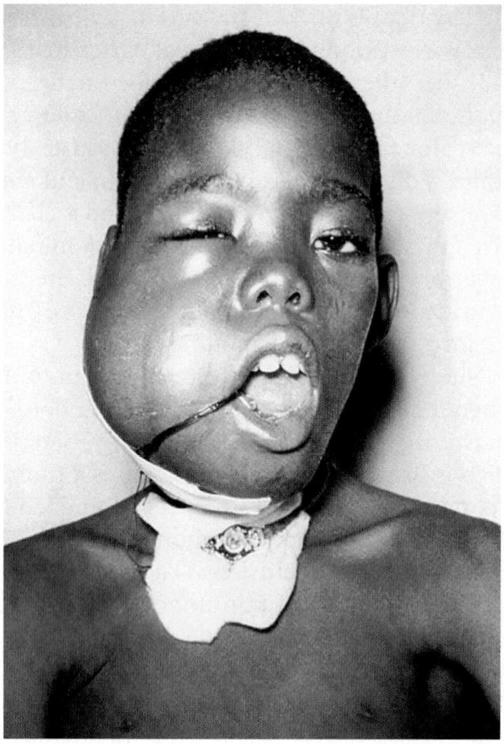

FIGURE 10-10 This Nigerian child has a large facial tumor due to malignant Burkitt's lymphoma. The tumor can grow to this size in a mere four to six weeks.

If 95% of the population in developing countries is infected with EBV, why doesn't everyone get Burkitt's lymphoma? Burkitt's lymphoma seems to happen most often in a person who first has a condition that weakens his/her immune system, such as chronic malaria or AIDS. Children in central Africa often suffer from the aforementioned infections. Then, when these children come down with Epstein-Barr virus, Burkitt's lymphoma occurs. EBV persistently infects B lymphocytes (resulting in a latent infection), the lymphocytes of the immune system that produce antibodies. Burkitt's lymphoma is a solid tumor of B lymphocytes. It affects the jaws and very rapidly spreads to the soft tissues and the parotid glands. When Burkitt's lymphoma tumors are analyzed, genetic aberrations such as chromosomal translocations involving chromosomes 8 and either 14, 22, or 2 occur. These translocations move the *c-myc* gene near an immunoglobin heavy chain or light chain, resulting in abnormal regulation of the *c-myc* gene and increased tumorigenicity of the cells. EBV DNA can be isolated from tumors, and viral-specific antigens can be detected in the tumors. Infection with EBV induces a lymphoma-like disease in new world primates. *In vitro* EBV immortalization of B-cells further supports the link. The mechanism by which the virus transforms cells is still uncertain. The transforming gene(s) have not been precisely identified. Possible candidates are listed in Table 10-4.

Kaposi's Sarcoma Virus

Kaposi's sarcoma was known as a rare cancer before the AIDS epidemic. It was first described in 1872 by the Hungarian dermatologist Moritz Kaposi. It is referred to today as **classic Kaposi's sarcoma** and occurs most often in elderly men, with a ratio of 10 to 15 males to 1 female. The average age of onset is between 50 and 70 years in men of Mediterranean, Middle Eastern, or Eastern European ethnic decent. Single or multiple lesions are usually located in the lower extremities, especially the ankle and soles of the feet (**FIGURE 10-11**).

In 1981, the rapid and disseminated form of Kaposi's sarcoma in young homosexual or bisexual men was first reported as part of a new epidemic now known as AIDS. Kaposi's sarcoma has reached epidemic proportions in parts of Africa and the Western world. It continues to pose a problem in AIDS patients receiving **highly active antiretroviral therapy** (**HAART**) and transplant patients who must undergo immunosuppressive treatment to prevent organ rejection. The lifetime risk for Kaposi's sarcoma in homosexual male AIDS patients during their lifetime is 50%.

Patrick Moore at the Columbia School of Public Health and Yuan Chang at Columbia College of Physicians and Surgeons identified Kaposi's sarcoma virus, also called human herpesvirus-8 (HHV-8), in 1994 as the infectious cause of Kaposi's sarcoma. DNA sequences of the viral genome were

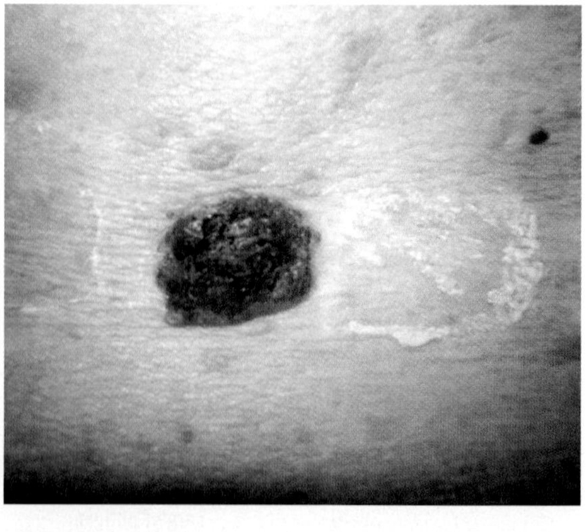

FIGURE 10-11 A Kaposi's lesion is visible in the center of the photograph. The lesion is typically red, purple, or brown and presents as a patch, plaque, or nodular skin.

detected via polymerase chain reaction (PCR) in biopsy material (Kaposi's lesions) from patients with classic, transplant-related, and AIDS-associated Kaposi's sarcoma, and was absent from skin tissue from patients who did not have Kaposi's sarcoma.

Less than 3% of infected cells in Kaposi's sarcoma lesions display evidence of lytic KSHV infection. Transcripts of the latent viral genes are found in most Kaposi's sarcoma tumor cells. The viral genome contains sequences that encode a D type cyclin, an interleukin-6 homolog, a *Bcl2* homolog, and a number of other genes potentially implicated in the growth deregulation that may be relevant to its proposed role as a transforming virus. Currently there are clinical trials involving antivirals purposely directed against Kaposi's sarcoma virus (e.g., ganciclovir and foscarnet) in progress. These antivirals target mainly lytic virus and more than likely will be ineffective against latent virus infection found in most Kaposi's sarcoma lesions. Alternatively, RNA interference and gene therapy have been proposed as a future management option of Kaposi's sarcoma.

Hepatitis B Virus (HBV)- and Hepatitis C Virus (HCV)-Related Hepatocellular (Liver) Cancer (HCC)

Of all the hepatitis viruses, only HBV and HCV cause chronic hepatitis, which can progress to cirrhosis and HCC. HBV and HCV are biologically very different viruses. HBV contains a double-stranded DNA (dsDNA) genome that replicates by reverse transcription of an RNA pregenome. HCV is an RNA virus that replicates in the cytoplasm of the cell. (The replication of HBV and HCV is discussed in Chapter 17.)

Both viruses mainly infect hepatocytes of the liver, although HCV also infects B lymphocytes. Despite their different life cycles and genes, they share common characteristics in the mechanisms of the chronic liver disease they cause and their association with cirrhosis, the need for a liver transplant, and HCC (**FIGURE 10-12**). HCC tumors in patients infected with HBV usually harbor integrated viral DNA. HCV contains an RNA genome that does not contain an obvious oncogene and does not integrate into host genomes. Its

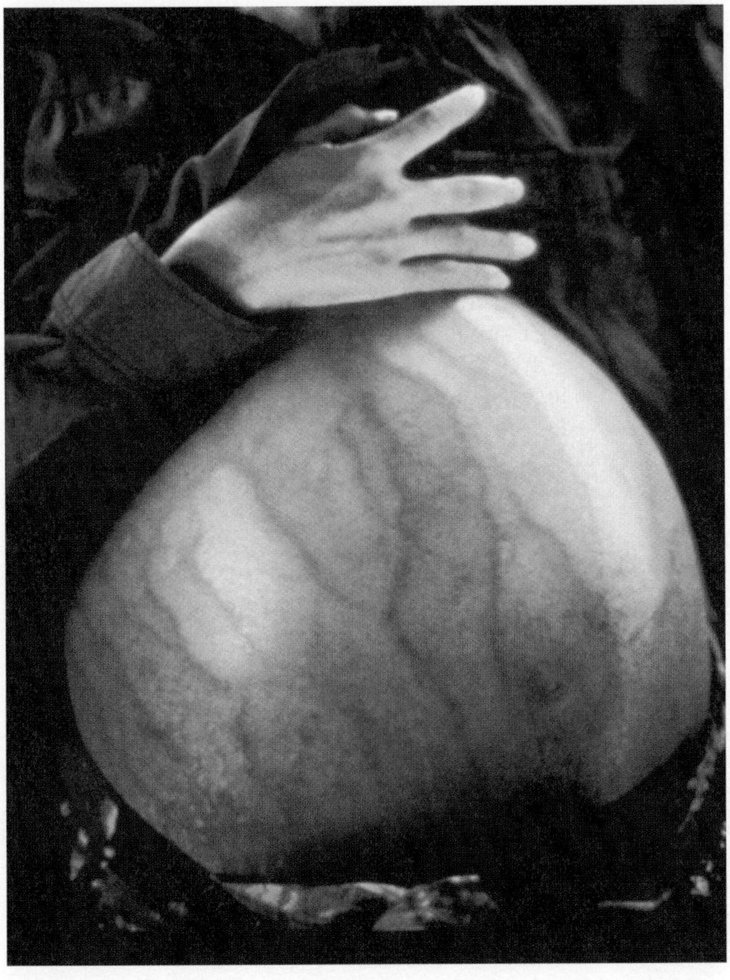

FIGURE 10-12 This Cambodian woman has a distended abdomen due to a hepatoma resulting from chronic hepatitis B infection. The incidence of HCC in patients with a hepatitis B infection is 12 to 300 times greater than in people not infected with HBV. The HBV DNA is incorporated into the hepatocytic DNA during the disease pathogenic process. Note how the tumor commands its own blood supply.

mechanism of oncogenesis is not clear. Recent studies suggest that HCV causes chromosomal instability. HCC tumors in HCV patients have a five- to tenfold increase in mutation frequency of several cellular genes, such as the immunoglobulin heavy chain, c-*bcl*-6, p53, and certain β-catenins.

The routes of transmission by HBV and HCV are similar. Transmission occurs when blood or body fluids of infected persons enter the body of a person who is not immune. Both are spread when sharing needles with an infected person or from an infected mother to the baby during birth. HBV can be transmitted sexually, too; however, there is no evidence of sexual transmission of HCV among monogamous couples. The rate of progression from chronic hepatitis through cirrhosis to cancer in hepatitis C is dramatically higher than that seen with the progression of HBV-induced cirrhosis to cancer. HCC is usually not detected until it is in an advanced stage. Hence, liver cancer is usually fatal within a year of diagnosis.

Liver cancer is the fifth most common cancer in the world. The hepatitis B vaccine is the first anticancer vaccine. Studies have shown that hepatitis B vaccination programs caused the number of hepatitis B carriers to be decreased substantially in some communities; for example, a 10-year study (1984–1994) in Taiwan found that the use of the hepatitis B vaccine reduced the carrier state in children from 9.8% in 1984 to less than 1.3% in 1994. It is anticipated that this significant decrease will be linked to a lower incidence of liver cancer in children. (The hepatitis B vaccine will be discussed in greater detail in Chapter 17.) There is no vaccine available for HCV. Donated blood is screened for the presence of HBV and HCV in both the United States and Canada. Most blood centers in southeast Asia screen donated blood for HBV, and even fewer countries are undertaking HCV screening.

Human Papillomavirus (HPV)

HPV infections are common among sexually active adults and adolescents. Genital HPV infection is one of the most prevalent sexually transmitted diseases of the world today. In an alarming study of college-age women from a state university in New Brunswick, New Jersey, 43% converted from HPV-negative to HPV-positive during their three-year period of study. High-risk types of HPVs cause cervical cancer. Cervical cancer is a major cause of death among women in developing countries and is the third most common cancer among women worldwide.

Papanicolaou (Pap) smear screening programs have reduced cervical cancer mortalities. Between 1947 and 1984 there was a 70% decline in the mortality rate in the United States due to cervical cancer, a reduction that is directly attributed to early detection and removal of the HPV-infected, premalignant tissue. Routine cervical cytological screening in the United States results in the treatment of at least 750,000 women each year for cellular abnormalities suspected to represent possible precancerous lesions.

Pinpointing the cause of cervical cancer has been of interest to epidemiologists and physicians for over 160 years. In 1842, Italian surgeon and amateur epidemiologist D. Rigoni-Stern reported that cancer was about five times higher in Catholic nuns than in other women because of an excess of breast cancer in nuns. He also reported four deaths of nuns from uterine cancer. At the time, cervical cancer was not distinguishable from uterine cancer.

Rigoni-Stern studied the death certificates of 74,184 women who died, and found that 1288 of the women were nuns. Ironically, this study was gradually embellished as a nun's tale with invented details. Subsequent authors began reporting that cervical cancer was "rare in nuns and common in prostitutes," suggesting a connection between intercourse and cervical cancer. The idea that cervical cancer was rare among nuns became dogma and most papers regarding the epidemiology of cervical cancer mentioned Rigoni-Stern's study. Subsequent studies followed, but careful review of the literature did not support the dogma that cervical cancer is rare in nuns.

In 1986, Ronald Ostrow and his colleagues reported in the journal *Science* the detection of papillomavirus DNA in human semen. Ostrow's report supported the debate that sexual transmission of HPV DNA could occur via semen. Over the past 20 years, there has been an explosion of research with focused interest on the specific molecular biology of HPVs associated with genital lesions.

Papillomaviruses Are Traditionally Described as Types

Over 120 papillomavirus types have been completely described. The most intensely studied host is humans, but papillomaviruses have been detected from most mammals and birds. Most papillomaviruses cause benign **papillomas** or **warts** in the skin (especially the hands, soles of the feet, and genitals) or mucous membranes (**FIGURE 10-13**). HPV infection occurs when the skin is damaged in some way, providing the virus with a means of entry.

Nearly 120 HPVs have been described based on the isolation of complete genomes. A database established from 30 years of sequencing information of thousands of papillomavirus isolates has

VIRUS FILE 10-2 Mouth and Throat Cancer and Oral Sex

Approximately 1 in 10,000 people develop oral cancers each year. In 90% of the cases, tobacco, poor diet, poor oral hygiene, and alcohol consumption attribute to the cause of these cancers. The International Agency for Research on Cancer conducted a case-control study that involved the recruitment of patients who had cancer of the oral cavity (tongue, lips, gum, palate, and other parts of the mouth) or the oropharynx (the part of the pharynx between the soft palate and the epiglottis), including the tonsils. The study was conducted in nine countries (Italy, Spain, Northern Ireland, Poland, India, Cuba, Canada, Australia, and Sudan) from 1996 to 1999. Cancer and control (i.e., cancer-free) patients were recruited at participating general and cancer hospitals. Control patients did not have a history of oral cancer and were matched at the same hospital locations as the cancer patients by age and gender. A total of 1670 cancer patients and 1732 healthy control subjects participated in the study.

Biopsy specimens and/or exfoliated cells from control subjects were subjected to PCR analysis using primers specific to 14 HPV high-risk types and 6 low-risk types. Plasma from patients was collected and tested for antibodies against HPV-16 E6 and E7 proteins in an ELISA assay. Each participant was administered a standardized questionnaire by a specially trained interviewer. The survey included questions regarding demographics, education, occupation, use of tobacco and alcohol, dietary habits, marital status, disease history, cancer history, oral cavity health, and sexual history (number of lifetime sexual partners, visits to prostitutes, frequency of oral sex).

HPV DNA was detected in 4% of oral cavity biopsy samples and 18% of oropharynx and tonsil biopsy samples. HPV-16 DNA (the strain most often associated with cervical cancer) was found in 95% of the biopsies that tested positive by PCR for HPV DNA. HPV DNA was detected less frequently in tobacco smokers and chewers and more frequently among cancer patients who reported more than one sexual partner or practiced oral sex. Antibodies to HPV-16 proteins were associated with risk for cancers of the oral cavity and the oropharynx. Cancer patients were three times more likely to have antibodies against HPV than healthy control subjects.

The results demonstrate the connection between HPV and oral cancer for which the disease cannot be blamed on decades of smoking and drinking because the patients were too young. New studies also suggest that HPV is now thought to be the main cause of throat cancer in people under the age of 50 years old.

HPV infections are common. About 30% of 25-year-old women in the United States are infected with HPV, and it is thought that about 10% of infections involve high-risk types of HPVs. Recent studies indicate that 50% of males are infected with HPV. Gardasil, which protects against some but not all strains of HPV has been approved for girls and women. Not only will it prevent many cervical and genital cancers, but possibly cancers of the oral cavity, oropharynx, and throat.

Reference

Giuliano, A. R., et al. 2011. "Incidence and clearance of genital human papillomavirus infection in men (HIM): A cohort study." *Lancet* 377(9769):932–940.

Herrero, R., et al. 2003. "Human papillomavirus and oral cancers: The International Agency for Research on Cancer Multicenter Study." *J Nat Cancer Inst* 95(23):1772–1783.

Hocking, J. S., et al. 2011. "Head and neck cancer in Australia between 1982 and 2005 show increasing incidence of potentially HPV-associated oropharyngeal cancers." *Br J Cancer* 104:886–891.

Partridge, J. M., et al. 2007. "Genital human papillomavirus infection in men: Incidence and risk factors in a cohort of university students." *J Infect Dis* 196(8):1128–1136.

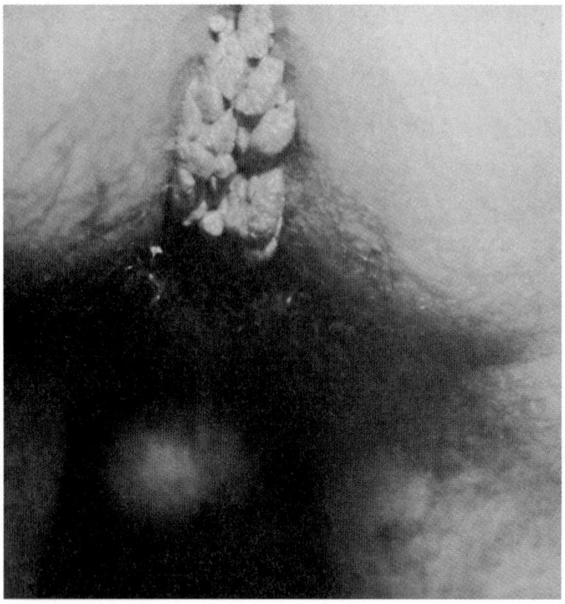

a

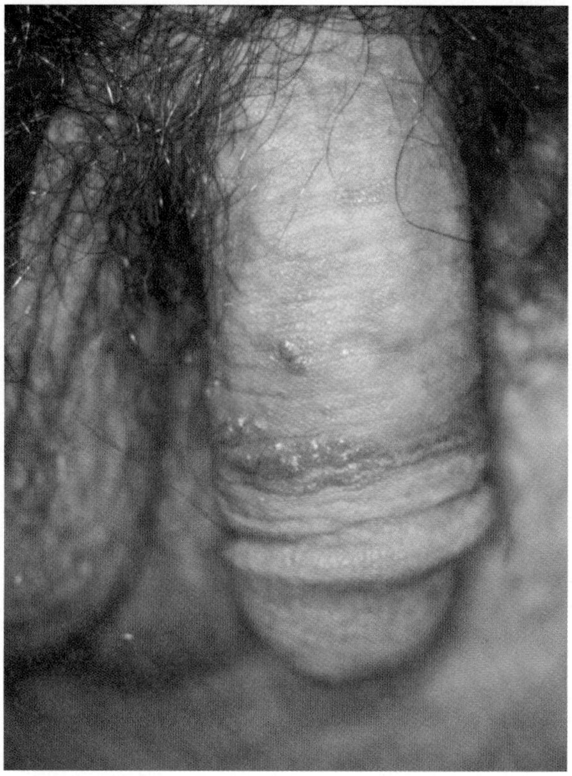

b

FIGURE 10-13 (a) Genital warts are a sexually transmitted disease caused by HPVs. This man has venereal warts in the anal region of the perineum. **(b)** This patient has soft, wart-like growths on the penis, shown 12 hours after podophyllin application.

been used to generate a classification system for papillomaviruses. The HPVs have been divided into low-, intermediate-, and high-risk types. High-risk types of HPV are likely to be responsible for a high proportion of carcinomas of the cervix, vulva, vagina, anus, and penis (e.g., types 16 and 18). There is a 4- to 20-year latent period between infection and development of cancer. "Low-risk" types are benign from an oncologist's point of view (e.g., types 6 and 11). The "intermediate-risk" type is used to distinguish HPVs that are frequently found in precancerous lesions but are less often represented in cancers (e.g., types 31, 33, 51, 52, and 83).

Papillomavirus Particle Structure and Genome

Papillomaviruses are small, nonenveloped, icosahedral-shaped dsDNA viruses (**FIGURE 10-14**). The particles are 52 to 55 nm in diameter. Virus particles contain dsDNA in a circular form that is approximately 8000 base pairs in length (**FIGURE 10-15**).

HPVs infect stratifying basal epithelial cells. They cannot affect the dermis because those cells are not metabolically active. Viral particles infect the host by entering the basal epithelial cells through a break in the skin. Genome replication, nucleocapsid formation, and virion maturation occur in the nucleus of the epithelial cells (**FIGURE 10-16**).

Cervical Cancer: Oncogenesis

HPV-16 (50%), HPV-18 (15%), HPV-45 (8%), and HPV-31 (5%) DNA is found in nearly all cervical cancer cells using PCR for detection. The remaining 22% of the samples contain other high-risk HPV types. **FIGURE 10-17** illustrates the differences in HPV-16 DNA in benign versus malignant tumors. In Figure 10-17a, the HPV genome is maintained as a circular extrachromosomal episome in benign tumors. RNAs transcribed from the episome contain a destabilizing sequence (59AUUUA39). In Figure 10-17b, the viral genome in cervical cancer cells is integrated into the host chromosomal DNA (purple). The viral genome is interrupted upstream of the *E6*/*E7* open reading frames. The mRNAs produced from this DNA do not contain the destabilizing sequence. The integrated *E6*/*E7* early genes are overexpressed in cancer cells. The E6 and E7 proteins inactivate the products of tumor suppressor gene products p53 and pRb. This in turn leads to disruption of the normal cell cycle and chromosomal abnormalities.

Even though most women with cervical cancer are over 45 years of age when diagnosed, HPV infection and disease pathogenesis are known to begin at the onset of sexual activity. Pap smear screening is central to treating cervical cancer.

VIRUS FILE 10-3 Dede Kaswara's Battle With Warts

In 2007, a Discovery Channel documentary entitled, *Tree Man, Search for a Cure,* aired on television. The program was about an Indonesian fisherman, Dede Kaswara, who was known as the "the tree man of Java" in a traveling freak show. Dede had cauliflower-like tumors all over his body. His hands and feet looked like branches of trees growing from the surface of his skin. Dr. Anthony Gaspari, an American dermatologist who saw the program was curious about the growths. He contacted the journalist who worked on the story and made arrangements to meet with Dede and biopsy the growths that covered his body (**FIGURE VF 10-1**). Dede was 39 years old. His quality of life was very affected by the growths. He could not bathe or take care of himself. His brother-in-law took care of him. His wife left him and his sister was taking care of his children. Dede was isolated and depressed.

FIGURE VF 10-1 Dede's warts in June, 2007, before treatment.

(continued)

VIRUS FILE 10-3 Dede Kaswara's Battle with Warts (continued)

After questioning Dede, Gaspari learned that the warts first started to appear on his knee after a cut on his knee at the age of 15. At the time he was a construction worker and fisherman. He had to stop working in his 20s. His affliction went untreated for nearly a decade. His warts were growing in size about 5 mm per month. Gaspari surgically removed three of the tree-like warts and drew blood samples from Dede, his parents, sister, and children. None of Dede's family members, neighbors, or friends suffered from a condition similar to his.

Gaspari returned to the United States and studied Dede's biopsied warts, along with other researchers interested in the case. They determined that Dede was suffering from a rare human papillomavirus type 2 (HPV-2) infection based on HPV typing by polymerase chain reaction and sequencing. He was also surprised to discover that Dede suffered from a severe immunodeficiency known as chronic CD4+ T-cell **lymphocytopenia**. Someone suffering from lymphocytopenia does not have enough lymphocytes to fight infections. Lymphocytes are a subset of white blood cells. His first thoughts were that Dede may have AIDS, but he had no risk factors for HIV infection and tested negative for HIV-1 and HIV-2. A team of doctors from the United States and Indonesia collaborated on Dede's case.

They determined that he suffered from a severe immunodeficiency that allowed an untreated HPV-2 infection to cause extreme disfigurement (giant cutaneous horns on his upper and lower extremities). His warty condition, called **generalized verrucosis**, is rare. Dede was treated with acitretin (unsuccessfully), interferon β, and cidofovir (which was too toxic for him). He has undergone a series of surgical procedures. A bone saw was used to remove the cutaneous horns from his hands. Many verrucous lesions on his face and trunk were excised. The surgical procedures are an ongoing treatment to remove recurrent warts.

This rare condition was also reported in a Chinese patient who was eventually treated successfully with radiation therapy to his hands and feet. The medical research team compared the HPV-2 isolates from Dede to the Chinese patient's isolate. The HPV-2 *E2* gene codes for a transcriptional regulator that drives the transcription of the HPV genome. The *E2* gene of Dede's HPV-2 isolates contained missense mutations that were not present in reference strains of HPV-2 or the Chinese patient's HPV-2 *E2* region of the viral genome. Missense mutations are point mutations in which a single nucleotide is changed, resulting in a codon that codes for a different amino acid. Researchers speculate that these distinct nonsense mutations may be the reason why Dede's HPV-2 infection is so aggressive.

References

Gober, M. D., et al. 2007. "Novel homozygous frameshift mutation of *EVER1* gene in an epidermodysplasia verruciformis patient." *J Invest Dermatol* 127(4):817–820.

Lei, Y.-J., et al. 2007. "HPV-2 isolates from patients with huge verruca vulgaris possess stronger promoter activities." *Intervirology* 50(5):353–360.

Wang, C. W. W., et al. 2007. "Multiple cutaneous horns overlying varruca vulgaris induced by human papillomavirus type 2: A case report." *Br J Dermatol* 156:760–762.

Wang, C., et al. 2007. "Detection of HPV-2 and identification of novel mutations by whole genome sequencing from biopsies of two patients with multiple cutaneous horns." *J Clin Virol* 29:334–342.

University of Maryland dermatologist featured in documentary on Indonesian "Treeman." 2009. Retrieved March 4, 2011, from http://www.umm.edu/dermatology/treeman.htm

My Shocking Story–Half Man Half Tree Video. 2008. Discovery Channel. Retrieved March 4, 2011, from http://www.yourdiscovery.com/video/my-shocking-story-half-man-half-tree/

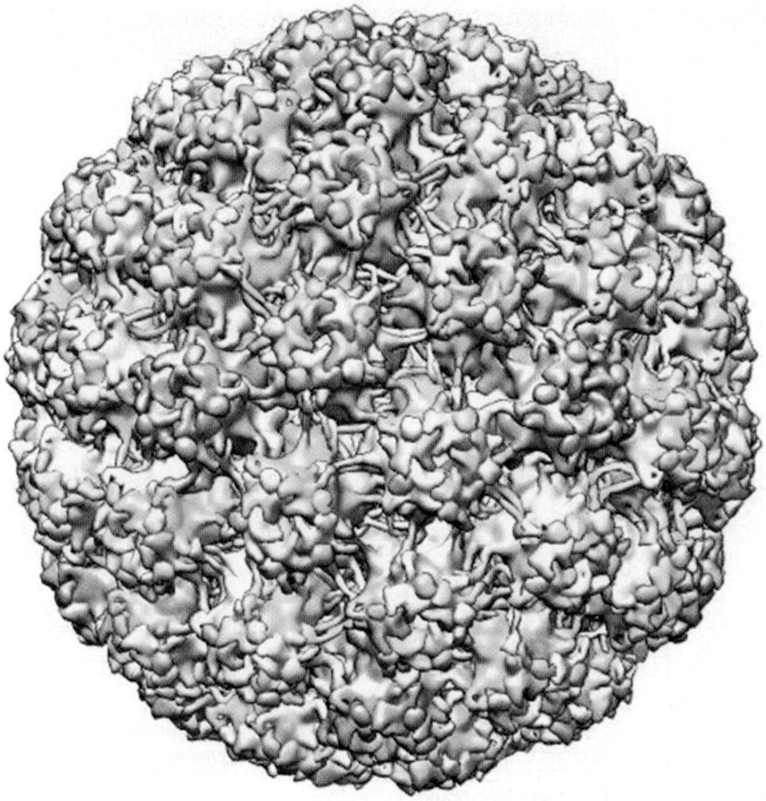

FIGURE 10-14 Computerized three-dimensional reconstruction of a bovine papillomavirus particle. HPV-16 VLP particles are similar. A VLP is an empty particle or shell. VLPs do not contain viral DNA and cannot cause disease.

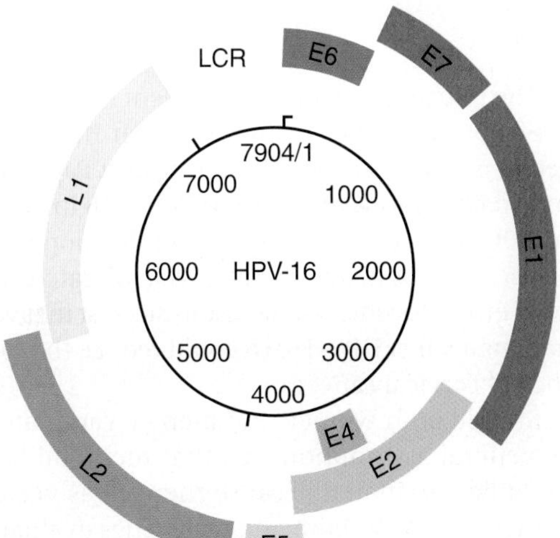

FIGURE 10-15 Genomic map of HPV-16. The circular dsDNA is 7904 base pairs in length. Transcription occurs in a clockwise manner. The opening reading frames are designated E1–E7, L1, and L2. The L1 and L2 gene products make up the capsid or structural proteins of the virus. E5, E6, and E7 code for transforming proteins of HPV-16 (see Table 10-4). Adapted from Cann, A., "Papovaviruses," Course BS335: *Virology*.

Despite widespread screening in the United States, approximately 4200 deaths due to invasive cervical carcinoma were reported in 2010. About 35% of invasive cervical cancers and 57% of deaths occur in the United States in women over the age of 55.

HPV Vaccines

Cervical cancer is the third most common cancer in women worldwide. There are an estimated 555,000 new cases and 310,000 deaths every year, and almost 85% of these cases occur in developing countries. Efforts to develop an HPV vaccine that targeted high-risk types of HPVs began in the early 1990s. Phases II and III vaccine clinical trials showed positive results against HPV-16 and HPV-18. A 3-year study to test the safety, efficacy, and immunogenicity of a HPV vaccine involving 1113 North American or Brazilian women between the ages of 15 and 25 years was conducted by Diane Harper and colleagues. The women participating had the following traits:

- They had 2–5 sexual partners.
- Most began sexual activity between the ages of 15 and 19.
- Approximately half were smokers.
- They tested seronegative for HPV-16 and -18.

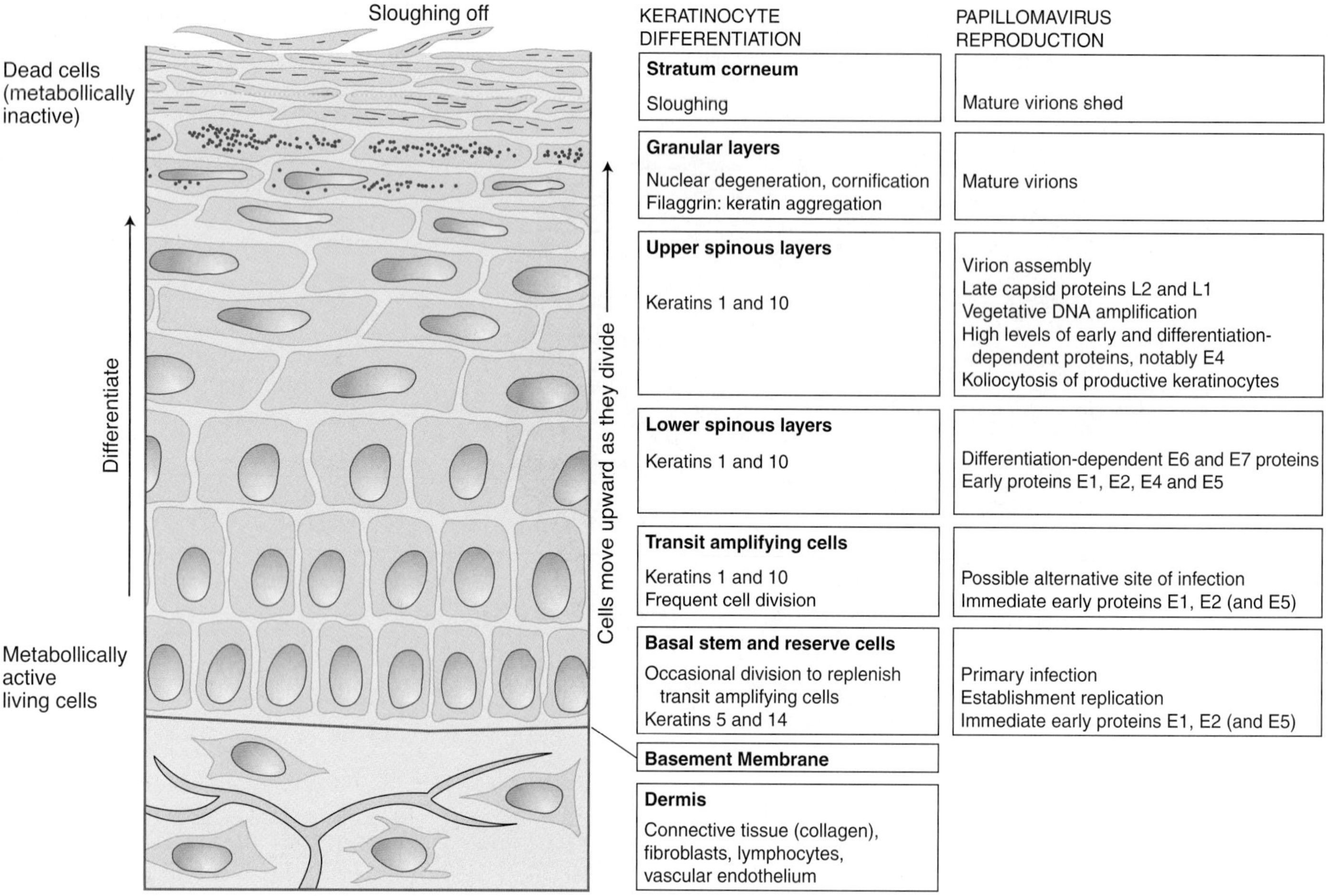

FIGURE 10-16 Papillomaviruses enter through a break in the skin. The primary site for replication is in the basal and stem cell reserves. These cells are metabolically active and support viral replication. HPVs cannot yet be grown in cell culture because the viral life cycle is tightly linked to epithelial cell differentiation. DNA replication can occur in less differentiated layers of the epidermis, and late protein synthesis and viral packaging can only take place in more differentiated epithelial layers. With the exception of HPV-1, which causes plantar and common warts, most HPV-induced lesions rarely produce large quantities of virus particles. Adapted from Fields, B. N., et al. *Fields Virology*, Fourth Edition. 2 vols. Lippincott Williams & Wilkins, 2001.

- They tested negative by PCR for HPV-16 and -18 DNA.
- They had no history of abnormal Pap smears.
- They were never treated for genital warts or cervical cancer.

The women were immunized at a 0-, 1-, and 6-month schedule with a placebo or bivalent vaccine containing HPV-16/18 viruslike particles (VLPs). VLPs mimic the true structure of HPV virions but do not cause an infection (**FIGURE 10-18**). All women were required to report for follow-up appointments and Pap smears at 18 months and 27 months postvaccination. Their cervical samples were analyzed via PCR for HPV DNA. In addition to Pap smears, blood was drawn from the patients to assess seroconversion. Women also recorded their symptoms or adverse reactions for the first week after each injected immunization dose.

The researchers found the bivalent HPV vaccine to be 100% effective against persistent HPV-16/18 infections. It was 91.6% effective against Pap smear abnormalities associated with HPV-16/18 infection and 100% effective against lesion development. The vaccine was generally safe, well tolerated, and highly immunogenic. These results suggest that vaccinations against HPV-16/18 could reduce the incidence of cervical cancer.

Should both women and men be vaccinated? The general assumption was that men and boys should be vaccinated because they act as vectors for infection. Al V. Taira and colleagues evaluated the benefit and cost-effectiveness of adopting a vaccine strategy for both sexes. Their results, which were published in the Centers for Disease Control and Prevention's (CDC's) *EID* journal, suggested that vaccinating a cohort of 12-year-old females against HPV-16/18 would reduce cervical cancer cases by 61.8%. If male participants were

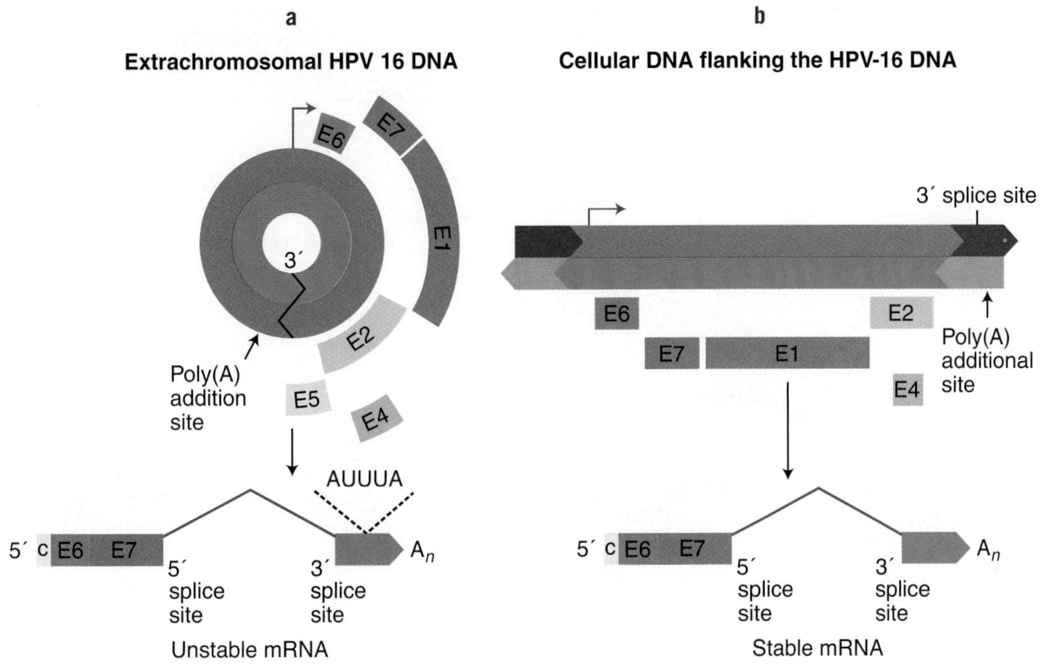

FIGURE 10-17 (a) Benign tumors contain episomal forms of HPV-16 DNA. **(b)** Malignant cells contain integrated HPV-16 DNA. The integrated DNA contains the *E6/E7* HPV genes that are involved in oncogenesis. Adapted from Flint, S. J., et al. *Principles of Virology: Molecular Biology, Pathogenesis, and Control of Animal Viruses*, Second Edition. ASM Press, 2003.

vaccinated as well it would reduce cervical cancer cases by an additional 2.2%. Thus, female vaccination was generally considered cost-effective, whereas male vaccination was not. There are two HPV vaccines available to protect against the types of HPV that causes most cervical cancers: Gardasil and Cervarix. A new international study that included over 4000 sexually active males between the ages of 16 and 26 originating from 18 different countries participated in a controlled clinical trial. One group of males was free of HPV infection at the start of the trial and was vaccinated with Gardasil. The other group was not free of HPV infection and was given either the Gardasil vaccine or placebo. The study proceeded for three years. Over 90% of the males in the first group did not get HPV-related genital warts. In the second group, only about 0.5% developed genital warts. This raises the question of whether or not all boys and young men should be vaccinated.

A survey of 278 parents conducted by Gregory Zimet and colleagues asked if they were willing to vaccinate their adolescent children against organisms that caused sexually transmitted diseases (STDs). Their results, which indicated that parents were receptive to the idea of vaccinating their children against STDs, were reported in the February 2005 issue of *Archives of Pediatric and Adolescent Medicine.*

An inactivated quadrivalent HPV vaccine manufactured by Merck, Gardasil, was licensed by the FDA in 2006. It protects against infection by four HPV types: 6, 11, 16, and 18. The FDA approved Gardasil for use in females as a prevention of cervical cancer, and some vulvar and vaginal cancers, caused by HPV types 16 and 18 and for use in males (ages 9–26) for the prevention of genital warts caused by HPV types 6 and 11. Gardasil is given through a series of three injections into muscle tissue over a 6-month period. The CDC highly recommends that girls aged 11 to 12 (in general, before their first sexual contact) be vaccinated.

A bivalent vaccine HPV vaccine, Cervarix, produced by GlaxoSmithKline, was licensed in Australia, the Philippines, and the European Union in 2007. The vaccine protects against HPV 16 and 18. In 2009, it was approved in the United States by the FDA for use in females ages 10–25 for the prevention of cervical cancer caused by HPV types 16 and 18. The vaccine is also given in three doses over a 6-month period. About 30% of cervical cancers will not be prevented by these vaccines. In the case of Gardasil, 10% of genital warts will not be prevented by the vaccine. The CDC and FDA continue to monitor the safety of HPV vaccines. As of February 14th, 2011, about 33 million doses of Gardasil were distributed in the U.S. Since February 14, 2011, the Vaccine Adverse Event Reporting System received 18,354 reports of adverse effects following Gardasil vaccination in the United States. Of these, 92% of the reports were considered nonserious and 8% were considered serious (life threatening or resulted in a persistent disability, congenital birth defect, hospitalization, or prolonged hospitalization). For updated information, see http://www.cdc.gov/vaccines/pubs/vis/

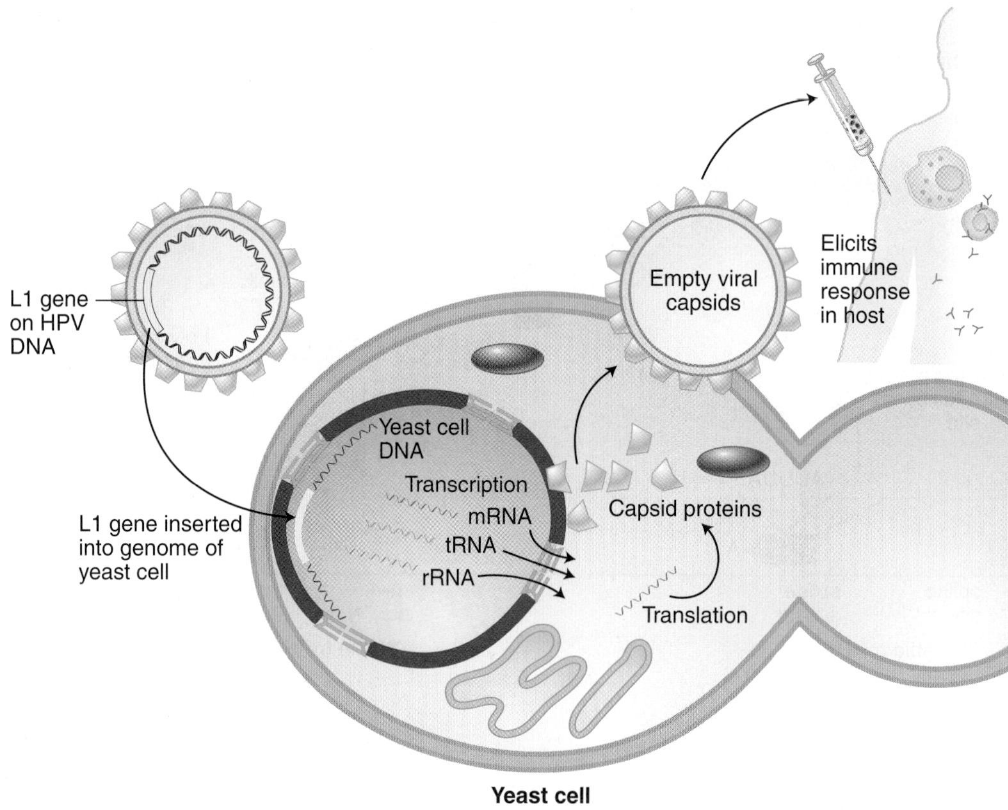

FIGURE 10-18 The bivalent HPV-16 and HPV-18 vaccine was made by expressing the HPV-16 and HPV-18 *L1* genes in *Saccharomyces cerevisiae* (yeast) cells. The *L1* gene product is produced in high concentrations and self-assembles into a viruslike particle (VLP). It resembles the HPV viral particle but does not contain the DNA genome. It cannot infect human cells and cause papillomas or cancer. The bivalent vaccine contained both HPV-16 and HPV-18 VLPs. The vaccine was injected intramuscularly into the arm.

10.6 Animal DNA Tumor Viruses

Adenoviruses and some of the polyomaviruses (such as SV-40) induce tumors in experimental animal systems but not in humans. These viruses are DNA tumor viruses. They transform cells at a frequency of less than 1 in 100,000 infected cells.

Adenoviruses

Adenoviruses were isolated from human adenoids of children by Rowe and colleagues in 1953. At the time Rowe was searching for different tissues in which to propagate polioviruses. Soon it was discovered that adenoviruses could be isolated from every species of mammals, birds, and amphibians, and it became evident that adenoviruses could persist in lymphoid tissues for many years. Approximately 50% to 80% of tonsils that are surgically removed are infected with adenoviruses.

Certain human adenoviruses cause malignant tumors in baby rodents, such as hamsters and mice. Two adenovirus genes, *E1A* and *E1B*, are responsible for oncogenic transformation of rodent cell lines. *E1A* inactivates the tumor suppressor gene product pRb, and *E1B* inactivates the tumor suppressor protein p53. Despite the potent oncogenic properties of some human adenoviruses in animals and tissue culture cells, the virus has not been linked to any human cancers. Subsequently, it was discovered that adenoviruses cause respiratory tract, GI tract, and eye infections, including highly contagious conjunctivitis ("pink eye") and possibly obesity (see **VIRUS FILE 10-4**) in humans. Respiratory epidemics of adenovirus are often prevalent on army bases. Adenovirus infections are also common in the immune-compromised, such as AIDS patients.

By 2010, 56 human adenoviruses had been identified. Most individuals are infected with one or more types of adenoviruses before the age of 15. Direct contact, respiratory droplets, or ingestion spreads adenovirus infections. Most infected persons are asymptomatic. There is no specific antiviral treatment for those individuals experiencing symptoms. In rare cases, ribavirin is used to treat adenovirus infections.

Adenoviruses are 80 nm in diameter and of icosahedral symmetry. The capsid is composed of three different major proteins—hexon, penton-base, and a knobbed fiber—in addition to a number of minor proteins (**FIGURE 10-19**). The viral genome is a linear, dsDNA that is approximately 36 to 38 kilobase pairs in length for mammalian adenoviruses. The genome contains inverted repeats at its ends. Adenoviruses are a prime candidate as gene-therapy vectors because they cause mild diseases in humans and they deliver their genome to the nucleus where it can replicate with high efficiency. Adenoviruses can infect a wide variety of tissues such as lung, brain, pancreas, thyroid, heart, and skeletal muscle. Over the past decade, adenoviruses have been genetically manipulated to manage cancers (see Section 10.7), cardiovascular disease, genetic disorders, and eye diseases such as glaucoma.

Glaucoma is a blinding eye disease characterized by abnormally high intraocular fluid pressure of the eye. It can result in a damaged optic disk, hardening of the eyeball, and complete loss of vision. The goal of the glaucoma filtration procedure is to create a new passageway by which aqueous fluid inside the eye can escape, thereby lowering the intraocular pressure. A complication of this surgery is side effects such as controlling unwanted wound healing after the glaucoma filtrations surgery. Adenovirus-based vectors are potential vehicles as an accompanying therapy to glaucoma filtration surgery. A general schematic of adenovirus therapy is shown in **FIGURE 10-20**.

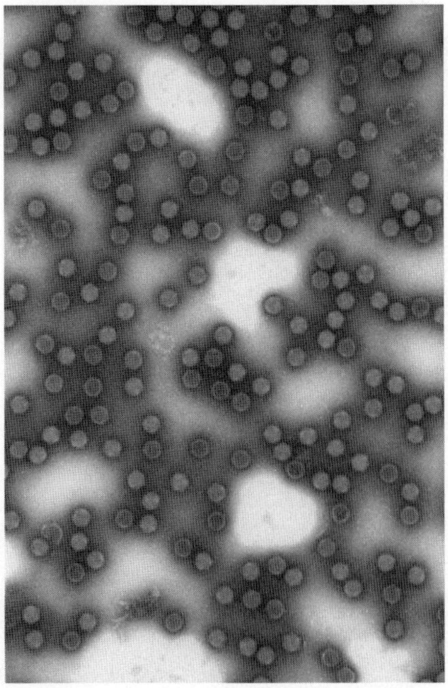

FIGURE 10-19 Transmission electron micrograph of adenoviruses in patient sample showing the penton fibers. Magnification is 40,000×.

Other developments in adenovirus gene therapy are directed at managing metastatic thyroid cancer, pancreatic adenocarcinoma, late-stage prostate cancer, brain gliomas, and hepatocellular carcinoma (HCC). In addition to cancer therapy, adenovirus-based vectors are promising vehicles for cystic fibrosis gene therapy.

Upon closer scrutiny of adenovirus gene therapy systems, it has become evident that some technical problems and potential side effects in the patient do exist; for example, the adenovirus genes do not integrate into the host chromosome and, therefore, the therapeutic genes do not persist for very long in cells. To overcome this, an improved adeno-transposon vector aimed toward efficiently integrating into the host chromosome has been developed.

When high doses of the vector are used, it causes an acute immune response. The first victim of an acute immune response was Jesse Gelsinger (see Chapter 9). During adenovirus infections, there is a well-orchestrated immune response to suppress the virus during the course of infection. This response is also effective against the vector. It has become increasingly important to unravel the complex molecular events involved in mounting the host defenses.

Simian Virus-40 (SV-40)

In 1960 simian vacuolating virus-40 (SV-40) was isolated from primary African green monkey kidney cells by Sweet and Hilleman during safety testing of the poliovirus vaccine. They observed that the SV-40 infected cells showed distinctive morphological changes (vacuolization). Prior to this SV-40 escaped detection because rhesus or cynomolgus monkey kidney cells were used to prepare the vaccine. SV-40 did not cause any observable cytopathic effects in the rhesus or cynomolgus monkey kidney cells and was a frequent contaminant of the rhesus macaque kidney cell culture cells (**VIRUS FILE 10-5**).

SV-40 is a polyomavirus. Two other polyomaviruses have been found in humans: Jamestown Canyon virus (JCV) and BK virus (BKV). Both JCV and BKV cause tumors in animals and in about 5% of AIDS patients. The focus in this section will be on SV-40.

Polyomaviruses have small, simple circular dsDNA genomes (e.g., the SV-40 genome is 5.2 kb in length). The SV-40 viral particle is nonenveloped, measures 45 nm in diameter, and consists of three different capsid proteins (**FIGURE 10-21a**). SV-40 enters cells by binding to the major histocompatibility complex class I molecules (MHC; see Chapter 7). The virion enters through endocytosis

VIRUS FILE 10-4 The Obesity Viruses: SMAM-1 and Adenovirus 36

During the past 20 years there has been a dramatic increase in obesity in the United States. In 1991, four U.S. states had obesity rates of 15% to 19% and no states had rates at or above 20%. In 2003, 15 states had rates of 15% to 19%, 31 states had rates of 20% to 24%, and four states had rates of more than 25%. Today 30% of U.S. adults 20 years of age and older—over 60 million people—are obese. The increasing obesity trends among U.S. adults raises concern because being overweight or obese increases the risk of many diseases and health conditions, such as hypertension, Type 2 diabetes, stroke, heart disease, sleep apnea, respiratory problems, and certain cancers (e.g., breast, colon, and endometrial).

It is likely that there are many causes of obesity. Animal obesity has been attributed to nutritional, environmental, genetic, seasonal, endocrine, pharmacological, or viral origin. The focus of obesity studies has been on the genetic and behavioral components of obesity.

An infectious cause of obesity has received little attention. Over the past 20 years, though, six different infectious agents have been reported to cause obesity in animals (**TABLE VF 10-1**). It is not known whether these six pathogens contribute to obesity in humans. Two of them, however—SMAM-1 (a chicken adenovirus) and human adenovirus 36—have been associated with human obesity. For ethical reasons, the definitive experiment of injecting humans with these adenoviruses to determine a causal role for this virus in human obesity is not possible.

SMAM-1 is an avian adenovirus that caused high mortality in poultry during epidemics in India in the 1980s. Adenovirus 36 can infect chickens and humans and induce adiposity in chickens. The virus was first isolated in 1978 from a diabetic child in Germany. Obese and nonobese volunteers from Wisconsin, Florida, and New York provided serum samples, which were then screened for the presence of SMAM-1 or adenovirus 36 antibodies. Results of these human studies showed that 5% of nonobese individuals possessed adenovirus 36 antibodies, whereas 30% of the obese individuals were antibody-positive. In another study, 20% of 52 obese individuals screened tested positive for

TABLE VF 10-1 Obesity Observed in Animals Injected with Viruses or Prion Agents

Virus/Infectious Agent	Obesity Observed in These Animal Model(s) after Injection with the Virus	Obesity Observations	Obesity Mechanism
Canine distemper virus	Mice Chickens	Body weight; fat cell size and number increases; 26% of surviving mice are obese	Downregulation of leptin receptor
Rous-associated virus-7	Chickens	Fat deposition around crop and abdominal fat pads of birds	No proposed molecular mechanism
Borna disease virus	Rats	Rats that survive are obese	Hypothalamic damage
Scrapie agent	Mice Hamsters	Obesity observed in hamsters	Hypothalamic-pituitary adrenal problem
SMAM-1 avian adenovirus	Chickens	Develop excessive visceral fat	No proposed molecular mechanism
Human adenovirus-36	Chickens Mice Monkeys	Increased adipose tissue	No proposed molecular mechanism

FIGURE VF 10-2 Exercise plays an important role in weight loss and maintenance.

SMAM-1 antibodies. These data show an association of SMAM-1 and adenovirus 36 antibodies with human obesity, but they do not establish a causal relationship.

One U.S. firm has developed the technology to test humans for adenovirus 36 antibodies. A positive test means that a person has been exposed to adenovirus 36 and the infection may contribute to difficulties in maintaining a healthy weight. Nonobese individuals who test positive for antibodies may be at risk for developing obesity. They should develop proper eating and exercise habits (**FIGURE VF 10-2**). An adequate understanding of viral infection as an etiological factor is needed for better management of obesity. For certain individuals, it may be possible one day to prevent or treat obesity of infectious origin.

References

Atkinson, R. L., et al. 2005. "Human adenovirus-36 is associated with increased body weight and paradoxical reduction of serum lipids." *Int J Obesity* 29:281–286.

Dhurandhar, N. V. 2001. "Infectobesity: Obesity of infectious origin." *J Nutr* 131(10):2794S–2797S.

Dhurandhar, N. V., et al. 2000. "Increased adiposity in animals due to a human virus." *Int J Obesity* 24:989–996.

Dhurandhar, N. V., et al. 2002. "Human adenovirus Ad-36 promotes weight gain in male rhesus and marmoset monkeys." *J Nutr* 132:3155–3160.

Gabbert, C., et al. 2010. "Adenovirus 3 and obesity in children and adolescents." *Pediatrics* 126(4):721–726.

Na, H. N., et al. 2010. "Association between human adenovirus-36 and lipid disorders in Korean schoolchildren." *Int J Obes (Lond)* 34(1):89–93.

Zinn, A. R. 2010. "Unconventional wisdom about the obesity epidemic." *Am J Med Sci* 340(6):481–491.

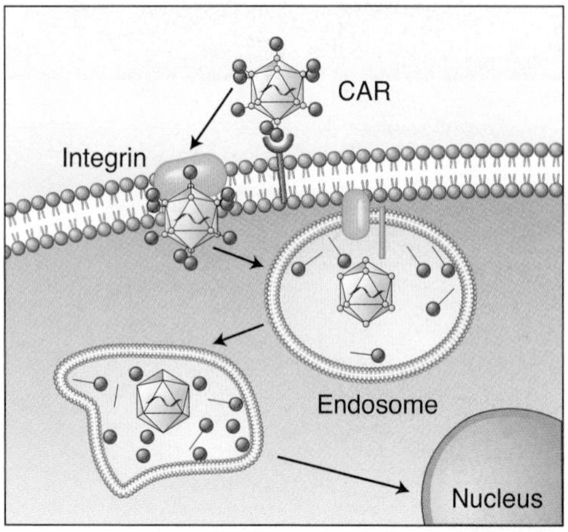

FIGURE 10-20 Adenovirus gene therapy. Adenoviruses fuse with the host membrane by binding to CAR (a cell surface *c*oxsackie-*a*denovirus cell surface *r*eceptor) and internalize into endosomes using cellular integrins. At some stage thereafter, the viral components are released and they are targeted to the nucleus of the cell. Adapted from Baker, A. H., *Heart and Metabolism* 18 (2002): 28–35.

mediated via invaginations called caveolae ("little caves"). Caveolin proteins are integral membrane proteins present within the caveolae that facilitate endocytosis. The virion is transported to the endoplasmic reticulum and enters it, and subsequently the viral genome enters the nucleus of the cell.

The SV-40 genome consists of two sets of genes: the early expressed genes, which encode viral gene products essential for viral replication such as large and small T antigens, and a 17KT protein. A set of late genes encodes the structural proteins necessary for viral assembly. This includes the three capsid proteins (VP1, VP2, VP3) and a maturation protein called agnoprotein. The genome also contains regulatory sequences that are not translated, such as an origin of replication (ORI), promoters, and enhancer sequences for replication. DNA replication is bidirectional, starting at the ORI (**FIGURES 10-21b** and **22**).

SV-40 is of monkey origin. Its natural host is the Asian macaque (*Macaca mulatta*). It infects several species of monkeys in captivity but typically does not cause symptoms or disease in them. The virus is presumably spread via contaminated urine. It is also able to infect and cause sarcomas in hamsters. For this reason, much attention has focused on the viral oncogenes of SV-40: large and small T antigens.

Large T antigen is a large, multifunctional protein that is involved in a number of cellular processes including transcriptional activation and repression, differentiation, and the stimulation of the cell cycle. Large T antigen interacts with the cellular tumor suppressor p53 and pRb, whereas small T antigen interacts with a cellular phosphatase, PP2A. In doing so, these viral proteins disrupt the p53 and pRb cellular pathways, inactivating cell cycle control. This in turn can lead to cellular transformation, the blocking of apoptosis, and the stimulation of cell division.

Large T antigen is controlled by phosphorylation. In the life cycle of SV-40, it functions as a replicative helicase (unwinding the genome at the ORI and promoter). It also recruits DNA replication proteins such as topoisomerase, replication protein A, and DNA polymerases to the ORI and promoter site of replication. It indirectly promotes the activation of late transcription and plays a role in viral maturation by influencing the phosphorylation of capsid proteins. Small T antigen plays a role in the lytic life cycle of the virus.

10.7 Oncolytic Viruses

Cancer Therapy: The Last 150 Years

Surgery

Up to the early 1900s, the only cancer therapy option was the surgical removal of tumors. Surgery did not progress until **anesthesia** was introduced in 1846. Anesthesia is a procedure in which the patient is administered a drug (anesthetic) to facilitate surgery. It reduces the pain of the patient and can render them unconscious during the operation. Ether was the first drug used as an anesthetic. It took time for surgery to evolve. It had to become more aseptic and techniques were static at this time. There were some reports of cancers cured by surgery. If a tumor was quickly diagnosed and easily accessible for removal, the patient had a higher likelihood of surviving. For those patients for whom this was not the case, prognosis was bleak.

Radiation Therapy

X-rays were discovered by Wilhelm Rontgen while experimenting with a cathode-ray tube in 1895. Marie Curie discovered the radioactive element radium in 1898. Radium is a source of gamma (γ) rays. X-rays and γ rays cause biological damage indirectly. They release their energy by colliding with cells. This produces fast-moving electrons causing biological damage to tissues leading to cell death. Through happenstance, it was discovered that tissues with the highest mitotic activity—such as rapidly dividing cancer cells—were the most radiosensitive. Subsequently their application to treat cancer was fervent. Engineers built powerful X-ray sources. Early radiotherapy consisted of a single massive dose of radiation, lasting about an hour. Side effects were severe. Radiotherapy

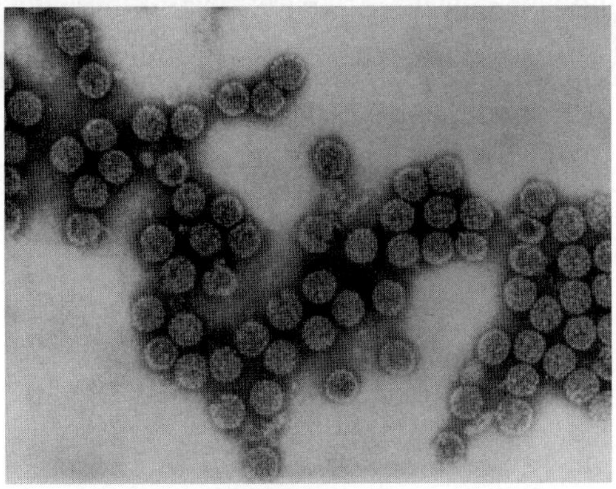

a

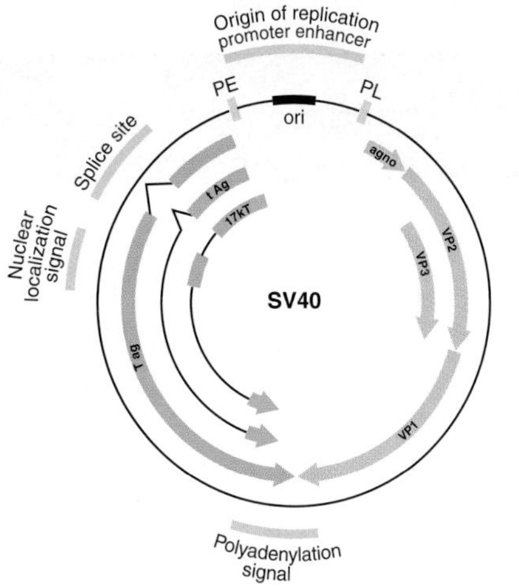

b

FIGURE 10-21 (a) Specimen containing SV-40 particles, which are nonenveloped. The specimen contains three capsid proteins, VP1-3. **(b)** The SV40 genome contains five regions: a control region or nontranslated region that contains the origin of replication, early and late promoters and enhancers, the early transcription region which codes for large T antigen, small T antigen and a smaller product termed 17kT; the late transcription regions which codes for the structural proteins VP1, VP2, and VP3 on overlapping reading frames and a smaller maturation protein called agnoprotein; the polyadenylation signal for mRNAs and regions containing a nuclear localization signal an intron splice site. The arrowheads point in the direction of transcription. The late mRNAs are expressed after viral DNA replication has occurred. The viral and late transcripts are produced from opposite strands on the viral DNA. Early and late transcription occurs bidirectionally near the origin of replication. The cellular RNA polymerase II transcribes the viral mRNAs. Only a few hundred early transcripts are produced per cell in contrast to the several hundred thousand copies per cell of late viral mRNA transcripts. When enough large T antigen is accumulated in the cell, it binds to the viral DNA and represses early transcription, indirectly activating late transcription of SV40 mRNAs. Differential splicing leads to 2 mRNAs. Adapted from Poulin, D. L., and DeCaprio, J. A. 2010. "Is There a Role for SV40 in Human Cancer." *Journal of Clinical Oncology* 24(26): 4356–4365.

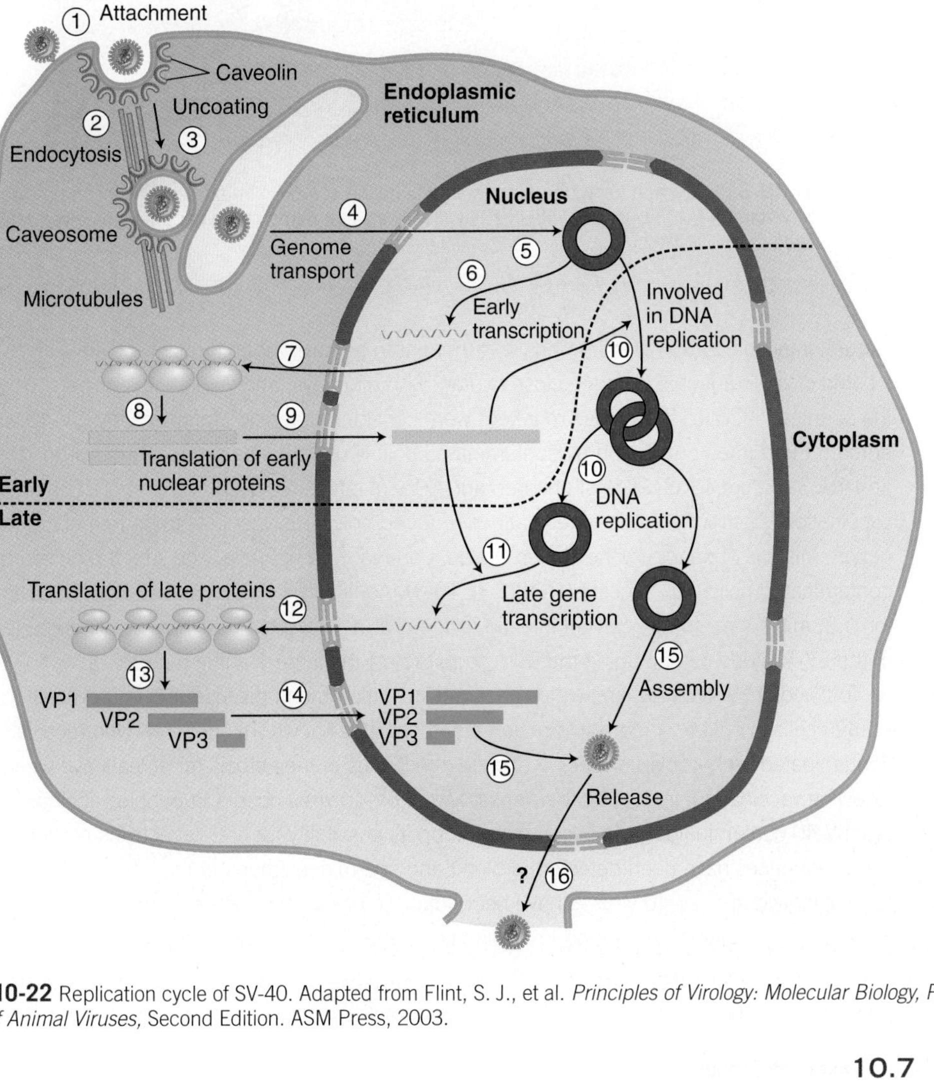

FIGURE 10-22 Replication cycle of SV-40. Adapted from Flint, S. J., et al. *Principles of Virology: Molecular Biology, Pathogenesis, and Control of Animal Viruses,* Second Edition. ASM Press, 2003.

VIRUS FILE 10-5 The SV-40 Controversy: Passenger or Emerging Pathogen? Will an SV-40 Large T Antigen Vaccine Become Routine?

FIGURE VF 10-3 Photograph showing people standing in line at a polio immunization station outside a local grocery store in Columbus, Georgia, in 1961. It is likely that many of the vaccinees received a SV-40 contaminated poliovirus vaccine.

Natural infection by SV-40 in humans was thought to be rare. Inhabitants of Indian villages located close to a jungle, living in close contact with monkeys, and zookeepers or persons working at animal facilities that house monkeys were considered at risk. Unfortunately, SV-40 virus entered the human population by contaminated poliovirus vaccines produced between 1955 and 1963. Before 1963, 90% of children and 60% of adults received one or more contaminated vaccinations from which federal regulations required poliovirus vaccines to be free of contamination. Millions of people in the United States, Canada, Europe, Asia, and Africa received the contaminated poliovirus vaccine (**FIGURE VF 10-3**). Adenovirus vaccines used from 1961 to 1965 for U.S. military personnel carried infectious adenovirus SV-40 hybrids that contained sequences of the SV-40 genome, including the DNA sequence of the large T antigen gene.

To this day SV-40 is not known to cause any specific human disease. It can, however, infect human cells in culture and transform some of them. It is known that children who received the contaminated oral poliovirus vaccine excreted poliovirus in their stools for at least five weeks following vaccination, indicating that replication of SV-40 virus occurs in people vaccinated with oral SV-40 contaminated vaccines and opens up its possibility for oral–fecal transmission. Viral DNA sequences have been detected in blood and sperm specimens from cancer patients and normal individuals. SV-40 viruses have been found in urine and sewage samples. Antibodies to SV-40 capsid proteins were found in human sera, and SV-40 sequences have been detected in

VIRUS FILE 10-5 The SV-40 Controversy: Passenger or Emerging Pathogen? Will an SV-40 Large T Antigen Vaccine Become Routine? (continued)

people who are too old or too young to have been associated with SV-40 contaminated poliovirus vaccines—suggesting horizontal transmission (person-to-person).

Is there an association between SV-40 and human diseases? There is considerable debate in the scientific community. Numerous reports discuss the association between SV-40 in humans who have inflammatory kidney diseases or rare tumors such as malignant mesothelioma, non-Hodgkin's lymphoma, and brain and bone tumors. SV-40 sequences are usually detected in an episomal state in tumors and rarely as integrated into the chromosome of tumor cells. Infectious SV-40 virus particles have been isolated from brain tumors. When SV-40 was injected into hamsters, the same type of cancers are produced as in humans (such as lymphomas, bone cancer, and malignant mesothelioma).

The National Cancer Institute (NCI) conducted studies to determine the possible association between SV-40 exposure and non-Hodgkin's lymphoma. In 2004, they reported no statistical association between the virus and non-Hodgkin's lymphoma. NCI researchers also evaluated the cancer risk in 54,796 U.S. children whose mothers received contaminated poliovirus vaccines during pregnancy. The scientists hypothesized that their children would be at increased risk for SV-40 associated cancers because of SV-40 mother-to-child transmission. They also studied 254 zoo workers exposed to monkeys at work. All of the NCI studies published in 2004 concluded that there is no substantial evidence to indicate that SV-40 causes cancer in humans.

Scientists who believe that SV-40 is associated with rare cancers and kidney disease suggest that the mechanism of tumorigenesis in humans is related to the two viral oncoproteins: the large and small T antigens. The possibility that SV-40 may be a cofactor in the cause of some human tumors has stimulated the preparation of a vaccine against large T antigen. Such a vaccine may serve as a useful intervention against SV-40 associated cancers in the future.

References

Barbanti-Brodano, G. L., et al. 2004. "Simian virus 40 infection in humans and association with human diseases: Results and hypotheses." *Virology* 318(1):1–9.

Engels, E. A., and Chen, J. 2004. "Poliovirus vaccination during pregnancy, maternal seroconversion to simian virus 40, and risk of childhood cancer." *Am J Epidemiol* 160(4):306–316.

Engels, E. A., Viscidi, R. P., et al. 2004. "Case-control study of simian virus 40 and non-Hodgkin lymphoma in the United States." *J Natl Cancer Inst* 96(18):1368–1374.

Engels, E. A., et al. 2004. "Serological evidence for exposure to simian virus 40 in North American zoo workers." *J Infect Dis* 190(12):2065–2069.

Gazdar, A. F., Butel, J. S., and Carbone, M. 2002. "SV40 and human tumours: Myth, association or causality?" *Nat Rev* 2:957–964.

Gee, G. V., et al. 2010. "SV-40 associated miRNAs are not detectable in mesotheliomas." *Br J Cancer* 103(6):885–888.

Vilchez, R. A., and Butel, J. S. 2004. "Emergent human pathogen simian virus 40 and its role in cancer." *Clin Microbiol Rev* 17(3):495–508.

evolved from a horribly inexact science to a targeted approach in which tumors were irradiated specifically. It was the preferred method of treatment for many cancers. By the 1940s, despite technical advances and some promising results, radiation had done little to improve long-term survival. Surgery, which had been placed on the backburner, was once again being advocated, as was chemotherapy.

Chemotherapy

Chemotherapy is the use of a chemical agent to selectively kill cancerous cells. The first report of a chemotherapy clinical trial for cancer treatment was published in the *Journal of American Medical Association* (*JAMA*) in 1946. Two of the lead authors of the study were Louis Goodman (a physician and pharmacologist) and Alfred Gilman (pharmacologist and Major of the Medical Corps, U.S. Army). They were recruited by the U.S. Department of Defense to research potential applications of chemical warfare agents. Their study was based on an observation of soldiers accidently exposed to **mustard gas** (a chemical weapon used during WWI) during a military operation. Autopsies of some of the victims indicated a suppression of white blood cells in the body. White blood cells divide rapidly, as do cancerous cells.

Their logic was that mustard gas—chemical name, methyl-bis(β-chloroethyl)amine hydrochloride—could have a similar effect on cancer cells.

In their study, 67 patients with advanced cancers of white blood cells such as Hodgkin's disease, lymphosarcoma, and leukemia, and a few miscellaneous cancers were injected intravenously with the mustard gas compound. Most of the patients had undergone radiation therapy but were no longer responding positively to it. Patients were injected with nitrogen mustards rather than inhaling it because mustard gas is a **vesicant**, causing blistering (or vesicles) and burning of the skin equivalent to second- and third-degree burns that could become life-threatening (**FIGURE 10-23**). It causes respiratory symptoms such as coughing, bronchitis, and long-term respiratory problems. Eyelids will swell, causing the exposed individual to blink a lot. Exposure to large amounts of mustard gas can cause death.

Results of the 1946 study were mixed. The most impressive responses occurred in patients who had Hodgkin's disease and lymphosarcoma. Of the 27 Hodgkin's disease patients, nearly everyone benefited from the treatment. Some had dramatic improvements, including partial or complete disappearance of Hodgkin's tumor masses and remission. Eight of 13 patients with lymphosarcoma experienced remissions of three weeks to months. All of the patients were resistant to radiation therapy. When treatment was repeated, it became less effective and the remissions became shorter. The nitrogen mustard therapy caused pain and swellings that were tender at the injection site. Nausea and vomiting usually occurred within one to three hours of injection. Other toxic effects were considered an extension of the therapeutic effects. During this time, heavy metals and nitrogen mustards were used, but overall had no striking effects.

During the 1940s, Dr. Sidney Farber, an Assistant Professor of Pathology at Harvard Medical School, performed autopsies on about 200 children who had acute leukemia. He observed that the children who had been treated with folic acid experienced an "acceleration phenomenon" of the leukemia process unlike anything he had seen before. He rationalized that drugs that were antagonists of folic acid could be a potential treatment for leukemia (suppressing the leukemia process). He followed through with this idea. A total of 16 infants or children suffering from acute leukemia were injected intramuscularly with **aminopterin**. The typical treatment regimen was doses in the 0.5- to 1-milligram range of aminopterin administered three times a week for at least a month. Aminopterin was the most powerful folic acid antagonist available at that time. Many of the children were moribund before the experimental treatment was administered. Ten of the 16 leukemia patients responded well. Six did not respond, and four of those died at the time of publication. Five of the ten who responded well clinically were described as case reports. They had marked improvement, including the return of white blood cell counts to normal; increased appetite; the spleen, liver, and lymph nodes were no longer enlarged; and temporary remissions with daily injections. Farber's approach was the first chemotherapy to bring about regular remissions for leukemia. It spurred a quest for additional chemical compounds for the treatment of cancer. Steroids such as cortisone and prednisone were produced. Subsequently nucleic acid synthesis antagonists were discovered (**FIGURE 10-24**).

In the 1950s a small study (six female participants) determined that methotrexate caused dramatic tumor regression in all six women suffering from choriocarcinoma (an aggressive solid tumor of the uterus). Five of the six women experienced clinical remission. One female participant died. Death was attributed to methotrexate toxicity. Like aminopterin, methotrexate is also a folic acid antagonist. It was the first study to show that a solid tumor could be cured. All other successes were treatments for leukemia, including the effectiveness of combinations of chemotherapeutic agents to increase duration and remissions in leukemia patients. Even though significant strides were achieved in cancer therapy in the early 1900s, there were no cures applicable to all cancers at all stages. Tumors could be surgically removed, but high-energy beams of radiation or intravenous injections of chemicals, or a combination thereof, proved to be insufficient. **FIGURE 10-25** is a timeline of some of the milestones in cancer therapy.

Immunotherapy

Cancer immunotherapy is designed to elicit or induce the immune system to recognize and eliminate malignant tumors. Attitudes fluctuated back and forth for about 100 years as to whether or not immunotherapy was plausible. William Coley, a New York surgeon, was credited with the first immunotherapy large-scale experiments. His experiments began in the 1890s. His interest in developing immunotherapy occurred after the loss of his first cancer patient—a young girl with a sarcoma on her right arm. Dr. Coley surgically removed her tumor but the cancer metastasized and she died shortly thereafter. Coley began studying medical records of other cancer patients at the hospital and reviewing publications. He found reports of observations of cancer patients who went into remission after a group A streptococcal infection

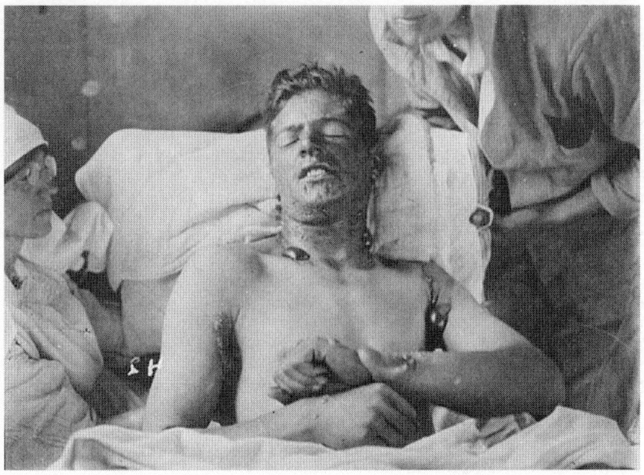

FIGURE 10-23 World War I Canadian solider with mustard gas burns.

of the skin. Coley began deliberately injecting cancer patients with *Streptococcus*.

In a pre-antibiotic era, his experiments were problematic. He had difficulty causing an infection in some patients even after repeated injections and had limited success. He changed the preparation of bacteria. He began injecting cancer patients with a combination of heat-killed streptococci and heat-killed *Serratia marcescens*. The combination of heat-killed bacteria was called "**Coley's toxins**." This combination gave William Coley most of his clinical success. Coley's therapy was harsh. Patients experienced a raging fever for 12 to 24 hours after injection of Coley's toxins. Injections were carried out daily or every other day, for several weeks and sometimes months. The highest success rate occurred in patients who had inoperable sarcomas. In 43 years, Coley treated nearly 900 cancer patients at about a 10% cure rate.

Why were some patients "cured"? It wasn't clear at the time, but later research suggests that the lipopolysaccharide (LPS) found in the outer membrane of the *Serratia marcescens* bacterium acts as an endotoxin that induces a strong immune response.

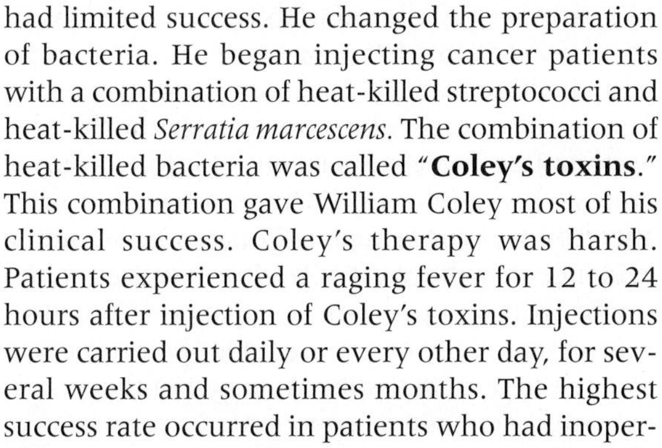

FIGURE 10-24 An assistant working at the National Cancer Institute chemotherapy testing program in 1950.

In this case, it induced a serum factor later purified and named **tumor necrosis factor**. Tumor necrosis factor is a multifunctional cytokine (refer to Chapter 7, Section 7.1) that causes death (necrosis) of tumor cells. A few later reports say that interleukin-12 is also a contributor to tumor regression.

Coley's successes were not widely accepted by the scientific and medical community at the time, probably due to the high fevers caused by the treatments and perceived low cure rates. From 1900 to 1995, opinions fluctuated for and against cancer immunotherapy within the medical and scientific community. By the 1990s, opinions were tipped in favor of cancer immunotherapy. The current consensus is that there is now overwhelming evidence that both the innate and adaptive immune defenses can recognize and eliminate tumors.

Virotherapy

Even before viruses were discovered or visualized by the electron microscope, physicians reported observations of malignant tumors regressing during a natural virus infection. Most of the patients who experienced the short-lived remissions coinciding with a viral infection had cancers that caused immune suppression such as leukemia or lymphomas. One of the most cited examples is a 1904 paper by Dr. George Dock. He described a 42-year-old woman with leukemia that went into remission after a presumed influenza infection. The woman's greatly enlarged spleen and liver shrunk to a nearly normal size and her leukocyte count dropped more than 70-fold after infection. This was reported before it was determined that *influenza* was caused by a viral infection. In another case published in 1953, a 4-year-old boy suffering from leukemia experienced a similar remission after the onset of *chickenpox* (chickenpox is caused by a herpesvirus; see Chapter 15). During the 1970s, more reports of remissions coinciding with *measles* infection were described in patients with Hodgkin's disease, leukemia, and Burkitt's lymphoma.

Four historically significant clinical **virotherapy** trials were carried out from 1949 to 1974. Virotherapy (also called oncolytic virotherapy) is an experimental form of cancer therapy in which viruses are used to target and destroy cancer cells without harming healthy cells. These were desperate times for patients with advanced stages of cancer not responsive to any method of treatment: surgery, radiation, or chemotherapy. Volunteers for clinical trials were usually made aware of the inherent risks in the unproven methods or procedures undertaken. The protocols of these early trials would not meet today's safety

FIGURE 10-25 Timeline of early milestones in cancer therapy. Adapted from Kelly, E., and Russell, S. J. 2007. "History of oncolytic viruses: Genesis to genetic engineering." *Molec Ther* 15(4):651–659.

1846	First public demonstration of surgery to remove a tumor using anesthesia (ether)
1896	Radiation therapy for clinical use
1946	Nitrogen mustards introduced as chemotherapeutic agent
1948	Folic acid antagonists used for leukemia treatment
1958	Methotrexate cured the first solid tumor of the uterus
1965	Combination chemotherapy improves frequency and duration of remission in children with leukemia
1968	Stereotactic radiosurgery developed to localize focused high-dose X-ray beams to areas of the brain where tumors or abnormalities occur
2005	First oncolytic virus approved for use

and ethical requirements. The production of viruses administered to patients was not regulated or quality-controlled. In some cases, viruses were extracted from a previous cancer patient and passed to the next patient. Some investigators injected human cancer cells into chronically ill patients and prisoners without informed consent. This led to outrage and legal proceedings and, in part, is the reason why today's clinical trials must adhere to regulatory laws.

Hepatitis viruses were among the first to be used for virotherapy. In 1949, researchers at the University College of Medicine in Columbus, Ohio, observed remissions of two patients afflicted with Hodgkin's disease after the onset of viral hepatitis. Based on this observation, 22 Hodgkin's disease volunteers were given parenteral injections of unpurified human serum and tissue extracts containing hepatitis virus. The source of serum and tissue extracts was patients suffering from "infectious hepatitis" (likely a self-limiting hepatitis A virus infection) or from serum hepatitis, most likely the result of hepatitis B infection (see Chapter 17). Fourteen of the 22 volunteers developed hepatitis. Seven out of the 22 volunteers improved for at least a month. Improvement meant a disappearance of signs and symptoms of Hodgkin's disease. At least one volunteer died as a direct result of contracting viral hepatitis.

During the 1950s, infections caused by mosquito-borne viruses were common in the United States and abroad. It is not surprising that physicians began investigating these viruses and applying them toward virotherapy. The Egypt 101 virus isolate of West Nile virus was used in more than 150 virotherapy clinical trials against a wide variety of cancers. In a 1952 report, 34 volunteers presented with advanced unresponsive cancers were injected intramuscularly or intravenously with bacteria-free crude suspensions of infected mouse brains, chick embryos, or human tumor tissues. **Viremia** was confirmed in 27 patients, including evidence that the West Nile virus localized to the tumor tissues in patients. In 5 patients, the virus was preferentially concentrated in their tumors in contrast to normal tissues. At least 4 of the 34 volunteer cancer patients experienced a regression of tumor growth. Some patients experienced encephalitis. Patient safety was a concern. Because it was difficult to control neurotoxicity caused by the inoculations and tumor regressions were rare, the use of flaviviruses were abandoned. Instead, virotherapy shifted toward adenoviruses, herpesviruses, paramyxoviruses, picornaviruses, and poxviruses.

Adenoviruses rapidly made their way to the bedside of cancer patients because of modest side effects. Occasionally adenoviruses caused eye inflammation, but the virus inoculations did not cause encephalitis, making it a safer oncolytic virus for clinical trials. A study in 1956 led by J. Georglades utilized adenoviruses in a clinical trial to treat women suffering from advanced cervical cancer. Tissue culture supernatants containing adenoviruses were injected either intra-arterially, intratumorally, or intravenously into cancer patients. The researchers reported areas of necrosis (death) in tumors in 20 of 30 patients within 10 days of adenovirus injections. The oncolytic effects were striking and generalized to the tumor (also called **oncotropism**), but the adenoviruses were cleared rapidly by the body's immune system. Patients did not survive significantly longer. More than half of the patients died within a few months of the beginning of the clinical trial, all from complications of the initial cancer. The researchers discovered anti-adenovirus antibodies present in patient serum. The adenoviruses were eliminated by the body's immune system creating a roadblock in virotherapy. Enthusiasm and application of adenovirus treatments quickly dissipated. Focus shifted toward enteroviruses, including poliovirus and coxsackie B3, vesicular stomatitis virus, and vaccinia virus, in human trials. However, no winners equal to that of prior virotherapy successes emerged.

During the 1970s, the spotlight shifted toward mumps virus as an immunotherapy agent and not as an oncolytic virotherapy agent. In what would be considered a very large clinical trial for its time, Teruo Asada from Osaka University, Japan, reported complete tumor regression in 37 of 90 terminally ill cancer patients. The majority of patients suffered from gastric, lung, or uterine carcinomas. Asada's source of mumps virus for experimentation was either mumps virus purified from saliva by filtration or cell culture supernatants containing mumps viruses. Most preparations also contained glycerin, antibiotics, and an antifungal drug. Dr. Asada administered the mumps virus solution either topically, intravenously, or by inhalation using several types of applications. He applied the mumps solution to the surface of the tumor or scarified the tumor and then applied it. The mumps virus solution was also injected into tumors. Patients were instructed to ingest the virus solution or eat bread soaked with mumps virus. A tampon was stringed and soaked into the mumps virus solution and then inserted and pulled into and out of the throat of those patients with esophageal cancer, allowing the virally soaked tampon to make contact with esophageal lesions. The mumps solution was also injected into the rectum through the anus. Mumps virus was also administered intravenously and inhaled using a **nebulizer**.

Results were promising: 41% of patients went into complete remission or tumors shrunk more than 50%; tumor growth was reduced at least 50% in 47% of patients; and 12% of patients had no response to the mumps virus treatments. Seven patients (8%) experienced side effects. These were some of the best results reported in the literature at this time. More studies followed, but subsequent results were lackluster by comparison. **TABLE 10-5** is a summary of these pioneering virotherapy clinical trials to treat cancer patients. Most known oncolytic viruses can now be engineered to eliminate their pathogenicity, making them safe to use without destroying their oncolytic potency. With these "designer viruses" under development comes a sense of optimism in the field.

Virotherapy in the 21st Century: Hope for the Future

Even today, with most advanced cancers, radiation and chemotherapy treatments are toxic, and cancer remission is brief because of the emergence of chemo-resistant cells. There remains an urgent need for innovative strategies to treat aggressive metastatic cancers that are incurable. Improved virotherapy is on the horizon. Shanghai Sunway Biotech (Shanghai, China) was granted the first marketing approval for an oncolytic virus in November 2005. The virus was a genetically modified human type-5 adenovirus (named H101). Approval was granted because of high success rates in chemotherapy patients who received intratumoral adenovirus injections into incurable head and neck tumors compared to those cancer patients treated with chemotherapy alone. Approval in this case was for the treatment of nasopharyngeal carcinoma. Examples of additional oncolytic viruses currently in phase I and II clinical trials in different countries are listed in **TABLE 10-6**. There is a growth spurt of publications about oncolytic virotherapy. It has only been since the start of the 21st century that researchers began gaining a more comprehensive understanding of the molecular biology of tumorigenesis to be able to develop oncolytic virotherapy.

For viruses to be more widely approved and used as anticancer agents, they will have to meet stringent safety and efficacy standards and be amenable to study/monitor in human tissues. What features make an attractive therapeutic oncolytic virus? The following list must be considered when optimizing oncolytic viruses for cancer treatment:

1. The oncolytic virus should not be a human pathogen. This reduces the chance of preexisting immunity that would inhibit its therapeutic effectiveness.
2. The oncolytic virus should not cause toxicity or severe side effects to normal tissues and cells.
3. The oncolytic virus can be genetically manipulated to facilitate the introduction of therapeutic or suicide genes and/or genes used to monitor its viral spread through the tumor or cancerous mass/tissues.
4. The oncolytic virus has a rapid life cycle such that it can replicate in cancer cells, lyse the cancer cells, and spread to other cancer cells quickly.

TABLE 10-5 Historically Significant Human Clinical Cancer Therapy Trials

Year	Type of Cancer	Number of Patients	Oncolytic Virus	Administration and Preparation of Oncolytic Virus	Outcome	Side Effects
1949	Hodgkin's Disease	22	Hepatitis A and B	Parenteral injections of unpurified human serum and tissue extracts containing hepatitis virus	14/22 developed hepatitis; 7/22 improved clinically; 4/22 tumors reduced in size	Malaise, fever, 1 confirmed death
1952	Incurable, advanced cancer	34	West Nile Virus (Egypt 101)	Intravenous, intramuscular injections of bacteria-free crude suspensions of infected mouse brains, chick embryos, or human tumor tissues	27/34 infected with Egypt 101;14/34 virus found in tumors (**oncotropism**); 4/34 transient tumor regression	Fever, malaise, encephalitis
1956	Cervical cancer	30	Adenovirus	Intra-arterial, intratumoral, or intravenous injections of tissue culture supernatants containing adenoviruses	26/30 had localized necrosis of tumors	Infrequent fever, malaise, vaginal bleeding
1974	Miscellaneous advanced cancers: majority of lung, gastric, uterine	90	Wild-type mumps virus	*External:* post-scarification. *Internal:* Intravenous, intratumoral, oral, inhalation, rectal; mumps virus purified from human saliva by filtration or cell culture supernatants containing mumps viruses	37/90 complete regression or >50% decrease in tumor size; 52/90 tumors decreased <50%; 11/90 unresponsive	7/90 adverse reactions: bleeding, fever

Source: Adapted from Kelly, E., Russell, S. J. 2007. "History of oncolytic viruses: Genesis to genetic engineering." *Molec Ther* 15(4):651–659.

5. The oncolytic virus can be administered systemically (e.g., intravenous injection is the preferred route because many patients have metastatic disease).
6. The oncolytic virus can eradicate the tumor and also establish anti-tumor immunity to contain metastases.
7. The oncolytic virus does not enter the nucleus of its target cell or cannot recombine with the host cell genome to minimize the risk of virus–host genetic recombination events.
8. The oncolytic virus naturally replicates specifically in tumor tissue but not normal tissues or cells.
9. The oncolytic virus is susceptible to an antiviral drug.
10. The oncolytic virus is well characterized in terms of viral genome and protein function.
11. The viral mechanism of oncolytic action and tumor specificity is well characterized.

The diverse arsenal of oncolytic viruses with proven ability to replicate selectively in tumor tissue under development is ever expanding. The majority of them have been genetically modified in some way to reduce their pathogenicity in humans without destroying their oncolytic potency (**TABLE 10-7**). Some animal viruses lack pathogenicity in humans but are capable of destroying human tumor tissue (Table 10-7). *The main problem with using animal viruses for human cancer therapy is the omnipresent risk of virus evolution giving rise to a new human pathogen able to spread from the patient to non-immune contacts.*

Intelligent Design of Oncolytic Viruses

It is believed that oncolytic viruses cause tumor cells to die in four main ways:

1. As a consequence of infecting and replicating in cancer cells.
2. By inducing apoptosis.
3. By causing the cancer cell to undergo lysis and the expulsion of numerous progeny virus particles during the course of viral infection.
4. By a more complicated mechanism such as stimulating the host's immune system. In general, the immune system plays a natural role in surveying and recognizing tumor-associated antigens and in eliminating neoplastic cells prior to tumor development.

TABLE 10-6 Examples of Open and Planned Oncolytic Virotherapy Pediatric Phase I Clinical Trials

Virus Type	Virus Name	Genetic Modification of Virus	Age of Participant	Tumor Type(s)	Source of Virus	Participating Institutions
Herpes simplex virus-1	HSV 1716, Seprehvir	RL1 (γ34.5) deletion	13–30	Non-central nervous system (CNS) solid tumors and glioblastoma tumors >3 ml tumor volume	Crusade Laboratories, Glasgow, UK	Cincinnati Children's Hospital Medical Center, Cincinnati, Ohio, USA
Reovirus	Reolysin	None	3–21	Solid tumors	Oncolytic Biotech, Calgary, Alberta, Canada	Children's Oncology Group
Seneca Valley virus	SVV-001, JX-594	None	3–21	Neuroblastoma, rhabdomyosarcoma, Wilm's tumor, retinoblastoma, adrenocortical carcinoma, carcinoid tumors	Neotropix via Malvern, Pennsylvania, USA, via the National Cancer Institute (NCI)	Children's Oncology Group
Vaccinia	JX-594	Thymidine kinase gene deletion; GM-CSF insertion	2–21	Non-CNS solid tumors >1 cm	Jennerex Biotherapeutics, San Francisco, CA, USA	Cincinnati Children's Hospital Medical Center, Cincinnati, Ohio, USA, and Texas Children's Hospital

Source: Adapted from Hammill, A. M. et al., 2010. "Oncolytic virotherapy reaches adolescence." *Pediatr Blood Cancer* 55(7):1253–1263.

The safety of oncolytic viruses will always be the primary factor in clinical progress. Oncolytic viruses that are promising cancer therapy candidates are usually designed based on at least one of four broad categories of tumor targeting mechanisms to date (**FIGURE 10-26**). The mechanisms target transcription, attachment, interferon (IFN) signaling, and apoptosis activities of cells. Each mechanism of targeting will be explained further in relationship to viral replication cycles inside of normal in comparison to cancer cells. Table 10.7 also lists the mechanisms exploited by oncolytic viruses under development, some of which are already being used in preclinical (animal) and clinical (human) trials. It is almost certain that there are other defective key cellular pathways that will be exploited by designed oncolytic viruses through careful and creative strategies for achieving tumor-specific viral replication.

Targeting Transcription

Certain types of viruses can be genetically engineered with a **tissue-specific promoter** in their genomes to regulate the expression of the viral genome whose products are essential for viral production. For example, the *E1B* and *E1A* genes of adenovirus CG7870 are placed under the control of prostate tumor-specific promoters, **p**rostate **s**pecific **a**ntigen (PSA), and rat prostatic basic protein (known as probasin), targeting the virus to replicate in cancerous prostate tissues and not normal tissues (Figure 10-26a).

Targeting Attachment

Cancer cells have acquired mutations that are beneficial to the growth of a tumor. Many of these cancer-specific mutations result in the overexpression of normal or mutated proteins on the cell surface that are expressed at low levels, or absent, on neighboring normal cells. *Viruses can be adapted to use receptors that are more abundantly or exclusively expressed on tumor cell surfaces than on normal cells.* For example, integrin $\alpha_2\beta_1$ is expressed abundantly on the surface of ovarian and prostate cancer cells. Echovirus Type 1 has exploited integrin $\alpha_2\beta_1$ as a receptor for cell attachment, allowing the virus to enter cancer cells and a subsequent lytic infection. Coxsackie virus A21 utilizes the intercellular adhesion molecule 1 (ICAM-1) and/or decay-accelerating factor (DAF) for entry. Both ICAM-1 and DAF are overexpressed on the surface of a variety of cancer cells, including cells of melanoma, breast cancer, prostate cancer, and multiple myeloma.

A number of approaches to mediate the viral entry of measles viruses are underway. For example, the measles *H* gene that codes for the viral attachment protein was genetically manipulated to redirect the virus so that it no longer binds to its

TABLE 10-7	Examples of Oncolytic Viruses and Their Mechanisms of Tumor Selectivity			
Virus Family	**Genome**	**Virus Name**	**Genetic Modification of Virus**	**Mechanism of Tumor Targeting**
Adenoviridae	dsDNA 36–38 kb	Adenovirus HB101	*E1B* and *E3* genes deleted	The virus is engineered to disable viral proteins that prevent apoptosis. Infected cells are impaired in their ability to undergo apoptosis. The virus can only replicate and produce new viruses and spread only in cancer cells.
		Adenovirus Onyx-015	*E1B* gene deleted	The virus is engineered to disable viral proteins that prevent apoptosis. Infected cells are impaired in their ability to undergo apoptosis. The virus can only replicate and produce new viruses and spread only in cancer cells.
		Adenovirus CG7870	*E1B* gene under the control of PSA promoter (**p**rostate **s**pecific **a**ntigen); *E1A* gene is under control of rat prostatic basic protein (probasin)	The *E1B* and *E1A* virus genes are engineered to be under control of prostate tumor-specific promoters, such that the virus only replicates in prostate tissues.
Poxviridae	dsDNA 130–280 kb	Vaccinia JX594-GM-CSF	Thymidine kinase gene deleted	The virus is engineered to disable viral proteins that antagonize the cellular interferon response. Cancer cells will be impaired in their ability to release or respond to interferon. It also expresses genes for granulocyte macrophage colony-stimulating factor (GM-CSF). GM-CSF is a cytokine that stimulates white blood cells.
		Vaccinia vSP	*SP1* and *SP2* genes deleted	The *SP1* and *SP2*, anti-apoptosis genes of the vaccinia virus, are deleted. The virus replicates well in cancer cells but not in normal cells.
		Myxoma virus	*Akt* and *ERK* genes added	Myxoma virus requires activated Akt (a host protein kinase) and ERK (**e**xtracellular signal **r**egulated **k**inase) to grow in cancer cells.
Herpesviridae	dsDNA 120–200 kb	Herpes simplex virus NV1020	γ*34.5* gene deleted	The γ*34.5* gene codes for a neurovirulence factor. Viruses deleted of this gene are not neurovirulent and do not replicate in normal brain cells but can replicate in cancer cells.
		Herpes simplex virus G207	γ*34.5* and *ICP6* deleted	The γ*34.5* and *ICP6* genes code for neurovirulence factors. Viruses deleted of this gene are not neurovirulent and do not replicate in nondividing cells but can replicate in cancerous (dividing) brain cells.
Reoviridae	dsRNA 22–27 kb	Reovirus (Reolysin)	None	Replicates in cancer cells with a mutant or activated *RAS*, inactive *PKR* (*PKR* is part of the interferon [IFN] response). Normal cells express wildtype *RAS* that activates *PKR*, leading to an abortive reovirus infection. Cancer cells lack the IFN response, allowing the virus to replicate well in them.
Retroviridae	+ssRNA 8 kb	Murine leukemina virus (MLV)	ACE-CD (this is the name of the retroviral vector; the 5′U3 region is replaced by a cytomegalovirus promoter)	Cannot replicate in quiescent normal cells but can be targeted to dividing cancer cells.

(continued)

TABLE 10-7 Examples of Oncolytic Viruses and Their Mechanisms of Tumor Selectivity (continued)

Virus Family	Genome	Virus Name	Genetic Modification of Virus	Mechanism of Tumor Targeting
Rhabdoviridae	–ssRNA 11–12 kb	Vesicular stomatitis virus (VSV)	*M51* deleted	VSV with *M51* deletion (matrix protein mutant) renders the tumor sensitive to the effects of interferon (IFN). Therefore, it cannot replicate in normal cells (that have a functional IFN response system), but replication in tumor cells with a defective IFN response system is unaffected.
			Addition of *IFN*-α/β gene	The modified VSV is very sensitive to the effects of IFN, blocking their replication in normal cells. However, cancer cells lack the IFN response, allowing VSV replication to occur in them. The addition of the *IFN*-α/β gene ensures that the VSV cannot replicate in normal cells, but it can selectively replicate in cancer cells (making it a safer therapeutic oncolytic virus). VSV has high replicate rates in cancer cells.
Orthomyxoviridae	–ssRNA 13.6 kb	Influenza A	*NS1* gene deletion	*NS1*'s major function is to counteract or prevent a *PKR*-mediated IFN response. *NS1* deletion allows the virus to replicate in cancer cells deficient in interferon (IFN) pathway.
Picornaviridae	+ssRNA 7.2–8.4 kb	Poliovirus	PV1 (RIPO). The internal ribosomal entry site (IRES) of poliovirus is replaced with an IRES from human rhinovirus serotype 2 (HRV2)	The virus is no longer neurovirulent with the HRV2 IRES. The virus can selectively destroy cancer cells, leaving neuronal cells untouched.
		Coxsackie virus A21	None	The virus binds to intercellular adhesion molecule-1 (ICAM-1) and decay accelerating-factor (DAF) for entry. A variety of cancer cells overexpress ICAM-1 and DAF. CVA21 binds to ICAM-1 and DAF, causing lysis of the cancer cells.
		Echovirus Type 1	None	The virus binds to integrin $\alpha_2\beta_1$ to enter cells. In cancer cells integrin $\alpha_2\beta_1$ 1 is overexpressed. The virus binds to its receptor, integrin $\alpha_2\beta_1$, enters, and causes a lytic infection of cancer cells.
		Seneca Valley virus	None	This virus replicates and causes a lytic infection specifically in cancer cells with neuroendocrine characteristics. The virus causes no harmful effects in normal cells. Exact mechanism of tumor specificity is unknown.
Togaviridae	+ssRNA 12 kb	Sindbis virus	None	The virus binds to laminin to enter cells. In cancer cells, laminin is overexpressed. The virus binds to its receptor, laminin, enters, and causes a lytic infection of cancer cells.
Parvoviridae	ssDNA 4–6 kb	Parvovirus H1	P4 promoter from parvovirus minute virus regulates transcription	This virus induces heat shock protein-72 (hsp-72) in human tumor cells but not normal cells.
Paramyxoviridae	–ssRNA 16–20 kb	Measles virus-CEA	*CEA* (carcinoembryonic antigen) gene added to monitor viral replication in cancer cells	The virus normally binds to CD46 (complement regulatory protein) and signaling lymphocyte activation molecule (SLAM) to enter cells. The *H* (attachment) and *F* (fusion) genes of the virus have been altered causing the virus to exploit alternative receptors present on the outside of cancer cells.
		Newcastle disease virus PV701	Naturally attenuated	Exploits the IFN defects in cancer cells. Normal cells almost unaffected.

Source: Adapted from Russel, S. J., Peng, K. W. 2007. "Viruses as anticancer drugs." *Trends Pharmacol Sci* 28(7):326–333.

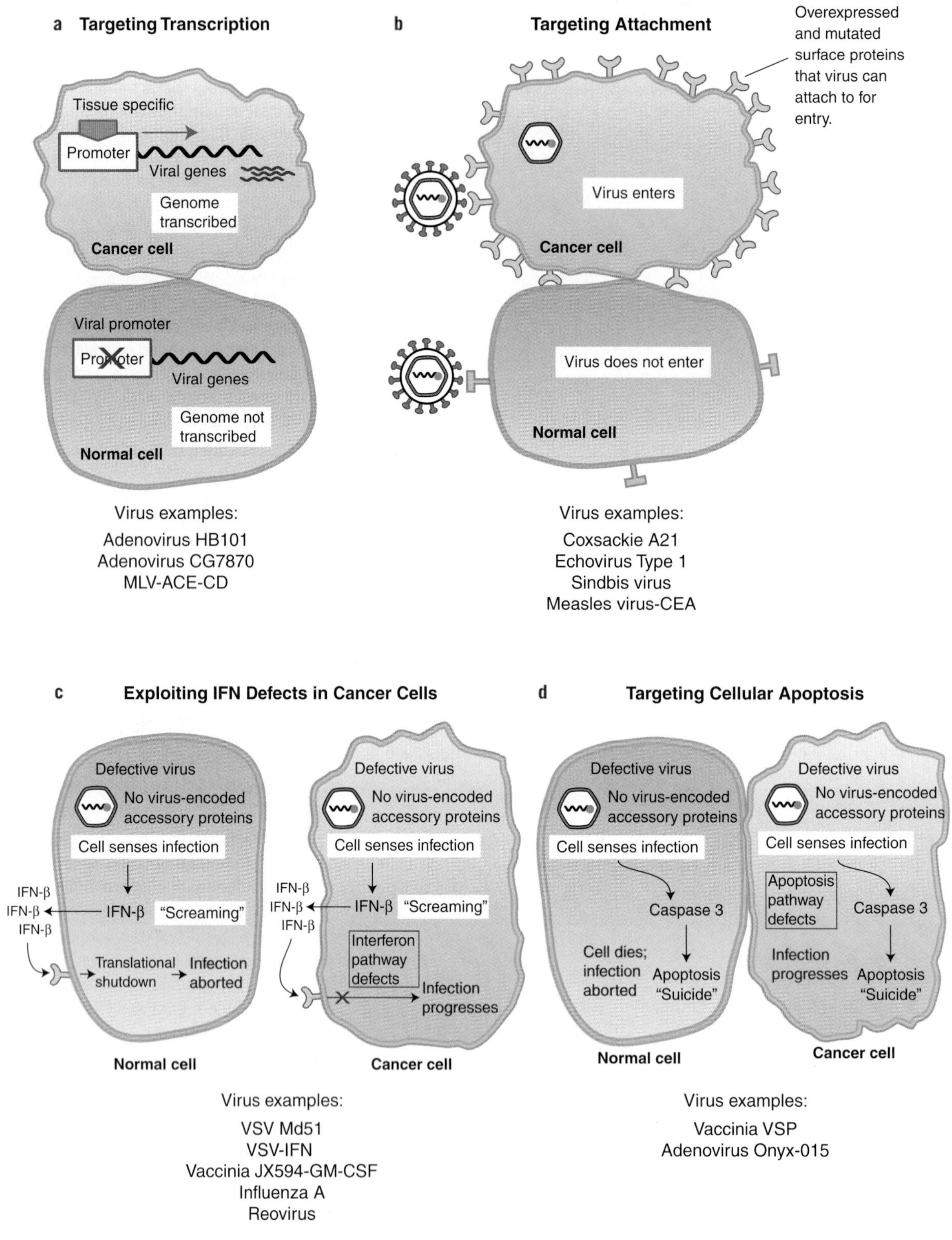

FIGURE 10-26 Mechanisms of tumor targeting by oncolytic viruses**. (a)** Oncolytic viruses genetically engineered with a **tissue-specific promoter** in their genomes to regulate the expression of the viral genome whose products are essential for viral production. The viral genome is transcribed in cancer cells but not normal cells. **(b)** Viruses can be adapted to use receptors that are more abundantly or exclusively expressed on tumor cell surfaces than on normal cells. **(c)** Oncolytic viruses disable viral proteins that antagonize the cellular IFN response, making them very sensitive to the effects of IFN. Normal cells then release IFN upon infection, causing neighboring cells to shut off translation. The oncolytic viruses infect the cancer cells that are impaired in their ability to release or respond normally to interferon, allowing the virus to generate progeny and spread **only in cancer cells**. **(d)** Tumor suppressor p53 or Rb proteins in tumor cells are often nonfunctional, blocking apoptosis and cell cycle arrest in them. Oncolytic viruses do not contain apoptosis inhibitor proteins, allowing the viruses to replicate efficiently in cancer cells without fear of triggering premature apoptosis. Adapted from Russel, S. J., and Peng, K. W. 2007. "Viruses as anticancer drugs." *Trends Pharmacol Sci* 28(7):326–333.

natural host cell receptors, complement regulatory protein (CD46), and signaling lymphocyte activation molecule (SLAM). Sequences that code for the ligands of cellular growth factors were inserted into the 3′ end of the gene (resulting in a different C terminus of the gene product). Alternatively, instead of binding to CD46 or SLAM, the virus *H* gene product attaches to tumor-associated receptors such as growth factor receptors (Figure 10-26b).

Exploiting the Defects of the Interferon (IFN) Signaling Pathways in Tumor Cells

In contrast to normal cells, **cancer cells possess defective IFN signaling pathways** (refer to Chapter 7, Section 7.2, for a review of IFN pathways), making them susceptible/permissive to virus infection even after exposure to interferons (**FIGURE 10-27**). Researchers have engineered oncolytic viruses to disable viral proteins that antagonize the cellular IFN response, making them very sensitive to the effects of IFN. Normal cells then release IFN upon infection, causing neighboring cells to shut off translation. The oncolytic viruses infect the cancer cells that are impaired in their ability to release or respond normally to interferon, allowing the virus to generate progeny and spread **only in cancer cells**.

Toward this end, vesicular stomatitis virus (VSV) was engineered to take advantage of the defective interferon pathway in cancer cells. The *M* (matrix) gene product of VSV blocks the synthesis and release of active IFNs. By deleting the *M* gene or inserting the *IFN*-β gene into the genome of VSV, it becomes sensitive to the effects of IFN but is still able to replicate in cancer cells that are defective in their ability to produce and respond to IFN. The engineered VSV cannot replicate in normal cells but it can selectively replicate in cancer cells, making it a safer therapeutic virus (Figure 10-26c).

Targeting Cellular Apoptosis

Normal cells infected by oncolytic viruses undergo rapid lysis mediated by p53-induced apoptosis or Rb-induced cell cycle arrest. Both p53 and Rb function as tumor suppressor proteins in normal cells. In contrast, the p53 or Rb proteins in tumor cells are often nonfunctional, blocking apoptosis (refer to Chapter 7, Section 7.2, for a review of apoptosis pathways) and cell cycle arrest in them (Figure 10-27). The *E1A* and *E1B* gene products of adenoviruses prevent apoptosis in infected cells by inhibiting p53 and Rb. The *E1A* and *E2B* genes of adenovirus HB101 were deleted from its genome, allowing the virus to be able to replicate efficiently in cancer cells without fear of triggering premature apoptosis (Figure 10-26d).

Challenges of Virotherapy

Immunity

Researchers agree that both the adaptive and innate immune responses can significantly impact the effectiveness of oncolytic virus therapeutics. *Viruses are vulnerable to antiviral host defenses, in that they are usually eliminated from the body before they have had a chance to cause substantial damage to the tumor.* Viruses encounter resistance from innate and adaptive immune responses, including activation of natural killer cells, release of interferons, amplification of antigen-specific T_C lymphocytes, and the secretion of high-affinity antiviral antibodies. These host responses are a double-edged sword. On the one hand, the immune responses are an impediment to the delivery and/or spread of oncolytic viruses. On the other hand, viral stimulation of the adaptive immune response seems to activate anti-tumor immune surveillance systems, increasing the effectiveness of oncolytic virotherapy.

The adaptive immune response, which is of serious concern for the delivery and spread of oncolytic viruses, is triggered by free circulating, virion-associated, or cell-associated viral proteins. The viral products are recognized by specific immunoglobulin surface receptors present on B lymphocytes, causing the eventual activation of B cells and production of antiviral antibodies. These antibodies might be neutralizing (preventing the virus from binding to cellular receptors followed by host cell entry) or could facilitate complement-mediated lysis of virally infected cells. Also, viral proteins are presented to T_H cells as peptides in the context of MHC molecules that activate immunity, including the production of cytokines or the antigen-restricted lysis of virus-infected cells by T_C cells (refer to Chapter 7 for a review of the immune system).

Older clinical trials suggest that oncolytic viruses can be effective when the immune system is suppressed. More recent clinical and preclinical trials suggest the presence of neutralizing antibodies against viral antigens significantly reduces the therapeutic efficacy of a number of oncolytic adenoviruses, vesicular stomatitis viruses, and reoviruses. Contrary to these results, when measles viruses and herpesviruses were injected into tumors, both types of viruses retained their therapeutic activity in spite of a robust neutralizing antibody response. It seems that antivirus immunity can interfere with the therapeutic efficacy of some oncolytic viruses and not others. Clinical trials are in progress in which oncolytic viruses are administered in combinations with chemotherapy and immunosuppressive drugs. Oncolytic virus therapy is in a maturation phase. Design of the next generation of oncolytic viruses should be based

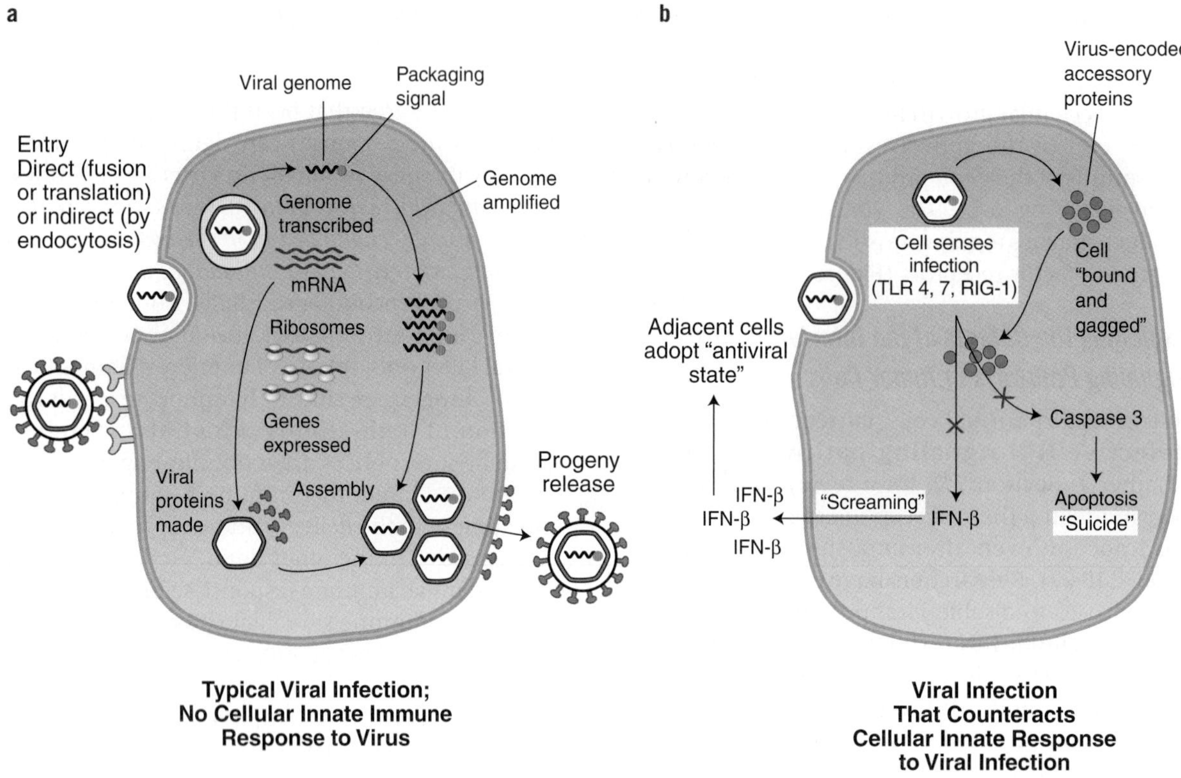

FIGURE 10-27 Virus replication in cells that do not suppress viral infection via IFN and apoptosis innate immune responses **(a)** and cells that produce IFNs (via cellular induction of TLR: Toll-like receptors; RIG-1, retinoic-acid–inducible protein pathways) and in which apoptosis is induced but the invading virus can counter these responses with accessory proteins **(b)**. Adapted from Russel, S. J., Peng, K. W. 2007. "Viruses as anticancer drugs." *Trends in Pharmacol Sci* 28(7):326–333.

on the current knowledge of virology, immunology, and cancer biology.

Delivery

Oncolytic viruses replicating in tumor cells will spread newly formed viruses to nearby and distant tumor cells throughout the tumor, causing oncolysis and tumor regression. The **burst size**, or number of viral progeny released per a single infected cell, can range from 1 to 100,000. The route of oncolytic virus administration should be flexible. For example, a patient suffering from a metastatic cancer, requires that the oncolytic viruses would be delivered intravenously or intra-arterially, allowing oncolytic viruses to spread systematically. When oncolytic viruses are given intravenously, within minutes, the liver absorbs most of the viruses. Viruses that escape the liver enter circulation, where they may be quickly neutralized through absorption by blood cells, through complement cascade or by neutralizing antibodies, particularly if the patient has preexisting immunity. To gain access to the tumor, oncolytic viruses must leave the circulation by traversing a leaky vascular system and the **tumor microenvironment**, which has multiple barriers to efficient replication of viruses, such as the creation of an anoxic and necrotic environment, tumor vasculature, and extracellular matrix that is laden with leukocytes, interferons, cytokines, and other immune mediators. The tumor microenvironment consists of the normal cells, molecules, and blood vessels that surround and feed a tumor A tumor can change its microenvironment, and a microenvironment can affect how a tumor grows and spreads.

For a patient who has a significant tumor in an organ and systemic spread, therapy may be administered intravenously and intratumorally. Oncolytic viral spread in solid tumors *in vivo* is often limited. This could be due to the size of the virus, the cellular mixture of the tumor mass (which will consist of cancer and up to 50% normal cells), or low expression of receptors for viral infection. Tumor cell populations are also heterogeneous in nature. Oncolytic viruses may be able to infect and kill some cancer cells in a tumor better than others. **FIGURE 10-28** illustrates a model of how a combination of oncolytic viruses may be used as a beneficial cancer treatment within the tumor microenvironment. A single direct injection into the tumor may be insufficient. The most effective method is yet to be determined. A study led by G. William Demers demonstrated that intravenous injection of adenoviruses required 1000

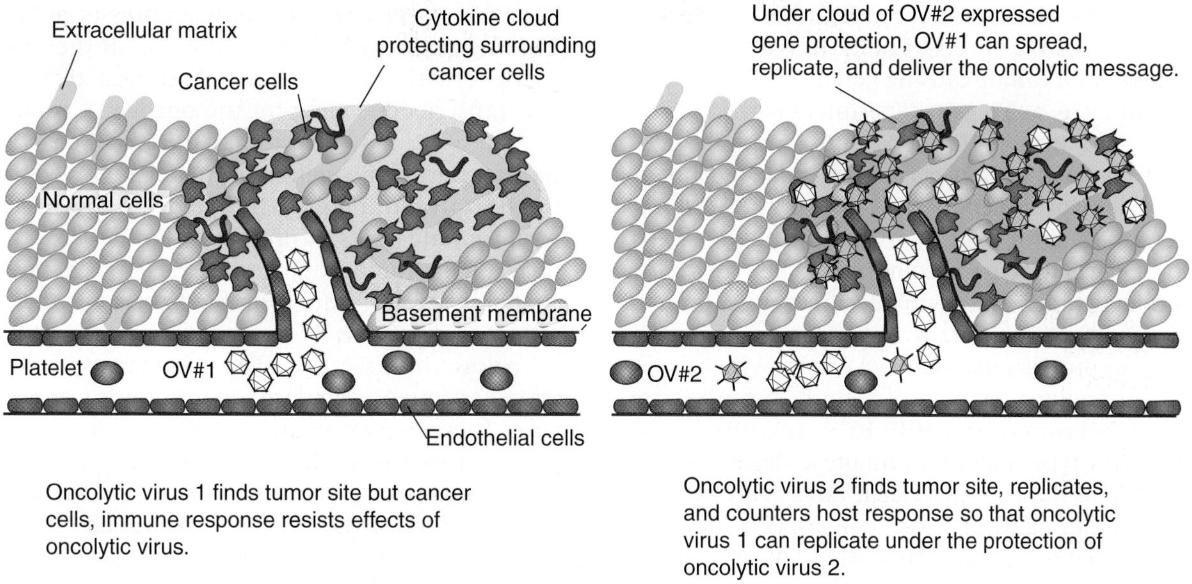
Extracellular matrix
Cancer cells
Cytokine cloud protecting surrounding cancer cells
Normal cells
Basement membrane
Platelet
OV#1
Endothelial cells
Oncolytic virus 1 finds tumor site but cancer cells, immune response resists effects of oncolytic virus.
Under cloud of OV#2 expressed gene protection, OV#1 can spread, replicate, and deliver the oncolytic message.
OV#2
Oncolytic virus 2 finds tumor site, replicates, and counters host response so that oncolytic virus 1 can replicate under the protection of oncolytic virus 2.

FIGURE 10-28 Models on how oncolytic virus combinations can be beneficial to reach the best efficacy to treat cancer. **(a)** Oncolytic virus 1 (OV1) is injected into patient and can find the tumor site and start to replicate, but the cancer cells (red) induce an immune response, secreting a yellow cloud of cytokines (e.g., IFN) that alter neighboring cancer cells to become resistant to infection. **(b)** Oncolytic virus 2 (OV2) can find the tumor and start its replication. OV2 contains a gene that codes for a gene product (green cloud) that can counter the cellular response, allowing it to replicate in the cancer cells. OV1 can now replicate and spread via OV2's protection. OV2 can only replicate in cancer cells, and therefore only OV1 can replicate in cancer cells, allowing the oncolysis to occur in cancer and not normal cells. Adapted from Le Boeuf, F., and Bell, J. 2010. "United virus: The oncolytic tag-team against cancer." *Cytokine Growth Factor Rev* 21:205–211.

times the viral load necessary to achieve the desired results as with an intratumoral injection. It appears that efficient delivery of oncolytic viruses to the tumor site is influenced by a number of parameters such as tumor size, tumor site, tumor type, and immune responses. Optimizing the delivery of oncolytic viruses to the tumor site is critical for anti-tumor efficacy.

Biomarkers to Track Progress

Where do oncolytic viruses go and how many reach their target site? How many progeny viruses are released per infected cell and, when they are released, where do they go? How fast and far do the oncolytic viruses spread in the body? Within the tumor? Answers to these important questions are typically obtained through direct analyses of tissues harvested from multiple sites at multiple times. Biomarker genes incorporated into the genome of oncolytic viruses will facilitate noninvasive monitoring of viral replication and production of viral progeny in humans. New monitoring strategies are expected to enhance the quality of pharmacological information, providing a rationale for intelligent protocol modifications that will accelerate clinical success.

Cancer has become one of the leading causes of death in Western society. Current cancer therapies like radiation therapy and chemotherapy often have only a limited effect. The future of virotherapy for cancer treatment is bright. There are many possibilities when one considers the different types of virus species with diverse mechanisms of action. To get there, however, clinical validation is needed.

Summary

Cancer develops when cells in a part of the body begin to grow out of control. Normal body cells grow, divide, and die in an orderly fashion. During the early years of a person's life, normal cells divide more rapidly until the person becomes an adult. After that, cells in most parts of the body divide only to replace worn-out or dying cells and to repair injuries. Humans have suffered from cancer throughout history. Hippocrates was the first to describe tumors and coin the term carcinoma.

During the late 1700s, connections between occupation, environment, and the onset of cancer were drawn. Early cancer biology research began in the 1900s. The discovery that a filterable agent (the retrovirus Rous sarcoma virus) caused a leukemia common in chickens suggested that cancer could be an "infectious disease." As molecular biology began to advance, evidence from several laboratories demonstrated that RSV contained a gene—*src*—that was involved in the oncogenic transformation of cells. The term oncogene was coined for a tumor-causing gene.

In 1976, an elegant experiment performed by Michael J. Bishop, Harold Varmus, and colleagues showed that the *src* oncogene had a cellular counterpart. They proposed that viruses hijacked cellular genes. Today it is known that oncogenes exist throughout the chromosome of humans. Their gene products function as cytokines, receptors, protein kinases, G-proteins, transcription factors, and other nuclear proteins that regulate cell growth.

Approximately 20% of all human cancers are associated with viruses. HBV, HCV, HPV, EBV, KSHV, and HTLV types 1 and 2 contribute to cancers. Eighty percent of the viral-associated cancers are cervical cancers caused by high-risk types of HPV and liver cancer that is caused by HBV and HCV.

Transformed cells undergo phenotypic, genetic, and metabolic changes when grown in culture. The cells become immortalized, undergoing a loss of cell cycle control. Cell surfaces may change in membrane structure and function, and tumor-associated carbohydrate antigens are expressed. The cells display the loss of the need for adhesion, resulting in anchorage-independent growth. When transformed cells are injected into immune deficient nude mice, they may produce tumors.

When cancer cells of tumors are observed *in vivo*, their properties are somewhat similar to that seen in transformed cells in culture. These cells show an increase in telomerase activity and increased levels of enzymes involved in nucleic acid synthesis and lytic enzymes. The DNA of the cells exhibit altered methylation patterns. The cells divide out of control in part due to the loss of tumor suppressor gene function. The cancer cells escape the immunosurveillance response. Tumors arise from a series of events that lead to greater loss of the regulation of cell division.

DNA or RNA tumor viruses are associated with cancer. Retroviruses are RNA tumor viruses that encode reverse transcriptase, an enzyme that is involved in the complex process that converts RNA into DNA. Retroviruses contain at least three genes—*gag*, *pol*, and *env*—in addition to repetitive sequences (LTRs) located at the ends of the genome. Some retroviruses contain an additional gene, an oncogene referred to as *v-onc*. *V-onc* genes are capable of inducing cellular transformation. They are not essential for viral replication. Instead, the *v-onc* gene was hijacked from the genome of its host during the course of evolution.

Retroviruses can cause cancer via the integration of a provirus into the host's genome if it integrates within or near a cellular oncogene (*c-onc*). Provirus integration may result in the insertional inactivation of a cellular tumor suppressor gene, resulting in uncontrolled cell division. If the retrovirus contains a *v-onc*, it may become activated and transcribed, coding for the oncogenic product. The *v-onc* may be a mutated form of a *c-onc* that interferes with cellular signaling involved in growth, promoting growth and cell division at inappropriate times during the cell cycle. Lists of *c-onc*, cellular tumor suppressor genes, *v-onc* genes, and their functions are found in Tables 10-2, -3, and -4, respectively.

Approximately 8% of the human genome contains sequence similarity to infectious retroviruses. These sequences are termed human endogenous retroviruses (HERVs). HERVs contain the easily recognized *gag*, *pol*, *env*, and LTRs. Despite intense efforts, there is little direct evidence to support the claim that HERVs are etiological cofactors in chronic diseases such as cancer. Two human retroviruses have been implicated in human cancers: HTLV-1 causes adult T-cell leukemia, and HTLV-2 is associated with T-cell malignancies.

DNA tumor viruses are a diverse group of viruses with different structures, genome organization, and replication strategies. Some induce tumors in humans, such as EBV, KSHV, HPV, and HBV. Others induce tumors in experimental animals, such as adenovirus and SV-40. Most of the oncogenes of DNA tumor viruses are essential viral replication genes in contrast to the oncogenes of retroviruses that are cellular counterparts. DNA tumor viruses inactivate tumor suppressor gene products of the host such as p53 and pRb. The oncogenic potential of DNA tumor viruses is relatively low. In general, most cancers caused by DNA tumor viruses occur when cells are persistently infected for years.

Burkitt's lymphoma is an aggressive childhood B-cell lymphoma that is prevalent in central Africa. It is characterized by a rapidly growing tumor in the jaw bones, orbital region, or abdomen. Burkitt's lymphoma is intimately associated with EBV infection. The precise mechanism by which the virus transforms B cells is still uncertain.

Kaposi's sarcoma is a rare skin cancer that is caused by KSHV (also known as human herpesvirus 8). The mRNAs of latent viral genes are found in Kaposi's sarcoma tumor cells. KSHV has a large genome that codes for more than 80 genes. Of these 80 genes, three of them encode cellular homologs involved in growth regulation and may play a role in cellular transformation.

Hepatocellular carcinoma (HCC or liver cancer) is the fifth most common cancer in the world. HBV (a dsDNA virus) and HCV (an ssRNA virus) both cause chronic hepatitis that can progress to cirrhosis and HCC. Despite their genome and replication strategy differences, they share common characteristics in the mechanisms of chronic liver diseases and their association with HCC.

Human papillomavirus (HPV) infections are common among sexually active adults and adolescents. High-risk types of HPVs cause cancer of the cervix, vulva, vagina, anus, and penis. HPV is a small DNA tumor virus that infects basal epithelial cells through a break in the skin. It takes years for HPV lesions to become cancerous. Benign papillomas contain an episomal form of the viral genome. Malignant HPV papillomas contain integrated viral E6/E7 oncogenes into the host chromosome.

Adenoviruses and some of the polyomaviruses (e.g., SV-40) cause tumors in experimental animals such as hamsters or other rodents. Adenovirus causes mild diseases of the respiratory tract, gastrointestinal tract, and conjunctivitis ("pink eye") in humans. Adenoviruses can cause severe respiratory epidemics in military barracks. From the early 1960s until 1995, U.S. military recruits were vaccinated against adenovirus types that caused febrile (fever) respiratory infections. Adenoviruses have been prime candidates as gene therapy vectors because they can package many additional genes. The major technical issue that needs to be addressed is the control of an acute immune response against the viral vector. If this can be achieved, it will improve its effectiveness as a therapeutic and reduce side effects.

SV-40 is a small dsDNA virus that was discovered as a contaminant of African green monkey kidney cells in culture. It causes tumors in hamsters that correlate to rare tumors of the lung, brain, bone, and lymphomas in humans. The SV-40 genome contains two sets of genes: those expressed early or those express late after infection. The oncogenic properties of SV-40 have been mapped to the large and small T antigen genes. Large T antigen disrupts the function of cellular p53 and pRb tumor suppressor genes. Small T antigen interacts with a cellular phosphatase PP2A.

Cancer is a multistep process. Viruses are thought to be co-carcinogens in the development of human tumors. The relative simplicity of the viral genome (compared to the enormous complexity of the cellular genetic material) has permitted the identification of genes involved in the genesis of human cancer.

Even before viruses were discovered or visualized by the electron microscope, physicians reported observations of malignant tumors regressing during a natural virus infection. Most of the patients who experienced the short-lived remissions coinciding with a viral infection had cancers that caused immune suppression such as leukemia or lymphomas. Physicians began to carry out significant clinical trials based on these observations. In the 1940s, terminally ill cancer patients volunteered for the first virotherapy clinical trials. Virotherapy, also called oncolytic virotherapy, is an experimental form of cancer therapy in which viruses are used to target and destroy cancer cells without harming healthy cells. Virotherapy trials slowed down during the 1970s to 1990s. As technology improved, along with a better understanding of tumor biology, virotherapy to treat cancer was re-energized. Improved oncolytic virotherapy strategies are on the horizon. Today there is a diverse arsenal of oncolytic viruses with proven ability to replicate selectively in tumor tissue, and more are being developed. The majority of them have been genetically modified in some way to reduce their pathogenicity in humans without destroying their oncolytic potency. Oncolytic viruses will become a major new weapon in the physician's arsenal to combat cancer.

CASE STUDY QUESTIONS

These questions relate to the Case Study presented at the beginning of the chapter.

1. Could a DNA tumor virus be causing John's signs and symptoms? If so, which one(s)?
2. If the biopsy is negative but John does have an infection with a tumor virus, what types of treatment might John undergo? (For hints, refer to the website http://www.cdc.gov and/or to Chapter 17.)
3. Is John infectious? Will Tom need to take special precautions when visiting his grandfather? (For hints refer to the website http://www.cdc.gov and/or to Chapter 17.)
4. Is there a vaccine that could have prevented this? Elaborate upon your answer.
5. How did John become infected? In other words, what was the likely mode of transmission of this viral infection?

CASE STUDY 1: CANCER THERAPY

I Am Legend (2007) is a post-apocalyptic science fiction film. The movie opens with a TV news broadcast of a reporter interviewing Dr. Alice Krippen. Dr. Krippen explains that she has engineered a measles virus (Krippen Virus or K.V.) to be helpful rather than harmful. This measles virus was engineered to cure cancer. According to Dr. Krippen, 10,009 human subjects had been infected with the engineered measles virus during clinical trials. All of them were cured of cancer.

1. Given your new knowledge of oncolytic viruses, the idea of engineering measles virus against cancer isn't just science fiction. Compared to standard forms of cancer therapy like chemotherapy and radiation, why are oncolytic viruses less toxic to the patient?
2. List the characteristics of an "ideal" replicating oncolytic virus.

Then the film jumps ahead to a scene in New York City, three years after the interview with Dr. Krippen. Military virologist Robert Neville is the last surviving uninfected healthy human in New York City. The Krippen Virus had mutated into a lethal strain, killing 90% of the human population. There are infected survivors in the city, who are called "Darkseekers." Because of their sensitivity to ultraviolet light, they hide in dark places and inside of buildings during the day. These Darkseekers are aggressive and have some basic problem-solving intelligence. They are cannibals and hunt infected humans.

3. Are there viruses that turn humans into Darkseekers? What types of characteristics would a virus have to create a Darkseeker (for example, cellular tropism, molecular effects, or pathogenesis)?

Neville spends time performing experiments on rats. He traps a Darkseeker in attempts to treat or cure the disease with serum (which he referred to as a "vaccine").

4. Is serum used as a vaccine? Is serum used to cure individuals of viral infections? Explain your answer.
5. Most adults have preexisting immunity to measles virus through vaccination. Discuss how this may be a roadblock and/or a safety feature in the use of a modified measles virus as a cancer therapy.
6. Using oncolytic viruses to treat cancer is not a new idea. There were numerous attempts in the 1950s and 1960s using oncolytic viruses to treat tumors but researchers abandoned the field. Speculate why its use was abandoned and why there is renewed interest today.

Resources

Primary Literature and Reviews

Alisjahbana, B., et al., 2010. "Disfiguring generalized verrucosis in an Indonesian man with idiopathic CD4 lymphopenia." *Arch Dermatol* 146(1):69–73.

Asada, T. 1974. "Treatment of human cancer with mumps virus." *Cancer* 34(6):1907–1928.

Barbanti-Brodano, G., et al. 2004. "Simian virus 40 infection in humans and association with human diseases: Results and hypotheses." *Virology* 318(1):1–9.

Bernier, J., et al. 2004. "Radiation oncology: A century of achievements." *Nat Rev Cancer* 4: 737–747.

Block, T. M., et al. 2003. "Molecular viral oncology of hepatocellular carcinoma." *Oncogene* 22(33):5093–5107.

Boch, F. X., et al. 2004. "Primary liver cancer: Worldwide incidence and trends." *Gastroenterology* 127(5 Suppl 1): S5–S16.

Breitbach, C. J., et al. 2010. "Navigating the clinical development landscape for oncolytic viruses and other cancer therapeutics: No shortcuts on the road to approval." *Cytokine Growth Factor Rev* 21:85–89.

Bridle, B. W., et al. 2010. "Combining oncolytic virotherapy and tumour vaccination." *Cytokine Growth Factor Rev* 21: 143–148.

Chang, C. H., et al. 2011. "Prevalence of human herpesvirus 8 DNA in peripheral blood mononuclear cells of the acute and chronic leukemia patients in Taiwan." *FEMS Immunol Med Microbiol* 61(3):356–358.

Chen, H. L. 1996. "Seroepidemiology of hepatitis B virus infection in children: Ten years of mass vaccination in Taiwan." *JAMA* 276(11):906–908.

Cheng-Liang, L., et al. 2003. "Trends in the incidence of hepatocellular carcinoma in boys and girls in Taiwan after large-scale hepatitis B vaccination." *Cancer Epidemiol Biomarkers Prev* 12:57–59.

Centers for Disease Control and Prevention (CDC). 2007. "Quadrivalent human papillomavirus vaccine." *MMWR* 56:1–24.

Comins, C., et al. 2008. "Reovirus: Viral therapy for cancer 'as nature intended.'" *Clin Oncol* 20:548–554.

Costas, J. 2002. "Characterization of the intragenomic spread of the human endogenous retrovirus family HERV-W." *Mol Biol Evol* 19(4):526–533.

De Martel, C., and Franceschi, S. 2009. "Infections and cancer: Established associations and new hypotheses." *Crit Rev Oncol Hematol* 70:183–194.

Dock, G. 1904. "The influence of complicating diseases upon leukemia." *Am J Med Sci* 127(4):563–592.

Engels, E. A., et al. 2004. "Case-control study of simian virus 40 and non-Hodgkin lymphoma in the United States." *J Natl Cancer Inst* 96(18):1368–1374.

Engels, E. A., and Chen, J. 2004. "Poliovirus vaccination during pregnancy, maternal seroconversion to simian virus 40, and risk of childhood cancer." *Am J Epidemiol* 160(4):306–316.

Engels, E. A., et al. 2004. "Serological evidence for exposure to simian virus 40 in North American zoo workers." *J Infect Dis* 190(12): 2065–2069.

Erlwein, O., and McClure, M. O. 2010. "Progress and prospects: Foamy virus vectors enter a new age." *Gene Therapy* 17:1423–1429.

Everts, B., and van der Poel, H. G. 2005. "Replication-selective oncolytic viruses in the treatment of cancer." *Cancer Gene Ther* 12:141–161.

Farber, S., et al. 1948. "Temporary remissions in acute leukemia in children produced by folic acid antagonist, 4-aminopteroyl-glutamic acid (aminopterin)" *N Engl J Med* 238(23):786–793.

Fiander, A. 2011. "The prevention of cervical cancer in Africa." *Womens Health (Lond)* 7(1);121–132.

Fielding, A. K. 2005. "Measles as a potential oncolytic virus." *Rev Med Virol* 15:135–142.

Frei, E., et al. 1965. "The effectiveness of combinations of antileukemic agents in inducing and maintaining remission in children with acute leukemia." *Blood* 26(5): 642–656.

Gazdar, A. F., Butel, J. S., and Carbone, M. 2002. "SV40 and human tumours: Myth, association or causality?" *Nat Rev* 2:957–964.

Georgiades, J., et al. 1959. "Research on the oncolytic effect of APC viruses in cancer of the cervix uteri." *Biul Inst Med Morsk Gdansk* 10:49–57.

Goodman, L. S., et al. 1946. "Nitrogen mustard therapy." *JAMA* 132:126–132.

Griffiths, D. J. 2001. "Endogenous retroviruses in the human genome sequence." *Genome Biol* 2(6):1017.1–1017.5.

Griffiths, M. 1991. "Nuns, virgins, and spinsters': Rigoni-Stern and cervical cancer revisited." *Br J Obstet Gynaecol* 98:797–802.

Giuliano, A. R., et al. 2011. "Efficacy of quadrivalent HPV vaccine against HPV infection and disease in males." *N Engl J Med* 364:401–411.

Hammill, A. M., et al. 2010. "Oncolytic virotherapy reaches adolescence." *Pediatr Blood Cancer* 55(7):1253–1263.

Harper, D. M., et al. 2004. "Efficacy of a bivalent L1 virus-like particle vaccine in prevention of infection with human papillomavirus types 16 and 18 in young women: A randomized controlled trial." *Lancet* 364:1757–1765.

Harrington, K. J., et al. 2010. "Clinical trials with oncolytic reoviruses: Moving beyond phase I into combinations with standard therapeutics." *Cytokine Growth Factor Rev* 21:91–98.

Hartkopf, A. D., et al. 2011. "Oncolytic virotherapy of gynecologic malignancies." Gynecol Oncol 120(2): 302–310.

Herrero, R., et al. 2003. "Human papillomavirus and oral cancer: The International Agency for Research on Cancer Multicenter Study." *J Natl Cancer Inst* 95(23): 1772–1783.

Hoster, H. A., et al. 1949. "Studies in Hodgkin's syndrome IX. The association of 'viral' hepatitis and Hodgkin's disease (a preliminary report)." *Cancer Res* 9:473–480.

Jeang, K. T. 2010. "HTLV-1 and adult T-cell leukemia: Insights into viral transformation of cells 30 years after virus discovery." *J Formos Med Assoc* 109(10):688–693.

Kapadia, R., and Coffey, M. C. 2010. "The use of immunohistochemistry to determine oncolytic reovirus distribution and replication in human tumors." *Methods* 52: 301–306.

Karlsson, H., et al. 2001. "Retroviral RNA identified in the cerebrospinal fluids and brains of individuals with schizophrenia." *PNAS USA* 98(8):4634–4639.

Kaul, S. A., et al. 2010. "Tumor viruses and cancer biology: Modulating signaling pathways for therapeutic intervention." *Cancer Biol Ther* 10(10):961–978.

Kelly, E., and Russell, S. J. 2007. "History of oncolytic viruses: Genesis to genetic engineering." *Mol Ther* 15(4):651–659.

Kirn, D. H. 2006. "The end of the beginning: Oncolytic virotherapy achieves clinical proof-of-concept." *Mol Ther* 13(2):237–238.

Le Boeuf, F., and Bell, J. C. 2010. "United virus: The oncolytic tag-team against cancer!" *Cytokine Growth Factor Rev* 21:205–211.

Lerner, B. H. 2004. "Sins of omission—Cancer research without informed consent." *N Engl J Med* 351(7): 628–630.

Lewis, D. A. 2001. "Retroviruses and the pathogenesis of schizophrenia." *PNAS USA* 98(8):4293–4294.

Li, M. C., et al. 1958. "Therapy of choriocarcinoma and related trophoblastic tumors with folic acid and purine antagonists." *N Engl J Med* 259(2):66–74.

Liu, T. C., and Kirn, D. 2008. "Gene therapy progress and prospects cancer: Oncolytic viruses." *Gene Ther* 15: 877–884.

Lopez-Rios, F., et al. 2004. "Evidence against a role for SV40 infection in human mesotheliomas and high risk of false-positive PCR results owing to presence of SV-40 common sequences in common laboratory plasmids." *Lancet* 364:1157–1166.

Lu, W., et al. 2004. "Intra-tumor injection of HB101, a recombinant adenovirus, in combination with chemotherapy in patients with advanced cancers: A pilot phase II clinical trial." *World J Gastroenterol* 10(24): 3634–3638.

Machida, K., et al. 2004. "Hepatitis C virus induces a mutator phenotype: Enhanced mutations of immunoglobin and proto-oncogenes." *PNAS USA* 101(12): 4262–4267.

Meiering, C. D., and Linial, M. L. 2001. "Historical perspective of foamy virus epidemiology and infection." *Clin Microbiol Rev* 14(1):165–176.

Meyers, C., and Laimins, L. A. 1994. "*In vitro* systems for the study and propagation of human papillomaviruses." *Curr Top Microbiol Immunol* 186:199–215.

Munger, K., et al. 2004. "Mechanisms of human papillomavirus-induced oncogenesis." *J Virol* 78(21): 11451–11460.

Nakamura, A., et al. 2003. "Human endogenous retroviruses with transcriptional potential in the brain." *J Hum Genet* 48(11):575–581.

Ostrow, R. S., et al. 1986. "Detection of papillomavirus DNA in human semen." *Science* 23:731–733.

Parish, C. R., 2003. "Cancer immunotherapy: The past, the present and the future." *Immunol Cell Biol* 81: 106–113.

Petrizzo, B. L., et al. 2011. "Translating tumor antigens into cancer vaccines." *Clin Vaccine Immunol* 18(1):23–34.

Poiesz, B. J., Poiesz, M. J., and Choi, D. 2003. "The human T-cell lymphoma/leukemia viruses." *Cancer Invest* 21(2):253–277.

Pomfret, T. C., et al. 2011. "Quadrivalent human papillovirus (HPV) vaccine: A review of safety, efficacy, and pharmacoeconomics." *J Clin Pharm Ther* 36(1):1–9.

Quetglas, J. I., et al. 2010. "Alphavirus vectors for cancer therapy." *Virus Res* 153:179–196.

Rommelaere, J., et al. 2010. "Oncolytic parvoviruses as cancer therapeutics." *Cytokines Growth Factor Rev* 21: 185–195.

Rui-Mei, L., et al. 2002. "Molecular identification of SV40 infection in human subjects and possible association with kidney disease." *J Am Soc Nephrol* 13:2320–2330.

Russel, S. J., and Peng, K. W. 2007. "Viruses as anticancer drugs." *Trends Pharmacol Sci* 28(7):326–333.

Sadagopan, S., et al. 2011. "Kaposi's sarcoma-associated herpesvirus-induced angiogenin plays roles in latency via the phospholipase Cγ pathway: Blocking angiogenin inhibits latent gene expression and induces the lytic cycle." *J Virol* 85(6):2666–2685.

Skrabanek, P. 1988. "Cervical cancer in nuns and prostitutes: A plea for scientific continence." *J Clin Epidemiol* 41(6):577–582.

Southman, C. M., and Moore, A. E. 1952. "Clinical studies of viruses as antineoplastic agents, with particular reference to Egypt 101 virus." *Cancer* 5:1025–1034.

Southman, C. M., et al. 1957. "Homotransplantation of human cell lines." *Science* 125:158–160.

Southman, C. M., and Moore, A. E. 1958. "Induced immunity to cancer cell homografts in man." *Ann NY Acad Sci* 73:635–653.

Stanford, M. M., et al., 2010. "Novel oncolytic viruses: Riding high on the next wave?" *Cytokine Growth Factor Rev* 21:177–183.

Starnes, C. O. 1992. "Coley's toxins in perspective." *Nature* 357–358.

Tsung, K., and Norton, J. A. 2006. "Lessons from Coley's toxin." *Surg Oncol* 15:25–28.

Vilchez, R. A., and Butel, J. S. 2004. "Emergent human pathogen simian virus 40 and its role in cancer." *Clin Microbiol Rev* 17(3):495–508.

Zemp, F. J., et al. 2010. "Oncolytic viruses as experimental treatments for malignant gliomas: Using a scourge to treat a devil." *Cytokine Growth Factor Rev* 21:103–117.

Zimet, G., et al. 2005. "Parental attitudes about sexually transmitted infection vaccination for their adolescent children." *Arch Pediatr Adolesc Med* 159:132–137.

Zimmer, C. 2001. "Do chronic diseases have an infectious root?" *Science* 293:1974–1977.

Popular Press

Bookchin, D., and Schumacher, J. 2004. *The Virus and the Vaccine: The True Story of a Cancer-Causing Monkey Virus, Contaminated Polio Vaccine, and the Millions Exposed.* New York: St. Martin's Press.

Crawford, D. H. 2005. "An Introduction to Viruses and Cancer." *Microbiol Today* August 5:110–112.

Ewald, P. W. 2002. *Plague Time: The New Germ Theory of Disease.* New York: Anchor Books.

Gallo, R. 1991. *Virus Hunting, Aids, Cancer & The Human Retrovirus: A Story of Scientific Discovery.* New York: Basic Books.

Higley, T. L. 2005. *Retrovirus.* Uhrichsville, OH: Barbour. (Fiction)

Li, J. J. 2006. *Laughing Gas, Viagra, and Lipitor: The Human Stories Behind the Drugs We Use.* New York: Oxford University Press.

Matheson, R. 2007. *I am Legend.* New York, Tor Books. (Fiction)

Mukherjee, S. 2010. *The Emperor of All Maladies: A Biography of Cancer.* New York, Scribner.

Rosenberg, S. A., and Barry, J. M. 1992. *The Transformed Cell: Unlocking the Mysteries of Cancer.* New York: Avon Books.

Skloot, R. 2010. *The Immortal Life of Henrietta Lacks.* New York: Crown.

Weinberg, R. A. 1999. *One Renegade Cell (Science Masters).* New York: Basic Books.

Video Productions

Bad Blood. 2010. Necessary Films.

I am Legend. 2007. Warner Brothers. (Fiction)

NewsHour Medical Ethics and Issues. 2008. Films for the Humanities.

Understanding Hepatitis. 2007. Films for the Humanities.

My Shocking Story—Half Man Half Tree. 2007. Discovery Channel

Waging War on Cancer. 2006. Films for the Humanities.

History of Cancer. 2006. Films for the Humanities.

New Directions: The Fight Against Cancer. 2004. Films for Humanities.

Voices: Cancer Patients Speak Out. 2004. Films for the Humanities.

Cancer Story. 2004. Films for the Humanities.

Trial and Error: The Rise and Fall of Gene Therapy. 2003. BBCW Production.

Fighting Hepatitis C. 2001. Films for the Humanities.

Sexually Transmitted Diseases: The Silent Epidemic. 2000. Films for the Humanities.

Great Minds of Science: Viruses. 1997. Films for the Humanities.

Hepatitis B: The Enemy Within. 1996. Films for the Humanities.

Hepatitis C: The Silent Scourge. 1995. Films for the Humanities.

eLearning

go.jblearning.com/shors2

The site features eLearning, an online review area that provides quizzes and other tools to help you study for your class. You can also follow useful links for in-depth information, or just find out the latest virology and microbiology news.

CHAPTER 11

Poliovirus and Other Enteroviruses

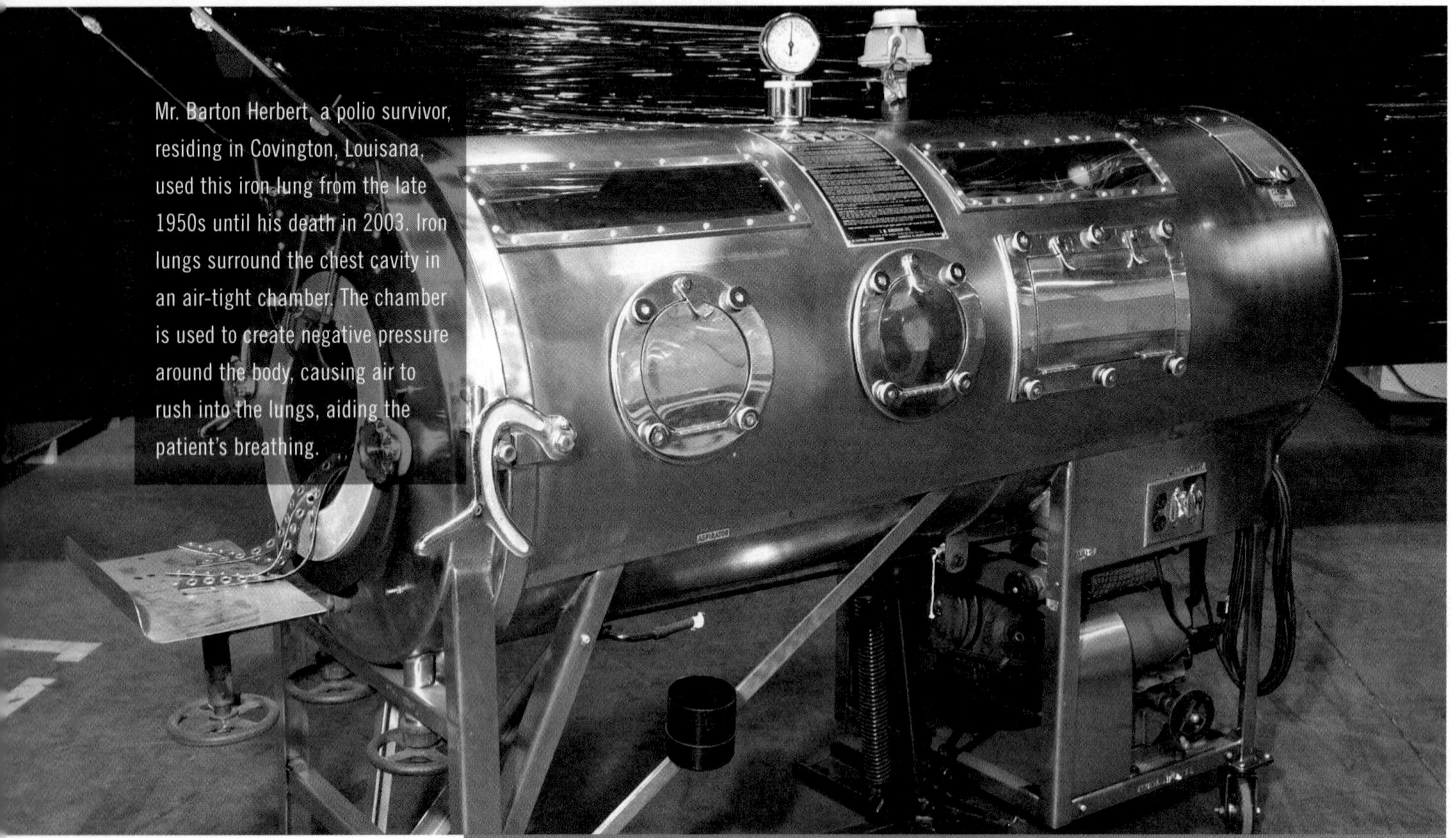

Mr. Barton Herbert, a polio survivor, residing in Covington, Louisana, used this iron lung from the late 1950s until his death in 2003. Iron lungs surround the chest cavity in an air-tight chamber. The chamber is used to create negative pressure around the body, causing air to rush into the lungs, aiding the patient's breathing.

OUTLINE

"The fight against infantile paralysis is a fight to the finish, and the terms are unconditional surrender."

Franklin Delano Roosevelt, 32nd U.S. President and polio survivor, 1944

CASE STUDY: POLIOMYELITIS

During July 2005, a seven-month-old unvaccinated infant girl in an Amish community was hospitalized in a central Minnesota hospital because she was failing to thrive. The child suffered from diarrhea and recurrent infections and was diagnosed as having a severely compromised immune system. She was placed in isolation and evaluated for a bone marrow transplant. Meanwhile, the Minnesota Department of Health (MDH) began an investigation that isolated an enterovirus from the infant's stool sample and identified it as a poliovirus type 1 vaccine-derived poliovirus Sabin strain. The isolate's genomic sequence, however, was determined to be 2.3% different from the parent Sabin poliovirus type 1 strain, and it was estimated that it may have been replicating for about two years in an immune-compromised individual. Oral poliovirus vaccine (OPV) is still widely used in most countries but has not been used in the United States since 2000 and in Canada since 1997.

References

Bahta, L., et al. 2005. "Poliovirus infections in four unvaccinated children—Minnesota, August–October 2005." *MMWR* 54(Dispatch):1–3.

Roberts, L. 2005. "Minnesota polio case stumps experts." *Science* 310:213.

11.1 Brief Overview of Enteroviruses

Enteroviruses, which are RNA viruses of the *Picornaviridae* family, are small, naked icosahedral viruses that multiply in the mucosa of the gut. These viruses are ubiquitous pathogens that are transmitted from person to person by an oral–fecal route. The prefix *entero* comes from the Latin word meaning intestine. In the United States it is estimated that enteroviruses cause five to ten million symptomatic infections, resulting in 30,000 to 50,000 hospitalizations annually. Most infections occur during childhood and they usually produce lifelong immunity.

There are at least 70 distinct types of human enteroviruses, and over 20 recognizable clinical syndromes have been associated with enterovirus infections. **TABLE 11-1** provides examples of these clinical manifestations and the common names of the enteroviruses with which they are associated. In general, most infections cause a mild, self-limiting disease; however, they can produce severe and sometimes fatal illnesses such as **meningitis, encephalitis, myocarditis,** and **poliomyelitis**. When serious conditions do occur, it is the result of a spillover viremia leading to secondary infections of nongastrointestinal cells. This is the case with poliomyelitis, which involves the destruction of spinal cord cells, or carditis, a condition in which the myocardial (heart) cells are adversely affected. Unlike the majority of infectious agents that cause gastroenteritis, enteroviruses are usually shed for prolonged periods (as long as six weeks) in feces. Poliovirus is the most thoroughly studied enterovirus prototype for the molecular virologist and has provided important insight into picornavirus biology.

11.2 The History of Polio

The discovery of the 18th dynasty (1500 B.C.) Egyptian stele described in Chapter 1, which shows a priest with a shriveled leg characteristic of the flaccid (i.e., limp, weak, and lacking muscle tone) paralysis or "foot drop" typical of poliomyelitis, is evidence that this disease has a long history. Paralytic disease did not appear to be a significant problem until outbreaks began in northern Europe in the late 1800s, when reports of polio epidemics of a notable size began to appear intermittently (**FIGURE 11-1**). From that time on, the disease was reported seasonally (peaking in the summer months in temperate climates) and epidemics occurred more often in industrialized countries. In 1916, the United States faced one of the worst polio epidemics of the 20th century, when polio killed 6000 people and paralyzed an additional 27,000. New York City was faced with so many cases that city officials responded by flooding the streets with 54 million gallons of water a day. They thought that better sanitation in the community would defeat polio because increased sanitation had defeated diseases like typhoid, dysentery, and tuberculosis. Unfortunately, flooding the streets did not have any affect on polio cases (**FIGURE 11-2**). Quarantine was also advised—in some cases, an entire town

TABLE 11-1	Human Enteroviruses and Their Clinical Syndromes
Virus	**Syndrome**
Polioviruses, types 1–3	Paralysis (complete to partial muscle weakness) Aseptic meningitis Summer "fever" illness
Coxsackieviruses, Group A, types 1–24	Herpangina (mouth blisters or sores) Hand-foot-and-mouth disease (A10, A16) "Common cold" Paralysis Aseptic meningitis Exanthema (rash) Hepatitis Infantile diarrhea Pneumonitis (allergic alveolitis) of infants Lymphatic or nodular pharyngitis (sore throat) Acute hemorrhagic conjunctivitis (type A24 variant)
Coxsackieviruses, Group B, types 1–6	Pericarditis, myocarditis Paralysis (infrequent) Macular skin rash Hepatitis Summer "fever" illness Upper respiratory illness and pneumonia Aseptic meningitis Pleurodynia (pain between the ribs or in the chest wall area) Meningoencephalitis and myocarditis in children
Echoviruses, types 1–33	Meningitis Paralysis Encephalitis, ataxia (lack of coordination or muscle control) or Guillain-Barré syndrome Exanthema (rash) Respiratory illness Diarrhea Pericarditis and myocarditis Liver disturbance
Enteroviruses, types 68–71*	Pneumonia and bronchiolitis Acute hemorrhagic conjunctivitis (type 70) Paralysis (types 70, 71) Meningoencephaltitis (types 70, 71) Hand-foot-and-mouth disease (type 71)
Enterovirus 72 (hepatitis A)	Hepatitis

*Since 1969, new enterovirus types have been assigned type numbers rather than being classified as coxsackieviruses or echoviruses. The common names of the previously identified enteroviruses have been retained. (*Source*: Adapted from Melnick, J. L. 1996. "Enteroviruses: Polioviruses, coxsackieviruses, echoviruses, and newer enteroviruses." In: Fields, B. N., Knipe, D. M., Howley, P. M., et al., eds. 1996. *Fields Virology*, Third Edition. Philadelphia: Lippincott-Raven, pp. 655–712.)

was quarantined—and many communities that bordered New York City forbade nonresidents from entering. By 1953, the incidence of paralytic polio in the United States was more than 20 per 100,000. Even though this rate was not as high as that of other common diseases such as measles, polio became a public concern for several reasons:

- It was a disease of mysterious, seasonal appearance.
- It could paralyze respiratory muscles.
- It had disfiguring, crippling, and sometimes fatal effect.

In 1909, before the era of tissue culture experimentation, Karl Landsteiner and Erwin Popper reproduced poliomyelitis disease in two rhesus monkeys. To do this, they created a filtrate from ground-up central nervous system (CNS) tissue that had been removed from a nine-year-old boy who had died from polio. They injected this filtrate into the peritoneum of the monkeys. The monkeys became sick and, based upon microscopic examination of the monkeys' spinal cords, Landsteiner and Popper showed that a filterable agent caused the poliomyelitis. For the next 40 years, poliovirus

1789	British physician Michael Underwood describes polio
1887	Medin describes polio outbreak in Stockholm, Sweden
1894	First major outbreak in U.S. (132 cases in Vermont)
1908	K. Landsteiner and E. Popper showed that a filterable agent caused poliomyelitis in monkey animal model
1916	Large outbreak of polio in the U.S. Over 9000 cases reported in NYC alone; disease killed 6000 and paralyzed 27,000
1921	FDR contracts polio and suffers severe paralysis
1928	Philip Drinker and Louis Shaw invent the Drinker Respirator (iron lung)
1932	FDR elected President of the U.S.
1934	Major outbreak of polio in Los Angeles: ~2500 cases from May to November
1942	The first Sister Kenny Institute opens in Minneapolis
1945	WWII ends; many U.S. polio outbreaks result in 20,000 cases/year (1945–1949)
1952	58,000 cases of polio in the U.S., Salk killed vaccine testing begins
1953	35,000 cases of polio in the U.S.
1954	Salk killed vaccine field trial sponsored by the National Foundation for Infantile Paralysis
1955	Nationwide vaccination program begins
1958–1959	Field trials of Sabin oral live vaccine
1961	Sabin vaccine approved for use
1964	Only 121 cases of polio in the U.S.
1979	Multistate polio outbreaks in Amish communities in the U.S.
1981	*Time* magazine reports post-polio syndrome among survivors of polio
1988	WHO passes resolution to eradicate polio by the year 2000
1994	China immunizes 80 million children against polio
1995	India immunizes more than 87 million children against polio
1996	CDC announces change in polio vaccination in the U.S.: two doses of inactivated polio vaccine (IPV) followed by two doses of oral polio vaccine (OPV)
1999	Number of world polio cases falls to ~7000
2000	OPV vaccination discontinued in the U.S.
2001	575 million children vaccinated in 94 countries
2005	Polio-free areas experience new outbreaks from Nigeria to Sudan
2005	Polio outbreak in Amish community in Minnesota
2006	Polio outbreak in Namibia (affected primarily young adults born before 1990). Namibia was declared polio-free in 2001.
2010	Highly lethal poliovirus outbreak in Congo-Brazzaville (42% fatal in males ages 15–25).
2011	Outbreak of polio war zone at Afghanistan/Pakistan border.

FIGURE 11-1 Timeline of the history of poliomyelitis.

research was limited because it was dependent upon animal models and formalin-treated nervous system tissue as a source of a vaccine for field trials, and appropriate tests for the safety and efficacy of the nervous tissue vaccines were lacking. Results obtained were inadequate to justify human vaccine trials. It was not until the advent of tissue culture techniques that the development of a polio vaccine became realistic.

In 1949 John Enders, Fredrick Robbins, and Thomas Weller published a paper in *Science* that described the successful cultivation of the Lansing strain of poliovirus in human embryonic, nonnervous tissues. Soon, poliovirus was propagated in monkey kidney cells and a number of laboratories began to work toward the development of a polio vaccine. (Polio vaccines are discussed in detail later in this chapter.)

11.3 Clinical Features of Poliomyelitis

Poliomyelitis, or polio, is rare today because of vaccination efforts. The last cases of polio in the United States were unvaccinated individuals in communities that were not vaccinating: 1972

FIGURE 11-2 New York City was the epicenter of the first major polio outbreak in the United States in 1916. This photograph shows a New York City Sanitation worker sweeping the streets (circa 1910).

(Christian Scientists), 1979 and, 2005 (members of Amish communities). After the 1997 transition to all-inactivated poliovirus vaccine (IPV) schedules for routine vaccination of children in the United States, vaccine-associated cases of polio were dramatically reduced. In 2000 and 2009, individuals contracted the vaccine-derived poliovirus from someone who had received the live oral poliovirus vaccine (OPV). The patients had weakened immune systems and multiple health problems.

The mouth is the portal of entry for polioviruses. Person-to-person poliovirus infections are transmitted via an oral–fecal route or, less frequently, an oral–oral route. Direct contact with feces occurs with activities such as diaper changing. Indirect transmission is seen with poor sanitary conditions and may occur by numerous routes including contaminated water, food, and fomites. Infants, particularly those in diapers, appear to be the most efficient transmitters of infection. The incubation period for poliomyelitis is 6 to 20 days but can range from 3 to 35 days. Poliovirus may be present in the stool for 3 to 6 weeks and for approximately 2 weeks in saliva.

The course of infection is variable, and up to 95% of all poliovirus infections are unapparent or asymptomatic. Infected persons who are asymptomatic shed virus in stool and are able to transmit the virus to others. About 4% to 8% of poliovirus infections cause minor illness with symptoms of malaise, gastrointestinal disturbances, fever, influenza-like illness, and sore throat. A complete recovery occurs within a week. In 1% to 2% of infected persons, following a minor illness, symptoms of stiffness of the neck, back, and/or legs occur. These symptoms persist for two to ten days followed by a complete recovery. Less than 1% of all polio infections cause a major illness, resulting in clinical signs involving the CNS, such as **flaccid paralysis**. These severe symptoms depend on the strain and type of virus. There are three poliovirus serotypes (called types 1, 2, and 3). Infection with one serotype does not confer immunity against a different serotype. Infection with serotype 1 is most likely to cause paralysis, and serotype 2 is the least likely to cause paralysis.

The term *poliomyelitis* is derived from the Greek words ***polios*** (gray), ***myelos*** (refers to marrow), and ***itis*** (inflammation). The severe form of the disease involves the inflammation and sometimes destruction of neurons located in the gray matter of the anterior horn of the spinal cord. Poliovirus infects motor neurons, not sensory neurons. Recovery may take up to two years and may be incomplete; weakened muscles may be a permanent result. In paralytic poliomyelitis the inflammation is severe enough to destroy the motor nerves and the resulting paralysis is permanent. The muscles that are commonly weakened by polio are shown in **FIGURE 11-3**.

The major illness is divided into three forms:

- spinal paralysis
- bulbar
- bulbospinal

Spinal poliomyelitis is more common than bulbar poliomyelitis and in general is characterized by asymmetric paralysis (occurring on one side of the body). Invasion of the lower spinal cord by poliovirus results in paralysis of the legs, known as spinal poliomyelitis. Invasion of the upper spinal cord and medulla by poliovirus results in **bulbar poliomyelitis** and paralysis of breathing. It is less common and results in the weakness of muscles, impairing the ability to talk or swallow, and the patient requires an iron lung or respirator. **Bulbospinal poliomyelitis** is a combination of bulbar and spinal paralysis.

People are the most infectious immediately before and one to two weeks after the onset of paralytic disease. About 80% of individuals who suffer from severe symptoms will experience permanent paralysis, 10% will die, and 10% will make a full recovery.

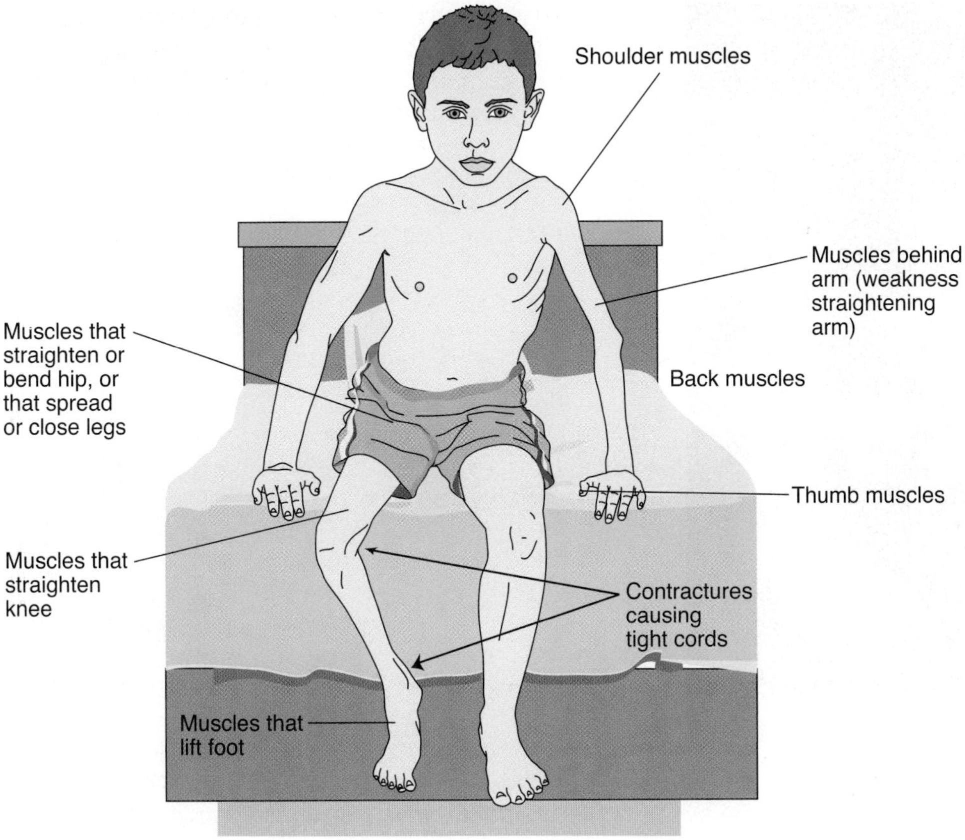

FIGURE 11-3 Muscles commonly weakened by poliomyelitis. Note a contracture in this context is a shortening of the muscle caused by paralysis of the antagonist muscle, resulting in distortion or deformity of a body joint. Adapted from Werner, J. *Disabled Village Children: A Guide for Community Health Workers, Rehabilitation Workers, and Families,* Second Edition. The Hesperian Foundation, 1999.

Post-Polio Syndrome

Post-poliomyelitis syndrome (**PPS**) occurs in a large proportion of individuals who recovered from paralytic poliomyelitis. The time between acute poliomyelitis and PPS ranges from 8 to 71 years with a mean interval of 36 years. The onset of symptoms is usually insidious. The most common three symptoms are new weakness in muscles or limbs involved at the time of acute poliomyelitis, fatigue, and pain in the muscles and joints. Fatigue is worsened during physical activity. Other less frequent symptoms are difficulty breathing, sleep abnormalities, joint deformities, atrophy and twitching, speech problems, difficulty swallowing, and cold intolerance. PPS can result in disabilities and handicaps. Mobility, especially stair climbing, is most frequently affected.

The cause of PPS remains unclear. The most widely accepted hypothesis suggests that the motor neurons infected with poliovirus degenerate and die. Surviving neurons sprout new nerve endings to restore muscle function. In PPS, the muscle fibers of the surviving motor neurons slowly deteriorate over time. Eventually nerve endings are destroyed and permanent weakness occurs (**FIGURE 11-4**). Effective management of PPS requires an interdisciplinary approach using a team of physicians and healthcare personnel. The approach must be symptom-specific (e.g., management of pain, weakness, and fatigue). Experimental therapies such as the use of an interferon (antiviral), prednisone (anti-inflammatory drug), low doses of pyridostigmine (nerve gas antidote), and insulin-like growth factor (to regenerate nerve endings) have not shown significant improvement in PPS symptoms.

11.4 Classification and Structure of Poliovirus

Polioviruses are small (~30 nm in diameter), acid-stable (surviving pH of 3.0 or lower), nonenveloped, icosahedral-shaped viruses that belong to the genus Enterovirus and family *Picornaviridae* (from the Greek *pico,* for "small"). In 1985, poliovirus type 1 was one of the first human viruses whose three-dimensional structure was solved in atomic detail by X-ray crystallography. The poliovirus virion consists of four distinct capsid proteins designated VP1, VP2, VP3, and VP4. The basic building block (protomer) of the capsid consists of one copy of VP1–VP4. The capsid is made up of 60 protomers with icosahedral symmetry to form a spherical particle. The shell of the virus is formed by VP1 to VP3, and VP4 lies on its inner surface. VP1, VP2, and VP3 do not share sequence homology, yet they all form the same topology. There are

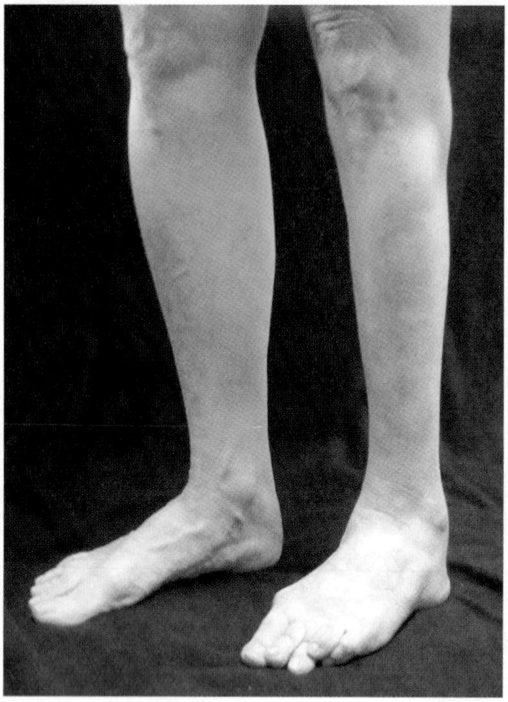

a

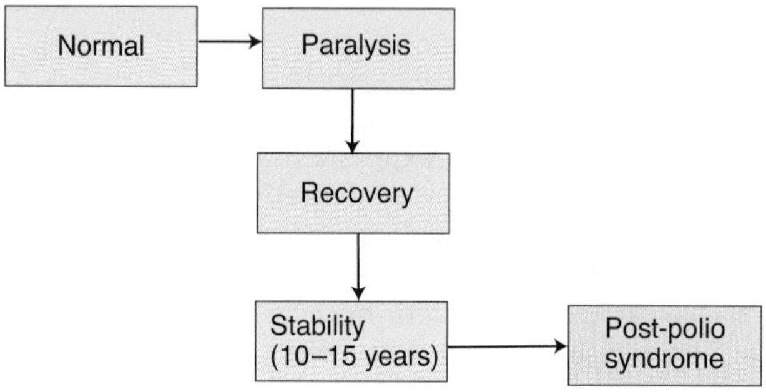

Normal
Paralysis
Recovery
Stability
(10–15 years)
Post-polio
syndrome

b

FIGURE 11-4 (a) Asymmetric atrophy of left leg in patient with PPS. **(b)** Clinical course of PPS. Overlapping boxes represent times when the motor neurons are becoming less stable. **(c)** Hypothetical model for the cause of PPS. It shows the changes in motor neurons before, during, and after poliovirus infection in individuals who experienced the paralytic form of the disease. Adapted from Raut, A., "After the storm," *Meducator* 2 (2003).

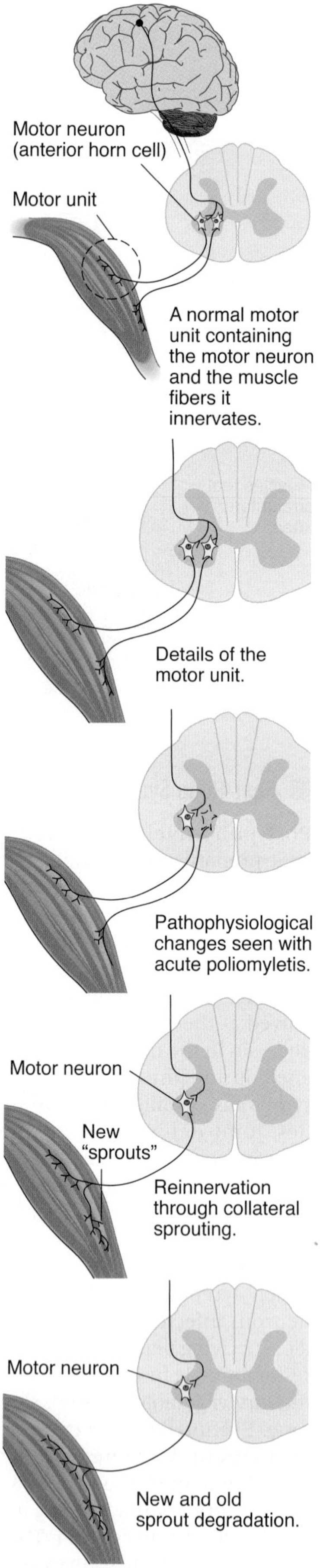

Motor neuron
(anterior horn cell)
Motor unit
A normal motor unit containing the motor neuron and the muscle fibers it innervates.
Details of the motor unit.
Pathophysiological changes seen with acute poliomyletis.
Motor neuron
New "sprouts"
Reinnervation through collateral sprouting.
Motor neuron
New and old sprout degradation.

c

TABLE 11-2 Origin of the Three Original Poliovirus Strains

Type	Strain	Source/Geographic Origin
1	Brunhilde	Poliovirus isolated from spinal cord of a chimpanzee, Brunhilde, who was inoculated in 1939 with stool specimens from seven paralytic patients in Baltimore, MD.
2	Lansing	From patient who died of paralytic polio in Lansing, MI, in 1938.
3	Leon	Brain and spinal cord specimen from 11-year-old boy named Leon, who died from paralytic polio in Los Angeles, CA, in 1937.

three antigenic types (serotypes 1, 2, and 3) of poliovirus. All polioviruses within a serotype can be neutralized by type-specific antisera. **TABLE 11-2** lists the three original poliovirus strains isolated by American scientists in the 1950s along with the name of a strain and its origin. Today they are known as types 1, 2, and 3. Albert Sabin attenuated three different serotypes for use as an oral live trivalent vaccine. The genome of poliovirus consists of one infectious (+) sense single-stranded RNA (ssRNA) molecule that is 7441 nucleotides in length. It will be described in further detail later in this chapter in the discussion of viral replication.

Stability of Enteroviruses in the Environment

Enteroviruses are able to pass through the stomach and the small intestine because they are resistant to pH levels less than three as well as to several proteases. Enterovirus particles are nonenveloped, making them resistant to detergents, 70% alcohol, and other lipid solvents such as ether and chloroform. They are also resistant to disinfectants such as 5% Lysol and 1% quaternary ammonium compounds. Enteroviruses are usually present in organic matter such as feces; therefore, standard **inactivation protocols** may require prolonged contact time. **Inactivated** viruses are dead. In other words, the virus is not destroyed but it is no longer able to infect cells. Inactivation protocols are developed to determine what procedures are necessary to decontaminate or completely disinfect surfaces where viruses are present.

Chemical inactivation of enteroviruses can be achieved with chlorine, hydrochloric acid, and aldehydes. Heating at 50°C (122°F) for one hour in the absence of calcium and magnesium will also inactivate enteroviruses. Viruses may be stable several days to several weeks at 4°C (39.2°F).

11.5 Laboratory Diagnosis of Poliovirus Infections

In the past, diagnosis was based on the recovery of poliovirus from stool and sometimes throat samples but not from cerebral spinal fluid (CSF). Poliovirus grows rapidly (the life cycle is complete within a few days) in any human or monkey cell lines and shows good cytopathic effects. The serotype of the isolate is identified using neutralization assays. The nucleotide sequences of all wild-type and vaccine-type strains have been identified, and as a result any poliovirus isolated from a person with flaccid paralysis is tested further (e.g., genomic sequencing) to determine if the infection is a result of an infection with either vaccine-derived (e.g., a reverted Sabin poliovirus strain) or wild-type virus. Improved laboratory diagnosis of poliomyelitis is an important part of the World Health Organization (WHO) initiative for global eradication of poliomyelitis.

11.6 Cellular Pathogenesis

Humans and nonhuman primates are the only natural hosts of poliovirus. There are no animal reservoirs. Nonhuman primates have been used as models of human conditions. Mice or any other affordable experimental animals were found to be resistant to poliovirus infection, presumably due to the lack of the poliovirus receptor. Natural infection occurs via ingestion of polioviruses.

Once ingested, polioviruses invade two lymphoid tissues: the tonsils and the Peyer's patches (lymph nodes in the walls of the intestines near the junction of the ileum and colon). The virus spreads via the regional lymph nodes to the blood. A major viremia causes the minor illness symptoms in infected individuals (e.g., sore throat, headache, and fever). **FIGURE 11-5** shows how polioviruses spread through the bloodstream. In a very small percentage of patients, the CNS is involved. The polioviruses are carried via the bloodstream to the anterior horn cells of the spinal cord in which the virus replicates, resulting in lesions that are widely distributed throughout the spinal cord and parts of the brain. The hallmark of poliomyelitis is the selective destruction of motor neurons that leads to paralysis and in severe cases, respiratory arrest, and death. The molecular mechanism by which poliovirus infection causes poliomyelitis is

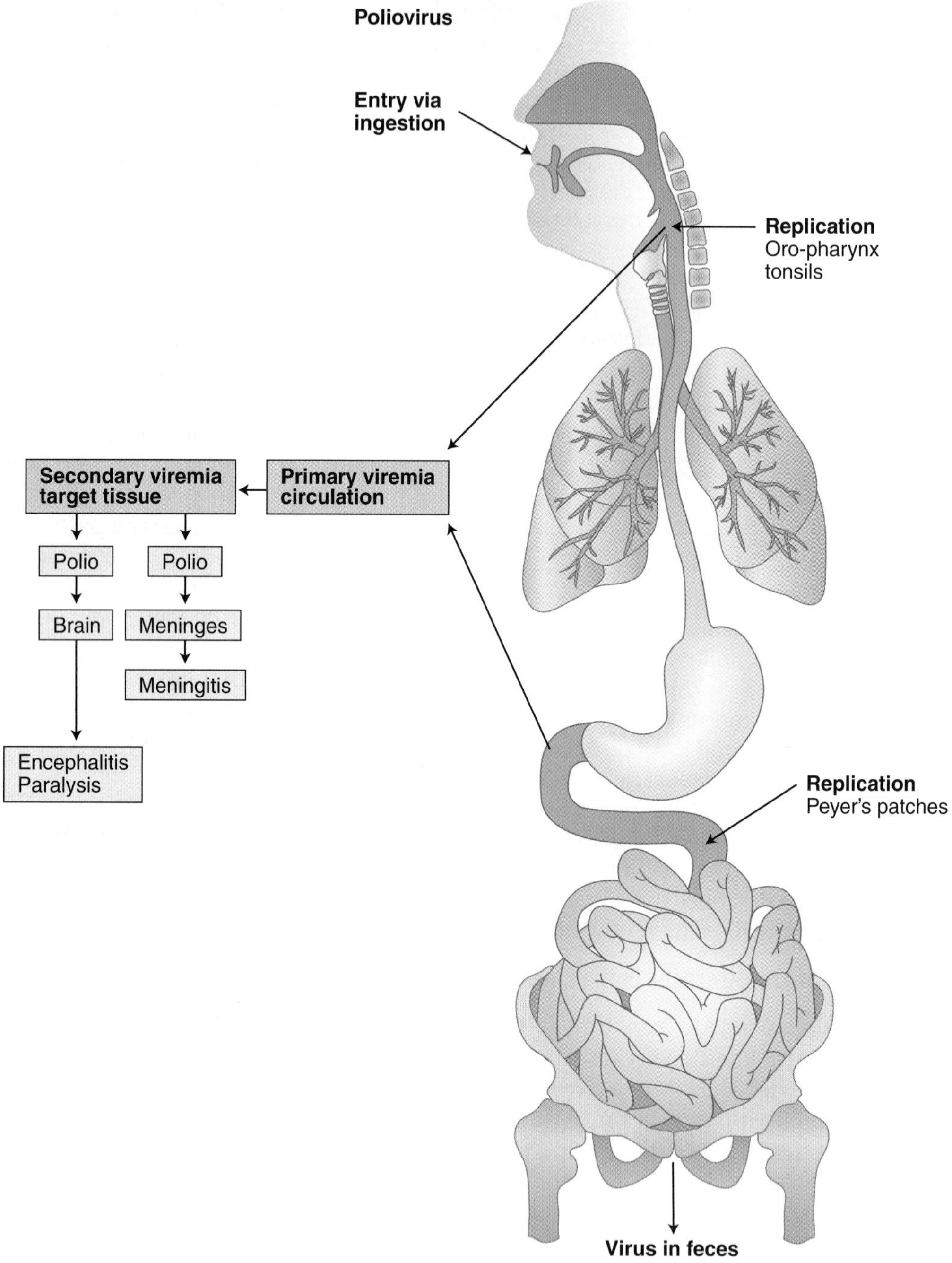

FIGURE 11-5 Poliovirus pathogenesis. Adapted from University of South Carolina, School of Medicine. "Virology: Picornaviruses, Part One." Microbiology and Immunology. University of South Carolina, 2008. http://pathmicro.med.sc.edu/virol/picorna.htm.

still poorly understood. This is remarkable considering that poliovirus is one of the most thoroughly investigated viruses of all time.

11.7 Viral Replication

The prototype of the genus Enterovirus of the *Picornaviridae* family is poliovirus, the causative agent of poliomyelitis. The first step toward identifying host receptors for poliovirus came with the isolation of monoclonal antibodies that block binding of poliovirus to cells. As mentioned earlier, polioviruses could not infect mouse cells because they lacked the correct cellular receptor. To bypass the receptor-binding step, the RNA genome of poliovirus was transfected into mouse cells (receptor-negative cells). It resulted in the production of a single cycle of infectious viruses. When mouse cells were transformed with human genomic DNA from HeLa cells, the mouse cells became susceptible to poliovirus infection. An antireceptor antibody blocked the infection. The results suggested that the poliovirus receptor gene had been introduced into mouse cells, resulting in the expression of the receptor and susceptibility to poliovirus infection. The **poliovirus receptor** (**PVR**, also known as CD155) is a cell surface integral membrane protein and a member of the immunoglobulin (Ig) superfamily of proteins. The function of CD155 is that of an activating ligand that stimulates natural killer cells.

The interaction of poliovirus with its receptor leads to a conformational change in the capsid. VP4, an internal capsid protein, detaches from the capsid. The capsid swells and a pore is formed in the cell membrane through which the RNA proceeds. (It is not known precisely where the uncoating event occurs—at the cell membrane or from within an endosome.) After the virion RNA is released into the cytoplasm of the cell, it serves as a messenger RNA (mRNA) that is translated into a single, highly autocatalytic polyprotein whose proteolytic cleavage products serve as capsid precursors and replication proteins.

The poliovirus genome is made up of a positive sense ssRNA and both ends of the genome are modified. The 5′ end of the poliovirus genome forms a cloverleaf or tRNA-like structure that plays a key role in the replicative process of the RNA molecule (**FIGURE 11-6**). A small peptide, VPg ("*v*iral *p*rotein linked to the *g*enome") is attached to the 5′ cloverleaf end of the viral genome. VPg is essential for the initiation of viral replication. It was discovered 30 years ago and is conserved in all picornaviruses, including poliovirus, hepatitis A virus and group A coxsackieviruses. VPgs are linked to the 5′ terminal UMP molecule of the genomic RNA by phosphodiester bonds to the hydroxyl group of a tyrosine residue found in VPg. Free VPg peptides, uridylylated at the same reactive tyrosine residue, are found in the cytoplasm of infected cells and serve as primers for viral replication.

Farther downstream of the clover-leaf structure is an **untranslated region** (**UTR**) that promotes the initiation of translation by its internal ribosomal entry site (IRES). The IRES element contains extensive regions of secondary stem-loop structures. The 3′ end of the poliovirus genome is polyadenylated (Figure 11-6).

Poliovirus proteins are synthesized by the translation of a single, long, open reading frame on the viral +ssRNA genome. This is followed by a series of cleavages. Viral proteases cleave themselves (**autocleavage**) out and cleave the polyprotein into 11 separate gene products involved in replication and packaging. The viral protease 2A cleaves the p220 subunit of the cap binding complex (eIF-4G), making host cell mRNA unrecognizable to ribosomes. This results in the shutdown of host protein synthesis but frees the ribosomes to translate viral RNA. Viral mRNA is translated via cap-independent translation: The 5′ UTR contains the IRES, which serves as a ribosome docking site for the 40S subunit of ribosomes. IRES elements were first discovered in poliovirus, encephalomyocarditis, and other picornaviruses. IRES-dependent translation is now known as a novel mechanism of eukaryotic cap-independent translation. An increasing number have been discovered not only in viral genomes but also in the 5′ UTRs of many cellular mRNAs. **FIGURE 11-7** shows the genomic

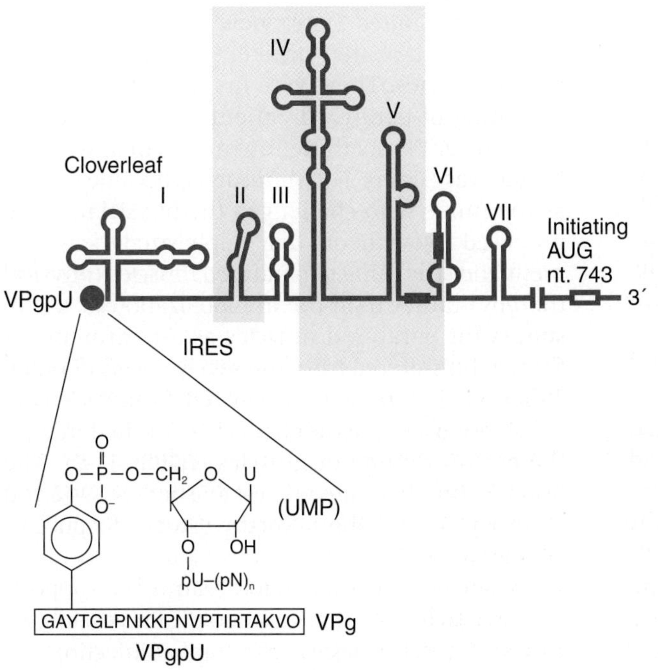

FIGURE 11-6 Schematic diagram showing the 5′ and 3′ nontranslated regions of the poliovirus genome. VPg (●) is attached to the 5′ cloverleaf end of the viral genome. VPg is linked to the 5′ terminal UMP molecule of the genomic RNA by phosphodiester bonds to the hydroxyl group of a tyrosine residue found in VPg. The 5′ end also contains a structural **IRES** element. Adapted from Rotbart, H. A., ed. *Human Enterovirus Infections.* ASM Press, 1995.

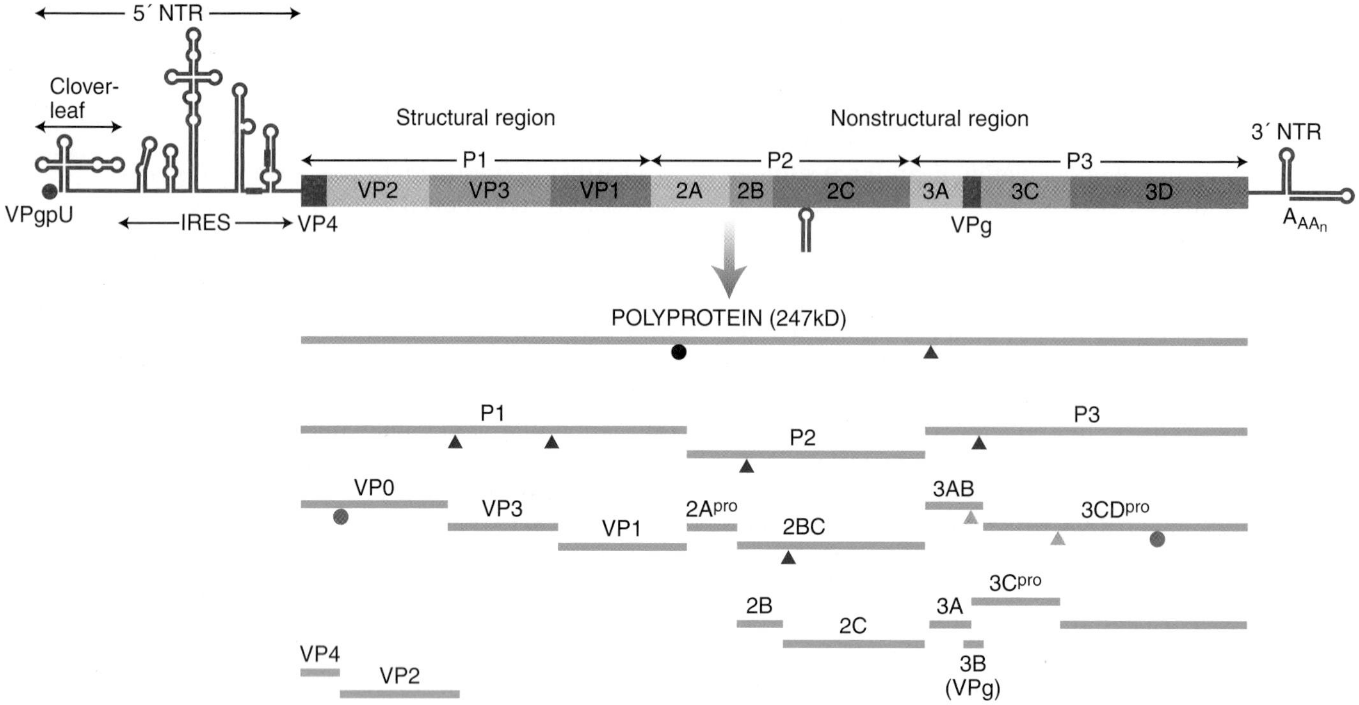

FIGURE 11-7 Genomic organization and proteolytic processing of poliovirus. Reproduced from *Virus Research*, Vol. 111, Mueller, S., Wimmer, E., and Cello, J., "Poliovirus and poliomyelitis: A tale of guts, brains, and an accidental event," pp. 175–193, Copyright 2005, with permission from Elsevier.

organization and proteolytic processing of the poliovirus polyprotein. The single open reading frame is translated into a polyprotein of 247 kilodaltons that is processed into three precursor proteins by virally encoded proteases named 2A, 3C, and 3D. The precursor proteins are further processed by virally encoded proteases (**Figure 11-7**). The cleavage results in four structural and seven nonstructural proteins.

RNA replication occurs entirely in the cytoplasm of the infected cell. Host cells lack the necessary cellular machinery to replicate the viral RNA genome. Poliovirus uses a viral RNA-dependent RNA polymerase to produce a −ssRNA strand from the packaged genomic +ssRNA template (mRNA) that was used for protein synthesis. The viral VPg protein attached to uridine serves as a primer. The first round produces a single antisense molecule. The antisense molecule is then used to produce (+) sense copies of the original genome that can be packaged into the viral capsids prior to viral release.

After the viral genome has been translated to produce the necessary proteins and replicated its genome, the capsid proteins self-assemble into a structure of icosahedral symmetry that contains 60 copies of each viral capsid protein (VP1, 2, 3, and 4). The newly synthesized +ssRNA genome enters the incomplete capsid and is secured inside once the viral proteases finish their cleavage activity. When the genome is secured and the virion is mature, as many as 100,000 virions may be released during cell lysis of a single infected cell. **FIGURE 11-8** shows the life cycle of poliovirus.

11.8 Treatment

During the era when polio outbreaks were common in the United States, new therapies were developed, such as the "drinker respirators" or iron lung machines. They were first used to assist the breathing of paralyzed patients in the 1930s. In 1940, Sister Elizabeth Kenny (nurses then were referred to as "sisters") of Minneapolis became known as the nurse who challenged the physicians. She proposed a new theory that emphasized the movement and stretching of paralyzed muscles (physical therapy) rather than the immobilization or use of splints for paralyzed muscles as a treatment for those who suffered from the severe forms of polio. Other types of therapy used by Kenny included the use of hot packs and hot baths in conjunction with the active training of muscles (**FIGURE 11-9**). The Sister Kenny Institute was established in 1942 and is now part of Abbott Northwestern Hospital in Minneapolis.

There is no cure for polio. Treatment is supportive care, including physical therapy. Prevention via vaccination is the best way to combat infection. No

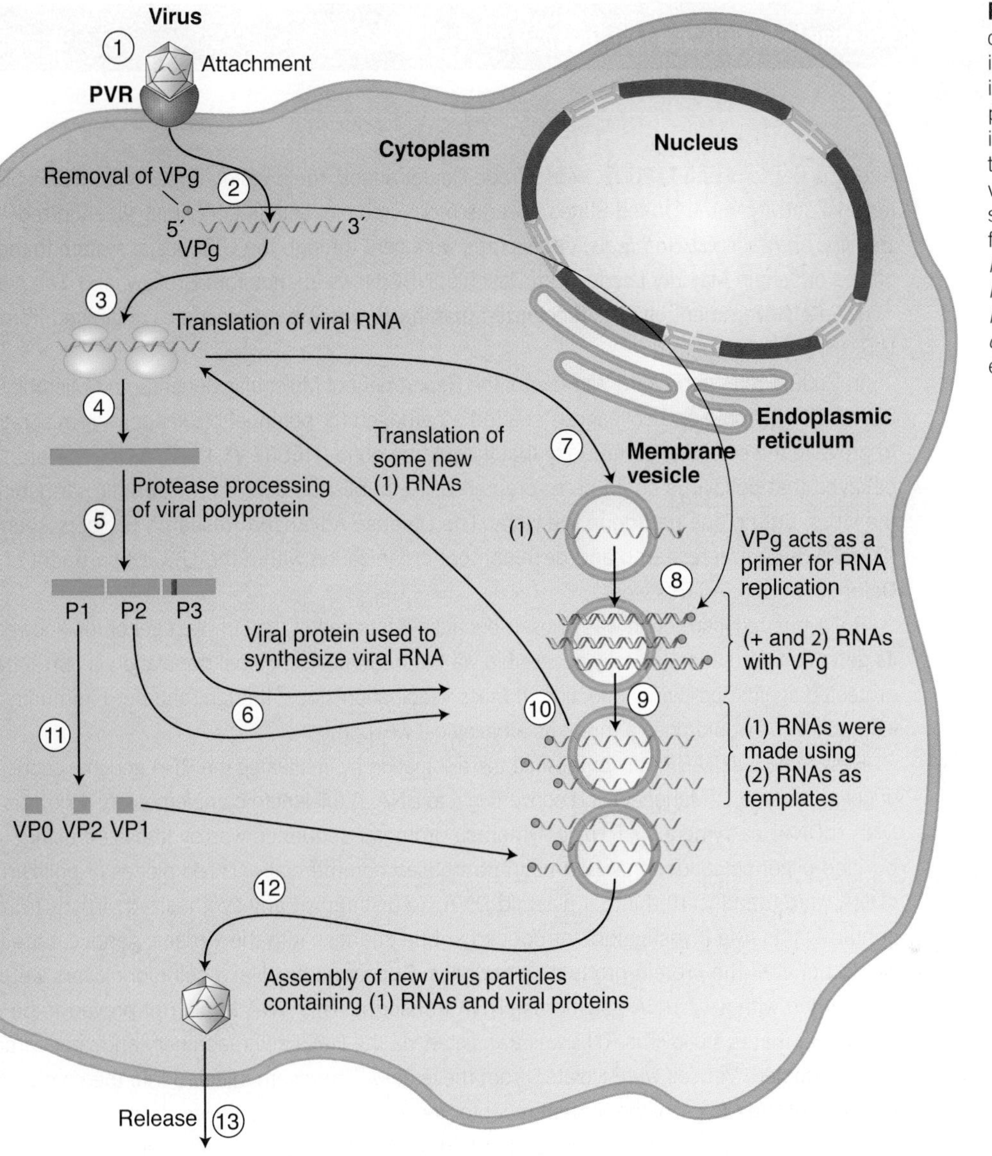

Note: This diagram only shows the initial processing of viral polyprotein in step 5.

FIGURE 11-8 The life cycle of polioviruses occurs solely in the cytoplasm of the infected cell. Note that the proteins that participate in RNA synthesis are transported to membrane vesicles where RNA synthesis occurs. Adapted from Flint, S. J., et al. *Principles of Virology: Molecular Biology, Pathogenesis, and Control of Animal Viruses*, Second edition. ASM Press, 2003.

antiviral chemotherapeutic agents exist. The development of antipoliovirus drugs has seemed unnecessary because the disease is near eradication. No poliovirus antiviral is likely to profit a company.

Only recently have the WHO and CDC called for the development of a safe antiviral that would be used to prevent and treat poliovirus infections. The driving factor is vaccine-driven outbreaks and the potential for post-eradication outbreaks.

11.9 Prevention

Poliomyelitis was the most widely known disease of the 20th century until AIDS. Scientists set out on a mission to learn more about the disease, gradually allowing them to develop more effective treatment, rehabilitation techniques, and prevention.

Chemicals and Gamma Globulin

During the 1940s, researchers observed that if certain species of monkeys were treated in their nose with chemicals such as picric acid, sodium alum, and zinc sulfate and then challenged with poliovirus, the chemicals seemed to block viral infection of the nasal mucosa. When this treatment was tried in humans, the results were not promising and this technique was quickly abandoned. The use of convalescent serum for poliomyelitis was recommended as early as 1915. In 1952, passive immunity attempts to inject 54,772 children who were at risk of contracting polio with convalescent

VIRUS FILE 11-1 Creating Poliovirus in a Test Tube

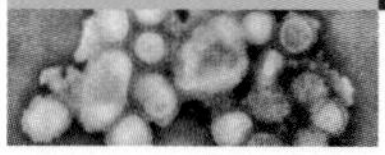

After the September 11, 2001, World Trade Center attack, the ongoing threat of domestic bioterrorist activity in the United States became more credible. Letters containing weaponized *Bacillus anthracis* (*B. anthracis*) endospores were sent through the U.S. postal system to the offices of Senate Majority Leader Tom Daschle, NBC News anchor Tom Brokaw, and *The New York Post*. Government officials recognized that the *B. anthracis* endospores originated from the U.S. biodefense program.

In 2002, Professor Eckard Wimmer of the Department of Molecular Genetics and Microbiology, SUNY Stony Brook, and colleagues created a pathogen (or potential bioweapon) from scratch to prove to the scientific community that it could be done (**FIGURE VF 11-1**). Many researchers believed that poliovirus could be re-created using published data on the genetic sequence of the virus, but no one had done the work. The Defense Advanced Research Projects Agency (DARPA, a central research and development organization within the U.S. Department of Defense) supported this work.

Poliovirus reconstruction was chosen because it has one of the simplest genomes known and its genome information is easily accessible via the Internet. The global population is currently protected against poliovirus infection. If mass vaccination stops, though, and herd immunity is lost, poliovirus would pose a threat as a biological weapon.

Researchers at DARPA accomplished the re-creation by first taking the RNA genome sequence of poliovirus type 1 (Mahoney) and converting it to DNA. A full-length complementary poliovirus DNA (cDNA) carrying a T7 RNA polymerase promoter sequence was synthesized using purified oligonucleotides that had been purchased commercially. Three pieces of poliovirus cDNA were then inserted into a plasmid DNA. To distinguish any synthesized viruses from lab contaminants, the investigators introduced subtle changes into the viruses' genetic code that would not alter the protein products it encodes. Subsequently, HeLa cell monolayers were transfected with a T7 RNA polymerase RNA transcript. This RNA transcript proved to be infectious; that is, poliovirus RNA was translated via the HeLa cell machinery and viral particles self-assembled. Viruses were isolated from these cells. They were injected into the cerebrum of mice engineered to carry the human receptor for poliovirus. The mice were examined daily for paralysis and/or death. The synthetic virus had the biochemical and pathogenic characteristics of poliovirus, but researchers found that the newly created viruses were much less virulent than the typical laboratory strain. As a result, higher doses than usual of the synthetic poliovirus were needed to kill the mice.

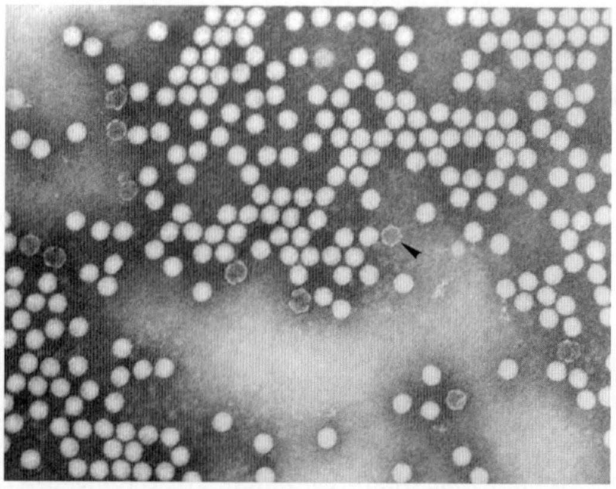

FIGURE VF 11-1 Transmission electron micrograph of poliovirus type 1. The nonenveloped virions are ~30 nm in diameter and have icosahedral symmetry.

VIRUS FILE 11-1 **Creating Poliovirus in a Test Tube (continued)**

This poliovirus project took nearly three years to complete in a highly sophisticated research laboratory using the most advanced DNA synthesis technology and could not have been completed in a substandard facility. Current technology is not sophisticated enough to re-create more complex viruses such as smallpox and Ebola virus.

Now that the genome has been recreated, what are the implications of these experiments? Should this work have been done at all?

References

Cello, J., Paul, A. V., and Wimmer, E. 2002. "Chemical synthesis of poliovirus cDNA: Generation of infectious virus in the absence of natural template." *Science* 97(5583):1016–1018.

Racaniella, V. R. 2000. "It is too early to stop polio vaccination." *Bull WHO* 78(3):359–360.

serum had no apparent effect on the incidence or severity of paralysis, probably because this immunity was short-lived. Another 235,000 children were inoculated with gamma globulin the following year with the same results. Gamma globulin failed as a prophylactic measure to prevent poliovirus infection in humans and the supply of serum was no longer offered.

Inactivated Vaccines

As soon as tissue culture techniques became available, a number of medical research laboratories began to develop a vaccine against poliovirus. Jonas Salk's laboratory began working on a formalin-inactivated vaccine, whereas Albert Milzer and colleagues pursued ultraviolet irradiation as a means of virus inactivation. Albert Sabin and his research team took the approach of a live-attenuated vaccine.

Milzer's vaccine was immunogenic in humans but was never adopted for general use. Salk continued work on the formalinized vaccine that had been prepared by propagating the virus in monkey kidney cells. By 1953, preliminary tests of Salk's inactivated vaccine on children and adolescent volunteers at the D. T. Watson Home for Crippled Children and the Polk School for the Retarded and Feeble-Minded were conducted. (During this

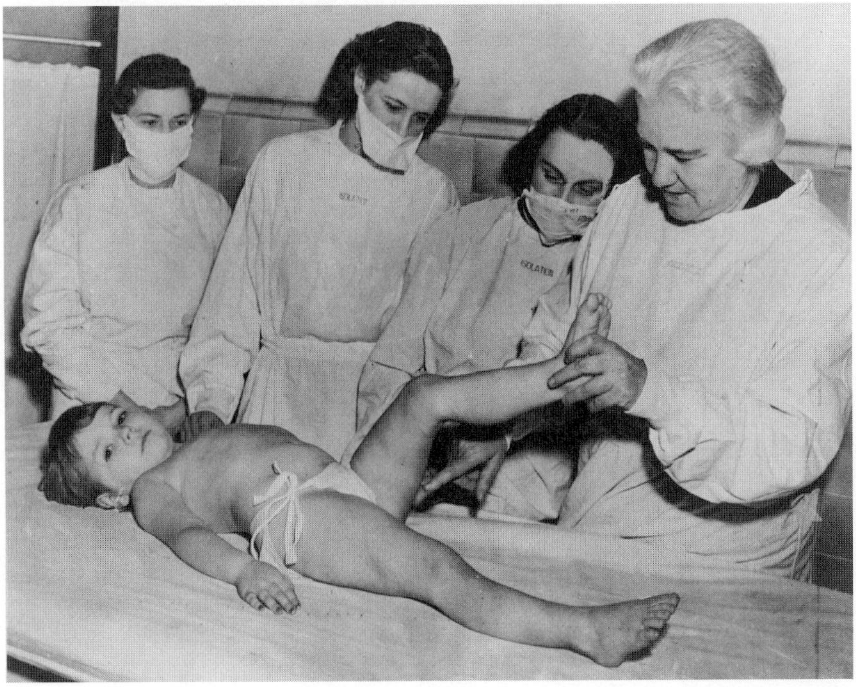

FIGURE 11-9 Sister Kenny demonstrates therapy techniques at the Sister Kenny Institute, 1942.

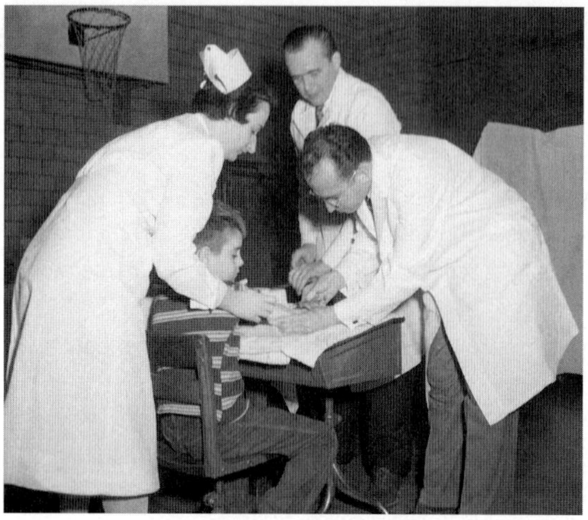

FIGURE 11-10 Jonas Salk vaccinating a boy at Arsenal Elementary School, located in Pittsburgh, Pennsylvania (Feb. 23, 1954).

time, this type of human testing was a standard practice.) Results were favorable and the largest experiment of its time was set up. A mass vaccination trial, including placebo controls, was sponsored by the March of Dimes (then known as the National Foundation for Infantile Paralysis). A total of 1,829,916 children from all parts of the United States, Canada, and Finland were vaccinated (**FIGURE 11-10**). Results of this inactivated poliovirus vaccine (IPV) were presented by manufacturers in 1955. The vaccine was said to be safe and 70% effective (booster shots would be needed to increase its efficacy). Within days, the vaccine was licensed for use.

The Cutter Episode

Not long after the vaccine was licensed and readily available, cases of paralytic poliomyelitis were reported in children who had received the Salk vaccine. An epidemiological investigation revealed that all of these vaccinees were given a vaccine produced by the same manufacturer—Cutter. Live virus was isolated from the vaccine lots, indicating that the inactivation process was not complete. A total of 260 cases were reported; 94 infections were determined to be from the vaccination, 126 were through family contacts, and 40 were through community contacts. Of these 260 cases, 192 were paralytic. There were no deaths.

Amazingly, the "Cutter Incident" did not change public confidence in the vaccine. As a result of this incident, though, new requirements were introduced for safety testing of the vaccine. A surveillance unit was created at the Centers for Disease Control and Prevention to maintain and scrutinize all vaccination programs.

Live, Attenuated Poliovirus Vaccines

Attenuated (as opposed to inactivated) vaccines were more appealing to medical researchers because it was assumed that an active infection came closest to reproducing the natural situation. The idea is to fool the body's immune defense system into thinking it is under attack by a pathogenic form of the virus, thus causing it to produce the defense mechanisms (e.g., antibodies) that will fight off the poliovirus if it is encountered in the future. It was expected that attenuated viruses would produce longer-lasting immunity. The major concern related to the creation of attenuated virus vaccines was the issue regarding the genetic stability of the poliovirus vaccine strains. In other words, would they revert to neurovirulence after multiplying in the human host? Three research groups worked to develop a live, attenuated poliovirus vaccine during the early 1950s. Sabin's group obtained three different poliovirus strains and passaged them in a variety of ways: monkey testicular cells or intracerebral passages in rhesus monkeys or passaged by feeding chimpanzees and collecting excreted strains, and/or by cynomolgus monkey kidney tissue culture (MKTC) cells. Attenuation/passage histories of the three Sabin vaccine seeds now in use are listed in **TABLES 11-3**, **11-4**, and **11-5**. **FIGURE 11-11** is a schematic illustrating vaccine development by Sabin's team. The definitive test used to determine if the vaccine strains would revert to virulence was to

TABLE 11-3 Poliovirus Type 1, Sabin Strain, Passage History

Year	Manipulation
1941	Francis and Mack: Isolation of Mahoney strain from pooled feces of 3 healthy children in Cleveland, Ohio. Salk: 14 rhesus monkey passages and 2 *in vitro* testicular tissue culture passages
1953	Li and Schaefer: 11 monkey kidney tissue culture (MKTC) passages. Additional tissue culture passages in monkey kidney and skin
1954	Sabin: 5 passages in cynomolgus MKTC 3 single plaque passages Selection by neurovirulence testing
1956	Sabin: 2 passages in cynomolgus MKTC
1956	Merck, Sharp & Dohme: 1 passage in rhesus MKTC

Source: Compiled from Sabin, A. B., Boulger, L. R. 1973. "History of sabin attenuated poliovirus oral live vaccine strains." *J Biol Standard* 1:115–118. and Sutter, R. W., et al. "Poliovirus vaccine-live." In *Vaccines*, Fifth Edition. S. Plotkin, W. Orenstein, and P. Offit, editors. Saunders, Elsevier, 2008.

TABLE 11-4	Poliovirus Type 2, Sabin Strain, Passage History
Year	**Manipulation**
Unknown	Fox and Gelfand: P712 strain isolated from feces from healthy children in Louisiana.
1954	Sabin: 4 passages in cynomolgus MKTC 3 serial passages of plaque isolates Selection by neurovirulence testing in cynomolgus monkeys inoculated intraspinally Fed to chimpanzees and excreted strains with the least neurovirulence underwent 3 single plaque passages
1956	Sabin: 2 passages in cynomolgus MKTC
1956	Merck, Sharp & Dohme: 1 passage in rhesus MKTC

Source: Compiled from Sabin, A. B., Boulger, L. R., 1973. "History of sabin attenuated poliovirus oral live vaccine strains." *J Biol Standard* 1:115–118. and Sutter, R. W., et al. "Poliovirus vaccine-live." In *Vaccines*, Fifth Edition. S. Plotkin, W. Orenstein, and P. Offit, editors, Saunders, Elsevier, 2008.

inject the attenuated strains intracerebrally or directly into the spinal cord of monkeys. Attenuated strains that passed the definitive test were used as vaccine candidates for field trials.

It was difficult to conduct field trials in the United States because the Salk vaccine had already

TABLE 11-5	Poliovirus Type 3, Sabin Strain, Passage History
Year	**Manipulation**
1937	Kessel and Stimpert: Leon strain isolated from brain stem and spinal cord of an 11-year-old boy who died of bulbo-spinal poliomyelitis in Los Angeles. 20 intracerebral passages in rhesus monkeys
1952	Melnick: 8 passages in rhesus monkey testicular tissue culture
1953	Sabin: 3 passages in cynomolgus MKTC 30 rapid (24 hour) passages at low dilution in cynomolgus MKTC 3 terminal dilution passages 1 low dilution pass 9 plaques isolated, single plaques passed 3 times Selection by neurovirulence testing in cynomolgus monkeys
1956	Sabin: 3 passages in cynomolgus MKTC
1956	Merck, Sharp & Dohme: 1 passage in rhesus MKTC

Source: Compiled from Sabin, A. B., Boulger, L. R., 1973. "History of sabin attenuated poliovirus oral live vaccine strains." *J Biol Standard* 1:115–118. and Sutter, R. W., et al., "Poliovirus vaccine-live." In *Vaccines*, Fifth Edition. S. Plotkin, W. Orenstein, and P. Offit, editors. Saunders, Elsevier, 2008.

been licensed and was being widely used. In 1958, the first large-scale field trial, in which 200,000 children were vaccinated with the attenuated Sabin serotype 2 poliovirus vaccine, was conducted in Singapore. The same year, laboratory protocols to grow the vaccine strains and produce enough vaccine for 300,000 children was provided to Professor Mikhail Petrovich Chumakov, who was the director of the Poliomyeltitis Research Institute in Moscow. Chumakov quickly manufactured more of this vaccine using the seed strains provided by Sabin. In just over a year, 15 million Russians were vaccinated without any untoward effects and with evidence of effectiveness. By 1960, 100 million participants in Russia and eastern European countries were vaccinated. This provided confidence that the attenuated viruses would not revert to being neurovirulent. After these successful field trials, the United States granted licensure of the Sabin live poliovirus vaccine.

The results were that within ten years, two effective and safe vaccines were available for general use. The live vaccine developed by Sabin included all three attenuated viruses (types 1, 2, and 3). The vaccine was given by mouth in the form of a few drops of liquid; hence, it was known as the oral poliovirus vaccine (OPV). The incidence of poliomyelitis in the United States dropped dramatically after the administration of IPV and OPV (**FIGURE 11-12**). People waited in long lines for the oral vaccine against poliovirus (**FIGURE 11-13**).

During the early years of IPV and OPV production, some monkey kidney cell cultures used to produce the vaccine strains were contaminated with simian virus-40 (SV-40). As detailed in Chapter 10, SV-40 causes tumors in hamsters and transform cells in culture. Vaccinees developed antibodies against SV-40 and the virus could be isolated from the feces of some vaccinees. Long-term studies were conducted to determine if individuals receiving the contaminated vaccines would be at higher risk for cancer (refer to Chapter 10). The composition and immunity of the current IPV and OPV is shown in **TABLE 11-6**. From 1996 to 1999, the CDC recommended sequential use of IPV/OPV. Both vaccines induce immunity to poliovirus, but OPV is better at preventing the disease from spreading to other people. It was recommended that the IPV would be administered at age 2 and 4 months followed by OPV at age 12–18 months and 4–6 years. For a few people (about 1 in 2.4 million), OPV actually causes poliomyelitis, a situation termed **vaccine-associated paralytic polio (VAPP)**. In 2000, to reduce VAPP, the exclusive use of IPV was recommended and OPV is no longer readily available in the United States. Children get four doses of IPV at 2 and 4 months, a dose at 6–18 months, and a

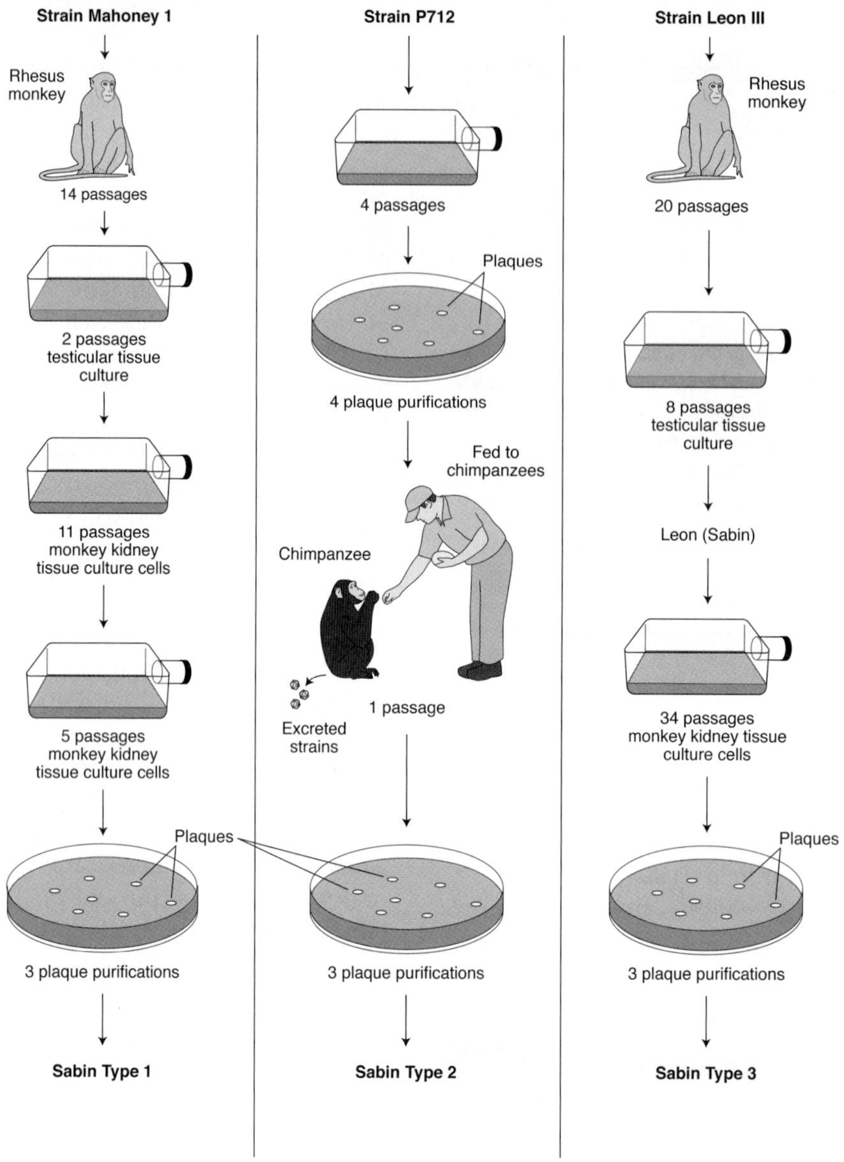

FIGURE 11-11 Schematic of the passage history used to develop the Sabin poliovirus vaccine. Adapted from Minor, P. D. 2004. "Polio eradication, cessation of vaccination and re-emergence of disease." *Nat Rev Microbiol* 2:473–482.

FIGURE 11-12 Incidence of poliomyelitis in the United States from 1941 to 1971. The blue line represents paralytic polio only; the solid line contains both paralytic and nonparalytic forms of poliomyelitis. Adapted from D. S. Younger, ed. *Motor Disorders.* Lippincott Williams & Wilkins, 1999.

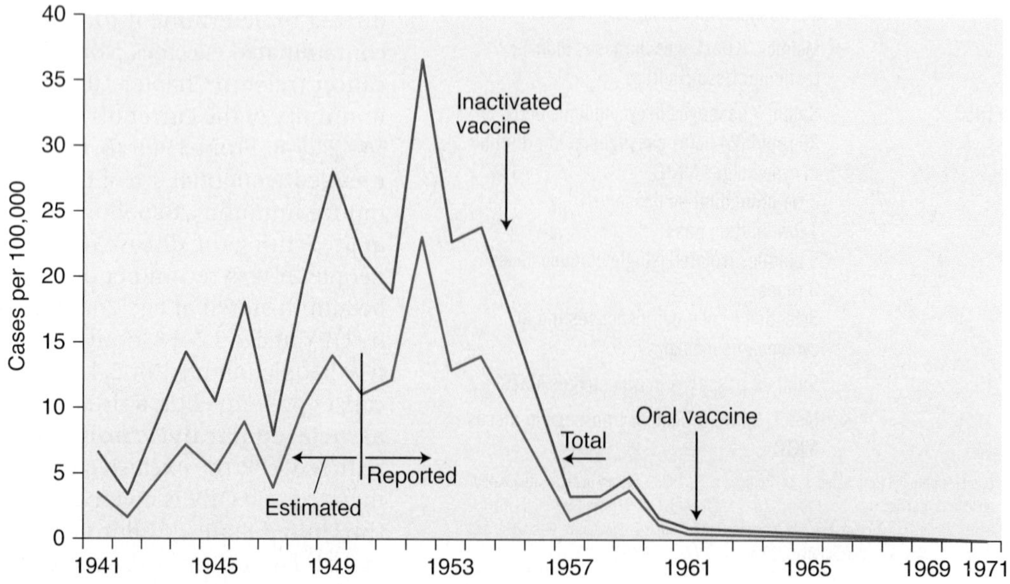

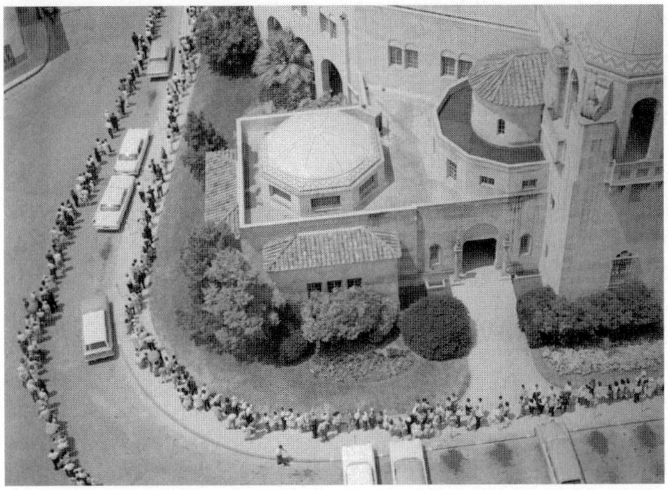

FIGURE 11-13 An aerial view of a crowd surrounding a massive city auditorium in San Antonio, TX, awaiting polio immunization, 1962.

booster dose at 4–6 years. This has helped to eliminate VAPP.

11.10 Poliovirus Eradication Is Unfinished Business

In 1988, the World Health Assembly, the directing council of the WHO, passed a resolution to globally eradicate poliomyelitis by the year 2000. This was considered possible for several reasons, all of which are still true:

- Two vaccines that induce immunity were available.
- There was no animal reservoir.
- There were only three serotypes of wild-type viruses, all of which were genetically stable.
- The OPV was inexpensive and easy to administer in mass vaccination campaigns.

Several potential roadblocks to global eradication were identified, including:

- Poliovirus is contagious, particularly by an oral–fecal route.
- IPV is inefficient in preventing the spread of virus.
- The use of OPV in tropical countries has been problematic.
- There are tensions in many countries between those who favor mass vaccination and those who do not.
- Verification of success is difficult as many poliovirus infections are not clinically obvious. (Only about 1 in 100 people who are infected with poliovirus suffer from paralysis.)

Although the goal of eradicating the poliomyelitis was not realized by 2000, progress toward it has been significant. Between 1988 and 2003, the number of polio-endemic countries was reduced from 125 to 6 (Afghanistan, Egypt, India, Niger, Nigeria, and Pakistan). The number of wild-type poliovirus infections reported annually decreased from

TABLE 11-6 Inactivated and Oral Polio Vaccines Composition and Protective Capabilities

Vaccine Composition	Vaccine Efficacy
Inactivated polio vaccine	
• Contains three serotypes of vaccine virus	• Highly effective in producing immunity to poliovirus
• Grown on monkey kidney (Vero) cells	• >90% immune after two doses
• Inactivated with formaldehyde	• >99% immune after three doses
• Contains 2-phenoxyethanol, neomycin, streptomycin, polymyxin B	• Duration of immunity not known with certainty
Oral polio vaccine	
• Contains three serotypes of vaccine virus	• Highly effective in producing immunity to poliovirus
• Grown on monkey kidney (Vero) cells	• >50% immune after one dose
• Contains neomycin and streptomycin	• >95% immune after three doses
• Shed in stool for up to six weeks following vaccination	• Immunity probably lifelong

350,000 in 1988 to roughly 1,350 in 2005. As of 2010, four polio endemic countries remain: Nigeria, India, Afghanistan, and Pakistan. Any other countries reporting cases were "imported" from endemic regions. Northern Nigeria continues to account for the majority of global cases. Increased cases of polio in Nigeria were linked to a boycott of poliovirus vaccinations by leaders in several Nigerian states who falsely claimed the vaccine causes sterility and spreads AIDS. Countries that were declared polio-free have now experienced new outbreaks or reinfections. Twenty-six previously polio-free countries have been reinfected since 2003. Of these, nine continued to report cases in 2006 (Angola, Cameroon, Chad, Kenya, Ethiopia, Bangladesh, Nepal, Namibia, and Yemen). Some scientists hypothesize that reinfections in 2005 were due to polio-infected Muslims who spread the virus while making their pilgrimage to Mecca (the Hajj, Saudi Arabia), which took place in January 2005. In response to these outbreaks, the WHO has intensified efforts in surrounding countries and created educational programs to reinforce the importance and safety of vaccination. Saudi Arabian health officials recommended that all Hajj travelers from polio-endemic countries be vaccinated regardless of their immune status.

In 2009, a total of 1606 cases were reported in 23 countries. The majority of cases (1256) were in the four endemic countries (**FIGURE 11-14**). The largest number was in India (741) and Nigeria (388). At the beginning of 2010, there was a 41% reduction in worldwide cases reported compared to the same time period in 2009.

A very deadly polio outbreak began during October 2010 in the Republic of Congo (also known as Congo-Brazzaville). Routine vaccination in the Republic of Congo had kept it polio-free since 2000. The new outbreak, caused by wild poliovirus type 1, was killing 42% of infected individuals, most of whom were young men between the ages of 15 and 25. Because polio usually strikes children under the age of 5, healthcare workers originally searched for another cause of this epidemic. Few fecal samples were collected at the onset of the outbreak. The mortality rate was also disturbing. Poliovirus infections kill up to 10% of individuals *in developing countries.* At the time of this writing, WHO and CDC epidemiologists are investigating whether a simultaneous outbreak of another deadly virus with a different route of transmission is occurring. This explosive outbreak raised questions. Can poliovirus be eradicated? Where will the next outbreak of poliovirus occur in polio-free parts of the world? The best strategy appears to be long-term disease prevention via vaccination. Poliovirus will likely be impossible to eradicate.

11.11 Other Enteroviruses (Nonpolio)

Historically, poliovirus has been the most significant enterovirus. There are, however, at least 70 viruses within the Enterovirus genus that are known to infect humans.

Myocarditis and Dilated Cardiomyopathy

Over the past 50 years, extensive research has addressed the role of enteroviruses in myocarditis and **dilated cardiomyopathy** (**DCM**). Myocarditis is the inflammation or degeneration of the myocardium (muscular tissue of the heart). In some cases, it can lead to congestive heart failure requiring hospitalization, heart failure medications, or a heart transplant. Myocarditis is often asymptomatic. About 10% of patients experience fever, chest pains, shortness of breath, heart palpitations (the sensation of a skipped heartbeat), and fatigue.

DCM is a disease of the heart muscle that causes the heart to become enlarged and to pump less effectively. As a result, the heart becomes thin, weak, or floppy, and as a result it is unable to pump blood efficiently around the body. This causes fluid to build up in the lungs, causing congestion and the feeling of breathlessness.

DCM is a serious condition. It has a 50% mortality rate in the two years following a diagnosis and may account for nearly half of all patients requiring a heart transplant. Viruses, bacteria, and protozoan have all been shown to infect the myocardium, resulting in myocarditis. The enteroviruses associated with myocarditis and DCM are the Coxsackie group B viruses. They can replicate in the heart but are difficult to isolate in an infectious form during chronic disease. Myocardial damage is explained by direct viral damage and/or the consequences of the immune response.

The WHO has reported that approximately 1.5% of all enteroviral infections are associated with cardiovascular signs and symptoms. Over a person's lifetime, repeated exposure to enteroviruses increases the likelihood of acquiring an enteroviral infection associated with cardiac symptoms. It has been suggested that 70% of the general population has been exposed to cardiotropic viruses and that half of this group experiences an episode of acute viral myocarditis. There are several factors that influence the incidence of enterovirus-associated heart disease:

- Host age
- Host gender
- Host race and geographical location

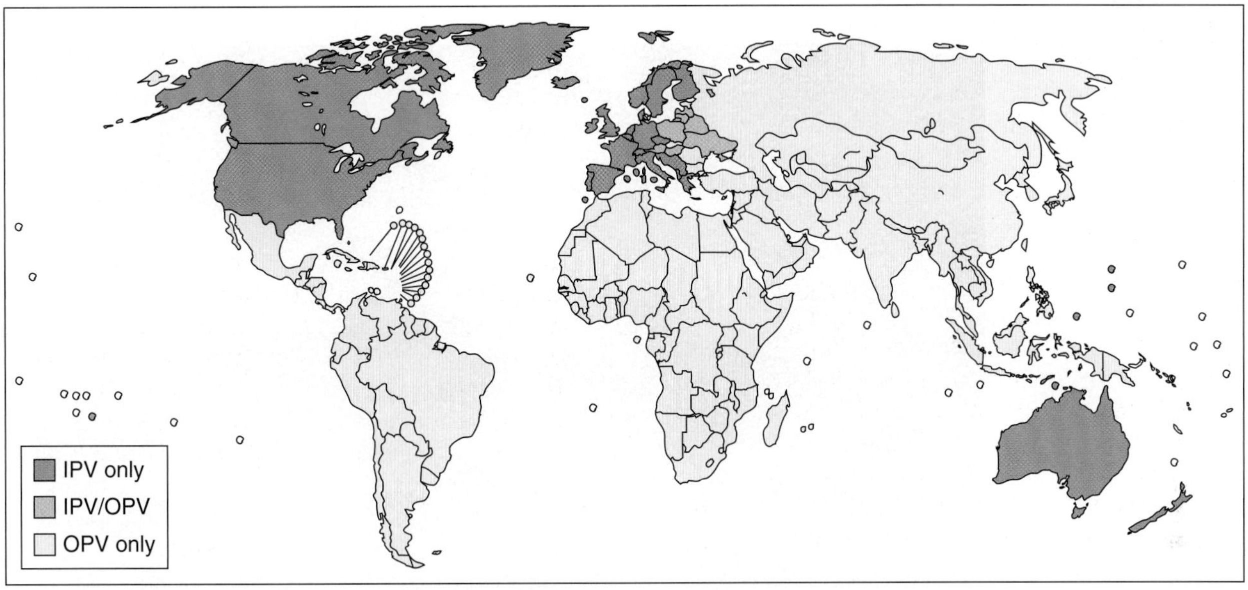

a

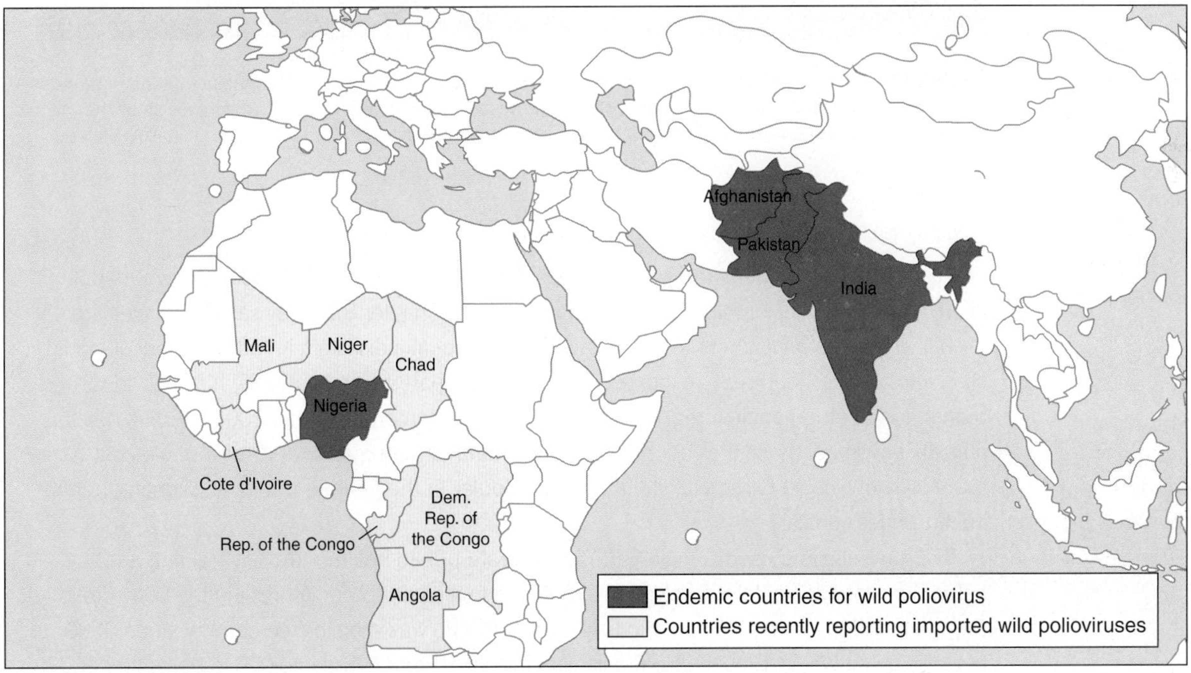

b

FIGURE 11-14 (a) Wild poliovirus pockets in the world, 2005. Data are from the WHO as of October 25, 2005. Courtesy of the World Health Organization. **(b)** Polio-affected countries as of November 10, 2010. Travelers are recommended to get a polio immunization (booster) before traveling to these areas. The four endemic areas (in red) are Nigeria, Pakistan, Afghanistan, and India. Countries in yellow represent imported cases of polio. Courtesy of the World Health Organization.

VIRUS FILE 11-2 Using Google Earth to Track Poliovirus Down the Congo River

FIGURE VF 11-2 Google Earth map image of a portion of the Congo River near Mbandaka. The main islands are represented by the yellow push pins. Red dots represent seasonal villages. The white lines show the limits of the upper and lower limits of the district area of responsibility. © 2009 Kamadjeu; licensee BioMed Central Ltd.

The use of maps to understand geographic patterns and disease dates to John Snow's mapping of an 1854 cholera epidemic that occurred in England (Chapter 8). Today satellite images are used in disaster management (e.g., tsunami and earthquake disasters in Indonesia and Japan; Haiti earthquake), outbreak response, emergency preparedness to identify hazard zones, and humanitarian emergencies (e.g., refugee camps in Sudan and Chad in 2004). Public health officials have had challenges accessing satellite data of high quality and reasonable cost. The availability of Google Earth satellite images is popular in the general public and among public health professionals.

CDC investigators used Google Earth to generate public health maps for training and to help understand the entities involved in a response to a recent polio outbreak in the Democratic Republic of Congo. The Democratic Republic of Congo was a polio-free country since 2000 because of the WHO's Global Polio Eradication Initiative. From February 2006 to December 2007, 54 cases of poliomyelitis caused by wild-type poliovirus type 1 were reported in 4 provinces in the Democratic Republic of Congo. Mass immunization campaigns were launched to interrupt the circulation of polioviruses and stop the spread to neighboring provinces and countries. The campaign targeted the vaccination of children aged 9 months up to 5 year olds living in affected and high-risk districts.

Little progress was made. Geospatial distribution of polio cases showed that the outbreak seemed to follow the course of the Congo River. The Congo River is the second longest river in Africa. It is

VIRUS FILE 11-2 **Using Google Earth to Track Poliovirus Down the Congo River (continued)**

2700 km (2920 miles) long. Investigators suspected that the river could play a role in the spread of the outbreak to neighboring districts. A **River Strategy** was adopted to target populations along the river. *Detailed maps of entire sections of the river were not known or available.* For this reason, vaccination services and plans to live on the islands or tributaries of the river were not included in the initial planning. The River Strategy involved the use of Google Earth to develop a better plan for allocating resources (e.g., fuel, rafts, canoes, and outboard motors) and dispatch vaccination teams along the river (**FIGURE VF-2**). Google Earth provided high-quality and free satellite imagery with a flexible platform set apart from traditional geographic information systems. This allowed the creation of high-quality maps for the response teams to vaccinate children living in seasonal villages and permanent villages on islands or stilts along the Congo river. With the improved planning, children previously missed by routine immunization were identified and vaccinated. Google Earth was successfully implemented as a planning tool to contain this polio outbreak in the Democratic Republic of Congo.

References

Kamadjeu, R. 2009. "Tracking the polio virus down the Congo River: A case study on the use of Google Earth in public health planning and mapping." *Int J Health Geograp* 8:4 doi10.1186/1476-072X-8-4.

Google Earth Home Page: http://www.google.com/earth/index.html

Children under six months of age are highly susceptible to Coxsackie virus myocarditis. During late childhood and adolescence, the incidence of viral myocarditis rises. Patients between 20 and 39 years of age account for 52% all viral myocarditis cases. The incidence of Coxsackie myocarditis and DCM is significantly higher in males than females. Depending on the study, the male-to-female ratios range from 2:1 to 4:1. In the United States the mortality rate in males is twice that of females, and myocarditis for blacks is 2.4 times higher than in whites. DCM seems to be higher in underdeveloped and tropical countries.

Respiratory Enteroviruses

Of all of the clinical syndromes reported to be associated with enteroviral infections, respiratory disease accounts for 14% to 21% of them. Respiratory illnesses are the most frequent and universal illnesses suffered by children and adults. On average, children experience two to seven episodes and adults experience one to three episodes of respiratory illnesses per year, most of which are caused by viruses. A variety of viruses cause respiratory illnesses, such as respiratory syncytial virus (RSV), parainfluenza viruses, influenza viruses, adenoviruses, rhinoviruses, and enteroviruses. Worldwide, enteroviruses account for 2% to 15% of all viruses that cause upper- and lower-respiratory tract infections. Respiratory diseases caused by enteroviruses are mild and of a self-limiting nature and, thus, treatment is supportive.

Enterovirus 71: A Reemerging Viral Pathogen?

The majority of enteroviruses cause asymptomatic, mild, or self-limiting infections in children. Enteroviruses have been placed into subgroups based on differences in the range of hosts and pathogenicity:

- Polioviruses
- Group A Coxsackie viruses
- Group B Coxsackie viruses
- Echoviruses

Since 1970, newly identified enteroviruses have been allocated sequential numbers (68–71), of which enterovirus 71 is the most acknowledged. It was first isolated in California in 1969 and is a frequent cause of **hand, foot, and mouth disease** (**HFMD**) epidemics associated with severe neurological complications in a small proportion of cases (**FIGURE 11-15**). There has been a significant increase in enterovirus 71 epidemic activity throughout the Asia-Pacific region

FIGURE 11-15 Cutaneous lesions of the fingers and palms are present in two thirds of individuals with HFMD. Herpetic-like lesions may also be present inside the mouth and on the soles of the feet.

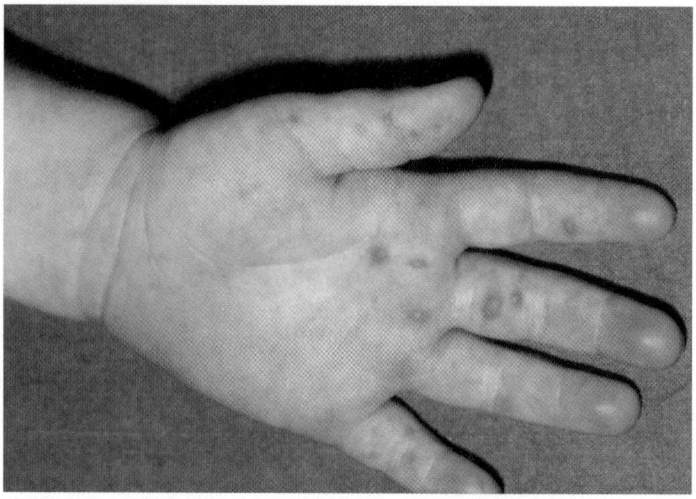

since 1997. The large recent outbreaks that occurred in Taiwan, Bulgaria, Hungary, and Malaysia have raised important questions about the biology of the virus and the factors underlying the widespread emergence of enterovirus 71 infections.

Two disease patterns have been observed: small outbreaks associated with an occasional patient death and unusual severe outbreaks in which a significant number of individuals experience brain stem encephalitis associated with pulmonary edema and hemorrhage, resulting in death. Over 90% of fatal cases were in children less than five years of age who died within a day of hospital admission. There has been a great deal of interest in trying to explain the spread and mysterious virulence of these severe enterovirus 71 infections in molecular and genetic terms. This viral agent has been changing genetically from the strains isolated in the 1970s. It is not clear whether the genetic changes in the genome explain the genesis of the enterovirus 71 pulmonary edema.

Enterovirus 71 joins the ranks of the increasing numbers of microbes of emerging or reemerging infectious diseases. Poliovirus infections are controlled in most parts of the world through the efforts of the WHO, and as a result attention is being turned to the nonpolio enterovirus infections, particularly enterovirus 71.

Summary

The enteroviruses are small (30 nm), naked viruses that contain infectious ssRNA of positive polarity. They are ubiquitous pathogens that belong to the family Picornaviridae. An estimated 10 to 15 million symptomatic infections are caused by enteroviruses, resulting in 30,000 to 50,000 hospitalizations in the United States per year. Although most enterovirus infections are relatively mild and result in the full recovery of the patient, these viruses can also cause severe illnesses such as poliomyelitis, meningitis, encephalitis, and myocarditis. Poliovirus is the most well-known and studied enterovirus.

Poliovirus infections in the United States suddenly emerged in epidemic form at the turn of the 20th century, causing paralysis and death, mainly in children. Poliomyelitis became a public concern because it was a disease of mysterious seasonal appearance (peaking in the summer months in temperate climates), had a disfiguring and/or crippling nature, and could paralyze respiratory muscles (iron lungs, which were invented in 1928, were used to assist breathing). The viral etiology of poliomyelitis was discovered in 1909.

During the 1940s and 1950s, poliomyelitis patients underwent treatments developed by Sister Elizabeth Kenny that emphasized physical therapy instead of immobilization of the affected limbs. The isolation, genetic studies, and characterization of polioviruses did not begin until the 1950s. The growth of polioviruses in cultured mammalian cells radically changed the field of medical virology. These advanced techniques allowed researchers to isolate, grow, and characterize the biology of polioviruses. Most importantly, the techniques led to the development of two successful vaccines: Jonas Salk's inactivated poliovirus vaccine (IPV) and Albert Sabin's live, attenuated oral vaccine (OPV). These vaccines were licensed for use in the United States in 1955 and 1961, respectively.

Poliomyelitis is rare today. The virus is spread via an oral–fecal route of transmission. Up to 95% of poliovirus infections are asymptomatic. About 4% to 8% of infected individuals experience minor illness, and less than 1% of all poliovirus infections cause major illness, resulting in clinical signs or symptoms involving the CNS such as flaccid paralysis. Laboratory diagnosis today is focused on isolating the poliovirus from the patient and determining its serotype using neutralization assays. Genomic sequencing is used to determine if the viral isolate is a wild- or vaccine-type strain. Post-polio syndrome occurs in a large proportion of individuals who recovered from paralytic poliomyelitis. Its onset is insidious. Common symptoms include muscle weakness, fatigue, and pain in the muscles and joints. Treatment consists of managing symptoms. Global eradication of poliomyelitis is in the home stretch. There remain small pockets of polio in impoverished nations experiencing civil unrest, population displacements, famines, and national disasters. International political will and commitment is necessary to achieve eradication.

Research regarding the molecular biology of polioviruses continues. Virus properties include its absence of a lipid envelope, stability over a pH range of 3 to 10, and resistance to alcohol and other solvents. The three-dimensional structure of polioviruses was deciphered in 1985. The capsid consists of four different proteins (VP1-4) that make up the icosahedral shell of the virus particle. There are three serotypes of poliovirus that are distinguishable via neutralization of cytopathic effects in cell cultures by specific antisera. The host range of polioviruses is limited to primates and primate cell cultures. Nonprimate cell cultures are resistant to poliovirus infection because they lack the poliovirus receptor (PVR).

The monocistronic +ssRNA poliovirus genome of poliovirus is 3', polyadenylated and covalently linked to a virus-encoded oligopeptide (VPg) at its 5', terminus. Upon entry into the cytoplasm, the genome functions as monocistronic mRNA consisting of a single open reading frame. It is translated by a cap-independent mechanism (using the IRES) to produce a large precursor polyprotein that is proteolytically processed by virus-encoded proteases to yield structural and nonstructural proteins. Assembled polioviruses lyse the host cell upon release of viral progeny.

Nonpolio enteroviruses that cause acute illnesses, particularly in infants and children, have been studied with increasing frequency over the last 30 years. They have been associated with occasional outbreaks in which a larger-than-usual number of patients develop clinical disease, sometimes with fatal consequences. This chapter has discussed nonpolio enteroviruses that cause respiratory tract illnesses, myocarditis, and HFMD. The future likely will focus on the intervention of the more serious illnesses caused by nonpolio enteroviruses, either through the development of antiviral therapies or vaccines. It is an important and exciting area for future study.

CASE STUDY QUESTIONS

These questions relate to the Case Study presented at the beginning of the chapter.

1. Who is likely the original source of this virus?
2. An investigation was needed to identify potential contacts with the infant. Create a list of individuals who should be interviewed.
3. Who is most at risk for contracting vaccine-derived poliomyelitis?

CASE STUDY 1: ECHOVIRUS-4

A group of 17 young Italians from Turin (Italy) traveled to Krishnanagar, a town about 80 km (31 miles) from Calcutta, India. The group consisted of 11 women and 6 men, all between the ages of 18 and 32. Before the trip, the group members received tetanus, hepatitis A, hepatitis B, and typhoid fever vaccines. All of them were taking prescribed antimalarial medication. They spent two weeks staying at a hostel and touring the city, enjoying the historical sites and cuisine.

After returning to Italy, 8 of them became ill within 2 to 3 days of their return. They developed the following signs and symptoms:

- 8/8 experienced a fever greater than 99.5°F (37.5°C)
- 8/8 had a headache
- 2/8 had a stiff neck
- 1/8 experienced vomiting
- 1/8 had a sore throat

Seven of the sick individuals were hospitalized. Lumbar punctures were performed on 6 out of 8 them. All of the cerebrospinal fluid specimens from the lumbar puncture had increases in the proportion of lymphocytes indicative of a viral infection. Real-time polymerase chain reaction on the 5′ untranslated region of the viral genomes tested positive for enteroviruses. Echovirus-4 was isolated from 4 out of 6 stool samples and 1 out of 6 throat swabs collected from the travelers. The duration of their symptoms and hospitalization stay ranged from 3 to 5 days. All of the travelers became ill at the same time. No family members or healthcare staff became infected.

1. What is the most frequent travel-related illness among individuals going from industrialized countries to developing regions?
2. Do you think all of the sick travelers were infected by the same source? Explain your answer.
3. What is the suspected source of contamination in this case? (Hint: See Gobbi, F., et al. 2010 in the References)
4. What are possible routes of transmission?
5. The most frequently isolated serotypes of echovirus in Europe are 30, 23, and 6. Research the scientific literature using PubMed or ScienceDirect. In what countries is echovirus-4 prevalent?
6. What dietary recommendations would you follow if you were to travel to certain parts of India? Gambia? Thailand? Brazil? Kenya? Nepal?
7. How do echoviruses differ at the molecular level from coxsackie viruses? From polioviruses? From other enteroviruses?
8. Do all members of the Picornaviridae family cause the same diseases? Elaborate upon your answer.

References

Gobbi, F., et al. 2010. "Echovirus-4 meningitis outbreak imported from India." *J Travel Med* 17(1):66–68.

Wichman, R. J., et al. 2005. "Risk and spectrum of diseases in travelers to popular tourist destinations." *J Travel Med* 12(5):248–253.

Resources

Primary Literature and Reviews

Belnap, D. M., et al. 2000. "Three-dimensional structure of poliovirus receptor bound to poliovirus." *PNAS USA* 97(1):73–78.

Chia, J. K. S. 2005. "The role of enterovirus in chronic fatigue syndrome." *J Clin Pathol* 58:1126–1132.

Couzin, J. 2006. "Report concludes polio drugs are needed—After disease is eradicated." *Science* 311:1539.

Enserink, M. 2010. "What's Next for disease?" *Science* 330:1736–1739.

Frisk, G. 2001. "Mechanisms of chronic enteroviral persistence in tissue." *Curr Opin Infect Dis* 14(3):251–256.

Hellen, C. U. T., and Sarnow, P. 2001. "Internal ribosome entry sites in eukaryotic mRNA molecules." *Genes Dev* 15(13):1593–1612.

Ho, M. 2000. "Enterovirus 71: The virus, its infections and outbreaks." *J Microbiol Immunol Infect* 33:205–216.

Howard, R. S. 2005. "Poliomyelitis and the postpolio syndrome." *BMJ* 330:1314–1318.

Kamadjeu, R., et al. 2009. "Tracking the polio virus down the Congo River: A case study on the use of Google Earth in public health planning and mapping." *Int J Health Geogr* 8:4 doi:1186/1476-072X-8-4.

Lukashev, A. N. 2005. "Role of recombination in evolution of enteroviruses." *Rev Med Virol* 15:157–167.

Minor, P. D. 2004. "Polio eradication, cessation of vaccination and re-emergence of disease." *Nat Rev Microbiol* 2:473–482.

Mueller, S., Wimmer, E., and Cello, J. 2005. "Poliovirus and poliomyelitis: A tale of guts, brains, and an accidental event." *Virus Res* 111:175–193.

Orent, W. 2000. "The end of polio?" *The Sciences* March/April, pp. 25–31.

Palacios, G., and Oberste, M. S. 2005. "Enteroviruses as agents of emerging infectious diseases." *J Neurovirol* 11:424–433.

Roberts, L. 2010. "Polio outbreak breaks the rules." *Science* 330:1730–1731.

Robbins, F. C., and Enders, J. F. 1950. "Tissue culture techniques in the study of animal viruses." *Am J Med Sci* 220(3):316–338.

Rotbart, H. A., ed. 1995. *Human Enterovirus Infections.* Washington: ASM Press.

Sabin, A. B., and Boulger, L. R. 1973. "History of Sabin attenuated poliovirus oral live vaccine strains." *J Biol Stand* 1:115–118.

Sabin, A. B. 1987. "Role of my cooperation with Soviet scientists in the elimination of polio: Possible lessons for relations between the U.S.A. and the U.S.S.R." *Perspect Biol Med* 31(1):57–64.

Schein, C. H., et al. 2006. "NMR structure of the viral peptide linked to the genome (VPg) of poliovirus." *Peptides* 27(7):1676–1684.

Solomon, T. 2003. "Exotic and emerging viral encephalitides." *Curr Opin Neurol* 16:411–418.

Stalkup, J. R., and Chilukuri, S. 2002 "Enterovirus infections: A review of clinical presentation, diagnosis, and treatment." *Dermatol Clin* 20:217–223.

Trojan, D. A., and Cashman, N. R. 2005. "Post-poliomyelitis syndrome." *Muscle Nerve* 31:6–19.

Vaccines and Biologicals Dept., World Health Organization, Geneva, Switzerland. Div. of Viral and Rickettsial Diseases, National Center for Infectious Diseases; Global Immunization Div., National Immunization Program, CDC2004. "Progress toward global eradication of poliomyelitis, January 2003–April 2004." *MMWR* 53(24): 532–535.

Whitton, J. L., and Feuer, R. 2004. "Myocarditis, microbes and autoimmunity." *Autoimmunity* 37(5):375–386.

World Health Organization (WHO) Sudan Office, Khartoum, Sudan, et al. 2005. "Progress toward poliomyelitis eradication—Poliomyelitis outbreak in Sudan, 2004." *MMWR* 54(4):97–99.

World Health Organization (WHO) Namibia Office, Windhoek, et al., 2006. "Outbreak of polio in adults—Namibia, 2006." *MMWR* 55(44):1198–1201.

World Health Organization (WHO). 1992. "New approaches to poliovirus diagnosis using laboratory techniques: Memorandum from a WHO meeting." *Bull WHO* 70(1): 27–33.

Popular Press

Allen, A. 2007. *Vaccine: The Controversial Story of Medicine's Greatest Lifesaver.* New York: W. W. Norton.

Bookchin, D., and Schumacher, J. 2004. *The Virus and the Vaccine: The True Story of a Cancer-Causing Monkey Virus, Contaminated Polio Vaccine and the Millions of Americans Exposed.* New York: St. Martin's Press.

Clarke, M. 1953. "How they're closing in on polio." *Popular Science* May Issue.

Gilden Seavey, N., Smith, J. S., and Wagner, P. 1998. *A Paralyzing Fear: The Triumph Over Polio in America.* New York: TV Books.

Gould, T. 1995. *A Summer Plague: Polio and Its Survivors.* New Haven, CT: Yale University Press.

Halstead, L. S. 1998. "Post-polio syndrome." *Sci Am* 278(4): 42–47.

Kehret, P. 2006. *Small Steps: The Year I Got Polio,* Morton Grove, IL: Albert Whitman.

Oshinsky, D. M. 2005. *Polio: An American Story, The Crusade That Mobilized the Nation Against the 20th Century's Most Feared Disease.* New York: Oxford University Press.

Pipes, D. *The Hidden Hand: Middle East Fears of Conspiracy.* Hampshire, England: Palgrave Macmillan.

Rogers, N. 1992. *Dirt and Disease: Polio Before FDR.* Newark, NJ: Rutgers University Press.

Video Productions

The Final Inch: In the Fight to Eliminate Polio, the Last Step Is the Biggest. 2009. Vermilion Films.

The Polio Crusade. 2009. PBS Home Video.

Historic Polio Vaccine Films (1956–1961). 2007. Quality Information Publishers.

Warm Springs. 2005. HBO production.

FDR, A Presidency Revealed. 2005. The History Channel

RX for Survival: A Global Health Challenge: The Polio Campaign in India, The Campaign Reaches Its Climax. 2005. PBS.

In Search of the Polio Vaccine. 1998. The History Channel.

Breathing Lessons. 1996. Fanlight Productions.

A Paralyzing Fear: The Story of Polio in America. 1988. PBS Home Video.

eLearning

go.jblearning.com/shors2

The site features eLearning, an online review area that provides quizzes and other tools to help you study for your class. You can also follow useful links for in-depth information, or just find out the latest virology and microbiology news.

CHAPTER 12

Influenza Viruses

Wild waterfowl such as these geese and ducks are the largest natural reservoir for influenza viruses. In these birds, infections are generally localized to the intestinal tract, and high concentrations of viruses are shed in the feces without causing disease.

OUTLINE

"The duck is the Trojan horse."

Dr. Robert G. Webster, Influenza Researcher, St. Jude Children's Research Hospital

OUTLINE (continued)

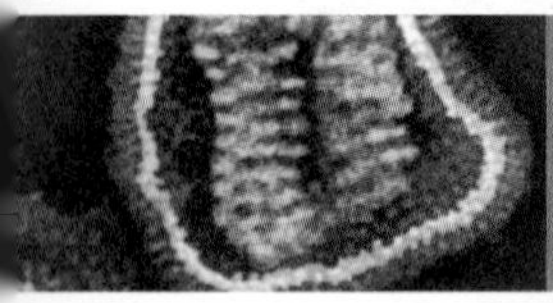

CASE STUDY: AVIAN INFLUENZA

Salesman Walter Smith began to travel extensively throughout Eastern Asia for his job. After business hours, he would sightsee. One day, as he walked the streets of Hanoi, Vietnam, he observed poultry for sale everywhere, especially chickens. There were prized fighting cocks displayed on a grassy median; their owners would suck the blood or phlegm out of the bird's mouths following a fight.

Walter was fascinated by the local traditions and loved exotic cuisine. As noon approached, he peered at the menu outside of a crowded restaurant and saw a dish on it that he had never eaten before: duck blood pudding, a delicacy. Intrigued, Walter decided to try it and found it delicious. He had two helpings and then returned to the hotel for the evening. He spent the next three days touring more of the city. On the fourth day, when he awoke in his hotel room, Walter felt feverish. He had a sore throat and his entire body ached. Over the course of the next two days he developed a bad cough. His illness continued to worsen and Walter went to the hospital. Upon his arrival, he saw the waitress who had waited on him at the restaurant where he had eaten the duck blood pudding. She had similar symptoms. Within the next hour, several chefs and more individuals who had dined at the restaurant arrived with the same symptoms. The doctors quickly noticed a pattern of illness and began setting up a ward to quarantine individuals related to this new outbreak of respiratory illness. Later, it was confirmed that this was an avian influenza outbreak.

12.1 History

Influenza is not a new disease. Historians have described outbreaks of a highly contagious respiratory illness followed by pneumonia as early as 400 B.C. The word *influenza* was derived from the medieval Latin word *influentia* because it was believed diseases were caused by bad heavenly fluid or "influence." Thus, *influenza* was used as a general term; for example, scarlet fever was referred to as *influenza di febbre scarlattina*; catarrhal fever (a very bad head cold involving copious mucous production) was called *influenza di catarro*. The first word in the phrase became the most significant. The French term *la grippe* ("the seizure") became the most popular term during the early 1900s. In English, diseases correlated with most respiratory tract infections became known as ***influenza***, later abbreviated to the modern *flu*.

There have been major influenza epidemics (localized outbreaks) and pandemics (worldwide epidemics). Pandemics occurred in 1918 (Spanish influenza), 1957 (Asian flu), 1968 (Hong Kong flu), 1977 (swine flu), and 2009 (H1N1 swine flu). They occur when the influenza virus mutates dramatically. These pandemics resulted in high morbidity and mortality.

Today, millions of Americans contract the flu each year. The CDC used to report that during the average season there were approximately 36,000 deaths nationwide and many more hospitalizations. Today, the CDC states that the range of annual deaths is too wide to continue using a single number. Based on 31 seasons (1976–2007), the average is closer to 23,600 with a range of approximately 3300 to 49,000 influenza deaths per year. The elderly, pregnant women, very young children and otherwise immune-compromised individuals are at the highest risk of developing serious flu complications. This group accounts for >90% of influenza-related deaths. Influenza is the seventh leading cause of death in the United States. Many virologists think that another influenza pandemic that could kill millions of humans is inevitable.

12.2 Epidemiology

At any given time of the year, one or two strains of influenza virus dominate a particular region of the world. In temperate or colder regions influenza illness usually occurs during the winter, but in tropical regions, influenza occurs throughout the year. Influenza epidemics can become unmanageable at alarming speeds for several reasons:

- The incubation period (the time between infection and the first appearance of symptoms) is extremely short (1–4 days).
- Copious numbers of infectious virus particles are shed in droplets discharged by sneezing or coughing. One droplet can contain 100,000 to 1,000,000 virus particles.
- Many symptomatic individuals do not stay at home. Instead, they continue their normal activities, spreading the illness to many contacts in the process.
- Individuals lack immunity to the strain of influenza circulating in the area at the time of the epidemic.

Schools are a primary source for sharing influenza viruses. Children bring the viruses home. Parents who contract the flu then disseminate the infection into the workplace. The best indicator of the scale of an epidemic is an escalation in absenteeism from schools and the workplace that is accompanied by an increase in hospital admissions and deaths, especially among the elderly. Epidemiologists measure the increase in mortality during an influenza epidemic by comparing it to the average number of deaths during comparable winters without an influenza epidemic. The number of "excess deaths" helps to determine epidemic years.

In the United States, the Spanish flu killed 675,000 Americans, with a 4.39/1000 national average. Death rates peaked in October of 1918 (**FIGURE 12-1a**). This epidemic was unique in that it killed healthy adults ages 20 to 40 (Figure 12-1b) in addition to children and elderly, who are usually within the high-risk group. This epidemic decreased life expectancy in the United States by 11 years (Figure 12-1c).

The 1957–1958 Asian flu caused 70,000 deaths, and the 1968–1969 Hong Kong flu resulted in 34,000 deaths in the United States. All of these deaths were caused by influenza Type A viruses (see Section 12.4, Classification of Influenza Viruses). Each flu season is unique, but it is estimated that approximately 10% to 20% of United States residents get the flu, and an average of 114,000 persons are hospitalized for flu-related complications during each season.

12.3 Clinical Features

Influenza can be hard to diagnose because there are several other diseases that resemble it. Accurate diagnosis is based on the patient's history and phys-

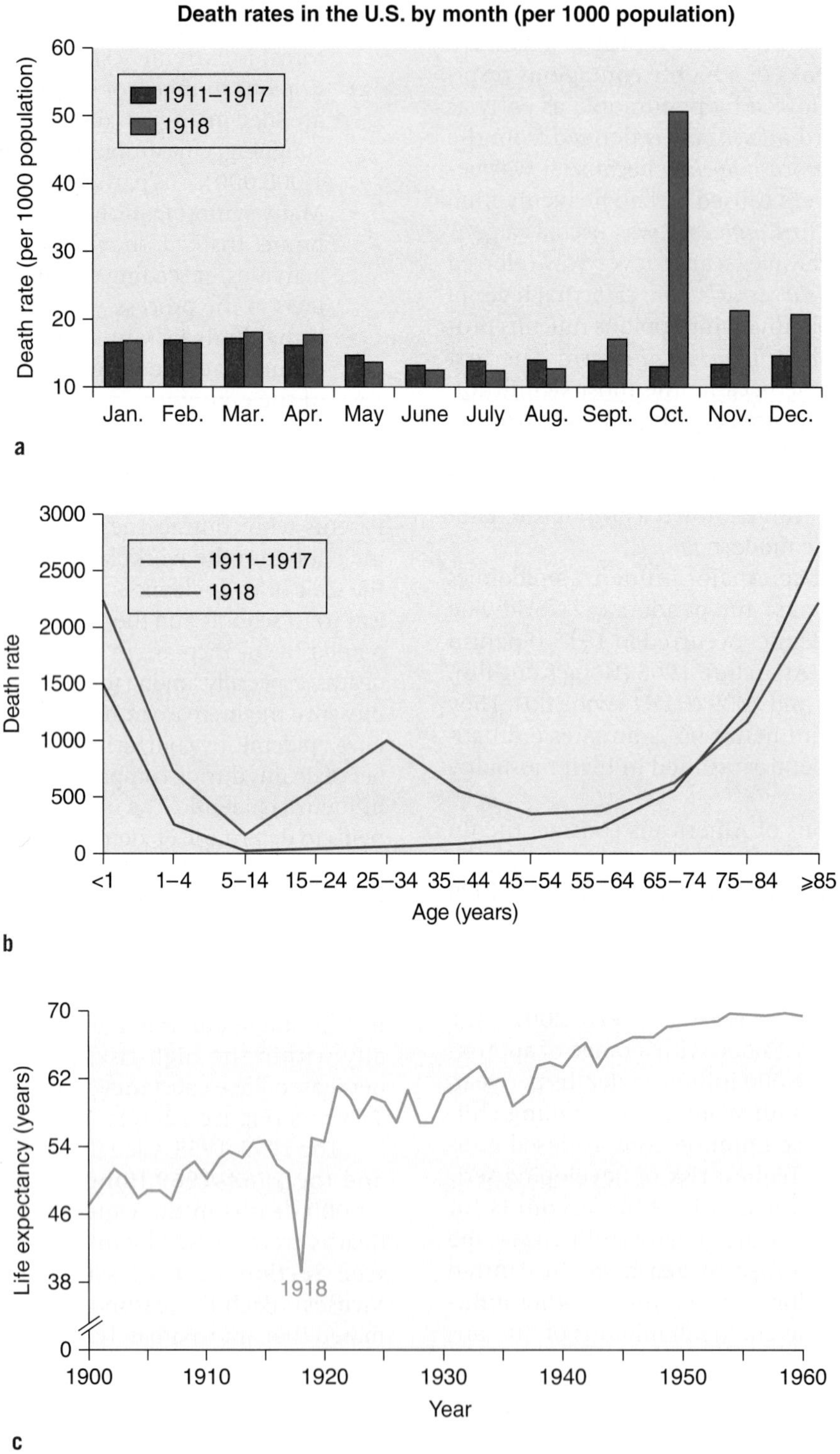

FIGURE 12-1 (a) During the Spanish flu pandemic of 1918, the month of October had the highest number of Spanish flu deaths. Adapted from PBS, "Influenza 1918," *American Experience,* http://www.pbs.org/wgbh/amex/influenza/maps. **(b)** This graph compares the age distribution of influenza deaths in 1918 and 1911–1917. Influenza pandemics result in a **"U-shaped" curve** with mortality peaks in the very young and elderly. The Spanish flu was unusual in that it also killed those in the prime of their lives, resulting in a characteristic **"W-shaped" curve**. Adapted from J. K. Taubenberger, *ASM News* 65 (1999): 473–478. **(c)** U.S. life expectancy dropped in 1918. Adapted from J. K. Taubenberger, *ASM News* 65 (1999): 473–478.

VIRUS FILE 12-1 Retrospective Studies

FIGURE VF 12-1 Collage of 1918 influenza artifacts found at antique stores originally used to treat influenza during this period in the Oshkosh area. Headlines, advertisements, and photographs from the Oshkosh Daily Northwestern newspaper are also shown.

*A **retrospective study** is defined as a study that looks backward in time. Based on this definition, determine by means of a retrospective study the impact the 1918 Spanish flu had on the community you currently reside in. An example by students at the University of Wisconsin–Oshkosh follows.*

In the spring semester of 2001, three very different classes at the University of Wisconsin-Oshkosh joined together to create the "Flu Project" as part of a multidisciplinary approach to learning about the types of research relevant to the study of epidemiology, archaeology, and gerontology. As a result of participating in the project through their individual courses, 70 students (35 virology students, 5 archaeology students, and 30 psychology students) were involved. The students participated in data collection and analysis about the spread of a deadly disease, the memorialization of those who died, and the memories of those who survived (**FIGURE VF 12-1**).

Undergraduates in virology class learned about epidemiology by studying the death records of those who died of the flu and related pneumonia in Winnebago County during the months of the epidemic. These students discovered where victims of the flu lived and the condition of their lives that may have contributed to the contagion and its deadliness. The psychology students interviewed flu survivors about their experiences and attitudes toward infectious diseases then and now. The archaeology students studied the headstones and graveyards to determine if burial practices changed during times of crisis (e.g., many bodies buried in a short time period). The students also learned the value of record keeping. Their biggest surprise was the inaccuracies in reporting the numbers of flu cases at the actual time of the epidemic. The students now have put together a more accurate picture of this historical event and have contributed to the knowledge base about it.

References

Shors, T., and McFadden, S. H. 2009. "Facilitated learning through interdisciplinary undergraduate research involving retrospective epidemiological studies and memories of older adults." *CURQ* 30(2):36–41.

Shors, T., and McFadden, S. H. 2009. "1918 Influenza: A Winnebago County, Wisconsin perspective." *Clin Med Res* 4:147–156.

ical examination in a clinical setting. Sometimes laboratory testing is performed before treatment is initiated.

Uncomplicated Infection

There is a tendency for patients—and even some doctors—to label all respiratory ailments as the flu. In reality, influenza is a distinct clinical entity that begins after a two- to three-day incubation period. The illness starts with an unexpected onset of symptoms: headache; aching in the limbs and back; fever; **malaise** (a vague feeling of physical discomfort or uneasiness); a dry cough; a dry, tickling throat, which can lead to the voice becoming husky and even lost; a sore throat; and **myalgia** (pain in muscles). Many patients can pinpoint the hour that they started feeling ill.

In general, most individuals feel weak and remain in bed. Their temperature will be high and continuous (100°F to 103°F for three days). Cough and weakness can persist for another week or more. Eyes are watery and burning and can even be painful upon movement. The nose may be runny or blocked with discharge. On examination by a physician, the patient's eyes appear red and his or her cheeks may be flushed. A chest x-ray will be normal unless a secondary bacterial infection has taken place. Usually the infection resolves itself within seven days, although patients often complain of feeling listless for a few weeks after the infection.

Many individuals think that a cold and the flu are caused by the same virus. They are not. The common cold, or acute viral rhinitis, is a much milder disease that can be caused by one of over 100 different types of rhinoviruses and is not lethal. The differences between a cold and the flu are listed in **TABLE 12-1**.

Complicated Infection

Complications depend on the age of the patient. Young children may develop croup (laryngitis with a hoarse, barking cough and difficult breathing), secondary bacterial pneumonia, and middle ear infections. Elderly are particularly vulnerable to life-threatening pneumonia caused by secondary bacterial infections (usually due to *Staphylococcus aureus*, *Streptococcus pneumoniae*, or *Haemophilus influenzae*). Preexisting chronic conditions such as congestive heart failure also can be exacerbated.

In addition to the elderly, anyone suffering from chronic conditions affecting the pulmonary (lung), cardiac (heart), renal (kidney), hepatic (liver), or endocrine systems is at risk. Modern medicine has kept many children alive who have immune deficiencies or diseases such as cystic fibrosis, as well as saved patients of any age with organ transplants, cancer, or acquired immunodeficiency syndrome (AIDS). This has increased the number of individuals at risk of death during an influenza epidemic.

Reye Syndrome

Besides secondary bacterial infections, there is a rare condition called **Reye syndrome** that is sometimes associated with influenza (or other viral illnesses such as chickenpox) in children. Studies have shown that using aspirin or salicylate-containing medications to treat viral illnesses increases the risk of developing Reye syndrome. It affects all organs of the body but is most harmful to the brain and the liver, causing an acute increase of pressure within

TABLE 12-1 Cold or Flu?

Symptoms	Cold	Flu
Fever	Rare in adults and older children, but can be as high at 102°F in infants and small children	High (100 to 104°F), lasting 3–4 days.
Headache	Rare	Prominent, sudden onset
General muscle aches and pain	Slight/mild	Common, often severe
Sore throat	Sometimes	Common
Cough, chest discomfort	Mild hacking cough	Common, can become severe
Sneezing	Often	Sometimes
Runny or stuffy nose	Often	Sometimes
Fatigue, weakness	Mild	Often extreme, lasting 2–3 weeks
Extreme exhaustion	Never	Sudden onset; can be severe
Complications	Sinus congestion or earache	Bronchitis, pneumonia
Prevention	None	Annual vaccination, antiviral drug
Treatment	Temporary relief of symptoms	Antiviral drug within 24–48 hours after onset of symptoms

the brain and massive accumulations of fat in the liver and other organs.

Symptoms of Reye syndrome include persistent or recurrent vomiting, listlessness, personality changes (i.e., irritability or combativeness), disorientation or confusion, delirium, convulsions, and loss of consciousness. There is no cure for Reye syndrome. Successful management depends on early diagnosis.

Anthrax, Influenza, or SARS?

After the terrorist attacks on the United States of September 11, 2001, clinicians were concerned about how to evaluate persons who may be at increased risk for inhalational anthrax but who have symptoms of an influenza-like illness. By November 2001, the Centers for Disease Control and Prevention (CDC) published a table listing the signs and symptoms of ten cases of inhalation anthrax compared to laboratory-confirmed influenza. It is important to be able to distinguish among respiratory infections—especially during the flu season—so that they can be treated appropriately (**TABLE 12-2**).

12.4 Classification of Influenza Viruses

True influenza is caused by any of several genera of the *Orthomyxoviridae* family of viruses. *Orthos* is the Greek term meaning standard or correct and *myxa* means mucus. This family contains three influenza genera categorized as influenza type A, type B, and type C viruses. All three types of influenza viruses can cause similar symptoms in humans. Infection with one type of influenza, however, does not confer immunity to the other types. **TABLE 12-3** compares the epidemiological, clinical, genomic, and structural differences among the three virus types. All three share certain structural features, including segmented genomes (single-stranded RNA [ssRNA] of negative polarity), envelope glycoproteins critical for virus entry and exit of cells, and a host-derived envelope.

12.5 Laboratory Diagnosis

Most physicians use a cluster of signs to diagnose influenza, backed up by the knowledge that an outbreak is in progress. Influenza viruses, common cold viruses, and bacteria can cause similar upper respiratory tract symptoms. The intensity, severity, and frequency of the symptoms vary. **FIGURE 12-2** illustrates the natural course of influenza infection. If a tentative diagnosis of influenza has been made, laboratory testing may be an appropriate confirmatory step.

Office-Based Rapid Viral Tests

There are at least six commercially available, office-based techniques for detecting influenza A and/or B. All of these tests can be completed within 10 to 20 minutes. Rapid influenza tests are

TABLE 12-2 Symptoms and Signs of Inhalational Anthrax vs. Laboratory-Confirmed Influenza and SARS

Symptoms/Signs	Inhalational Anthrax (n = 10)*	Laboratory-Confirmed Influenza*	SARS (n = 62)**
Fever or chills	100%	83%–90%	100%
Fatigue/malaise	100%	75%–94%	100%
Nonproductive cough	90%	84%–93%	90%
Shortness of breath or chest discomfort/pain	80%	6%	72%
Headache	50%	84%–91%	85%
Sore throat	20%	64%–84%	12%
Runny nose	10%	79%	10%
Nausea or vomiting	80%	12%	32%
Abdominal pain	30%	22%	NA
Myalgias (muscle pain)	50%	67%–94%	87%

The CDC reported the signs and symptoms of ten cases of inhalational anthrax that occurred after the tragic events of September 11, 2001, and compared it to laboratory-confirmed influenza (*data from *MMWR*, 2001, 50[44]:984–986). The inhalational anthrax cases began appearing just before flu season, raising diagnostic concerns for healthcare providers. Many persons who were exposed to inhalational anthrax in the bioterrorism-related anthrax attacks completed a 60-day course of antimicrobial prophylaxis. From February to May 2003, researchers investigated the clinical signs, symptoms, and course of all probable SARS cases defined by WHO criteria during the Vietnam outbreak (**data from Thu Vu, H., et al. 2004. *EID* 10[2]:334–338).

NA = Not available.

TABLE 12-3	Differences Between Influenza Types A, B, and C Viruses		
Feature	**Influenza A**	**Influenza B**	**Influenza C**
Host Range	Humans, pigs, horses, birds, marine mammals	Humans only	Humans and pigs
Epidemiology	Antigenic shift and drift	Antigenic drift only	Antigenic drift only
Clinical Features	May cause pandemics with significant mortalities in affected young people	Severe disease, generally confined to elderly or high-risk, pandemics not seen	Mild disease, common in children, without seasonality
Genome	8 gene segments	8 gene segments	7 gene segments
Structure	10 viral proteins M2 unique	11 viral proteins NB unique	9 viral proteins HEF unique

cheaper than virus isolation or PCR testing: ~\$5 for first 1000; <\$10 thereafter. These rapid tests are either enzyme-linked immunosorbent assay (ELISA) or immunofluorescent assays (for review of techniques see Chapter 5). The BD Directigen and QuickVue tests distinguish between influenza A and B viral antigens. If the chosen test is positive, antiviral therapy may be considered if symptoms have been present for less than 48 hours. A negative test result may support the decision to withhold antiviral drug therapy. Although rapid tests are relatively easy to perform, the CDC estimates that up to 30% of samples may produce negative results, yet be positive by viral culture methods. Rapid tests may also produce false-positive results.

Other Tests

Viral culture is the most accurate method for identifying specific viral strains. Virus can be recovered from nasal swab specimens, nasal washes, or com-

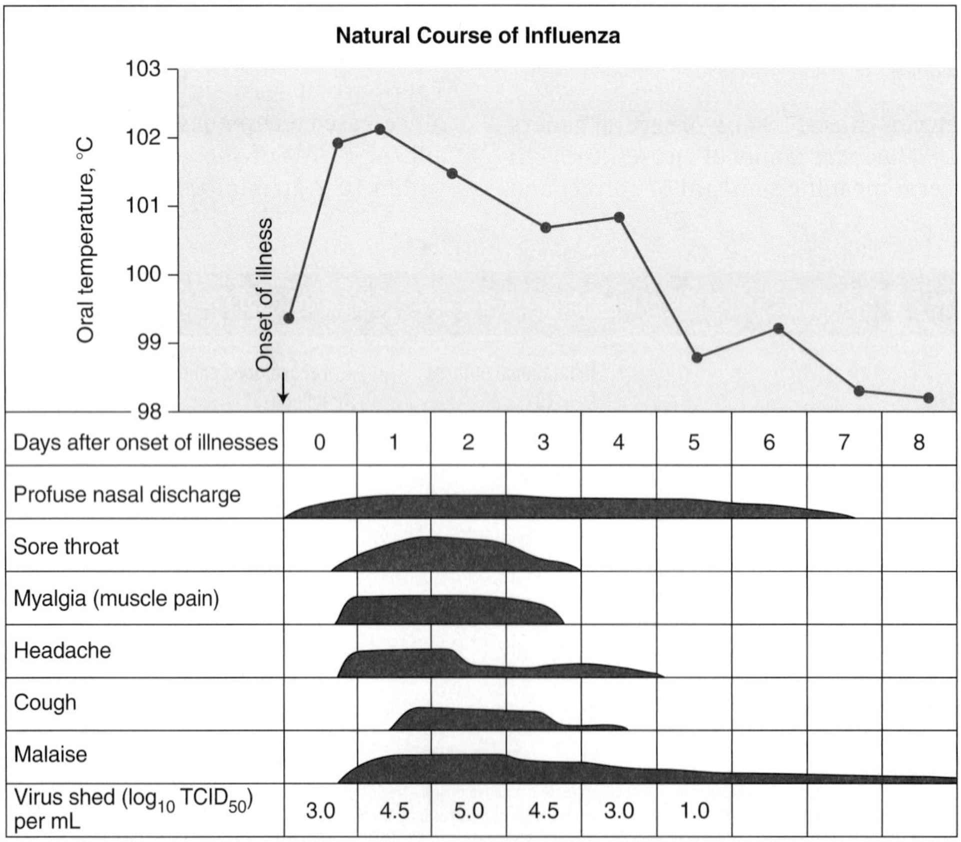

FIGURE 12-2 The natural course of influenza in an otherwise healthy 28-year-old male patient. Many symptoms overlap throughout the duration of influenza. Virus particles are shed during the first five days of the infection. Adapted from Dolin, R. 1976. *American Family Physician* 14:74.

bined nose-and-throat swab specimens. It can also be isolated directly from sputum samples. Samples are placed in special transport containers, labeled, and transported properly to the viral diagnostic lab as soon as possible.

The specimens are inoculated onto rhesus or cynomolgus monkey kidney cells or Madin-Darby canine kidney (MDCK) cell cultures where the virus is detected by cytopathic effect or other assays. Results are not available for two to ten days. Viral culture is also costly (about $75) and requires special laboratory procedures as well as skilled technicians to carry out these procedures. In sophisticated laboratories, nucleic acid techniques (polymerase chain reaction [PCR]), electron microscopy, cytology, and histology may be used to diagnose viral illness.

Serology Testing

Sometimes in clinical practice a serum sample is collected from the patient when the patient comes in for diagnosis and once again 10 to 20 days later, after the patient has recovered (**convalescent serum**). The serum samples are analyzed for an increase in antibody titer. A fourfold or greater increase in antibody titer is considered diagnostic of infection. This information is collected too late and therefore does not affect clinical decisions; however, it does confirm evidence of the infection.

"Flu Chips"

Several research teams are currently developing "flu chips" or **microarrays** as a way to quickly identify (within an hour or less) and screen for influenza (types A, B, and C), avian influenza, severe acute respiratory syndrome–coronavirus (SARS-CoV), respiratory syncytial virus (RSV), and other airborne bioterrorist pathogens such as *Bacillus anthracis* (refer to Chapter 5, Section 5.2, Microarrays and Protein Arrays: A New Option in Viral Diagnostics). Several teams are developing diagnostic microarrays that address the needs of developing nations by creating an inexpensive and field-portable test kit for influenza A respiratory tract infections.

When a U.S. public health emergency was declared in December 2009, the FDA approved the temporary use of a resequencing influenza A microarray detection panel for clinical respiratory symptoms. The microarray panel was approved for use to diagnose infections caused by the 2009 H1N1 influenza A virus through April 26, 2010.

Glycan microarray analysis focused on detecting the virus-receptor specificity of new influenza strains, especially H5N1 is being used by research laboratories. Influenza viruses use host glycans that contain sialic acid moieties as cell-surface receptors. These glycans vary in structure from species to species. The subtle difference in recognizing glycan structures can rapidly assess whether influenza strains from avian or other species will easily infect humans. Carbohydrate microarrays were used to analyze virus receptor specificity of fatal cases of the pandemic H1N1 influenza A virus.

12.6 Cellular Pathogenesis

Influenza is spread person-to-person via droplets containing the virus. Large numbers of virus are present in respiratory secretions of infected persons. The virus is dispersed in aerosol form by sneezing, coughing, and talking. Influenza infection is acquired by the respiratory route through the inhalation of aerosoled virus particles. Once the virus enters the respiratory tract, it can attach to and penetrate ciliated columnar epithelial cells lining the sinuses and airways (**FIGURE 12-3a**).

The primary site of infection occurs at the tracheobronchial tree, but the nasopharynx is also involved (Figure 12-3b). After the virus attaches, is adsorbed, and replication begins, the cilia are destroyed, releasing progeny virus that spread to nearby cells. Hence, the cleaning system in the lungs does not work well. As the numbers of cilia are reduced, there are fewer to clean the lungs and more mucus stays in the airways, clogging them and making the individual cough. The loss of epithelial cells and destruction of cilia can contribute to secondary bacterial invasion, manifested as sinusitis, otitis (inflammation of the ear), and **pneumonia** (infection of the alveoli and surrounding lung). Infectious virus can be isolated for one to seven days, with the peak of the released virus occurring on the fourth or fifth day after infection (**FIGURE 12-4**).

12.7 Immunity

Infection with influenza virus results in the development of antibodies against the H and N envelope glycoproteins of the virus. Anti-H antibodies neutralize the virus, preventing attachment/viral infectivity. Antibodies against N do not neutralize the virus, but they do reduce release of virus from infected cells. If the individual has been infected within the past couple of years by a closely related strain of influenza H subtype, anti-H IgA class antibodies may intercept and neutralize infecting virions. IgG antibodies may play a secondary role in neutralizing the virus, providing protection in the lung.

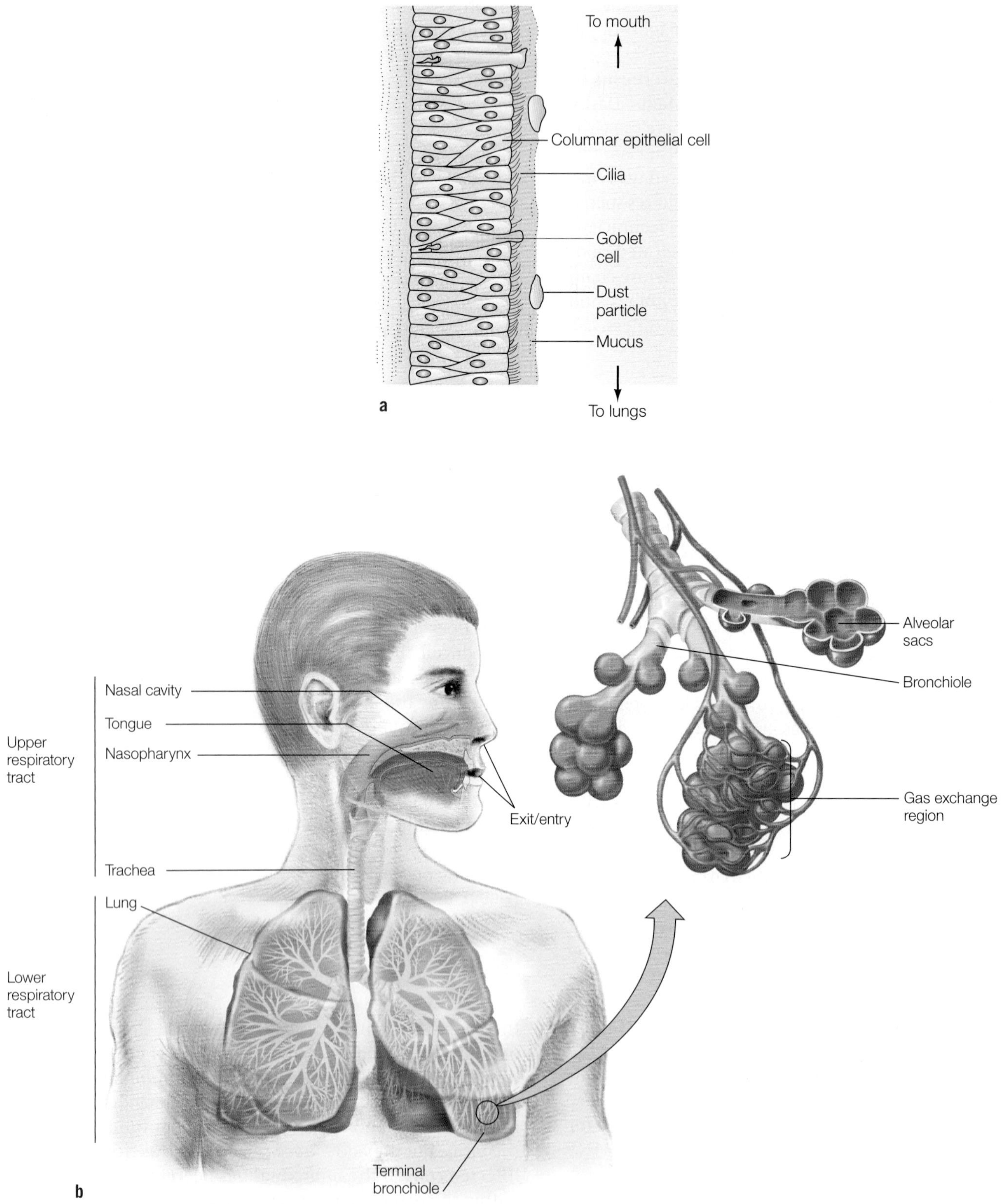

FIGURE 12-3 (a) The cilia are attached to columnar epithelial cells. Cilia sweep mucus produced in goblet cells as well as mucus coming from deeper glands within the lungs and the particulate matter trapped in the mucus. The presence of cilia aids the removal of mucus materials from lower parts of the airway to the throat for disposal. **(b)** Exit and primary sites of respiratory infections.

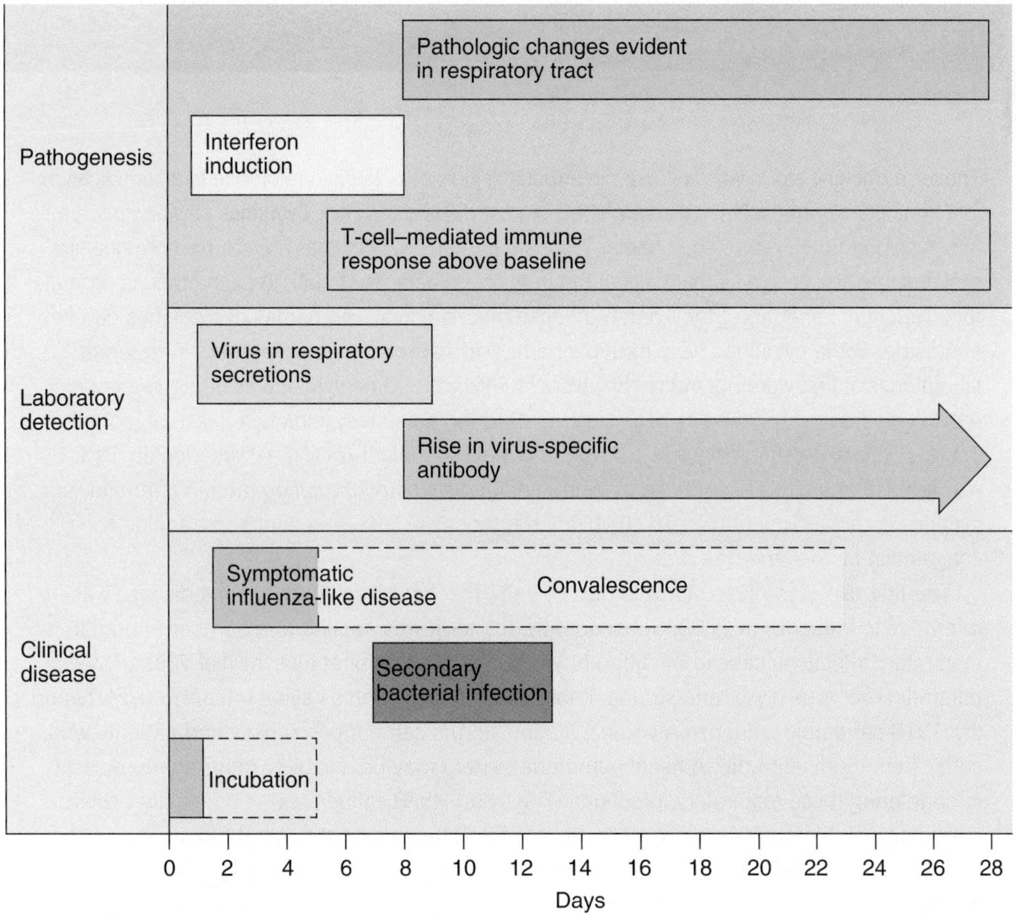

FIGURE 12-4 Time course of influenza A infection. Flu-like symptoms occur early. Later, pneumonia may result from a secondary bacterial infection in certain individuals. Adapted from Murray, P. R., et al. *Medical Microbiology*, Third Edition. Mosby Elsevier Health Science, 1998.

If preexisting antibodies are inadequate, the individual must mount an antibody response against the virus. Besides an antibody response, other arms of the immune system play a role in recovery. These are activated macrophages, natural killer T cells, and some cytokines such as interferon.

A **cytokine storm** or **systemic inflammatory response syndrome** (**SIRS**), has been suggested as an explanation for the devastating nature of the 1918 flu. From a clinical perspective, a cytokine storm is defined as an immune system that is overreacting toward the pathogen. In this situation, cytokine signaling immune cells such as macrophages and T-cells travel to the site of infection, causing damage to the body and thus the failure of multiple organs (**VIRUS FILE 12-2**). This is a rare event. Patients who suffer from them usually die. The 1918 H1N1 strain has been re-created using published sequences and reverse genetics (see Virus File 12-2). The re-created virus was highly virulent in macaques and mice. Scientists determined that the following chemokines or cytokines were present at elevated levels in the bronchi of macaques infected with the 1918 virus: interleukin-6 (IL-6), interleukin-8 (IL-8), monocyte chemotactic protein-1 (MCP-1 or CCL2), and CCL5 (RANTES). The H5N1 influenza strain also causes SIRS and is 50% lethal in humans.

12.8 Viral Replication

Properties of the Influenza A Virus Particle

Influenza particles are 80 to 120 nm in diameter. Influenza strains may be filamentous in a form of helical symmetry immediately after isolation (**FIGURE 12-5a**), but after several passages in cell cultures in the laboratory, the particle morphology becomes spherical (Figure 12-5b). Electron micrographs reveal influenza type A to be enveloped viruses, covered with approximately 500 projecting spikes. Approximately 80% of these rod-shaped glycoprotein spikes are **hemagglutinin** (**H**) antigen (a trimer), and 20% are mushroom-shaped **neuraminidase antigen** (N; a tetramer) (Figure 12-5c). The H is a **transmembrane protein**, whereas N is an **integral membrane protein**. The inside of the particle is lined by a **matrix 1** (**M1**) protein.

VIRUS FILE 12-2 The Perfect Storm of Cytokines That Can Kill You

The term cytokine storm was first used in a medical journal in 1993 in reference to a complication in a bone marrow transplant called graft-versus-host disease (GVHD). **Cytokines** play an important role in the immune system (see Chapter 7, Innate Immunity). Cytokines are soluble hormone-like proteins that act as signals that cells of the immune system use to talk to each other, coordinate their functions, and target their harsh defense mechanisms. The names of cytokines can be confusing. Some cytokines have multiple names, in part because researchers in different laboratories at first were not aware they were studying the same cytokine as other researchers. A cytokine storm response has been used to describe a fierce systematic release of more than 150 *pro-inflammatory cytokines* such as tumor necrosis factor-α (TNF-α), interleukin-1 (IL-1), and interleukin-6 (IL-6) as well as *cytokine antagonists of inflammation* (anti-inflammatory cytokines) such as interleukin 10 (IL-10) and interleukin 1 (IL-1) receptor in addition to coagulation factors and free radicals.

The first use of cytokine storm to describe the mechanism of an infectious disease was in reference to influenza in 2003. Subsequently, the term was applied to a number of human respiratory infections caused by different viruses such as respiratory syncytial virus (RSV), parainfluenza virus (PIV), and strains of influenza viruses such as avian influenza (H5N1) and the 1918 pandemic influenza A virus. Cytokine storms can happen rapidly and patients who suffer from them often die. A healthy immune system may be a liability rather than an asset in countering these respiratory infections. The avian H5N1 influenza A is unusually virulent in humans (in March 2011 there were a total of 316 human deaths out of 534 cases; 59% mortality rate). H5N1 causes a cytokine storm affecting the lungs, resulting in subsequent damage to the alveoli and lung tissue. Without prompt medical care, patients develop alveolar hemorrhaging, acute respiratory distress because pulmonary edema blocks off airways, and multi-organ failure, which leads to death. Research is in progress to study the cytokine profiles of different viral respiratory tract infections that trigger a cytokine storm. A short scenario of what happens in a H5N1 influenza A cytokine storm is as follows:

1. H5N1 influenza A viruses attach to and replicate in the pneumocytes of the alveolar epithelium and macrophages located in alveoli of the lower respiratory tract. (Note: Seasonal influenza viruses prefer cellular receptors located in the upper respiratory tract of the lung.)
2. Virus infection stimulates the infected and uninfected epithelial pneumocytes to release cytokines and chemokines.
3. Infected macrophages secrete cox-2 (an enzyme that causes inflammation, pain, and fever) and proinflammatory cytokines and chemokines (e.g., IL-6, IL-8, RANTES [CCL5], macrophage inflammatory protein [MIP], monocyte chemotactic protein [MCP], CCL8, TNF-α, type I interferons [α and β], CXCL9, CXCL10 [IP-10] and CXCL11).
4. Leukocytes such as macrophages, dendritic cells, and lymphocytes are drawn to and activated by the cytokines and chemokines into the alveolar spaces.
5. The dendritic cells and macrophages in the alveolar space present viral peptides to CD8+ cytotoxic T cells (MHC 1), leading to the proliferation of influenza-specific CD8+ cytotoxic T cells in the lung.
6. CD8+ cytotoxic T cells play an important role in controlling the viral infection, but in excessive numbers, might contribute to tissue damage.

The pro-inflammatory response triggered by the cascades of cytokines and chemokines include apoptosis, tissue damage and necrosis, and dilation of blood vessels. **FIGURE VF 12-2** is an illustration of the cytokine storm response in the pathogenesis of lung damage in an H5N1 infection.

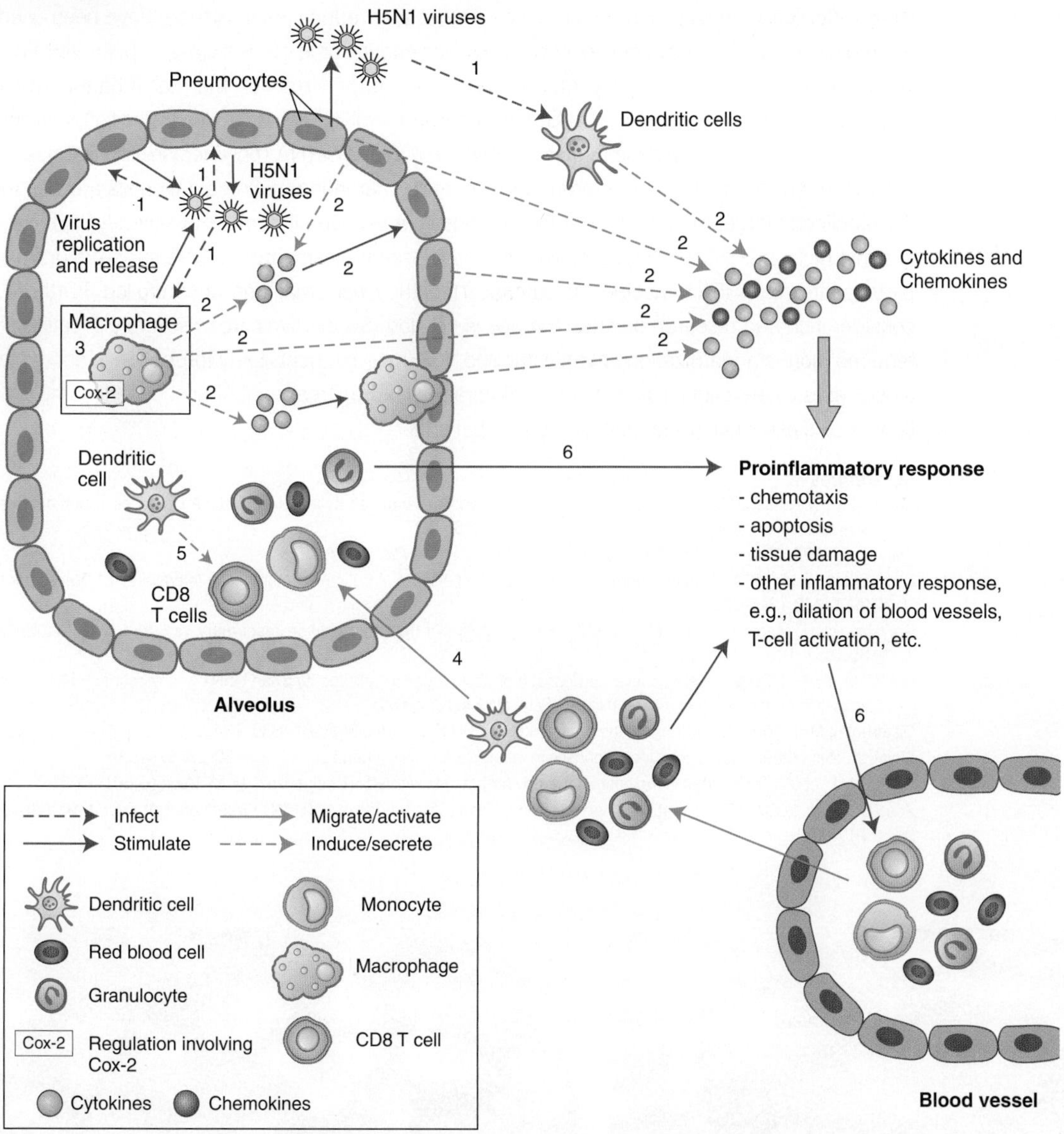

FIGURE VF 12-2 Model for the mechanism of the cytokine storm evoked by highly avian influenza viruses. The viruses infect pneumocytes and macrophages in the alveoli of the lower respiratory tract. The macrophage becomes activated and presents a viral peptide on its outside surface. In turn, this activates a T cell, resulting in an uncontrolled response of proinflammatory chemokines and cytokines. The blood vessels of the lungs become dilated and filled with leukocytes, resulting in destruction and necrosis of lung tissues. Adapted from Sriyal, J., et al. 2009. "Innate immune responses to influenza A H5N1: Friend or foe?" *Cell* 30(12):574–584.

Treatment of H5N1 is challenging for physicians. The viral neuraminidase inhibitor oseltamivir (Tamiflu) is the drug of choice possibly at a higher dose (150 mg twice daily) and longer duration (10 days). Recommended dose for seasonal influenza cases is 75 mg twice daily for 5 days. Treatment early in the clinical course of infection is critical to survival. Influenza expert Robert Webster warns that it is not wise to rely on monotherapy and believes that a combination chemotherapy approach will safeguard against the emergence of drug-resistant

(continued)

strains. **Corticosteroids**, a class of steroid hormones that reduce inflammation, have been used in some patients, but they have not shown any benefits and in some cases may have been harmful. Drug development to treat H5N1 infections is targeting the virus and the host innate immune response. New antiviral drugs are in the pipeline: cyanovirin-N (hemagglutinin inhibitor), peramivir and long-acting neuraminidase inhibitor (LANI), T-705 and siRNA (polymerase inhibitors), and antibody-mediated therapy. Interferon inducers and other inflammation antagonists are also in the pipeline to target the innate immune responses. A very small number of severely ill H5N1 patients recovered after being administered convalescent plasma from H5N1 survivors or from participants in an H5N1 vaccine clinical trial. The study was small and uncontrolled. Further consideration and research on such therapy is needed. Supportive care of severely ill patients requires more standardization of ventilator and fluid management. Early intubation is critical in those cases where clinical signs point to respiratory failure. Treatment of nosocomial complications is also essential in the management of patients.

References

Chan, M. C. W., et al. 2005. "Proinflammatory cytokine responses induced by influenza A (H5N1) viruses in primary human alveolar and bronchial epithelial cells." *Respir Res* 6:135–137.

Clark, I. A. 2007. "The advent of the cytokine storm." *Immunol Cell Biol* 85:271–273.

Ferrara, J. L., et al. 1993. "Cytokine storm of graft-versus-host disease: A critical effector role for interleukin-1." *Transplant Proc* 25(1 Pt 2):1216–1217.

Grayson, M. H., and Holtzman, M. J. 2007. "Emerging role of dendritic cells in respiratory viral infections." *J Mol Med* 85:1057–1068.

Lee, S. M. Y., et al. 2009. "Systems-level comparison of host-responses elicited by avian H5N1 and seasonal H1N1 influenza viruses in primary human macrophages." *PLoS ONE* 4(12):e8072.

Olsterholm, M. T. 2005. "Preparing for the next pandemic." *N Engl J Med* 352(18):1839–1842.

Sriyal, J., et al. 2009. "Innate immune responses to influenza A H5N1: Friend or foe?" *Cell* 30(12):574–584.

White, N. J., et al. 2009. "What is the optimal therapy for patients with H5N1 influenza?" *PLoS Med* 6(6):e1000091.

Zou, B., et al. 2007. "Treatment with convalescent plasma for influenza A (H5N1) infection." *N Engl J Med* 357(14): 1450–1451.

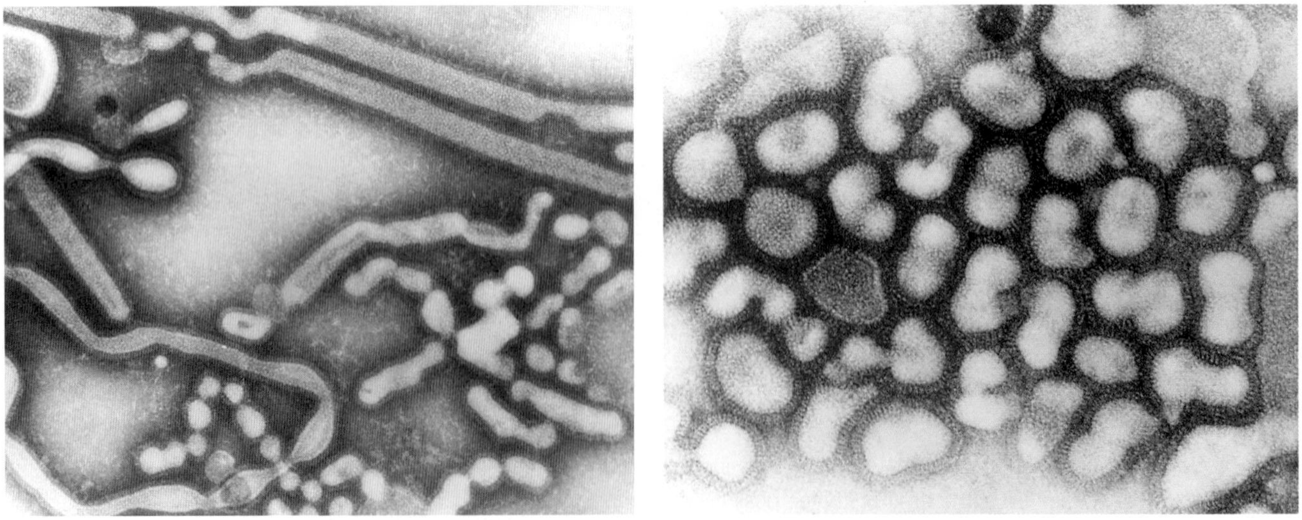

a b

FIGURE 12-5 Electron micrographs of influenza A spherical particles. Note the projecting spikes located on the outside of the viral particle. **(a)** When influenza viruses are isolated directly from patients, they are filamentous shaped as shown in the electron micrograph. **(b)** After several passages in the laboratory, the particle morphology will become spherical shaped.

(continued)

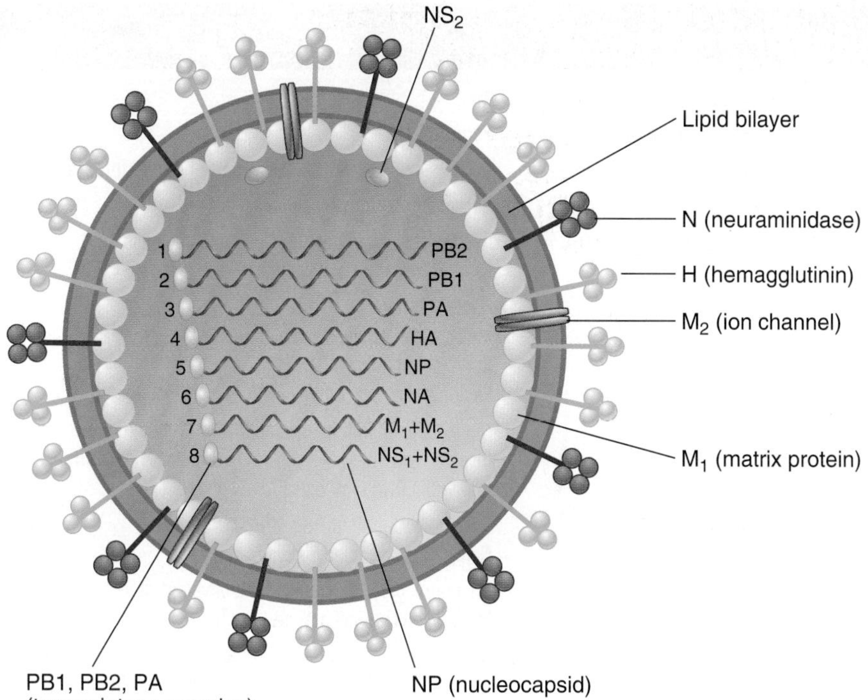

c

FIGURE 12-5 (Continued) Electron micrographs of influenza A spherical particles. Note the projecting spikes located on the outside of the viral particle. **(c)** Schematic diagram of influenza A virus structure's H and N spikes. The H is a trimer of identical subunits. N is a tetramer of identical subunits D. The ratio of H to N varies, but it ranges from 4:1 to 5:1.

In addition to the N and H proteins found on the surface of the virus, there are also small amounts of **M2 ion channel** integral membrane protein present in influenza A virus particles. M2 is not present in influenza B virions. M2 ion channel protein is inserted into the lipid bilayer membrane that surrounds the viral genetic material. Directly below the lipid membrane of the virus particle is a layer of viral matrix protein termed M1. It is the most abundant virion protein. Inside the virus are eight segments of ssRNA of negative polarity ranging from 934 to 2341 nucleotides in length (**TABLE 12-4**). The coding assignments for each segment are depicted in the schematic diagram of the influenza A particle in **FIGURE 12-6**. Each segment contains a transcriptase complex that consists of three additional viral polymerase proteins: PB1, PB2, and PA. Small amounts of NS2 are present in purified virions. NS1 is only found in infected cells.

Virus Nomenclature

The standard naming for influenza viruses is:

- influenza type/
- species isolated from/ (unless human)
- place of isolation/
- strain designation/
- year isolation/
- H#N# subtypes

For example, influenza A isolated from the first patient isolate in Hong Kong in 1968 is given the name A/Hong Kong/1/68(H3N2). If the influenza isolate has been isolated from another host, such as birds or animals, the host name is included (e.g., A/swine/Iowa/15/30[H1N1]). If the strain is a human strain, the host origin is not included in the name (e.g., A/Sing/1/57). Influenza type A is the most frequent cause of influenza and will be the focus of this chapter.

There are 16 antigenically different types of Hs, labeled H1–H16 subtypes, and 9 distinct Ns (N1–N9 subtypes). All subtypes differ by more than 30% amino acid sequence homology. All H and N subtypes infect waterfowl (e.g., ducks). Six subtypes of H (H1, H2, H3, H5, H7, and H9) and two subtypes of N (N1 and N2) have been recovered from humans (**TABLE 12-5**). All but the H5, H7, and H9 subtypes spread human to human efficiently. Besides humans and birds, other H and N subtypes are found in pigs, seals, and horses. Cross-reactivity among the different subtypes does not occur. In other words, if you are immune to influenza H1N1, you will not have any immunity against H2N2. The antigenic subtype differences are critical for vaccine preparation.

Genome Organization

Influenza A virus contains eight segments of ssRNA of negative polarity (see Table 12-4 for coding assignments). The complete nucleotide sequence of influenza A virus PR/8/34 and many RNA segments of other subtypes were completed in 1982.

TABLE 12-4 Influenza A Gene Coding Assignments

RNA Segment	Length (nt)	Encoded Polypeptide (aa)	Function
1	2341	PB1 757	Catalyzes nucleotide addition (part of transcriptase complex)
2	2233	PA 716	Part of transcriptase complex
3	2341	PB2 759	Recognizes cap of host mRNAs, endonuclease (part of transcriptase complex)
4	2073	HA 566	Hemagglutinin Major surface glycoprotein Sialic acid binding Fusion at low pH
5	1565	NP 498	Nucleocapsid protein, interacts with RNA segments Involved in switching of mRNA to template RNA synthesis and virion RNA synthesis
6	1413	NA 454	Surface glycoprotein Neuraminidase (cleaves sialic acid)
7	1027 spliced	M1 252 M2 97	M1 matrix protein inside of lipid envelope of virion M2 has ion channel activity
8	934 spliced	NS1 230 NS2 121	NS1, nonstructural protein (not detected in virions) inhibits export of poly(A)′ dmRNAs and pre-mRNA splicing, IFN antagonist NS2 plays a role in export of ribonucleoproteins (RNPs) out of the nucleus

Segments 1 to 6 code for only one viral protein; segments 7 and 8 code for two.

The Virus Life Cycle

Virus Adsorption and Entry

The influenza virus H attaches to sialic acid residues that are present on glycoproteins or glycolipids of the ciliated columnar epithelial cells lining the sinuses and airways. Next, the bound virus is endocytosed by the cell. Virions enter the cell within the endosomal vesicle (**FIGURE 12-7a**). Inside of the endosome, the virus is exposed to a relatively low pH (the pH drops from 7 to 5). The pH change triggers fusion between the viral lipid membrane and the host cell endosomal membrane (Figure 12-7b). The low pH-dependent fusion event is mediated by the H protein. The membrane fusion step is a

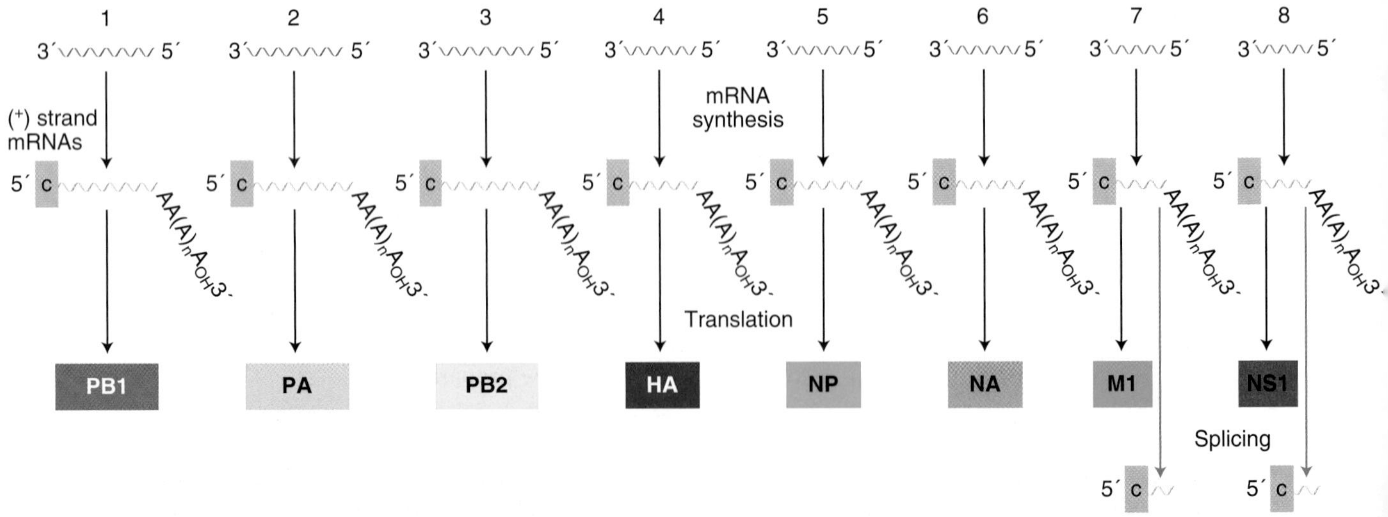

FIGURE 12-6 There are eight influenza A RNA segments, coding for ten influenza protein products. Gene segments 7 and 8 code for more than one protein on overlapping reading frames and the viral RNAs undergo a splicing event. Adapted from Flint, S. J., et al. *Principles of Virology: Molecular Biology, Pathogenesis, and Control of Animal Viruses*, Second Edition. ASM Press, 2003.

TABLE 12-5 Hemagglutinin (H) and Neuraminidase (N) Subtypes of Influenza A Viruses Isolated from Humans, Lower Mammals, and Birds

Subtype	Human	Swine	Horse	Bird
Hemagglutinin				
H1	√	√		√
H2	√			√
H3	√	√	√	√
H4				√
H5	√			√
H6				√
H7	√		√	√
H8				√
H9	√			√
H10				√
H11				√
H12				√
H13				√
H14				√
H15				√
H16				√
Neuraminidase				
N1	√	√		√
N2	√			√
N3		√		√
N4				√
N5				√
N6				√
N7			√	√
N8			√	√
N9				√

Source: Adapted from Levine, A. J. *Viruses*. W. H. Freeman and Company, 1992.

vital step in the life cycle of the virus. The H protein undergoes an irreversible conformational change that is triggered by the low pH of the endosome. The H protein is composed of two polypeptides, designated H1 and H2. Full-length H protein is referred to as H0. The H1 and H2 polypeptides are linked by a disulfide (S-S) bond between two cysteines. The H1 portion of H0 contains sialic acid-binding sites for attachment to the host cell. The H2 portion of H0 forms a membrane-spanning anchor called a fusion peptide, which is directly involved in the fusion mechanism (Figure 12-7b). The H2 peptide interacts with the host membrane after the conformational change is induced in H0 by low pH.

Influenza virus is not infectious unless the H protein is cleaved by cellular proteases. If the H protein contains a furin cleavage site, it undergoes proteolytic cleavage by a cellular protease inside of the trans-Golgi network. If it does not contain a furin cleavage site, the H0 precursor protein is cleaved into H1 and H2 by cellular proteases outside of the host cell. This limits their spread in hosts to tissues where the appropriate proteases are encountered. (Figure 12-7c shows a schematic of the cleavage site within the protein.) A general rule has been observed: The Hs of pathogenic influenza viruses are cleaved intracellularly by proteases and have the capacity to infect various cell types and cause systemic infections. The Hs of influenza virus that are cleaved by extracellular trypsin-like proteases are generally apathogenic, limiting their spread in hosts to tissues that contain the appropriate proteases. In some cases of influenza, the proteases of coinfecting bacteria may play a role in virus activation and cause an increase in the severity of the disease.

Uncoating Step

Uncoating takes place in the host cell endosome. Once the viral membrane fuses with the endosomal membrane of the host cell, the viral M2 transmembrane protein forms a transmembrane H^+ ion channel or small pore in the viral envelope. The M2 ion channel protein allows H^+ ions to penetrate the virion, weakening the interaction of the viral M1 matrix protein from the viral RNA, NP, and transcriptase complex (RNP). The RNPs are released into the cytoplasm and exported to the nucleus. The antiviral drugs **amantadine** (sold as Symmetrel) and **rimantidine** (sold as Flumadine) block the M2 ion channel function, resulting in the incomplete release of the viral RNPs, interfering with uncoating (**FIGURE 12-8**).

mRNA Synthesis and Replication of Virion RNA

Influenza virus genome replication occurs in the nucleus of the infected cell. This is an unusual feature for an RNA virus because most RNA viruses replicate their viral genomes in the cytoplasm of the infected cell. Following attachment, entry, and uncoating, the viral RNPs are transported to the nucleus and mRNA synthesis begins. During synthesis of mRNA, influenza engages in a process termed **cap-snatching**. The viral PB2 protein binds to the cap structures of the host mRNA in the nucleus of the cell. The cap, together with 10 to 13 nucleotides from the 5′ end of the host mRNA, is removed and PB1 and PA of the RNP complex uses this short string of nucleotides as a primer to

FIGURE 12-7 **(a)** Influenza virus attachment and entry. Adapted from Fields, B. N., et al. *Fields Virology,* Fourth Edition. 2 vols. Lippincott Williams & Wilkins, 2001. **(b)** Conformational changes in influenza H protein trigger membrane fusion. Adapted from Bergstrom, B. P., "Protein sorting: organelle biogenesis and protein secretion," *BIOL336: Cell Physiology,* http://www.muskingum.edu/~brianb/CellPhys/Lect11. **(c)** Influenza A H0 must be cleaved into H1 and H2 in order for the virus to be infectious. The sequence of the cleavage site determines which proteases (intracellular or extracellular) can cleave H0. Adapted from Acheson, N. H. *Fundamentals of Molecular Virology.* John Wiley & Sons, Inc., 2006.

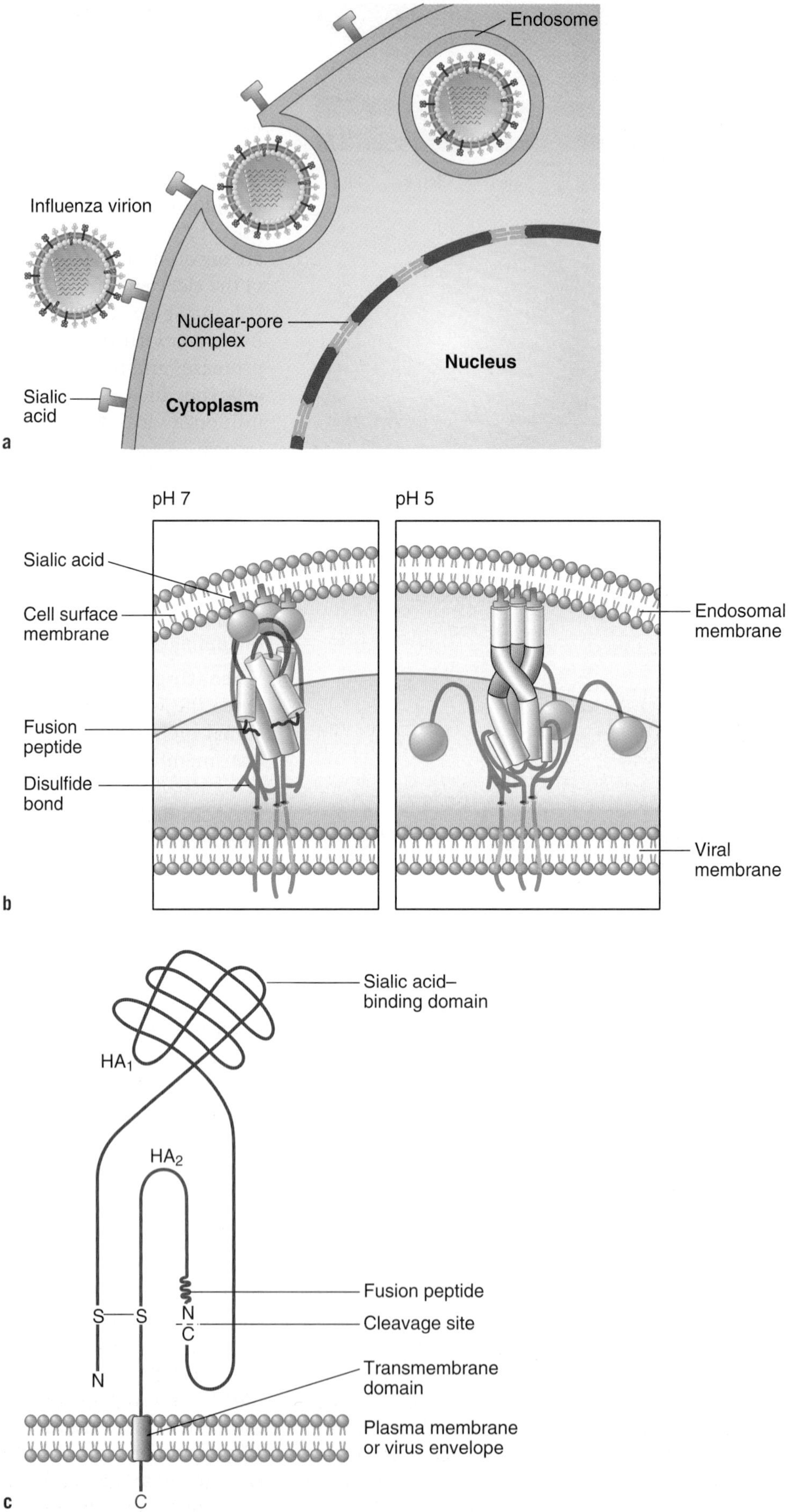

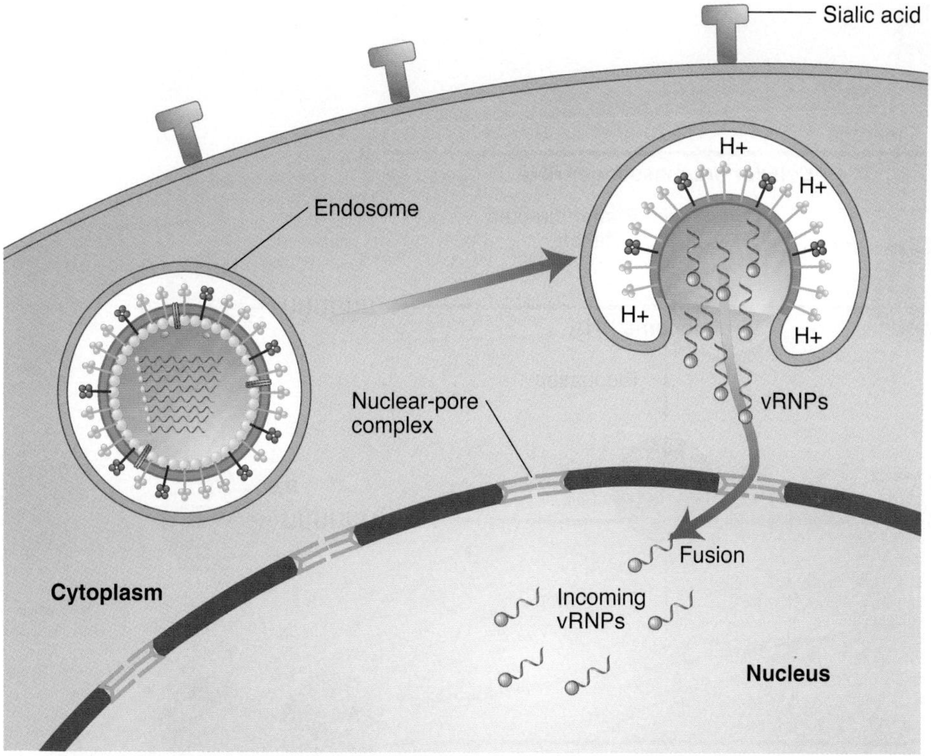

FIGURE 12-8 Fusion and uncoating of influenza A viruses. M2 facilitates the movement of protons from the endosome into the inside of the virion. The ribonucleoproteins are liberated to the cytoplasm and then transported to the host cell nucleus. Two drugs, amantadine and rimantidine, block the M2 ion channel activity, preventing viral uncoating. Adapted from Fields, B. N., et al. *Fields Virology,* Fourth Edition. 2 vols. Lippincott Williams & Wilkins, 2001.

initiate transcription of viral RNA (vRNA) by extending the primer (**FIGURE 12-9a**). For this reason, influenza viruses can only replicate in physiologically active cells containing a functional DNA-dependent RNA polymerase (RNA polymerase II) even though this enzyme does not transcribe the virus genome.

Transcription of the viral mRNA is terminated near the end of the genome segment at a point on the template vRNA that contains a run of uracil residues 17 to 22 residues from its 5′ end. These are copied into A residues in the mRNA, generating a poly A tail before the mRNA dissociates from the vRNA template. Synthesis of the antigenome positive-sense RNA does not use a primer and generates a complete complementary copy of the template RNA, termed vcRNA (Figure 12-9b). Two types of viral RNAs are generated: a positive sense viral mRNA and a positive sense viral anti genome RNA. **TABLE 12-6** distinguishes the difference between these two types of viral RNAs.

Viral mRNAs Are Not Cannabilized for Their 5′ Caps

Interestingly, the viral polymerase proteins selectively "snatch" caps from cellular pre-mRNAs but not viral mRNAs. This selectivity is essential because if the 5′ ends of the viral mRNAs were removed and utilized as primers, overall synthesis and translation of viral mRNAs would not occur. How has this selectivity evolved? While studying the splicing mechanism of RNA segments 7 and 8 of influenza, researchers discovered that the viral polymerase complex binds to the specific sequence 5′-AGCAAAAGCAGG-3′ that is found in all mRNAs complementary to nucleotides 1 to 12 of the 3′ end of each vRNA segment. This protects the 5′ ends of all viral mRNAs but not cellular pre-mRNAs from cleavage by the polymerase complex (Figure 12-9b).

Influenza Exploits Host Nuclear Splicing Machinery

After the "snatched" cap and associated nucleotides are used as a primer to transcribe each of the eight negative-sense RNA genome segments, six newly transcribed viral mRNAs (segments 1–6) are exported to the cytoplasm and translated immediately into viral proteins. Two of the primary viral transcripts (segments 7 and 8) are each spliced through the exploitation of the host splice machinery into at least two viral mRNAs. **FIGURE 12-10** is a schematic illustrating the different viral mRNAs generated from influenza A segments 7 and 8. In both cases, the splicing events remove a large portion of the first open reading frame (ORF), leaving the AUG initiation codon in place. The first alternative ORF is fused to the initiation codons of the M1 and NS1 ORFs. The newly generated ORFs direct the synthesis of novel proteins M2, NS2, and M3. All of these viral mRNAs code for nonstructural proteins. M3, however, has not been detected in infected cells and its role in the infectious cycle is not known.

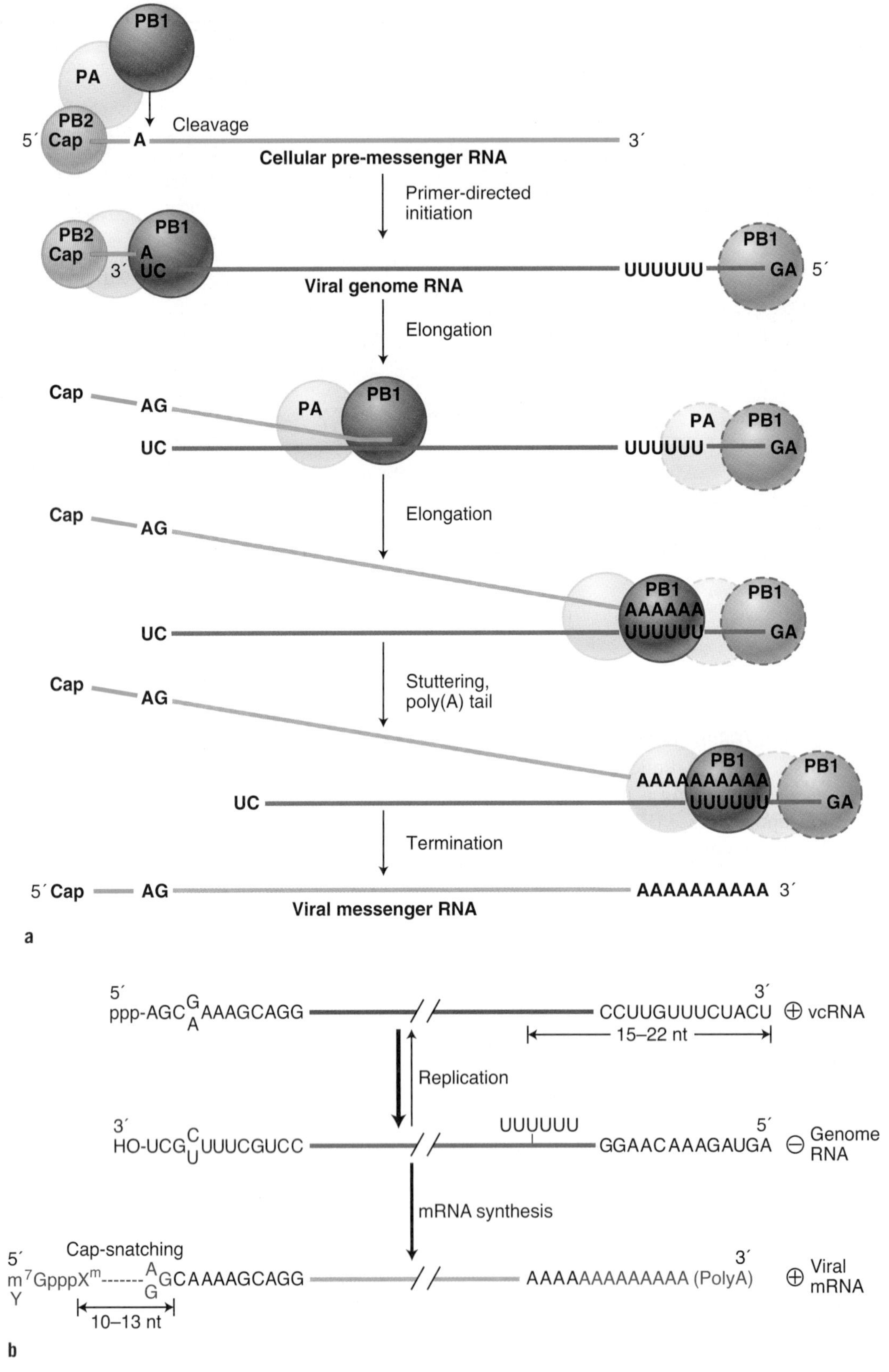

FIGURE 12-9 (a) Cap-snatching mechanism of influenza A viruses. Three influenza proteins are involved in stealing the cap of host mRNAs: PB2, PB1, and PA. The cap, along with 10 to 13 nucleotides from the 5′ end of the cellular mRNA, is removed from the host mRNA by PB1, and the RNA from the 59 end of the cleaved cellular mRNA is used as a primer to initiate transcription of viral RNAs. Adapted from Acheson, N. H. *Fundamentals of Molecular Virology.* John Wiley & Sons, 2006. **(b)** Schematic diagram distinguishing the differences between influenza virion (–) genome RNA; viral (+) mRNA containing the snatched cellular mRNA cap structure; and (+) viral complementary RNA or template RNA that is a complete copy of the genome RNA segment. Adapted from Strauss, E. G., and Strauss, J. H. *Viruses and Human Diseases.* Academic Press, 2001.

TABLE 12-6 The Differences Between the Influenza Virus mRNA and Antigenome

+ vmRNA	+ Antigenome RNA (vcRNA)
Shorter than template genome segment	Exact copy of the genome segment
Contains 3′ poly (A) tail	No 3′ poly (A) tail
Contains 5′ cap	No 5′ cap
Synthesis is insensitive to inhibitors of protein synthesis	Synthesis requires continuous viral protein synthesis

NS1 Prevents Nuclear Export of Cellular Pre-mRNAs, Facilitating Cap-Snatching

Two cellular proteins—cleavage and polyadenylation specificity factor (CPSF) and poly(A)-binding protein II (PABII)—are responsible for processing the 3′ end of cellular mRNAs. Influenza A virus NS1 protein, the most abundant viral protein in virally infected cells, interacts with CPSF and PABII, thereby inhibiting their function. As a result, cellular pre-mRNAs are not cleaved and contain very short 3′ poly (A) tails that accumulate in the nucleus in the infected cell.

These newly synthesized cellular pre-mRNAs are trapped in the nucleus of the infected cell and are almost completely degraded. Retaining cellular pre-mRNAs in the nucleus, though, facilitates the removal of the pre-mRNA 5′ cap by the influenza PB2 protein. The host mRNAs made after infection do not survive very long in the cell and little or no cellular proteins will be synthesized in influenza A-infected cells.

Even though cellular pre-mRNAs are not exported to the nucleus, viral mRNAs can be exported and are not affected by this because the influenza viral transcriptase adds the poly (A) tails onto the 3′ ends of viral mRNAs. The cellular CPSF

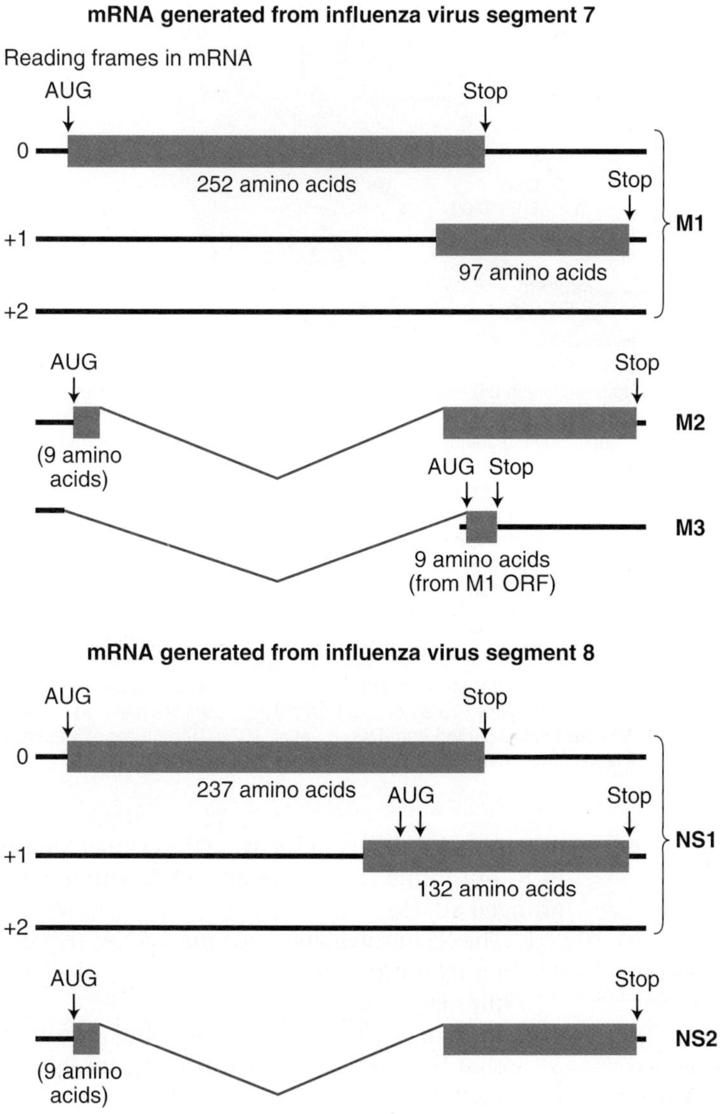

FIGURE 12-10 Splicing of RNA segments 7 and 8 of influenza A viruses. Shaded areas represent coding regions in either the 0, +1, or +2 open reading frames (ORFs). Thin lines represent untranslated regions. Introns are shown by the V-shaped lines. The first ORF codes for unspliced M1 and NS1 proteins. No evidence has been obtained that the M3 mRNA is detected in infected cells. Note that the M2 and NS2 proteins share common amino terminal sequences to the M1 or NS1 protein products, respectively.

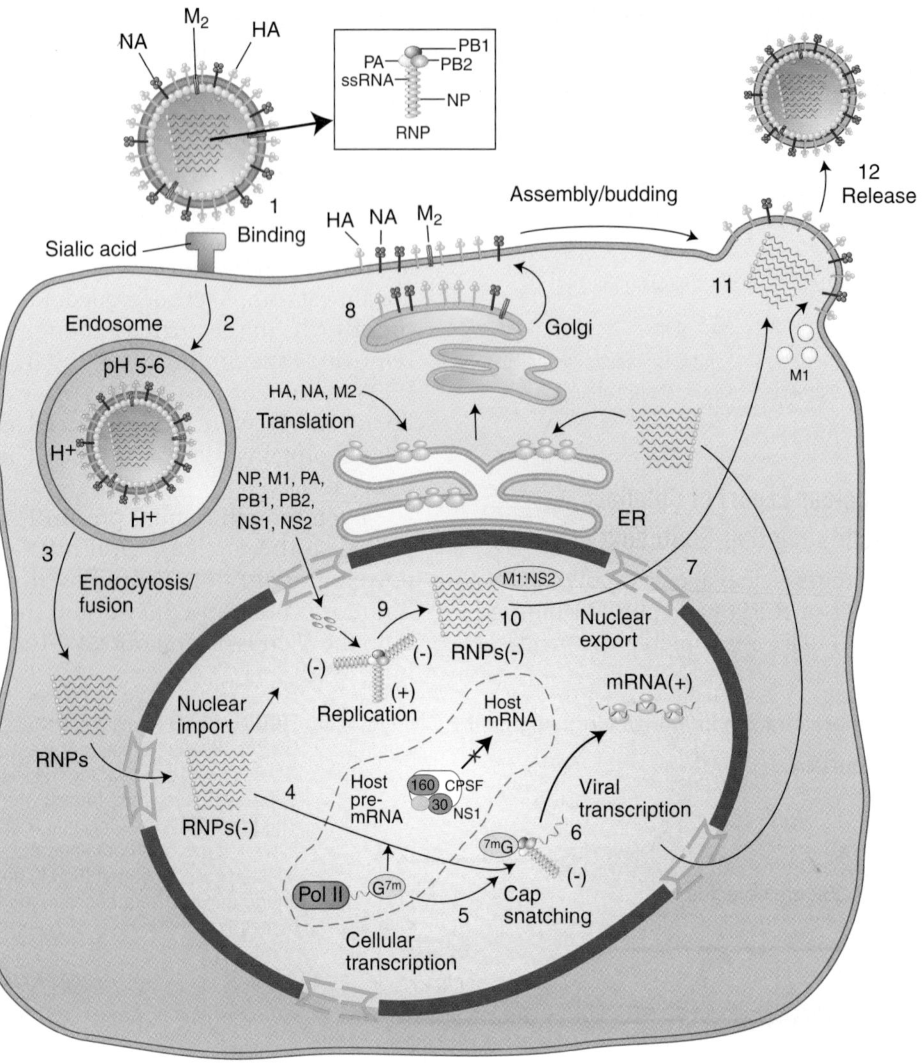

FIGURE 12-11 Life cycle of influenza A virus. Step 1: Influenza A hemagglutinin (H) attaches to host cell receptor (sialic acid). Step 2: The virus enters the host cell via endocytosis within an endocytic vesicle. Step 3: The M2 ion channel of the virus allows H^+ ions to penetrate the virus. The low pH of the endosome triggers a conformational change in H, resulting in fusion of the viral and endosomal membranes. Step 4: As the H^+'s enter through the M2 ion channel, the viral ribonucleoprotein complexes (vRNPs) are dissociated from M1 matrix proteins and released into the cytoplasm where they travel through the nuclear pores of the nucleus. Step 5: In the nucleus, the viral polymerase initiates viral mRNA synthesis with the 5′ caps cleaved (snatched) from the host pre-mRNAs. Step 6: The influenza PB2 binds to the 5′ cap of host pre-mRNAs, and the viral PA endonuclease cleaves the host pre-mRNA 10–13 nucleotides downstream of the 5′ cap of the host pre-mRNA. Step 7: Viral transcription is initiated. Step 8: Viral mRNAs are transported into the cytoplasm for translation into viral proteins. Step 9: The viral surface proteins—H, N, and M2—are processed in the endoplasmic reticulum (ER), glycosylated in the Golgi apparatus, and transported to the host cell membrane. Step 10: The NS1 protein of influenza A virus suppresses host mRNA synthesis by inhibiting 3′ end processing of host pre-mRNAs. Consequently, host mRNAs, including interferon β mRNAs are blocked from synthesis, but viral mRNAs are unaffected by NS1's activity. Besides synthesizing viral mRNAs, the viral polymerase is involved in the replication of vRNAs (–) vRNAà (+) vRNAà (–) vRNA. Step 11: The final (–) vRNPs are complexed with M1 and NS2 and exported out of the nucleus to the site of viral particle assembly. New viruses bud out. Step 12: Viruses are released as the N of newly formed viruses cleaves sialic acid residues present on the host cell. Adapted from Das, K., et al. 2010. "Structures of influenza A proteins and insights into antiviral drug targets." *Nat Structur & Molec Biol* 17(5):530–538.

and PABII proteins are not required for processing the 3′ ends of viral mRNAs (**FIGURE 12-11**).

Translational Control Mechanisms

Influenza A suppresses the interferon system in host cells (see the discussion on host resistance, including the interferon response, in Chapter 7). During influenza virus replication, double-stranded RNA (dsRNA) intermediates are formed. These dsRNA molecules would activate the cellular kinase PKR, but influenza counteracts PKR with a two-pronged attack:

1. The NS1 protein binds to viral dsRNA, sequestering these molecules, and thereby blocking the activation of PKR.
2. Influenza activates a 58-kDa cellular protein that interacts with PKR directly, inhibiting PKR in influenza-infected cells.

Virion Maturation, Assembly, and Release

After the capped viral mRNAs are exported to the cell cytoplasm, they are translated by host ribosomes. The H, N, and M2 glycoproteins are synthesized on membrane-bound ribosomes. Following synthesis these three membrane proteins enter the endoplasmic reticulum, where they are folded and glycosylated. Subsequently, they are transported to the Golgi network and the cell surface, where assembly takes place. At least one copy of each genome segment must be packaged into a new virion in order for it to be infectious.

The influenza virions are assembled at the membrane surface of the host cell and released by a budding process. H anchors the virus to the cell by binding to sialic acid-containing receptors on the cell surface. The enzymatic activity of the N removes the sialic acid from the carbohydrates present on the viral glycoproteins to prevent the viral particles from aggregating, thus preventing the clumping or aggregation of viruses to the cell surface; this in turn facilitates the viral release from H found on the host cell's surface. It is also believed that N may aid the virus at the beginning of its life cycle by helping its transport through the mucin layer present in the respiratory tract so that the virus can find its way to the target epithelial cells. Zamanivir (Relenza) and oseltamivir phosphate (Tamiflu) are antiviral compounds that inhibit the function of N (Figure 12-11).

12.9 Genetic Variation

Mutations are common during viral replication. Such mutations are more frequent in RNA viruses because their RNA-dependent RNA polymerases lack proofreading and correction ability. RNA polymerases are at least 1000 to 10,000 times more prone to error than human or viral DNA polymerases (which have proofreading and editing ability). The mutation rate of influenza is approximately 1.5×10^{-5} mutations per nucleotide per infectious cycle. This results in about two or three amino acid changes in the H protein each year. This rate is high compared to mammalian DNA genomes (10^{-8} to 10^{-11} per incorporated nucleotide).

Often there are "hot spots" for mutation in the viral genome. Influenza consists of an enveloped particle that contains two very important surface glycoproteins: H and N. The *H* and *N* genes contain the hot spots that code for antigenic sites recognized by virus-neutralizing antibodies. Epidemics and pandemics occur due to changes in the antigenic structure of the H and N proteins of the influenza virus. Two processes—antigenic drift and antigenic shift—mediate the genetic variation that causes new strains of influenza to appear (**FIGURE 12-12**).

Antigenic Drift

Antigenic drift is the small changes in the *H* and/or *N* genes that happen over time. It produces new virus strains that may not be recognized by the body's immune system, necessitating changes in influenza vaccines. When a person is infected with a particular influenza virus strain, the individual develops antibodies against that virus. As newer virus strains appear, the antibodies against the older strains no longer recognize the newer virus, and reinfection can occur. This is one of the main reasons why people can get influenza more than once. Antigenic drift can be responsible for localized outbreaks. In most years, one or two of the three virus strains in the influenza vaccine are updated with the changes in the circulating influenza viruses (Figure 12-12a and b). After the appearance of a new subtype, antigenic differences between isolates can be detected within a few years by using ferret antisera for analysis via hemagglutination inhibition assays.

Antigenic Shift

Antigenic shift is responsible for potentially pandemic strains of influenza A viruses: the rapid spread of novel strains of influenza A virus to all areas of the world. When antigenic shifts occur, large numbers of people have no antibodies to protect them against the virus, causing a pandemic. These shifts result in an unusually high number of cases and deaths for approximately two to three years. Antigenic shift does not occur with influenza B viruses because influenza A appears to infect humans and animals (birds, swine, horses, and seals), whereas influenza B only infects humans.

Antigenic shift usually occurs when a pig lung cell is infected with two different influenza A viruses: one of nonhuman origin (usually an aquatic bird) and the other a human strain (Figure 12-12c and d). Normally, influenza A viruses from birds grow poorly in humans and vice versa. **Reassortments** between human and bird influenza viruses, however, can produce a new strain of influenza virus that can infect and replicate well in humans. The virus is able to reassort its genes, swapping parts and whole segments of the virus in coinfected cells; for example, reassortment of influenza A viruses results in a new H#N# strain plus 256 other possible virus combinations. The pig and duck serves as a "mixing" or "reassortment" vessel (**FIGURE 12-13**).

Figure 12-13 is an illustration showing the different **reservoirs** of influenza A viruses (refer to

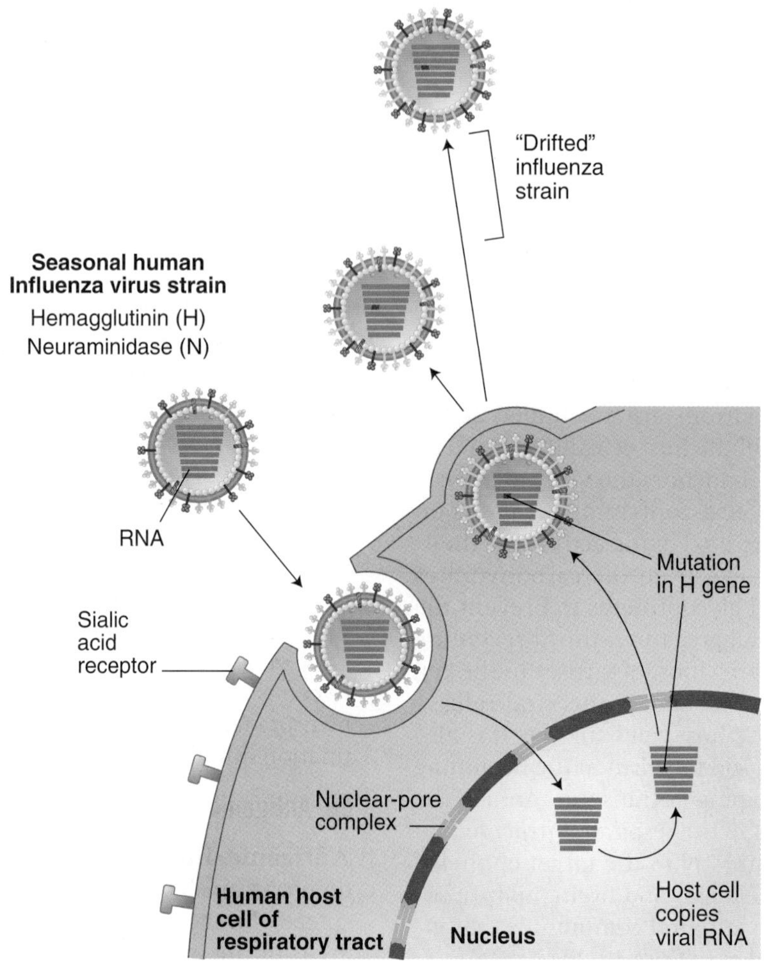

a

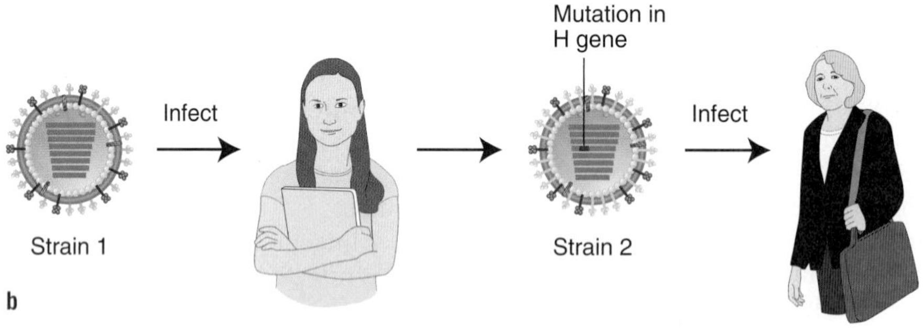

b

FIGURE 12-12 (a) Antigenic drift. Illustration at the cellular level. During replication of the viral RNA, mutation(s) occur. Antigenic drift is the gradual accumulation of new epitopes on the H protein (and, to a lesser degree, the N protein). Adapted from Branswell, H. 2011. "Flu factories: The next pandemic virus may be circulating on U.S. pig farms, but health officials are struggling to see past the front gate." *Sci Am* January:47–51. **(b)** Antigenic drift illustration at the host level.

(continued)

Chapter 8, Epidemiology, terms). Wild waterfowl (e.g., ducks) are the largest reservoir of influenza A viruses. Infections in wild waterfowl are usually asymptomatic. The influenza A viruses replicate in the intestinal tract and are transmitted by an oral–fecal route. The evolution rates of influenza A viruses in the wild aquatic birds is very low. However, upon transmission to domesticated birds (e.g., chickens or turkeys), rapid evolution occurs. The replication of duck influenza A viruses in chickens usually occurs in the respiratory and gastrointestinal tract and causes mild or no symptoms. Replication of influenza A viruses of H5 or H7 subtype, however, can lead to the emergence of highly pathogenic influenza viruses that cause devastating **epizootics**. An epizootic is an epi-

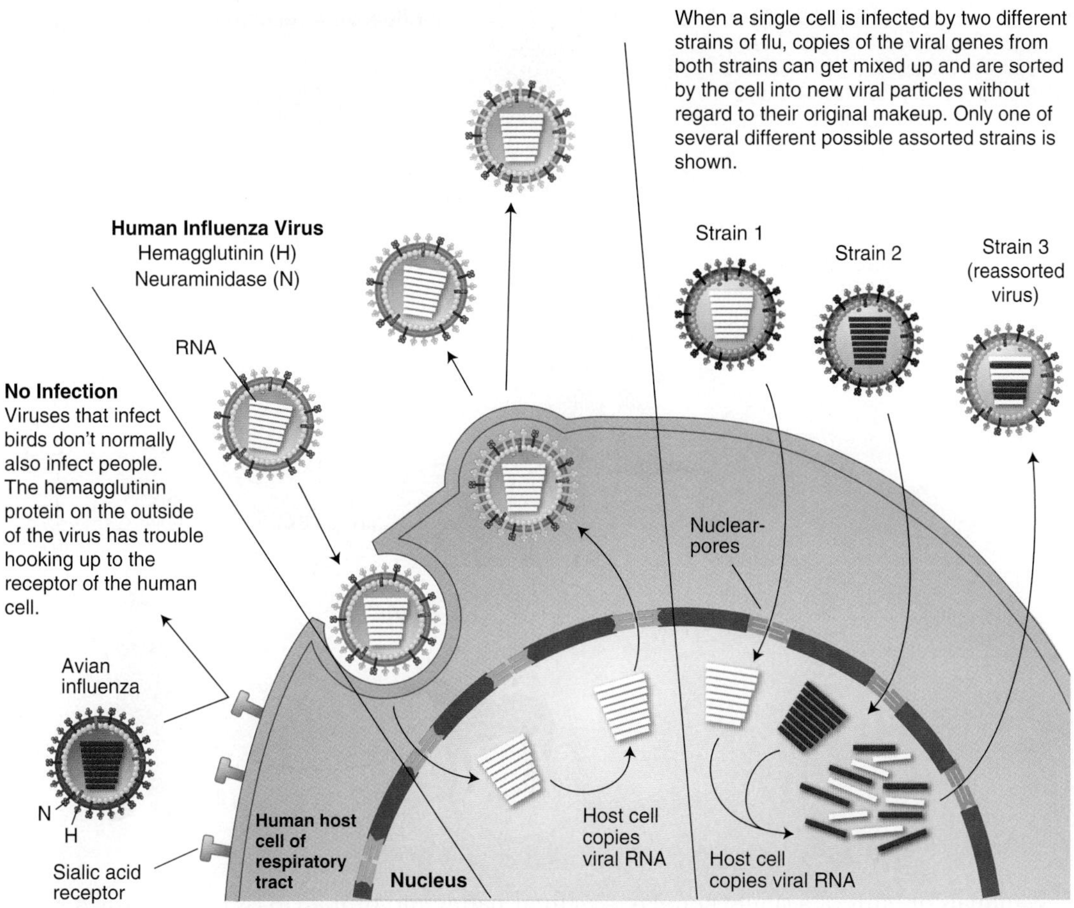
When a single cell is infected by two different strains of flu, copies of the viral genes from both strains can get mixed up and are sorted by the cell into new viral particles without regard to their original makeup. Only one of several different possible assorted strains is shown.
Human Influenza Virus
Hemagglutinin (H)
Neuraminidase (N)
RNA
No Infection
Viruses that infect birds don't normally also infect people. The hemagglutinin protein on the outside of the virus has trouble hooking up to the receptor of the human cell.
Strain 1
Strain 2
Strain 3 (reassorted virus)
Nuclear-pores
Avian influenza
N
H
Sialic acid receptor
Human host cell of respiratory tract
Nucleus
Host cell copies viral RNA
Host cell copies viral RNA

c

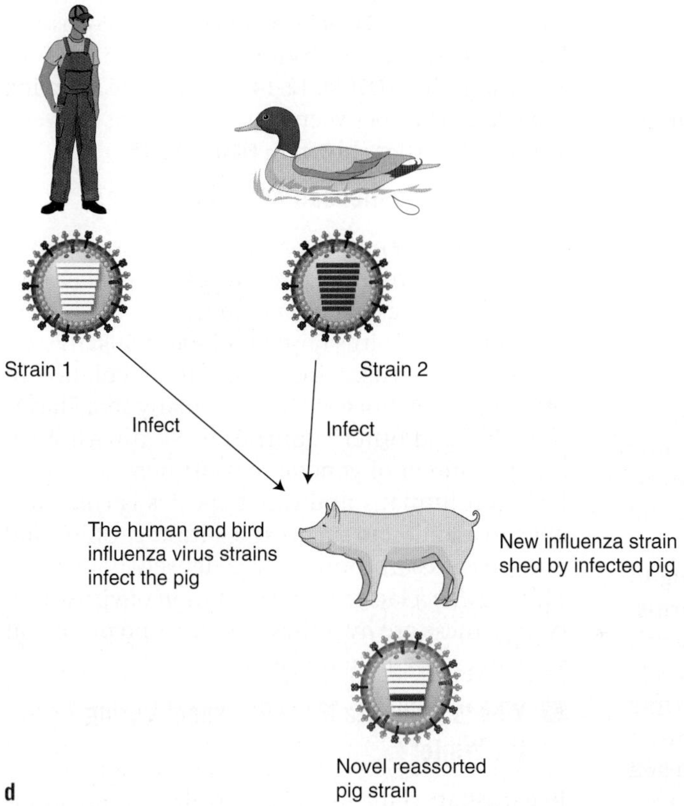
Strain 1
Strain 2
Infect
Infect
The human and bird influenza virus strains infect the pig
New influenza strain shed by infected pig
Novel reassorted pig strain

d

FIGURE 12-12 (Continued) **(c)** Antigenic shift. Illustration at the cellular level. Antigenic shift occurs when the influenza A virus acquires a new *H* or *N* gene. When shifts occur, large numbers of individuals have no antibodies to protect them from virus infection. Adapted from Branswell, H. (2011) "Flu factories: The next pandemic virus may be circulating on U.S. pig farms, but health officials are struggling to see past the front gate." *Sci Am* January:47–51. **(d)** Antigenic shift illustration at the host level.

FIGURE 12-13 The reservoir of influenza A viruses. All 16 HA and 9 NA subtypes circulate in wild waterfowl like ducks. They are the largest reservoir of influenza A viruses. Transmission of influenza viruses has been demonstrated between humans and pigs. In 1997, an avian influenza A strain directly infected a human, changing the current tradition of thinking that all influenza viruses would first go through a pig before transmission to humans. Adapted from Fields, B. N., et al. *Fields Virology*, Fourth Edition. 2 vols. Lippincott Williams & Wilkins, 2001.

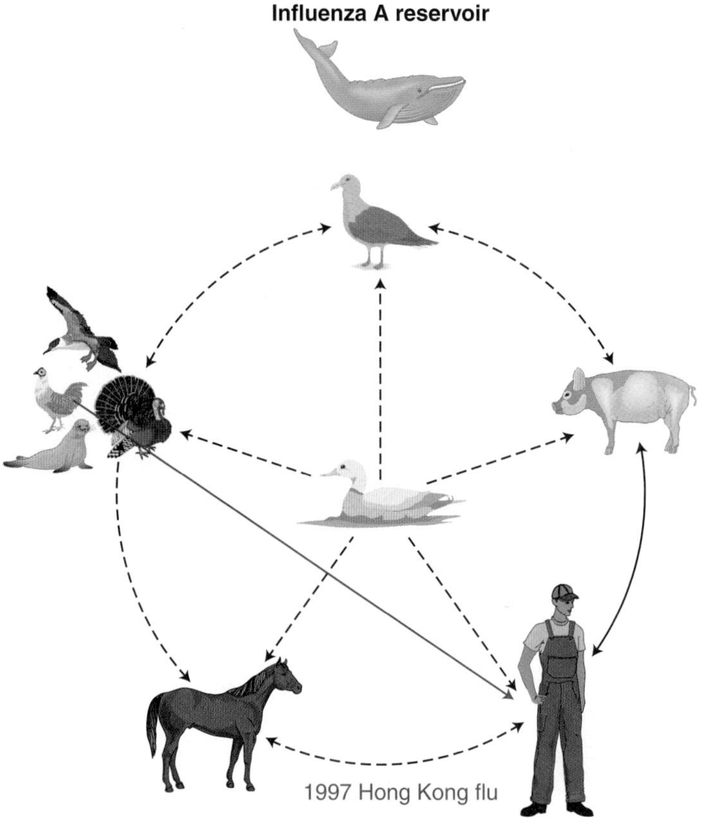

demic among animals or birds of a single population within a particular region.

The Swine-Origin H1N1 Influenza Virus

The outbreak of the 2009 influenza pandemic started in the Mexican town of La Gloria, Veracruz, in mid-February. By early April, Mexican public health officials were investigating high numbers of pneumonia/influenza-like illnesses. They informed the Pan American Health Organization (PAHO) and World Health Organization (WHO) of a possible pandemic. In April, the CDC collected two specimens from patients in southern California. The viruses were isolated from patient specimens, and further investigation revealed that the Mexican and Californian cases were caused by similar influenza A viruses. Data on the genetic composition became available soon after virus isolation. The CDC developed a RT-PCR diagnostic test kit to detect the 2009 H1N1 pandemic influenza A strain.

The H1N1 influenza A strain was alarming to experts because it was genetically traced to a triple reassortment virus of avian, human, and swine origin: North American avian influenza virus + classical swine influenza virus + a human (H3N2) virus that was circulating in the 1990s. Subsequently, a Eurasian avian-like swine influenza virus circulated within the classical pig population, resulting in a new reassortment virus that infected humans. The final pandemic human H1N1 influenza A virus contains genes of the following origins: PB2-North American avian, PB1-Human H2N2, PA-North American avian, **H1-classical swine**, NP-classical swine, **N1-Eurasian avian-like swine**, and NS classical swine (**FIGURE 12-14**). A timeline showing the relationship between pandemic strains between 1900 and 2010 is shown in **FIGURE 12-15**.

Is China the Incubator for Flu Viruses?

Why does it seem that pandemic strains originate in southern China? Some suggest this is because there is a close association of humans with domestic animals and birds raised for food (**FIGURE 12-16**). Influenza A viruses are constantly circulating in birds, pigs, and horses. There are many small farms in China and other countries in Southeast Asia. The likelihood of genetic reassortment of viruses between humans and other species is considerable. In addition to small-scale farming, China and Southeast Asian countries are densely populated. This may be a factor in the evolution of viruses. Of course, these are hypotheses; there is no proof that epidemics originate this way.

Why Is Influenza More Prevalent During the Winter?

In temperate climates such as North America, influenza generally affects people from November to

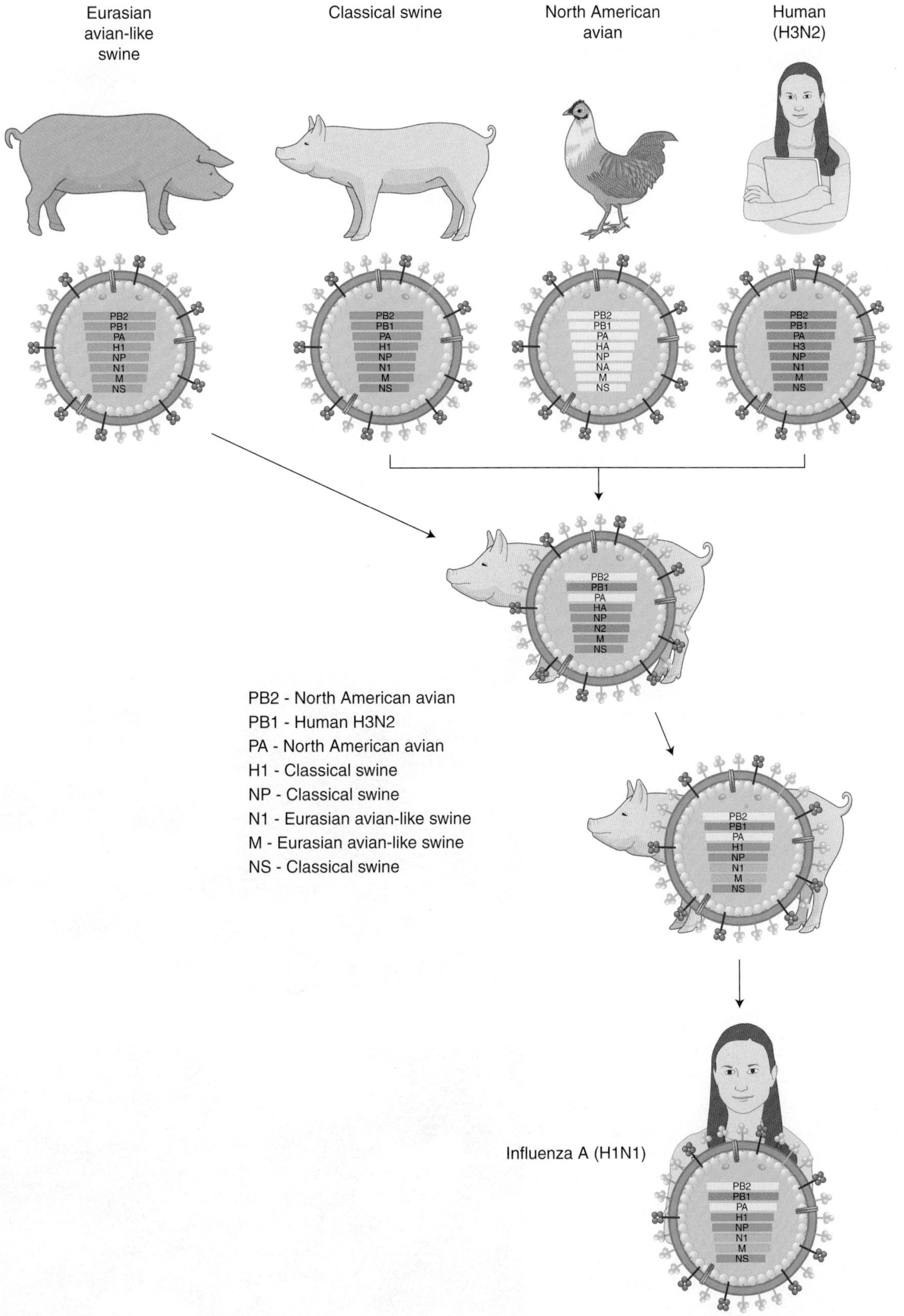

FIGURE 12-14 Genesis of the 2009 H1N1 swine influenza A. A triple reassortment between classical swine influenza A viruses, human H3N2, North American avian influenza A viruses occurred some time during the late 1990s. Reassortments of H3N2 and H1N2 swine influenza A viruses circulated in North American pig populations. A triple reassortment classical swine influenza A viruses reassorted with a Eurasian avian-like swine virus, resulting in an H1N1 swine influenza A virus circulating in humans in 2009. Neumann, G., et al. 2009. "Emergence and Pandemic Potential of Swine-Origin H1N1 Influenza Virus." *Nature* 459:931–939.

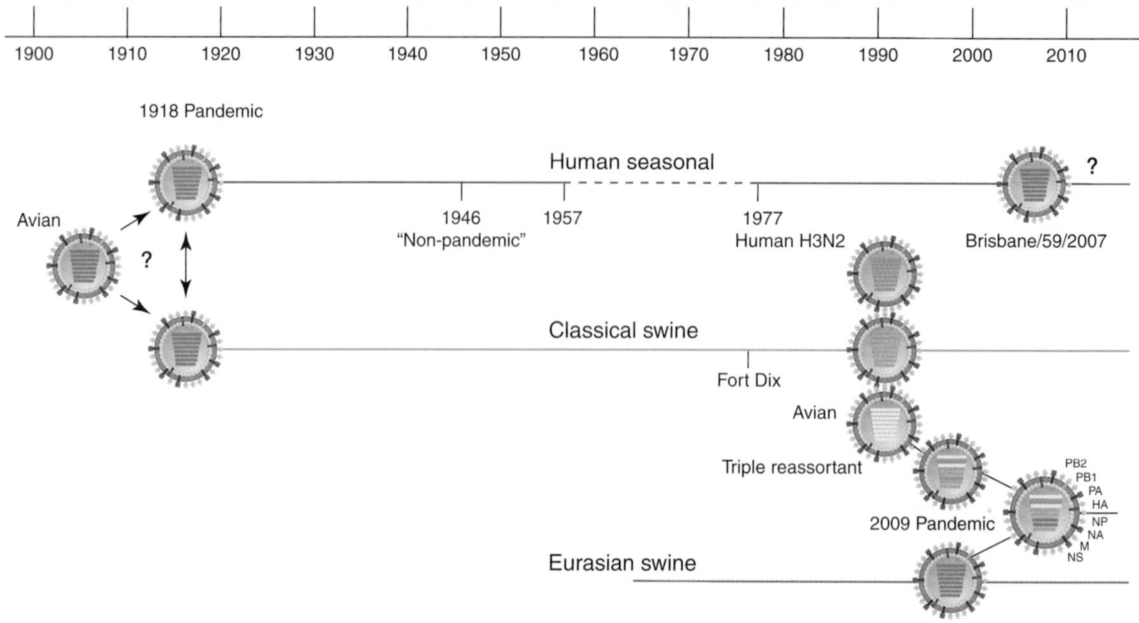

FIGURE 12-15 Timeline showing the relationship of H1N1 human and swine influenza A viruses. The 1918 influenza A pandemic virus gave rise to H1N1 viruses that caused seasonal epidemics that disappeared in 1957 and reemerged in 1977. In the late 1990s, the triple reassortment virus emerged in pigs, which contained PB2 and PA of an avian virus, PB1 and N of a human H3N2 virus, and the other four genes of classical swine virus. The 2009 H1N1 pandemic virus is a triple reassortment with the N and M genes from the Eurasian avian-like H1N1 lineage that emerged around 1970 in pigs. A nonpandemic 1946 strain and a limited outbreak of classical swine virus in soldiers at Fort Dix (United States) in 1976 are also indicated on the timeline. Adapted from Klenk, H. D., et al., 2011. "Molecular mechanisms of interspecies transmission and pathogenicity of influenza viruses: Lessons from the 2009 pandemic." *Bioessays* 33(3):180–188.

FIGURE 12-16 **(a)** Farmers migrating in China, bringing their birds and animals with them. Humans working in close association with birds and mammals may facilitate influenza A evolution and genetic variation. **(b)** It is not uncommon for farmers in southeast Asia to live with their birds and animals; for example, these chickens live with a family that works on a small farm.

a

b

March, whereas influenza infections occur throughout the year in tropical climates. A possible explanation for the high level of influenza activity during the winter is that the gathering of people indoors facilitates the transmission of the virus and viruses survive longer due to the dry indoor air. A 2007 study by Anice Lowen and Peter Palese's research teams demonstrated that the aerosol spread of influenza virus was dependent upon humidity and temperature. The virus was best transmitted at low humidity (20%) and colder temperatures (5°C or 41°F). They provide supporting evidence that weather conditions play a role in influenza virus transmission. Findings from a 2010 study confirmed that humidity drives seasonal variation of influenza transmission in temperate regions.

12.10 Pandemics in History

During the 20th century, three influenza A pandemics have occurred: the Spanish flu of 1918–19, the Asian flu of 1957–58, and the Hong Kong flu of 1968–69 (**TABLE 12-7**).

The 1918 Spanish Flu

The 1918 Spanish influenza killed more people in 25 weeks than AIDS has killed in 25 years. It killed more people in a year than the plagues of the Middle Ages killed in a century. Estimated deaths range between 20 and 50 million, including 675,000 Americans. In New York City alone its impact resulted in 21,000 children orphaned by influenza. Seven times as many people died of the 1918 influenza than in World War I. True numbers will never be known because many places were overwhelmed by the flu and did not keep mortality statistics (see Section 1.3, Viruses in History: Great Epidemics). This pandemic spread faster than any plague in history, charging across America in a mere seven days. It took only three months to sweep around the world. In Alaska, 60% of the Inuit population was wiped out. Islands in the South Pacific (where respiratory illness is uncommon and not life-threatening) lost 20% of their population. The Spanish flu was 25 times more deadly than previous influenza viruses. It killed 2.5% of its victims; normally, only 0.1% of those who get the flu die.

TABLE 12-7 Pandemics in History

Dates	U.S. Deaths	Influenza Strain
1918–1919 "Spanish flu"	675,000 (>20 million worldwide)	H1N1
1957–1958 "Asian flu"	70,000	H2N2 A/Singapore/57
1968–1969 "Hong Kong flu"	33,800	H3N2 A/Aichi/68
2009	7,500–12,000*	H1N1 A/Mexico/09

*Conservative range estimate based on CDC survey.

Why was this strain of influenza called the Spanish flu and what made it unique? Historians suggest that the name Spanish flu stuck because Spain did not censure its news reports of influenza outbreaks during WWI, whereas most European countries did.

At nearly the same time, in March 1918, influenza struck an Army Camp in Kansas. This strain of influenza was carried in the lungs of soldiers in the 15th U.S. Cavalry headed for service in Europe (**FIGURE 12-17**). Its effects were unique for several reasons:

- It killed 20- to 40-year-old adults (i.e., adults in the prime of their lives).
- It killed quickly (in 2–3 days).
- Infected individuals suffered from hemorrhagic symptoms.

This deadly flu was a pathologist's nightmare. Physicians recorded,

> It began undramatically (after a two-day incubation period) with a cough. Next, there was pain behind the eyes, in the ears and in the lumbar region. Soon a drowsy numbness invaded the body, and a fever set in; often the temperature soared to 104°F. The pulse was thready and unstable; the victim's tongue was thickly coated . . . every mortal fiber ached indescribably—the throat, the head, the nasopharynx. . . . why should the sickness affect so many organs of the body normally untouched? . . . most often the disease resembled encephalitis, with the patient lapsing into a coma . . . a dilation of the heart but even of fatty degeneration . . . a cough so intense that it ruptured the muscles of a soldier's rectum . . . retention of urine . . . puffy faces and swollen ankles of acute nephritis . . . the lungs were the organs most vitally affected . . . a patient's face so contorted in death that even close friends couldn't recognize him . . . and autopsy surgeons were encountering what one doctor termed 'a pathological nightmare'; lungs up to six times their normal weight, looking 'like melted red currant jelly.' (From Richard Collier, 1974. *The Plague of the Spanish Lady: The Influenza Pandemic of 1918–1919.* New York: Atheneum Publishing.)

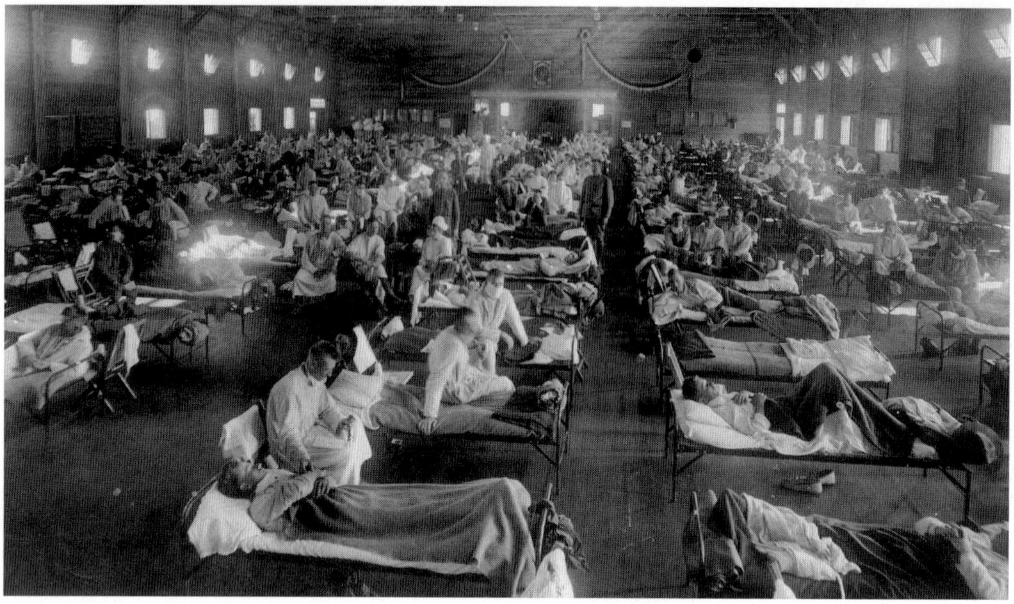

FIGURE 12-17 Many young servicemen died in U.S. Army Hospitals. This Emergency Hospital was located at Camp Funston, Kansas.

As a result of the Spanish flu making its appearance during WWI, some people speculated that Germany had put influenza into Bayer aspirin to use as germ warfare. The U.S. Public Health Service Hygienic Laboratory was directed to examine 200 tablets of aspirin imported from Germany to determine if they contained organisms responsible for the spread of the Spanish flu. There was no evidence to link organisms to the tablets and this speculation ended when the Germans fell ill, too.

In the United States many emergency hospitals were set up while scientists scrambled to determine what was causing the flu. Many thought it was caused by a bacterium, and a vaccine was incorrectly produced by researchers at the Mayo Clinic against a bacterium (Pfeiffer's bacillus; see paper by J. M. Eyler listed in primary literature), and distributed gratis. Public health officials banned public gatherings like funerals. Schools were closed for nearly two months.

People concocted medicines or home remedies to prevent the flu. Physicians predominately prescribed heroin hydrochloride, codeine sulfate, cocaine hydrochloridum, opium, morphine sulfate, elixer terpin hydrate (a concoction of turpentine, alcohol, and nitric acid), and paregoric elixir (made with powdered opium, benzoic acid, camphor, oil of anise, glycerin diluted alcohol, and morphine). Few pharmacists would sell more than an ounce of these drugs because of their habit-forming potential. Many remedies contained a large percentage of alcohol.

Influenza permeated every aspect of life, even infiltrating children's songs. The following rhyme was popular during the 1918 pandemic:

I had a little bird
and its name was Enza
I opened the window
and influenza.

Why Was the 1918 Spanish Flu So Deadly?

The mystery of why the 1918 Spanish flu was so deadly has intrigued researchers for a long time. As a young graduate student in 1951, Johan Hultin (**FIGURE 12-18**) traveled to Brevig Mission, Alaska, to find traces of tissue containing the killer influenza virus in a mass grave. In 1918, 72 of the 80 residents at Brevig Mission had died within a five-day period; the only survivors were children. After his first expedition, Hultin expected to find the virus "alive," preserved by the permafrost. Much to his surprise, he could not grow the virus when he returned with samples to the lab.

In 1997, Jeffrey Taubenberger's research team at the Armed Forces Institute of Pathology in Washington, DC, analyzed formalin-fixed, paraffin-embedded autopsy tissue specimens from U.S. servicemen who had experienced massive pulmonary edema with acute bronchopneumonia. After hearing news of Taubenberger's research, Hultin, then 72 years old, was inspired to repeat his expedition to the Brevig Mission grave. He exhumed the bodies of four of the 1918 flu victims, removed samples of the victims' frozen lung tissues, placed the samples in a fixative, and then provided these samples to Taubenberger for analysis. Taubenberger's team was able to sequence the coding regions of nine RNA fragments of the 1918 influenza virus from one of the Brevig Mission's victims. These RNA sequences were

FIGURE 12-18 The mystery of the 1918 killer flu intrigued Johan Hultin. As a young graduate student in 1951, Hultin made his first trip to find ice-bound traces of tissue containing the killer Influenza virus in an Alaskan mass grave. He expected to find the virus "alive." Much to his surprise, he could not grow the virus when he returned with samples to the lab. In 1997 (photo), he returned to the Alaskan mass grave and was able to recover influenza-infected lung tissues. He provided the samples to Jeffrey Taubenberger, and his team of researchers were able to obtain sequence information that was a start in solving a century-old riddle.

located in the coding regions of the H, N, NP, M1, and M2 proteins.

Why the 1918 flu killed so quickly and spread so easily continues to be a mystery. Analyses of antibody titers of 1918 flu survivors from the late 1930s suggest that the 1918 strain was an H1N1 subtype of influenza A virus, closely related to what is now known as "classic swine" influenza virus. As genetic technology has improved, we have learned more about this 1918 strain. Recently published sequences of the *H* and *N* genes suggest the 1918 virus was from a birdlike influenza virus that had not circulated in pigs or humans in the decades prior to this pandemic.

The *H* and *N* genes of the 1918 virus have been studied extensively. These two genes code for the major surface antigens of the virus and necessitate a new vaccine each year. This might suggest these genes would account for the virulence of the 1918 virus. In 2005, the 1918 Spanish influenza virus was reconstructed by a team of premiere researchers led by Terrence M. Tumpey and using the availability of the complete 1918 influenza gene coding sequence. The team used reverse genetics (**VIRUS FILE 12-3**) to generate the influenza virus in an effort to understand the molecular basis of its virulence. The 1918 virus was created and handled in an enhanced Biosafety Level 3 laboratory (**FIGURE 12-19**).

The reconstructed 1918 virus replicated in the absence of trypsin, caused death in mice and embryonated chicken eggs, and grew to high titers in bronchial epithelial cells that line the airway in mice. Its ability to replicate in MDCK cells without the addition of trypsin was unanticipated.

It is standard protocol to supplement cell cultures with trypsin to facilitate influenza infection. Trypsin expedites cleavage of H_0 into H_1/H_2, activating virus infectivity. This protocol adaptation was important in growing previously isolated human and avian influenza viruses. The reconstructed strain was extremely virulent in mice. The infected mice lost 13% of their body weight in two days after infection with the 1918 virus. The mice contained 39,000 times more virus particles in mouse lung tissues four days after infection than other contemporary H1N1 influenza viruses. All mice died within six days of infection. Nothing strikingly different, though, was observed in the 1918 *H* or *N* gene sequences until 2007.

In 2007, a team led by Yoshihiro Kawaoka of the University of Wisconsin–Madison infected macaques with the reconstructed 1918 influenza virus. These experiments were carried out in Biosafety Level 4 laboratories of the Public Health Agency of Canada in Winnipeg. Infected macaques died of acute respiratory distress typical of a cytokine storm (see Virus File 12-2). Concurrently, Tumpey's group at the CDC, in collaboration with colleagues at the Mount Sinai School of Medicine in New York, created 1918 viruses that contained point mutations in the *H* gene that altered the receptor-binding preference of the virus.

Contemporary human influenza viruses (which readily infect ferrets—the best animal model for human influenza viruses) bind to host cells that contain sialic acid bound to a galactose through a linkage called α-2,6. Avian influenza viruses bind to intestinal host cells containing sialic acid bound to a galactose through an α-2,3 linkage. Based on this idea, Tumpey's group mutated the *H* gene of the 1918 virus in such a way that it would bind α-2,6 or α-2,3 linkages. In doing so, their goal was to create an influenza strain that was more avian-like and unable to be transmitted to humans.

Ferrets were inoculated intranasally with the mutated 1918 viruses or with the original recreated 1918 virus. All three viruses created caused severe disease in ferrets. The 1918 virus that contained a double mutation that allowed the virus to bind to both α-2,6 and α-2,3 receptor was highly lethal and replication efficient in the upper respiratory tract of ferrets but was more avianlike in that the virus was unable to transmit between ferrets. The upper respiratory tracts of ferrets and

VIRUS FILE 12-3 Preparing a Pandemic Influenza Vaccine: Reverse Genetics or a Universal Vaccine?

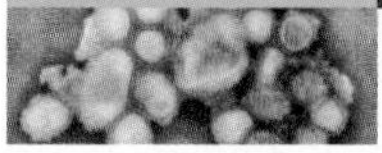

The World Health Organization has warned the public that another influenza pandemic is inevitable and possibly imminent. Influenza pandemics occur when a completely new influenza virus appears in a human population that does not have either partial or full immunity, resulting in several simultaneous epidemics in different parts of the world. The new virus has the capability of spreading from person to person, resulting in widespread illness and an enormous number of deaths. Given the high level of global travel today, a pandemic virus would spread rapidly, leaving little time to prepare. It would take several months before a vaccine became available. Preparing for the next influenza pandemic will provide benefits now. A new pandemic virus is predicted to be related to the emerging H5N1 avian influenza virus that first appeared in Hong Kong in 1997 or a similar H5N1 variant that emerged in 2003. It also could be similar to the 1918 H1N1 strain.

Creating a commercial vaccine against a pandemic strain of influenza poses challenges. Annual influenza vaccines are produced commercially from viruses grown in embryonated chicken eggs (**FIGURE VF 12-3**).

In response to a WHO pandemic alert regarding the 2003 variant bird strain, a team of researchers has prepared a candidate vaccine. The A/Hong Kong/213/03(H5N1) strain was isolated from killed embryonated chicken eggs. Therefore, traditional methods could not be used to produce a candidate vaccine. Instead, reverse genetics was used to generate a vaccine strain. Briefly, sequences coding for the polybasic amino acids located at the hemagglutinin cleavage site associated with high virulence were removed from the *H5* gene of the pathogenic avian strain. This resulted in a substantially attenuated viral strain. The removed cleavage site was then replaced with sequences from the HA site of an avirulent teal strain of influenza A. The candidate virus could be grown to high titers in eggs and did not cause embryo death, yet it was antigenically similar to the original H5N1 pathogenic strain. The researchers were able to produce a candidate vaccine virus through the usefulness of reverse genetics in less than four weeks from the time of virus isolation (Figure VF 12-3).

A "Universal" Human Influenza A Vaccine?

Influenza hemagglutinin and neuraminidase are large glycoproteins that protrude outside of the envelope of the virus. They are highly immunogenic and are the obvious choices as the main antigens in today's influenza vaccines. They have one large drawback, though: they drift and shift. This is why new vaccines are produced each year in an effort to "match" the vaccine and circulating strains of influenza. Influenza epidemics cause between 500,000 and 1 million deaths worldwide per year. The epidemiological models from the CDC predict that a new pandemic will cause 2 to 7.4 million deaths globally. A universal vaccine offers a solution that would counter the looming threat of an influenza pandemic because it would protect effectively against all strains of influenza viruses affecting humans. It would provide long-lasting immunity and be inexpensive to produce.

Researchers are working on a universal vaccine based on the use of the M2 protein of influenza virus. Like the hemagglutinin and neuraminidase proteins, the M2 protein is an integral membrane protein (**FIGURE VF 12-4**). The M2 protein is much smaller, though. Only 24 of the amino acids of the protein are exposed at the membrane's surface. These 24 amino acids are amazingly well conserved; for example, amino acid sequences of the M2 proteins from the first human influenza A virus isolated in 1933 are exactly the same as present influenza A strains. The M2 protein is abundantly expressed at the cell surface of influenza A-infected cells and is also part of the virus particle but in much smaller quantities (14 to 68 proteins/virus particle).

HOW THIS FLU VIRUS WORKS

The virus has two proteins on its surface called hemagglutinin, or H, and neuraminidase, or N.

The H protein connects to receptors on the outside of human cells, allowing the virus to meld with the cell and replicate.

The N protein allows the viruses to rip free from the cell, killing it. The viruses then march on to infect other cells.

H protein　N protein

Cell

Nucleus

Virus

HOW THE VACCINE IS MADE

A flu virus has eight gene segments. Scientists extract the two genes that are responsible for the H and N proteins. The virulent sections of the H gene are removed.

Six gene segments are taken from another flu strain, which grows well in eggs.

Dangerous avian flu virus

Section removed

Another flu virus

These genes are inserted into DNA loops...

H gene　N gene

DNA plasmids

PB2　PB1　PA　NP　M1 + M2　NS1 + NS2

...that are inserted into an animal cell line where they combine to produce a new strain, a seed virus.

The seed virus is inserted into chicken eggs so that it can multiply.

Chicken eggs

New flu strain (vaccine)

The inactivated strain is injected into humans, prompting the creation of swarms of antibodies that recognize the H proteins and neutralize the virus.

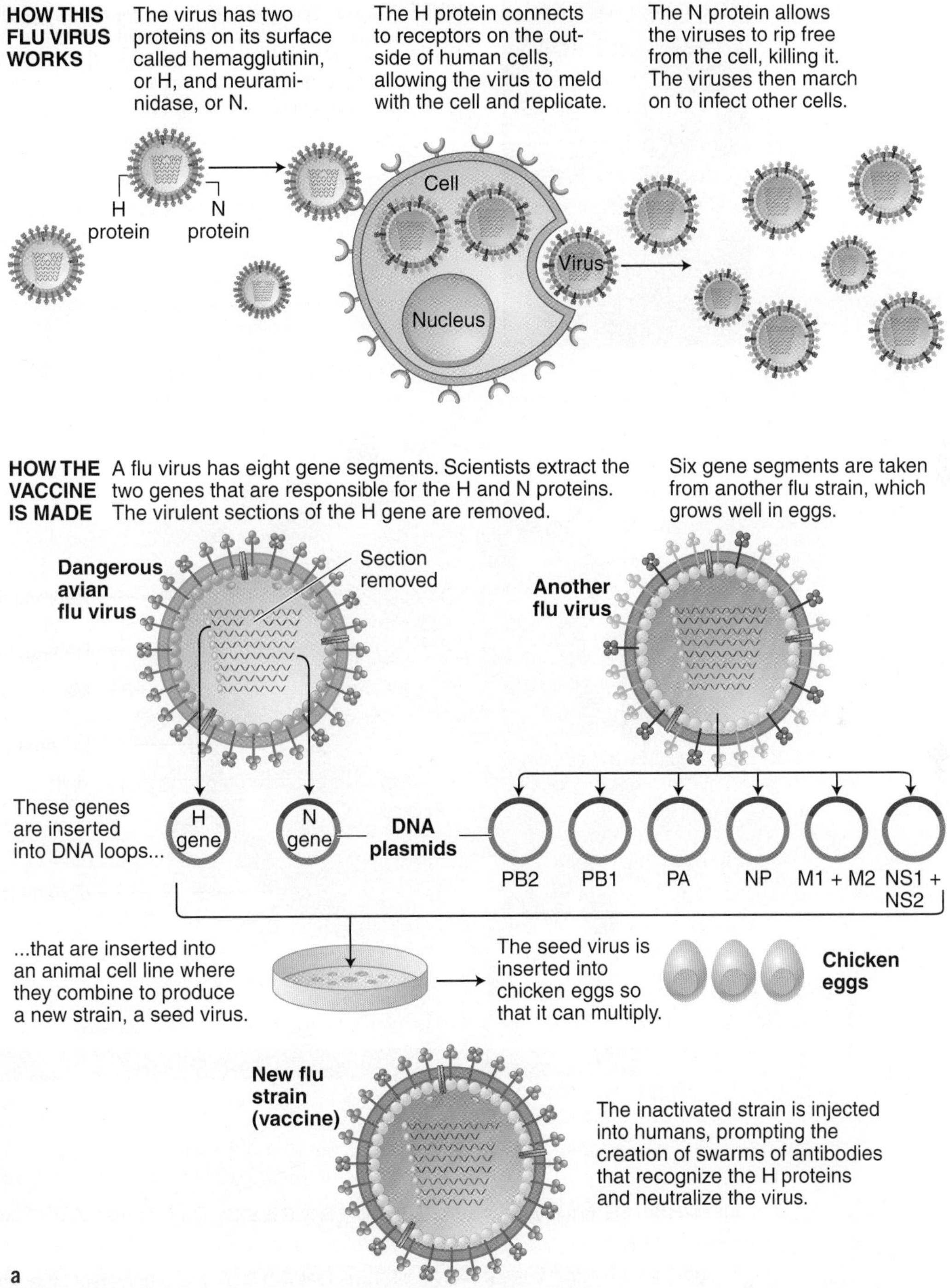

a

FIGURE VF 12-3 (a) Schematic of reverse genetics technology. Adapted from National Institute of Allergy and Infectious Diseases, National Institutes of Health, How the Flu Virus Changes, http://www3.niaid.nih.gov/healthscience/healthtopics/Flu/Research/ongoingResearch/FluVirusChanges.

(continued)

b

FIGURE VF 12-3 (Continued) **(b)** Robert Webster, is an influenza expert who helped to discover that live poultry markets in Hong Kong were the origin of the H5N1 influenza A virus that spread directly from chickens to humans in 1997. His laboratory has applied reverse genetics toward the rapid development of influenza vaccines.

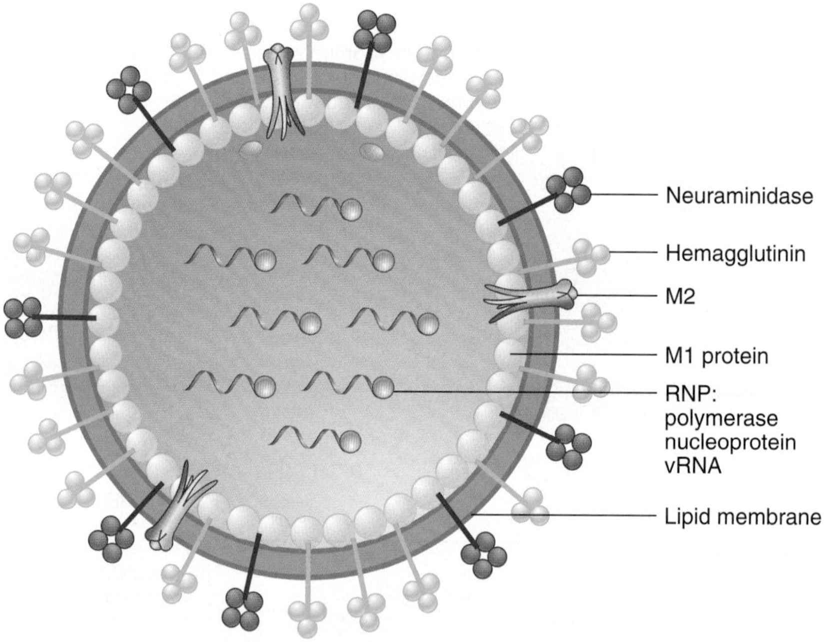

a

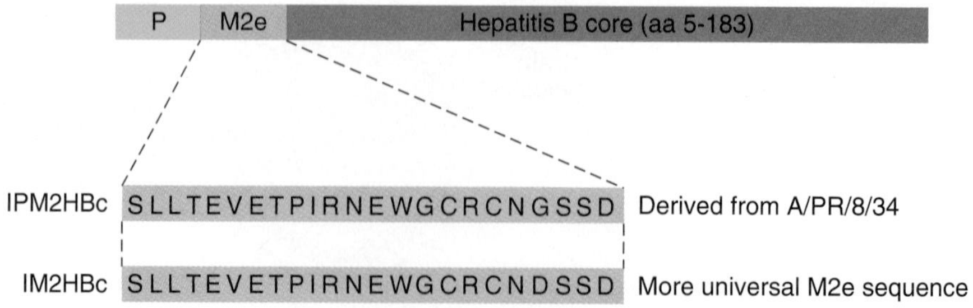

b

FIGURE VF 12-4 (a) Influenza A structure. Note the three influenza proteins that span the lipid membrane of the virus particle: hemaglutinin, neuraminidase, and M2. Adapted from Neirynck, S., et al. "A universal influenza A vaccine based on the extracellular domain of the M2 protein." *Nat Med* 5 (1999):1157–1163. **(b)** Schematic showing a promoter fused to a fusion gene consisting of M2e and HBc coding sequences. Adapted from Neirynck, S., et al. "A universal influenza A vaccine based on the extracellular domain of the M2 protein." *Nat Med* 5 (1999):1157–1163.

VIRUS FILE 12-3 Preparing a Pandemic Influenza Vaccine: Reverse Genetics or a Universal Vaccine? (continued)

Antibodies against the M2 protein are found in the sera of patients recovering from influenza, but data suggest it is weakly immunogenic (e.g., monoclonal antibodies directed against the extracellular domain of the M2 protein reduce cell-to-cell spread of the virus but don't block infectivity of the virus *in vitro*).

Researchers proceeded to create a chimerical protein consisting of the M2 extracellular domain (M2e) fused to the hepatitis B core antigen protein (HBc). When HBc is expressed in *Escherichia coli,* it self-assembles into an empty virus particle that is indistinguishable from hepatitis B particles isolated from the liver of hepatitis B patients (but does not cause hepatitis B). When the new M2e hepatitis B core fusion proteins were produced in *E. coli,* similar self-assembled particles were formed. The particles contained 240 M2e copies per self-assembled particle. Mice infected with a lethal influenza challenge were administered M2eHBc particles. After immunization, 90% to 100% of the mice were protected against the lethal virus challenge. The results of this *in vivo* model are encouraging. The M2eHBc vaccine is inexpensive and safe to produce. Perhaps a combination of old and new approaches will be required to curtail all pandemic influenza strains. Recent Phase I clinical trials demonstrated that the M2-based vaccines are safe and immunogenic in humans.

References

Du, L., et al. 2010. "Research and development of universal influenza vaccines." *Microbes Infect* 12:280–286.

Fiers, W., et al. 2004. "A universal human influenza A vaccine." *Virus Res* 103:173–176.

Kilbourne, E. D. 1999. "What are the prospects for a universal influenza vaccine?" *Nat Med* 5:1119–1120.

Neirynick, S., et al. 1999. "A universal influenza A vaccine based on the extracellular domain of the M2 protein." *Nat Med* 5:1157–1163.

Schodel, F., et al. 1994. "Hepatitis B virus core particles as a vaccine carrier moiety." *Int Rev Immunol* 11(2):153–165.

Schotsaert, M., et al. 2009. "Universal M2 ectodomain-based influenza A vaccines: Preclinical and clinical developments." *Expert Rev Vaccines* 8:499–508.

Webby, R. J., et al. 2004. "Responsiveness to a pandemic alert: Use of reverse genetics for rapid development of influenza vaccines." *Lancet* 363:1099–1103.

See also http://who.int/csr/disease/influenza/pandemic/en.

humans contain receptors that have α-2,6 linkages that are abundant. They also have α-2,3 linkages that are found deep in the lungs, where the virus may not be released as easily. Ferrets infected with the avianlike 1918 virus did not sneeze, which could be why transmission was reduced in ferrets. Several groups are now trying to create 1918 viruses that are more humanlike.

The 1957 Asian Flu

Although the 1957–1958 Asian flu was not as devastating as the 1918 Spanish flu, a significant

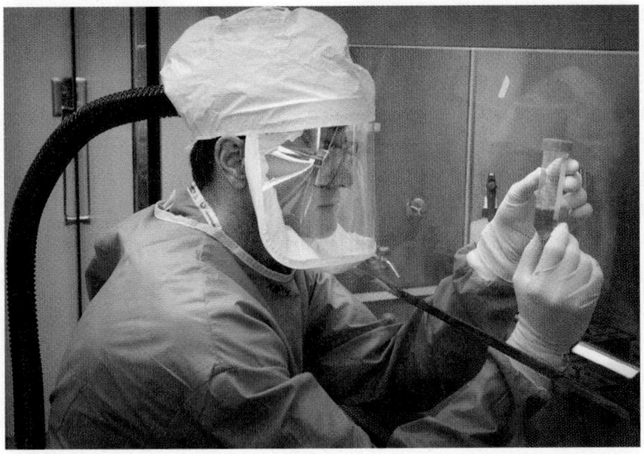

FIGURE 12-19 CDC researcher Dr. Terrence Tumpey working with the reconstructed 1918 pandemic influenza virus in a Biosafety Level-3 enhanced laboratory.

number of people—about 70,000—died in the United States alone. The Asian flu was first identified in February of 1957 in northern China and spread to the United States by that June.

Fortunately, the advances in technology that occurred in 1933 (Andrewes, Smith, and Laidlaw's isolation of the first human influenza virus) and in 1940 (Sir Frank Macfarlane Burnet's cultivation of influenza virus in embryonated chicken eggs) created important capabilities for influenza isolation and vaccine development. The 1957 strain of Asian influenza was rapidly identified and a vaccine was made available by August 1957.

The 1968 Hong Kong Flu

The Hong Kong influenza pandemic was first detected in Hong Kong during the early months of 1968 and reached the United States by September. Deaths in the United States (the victims were usually the elderly) peaked in December 1968 and January 1969. This pandemic was the mildest pandemic in the 20th century, killing only 33,800 people in the United States.

The 2009 H1N1 Swine Flu

The events surrounding the 2009 pandemic of swine-origin H1N1 influenza virus was described in Chapter 1 (Section 1.4, Recent Viral Outbreaks). *Contrary to what many experts expected, the first pandemic of the 21st century started in North America and not in Southeast Asia. The influenza virus was of swine instead of avian origin.* In late April, Mexico became the epicenter of the first influenza pandemic of the 21st century. The novel swine influenza A (H1N1) virus strain was rapidly identified. Mexican authorities implemented aggressive social interventions and deployed antiviral drugs to treat cases and contacts. From the beginning of the outbreak through December 5, 2009, 208 countries reported cases.

Based on conservative estimates, the number of pandemic H1N1 flu deaths in the United States ranged from 7500–12,000—less than half the number caused annually by seasonal influenza A and B viruses. Most conservative estimates are based on official medical records reporting influenza as the cause of death. However, many influenza deaths are not recorded this way because there are often complicating conditions such as heart disease or diabetes. For this reason, the number of influenza mortalities is likely underestimated. **FIGURE 12-20** compares 2009 pandemic deaths to seasonal influenza deaths and deaths from influenza during the 1968 and 1957 pandemics.

The New York City Health Department investigated an H1N1 outbreak at a high school in Queens from April 24 through May 8, 2009. A total of 124 students and employees of the high school

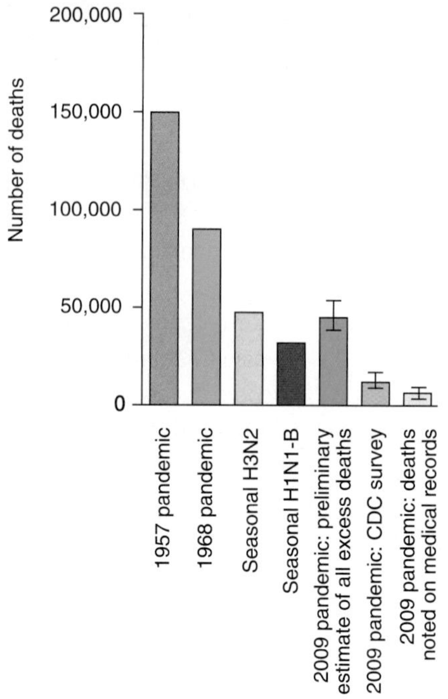

FIGURE 12-20 Estimated deaths during U.S. influenza pandemics. Courtesy of the CDC.

were confirmed by laboratory real-time-PCR assays to be infected with the 2009 H1N1 influenza A strain. Students and employees of the high school were asked to provide detailed information by participating in an online questionnaire. From their study, they were able to estimate the incubation period of 1.4 days. The outbreak was also linked with travel to Mexico.

A striking observation of the 2009 pandemic was that more than 75% of individuals infected were younger than 30 years of age. The peak age group affected most by H1N1 was 10–19 year olds (**FIGURE 12-21**). Less than 3% of cases occurred in persons over the age of 65. Epidemiologists analyzed serum samples collected from people in England taken before the start of the pandemic. These are referred to as **seroprevalence studies**. Older people had stronger antibody reactions toward H1N1 antigens than young people. A plausible explanation for this is that adults born before 1957 still have some cross-reacting neutralizing antibodies because of exposure to seasonal H1N1 viruses that had similarities to the new 2009 H1N1 strain. The 2009 H1N1 strain dominated other seasonal influenza strains circulating during the pandemic (**FIGURE 12-22**). Elderly people were spared of exposure to the usual influenza seasonal strains such as A/H3N2.

A study led by Wrammert examined antibodies from nine people who survived infection with the 2009 pandemic H1N1 strain (A/California/04/2009). The patients experienced mild to severe

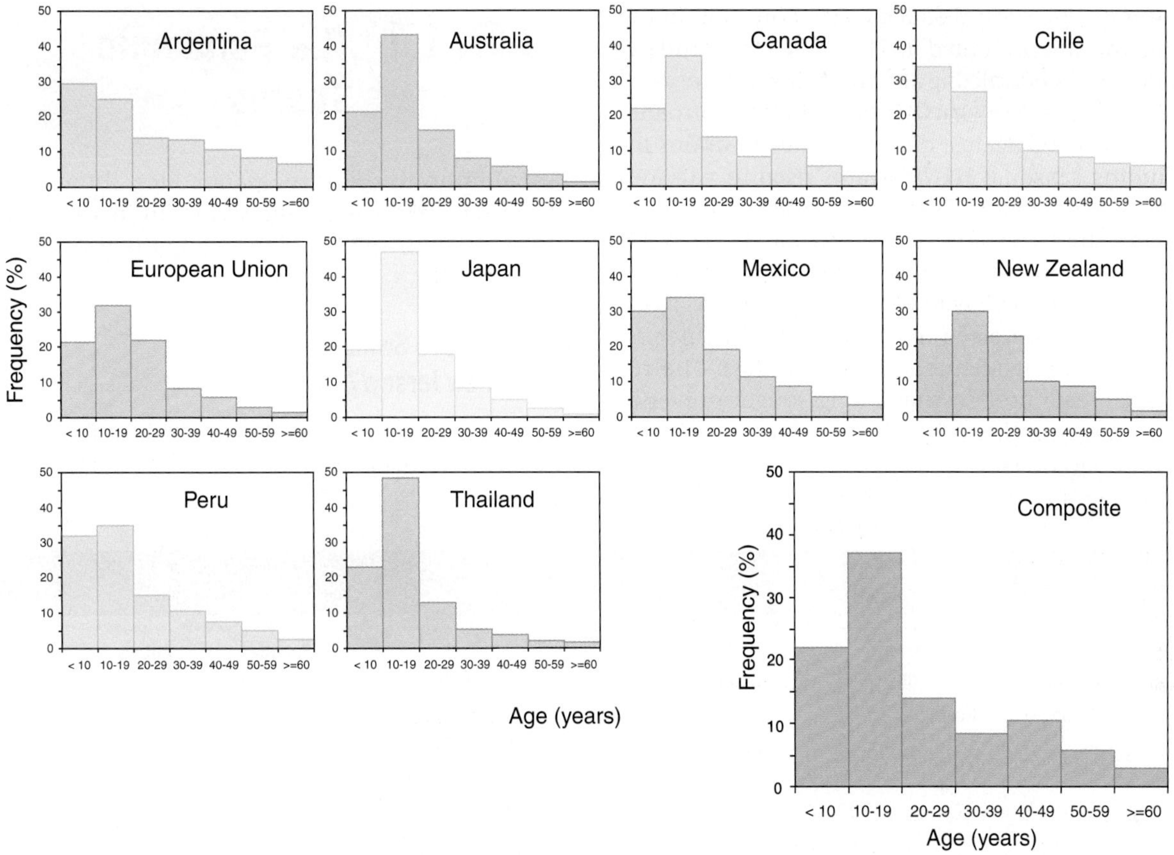

FIGURE 12-21 Age groups most affected by H1N1 in nine countries and the European Union on five continents through July 2009. Adapted from Reichert, T., et al. 2010. "Does glycosylation as a modifier of original antigenic sin explain the case age distribution and unusual toxicity in pandemic novel H1N1 influenza?" *BMC Infect Dis* 10:5.

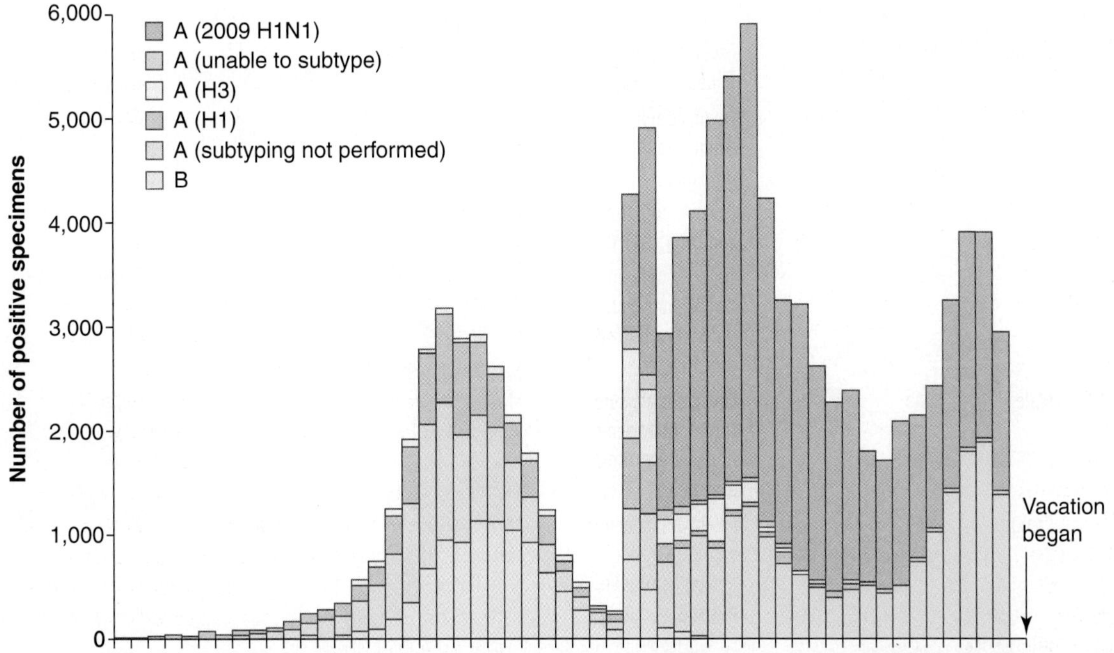

FIGURE 12-22 H1N1 dominated other influenza strains during the 2009 H1N1 influenza pandemic, gradually replacing the existing seasonal influenza strains. Courtesy of the CDC.

disease. The clinical data collected on eight of the patients are presented in **TABLE 12-8**. All of the patients had neutralizing antibodies against the 2009 H1N1 influenza virus. Their antibodies broadly cross-reacted against diverse influenza strains including seasonal H1N1 strains used in vaccines (e.g., A/Brisbane/59/2007 and others), and pandemic strains H1N1 (A/Brevig Mission/1/1918) and H5N1 (A/Indonesia/05/2005) by ELISA assays (see Chapter 5 to review ELISA assays). The survivors' antibodies protected mice intentionally infected with lethal pandemic influenza strains. The research team proffers that a long-sought universal influenza vaccine might be possible (see section 12.13 Vaccines).

12.11 The Pandemic Scares

In addition to the three pandemics, there have been several "scares": the swine flu in 1976, the Russian flu in 1977, and the Hong Kong avian flu in 1997.

The 1976 Swine Flu Scare: A/New Jersey/76/H1N1

In February 1976, several cadets at Fort Dix, New Jersey, were hospitalized with flu symptoms.

TABLE 12-8 Clinical Data for Pandemic H1N1 Influenza Survivors

Age	Gender	Pre-Existing Conditions	Initial Symptoms	Hospital Course	Sample Collection	Antiviral Treatment
30	Female	None	Fever, cough, labored breathing (**dyspnea**)	Acute respiratory distress syndrome, bacterial pneumonia, embolism, prolonged ventilator support, **tracheostomy***, discharged after 12 months	Day 31	Tamiflu
37	Male	High blood pressure (hypertension), interstitial lung disease	Fever, cough, shortness of breath, nausea, vomiting	Pneumonia, acute sinusitis, acute kidney failure, discharged after 8 days	Day 18	Tamiflu and Relenza
38	Female	None	Fever, cough, body aches	N/A	Day 15	None
21	Male	Congenital heart disease; repaired tetralogy of Fallot	Fever, cough, sore throat, nausea, diarrhea	N/a	Day 9	Tamiflu
24	Male	None	Fever, cough, nausea, vomiting, diarrhea	N/A	Day 11	Tamiflu
25	Male	None	Fever, cough, sore throat, body aches, nausea, vomiting, diarrhea	N/A	Day 9	Tamiflu
26	Male	None	Fever, cough, sore throat, body aches, nausea, vomiting, diarrhea	N/A	Day 9	None
45	Female	None	Fever, chills, cough, sore throat, body aches, headache, nausea, vomiting	N/A	Day 9	None

*A tube is inserted into the trachea to provide a passage for air because the pharynx is obstructed by inflammation during infection.

N/A= not applicable.

Source: Adapted from Wrammert, J., et al., 2011. "Broadly cross-reactive antibodies dominate the human B cell response against 2009 pandemic H1N1 influenza virus infection." *J Exp Med* 208:181–193.

Within 24 hours, a young private named David Lewis had died. Subsequently, the CDC investigated the outbreak. Nineteen throat washings were analyzed. Eleven of the washings were influenza A (mostly H3N2). Four of the samples were confirmed to contain H1N1 influenza A and thought to be close to the swine influenza virus responsible for the 1918–19 pandemic. Fear of influenza deaths in numbers similar to 1918 prompted the federal government's recommendation to vaccinate all Americans.

In all, 150 million doses of vaccine were prepared in the United States and 46 million doses were administered within a few months. The vaccine needed to be produced fast, and because of this the manufacturer's insurance carriers stated they would not indemnify such a hastily produced vaccine. Congress passed a liability protection bill. Early problems were reported with the vaccine, including cases of Guillain-Barré syndrome (GBS) and several deaths. GBS is an illness characterized by fever, nerve damage, and muscle weakness. The body damages its own nerve cells, resulting in muscle weakness and sometimes paralysis. It can last from weeks to months and 5% to 6% of individuals with GBS die. The vaccine campaign was suspended in the late fall of 1976 after the feared swine flu epidemic had not appeared anywhere else in the world and due to safety concerns associated with the vaccine.

A total of 532 swine flu vaccinees developed GBS within a 10-week period following vaccination, and 32 died. The government paid more than $90 million on claims in these cases. Research on the virus later showed that if it had spread, it would probably have been much less deadly than the Spanish flu.

The 1977 Russian Flu Scare: A/USSR/77/H1N1

In May 1977, influenza H1N1 was isolated in northern China. It spread rapidly and caused an epidemic in children and young adults worldwide. This virus was similar to the H1N1 strain isolated prior to 1957. Individuals born before 1957 were likely exposed to this strain and developed some immunity against the virus. This may be the reason why primarily children and younger people became ill from this H1N1 strain. The virus had spread around the world by January 1978, including to the United States. It has been suggested that the 1977 Russian H1N1 strain was not a result of genetic reassortment between a bird and human flu. Some believe that a frozen stock of older virus may account for this mild pandemic 27 years later, but there is no actual proof for this theory.

The 1997 Avian Flu Scare: H5N1

In May 1997, a 3-year-old boy in Hong Kong became ill with influenza-like symptoms and died 12 days later. The diagnostic laboratory isolated an influenza A virus but could not identify its H subtype with any reagents available to them. In August, it was confirmed that this was a H5N1 influenza A virus. H5 was never known to infect humans and was limited to birds. This "jump" of an avian strain directly to humans (bypassing the pig mixing vessel) had never happened before. It was also discovered that this A/HongKong/258/97/H5N1 strain was responsible for sudden die-offs of chickens on three rural chicken farms during March of 1997 in Hong Kong.

It is believed that the emergence of the H5N1 influenza in Hong Kong in 1997 occurred when nonpathogenic H5N1 influenza spread from migrating shorebirds to ducks via fecal contamination of water. The virus was transmitted to chickens and then became established in live bird markets. During transmission between different species, the virus became very pathogenic in chickens and was occasionally transmitted to humans from chickens in the markets.

In November 1997, new human cases caused by the H5 strain occurred. By December 1997, there were 18 confirmed cases, of which 6 were fatal. Investigations by local authorities, the WHO, and the CDC continued. They were able to confirm that all but one of these patients had been near live chickens (e.g., in marketplaces) days before the onset of symptoms. On December 18, 1997, veterinary authorities began to slaughter chickens in wholesale facilities and importation of chickens from neighboring areas was stopped. A total of 1.6 million chickens were culled (**FIGURE 12-23**), and as a result of this fast intervention the outbreak resulted in much fewer casualties than previous outbreaks. These avian flu cases occurred during the "flu season," and as a result health officials were concerned that human and avian strains would reassort, leading to a potential pandemic with efficient person to person spread of the virus.

Lessons Learned from the 1997 Hong Kong Flu

Even though there were only 18 cases of the 1997 Hong Kong flu, it was highly lethal. Six of the 18 individuals died, resulting in a 33% fatality rate—even more deadly than the 1918 Spanish flu, which had a 4% average fatality rate. This Hong Kong strain also caused an 87.5% to 100% fatality rate in experimentally inoculated White Plymouth Rock and White Leghorn chickens.

FIGURE 12-23 A total of 1.6 million chickens were culled to curb the Hong Kong Flu in 1997.

The 1997 avian influenza A H5N1 virus was isolated from a child with fatal respiratory illness. The *H* gene sequence of the A/Hong Kong/156/97H5N1 virus was determined. All *H* gene products of the human influenza A viruses, including the 1918 strain, have a single basic arginine residue at the cleavage site between H1 and H2. Interestingly, the H protein of the avian Hong Kong virus contains an insertion of multiple basic residues upstream of the cleavage site that has been seen in highly pathogenic avian strains of influenza A virus. The addition of these residues increases the cleavage site's accessibility to proteases that can cleave the H_0 into H_1 and H_2. Hence, a single insertion (mutation) could enable the virus to spread much more efficiently and systematically to the brain, heart, and blood vessels. Other gene segments of this virus require further analysis to determine if they may also contribute to its lethality.

Avian strains continue to plague eastern Asia (especially Vietnam, Thailand, Cambodia, and Indonesia) and Russia, causing huge economic losses within the poultry industry. Most cases of H5N1 infection in humans are thought to have occurred from direct contact with infected poultry. It is highly lethal in humans—with as high as 50% mortality (**FIGURE 12-24a**). The most important control measures are rapid culling of infected or exposed birds, proper disposal of carcasses, quarantining, and rigorous disinfection of farms. In 2005, the significant spread of the virus in birds became more apparent, but it is poorly understood. Scientists are increasingly convinced that at least some migratory birds are now carrying the H5N1 virus to poultry flocks that lie along migratory routes (Figure 12-24b). There have been die-offs of migratory birds infected with a highly pathogenic H5N1 strain. It is still a rare disease in humans, but one that should be closely monitored, particularly because of the potential of this virus to evolve in ways that could start a pandemic.

12.12 Antivirals for Influenza Treatment

Scientists continue to search for molecules that will selectively inhibit virus-directed events without interfering with normal cellular activities. Fortunately, there are currently four licensed antiviral drugs available for influenza. These drugs, however, are of little value if prescribed greater than or equal to 48 hours after the onset of illness. There are currently two types of influenza antivirals available: M2 inhibitors and N inhibitors.

M2 Inhibitors

The first drugs effective against influenza A virus were M2 inhibitors. They were discovered through massive screening of compounds. Amantadine (Symmetrel) and rimantidine (Flumadine) block the M2 ion channel, preventing uncoating of the

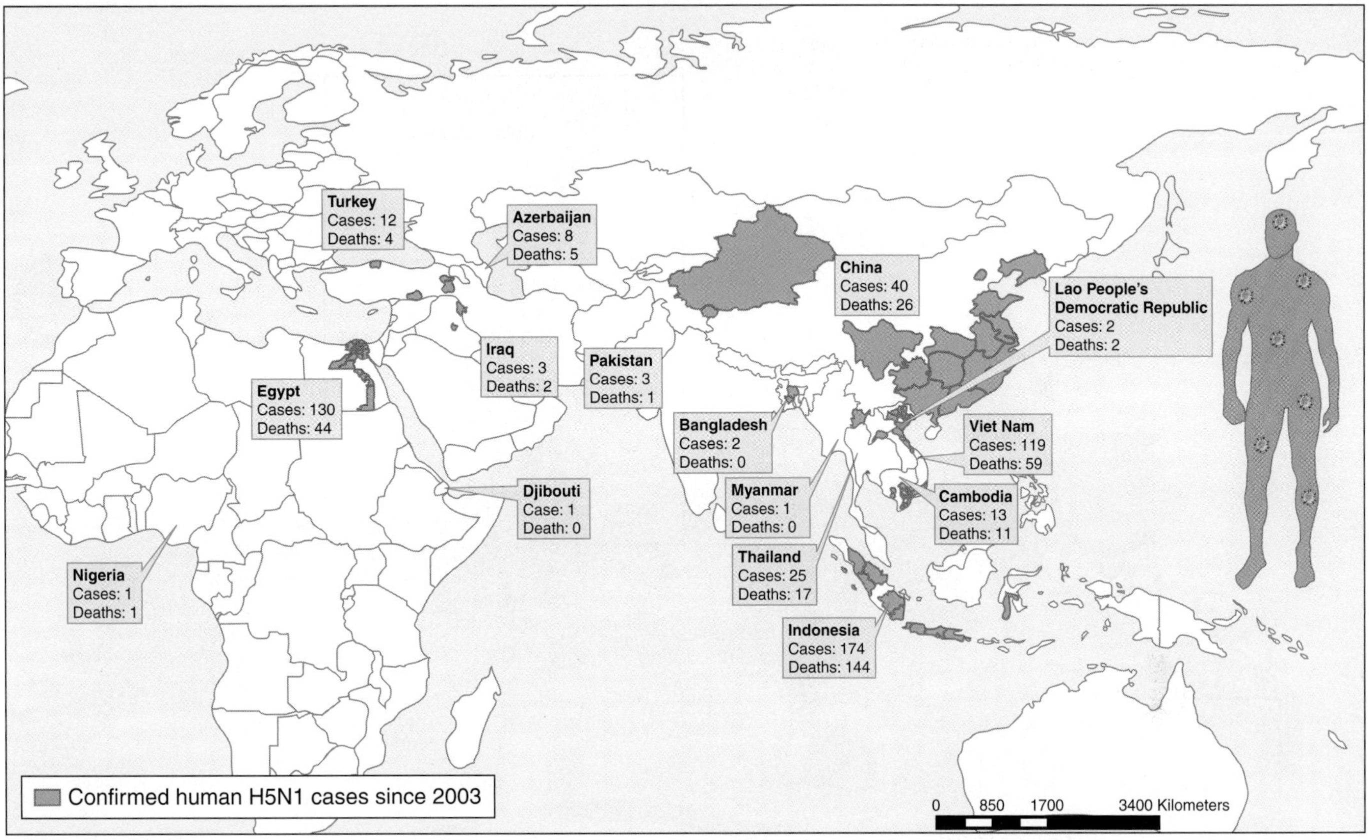

a

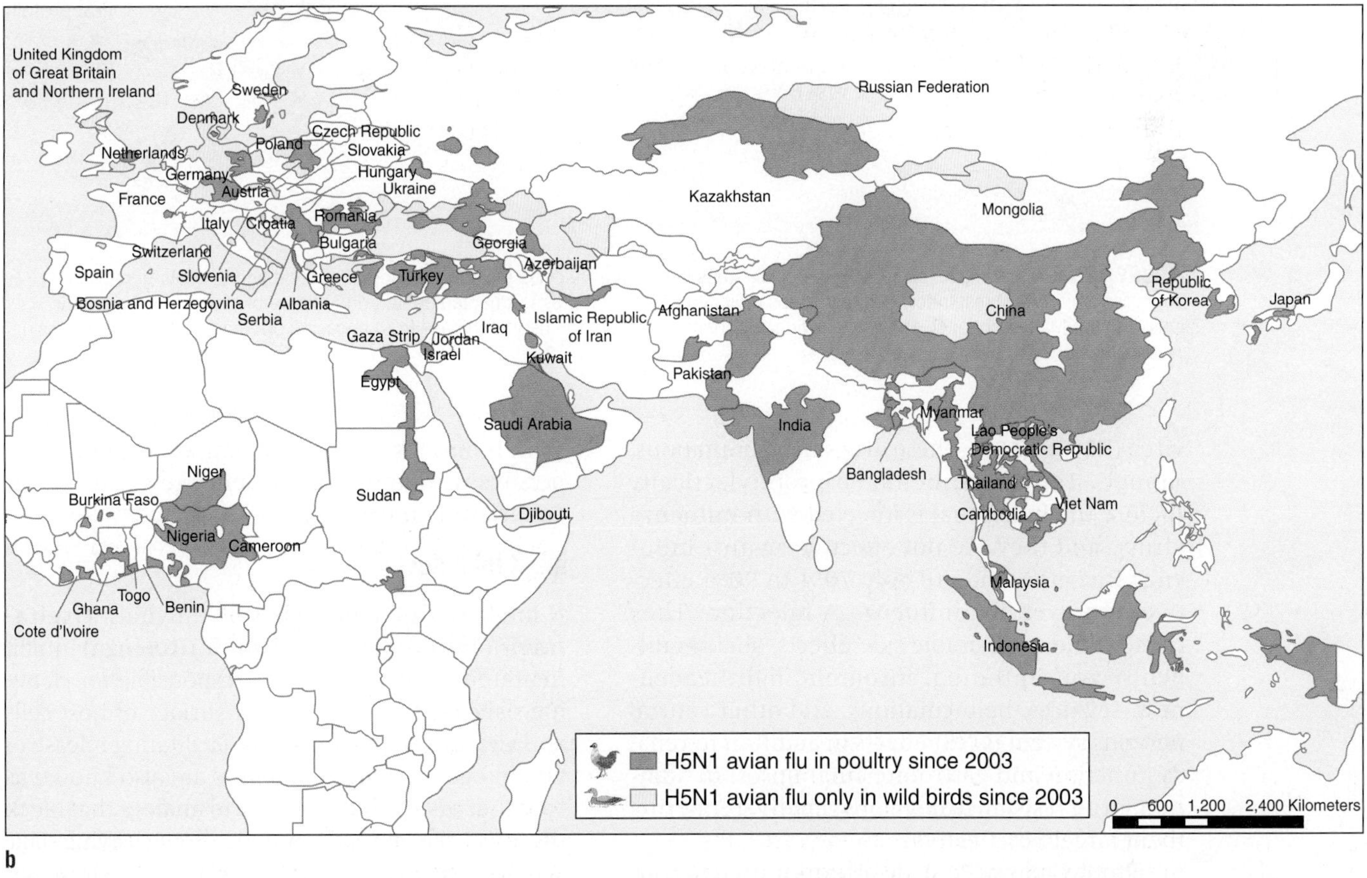

b

FIGURE 12-24 (a) Areas reporting human cases of H5N1 avian influenza January 2003-March 2007. Adapted from World Health Organization, 2011. **(b)** Areas reporting the occurrence of H5N1 avian influenza in poultry and wild birds January 2003–March 2007. Adapted from World Health Organization, 2008.

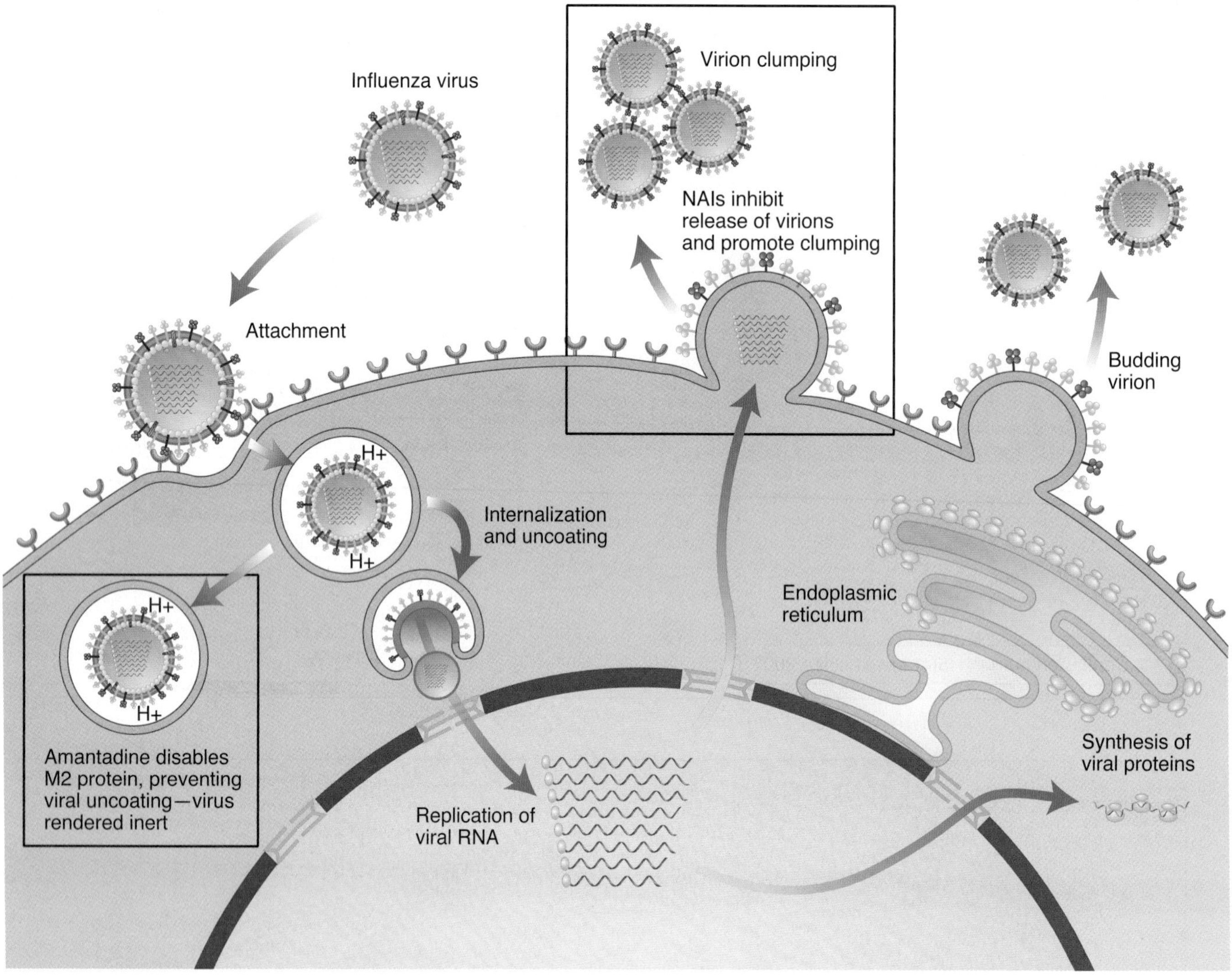

FIGURE 12-25 A schematic diagram representing steps in the viral life cycle (attachment, internalization and uncoating, replication, and exit from the host respiratory cell) and steps inhibited by antiviral drugs. Amantadine and rimantidine block viral uncoating by inhibiting the M2 ion channel, and oseltamivir and zanamivir prevent the neuraminidase from releasing budding viruses and dispersing virus particles.

virus (**FIGURE 12-25**). These drugs have limitations, though. They must be taken prophylactically (before the individual is infected with influenza virus), and they are not effective against influenza B viruses and are only 70% to 90% effective in preventing influenza A infection. They cause some undesirable side effects, such as difficulty concentrating, insomnia, lightheadedness, seizures, hallucinations, and other central nervous system (CNS) effects in addition to renal dysfunction and gastrointestinal upset. In addition, drug resistance develops rapidly, rendering them largely ineffective.

Prophylaxis with M2 inhibitors is used to control influenza A outbreaks in chronic care facilities. It is given to all residents and unvaccinated staff. It may also be used as adjunctive therapy in persons with immune deficiency (e.g., HIV or organ transplant patients).

N Inhibitors

N inhibitors are a new class of antivirals. **Oseltamivir (Tamiflu)** and **zanamivir (Relenza)** inhibit neuraminidase, the enzyme responsible for cleaving sialic acid residues on the surface of host cells and viral envelopes and, thus, facilitating release of viral progeny. N drug inhibitors are also known as "plug" drugs. They are sialic acid analogs that block the active site of N, preventing it from cleaving sialic acid residues present on the surface of the host and virus. The result is a reduction in the number of viral particles released to infect other cells.

Oseltamivir is available in pill form and zanamivir must be inhaled. N inhibitors must be initiated within 48 hours of onset of symptoms to be effective. They are effective against both influenza A and B viruses. Some patients, though, may not receive any benefit if the treatment is started after 36 hours of onset of symptoms. Both drugs have some side effects: Oseltamivir can cause nausea, vomiting, vertigo, and bronchitis. Zanamivir can cause nasal and throat irritation, headache, gastrointestinal upset, bronchitis, and cough (especially with individuals suffering asthma or chronic obstruction pulmonary disease [COPD]). The standard outpatient treatment for 2009 H1N1 influenza A was Tamiflu. The Food and Drug Administration (FDA) gave Tamiflu an Emergency Use Authorization on October 30, 2009, to extend its use to children younger than 12 months.

Peramivir is also a neuraminidase inhibitor. It is an unapproved drug in phase 3 clinical trials. It received an Emergency Use Authorization to treat hospitalized patients with known or suspected 2009 H1N1. It is administered intravenously, providing rapid drug delivery at high levels. Its use was limited to patients hospitalized and not responding to oral or inhaled antiviral drugs or for whom intravenous administration was the only feasible route of administration.

12.13 Vaccines

Immunization with a viral vaccine is the most effective method of preventing influenza. In the United States, persons at high risk and some healthcare workers are vaccinated each year in October and November, prior to the December to January flu season, with an inactivated trivalent vaccine.

Viruses for the vaccine are grown in eggs. A nasal-spray flu vaccine containing live attenuated viruses instead of killed viruses (sometimes called LAIV, for live attenuated influenza vaccine) was licensed in 2003. LAIV is only approved for healthy people between the ages of 5 and 49 years.

Vaccine Composition

Vaccine composition is based upon a prediction of which influenza strains will be circulating in the United States during the flu season. Each vaccine consists of three inactivated (killed) viruses. Two of the viruses are influenza A virus strains and the third is a strain of influenza B virus; for example, the 2006–07 vaccine contained A/New Caledonia/20/99, A/Wisconsin/67/2005, and B/Malaysia/2506/2004 viruses. This recommendation was based on antigenic analyses of recently isolated influenza viruses, epidemiologic data, and post-vaccination serologic studies in humans.

Effectiveness of the Vaccine

The effectiveness of the influenza vaccine depends upon several factors:

- the age of the vaccine recipient
- immunocompetence of the vaccine recipient
- the degree of similarity between the viruses in the vaccine and those in circulation

Target Groups for Vaccination

The following list consists of vaccination target groups. It includes persons at increased risk for complications, those persons who interact often with individuals who are at risk, and persons who can transmit influenza to those at high risk.

- Persons aged 50 or older
- Residents of nursing homes or other chronic-care facilities who house individuals with chronic medical conditions
- Adults and children with chronic disorders such as asthma, pulmonary or cardiovascular problems, and immunosuppression
- Children who are receiving long-term aspirin therapy
- Women in their second or third trimester of pregnancy during the influenza season
- Physicians, nurses, and medical emergency response workers
- Employees of nursing homes and chronic-care facilities
- Household members of persons at high risk
- Travelers to the tropics
- Travelers to the southern hemisphere during April to September
- Travelers with organized tour groups at any time of year

12.14 International Influenza Surveillance

The WHO Influenza Program was created in 1946 as an international center to collect and distribute information, coordinate laboratory work on influenza, and train laboratory workers. Today, WHO's program for global surveillance of influenza now maintains 110 National Influenza Centers in 83 countries and 4 WHO Collaborating Centers for Virus Reference and Research (in Atlanta, USA;

London, UK; Melbourne, Australia; and Tokyo, Japan).

Influenza Sentinel Physicians in the United States

Every year, influenza sentinel physicians conduct surveillances for influenza-like illness (ILI) in collaboration with their state health departments and the CDC. Data provided by sentinel physicians are combined with other influenza surveillance data to provide a national picture of influenza virus and ILI activity in the United States.

What Data Do Sentinel Physicians Collect? How and to Whom Are the Data Reported?

During influenza season (October through mid-May), sentinel physicians report the total number of patient visits each week and the number of patient visits for influenza-like illness by age group (0–4 years, 5–24 years, 25–64 years, and >65 years). These data are forwarded once a week to a central data repository at the CDC. Influenza viruses are constantly evolving and cause substantial morbidity and mortality almost every year. The data the sentinel physicians provide are critical for monitoring the impact of influenza and, when combined with other influenza surveillance data, can be used to guide prevention and control activities, vaccine strain selection, and patient care.

Will There Be Another Killer Flu?

It is impossible to predict when another pandemic might occur. Shifts occur when novel strains of the virus appear, to which the general population will have no immunity. If an influenza virus similar to the 1918 strain appears and behaves as it did in 1918, even taking into account the medical advances today, there will be unparalleled tolls of cases and deaths. Air travel will hasten the spread of the novel virus, decreasing time available to prepare for interventions. Healthcare systems could be rapidly overburdened, economies strained, and social order disrupted. Although it is not feasible to halt the spread of the pandemic virus, it should be possible to minimize the consequences by preparing for the challenge in advance. During the 2009 pandemic, international cooperation was critical. The mass media played a vital role communicating to the public and healthcare professionals. Preparedness plans were very important to confront the pandemic.

Pandemic Planning

The WHO has established a National Pandemic Planning Committee charged with developing strategies appropriate for their countries in advance of the next pandemic. Currently there are four WHO Collaborating Centers that are based in Atlanta, London, Melbourne, and Tokyo. These centers maintain repositories of different virus strains and also develop reagents and technologies for rapid diagnosis of new strains. The Centers have bio-containment facilities run by trained personnel to conduct studies on possible pandemic influenza viruses.

12.15 Lessons Learned from the Severe Acute Respiratory Syndrome (SARS) Outbreak

The 2002–2003 SARS-CoV outbreak serves as a reminder that we need to be prepared to confront unexpected viruses. SARS-CoV was first described on February 26, 2003. It was identified as a new disease by WHO physician Dr. Carlo Urbani, who diagnosed it in a 48-year-old businessman who had traveled from the Guangdong province of China, through Hong Kong, to Hanoi, Vietnam. The businessman died from the illness, and Dr. Urbani subsequently died from SARS on March 29, 2003, at the age of 46.

SARS-CoV began to spread, and within six weeks of its discovery it had infected thousands of people in 16 countries around the world. Its case–fatality rate proportion was 3.9%. Schools closed throughout Hong Kong, Canada, and Singapore. National economies were affected. The WHO and the CDC had identified SARS as a global health threat and issued an unprecedented travel advisory. SARS is a serious form of atypical pneumonia accompanied by fever. It results in acute respiratory distress and sometimes death, even in individuals regarded as young, fit, and healthy. The disease is very similar to influenza.

Initial diagnosis was based on patient travel history and is a dramatic example of how quickly world travel can spread a disease. It is also an example of how quickly a networked health system can respond to an emerging threat. This crisis was responded to by utilizing sanitation and surveillance techniques (**FIGURE 12-26**). Both China's Minister of Health and the Mayor of Beijing were

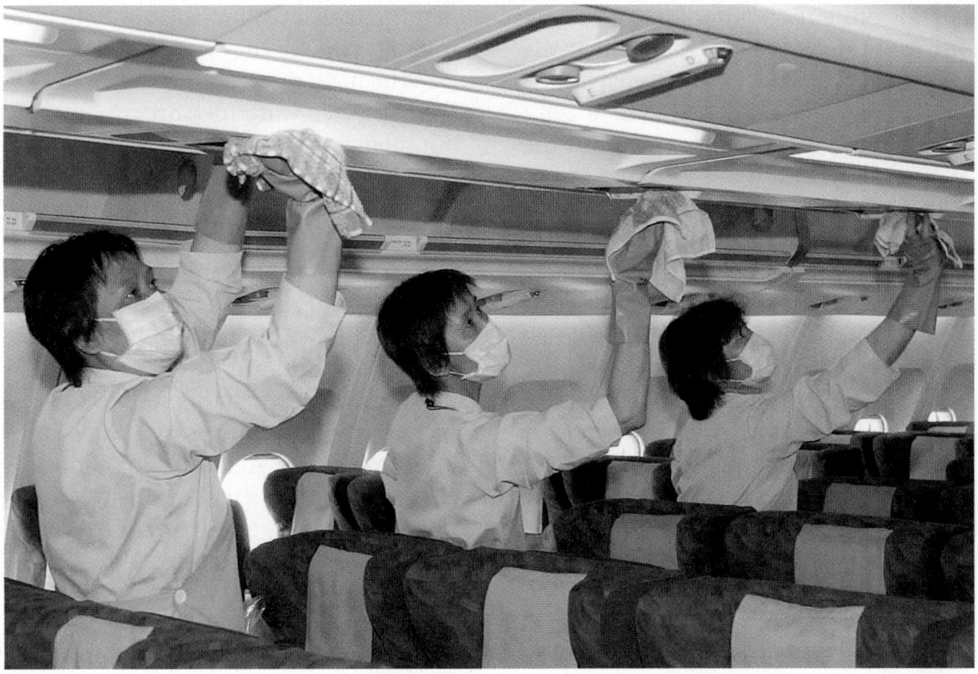

FIGURE 12-26 Airport maintenance personnel in protective clothing disinfecting airliner cabins to curb the spread of SARS.

fired for underreporting SARS cases. The identification of SARS-CoV represents one of the fastest identifications of a new pathogen in history. The experience with SARS is a reaffirmation of the need to respond to new infectious viral diseases.

Summary

Influenza viruses have been afflicting humans for centuries. Major influenza pandemics occurred in 1918, 1957, 1968, and 2009. Influenza can spread at alarming speeds because of its short incubation period; the large number of infectious virus particles shed in droplets discharged by sneezing, coughing, or talking; and the fact that many symptomatic individuals continue their normal social activities, spreading illness to many contacts. Pandemics occur when individuals lack immunity to the strain of influenza circulating in a particular area.

Symptoms of influenza include the sudden onset of headache, fever, aching in the limbs and back, malaise, dry cough, weakness, myalgia, and sore throat. Secondary bacterial pneumonia can be a life-threatening complication of influenza in young children, the elderly, and other high-risk groups such as AIDS patients. Some influenza symptoms overlap with other respiratory infections, such as anthrax and SARS. Rapid diagnostics is essential to proper treatment.

There are three types of influenza viruses—A, B, and C—of which A is the most prevalent. All three types can cause similar symptoms in humans. Diagnosis is based on symptoms and laboratory testing may be an appropriate confirmatory step. Several rapid ELISA tests are commercially available. Viral culture is the most accurate method for identifying specific viral strains. Several groups are developing microarrays or "flu chips" to distinguish among the different respiratory pathogens.

Influenza is spread person-to-person via droplet transmission. Inhaled virus particles enter the respiratory tract, adhering to and penetrating ciliated columnar epithelial cells that line the sinuses and airways. During infection, the cilia are destroyed and the cleaning system of the lungs does not work well. The highest number of infectious virus particles can be isolated on the fourth or fifth day after infection.

Influenza is an enveloped virus. Three proteins protrude from the viral lipid membrane: hemagglutinin (H), neuraminidase (N), and M2. There are 16 antigenically different types of Hs and 9 distinct Ns. All H and N types infect waterfowl (e.g., ducks, gulls, geese, teals, and terns). The antigenic type differences are critical for vaccine preparation. Only antibodies against the H protein will neutralize the virus. The H protein plays several roles in the life cycle of the virus. To gain entry, the H protein of the influenza virus particle binds to sialic acid present on the lining of the respiratory tract. It also mediates fusion of the membrane of the viral particle with the membrane of the cell to form endosomes. The H protein must be cleaved into two separate proteins in order for the virus to be infectious. Neuraminidase plays a role at the end of the viral life cycle. The N protein cleaves the sialic acid residues located on the outside of cells, facilitating the release of virus particles from the cell surface. This allows the virus to spread from cell to cell. The M2 channels the flow of H^+ ions from the lumen of the endosome into the interior of the virus particle. This results in the uncoating of the viral genome into the cell's cytoplasm. The eight segments of –ssRNAs of influenza A are transported to the cell nucleus. Transcription and replication occur in the nucleus. This is an unusual feature for an RNA virus because most RNA viruses replicate their viral genomes in the cytoplasm of the cell.

The eight RNA segments of influenza A code for at least ten gene products. Besides the segments encoding H, N, and M2, the other gene products code for structural proteins that function in a host mRNA cap-snatching mechanism or as part of a transcriptase complex involved in RNA synthesis. The remaining proteins are nonstructural proteins involved in viral RNA splicing and export or as an interferon antagonist.

Genetic variability is common during influenza replication. Influenza A undergoes antigenic drift and shift. Antigenic drift occurs when the viral genome acquires nucleotide point mutations over time. In antigenic shift, whole RNA segments are swapped between two different influenza-A virus strains within the same infected cell (usually pigs). The pig is considered the reassortment vessel for influenza A. Antigenic shift is responsible for pandemic strains of influenza A and their rapid spread to other parts of the world.

The main reservoir of influenza A is wild waterfowl. Infected waterfowl are usually asymptomatic but shed virus particles in their feces. In Hong Kong in 1997, an H5N1 avian influenza A virus spread from migrating shorebirds to ducks via fecal contamination of the water. The virus then jumped to chickens and became established in live bird markets. Humans in close contact with the birds became infected. This "jump" of an avian strain directly to humans (bypassing the pig-mixing vessel) had never happened before. Millions of chickens were slaughtered to contain the new influenza strain. Avian strains continue to plague eastern Asia and Russia, causing huge economic losses within the poultry industry and a modest number of human deaths.

Scientists continue to investigate why the 1918 influenza pandemic was so deadly. Today antivirals that inhibit N (oseltamivir and zanamivir) and M2 (amantidine and rimantidine) are available. Inactivated vaccine cocktails containing two influenza A strains and one influenza B strain are used to vaccinate high-risk groups each year before the flu season. The WHO has established a National Pandemic Planning Committee charged with developing strategies in advance of a forecast pandemic. Strategies include epidemiological surveillance, stockpiling antivirals, and vaccine development using reverse genetics or other methods. A unified global effort is central to mitigating the next major influenza pandemic.

CASE STUDY QUESTIONS

These questions relate to the Case Study presented at the beginning of the chapter.

1. Which countries are currently affected by a highly pathogenic avian influenza strain that infects humans?
2. What are the control measures for birds?
3. How do outbreaks of avian influenza spread within a country?
4. How are humans infected?
5. In this case, list possible route(s) of avian flu transmission.
6. List good hygiene practices to avoid spreading the virus through food.
7. Will an annual flu shot protect humans from avian influenza?

References

ProMED-mail. 2005. "Avian influenza, human—East Asia." 07 Mar;20050307.0682.
ProMED-mail. 2005. "Avian influenza—Eastern Asia: Viet Nam." 08 Feb;20050208.0426.

CASE STUDY 1: INFLUENZA TESTING

Your college roommate Sharon is a pre-nursing major who volunteers at a local senior citizens home on weekends. Sharon spent the prior weekend at home with her parents and her three-year-old brother James. When Sharon came back from the weekend, her parents called to tell her that James was vomiting, had a fever, and had a dry cough. They were en route to the hospital with James when Sharon recalled that, during her visit, James seemed lethargic and was not his usual self.

Other pertinent information to this case is that James attends a daycare facility three times per week where he had multiple contacts with sick children and his illness occurred in December.

On physical examination, James had a temperature of 101°F and was very weak. A nasopharyngeal aspirate specimen was collected and sent to the laboratory for rapid diagnostic tests assaying for influenza A and respiratory syncytial virus. The test results are shown in **FIGURE CS 12-1**.

1. What agent is causing the infection?
2. What symptoms does James have that are consistent with his illness?
3. What other agents could have similar symptoms?
4. Would antiviral therapies be effective for James?
5. Will antiviral therapies be effective for uninfected family members and contacts?
6. How would the diagnosis be confirmed?
7. When was James contagious?
8. How was the virus transmitted?

FIGURE CS 12-1 An influenza test where "C" represents a positive control result and "T" represents the test result. The left well is an influenza A test and the right well is an influenza B test.

9. Which family members are at greater risk for serious disease and why?

Two days later, Sharon developed a headache and sore throat. En route to volunteering at the senior citizens home, she went to the campus health center to find out if she contracted the same illness as James. This indeed was the case.

10. Create a list of Sharon's daily contacts. Are any of these contacts at high risk?

A week later Sharon felt much better, but her coursework had suffered during this time. She began studying and sleeping less to catch up in her courses. A few days later, she started to feel ill again. She developed chills and a cough producing green mucus.

11. What is the likely cause of this second illness?

CASE STUDY 2: RESPIRATORY ILLNESS

It is October and you strike up a conversation about the flu shot with Dan Jones, a 65-year-old member of the university custodial staff. You recommend to Dan that he get a flu shot. Dan, who is very healthy, states that he heard the shot gives you the flu, but, because he trusts your judgment he states that he will think about getting it. After much thought, Dan decided to get a flu shot, but he didn't make it to the vaccine clinic until mid-December. Two days after getting his flu shot, he began exhibiting flu-like symptoms. He was diagnosed with influenza A.

1. How is it possible that Dan fell ill with influenza A?
2. Why is influenza so difficult to control even when there is a national vaccination program?

CASE STUDY 3: H5N1 AVIAN INFLUENZA

Buckley Smith and George Washburn are two students taking a virology class. Both of them grew up on poultry farms near Omaha, Nebraska. They were interested in avian influenza and knew that the H5N1 avian influenza A virus is of a growing health concern. Human-to-human transmission is rare but the virus is transmitted rapidly from chicken-to-chicken on poultry farms. Smith and Washburn decided to research more about how avian influenza outbreaks on poultry farms are contained. They were surprised to learn that there were a number of disposal options for infected chicken carcasses including burial, rendering, composting, incineration, lactic acid fermentation, alkaline hydrolysis, and anaerobic digestion and burial.

1. Research each of the disposal options listed above at the website of materials located at the K-State Research Exchange: "Carcass Disposal: A Comprehensive Review." Retrieved March 18, 2011, from http://krex.k-state.edu/dspace/handle/2097/662. Explain each method. What are the advantages and disadvantages of each method?

The U.S. Environmental Protection Agency recommends three management options: land disposal by on-site burial, composting, and off-site landfills.

2. Why do you think landfilling and composting is an appropriate method for disposal of carcasses infected with avian influenza virus?
3. Mass disposal of carcasses is not a new practice. List other dates and locations of epizootics in which millions of animal carcasses were disposed. (Hint: There is an example in Chapter 1 and other examples in the Review website.)
4. In 1984, an outbreak of avian influenza in Virginia resulted in the disposal of 5700 tons of chicken carcasses. Describe a standardized burial trench used to bury approximately 800 pounds of poultry carcasses. (Hint: See Review on Burial located at the above website.)
5. What types of gases are emitted from the decomposition of animal carcasses in a landfill? Are the gas emissions hazardous? Explain your answer.
6. What types of monitoring or what testing is recommended to ensure public and private water supplies are safe?
7. List bacterial and viral pathogens or other infectious agents that might survive burial and subsequent dissemination from actual carcass burial sites.
8. The study cited in the Reference was performed to determine the survival of avian influenza (H6N2) after landfill leachate disposal. Summarize the results of Graiver et al.'s work. Be sure to include how long the waste was infectious, and how pH, temperature, and the leachate affected survival times of the avian influenza viruses.

Reference

Graiver, D. A., et al. 2009. "Survival of the avian influenza virus (H6N2) after land disposal." *Environ Sci Technol* 43(11):4063–4067.

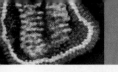

CASE STUDY 4: SICK DOGS AND CATS

A veterinarian working at the Khao Khiew Zoo near Bangkok, Thailand, found three dead birds, a sick white tiger, and a dead clouded leopard. The tiger and leopard were fed chicken carcasses. The veterinarian read recent reports of 3 pet house cats that died of avian influenza and 39 reports of direct bird-to-human transmission cases of avian influenza. He made calls to local animal health officials who told him that a virologist was investigating the deaths of 629 village dogs and 111 cats in the Suphan Buri District of central Thailand. Of these, 8 cats and 160 dogs had antibodies against avian H5N1 influenza A. Later, the veterinarian discovered that the chicken carcasses fed to the tiger and leopard were infected with avian H5N1 influenza virus.

1. What do the virologist's findings indicate?
2. Companion animals such as dogs and cats often come in close contact with humans and domestic animals (e.g., horses, pigs, poultry on a farm). Is it possible that cats and dogs may play a role in transmitting H5N1 to humans? Explain your answer.
3. How might a dog or cat become infected with an avian H5N1 influenza A virus?
4. House cats have been intentionally infected with chicken H5N1 influenza viruses and an H7N7 influenza virus isolated from a fatal human influenza case. The cats developed respiratory disease. If you owned a house cat and lived in an H5N1-affected area, how might you protect your cat from infection?
5. What other mammals besides dogs and cats are susceptible to avian H5N1 influenza infection and may be able to contribute to the spread of influenza? (Note: You may need to look up one or more of the References listed below or search for additional primary literature.)
6. H5N1 avian influenza virus has crossed species barriers. What other viruses cross species barriers?
7. Why do virologists raise concerns when a virus crosses the species barrier?
8. Researchers in the United Kingdom have created transgenic chickens that are resistant to infection by avian influenza virus. What was their strategy? How were the chickens genetically modified? (Hint: See Lydall, J., et al., in the References below.)

References

Boon, A. C. M., et al. 2007. "Role of terrestrial wild birds in ecology of influenza A virus (H5N1)." *EID* 13(11):1720–1723.

Butler, D. 2006. "Thai dogs carry bird-flu virus, but will they spread it?" *Nature* 439(7078):773.

Kulken, T., et al. 2004. "Avian H5N1 influenza in cats." *Science* 306:241.

Lyall, J., et al. 2011. "Suppression of avian influenza transmission in genetically modified chickens." *Science* 331(6014):223–226.

Maas, R., et al. 2007. "Avian influenza (H5N1) susceptibility and receptors in dogs." *EID* 13(8): 1219–1221.

Riedel, S. 2006. "Crossing the species barrier: The threat of an avian influenza pandemic." *Proc (Bayl Univ Med Cent)* 19(1):16–20.

Said, A. W. A., et al. 2010. "A sero-survey of subtype H3 influenza A virus infection in dogs and cats in Japan." *J Vet Med Sci* Dec 7, Epub ahead of print, PMID 21150133.

van Riel, D., et al. 2010. "Highly pathogenic influenza H7N7 isolated from a fatal human case causes respiratory disease in cats but does not spread systematically." *Am J Pathol* 177(5):2185–2190.

Resources

Primary Literature and Reviews

Abbott, A. 2001. "The flu HQ." *Nature* 414:10–11.

Amorim, M. J., and Digard, P. 2006. "Influenza A virus and the cell nucleus." *Vaccine* 24(44-46):6651–6655.

Auewarakul, P., et al. 2007. "An avian influenza H5N1 virus that binds to human-type receptor." *Virology* 81:9950–9955.

Basler, C. F., et al. 2001. "Sequence of the 1918 pandemic influenza virus nonstructural gene (NS) segment and characterization of recombinant viruses bearing the 1918 NS genes." *PNAS USA* 98:2746–2751.

Boffey, P. M. 1976. "Swine flu vaccination campaign: The scientific controversy mounts." *Science* 193:559–563.

Burnet, F. M. 1935. "Propagation of the virus of epidemic influenza on the developing egg." *Med J Aust* 2: 687–689.

Butler, D. 2010. "Portrait of a year-old pandemic." *Nature* 464:1112–1113.

Centers for disease control and prevention (CDC). 2001. "Notice to readers: considerations for distinguishing influenza-like illness from inhalational anthrax." *MMWR* 50:984–986.

Centers for disease control and prevention (CDC). 2002. "Prevention and control of influenza: recommendations of the advisory committee on immunization practices." *MMWR* 51(No. RR-3):1–31.

Centers for disease control and prevention (CDC). 2003. "Severe acute respiratory syndrome (SARS) and coronavirus testing—United States." *MMWR* 52: 297–302.

Chen, H., et al. 2005. "H5N1 virus outbreak in migratory waterfowl." *Nature* 436:191–192.

Chen, W., et al. 2001. "A novel influenza A virus mitochondrial protein induces cell death." *Nat Med* 7: 1306–1312.

Childs, R. A., et al. 2009. "Receptor-binding specificity of pandemic influenza A (H1N1) 2009 virus determined by carbohydrate microarray." *Nat Biotechnol* 27:797–799.

Cross, K. J., Burleigh, L. M., and Steinhauer, D. A. 2001. "Mechanisms of cell entry by influenza virus." *Expert Rev Mol Med* 6:1–18.

Cyranoski, D. 2005. "Deadly flu virus can be sent through the mail." *Nature* 437:1077.

Dankbar, D. M., et al. 2007. "Diagnostic microarray for influenza B viruses." *Anal Chem* 79(5):2084–2090.

Dawson, E. D., et al. 2007. "Identification of A/H5N1 influenza viruses using a single gene diagnostic microarray." *Anal Chem* 79(1):378–384.

Dawson, E. D., et al. 2006. "MChip: A tool for influenza surveillance." *Anal Chem* 78(22):7610–7615.

Enserink, M. 2007. "From two mutations, an important clue about spanish flu." *Science* 315:582–583.

Eyler, J. M. 2009. "The fog of research: Influenza vaccine trials during the 1918–19 pandemic." *J Hist Med* 64(4): 401–428.

Fouchier, R. A., et al. 2005. "Characterization of a novel influenza A virus hemagglutinin subtype (H16) obtained from black-headed gulls." *Virology* 79(5): 2814–2822.

Garman, E., Laver, G. 2004. "Controlling influenza by inhibiting the virus's neuraminidase." *Curr Drug Targets* 5:119–136.

Gernhart, G. 1999. "A forgotten enemy: PHS's fight against the 1918 influenza pandemic." *Pub Health Chron* 114: 559–561.

Ghedin, E., et al. 2005. "Large-scale sequencing of human influenza reveals the dynamic nature of viral genome evolution." *Nature* 437:1162–1166.

Gibbs, M. J., Armstrong, J. S., and Gibbs, A. J. 2001. "Recombination in the hemagglutinin gene of the 1918 Spanish flu." *Science* 293:1842–1845.

Hatta, M., Gao, P., Halfmann, P., and Kawaoka, Y. 2001. "Molecular basis for high virulence of Hong Kong H5N1 influenza A viruses." *Science* 293:1840–1842.

Hatta, M., and Kawaoka, Y. 2002. "The continued pandemic threat posed by avian influenza viruses in Hong Kong." *Trends Microbiol* 10(7):340–344.

Hoffman, E., et al. 2000. "A DNA transfection system for generation of influenza A virus from eight plasmids." *PNAS USA* 97:6108–6113.

Hollingsworth, D., Ferguson, N. M., and Anderson, R. M. 2007. "Frequent travelers and rate of spread of epidemics." *EID* 13(9):1288–1294.

Holmes, E., Taubenberger, J. K., and Grenfell, B. T. 2005. "Heading off an influenza pandemic." *Science* 309(5737):989.

Kash, J. C., et al. 2006. "Genomic analysis of increased host immune and cell death responses induced by 1918 influenza virus." *Nature* 443:578–581.

Klenk, H. D., et al. 2011. "Molecular mechanisms of interspecies transmission and pathogenicity of influenza viruses: Lessons from the 2009 pandemic." *Bioessays* 33(3):180–188.

Kobasa, D., et al. 2007. "Aberrant innate immune response in lethal infection of macaques with the 1918 influenza virus." *Nature* 445(7125):267–268.

Kuiken, T., et al. 2004. "Avian H5N1 influenza in cats." *Science* 306:241.

Lessler, J., et al. 2009. "Outbreak of 2009 pandemic influenza A (H1N1) at a New York City school." *N Engl J Med* 361(27):2628–2636.

Lin, B., et al. 2007. "Using a resequencing microarray as a multiple respiratory pathogen assay." *J Clin Microbiol* 45(2):443–452.

Liu, Q., et al. 2007. "Microarray-in-a-tube for detection of multiple viruses." *Clin Chem* 53:188–194.

Liu, Y., et al. 2010. "Altered receptor specificity and cell tropism of D222G hemagglutinin mutants isolated from fatal cases of pandemic A (H1N1) 2009 influenza virus." *J Virol* 84(22):12069–12074.

Longini, I. M., et al. 2005. "Containing pandemic influenza at the source." *Science* 309:1083–1087.

Loo, Y.-M., and Gale Jr., M. 2007. "Fatal immunity and the 1918 virus." *Nature* 445:267–268.

Lowen, A. C., et al. 2007. "Influenza virus transmission is dependent on relative humidity and temperature." *PloS Pathog* 3(10):1470–1476.

Mehlmann, M., et al. 2007. "Comparison of the MChip to viral culture, reverse transcription PCR and the QuickVue® influenza A+B test for rapid diagnosis of influenza." *J Clin Microbiol* 45(4):1234–1237.

Montalto, N. J. 2003. "An office-based approach to influenza: Clinical diagnosis and laboratory testing." *Am Fam Phys* 67:111–118.

Moscora, A. 2005. "Neuraminidase inhibitors for influenza." *N Engl J Med* 353(13):1363–1373.

Moser, T. 2010. "2009 H1N1 influenza pandemic." *J Pediatr Health Care* 24(4):258–266.

Munier, S., et al. 2010. "A genetically engineered waterfowl influenza virus with a deletion in the stalk of the neuraminidase has increased virulence for chickens." *J Virol* 84(2):940–952.

Nayak, D. P., Hui, E. K., and Barman, S. 2004. "Assembly and budding of influenza virus." *Virus Res* 106(2): 147–165.

Neumann, G., Whitt, M. A., and Kawaoka, Y. 2002. "A decade after the generation of a negative-sense RNA virus from cloned cDNA—What have we learned?" *J Gen Virol* 83:2635–2662.

Neumann, G., et al. 2009. "Emergence and pandemic potential of swine-origin H1N1 influenza virus." *Nature* 459:931–939.

Neumann, G., Kawaoka, Y. 2011. "The first influenza pandemic of the new millennium." *Influenza Other Respi Viruses* 5(3):157–166.

Obenauer, J. C., et al. 2006. "Large scale sequence analysis of avian influenza isolates." *Science* 311(5767): 1576–1580.

Osterholm, M. T. 2005. "Preparing for the next pandemic." *N Engl J Med* 352(18):1839–1842.

Oxford, J. S., Novelli, P., Sefton, A., and Lambkin, R. 2002. "New millennium antivirals against pandemic and epidemic influenza: The neuraminidase inhibitors." *Antivir Chem Chemother* 13:205–217.

Plotkin, S. A., and Orenstein, W. A., eds. 1999. *Vaccines*, 3rd ed. Philadelphia: W. B. Saunders.

Reid, A. H., Taubenberger, J. K., and Fanning, T. G. 2004. "Evidence of an absence: The genetic origins of the 1918 pandemic influenza virus." *Nat Rev Microbiol* 2: 909–914.

Richman, D. D., Whitley, R. J., and Hayden, F. G., eds. 2002. *Clinical Virology,* 2nd ed. Washington: ASM Press.

Schnitzler, S. U., and Schnitzler, P. 2009. "An update on swine-origin influenza virus A/H1N1: A review." *Virus Genes* 39(3):279–292.

Schultz-Cherry, S., and Jones, J. C. 2010. "Influenza vaccines: The good, the bad, and the eggs." *Adv Virus Res* 77:63–84.

Shaman, J., et al. 2010. "Absolute humidity and the seasonal onset of influenza in the continental United States." *PLoS Biol* 8(2):e1000316.

Sharp, P. A. 2005. "1918 flu and responsible science." *Science* 310:17.

Shih, S.-R., and Krug, R. M. 1996. "Surprising function of the three influenza viral polymerase proteins: Selective protection of viral mRNAs against the cap—snatching reaction catalyzed by the same polymerase proteins." *Virology* 226:430–435.

Smee, D. F., et al. 2010. "Combinations of oseltamivir and peramivir for the treatment of influenza A (H1N1) virus infections in cell culture and mice." *Antiviral Res* 88:38–44.

Smith, K. 2007. "Concern as revived 1918 flu virus kills monkeys." *Nature* 445:237.

Sriyal, J., et al. 2009. "Innate immune responses to influenza A H5N1: Friend or foe?" *Cell* 30(12):574–584.

Steinhauer, D. A. 1999. "Role of hemagglutinin cleavage for the pathogenicity of influenza virus." *Virology* 258:1–20.

Stevens, J., et al. 2006. "Glycan microarray technologies: Tools to survey host specificity of influenza viruses." *Nat Rev Microbiol* 4:857–864.

Stiver, G. 2003. "The treatment of influenza with antiviral drugs." *CMAJ* 168(1):49–56.

Subbarao, K., et al. 1998. "Characterization of an avian influenza A (H5N1) virus isolated from a child with a fatal respiratory illness." *Science* 279:393–396.

Taubenberger, J. K., et al. 1997. "Initial characterization of the 1918 Spanish influenza virus." *Science* 275: 1793–1796.

Taubenberger, J. K., et al. 2005. "Characterization of the 1918 influenza virus polymerase genes." *Nature* 437: 889–893.

Taubenberger, J. K., and Morens, D. M. 2006. "1918 influenza: The mother of all pandemics." *EID* 12(1): 15–22.

Thomas, J. K., and Noppenberger, J. 2007. "Avian influenza: A review." *Am J Health Syst Pharm* 64(2):149–165.

Townsend, M. B., et al. 2006. "Experimental evaluation of the FluChip diagnostic microarray for influenza virus surveillance." *J Clin Microbiol* 44(8):2863–2871.

Tumpey, T. M., et al. 2002. "Existing antivirals are effective against influenza viruses with genes from the 1918 pandemic virus." *PNAS USA* 99:13849–13854.

Tumpey, T. M., et al. 2005. "Characterization of the reconstructed 1918 Spanish influenza pandemic virus." *Science* 310:77–80.

Tumpey, T. M., et al. 2007. "A two-amino acid change in the hemagglutinin of the 1918 influenza virus abolishes transmission." *Science* 315:655–659.

Von Bubnoff, A. 2005. "The 1918 flu virus is resurrected." *Nature* 437:794–795.

Vu, H. T., et al. "Clinical description of a completed outbreak of SARS in Vietnam, February–May 2003." *EID* 10(2):334–338.

Walsh, K. B. et al. 2011. "Suppression of cytokine storm with a spingosine analog provides protection againsts pathogenic influenza virus." *PNAS.*

Watanabe, T., Kawaoka, Y. 2011. "Pathogenesis of the 1918 influenza virus." *PLoS Pathogens* 7(1):e1001218.

Webster, R. G. 1998. "Influenza: An emerging disease." *EID* 4:436–441.

Webster, R. G., et al. 2006. "H5N1 outbreaks and enzootic influenza." *EID* 12(1):3–8.

World Health Organization (WHO). 1980. "A revision of the system of nomenclature for influenza viruses: A WHO memorandum." *Bull WHO* 58(4):585–591.

Wrammert, J., et al. 2011. "Broadly cross-reactive antibodies dominate the human B cell response against 2009 pandemic H1N1 influenza virus infection." *J Exp Med* 208(1):181–194.

Writing committee of the WHO consultation on clinical aspects of pandemic (H1N1) virus infection. 2010. "Clinical aspects of pandemic 2009 influenza A (H1N1) virus infection." *N Engl J Med* 362(18):1708–1719.

Yang, Y., et al. 2007. "Detecting human-to-human transmission of avian influenza A (H5N1)." *EID* 13(9): 1348–1353.

Yewdell, J., and Garcia-Sastre, A. 2002. "Influenza virus still surprises." *Curr Opin Micro* 5:414–418.

Popular Press

Barry, J. 2004. *The Great Influenza: The Epic Story of the Deadliest Plague in History.* New York: Viking Adult.

Branswell, H. 2011. "Flu factories: The next pandemic virus may be circulating on U.S. pig farms, but health officials are struggling to see past the front gate." *Sci Am* January:47–51.

Collier, R. 1996. *The Plague of the Spanish Lady: The Influenza Pandemic of 1918–1919.* London: Allison & Busby.

Crosby, A. W. 1989. *America's Forgotten Pandemic. The Influenza of 1918.* Cambridge, England: University Press.

Davies, P. 2000. *Devil's Flu: The World's Deadliest Influenza Epidemic and the Scientific Hunt for the Virus That Caused It.* New York: Henry Holt.

De Serres, G., et al. 2010. "Contagious period for pandemic (H1N1) 2009." *EID* 16(5):783–788.

Fanning, P. J. 2010. *Influenza and Inequality: One Town's Tragic Response to the Great Epidemic of 1918.* Amherst, MA: University of Massachusetts Press.

Iezzoni, L. 2000. *Influenza 1918: The Worst Epidemic in American History.* New York: TV Books.

Johnson, N. 2006. *Britain and the 1918–1919 Influenza Pandemic: A Dark Epilogue (Routledge Studies in the Social History of Medicine).* London: Routledge.

Klenk, H. D., et al. 2010. "Molecular mechanisms of interspecies transmission and pathogenicity of influenza viruses: Lessons learned from the 2009 pandemic." *Bioessays* 33(3):180–188.

Kolata, G. 1999. *Flu: The Story of the Great Influenza Pandemic of 1918 and the Search for the Virus That Caused It.* New York: Farrar, Straus, & Giroux.

Munier, S., et al. 2010. "A genetically engineered waterfowl influenza virus with a deletion in the stalk of the neuraminidase has increased virulence for chickens." *J Virol* 84(2):940–952.

Normand, A. 2006. *The Return of the Spanish Lady: The 1918 Influenza Virus is Back.* Bloomington, IN: Authorhouse.

North, S. 2006. *Another Place to Die.* Morrisville, NC: Lulu Enterprises.

Quammen, D. 2007. "Deadly contact: How animals and humans exchange disease." *National Geographic* October: 78–105.

Quinn, R. 2008. *Flu: A Social History of Influenza.* New Holland Publishers.

Rice, G. 2005. *Black November: The 1918 Influenza Pandemic in New Zealand.* Christ-church, New Zealand: Canterbury University Press.

Video Productions

Contagion. 2011. Warner Brothers.

Killer Virus: Hunt for the Next Plague. 2009. Discovery Channel.

H1N1. Aired November 1, 2009. 60 Minutes, CBS Broadcasting Inc.

H1N1. Aired October 18, 2009. 60 Minutes, CBS Broadcasting Inc.

Wild Goose Chase. 2007. Mike Rowe Dirty Jobs, Discovery Channel.

Spanish Flu: H5N1's Deadly Ancestor. 2007. Films for the Humanities.

NOVA: Epidemic—Ebola Aids Bird Flu & Typhoid. 2007. WGBH Boston.

Spanish Flu: H5N1's Deadly Ancestor. 2007. Films for the Humanities.

Mega Disasters: Alien Infection. 2007. History Channel, A&E Production.

60 Minutes—Chasing the Flu. 2006. CBS Broadcasting Inc.

Influenza. 2006. Films for the Humanities & Sciences.

Outbreak in America: When the Flu Pandemic Hits Home. 2006. Films for the Humanities.

Outbreak in Asia: How a Pandemic Is Born. 2006. Films for the Humanities.

Influenza, An American Experience. DVD released 2006. Robert Kenner Production, PBS.

Charlie Rose with Sir Paul Nurse, Michael Leavitt, Julie Gerberding, Harvey Fineberg, Peter Palese, David Nabarro, and Laurie Garrett. 2006. Charlie Rose, Inc. Studio.

Fatal Contact: Bird Flu in America. 2006. Sony Pictures.

H5N1: Killer Flu. 2005. Films for the Humanities.

Ready or Not: The Next Big One. Aired September 15, 2005. ABC News Primetime.

R_x for Survival: Bird Flu—How Safe Are We? 2005. WGBH Boston.

1918. 2004. Directed by Ken Harrison. Image Entertainment.

Killer Flu. 2004. Discovery Channel.

Secrets of the Dead: Killer Flu. 2004. PBS. Bill Lyons Production.

History Undercover: The Doomsday Flu. 2001. History Channel Production.

Investigative Reports: Hunt for the Killer Flu. 2000. A & E Production.

eLearning

go.jblearning.com/shors2

The site features eLearning, an online review area that provides quizzes and other tools to help you study for your class. You can also follow useful links for in-depth information, or just find out the latest virology and microbiology news.

CHAPTER 13

Rabies

What do Edgar Allen Poe, Louis Pasteur, and Old Yeller have in common?

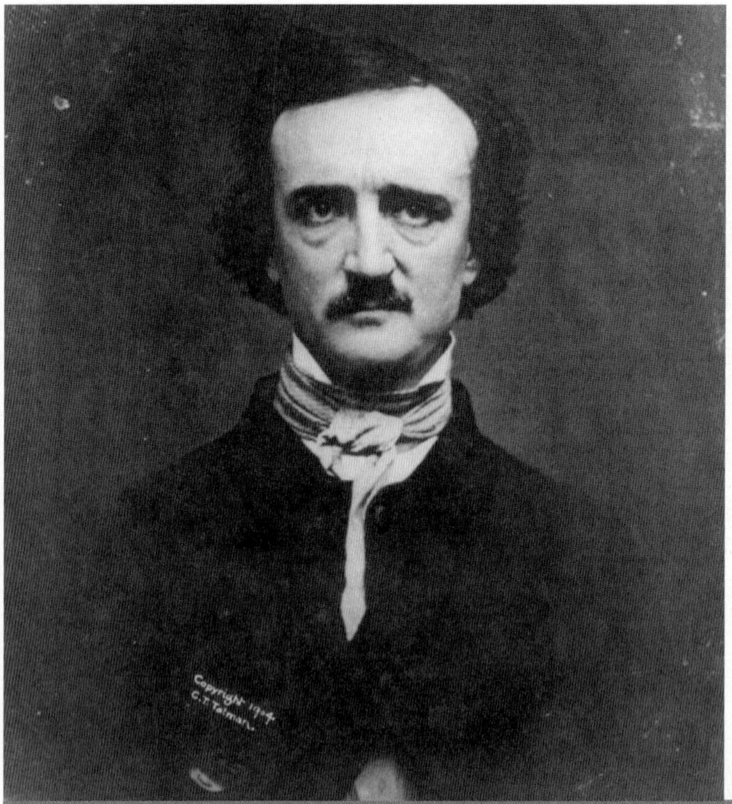

"TRUE!—nervous—very, very dreadfully nervous I had been and am; but why will you say that I am mad?"

Edgar Allan Poe, The Tell-Tale Heart

OUTLINE

CASE STUDY: POE'S MYSTERIOUS DEATH

In 1996 a group of physicians at the University of Maryland's College of Medicine gathered for their weekly clinical pathologic conference. Each physician was given an anonymous, complex case that was presented without diagnosis. Their role in the exercise was to discuss the patient's condition and course of treatment. One of the participants, Dr. R. Micheal Benitez, was (blindly) provided with the case of Edgar Allan Poe, which until that time had remained a mystery.

The basics of Poe's case were that he was found semiconscious on October 3, 1849, on a wooden plank that lay across two barrels located outside of Ryan's saloon on Lombard Street in Baltimore. He was wearing someone else's clothes, having been robbed of his suit. Poe had been traveling from Virginia to Baltimore on business. Finding him in this predicament was probably not a huge surprise, as he had a history of alcohol and opiate abuse. He was taken to Washington College Hospital (later renamed Church Hospital), and his final four days were described in the medical record as follows:

- When Poe arrived at the hospital, he was unresponsive.
- At 3 A.M. he regained consciousness. He began to perspire heavily, was delirious, and had tremors and visual hallucinations for the next 28 hours.
- On the third day, he was calm, alert, and lucid. The tremors had disappeared.
- When queried, Poe said he had a headache, some abdominal discomfort, and felt terrible in general.
- He was transferred to a ward room because his condition had vastly improved.
- Physicians offered him alcohol (to relieve possible alcohol-withdrawal symptoms), but he refused to drink it. He was hydrophobic and swallowed a little bit of water with great difficulty.
- That evening Poe relapsed into a delirious state, becoming violent.
- He was restrained and lapsed into a coma until his death.
- Poe died on October 7, 1849, his fourth day of hospitalization. The Baltimore Commissioner of Health recorded "congestion of the brain" as his cause of death on the death certificate.
- There was no autopsy.

Based upon the historical evidence, Benitez concluded that rabies was the likely cause of Poe's death and that he may have contracted rabies from one of his many pets. Benitez had analyzed other possible causes for delirium, such as trauma, epilepsy, cancer, and alcohol and opiate withdrawal, but none of these maladies were consistent with Poe's symptoms, and there were accounts that Poe had not been drinking for six months prior to his death. At the conference Benitez commented, "It is also unusual for a patient to recover for a brief time and then worsen and die." Poe's symptoms were more consistent with a central nervous system (CNS) infection, such as that of rabies, rather than drug or alcohol abuse. The average survival after the onset of rabies symptoms at that period was four days—the same number of days Poe had been hospitalized before his death.

Benitez ruled out other potential reasons for Poe's delirium, such as metabolic, endocrine, nutritional, poisoning, or infectious diseases that could cause delirium (e.g., yellow fever or malaria). Poe's hydrophobic symptoms are a hallmark of this case. We may never know if this analysis is correct; however, Benitez's theory is plausible.

13.1 History of Rabies

Rabies, which was known as **hydrophobia** (fear of water) in the 1800s, is a terrifying disease. It affects the CNS, resulting in delirium and hydrophobia, and causes aggression or stupor.

French scientist Louis Pasteur (1822–1895) made numerous contributions to the field of microbiology. He was trained as a chemist and then applied his training to microbiology applications. He isolated and identified microorganisms responsible for the fermentation of wine, beer, and vinegar. His most important discovery, however, became the foundation for the science of microbiology. His "germ theory of disease," which proposes that infectious diseases are caused by germs, has become a cornerstone of modern medicine.

Pasteur's pioneering work regarding the development of vaccines against chicken cholera, anthrax, and swine erysipelas (a superficial bacterial infection that can spread to the lymphatics) allowed him to apply his concepts toward the treatment and prevention of rabies.

Pasteur and his colleague Emile Roux realized that conquering rabies would be a significant achievement for science and of great benefit to the public at large. They determined that the rabies agent existed in the spinal cord and began to make an inactivated vaccine by drying strips of spinal cords that were removed from rabid rabbits. After four years of intensive research, they created a vaccine consisting of dried spinal cord tissue injected under the skin of dogs that was used to induce immunity successfully in 1885. Fifty dogs vaccinated and challenged with rabies survived.

Pasteur was, however, cautious in using the rabies vaccine on humans, and insisted that additional years of research were needed before the treatment could be applied to humans. The press had picked up quickly on his successful vaccination of dogs, though, and on July 6, 1886, Pasteur's first human subject arrived at his laboratory.

Two days earlier, nine-year-old Joseph Meister had been mauled and bitten 14 times by a rabid dog. Meister's mother appealed to Pasteur for the rabies treatment. After consulting his colleagues Pasteur treated the boy, inoculating him 13 times over a 15-day period with the dried spinal cord suspension. The child survived and remained in fine health. Months later, another victim showed up at Pasteur's lab, followed by still others.

Pasteur's success made him a hero and a legend. In 1887 the Pasteur Institute was founded to treat rabies victims. Today it is widely respected for its research in the field of microbiology. More recent discoveries at the Pasteur Institute include the discovery of HIV-1 and HIV-2 (1983 and 1985), a vaccine against *Helicobacter pylori* (the cause of many peptic ulcers and gastric cancers; 1991), the complete genome sequence of *Mycobacterium tuberculosis* (the causative agent of tuberculosis; 1998), and the discovery of the nicotine receptor responsible for the pain-relieving activity of nicotine (1999). The Institute's website is http://www.pasteur.fr/english.html.

13.2 Epidemiology

Rabies is one of the oldest known viral diseases. The word rabies originated in 3000 B.C. in Sanskrit, the official language of India. *Rabhas* ("rabies") means "to do violence." In Latin the word rabies is derived from *rabere*, which means "to rave." The Italian scholar Girolamo Fracastoro (1478–1553) first described the rabies disease in 1546 (350 years before Louis Pasteur and Emile Roux developed the rabies vaccine). His description of classic rabies symptoms was as follows:

> *The patient can neither stand nor lie down, like a mad man he flings himself hither and thither, tears his flesh with his hands, and feels intolerable thirst. This is the most distressing symptom, for he so shrinks from water and all liquids that he would rather die than drink or be brought near to water. It is then they bite other persons, foam at the mouth, their eyes look twisted, and finally they are exhausted and painfully breathe their last.*

In the 19th century, canine or street rabies was a scourge everywhere, especially in Europe. Rabies victims killed themselves or were killed after a rabid dog bite. The image of the mad dog has symbolized man's fear of rabies for centuries (**FIGURE 13-1**). Even though there are vaccines available today, rabies remains a significant wildlife management and public health challenge. Education is essential in preventing this disease.

Rabies epidemiology is extremely complex and differs greatly from one continent to another. Rabies surveillance programs are used to address the following questions:

- What animals have rabies and in what regions of a country?
- How many people get rabies and from what animals did they contract it?
- What are the best strategies to prevent rabies in humans and animals?

Rabies in Animals

Rabies is a disease of mammals. Rabies in the United States is primarily a disease that affects wildlife populations. Worldwide, it is particularly a disease in wild dogs, coyotes, foxes, raccoons, skunks, mongooses, and bats. The disease is less frequent in domesticated dogs, cats, cows, horses, pigs, sheep, and goats. **FIGURE 13-2** is a composite showing the numbers of animal reservoirs in the world in 2003. Between 1000 and 5000 animal cases were reported in the former Soviet Union, parts of Asia, South America, and Europe in the year 2000.

Control of canine (dog) rabies in Asia, Africa, parts of South America, and Mexico has proved difficult. Foxes are an important rabies vector in Europe, Canada, Alaska, and the former Soviet Union. The raccoon dog, which is native from Siberia to Japan and in northern India, is also an important rabies reservoir in Europe. Raccoon dogs were introduced into Europe because of the commercial value of their long, thick fur and have now spread across Europe.

Mongooses, which were introduced into the Caribbean islands for rat control, are now an important rabies reservoir there as well as in their native southern Africa. The vampire bat is a threat

FIGURE 13-1 This 1826 cartoon depicts citizens equipped with weapons to protect themselves and remove the rabid dog roaming the streets of London, England. A street peddler falls to avoid the dog while women run into hiding.

to cattle and humans in Mexico and parts of South America. Insectivorous bats in North America are widely infected with the rabies virus.

Rabies has been extensively studied in Thailand. These studies have shown that dogs account for 95% of rabies cases, cats account for 3% of the cases, and rats, rabbits, civet cats, monkeys, tigers, squirrels, and other animals are responsible for the remaining 2% of cases—evidence that rabies virus is highly infectious among mammals.

Before 1992, Ontario, Canada, was known as the rabies capital of North America. Since the imple-

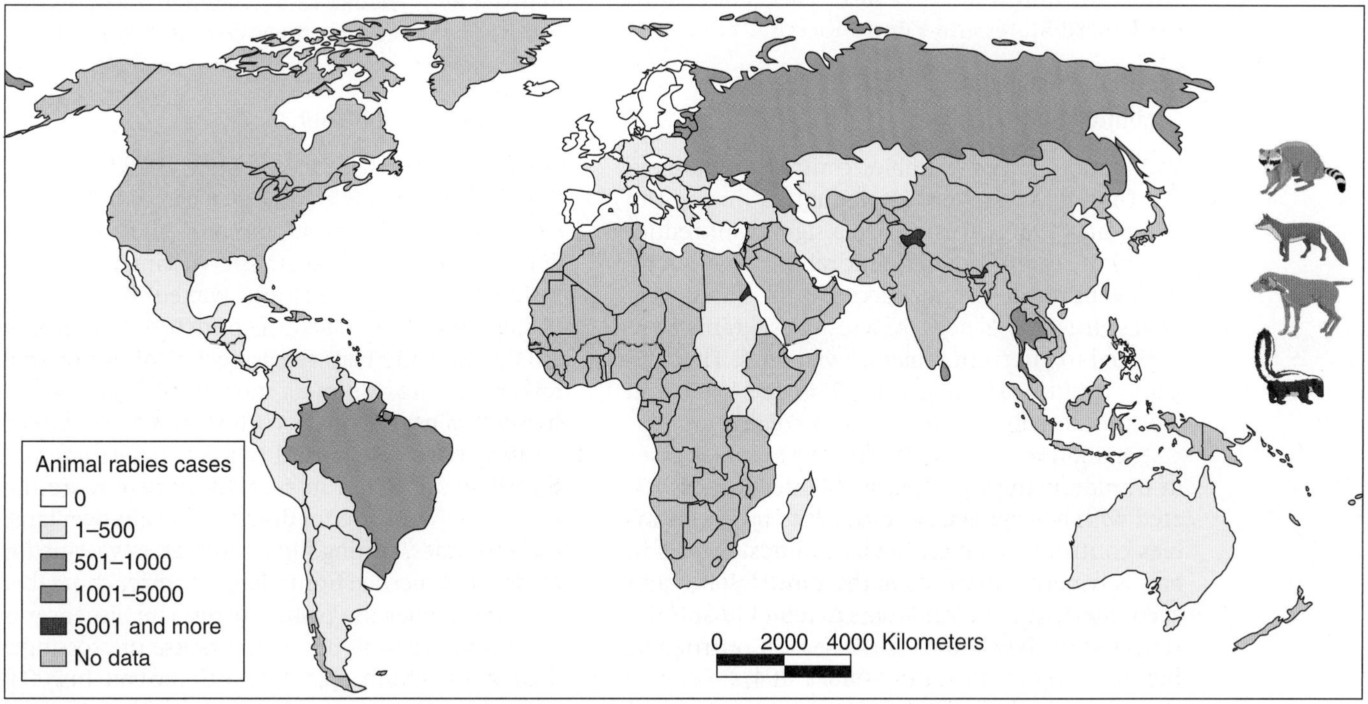

FIGURE 13-2 Rabies in animals: indicator status worldwide in 2000. Adapted from World Health Organization, 2005

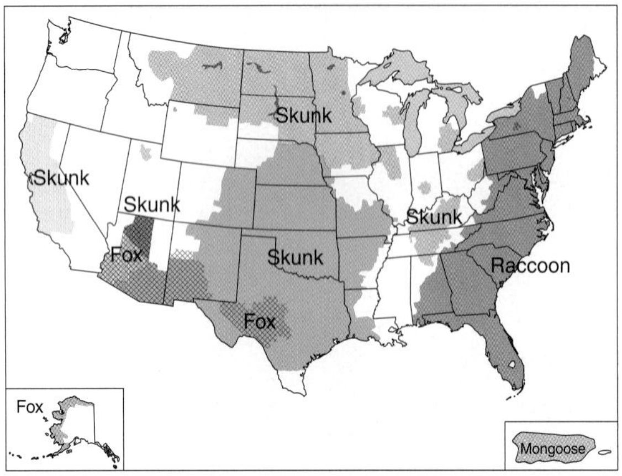

FIGURE 13-3 Distribution of Major Terrestrial Reservoirs of Rabies in the United States. Adapted from the CDC, "Terrestrial Rabies Reservoirs in the United States, 2009," http://www.cdc.gov/rabies/resources/publications/2009-surveillance/reservoirs.html.

mentation of rabies control programs, though, the number of wildlife rabies cases has been reduced by 95%. The programs focused on the vaccination of arctic foxes and mid-Atlantic raccoons in addition to an educational focus regarding various bat strains.

The distribution of wild animal rabies in the United States in 2009 is shown in **FIGURE 13-3**. Wildlife in the United States reported to be the most commonly infected with rabies are raccoons, skunks, foxes, coyotes, and bats. The spatial boundaries of wildlife rabies are temporally dynamic. **TABLE 13-1** displays the breakdown of wildlife, domestic, and human rabies cases reported in the United States and Puerto Rico in 2001.

TABLE 13-1 Cases of Rabies in the United States and Puerto Rico Reported to the CDC in 2006

Domestic Animals	Number of Rabies Cases
Dogs	79
Cats	318
Cattle	82
Horses/mules	53
Sheep/goats	11
Other domestics*	4
Total of all domestic animals	**547**
Wild Animals	**Number of Rabies Cases**
Skunks	1,494
Raccoons	2,615
Foxes	427
Bats	1,692
Rodents/lagomorphs**	44
Other wild animals***	121
Total wild animals	**6,393**
Humans	3

*Other domestics include ferrets and llamas.

**Rodents and lagomorphs include groundhogs, beavers, chipmunks, rabbits, badgers, bobcats, deer, opossum, mongooses, otters, coyotes, and ringtails.

***Other wild animals includes bobcats, cougars, deer, opossum, otters, coyotes, fishers, mongooses, and ringtails.

Rabies Management Programs in the United States

Vaccination and other rabies-control programs in the United States were implemented during the 1940s and 1950s. This caused a significant reduction in the number of domestic rabies cases, but by the 1960s it still had not succeeded in eliminating rabies virus in wild dogs. As a result programs were initiated to interrupt rabies in wild dogs. The programs continued through the 1970s and 1980s and did substantially reduce the spread of dog rabies.

A small but increasing number of rabies cases in human beings resulted from infection associated with bats, which are difficult to target for rabies control by conventional methods. Rabies in bats is widely distributed in the United States and accounted for 17% of all cases of rabies in animals reported to the Centers for Disease Control and Prevention (CDC) in 2009 (**FIGURE 13-4**).

The Wildlife Services (WS) program—part of the U.S. Department of Agriculture—is responsible for providing assistance in wildlife disease management (**FIGURE 13-5**). In 2004, 15 states distributed oral vaccines for raccoons; Texas distributed baits for gray fox and coyote as well.

Oral Rabies Vaccine

The oral vaccine used to vaccinate wildlife populations is combined with an outer bait matrix that is composed of fishmeal (for raccoons and coyotes) or dog food (for gray foxes). The vaccine is inserted inside of a small packet and waxed into the bait (**FIGURE 13-6**). The packet is punctured when the animal bites the bait, releasing 1.5 mL of the oral rabies vaccine. Animals mount an immune response against the rabies virus within 2 to 3 weeks.

Baits are distributed by hand in urban and suburban areas, but in rural areas planes are the most effective means for distributing baits over large vaccine zones. During flight, oral rabies vaccine for wildlife is dispensed from a bait dispenser chute that is located under the plane's wing. The dispenser is turned off when flying over a house or crossing a road to avoid human contact with the bait. In 2003, more than 10 million baits were distributed in the United States and Canada.

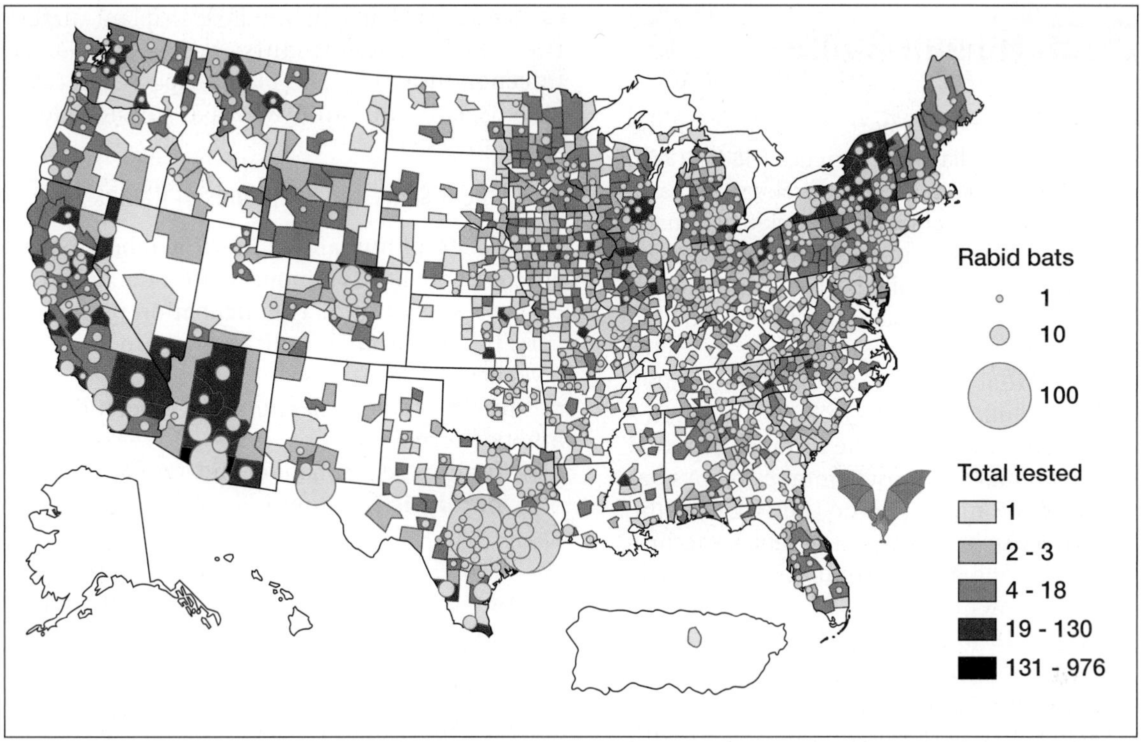

FIGURE 13-4 Distribution of bat rabies in the United States in 2009. Bat rabies has been reported in 47 of the 48 contiguous states. Adapted from the CDC, "Rabid animals reported in the United States during 2009," http://www.cdc.gov/rabies/resources/publications/2009-surveillance/rabid-animals.html.

FIGURE 13-5 Biologists collect blood and other biological samples to evaluate the effectiveness of the oral rabies vaccine before releasing the animals back into the wild at the site of capture.

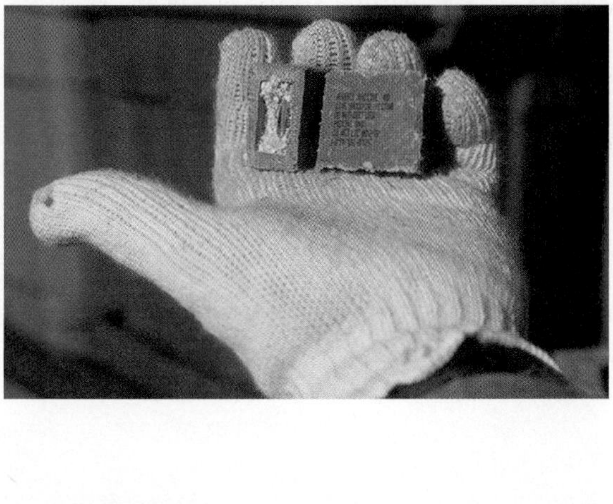

FIGURE 13-6 Merial Raboral V-RG rabies vaccine bait. The bait consists of a fishmeal polymer cube that is hollow. A plastic packet containing the vaccine is inserted into the hollow area of the bait and sealed with wax**.** Planes are the most effective means for distributing baits over large-scale oral rabies vaccine zones. During flight, oral rabies vaccine for wildlife is dispensed from a chute located under the aircraft's wing.

13.3 Human Rabies

Human rabies has been reported in all parts of the world except Australia and Antarctica (**FIGURE 13-7**). The dog is the global reservoir of rabies, and as a result the majority of rabies cases outside of Canada and the United States occur in countries where dog rabies is not well controlled. The World Health Organization (WHO) estimates that over 10 million humans are vaccinated postexposure to rabies annually and that there were more than 55,000 rabies-related human deaths in 2004. This number represents a fraction of the number of actual cases. Human rabies is most common in boys between the ages 5 and 14. In the United States the highest incidence of postexposure rabies vaccination occurs among boys in rural areas, primarily during the summer months. Many baby-boomers recall the 1957 Disney movie *Old Yeller,* which was based on a novel by Fredrick B. Gipson. The story takes place in Texas in 1869. Two boys, Travis and Arliss Coates, live on a remote homestead with their mother. The boys befriend a stray yellow mongrel dog they name Old Yeller. The dog proves his loyalty by saving Mrs. Coates from a rabid wolf but contracts rabies itself. The viewers watch the terrifying, raving transformation of the boys' beloved pet. In the end, Travis is forced to shoot his dog (thus traumatizing several generations of movie-watching children).

Since 1990 over 92% of human rabies cases in the United States were associated with bat bites. Human-to-human transmission by bite is extremely rare. Besides bat bites, there have been cases associated with direct implantation of infected organs during human transplant surgery (i.e., corneal transplants; see Section 6.1). Infection by aerosol has been suspected in caves heavily contaminated with bat guano and in laboratories.

Clinical Signs and Symptoms of Rabies

About 75% of humans contracting rabies experience encephalitis symptoms within the first 20 to 60 days of exposure. The incubation period can vary, though, from as little as 4 days to as long as 6 years (1%–3% of cases).

Human rabies infections have been divided into two forms:

- **furious** (encephalitic) and
- **paralytic** (or dumb).

With either form, the average course of the disease runs from 2 to 14 days before coma supervenes. Death occurs an average of 18 days after the onset of symptoms. Symptoms during the primary or prodromal period of rabies are nonspecific and are similar to those of other viral infections:

- headache
- malaise
- fever
- anorexia
- nausea
- vomiting

Secondary symptoms of furious rabies are:

- hydrophobia
- difficulty swallowing
- agitation
- anxiety
- hallucinations

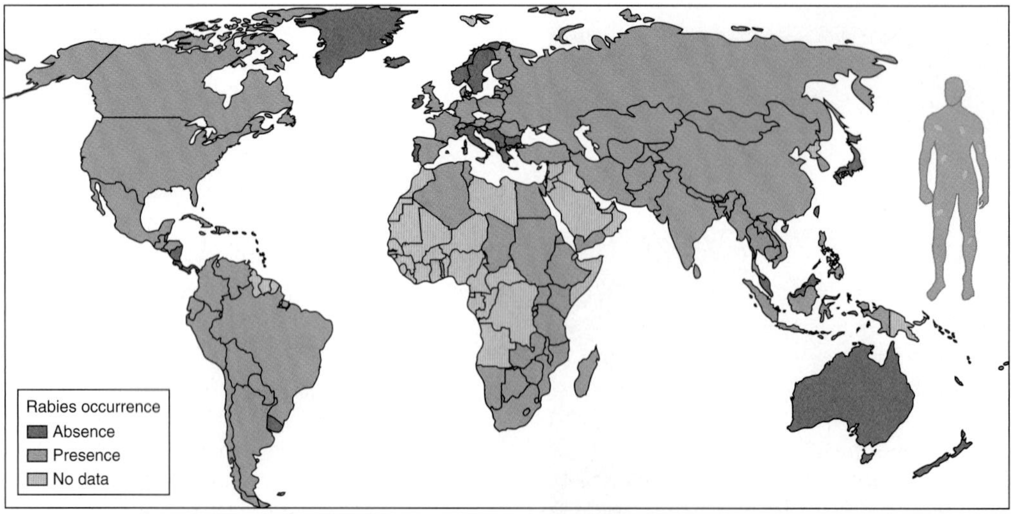

FIGURE 13-7 Total number of human rabies deaths in 2005, including imported cases in humans and animals. Adapted from the World Health Organization, 2005.

FIGURE 13-8 Here are two dogs afflicted with dumb rabies, manifested as depression, and an attempt at self-imposed isolation. Domesticated animals afflicted with dumb rabies may become increasingly depressed and try to hide in isolated places, whereas wild animals seem to lose their fear of human beings, often appearing unusually friendly.

- hypersalivation
- bizarre behavior
- biting
- jerky and violent contractions of the diaphragm

Secondary symptoms of paralytic rabies include:

- lack of hydrophobia
- lack of hyperactivity
- lack of seizures
- weakness and ascending paralysis

These symptoms also apply to animals. Animals that are predators show signs of furious rabies, whereas nonpredators display paralytic or dumb rabies symptoms (**FIGURE 13-8**).

Rabies Diagnosis: Rabies Diagnostic Testing in Animals

All individuals involved in rabies testing are required to receive preexposure immunization followed by regular serologic tests and booster immunizations when necessary. In addition, all laboratory workers must participate in national rabies proficiency testing, available through the Wisconsin State Laboratory of Hygiene in Madison, Wisconsin.

In the United States, direct fluorescent antibody test (dFA) is the gold standard for rabies diagnosis in animals. Animals that are behaving abnormally or show signs consistent with rabies should be tested when human or animal exposures have occurred. The test is performed postmortem.

Rabies virus is present in nervous tissues and, thus, the brain is the ideal test tissue (not blood, as is the case with many other viruses). Two or three areas of the animal's brain are tested: the brain stem (medulla), which merges with the spinal cord, and the cerebellum or the hippocampus. A piece of each tissue is removed and placed on an impression slide. The tissues are then air dried and fixed in acetone before being incubated with labeled antibodies (antirabies N protein labeled with a fluorescent dye [fluorescein-isothiocynate, or FITC]). After incubation, the slides are washed and viewed under a fluorescence microscope in a dark room. dFA testing is rapid (30 minutes to less than 4 hours). The results may save a patient from the unnecessary physical, emotional, and financial burdens if the animal is not rabid.

Rabies Diagnosis in Humans

Laboratory tests for rabies are rarely conducted in developing countries. There are several routine tests performed in the United States on humans antemortem (before the death of the patient). Saliva, tears, and cerebral spinal fluid (CSF) samples are used for virus isolation in tissue culture, or reverse-transcriptase polymerase chain reaction (RT-PCR) is used to detect viral RNA. Serum and CSF are immediately tested for antibodies to rabies virus in unvaccinated patients. Rabies antibodies usually appear during the second week of the illness. Skin punch biopsies at the nape of the neck are examined for rabies antigen that may be present in the cutaneous nerves at the base of the hair follicles.

Postmortem dFA tests are performed on the brain of the victim. Long-needle biopsies on two or more samples of the brain (e.g., the brain stem and cerebellum) are tested by RT-PCR, virus isolation in cell culture, or suckling-mouse inoculation. Further histologic, immunocytochemistry, and electron microscopic (**FIGURE 13-9**) examination of autopsy tissues may also be performed.

Refresher What Is Encephalitis?

Encephalitis occurs when brain tissue becomes infected and inflamed. Viruses are often the cause. Encephalitic symptoms include fever, headache, stiff neck, nausea, and vomiting. Symptoms worsen to seizures, paralysis, blurred vision, delirium, and coma. Several different viruses can cause encephalitis, such as rabies, West Nile virus, and herpes simplex virus.

Differential diagnosis is a systematic method used by physicians to diagnose a disorder; for example, rabies symptoms are similar to other viral encephalitis disorders such as herpes simplex encephalitis. A physician must determine which virus is causing encephalitis (**FIGURE RB 13-1**).

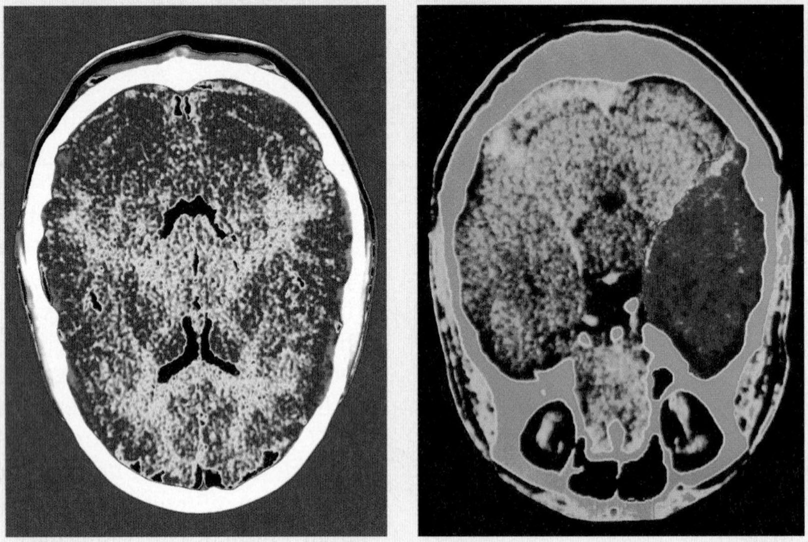

FIGURE RB 13-1 Computed tomographic (CT) scans of the human brain: Healthy brain versus brain with herpes encephalitis. Encephalitis results in the inflammation of the encephalon (red spot).

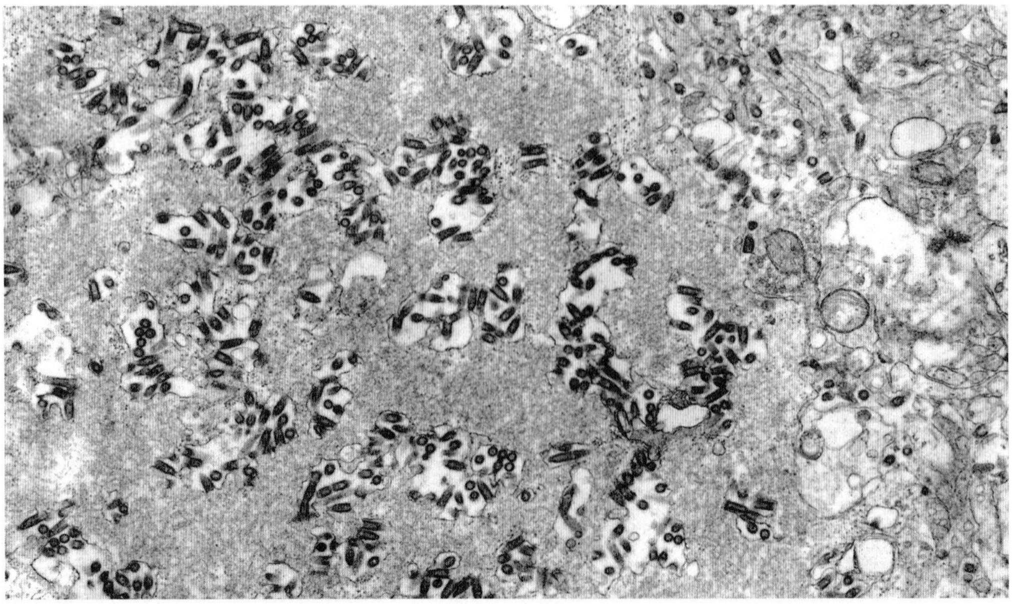

FIGURE 13-9 Rabies infection in the brain. This micrograph covers just part of the cytoplasm of an infected neuron. Two hallmarks of rabies virus infection are seen: there is minimal damage to the structure of infected neurons even though the extent of the infection is dramatic, and large numbers of bullet-shaped virions accumulate as a result of budding upon the endoplasmic reticulum membranes of these cells. Magnification approximately 25,000×.

Survivors

Only seven individuals who had received pre- or postexposure rabies prophylaxis (passive antibody, immune globulin, or vaccine) have survived. **TABLE 13-2** lists the individuals who have recovered from rabies encephalitis. Half of the survivors had severe, permanent neurological disorders.

Despite efforts to treat patients with steroids, interferon, and other antivirals, there is no therapy of proven value. Heavy sedation pain medication is used to relieve the patient of agonizing symptoms. Supportive care has prolonged life but there is no expectation of survival in unvaccinated patients (**FIGURE 13-10**). The first individual who survived rabies without receiving immune prophylaxis was a 15-year-old girl from Fond du Lac, Wisconsin, in 2004 (**VIRUS FILE 13-1**). She was treated with ketamine, midazolam, and phenobarbital to induce coma and two antivirals: ribavirin and amantidine. After 31 days she was removed from an isolation ward and was discharged to her home after 76 days. Five months after hospitalization, she was alert and communicative but with uncontrolled movements (choreothetosis), dysarthria (a speech condition), and an unsteady gait. Physicians who followed up on her progress after 27 months recorded some fluctuating gait and speech difficulties. She has had no problems with daily activities, including driving a car. She graduated from college and is currently employed. **TABLE 13-3** lists some of the patients treated by the Milwaukee protocol or variations of it.

Pasteur described "spontaneous recovery" in a dog that showed signs of rabies, and there have been a few other reports of laboratory animals recovering from rabies infections; for example, in 1980, Fekadu and Baer reported the recovery of two beagles that had been infected with an Ethiopian strain of street (wild) rabies virus.

Pathogenesis

The most common way that rabies virus enters the body is through a break in the skin. The virus replicates in muscle cells surrounding the wound. At some point during the incubation period, the virus attaches to nicotinic acetylcholine receptors of the peripheral nerve cells of the neuromuscular junction. The virus journeys within the axons of the nerves at a rate of 0.5 to 15 inches per day by retrograde axonal transport (**FIGURE 13-11**). It may take days or weeks before the virus reaches the spinal cord motor neurons of the CNS. This is why postexposure prophylaxis is possible. Timely administration of antirabies immunoglobulin or vaccine prevents the spread of the virus to the CNS, thereby stopping the disease.

During the course of infection, the virus replicates in the motor neurons of the spinal cord as it makes its rapid ascent to the brain. While in the neurons, the virus is hidden from the body's immune surveillance. The virus spreads to many tissues via peripheral nerves (e.g., the cardiac muscle, adrenal gland, kidney retina, pancreas, cornea, nerves surrounding skin follicles, and salivary glands). Budding of infectious particles from infected cells takes place predominately in the salivary glands, excreting virus particles that are transmissible to other mammals. Upon biopsy or autopsy, there is little evidence of brain damage even though fluorescent antibody labeling and electron microscopic analysis of brain tissues shows that almost all of the neurons are infected. The neural dysfunction of this disease may be mediated through the inhibition of neurotransmitters.

Immunity

There are no antibody responses detected during the incubation period of rabies virus infection. Once neurological symptoms develop, antibodies appear in serum and rarely in the CSF suggesting limited penetration into the CNS where neutralizing antibodies to the virus are needed the most. The demonstration of antibodies is only useful during the late stages of the disease. Some patients mount a cell-mediated response. Those patients experience the encephalitic (furious) form of the disease. Rabies virus infection can inhibit the innate immune response, especially by inhibiting type I and II interferon signaling. The expression of the rabies virus phosphoprotein is critical for the prevention of type I and II interferon responses. The induction of apoptosis by rabies virus has been a controversial topic, but increasing evidence suggests that pathogenic rabies virus strains do not induce apoptosis.

13.4 Management of Human Rabies

Preexposure Prophylaxis

All high-risk individuals should be vaccinated against rabies virus. Approximately 40,000 individuals in the United States receive preexposure vaccinations. Examples of high-risk individuals are:

- veterinarians
- animal handlers, such as zookeepers
- dog catchers
- mail carriers
- speleologists and spelunkers

TABLE 13-2 Human Rabies Survivors

Year	Age (yrs)	Gender/ Country	Exposure	Prophylaxis	Incubation Period	Symptoms	Recovery
1970	6	Male/Ohio, USA	Left thumb bite by rabid bat	Duck embryo vaccine the next day and 13 additional doses	20 days	Encephalitis, seizures, paralysis, coma	**Complete recovery** in 6 months
1972	45	Female/ Argentina	Bite by rabid dog	Suckling-mousebrain rabies vaccine 10 days after bite	21 days	Headache, loss of motor strength, tremors, cerebellar dysfunction	**Complete recovery** 13 months after onset of rabies symptoms; relapsed twice after vaccine boosters
1977	32	Male/ New York, USA	Inhaled aerosol of laboratory strain of rabies virus (SAD strain)	Preexposure duck-embryo rabies vaccine*	14 days	Fever, nausea, encephalitis, respiratory arrest	**Partial recovery**; dementia, personality disorder
1985	10	Male/India	Stray dog bite, lower left leg	Semple's vaccine (phenol-inactivated nerve tissue)	120 days	Fever, pain, hydrophobia, anxiety, photophobia, aerophobia	**Complete recovery**; case was severe, required 16 days of respiratory support
1986	18	Male/India	Stray dog bite, lower left leg	Semple's vaccine (phenol-inactivated nerve tissue)	60 days	Fever, pain, hydrophobia, anxiety, photophobia, aerophobia, alternating drowsiness with violent behavior	**Complete recovery**; patient also had cardiac irregularities and needed oxygen for 3 days
1992	9	Male/ Mexico	Bite on forehead, nose, and left cheek, by rabid dog	Vero-cell vaccine the next day and at days 3, 7, 14, and 30	15 days	Encephalitis, fever, convulsions, coma, paralysis	**Partial recovery**; blind and deaf, neurological dysfunction, died 34 months later
2000	6	Female/ India	Bites on hands and face by rabid dog	Three doses of purified chick embryo at days 0, 3, and 7; no wound treatment; 1 dose of human diploid cell vaccine on day 2 after hospital admission during symptoms	16 days	Encephalitis, fever, difficulty swallowing, photophobia, visual hallucinations, hypersalivation, frothing at the mouth, profuse sweating	**Partial recovery**; child was comatose for 3 months; responding to simple verbal commands and oral feeding at the end of 5 months; rigidity and tremors of limbs; discharged after 6 months with follow-ups every 3 months, which indicated little improvement
2004	15	Female/ Wisconsin, USA	Bite by rabid bat	**No vaccine or immune globin**; instead patient was given ketamine, midazolam, ribavirin, and amantidine after onset of symptoms	21 days	Double vision, fatigue, tingling and numbness of left hand, nausea and vomiting, blurred vision, nerve palsy, muscular twitching, tremors, fever, hypersalivation	**Partial recovery**; was able to return to school. Some long-term effects.

*Six months prior to this episode, the individual had a neutralizing antibody titer of 1:32.

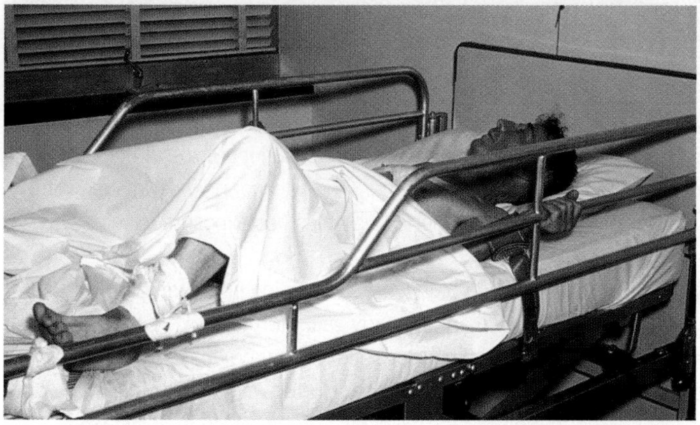

FIGURE 13-10 In 1958, this patient was presented with early, though progressive symptoms due to what was confirmed as rabies.

- trappers and hunters in areas where animal rabies is enzootic
- certain laboratory workers, such as researchers who work with live rabies virus or are involved in the production of rabies vaccines
- individuals who come in frequent contact with potentially rabid animals
- Peace Corps workers
- international travelers who go to countries that have endemic dog rabies and lack immediate access to medical care

The shot schedule for preexposure vaccination is given in three 1-mL shots injected intramuscularly into the forearm or shoulder on days 0, 7, and 21 or 28. Individuals who work with live rabies virus through research or vaccine production must be tested every six months for rabies antibodies. A booster will be administered if necessary. The vaccine usually takes 7 to 14 days to induce immunity, which lasts approximately two years.

Wound Care

It was common practice until the middle of the 20th century to cauterize wounds caused by rabid animals. Today wounds are washed thoroughly with a solution of 20% soap, 70% ethanol, or 2% benzalkonium chloride and are irrigated with a virucidal agent such as povidone-iodine. All bite victims should seek medical attention and should supply the medical attendant with the following information:

- the animal involved
- whether the bite was provoked or unprovoked
- the vaccination status of the animal, if known
- whether or not the animal can be safely captured for rabies testing
- the geographic location of the incident

If the animal showed signs of rabies and can be captured, the animal is killed and the head is shipped to a qualified laboratory for testing.

Postexposure Prophylaxis

Immune Globulin

Unvaccinated individuals in the United States who have experienced a rabid animal bite are injected with human rabies immune globulin (HRIG, Hyperab) on day zero, along with a shot of vaccine. Rabies immune globin consists of antibodies from blood donors given rabies vaccine. It provides passive immunity until active antibodies are produced from the vaccine. It is injected at the bite site and in the buttocks. HRIG is expensive and is not available in all countries. Many countries use rabies immune globin of equine (horse) origin (ERIG). It is considerably cheaper and current preparations are highly purified and safe.

Postexposure Vaccination

There has been a pressing need for a safe and effective vaccine throughout history. Rabies vaccines can be divided into three different categories: nerve tissue vaccines, avian embryo vaccines, and cell culture vaccines. **TABLE 13-4** lists important past and current vaccines for humans.

Early vaccines made from nerve tissues produced serious adverse side effects, such as Guillain Barré-like syndrome, paralysis, autoimmune CNS disease, and meningoencephalitis (inflammation of the brain and meninges). The incidence of neurological effects produced by the Semple-type vaccine is as high as 1 in 200 in recipients, with a lethality of up to 14%. Regrettably, nerve tissue vaccines are still being used for rabies prophylaxis in Asia, South America, and Africa.

Vaccines have been continuously improved over the years. Many people remember the 1950s

VIRUS FILE 13-1 Why Did Jeanna Giese Survive?

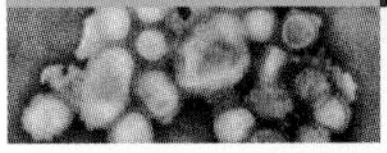

Fifteen-year-old Jeanna Giese was attending a church service in September, 2004, when a bat struck an interior window of the church. She picked up the bat and released it outside of the building. While handling the bat, she was bitten on her left index finger. The bite wound was washed with hydrogen peroxide, but she did not receive medical treatment. No rabies postexposure vaccine was administered.

About one month after the bat bite, she complained of fatigue and tingling in her left hand. Over the course of the next three days, she developed an unsteady gait, blurred and double vision, nausea, and vomiting. She was referred to a neurologist. A magnetic resonance image (MRI) of her brain was normal. By the fifth day, she had a fever. Her speech was slurred and she experienced tremors in her left arm. She was transferred to the Children's Hospital of Wisconsin in Milwaukee. Giese began salivating and became hydrophobic. A history of the bat bite was obtained. Samples of her serum, cerebral spinal fluid (CSF), skin biopsies, and saliva were sent to the CDC for the diagnosis of rabies. The CDC confirmed the presence of rabies-specific antibodies in her CSF and serum. A team of eight specialists developed a new protocol to treat Giese.

Jeanna Giese's parents were counseled about her diagnosis and prognosis (death). They were offered hospice care and an aggressive treatment that had never been tested. The treatment was based on a combined strategy that uses **anti-excitatory** and antiviral drugs, along with supportive care. The parents opted for the latter treatment.

Geise was intubated and put into a drug-induced coma to suppress brain activity. She was treated with a cocktail of ketamine, midazolam, and phenobarbitol. She also received two antivirals: **ribavirin** and **amantadine**. (Neither rabies vaccine nor rabies immune globulin was administered to Giese because she had already developed an antibody response toward the rabies virus.) During her comatose state, she was continuously monitored for anti-rabies antibodies and detection of rabies nucleic acid in saliva. After seven days, her coma (anti-excitatory) drugs were tapered and the antivirals were discontinued. She became increasingly alert. She was removed from isolation on the 31st day of hospitalization because her titers of neutralizing antirabies antibodies from her CSF and blood were high. No virus or viral nucleic acids were detected in her saliva. It was assumed that she had cleared the virus.

She was discharged from the hospital on the 76th day after rehabilitation (**FIGURE VF 13-1**). The total cost of her treatment was approximately $800,000. She graduated from high school in 2007. Two years after the bat bite, she was still experiencing some numbness of the bitten finger, an alteration in the tone of her left arm, and a wider gait when she runs. In 2011, with the exception of slower speech, she continues to improve. She remains an animal lover. She drives a car, has graduated from college, and works at a pet daycare, boarding, and grooming facility.

Jeanna Giese made history as being the only survivor of rabies who did not receive any postexposure prophylaxis. We do not know why Jeanna Giese survived rabies. It is not known whether her recovery was related to an infection by an attenuated bat rabies variant or her therapy. This hypothesis is difficult to test because no virus was recovered from Jeanna Giese's bodily fluids.

Giese's recovery drew global interest. It prompted the use of the "Milwaukee protocol" to treat at least 33 other patients (located in the U.S. [Texas, California, Indiana], Dominican Republic, The Netherlands, Brazil, Columbia, Peru, Canada [Alberta], Thailand, Equitorial

a

FIGURE VF 13-1 (a) Jeanne Giese, the world's first survivor of rabies infection.

(continued)

VIRUS FILE 13-1 Why Did Jeanna Giese Survive? (continued)

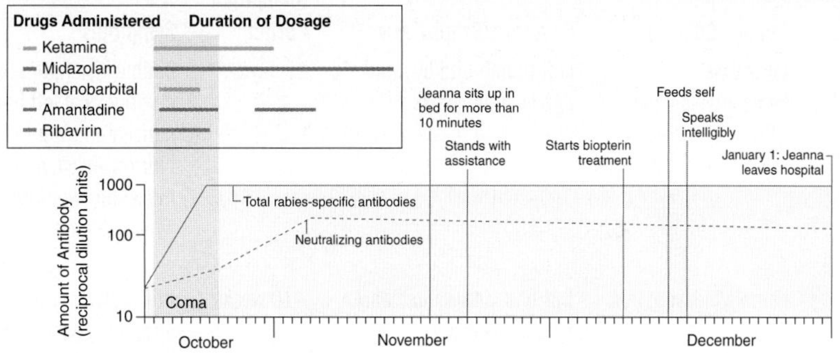

b

FIGURE VF 13-1 (Continued) **(b)** Timeline and treatment during Giese's hospital stay. It includes the incubation time after infection until two months after her discharge from the hospital. Adapted from Willoughby, R. E., Jr., et al., 2007. "A cure for rabies?" *Sci Am* March:89–95.

New Guinea, Qatar, Germany, Ireland, and India). Five additional patients at the time of this writing have survived besides the index patient (Geise). These survivors underwent modifications of the Milwaukee protocol. The original protocol and new versions can be found at: http://www.mcw.edu/rabies. Protocol modifications include the omission of ribavirin and the supplementation of tetrahydrobiopterin (also known as BH_4).

On a global scale, about 55,000 cases of human rabies occur each year. It is the number one fatal infectious disease in China due to rabid dog bites. Perhaps the best way to do follow-up research is to apply the Milwaukee protocol to rabid animals to systematically determine what parts of the protocol are critical in defeating rabies: the induced coma or the antivirals. Unfortunately, to date, veterinary schools have not permitted these studies because officials have raised concerns about treating rabies animals in their intensive care units.

References

Hemachudha, T., et al. 2006. "Failure of therapeutic coma and ketamine for therapy of human rabies." *J Neurovirol* 12:407–409.

Hunter, M., et al. 2010. "Immunological correlates in human rabies treated with therapeutic coma." *J Med Virol* 82:1255–1265.

McDermid, R. C., et al. 2008. "Human rabies encephalitis following bat exposure: Failure of therapeutic coma" *CMAJ* 178(5):557–561.

Rubin, J., et al. 2009. "Applying the Milwaukee protocol to treat canine rabies in Equatorial Guinea." *Scand J Infect Dis* 41:372–380.

Van Thiel, P-P A. M., et al. 2009. "Fatal human rabies due to Duvenhage virus from a bat in Kenya: Failure of treatment with coma-induction, ketamine, and antiviral drugs." *PLoS Negl Trop Dis* 3(7):e428.

Wilde, H. et al., 2008. "Viewpoint: Management of human rabies." *Trans R Soc Trop Med Hyg* 102:979–982.

Willoughby, R. E., Jr. 2007. "A cure for rabies?" *Sci Am* April issue:89–95.

Willoughby, R. E., Jr., et al. 2005. "Survival after treatment of rabies with induction of coma." *N Engl J Med* 35(24):2508–2514.

Willoughby, R. W., et al. 2007. "Long-term follow-up after treatment of rabies by induction of coma." *N Engl J Med* 357(9):945–946.

Willoughby, R. W., et al. 2008. "Generalized cranial artery spasm in human rabies." *Dev Biol (Basel)* 131:367–375.

Willoughby, R. E., Jr. 2009. "Are we getting closer to the treatment of rabies?" *Future Virol* 4(6):563–570.

Willoughby, R. E., Jr., et al. 2009. "Tetrahydrobiopterin deficiency in human rabies." *J Inherit Dis* 32:65–72.

Willoughby, R. E., Jr. 2009. "Early death and the contradiction of vaccine during treatment of rabies." *Vaccine* 27:7173–7177.

human rabies vaccine made from duck embryos, which required 21 painful shots in the abdomen.

Three vaccines are currently available in the United States: HDCV, RVA, and PCECV. Unvaccinated individuals who are exposed to rabies receive one shot of HRIG along with the administration of one dose of an available vaccine immediately after exposure (day 0). Subsequently, the individual should receive additional vaccine injections at days 3, 7, 14, and 28. Injections must be administered intramuscularly in the deltoid area.

13.5 The Rabies Virus Life Cycle

Classification of Virus

Rabies virus is classified as follows:

Order: Mononegavirales
Family: Rhabdoviridae
Genus: Lyssavirus

TABLE 13-3 Human Rabies Cases Treated with the Milwaukee Protocol*

Year	Age (yrs)	Gender/Country	Exposure/Prophylaxis	Incubation Period	Symptoms	Recovery
2004	15	INDEX CASE Female/Wisconsin, USA	Left thumb bite by rabid bat/no prophylaxis	21 days	Double vision, tingling and numbness of left hand, nausea and vomiting, blurred vision, nerve palsy, muscular twitching, fever, hypersalivation	Nearly full recovery after a year of extensive rehabilitation, survivor was able to attend college and drive a car; minor speech problems
2006	10	Female/Indiana, USA	Bat bite, details unclear/ no prophylaxis	~10 weeks	Irritability, slurred speech, difficulty swallowing and breathing, hypersalivation	*Milwaukee protocol implemented on the 5th day of hospitalization. Died on the 26th day of hospitalization
2006	11	Male/California, USA	Dog bite in Philippines, details unclear/ no prophylaxis	~2 years	Sore throat, fever, chest tightness, insomnia, agitation, irregular lip and mouth movements, hallucinations, hydrophobia, hypersalivation	*Milwaukee protocol started on the first day of hospitalization. Life support withdrawn on the 24th day of hospitalization
2007	5	Male, Equatorial New Guinea	Bite on neck by rabid dog/ none	5 weeks	Hydrophobia, aerophobia, agitation, anxiety, phonophobia, unsteady gait, drooling, combativeness	Survived 18 days after symptoms began; died of complications related to malnutrition
2008	15	Male, Brazil, South America	Bite by rabid vampire bat/ yes—vaccination after bit (4 of 5 shots administered)	1 month	Tremors, anxiey, agitation, hypersalivation	Partial recovery, discharged from hospital with good cognitive function, speaking, some motor limitations
2008	73	Male, Edmonton, Canada	Bat bite on a farm, left shoulder/immune globulin given but no vaccination	6 months	Irritability, lethargy, hypersalivation	*Milwaukee protocol started 15 days after onset of symptoms, death 65 days after onset of symptoms
2011	8	Female, California, USA	Scratch by a rabid feral cat		Paralysis, could not swallow, strong immune response to rabies virus	Milwaukee protocol started after onset of severe symptoms; expected to make a full or nearly full recovery

*The Milwaukee protocol involves induction of coma with ketamine, midazolam and phenobarbital, and antiviral therapy with ribavirin and amantadine. In some cases the protocol was modified.

The order Mononegavirales represents a group of viruses that contain nonsegmented single-stranded RNA (ssRNA) genomes of negative sense. ***Rhabdo*** is the Greek term for "rod-shaped;" ***lyssa*** in Greek means "frenzy or madness." Rabies was the first of seven *Lyssavirus* genotypes identified. All but one of them causes fatal encephalitis in humans. Most of these viruses have been found in bats; perhaps rabies virus is really a bat virus that has crossed species. New *Lyssaviruses* continue to be discovered.

Properties of the Rabies Virus Particle

Rabies virus particles and other members of the Rhabdovirus family have a unique bullet-shaped appearance, with one end flattened and the other rounded. The average length of infectious or mature rabies virus particles is 180 nm, and the average diameter is 75 nm. The surface of the virus particle is covered with approximately 400 tightly arranged glycoprotein spikes. The spikes are an average 7 nm in length by 3 nm in diameter. They

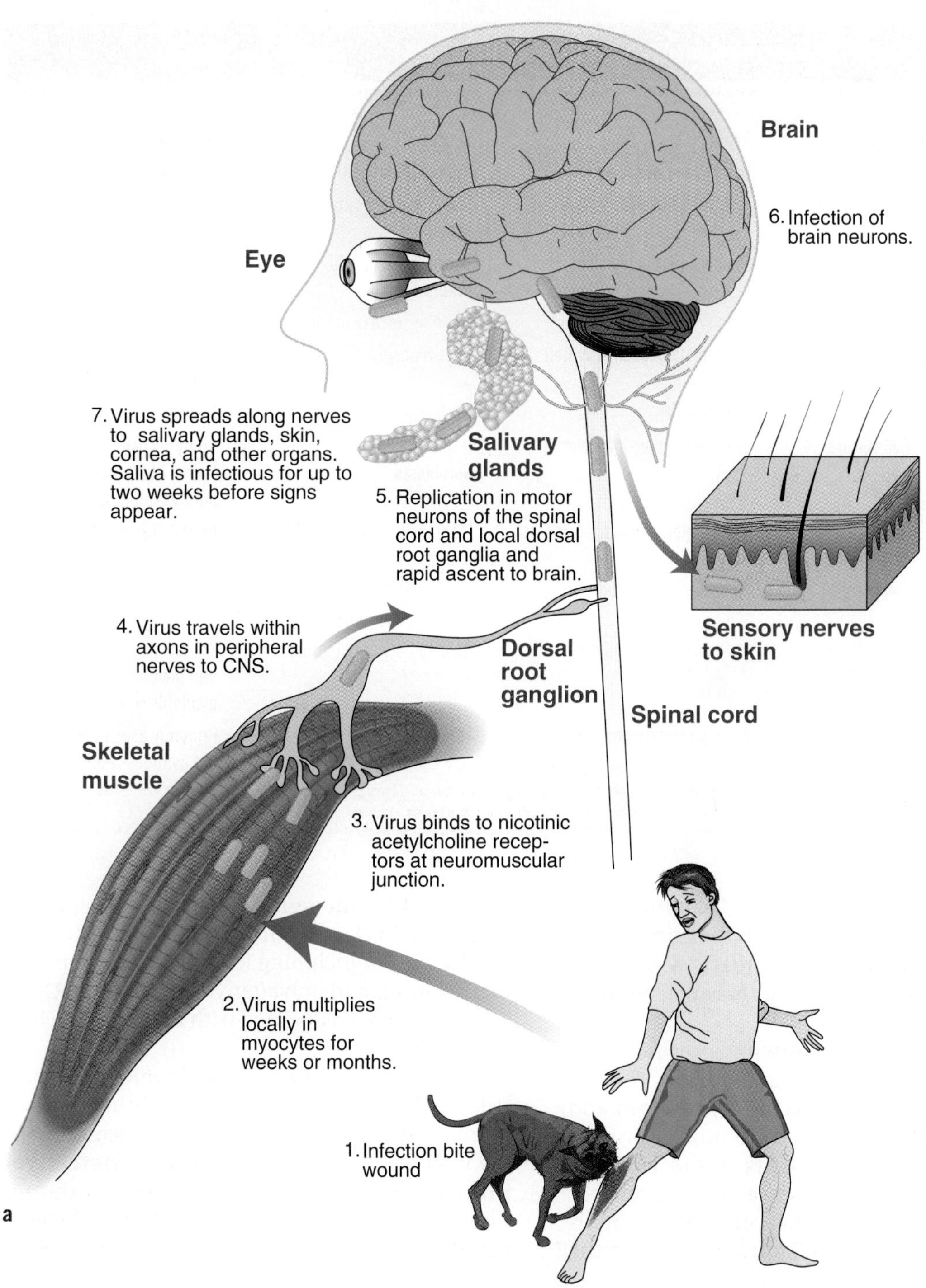

Brain
6. Infection of brain neurons.
Eye
7. Virus spreads along nerves to salivary glands, skin, cornea, and other organs. Saliva is infectious for up to two weeks before signs appear.
Salivary glands
5. Replication in motor neurons of the spinal cord and local dorsal root ganglia and rapid ascent to brain.
4. Virus travels within axons in peripheral nerves to CNS.
Dorsal root ganglion
Sensory nerves to skin
Spinal cord
Skeletal muscle
3. Virus binds to nicotinic acetylcholine receptors at neuromuscular junction.
2. Virus multiplies locally in myocytes for weeks or months.
1. Infection bite wound

a

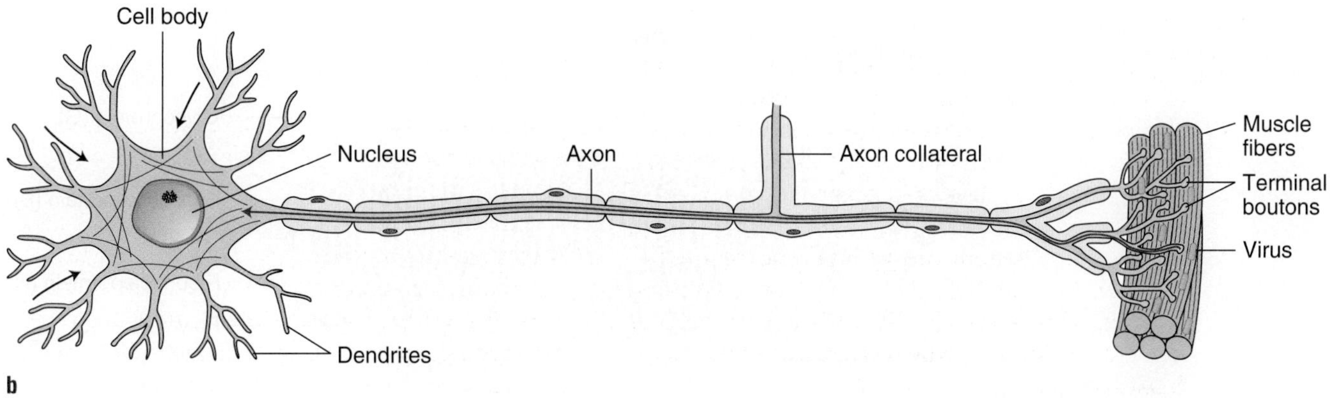

Cell body
Nucleus
Axon
Axon collateral
Muscle fibers
Terminal boutons
Virus
Dendrites

b

FIGURE 13-11 (a) Sequential steps of rabies transmission and pathogenesis after a dog bite in a lower limb of a human. **(b)** Rabies virus attaches to nicotinic acetylcholine receptors of nerve cells and journeys within the axons of the nerves. Adapted from Jackson, A. C., and Wunner, W. H., eds. *Rabies*. Academic Press, 2002.

TABLE 13-4 Important Past and Current Rabies Vaccines for Humans

Vaccine Name	Category	Type	Source	Where Used
Pasteur	Nerve tissue	Inactivated by drying	Rabbit spinal cord	No longer used
Fermi	Nerve tissue	Phenolized live virus	Sheep, goat, or rabbit brains	No longer used
Semple	Nerve tissue	Phenol-inactivated virus	Sheep, goat, or rabbit brains	Asia, Africa
Fuenzalida	Nerve tissue	Inactivated	Suckling-mouse brain	South America
PDEV	Avian	β-Propiolactone inactivated	Duck embryo	Primarily in Europe but available worldwide
DEV	Avian	Inactivated	Duck embryo	No longer used
HDCV	Cell culture	β-Propiolactone inactivated	Human-cultured fibroblasts	Primarily used in the United States and Europe but available worldwide
RVA	Cell culture	β-Propiolactone inactivated	Fetal rhesus lung cell culture	United States
PHK	Cell culture	Formalin inactivated	Primary Syrian hamster kidney cell culture	China, Russia
PCECV	Cell culture	β-Propiolactone inactivated	Chick embryo cell culture	Primarily used in Germany and the United States but available worldwide
PVRV	Cell culture	β-Propiolactone inactivated	Vero cell line	Primarily used in France but available worldwide

Abbreviations for vaccines: PDEV = purified duck embryo vaccine; DEV = duck embryo vaccine; HDCV = human diploid cell culture vaccine; RVA = rabies vaccine adsorbed; PHK = primary hamster kidney; PCECV = purified chick embryo cell culture vaccine; PVRV = purified Vero rabies vaccine.

are composed of trimers of viral glycoprotein (G) (**FIGURE 13-12**). The glycoprotein is the primary surface antigen that induces and reacts with virus-neutralizing antibodies. It is also associated with receptor activities.

Internal to the glycoprotein spikes is a host-derived envelope or membrane bilayer. The envelope is lined with a **matrix (M) protein** and contains the helical-shaped ribonucleoprotein core (RNP). The RNP consists of the viral genomic RNA that is tightly encased with nucleocapsid (N) proteins. It also contains two additional proteins, the phosphoprotein (P) and the catalytic RNA-dependent RNA polymerase or large protein (**L protein**). The L protein has several functions, including RNA synthesis, capping, methylation, and polyadenylation of viral RNAs. The M protein is associated with both the envelope and the RNP.

The standard bullet-shaped rabies virus particles are designated **B particles**. Other shorter, often cone-shaped, particles are coproduced during infection. They are noninfectious **defective-interfering (DI)** particles. They contain shorter (defective) RNA genomes. DI particles can be separated from B particles by centrifugation in a sucrose

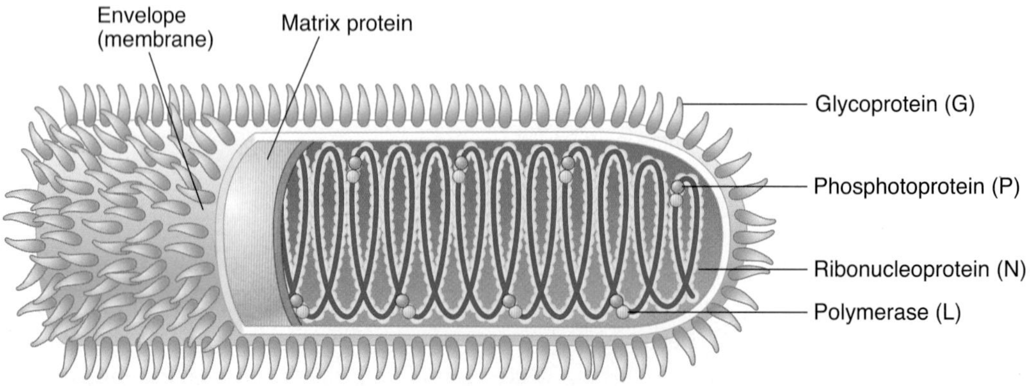

FIGURE 13-12 Longitudinal section of rabies virus. Adapted from the CDC.

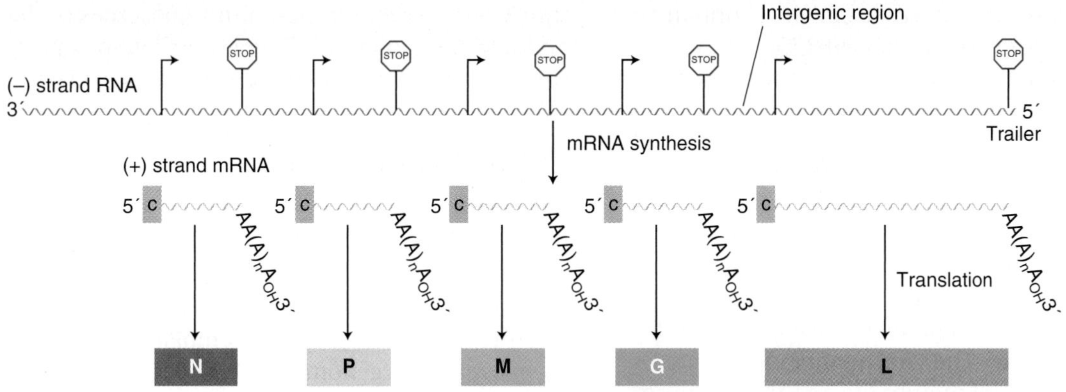

FIGURE 13-13 Genome organization of rabies virus. Adapted from Flint, S. J., et al. *Principles of Virology: Molecular Biology, Pathogenesis, and Control of Animal Viruses*, Second Edition. ASM Press, 2003.

density gradient. DI particles are readily generated in cell cultures infected with laboratory-adapted (fixed) strains of rabies virus. Rabies DI particles have been shown to be noninfectious in cell culture and in animals.

Viral Replication: Genome Organization

The rabies virus genome consists of an 11,932-nucleotide nonsegmented ssRNA genome of negative-sense polarity. The 3′ end of the genome contains a noncoding leader sequence. Downstream of the leader sequence are the five structural genes of the rabies virus genome: *N, P, M, G,* and *L*, followed by a noncoding trailer sequence at the 5′ end (**FIGURE 13-13**). The rabies virus gene functions are listed in **TABLE 13-5.** Each gene is separated by a short intergenic sequence. There are two nucleotides separating the *N* and *P* genes, five nucleotides separating the *P* and *M* genes, and five nucleotides separating the *M* and *G* genes. There is a larger intergenic sequence of 423 nucleotides between the *G* and *L* genes of rabies virus. It is long enough to represent a potential gene, but there is no detectable protein found that represents this coding region. It has been given the designation of a remnant gene or pseudogene (Ψ).

TABLE 13-5 Rabies Virus Gene Coding Assignments

Gene	Length	Function of Gene Product
N	1424	Coat protein (nucleocapsid); part of the RNP core
P	991	Polymerase-associated phosphoprotein (NS); part of the RNP core
M	805	Matrix protein; part of the RNP core
G	1675	Glycoprotein (G-protein), receptor-binding protein, major surface antigen
L	6475	Polyprotein with RNA replicase, 5′ capping, methylation, and 3′ polyadenylation activity

Attachment

The glycoprotein of rabies virus is of major importance immunologically and has been studied extensively for this reason. It is capable of inducing and reacting with virus-neutralizing antibodies. The rabies virus surface glycoprotein is responsible for binding (attachment) to its host receptor(s) and mediating pH-dependent membrane fusion with a host cell.

The rabies glycoprotein contains significant sequence homology to snake venom neurotoxins. Snake venom neurotoxins tightly bind nicotinic acetylcholine receptors and block their activity. Nicotinic acetylcholine receptors were first identified as a potential receptor for rabies virus because the amino acid sequence of rabies virus glycoprotein that is similar to snake neurotoxins was the region correlated with receptor binding. Nicotinic acetylcholine receptors are located at the postsynaptic muscle membrane but not at presynaptic nerve membranes. Other research suggests that rabies viruses initially replicate in muscle cells. In either situation, rabies virus enters neurons using neuronal cell adhesion molecule (NCAM) and another unknown receptor. Different mechanisms have been proposed for how rabies viruses get transported through the axon to the cell body of a neuron (Figure 13-13). None of the receptors discovered to date are essential for viral replication in cell cultures (*in vitro* experiments).

A single point mutation in the rabies virus glycoprotein gene results in an apathogenic (non-neurotropic) variant virus. If amino acid 333 (arginine or lysine) is changed to isoleucine or glutamine of the 505 amino acid protein, the virus is rendered avirulent. The spread of the virus in the CNS is reduced, making the virus significantly less neuroinvasive. Virulent strains of virus contain an arginine or lysine at amino acid 333. It has been postulated that rabies viruses may use more than one receptor or an additional **co-receptor**. More research is needed to understand the function of rabies virus receptors for viral attachment, entry, spread, and pathogenicity.

Rabies virus particles can be found in vesicles or clathrin-coated pits and at later time points of infection, in lysosomes. The virus enters nerve terminals via low pH-dependent fusion with endosomal membranes. The viral membrane then fuses with an endosomal membrane and the glycoprotein undergoes a conformational change at a low pH. The acidic interior of the endosome allows the viral nucleocapsid to escape into the cytoplasm. The M-protein dissociates from the RNP during this uncoating step.

There is no protein synthesis machinery available for the rabies virus in axons of neurons. For this reason, rabies virus must reach the neuronal cell body for replication, transcription, and protein synthesis. The mechanisms for the transport of rabies virus or capsids through the axon to the cell body are not completely understood.

Transcription of Viral Genes

During uncoating, the RNP is released into the host cell cytoplasm. The viral M protein dissociates from the nucleocapsid. This step is necessary for viral transcription to occur.

Direct translation of viral proteins via a viral mRNA cannot occur because the viral genome is negative sense RNA. The viral L polymerase and P proteins (which were packaged inside of the particle and therefore brought into the cell via the virion) begin to transcribe the viral genomic RNA within the cytoplasm of the cell. Transcription begins at the 3′ end of the viral RNA genome. The 5′ end-capped and polyadenylated mRNAs that encode the viral proteins are synthesized (Figure 13-13). The most popular model for transcription of rhabdoviruses is the stop-start model in which the viral polymerase stops transcription at a conserved signal sequence and ignores the intergenic region of 2–24 nucleotides and restarts at the transcription signal sequence. It transcribes the RNA sequentially (e.g., the leader sequence, *N, P, M, G,* and *L* genes). A separate positive-sense RNA transcript is generated for each viral gene. Each viral transcript is capped and polyadenylated by the viral L-protein.

Viral Protein Synthesis

The viral messenger RNAs (mRNAs) resemble cellular mRNAs and are translated by cellular ribosomes. The N, P, M, and L viral mRNAs are translated by free ribosomes, whereas the G-mRNA is translated on membrane-bound ribosomes of the endoplasmic reticulum. The glycoprotein is post-translationally modified (glycosylated) via the Golgi apparatus.

Viral Genome Replication

When sufficient N is available, the synthesis of full-length anti-genome begins instead of the production of individual viral mRNAs, switching from transcription to replication. The anti-genome or replicative intermediate RNA serves as a template for progeny (negative-sense) genomic RNA, which will be packaged into the viral particle. During infection, an excessive amount of genome to anti-genome is synthesized. Fifty copies of genomic RNAs are produced for each anti-genome produced.

Virus Assembly and Release

Rabies virus has evolved regulatory mechanisms that produce viral gene products at the optimum amounts necessary for efficient viral production but low enough not to be recognized by the immune system or to interfere with vital functions of its host cells. As soon as enough genomic negative-sense RNA, N, P, and L proteins have accumulated within the infected cells, rabies virus nucleocapsids are formed and virus assembly begins. The cytoplasmic M protein combines with the RNP complexes. The mature infectious bullet-shaped particles acquire the cellular lipid bilayer as the RNP-M buds through the host cell plasma membrane. **FIGURE 13-14** shows a complete rabies virus life cycle.

13.6 Genetic Variation

Fixed Versus Street Strains

The term *fixed* strain refers to strains of rabies virus used in research laboratories. *Street* (wild) strains are those strains isolated from patients or animals that have experienced the disease via a bite from a rabid animal or other means (e.g., via infected organ during a transplantation procedure). The rabies virus virion-associated RNA replicase/transcriptase (L protein) does not possess RNA repair/proofreading capabilities. Consequently, it is error prone. Mutations are introduced at a relatively high frequency into the rabies virus RNA genome. The mutation rate has been estimated to be in the order of 10^{-4} to 10^{-5} per nucleotide per cycle. Mutations may affect the survival of the infected cell or mammalian host. L protein may affect the tropism of the virus for neurons and be involved in viral persistence.

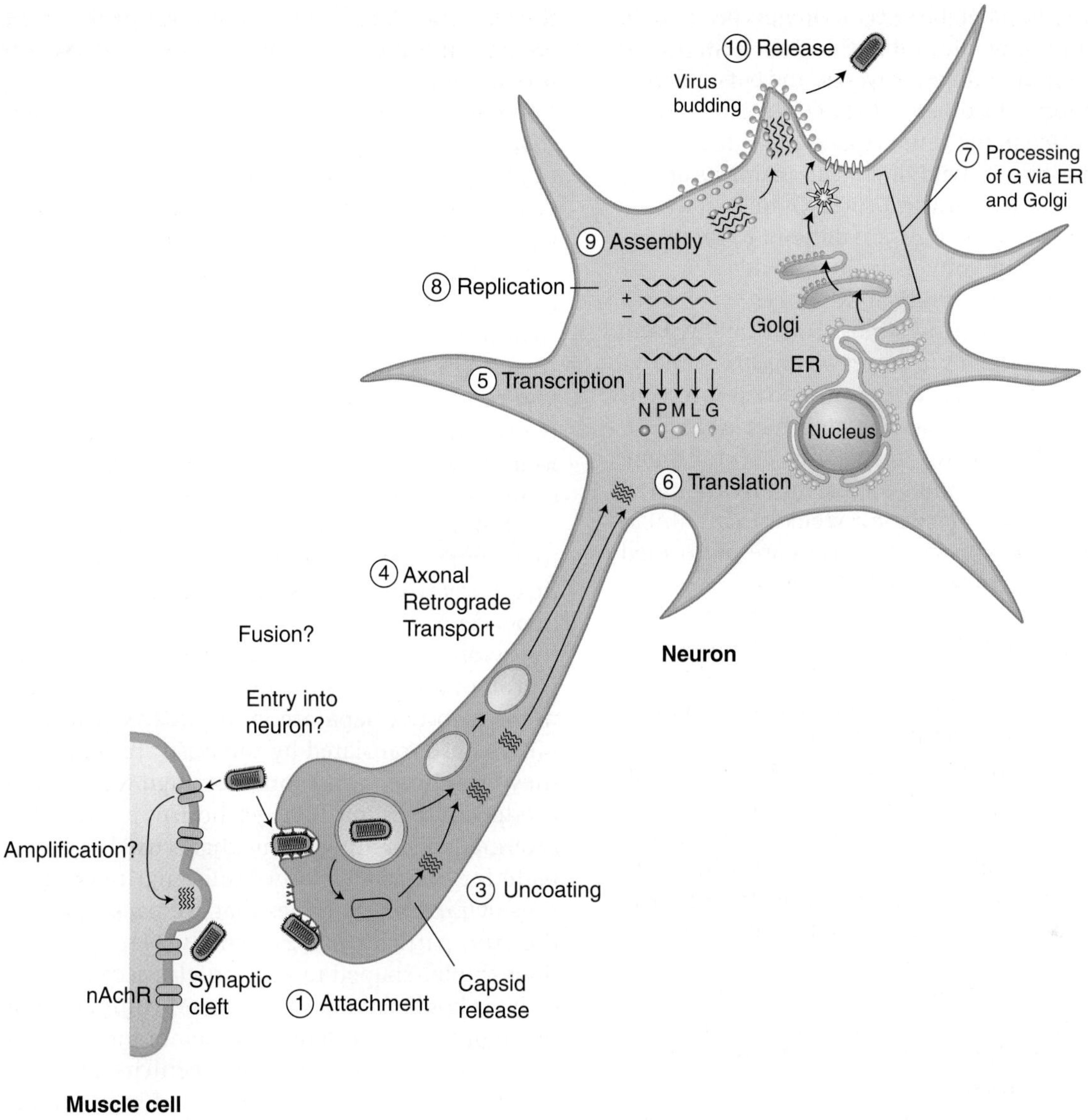

FIGURE 13-14 Life cycle of rabies virus.

Summary

Rabies is one of the oldest and most feared diseases. The disease is caused by the rabies virus. It is transmitted through the saliva of an infected mammal by an unprovoked bite. There are two forms of the disease: furious and dumb. The classic symptoms of furious rabies include hydrophobia, hypersalivation, delirium, tremors, and other encephalitis-like symptoms. The dumb form of rabies symptoms lack hydrophobia, seizures, and hyperactivity; instead, victims experience weakness and ascending paralysis.

Louis Pasteur and Emile Roux developed the first rabies vaccine during the late 1800s. It was prepared by drying the spinal cords of rabid rabbits. The vaccine was successfully tested on dogs; the first human to receive a postexposure vaccination was nine-year-old Joseph Meister in 1886. Meister had been severely bitten and mauled by a rabid dog. He made a full recovery after being inoculated with Pasteur's crude nerve tissue vaccine 13 times over a 15-day period. Today postexposure vaccination is a standard practice used to prevent rabies in individuals who have been bitten or scratched by a rabid mammal.

Rabies epidemiology is complex. The disease remains a significant wildlife management issue and public health challenge. Education plays a vital role in preventing this disease.

Rabies is a disease of both domestic and wild animals. Canine rabies continues to be a large problem in developing countries. Foxes and raccoon dogs are important rabies vectors in Europe. Wildlife most commonly infected in the United States are raccoons, skunks, foxes, coyotes, and bats. Wildlife vaccination and other rabies control measures implemented in the United States has substantially reduced rabies in wild dogs. The majority of human rabies cases in the United States today are caused by bat bites. Bats are difficult to target for rabies control by conventional methods.

Despite rapid medical advances, rabies remains a disease with an almost 100% fatality rate. Diagnosis of animals involves postmortem analysis of brain tissues via direct antibody fluorescent test. Diagnosis in humans antemortem is based on RT-PCR tests on saliva, tears, and cerebral spinal fluid. Rabies antibodies appear in unvaccinated patients during the second week of the illness and may aid diagnosis. Autopsies are performed on humans to confirm rabies cases.

There are no cures or antiviral drugs effective against this disease. Once symptoms begin, treatment is supportive. To date there is only one documented survivor of a rabies patient who was treated with a novel therapeutic regimen after the onset of rabies symptoms. This patient did not receive a postexposure vaccination. Preexposure vaccination of high-risk individuals and postexposure rabies vaccination and immune globulin are used today to prevent the disease. Three categories of vaccines are used: nerve tissue vaccines, avian embryo vaccines, and cell culture vaccines. Nerve tissue vaccines often cause vaccine complications. These vaccines are inexpensive to produce and are still used in developing countries.

Postexposure vaccination soon after infection is successful in preventing rabies because the virus infects muscle cells for weeks before it travels to the CNS. The vaccine elicits an immune response that prevents the virus from spreading throughout the CNS. Once the virus travels to the CNS, it spreads through the peripheral nerves to the salivary glands, skin, mucosal surfaces of the gut, and other organs. Saliva is infectious for several days and up to two weeks before symptoms occur.

The rabies virus is a unique enveloped, bullet-shaped rhabdovirus. It contains a single-stranded negative-sense RNA genome. The surface of the virus is covered with trimers of glycoprotein (G) spikes. The virus contains a helical-shaped ribonucleoprotein core (RNP) that consists of the viral genomic RNA, which is tightly encased with nucleocapsid (N) proteins, a few copies of phosphoprotein (P), an RNA-dependent RNA transcriptase, and replicase (L) protein. A matrix (M) protein is also part of the RNP core. This lethal virus contains a mere five genes in this order: 3′ N, P, M, G, and L 5′.

Rabies virus infects nerve cells by binding to nicotinic acetylcholine receptors. The viral particle fuses with the host membrane via low pH-dependent fusion. The virus replicates solely in the cytoplasm of its host cell. After uncoating, the L protein transcribes the negative-sense RNA genome into five separate viral mRNAs. The viral mRNAs are translated by the hosts' protein synthesis machinery. As soon as enough N protein is produced, the L protein synthesizes a viral RNA intermediate or antigenome that is used as a template for the replication of full-length negative-sense viral RNA genomes that are packaged into the virus particle during assembly. Mature, infectious, bullet-shaped rabies particles acquire their cellular lipid bilayer as the RNP-M complex buds through the host cell plasma membrane.

Many aspects of the pathogenicity of rabies are still unknown. New patterns of rabies infection present a problem for both epidemiologists and virologists. The future of rabies control is a matter of continued research and prevention efforts. Pasteur said it best:

> *Let me tell you the secret that has led me to my goal: my strength lies solely in my tenacity.*
>
> Louis Pasteur (1822–1895)

CASE STUDY 1: A RABID COW NAMED MILLIE

James Welles is a 20-year-old college student who grew up on a dairy farm in Wisconsin. Every summer he returns home to work on the farm. One summer day, while walking through the cow pasture, James notices a raccoon scurrying in circles around an oak tree. Suddenly, he is distracted by his favorite Holstein cow Millie, who is standing alongside a fence. James was worried about Millie because lately she didn't have her usual appetite.

As Millie unsteadily stood next to the fence, she appeared to be in a stupor and foaming at the mouth. James was concerned and later discussed Millie's behavior with his father, who immediately called the local veterinarian. Based on Millie's loss of appetite, the veterinarian first thought that the cow had an intestinal obstruction. A few days later, though, the cow became aggressive and uncoordinated. Her health declined rapidly and she died. The veterinarian decided that a necropsy was needed to determine the cause of death.

1. What diagnostic tests should be performed to determine if Millie had rabies?
2. Should James and his parents receive post-exposure rabies vaccines? Why or why not?

Millie had been milked 12 times during the week before her death. Her milk was pooled with the milk from the other dairy cows on the farm. A portion of the milk was not pasteurized and was distributed for human consumption.

3. Should individuals who drank the unpasteurized milk receive postexposure prophylaxis? Why or why not?
4. Speculate how Millie may have contracted rabies. How would you test your hypothesis?

Note: The CDC reported similar cases of cow rabies in *MMWR* March 26, 1999, Vol. 48, No. 11, pp. 228–229.

CASE STUDY 2: TEXAS TEEN RABIES CASE

During February 2009, a 17-year-old girl in Texas began experiencing headaches. Her headaches continued for two weeks. Her mother took her to the emergency room of a local hospital. When she arrived, she still had a severe headache and neck pain. She felt dizzy and was photophobic. Her face and forearms tingled. Doctors examined her carefully. A lumbar puncture revealed a high white blood cell count of $163/mm^3$ (normal range is 0–5 mm^3) present in her cerebral spinal fluid (CSF). Her lymphocyte counts were also high at 97% (normal range is 40%–80%). She was treated with ceftriaxone and dexamethasone. Ceftriaxone is a broad spectrum antibiotic used to treat Gram-positive and Gram-negative infections. Dexamethasone is a steroid that has anti-inflammatory and immunosuppressant properties. No bacteria were cultured from the CSF. Therefore, her medications were discontinued. After a three-day hospital stay, her symptoms resolved on their own, and she was discharged from the hospital on February 28.

On March 6, she returned to the hospital. Her headaches returned and were intensified. She had similar symptoms: photophobia, along with vomiting, neck, back, and muscle pain. Another lumbar puncture was performed. This time it revealed 160 mg/dL of protein (normal levels are 15–45 mg/dL), and her white blood cell count increased to $185/mm^3$, with 95% of the white blood cells being lymphocytes. She was transferred to a Children's Hospital the same day.

Upon admission to the hospital, she had an itchy papular rash on her arms and back. She was photophobic. The teenager was diagnosed with infectious encephalitis and treated with intravenous acyclovir (used to treat herpesvirus infections), ceftriaxone (a broad spectrum antibiotic), and four antibiotics used to treat *Mycobacterium tuberculosis* infections (ethambutol, isoniazid, pyrazinamide, and rifampin). On the fourth day of hospitalization, she lost sensation and strength in her right arm and leg. She became combative and agitated. A team of doctors were

(continued)

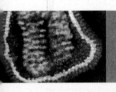

involved in an extensive workup to determine the cause of the girl's encephalitis, including a travel and medical history.

At this time, it was revealed that the girl had come into contact with flying bats in a cave while on a camping trip two months before her headaches began. She didn't recall any bat bites or scratches. Immediately tests were performed on serum and CSF, saliva, and nuchal skin biopsies for the presence of rabies virus RNA, antigens, or rabies antibodies. Four serum samples and CSF samples tested positive for antibodies toward rabies virus. On March 14, after the positive rabies test results, she was vaccinated with one dose of rabies vaccine and immunoglobulin. Her symptoms resolved and she was discharged on March 22. She returned to the emergency room again on March 29 with a recurring headache. She left before a lumbar puncture could be performed. On April 3, she returned to the emergency room while experiencing vomiting and a severe headache. Again, her headache resolved and she did not return for follow-up as an outpatient.

1. Would family members or the boyfriend of the girl meet the criteria for rabies post-exposure prophylaxis? Why or why not?
2. Rabies virus can cross react serologically in laboratory tests with other members of the *Lyssavirus* genus. Research and create a list of other Lyssaviruses, including the countries endemic to infections caused by them.
3. Could any of the Lyssaviruses you listed above be the cause of the teenager's encephalitis? Explain your answer.
4. Doctors also considered that Kern Canyon Virus (KCV) could be the cause of her condition. What is KCV? What are the characteristics of this virus? What symptoms does it cause? Where is it typically found? Why was KCV ruled out?
5. During diagnostic testing, what other etiological agents were suspected to cause her symptoms? (Hint: See Table 2 from Murphy, J., et al., 2010, in the References.)
6. What was the final assessment of her illness by the team of doctors who treated her? (Hint: Read the paper by Murphy, J. et al., 2010.)
7. How does this case differ from Jeanna Geisse's and other rabies survivors?
8. Research the primary literature. Are there documented cases of human-to-human transmission of rabies in a healthcare setting? List your resources.
9. Can lyssaviruses be spread by droplet transmission? Summarize the research experiments conducted in the paper by Johnson, N., et al., 2006, in the References.

References

Johnson, N., et al. 2006. "Airborne transmission of lyssaviruses." *J Med Microbiol* 55:785–790.

Murphy, J., et al. 2010. "Presumptive abortive human rabies—Texas, 2009." *MMWR* 59(7):185–190.

CASE STUDY 3: RABID DOGS

Canadian Olympic athletes were preparing for the 29th Olympic Summer Games that were to be held August 8–24 in Beijing, China, in 2008. Some of the athletes were concerned about rabies in China, because it is a serious public health burden there. Approximately 110,075 human deaths from rabies occurred in China between 1950 and 2008. Rabies remains a neglected disease. In 2006, 140,000 people in Beijing reported animal bites (usually caused by dogs). In 2007, rabies was the number one fatal infectious disease in China in humans. Rabies immunoglobulin is not available for post-exposure treatment, except in Hong Kong. Only

3% of dogs are properly vaccinated in China. (Hint: Refer to Shaw et al., 2007, in the References to answer the following questions.)

1. How would you educate a traveler to China on the risks of rabies?
2. What actions should a traveler take if an exposure (e.g., dog bite) does occur?
3. Should you be vaccinated against rabies before traveling to China?
4. What are the dog policies used to combat rabies in Beijing?
5. What are the dog policies used to combat rabies in Shanghai?
6. What are the dog polices to combat rabies in rural China?

In 2009, a 42-year-old physician traveling to India was bitten by a rabid dog. He returned to Virginia but died of rabies. Research the case study by Bullock, 2010 (in the References).

7. Summarize the patient's exposure history, clinical course, and treatment, and describe the efforts to identify close contacts requiring post-exposure prophylaxis (PEF).

References

Bullock, B., et al., 2010. "Human rabies—Virginia, 2009." *MMWR* 59(38):1236–1238.

Shaw, M. T. M., et al. 2007. "Traveling to China for the Beijing 2008 Olympic and Paralympic Games." *Travel Med Infect Dis* 5:365–373.

Resources

Primary Literature and Reviews

Adedeji, A. O., et al. 2010. "An overview of rabies—History, epidemiology, control and possible elimination." *Afr J Microbiol Res* 4(22):2327–2338.

Alverez, L., et al. 1994. "Partial recovery from rabies in a nine-year-old boy." *Pediatr Infect Dis J* 13(12): 1154–1155.

Benitez, R. M. 1996. "A 39-year-old man with mental status change." *Md Med J* 45(9):765–769.

Blanton, J. D., et al. 2010. "Rabies surveillance in the United States during 2009." *JAVMA* 237(6):646–657.

Centers for Disease Control and Prevention. (CDC). 1972. "Human rabies—Texas." *MMWR* 21(14):113–114.

Centers for Disease Control and Prevention (CDC). 1977. "Rabies in a laboratory worker—New York." *MMWR* 26(21):182–183.

Centers for Disease Control and Prevention. 1977. "Follow-up on rabies—New York." *MMWR* 26(31): 249–250.

Centers for Disease Control and Prevention (CDC). 1999. "Human rabies prevention—United States, 1999 Recommendations of the Advisory Committee on Immunization Practices (ACIP)." 48(RR-1):1–17.

Centers for Disease Control and Prevention (CDC). 2004. "Update: Investigation of rabies infections in organ donor transplant recipients—Alabama, Arkansas, Oklahoma, and Texas, 2004." *MMWR* 53(27): 615–616.

Christenson, J. C., et al. 2007. "Human rabies—Indiana and California, 2006." *MMWR* 56(15):361–365.

Cohen, J., and Powderly, W. G., eds. 2004. *Infectious Diseases*, 2nd ed. New York: Mosby.

Crepin, P., et al. 1998. "Intravitam diagnosis of human rabies by PCR using saliva and cerebrospinal fluid." *J Clin Microbiol* 36(4):1117–1121.

Davis, A. D., Rudd, R. J., and Bowen, R. A. 2007. "Effects of aerosolized rabies virus exposure on bats and mice." *J Infect Dis* 195(8):1144–1150.

Dietzschold, B., et al. 1985. "Differences in cell-to-cell spread of pathogenic and apathogenic rabies virus in vivo and in vitro." *J Virol* 56(1):12–18.

Dietzchold, B., and Koprowski, H. 2004. "Rabies transmission from organ transplants in the USA." *Lancet* 364:648–649.

Dixon, B. 2007. "Watch your bats." *Lancet Infect Dis* 7(1):8.

Dutta, J. K., and Dutta, T. K. 1994. "Treatment of clinical rabies in man: Drug therapy and other measures." *Clin Pharmacol Ther* 32(11):594–597.

Emmons, W., et al. 1973. "A case of rabies with prolonged survival." *Intervirology*:60–72.

Fekadu, M., and Baer, G. M. 1980. "Recovery from clinical rabies of 2 dogs inoculated with a rabies virus strain from Ethiopia." *Am J Vet Res* 41(10):1632–1634.

Fekadu, M., et al. 1996. "Possible human-to-human transmission of rabies in Ethiopia." *Ethiop Med J* 34: 123–127.

Gautret, P., et al. 2007. "Rabies preexposure vaccination in travelers." *J Travel Med* 14(2):136.

Gode, G. R., et al. 1988. "Treatment of 54 clinically diagnosed rabies patients with two survivals." *Indian J Med Res* 88:564–566.

Goldrick, B. A. 2005. "Human rabies: Unusual cases shine the spotlight on an old disease." *Am J Nursing* 105(5): 31–34.

Hemachudha, T., et al. 2006. "Failure of therapeutic coma and ketamine for therapy of human rabies." *J Neurovirol* 12:407–409.

Hunter, M., et al. 2010. "Immunological correlates in human rabies treated with therapeutic coma." *J Med Virol* 82:1255–1265.

Jackson, A. C. 2006. "Rabies: New insights into pathogenesis and treatment." *Curr Opin Neurol* 19: 267–270.

Jackson, A. C., and Wunner, W. H., eds. 2006. *Rabies San Diego.* Elsevier Science.
Johnson, N., et al. 2010. "The immune response to rabies virus infection and vaccination." *Vaccine* 28: 3896–3901.
Lemos-Filho, L., and Fries, B. 2006. "An encounter with rabies in New York City." *Clin Infect Dis* 43(11):1492.
Madhusudana, S. N., et al. 2002. "Partial recovery from rabies in a six-year-old girl." *Int J Infect Dis* 6:85–86.
McColl, K. A., et al. 1993. "Polymerase chain reaction and other laboratory techniques in the diagnosis of long incubation rabies in Australia." *Aust Vet J* 70(3): 84–89.
McDermid, R. C., et al. 2008. "Human rabies encephalitis following bat exposure: Failure of therapeutic coma" *CMAJ* 178(5):557–561.
McGuill, M., et al. 1999. "Mass treatment of humans who drank unpasteurized milk from rabid cows—Massachusetts, 1996–1998." *MMWR* 48(11):228–229.
Meng, S., et al. 2010. "Transmission dynamics of rabies in China over the last 40 years: 1969–2009." *J Clin Virol* 49:47–52.
Morgan, M., and Palmer, J. 2007. "Clinical review: Dog bites." *BMJ* 334:413–417.
Nicolle, L. 2002. "Rabies: Still with us." *Can J Infect Dis* 13(2):83–84.
Plotkin, S. A. 2000. "Rabies." *Clin Infect Dis* 30:4–12.
Porras, C., et al. 1976. "Recovery from rabies in man." *Ann Intern Med* 85:44–48.
ProMED-mail. Rabies, human—USA (Texas). ProMED-mail 2006;13 May:20060513.1360. http://www.promedmail.org. Accessed March 21, 2011.
ProMED-mail. Rabies, human—Canine—China (Hunan). ProMED-mail 2007;05 March:20070305.0782. http://www.promedmail.org. Accessed March 21, 2011.
Rubin, J., et al. 2009. "Applying the Milwaukee protocol to treat canine rabies in Equatorial Guinea." *Scand J Infect Dis* 41:372–380.
Rupprecht, C. E., Willoughby, R., and Slate, D. 2006. "Current and future trends in the prevention, treatment and control of rabies." *Expert Rev Anti Infect Ther* 4(6):1021–1038.
Qanungo, K. R., et al. 2004. "Two RNA polymerase complexes from vesicular stomatitis virus-infected cells that carry out transcription and replication of genome RNA." *PNAS* 101(16):5952–5957.
Salinas, S., et al. 2010. "A hitchhiker's guide to the nervous system: The complex journey of viruses and toxins." *Nat Rev Microbiol* 8:645–655.
Schnell, M. J., et al. 2010. "The cell biology of rabies virus: Using stealth to reach the brain." *Nat Rev Microbiol* 8(51):51–61.
Smith, J. S. Fishbein, D. B., and Rupprecht, C. E. 1991. "Unexplained rabies in three immigrants in the United States." *N Engl J Med* 324(4):205–211.
University of Alabama at Birmingham Hospital, et al. 2004. "Investigation of rabies infections in organ donor transplant recipients—Alabama, Arkansas, Oklahoma, and Texas, 2004." *MMWR* 53(26):586–589.
Warrell, M. J., and Warrell, D. A. 2004. "Rabies and other lyssavirus diseases." *Lancet* 363:959–969.
Whelan, S. P., and Wertz, G. W. 2002. "Transcription and replication initiate at separate sitees on the vesicular stomatitis genome." *PNAS* 99(14):9178–9183.
Wiktor, T. J., et al. 1977. "Induction and biological properties of defective interfering particles of rabies virus." *Virology* 21(2):626–635.
Wilde, H., et al. 2008. "Viewpoint: Management of human rabies." *Trans R Soc Trop Med Hygiene* 102:979–982.
Willoughby, R. E., et al. 2004. "Recovery of a patient from clinical rabies—Wisconsin, 2004." *MMWR* 53(50): 1171–1173.
Willoughby, R. E., Jr., et al. 2005. "Survival after treatment of rabies with induction of coma." *N Engl J Med* 352(24):2508–2514.
Willoughby, R. W., et al. 2008. "Generalized cranial artery spasm in human rabies." *Dev Biol (Basel)* 131: 367–375.
Willoughby, R. E., Jr. 2009. "Are we getting closer to the treatment of rabies?" *Future Virol* 4(6):563–570.
Willoughby, R. E., Jr., et al. 2009. "Tetrahydrobiopterin deficiency in human rabies." *J Inherit Dis* 32:65–72.
Willoughby, R. E., Jr. 2009. "Early death and the contradiction of vaccine during treatment of rabies." *Vaccine* 27: 7173–7177.
Winkler, W. G., et al. 1973. "Airborne rabies transmission in a laboratory worker." *JAMA* 226(10):1219–1221.
Wu, X., et al. 2009. "Reemerging rabies and lack of systemic surveillance in People's Republic of China." *EID* 15(8):1159–1164.

Popular Press

Cockrum, E. L. 1997. *Rabies, Lyme Disease, Hanta Virus: And Other Animal-Borne Human Diseases in the United States and Canada.* Tucson, AZ: Fisher Books.
Finley, D. 1998. *Mad Dogs: The New Rabies Plague.* College Station, TX: Texas A & M University Press.
Gipson, F., and Poison, S. 2001. *Old Yeller.* New York: Harper Perennial Modern Classics, reissue edition.
Lite, J. 2008. "Medical mystery: Only one person has survived rabies without vaccine—But how?" *Sci Am* 8:4. Retrieved March 21, 2011, from http://www.scientificamerican.com/article.cfm?id=jeanna-giese-rabies-survivor
Walsh, J. E. 2000. *Midnight Dreary: The Mysterious Death of Edgar Allan Poe.* New York: St. Martin's Minotaur.
Willoughby, R. E., Jr. 2007. "A cure for rabies?" *Sci Am* 4:89–95.

Video Productions

Louis Pasteur. 2007. Films for the Humanities.
Rabies. 2006. The American Veterinary Medical Association.
The Girl Who Survived Rabies. 2005. BBC Documentary.
Rabid. 1977. Somerville House. Film directed by David Cronenberg.
Old Yeller. 1957. Walt Disney Production.

eLearning

go.jblearning.com/shors2

The site features eLearning, an online review area that provides quizzes and other tools to help you study for your class. You can also follow useful links for in-depth information, or just find out the latest virology and microbiology news.

CHAPTER 14

Poxviruses

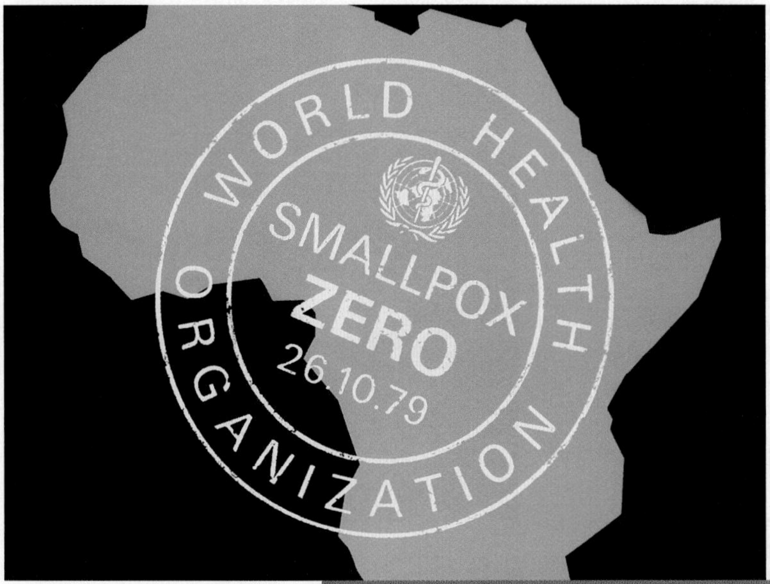

This poster was produced in six different languages by the World Health Organization. It is celebrating the eradication of smallpox from the Horn of Africa on October 26, 1979 (exactly two years after the last naturally occurring case that occurred in Somalia). Reproduced from Fenner, F., et al. *Smallpox and Its Eradication.* World Health Organization, 1988.

"Could it not be contrived to send the Small Pox among those disaffected tribes of Indians? We must on this occasion use every stratagem in our power to reduce them."

Lord Jeffrey Amherst, Commanding General of British forces in North America during the French and Indian War, July 1763

OUTLINE

CASE STUDY: SQUIRRELPOX

Aiden Robinson and Michael Smith were college students participating in a continuing education and extension program that involved traveling to Scotland and England. Aiden and Michael are biology students. Their study abroad included a biology component in which they utilized a capture-recapture model to estimate the number of squirrels present in different locations throughout England and Scotland. Aiden and Michael began live-trapping squirrels, tagging the squirrels, and releasing them back into their habitat. It didn't take either one of them very long to notice that 95% of the squirrels they trapped were an invasive species of grey squirrel (*Sciurus carolinensis*). They trapped only a handful of native Eurasian red squirrels (*Sciurus vulgaris*; **FIGURE CS 14-1**). They wondered if the grey squirrel invasion could cause the red squirrel to become extinct. Aiden and Michael created a list of explanations for the displacement of red squirrels, including diseases such as squirrelpox. Squirrelpox virus (SQPV) causes high mortalities in red squirrels but appears to be nonpathogenic in grey squirrels.

FIGURE CS 14-1 The Eurasian red squirrel (*Sciurus vulgaris*).

14.1 History

Smallpox is a human infectious disease caused by variola virus (family *Poxviridae*) and was a worldwide scourge for thousands of years. Dr. Michael T. Olsterholm of the University of Minnesota School of Public Health has called smallpox "the lion king of infectious diseases." Smallpox killed approximately 500 million people in the 20th century—more than the 320 million deaths caused by wars, the 1918 Spanish flu pandemic, and acquired immune deficiency syndrome (AIDS) combined. Those who survived smallpox were often left blind from corneal ulcerations and badly scarred by pockmarks. Famous victims and survivors of smallpox are listed in **TABLE 14-1**.

It has been speculated that smallpox emerged sometime after the first agricultural settlements, in about 10,000 B.C. The first evidence of smallpox comes from mummified remains of the Egyptian pharaoh Ramses V. Written descriptions of the disease appeared in China (340 A.D.) and southwestern Asia (910 AD). Smallpox was carried by explorers and traders from the Old World to the New World, where the population had no exposure or immunity to smallpox before the Europeans introduced it in the 16th century. In 1519, Spanish conquistador Hernando Cortez and his followers landed in Mexico, in the heart of the Aztec empire, followed soon after by a second group of Spanish explorers. In the second group was an African slave who was afflicted with smallpox. In the ensuing months, smallpox spread throughout the Aztec empire. Over time, the Aztec and Inca empires, and eventually the Native American Indians, were devastated by outbreaks of smallpox. Smallpox ravaged Europe in the 17th and 18th centuries as well, afflicting up to 90% of the children in Britain alone. It was so pervasive that it was considered unusual if someone did not have **pox scars** on their face. In the United States, smallpox epidemics involving all age groups occurred frequently during the 18th century. Small hospitals referred to as "pest-houses" were used to isolate smallpox patients and their contacts during an outbreak.

TABLE 14-1 Famous Individuals Who Suffered from Smallpox

Individual	Age Contracted	Year Contracted	Outcome
Pharaoh Ramses V	35	1157 B.C.	Died (see Chapter 1)
Mary II of England	32	1694	Died
Peter II, Emperor of Russia	15	1730	Died
Louis XV, King of France	64	1774	Died
Wolfgang Amadeus Mozart	11	1767	Severe case, survived
George Washington	19	1751	Severe case, survived
Abraham Lincoln	54	1863	Mild case, survived
Joseph Stalin	7	1886	Severe case, survived

14.2 Clinical Features of Human Poxviruses

Smallpox (Variola)

During the World Health Organization's (WHO's) eradication program, smallpox recognition cards were widely used by workers searching for cases in remote areas of India and Africa, where the last pockets of smallpox existed (**FIGURE 14-1**). Ultimately, the most effective method used to ensure prompt reporting was the offer of a reward, which was given to both the health worker who investigated the case and the person who reported it. At the beginning of 1974, a reward of 50 rupees (US$6.25) was offered per case in Indian states.

By the end of 1974, the reward was increased to 100 rupees (US$12.50) and by July 1975, the reward increased to 1000 rupees (US$125.00). Workers were sometimes paid as little as 10 rupees or less per day by their employers; hence, this reward was a significant amount. The reward system was also used in Bangladesh and Somalia. In 1978, the WHO increased the award to US$1000 for reporting a confirmed case.

Variola major and **variola minor** are two strains of the same virus (variola) but cause different diseases. Variola major is more common and causes a severe disease, with a more extensive rash and a higher fever. Variola minor causes a milder disease that has a 1% to 2% fatality rate in unvaccinated individuals. There are four types of variola major smallpox:

1. **ordinary** (the most frequent type, accounting for 89% or more of cases and with a 30% mortality rate);
2. **vaccine modified** (mild, occurring in 2.1% of previously vaccinated persons, and not lethal);
3. **flat or malignant pox** (rare—only 6.7% of cases—and very severe, with a 96% mortality rate);
4. and **hemorrhagic** (rare—2.4% of cases—and very severe, with a 96% mortality rate).

Historically, variola major has an overall fatality rate of about 30%; however, flat and hemorrhagic smallpox usually are fatal. Variola minor is a less common presentation of smallpox, and a much less severe disease, with death rates historically of 1% or less (**FIGURE 14-2**).

The average incubation period of smallpox is 12 to 14 days. During this time, people are not contagious. The first symptoms of smallpox include a fever of 101°F to 104°F, a splitting headache, and a severe backache. Vomiting occurs in about half of all infected individuals, and about 10% of individuals experience diarrhea. Delirium and convulsions may occur in a small percentage of individuals (7% to 15%), usually children. During this 2- to 4-day prodromal phase, the individual is too sick to carry on normal daily activities. The person may be contagious during this stage of infection. By the second or third day, the individual is feeling better. As the fever declines, a macular rash (small red spots) appears on the tongue and in the mouth (oral mucous membranes). The rash becomes papular (raised) and vesicular (blistery). It harbors large quantities of virus. As the vesicles rupture, large numbers of virus are liberated into the saliva. At this time, the person is most contagious. The lesions can also occur along the respiratory tract and some individuals may experience a sore throat. Around the same time that the vesicles break, a skin rash appears: first on the face, then spreading to the arms and legs, and finally to the hands and feet, all within a 24-hour period. This rash is described as being a **centrifugal** rash (it mainly covers the face and extremities as opposed to the entire body) and is important in the diagnosis of smallpox.

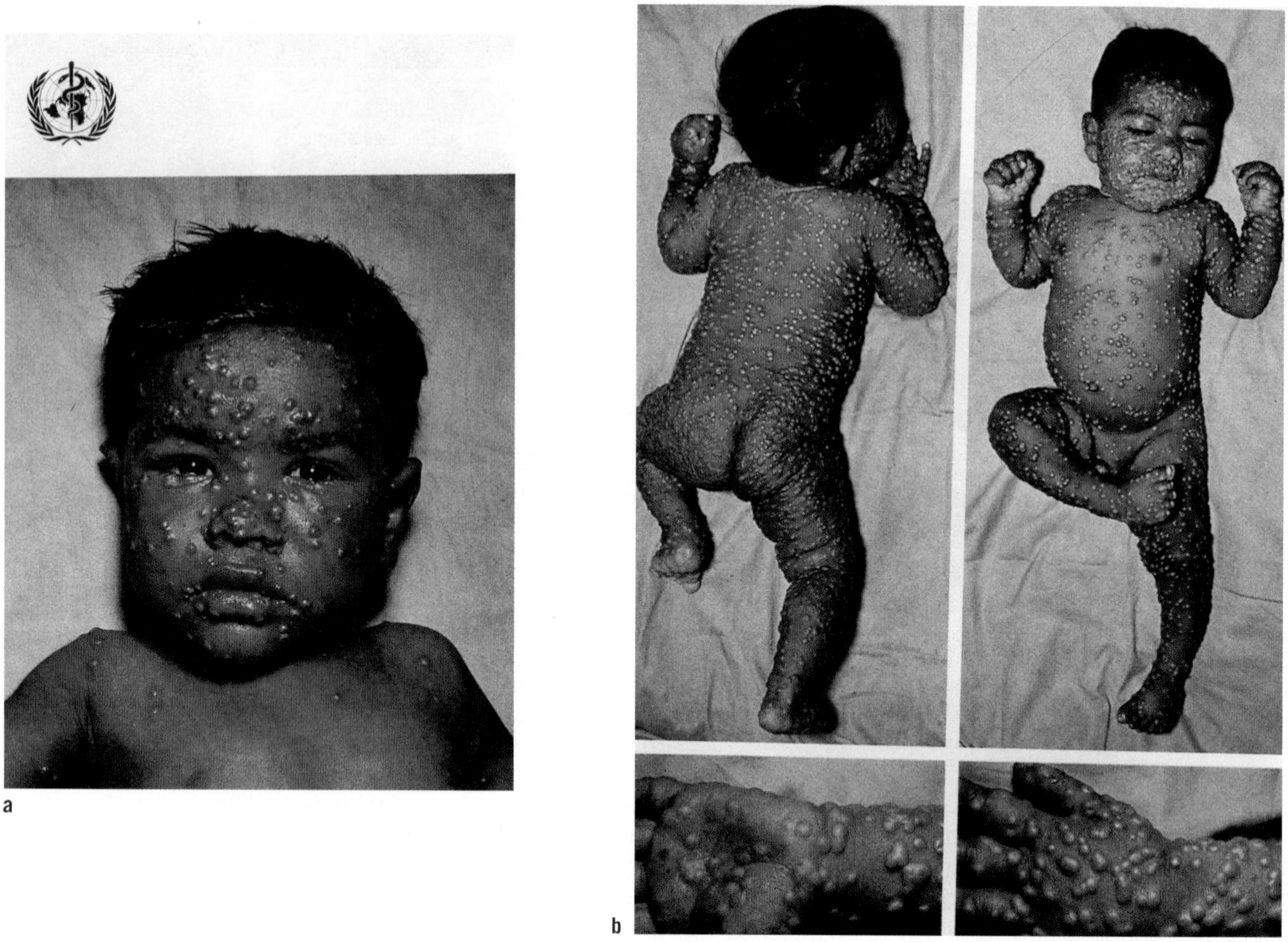
a
b

FIGURE 14-1 (a) Front of the WHO recognition smallpox card used during the eradication program. The cards were shown by workers to inquire whether anyone had seen a person with a similar rash. **(b)** Reverse side of the smallpox recognition card.

It is a raised rash that becomes vesicular, containing an opaque and turbid fluid by the fourth or fifth day. The vesicles mature into pustules by the seventh day. By the second week, the rash begins to crust over and the scabs separate by days 22 to 27. In fatal cases, death occurs between the tenth and sixteenth day of illness. The communicability of the disease extends from the onset of the rash to the complete separation of all scabs.

Monkeypox

Human monkeypox virus infections are rare. Before 1970, monkeypox was recognized as a disease of animals, normally occurring in the rainforests of central and western Africa. The monkeypox virus was first isolated from infected cynomolgus monkeys in 1958. Between 1970 and 1986, the first human cases of monkeypox were reported from western African and Congo Basin countries of Africa as smallpox disappeared. The smallpox vaccine protects or mitigates the clinical manifestations of monkeypox.

The first outbreaks of **human-to-human transmission** of monkeypox occurred in 13 villages in the Kasai Oriental, Zaire, from 1996 to 1997. The outbreaks were investigated by the Zaire Ministry of Health and the WHO. It has been postulated that lack of vaccination and a concurrent epidemic in a large number of animals allowed the monkeypox virus to cross the species barrier into humans. In June 2003, the first cases of monkeypox were reported among several people in a multistate outbreak in the United States who had had contact with infected pet prairie dogs and other exotic pets. The case-fatality rate of monkeypox ranges from 1% to 14%.

The signs and symptoms of monkeypox are similar to ordinary or modified smallpox but milder. Patients experience a fever, respiratory symptoms, and vesicular rash similar to smallpox victims (**FIGURE 14-3**). One difference is that **lymphadenopathy** (a chronic, abnormal enlargement of the lymph nodes) is more prominent in monkeypox cases during the early stages of the disease.

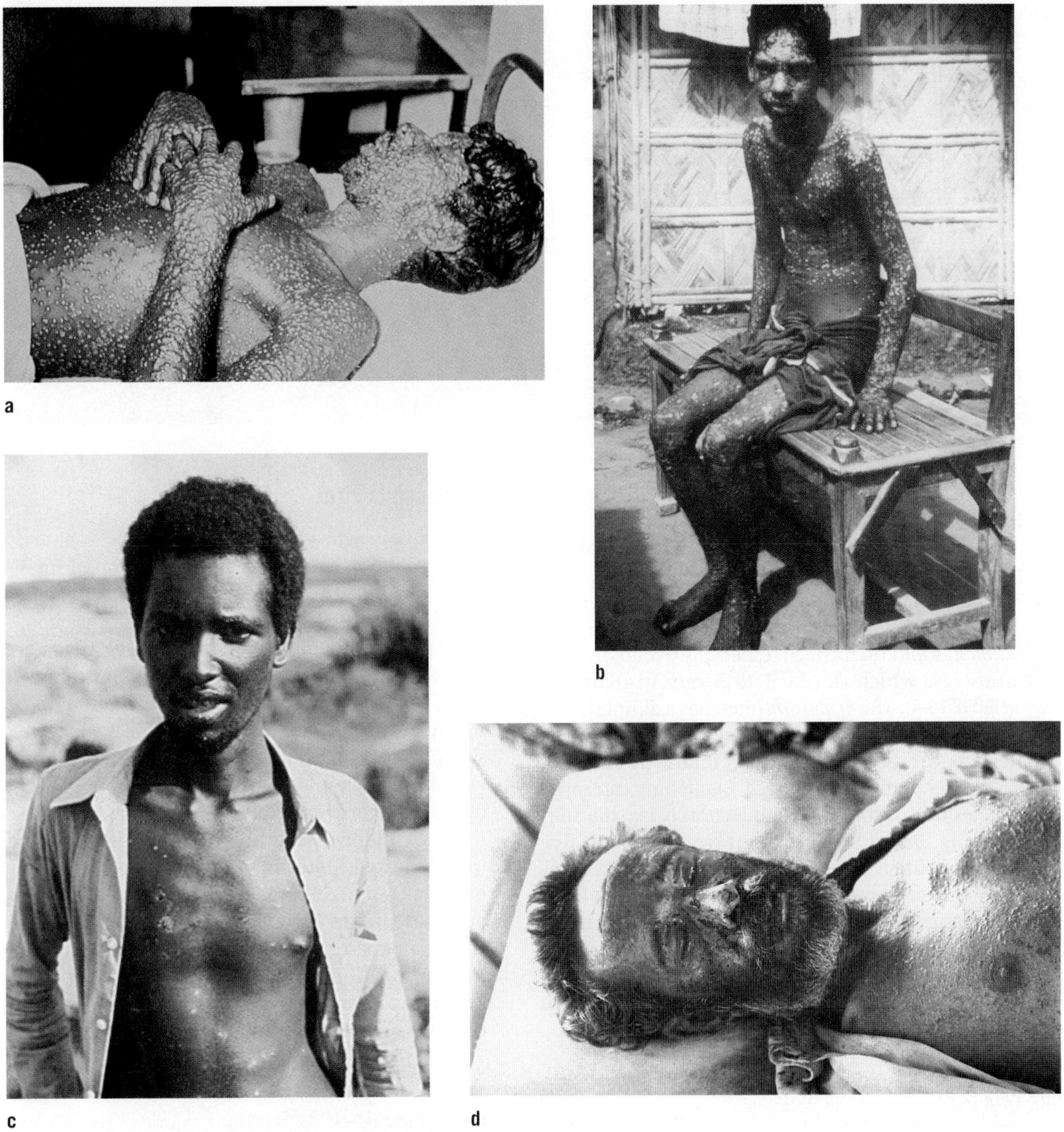

FIGURE 14-2 (a) This man is suffering from an ordinary variola smallpox infection. He is a patient at the Military Hospital No. 45, an isolation hospital at Gungalin, Australia. **(b)** The usual centrifugal pattern of smallpox maculopapular rash (less lesions on the trunk and more on his face and extremities) is revealed by this photograph of a Bangladesh boy (1973). **(c)** This man is the last known person in the world to have natural smallpox of any kind. At 23 years old, Ali Maow Maalin suffered from variola minor. He lived in Merka, Somalia (1977). **(d)** This Bangladesh man suffered from the most severe form of smallpox (hemorrhagic). He likely died from the viral illness (1975).

Molluscum Contagiosum

Molluscum contagiosum is a common poxvirus infection of worldwide distribution that can become a public health problem in areas of low sanitary and hygienic conditions. It represents 1% of all skin infections. The virus is transmitted by direct contact, including sexual contact, or more commonly by a consequence of indirect contact through **fomites** such as sharing towels or from swimming pools. It can be transmitted rapidly among children in daycare centers and kindergartens. Today the disease is becoming a significant opportunistic infection in patients with human immunodeficiency

FIGURE 14-3 Mother and daughter crusted primary sites of monkeypox infection.

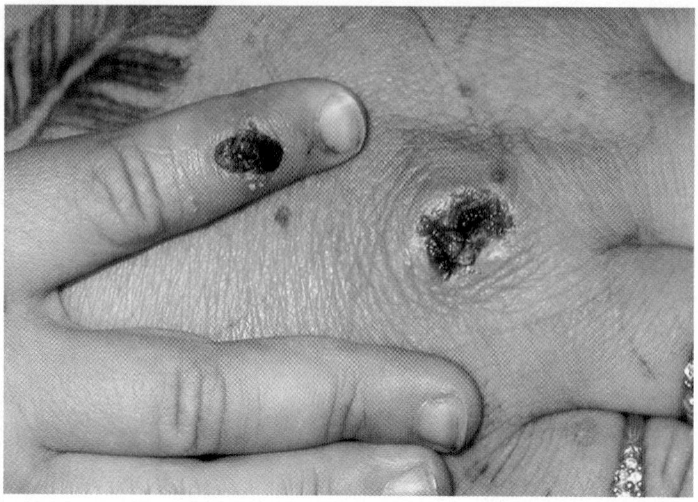

virus (HIV) infections, especially those with severely depressed CD4+ T lymphocyte numbers. HIV patients may experience persistent and fulminating Molluscum contagiosum.

The incubation period is two to eight weeks. The virus causes pink, pearl-like lesions (referred to as *molluscum*) on the face (especially eyelids), arms, and legs, which can be 1 to 5 mm in diameter (**FIGURE 14-4**). The *molluscum* often has a dimple in the center. If the infection is sexually transmitted, the lesions can be seen on the genitals, thighs, and the lower part of the stomach. Scratching can cause the lesions to spread. The infection is usually self-limiting. It can take between six months to five years for the lesions to disappear. There is continued debate about whether the lesions should be treated (e.g., by surgical removal, scraping, or freezing) or allowed to resolve spontaneously. Secondary bacterial infections are a complication of Molluscum contagiosum lesions. Molluscum contagiosum can serve as a marker for severe immunodeficiency.

Vaccinia Virus

Vaccination with **vaccinia virus** has been directly responsible for the successful eradication of smallpox (variola). Infection with this virus provides adequate cross-protective immunity against smallpox and monkeypox viruses. Although the exact origins of vaccinia virus are a mystery, genetically vaccinia virus is most similar to buffalopox. Inoculation with vaccinia virus produces a localized skin infection. In persons who are immunocompromised, vaccinia may disseminate and cause severe disease. Adverse reactions have been rare because routine childhood immunization was discontinued in the United States in 1972. Recent mysterious outbreaks of an emergent poxvirus from humans and cattle in Rio de Janeiro State, Brazil, may represent a persistent vaccinia-like virus infection in nature. A 2008 to 2010 study of vaccinia virus infections occurring on 56 dairy farms located in the Amazon biome of Brazil concluded that animal

FIGURE 14-4 Molluscum contagiosum lesions.

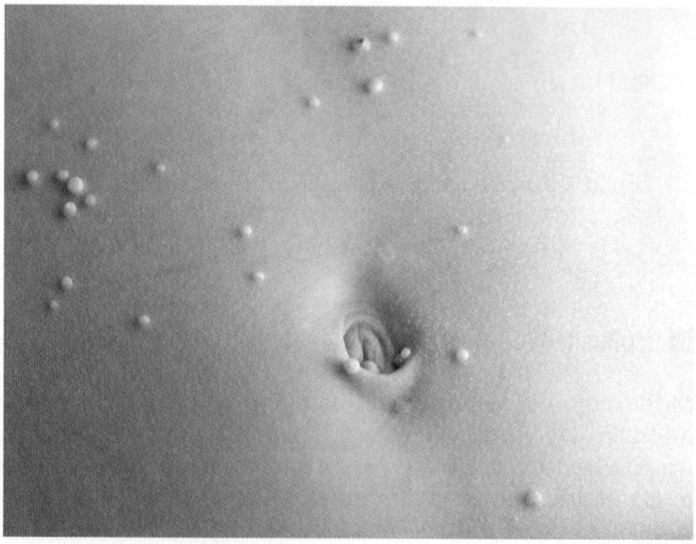

movement on farms located along a 41 km highway was the major cause of vaccinia (Cantagalo strain) infections in dairy cows and workers. Cattle trading at auctions launched the spread of the virus. Migration of infected dairy cows and workers was also associated with spread of the vaccinia viruses.

14.3 Laboratory Diagnosis of Poxvirus Infections

In the United States, the last natural case of smallpox occurred in 1949 and routine smallpox vaccination of children was discontinued in 1972. Very few U.S. physicians have seen an actual case of smallpox. Variola major is a potential biological weapon today, though, so the Centers for Disease Control and Prevention (CDC) has prepared information and instructions for physicians in case of a smallpox emergency. The CDC has defined a smallpox clinical case definition as, "an illness with acute onset of fever greater than or equal to 101°F (38.3°C) followed by a rash characterized by firm, deep-seated vesicles or pustules in the same stage of development without other apparent cause."

Typically, diagnosis of poxvirus infections is made based on the appearance of the lesions. Smallpox produces a centrifugal rash, as compared to a chickenpox rash that is more concentrated on the trunk of the body (**FIGURE 14-5**). Preliminary tests on specimens that include scrapings of the pustules, blood, and scabs may be done by state health departmental laboratories. If smallpox or another poxvirus is suspected, scabs are also collected and analyzed by the CDC for definitive testing via electron microscopy to visualize poxvirus particles. Other methods such as virus isolation, histology, IgM and IgG enzyme-linked immunosorbent assay (ELISA), or polymerase chain reaction (PCR) techniques, followed by restriction fragment-length polymorphism (RFLP) or DNA sequencing analysis, may be used to detect the causative agent.

Electron microscopy has played a major role in viral diagnosis in the past. If available, it can be a first-line method for laboratory diagnosis of poxvirus infections and may provide one of the first clues to the cause of the unknown rash illness. PCR-RFLP analysis is the only way to accurately distinguish between variola, monkeypox, and vaccinia virus infections.

14.4 Cellular Pathogenesis

Despite the historical significance of smallpox, very little is known about the pathogenesis and virulence of variola virus. The knowledge we have

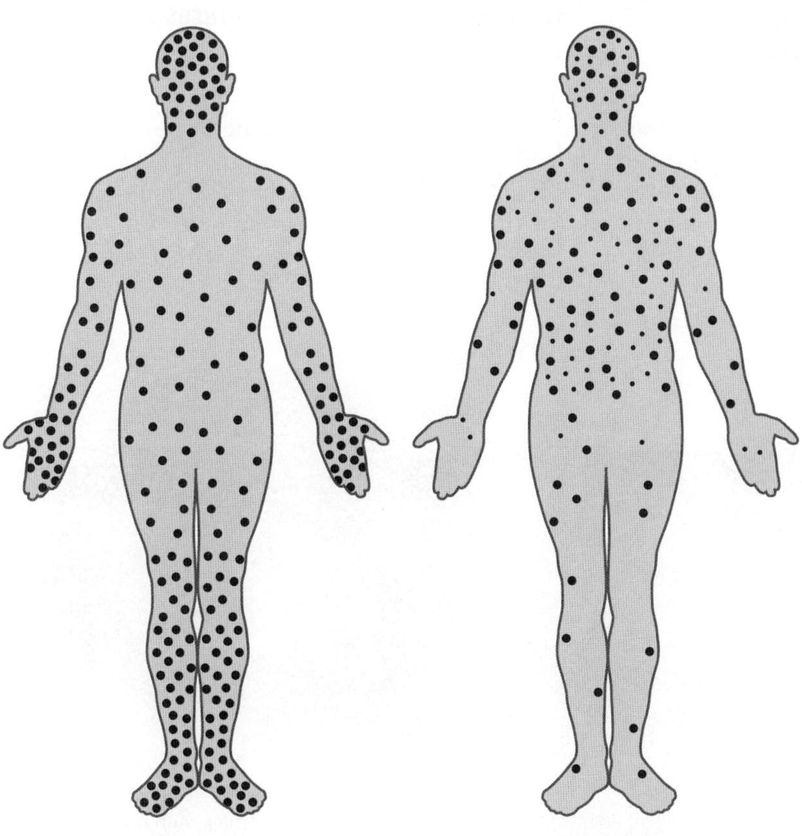

FIGURE 14-5 Rash distribution of smallpox vs. chickenpox. Adapted from World Health Organization, *Diagnosis of Smallpox Slide Set*, http://www.who.int/emc/diseases/smallpox/slideset.

comes from an era that predates modern advances in molecular virology and immunology. Virtually no modern molecular biology has been applied to the study of live variola virus. Today, variola virus is stored safely in two international repositories: the State Research Center of Virology and Biotechnology (Vector) near Novosibirsk in Russia and at the CDC in Atlanta, Georgia. The virus can only be safely worked with in a BSL-4 (Biosafety Level-4) maximum containment facility, sometimes referred to as a "hot lab." There is concern, though, that smallpox could reemerge in modern society through the use of variola virus as a bioweapon. To improve our understanding of variola pathogenesis, further research is needed to develop rapid diagnostics, antivirals, and a protective but less **reactogenic** and safer vaccine to treat those exposed to variola virus. To address these needs, research proposals with variola virus have been approved by the WHO. These proposals include the development of an animal model (e.g., the use of cynomolgus monkeys to attempt to repeat some of the pathologic features of human smallpox) because it is impractical and unethical to safely do smallpox experiments in humans. In nature, variola is uniquely restricted to humans, which is why surrogate primate models are needed to test vaccines and therapeutics.

14.5 Naming and Structure of Poxviruses

The name **variola** is derived from the Latin *varius* (spotted) or *varus* (pimple). The word *poc* or *pocca*—a bag or pouch—describes an **exanthematous** disease (a disease accompanied by skin eruption), such as smallpox, and English accounts began to use the word *pockes*. When syphilis appeared in Europe in the late 15th century, writers began to use the prefix *small* to distinguish variola, the smallpox, from syphilis, the great pox (which is caused by the spirochete bacterium *Treponema pallidum*). Vaccinia virus, the "vaccine virus," is a relative of variola virus. It is the prototype of poxviruses. Information regarding poxvirus structure and replication is based on laboratory experiments involving cell cultures infected with vaccinia virus.

Poxviruses are the largest of all animal viruses. They can be visualized by light microscopy. Virions are **enveloped** rectangular or "brick-shaped," ranging from 270 to 350 nm on average. They contain **linear double-stranded DNA (dsDNA)** genomes with closed ends that vary from 130 to 230 kilobase pairs (kbp) in length. Ends of the genome have inverted terminal repeat (ITR) sequences. Clinical specimens contain two different morphologies of virions: the intact M (or "mulberry") form mainly found in the vesicular fluid, and the C (or "capsule") form that is associated with dried scabs. Internally, poxviruses have a nucleoid and two **lateral bodies** surrounded by an outer membrane, as revealed by staining thin sections of infected cells using electron microscopy (**FIGURE 14-6**). The lateral bodies contain various enzymes essential for viral replication. Inside of the cell, the virion often has a double membrane.

Broad Host Range of Poxviruses

Naturally occurring poxvirus infections affect humans and many species of animals and insects. Most poxviruses were named after the animal from which they were originally isolated (e.g., cowpox and camelpox), but the main reservoirs for the viruses may be rodents or other species (**TABLE 14-2**). The one large *Poxviridae* family con-

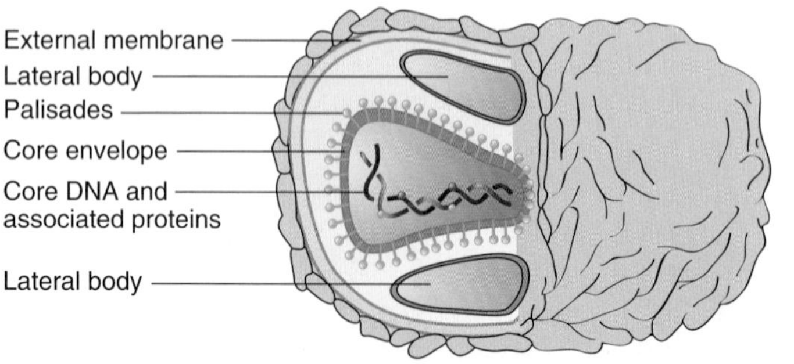

a

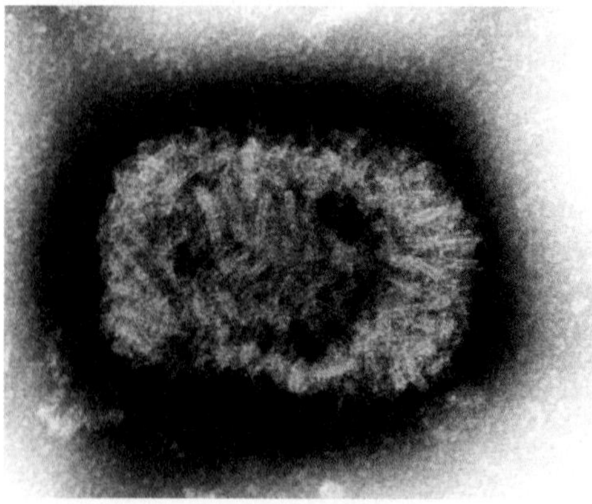

b

FIGURE 14-6 (a) Schematic drawing of a typical poxvirus virion. Adapted from an illustration by inDepthLearning, Ann Arbor, Michigan. **(b)** Electron micrograph of variola virus particles (310,000 ×, negative stain).

TABLE 14-2 Human Poxvirus Pathogens and Their Natural Hosts

Virus	Natural Host*/Reservoir	Disease	Geographical Distribution
Variola major and minor viruses	Humans*	Smallpox, rash, generalized disease	Previously worldwide; eradication in humans in 1977
Monkeypox virus	Squirrels,* monkeys, rabbits, mice, rats, humans	Monkeypox rash, generalized disease	Western and central Africa
Molluscum contagiosum virus	Humans*	Molluscum contagiosum, skin nodules	Worldwide
Orf virus	Sheep,* dogs, goats, humans, chamois, reindeer, musk ox, Himalayan tahr, Steenbok, alpaca	Localized skin lesions	Worldwide
Pseudocowpox virus	European cattle,* humans	Milker's nodules	Worldwide
Bovine papular stomatitis virus	European cattle,* humans	Localized skin lesions	Worldwide
Cowpox virus	Rodents,* cats, cows, zoo animals, humans	Localized pustular lesions	Western Asia, Europe
Vaccinia virus	Natural host unknown (buffalo** and cattle?), humans	Buffalopox, localized pustular lesions	Asia, India, Brazil, and laboratory
Tanapox	Monkeys (?), rodents (?), humans	Tanapox, localized nodular lesions	Eastern and central Africa
Yaba monkey tumor poxvirus	Monkeys (?), humans	Localized nodular skin lesions	Western Africa

*Natural host.

**During the WHO smallpox eradication program, water buffalo in India and cattle in Brazil were infected with the local vaccine strain of vaccinia virus, which apparently persists in these animals and occasionally infects humans.

Source: Adapted from White, D. O., and Fenner, F. J., *Medical Virology*, Fourth Edition, San Diego: Academic Press.

TABLE 14-3 Members of the *Poxviridae* Family That Infect Vertebrates (Subfamily *Chordopoxvirinae*)

Genus	Member Viruses
Avipoxvirus	Canarypox, fowlpox, pigeonpox, juncopox, quailpox, turkeypox, starlingpox, sparrowpox, peacockpox, penguinpox
Capripoxvirus	Goatpox, sheeppox, lumpy skin disease
Leporipoxvirus	Hare fibroma, myxoma, squirrel fibroma, rabbit fibroma
Molluscipoxvirus	**Molluscum contagiosum virus***
Orthopoxvirus	**Variola (smallpox),* vaccinia, monkeypox,*** buffalopox, camelpox, **cowpox,*** ectromelia (mousepox), volepox, raccoonpox, skunkpox, elephantpox
Parapoxvirus	**Pseudocowpox,* orf,* bovine papular stomatitis virus,*** sealpox, parapox of deer
Suipoxvirus	Swinepox
Yatapoxvirus	**Tanapox,* Yaba monkey tumor***

*Human pathogens

tains two subfamilies: *Chordopoxvirinae,* which infect vertebrates, and *Entomopoxvirinae,* which infect insect hosts. This chapter focuses on members of the *Poxviridae* family that infects humans: smallpox (variola major and minor), Molluscum contagiosum virus (MCV), monkeypox (a recently emerged virus in the United States in imported wild rodents from Africa), and vaccinia (the vaccine strain against smallpox). With the exceptions of the "extinct" variola virus and MCV, all other poxvirus infections of humans today are **zoonoses** (**TABLE 14-3**).

14.6 Vaccinia Virus Replication

Vaccinia virus is the prototype of poxviruses. It can be propagated in a wide host range of tissue culture cells (e.g., monkey, rabbit, hamster, and mouse). Vaccinia virus is like the *Escherichia coli* of a bacteriology laboratory: It can be grown to high titers in a short period of time. Within 12 to 24 hours, 1×10^8 or 10^9 infectious viruses/ml of cell culture medium can be produced. Infectious vaccinia virus particles

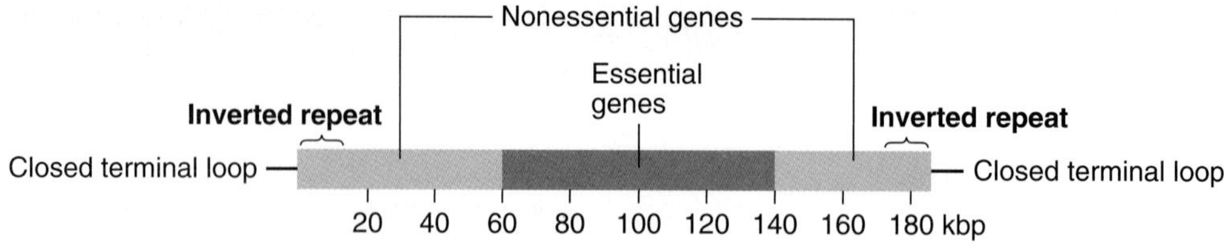

FIGURE 14-7 Typical poxvirus genome structure. Adapted from Cann, A. J., "Poxviruses," *MicrobiologyBytes: Virology*, http://www.microbiologybytes.com/virology/Poxviruses.html

are large (300–400 nm in diameter) and brick-shaped. Lipoprotein membranes surround a complex core structure that contains a 200,000 base pair (bp) linear, dsDNA genome that encodes at least 200 genes. The ends of the genome consist of a terminal hairpin loop (no free ends) with inverted sequences. The complete genome sequences of more than 40 poxviruses are available via the **Bioinformatics Resource Center** (http://www.poxvirus.org). Most of the **essential genes** are located in the central part of the genome, whereas **nonessential genes** (i.e., not essential for growth in cell cultures) are located at the ends of the genome (**FIGURE 14-7**). Approximately 90 essential genes are highly conserved among various poxviruses. Their functions are essential for viral replication and **morphogenesis**.

The entry of poxviruses into cells is complicated by the existence of multiple forms of infectious particles. One form contains two **extracellular enveloped virion** (**EEV**) membranes. The membranes of EEVs are obtained from the Golgi apparatus and the plasma membrane of the host cell. The second form is an **intracellular mature virion** (**IMV**), which only contains the inner membrane derived by a Golgi wrapping event. IMVs are released upon cell lysis. Each infectious form is thought to enter cells by different mechanisms. Vaccinia virus can enter almost every cell line tested. This means the receptor for vaccinia virus entry is highly conserved and ubiquitous in nature or that the virus can enter via more than one receptor. So far, some studies have shown that the A27L virion protein of vaccinia interacts with cell surface glycosaminoglycans (GAGs). GAGs are ubiquitously expressed on many different cell surfaces. The mechanism for cell entry and penetration is complex and may involve more than one mechanism.

During uncoating the outer membrane of the EEV is removed and the particle enters the cell, where it loses the second membrane and the viral core passes into the cytoplasm. *One of the unique hallmarks of poxviruses is that these viruses have acquired all of the functions necessary for genome replication in the cytoplasm even though they have DNA genomes.* Poxvirus gene expression and genome replication can occur in enucleated cells, but maturation of particles is blocked. Viral enzymes associated with the core carry out the three stages of gene expression: **early**, **intermediate**, and **late gene expression**. The three distinct classes of viral messenger RNAs (mRNAs) are transcribed from genes containing promoters and other sequence elements that provide the basis for their programmed order of expression. No splicing occurs in poxvirus mRNAs.

A complete early transcription system is synthesized late during infection and is packaged into the viral core. This viral transcriptional machinery consists of a DNA-dependent RNA polymerase, a transcription factor (VETF), capping and methylating enzymes, and a poly(A) polymerase. The machinery has the capability to synthesize mRNAs that are recognized and translated by the cell's protein synthesis machinery. The early mRNAs encode enzymes and factors needed for transcription of the intermediate class of genes that in turn encode enzymes for late gene expression. Late expression occurs after genome replication. The vaccinia virus D10 gene encodes a decapping enzyme that may accelerate viral mRNA turnover and help eliminate competing host mRNAs from being translated by the translational machinery. This would allow for stage-specific synthesis of viral proteins.

After the late structural proteins are synthesized, infectious progeny particles are assembled. Assembly involves interactions with the cytoskeleton (e.g., actin-binding proteins). The particles are wrapped with a Golgi-derived membrane and transported to the periphery of the cell. The EEV form picks up an additional membrane as it is released from the cell, completing the virus replication cycle. EEVs can initiate infection, mediating cell-to-cell spread (**FIGURE 14-8**).

14.7 Poxviruses and Immune Evasion

Poxviruses do not undergo a latent stage by integrating viral DNA into its host chromosome. Nor do they cause persistent infections that would make

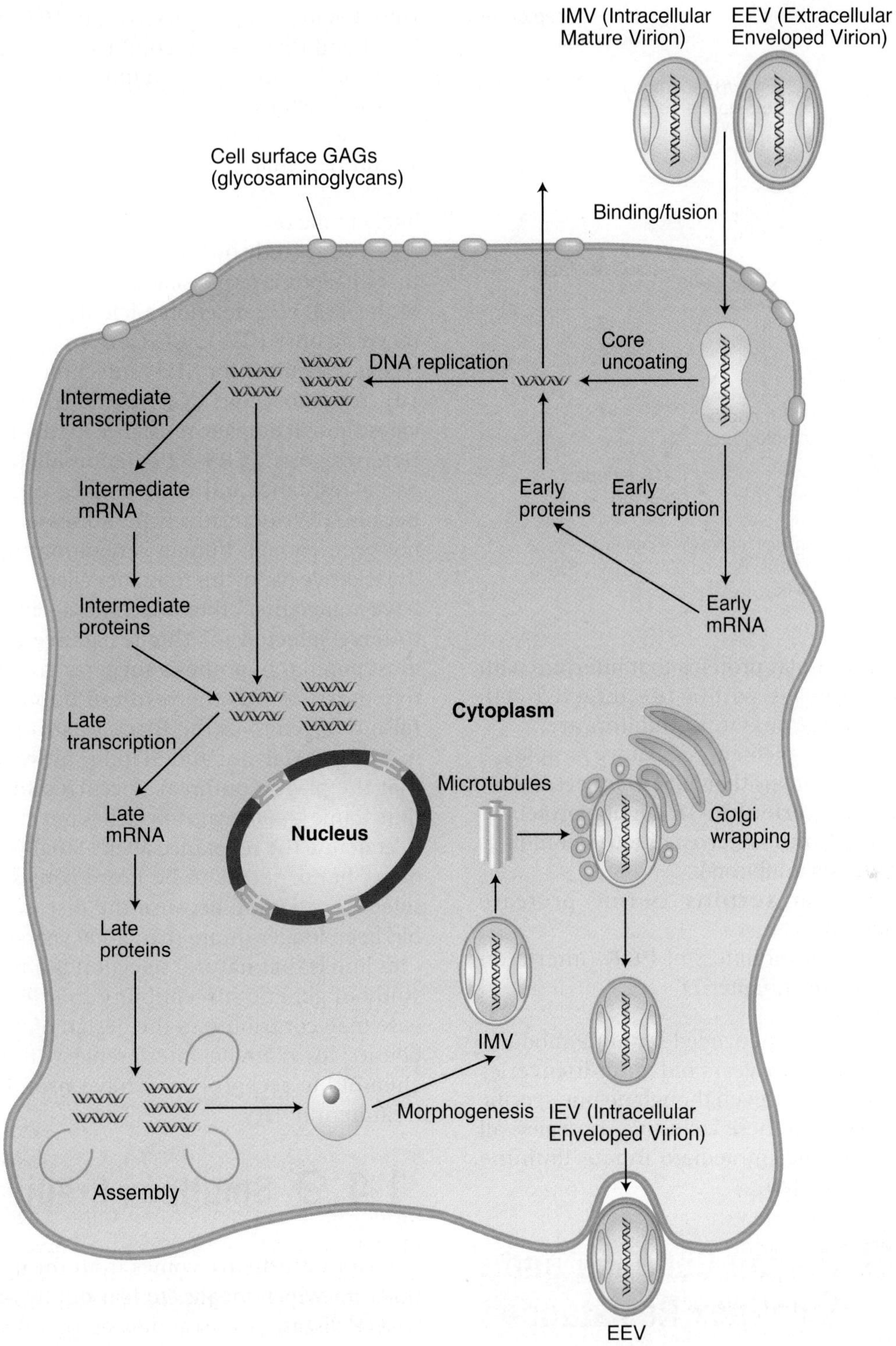

FIGURE 14-8 Life cycle of vaccinia virus. Adapted from McFadden, G., *Nat. Rev. Microbiol.* 3 (2005): 201–213.

them invisible to the immune system. Instead, poxviruses use a number of viral proteins to evade host innate (nonspecific) immune responses. They produce **viroceptors** and **virokines**. Viroceptors are altered cellular receptors that have lost their transmembrane anchor sequences. Consequently, these unanchored viral proteins are secreted from infected cells, where they sequester ligands onto the receptor portion of the protein. Virokines are secreted viral proteins that resemble host **cytokines**. The majority of poxvirus virokines and viroreceptors were discovered by investigators because of the deduced sequence homologies to their cellular counterparts. Besides secreted viral inhibitors, poxviruses also

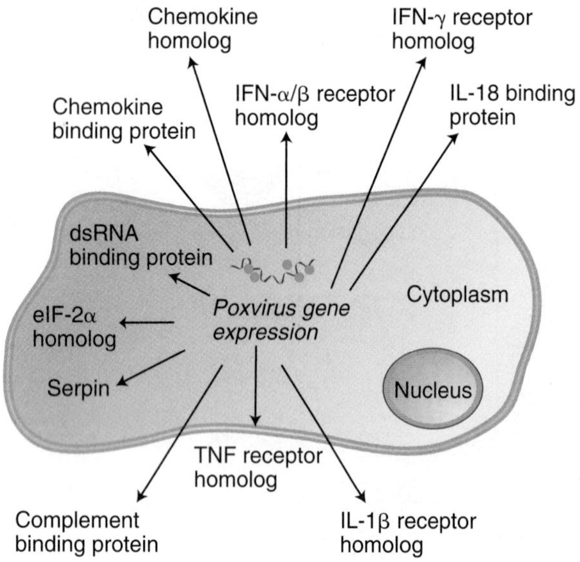

FIGURE 14-9 Representation of poxvirus immune evasion proteins. Adapted from Moss, B., and Shisler, J. L., *Semin Immunol* 13 (2001): 59–66.

produce intracellular proteins that interfere with signaling pathways within the infected cell. Examples of these poxvirus inhibitors are:

- Secreted complement regulatory proteins
- Secreted proteins that bind to **interferons**
- Secreted **interleukin**-18 binding proteins
- Secreted **tumor necrosis factor** homologs (**apoptosis** inhibitors)
- Intracellular **serpins** (serine protease inhibitors)
- Intracellular inhibitors of **PKR** (interferon pathway; see Chapter 7)

The field of poxvirus-encoded immune modulators is more than 25 years old, and discoveries continue in this area. Even though no one specific poxvirus uses all of these collective strategies, all seem to target the immediate innate immune response (**FIGURE 14-9**).

14.8 Human Genetics and Smallpox Resistance

In order for an infectious disease to cause selective pressure on host genetics, it would need to have a significant effect on morbidity and mortality, before reproductive age, over a long time. Smallpox was endemic throughout much of the world for more than 2000 years. In populations where epidemics occurred frequently, smallpox was a childhood disease, killing as many as 30% of its victims. Because human populations were exposed to the virus for generations, resistant genes should be common in smallpox survivors. In 1965 and 1966, Vogel and Chakravarti conducted a comparative study of 415 unvaccinated Indian smallpox survivors and victims. They discovered that people who were blood type A were seven times more likely to contract smallpox, three times more likely to develop a severe case of the disease, and twice as likely to die of it.

A 2003 study by Galvani and Slatkin suggested that smallpox is responsible for the genetic selection of the CCR5-32 deletion allele that confers resistance against HIV-1. CCR5 is a chemokine co-receptor required for HIV-1 infection (see Chapter 16). A homozygous CCR5-32 deletion allele provides almost complete resistance to HIV-1 infection. Heterozygous CCR5-32 deletion allele provides partial resistance and a slower disease progression. Because HIV is a relatively new disease, there have not been enough human generations or time for the selective rise of this resistance allele. Researchers have argued that other infectious agents are likely to have selected for this resistance allele. The most popular hypothesis suggests that this selective pressure was the result of bubonic plague (also referred to as the Black Death) pandemics in Europe. Galvani and Slatkin, however, argue that the plague outbreaks occurred in a shorter time frame—400 years—which is not enough time to generate the resistance allele. Smallpox, on the other hand, seems to be more consistent with selective pressure because the disease selection has been longer (more than 2000 years). The bottom line is that natural selection is a reality. The study of genetic susceptibility to infectious disease may contribute to the design of new therapeutics; for example, interference with the CCR5 chemokine receptor may have promise in the treatment of HIV.

14.9 Smallpox Eradication

The word **eradicate** comes from the Latin word *eradicare,* which means "to tear out by the roots." Several diseases were candidates for a global eradication program, such as hookworm, yellow fever, and the eradication of mosquito vectors that carry the malarial parasite and other disease-causing organisms. Smallpox is the only human disease that has officially been eradicated.

There were several reasons why it was possible to eradicate smallpox:

- Smallpox has a narrow host range (variola only infects humans).
- There is no long-term carrier state in humans.

a b

FIGURE 14-10 (a) WHO smallpox eradication workers faced rugged terrain as they rode four-wheel drive vehicles through Bangladesh countryside in an attempt to eradicate smallpox from the country. Off-road Land Rovers enabled the country's eradication program to succeed, giving public health practitioners access to even the most inaccessible rural villages (1975). (b) Nigerian child being immunized during the Smallpox Eradication and Measles Control Program of West Africa (1960).

- There is no animal reservoir.
- A highly effective and inexpensive freeze-dried vaccine was available.
- Surveillance of the disease was easy (e.g., observation of centrifugal rash).
- The WHO created a program to eradicate it.

In 1967, the WHO launched a **mass-vaccination strategy** to vaccinate 100% of the human population. At that time, approximately 10 to 15 million cases of smallpox occurred annually and the disease was endemic in 30 countries. Over time, the strategy gradually changed to **containment vaccination** or **surveillance containment** around newly discovered cases or outbreaks. This new strategy evolved after countries like Nigeria and India experienced outbreaks originating in areas where religious groups resisted vaccination. To prevent spread of these outbreaks to populated areas, healthcare workers went on house-to-house searches to control local outbreaks. This was so successful in preventing spread of the disease that controlling outbreaks through variations of surveillance containment was adopted (**FIGURE 14-10**). Evaluation and assessment procedures constantly evolved in response to new experiences and lessons learned in the field. Important innovations, such as the use of **smallpox recognition cards** (Figure 14-1), **rewards**, the **bifurcated needle** (**FIGURE 14-11**), **rumor registers**, and **containment books** were used in the field. Every case of fever with rash was recorded, monitored, and treated as smallpox unless proven otherwise. Four watch guards were placed at infected homes and all villages within ten miles of a case of known or suspected smallpox were searched and vaccinations were carried out. In 1977, the last natural case of smallpox was reported in Somalia. Smallpox was the first and only infectious disease to be eradicated.

14.10 Recombinant Vaccinia Viruses as Research Tools and Vaccines

Research on poxviruses did not languish after the eradication of smallpox. There was an explosion of interest in vaccinia virus, the prototype of poxviruses, in the 1980s. **Recombinant DNA technology** was applied to vaccinia virus. Inserting foreign

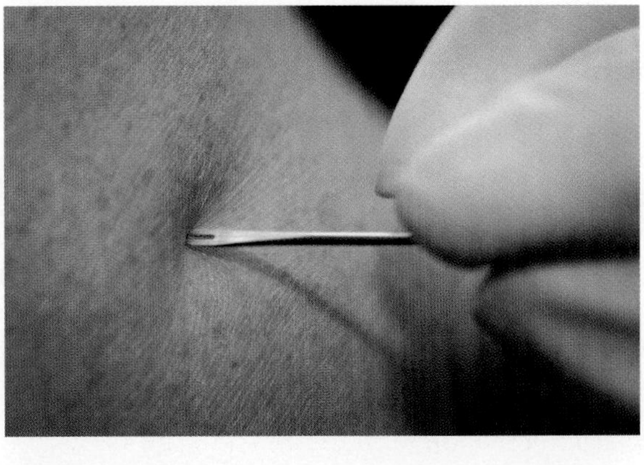

FIGURE 14-11 The bifurcated needle replaced the jet-injector to administer the smallpox vaccine. The jet injector could do over 1000 vaccinations in an hour, but its use was not practical and was too expensive to use for house-to-house vaccination. By 1970, the bifurcated needle replaced traditional methods and was used everywhere.

genes into vaccinia virus has provided a way to study the expression of these foreign genes within the cytoplasm of mammalian cells. These new or **recombinant vaccinia viruses** have multiple applications in research, gene therapy, and vaccinology; for example, a successful recombinant vaccinia virus rabies vaccine has been used to eliminate rabies in western Europe and the United States (see Chapter 13). The construction of a recombinant vaccinia virus is illustrated in **FIGURE 14-12**. Briefly, the foreign gene of interest (e.g., rabies glycoprotein gene) is inserted into a **plasmid DNA** using restriction enzymes and T4 DNA ligase. The gene is inserted downstream of a vaccinia virus promoter and in the middle of a nonessential vaccinia virus gene such as the **thymidine kinase** (***TK***) gene. The presence of the *TK* gene allows for the homologous recombination incorporation of the foreign gene into the vaccinia genome. The engineered plasmid DNA is transfected into TK–tissue culture cells (cells that do not contain a *TK* gene) simultaneously infected with vaccinia viruses. Absence of TK activity in host cells and disruption of the *TK* gene in vaccinia virus renders host cells resistant to toxic effects of 5-bromodeoxyuridine (BUdR). This selection enriches for cells that carry recombinant vaccinia virus.

Thymidine kinase is an essential *cellular* enzyme that is expressed during cell division activity. This enzyme allows a cell to utilize an alternate metabolic pathway for incorporating thymidine into DNA. BUdR is a mutagenic nucleoside that substitutes for thymidine in DNA, acting as a chain terminator. It causes breaks in the chromosomes of cells. It is used as an anticancer drug and is a potential antiviral drug. In the practice of creating vaccinia virus recombinants, a TK selection system has been developed. Only those cells and viruses that do not contain a functional *TK* gene can survive the effects of BUdR. Cells or viruses with an

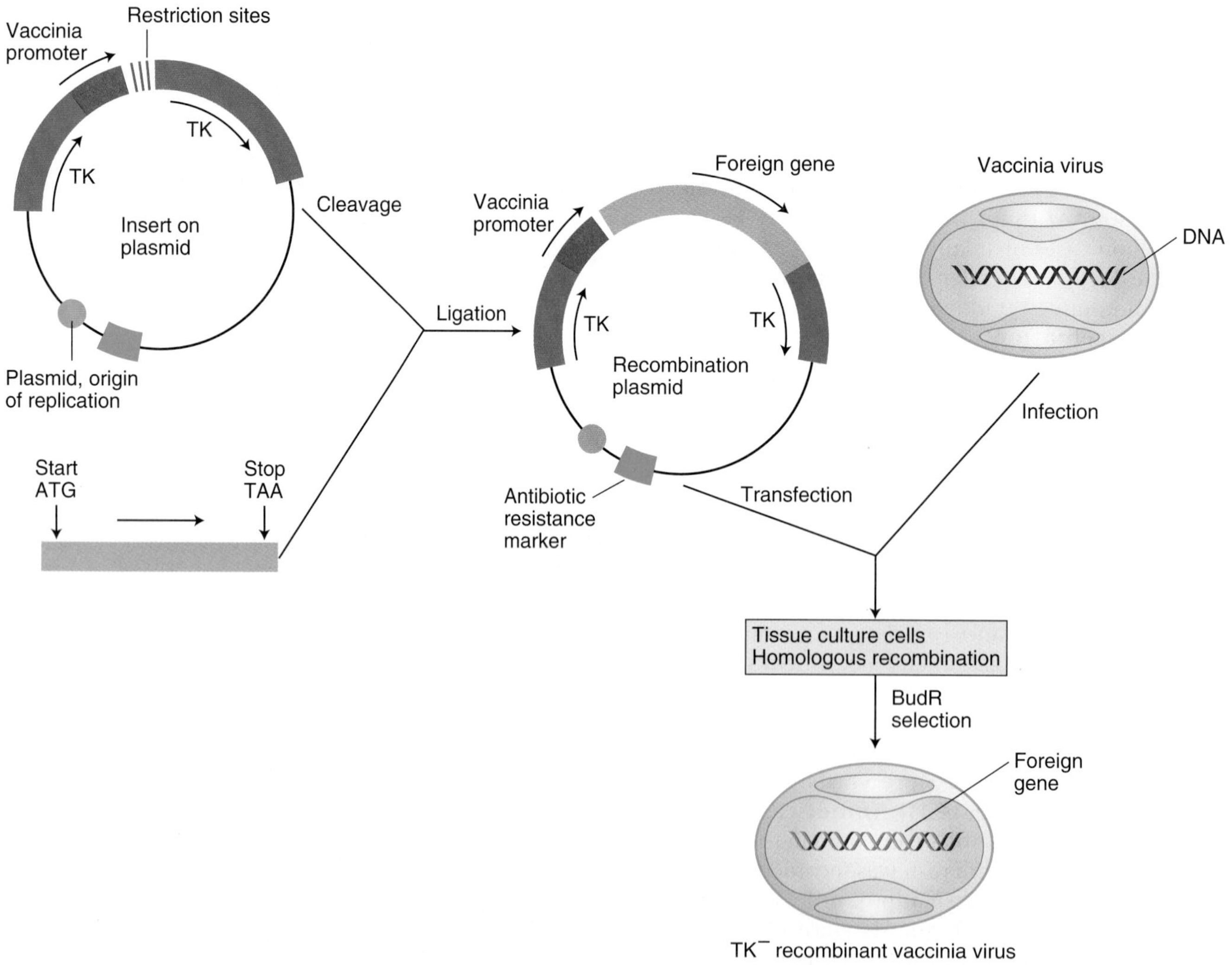

FIGURE 14-12 Flowchart depicting the creation of recombinant vaccinia viruses. Adapted from Mooney, H. A., and Bernardi, G. *Introduction of Genetically Modified Organisms into the Environment.* John Wiley & Sons, 1990.

intact *TK* gene will incorporate BUdR, resulting in genetic mutations and likely cell death and/or viral inactivation.

One of the major advantages of using vaccinia virus is the fact that it has a large dsDNA virus that can package more than 200 genes, including foreign genes. RNA viruses have smaller genomes and a packaging size limit. RNA viruses are also more difficult to genetically engineer. Foreign genes can be introduced into nonessential parts (e.g., the *TK* gene) of the vaccinia genome via homologous recombination. Hence, vaccinia viruses are able to express one or more unrelated genes in a cell culture system. This makes it a very attractive tool for molecular biologists. One very important application of vaccinia virus is using its recombinants as new **live vaccines** against infectious agents of humans and animals.

A disadvantage of live vaccinia virus is the complications that can follow vaccination. To overcome this, attenuated strains of vaccinia virus have been generated. The modification of these attenuated strains will result in the production of safer live vaccines. It is recommended that laboratory workers

VIRUS FILE 14-1 Deliberate Use of Myxoma Poxviruses to Control Australian Wild Rabbits

Thomas Austin, a resident of Victoria in southern Australia, loved to hunt rabbits during his trips to England in the mid-1800s. To promote his hobby, he asked his brother in England to ship him some rabbits. In 1859, Austin released 24 wild European rabbits that had been trapped and shipped by his brother onto his farmland in southern Australia.

Rabbits, of course, are prolific breeders. They quickly populated Austin's farm and his hunting tallies were published in the Victorian newspapers. In 1867, a newspaper reported he had killed 14,362 rabbits on his property for the year.

Austin didn't realize the rabbits were spreading throughout Australia. He had introduced a serious mammalian pest: an **invasive species**. The rabbits out-bred, out-ate, and out-competed other native Australian species. They multiplied so fast that two million rabbits could be shot or trapped each year without any noticeable effect on the population size. By 1900, the rabbit population was firmly established throughout all of Australia. Serious erosion problems occurred because the rabbits would consume native plants that would otherwise retain the soil. Some plant species became extinct. From 1901 to 1907, the western Australian government built a rabbit-proof fence that ran 1833 kilometers between Cape Keraudren in the north and Starvation Boat Harbor in the south to prevent their spread. The rabbits managed to get past the fence because kangaroos, emus, and wombats made holes under the fences. Further complicating matters, storms and drifting sand damaged the fences, allowing the rabbits to penetrate the fence, and they were well beyond the fenceline before it was completed. By the 1940s, there were hundreds of millions of rabbits, causing massive agricultural and environmental damage wherever they were found (**FIGURE VF 14-1**). The invasion of Australia by rabbits remains one of the fastest recorded for any mammal species.

Conventional control measures such as hunting and poisoning were applied to reduce rabbit populations. Hunting was labor intensive and limited to certain geographic areas. The disadvantage of poisoning was that the rabbit could not be used as food for either humans or predators thereafter. Meanwhile, **myxomatosis** was causing mild benign lesions (myxomas) in the South American wild rabbit population. A poxvirus (**myxoma virus**) was isolated and determined to be the culprit of myxomatosis. This viral disease was transmitted via a bite of an infected blood-feeding flea or mosquito. When European rabbits were brought into a South American animal laboratory, the myxoma virus caused a highly lethal infection in the European rabbits. It was a vastly different disease than observed in the South American rabbits. The rabbits became bedraggled with eyes almost totally swollen shut and lots of discharge coming out of their eyes and noses. The ears and heads were swollen. Myxomatous nodules were found on the eyelids, face, ears, and sometimes the rest of the body. The myxoma virus was shed in discharges and

(continued)

FIGURE VF 14-1 High densities of rabbits surrounding a waterhole in Canberra, Australia, during the 1963 drought. It depicts the problem of rabbit control prior to the introduction of myxomatosis.

the disease could be spread by close contact with infected rabbits. Myxomatosis killed 99.5% of the European rabbits about 10 to 12 days after infection.

Investigators in England began experiments in the 1930s to determine if myxoma virus infected any other common animals. They discovered the virus was species-specific. Fifteen years later (~1950), the virulent myxoma virus was introduced by Australian farmers via inoculation campaigns into wild rabbit populations. It was initially highly effective in controlling plague proportions of rabbits, but the impact was reduced when the virus and rabbit host adapted. The myxoma virus adapted into a benign strain and genetically resistant rabbits rebounded.

In 1993, the Australian government approved the genetic modification of myxoma virus. Scientists proposed to introduce genes that code for rabbit sperm and egg proteins into myxoma virus. In doing so, female rabbits infected with the recombinant virus would generate an antibody response toward the viral and sperm or egg antigens. The sperm and egg proteins would behave like foreign invaders, triggering an autoimmune response. This **immunocontraceptive vaccine** delivered by myxoma virus would neutralize the antigens, thereby inducing infertility in female rabbits. Recombinant myxoma viruses expressing reproductive rabbit antigens have been constructed. To date, the engineered viruses are being studied in the laboratory. This strategy for controlling wild rabbit populations is based on the control of fertility rather than mortality.

References

Fenner, F., and Fantini, B. 1999. *Biological Control of Vertebrate Pests: The History of Myxomatosis: An Experiment in Evolution.* Oxfordshire, England: CABI Publishing.

Gu, W., et al. 2004. "Immune response in rabbit ovaries following infection of a recombinant myxoma virus expressing rabbit zona pellucida protein B." *Virology* 318(2):516–523.

Hayes, R. A., and Richardson, B. J. 2001. "Biological control of the rabbit in Australia: Lessons not learned?" *Trends Microbiol* 9(9):459–460.

Holland, M. K., and Jackson, R. J. 1994. "Virus-vectored immunocontraception for control of wild rabbits: Identification of target antigens and construction of recombinant viruses." *Reprod Fertil Dev* 6:631–642.

Mackenzie, S. M., et al. 2006. "Immunocontraceptive effects on female rabbits infected with recombinant myxoma virus expressing rabbit ZP2 or ZP3." *Biol Reprod* 74:511–521.

Morell, V. 1993. "Australian pest control by virus causes concern." *Science* 261:683–684.

who directly handle cultures or animals infected with vaccinia virus be vaccinated. Laboratory workers using highly attenuated strains of vaccinia virus, such as the modified vaccinia Ankara (MVA) virus, the New York vaccinia (NYVAC) virus, or avipox viruses such as ALVAC (attenuated canarypox) and TROVAC (attenuated fowlpox) do not require vaccination. These attenuated viruses are called **suicide vectors**. All of them undergo **abortive replication** in mammalian cells. In other words, these viruses can infect cells, but there is a block in their life cycle such that early viral proteins are produced but late proteins (and therefore full infectious particles) are not produced in mammalian cells. **These vectors can express high levels of a foreign antigen without being pathogenic to a host**. To engineer and work with these attenuated viruses, they can be propagated in chick embryo fibroblasts (CEFs) or sometimes baby hamster kidney (BHK-21) cells. All of these attenuated poxviruses have a very restricted host range. Recombinant vaccinia or other attenuated poxvirus vectors have been genetically engineered to contain genes from a wide variety of infectious agents including HIV and severe acute respiratory syndrome-coronavirus (SARS-CoV).

14.11 Prevention: Vaccines

Smallpox Vaccine History

Attempts to prevent the disease that caused "the speckled monster" date to practices in China beginning in the late 17th century. Two such practices were to inhale or plug the nose with powdered smallpox scabs, or to put the underwear of an infected child on a healthy child for several days. **Variolation** was practiced by Hindus in India during the 16th century. It is the intentional inoculation of dried smallpox scabs into the skin of an uninfected individual, causing a mild form of the disease and immunity against subsequent exposure to variola virus. The practice spread from central Asia, east to China, and then west to Africa and the Ottoman Empire. This practice involved the direct inoculation of smallpox into susceptible persons, so it is not surprising that 2% to 3% of variolated people died, as compared to 30% who died when they contracted smallpox naturally.

Lady Mary Wortley Montague (1689–1762), the wife of the British ambassador to Turkey, learned about variolation (referred to as *engrafting*) during her travels to Constantinople (now Istanbul) in 1717, when she observed Muslims using this technique. Montague expressed interest regarding this practice because she had survived smallpox in 1715. The disease left her without eyelashes and a badly pockmarked face; her brother had died from it. After returning to London, Montague had the embassy surgeon, Charles Maitland, engraft her 4-year-old daughter in 1721. Maitland gained permission to conduct a study in which six prisoners underwent the variolation procedure. The prisoners survived after exposure to two children with smallpox. They had a milder form of the disease and developed immunity. This study was witnessed by members of the Royal Society of London, and based on the results these inoculations began the acceptable medical practice of variolation in England.

Edward Jenner (1749–1823), a physician in England, is credited with **cowpox vaccination**, the first attempt to control smallpox without transmitting the disease. As mentioned in Chapter 1, Section 1.1, Jenner experimented with the folklore that milkmaids who contracted cowpox "did not take the smallpox," and in 1796 he variolated the 8-year-old son of a local farmer with fluid from the cowpox pustules from the hand of a local dairymaid. The boy developed a slight fever and a few lesions but remained unscathed for the most part. A few months later, the boy was injected with smallpox again and still failed to develop the disease. Jenner continued these experiments and published his findings in *Inquiry into the Cause and Effects of the Variolae Vaccinae* in 1798. The immediate reaction to Jenner's work was ridicule. Critics, especially the clergy, claimed it was repulsive and ungodly to inoculate someone with material from a diseased animal. A satirical cartoon of 1802 showed people who had been vaccinated sprouting cow heads. Despite these initial reactions, Jenner did go on to become famous, and the word vaccination was coined.

In the 1800s, **arm-to-arm passage** (or **arm-to-arm vaccination**) became a practice to prevent smallpox. The fluid or *lymph* taken from a vesicle or pustule from a previously vaccinated person was passaged arm-to-arm between humans. Occasionally, the vaccine was lost and further strains were obtained from cows or horses. Arm-to-arm passage was not a safe practice. Sometimes the vaccine became contaminated with variola major or minor, which sometimes resulted in the inadvertent transmission of other infectious diseases such as syphilis to tuberculosis. At least one recorded epidemic of hepatitis B was contributed to arm-to-arm passage. In 1898, the Vaccination Act banned arm-to-arm vaccination.

As early as 1876, the New York City health department constructed a laboratory to prepare smallpox vaccine. The laboratory was in a three-story building that included a stable on the ground floor containing stanchions to house heifers (**vaccinifers**) for vaccine production and an operating room and a sterilizing room on other floors. Cowpox lesions (referred to as **seed lymph**) from infected calves were smeared into superficial linear bloody incisions made with sharp knives on the shaved, washed, and sterilized skin of the belly or flank of uninfected calves. Vesicles appeared within the scarified incisions in a few days. Four days later after the appearance of lesions, the skin was scraped with a currette and the **vaccine (calf) pulp** was collected in a jar. Glycerin was added to prevent microbial contamination and act as a preservative. The pulp was emulsified and clarified via centrifugation to yield **vaccine lymph**. This early vaccine product was very crude and was contaminated with fragments of cells and bacteria.

By 1902, smallpox vaccination was practiced throughout industrialized countries. Little is known about the origins or history of the viruses present in the seed lymph used for vaccination until the 1960s. Vaccine technology was not consistent among manufacturers.

Viruses were passed through a variety of vaccinifers such as cows, sheep, water buffalo, and sometimes rabbits, horses, and even humans. Seed lymph was obtained from "smallpox direct," cowpox, horsepox, sheeppox, goatpox, or even vaccinia in the human body. As protocols became standardized, vaccines produced in animal skin were tested for the presence of specific pathogenic microbes. These specifications allowed for the presence of a low concentration of nonpathogenic bacteria and fungi. National surveys in the United States revealed that 266 per million vaccinees experienced life-threatening complications after a primary vaccination.

By 1967, the WHO recommended the use of a seed lot containing the Lister or the New York City Board of Health strain (NYCBOH) grown by Wyeth Laboratories for vaccine production. These two were the least pathogenic of the effective vaccines. **FIGURE 14-13** shows the preparation of freeze-dried smallpox vaccine from virus grown on the skin of a calf around 1980.

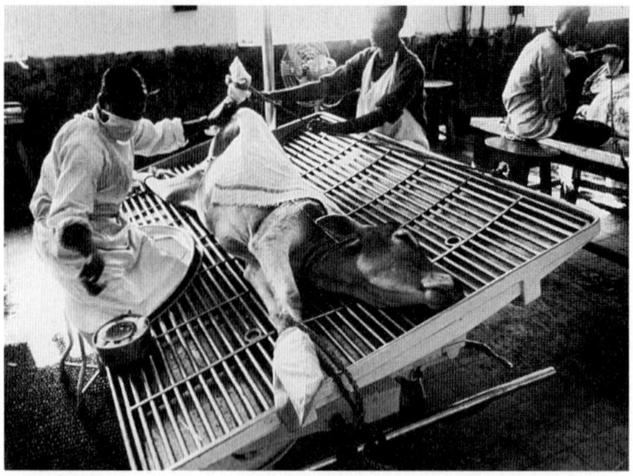

FIGURE 14-13 Two laboratory workers preparing smallpox vaccine (1980). The calf is bound to a grated table. Vaccinia virus was introduced into the scarified areas of skin.

Mass smallpox vaccination in the United States was discontinued in 1972. The country created a strategic national **stockpile of smallpox vaccine** available for biodefense consisting of the Wyeth Dryvax produced from 1980–1982 and the Connaught (now Aventis-Pasteur) vaccine WetVax produced in the 1950s. There was enough vaccine to vaccinate everyone in the country in the event of an emergency. In 2008, the Dryvax smallpox vaccine prepared from calf lymph was replaced by **ACAM2000**, a more modern product manufactured by Acambis.

The vaccine stockpiles of the European Union and the WHO were also produced from calf-lymph. These vaccines have not been tested by modern methods for the presence of mycoplasma or bovine viruses. The WHO has agreed to develop safer, tissue culture-based vaccines that contain **attenuated** strains of vaccinia virus.

Smallpox Vaccination Today

Currently the only individuals in the United States vaccinated against smallpox are laboratory workers who do research with poxviruses, healthcare workers, and members of the military. After the events of September 11, 2001, the Bush administration called for a half-million healthcare workers to be vaccinated against smallpox. By 2004, only 40,000 had complied.

The Dryvax Vaccine

The smallpox vaccine available in the United States used through 2007 was the Dryvax, live-attenuated preparation of vaccinia virus. Dryvax was cultured in the skin of calves and freeze-dried. It was originally created in the late 1800s by American Home Products, which later became Wyeth Laboratories. Wyeth ended production of the vaccine in 1978.

The Dryvax vaccine was a mixed population of vaccinia strains. It contained trace amounts of the antibiotics polymyxin B, streptomycin, tetracycline, and neomycin. The vaccine was reconstituted in 50% glycerin and a small amount of phenol as a preservative. The vaccine was given with multiple punctures of a bifurcated (two-pronged) needle into the deltoid muscle of the arm: 3 times for a

VIRUS FILE 14-2 Farmer Jesty and the Importance of Self-Promotion

". . . in science credit goes to the man who convinces the world, not the man to whom the idea first occurs."

Francis Galton

Twenty-two years before Edward Jenner was credited with cowpox vaccination, Benjamin Jesty, a farmer in the Village of Yetminster in Dorset, United Kingdom, observed dairymaids with unblemished complexions. They contracted cowpox but were never afflicted with smallpox even after nursing relatives suffering from the disease. Jesty himself had also contracted cowpox while working with cattle as a young man. He was intelligent, innovative, flourishing financially, and a prominent supporter of the community during a time when the approach to farming was evolving. Crop rotation was introduced. Seed quality was improved and new crops were being grown on a commercial scale. Cattle were being bred for the production of leather and meat, rather than just for dairy products.

Smallpox was widespread throughout Europe, Africa, and the Americas. It was a major endemic disease everywhere in the world except Australia. There were large outbreaks of smallpox in England at a time when 80% of deaths from the infection were children younger than five years of age. In his book entitled, *History of England from the Accession of James the Second,* T. B. Macaulay wrote,

> *"The smallpox was always present, filling the churchyard with corpses, tormenting constant fears of whom it had not yet stricken, leaving on those whose lives it spared the hideous traces of its power, turning the babe into a changeling at which the mother shuddered, and making the eyes and cheeks of the betrothed maidens objects of horror to the lover."*

Jesty was not alone in his fear of smallpox. In 1774, smallpox was making its way through the farming community. He was determined to spare his family from smallpox. He had the idea that substituting cowpox material was a safer form of inoculation than using smallpox material. He discussed this idea with his wife and at least one other person. He carried out his idea by taking his wife and two sons, three-year-old Robert and two-year-old Benjamin, to a nearby herd of grazing cows that he knew had symptoms of cowpox. He took a stocking needle and transferred material from the cowpox lesions present on the cow teats, and punctured the skin below his wife's elbow. Stocking needles were used to knit knee-length stockings, the ends were slim to a point and ideal for piercing the skin. He repeated the procedure with his sons. His wife's arm became inflamed. She developed a fever but recovered in due course. Word spread throughout Dorset. As word got out about his "procedure," members of the medical, farming, and religious communities scorned him. People were superstitious. Despite the negative reaction in the community, his wife and sons never suffered from smallpox even after repeated exposures during smallpox epidemics. The boys were deliberately exposed to smallpox by Jesty at least once.

Farmer Jesty's vaccination technique was performed 22 years before Jenner is credited with cowpox vaccination. He did not record his vaccinations. He made no attempt to publicize his technique until he heard of Jenner's first award. His main motive was the well-being of his family, and he didn't extend his technique to the greater community. Edward Jenner, on the other hand, had a reputation as a cautious investigator and prolific author. His publications and self-promotion encouraged the widespread adoption of his vaccination procedure for the masses. Jenner is credited with smallpox vaccination and the development of vaccination in general.

(continued)

VIRUS FILE 14-2 Farmer Jesty and the Importance of Self-Promotion (continued)

He received the acclaim of the medical profession. The Original Vaccine Pock Institute awarded Jesty a testimonial scroll, 15 guineas expenses, and a pair of gold mounted lancets and arranged for his portrait to be painted by the artist Michael W. Sharp.

Others like Benjamin Jesty, Dr. Zabdiel Boylston (1679–1766), and Reverend Cotton Mather (1663–1728) utilized similar inoculation practices in the New World. Boylston and Mather were influential but not as relentless in promoting how vaccination changed the way medicine was practiced. Jenner certainly was a hero, but he is one of many. Breakthroughs in medicine often result from countless small steps rather than one giant leap. Farmer Jesty died seven years before Jenner. He wrote his own epitaph. It was his way of publishing notice of his vaccinations posthumously (**FIGURE VF 14-2**).

FIGURE VF 14-2 Benjamin Jesty's tombstone, which is located in a St. Nicholas of Myra Church cemetery at Worth Matravers, UK.

References

Gross, C. P., and Sepkowitz, K. A. 1998. "The myth of the medical breakthrough: Smallpox, vaccination, and Jenner reconsidered." *Int J Infect Dis* 3(1):54–60.

Hammarsten, J. F., et al. 1979. "Who discovered smallpox vaccination? Edward Jenner or Benjamin Jesty?" *Trans Am Clin Climatol Assoc* 90:44–55.

Pead, P. J. 2003. "Benjamin Jesty: New light in the dawn of vaccination." *Lancet* 362:2104–2109.

Pead, P. J. 2006. *Vaccination rediscovered: New light in the dawn of man's quest for immunity.* Chichester, West Sussex, England: Timefile Books.

Pead, P. J. 2006. "Portrait: Benjamin Jesty: The first vaccinator." *Lancet* 368:2202.

primary vaccination and 15 times for a revaccination. The virus was live, so it could spread to other parts of the body by inadvertent scratching or other contact. Thus the vaccination site was covered and the dressing changed every one to three days. Contaminated dressings were handled as infectious waste. **FIGURE 14-14** shows a typical smallpox reaction. Immunity lasted approximately three years.

Dryvax Vaccine Complications

Data from the 1960s suggest that 1000 of every 1 million vaccinees have a serious reaction to the vaccine. About one in 1 million of primary vaccinees will die, and one in 4 million revaccinees will die. Rates of adverse reactions and deaths could vary considerably today. Serious side effects have occurred in people with skin conditions such as eczema or atopic dermatitis, as well as in people with weakened immune systems or people with the following risk factors: heart disease, high blood pressure, diabetes, and high blood cholesterol. Severe complications that may follow a primary or revaccination include **encephalitis**, **vaccinia necrosum** (**FIGURE 14-15a**), **eczema vaccinatum** (Figure 14-15b), **encephalomyelitis**, **encephalopathy**, and **myopericarditis**

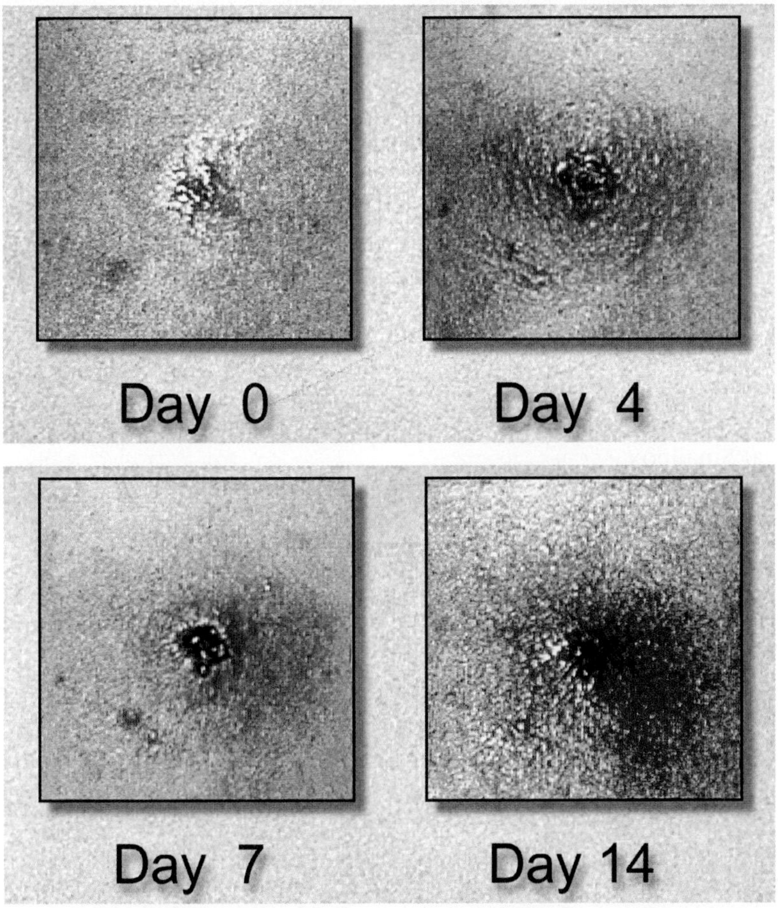

FIGURE 14-14 Expected vaccine site reaction and progression following a smallpox vaccination. The pustule dries from the center outward and forms a scab that separates 14–21 days after vaccination, leaving a pitted scar.

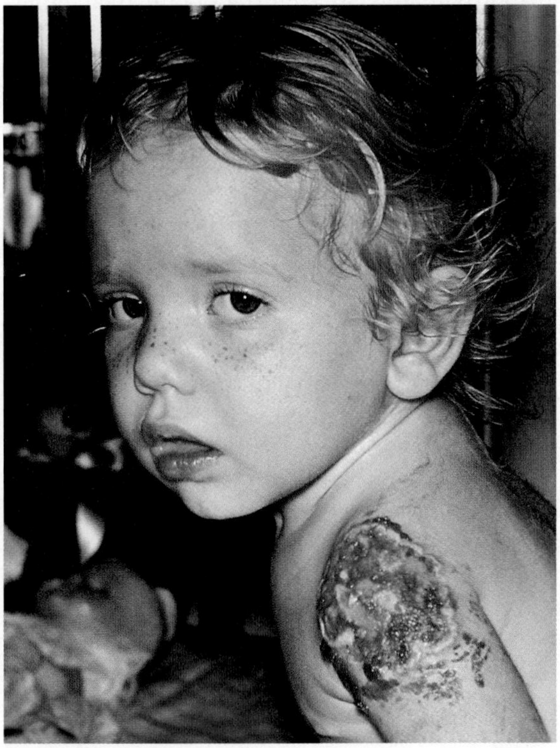

a

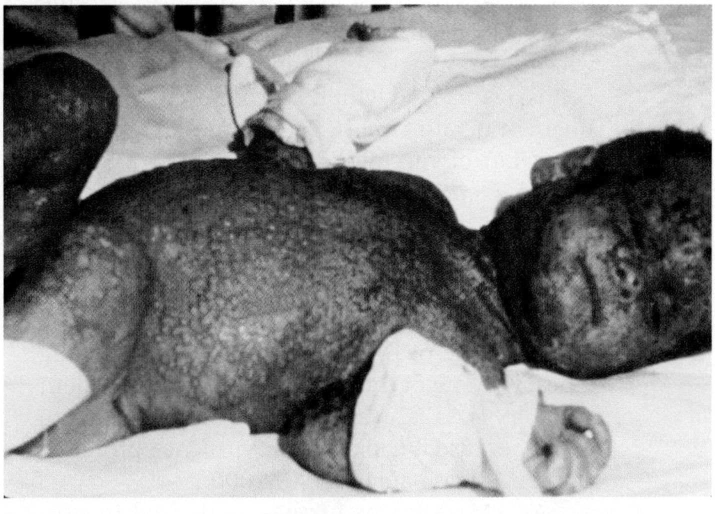

b

FIGURE 14-15 (a) This young child is suffering from a necrotic vaccination site. It is one of the most severe complications of the smallpox vaccine and is almost always life-threatening. This child required a skin graft to correct the necrosis. Those most susceptible to this condition are the immunocompromised. **(b)** This eight-month-old boy developed eczema vaccinatum after acquiring a vaccinia infection from a recently vaccinated sibling. Eczema vaccinatum occurs in people with atopic dermatitis who come in contact with vaccinia virus.

(**heart inflammation**). Adverse reactions may be treated with **vaccinia immunoglobulin** and **cidofovir**. Any one of these adverse effects may result in permanent disability and/or death. About two-thirds of complications after smallpox vaccination might have been preventable and avoided with better screening during the smallpox eradication era. Adverse reactions in the United States might be higher today than previous reports because there is a much higher proportion of people at risk for complications as a result of organ transplantation, cancer, cancer therapy, radiation, immune-modulating medications, eczema, atopic dermatitis, and other illnesses such as AIDS. The incidence of adverse cardiac events related to smallpox vaccinations administered during the National Smallpox Vaccination Program (NSVP) in 2002 received widespread attention. Safer vaccines were needed in case of an emergency bioterrorism event that would require mass vaccination. For this reason, the Dryvax vaccine was replaced with ACAM2000, a cell culture–derived smallpox vaccine.

The ACAM2000 Vaccine

The advances in smallpox vaccine production technology were made possible because cell culture-based virus propagation could replace growth of vaccinia viruses on the skin of cows. Growth in cell culture allowed for many improvements in the quality of vaccine product, such as eliminating possible contaminants and reducing lot-to-lot variation. ACAM2000 is a live vaccinia virus vaccine derived from the plaque purification of Dryvax vaccine strains (Wyeth Laboratories, Murietta, Pennsylvania, calf lymph vaccine, New York City Board of Health Strain). Six isolated clones (name CL1-CL6) of vaccinia from a pool of 30 vials of Dryvax were plaque purified and passaged five times in a human embryonic lung fibroblast cell line (MRC-5 cells) and subsequently tested for virulence. The goal of vaccine production was to produce a vaccine that would meet or exceed the safety profile of Dryvax without compromising efficacy. Some of the isolates were more virulent than Dryvaz. CL2 was chosen as the new vaccine strain because it had a similar virulence profile to Dryvax. It was named ACAM1000.

ACAM1000 was further evaluated in preclinical trials (animal studies). CL2 was at least as safe, if not safer, than Dryvax in animal models. ACAM1000 proceeded to human clinical trials with similar results to the Dryvax vaccine. Because of the success of the ACAM1000 vaccine in preclinical and clinical tests, along with the growing concern of terrorist threats after the attacks of September of 2001, the United States government increased their contract to Acambis to 209 million vaccine doses. In order to produce such a large quantity of vaccine with as few passages of ACAM1000 possible, Acambis partnered with Baxter BioScience (located in Deerfield, Illinois). They inoculated african green monkey kidney (Vero) cells in large 1200-liter bioreactors with ACAM1000. The resulting virus stock was renamed ACAM2000.

Cellular debris was removed by a large-pore filtration step and the cellular genomic DNA was digested. The vaccine lots were tested to be free of bacterial or fungal contamination and for human and bovine pathogens. ACAM2000 is the new U.S. stockpile of smallpox vaccine. There are over 200 million doses of ACAM2000. ACAM2000 is used to vaccinate all Department of Defense personnel. There remain concerns related to cardiac complications. There is still a need for a safer smallpox vaccine for the general population and especially for the immunocompromised, infants, and others with preexisting conditions, for whom ACAM2000 cannot be given.

14.12 Orthopoxvirus Antivirals

The discontinued use of smallpox mass-vaccination with vaccinia virus has resulted in the lack of herd immunity against variola and monkeypox viruses. Infection with variola major is associated with an approximately 30% mortality rate. In an emergency today (e.g., intentional use of variola major in a bioterrorism event), immunocompromised individuals or other high-risk groups cannot be inoculated with vaccinia virus because of the complications associated with the smallpox vaccine. There is no licensed drug treatment available for clinical smallpox; thus, the focus has been on the development of antiviral strategies against variola as an alternative or additional strategy for the prevention or postexposure treatment of smallpox and/or monkeypox. **Cidofovir** (also called **HPMPC** or **Vistide**) has been used to treat Molluscum contagiosum in AIDS patients through intravenous injection. It has also been used as a topical treatment (1% cream) for orf infections. It causes the pox-induced lesions to regress after two to four months of therapy. Cidofovir has been licensed for the treatment of cytomegalovirus retinitis in AIDS patients as well. Cytomegalovirus is a large herpesvirus that contains a dsDNA genome (see Chapter 15). Cidofovir is a nucleoside/nucleotide analog that interferes with virally encoded DNA polymerases during viral replication. It is currently being tested as an antiviral treatment after lethal monkeypox infection of cynomolgus monkeys. As described

TABLE 14-4 Antivirals in the Pipeline Against Smallpox

Drug	Viral Target/Mechanism
Cidofovir (S-HPMPC)	Viral DNA polymerase inhibitor
CMX-001	Viral DNA polymerase inhibitor
N-methanocarbathymidine (N-MCT)	Viral DNA polymerase inhibitor
Neplanocin A	Viral S-adenosyl-L-homocysteine hydrolyase inhibitor
WYCRCK and KCCRCK peptides Other small unnamed molecules*	Viral Type 1 topoisomerase inhibitor
ST-246	Inhibition of extracellular virus formation
Gleevec CI-1033 Aurintricarboxylic acid (ATA)	Host signal transduction cascades (e.g., EGF receptor or EGFR kinase or ERK phosphorylation)

*Bond, A., et al. 2006. "Novel and specific inhibitors of a poxvirus type 1 topoisomerase." *Mol Pharm* 69(2):547–557.

earlier in this chapter, monkeypox causes a disease similar to human smallpox and this animal model is being used to measure the effectiveness of cidofovir as a preventative or postexposure antiviral therapy.

An ideal antiorthopoxvirus drug has yet to be discovered. The best antiviral drugs target essential viral enzymes (**TABLE 14-4**). Several viral gene candidates play essential roles in poxvirus replication. Potential targets include viral genes involved in DNA replication and recombination, mRNA synthesis and modification, and protein modification and virus assembly. Selected poxvirus enzyme targets are described in detail by Yuk-Ching Tse-Dinh (see References). Ultimately, the development of antiviral drugs against smallpox could deter terrorists from releasing variola because its effects would be diminished. Likewise, new antiorthopoxvirus therapies are likely to be useful in the treatment of other human orthopoxvirus infections.

14.13 Variola Virus in the Laboratory

There have been two fully documented laboratory-associated outbreaks of smallpox. Both outbreaks occurred in the United Kingdom after smallpox was no longer endemic there. The first outbreak occurred at the London School of Hygiene and Tropical Medicine in 1973. A 23-year-old female assistant in a mycology laboratory contracted smallpox after observing a friend inoculating chorioallantoic membranes of eggs with variola viruses at an open bench in a poxvirus laboratory in the same building. This laboratory accident resulted in four cases of smallpox, of which two were fatal. It was concluded that the young assistant inhaled aerosoled virus. At the time, variola was not studied within the setting of a BSL-4 laboratory built for dangerous pathogens.

The second laboratory-associated outbreak generated a great deal of publicity in the media. Janet Parker, a 40-year-old medical photographer in the Anatomy Department of the Medical School of the University of Birmingham, came down with smallpox in August 1978. After the appearance of the rash and a diagnosis was confirmed, she was placed in an isolation hospital. Health authorities identified 290 contacts with Mrs. Parker during her illness. These included her mother and other relatives and colleagues. All were vaccinated and placed under surveillance at home. It was concluded that Mrs. Parker contracted smallpox by an airborne route, either through a duct or telephone room while visiting the medical microbiology department where she regularly entered the animal pox room. Mrs. Parker died; her mother contracted smallpox but survived. The tragedy continued with the suicide of Professor Henry Bedson, the chief of the poxvirus laboratory. As a result of this outbreak, close attention was paid toward the handling of variola stocks in laboratories. As mentioned earlier in the chapter, today variola is a BSL-4 pathogen that is stored under high security at two repositories in the world. Any research studies using variola must be conducted in these two facilities.

14.14 The Variola Destruction Debate

Since the eradication of smallpox, it has been repeatedly proposed to destroy all of the last remaining stocks of variola virus that are stored in the two repositories. The destruction has been postponed indefinitely, though. New research proposals to develop diagnostics, safer vaccines, and antivirals against variola have been approved by the WHO's advisory committee. This research has been driven forward because of the concerns of bioterrorism.

All approved research must be conducted at the two laboratories where the viruses are stored. These plans have been aggressively protested by at least two advocacy groups: the Sunshine Project of Austin, Texas, and the Third World Network, with headquarters in Penang, Malaysia. In 2005 the U.S. Congress passed an amendment, later signed into law by President George W. Bush, which is part of the **Intelligence Reform and Terrorism Prevention Act**. The Act bans the synthesis of variola virus from scratch (e.g., using a DNA synthesizer and sequence information from a database such as the **Poxvirus Bioinformatics Resource Center**, which is available via the Internet) but does not prohibit researchers from engineering similar poxviruses to contain variola genes. In May 2011, the World Health Assembly voted 27-7 in favor of keeping the stockpiles of the poxvirus that causes smallpox to study for another three years. They will vote again in 2014.

14.15 Bioterrorism and Biowarfare

The use of variola major as a biological weapon has a long history. The most documented incident was the intended use of variola major virus as germ warfare against the Native Americans during the French and Indian Wars (1754–1763). In the 18th century, the British fought France and its Indian allies for possession of what was to become Canada during the French and Indian Wars. In 1763, while in Fort Pitt on the Pennsylvania frontier, British captain, Simeon Ecuyer gave two blankets and a handkerchief, taken from smallpox patients in the Fort's infirmary, to the Delaware Indians surrounding the fort at a peace-making parley. It was his hope to induce a smallpox epidemic among the Indians and defeat them.

Another example of intentional transmission of variola major occurred much later, in the 1860s. Following the Civil War, there was an unprecedented increase in student enrollment in the medical colleges of Ohio, resulting in a high demand for cadavers. Physicians and their students needed to renew their own knowledge of anatomy or improve their surgical techniques. Professional grave robbers, which were also called body snatchers or resurrectionists, provided services to meet the demands. William Cunningham (1807–1871) was a professional grave robber for the Medical College of Ohio in Cincinnati when several anatomy students played a trick on him. In an act of revenge, he delivered a corpse of a recently buried smallpox victim to the anatomy dissecting room and a number of anatomy students were infected.

Smallpox as a bioterrorism weapon coincided with its declared eradication in 1980. The World Health Assembly recommended that all countries stop vaccination and that smallpox virus stocks be destroyed or transferred to one of the two specific reference laboratories. Countries complied; however, viral stocks in Russia were moved to a biological weapons development facility known as Vector. The Russians began producing upwards of 20 tons of variola virus for the use in bombs and intercontinental ballistic missiles. Much research focus was placed on optimizing virus viability during and after dissemination of a bioweaponized bomb or missile. Funding in the 1990s declined for the Russian laboratories and many scientists left to work in other laboratories. The fate of the viral stocks is uncertain.

The bottom line is that acts of bioterrorism occurred in the past and they will inevitably occur in the future. These types of acts are intended to generate panic and fear through their unpredictable manner.

Ordinary citizens are much more susceptible to bioweapons than are military targets. Military personnel are vaccinated against potential biological agents. In contrast, civilian populations will require rapid diagnosis and antimicrobial treatment and vaccination if available. Preparation is essential in order to protect ordinary citizens. This preparation takes two forms:

1. Intelligence and law enforcement activities.
2. Public health activities to prepare and respond to such acts to lessen their impact.

Is the use of variola major as a bioweapon a realistic threat today? What would happen if a terrorist infected him- or herself intentionally with variola virus and walked through a crowded subway or shopping mall? Would this event cause a massive outbreak? In 1947, a situation somewhat like this happened in New York City. An American citizen, Eugene LeBar, contracted smallpox while in Mexico. Shortly after arriving in New York City, he developed a rash characteristic of hemorrhagic smallpox. He was hospitalized but not diagnosed until two more cases occurred. The new cases were individuals who had had contact with LeBar during his hospitalization. LeBar subsequently died. Within three weeks, 5.35 million New York City residents were vaccinated. The rapid vaccination response curbed this epidemic to a total of 2 deaths and 12 cases in less than a month.

What if a situation similar to what happened in 1947 occurred today? Smallpox has been eradicated globally and vaccination against variola ceased in the 1970s in most countries. As a result, today's human population is immunologically naive and susceptible to infection by variola virus.

The fact that smallpox has been eradicated with a vaccine is a good indication that we can contain and control this disease (via vaccination and isolation of patients). Other things to consider are that it requires a sophisticated biosafety laboratory to grow the virus. Terrorists who consider using the virus as a weapon would have to either intentionally infect him- or herself or produce large volumes of virus that would be aerosoled into the human population, a method that has never been done. If a terrorist did manage to inoculate him- or herself with variola major, it would be very difficult to disseminate the virus because the individual would be very ill during the most contagious time period of the disease. From the historical record, most individuals who contracted smallpox did so when they were caring for an infected person (i.e., they were in very close contact). It is also known that if a person is vaccinated within three to four days of exposure to the virus the disease will either be prevented or greatly ameliorated.

Preparing is the key to preventing the return of smallpox. There is currently a stockpile of vaccine to cover the entire U.S. population of healthy persons. Military staff members are vaccinated, and healthcare and emergency workers may volunteer to be vaccinated. Unfortunately, the current vaccine cannot be given to immunocompromised or other high-risk individuals. At the time of this writing, 46% of all hospitals in the United States have written hospital terrorism preparedness emergency response plans. In addition, staged drills for biological attacks and severe epidemics are conducted at least once a year. Much work remains; however, through public education, surveillance, and continual national and international preparedness, the vulnerability to **weapons of mass destruction** will be significantly reduced.

Summary

. . . the confluent smallpox—invented perhaps as the cruelest remedy for human vanity . . .

Lawrence Durrell (1912–1990), playwright

Smallpox was one of the most devastating infectious diseases in human history and is the only human disease to be intentionally eradicated from the world's human population. It has contributed such words as variolation, inoculation, vaccination, immunity, and bioterrorism to our language.

Eradication of smallpox was possible for several reasons:

- Smallpox has a narrow host range (variola only infects humans).
- There is no long-term carrier state in humans.
- There is no animal reservoir.
- A highly effective and inexpensive freeze-dried vaccine was available.
- Surveillance of the disease was easy (e.g., by observation of centrifugal rash).
- The WHO created a coordinated program to eradicate it.

In addition to smallpox, there are other diseases caused by **poxviruses** that afflict many species of animals and insects (e.g., camelpox, canarypox, skunkpox). Poxviruses infect a diversity of hosts, but each poxvirus is very species-specific; for example, rabbitpox only infects rabbits and smallpox only infects humans.

Smallpox is caused by variola major or variola minor, viruses that belong to the *Poxviridae* family. Variola major is the severe and most common form of smallpox. It can cause ordinary, vaccine-modified, flat, or malignant pox and hemorrhagic smallpox. The key clinical features used to diagnose smallpox are recognition of a high fever and the presence of a centrifugal rash. If a smallpox case were suspected today, pustules would be analyzed by experts at the CDC for definitive testing via electron microscopy and other molecular methods such as ELISAs or PCR techniques. Other human poxvirus infections are caused by monkeypox, Molluscum contagiosum, and vaccinia viruses.

Very little is known about the pathogenesis of smallpox because modern molecular biology has not been applied to the study of live variola virus. This has occurred because smallpox was eradicated before advanced molecular techniques were developed. Variola virus strains are stored under high security in two international repositories located in the United States and Russia. The WHO has approved research proposals that involve the development of an animal model to study variola pathogenesis. Poxvirus research is now focused on the development of rapid diagnostics, antivirals, and a safer vaccine against variola.

The majority of molecular biology we know regarding poxviruses comes from research on vaccinia virus. Vaccinia is the vaccine strain against smallpox and is the prototype of poxviruses. Genetically, its closest relative is buffalopox. Vaccinia virus readily grows in almost all cell lines tested. It is a large, brick-shaped virus that has at least two infectious forms produced during its life cycle. The extracellular enveloped virion (EEV) contains two membranes wrapped around its large linear dsDNA genome, whereas the intracellular mature virion (IMV) contains one membrane derived from Golgi wrapping. No specific host-cell receptors for viral entry have been identified. One of the unique hallmarks of poxviruses is that they have acquired all of the functions necessary for genome replication in the cytoplasm even though they possess DNA genomes. A complete early transcription system is synthesized late during infection and is packaged into the viral core. This viral transcriptional machinery consists of a DNA-dependent RNA polymerase, a transcription factor (VETF), capping and methylating enzymes, and a poly(A) polymerase. The virus has the capability to synthesize mRNAs that are recognized and translated by the cell's protein synthesis machinery. Poxvirus expression is sequential and temporally regulated (early, intermediate, and late mRNAs).

Poxviruses produce a number of immune evasion proteins such as viroceptors, virokines, interferon resistance proteins, and serpins. Even though no one specific poxvirus produces all of these collectively, all seem to target the nonspecific immune response.

Research on poxviruses did not languish after its eradication. Recombinant vaccinia viruses have been used as research tools and in the development of new live vaccines against a variety of infectious agents. Most notable is the successful use of a recombinant vaccinia rabies vaccine to eliminate rabies in wild raccoon and fox populations in western Europe and in the United States.

Mass vaccination against smallpox stopped in 1972. Today, some military and emergency health-care workers have received the vaccine. This has been prompted by the threat of the use of variola as a bioterrorist agent. The current live vaccinia virus vaccine has caused complications in vaccinated individuals with skin conditions, weakened immune systems, heart disease, or other chronic conditions. Research is now focusing on the development of a safer, tissue culture-based vaccine that contains attenuated strains of vaccinia virus. A second focus has been placed on the development of rapid diagnostics and antiviral strategies as an alternative or additional strategy for the prevention or postexposure treatment of smallpox and/or monkeypox.

The anthrax attacks in the United States in the fall of 2001 called attention to the need for coordinated emergency planning and the stockpiling of antibiotics or antivirals and vaccines in case of a widespread bioterrorist attack. All levels of communities (national, state, county, city, neighborhood) are affected by these concerns and need to prepare. Public education and continual surveillance are needed to reduce the vulnerability to biological weapons such as that of a smallpox outbreak.

CASE STUDY QUESTIONS

These questions relate to the Case Study presented at the beginning of the chapter.

1. How would you prove that the grey squirrels displaced the red squirrels by habitat effects? By competition? By predation? Or as a result of a pathogen(s) carried by the grey squirrel?
2. If the grey squirrels are the reservoir of SQPV, how is the virus transmitted to red squirrels?
3. How would you determine if the red squirrels trapped had been infected with SQPV and survived?
4. How would you determine if the red squirrels had never been infected with SQPV?
5. Create a plan to effectively manage and protect the remaining red squirrel populations.
6. What are the signs and symptoms of squirrel pox?
7. Assuming you could find some infected/sick red squirrels, how would you go about isolating SQPV from the red squirrels?

References

Gurnell, J., et al. 2006. "Squirrel poxvirus: Landscape scale strategies for managing disease threat." *Biol Conserv* 131:287–295.

Sainsbury, A. W., et al. 2008. "Poxviral disease in red squirrels *Sciurus vulgaris* in the UK: spatial and temporal trends of an emerging threat." *Ecohealth* 5(3):305–316.

Thomas, K., et al. 2003. "A novel poxvirus lethal to red squirrels (*Sciurus vulgaris*)." *J Gen Virol* 84: 3337–3341.

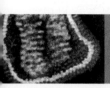

CASE STUDY 1: SQUIRRELPOX VACCINATION

In 1985 Bob Miller, an 18-year-old National Guard member, received his first vaccination against smallpox at a U.S. Army facility. The vaccination gave him a lot of discomfort. His friend Jennifer applied compresses to the vaccination site on his arm to relieve some of the irritation and soreness caused by the inoculation. Two weeks later, Jennifer developed a slight fever and she felt very tired. A painful ulcer formed on her upper lip. It looked similar to Bob's vaccination lesion. The thought never crossed her mind that she could have contracted a poxvirus infection from Bob. Jennifer had been vaccinated against smallpox as a child even though she didn't have a scar on her arm from the inoculation. She assumed she was immune to a smallpox infection. Jennifer's mother encouraged her to go to the local clinic to determine what was causing the lesion. A physician at the clinic referred Jennifer to a dermatologist.

1. What does the vaccine against smallpox consist of? Is it a live or killed vaccine?
2. Could Jennifer be suffering from a poxvirus infection? Explain your answer. How are poxvirus infections contracted?
3. What tests might the dermatologist use to determine the cause of Jennifer's lip ulcer?
4. Would an intramuscular shot of vaccinia immune globulin be considered a viable treatment of her skin lesion?
5. Are Bob and Jennifer contagious? Explain your answer.
6. Should an investigation be conducted to determine whether Jennifer or Bob had transmitted disease to further contacts, such as family members?

References

Centers for Disease Control and Prevention (CDC). 1981. "Vaccinia outbreak—Newfoundland." *Morbid Mortal Wkly Rep MMWR* 30(36):453–455.

Centers for Disease Control and Prevention (CDC). 1982. "Epidemiologic notes and reports disseminated vaccinia infection in a college student—Tennessee." *Morbid Mortal Wkly Rep MMWR* 31(50):682–683.

Centers for Disease Control and Prevention (CDC). 1983. "Vaccinia outbreak—Nevada." *Morbid Mortal Wkly Rep MMWR* 32(31):403–404.

Centers for Disease Control and Prevention (CDC). 1984. "Contact spread of vaccinia from a recently vaccinated marine—Louisiana." *Morbid Mortal Wkly Rep MMWR* 33(3):37–38.

Centers for Disease Control and Prevention (CDC). 1985. "Epidemiologic notes and reports contact spread of vaccinia from a National Guard vaccinee—Wisconsin." *Morbid Mortal Wkly Rep MMWR* 34(13): 182–183.

CASE STUDY 2: ENVELOPE OF SMALLPOX SCABS

On March 31, 2003, a librarian at the College of Sante Fe's Fogelson Library discovered a small yellow envelope labelled: "Scabs from vaccination of W. B. Yarrington's children." It was tucked inside of an 1888 Civil War medicine book that had been written by Dr. W. D. Kelly, a physician who had done work on childhood vaccinations during the late 1800s. The librarian decided not to open the envelope and contacted the Collections Manager of the National Museum of Civil War Medicine via email.

She queried whether the museum would want the envelope and if the scabs could be dangerous. This started a number of communications involving the museum director, the FBI, staff at Walter Reed Army Medical Center, and the CDC. The 115-year-old scabs were from smallpox victims or were used to vaccinate/inoculate healthy individuals. During the late 1800s pus or bits of scabs from smallpox patients who suffered from a mild case of smallpox were implanted into the skin of healthy individuals in order to cause a mild illness, resulting in protective immunity. Within two days, the FBI picked up the envelope of scabs and delivered it to the CDC for a battery of tests.

1. What biosafety precautions should be taken when handling this material?
2. Assuming that the scabs do not yield live infectious virus, what kind of viral characterization can be done on this material? What methods or techniques might be used to analyze these scabs? What would these methods detect and tell us about the virus?
3. If the desiccated scabs did contain live viruses that could be grown in the laboratory, could this virus provide insight into the evolution of smallpox vaccines in the United States? Explain your answer.

Poxviruses in general are very hardy in the environment. Samples of variola virus freeze-dried in the laboratory have been revived after storage for 20 years. Study of the long-term survival of variola virus under natural conditions is lacking. Scientists have questioned how long variola virus is stable in victims' graves, clothing, and possessions or tissues. When the mummy of the Egyptian Pharaoh Ramses V was examined by Donald Hopkins in 1979, dried pustules covered mainly the face, neck, shoulders, and arms. There was no rash on the chest. The rash appeared to be characteristic of smallpox. Hopkins raised the issue of whether or not scabs and shrouds of smallpox victims posed an unrecognized threat to archaeologists and anthropologists in the field. Pieces of skin from the pharaoh were analyzed by electron microscopy but did not reveal virus particles. Another well-preserved mummy, this time of a two-year-old Italian boy who died from a severe form of smallpox, was discovered in 1986. The tissue remains of this 16th century mummy were examined by immunoelectron microscopy, which positively identified variola virus in the tissues. This was the first time that virus had been identified from such ancient tissues. Both of these mummies were preserved in a dry, cool crypt.

To determine the potential for transmission resulting from contact with old scabs, researchers conducted experiments in 1954 to study the survival of variola virus under normal conditions. Scabs were taken from three smallpox patients. The scabs were placed in unsealed envelopes and stored in a laboratory cupboard that was kept between 15°C and 30°C, with humidity varying from 35% to 98%. Over the next 12 years, scientists tested the scabs for live virus particles. The study was terminated after 13 years because they ran out of scabs for testing. They concluded that the virus does indeed survive under cool and dry environmental conditions.

4. As cities continue to grow, cemeteries are relocated. Are excavators involved in the relocation of smallpox victims in danger of being infected? (Refer to Joseph Kennedy's 1994 paper, "The archeological recovery of smallpox victims in Hawaii: Scientific investigation or public health threat?" *Perspect Biol Med* 37(4):499–509.)
5. Smallpox has been eradicated worldwide and thus no herd immunity exists against this disease. Should it be recommended that archaeologists and excavators be vaccinated as a precaution in special circumstances where victims interred may be disturbed? Why or why not?
6. What physical factors are most directly related to the viability of variola major?
7. Would you predict smallpox victims buried in permafrost to be potentially infectious? Explain your answer.
8. What precautions, if any, would you take in exhuming interred smallpox victims?

References

Ambrose, C. T. 2005. "Osler and the infected letter." *EID* 11(5):689–693.

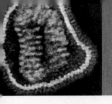

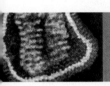

CASE STUDY 2: ENVELOPE OF SMALLPOX SCABS (continued)

Fornaciari, G., and Marchetti, A. 1986. "Italian smallpox of the sixteenth century." *Lancet* 2:1469–1470.

Fornaciari, G., and Marchetti, A. 1986. "Intact Smallpox Virus Particles in an Italian Mummy of Sixteenth Century." *Lancet* 2(857):625.

Maccallum, F. O., and McDonald, J. R. 1957. "Survival of variola virus in raw cotton." *Bull World Health Organ* 16(2):247–254.

Meers, P. D. 1985. "Smallpox still entombed?" *Lancet* 1(8437):1103.

ProMED-mail. 1997. "Smallpox, human remains, likelihood." 1997; 11Mar;19970311.0546.

ProMED-mail. 2004. "Smallpox, 115-year old scabs found, USA." *ProMED mail* 2004; 2 Jan;20040102.0016.

Stone, R. 2002. "Is live smallpox lurking in the arctic?" *Science* 295:2002.

Zuckerman, A. J. 1984. "Palaeontology of smallpox." *Lancet* 22(29):1454.

CASE STUDY 3: LABORATORY-ACQUIRED COWPOX

Rhonda Meskonic was a student working in a poxvirus research lab at a university. She was in charge of autoclaving biohazardous waste collected during experiments in the research laboratory. Basic research in the lab was conducted using strains of poxviruses that were nonpathogenic to humans. Rhonda followed proper laboratory procedures such as wearing gloves and a laboratory coat while she was working in the laboratory.

One day Rhonda developed a painful, ulcerated lesion on her finger in the same spot in which she had experienced a paper cut a few days before the lesion erupted. The lesion did not go away on its own after several weeks. The lesion was a very distinct pustule. Rhonda went to a doctor who examined the lesion and referred her to a dermatologist. Given the fact that she worked in a poxvirus research lab, the doctor considered the possibility of the pustule being caused by a poxvirus infection. The dermatologist biopsied the lesion and it was sent to the CDC for further investigation. Real time PCR of biopsied tissue showed that it tested positive for a non-variola poxvirus. The sample did not contain vaccinia virus but it did contain cowpox DNA. The student had never worked with cowpox. There were modified cowpox strains used for research stored in freezers but no researchers in the lab had worked with cowpox for five years. Laboratory equipment, benches, and other surfaces were swabbed for environmental testing. Cowpox DNA was found in environmental swabs of several surfaces in the laboratory and a freezer room. No live virus was recovered from the swabs. Cowpox was detected in several viral stocks and two cell lines used in the laboratory. Doctors concluded that Rhonda was inadvertently infected with a recombinant (genetically modified) cowpox strain while handling chemicals and contaminated samples in the laboratory.

1. Where in the world are human cowpox infections endemic today?
2. Why is it more difficult to identify infections in a laboratory worker infected with a research or genetically modified strain of virus as compared to strains found in natural circulation?
3. Why didn't the student worker get vaccinated before she started working in the laboratory?
4. More than 500,000 people work in laboratories in the United States. Estimates are that 3 out of every 1000 laboratory workers become infected each year. What are the most common infections according to the primary literature?
5. Cowpox is an emerging zoonosis. In 2009, there were more than 33 human cases of cowpox in France and Germany. What was the potential common source of infection? (*Hint:* See Table 14-2 and look up some of the references listed next.)

References

Dina, J., et al. 2011. "Genital ulcerations D to a cowpox V: A misleading diagnosis of herpes." *J Clin Virol* 50:345–347.

Essbauer, S., et al. 2010. "Zoonotic poxviruses." *Vet Microbiol* 140(3–4):229–336.

Haenssle, H. A., et al. 2006. "Orthopoxvirus infection transmitted by a domestic cat." *Am Acad Dermatol* 54:S1–S4.

(continued)

ProMED mail. Cowpox, Human—USA: (Georgia), Laboratory Infection. ProMED mail 2011; 9 Feb:20110209.0444.

Reardon, S. 2011. "First U.S. cowpox infection: Acquired from lab contamination." *ScienceInsider.* Accessed May 10, 2011, from http://news.sciencemag.org/scienceinsider/2011/02/first-us-cowpox-infection acquired.html

Sewell, D. L. 1995. "Laboratory-associated infections and biosafety." *Clin Microbiol Rev* 8(3):389–405.

Singh, K. 2009. "Laboratory-acquired infections." *Clin Infect Dis* 49(1):142–147.

Wiodaver, C. G., Palumbo, G. J., Waner, J. L. 2004. "Laboratory-acquired vaccinia infection." *J Clin Virol* 29(3):167–170.

Resources

Primary Literature and Reviews

Artenstein, A. W. 2008. "New generation smallpox vaccines: A review of preclinical and clinical data." *Rev Med Virol* 18:217–231.

Bahar, M. W., et al. 2011. "How vaccinia evolved to subvert the host immune response." *J Struct Biol* doi:10.1016/j.jsb.2011.03.010.

Bardell, D. 1976. "Smallpox during the American War of Independence." *ASM News* 42(9):526–530.

Besser, J. M., Crouch, N. A., and Sullivan, M. 2003. "Laboratory diagnosis to differentiate smallpox, vaccinia, and other vesicular/pustular illnesses." *J Lab Clin Med* 142:246–251.

Bhattacharjee, Y. 2006. "U.S. panel wants security rules applied to genomes, not pathogens." *Science* 314:743.

Bisht, H., et al. 2004. "Severe acute respiratory syndrome coronavirus spike protein expressed by attenuated vaccinia virus protectively immunizes mice." *PNAS USA* 101:6641–6646.

Blenden R. J., et al. 2003. "The public and the smallpox threat." *N Engl J Med* 348:426–432.

Block, S. M. 2001. "The growing threat of biological weapons." *Am Sci* 89:28–37.

Breman, J. G., Arita, M. D., and Fenner, F. 2003. "Preventing the return of smallpox." *N Engl J Med* 348(5):463–466.

Bugert, J. J., and Darai, G. 2000. "Poxvirus homologues of cellular genes." *Virus Genes* 21(1–2):111–133.

Bugert, J. J., Melquiot, N., and Kehm, R. 2001. "Molluscum contagiosum virus expresses late genes in primary human fibroblasts but does not produce infectious progeny." *Virus Genes* 22(1):27–33.

Buller, R. M., and Palumbo, G. J. 1991. "Poxvirus pathogenesis." *Microbiol Rev* 55(1):80–122.

Bungum, T., and Day, C. 2006. "Smallpox-related knowledge and beliefs among recent college graduates." *J Am Coll Health* 55(3):181–183.

Centers for disease control and prevention (CDC). 1997. "Human monkeypox—Kasai Oriental, Zaire, 1996–1997." *MMWR* 46(14):304–307.

Centers for disease control and prevention (CDC). 2003. "Update: Multistate outbreak of monkeypox—Illinois, Indianna, Kansas, Ohio, and Wisconsin, 2003." *MMWR* 52(27):642–646.

Centers for disease control and prevention (CDC). 2003. "Smallpox vaccination and adverse reactions. Guidance for clinicians." *MMWR* 2003(52):1–29.

Centers for disease control and prevention (CDC). 2003. "Recommendations for using smallpox vaccine in a pre-event vaccination program." 2003. *MMWR* 52:RR–7.

Condit, R. C., Moussatche, N., and Traktman, P. 2006. "In a nutshell: Structure and assembly of the vaccinia virion." *Adv Virus Res* 66:31–124.

Croft, D. R., et al. 2007. "Occupational risks during a monkeypox outbreak, Wisconsin, 2003." *EID* 13(8): 1150–1157.

Damaso, C. R., et al. 2000. "An emergent poxvirus from humans and cattle in Rio de Janeiro State: Cantagalo virus may derive from brazilian smallpox vaccine." *Virology* 277:439–449.

Davies, H. 2007. "Ethical reflections on Edward Jenner's experimental treatment." *J Med Ethics* 33(3):174–176.

De Clercq, E., and Neyts, J. 2004. "Therapeutic potential of nucleoside/nucleotide analogues against poxvirus infections." *Rev Med Virol* 14:289–300.

DeWitt, S. Jr. 1978. "Victoria over variola." *ASM News* 44(12):639–644.

Diamond, J. 1990. "A pox upon our genes: Smallpox vanished twelve years ago, but its genetic legacy may still linger with us." *Nat His* 2:26–30.

Enserink, M. 2005. "Unnoticed amendment bans synthesis of smallpox virus." *Science* 307:1540–1541.

Enserink, M. 2005. "WHA gives yellow light for variola studies." *Science* 308:1235.

Esposito, J. J., et al. 2006. "Genome sequence diversity and clues to the evolution of variola (smallpox) virus." *Science* 313(5788):807–812.

Essbauer, S., et al. 2009. "Zoonotic poxviruses." *Vet Microbiol* 140:229–236.

Fenner, F., et al. 1988. "Smallpox and its eradication." Geneva, Switzerland: World Health Organization (WHO).

Fenner, F. 2000. "Adventures with Poxviruses of Vertebrates." *FEMS Microbiol Rev* 24:123–133.

Flavio, G., et al. 2002. "Characterization of a vaccinia-like virus isolated in a Brazilian forest." *J Gen Virol* 83:223–228.

Galvani, A. P., and Slatkin, M. 2003. "Evaluating plague and smallpox as historical selective pressures for the CCR5-Δ32 HIV-Resistance allele." *PNAS USA* 100(5):15276–15279.

Griffiths, P. D. 2002. "Benefits of bioterrorism." *Rev Med Virol* 12:131–132.
Guarner, J., et al. 2003. "Monkeypox transmission and pathogenesis in prairie dogs." *EID* 10(3):426–431.
Harrison, S. C. 2004. "Discovery of antivirals against smallpox." *PNAS USA* 101(31):11178–11192.
Houston, C. S., and Houston, S. 2000. "The first smallpox epidemic on the canadian plains in the fur-traders' words." *Can J Infect Dis* 11(2):112–115.
Huhn, G. D. 2005. "Clinical characteristics of human monkeypox and risk factors for severe disease." *CID* 41:1742–1751.
Jacobs, B. L., et al. 2009. "Vaccinia virus vaccines: Past, present and future." *Antiviral Res* 84:1–13.
Jahrling, P. B., et al. 2004. "Exploring the potential of variola virus infection of cynomolgus macaques as a model for human smallpox." *PNAS USA* 101(42):15196–15200.
Lane, H. C., La Montagne, J., Fauci, A. S. 2001. "Bioterrorism: A clear and present danger." *Nat Med* 7(12):1271–1273.
Langland, J. O., et al. 2006. "Suppression of proinflammatory signal transduction and gene expression by the dual nucleic acid binding domains of the vaccinia virus E3L proteins." *Virology* 80(20):10083–10095.
Levine, R. S., et al. 2007. "Ecological niche and geographic distribution of human monkeypox in Africa." *PLoS ONE* 1:e176. doi:10.1371/journal.pone.0000176.
Likos, A. M., et al. 2005. "A tale of two clades: Monkeypox viruses." *J Gen Virol* 86:2661–2672.
Maurer, D. M., and Harrington, B. 2003. "Smallpox vaccine: Contraindications, administration, and adverse reactions." *Amer Fam Phys* 68(5):889–896.
McFadden, G. 2004. "Smallpox: An ancient disease enters the modern era of virogenomics." *PNAS USA* 101(42): 14994–14995.
McFadden, G. 2005. "Poxvirus tropism." *Nat Rev Microbiol* 3(3):201–213.
Mcfee, R. B., Leikin, J. B., and Kiernan, K. 2004. "Preparing for an era of weapons of mass destruction (WMD)—Are we there yet? Why should all be concerned. Part II." *Vet Hum Toxicol* 46(6):347–351.
Medeiros-Silva, D. C., et al. 2010. "Clinical signs, diagnosis, and case reports of vaccinia, virus infections." *Braz J Infect Dis* 14(2):129–134.
Moss, B. 1996. "Genetically engineered poxviruses for recombinant gene expression, vaccination and safety." *PNAS USA* 93:11341–11348.
Moss, B., and Shisler, J. L. 2001. "Immunology 101 at Poxvirus U: Immune evasion genes." *Semin Immunol* 13:59–66.
Moss, B. 2011. "Smallpox vaccines: Targets of protective immunity." *Immunol Rev* 239(1):8–26.
Muraskin, W. 2007. "Expunging variola: The control and eradication of smallpox in India 1947–1977." *J Public Health Policy* 28(1):153–155.
Murphy, F. A., and Osburn, B. I. 2005. "Adventitious agents and smallpox vaccine in strategic national stockpile." *EID* 11(7):1086–1089.
Nalca, A., Zumbrun, E. E. 2010. "ACAM2000: The new smallpox vaccine for United States strategic national stockpile." *Drug, Design, Dev Ther* 4:71–79.
Nicas, M., et al. 2004. "The infectious dose of variola (smallpox) virus." *App Biosafety* 9(3):118–127.
Odom, M. R., et al. 2009. "Poxvirus protein evolution: Family wide assessment of possible horizontal gene transfer events." *Virus Res* 144:233–249.
Paran, N., Sutter, G. 2009. "Smallpox vaccines: New formulations and revised strategies for vaccination." *Hum Vaccin* 5(12):824–831.
Parrino, J., and Graham, B. S. 2006. "Smallpox vaccines: Past, present and future." *J Allergy Clin Immunol* 18(6): 1320–1326.
Parrish, S., and Moss, B. 2006. "Characterization of a vaccinia virus mutant with a deletion of the D10R gene encoding a putative negative regulator of gene expression." *J Virol* 80(2):553–561.
Parrish, S., Resch, W., and Moss, B. 2007. "Vaccinia virus D10 protein has mRNA decapping activity, providing a mechanism for control of host and viral gene expression." *PNAS USA* 104(7):2139–2144.
Perdiguero, B., and Esteban, M. 2009. "The interferon system and vaccinia virus evasion mechanisms." *J Interferon Cytokine Res* 29(9):581–598.
Perkus, M. E., Tartaglia, J., and Paoletti, E. 1995. "Poxvirus-based vaccine candidates for cancer, AIDS and other infectious diseases." *J Leukoc Biol* 58:1–13.
Pommerville, J. C. 2003. "Integrating the agents of bioterrorism into the general biology curriculum. I. A primer on bioterrorism." *Am Biol Teach* 64(9):649–656.
Pommerville, J. C. 2003. "Integrating the agents of bioterrorism into the general biology curriculum. II. Mode of action of biological agents." *Am Biol Teach* 65(1):13–23.
ProMED-mail. 2003. "Cowpox, bovine, human—Brazil (Sao Paulo)." *ProMED mail* 2003;11 Jan;20030111.0095.
ProMED-mail. 2007. "Smallpox vaccine, eczema vaccinatum–USA." *ProMED mail* 2007;18 Mar;20070318.0947.
Quixabera-Santos, J. C. et al. 2011. "Animal movement and establishment of vaccinia virus cantagalo strain in Amazon Biome, Brazil." *EID* 17(4):726–729.
Rimoin, A. W., et al. 2010. "Major increases in human monkeypox incidence 30 years after smallpox vaccination campaigns cease in the Democratic Republic of Congo." *PNAS* 107(37):16262–16267.
Rubins, K. H., et al. 2004. "The host response to smallpox: Analysis of the gene expression program in peripheral blood cells in a nonhuman primate model." *PNAS USA* 101(42):15190–15195.
Ruef, C. 2006. "Five Years After 9/11—Fearing smallpox, . . . and the vaccine against it." *Infection* 34(5): 239–240.
Segal, S., and Hill, A. V. S. 2003. "Genetic susceptibility to infectious disease." *Trends Microbiol* 11(9):445–448.
Sivapalasingam, S., et al. 2007. "Immunological memory after exposure to variola virus, monkeypox virus, and vaccinia virus." *J Infect Dis* 195(8):1151–1159.
Stittelaar, K. J., et al. 2006. "Antiviral treatment is more effective than smallpox vaccination upon lethal monkeypox virus infection." *Nature* 439(7077): 745–748.
Sulaiman, I. M., et al. 2007. "Gene chip resequencing of the smallpox virus genome can identify novel strains: A biodefense application." *J Clin Microbiol* 45(2): 358–363.
Tse-Dinh, Y-C 2008. "An update on the development of drugs against smallpox." *Curr Opin Invest Drugs* 9(8):865–870.
Van Rijn, K. 2006. "'Lo! The poor Indian!', Colonial responses to the 1862–63 smallpox epidemic in British Columbia and Vancouver Island." *Can Bull Hist* 23(2):541–560.
Vogel, F., Chakravartti, M. R. 1966. "ABo blood types and smallpox in a rural population of West Bengal and Bihar (India)." *Humangenetik* 3(2):166–180.

Wei, H., et al. 2009. "Coadministration of cidofovir and smallpox vaccine reduced vaccination side effects but interfered with vaccine-elicited immune responses and immunity to monkeypox." *J Virol* 83(2):1115–1125.

Wiser, I., Balicer, R. D., and Cohen, D. 2007. "An update on smallpox vaccine candidates and their role in bioterrorism related vaccination strategies." *Vaccine* 25(6):976–984.

Xiao, Y., et al. 2007. "A protein-based smallpox vaccine protects mice from vaccinia and ectromelia virus challenges when given as a prime and single boost." *Vaccine* 25(7):1214–1224.

Zilinskas, R. A. 1997. "Iraq's Biological Weapons: The Past as Future?" *JAMA* 278:418–424.

Popular Press

Alibek, K., and Handelman, S. 1999. *Biohazard: The Chilling True Story of the Largest Covert Biological Weapons Program in the World—Told from the Inside by the Man Who Ran It.* New York: Random House.

Allen, A. 2007. *Vaccine: The Controversial Story of Medicine's Greatest Lifesaver.* New York: W. W. Norton.

Carrel, J. L. 2004. *The Speckled Monster: A Historical Tale of Battling Smallpox.* New York: Plume.

Fenn, E. A. 2002. *Pox Americana.* New York: Hill and Wang.

Glynn, I., and Glynn, J. 2004. *The Life and Death of Smallpox.* Cambridge, England: Cambridge University Press.

Henderson, D. A. 1976. "The Eradication of Smallpox." *Sci Am* 235(4):25–34.

Henderson, D. A. 2009. *Smallpox—the Death of a Disease: The Inside Story of Eradicating a Worldwide Killer.* Amherst, New York: Prometheus Books.

Hopkins, D. R. 1983. *Princes and Peasants: Smallpox in History.* Chicago: University of Chicago Press.

Hopkins, D. R. 2002. *The Greatest Killer: Smallpox in History.* Chicago: University of Chicago Press.

Munn & Co., eds. "Vaccine virus—Its preparation and its use." *Sci Am* January 19, 1901, pp. 41–42.

The New York Times. April 16, 1947. "Curb of smallpox a 'miracle' says City Health Commissioner."

Pead, P. J. 2006. *Vaccination Rediscovered: New Light in the Dawn of Man's Quest for Immunity.* West Sussex, UK: Timefile Books.

Persson, S. 2010. *Smallpox, Syphilis and Salvation: Medical Breakthroughs that Changed the World.* New Zealand: Exisle Publishing.

Preston, R. 2002. *The Demon in the Freezer.* New York: Random House.

Rodriguez, A. M. 2006. *Edward Jenner: Conqueror of Smallpox (Great Minds of Science).* Berkley Heights, NJ: Enslow Publishers.

Roy, J. 2010. Smallpox Zero: *An Illustrated History of Smallpox and Its Eradication.* Emory University Press: Emory Global Health Institute.

Stutzenberger, F. 1999. "Smallpox: Biological weapon on the frontier." *Muzzleloader Magazine* 70–73.

Tucker, J. B. 2002. *Scourge: The Once and Future Threat of Smallpox,* New York: Grove Press.

Williams, G. 2010. *Angel of death: The story of smallpox.* United Kingdom: Palgrave Macmillion.

Zelicoff, A. P., and Bellomo, M. 2005. *Microbe: Are we ready for the next plague?* New York: Amacom Books.

Video Productions

Biotechnology and Your Health: Pharmaceutical Applications. 2009. Films for the Humanities.

The Invisible Enemy: Weaponized Smallpox. 2008. Films for the Humanities.

Edward Jenner: The Man Who Cured Smallpox. 2007. Arts Magic.

History's Mysteries: Smallpox. 2006. The History Channel.

Insidious Killers: Chemical and Biological Weapons. 2006. The History Channel.

Smallpox, made-for-TV movie. January 2, 2005. FX Network.

Anthrax, Smallpox Vaccinations and the Mark of the Beast, presented by Dr. Leonard Horowitz. 2005. UFO TV.

R_x for Survival: A Global Health Challenge: Deadly Diseases, Smallpox, and the End of Smallpox: The Last Case. 2005. PBS.

History Undercover: Clouds of Death. 2001. The History Channel.

History's Mysteries: Smallpox Deadly Again? 2001. The History Channel.

Plague War: A Report on the Biological Weapons Threat and How the Soviet Union Secretly Amassed an Arsenal of Bio-weapons. 1998. Frontline PBS.

Germ Genie: The Threat of Biological Warfare. 1994. Films for the Humanities.

Tales from the Gimi Hospital. 1988. Kino Video (fiction).

eLearning

go.jblearning.com/shors2

The site features eLearning, an online review area that provides quizzes and other tools to help you study for your class. You can also follow useful links for in-depth information, or just find out the latest virology and microbiology news.

CHAPTER 15

Herpesviruses

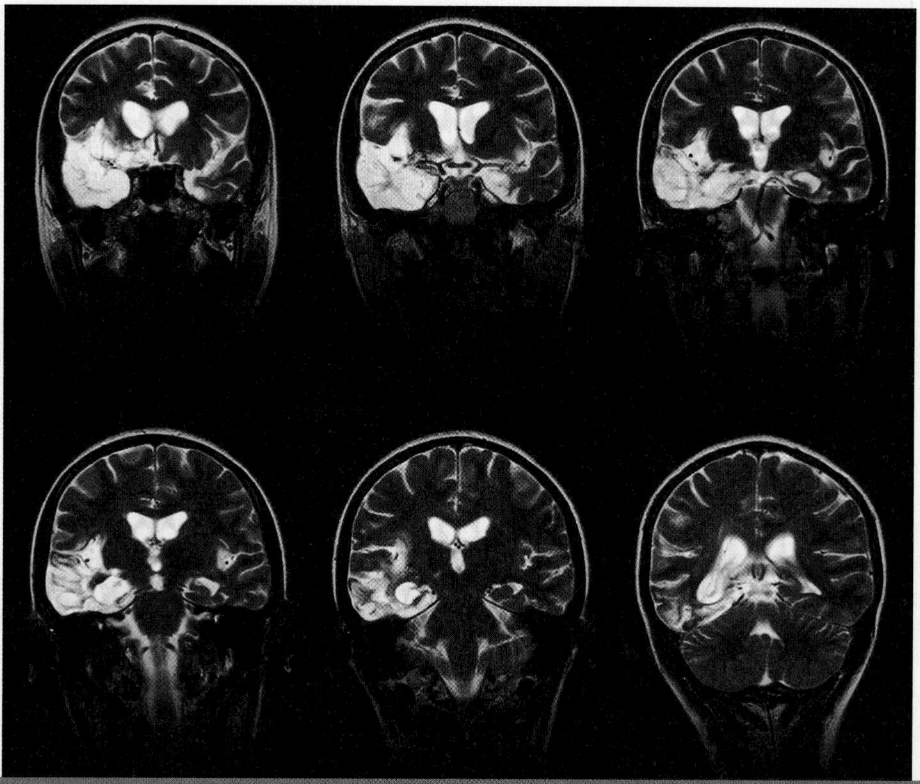

An MRI showing brain inflammation caused by an infection with the herpes simplex virus. The orange spot is a large lesion caused by the virus.

"O'er ladies' lips, who straight on kisses dream, Which oft the angry Mab with blisters plagues, Because their breaths with sweetmeats tainted are."

William Shakespeare, "Romeo and Juliet," Act 1, Scene IV

The blisters that Shakespeare refers to are in fact cold sores produced by the herpes simplex virus I (HSV-I).

OUTLINE

CASE STUDY: WRESTLERS WITH HERPES

During the summer of 1989, 175 male high school wrestlers attended a Minnesota wrestling camp. These young athletes participated in wrestling practices and competitions. During the final week of camp, many of the boys complained of sore throats, headaches, fever, and malaise. An inflamed skin rash that had the appearance of small, fluid-containing vesicles was observed on the head, neck, and arms of the wrestlers. It was particularly common in wrestlers who engaged in the lock-up position that places the face, neck, and arms of the opposing wrestlers in close contact. Some of the boys continued to wrestle for at least two days after the rash had appeared.

15.1 History

The Greek physician Hippocrates (570–510 B.C.) documented herpesvirus infections as an assortment of cutaneous lesions compatible with herpes simplex or varicella zoster lesions. Greek scholars coined the term ***herpes,*** which means "to creep or crawl," in reference to the spreading nature of herpetic lesions. It came from the word ***herpeton,*** an animal that goes on all fours. In the first century A.D., the emperor of Rome, Tiberius Claudius Nero, banned kissing in public ceremonies after a herpes simplex virus epidemic. Herpesvirus infections have existed for centuries and evidence suggests that these viruses infect virtually all species of vertebrates and at least one invertebrate (the oyster; see **VIRUS FILE 15-1**). They are among the most persistent and common viruses found in humans. Once individuals become infected by a herpesvirus, the infection remains in their body for life.

Herpesvirus lesions mysteriously appear and disappear because all herpesviruses are capable of establishing and maintaining a **latent state** of infection at a specific site within an immunocompetent host. A latent state of infection is one in which the herpesvirus remains present within the host but replicates slowly or not at all and produces few or no symptoms. Herpesviruses may remain silent for many years only to be reactivated later. Individuals can be infected with more than one herpesvirus during their lifetime. **TABLE 15-1** lists the nine known herpesvirus infections and the types of infections in humans.

TABLE 15-2 lists the seroprevalence of these infections based on healthy children and adult seroprevalence in the United States and developing countries. **Seroprevalence** is the number of persons in a population who test herpesvirus-positive based on serology (blood serum) specimens; this is often presented as a percent of the total specimens tested or as a rate per 100,000 persons tested. Despite certain similarities, infection with one member of the *Herpesviridae* family does not confer protection against infection or disease with the other members of the *Herpesviridae* family.

15.2 Clinical Signs and Symptoms of Human Herpesviruses

Herpes Simplex Virus Type 1 (HSV-1) and Herpes Simplex Virus Type 2 (HSV-2)

HSV-1 and HSV-2 have preference for infecting specific anatomic sites—usually the mouth or pharynx and genitals, respectively. HSV-1 reactivation occurs most frequently above the waist, compared to HSV-2 reactivation that typically occurs below the waist. The lesions of HSV-1 and -2 initially look similar. Both start with a clear vesicle containing infectious virus with red lesions at the base of the vesicle. The lesions may become encrusted and ulcers may develop. HSV-2 is associated with sexual transmission, whereas infections in children are usually associated with HSV-1.

HSV-1 can cause a number of herpetic diseases, including cold sores of the mouth, lesions on the lip, herpes keratitis of the eye (which is the leading cause of corneal blindness in the United States), herpes gladiatorum (transmitted during frequent contact in wrestling), and herpes rugbeiorum (seen in other contact sports like rugby). HSV-1 can cause genital herpes in about 10% of cases. HSV-1 reactivation is associated with the following factors:

- Immune suppression by cytotoxic drugs
- Sexual contact
- Physical and emotional stress
- Temperature changes (hot or cold, such as burning the mouth with hot coffee)

VIRUS FILE 15-1 Are Oyster Herpes Outbreaks a Symptom of Global Warming?

A newly described strain of herpesvirus known as Ostreid herpesvirus 1 (OsHV-1) μvar is spreading around the world, infecting the Pacific cupped oyster (*Crassostrea gigas*). Herpes-like virus infections causing mass mortalities have been reported worldwide in several other mollusk species such as clams, scallops, and abalone. The OsHV-1 μvar virus is more virulent than other strains identified in the past. It wipes out 80% of an oyster bed (**VF FIGURE 15-1**) within a week. The only outward sign that something is wrong is high mortalities in the oyster beds. There are no visible oyster symptoms. Polymerase chain reaction (PCR) of the viral DNA is the tool used to rapidly detect OsHV-1 infections in the aquaculture industry.

OsHV-1 μvar is causing high mortalities (up to 100%) in oyster hatcheries. Spat, oysters less than one year old, and juveniles (12- to 18-month-old oysters) are most affected. Adult oysters appear to be less sensitive to the disease than spat. Studies suggest adult oysters have the presences of a higher antiviral immune response. There is no cure once the virus has infected an area. The oyster herpesviruses cannot be transmitted to humans and there is no safety risk to consumers of oysters. The oyster herpesvirus does not affect native oysters.

In 2008 and 2009, outbreaks occurred in France, a leader in oyster production. It appears to have spread to England, Ireland, New Zealand, Australia, and Japan. OsHV-1 μvar has cost North Island oyster farms over 12 million New Zealand dollars (which is equivalent to 9 million U.S. dollars) in sales. It has caused Nelson hatchery to close, which had supplied 20% of New Zealand's stock oysters. Oysters are seen as one of the keys to growing aquaculture into a billion-dollar industry by 2025.

What caused the emergence of such a virulent oyster herpesvirus? OSHV-1 μvar infects the young oysters during the breeding season. Outbreaks of OsHV-1 μvar in the United Kingdom occurred when seawater temperatures underwent a sudden change, exceeding 61°F (16°C). The herpesvirus was dormant (in a state of latency) and reactivated by the environmental temperature change (see Section 15.4, Herpesvirus Life Cycle: Viral Replication and Latency) or other

FIGURE VF 15-1 Oyster beds located in Japan. Bay is at low tide.

(continued)

VIRUS FILE 15-1 Are Oyster Herpes Outbreaks a Symptom of Global Warming? (continued)

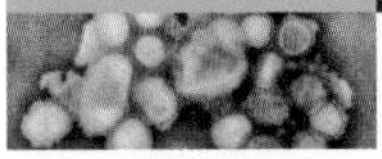

environmental factors. Some experts wonder if global warming plays a part in the emergence of this highly virulent herpesvirus. But other factors likely contributed to the outbreaks, such as the introduction of possibly infected spat, movements and mixing of oyster populations and age groups, reused equipment and gear from infected oyster beds, or water transferred from an infected area. Restrictions are placed on the movement of oysters into bays where farms were not affected by the virus.

References

Batista, F. M., et al. 2005. "Detection of ostreid herpesvirus-1 (OsHV-1) by PCR using a rapid and simple method of DNA E from oyster larvae." *Dis Aqua Org* 64:1–4.

Farley, C. A., et al. 1972. "Oyster herpes-type virus." *Science* 178:759–760.

Pepin, S. A., et al. 2008. "Detection of a particular ostreid herpesvirus I genotype associated with massive mortality outbreaks of Pacific oysters, *Crassostrea gigas,* in France in 2008." *Virus Res* 153(1):92–99.

ProMED-mail. Oyster herpesvirus—England, Ireland, Emerging, OIE. ProMED-mail. 17-Aug:20100817.2858.

ProMED-mail. Oyster herpesvirus—Netherlands. ProMED-mail 2010. 10-Sep: 20100919.32.59.

ProMED-mail. Oyster herpesvirus—New Zealand. ProMED-mail 2010. 03 Dec: 20101203.4337.

ProMED-mail. Oyster herpesvirus—New Zealand. ProMED-mail 2011;23 Feb: 20110223.0591.

Savin, K. W., et al. 2010. "A neurotropic herpesvirus infecting the gastropod, abalone, shares ancestry with oyster herpesvirus and herpesvirus associated with the amphioxus genome." *Virol J* 7:308.

Schikorski, D., et al. 2011. "Experimental ostreid herpesvirus 1 infection of the Pacific oyster *Crassostrea gigas:* Kinetics of virus DNA detection by qPCR in seawater and in oyster samples." *Virus Res* 155:28–34.

Schikorski, D., et al. 2011. "Experimental infection of Pacific oyster *Crassostrea gigas* spat by ostreid herpesvirus 1: Demonstration of oyster spat susceptibility." *Vet Res* 42:27.

Zell, R., et al. 2008. "Impact of global warming on viral diseases: What is the evidence?" *Curr Opin Biotech* 19:652–660.

TABLE 15-1 Human Herpesvirus Nomenclature and Clinical Syndromes

Acronym Name	Common Name	Disease in Healthy Persons	Disease in Immunosuppressed Persons
HHV-1 or HSV-1	Human herpesvirus 1 or herpes simplex virus 1	Cold sores (oral herpes), herpes keratitis (eye infection), mat herpes or herpes gladiatorum, herpes rugbeiorum, eczema herpeticum in children	Cold sores (oral herpes), herpes keratitis (eye infection), mat herpes or herpes gladiatorum, herpes rugbeiorum, eczema herpeticum in children, visceral infections (infections of internal organs)
HHV-2 or HSV-2	Human herpesvirus 2 or herpes simplex virus 2	Genital herpes, aseptic meningitis, neonatal herpes	Genital herpes, disseminated infection
HHV-3 or VZV	Human herpesvirus 3 or varicella zoster virus	Chickenpox, shingles	Disseminated infection
HHV-4 or EBV	Human herpesvirus 4 or Epstein-Barr virus	Mononucleosis, hepatitis, encephalitis	Burkitt lymphoma, oral hairy leukoplakia
HHV-5 or CMV	Human herpesvirus 5 or human cytomegalovirus	Mononucleosis, congenital infection	Hepatitis, retinitis, other visceral infections (infections of internal organs)
HHV-6	Human herpesvirus 6	Exanthem subitum or fever illness and seizures in infants and young children, encephalitis	Fever and rash, encephalitis, bone marrow suppression
HHV-7	Human herpesvirus 7	Exanthem subitum or fever illness and seizures in infants and young children, encephalitis	Encephalitis?
HHV-8 or KSVH	Human herpesvirus 8 or Kaposi's sarcoma-associated herpesvirus	Fever associated with a rash and mononucleosis?	Kaposi's sarcoma and B cell lymphoma, Castleman disease
Herpes B Virus	Monkey B virus	Mucocutaneous lesions, encephalitis	?

Source: Adapted from Straus, SE. Introduction to *Herpesviridae*. In: Mandell G. L., Kolin R., eds. *Principles and Practice of Infectious Diseases*, 6th ed. Philadelphia: Elsevier, Churchill Livingstone; 2005:1756–1762.

TABLE 15-2 Seroprevalence of Human Herpesviruses

Acronym	Common Name	Healthy Children Seroprevalence	Healthy Adults United States Seroprevalence	Healthy Adults Developing Countries Seroprevalence
HHV-1 or HSV-1	Human herpesvirus 1 or herpes simplex virus 1	20%–40%	50%–70%	50%–90%
HHV-2 or HSV-2	Human herpesvirus 2 or herpes simplex virus 2	0%–5%	20%–50%	20%–60%
HHV-3 or VZV	Human herpesvirus 3 or varicella zoster virus	50%–75%	85%–95%	50%–80%
HHV-4 or EBV	Human herpesvirus 4 or Epstein-Barr virus	10%–30%	80%–95%	50%–80%
HHV-5 or CMV	Human herpesvirus 5 or human cytomegalovirus	10%–30%	40%–70%	40%–80%
HHV-6	Human herpesvirus 6	80%–100%	60%–100%	60%–100%
HHV-7	Human herpesvirus 7	40%–80%	60%–100%	40%–100%
HHV-8 or KSVH	Human herpesvirus 8 or Kaposi's sarcoma-associated herpesvirus	<3%	<3%	10%–60%
Herpes B Virus	Herpes B virus	0	<<1%	<<1%

Source: Adapted from Straus SE. Introduction to *Herpesviridae.* In: Mandell G. L., Kolin R., eds. *Principles and Practice of Infectious Diseases*, 6th ed. Philadelphia: Elsevier, Churchill Livingstone; 2005:1756–1762.

- Too much ultraviolet light (e.g., sunburn)
- Menstruation
- Pregnancy
- Lactation
- Malnutrition
- Excessive fatigue

HSV-2 is responsible for 90% of genital herpes cases. Approximately 45 million persons in the United States suffer from genital herpes infections (**FIGURE 15-1**). New infections occur at a rate of about one million new cases per year. The majority of infections are unrecognized and therefore undiagnosed. Individuals with HSV-2 infections can shed virus even during asymptomatic periods. Safe-sex practices such as the use of condoms in combination with valacyclovir therapy reduce the spreading of HSV-2 infection during heterosexual intercourse by 75%. A pregnant woman who has an active case of genital herpesvirus infection should deliver the infant by caesarean section because HSV-2 can be transmitted to the infant during a vaginal delivery.

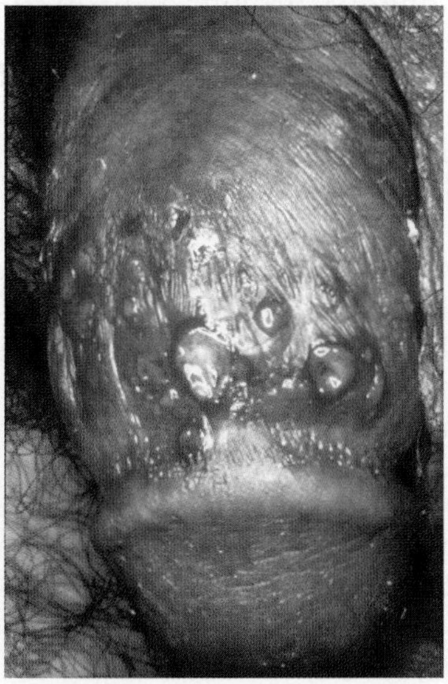

FIGURE 15-1 This male has a herpetic rash caused by HSV-2 on the shaft of his penis. The blisters break, leaving tender ulcers that may take two to four weeks to heal after the first time they appear.

Varicella Zoster Virus (VZV)

VZV causes two different clinical diseases: **chickenpox** and **shingles**. Chickenpox is a mild disease that usually afflicts children but can be severe in infants, adults, and persons with impaired immune systems.

There are several explanations for the origin of the name chickenpox. One of the most accepted suggestions is that the name is derived from a scribal error for "chick-peas," which was interpreted as chick peas, suggesting that the chickenpox vesicles resembled chick peas. Another idea is that the name came about because the rash looked as though the bill of a chicken had pecked the child.

Chickenpox is the only herpesvirus that spreads person to person by coughing or sneezing (airborne), making it highly contagious. Like other herpesviruses, it is also spread by close contact. Symptoms develop 10 to 21 days after contact with an infected person. When a person is initially infected with VZV, the virus infects the skin or mucosa of the respiratory tract and progresses through the blood and lymphatic system to the cells of the **reticuloendothelial system**. The reticuloendothelial system is a part of the immune system that consists of phagocytic cells (mainly monocytes and macrophages) located in reticular connective tissue. Their primary role is phagocytosis. The first sign of disease is often the itchy exanthematous rash that is characteristic of chickenpox (**FIGURE 15-2**). Some people experience a fever and malaise. The blisters dry and form scabs in four to five days. On average, a person has 300 to 400 lesions on the body during an attack. Adults are more susceptible to complications such as bacterial infection of the skin, swelling of the brain, and pneumonia. The Centers for Disease Control and Prevention's (CDC's) Advisory Committee on Immunization Practices recommends that all children be routinely vaccinated at 12 to 18 months of age and that all susceptible children receive the vaccine before their 13th birthday (see Section 15.6).

After a primary VZV infection, the virus remains dormant/latent in the dorsal root ganglia (neurons of the nerve roots) of all persons who had chickenpox. The virus can be reactivated later in life to cause an illness referred to as **shingles** or **herpes zoster**. VZV is most commonly reactivated after the age of 60 and the risk of reactivation increases with age. It is also more common and severe in immunocompromised patients (e.g., AIDS patients). The onset of shingles is characterized by severe pain, numbness, and itching, followed by a vesicular rash that forms in a three- to five-day period. The rash follows a nerve on one side of the body. In an otherwise healthy individual, the disease lasts 10 to 15 days, but it may take a month for the skin to return to normal. In an immunocompromised patient, lesion formation can take up to two weeks and scabbing may not take place until three to four weeks into the course of the disease. Chronic shingles may also occur in AIDS patients, in which case new lesions form in the absence of healing of existing lesions. This can be particularly debilitating.

An individual suffering from shingles is contagious to individuals who have not had chickenpox; thus, people who have not had chickenpox can be infected with VZV and contract chickenpox if they have close contact with the person who has shingles. One cannot, however, get shingles from someone afflicted with shingles. Shingles is caused only by VZV that has been dormant since an individual acquired chickenpox.

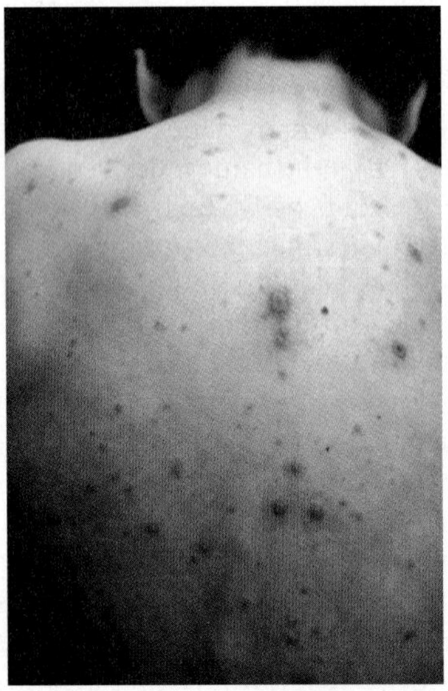

FIGURE 15-2 This child has the characteristic chickenpox rash. The lesions are blister-like and contain pus. They appear on the face, scalp, or trunk of the body.

Cytomegalovirus (CMV)

CMV infections are common in all human populations, ranging from 40% to 70% adults in the United States and nearly 40% to 80% in developing countries. Healthy individuals who become infected with CMV after birth experience few or no symptoms of the disease and no long-term health consequences of the disease. Once a person is infected, the virus is dormant within the person's body for life. For the majority of people, CMV is not a serious disease. CMV is, however, the most common **opportunistic pathogen** in the immunocompromised patient. It can cause a variety of diseases, such as:

- **congenital CMV syndrome** in **neonates** (newborn infants younger than four weeks old)
- infectious mononucleosis with prolonged fever and hepatitis in young adults
- pneumonia in bone marrow transplant recipients
- disease syndromes in lung, liver, kidney, and heart transplant recipients
- **retinitis** (an infection of the eyes) in AIDS patients

CMV is transmitted from person to person via close, intimate contact with a person excreting the virus in his or her saliva, urine, or other bodily fluids. It can be transmitted sexually or via breast milk, transplanted organs, and (rarely) blood transfusions. CMV is the most important cause of **congenital viral infection** (during childbirth) in the United States, where approximately 1% to 3% of women are infected during pregnancy. Developing unborn babies are at highest risk for developing complications of CMV infection. These complications include hearing loss, visual impairment, varying degrees of mental retardation, and motor problems. When a pregnant woman who has never had CMV infection becomes infected with it, there is the potential risk that after birth the infant may have CMV-related complications. In contrast, infants and children who contract CMV after birth experience few, if any, symptoms or complications.

Epstein-Barr Virus (EBV)

It is estimated that EBV causes 79% of **infectious mononucleosis** cases and that CMV causes the remaining 21%. The term infectious mononucleosis is used because the white blood cells or mononuclear cells increase dramatically within an infected individual. EBV is a common infection throughout the world. The vast majority of EBV infections occur in young children, and are asymptomatic. When EBV strikes individuals in their teens or 20s, it generally becomes symptomatic. Adults over the age of 35 are less likely to contract mononucleosis, but those who do often become very ill. In developing countries such as Africa, EBV infection is associated with Burkitt's lymphoma in children (see Chapter 10).

EBV is transmitted via intimate contact with saliva of an infected person (e.g., kissing, or sharing beverages or eating utensils). Hence, mononucleosis is sometimes referred to as "the kissing disease." The incubation period ranges from four to six weeks. Symptoms of infectious mononucleosis are:

- sore throat
- fever
- swollen lymph glands
- malaise
- exhaustion
- enlarged spleen (sometimes)
- enlarged liver (sometimes)
- heart problems (rare)
- central nervous system (CNS) problems (rare)

These symptoms usually resolve within one or two months. EBV remains latent in the throat and blood for the rest of the infected person's life.

HHV-6 and HHV-7

HHV-6 was isolated from T-cell cultures derived from the blood of AIDS patients, whereas HHV-7 was isolated from CD4+ T cells of a healthy person. Like HIV, both viruses can infect and kill CD4+ T cells. HHV-6 infects nearly all humans by two years of age. There are two distinctly different HHV-6 viruses: HHV-6A and HHV-6B. HHV-6A has rarely been associated with disease. HHV-6B causes **sixth disease** (also referred to as **exanthema subitum** or **roseola infantum**). Sixth disease is a mild disease of children. It begins with a sudden fever followed by a red rash. The name sixth disease comes from the fact that when childhood rash diseases were enumerated, it was the sixth listed. The diseases are:

1. Measles
2. Scarlet fever
3. Rubella (German measles)
4. Duke's disease
5. Fifth disease caused by parvovirus B19
6. Exanthum subitum (also called sixth disease or roseola infantum)

Both HHV-6 and HHV-7 have been associated with exanthum subitum and infantile fever and seizures. However, most cases are due to HHV-6. It is estimated that 5% to 25% of visits to emergency rooms for fever in infants are due to HHV-6. HHV-6 also causes approximately one third of fever seizures in children up to two years of age. HHV-6 may also play a role in multiple sclerosis (MS). HHV-6 reactivation may have the potential to trigger autoimmunity and tissue damage associated with MS lesion development (**VIRUS FILE 15-2**). HHV-7 shares 20% to 75% amino acid identity with HHV-6 in many of their viral proteins.

Kaposi's Sarcoma-Associated Herpesvirus (KSHV or HHV-8)

KSHV DNA sequences have been detected in all Kaposi's sarcomas (see Chapter 10). In 1972, Hungarian dermatologist Moritz Kaposi described an aggressive pigmented sarcoma of the skin. He reported three men who died within 16 months after the sarcomas appeared. Autopsy revealed that the sarcomas spread aggressively throughout the body. Kaposi's sarcoma became known as a skin cancer affecting elderly men of Mediterranean and eastern European descent. In 1980, Alvin Friedman-Kein reported on 50 homosexual men with Kaposi's sarcoma of the skin, lymph nodes, mucosa, and viscera. His report was a harbinger of the AIDS epidemic as Kaposi's sarcoma was one of the earliest manifestations of AIDS. The morbidity and mortality associated with AIDS-Kaposi's sarcoma promoted vigorous research efforts into its etiology, which eventually led to the discovery of KSHV in 1994.

A primary infection caused by KSHV has not been described. Most infections are likely asymptomatic or unrecognized. It appears that primary infection in an immunocompetent host is self-limiting whereas a primary infection in immunosuppressed persons can be severe and have life-threatening consequences.

Classical Kaposi's sarcoma predominantly involves the skin of the face (usually the nose), genitalia, oral cavity, and lower extremities. It causes little or no pain. Epidemic Kaposi's sarcoma, which refers to Kaposi's sarcoma in HIV-infected individuals, is usually aggressive, involving the skin, gastrointestinal tract, and respiratory tract. This

VIRUS FILE 15-2 Does Epstein-Barr Virus Play a Role in the Development of Multiple Sclerosis?

A key feature of herpesviruses is their close adaptation to their host. After a primary infection with herpesviruses, the infection becomes enigmatically silent in neurons or other cell types for the duration of one's life. Scientists have proposed that particular herpesviruses (e.g., Epstein-Barr virus and HHV-6) may play a role in debilitating neurological diseases such as multiple sclerosis (MS). MS commonly affects people between the ages of 20 and 40. More than 1.5 million people worldwide and at least 400,000 individuals in Europe alone are affected by MS. MS is second only to trauma as a cause of acquired disability in young adults in most Caucasian populations.

Epstein-Barr virus (EBV) has become a leading candidate as a trigger for MS in recent years. Emerging evidence demonstrates a strong association between EBV infection and several autoimmune diseases, including MS. The problem with all of these studies is the ubiquity of herpesviruses in humans. They "correlate" with everything, and uninfected control patients are difficult to find in significant numbers.

MS may be triggered by multiple factors, including genetics, viral infection, autoimmune reactions, and environmental conditions. Collaborative studies during recent years have determined that the genetic risk factors that predispose for MS are the HLA-DR and HLA-DQ alleles of

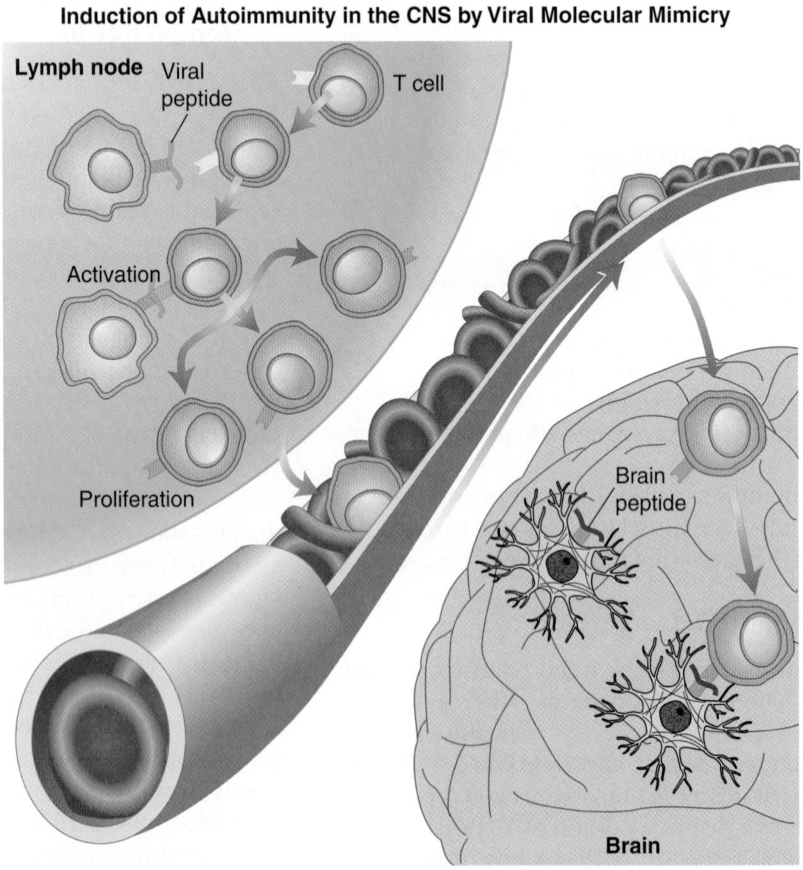

FIGURE VF 15-2 Model for autoimmune disease caused by viral molecular mimicry within the central nervous system—a situation where a viral peptide triggers autoimmunity against a cellular peptide with the same or very similar sequences. Adapted from Wucherpfennig, K. W., *Mechanism for the Activation of Self-Reactive Autoimmune Diseases.*

VIRUS FILE 15-2 Does Epstein-Barr Virus Play a Role in the Development of Multiple Sclerosis? (continued)

the HLA class II region on the human chromosome 6p21 (http://hla.alleles.org/). HLA, the acronym for human leukocyte antigen system, is the name of the major histocompatibility complex in humans (see Chapter 7). This is also the location mapped to the highest risk of conferring genes for most major **autoimmune diseases**, including MS, which is particularly associated with the DRB1*1501 allele that encodes HLA-DR2b. An autoimmune disease occurs when the body tissues are attacked by the body's own immune system. Evidence of autoimmunity is the presence of **autoantibodies**, antibodies directed against the person who produced them and T cells that are reactive with host antigens. MS is a relatively common disease in Europe, the United States, Canada, New Zealand, and parts of Australia. There are very low rates or a virtual absence of the disease near the equator. This geographic distribution can be attributed both to genetic and environmental influences.

EBV mainly infects resting B cells, although other cell types can be infected (e.g., T cells or epithelial cells). Primary infection occurs in young childhood and is usually asymptomatic, or during adolescence/early adulthood, in which case individuals develop infectious mononucleosis. Following infection with EBV, the virus exists as a lifelong latent viral presence within B cells, with periodic reactivation. EBV resides in memory B cells, which traffic into inflamed tissues. The presence of EBV could be a bystander of this phenomenon or a participant of an autoimmune disease.

Epidemiological studies have shown that individuals with a history of infectious mononucleosis in adolescence are at increased risk of developing MS (**VF FIGURE 15-2**).

References

Kakalacheva, K., et al. 2011. "Viral triggers of multiple sclerosis." *Biochem Biophys Acta* 1812:132–140.

Lunemann, J. D., and Munz, C. 2009. "EBV in MS: Guilty by association?" *Trends Immunol* 30(6):243–248.

Maghzi, A-H., et al. 2011. "Viral pathophysiology of multiple sclerosis: A role for Epstein-Barr virus infection?" *Pathophysiology* 18:13–20.

Nociti, V., et al. 2010. "Epstein Barr antibodies in serum and cerebrospinal fluid from multiple sclerosis, chronic inflammatory demyelinating polyradiculoneuropathy and amyotrophic lateral sclerosis." *J Neuroimmunol* 225:149–152.

Pender, M. P. 2011. "The essential role of Epstein-Barr virus in the pathogenesis of multiple sclerosis." *Neuroscientist* [E-pub ahead of print].

Pohl, D. 2009. "Epstein-Barr virus and multiple sclerosis." *J Neurol Sci* 286:62–64.

Villegas, E., et al. 2011. "Low intrathecal immune response of anti-EBNA-1 antibodies and EBV DNA from multiple sclerosis patients." *Diagn Microbiol Infect Dis* 70:85–90.

form is most common in the United States where it affects men who have sex with other men. Epidemic Kaposi's sarcoma also occurs in heterosexual HIV-infected individuals in Africa. The number of childhood cases of Kaposi's sarcoma has significantly increased in Africa with the AIDS epidemic. For example, in the early 1980s, Kaposi's sarcoma caused 0% to 2% of childhood cancers, but by 1992, it accounted for about 25% of childhood malignancies. The diagnosis of Kaposi's sarcoma is easily confirmed by a biopsy.

KHSV was first assumed to be transmitted sexually because of reports linking sexual risk factors such as the number of sexual partners with the numbers of cases. However, it is now known that KSHV is transmitted largely via saliva. KHSV is detectable in peripheral blood, suggesting that KSHV transmission by blood and blood products is possible. This is obviously of great concern particularly with regard to the safety of blood donations and transmission of KSHV via solid organ donation because transplant recipients require immune suppression to stay alive. Kaposi's sarcoma has been documented both in recipients of organs from KSHV-infected donors and infected recipients due to KSHV reactivation. This raises the question of whether donors, recipients, or both should be screened for KSHV antibodies.

Viral loads are performed by PCR of viral DNA present in plasma or peripheral blood mononuclear cells. Treatment of Kaposi's sarcoma in patients infected with HIV slows the progression of the disease but is not curative. Effective combination antiretroviral therapy (cART) prevents the development of Kaposi's sarcoma in patients with HIV. The incidence of Kaposi's sarcoma has decreased dramatically in countries such as the United States, where cART is broadly accessible. Yet, Kaposi's sarcoma still remains the second most common tumor in people with HIV/AIDS in the United States. Without cART, treatment options include topical chemotherapeutic agents, cryotherapy, irradiation, and laser treatment. Antiviral drugs such as ganciclovir, foscarnet, cidofovir, and adefovir (but not acyclovir) inhibit KSHV during lytic infections. These drugs are not effective during latent replication of KSHV. The majority of cells infected with KSHV are latently—not lytically—infected, which is the likely explanation for the lack of effectiveness of drug therapy. **TABLE 15-3** lists the systemic therapies used to treat Kaposi's sarcoma.

KSHV is also associated with two lymphoma disorders: primary effusion lymphoma and multicentric Castleman disease. Prognosis is poor. Like Kaposi's sarcoma, lymphoma cells are latently infected, making drug therapy ineffective. Castleman disease is almost always associated with HIV-infected individuals.

Herpes B (Herpesvirus Simiae or Monkey B Virus)

Herpes B most commonly infects macaques (a type of monkey used extensively in biochemical research; see **FIGURE 15-3**). Most macaques carry herpes B virus with no signs of overt disease. Humans have contracted herpes B infections through direct contact with macaques, such as a bite or scratch or indirect contact with a contaminated **fomite** such as a cage scratch. Fomites are inanimate objects that are capable of transmitting infectious pathogens from one individual to another (e.g., a handkerchief, drinking glass used by an infected person, diaper). The first documented case of human herpes B infection occurred in 1932 after a researcher was bitten by an apparently healthy rhesus macaque. The researcher died 15 days later of **encephalomyelitis** (inflammation of the brain and spinal cord).

The incubation period of herpes B in humans ranges from a few days to a month. Numbness and sometimes vesicles or ulcers occur at the exposure site. The infection causes influenza-like symptoms such as fever, muscle aches, malaise, and headache. Other variable symptoms include nausea, vomiting, abdominal pain, and hiccups. Neurological symptoms begin when the infection spreads to the central nervous system. Spread of infection to the nervous system is ominous. Most patients who show neurological symptoms die, even with antiviral therapy and supportive care. Symptoms vary depending upon what part of the spinal cord or brain is affected. As the virus spreads to the brain, ataxia (loss of coordination), agitations, seizures, confusion progressing to a coma, and progressive ascending paralysis occur. Deaths are often caused by respiratory failure associated with paralysis.

Only 40 cases of human herpes B infections have been reported. The majority of these cases occurred when large numbers of rhesus macaques were used in the production and test of poliovirus vaccines in the 1950s and '60s. Before available antiviral therapy, this virus was a serious zoonotic threat to people who came in contact with macaques. Full protective equipment is used when working with all macaques in a research setting. In 1989, the National Institute of Health's National Center for Research Resources began funding pathogen-free macaque colony development. This greatly reduced the numbers of cases of herpes B infections; however, herpes B virus is also prevalent in **free-ranging macaques** native to Southeast Asia. Free-ranging macaques are a tourist attraction and humans are at risk of contact with infected animals. Research has shown the risk of acquiring herpes B infections from macaques is low, but unfortunately the death rate is high.

15.3 Laboratory Diagnosis of Herpesvirus Infections

Herpesviruses have closely adapted to their host. When symptoms are observed, speed in using diagnostic procedures is important because the peak of shedding virus often occurs before the appearance of symptoms. The diagnostic method used varies for the different herpesvirus infections. The most definitive test is the presence of viable virus in a clinical specimen (e.g., oral and genital lesions, ocular samples, biopsies, and cerebrospinal fluid [CSF]) (**TABLE 15-4**). HSV-1 and VZV-infected cells in tissue culture show the presence of multinucleated giant cells, along with intranuclear inclusion bodies (**FIGURE 15-4**).

15.4 Herpesvirus Life Cycle

Virus Structure and Classes of Herpesviruses

Herpesviruses are large, enveloped, pleomorphic particles that are 150 to 300 nm in diameter when seen by electron microscopy. Herpesvirus virions

TABLE 15-3 Systemic Therapies for Kaposi's Sarcoma

Intervention Approach	Drug	Side Effects	Comments
Immunomodulation: combine antiretroviral therapy (cART) to suppress HIV replication	5 classes of drugs and >25 antivirals available	Depends on antivirals used	cART required for all HIV-associated Kaposi's sarcoma.
Immunomodulation (restore T-cell immunity)	Rapamycin	Edema, hypertension, elevated creatinine, hyperlipidemia	Recipients of organ transplants are susceptible to Kaposi's sarcoma as a result of treatment with immuno-suppressive drugs. Rapamycin, an immunosuppressive drug, may also have antitumor effects.
Immunomodulation, antiviral and **anti-angiogenic** effects (inhibits growth of new blood vessels; blood vessel growth plays a role in the growth of tumors)	Interferon α	Flu-like symptoms, fatigue, neutropenia, elevated transaminases, hypothyroidism	Rarely used.
Enhances T_H responses, upregulates interferon γ, may downregulate a KSHV-encoded protein, vGPCR	Interleukin 12	Neutropenia, elevated transaminases, depression	Not commercially available.
Immunomodulation and anti-angiogenic effects (inhibits growth of new blood vessels; blood vessel growth plays a role in the growth of tumors)	Thalidomide	Fatigue, depression, rash, neuropathy (disease or abnormality of the nervous system), somnolence	Typically not used. Phase II dosing involving Kaposi's sarcoma patients in progress.
Immunomodulation and anti-angiogenic effects (inhibits growth of new blood vessels; blood vessel growth plays a role in the growth of tumors)	Lenalidomide	Fewer neuropathic effects than thalidomide, cytopenias, edema, rash, infections, fatigue	Not generally used. Being tested in clinical trial on AIDS patients with Kaposi's sarcoma
Cytotoxic effects: causes the formation of free radicals resulting in DNA damage	Bleomycin	Nausea, vomiting, mucositis, neutropenia, hair loss, risk of cardiomyopathy	
Cytotoxic effects: disruption of microtubule function	Vincristine/ vinblastine	Hair loss, vomiting, nauseau, other allergic reactions	
Cytotoxic effects: topoisomerase II inhibitor	Etoposide	Neutropenia, anemia, alopecia (hair loss), small risk of developing acute myelogenous leukemia	It can be administered orally. Low doses applied in studies with AIDS-associated Kaposi's sarcoma.
Cytotoxic effects: liposomes accumulated within the vascularized Kaposi's sarcoma lesions	Pegylated liposomal doxorubicin	Anemia, thrombocytopenia, leukopenia, hand-foot syndrome	Standard therapy where available. FDA approved for use in Kaposi's sarcoma patients.
Cytotoxic effects: microtubule stabilizer	Paclitaxel	Hair loss, anemia, thrombo-cytopenia, neutropenia, neuropathy, myalgia infusion reactions	FDA approved for use in Kaposi's sarcoma patients. Often used as a second therapy in advanced cases.
Human monoclonal anti-vascular endothelial growth factor (VEGF) antibody	Bevacizumab	Hypertension, proteinuria, neuropathy	Being used in studies as a combina-tion therapy with pegylated liposomal doxorubicin.
Tyrosine kinase inhibitor that targets VEGFR1, VEGRF2, and PDGFR	Sorafenib	Rash, hand-foot syndrome, gastrointestinal symptoms, fatigue, elevated lipase	Patients currently accrued in Phase I study at the National Cancer Institute.

Source: Adapted from Uldrick, T. S., Whitby, D. 2011. "Update on KSHV epidemiology, Kaposi sarcoma pathogenesis, and treatment of Kaposi sarcoma." *Cancer Lett* 305(2):150–162.

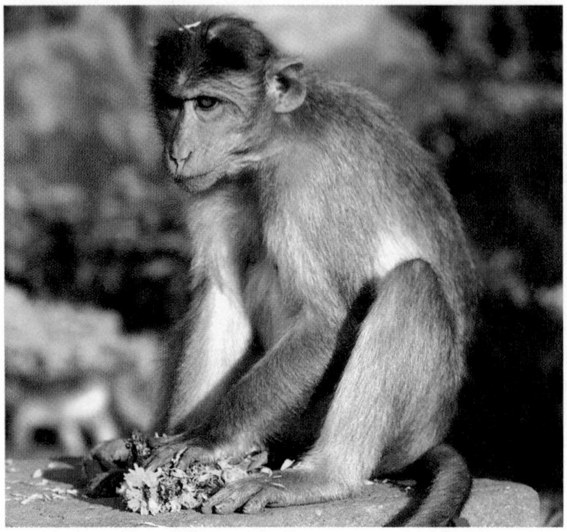

FIGURE 15-3 Photograph of a bonnet macaque (*Macaca radiata*). Herpes B infection is most commonly reported in the rhesus, cynomolgus, stumptail, pig-tailed, bonnet, Japanese, and Taiwan macaques.

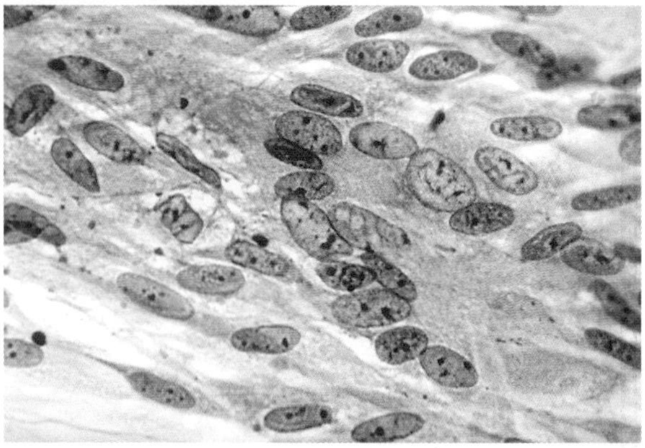

FIGURE 15-4 Photomicrograph showing intranuclear inclusions produced by VZV in tissue culture. Magnification 500×.

comprise more than 30 virally encoded proteins and also contain cellular components. The viral envelope is obtained through a **double envelopment process**. The envelope is fragile and easily damaged by heat, desiccation (drying out), 70% alcohol, or disruption by soap or detergents. The envelope must be intact for herpesviruses to be infectious. The envelope is extremely sensitive to damage; thus, the virus is usually transmitted through direct contact with mucosal surfaces or secretions of an infected person (e.g., the mouth or lips or on the genitals). Herpesviruses dry out and become damaged when exposed to air so cannot be transmitted by toilet seats or other inanimate objects—a common misperception.

Virions contain linear double-stranded DNA (dsDNA) genomes that range from 125 to 229 kilobase pairs in length. An icosahedral capsid surrounds the DNA genome and an amorphous proteinaceous **tegument** surrounds the nucleocapsid. The tegument consists of at least eight different viral proteins (**FIGURE 15-5**). There are eight herpesviruses that naturally infect humans. Middle-aged adults usually have antibodies to most of these, with the exception of HHV-8 (which is common in HIV+ men). HSV-2 is also not very common, unless an individual has had multiple sexual partners.

The *Herpesviridae* family is divided into three subfamilies: *Alphaherpesvirinae*, *Betaherpesvirinae*, and *Gammaherpesvirinae*. The subfamilies are distinguished from each other based on such properties as the type of cells in which viral latency is established and the characteristics of their growth in cell cultures (**TABLE 15-5**). Viruses are named according to their natural host species and numbered in

TABLE 15-4 Standard Diagnostic Tests of Herpesvirus Infections

Type of Infection	Method	General Availability
Skin and genital lesions, mucocutanious	Virus isolation	Variable
	Cytology	Yes
	Immunofluorescence or immunoperoxidase staining (detection of viral antigens)	Yes
	ELISA (detection of viral antigens)	Yes
	Shell vials	Variable
Keratitis	Virus isolation	Variable
	Immunofluorescence or immunoperoxidase staining (detection of viral antigens)	Yes
	Polymerase chain reaction of viral DNA	Variable
Encephalitis	Polymerase chain reaction of viral DNA	Variable
Previous infection	Serology increased titers of antibodies	Variable
	Serotype	Variable

Source: Adapted from Aurelian, L. "Herpes simplex viruses". In: Specter, S., Hodinka, R. L., Young, S. A., Wiedbrauk, D. L., eds. *Clinical Virology Manual,* Fourth Edition. Washington, DC: ASM Press; 2009:440.

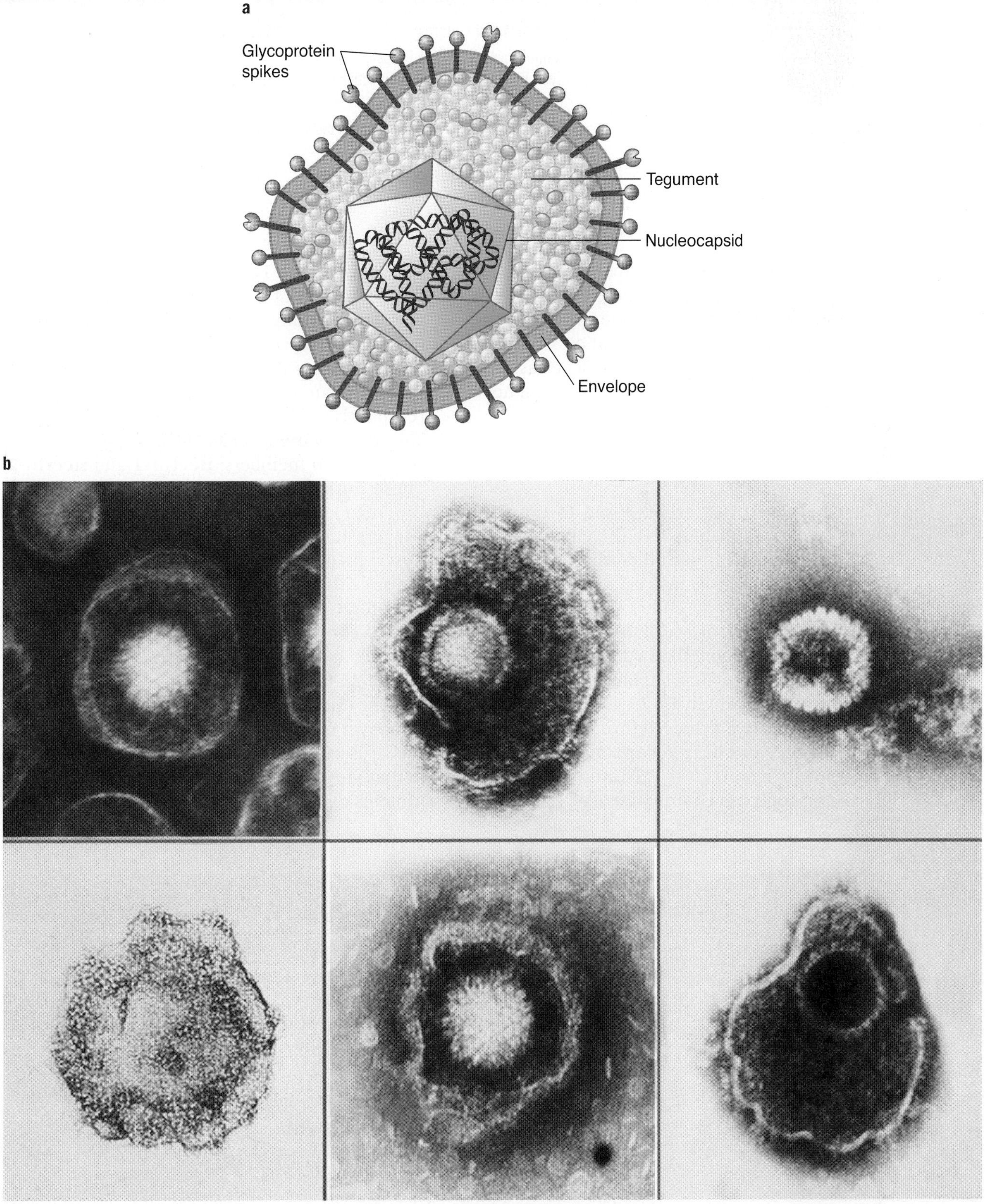

FIGURE 15-5 (a) Schematic of herpesvirus structure. There are more than a dozen different types of integral membrane proteins within the viral envelope. The tegument contains at least eight proteins that serve important functions after the virus has penetrated the host. **(b)** Electron micrographs of various viruses from the Herpesviridae family.

TABLE 15-5 Properties of the Subfamilies of the *Herpesviridae* Family

Subfamily	Examples	Growth in Cell Cultures	Latency
Alphaherpesvirinae	HSV-1, HSV-2, VZV, herpes B virus	Rapid growth and spread in cell cultures, short replication cycle, lytic infection in fibroblasts and epithelial cells	Established in sensory and cranial nerve ganglia (neurons)
Betaherpesvirinae	CMV, HSV-6, HSV-7	Slow infection, long replication cycle, cells form enlarged (cytomegalic) cells	Monocytes, macrophages, CD34+ cells, memory B cells
Gammaherpesvirinae	EBV, HHV-8	Limited to growth in lymphoblastoid cells.	Memory B or B cells

the order in which they were first identified; for example, herpes simplex virus type 1 (HSV-1) was the first human herpesvirus identified. In the future, it is likely that the classification scheme of the *Herpesviridae* family members will be based on genetic organization, such as repetitive sequences, selected genes, and gene clusters.

Herpesvirus Entry and Uncoating

When herpesviruses are exposed to mucosal membranes or skin abrasions, the virus attaches to cells of the epidermis and dermis of the skin. Entry of herpesviruses into cells depends upon multiple cell-surface receptors and multiple proteins located on the surface of the virion. The viral envelope contains more than a dozen viral integral membrane glycoproteins, but only five have been shown to participate in HSV entry: gB, gC, gD, gH, and gL. Entry is achieved through a **binding receptor** and an **entry receptor**. The viral gB or gC envelope proteins of HSV attach or dock to **glycosaminoglycan chains** present on cell surface **proteoglycans**. **Heparan sulfate** is the most common kind of glycosaminoglycan and is the preferred binding receptor. This facilitates binding of the viral gD protein to an entry receptor that is one of two different types of cellular receptors that are structurally unrelated. The first type of receptor is the **tumor necrosis factor receptor superfamily member 14 (TNFSR14)**. The second contains two members: **nectin-1** and **nectin-2**. These entry receptors are listed in **TABLE 15-6**. Binding to any one of these three entry receptors triggers fusion of the viral envelope with the host cell's plasma membrane. This fusion activity requires the action of a gB, gD, gH–gL, and gD receptor. Glycoproteins gB, gH, and gL are structurally conserved among all herpesviruses and probably have similar essential roles in viral entry (**FIGURE 15-6**).

The nucleocapsids of herpesviruses are covered by a layer of tegument proteins (Figure 15-5). As the virus uncoats and the nucleocapsid is released in the cytoplasm of the cell, some of the tegument proteins remain in the cytoplasm (e.g., the **virion**

TABLE 15-6 Human Herpesvirus Genome Sizes and Receptors

Human Herpesvirus	Genome Size kbp	Entry Receptor(s)
HSV-1 (HHV-1)	152	Nectin 1, nectin 2, tumor necrosis factor receptor superfamily member 14, 3-O-sulfotransferases
HSV-2 (HHV-2)	152	Nectin 1, nectin 2, tumor necrosis factor receptor superfamily member 14, 3-O-sulfotransferases
Varicella zoster (HHV-3)	125	Insulin degrading enzyme
Epstein Barr virus (HHV-4)	172	CD21, major histocompatibility complex class II
Cytomegalovirus (HHV-5)	229	Platelet-derived growth factor α epidermal growth factor receptor?
HHV-6	165	CD46
HHV-7	145	CD4?
Kaposi's sarcoma associated herpesvirus (HHV-8)	165	Integrin α, β, light chain of the human cystine/glutamate transporter system, dendritic cell-specific ICAM-3 grabbing nonintegrin
Herpes B virus	150	unknown

Source: Adapted from Straus, S. E. Introduction to *Herpesviridae*. In: Mandell G. I., Bennett, J. E., Dolin, R. eds. *Principles and Practice of Infectious Diseases*, Sixth Edition. Philadelphia: Elsevier, Churchill Livingstone; 2005:1756–1762.

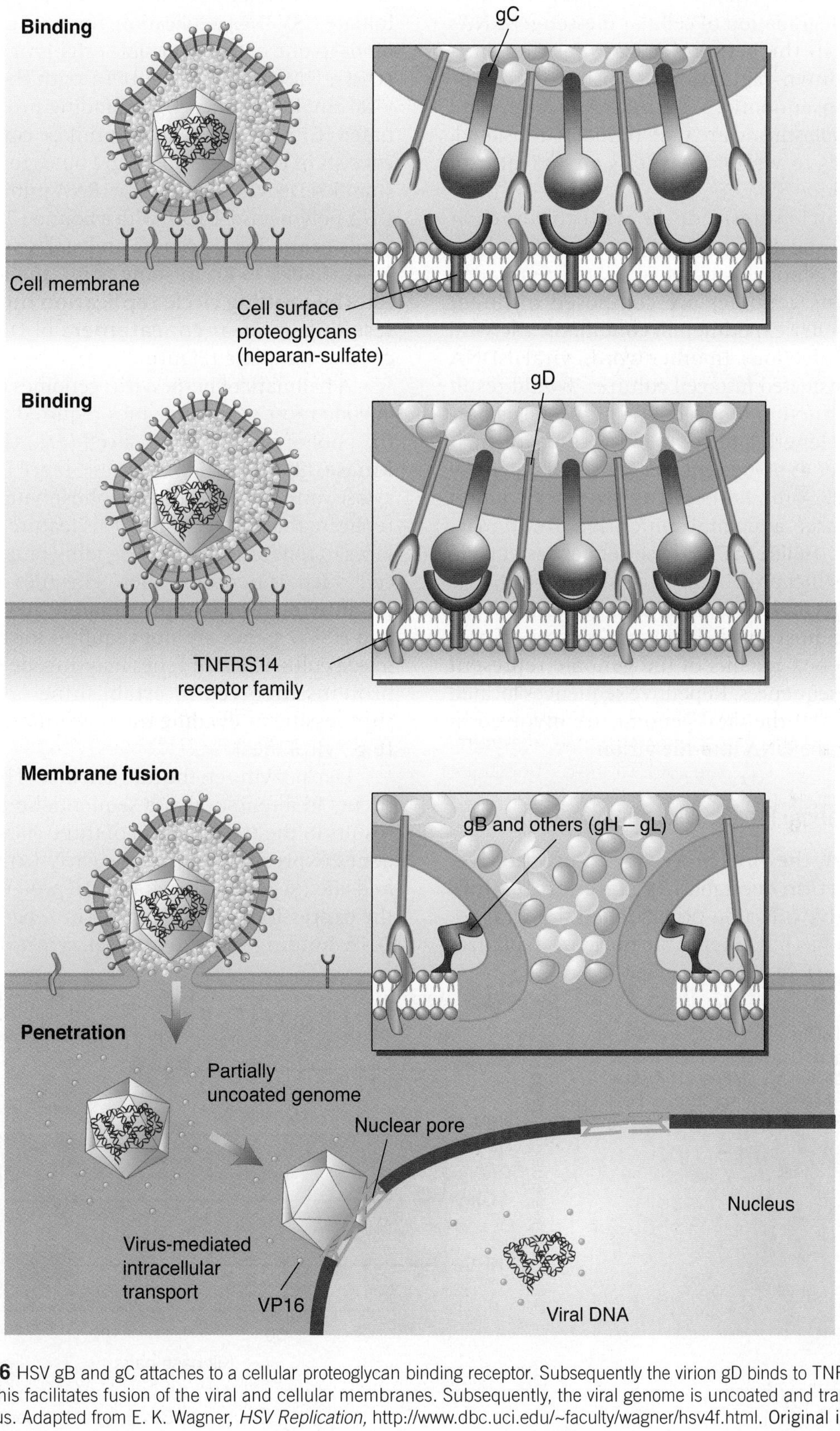

FIGURE 15-6 HSV gB and gC attaches to a cellular proteoglycan binding receptor. Subsequently the virion gD binds to TNFRS14 entry receptors. This facilitates fusion of the viral and cellular membranes. Subsequently, the viral genome is uncoated and transported to the nucleus. Adapted from E. K. Wagner, *HSV Replication,* http://www.dbc.uci.edu/~faculty/wagner/hsv4f.html. Original illustration by Imagecyte.com.

host shutoff protein, *vhs*), and other proteins such as the **VP16 protein** are transported to the nucleus. The viral *vhs* protein plays a role in the rapid degradation of cellular messenger RNAs (mRNAs) in the cytoplasm, causing the shutdown or down-regulation of host protein synthesis. Consequently, viral mRNAs accumulate in the cytoplasm and are preferentially translated. VP16 binds to viral DNA and is an activator of transcription.

The nucleocapsid docks with the nuclear membrane and delivers the genome to the nucleus, where the genome is transcribed and replicated. Herpesvirus genomes are composed of linear dsDNA. Under experimental conditions, the viral dsDNA is infectious. In other words, viral dsDNA alone, transfected into cell cultures, would result in the production of infectious herpesvirus particles. The length and complexity differ for each virus. The ends of the genomes consist of repetitive sequences. Some herpesvirus genomes contain internal large and small repetitive unique (U) sequences (**FIGURE 15-7**). These sequences encode over 50 distinct protein products such as DNA replication enzymes, glycoproteins important in viral host range, host defense proteins, and capsid proteins. The box regions of the genome represent repetitive sequences. Repetitive sequences located at the ends of the viral genome are involved in packaging the DNA into the virion.

Replication: Productive, Lytic Infection

Upon entry, the virus proceeds to a **productive, lytic infection** or establishes a **latent infection**. A productive infection occurs when the viral genome that has just entered the nucleus circularizes and is transcribed by the cellular DNA-dependent RNA polymerase and is regulated by viral- and cellular-encoded regulatory nuclear proteins. To initiate HSV DNA replication, the viral UL9 protein binds to one of three possible origins of replication (ORI). UL9 unwinds the DNA with the help of a viral single-stranded DNA binding protein designated ICP8. Next a helicase/primase complex that consists of UL5, UL8, and UL52 binds to the single-stranded DNA and synthesizes RNA primers. A viral DNA polymerase, UL30, with a bound UL42 protein binds to the RNA primers and starts synthesis of DNA (**FIGURE 15-8**). At some point, the DNA replicates by a **rolling circle replication mechanism**, resulting in linear **concatemers** of DNA from a circular template (Figure 15-8).

A hallmark of herpesvirus genomes is that they encode most of the enzymes required to increase the pool of nucleotides in a cell (e.g., **thymidine kinase**, ribonucleotide reductase, uracil DNA glycosylase, and deoxyuridine triphosphatase) and to replicate the DNA genome. This feature allows the virus to replicate in slowly dividing or nondividing cells such as neurons. These enzymes are targets for antiviral drugs. Approximately 50% of HSV-1 and HSV-2 genes are not required for replication in cell culture. These genes may encode regulatory proteins involved in establishing latency and suppressing or evading the host immune system (e.g., virokines).

Like poxviruses, the transcription of herpesvirus occurs in a temporal and sequential manner that results in the transcription of three classes of viral genes expressed: immediate-early (α), early (β), and late (γ) genes. The viral VP16 protein activates the expression of the α genes. The α genes encode DNA binding proteins that play a role in viral

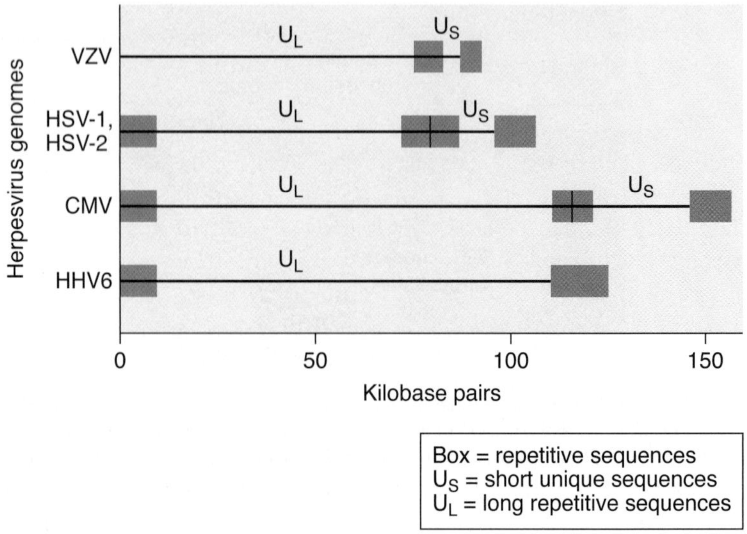

FIGURE 15-7 Comparison of the genomes of herpesviruses. The genome contains unique (U) sequences and repeat sequences (the boxes). Adapted from Murray, P. R., et al. *Medical Microbiology*, Third Edition. Mosby Elsevier Health Science, 1998.

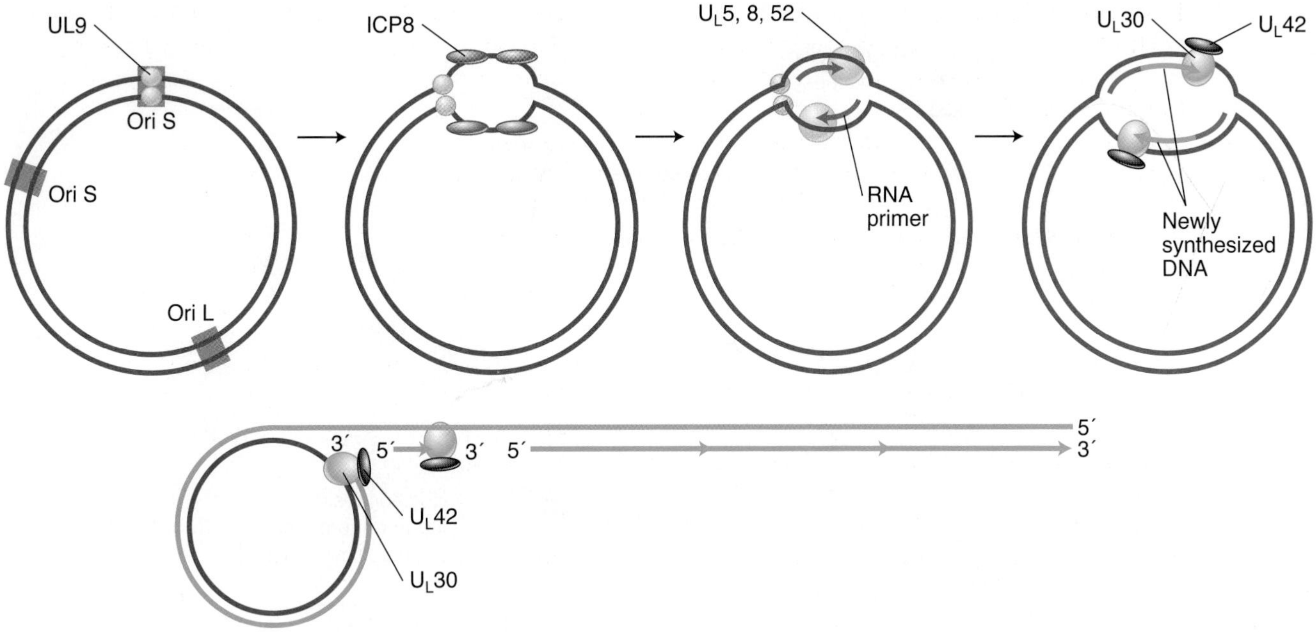

FIGURE 15-8 HSV replication initiates at one of three possible ORIs. Replication is bidirectional. RNA primers are created by a U_L5, U_L8, U_L52 complex. Newly synthesized DNA is generated by a viral DNA polymerase. U_L42 is bound to U_L30. U_L42 increases U_L30's ability to extend and detach from the growing strand of DNA from the template DNA. Adapted from Acheson, N. H. *Fundamentals of Molecular Virology.* John Wiley & Sons, Inc., 2006.

transcription of β genes. The β genes encode DNA replication and additional viral transcription factors. The γ genes encode the late structural proteins that are produced after viral genome replication has begun (**FIGURE 15-9**).

After the γ proteins are synthesized, they are localized into the infected cell nucleus, where capsid assembly occurs. Viral genomic DNA concatemers are cleaved into monomers.

The monomers are packaged into the capsids and virion assembly proceeds. As the tegument proteins wrap around the nucleocapsid, viral glycoproteins present in the inner nuclear membrane promote the budding of the virion through the nuclear membrane. This is the first or primary budding event. New research suggests that viral membrane formation is a **double envelopment process**. The first envelopment of the virus by the nuclear membrane translocates the nucleocapsid to the cytoplasm of the cell. The nucleocapsid then buds through Golgi-derived vesicles. Mature virus particles are released after fusion with the Golgi-vesicle membrane and the cellular plasma membrane, and cell lysis occurs. As a result, herpesviruses undergo a two-step envelope process: de-envelopment and re-envelopment (**FIGURE 15-10**). Death is a result of cellular injuries that occur as a result of viral replication and cellular responses to infection.

Viral Replication and Latency

Infected cells producing herpesvirus particles (a lytic infection) do not survive. Viruses benefit from their host cell remaining alive. HSV-1 and -2 have developed a survival strategy in immunocompetent hosts that centers on the establishment of lifelong **latency** in sensory neurons innervating the primary site of infection. After a productive primary infection, these herpesviruses enter sensory neuron axons and migrate along the axon to the cell body in a ganglion in the CNS (**FIGURE 15-11a**). In the sensory neuron cell body, the genome of the herpesvirus persists as an episome (not integrated) in the nucleus of the neuron. Episomes are circular DNAs that replicate independently of the cell's nucleus. During this latent stage, only a limited group of viral genes—the **latency-associated transcripts** (**LATs**)—are expressed. The role of LATs is still under debate. However, a number of microRNAs are now known to be expressed from the LATs. Scientists have suggested that LATs may prevent expression of viral genes via an antisense mechanism and thus may play a role in preventing apoptosis in neurons.

While latent, viral replication does not proceed further. No viral particles are detected. The virus hides from the immune system for months or even years until it is **reactivated** into a productive infection (Figure 15-11b). Once VZV, HSV-1, or

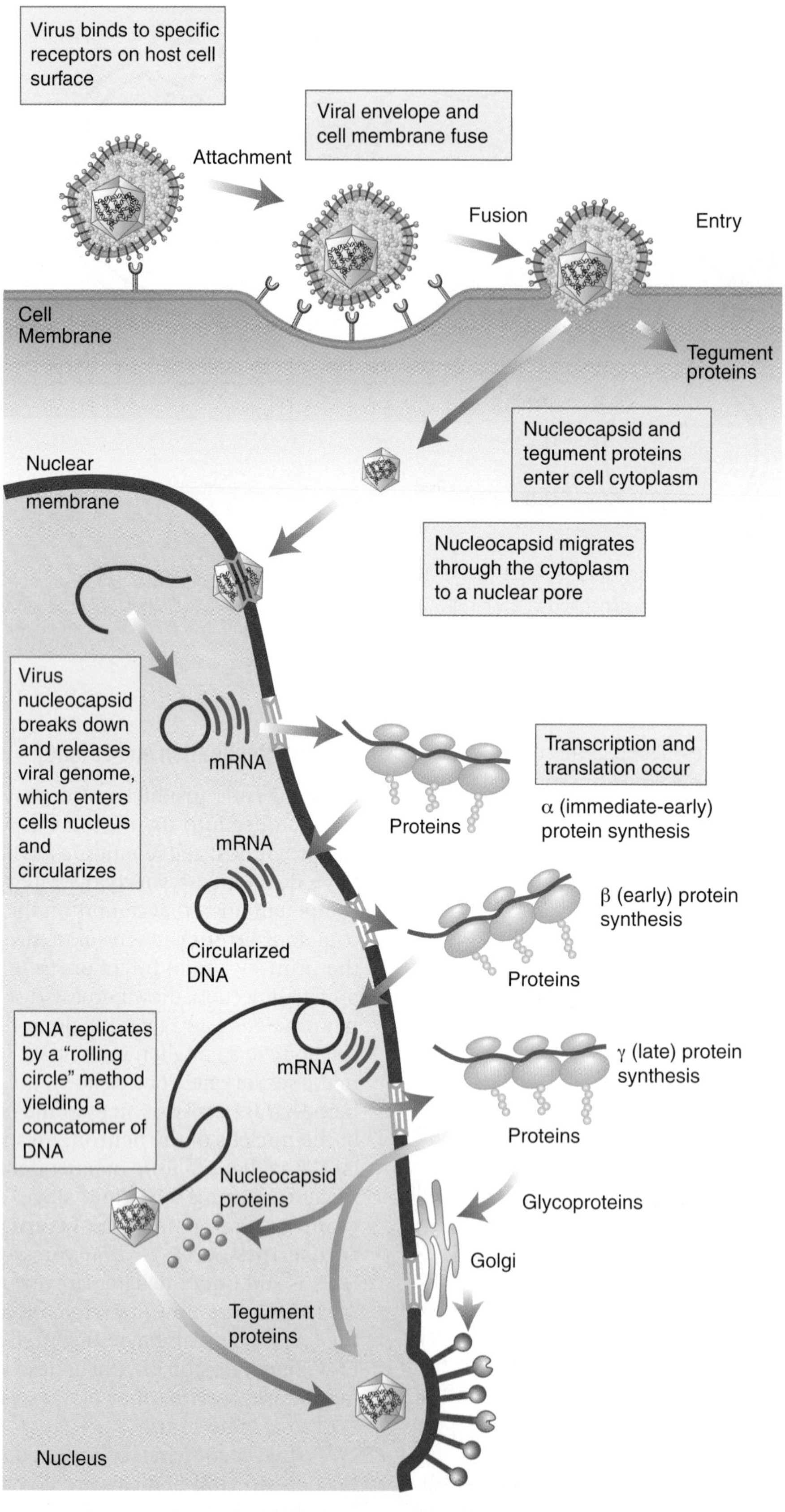

FIGURE 15-9 Upon entry, the genome circularizes and viral tegument proteins and cellular factors are involved in the transcription of the viral genome. Transcription and translation occur in three phases: (α) immediate-early, (β) early, and (γ) late. Adapted from Cohen, J., and Powderly, W. G. *Infectious Diseases,* Second Edition. Mosby, 2003.

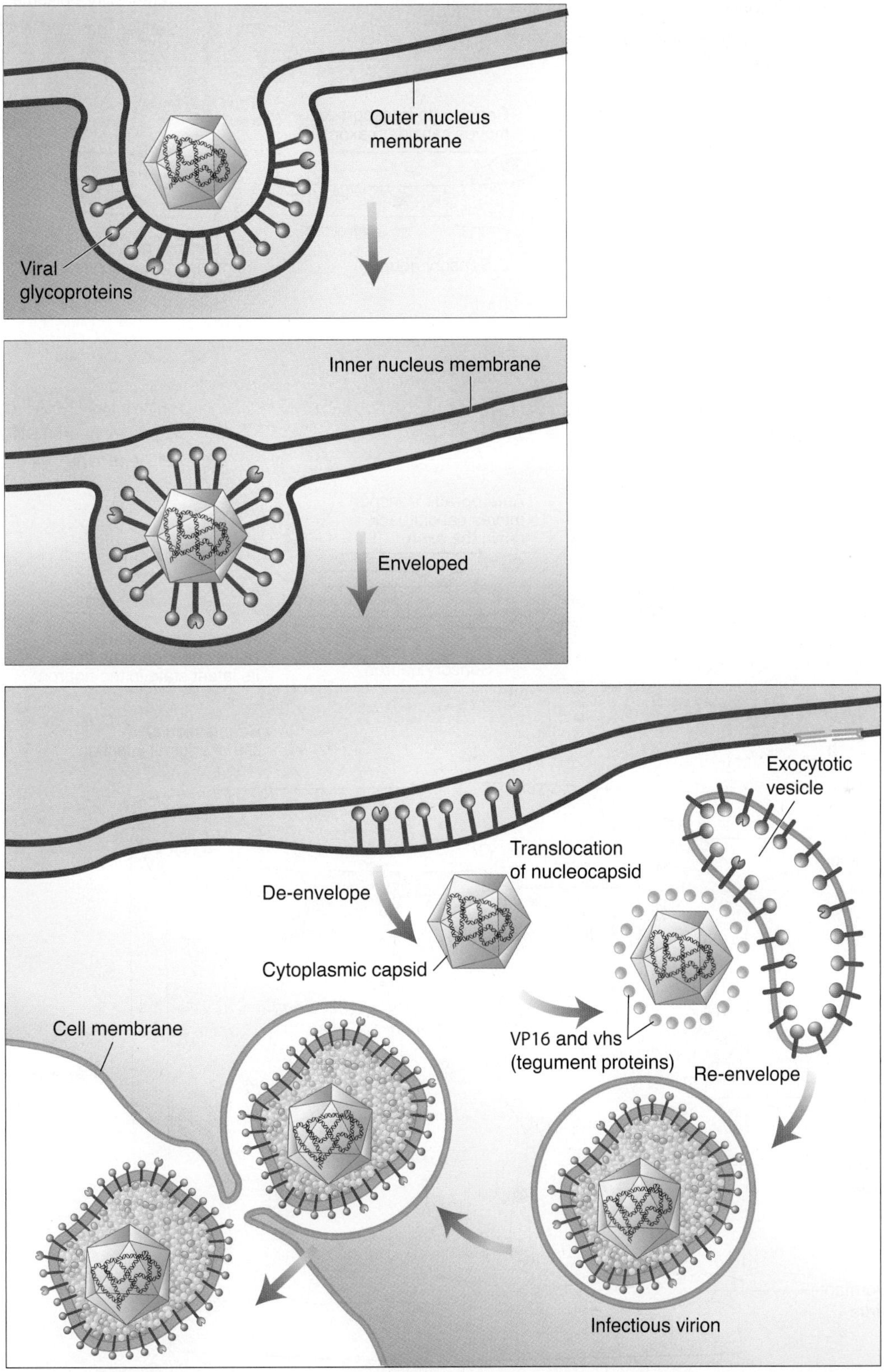

FIGURE 15-10 Herpesvirus **morphogenesis** requires de-envelopment–re-envelopment processes. The virion is enveloped by the nuclear membrane that translocates the nucleocapsid to the cytoplasm where it matures. The majority of tegument proteins (e.g., VP16, *vhs*) is added to the nucleocapsids and the final envelope is acquired by budding into the glycoprotein-containing Golgi-derived vesicles. Adapted from Wagner, E. K., *HSV Replication.* Original illustration by Imagecyte.com.

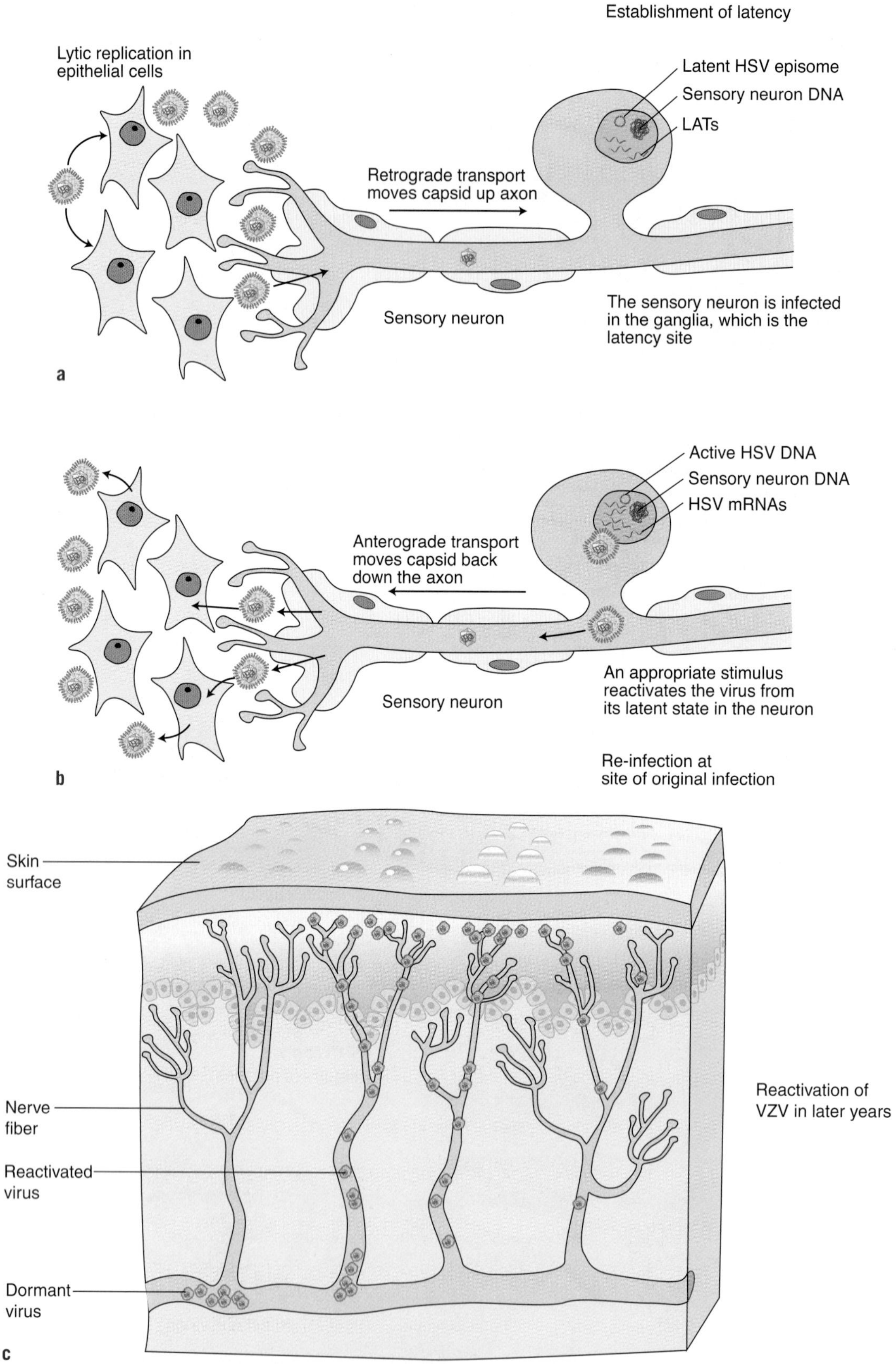

FIGURE 15-11 (a) HSV undergoes a productive infection in epithelial cells of the genital or oral mucosal surface. During latency, the virus travels up the axon to the neuronal cell body, where it remains dormant. Adapted from Taylor, T. J., et al., *Front Biosci* 7 (2002): d752–d764. **(b)** Once reactivated, the virus travels down the axon and establishes a productive, lytic infection. Adapted from Taylor, T. J., et al., *Front Biosci* 7 (2002): d752–d764. **(c)** Reactivation of VZV (chickenpox) in later years. A decrease in the body's immune system may contribute to an outbreak of shingles. In step (1) a cluster of small bumps begins to form and then (2) turns into blisters. The blisters will crust over (step 3) and eventually disappear (step 4). The typical course of shingles takes four to five weeks. Adapted from Zamula, E., *FDA Consumer Magazine* 35 (2001).

HSV-2 is reactivated, it travels the nerve pathway to the surface of the skin (Figure 15-11c). The molecular details of latency and reactivation are different for each group of herpesvirus, and it has proven difficult to dissect at the molecular level.

15.5 Antivirals/Treatment of Herpesvirus Infections

There are a number of antiviral drugs available for the treatment of active herpesvirus infections and prophylactic use (**TABLE 15-7**). Treatment is dependent upon the type, location of infection, and immune status of the patient. The most commonly prescribed anti-HSV drug is **acyclovir**. It is relatively nontoxic and has been used successfully for long-term prophylaxis. It can be administered topically to the skin, intravenously, orally, or topically to the eye. Most anti-HSV drugs are nucleotide analogs or inhibitors of the herpesvirus-encoded DNA polymerase.

Acyclovir, a guanosine analog (ACG), is converted 200 times more efficiently to ACG-monophosphate (ACG-P) by herpesvirus thymidine kinase (TK) than cellular thymidine kinases and other enzymes. After the viral TK converts ACG to ACG-P, a cellular guanosine monophosphate kinase phosphorylates (activates) ACG-P to ACG-triphosphate (ACG-PPP). Both the herpesvirus DNA polymerase and the cellular DNA polymerase incorporate ACG-PPP into the growing strand of herpesvirus or cellular DNA instead of GTP. This results in a **chain termination** reaction, preventing the elongation of DNA. The herpesvirus DNA polymerase is ten times more inhibited than the cellular DNA polymerase by the incorporation of ACG-PPP. The mechanism of action for acyclovir is shown in **FIGURE 15-12**. **Drug resistance** occurs when mutations occur in the herpesvirus *TK* gene, reducing the conversion of ACG to its active form. Less frequently, the viral DNA polymerase is mutated, usually resulting in a less virulent form of herpesvirus.

Some research suggested that diet is an important factor in keeping HSV outbreaks in remission. Foods high in L-arginine may cause herpesvirus outbreaks, whereas foods rich in L-lysine appear to reduce the occurrence, severity, and healing time for recurrent HSV infection. A few popular reports claimed that supplements of L-lysine were a useful alternative measure in controlling HSV outbreaks. Herpes simplex virus proteins synthesized in infected cells contain more arginine than lysine, methionine, phenylalanine, tyrosine, and isoleucine relative to leucine than proteins synthesized in uninfected cells. A favorable ratio of lysine to arginine suppresses viral replication and does not appear to cause harm to the host. Foods with a high lysine-to-arginine ratio are beef, chicken, fish, lamb, milk, cheese, beans, brewer's yeast, and most fruits and vegetables. Foods with a high arginine-to-lysine ratio are carob, chocolate, coconut, oats, whole wheat flour, white flour, gelatin, peanuts, soybeans, and wheat germ. That being said, research is conflicting on the benefits of high concentrations of L-lysine. The estimated average requirement for lysine is 31 mg/kg/day. Chronic use of higher doses than 3 g of lysine per day is not recommended until further safety guidelines have been determined by clinical studies. Diets too high in lysine or with a high lysine:arginine ratio have been reported to have a hypercholesterolemic effect in animal studies.

Varicella zoster immune globin and zoster immune plasma have been given to the immunocompromised or newborns at risk. They prevent the infection or minimizes the symptoms.

TABLE 15-7 FDA-Approved Antiherpesvirus Drugs

Drug	Chemical Class	Target Herpesvirus/Disease
Idoxuridine	Thymidine analog	Topical treatment of HSV-1 and HSV-2 keratoconjunctivitis
Vidarabine	Adenine arabinoside	Topical treatment of HSV-1 and HSV-2 keratoconjunctivitis
Trifluridine	Thymidine analog	Keratoconjunctivitis and keratitis caused by HSV-1 and HSV-2
Acyclovir	Guanosine analog	HSV-1, HSV-2 (oral form), VZV (intravenous or oral form)
Famciclovir	Guanosine analog	HSV-1, HSV-2, VZV (intravenous or oral)
Penciclovir	Guanosine analog	HSV-1 and HSV-2 (topical)
Valacyclovir	Guanosine analog	HSV-1, HSV-1, VZV
Foscarnet	Pyrophosphate analog	CMV, HSV-1, HSV-2
Cidofovir	Cytosine analog	HSV-1, HSV-2, CMV

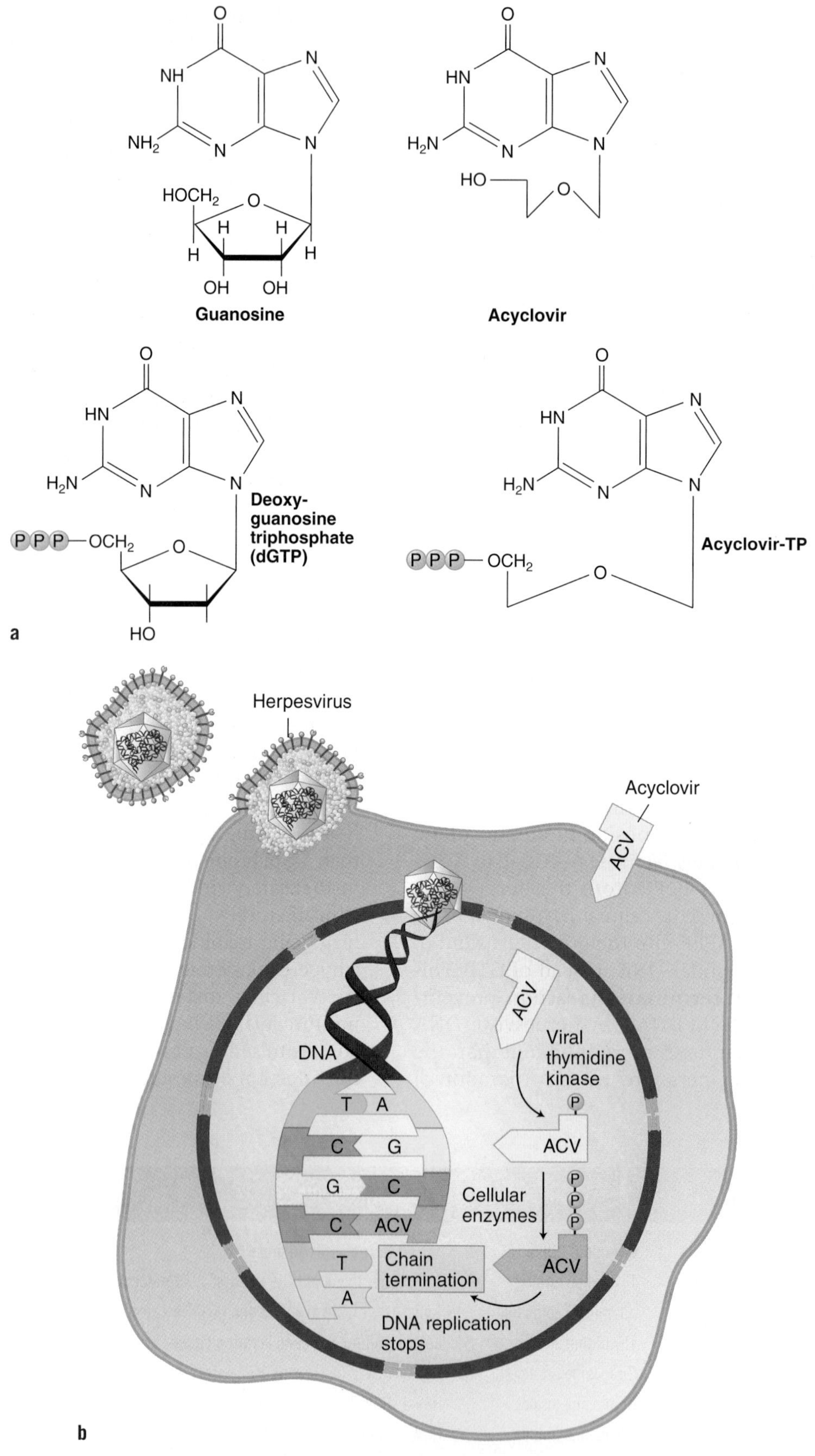

FIGURE 15-12 (a) Chemical structure of acyclovir and guanosine and their triphosphate forms. **(b)** Mechanism of action of the herpesvirus antiviral acyclovir (ACV). ACV has a high binding affinity for herpesvirus (but not the host) DNA polymerase. This leads to its enzyme inactivation. The incorporation of ACV-PPP into replicating DNA blocks chain elongation. Adapted from Cohen, J., and Powderly, W. G. *Infectious Diseases,* Second Edition. Mosby, 2003.

15.6 Chickenpox and the Development of Other Herpesvirus Vaccines

VZV was isolated from vesicular fluid of both chickenpox and herpes zoster lesions via cell culture by Thomas Weller and colleagues in 1952. This led to the development of a chickenpox vaccine by Takahashi and colleagues at Osaka University, Japan, in 1974. An Oka VZV strain was isolated from the fluid from vesicles of an otherwise healthy three-year-old Japanese boy (whose family name was Oka) who had a typical case of chickenpox. The virus was attenuated by passaging it 11 times in human embryonic lung fibroblasts at 34°C, followed by 12 passages in guinea pig fibroblasts at 37°C and 7 passages in human lung fibroblasts at 37°C. The entire genome of the Oka strain and its parent virus (P-Oka) has been sequenced. The Oka VZV strain has 42 mutations, resulting in 20 amino acid changes or base changes in the origin of DNA replication region. The exact mechanism of attenuation is still unclear.

Vaccination studies using the live, attenuated Oka strain were carried out in Japan in the 1970s. Early participants were immune-compromised children who had undergone steroid therapy for leukemia. The chickenpox vaccine (Varivax Oka/Merck) was licensed by the FDA for use in the United States in March 1995. No other live attenuated vaccines had been approved for licensure in the United States from 1969 to 1995. Current CDC recommendations are one dose of the vaccine between 12 and 18 months of age and a catch-up vaccination at any age after that if the child has never had chickenpox. Healthy individuals who have not been vaccinated until 13 years of age or older should get two doses of the vaccine four to eight weeks apart.

Clinical trials continue after licensure. Since 1995, millions of children have been vaccinated and there has been a 75% decline in the number of hospitalizations related to VZV infection.

One of the unanticipated results of chickenpox vaccination among immunized children is the growing number of varicella outbreaks attributed to **breakthrough varicella**, a wildtype virus that causes a mild form of chickenpox (an average maximum number 50 lesions or less) in children who have been immunized at least 42 days prior to their illness (**FIGURE 15-13**). According to the manufacturer of the vaccine, the breakthrough rate observed is 0.3% to 3.8% of vaccinees per year. Explanations

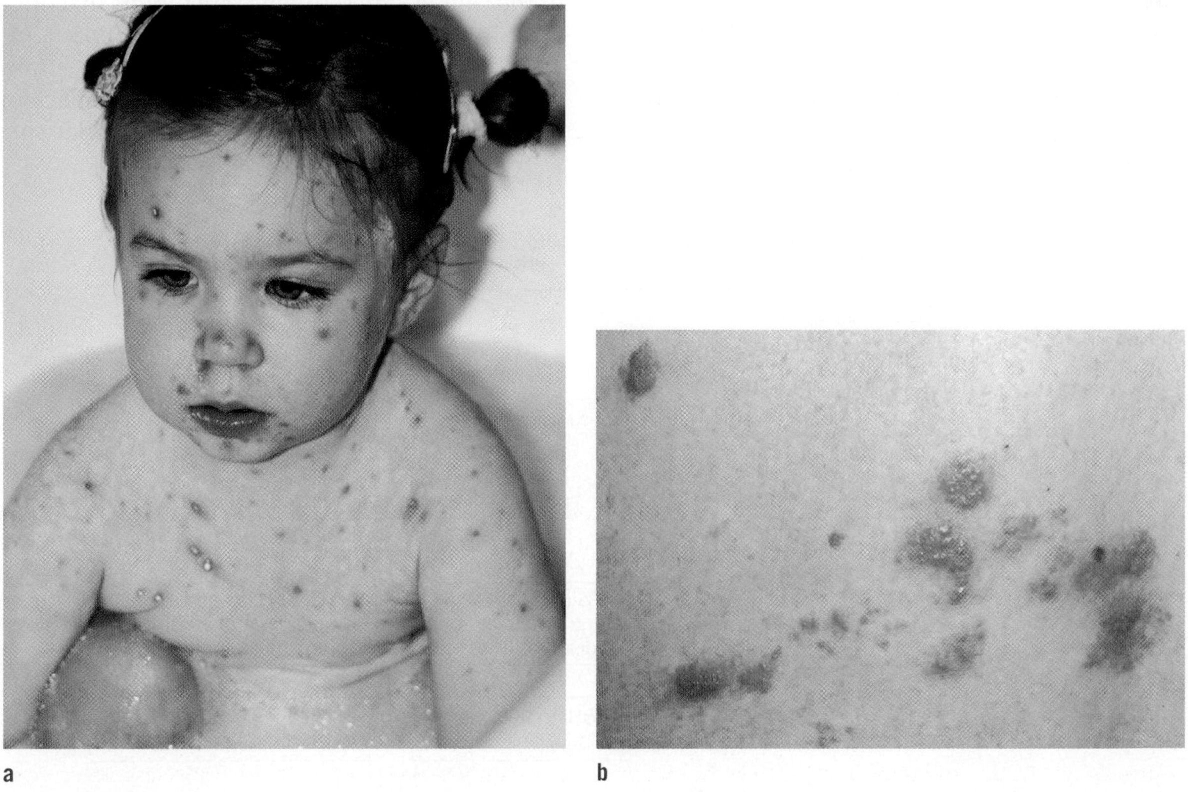

FIGURE 15-13 (a) An 18-month-old child with mild breakthrough chickenpox. **(b)** Shingles rash caused by VZV.

for breakthrough cases include a lessened immune response among the youngest vaccinees or the possibility of genetic variation among circulating VZV strains. Breakthrough cases appear to be more common in the United States than in Japan.

Will the Chickenpox Vaccine Prevent Herpes Zoster (Shingles)?

The Department of Veteran Affairs conducted a Shingles Prevention Study between 1998 and 2001 with a three-year follow-up (Cooperative Study No. 403). A total of 38,546 immune-competent adults age 60 or older participated in a randomized, double-blind, placebo-controlled clinical trial investigating the live attenuated Oka/Merck VZV vaccine. The participants received a one-time dose of the VZV vaccine or a placebo. More than 95% of the participants completed the trial. The results were reported in the *New England Journal of Medicine* in 2005. There were 11.2 cases of herpes zoster per 1000 persons in the placebo group and 5.42 cases per 1000 persons in the vaccine group. Thus, the incidence of shingles cases among the adult participants was reduced 51.3% in those who were vaccinated. A new vaccine called Zostavax (Merck) was approved by the FDA in 2006 to reduce the risk of shingles in people 60 years of age or older. Each dose contains about 14 times more live attenuated VZV than Varivax. In 2011, the FDA approved prescribing the Zostavax vaccine to persons 50 years of age or older. Licensure was granted after an exhaustive Merck study determined that vaccination with Zostavax reduced the risk of developing shingles by nearly 70% in adults aged 50 to 59.

Other Herpesvirus Vaccines in the Pipeline: Dream or Reality?

CMV infection is a major disease of immunocompromised patients and is the leading cause of inner ear (cochlea) hearing loss and nervous system damage in children in the United States (**FIGURE 15-14**). Developing a vaccine against CMV has been a top priority for the prevention of congenital CMV infection, but public awareness is still lacking. It is known that both cell-mediated immunity and antibodies are required to relieve the symptoms of CMV infection. Studies have shown that maternal antibodies reduce the risk of congenital infection in future generations by 69%. A recombinant canarypox vaccine containing CMV genes has been generated and is being investigated in clinical trials. It has been proposed that this combined vaccine might generate the cellular and antibody-mediated immunity needed for protection against CMV infections.

15.7 The Use of Genetically Engineered HSV to Treat Brain Tumors

Malignant **gliomas** affect 15 per 100,000 adults in the United States. Gliomas are tumors that originate from the brain or spinal cord. About 30% of CNS brain tumors are malignant gliomas. Brain gliomas are very aggressive and there is a lack of effective therapy. Current approaches include sur-

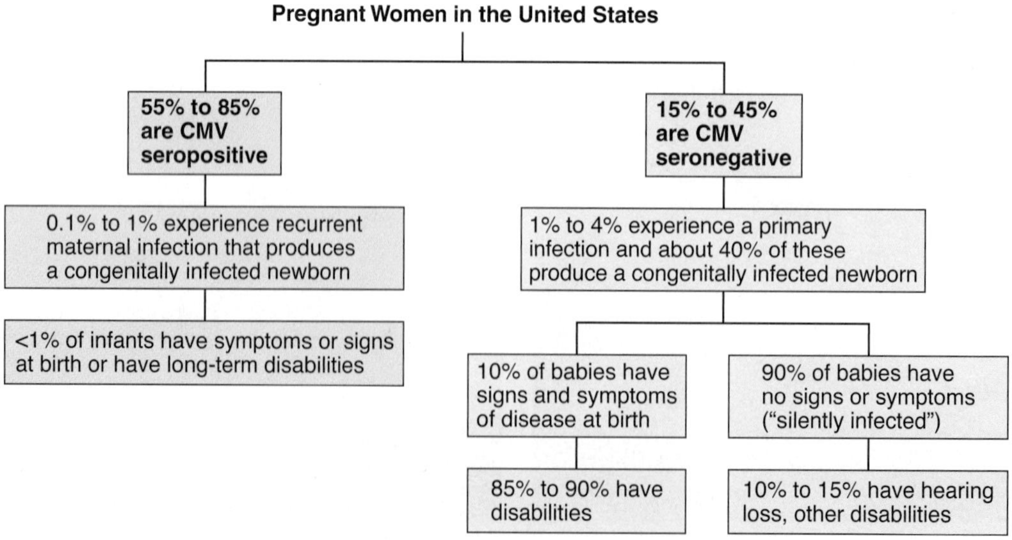
Pregnant Women in the United States
55% to 85% are CMV seropositive
15% to 45% are CMV seronegative
0.1% to 1% experience recurrent maternal infection that produces a congenitally infected newborn
1% to 4% experience a primary infection and about 40% of these produce a congenitally infected newborn
<1% of infants have symptoms or signs at birth or have long-term disabilities
10% of babies have signs and symptoms of disease at birth
90% of babies have no signs or symptoms ("silently infected")
85% to 90% have disabilities
10% to 15% have hearing loss, other disabilities

FIGURE 15-14 Neonatal CMV statistics in the United States. Adapted from National Congenital CMV Disease Registry, *What Everyone Should Know About CMV*, http://www.bcm.edu/web/pediatrics/documents/CMV_Brochure_eng.pdf.

gery, radiation, and chemotherapy. These methods extend the lives of patients an average of 14 weeks to one year. About 5.5% of individuals survive five years, but often these methods fail to kill all of the tumor cells. Patients with brain gliomas are ideal candidates for novel molecular-based therapies for several reasons:

- Metastases are rare.
- Magnetic resonance imaging (MRI) studies are available to monitor the outcome (**FIGURE 15-15**).
- Viruses provide a delivery technique that can target the tumor.

Previous strategies involved the use of engineered recombinant adenoviruses and retroviruses to transfer the HSV thymidine kinase gene into tumor cells followed by treatment of the patient with acyclovir. These strategies have had limited success. Genetically engineered HSV is an attractive vehicle because it can replicate in neurons and glia and the HSV DNA can be easily manipulated. Wildtype HSV is neurovirulent. Hence, the HSV vector must be engineered to be able to replicate in a dividing tumor but not in surrounding healthy cells of the CNS. Recombinant HSVs have been engineered that contain mutations in the viral thymidine kinase, DNA polymerase, ribonucleotide reductase, and the *γ134.5* gene. The *γ134.5* gene has been mapped to the neurovirulence phenotype of the virus. These mutations are necessary to create a replication-defective, safe strain of HSV gene therapy vector for treatment of brain gliomas and other neurological applications. To date, replication-defective HSV vectors have been optimized and plans are underway to carry out phase I clinical trials.

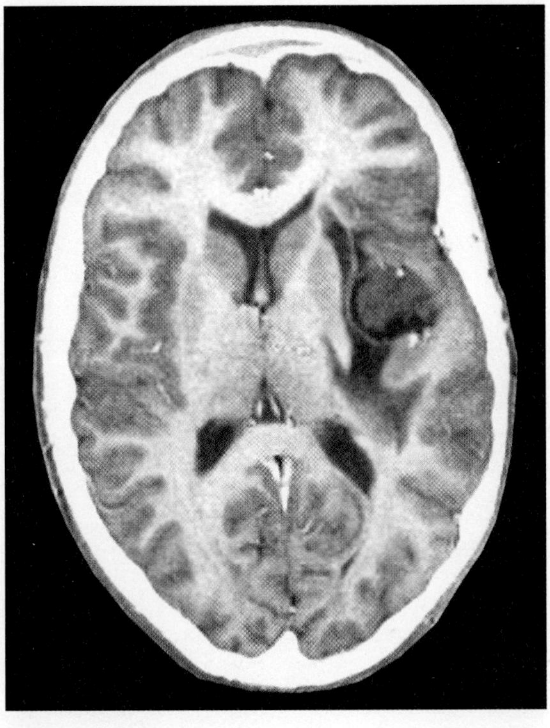

FIGURE 15-15 Color-enhanced cerebral MRI showing a brain glioma.

Summary

Herpesviruses are ubiquitous in vertebrates and have existed for hundreds of millions of years. Once individuals have become infected with a herpesvirus, the infection remains in their body for life. Herpesviruses have evolved an efficient survival strategy in which the virus can remain dormant (latent) for periods of time, hiding within the cells of the body by evading the host's immune system. While dormant, the virus is protected from therapeutic agents. For the α herpesviruses such as HSV-1, latent infections occur in the cell body of neurons, where viral DNA remains dormant within the cytoplasm or nucleus until activated to replicate and migrate along the neural axis to an epithelial cell surface. The β and γ herpesviruses are latent in other cell types. A variety of stimuli may activate a herpesvirus infection:

- immune suppression by cytotoxic drugs
- sexual contact
- physical and emotional stress
- temperature changes (hot or cold, such as burning the mouth with hot coffee)
- too much ultraviolet light (e.g., sunburn)
- menstruation
- pregnancy
- lactation
- malnutrition
- excessive fatigue

Herpesviruses are large, linear dsDNA enveloped viruses that replicate in the nucleus of the host cell. The herpesvirus obtains its envelope from the inner membrane of the nucleus as it buds out of the host cell's nucleus. The envelope of herpesviruses is very fragile and must be intact in order for the virus to be infectious.

There are nine members of the human *Herpesviridae* family that naturally infect humans:

- herpes simplex virus type 1 (HSV-1)
- herpes simplex virus type 2 (HSV-2)
- varicella zoster virus (VZV)
- Epstein-Barr virus
- cytomegalovirus
- human herpesvirus 6
- human herpesvirus 7
- human herpesvirus 8 (Kaposi's sarcoma-associated virus)

The *Herpesviridae* family is divided into three subfamilies (*Alphaherpesvirinae, Betaherpesvirinae,* and *Gammaherpesvirinae*), which are distinguished from each other based on the type of cells in which they establish latency.

HSV enters primarily through a break in the skin and intact mucous membranes. HSV-1 (oral herpes) occurs primarily above the waist, whereas HSV-2 (genital herpes) occurs primarily below the waist. HSV is lytic to human epithelial cells and latent in neural tissue. VZV causes a primary infection termed varicella (chickenpox) and is reactivated later in life as herpes zoster (shingles). Epstein-Barr virus has a tropism for B-lymphocytes. It causes infectious mononucleosis, Burkitt's lymphoma, and nasopharyngeal carcinoma. Cytomegalovirus infections usually occur early in childhood. A primary infection is usually mild or asymptomatic in people with normal immune systems. CMV causes a persistent, latent infection of T-lymphocytes. CMV infection is a major disease of immunocompromised patients and is the leading cause of inner ear (cochlea) hearing loss and nervous system damage in children in the United States. HHV-6 causes roseola, and HHV-8 is implicated in the development of Kaposi's sarcoma tumors. Other herpesviruses of interest are herpes gladiatorum, which is found among young wrestlers who pick up mat burns and other repeated abrasions. Herpes B (herpesvirus simiae) is a potential hazard to laboratory workers or zookeepers working with macaque monkeys.

The genomes of the human herpesviruses have been sequenced. Herpesvirus genomes vary remarkably in size and structure among the different herpesviruses. Hallmarks of herpesvirus replication include:

- Replication is in the nucleus.
- Sequential transcription and translation of immediate-early (α), early (β), and late (γ) genes produces α, β, and γ proteins.
- The viral envelope is obtained by a double envelopment process.
- A productive infection is cytocidal.
- They are able to establish latent infections persisting as an episome in the neurons or lymphocytes while only a subset of transcripts are expressed (LATs).
- Reactivation triggers replication and recurrent or continuous shedding of infectious virus.

Several antiviral drugs are available to treat herpesvirus infections. The most commonly prescribed drug is acyclovir. Its mechanism of action is the inhibition of the herpesvirus-encoded DNA polymerase. Another focus has been placed on prevention of herpesvirus infections, which is the approach of the approved medication **Varivax**, a live, attenuated Oka/Merck chickenpox vaccine, and the development of a cytomegalovirus vaccine to prevent congenital neonatal infections. Other medical research includes the use of replication-defective HSV viruses as a vehicle to destroy brain gliomas.

CASE STUDY QUESTIONS

These questions relate to the Case Study presented at the beginning of the chapter.

1. This outbreak was caused by a herpesvirus. Which one?
2. What tests were likely used to diagnose this infection?
3. How was the virus transmitted?
4. Speculate about how the afflicted wrestlers were treated.
5. How can this viral disease be prevented?

Reference

Centers for Disease Control and Prevention. 1990. "Epidemiologic notes and reports herpes gladiatorum in a high school wrestling camp–Minnesota." *Morbid Mortal Wkly Rep MMWR* 39(5):69–71.

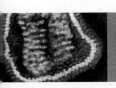

CASE STUDY 1: ELEPHANT HERPES

Asian and African elephants were used in transportation, labor, and ritual for thousands of years. They are the world's largest mammals, weighing 11,000 to 13,200 pounds. These elephants are giant herbivores that can tear down huge tree limbs with their muscular trunks. Asian elephants are endangered, and both Asian and African elephants are difficult to breed in captivity. Today, some Asian and African elephants live in national zoos throughout the world. Zoo spectators enjoy observing the elephants using their versatile trunks to squirt water over their backs, manipulate tiny objects, or blow dirt onto their backs during dust baths. Between 1995 and 2003, approximately 32 Asian and African zoo elephants were infected with a highly virulent new herpesvirus in the United States and Canada.

1. The herpesvirus that infected these zoo elephants caused unique pathological effects. List them. (*Hint:* Refer to the references below.)
2. What parts of the elephant were found to contain herpesvirus-like particles? What cell type do these viruses infect?
3. How does the virus spread among the zoo elephants?
4. Are wild elephants susceptible to the infection?
5. Can elephant herpesvirus infect people?

References

Richman, L. K., et al. 1999. "Novel endotheliotropic herpesviruses fatal for Asian and African elephants." *Science* 283:1171–1176.

Richman, L. K., et al. 2000. "Clinical and pathological findings of a newly recognized disease of elephants caused by endotheliotropic herpesviruses." *J Wildl Dis* 36(1):1–12.

Resources

Primary Literature and Reviews

Adler, S. P. 2005. "Congenital cytomegalovirus screening." *Pediatr Infect Dis J* 24(12):1105–1106.

Arbuckle, J. H., et al. 2010. "The latent human herpesvirus-6A genome specifically integrates in telomeres of human chromosomes *in vivo* and *in vitro*." *Proc Natl Acad Sci USA* 107(12):5563–5568.

Arvin, A. 2005. "Aging, immunity, and varicella-zoster virus." *N Engl J Med* 352(22):2266–2267.

Atanasiu, D., et al. 2006. "The stable 2-kb LAT intron of herpes simplex stimulates the expression of heat shock proteins and protects cells from stress." *Virology* 350(1): 26–33.

Baines, J. D., et al. 2007. "Electron tomography of nascent herpes simplex virus virions." *Virology* 81(6): 2726–2735.

Baker, M. L., et al. 2005. "Common ancestry of herpesvirus and tailed DNA bacteriophages." *Virology* 79(23): 14967–14970.

Bernstein, D. I., et al. 2002. "Effect of previous or simultaneous immunization with canarypox expressing cytomegalovirus (CMV) glycoprotein B (gB) on response to subunit gB vaccine plus MF59 in healthy CMV-seronegative adults." *J Infect Dis* 85 (5):686–690.

Burton, E. A., Fink, D. J., and Glorioso, J. C. 2005. "Replication-defective genomic HSV gene therapy vectors: Design, production and CNS applications." *Curr Opin Mol Ther* 7(4):326–336.

Cullen, R. R. 2001. "Journey to the center of the cell." *Cell* 105:697–700.

Davenport, D. S., et al. 1994. "Diagnosis and management of human B virus (herpes virus simiae) infections in Michigan." *Clin Infect Dis* 19(1):33–41.

Deepak, S., and Spear, P. G. 2001. "Herpesvirus and heparan sulfate: An intimate relationship in aid of viral entry." *J Clin Invest* 108:503–510.

Dove, A. 2006. "A long shot on cytomegalovirus." *Scientist* (Dec):40–45.

Efstathiou, S., and Preston, C. M. 2005. "Towards an understanding of the molecular basis of herpes simplex virus latency." *Virus Res* 111:108–119.

Fowler, K. B., Stagno, S., and Pass, R. F. 2003. "Maternal immunity and prevention of congenital cytomegalovirus infection." *JAMA* 289(8):1008–1011.

Gogev, S., et al. 2003. "Biosafety of herpesvirus vectors." *Curr Gene Ther* 3:597–611.

Goins, W. F., et al. 2004. "Delivery using herpes simplex virus: An overview." *Methods Mol Biol* 246:257–299.

Gomi, Y., et al. 2002. "Comparison of the complete DNA sequences of the oka varicella vaccine and its parental virus." *Virology* 76(22)11447–11459.

Griffith, R. S. 1981. "Relation of arginine-lysine antagonism to herpes simplex growth in tissue culture." *Chemotherapy* 27(3):209–213.

Griffith, R. S. 1987. "Success of L-lysine therapy in frequently recurrent herpes simplex infection." *Dermatologica* 175:183–190.

Grose, C. 2005. "Varicella vaccination of children in the United States: Assessment after the first decade 1995–2005." *J Clin Virol* 33(2):89–95.

Grunewald, K., et al. 2003. "Three-dimensional structure of herpes simplex virus from cryo-electron tomography." *Science* 302:1396–1398.

Huang, W.-C., et al. 2011. "Varicella breakthrough infection and vaccine effectiveness in Taiwan." *Vaccine* 29: 2756–2760.

Huff, J. L., and Barry, P. A. 2003. "B-virus (*Cercopithecine herpesviridae* 1) infection in humans and macaques: Potential for zoonotic disease." *EID* 9(2):246–250.

Jones, F. E., Smibert, C. A., and Smiley, J. R. 1995. "Mutational analysis of the herpes simplex virus virion host shutoff protein: Evidence that vhs functions in the absence of other viral proteins." *Virology* 69(8): 4863–4871.

Kakalacheva, K., et al. 2011. "Viral triggers of multiple sclerosis." *Biochem Biophys Acta* 1812(2):132–140.

Kent, J. R., et al. 2003. "Herpes simplex virus latency-associated transcript gene function." *J Neurovirol* 9(2):285–290.

Khanna, R., and Diamond, D. J. 2006. "Human cytomegalovirus vaccine: Time to look for alternative options." *Trends Mol Med* 12:27–33.

Marcason, W. 2003. "Will taking the amino acid supplement lysine prevent or treat the herpes simplex virus?" *J Am Diet Assoc* 103(3):351.

Marcellin, A. G., et al. 2007. "KSHV after an organ transplant: Should we screen?" *Curr Top Microbiol Immunol* 312: 245–262.

Mettenleiter, T. C. 2004. "Budding events in herpesvirus morphogenesis." *Virus Res* 106(2):167–180.

Mettenleiter, T. C. 2006. "Intriguing interplay between viral proteins during herpesvirus assembly or: The herpesvirus assembly puzzle." *Vet Microbiol* 113(3–4): 163–169.

Mettenleiter, T. C., Klupp, B. G., and Granzow, H. 2006. "Herpesvirus assembly: A tale of two mechanisms." *Curr Opin Microbiol* 9(4):423–429.

Mick, G. 2010. "Vaccination: A new option to reduce the burden of herpes zoster." *Expert Rev Vaccines* 9(3 Suppl): 31–35.

Oxman, M. N., et al. 2005. "A vaccine to prevent herpes zoster and postherpetic neuralgia in older adults." *N Engl J Med* 352(22):2271–2284.

Perfect, M. M., et al. 2005. "Use of complementary and alternative medicine for the treatment of genital herpes." *Herpes* 12(2):38–41.

Pfeffer, S., et al. 2005. "Identification of MicroRNAs of the herpesvirus family." *Nat Methods* 2(4):269–276.

Rainov, N. G., and Ren, H. 2003. "Gene therapy for human malignant brain tumors." *Cancer J* 9(3):180–188.

Ramachandran, S., and Kinchington, P. R. 2007. "Potential prophylactic and therapeutic vaccines for HSV infections." *Curr Pharm Des* 13(19):1965–1973.

Rouse, B. T., and Kaistha, S. D. 2006. "A tale of two α-herpesviruses: Lessons for vaccinologists." *Clin Infect Dis* 42:810–817.

Samols, M., and Renne, R. 2006. "Virus-encoded microRNAs: A new chapter in virus-host interactions." *Future Virol* 1(2):233–242.

Sanchez, V., and Spector, D. H. 2002. "CMV makes a timely exit." *Science* 297:778–779.

Schleiss, M. 2005. "Progress in cytomegalovirus vaccine development." *Herpes* 12(3):66–75.

Spear, P. G. 2004. "Herpes simplex virus: Receptors and ligands for cell entry." *Cell Microbiol* 6(5):401–410.

Tang, Q., and Maul, G. G. 2006. "Mouse cytomegalovirus crosses the species barrier with help from a few human cytomegalovirus proteins." *Virology* 80:7510–7521.

Thompson, R. L., and Sawtell, N. M. 2000. "HSV latency-associated transcript and neuronal apoptosis." *Science* 289(5485):1651.

Uldrick, T. S., Whitby, D. 2011. "Update on KSHV epidemiology, kaposi sarcoma pathogenesis, and treatment of kaposi sarcoma." *Cancer Lett* 305(2):150–162.

Vazquez, M., et al. 2004. "Effectiveness over time of varicella vaccine." *JAMA* 291(7):851–855.

Voumvourakis, K. I., et al. 2010. "Human herpesvirus 6 infection as a trigger of multiple sclerosis." *Mayo Clin Proc* 85(11):1023–1030.

Wagner, E. K., and Bloom, D. C. 1997. "Experimental investigation of herpes simplex virus latency." *Clin Microbiol Rev* 10(3):419–443.

Weller, T. H., and Stoddard, M. B. 1952. "Intranuclear inclusion bodies of human tissue inoculated with varicella vesicle fluid." *J Immunol* 68(3):311–319.

Wills, E., Scholtes, L., and Baines, J. D. 2006. "Herpes simplex virus 1 DNA packaging proteins encoded by UL6, UL15, UL28, and UL33 are located on the external surface of the viral capsid." *Virology* 80(21): 10894–10899.

Popular Press

Ewald, P. W. 2002. *Plague Time: The New Germ Theory of Disease.* New York: Anchor Books.

Regush, N. 2000. *The Virus Within, A Coming Epidemic: How Medical Detectives Are Tracking a Terrifying Virus That Hides in Almost All of Us.* New York: Dutton Books.

Video Productions

Reducing Water Pollution. 2010. Films for the Humanities & Sciences.

New Weapons to Fight Multiple Sclerosis. 2008. Films for the Humanities & Sciences.

Sexually Transmitted Infections. 2003. Films for the Humanities & Sciences.

Genital Herpes. 2001. Films for the Humanities & Sciences.

Hepatitis, Cytomegalovirus, and Epstein-Barr Virus. 1999. Films for the Humanities & Sciences.

Shingles: Treating Chronic Pain. 1999. Films for the Humanities & Sciences.

Ocean Resources. 1993. Films for the Humanities & Sciences.

eLearning

go.jblearning.com/shors2

The site features eLearning, an online review area that provides quizzes and other tools to help you study for your class. You can also follow useful links for in-depth information, or just find out the latest virology and microbiology news.

CHAPTER 16

Human Immunodeficiency Virus (HIV)

A group of people marching in a gay and lesbian parade while holding up large letters spelling out "Get Tested".

OUTLINE

"The AIDS epidemic is one of the greatest challenges to our society today. It is a fight that we cannot afford to lose."

Anthony S. Fauci, Director of the National Institute of Allergies and Infectious Diseases

OUTLINE (continued)

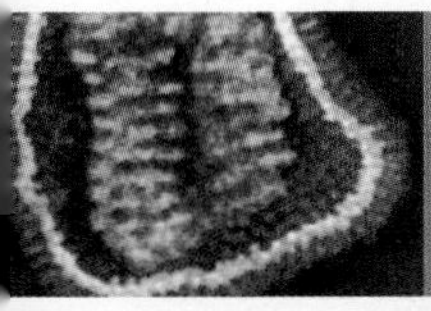

CASE STUDY: MAGIC JOHNSON AND HIV

Former Los Angeles Lakers basketball star Earvin "Magic" Johnson announced at a press conference on November 7, 1991, that he had HIV and retired from professional basketball. He acquired his infection through heterosexual transmission. At the time of his announcement his wife was pregnant, and both she and the fetus had tested negative. In his book entitled, *My Life,* Johnson admitted to having casual, unprotected sex with many women. After winning five NBA championship rings, an Olympic gold medal, and an NCAA championship, one of our national sports icons was faced with a life-threatening HIV infection.

Johnson's announcement increased the public's awareness and concern about HIV and the behavioral changes that would lead to reduced risk. The nonprofit Magic Johnson Foundation, founded in December of 1991, was charged with educating the public about HIV and with providing supportive HIV programs to address the social needs of inner-city children and young adults throughout the United States.

Johnson attempted brief professional comebacks from 1992–1993 and 1995–1996. During the 1992–1993 season, several players voiced their concerns. Basketball is a highly physical sport; thus, players may collide, crash, and inadvertently draw blood from one another while fighting for rebounds. Karl Malone of the Utah Jazz expressed his fears and opposition toward Johnson's first comeback. He pointed to the scars and scabs on his arms and said, "They can't tell me you're not at risk. And you can't tell me there's one guy in the NBA who hasn't thought about it." Other players, however, seemed very receptive to Johnson's return to basketball. Phoenix Suns player Charles Barkley added, "It's not like we are going to have unprotected sex with Magic on the floor." Later, Malone supported Magic's decision to come back for a second time in 1995.

As soon as Magic Johnson was diagnosed with HIV, he began antiretroviral therapy under the supervision of the Laker's team physician, Dr. Michael Mellman, and HIV expert Dr. David Ho. Within a few months, his viral load was reduced. In 1999, Johnson's wife, Cookie, said in an interview that her husband was healed. Johnson said there were no traces of virus in his bloodstream. Sixteen years after his announcement, Johnson is still alive and living normally. He remains a tireless crusader against acquired immunodeficiency syndrome (AIDS).

The 2010 UNAIDS Report on the Global AIDS Epidemic, published by the Joint United Nations Programme on HIV/AIDS, stated that *the HIV epidemic has turned the corner.* The epidemic has halted and has begun to reverse as the number of new HIV infections has fallen globally by 19% since the peak of the epidemic (1999). A world map representing the average number of adults and children estimated to be living with HIV as of the end of 2009 is shown in **FIGURE 16-1**.

The report includes the following additional scorecard updates:

- The number of people living with HIV in 2008 was 33.4 million.
- The number of people living with HIV in 2009 was 32.8 million.
- In 33 countries, HIV incidence has fallen by more than 25% between 2001 and 2009. Twenty-two of these countries are in sub-Saharan Africa. The biggest epidemics in sub-Saharan Africa—Ethiopia, Nigeria, South Africa, Zambia, and Zimbabwe—have either stabilized or are showing signs of decline.
- There are 15 million people in low- and middle-income countries estimated to be living with HIV who need antiretroviral treatment today. Of this group, 5.2 million have access to antiviral treatment, translating into fewer AIDS-related deaths.

  In 2009 alone, 1.2 million people received HIV antiretroviral therapy for the first time—an increase in the number of people receiving treatment of 30% in a single year.
- In 2009, an estimated 370,000 children contracted HIV during the perinatal and breast-feeding period, down from 500,000 in 2001. This is a 24% drop from five years earlier.
- *Several countries do not fit the overall trend. In seven countries—five of them in Eastern Europe and Central Asia—HIV incidence increased by more than 25% between 2001 and 2009.*
- Among young people in 15 of the most severely affected countries, HIV prevalence has fallen by more than 25% as they have adopted safer sexual practices.
- Slightly more than half of all people living with HIV are women and girls. In sub-Saharan Africa, more women than men are living with HIV; young women aged 15 to 24 years are as much as eight times more likely than men to be HIV positive. Protecting women and girls from HIV means protecting against gender-based violence and promoting economic independence from older men.
- Condom availability in places of need is increasing significantly, with 25.8 million female condoms provided through international and nongovernmental funding sources in 2009. Condom distribution increased by 10 million between 2008 and 2009.
- Trend analysis shows a general decline in the percentage of people who have had more than one sexual partner in the past year in sub-Saharan Africa.
- Fewer people are dying from AIDS-related causes. About 14.4 million life-years have been gained by antiretroviral therapy since 1996.

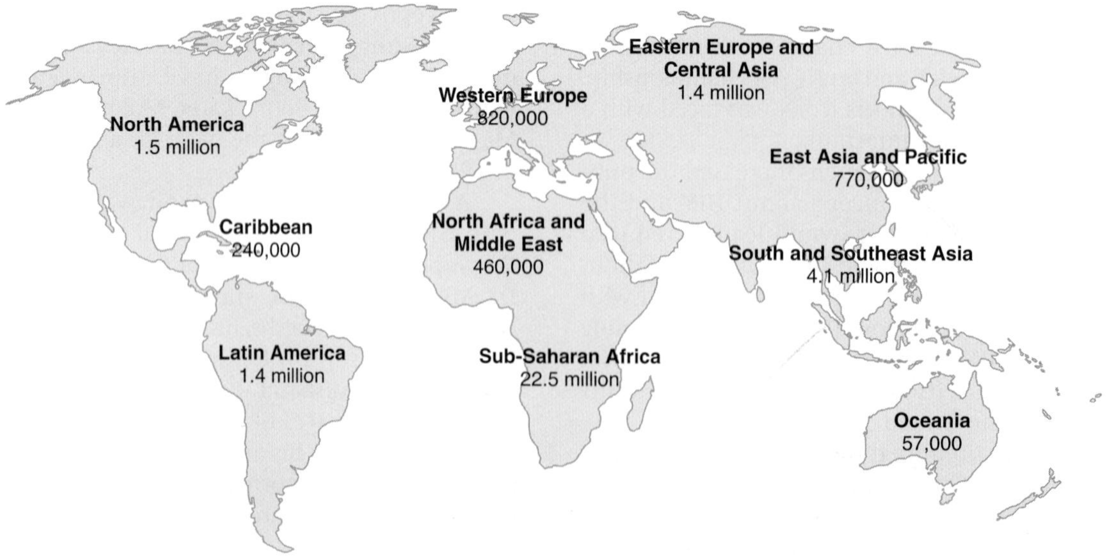

FIGURE 16-1 Regional average estimates of adults and children newly infected and living with HIV and AIDS-related deaths at the end of 2009. The total average of people living with HIV in 2009 was 32.8 million (30.9–34.7 million). Courtesy of UNAIDS.

16.1 The History of HIV

Gay Cancer, GRID, and AIDS (1980–1983)

During the early 1980s, physicians in New York and California made the first observations of a new infectious disease of viral origin. The following laboratory-confirmed infections were observed in young, homosexual men:

- The sudden appearance of a rare skin cancer called Kaposi's sarcoma, which became referred to as the **gay cancer**. Previously this cancer was only seen in elderly men of Mediterranean descent (**FIGURE 16-2a**).
- Atypical pneumonia caused by *Pneumocystis carinii*
- Cytomegalovirus infections
- Oral candidal mucosal infections (Figure 16-2b)

All of these observed **opportunistic infections** suggested that these individuals suffered from a type of cell-mediated immune suppression. In 1981, the Centers for Disease Control (CDC) named the syndrome **gay-related immune deficiency (GRID)**. As this gay men's health crisis was unraveled, physicians noted that many patients were also being treated for hepatitis B, herpes, and parasitic infections such as amebiasis and giardiasis. Other common trends among GRID patients included the use of drugs to heighten sexual experiences and unprotected, anonymous sex at bathhouses. The number of known deaths in the United States in 1981 was 234, and then climbed to 853 in 1982. At the same time, physicians reported that heterosexual individuals who received blood products or blood transfusions (e.g., hemophiliacs) began developing similar opportunistic infections. These latter cases were traced to the blood donors who were homosexual

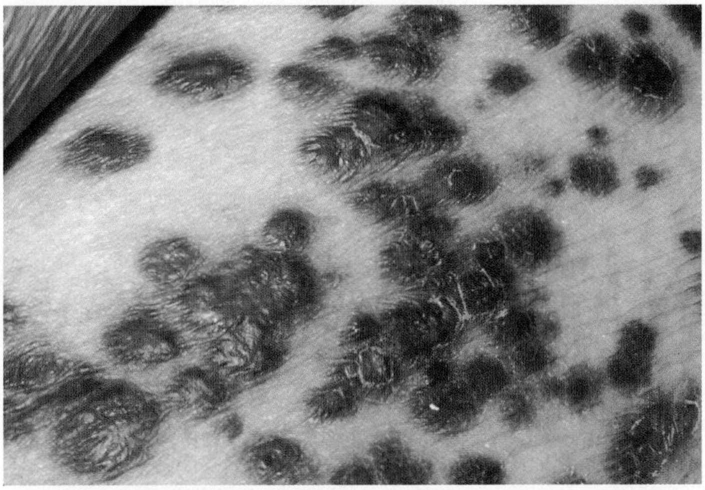

a

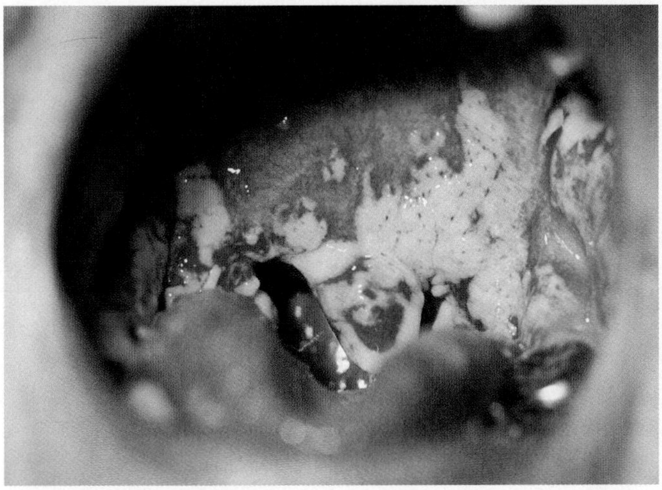

b

FIGURE 16-2 **(a)** These brown nodules represent Kaposi's sarcoma lesions commonly found in AIDS patients. **(b)** This AIDS patient has a chronic opportunistic oral candidiasis infection.

men dying of GRID. In 1982, the CDC renamed GRID to **acquired immune deficiency syndrome (AIDS)**. In 1983, the CDC warned blood banks of a possible problem with the national blood supply. The number of known deaths in the United States during 1983 was 2304. In 1985, film star Rock Hudson died from the complications of AIDS. He was the first major American celebrity whose AIDS diagnosis became public knowledge. U.S. President Ronald Reagan mentioned the term "AIDS" in public for the first time that same year. **TABLE 16-1** lists some of the famous people who are known to have or have had the virus known as HIV, including those who have died.

The Discovery of Human Immunodeficiency Virus (HIV)

The discovery of the cause of AIDS was published in the journal *Science* by two different research teams: Luc Montagnier's team at the Pasteur Institute in Paris (May 20, 1983) and Robert C. Gallo's group at the National Cancer Institute at the National Institutes of Health (NIH) in Bethesda,

TABLE 16-1 Famous HIV+ Individuals and Famous Individuals Who Have Died from AIDS-Related Causes

Individual	Occupation	HIV Status
Arthur Ashe	Tennis player and social activist	Infected in 1983 via blood transfusion during heart surgery; died in 1993.
John Curry	Figure skating champion	Died of AIDS-related illness in 1994.
Earvin "Magic" Johnson	Former Los Angeles Lakers point guard	Announced his HIV+ status on November 7, 1991.
Greg Louganis	Olympic diver	HIV+ in 1988; announced it in 1995.
Tommy Morrison	Former world boxing champion	Tested HIV+ in 1996, automatically retiring him from boxing.
Michael Bennett	Choreographer of the Broadway show *A Chorus Line*	Died in 1987.
Rock Hudson	Movie star	First major American celebrity whose AIDS diagnosis became public knowledge; died in 1985.
Liberace	Pianist	Died in 1987 of cytomegalovirus pneumonia due to AIDS.
Freddie Mercury	Lead singer of the band Queen	Died from AIDS-related causes in 1991.
Anthony Perkins	Actor; famous for playing Norman Bates in the movie *Psycho*	Died of AIDS-related causes in 1992.
Isaac Asimov	Science fiction author	Infected by a blood transfusion during heart surgery. Died in 1992.
Randy Shilts	American journalist and author of *And the Band Played On*	Died in 1994.
Ryan White	Teenager with hemophilia; AIDS activist	Contracted AIDS via blood products; died in 1990 at the age of 18.
Elizabeth Glazer	Wife of *Starsky and Hutch* actor Paul Michael Glazer	Infected via a blood transfusion. Her son and daughter were also infected. She started the Pediatric AIDS Foundation. She died in 1994; her daughter died in 1988. Her son, now a young adult, remains healthy.
Robert Reed	American actor, played Mike Brady in *The Brady Bunch*	Died of AIDS-related complications in 1992.
Tom Fogerty	American musician; played rhythm guitar in Creedence Clearwater Revival (CCR); elder brother of John Fogerty, the lead singer and guitar player in CCR	Died of AIDS related complications in 1990.
Tommy Sexton	Canadian comedian	Died of AIDS related complications in 1993.
John Holmes	American pornographic actor; one of the most famous male porn stars of all time	Died of AIDS related complications in 1988.
Bill Goldsworthy	Canadian-born player in the National Hockey League	Died of AIDS related complications in 1996.
Kurt Raab	West German stage and film actor, screenwriter, and playwright	Died of AIDS related complications in 1988.

Maryland (May 4, 1984). Both described a retrovirus isolated from cultured T lymphocytes derived from a lymph node biopsy of an AIDS patient with **lymphadenopathy** (swollen lymph nodes). Lymphadenopathy was considered a precursor syndrome to AIDS. The virus Gallo identified had originated from tissue samples sent to him by Montagnier, whose group suggested that the name of the virus be called Lymphadenopathy-Associated Virus (LAV). Gallo recommended the virus be named Human T-cell Lymphotrophic Virus Type III (HTLV III). In 1986, the name was changed from HTLV-III/LAV to the **human immunodeficiency virus** (**HIV**).

There has been controversy regarding which laboratory discovered HIV first. In 1990, the NIH's Office of Scientific Integrity began to investigate Gallo's HIV research program. The HIV discovery dispute was finally resolved in Washington, DC, in July 1994. At this time, it was acknowledged that the virus isolated at Gallo's laboratory was one of the first viruses isolated in 1983 at the Pasteur Institute. This dispute is the focus of John Crewdson's book, *Science Fictions: A Scientific Mystery, a Massive Cover-up, and the Dark Legacy of Robert Gallo.*

HIV-1 Versus HIV-2

The first HIV isolated by Montagnier's and Gallo's research teams were designated HIV-1. Shortly after the first HIV discovery, Montagnier's group isolated another strain, which they designated HIV-2. HIV-2 is rare in the United States but is endemic in Western Africa. HIV-2 is significantly less infectious and progresses more slowly to AIDS than HIV-1.

The Origin of HIV-1 and the AIDS Pandemic

Where did HIV originate? Has it been evolving for hundreds of years? The most widely accepted mode is the **hunter theory**, or extensions of it. Myths or controversial theories that have surfaced regarding the origin of HIV include:

1. house cat theory
2. hunter theory (or crossing the species barrier theory)
3. contaminated vaccine theory
4. colonization theory (or Heart of Darkness theory)
5. conspiracy theory

The House Cat Theory

Feline immunodeficiency virus (FIV) is a retrovirus that was discovered in 1987. It causes an infection of domestic cats that results in an immunodeficiency syndrome resembling HIV-1 infection in humans, referred to as feline AIDS, or it may cause feline leukemia. FIV has not been shown to cause disease in humans; however, HIV-1 and FIV are related retroviruses. A 2005 study published in the journal *AIDS* showed that cats vaccinated with the p24 viral HIV-1 protein appeared to be at least as well protected against FIV as those immunized with the FIV p24 protein currently used by veterinarians. This surprise finding suggests that the p24 protein of these two retroviruses is an important immunogen that is evolutionarily conserved. It may lead the way to the development of a protective vaccine against HIV-1 based on the cross-reactive regions of the p24 HIV-1 and FIV-proteins.

The Hunter Theory

This is the most commonly accepted theory in the scientific community. It is based on the premise that the AIDS pandemic is a **viral zoonosis**. A virus related to HIV-1 has been isolated from the common chimpanzee (*Pan troglodytes troglodytes*), which is native to Cameroon, Gabon, and the Democratic Republic of Congo (**FIGURE 16-3**). The virus is referred to as simian immunodeficiency virus (SIV_{cpz}). The genetic sequence of SIV_{cpz} and the early strains of HIV-1 are nearly identical. The conclusion thus far is that the chimpanzee is the natural reservoir for HIV-1. The belief is that AIDS arose when chimpanzee-to-human transmission of SIV_{cpz} occurred. The naturally infected chimps were killed and eaten and their blood entered cuts or wounds on the hunter, and subsequently the SIV_{cpz} strain adapted in the human host to become HIV-1. A 2004 study has shown that retrovirus transmission from primates to hunters continues to occur today (see Chapter 10). These infections are likely acquired through the butchering and consumption of contaminated monkey or ape meat (**bushmeat**). There have been calls to ban bushmeat hunting to prevent the transmission of simian viruses to humans.

A serological study on plasma samples used for malarial studies in Africa between 1959 and 1982 was published in 1998 in the journal *Nature*. It reported that the oldest sample that contained HIV-1 antibodies dated to 1959. This seropositive sample also contained viral HIV-1 genetic material that was sequenced. It is currently the oldest known case of HIV-1 infection. The sample is from an adult Bantu-speaking male who lived in Kinshasa, Democratic Republic of Congo (Figure 16-3). It has been hypothesized that this person became infected in rural Cameroon and then traveled by river to Kinshasa.

The Contaminated Polio Vaccine Theory

This is an extension of the hunter theory proposed by virologist Preston A. Marx. His theory posits that the initial spread of HIV-1 occurred in Africa

FIGURE 16-3 The oldest ancestor of HIV-1 to date is from the Democratic Republic of Congo. Kinshasa has the greatest genetic diversity of HIV-1 isolates, suggesting that the virus has been there longer than anywhere else. The initial HIV-2 epidemic is believed to have occurred during the war when Guinea-Bissau gained its independence from Portugal (1963–1974). Shaded areas represent the geographical distribution of chimpanzees and sooty mangabeys.

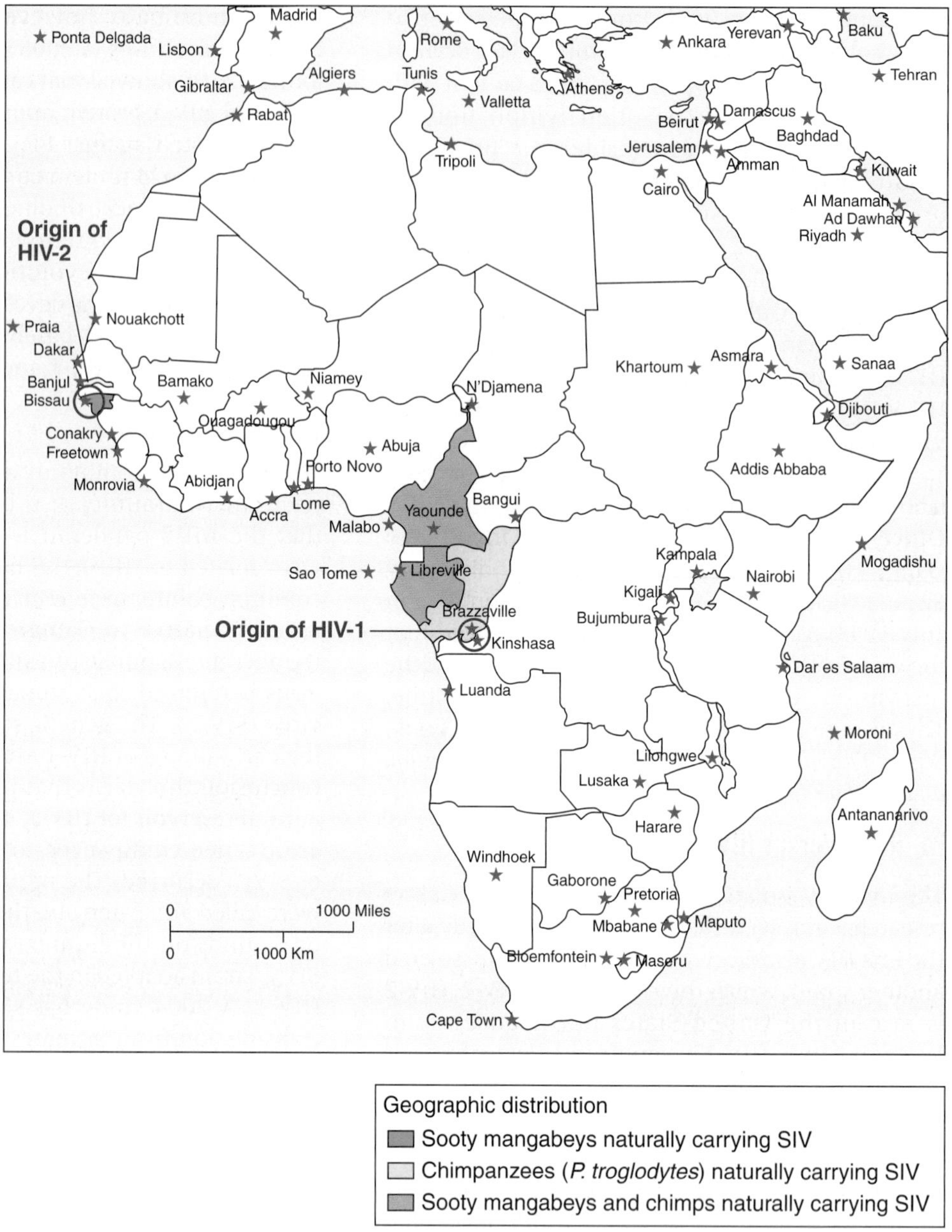

during the large-scale vaccination campaigns that began in the early 1950s. Disposable plastic syringes were introduced in the 1950s and were used to administer vaccines and medicines. This was a big change in medical practice; up until then, using huge quantities of traditional, metal syringes was too expensive in Africa. Multiple patients were inoculated with the same syringe without sterilization between patients. This allowed the rapid transfer of virus particles from one individual to another, creating the potential for the SIV_{cpz} to mutate and replicate in each new individual it entered.

Others believe in a slightly modified extension of this theory, suggesting that the oral polio vaccines were contaminated with SIV_{cpz} virus. Edward Hooper, author of *The River,* suggested that HIV could be traced to the oral polio vaccine referred to as Chat. The Chat vaccine was manufactured at the Wistar Institute in Philadelphia and tested in Africa. It was administered to a million people in the Democratic Republic of Congo and surrounding areas during the late 1950s. Hooper believed the vaccine was produced in kidney cells taken from chimpanzees that had been infected with SIV_{cpz}. Vaccinees received the contaminated vaccine and subsequently became infected with HIV-1. The Chat polio vaccines have since been exonerated. Original stocks prepared by the Wister Institute were tested for the presence of viral contamination. No HIV-1 or SIV_{cpz} or chimpanzee nucleic acid sequences were detected in the Chat vaccine. Furthermore, the vaccines made by the Wister Institute were created using the kidney cells of rhesus monkeys.

The Colonization Theory or Heart of Darkness Theory

In 2000, Jim Moore, a specialist in primate behavior, proposed another variant of the hunter theory. He pointed out that between 1880 and 1915 the colonial rule of French Equatorial Africa (also referred to as the Heart of Africa or the Heart of Darkness) was very harsh. Colonial authorities forced people to construct railroads or other projects. Many Africans were forced into labor camps and sent into the forest in search of valuable rubber for weeks at a time, year after year, and women were held hostage in unsanitary conditions and nearly starved. The physical demands were extreme. There was little time or energy left devoted to agriculture, leading to the increased reliance on bushmeat for food. These factors would have been sufficient to create poor health, including weakened immune systems. A stray and perhaps infected chimpanzee was likely a welcome source of food for laborers.

Moore believed that well-intentioned but undersupplied doctors routinely vaccinated the workers against smallpox with unsterile needles to keep them alive and working. A half a dozen syringes were used to draw blood to identify carriers and intravenously treat approximately 90,000 workers for sleeping sickness from 1917 to 1919. About 100,000 arm-to-arm passage smallpox inoculations were carried out prior to 1914 (refer to Chapter 14), and inoculation material with pox vesicles likely contained a high concentration of lymphocytes (the primary host cell of HIV-1). Many of the camps actively employed prostitutes, creating additional means of HIV/SIV_{cpz} transmission.

The transmission of pathogens was inevitable in the camps and disease was rampant. Anyone sick would not have stood out as being different because the laborers were present in an already disease-ridden population. It has been speculated that half of the labor workers died in the camps, and many of them would have died before the first symptoms of AIDS. No medical records for these camps exist.

The Conspiracy Theory

A significant number of individuals believe that HIV is a genetically engineered or man-made virus. During the 1980s, the Soviet press suggested that HIV was manufactured as part of an American biological warfare research program. Third World newspapers and publications run by Soviet front organizations picked up this disinformation. The Soviet reports were stopped in 1987.

Other individuals believe that HIV was genetically engineered to wipe out large numbers of black and homosexual people. Even though these theories cannot be proved without a doubt, the evidence they are based on is tenuous at best, as it ignores the clear link between SIV_{cpz} and HIV. It also ignores the fact that HIV antibodies have been detected in serum samples in people as far back as 1959—a time when genetic engineering technology was not available to "create" HIV. Genetic engineering didn't become a reality until 1977, when a man-made gene was used to manufacture human somatostatin in the bacterium *Escherichia coli.*

Where Did HIV-2 Originate?

Like HIV-1, HIV-2 entered the human population as the result of a zoonotic, or cross-species, transmission. It transferred from sooty mangabeys to humans during the first half of the 20th century in Western Africa. The sooty mangabey is an Old World monkey that is native to Guinea-Bissau, Gabon, and Cameroon. Sooty mangabeys are naturally infected with SIVsm that is genetically similar to HIV-2. They are commonly hunted for food and kept as pets in this region of the world. It is believed that the initial HIV-2 epidemic coincided with the independence war (1963–1974) in Guinea-Bissau, a former Portuguese colony. The war caused social disruption and a mass vaccination effort that may have brought forth a regional HIV-2 epidemic by increasing the number of people who were administered (HIV-2-contaminated) vaccine injections through unsterile reusable needles at the Canchungo clinic in Guinea-Bissau. HIV-tainted blood transfusions also occurred. Social disruption resulted in the increased movement of people and sexual activities such as rape and prostitution. The first reported cases of HIV-2 in Europe occurred among Portuguese soldiers returned from the independence war. Today, up to 10% of the general population in Guinea-Bissau is HIV-2 seropositive.

The AIDS Pandemic: Patient Zero Versus the Earliest AIDS Cases

The AIDS epidemic appeared suddenly and nearly simultaneously in several metropolitan areas of the United States. The greatest number of cases occurred among homosexual men living in Los Angeles, San Francisco, and New York City. In 1981, the CDC began an investigation of this epidemic, led by Dr. William Darrow and colleagues, to determine the risk factors involved in AIDS. The term **patient zero** refers to the first infected patient in an epidemiological investigation who is

TABLE 16-2 Earliest Cases of AIDS

Year of Death	Case	Location	Evidence
1959	Preserved blood sample from a male	Kinshasa, Democratic Republic of Congo, Africa	HIV and antibodies in blood sample of this individual who died in 1959.
1959	25-year-old male, former naval seaman	Manchester, England, UK	Unexplained immunodeficiency, pneumocystis, cytomegalovirus infection. PCR detected HIV proviral DNA. Findings were controversial.
1969	15-year-old male prostitute	St. Louis, Missouri, USA	Frozen tissue and serum samples contained HIV antibodies. Patient died of aggressive Kaposi's sarcoma.
1976	30-year-old Norwegian sailor and long-haul truck driver throughout Europe (mainly Germany)	Oslo, Norway	Had sex with African women, including prostitutes in Cameroon during the early 1960s. His wife and daughter both died from AIDS in 1977.
1977	47-year-old female Danish surgeon who had worked in a primitive hospital in Zaire from 1972 to 1975	Europe/Denmark	Died of AIDS-like illness; no obvious risk factors.
1984	31-year-old homosexual French-Canadian flight attendant	Quebec City, Canada	Kaposi's sarcoma; died of kidney failure caused by AIDS-related infections.

likely responsible for the spread of a particular infectious disease.

Randy Shilts, author of *And the Band Played On*, chronicled Darrow's investigation. It was Shilts who proposed that a homosexual Canadian flight attendant was likely patient zero. The flight attendant admitted to anonymous, unprotected sex with as many as 2500 partners, even after he developed Kaposi's sarcoma. As research continued, though, earlier cases of AIDS were reported (**TABLE 16-2**), disproving the patient zero theory.

16.2 HIV Transmission

HIV is present in blood, semen, vaginal fluids, breast milk, and, in very low levels, saliva and tears. The most common ways that HIV is transmitted are by means of

- anal or vaginal intercourse
- sharing of HIV-contaminated needles (e.g., by injection drug users)
- blood transfusions using infected blood or blood-clotting factors (this is now rare in the United States)
- accidental needlestick injuries (healthcare workers are at risk while working with HIV-positive patients)
- congenital AIDS (before or during birth or through breastfeeding)
- sharing of HIV-contaminated tattoo needles, razors, acupuncture needles, or ear-piercing implements

Rare Routes of HIV Transmission: Organ Transplants and Pre-Mastication

Organ donors are routinely screened for HIV infection in the United States. From 1994 to 2007, nearly 300,000 transplants were performed without any reported cases of HIV transmission. HIV transmission was always theoretically possible. The window between the time of HIV infection and the time of detectable HIV-specific antibodies by ELISA assays ranges from three to eight weeks. There is an eight- to ten-day window during the early period of infection in which HIV NAT tests do not work. Therefore, the ELISA test is less sensitive than NAT (nucleic acid/PCR testing; see Section 16.8 for laboratory testing of HIV). NAT detects HIV infection before antibodies develop and are detected by serology. HIV NAT is not consistently used for organ donor screening. The chances of an organ recipient being infected during that small window are remote but the risk of dying while on the waiting list is not incredibly small. Deceased donors are screened for HIV infection at the time of the donor's cardiac or brain death. The organ is transplanted typically within three to four hours into the recipient. There is no national policy for the type or timing of HIV screening tests used for living donors.

In 2009, the first documented case of HIV transmission through organ transplantation of an organ from a ***living*** **donor** occurred in a New York City hospital. A living donor is someone who willingly volunteers to give his/her kidney for the intent of transplantation. Living donors undergo comprehensive physicals and psychological evaluation. In this case, the living donor was evaluated and tested HIV negative ten weeks before surgery. The recipient tested HIV negative 12 days before transplant surgery by an ELISA test. The recipient had no history of sexually transmitted diseases, injection drug use, or other high-risk activities. Upon an extensive investigation, it was determined that the donor contracted HIV via sexual transmission during the ten-week period before the surgery. This was a rare case. This incident spurred new testing recommendations for living donors and the questioning of living donors about risky behavior (e.g., sexual history, intravenous drug use). The CDC recommends that living donors be rescreened by ELISA and NAT as close to the time of organ recovery as possible, but no longer than seven days before organ donation. Living donors should also be advised to avoid behaviors that put them at risk for contracting HIV before organ donation.

Three unusual cases of HIV transmission in African American children aged 9 months to 39 months were reported in *Pediatrics* in 2009. Researchers found compelling evidence to determine that the three children were infected with HIV from mothers or caregivers primarily through pre-chewed food. Pre-chewing food is also called **pre-mastication**. In developing countries, dental care is lacking and commercially available baby foods and blenders are not available. Mothers and caregivers pre-masticate or pre-chew foods for infants. In all three of these cases, the children tested negative for HIV infection after birth. The fathers of the children were also HIV negative. Two of the mothers were HIV positive. In one case, the mother was HIV negative but the child spent much time with a great aunt who was HIV infected. The mothers and caregiver fed the children pre-chewed food. In all three cases, it was suggested that a factor aiding HIV transmission was mouth bleeding in the HIV-positive aunt or mothers as well as the infants due to teething or infection. The caregiver and mothers were not adhering to HAART regimens, increasing their blood HIV levels and the likelihood of transmission. This was the first report of HIV transmission by pre-mastication of food. Albeit rare, HIV-infected caregivers or parents are advised against this practice.

Can Kissing and Oral Sex Transmit HIV?

Social kissing is not a high-risk behavior. The risk increases, though, with open-mouthed or "French" kissing because considerable amounts of saliva are exchanged. The risk increases if the uninfected person has canker sores or other lesions in his/her mouth. Oral sex also poses a similar risk to uninfected individuals, as HIV-infected T lymphocytes from semen may enter abrasions or lesions in the mouth.

Ways That HIV Is Not Transmitted: The Fragility of HIV in the Environment

HIV is an enveloped virus. Soap easily disrupts the virus, which also is highly susceptible to disruption through drying in the environment. It is important to understand that finding a small amount of HIV in a body fluid does not necessarily mean that HIV can be transmitted by that body fluid; for example, HIV has not been recovered from the sweat of HIV-infected persons. Contact with saliva, tears, or sweat has never been shown to result in transmission of HIV. HIV cannot be acquired via sharing a toilet seat used by an AIDS patient, nor can it be transmitted by hugging, handshaking, or contact with the sweat or mucus (e.g., coughing and sneezing) of an infected individual.

Mosquitoes or other insects cannot transmit HIV, and it is not expected that an HIV–mosquito connection will occur. Pathogens that are transmitted by mosquitoes multiply in high numbers in the mosquito or other blood-sucking host and then concentrate in its salivary glands. When the mosquito takes a blood meal, it passes the pathogen into the victim's blood via its saliva. HIV cannot replicate in mosquitoes; nor can it concentrate in any part of the mosquito so that it could leave the mosquito.

16.3 Prevention of HIV Infection

There is no "silver bullet" method to prevent HIV infection; however, there are a variety of practices and methods in development that have the potential to stem the HIV epidemic. Preventative HIV vaccines are discussed later in this chapter. The following practices are used to prevent HIV transmission:

- abstinence from sex
- barrier methods (e.g., condoms)
- microbiocides
- preexposure prophylaxis (PREP)
- postexposure prophylaxis (PEP)

Abstinence from sexual activity eliminates the risk of acquiring HIV. The use of male and female condoms effectively reduces the risk of HIV transmission. The Roman Catholic Church, which has banned contraceptives, has been considering the issue of using condoms as a barrier to infection among married couples when one of the partners is infected with HIV. Some priests hold the view that it would be a sin for an HIV-seropositive person to engage in sex without the use of a condom

VIRUS FILE 16-1 Transmission of HIV by Dental Procedure

In 1990, a young woman who had no risk factors for HIV infection tested seropositive. The patient's family and friends were interviewed extensively and it was determined that she was not an intravenous (IV) drug user. She did not have any tattoo procedures, body piercings, acupuncture treatments, or artificial insemination procedures. She did not have sex with an HIV-infected person or receive a blood transfusion or other blood factors. She had never been pregnant and did not have a history of acquiring sexually transmitted diseases. Both of her prior boyfriends were seronegative for HIV infection. She had had never worked in a healthcare or any other setting where she could have been exposed to HIV-infected blood or other body fluids. She had no risk for HIV transmission.

In 1988, though, she had had two teeth extracted by dentist Dr. David Acer. About three months prior to this, Dr. Acer was diagnosed with HIV and a T-cell count that was less than 200. He had developed AIDS-related symptoms and continued to treat patients after his HIV status was known. The CDC began an epidemiologic investigation that involved a review of the dentist's medical records and interviews of staff at the dental practice, including a review of the infection-control procedures at Dr. Acer's dental practice. Dr. Acer sent letters to his patients recommending them to be tested for HIV. Florida Department of Health and Rehabilitative Services (HRS) tested about 1100 patients for HIV infection. The CDC investigation revealed that four additional patients were likely infected while undergoing tooth extractions or other invasive dental procedures performed by Dr. Acer (**FIGURE VF 16-1**).

How did scientists prove that Dr. Acer transmitted HIV to his patients? The advent of techniques used to analyze viral genetic sequences proved to be the valuable tools needed to answer this question. Even though HIV has a high mutation rate, genomic sequencing was able to determine that the HIV strains from different individuals are genetically distinct. In this case, this technique was used to resolve whether Dr. Acer was responsible for the HIV transmission event to five of his patients. In this study, sequences of the HIV proviral envelope gene from the dentist, seven of his patients, and 35 HIV-positive individuals located in the same geographical area as the dental

FIGURE VF 16-1 Metallic dental instruments must be sterilized before their use.

VIRUS FILE 16-1 **Transmission of HIV by Dental Procedure (continued)**

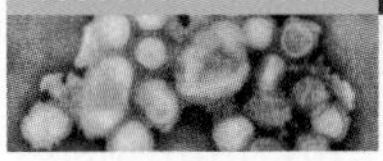

practice were amplified by polymerase chain reaction and sequenced. The proviral envelope gene (env) codes for gp120.

The sequences were analyzed by computer-based methods for genetic variation. The greater the variation, the easier it is to distinguish different strains of the virus. The study conducted showed that there was a high degree of **genetic relatedness or similarity** (within 3% variation) between the HIV strains of the five infected dental patients and Dr. Acer, suggesting that they became infected while undergoing care from a dentist with AIDS. Two other patients had high-risk behavioral exposures to HIV and genetically distant strains (varying by more than 13%), suggesting they contracted their HIV infections from other sources. The actual mode of transmission of the virus is still uncertain, but that does not mean it did not occur. What are some possibilities? What precautions are used in dental practices to prevent transmission of infectious diseases? Discuss potential injuries that may occur during invasive dental procedures.

References

Centers for Disease Control and Prevention (CDC). 1991. "Epidemiologic notes and reports update: Transmission of HIV infection during invasive dental procedure—Florida." *MMWR* 40(2):21–27, 33.

Ciesielski, C. A., et al. 1994. "The 1990 Florida dental investigation: The press and the science." *Ann Intern Med* 121(11): 886–888.

Hillis, D. M., and Huelsenbeck, J. P. 1994. "Support for dental HIV transmission." *Nature* 369(6475):24–25.

Myers, G. 1994. "Molecular investigation of HIV transmission." *Ann Intern Med* 121(11):889–890.

Ou, C.-Y., et al. 1992. "Molecular epidemiology of HIV transmission in a dental practice." *Science* 256:1165–1171.

Rom, M. C. 1997. *Fatal Extraction: The Story Behind the Florida Dentist Accused of Infecting His Patients with HIV and Poisoning Public Health.* San Francisco: Jossey-Bass.

because the act would ultimately kill the partner. Driving this debate is the fact that 165 million Roman Catholics live in Africa, the continent most affected by HIV and AIDS. If condom use is approved by the Roman Catholic Church, the approval offers a sign of hope for those suffering from AIDS, especially in Africa, where about 6600 people die every day of the disease.

Microbiocides are being developed to reduce sexual transmission of HIV to women. Formulations may be a gel, cream, suppository, or slow-release vaginal sponge that would be applied prior to sexual intercourse. The microbiocides could prevent HIV transmission through several different mechanisms of action, such as:

- **surfactants** (which disrupt the envelope of HIV)
- fusion inhibitors (which block attachment of HIV to the host-cell receptors and coreceptors)
- replication inhibitors (which inhibit HIV reverse transcriptase)

Microbiocides with only one mechanism of action will likely not offer complete protection against viral infection. As a result, a combination of microbiocides will increase their efficacy. There are nearly 30 microbiocide candidate products undergoing preclinical or early-phase clinical trials. An intriguing approach is the use of genetically engineered probiotic strains of *Lactobacillus*. These strains are involved in the production of yogurt and cheese. *Lactobacillus lactis* has been engineered to produce the antiviral agent **cyanovirin**. Cyanovirin is an HIV cell fusion inhibitor that was isolated from cultures of the blue green algae (cyanobacterium). *Lactobacillus jensenii* has been engineered to secrete CD4 proteins that bind to HIV gp120 that could inhibit viral entry. The idea is to use these bacteria to prevent vaginal and oral transmission. Will anti-HIV yogurt be a future microbiocide?

Preexposure prophylaxis (PREP) is the use of antiretroviral drugs by high-risk individuals prior to HIV exposure to reduce the risk of becoming infected. For this method to be desirable, the antiviral drugs must be effective and nontoxic, and must not cause resistant strains of HIV to develop. PREP trials testing tenofovir were provided to prostitutes in Cambodia, Cameroon, Malawi, and Nigeria, but local governments stopped the studies due to activist protests. Tenofovir is an HIV reverse transcriptase inhibitor. In PREP, it is used as an oral, once-daily, single-tablet regimen. New trials using Truvada are underway in Peru, Thailand,

Ghana, Botswana, and the United States. Truvada is a single-pill therapy that combines tenovir and emtricitabine reverse transcriptase inhibitors.

Postexposure prophylaxis (PEP) is providing antiretroviral drugs to lower the risk of HIV infection after an individual has had a possible exposure to the virus (e.g., through occupational, accidental needlestick injuries). PEP should be administered within 48 hours (not to exceed 72 hours) of a high-risk exposure to HIV. The sooner PEP is administered, the more effective it is. It is a four-week program in which two or three antiviral medications are provided several times per day. Some individuals may not be able to adhere to the regimen due to serious side effects of the drugs.

16.4 AIDS in Africa

A world map representing the incidence of HIV infection from 2001 to 2009 is shown in **FIGURE 16-4**. Africa, the world's second largest continent, has been hit harder by HIV/AIDS than any other region of the world. AIDS has erased the progress that was made in extending life expectancy. In the most heavily affected countries, life expectancy is below 51 years. This decline has been directly linked to HIV/AIDs. However, HIV incidence has fallen by more than 25% between 2001 and 2009 in 22 sub-Saharan countries. The largest epidemics in sub-Saharan Africa located in Ethiopia, Nigeria, South Africa, Zambia, and Zimbabwe have either stabilized or are showing signs of decline (Figure 16-4).

The future of Africa is threatened by a combination of ongoing crises that keep its people poor and its nations weak. The countries of Africa are suffering from unpayable debt, the AIDS epidemic, and trade policies that limit Africans from being able to sell their products at world prices in order to reduce their poverty status. Sub-Saharan Africa is the world's poorest region: 70% of its population lives on less than $2.00 per day. Approximately 200 million of the people in this region go hungry every day. Each day in Africa:

- 6600 people die of AIDS
- 8800 people are infected with the HIV virus
- 1400 newborn babies are infected with the HIV virus during birth or through breast-feeding

At least 12 million African children have lost one or both parents to AIDS. AIDS prevention campaigns can help to stop the spread of HIV, but access to antiretroviral drugs is a significant problem in Africa. Only a half a million of the 40.3 million HIV-seropositive people have access to antiretroviral drugs. This is because the drugs are too expensive and a sufficient number of trained healthcare workers to monitor and administer treatment are not available. The epidemic is putting strain on the healthcare sector and schools. This is a major concern because as the epidemic develops, there is a higher demand for care for those living with HIV and the number of healthcare workers affected. AIDS heavily affects schools, and education is vital. HIV/AIDS has dramatically affected labor and productivity, decimating the economy and social progress of Africa. The vast majority of HIV-infected people in Africa are between the ages of 15 and 49—the prime of their working lives. Factories, employers, schools, and hospitals have to train their staff to replace the workforce that has become too ill to work. Domestic and international support is needed for treatment and educational programs (including testing and support for those who are infected with the HIV virus) to further stop this global pandemic.

16.5 Central Asia and Eastern Europe: Hotspots in the Worldwide HIV Epidemic

Several regions and countries do not fit the overall trend of HIV epidemic decline or stabilization. HIV incidence increased by more than 25% in five countries in Eastern Europe and Central Asia between 2001 and 2009 in these areas (Figure 16-4). The Russian Federation and Ukraine are among Eastern European countries with the fastest growing numbers of cases of HIV. The primary route of transmission is intravenous drug use; however, sexual contact and mother-to-child transmissions have begun to increase. There are an estimated 1.5 to 2 million intravenous drug users in Russia alone. Detailed molecular analysis revealed that HIV strains isolated in Belarus, Russia, Kazakhstan, and other former Soviet Union countries evolved from southern Ukraine. HIV subtype A followed heroin trading routes into Eastern Europe. Subtype B is predominantly found in Western Europe (**FIGURE 16-5**).

In southern Ukraine, heroin was not popular during the mid-1990s. In Russia and other parts of Eastern Europe, people have been injecting **cher-**

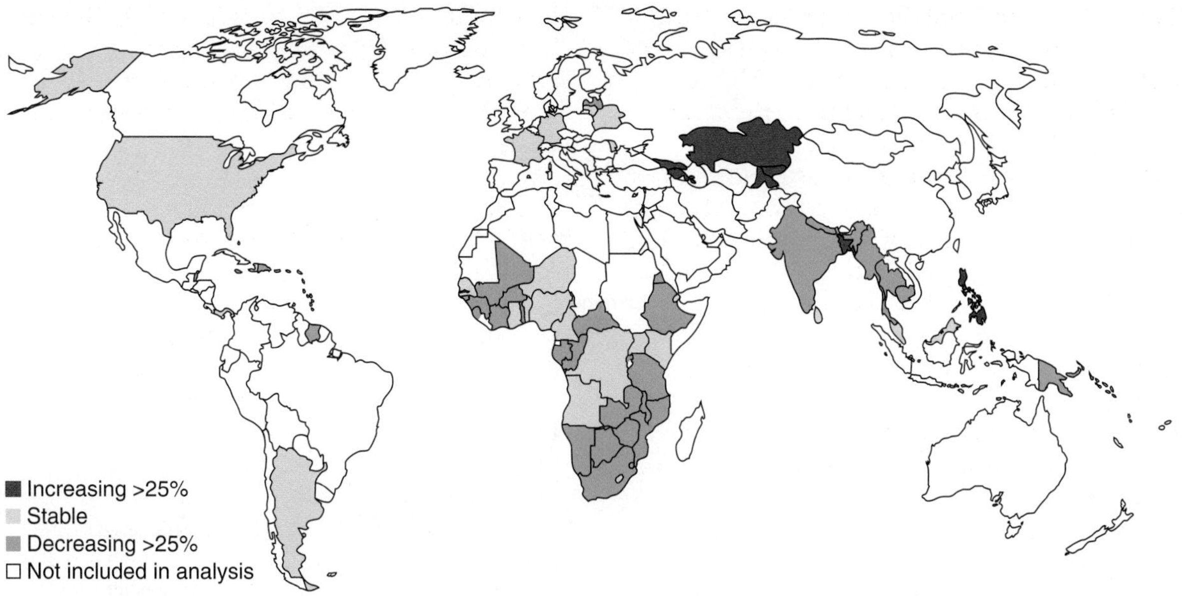

Increasing >25%

Armenia
Bangladesh
Georgia
Kazakhstan
Kyrgyzstan
Philippines
Tajikistan

Stable

Angola
Argentina
Belarus
Benin
Cameroon
Democratic Republic of the Congo
Djibouti
France
Germany
Ghana
Haiti
Kenya
Lesotho
Lithuania
Malaysia
Niger
Nigeria
Panama
Republic of Moldova
Senegal
Sri Lanka
Uganda
United States of America

Decreasing >25%

Belize
Botswana
Burkina Faso
Cambodia
Central African Republic
Congo
Côte d'Ivoire
Dominican Republic
Eritrea
Ethiopia
Gabon
Guinea
Guinea-Bissau
India
Jamaica
Latvia
Malawi
Mali
Mozambique
Myanmar
Namibia
Nepal
Papua New Guinea
Rwanda
Sierra Leone
South Africa
Suriname
Swaziland
Thailand
Togo
United Republic of Tanzania
Zambia
Zimbabwe

FIGURE 16-4 Changes in the incidence of HIV infection 2001–2009, selected countries. Note: Countries in Eastern Europe and Central Asia are HIV hotspots, with incidence increasing by greater than 25%. Adapted from *The 2010 UNAIDS Report on the Global AIDS Epidemic*. Geneva, Switzerland: Joint United Nations Programme on HIV/AIDS.

naya (a homemade heroin or opiate), methamphetamine (vindt), and methcathinone (jeff). Chernaya is made by adding water to ground poppies and alkalinized by sodium bicarbonate. The mixture is boiled and then the opiate alkaloids are extracted with solvents. Among the opiate alkaloids is morphine, which can be chemically converted to a heroin mixture that is reduced, boiled, and filtered. According to interview-based research, many dealers who prepare the chernaya sell prefilled

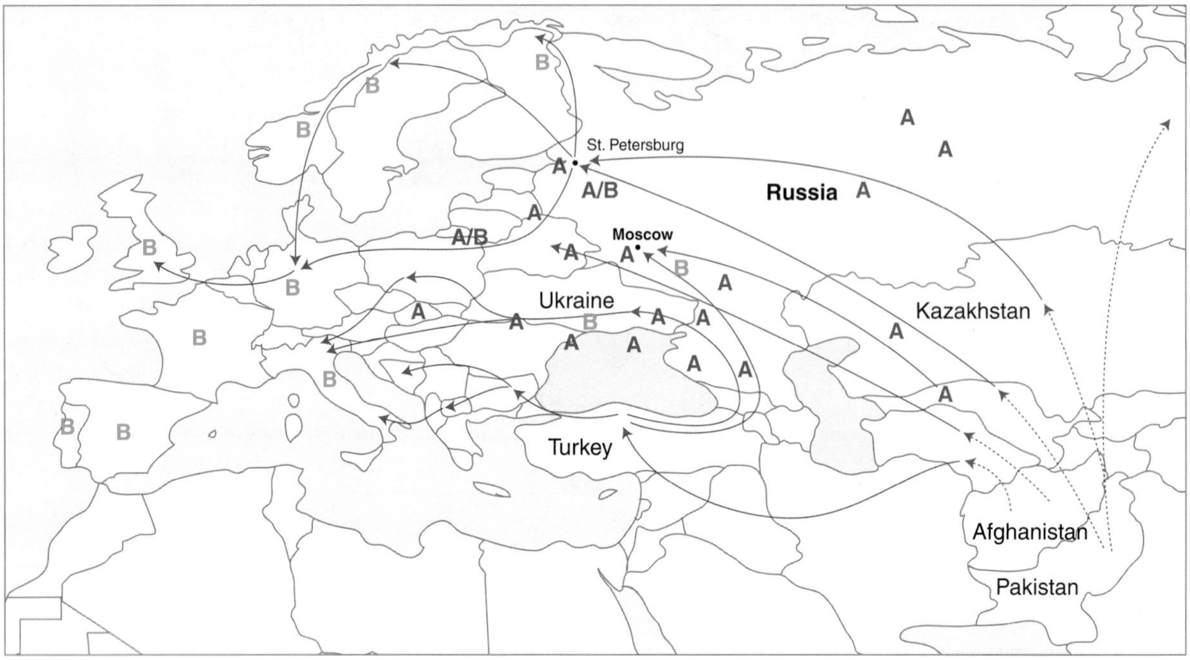

FIGURE 16-5 HIV subtype A followed homemade heroin drug trafficking routes into Eastern Europe. HIV subtype B predominates Western Europe. Adapted from Cohen, J. 2010. "Tracing the regional rise of HIV." *Science* 329:161.

used syringes picked off the street that are contaminated with blood (the blood may be from an HIV-infected injector). Heavy alcohol consumption is prevalent among intravenous drug users in Russia. It has been associated with sexual risk-taking and injection practices that facilitate HIV transmission. There is also a prevalence of sexual risk behaviors among non-intravenous drug users engaging in sex with intravenous drug users, spreading HIV beyond the intravenous drug user population. Methadone is used to treat heroin addicts. It reduces the cravings of heroin use and blocks the high associated with heroin use. Russia outlaws methadone. The Ukraine recently legalized it but struggles to meet the demand for it. Intervention to reduce HIV transmission will need to address alcohol consumption among non-intravenous drug users who engage in sexual activity with intravenous drug users, and drug use and the propensity of having sex while high on chernaya or methamphetamine.

16.6 AIDS in China

By the end of 2009, 740,000 adults and children were estimated to be living with HIV in China. About 48,000 new HIV infections are diagnosed each year. Of the 740,000 people diagnosed with HIV, 44.3% were infected through sexual transmission, 14.7% through homosexual transmission, 32.2% through intravenous drug use, 7.8% through commercial plasma donation and transfusion of infected blood and blood products, and 1% through mother-to-child transmission.

China is home to one of the largest injection drug user populations in the world. One reason for China's border provinces having large drug using populations is they are geographically located near large drug-producing and trafficking regions of Southeast and Central Asia. Yunnan and Guangxi have felt the hardest impact of the HIV/AIDS epidemic in China. Drug users make up 40% of the HIV individuals living in China. Preventing initiation of or continuation of high-risk sexual behaviors associated with drug use is needed to stem the Chinese HIV/AIDS epidemic.

A high degree of stigmatization of people living with HIV exists in the general population. A short summary by the Joint United Nations Programme on HIV/AIDS (UNAIDS) survey of 6000 people living in eight Chinese cities in 2008 revealed:

- 64.9% would be unwilling or strongly unwilling to live in the same room as an HIV-positive person.
- 41.3% are unwilling or strongly unwilling to work in the same place as an HIV-positive person.
- 47.8% would be unwilling or strongly unwilling to eat food in the same place as a person living with HIV.
- 63.6% would be unwilling or strongly unwilling to receive services (e.g., a haircut) from a person living with HIV.

Stigma and discrimination experienced by HIV-infected people are severe. Participants (41.7%) in the survey reported to have faced severe HIV-related discrimination. More than 76% of the survey participants said their family members had experienced discrimination as a result of their HIV status. Termination of pregnancy was recommended to pregnant women living with HIV.

16.7 General Epidemiology: HIV/AIDS in the United States Today

The number of individuals diagnosed with AIDS and the number of AIDS-related deaths in the United States have been declining since 1993 (**FIGURE 16-6a**). Highly active antiretroviral therapy (HAART) has slowed the progression of HIV in many infected individuals, resulting in a decline in the number of AIDS diagnosed cases and AIDS-related deaths.

A map of the United States that includes the estimated rates (per 100,000) for adolescents and adults (individuals 13 or older) *diagnosed with HIV infection* in 2008 is shown in Figure 16-6b. In 2008, the CDC estimated that in 2006 there were 56,300 people newly infected with HIV. Approximately 53% of these infections occurred in gay and bisexual men (Figure 16-4d). In 2009, the CDC estimated the number of diagnoses of HIV infection in the 40 states and 5 U.S.-dependent areas with confidential name-based HIV reporting was 42,959. The age group with the highest estimated number of diagnoses of HIV infection in 2009 was 20- to 24-year-olds, with 6237. At the end of 2008, the estimated number of individuals living with an HIV diagnosis in the 40 states and 5 U.S.-dependent areas was 682,668.

A map of the United States that includes the estimated rates (per 100,000) *diagnosed with AIDS infection* in 2008 is shown in Figure 16-4c. In 2009, the estimated number of people *diagnosed with AIDS* in the 40 states and 5 U.S.-dependent areas was 34,993. The age category with the highest number of estimated AIDS diagnoses in 2009 was 40- to 44-year-olds. The *cumulative* estimated number of individuals living with AIDS in the United States and dependent areas *from the beginning of the epidemic through 2009* is 1,142,714. According the *AIDS Scorecard* 2010 report, the incidence of HIV in the United States is stable.

Accurate surveillance is needed to monitor and assess the AIDS epidemic in the United States, where the statistics are collected by the CDC from each state. Not all states collect data in the same way, though. Most states use **confidential name-based reporting**, whereas some use other methods. In name-based reporting of AIDS cases, patient names are forwarded to the Health Department by healthcare providers when AIDS is diagnosed. Patient names are kept secure and confidential at the state health department, and no patient names are forwarded to the federal government. Name-based reporting for HIV infection is similar to name-based reporting for AIDS cases. These different methods in data collection pose a challenge when compiling national data. For the most up-to-date information on U.S. statistics and surveillance, visit http://www.cdc.gov/hiv/topics/surveillance/basic.htm.

A Risky Workplace? AIDS in the Adult Film Industry

There are over 200 pornographic film production companies in Los Angeles County, California, alone that employ at least 1200 workers who engage in work-related sexual contact. Production companies of heterosexual segments in general do not require condom use for any sexual act. Workers participate in voluntary programs that screen for HIV and other sexually transmitted diseases. At least 62 male and 6 female pornographic film performers died of AIDS-related illness from 1985 to 2002. It claimed the life of John Holmes, the most famous male porn star at the time. Holmes tested positive in 1985 and died in 1988. He reportedly had sex with more than 10,000 partners and continued to make porn movies until his death. In 2004, an outbreak of HIV was linked to a male actor who contracted the virus while working unprotected in Brazil. He then went on to infect at least three women who were working for an adult video company.

From 2009 to 2011, the **AIDS Healthcare Foundation** (AHF) filed workplace safety complaints against 16 California-based adult film companies including complaints against Larry Flynt's Hustler Video and Steve Hirch's Vivid Entertainment. AHF and other AIDS advocates are spearheading campaigns to improve the safety of workers by requiring adult film actors to use condoms. AHF has filed lawsuits against local Los Angeles county public health officials to enforce workplace safety regulations, lobbying for an overhaul of state workplace safety measures covering the two largest production centers located in California and Florida.

Testing programs cannot adequately prevent HIV transmission or other sexually transmitted diseases because individuals may transmit these diseases before their infection can be diagnosed.

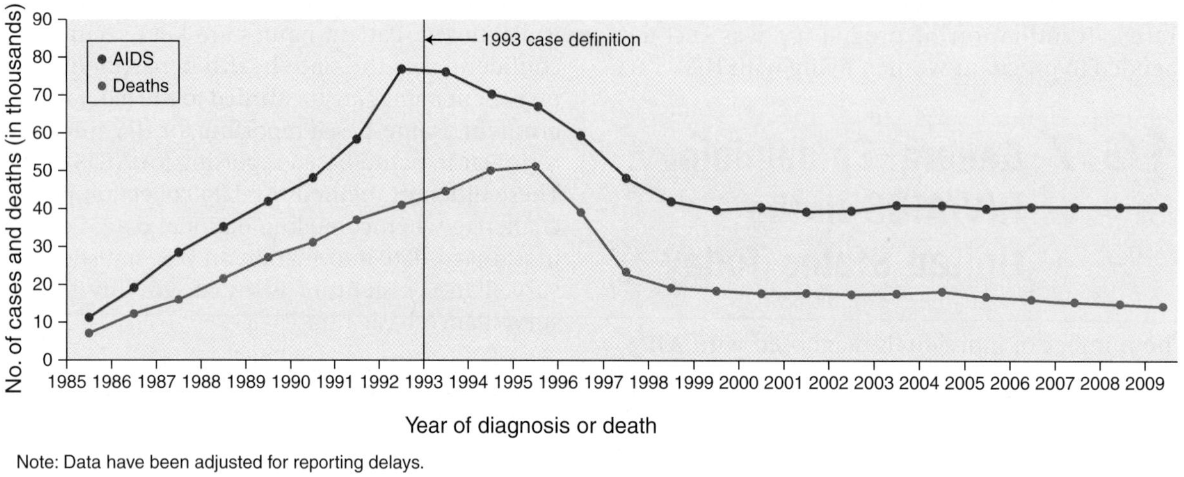

a

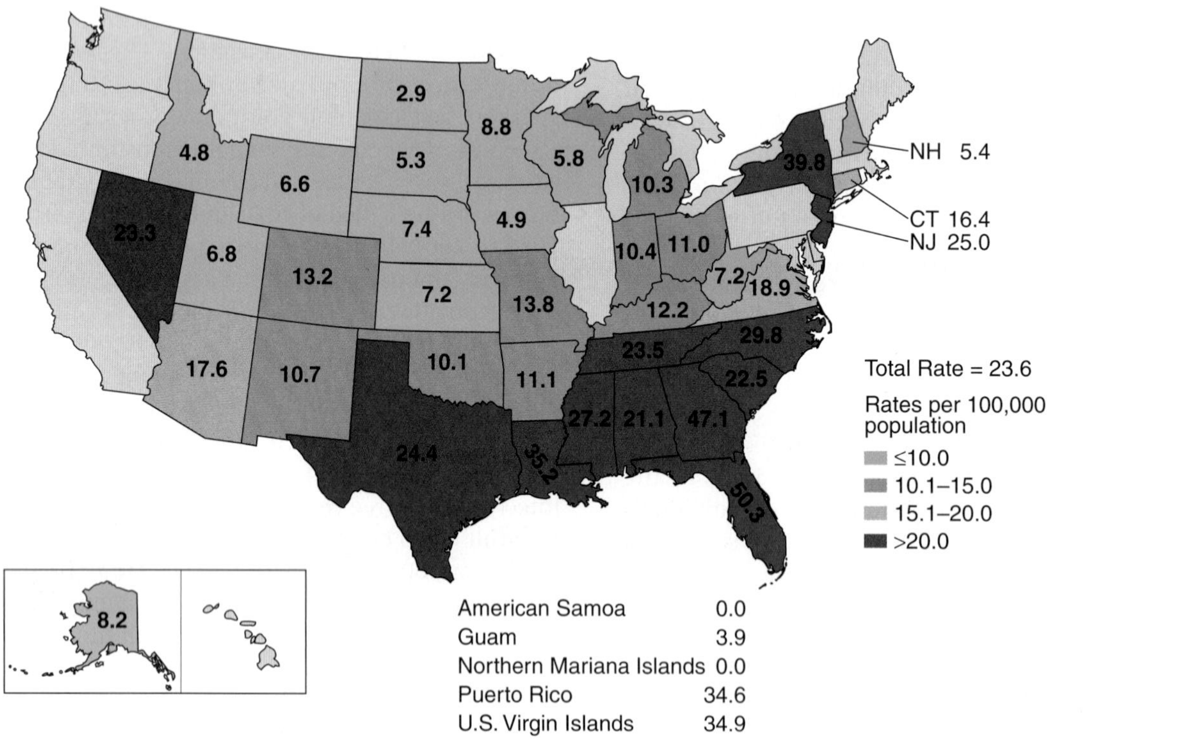

b

FIGURE 16-6 (a) Estimated AIDS diagnoses and deaths among adults and adolescents diagnosed with AIDS, 1985–2009, in the United States and dependent areas. Data are from the CDC, *HIV/AIDS Surveillance Report 2009.* **(b)** Estimated rates of diagnosis (per 100,000) of HIV infection for adults and adolescents in 2008 (37 states and 5 U.S.-dependent areas). Data include individuals with a diagnosis of HIV infection regardless of stage of disease at diagnosis. Data are from the CDC, *HIV/AIDS Surveillance Report 2009.*

(continued)

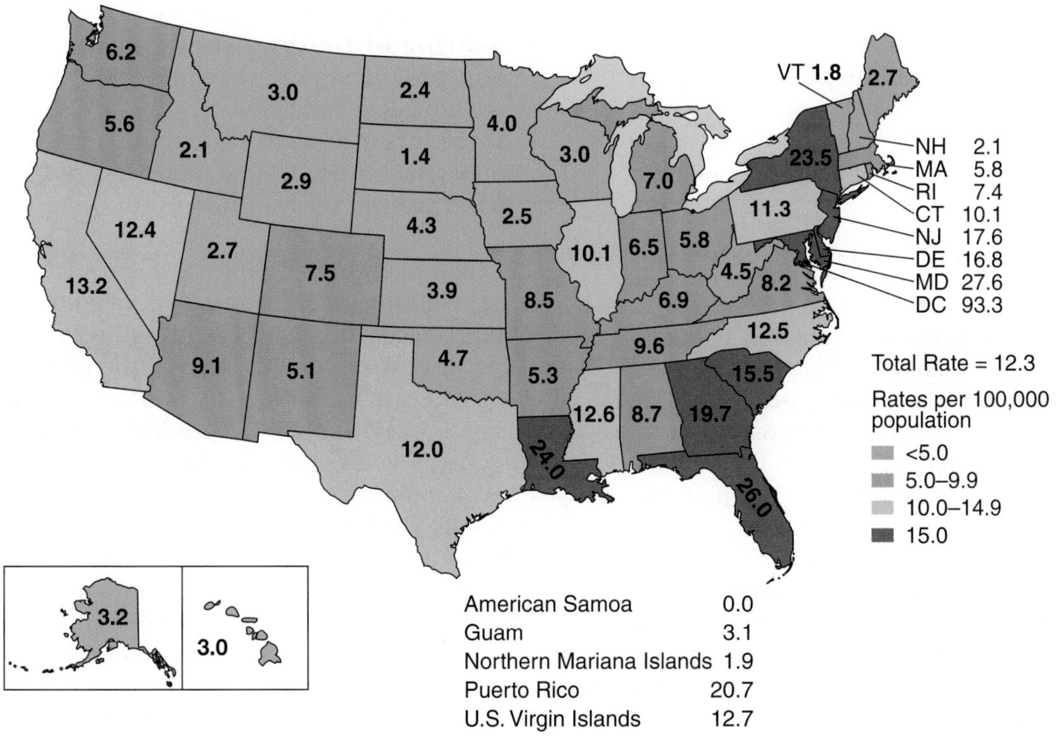

c

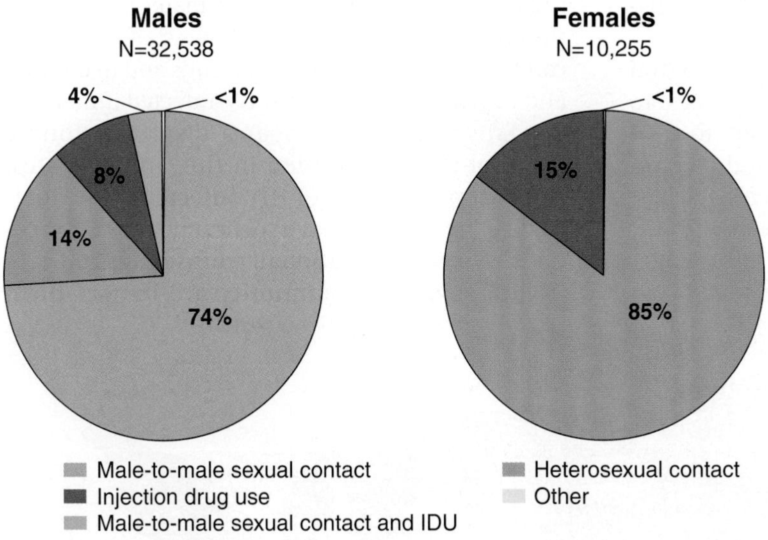

Note: Data include persons with a diagnosis of HIV infection regardless of their AIDS status at diagnosis. Data are from 33 states with confidential name-based HIV infection reporting since at least 2001. Data have been adjusted for reporting delays and cases without risk factor information and were proportionally redistributed.
*Heterosexual contact with a person known to have or to be at high risk for HIV infection.

d

FIGURE 16-6 (Continued) **(c)** Estimated rates of AIDS diagnosis (per 100,000) in 2008 (U.S. and dependent areas). Estimated numbers resulted from statistical adjustments that accounted for reporting delays, but not from incomplete reporting. Adapted from the CDC, *HIV/AIDS Surveillance Report 2008.* **(d)** Proportion of HIV infection among adults and adolescents by sex and transmission category in 2009. Data are from 40 states, and 5 U.S.-dependent areas. Heterosexual contact is defined as heterosexual contact with a person known to have, or be at high risk for HIV infection. The "Other" category includes hemophilia, blood transfusions, perinatal exposure, and risk factor not reported or identified. Data are from the CDC, *HIV/AIDS Surveillance Report 2008.*

For HIV infection, the period after infection until the virus is detectable by methods such as reverse transcriptase-polymerase chain reaction (RT-PCR) is called the **eclipse phase**, and the period until antibodies are detectable by serologic methods (i.e., enzyme-linked immunosorbent assay [ELISA]) is called the **window period**. The average eclipse phase is estimated to last 10 to 15 days but can be longer. Regulators and lawmakers have urged the industry to require that actors in adult films use condoms.

HIV/AIDS in U.S. Correctional Facilities

In January, 2010, there were 1,404,053 adults in state and federal prisons. HIV infection is five to seven times higher among prisoners than for the general population. Many prisoners are illegal drug users (e.g., in 2009, 50.7% of federal prisoners were incarcerated for drug-related offenses). The National Commission on AIDS stated in its 1991 report: "By choosing mass imprisonment as the federal and state governments' response to the use of drugs, we have created a de facto policy of incarcerating more and more individuals with HIV infection." Current epidemiology of HIV infection in state correctional institutions is shown in **FIGURE 16-7**. At the end of 2001, 3.2% of all female state prison inmates were HIV positive, compared to 2.0% of males.

There are numerous activities known to occur among prisoners that pose a risk for HIV transmission. These activities include:

- **Homosexual activity**—consensual sex and rape among male inmates without access to condoms. Many institutions object to condom distribution because the condoms can be used to hide contraband such as drugs. Some prisoners use ineffective makeshift devices such as the fingers from latex gloves and Saran Wrap to practice safer sex.
- **Intravenous drug use** with used syringes, pieces of pens, and light bulbs.
- **Tattooing** and **body piercing** with recycled sharpened and altered objects like staples, paper clips, and ballpoint pens.
- **Sharing of toothbrushes and shaving equipment** in facilities where these items are not adequately provided.
- **Incidents of violence**—fights that involve bites, lacerations, and bleeding of the participants.

HIV testing in U.S. prisons can be mandatory, routine, voluntary but encouraged, or only on demand. The courts have prevented local correctional facility policies on this issue. Prison inmates do have access to health care as a protected right. Treatment of HIV-seropositive prisoners has been criticized for years, though. According to the *Sourcebook of Criminal Justice Statistics*, an estimated 19,842 state and federal prisoners were known to be HIV positive. In 1995, 34% of all deaths in state prisons were AIDS-related. HAART and directly observed therapy (DOT) are available in some prison systems. Antiretrovirals account for a large portion of prison pharmaceutical budgets; hence, DOT (also known as "watch-take medication") assures that the patient ingests every dose dispensed to reduce costs. Some prisons attempt to segregate known HIV-infected inmates to "contain" the epidemic. This approach allows staff to be more careful around HIV-positive persons and to educate and provide expert care to HIV-infected inmates.

Jails and prisons bear a disproportionate share of the HIV burden in the United States. Approximately 25% of HIV-infected individuals in the United States have spent time in the correctional system. Educational, counseling, and care programs in the community are important to

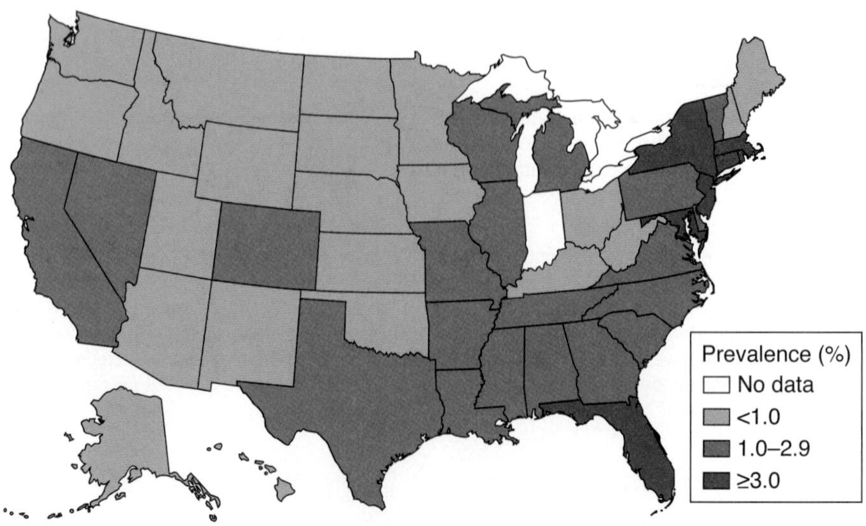

FIGURE 16-7 Known HIV cases in state correctional facilities shown as a percent of the total custody population in state prisons in 2002. These statistics are from *The Bureau of Justice Statistics Bulletin*, December 2004, NCJ 205333. Data are from U.S. Department of Justice, "HIV in prisons and jails, 2002," *Bureau of Justice Statistics Bulletin* (December 2004).

ensure that the HIV-infected patient has care both inside prison and after release. The "epidemic of imprisonment" contributes to the HIV epidemic in the United States.

AIDS and Seniors

AIDS among seniors is a growing problem. Approximately 24% of people living with HIV in the United States are over 50 years of age. Between 1990 and 2001, cases in seniors quintupled from 16,288 to 90,513. About 2.4% of these adults were infected with HIV by blood transfusions during surgeries from 1979 to 1985.

Surveys have shown that 24% to 30% of men and women ages 60–74 report that they are sexually active. This generation of individuals is living longer, healthier lives, but has not been targeted for sex education. Older women are more susceptible to HIV infection due to physical changes that occur with menopause, such as vaginal dryness and thinning of vaginal walls. This can lead to abrasions and tears that increase the risk of infection during unprotected sex. Older adults have not felt comfortable discussing their sexual histories and safe sex practices with partners and medical professionals, and they now represent a significant percentage of AIDS cases. Erectile dysfunction drugs, such as sildenafil (Viagra), tadalafil (Cialis), and vardenafil (Levitra), used by older adults, have been associated with higher risk of STDs, especially HIV transmission. **FIGURE 16-8** shows the 2005 statistics of persons by age group. In 2005, 15% of U.S. individuals over the age of 50 were diagnosed with HIV and 24% were diagnosed with AIDS. Thirty-five percent of the deaths in this age group were attributable to AIDS (Figure 16-6).

There are several barriers to diagnosis of HIV infection in older adults. The symptoms of other chronic illnesses resemble those of AIDS, such as:

- night sweats and hot flashes
- weight loss or fatigue
- respiratory problems
- skin rashes
- flulike symptoms or diarrhea
- chronic pain or numbness
- early signs of dementia

Healthcare providers rarely think to test for HIV in older adults. Many times the older adults are further along in the disease progression before they are diagnosed. There are treatment barriers. In some cases, older patients have been told they "have lived a long life" and will not be offered aggressive treatment for AIDS-related illnesses. This same age group usually has a marked decrease in income since diagnosis and may not have enough money to live on if they must give up work due to HIV.

The AIDS crisis among older adults needs specific outreach programs devoted to HIV information in general, safe sex practices, and testing. New media and marketing campaigns can raise awareness in older people while promoting respect and validation for the elderly as a group.

The first in-depth study on older adults with HIV (Research on Older Adults with HIV [ROAH]) was released in 2006. Key findings were also published in the book titled *Older Adults with HIV* (referenced in the Primary Literature list at the end of the chapter). Participants of ROAH were individuals who were diagnosed with HIV or AIDS, 50 years or older, and living in a community (i.e., not institutionalized). These participants were from five boroughs of New York City. They were surveyed about their health status, sexual orientation and behaviors, substance use, social networks, living arrangements, demographics, spirituality, stigma, and loneliness. Of the 1000 surveys sent to participants,

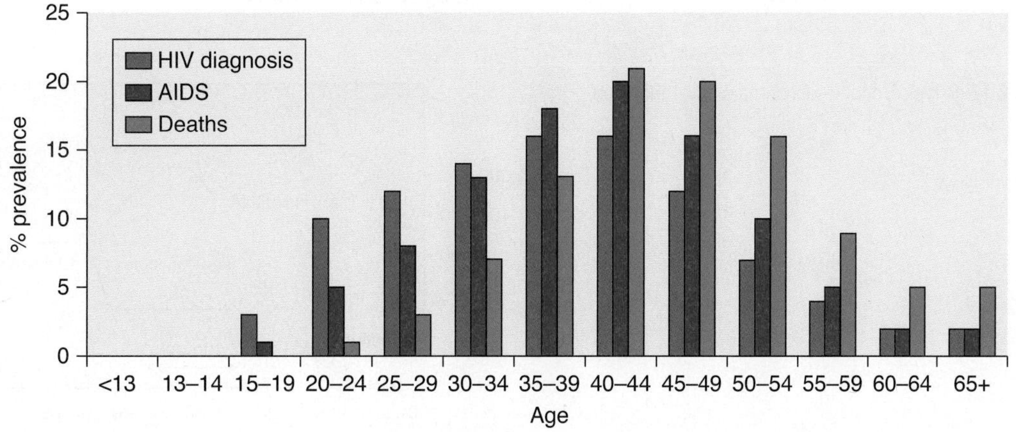

FIGURE 16-8 Estimated HIV and AIDS cases and deaths by age in United States for the year 2005. Data are from Avert, *United States Statistics Summary*, http://www.avert.org/statsum.htm.

914 responded. Of these, 70% were men, 29% women, and 1% identified as transgender.

Research suggests that by 2015, half of the people living with HIV in the United States will be over the age of 50. The ROAH study makes it clear that more resources are needed to design programs for older adults about the prevention and treatment of HIV/AIDS. The AIDS Community Research Initiative of America (ACRIA) published a special report and action plan on "Older Adults and HIV" in November, 2010. ACRIA urged the U.S. research agenda, the healthcare system, and aging and HIV service providers to respond to this emerging issue.

16.8 Clinical Symptoms

There are three phases of HIV-1 infection:

1. Primary HIV-1 infection
2. Chronic asymptomatic phase
3. AIDS

Primary HIV-1 Infection

Primary HIV infection is the period after infection but before the development of detectable antibodies against HIV-1 (**FIGURE 16-9**). The virus replicates rapidly during this window, which lasts a few weeks or months. The virus is present in high numbers in the blood, central nervous system, and lymphatic system and invades other tissues. The severity of a primary infection differs for each individual. About 50% of individuals do not experience any symptoms. **TABLE 16-3** contains a list of the signs and symptoms of primary HIV-1 infection for those people who do experience them. Their illness resembles an infection such as mononucleosis or influenza. These symptoms are generally referred to as acute retroviral syndrome (ARS). They typically occur two to six weeks after infection and last two to three weeks.

TABLE 16-3 Signs and Symptoms of Primary HIV-1 Infection

Signs and Symptoms	Frequency
Fever Fatigue Rash Muscle pain Sore throat Swelling of lymph nodes	Experienced by more than 50% of infected individuals
Headache Diarrhea Nausea and vomiting Night sweats Weight loss Thrush (oral candidiasis) Neurological symptoms (depression, dizziness, photophobia) Oral and genital ulcers Cough Enlargement of the spleen and liver	Experienced by 5% to 32% of infected individuals

The Chronic Asymptomatic Phase of Infection

During this phase, neither signs nor symptoms of the disease are present. This phase typically lasts an average of ten years. These HIV-1–infected persons are defined as **typical progressors**. The virus continues to replicate in the individual. As the levels of virus increase in the bloodstream, CD4+ T cell counts drop. When these counts get below 200 cells/μl of blood, the development of AIDS opportunistic infections occurs.

About 10% to 20% of HIV-1–infected individuals develop AIDS within five years of infection. These patients are called **rapid progressors**. About 5% to 15% of HIV-infected individuals remain free of AIDS for more than 15 years and are termed **slow progressors**. There is a subgroup of HIV-1–infected individuals who are **long-term non-progressors** (**FIGURE 16-10**). This 1% subgroup of

FIGURE 16-9 Typical course of a primary HIV-1 infection.

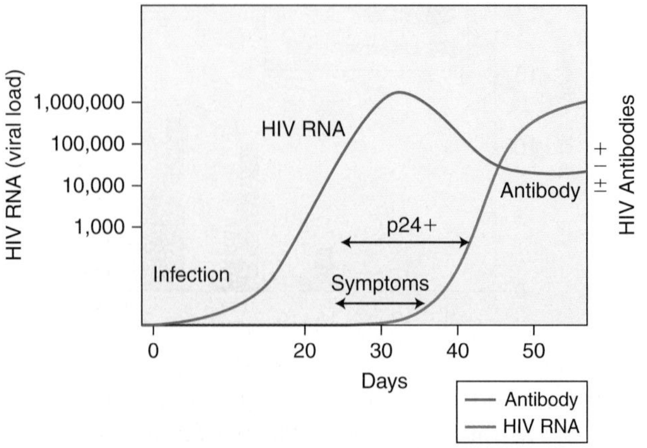

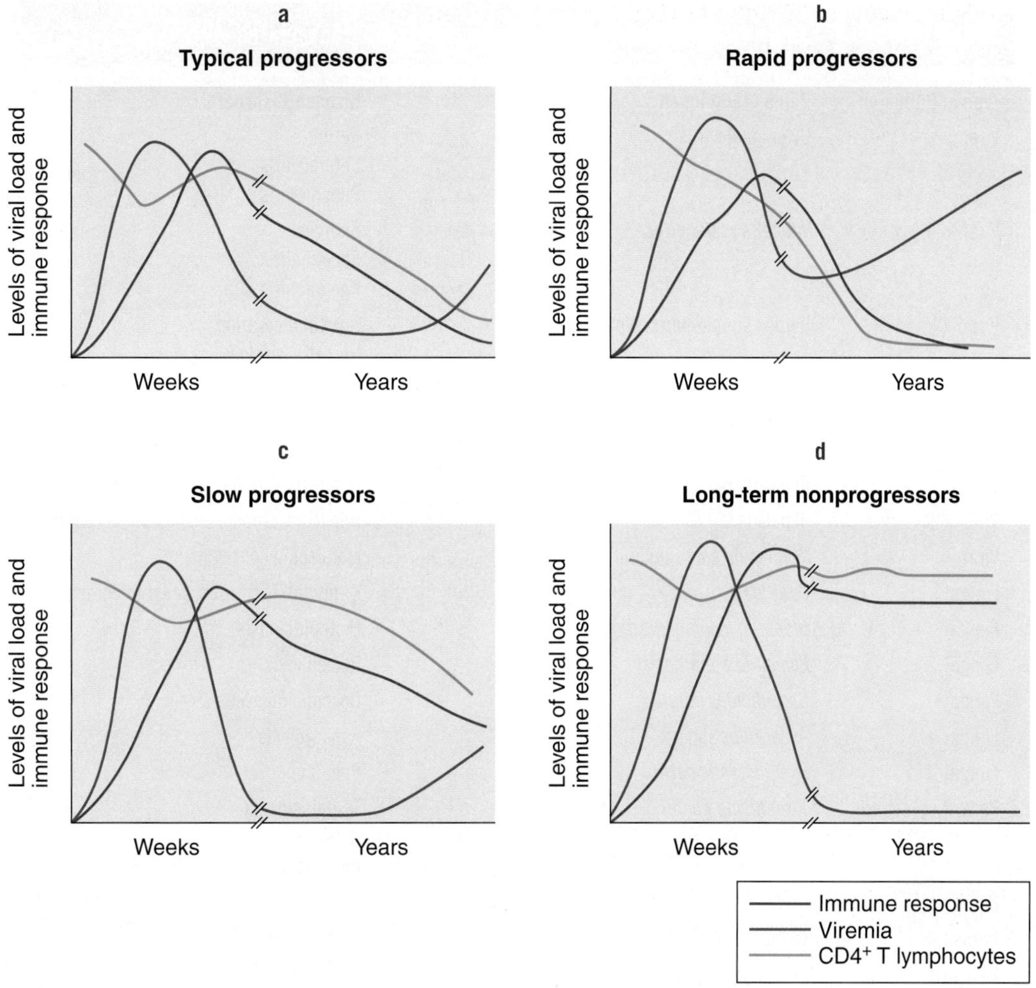

FIGURE 16-10 Graphs depicting the different types of progressors based on changes in viral load, CD4+ T cell counts, and antibody response during the natural course of HIV-1 infection. Adapted from J. Cohen and W. G. Powderly. *Infectious Diseases*, Second Edition. Mosby, 2003.

individuals have not previously taken or received antiretroviral therapy and show no sign of disease (e.g., normal CD4+ T cell counts and low levels of virus, which is, 500–1000 copies of RNA genomes/ml or <50 viruses/ml of plasma). The disease course of long-term nonprogressors is different due to genetic, immunologic, or other driving factors.

AIDS

Full-blown AIDS is the last phase of HIV-1 infection. Without antiretroviral therapy, the patient will die within two to three years. HIV does not kill the individual directly. Instead, it weakens the body's ability to fight other infections. As CD4+ T cell counts decrease below 50 cells/µl of blood, the number of opportunistic infections increases and there is a higher risk for death. HAART slows the rate of progression of this disease. Infections that rarely occur in people with competent immune systems are deadly to those with HIV. **TABLE 16-4** lists opportunistic infections commonly associated with AIDS. In addition to microbial infections and Kaposi's sarcoma, AIDS patients often suffer from **malignancies** such as systematic non-Hodgkin's lymphoma. **Wasting syndrome**, a loss of more than 10% of body weight due to fever or diarrhea for more than 30 days, is a common problem in many AIDS patients as well.

AIDS patients who are severely immunocompromised are managed with antimicrobial prophylactic regimens. Together with the use of antiretroviral therapy, this has led to longer survivorship and decreased mortality of patients. The development of drug resistance to the most commonly used antimicrobial agents is of major concern when long-term prophylaxis is used to manage patients. The U.S. Public Health Services and the Infectious Diseases Society of America have made recommendations for the prevention of opportunistic infections in AIDS patients. They recommend prophylaxis against the following disease-specific pathogens: *Pneumocystis carinii, Toxoplasma gondii, Mycobacterium tuberculosis, Mycobacterium avium, Streptococcus pneumoniae,* and *cytomegalovirus.*

TABLE 16-4	Examples of Opportunistic Infections Commonly Associated with AIDS	
Type of Pathogen	**Name of Pathogen**	**Disease/Disorder**
Viral	Cytomegalovirus	Retinitis Enterocolitis Pneumonitis
Viral	Varicella zoster virus	Shingles Encephalitis Pneumonitis
Viral	Herpes simplex virus types 1 and 2	Genital ulceration Herpetic whitlow Encephalitis
Viral	Papillomavirus	Common and genital warts Cervical and anal cancer
Viral	Hepatitis B	Hepatitis, chronic cirrhosis, liver cancer
Viral	Hepatitis C	Hepatitis, chronic cirrhosis, liver cancer
Viral	Molluscum contagiosum virus	Molluscum contagiosum
Fungal	*Candida albicans*	Candidiasis (thrush, yeast infection)
Fungal	*Histoplasma capsulatum*	Histoplasmosis
Fungal	*Penicillium marneffei*	Penicilliosis
Fungal	*Coccidioides immitis*	Coccidioidomycosis
Fungal	*Aspergillus* spp.	Aspergillosis
Fungal	*Pneumocystis carinii*	Pneumonia
Parasitic	*Toxoplasma gondii*	Toxoplasmosis Retinitis Pneumonitis
Parasitic	*Cryptosporidium parvum*	Cryptosporidiosis
Parasitic	*Isopora belli*	Isosporiasis
Parasitic	*Enterocytozoon, Encephalitozoon, Septata, Pleistophora, Nosema* spp.	Microsporidiosis
Bacterial	*Mycobacterium tuberculosis*	Tuberculosis
Bacterial	*Mycobacterium avium*	*Mycobacterium avium* complex
Bacterial	*Salmonella* sp., *Shigella flexneri, Clostridium difficile, Shigella, Campylobacter* sp.	Enterocolitis
Bacterial	*Rhodococcus equii, Haemophilus* sp.	Pneumonia
Bacterial	*Streptococcus pneumoniae, Moraxella catarrhalis, Haemophilus* sp., *Pseudomonas aeruginosa*	Sinusitis
Bacterial	*Bartonella henselae* and *Bartonella quintana*	Skin infections (subcutaneous nodules)

16.9 Laboratory Diagnosis of HIV

HIV tests are based on the detection of antibodies against HIV. HIV was isolated in 1983 and the first commercial antibody detection tests were available in 1985. Serologic assays are used to detect HIV antibodies in serum, whole blood, saliva, urine, and dried blood collected on filter paper. Most clinical laboratories screen for HIV-1 and HIV-2 antibodies by means of an ELISA, an assay that is widely used throughout the world. Initial tests were only able to detect IgG anti-HIV antibodies. Today all classes of anti-HIV antibodies can be detected with these assays. Positive ELISA test results should be confirmed by Western blot.

An ELISA test requires the drawing of a blood sample from a vein. It also requires two doctor visits: the first is for blood drawing and pretest counseling and the second is to receive the test results and additional counseling if needed. Healthcare workers may perform **rapid HIV tests**, which produce results within 5 to 30 minutes. Testing, counseling, and referrals can be done in one visit. Typically, rapid tests are used for the diagnosis of HIV infection in women during labor and delivery and to determine the infection status of patients following an accidental occupational needlestick injury. As for positive ELISA test results, positive rapid test results should be confirmed by Western blot.

The Federal Drug Administration (FDA) approved **HIV home testing systems** in 1996. These home tests refer to "home collection" that uses a simple "finger prick" of blood sample that has been applied to a filter paper and mailed in for laboratory testing. The confidential test results are obtained through a toll-free phone number and posttest counseling is provided by telephone when results are obtained. Consumers should be aware that there are several HIV home testing centers that are not approved and fraudulently marketed for home-use HIV testing. The only FDA-approved home test system is sold as "The Home Access HIV-1 Test System" or "The Home Access Express HIV-1 Test System."

The **Western blot** is the confirmatory test essential to exclude false-positive results. HIV proteins are separated by sodium dodecyl sulfate-polyacrylamide gel electrophoresis (SDS-PAGE) and transferred to a nitrocellulose membrane. The membrane is then reacted with test serum to determine which, if any, HIV proteins are recognized by patient antibodies (**FIGURE 16-11**; see Chapter 5 for more details regarding Western blot testing).

The CDC proposed guidelines for voluntary testing that would apply to every American aged 13 to 64. Approximately one fourth of the one million Americans infected with HIV don't know they are infected. This group is most responsible for HIV's spread. The CDC's plan was released in 2006 and is supported by the American Medical Association.

Genetic Variation/Groups

HIV mutates very readily. Hence, it is a highly variable virus that consists of many different strains.

FIGURE 16-11 CDC technician filling out an AIDS diagnostic report based on Western blot results.

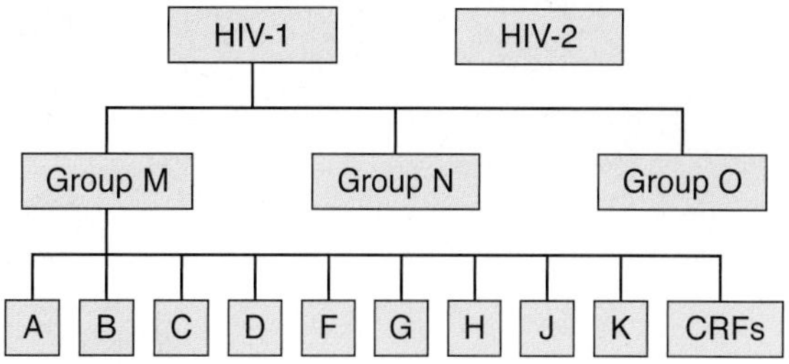

FIGURE 16-12 Schematic illustrating the classification of HIV strains into groups and further subtypes.

Researchers have divided HIV-1 into three distinct groups based on genetic similarities: the major group (M), the outlier group (O), and the new group (N). The three different groups may represent three different introductions of SIV into humans. The O group is restricted to west-central Africa and the N group was discovered in Cameroon in 1994. The first case of HIV-1 group O in the United States was identified in California. The patient was a native of Africa who came to the United States in 1994. Standard tests did not detect HIV infection during the initial evaluation by physicians.

The M group has been further divided into **clades** or **subtypes**: A, B, C, D, F, G, H, and K. Periodically, two viruses of different subtypes can infect the same cell in an individual, resulting in new progeny hybrid viruses with recombined genomes. These hybrid viruses often do not survive very long, but if they do infect more than one individual, they are referred to as **circulating recombinant forms** (**CRFs**). **FIGURE 16-12** is a diagram depicting the different classifications of HIV strains. It is almost certain that new genetic subtypes and CRFs will be discovered in the future. ELISA screening detects both HIV-1 and HIV-2. Routine ELISA screening and rapid tests detect all of the group M subtypes. Anyone who thinks they may have contracted the rare O subtype should seek expert medical advice.

16.10 HIV Virus Life Cycle

HIV Classification and Virion Structure

HIV is a member of the *Retroviridae* family and is placed in the *Lentivirus* genus (from the Latin word *lenti*, for "slow"). Lentiviruses have been implicated in slowly progressive, inflammatory diseases. They undergo long incubation periods and persistently infect monkeys or chimpanzees (e.g., SIV), sheep (e.g., Visna), horses (e.g., equine infectious anemia virus), cats (e.g., feline immunodeficiency virus), and goats (e.g., caprine arthritis-encephalitis virus).

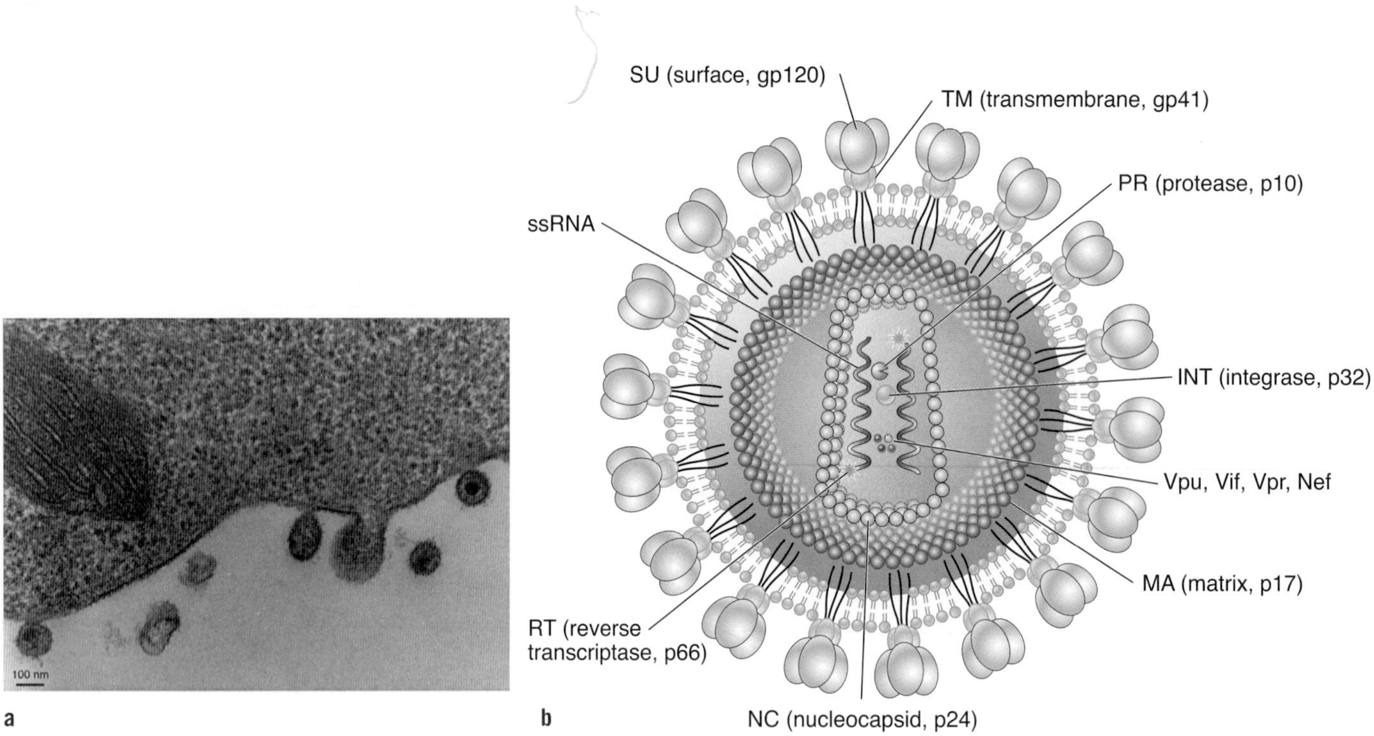

FIGURE 16-13 (a) Electron micrograph showing the presence of mature HIV virions in a tissue sample. Courtesy of Louisa Howard, Dartmouth College, Electron Microscope Facility. **(b)** Schematic of a mature HIV virion.

HIV particles are 100- to 120-nm icosahedron-shaped spheres surrounded by a membrane (**FIGURE 16-13a**). The membrane consists of protruding glycoproteins gp120 (the head) and gp41 (the spike). A matrix protein termed p17 is located underneath the bilipid layer. The capsid protein, p24, covers the two copies of the (+) single-stranded RNA (+ssRNA) viral genome. The HIV-1 genome is 9749 nucleotides in length. Reverse transcriptase (p64) and an integrase (p32) are bound to the viral genome (Figure 16-13b). A cellular $tRNA^{lys}$ is bound to the RNA genome that serves as a primer for the HIV reverse transcriptase. A nonbound protease, p10, is also found within the nucleocapsid of the virus particle.

HIV-1 Replication

HIV infects CD4+ T lymphocytes and will lyse them during a productive infection. Other cells harbor, replicate, and bud HIV without lysis, such as:

- natural killer cells
- CD8+ killer T cells
- macrophages
- cells of the nervous system (e.g., astrocytes, neurons, glial cells, and brain macrophages)
- dendritic cells (HIV concentrates on the cell surface)

During the early course of infection, HIV is macrophage tropic. Later in the course of infection, HIV is T-lymphocyte tropic. HIV enters cells via a cellular receptor and co-receptor. The major cellular receptor is CD4 that is abundantly present on T lymphocytes. The T lymphocyte co-receptor is **CXCR4** (also known as **fusin**). The macrophages' (often found in the skin and mucous membranes) co-receptor is **CCR5**. CD4 is present at low concentrations on macrophages. The rapid progression to AIDS has been associated with a switch in co-receptor preference from CCR5 to CXCR4. When the HIV virus attaches to a cell, the gp120 portion of the virus interacts with CD4. Subsequently, gp120 and gp41 undergo a conformational change, exposing new binding regions on these proteins, thus causing a co-receptor reaction. Subsequently, the virus anchors or embeds into the membrane of the CD4+ cell. The HIV gp41 protein changes into a coiled or hairpin-like shape that causes the virus and cell membrane to come together in proximity, allowing the membrane of the virus and host cell to fuse, resulting in entry and uncoating (**FIGURE 16-14**).

In the cytoplasm, the viral RNA is converted to DNA by the HIV-encoded **reverse transcriptase** (RT). Reverse transcriptase binds to the $tRNA^{lys}$ present on the viral genome and initiates DNA synthesis (see Chapter 10 for more information on retroviral genome replication and the functions of reverse transcriptase). When reverse transcription is completed, the final product consists of double-stranded DNA (dsDNA) that contains long terminal repeats (LTRs) located at the ends of the viral genome. The viral DNA is transported to the nucleus of the cell where the viral genome may be integrated into the host cell's

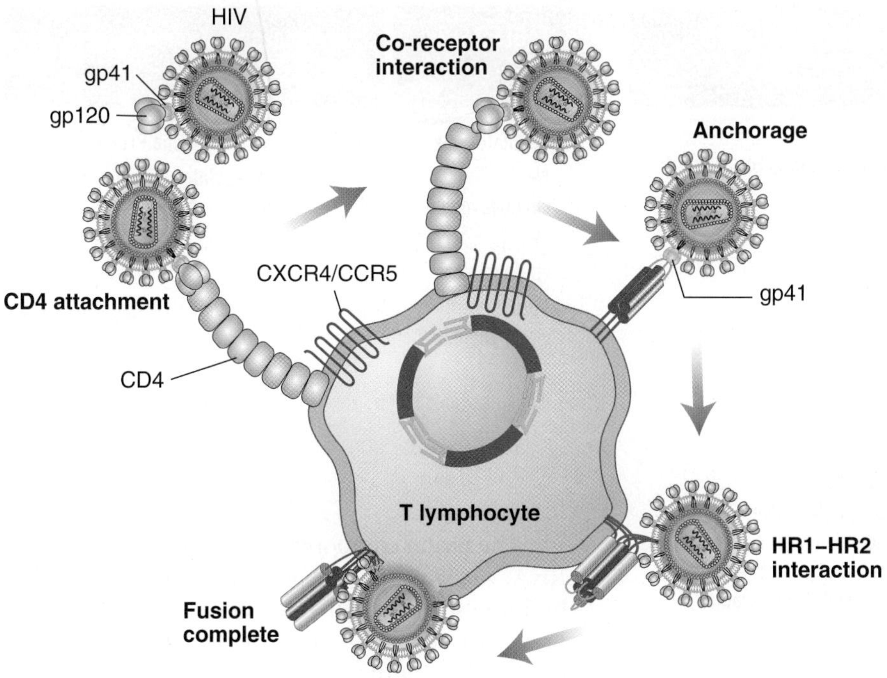

FIGURE 16-14 HIV enters cells by attaching to host cell CD4 receptors at a chemokine co-receptor such as CXCR4 or CCR5. Adapted from F. Hoffmann-La Roche Ltd., "Blockade of viral peptide helices prevents HIV from fusing with the host cell," *Roche Facets Clinical Trials.*

chromosomal DNA. Proviral genome integration is catalyzed by the integrase (p32) protein that was packaged into the original HIV particle. The HIV viral genome is more complicated than a typical retrovirus genome that consists of *gag, pol,* and *env* and that may also contain an *onc* gene (**FIGURE 16-15** and **TABLES 16-5** through **16-7**).

HIV-1 can **latently** infect cells. These latent HIV reservoirs are one of the main barriers to the elimination of HIV infection. During a productive infection, HIV-1 is reactivated and the completion of a lytic cell cycle occurs. HIV-1 utilizes the host cell's transcription and translational machinery. The proviral DNA is transcribed from a single promoter in the 5′ LTR into a 9 kb viral RNA primary transcript by the cellular RNA polymerase II. Tat and Rev are involved in the regulation of viral gene expression. RNA elements vital to replication and transcription are listed in **TABLE 16-8**. *tat* is a transcriptional transactivator that is essential for HIV-1 replication. It interacts with TAR. TAR is located at the 5′ end of all HIV RNAs. The binding of *tat* to TAR increases the transcription rate from the HIV LTR at least 1000-fold. Rev binds to the RRE to facilitate export of unspliced and incompletely spliced viral RNAs from the nucleus to the cytoplasm. The accessory proteins of HIV-1: *nef, vif, vpu,* and *vpr* play a role in HIV virulence (Table 16-7). Some full-length viral messenger RNAs (mRNAs) are packaged into virions following

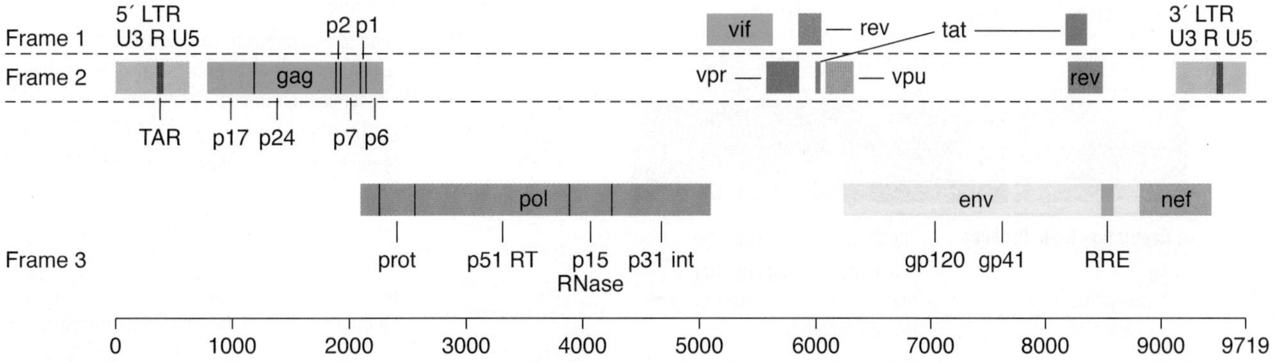

FIGURE 16-15 HIV-1 genome organization depicting open reading frames. The genome has extra open reading frames. A total of nine proteins are derived from *gag, pol,* and *env.* HIV makes six other proteins. HIV-2 contains a virion protein termed *vpx* that is not found in HIV-1 and SIVs. It is a 12-kd protein that is a gene homolog of HIV-1 *vpr.* The function of *vpx* is unknown. Modified from http://www.hiv.lanl.gov/content/sequence/HIV/MAP/landmark.html. Adapted from Sala, R. S. "Evolució I mutació del virus," *HIV.*

TABLE 16-5 HIV-1 Structural Gene Products and Their Functions

Structural Gene	Gene Product(s)	Function of Gene Product
Gag	p17	Matrix protein
	p24	Capsid protein
	p9	RNA-binding nucleocapsid protein
	p7	RNA-binding nucleocapsid protein
Pol	p10	Protease
	p15, 51	Reverse transcriptase RNase activity
	p32	Integrase
Env	gp120	Envelope glycoprotein (external), receptor binding
	gp41	Envelope glycoprotein (spike, transmembrane), receptor binding

transport to the cytoplasm, whereas other viral mRNAs are spliced within the nucleus to form mRNAs that are translated into the different viral proteins.

The ribosomes of the host cellular machinery translate the proviral mRNA into viral proteins. The *gag* gene and the *gag* and *pol* genes together are translated into large precursor polyproteins that are cleaved by a virus encoded protease. The full-length *env* protein, gp160, is transported through the *trans*-Golgi network where it is glycosylated and cleaved by a cellular protease into gp120 and gp41 to form mature envelope proteins that are targeted to the surface of the infected cell. The gp120 protein is attached to gp41 by non-covalent bonds. The genes of the smaller HIV proteins like tat overlap with the structural genes but are in different open reading frames. Their mRNAs are derived by alternative splicing of the structural gene mRNAs and translated accordingly.

TABLE 16-6 HIV-1 Regulatory Gene Products and Their Functions

Regulatory Gene Product	Function of Gene Product
Tat *Trans*-activator of transcription	RNA binding transactivating protein/enhances transcription of HIV full-length transcripts
Rev Regulator of virion protein expression	Facilitates the export of unspliced and incompletely spliced viral RNAs from the nucleus to the cytoplasm

TABLE 16-7 HIV-1 Accessory Gene Products and Their Functions

Regulatory Gene Product	Function of Gene Product
Nef negative regulatory factor	Downregulates the cell surface expression of CD4
Vpr viral protein R	Infectivity factor
Vpu viral protein U	Downregulates CD4 and enhances viral release
Vif viron infectivity factor	Required for the replication of HIV in lymphyocytes, macrophages and certain cell lines

The virion components of HIV-1 (e.g., the viral RNA, *gag, pol,* and *env* proteins) are assembled at budding sites located at the cellular plasma membrane. Two copies of full-length viral genomes that contain the cellular replication primer $tRNA^{lys}$ are packed into the virus particles. Other cellular factors may assist assembly of infectious viral particles. The *gag* and *gag-pol* proteins are cleaved by a HIV-specific protease to form mature virus particles. Our understanding of HIV-1 assembly, maturation, and release are just beginning to be characterized. Inhibitors are targeted at this HIV protease as anti-HIV chemotherapy. The life cycle of HIV-1 is shown in **FIGURE 16-16**.

HIV Cellular Pathogenesis

Ultimately, HIV causes AIDS by depleting the number of CD4+ T helper lymphocytes. This in turn results in the weakening of the immune system, allowing opportunistic infections to occur.

TABLE 16-8 HIV-1 RNA *Cis*-Acting Elements

RNA Element	Function
LTR	Long terminal repeat, contains promoter region, integration of proviral genome, binding site for cellular transcription factors
U3	Unique 3′ region, contains promoter region, binding site for cellular transcription factors
R	Terminal redundancy, contains promoter region, binding site for cellular transcription factors
U5	Unique 5′ region, contains promoter region, binding site for cellular transcription factors
TAR	Transactivation response element, activates transcription from LTR
RRE	Rev response element, involved in export of viral RNAs

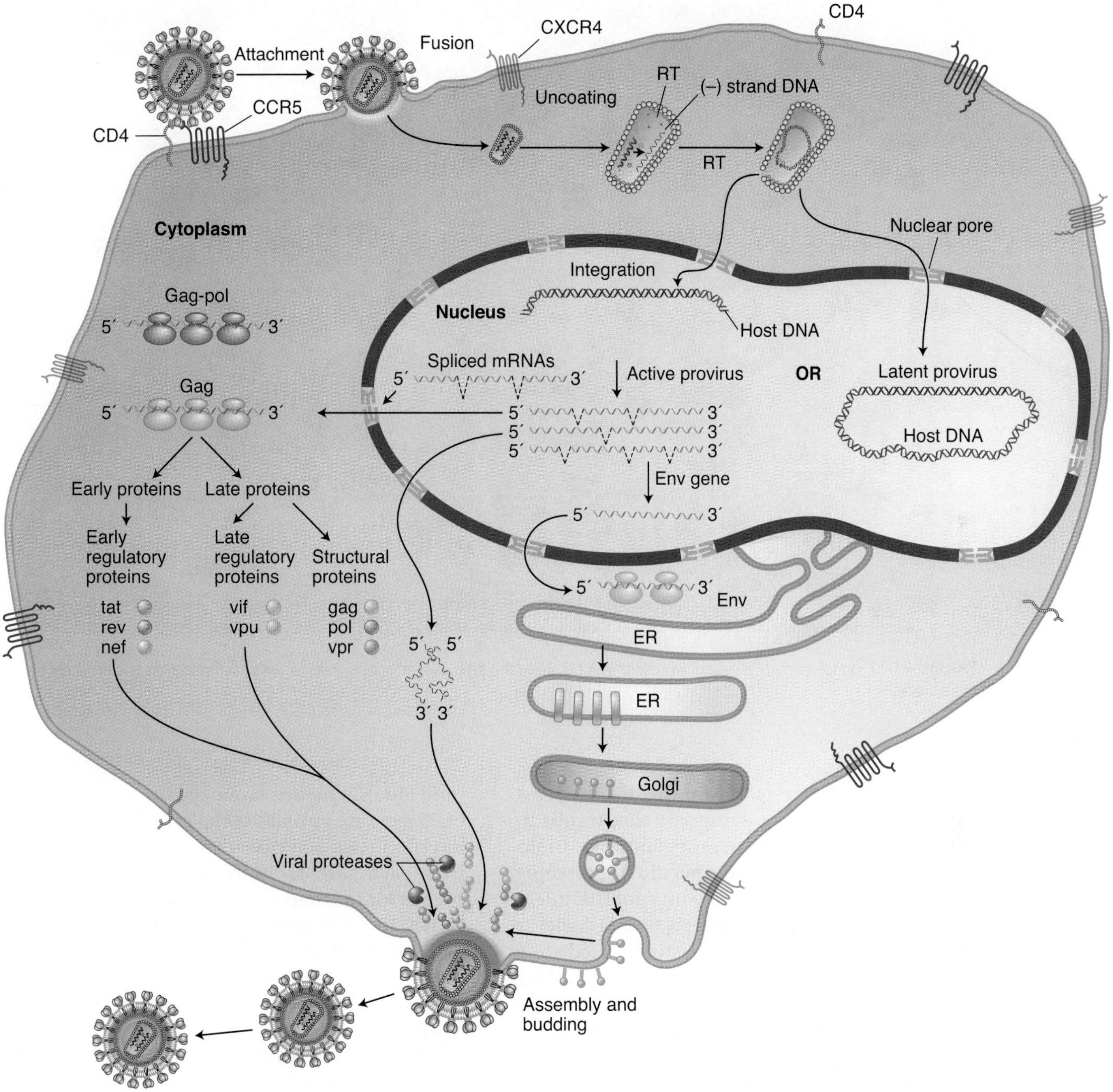

FIGURE 16-16 Life cycle of HIV-1. Adapted from Flint, S. J., et al. *Principles of Virology: Molecular Biology, Pathogenesis, and Control of Animal Viruses*, Second Edition. ASM Press, 2003.

Without T lymphocytes, essentially no immune response can occur. The body is unable to stop infections caused by other microbes or remove cancer cells. HIV causes **syncytia**, giant cells formed through the fusion of host cells (**FIGURE 16-17**). Viral *env* proteins are expressed on the surface of infected cells. These bind to uninfected cells, causing the plasma membrane to fuse, resulting in syncytial formation. Ultimately, these cells die prematurely.

AIDS dementia complex is a serious neurological complication that occurs in as many as 20% of infected individuals. Symptoms may affect cognitive reasoning (e.g., impaired short-term memory and concentration), behavior (e.g., apathy, withdrawal, irritability, depression, and personality changes), or motor performance (clumsiness, tremors, and leg weakness). When HIV infects monocytes and macrophages, it may travel to other organs, especially to the lungs and past the

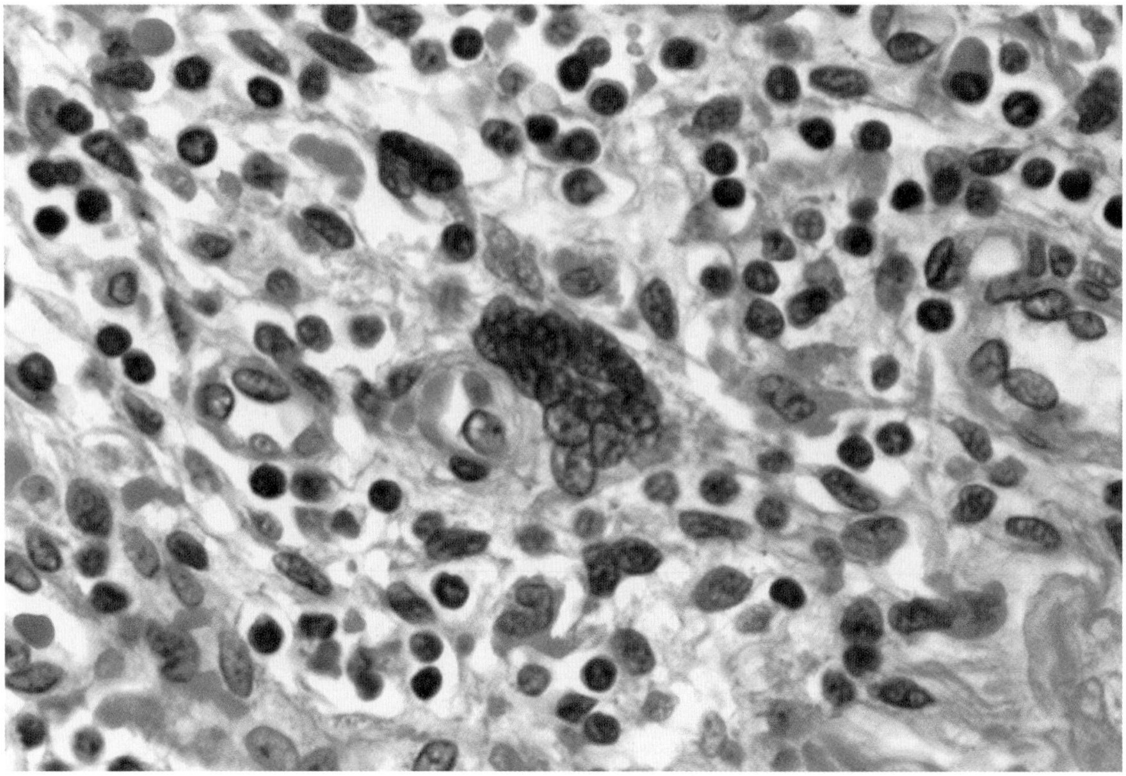

FIGURE 16-17 Lymph node of a patient who died of AIDS. Note the syncytial formation/multinucleated lymphocytes in the center of the photograph.

blood-brain barrier to the brain, triggering a cytokine and chemokine response that results in inflammation leading to neuronal injuries. Brain microglial nodules are nonspecific microscopic brain lesions often associated with fatal HIV infection. Current research is underway that looks at the roles of both viral and host factors in mediating HIV-induced neurological disease. Understanding HIV pathogenesis is essential to developing new approaches to AIDS therapy. What we know about HIV-1 biology and pathogenesis represents only a thin veneer on the surface of what needs to be known.

16.11 HIV Human Genetics/Resistance: The Smallpox Hypothesis

There are individuals who have had repeated exposures to HIV but either remain uninfected or, if infected, progress more slowly to full-blown AIDS. Recent studies have shown that there is a genetic component involved in host resistance. Individuals who are homozygous for a defective *CCR5* gene are virtually completely resistant to HIV infection (two defective copies are inherited, one from each parent). Individuals who are heterozygous for the defective *CCR5* gene progress more slowly to AIDS (only one defective *CCR5* inherited). In European populations, about 5% to 14% of individuals are heterozygous and 1% are homozygous in some populations, such as Caucasian North Americans. The allele is almost absent in African, Asian, Native American Indian, and Middle Eastern populations.

The CCR5 chemokine co-receptor is fundamental to establishing HIV infection. In the human host, secreted chemokines involved in inflammatory responses bind to chemokine receptors; for example, the following chemokines bind to the CCR5 receptor: RANTES, MIP-1α, and MIP-1β. These chemokines can block HIV infection if they are bound to CCR5. A defective *CCR5* gene that is missing 32 base pairs is a genetic **polymorphism** named CCR5Δ32. It codes for a protein that is severely truncated and cannot be detected at the cell surface. It blocks HIV entry because the viral gp120/gp41 cannot attach to it to infect cells. This defective co-receptor has been linked to the resistance to HIV infection (**FIGURE 16-18**).

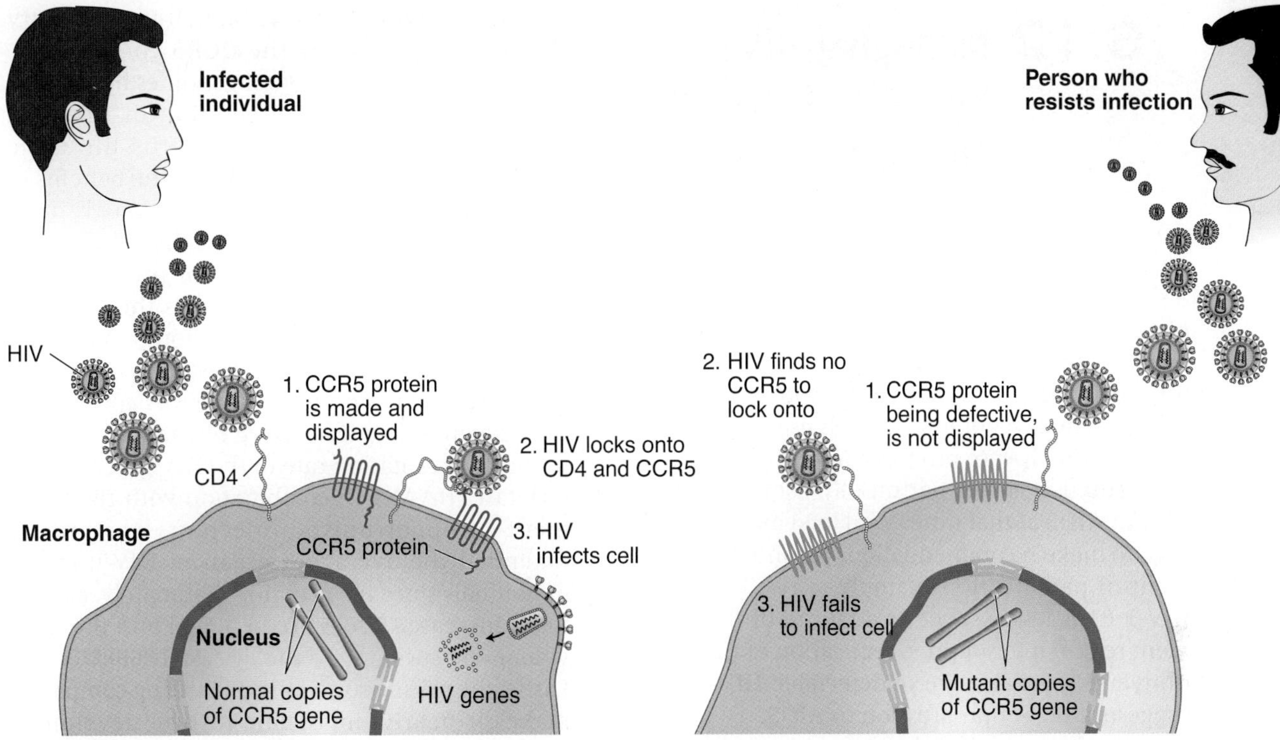

FIGURE 16-18 One mechanism for HIV resistance in humans. There are likely other genetic mutations that are also involved in resistance to HIV infection. Adapted from Coleman, F. "HIV," *Chem101: Chemistry,* http://www.wellesley.edu/Chemistry/Chem101/hiv/resist.gif.

The HIV resistance gene (CCR5Δ32) existed when the HIV pandemic began. Where did the gene come from? When did it arise? HIV is likely less than 100 years old. One hundred years is not sufficient enough time for **natural selection** to increase the frequency of a rare allele. In addition, the selective pressures caused by HIV should be much higher in Africa than in Europe and this is not the case. Scientists have estimated that selective pressures occurred in our history when deadly pathogens such as *Yersinia pestis* (causative agent of the bubonic plague or Black Death), *variola major* (smallpox), *Treponema pallidium* (syphilis), or *M. tuberculosis* killed millions of individuals approximately 600 to 700 years ago. Chemokine receptors play a central role in the immune response against many pathogens, particularly in the inflammatory response. Individuals who are CCR5Δ32 homozygous survive without immune system problems because of genomic redundancy of chemokine receptor functions; for example, CXCR4 or CCR2 compensate for CCR5.

The most popular theory among the genetic and medical literature is that the CCR5Δ32 allele led to increased survival during smallpox outbreaks that occurred in 14th century Europe. During this time, smallpox epidemics were ongoing, allowing for continuous natural selection. The defective allele was an unintended but welcomed consequence of HIV resistance. The frequency of the CCR5Δ32 variant in Europe is consistent with this scenario. Others speculated that the allele arose when the Black Death killed 25% to 40% of Europeans of all ages from 1346 to 1352 and 15% to 20% of Europe's population from 1665 to 1666. These were intermittent epidemics, though, and the bubonic plague virtually disappeared from Europe by 1750. Hence, there were no selective pressures to retain the rare allele. For this reason, the selective pressure of variola major epidemics more likely contributed to the higher frequency of CCR5Δ32 alleles. How would a CCR5Δ32 allele protect against variola infection?

Variola virus encodes a 35 kd secretory protein that binds to several CC-chemokine receptors. The variola C-terminal of the protein contains chemokine binding activity termed the smallpox virus-encoded chemokine receptor (SECRET) domain. During infection, the chemokine-binding protein is secreted and may act as a chemokine inhibitor, allowing the virus to evade the immune system, influencing smallpox pathogenesis. In turn, fortuitously, individuals who possess defective chemokine receptors would have been resistant to these effects by variola and HIV.

16.12 Managing HIV Patients: Antiviral Therapy

CD4+ T lymphocyte counts and HIV-1 RNA assays are used to track the progression of HIV infection. *A CD4+T lymphocyte count of less than 200 per µl of blood in an HIV-infected individual defines clinical AIDS.* **Virus load monitoring** is measured by the number of HIV-1 RNA genome copies per ml of plasma. The level of viral RNA is a powerful predictor of the disease progression to AIDS. Greater than 100,000 copies per ml of plasma will have clinical AIDS at that time. Viral load monitoring is used to make recommendations for antiviral treatment of patients; for example, a rise in plasma HIV-1 RNA levels suggests a failure of the treatment regimen to suppress replication of the virus. Conversely, an absence of detectable HIV-1 RNA suggests a slower progression to AIDS.

The first antiviral drug approved for HIV-1 treatment was zidovudine (AZT), a thymidine analog that inhibits reverse transcription of viral genomes. The benefits of this monotherapy were limited, though, because drug-resistant strains of the virus rapidly emerged. In 1996, virologist David Ho presented a mathematical model at the 12th World AIDS Conference in Vancouver. He postulated that highly active antiretroviral therapy might cure HIV infection within several years of treatment. HAART is based on administrating at least three different compounds to jointly block viral replication by inhibiting the viral protease and/or reverse transcriptase of the virus. It was hoped that the virus would be eliminated from the patient altogether. Unfortunately, there is a minority of cells that cannot be eliminated by chemotherapy.

Currently there are five classes of HIV inhibitors:

1. nonnucleoside or nucleoside **reverse transcriptase inhibitors** of the HIV reverse transcriptase
2. **protease inhibitors** that inhibit HIV maturation by blocking an HIV-specific protease, resulting in noninfectious virus
3. **fusion inhibitors** that block entry of HIV virions into a new host cell by blocking fusion of the viral gp41 envelope protein and host membrane
4. **integrase inhibitors** that block the HIV integrase activity, abolishing the integration of a DNA copy of the viral RNA genome into the host cell chromosome
5. **CCR5-blocking inhibitor**

The five classes of drugs are targeted at entry (blocking HIV gp41 or the CCR5 co-receptor), reverse transcription of viral genomes by the HIV reverse transcriptase, integration of the viral genome, and the end of the virus life cycle (**FIGURE 16-19**). From 1996 to 1997, the benefits of HAART were observed, but this therapy is not a cure. More drugs are commercially available to treat HIV (**TABLE 16-9**). The major problem with long-term antiretroviral therapy is the development of drug resistance and serious side effects that limit their usefulness. At least 15% of patients do not respond to protease inhibitors and experience severe side effects. Drug resistance is related to the high mutation rate of the HIV reverse transcriptase. HAART in combination with the treatment of opportunistic infections has greatly extended the average lifespan of HIV-infected individuals since its inception. Ultimately, as viral loads increase, T cells are depleted and the individual will succumb to the disease (**FIGURE 16-20**). Lastly, the pricing strategy used by drug companies makes it nearly impossible for the developing world to be able to afford HIV drugs; for example, the Roche drug Fuzeon is about $20,000/year/patient. HIV/AIDS advocacy groups are protesting these costs. The first CCR-5 blocking drug, Selzentry, was approved by the FDA for use in adult HIV patients in August 2007. The first integrase inhibitor, Isentress (raltegravir), was approved for use in HIV patients in October 2007.

16.13 Is an HIV Vaccine Possible?

Nearly 30 years after the discovery of HIV, there is still no vaccine or practical cure. It was proclaimed that Timothy Ray Brown was cured of an HIV infection in 2010 using an approach that is not practical or safe to cure most HIV patients (see Chapter 9, Virus File 9-1). An asymptomatic HIV patient harbors at least 1 million genetically distinct variants of HIV. An AIDS patient harbors 100 million genetically distinct variants of HIV. HIV variants can escape both antibody and T-cell immune responses. Will a vaccine ever be able to cope with HIV variability?

HIV causes a chronic infection in which there are reservoirs of virus in T cells, macrophages, and monocytes, where it may be integrated as a silent provirus. A vaccine needs to be able to prevent subsequent attacks by the same virus. In 2009, 33.3 million adults worldwide were living with HIV/AIDS. More than 85% of today's infections and deaths occur in developing countries where

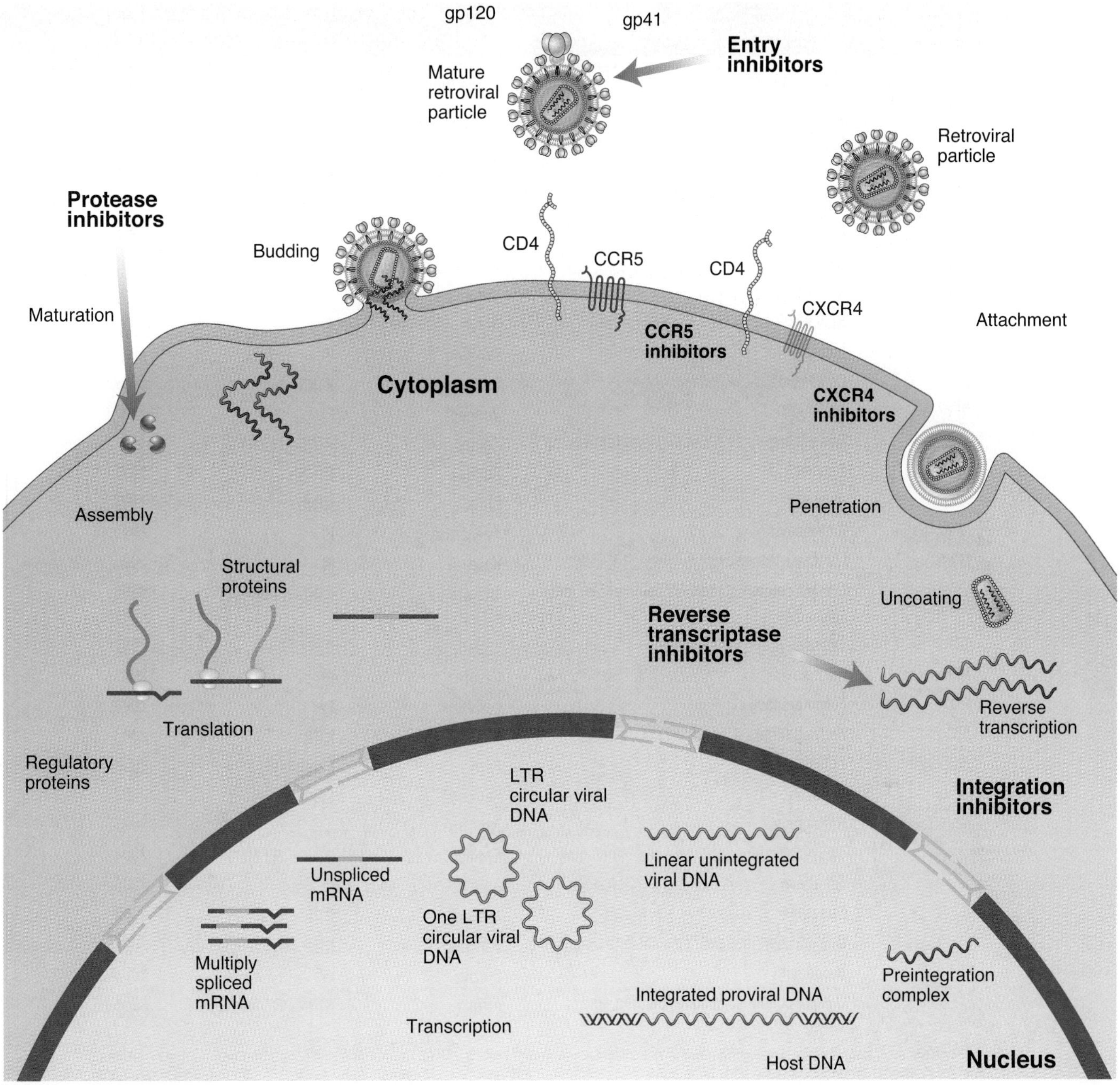

FIGURE 16-19 Antiviral drugs target different stages of the HIV-1 life cycle. Adapted from "Battling AIDS," *Nat Struct Biol* 6 (1999): 895–896.

significant health challenges continue to be a burden. HIV/AIDS threatens human welfare as well as social, political, and economic stability. Despite advances in HIV/AIDS research, the pandemic continues.

For the sake of simplicity, textbooks portray the HIV virion with a perfectly shaped enveloped virus containing gp120/gp41 spikes on its surface. The immune system faces a vastly different virus. About 1 in 1000 to 10,000 HIV particles are not defective and can reproduce in host cells. HIV particles carry numerous host cell-derived proteins within the HIV envelope. Many different serum proteins are found bound to the virion. Many of the original spikes have lost the gp120 subunits, leaving only the gp41 portion of gp160 remaining. Any remaining intact spikes may differ as much as 10% of amino acids between different HIV virions. This interferes with the affinity and neutralization of antibodies produced by the immune system to combat HIV (**FIGURE 16-21**). It makes the development of an HIV vaccine a complex research challenge. Promising vaccine candidates are advancing to clinical trials. Overall, a collective effort will be needed to slow the scourge of HIV/AIDS-worldwide.

TABLE 16-9	Approved Antiretroviral Therapy in 2010			
Abbreviation	**Generic Name**	**Trade Name**	**Drug Class***	**Year FDA Approval**
AZT, ZDV	Zidovudine, Lamivudine	Retrovir, Combivir	RTNI	1987
ddI	Didanosine	Videx	RTNI	1991
ddC	Zalcitabine	Hivid	RTNI	1992
d4T	Stavudine	Zerit	RTNI	1994
3TC	Lamivudine	Epivir	RTNI	1995
SQV	Saquinavir	Invirase, Fortovase	PI	1995
NVP	Nevirapine	Viramune	RTNNI	1996
RTV	Rionavir	Norvir	PI	1996
IDV	Indinavir	Crixivan	PI	1996
DLV	Delavirdine	Rescriptor	RTNNI	1997
NFV	Nelfinavir	Viracept	PI	1997
	One pill combines zidovudine and lamivudine	Combovir	RTNI + RTNI	1997
EFV	Efavirenz	Sustiva	RTNNI	1998
ABC	Abacavir	Ziagen	RTNI	1998
APV	Amprenavir	Agenerase	PI	1999
LPV/r	Lopinavir/Ritonavir	Kaletra	PI	2000
	One pill combines abacavir, lamivudine, and zidovudine	Trizivir	RTNI+RTNI+RTNI	2000
TDF	Tenovir	Viread	RTNI	2001
ATV	Atazanavir	Reyataz	PI	2003
FPV	Fosamprenavir	Lexiva	PI	2003
FTC	Emtricitabine	Emtriva	RTNI	2003
ENF, T-20	Enfuvirtide	Fuzeon	EI (gp41)	2003
	Maraviroc	Selzentry	EI (CCR5)	2007
	Raltegravir	Isentress	II	2007
	One pill combines abacavir and lamivudine	Epzicom	RTNI + RTNI	2004
	Tipranavir	Aptivus	PI	2005
	Etravirine	Intelence	RTNI	2008
	One pill combines emtricitabine and tenofovir	Truvada	RTNI + RTNI	2005
	Darunavir	Prezita	PI	2005
	One pill combines Sustiva, Emtriva, and Viread	Atripla	RTNNI+RTNI+RTNI	2006

*Reverse transcriptase nucleoside inhibitor (RTNI); reverse transcriptase non-nucleoside inhibitor (RTNNI); protease inhibitor (PI); entry inhibitor (EI); integrase inhibitor (II).

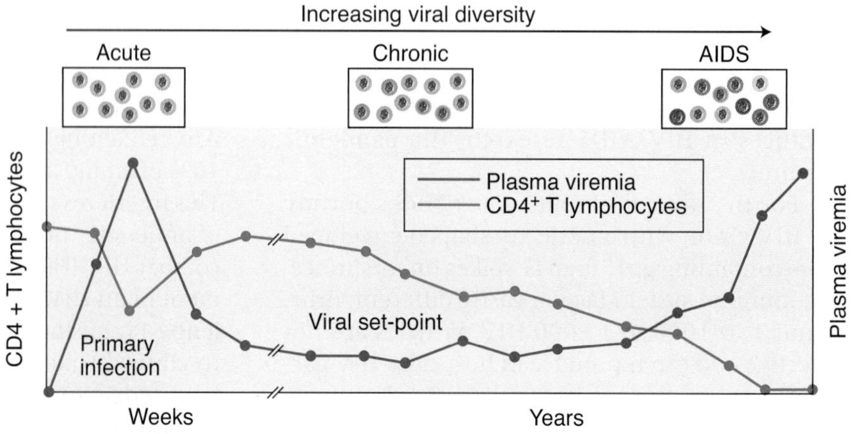

FIGURE 16-20 The progression of HIV infection based on T-cell counts and viral load. AIDS is defined as the stage when opportunistic infections and HIV-associated cancers occur. Note that as the infection proceeds, genetic diversity of HIV increases due to the error-prone reverse transcriptase. This situation challenges treatment of the patient. Adapted from Simon, V., and Ho, D. D. *Nat Rev Microbiol* 1 (2003): 181–190.

VIRUS FILE 16-2 Reasons Why HIV Is Invincible

SuperHIV!

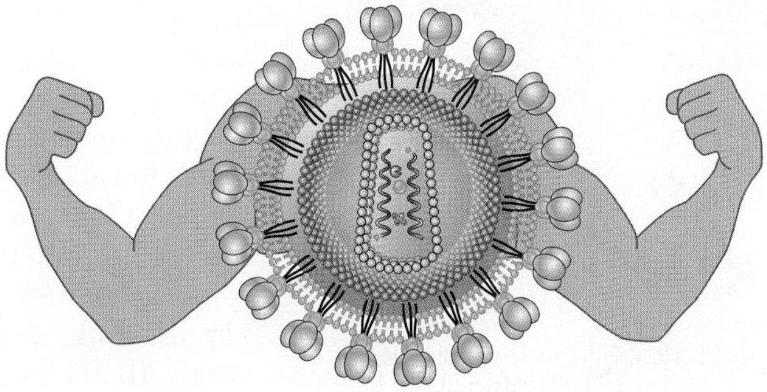

FIGURE VF 16-2 The high mutation rate of HIV makes infections difficult to treat and the development of a vaccine challenging.

At the time of this writing, more than 223,400 journal articles have been published on HIV research since the beginning of the AIDS pandemic. In his highly cited paper (see reference below), John M. Coffin outlined the reasons why HIV infection has been so challenging in the consideration of therapeutic strategies. Briefly listed are Coffin's theoretical insights into why HIV infection remains incurable and only modestly treatable despite international research efforts.

1. *HIV has an extraordinary mutation rate.* Reverse transcriptase does not have proofreading/error-correcting functions. Greater than 50% of viral genomes contain at least one error. Because thousands of viral generations of virus are present in an individual, this mutation rate may allow the virus to escape detection by the immune system and response to antiviral therapies. An individual who has been infected for ten years has viral genomes present in the body that are 3000 generations from the virus that initiated the infection. On average, the virus that is transmitted to someone else will be over 1000 generations removed from the initial infection.
2. The rapid mutation rate of the virus results in drug-resistant strains. *Antiviral drug resistance has been found for all compounds used as a therapy for HIV infection.* Hence, drug therapies only work for a short period of time due to the genetic diversity of HIV.
3. *Immunodeficiency is the result of the direct killing of infected cells either by HIV or the immune system.* Massive viral replication drives the pathogenic process that leads to immune degradation.

Reference

Coffin, J. M. 1995. "HIV population dynamics in vivo: Implications for genetic variation, pathogenesis, and therapy." *Science* 267:483–489.

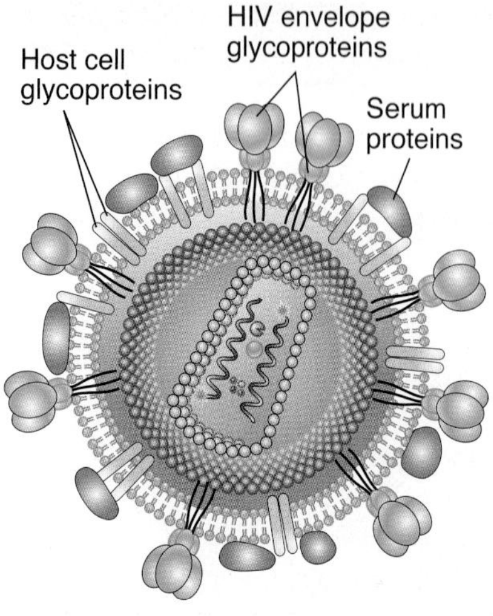

FIGURE 16-21 Schematic portrayal of a more realistic view of HIV virions found in an infected cell. The surfaces of HIV virions are not uniform, making it an even greater challenge for the immune system to clear the virus compared to other human viruses.

Summary

The AIDS pandemic began in the United States during the early 1980s. The first HIV-1 isolate was cultured in 1983 from the lymph node tissues of a patient. More than 223,400 medical research studies have been published since the first reported cases of what we now know as the acquired immunodeficiency syndrome (AIDS). AIDS affects men, women, and children in nearly every country in the world.

Several theories about the origin of HIV exist. The most widely accepted is the hunter theory or extensions of it. This theory is based on the premise that the AIDS pandemic is a viral zoonosis. A virus related to HIV-1 has been isolated from the common chimpanzee (*Pan troglodytes troglodytes*), which is native to Cameroon, Gabon, and the Democratic Republic of Congo. The virus is referred to as simian immunodeficiency virus (SIV_{cpz}). The genetic sequence of SIV_{cpz} and the early strains of HIV-1 are nearly identical. The belief is that AIDS arose when naturally infected chimps with SIV_{cpz} were killed and eaten and their blood entered cuts or wounds on the hunter. The SIV_{cpz} strain adapted in the human host to become HIV-1. Like HIV-1, HIV-2 entered the human population as the result of a zoonotic or cross-species transmission. It transferred from sooty mangabeys to humans during the first half of the 20th century in Western Africa.

In 33 countries, HIV incidence has fallen by more than 25% between 2001 and 2009. Twenty-two of these countries are in sub-Saharan Africa. Even the biggest epidemics in sub-Saharan African countries are stabilized or declining in new HIV cases. Several regions of the world do not fit this overall trend. Countries in Eastern Europe and Central Asian are HIV hotspots. The Russian Federation and Ukraine are among Eastern European countries with the fastest growing numbers of cases of HIV. The primary route of transmission is intravenous drug use; however, sexual contact and mother-to-child transmissions have begun to increase. Financial resources, treatment, and educational programs that include testing and support for those infected with HIV are especially needed to stop this global pandemic.

HIV is present in blood, semen, vaginal fluids, breast milk, and, in very low levels, saliva and tears. HIV is transmitted by:

- anal or vaginal intercourse
- sharing of HIV-contaminated needles (e.g., by injection drug users)
- blood transfusions using infected blood or blood-clotting factors (this is rare in the United States)
- organ transplants (rare)
- accidental needlestick injuries (healthcare workers are at risk while working with HIV-positive patients)
- congenital AIDS (before or during birth, or through breastfeeding)
- pre-mastication (rare)
- sharing of HIV-contaminated tattoo needles, razors, acupuncture needles, or ear-piercing implements.

There is no "silver bullet" method to prevent HIV infection. Abstinence from sex, barrier methods, microbiocides, and pre- and postexposure prophylaxis have the potential to stem HIV transmission. In the United States, HIV infection has hit the adult film industry, correctional facilities, and an increasing number of aging adults.

The clinical symptoms of HIV are divided into three phases:

1. primary HIV-1 infection
2. chronic asymptomatic phase
3. AIDS

Full-blown AIDS is the last phase of infection in which opportunistic infections increase and there is a higher risk for death. AIDS patients who are severely immunocompromised are managed with antimicrobial prophylactic regimens. Ultimately, HIV causes AIDS by depleting the number of CD4+ T helper lymphocytes. This results in the weakening of the immune system, which then is unable to combat infections and cancers associated with HIV infection. AIDS dementia is a serious neurological complication that occurs in as many as 20% of infected individuals.

HIV infection is diagnosed by serologic assays such as the ELISA, rapid HIV tests, and HIV home-testing systems. A Western blot must be used to confirm all positive test results. Commercial tests need to be updated on a continual basis to detect variant strains of HIV. HIV patients are managed with highly active antiretroviral therapy (HAART). Viral load testing and CD4+ T cell counts are used to make recommendations for antiviral treatment of patients. There are five classes of HIV inhibitors:

1. nonnucleoside or nucleoside reverse transcriptase inhibitors
2. protease inhibitors that inhibit HIV maturation, resulting in noninfectious virus
3. fusion inhibitors that bind gp41, blocking entry of HIV virions into a new host cell
4. Integrase inhibitors that block the HIV integrase activity, abolishing the integration of a DNA copy of the viral RNA genome into the host cell chromosome
5. CCR5-blocking inhibitor

HIV is a member of the *Retroviridae* family and is placed in the *Lentivirus* genus. HIV virions are enveloped with protruding glycoproteins gp120 (the head) and gp41 (the spike). Each virion contains two copies of +ssRNA genome that has a cellular $tRNA^{lys}$ bound to it, in addition to other bound viral proteins such as reverse transcriptase.

Viral entry is accomplished by binding CD4, a major cellular receptor present on the surface of T lymphocytes and a chemokine co-receptor (CXCR4 or fusin present on T cells and CCR5 present on macrophages). The rapid progression to AIDS has been associated with the switch in co-receptor preference from CCR5 to CXCR4. In the cytoplasm, the viral RNA genome is converted to DNA by the HIV-encoded reverse transcriptase. The viral DNA is transported to the nucleus where the viral genome may be integrated into the host cell's chromosome. The integrated HIV DNA is termed a provirus. The HIV viral genome is more complicated than that of a typical retrovirus. It codes for 15 proteins that serve as structural, regulatory, or accessory proteins during the life cycle of the virus.

HIV is an extraordinary virus. It replicates at an amazingly rapid speed and possesses a high mutation rate. The rapid mutation rate of the virus results in drug-resistant strains and strains that can evade the immune system. A very small percentage of individuals are resistant to HIV infection. It has been determined that individuals who are homozygous for a defective *CCR5* gene are virtually completely resistant to HIV infection. There are likely other genetic mutations that may also account for resistance to HIV infection. Despite over 30 years of medical research efforts, there is still no cure or vaccine for AIDS. Challenges remain in the area of therapeutic research. Continuous commitment and collaborative efforts may one day slow the scourge of HIV/AIDS in the world. Research efforts may shift toward long-term management of HIV infection rather than eradication of HIV from the body.

CASE STUDY QUESTIONS

These questions relate to the Case Study presented at the beginning of the chapter.

1. Magic Johnson's viral loads are undetectable. Is he cured, or can he be considered "virus-free"? Explain your answer.
2. If Magic Johnson were to discontinue HAART today, make a prediction as to what would happen to his viral loads and his CD4+ T cell counts. Explain your answer.
3. How could Magic Johnson have prevented his HIV infection?
4. List and explain the major routes of HIV transmission.
5. What is the difference between HIV and AIDS?
6. What does seropositive mean and how is a person tested for HIV?
7. What are some symptoms associated with the primary phase of HIV infection? What are the symptoms of AIDS?
8. The NBA has instituted safeguards for trainers and players. What are the precautions that are aimed to reduce the risk of contracting an HIV infection when treating open wounds on the basketball court?
9. If Magic Johnson were a less talented or charismatic NBA player, do you think the press and general public would have embraced his announcement and statements about HIV/AIDS in the same way? Discuss your answer.

References

Centers for Disease Control and Prevention. 1993. "Sexual risk behaviors of STD clinic patients before and after Earvin "Magic" Johnson's HIV-infection announcement—Maryland, 1991–1992." *MMWR* 42(3):45–48.

Johnson, E. "Magic," and Novak, W. 1992. *My Life.* New York: Random House.

Langer, L. M., et al. 1992. "Effect of Magic Johnson's HIV status on HIV-related attitudes and behaviors of an STD clinic population." *AIDS Educ Prev* 4(4):295–307.

Sigelman, C. K., Miller, A. B., and Derenowski, E. B. 1993. "Do you believe in Magic? The impact of "Magic" Johnson on adolescents' AIDS knowledge and attitudes." *AIDS Educ Prev* 5:2153–161.

Whalen, C. K., et al. 1994. "Preadolescents' perceptions of AIDS before and after Earvin Magic Johnson's announcement." *J Pediatr Psychol* 19(1):3–17; discussion 19–26.

Resources

Primary Literature and Reviews

Abbott, A. 2004. "Gut reaction. Consumers are stocking up on live yoghurts and fermented drinks that claim to improve health. But is there any science behind the marketing of these 'probiotic' products?" *Nature* 427:284–286.

Alejo, A., et al. 2006. "A chemokine-binding domain in the tumor necrosis factor receptor from variola (smallpox) virus." *Proc Natl Acad Sci USA* 103(15):5995–6000.

Allers, K., et al. 2011. "Evidence for the cure of HIV infection by CCR5Δ/Δ32 stem cell." *Blood* 117(10):2791–2799.

Ananda, S., et al. 2011. "HIV-negative drug addict diagnosed with AIDS and tuberculosis at autopsy: A case report and brief review of literature." *J Forensic Leg Med* 18(3):136–138.

Bailies, E., et al. 2003. "Hybrid origin of SIV in chimpanzees." *Science* 300:1713.

Bar, K. J., et al. 2010. "Wide variation in the multiplicity of HIV-1 infection among injection drug users." *J Virol* 84(12):6241–6247.

Barre-Sinoussi, F., et al. 1983. "Isolation of a T-lymphotropic retrovirus from a patient at a risk for acquired immune deficiency syndrome (AIDS)." *Science* 220(4599): 868–871.

Bellows, M. L., et al. 2010. "Discovery of entry inhibitors for HIV-1 via a new de novo protein design framework." *Biophys J* 99(100):3445–3453.

Bernard, M. A., et al. 2011. "HIV transmitted from a living donor—New York City, 2009." *MMWR* 60(10):297–301.

Berry, N., et al. 2001. "Analysis of oral polio vaccine CHAT stocks." *Nature* 410:1046–1047.

Blanchard, A., et al. 1998. "Molecular evidence for nosocomial transmission of human immunodeficiency virus from a surgeon to one of his patients." *Virology* 72(5):4537–4540.

Blancou, P., et al. 2001. "Polio vaccine samples not linked to AIDS." *Nature* 410:1045–1046.

Boyd, M. R., et al. 1997. "Discovery of cyanovirin-N, a novel human immunodeficiency virus-inactivating protein that binds viral surface envelope glycoprotein gp120: Potential applications to microbiocide development." *Antimicrob Agents Chemother* 41(7):1521–530.

Brennan, M., et al., eds. 2009. *Older Adults with HIV: An In-Depth Examination of an Emerging Population.* New York: Nova Science Publishers, Inc.

Britvan, L., et al. 1996. "Identification of HIV-1 group O infection—Los Angeles County, California, 1996." *MMWR* 45(26):561–565.

Buckhard, M., and Dean, G. A. 2003. "Transmission and immunopathogenesis of FIV in cats as a model for HIV." *Curr HIV Res* 1:15–29.

Bukrinskaya, A. G. 2004. "HIV-1 assembly and maturation." *Arch Virol* 149(6):1067–1082.

Burruano, L., et al. 2009. "HIV/AIDS epidemic in Eastern Europe: Recent developments in the Russian Federation and Ukraine among women." *Gender Med* 6(1): 277–289.

Busch, M. P., and Kleinman, S. H. 2000. "Nucleic acid amplification testing of blood donors for transfusion-transmitted infectious diseases." *Transfusion* 40: 143–159.

Centers for disease control and prevention (CDC). 2001. "Updated U.S. Public Health Service guidelines for the management of occupational exposures to HBV, HCV, and HIV and recommendations for postexposure prophylaxis." *MMWR* 50(RR11):1–42.

Chang, T. L.-Y., et al. 2003. "Inhibition of HIV infectivity by a natural human isolate of *Lactobacillus jensenii* engineered to express functional two-domain CD4." *Proc Natl Acad Sci USA* 100:11672–11677.

Chitnis, A., Rawls, D., and Moore, J. 2000. "Origin of HIV type 1 in Colonial French Equatorial Africa?" *AIDS Res Hum Retroviruses* 16(1):5–8.

Cholewinska, G. 2010. "Healthcare and antiretroviral treatment in HIV-infected detained persons at the penitentiary units." *HIV AIDS Rev* 7(4):10–12.

Citterio, P., and Rusconi, S. 2007. "Novel inhibitors of the early steps of the HIV-1 life cycle." *Expert Opin Investig Drugs.* 16(1):11–23.

Coffin, J. M. 1995. "HIV population dynamics *in vivo:* Implications for genetic variation, pathogenesis, and therapy." *Science* 267:483–489.

Cohen, J. 2006. "Novel attacks on HIV move closer to reality." *Science* 311:943.

Cohen, J. 2010. "Late for the epidemic HIV/AIDS in Eastern Europe." *Science* 329: 160–167.

Cohen, J. 2010. "Praised Russian prevention program faces loss of funds." *Science* 329:168.

Cohen, J. 2010. "Law enforcement and drug treatment: A culture clash." *Science* 329:169.

Cohen, J. 2010. "HIV/AIDS in Eastern Europe: HIV moves in on homeless youth." *Science* 329:170–171.

Cohen, J. 2010. "HIV/AIDS in Eastern Europe: Reducing HIV infection and abandonment of babies." *Science* 329:172.

Cohen, J. 2010. "HIV/AIDS investigators few and far between." *Science* 329:173.

Coleman, J. K., et al. 2005. "HIV-1 p24 vaccine protects cats against feline immunodeficiency virus infection." *AIDS* 19(14):1457–1466.

Delpech, V., Gahagan, J. 2009. "The global epidemiology of HIV." *Medicine* 37(7):317–320.

Deng, J., et al. 2007. "Design of second generation HIV-1 integrase inhibitors." *Curr Pharm Des* 13(2):129–141.

Dicko, M., et al. 2000. "Safety of immunization injections in africa: Not dimply a problem of logistics." *Bull World Health Organ* 178:163–169.

Douglas, J. M., et al. 1997. "Contraceptive practices before and after an intervention promoting condom use to prevent HIV infection and other sexually transmitted diseases among women—Selected U.S. sites, 1993–1995." *MMWR* 46(17):373–377.

Drucker, E., Alcabes, P. G., and Marx, P. A., 2001. "The injection century: Massive unsterile injections and the emergence of human pathogens." *Lancet* 358: 1989–1992.

Fiebig, E. W., et al. 2003. "Dynamics of HIV viremia and antibody seroconversion in plasma donors: Implications for diagnosis and staging of primary infection." *AIDS* 17:1871–1879.

Freed, E. O., and Mouland, A. J. 2006. "The cell biology of HIV-1 and other retro-viruses." *Retrovirology* 3:77.

Gallo, R. C. 2002. "The early years of HIV/AIDS." *Science* 298:1728–1730.

Gallo, R. C., and Montagnier, L. 2003. "The discovery of HIV as the cause of AIDS." *N Engl J Med* 349(24): 2283–2285.

Gallo, R. C., et al. 1984. "Frequent detection and isolation of cytopathic retroviruses (HTLV-III) from patients with AIDS and at risk for AIDS." *Science* 224(4648): 500–503.

Galvani, A. P., and Novembre, J. 2005. "The evolutionary history of the CCR5-Δ32 HIV-resistance mutation." *Microbes Infect* 7:302–309.

Galvani, A. P., and Slatkin, M. 2003. "Evaluating plague and smallpox as historical selective pressures for the CCR5-D32 HIV-resistance allele." *Proc Natl Acad Sci USA* 100(25):15276–15279.

Gao, F., et al. 1999. "Origin of HIV-1 in the chimpanzee *Pan troglodytes troglodytes.*" *Nature* 387:436–444.

Gaur, A. H., et al. 209. "Practice of feeding pre-masticated foods to infants: A potential risk factor for HIV transmission." *Pediatrics* 124:2.

Girard, F., et al. 2010. "Universal access in the fight against HIV/AIDS." *Science* 32:147–149.

Gottlieb, M. S., et al. 1981. "*Pneumocystis* pneumonia—Los Angeles." *MMWR* 30(21):250–252.

Hafeez, S., et al. 2011. "Infant feeding practice of pre-mastication: An anonymous survey among human immunodeficiency virus-infected mothers." *Arch Pediatr Adolesc Med* 165(1):92–93.

Hahn, B. H., et al. 2000. "AIDS as a zoonosis: Scientific and public health implications." *Science* 287:607–614.

Hall, H. I. 2008. "Estimation of HIV incidence in the United States." *JAMA* 300(5):520–529.

Hedrick, P. W., and Verrelli, B. C. 2006. "Ground truth for selection on CCR5-Delta32." *Trends Genet* 22(6): 293–296.

Heimer, R., et al. 2010. "Comparing sexual risks and patterns of alcohol and drug use between injection drug users (IDUs) and non-IDUs who report sexual partnerships with IDUs in St. Petersburg, Russia." *BMC Public Health* 10:676.

Holden, C. 1988. "Curbing Soviet disinformation." *Science* 242:665.

Hutter, G., et al. 2009. "Long-term control of HIV by CCR5 Delta32/Delta32 stem cell transplantation." *N Engl J Med* 360(7):692–698.

Jasny, B., et al. 2010. "HIV/AIDS: Eastern Europe." *Science* 329:159.

Jose, M., et al. 2005. "Stability of HCV, HIV-1 and HBV nucleic acids in plasma samples under long-term storage." *Biologicals* 33:9–16.

Keele, B. F., et al. 2006. "Chimpanzee reservoirs of pandemic and nonpandemic HIV-1." *Science* 303(5786): 523–528.

Kourtis, A. P. 2006. "Prevention of sexually transmitted human immunodeficiency virus (HIV) infection in adolescents." *Curr HIV Res* 4(2):209–219.

Kowala-Piaskowska, A., et al. 2010. "Woman living with HIV/AIDS and her child—epidemiological data." *HIV AIDS Rev* 9(3):72–78.

Kucirka, L. M., et al. 2011. "Risk of window period HIV infection in high infectious risk donors: Systematic review and meta-analysis." *Am J Transplant* (in press). DOI:10.1111/j.1600-6143.2010.03329.x

Laffoon, B., et al. 2010. "HIV transmission through transfusion—Missouri and Colorado, 2008." *MMWR* 59(41):1335–1339.

Lalani, A. S., et al. 1999. "Use of chemokine receptors by poxviruses." *Science* 286:1968–1971.

Lansky, A., et al. 2010. "Epidemiology of HIV in the United States." *J Acquir Immune Defic Syndr* 55(Suppl 2): S64–S68.

Laurent, C., et al. 2006. "Antiretroviral drug resistance and routine therapy, Cameroon." *EID* 12(6):1001–1004.

Lemey, P., et al. 2003. "Tracing the origin and history of the HIV-2 epidemic." *Proc Natl Acad Sci USA* 100(11): 6588–6592.

Leonard, J. T., and Roy, K. 2006. "The HIV entry inhibitors revisited." *Curr Med Chem* 13(8):911–934.

Letvin, N. L. 2009. "Moving forward in HIV vaccine development." *Science* 326:1196.

Levison, J., et al. 2011. "Think twice before recommending pre-masticated food as a source of infant nutrition." *Matern Child Nutr* 7(1):104.

Lewin, S. R., and Rouzioux, C. 2011. "HIV cure and eradication: How will we get from laboratory to effective clinical trials?" *AIDS* 25(7):885–897.

Liu, R., et al. 1996. "Homozygous defect in HIV-1 coreceptor accounts for resistance of some multiply-exposed individuals to HIV infection." *Cell* 86(3): 367–377.

Maruschak, L. M. 2004. "HIV in prisons and jails, 2002." *Bureau of Justice Statistics Bulletin,* Document NCJ 205333.

Marx, P. A., Alcabes, P. G., and Drucker, E. 2001. "Serial human passage of simian immunodeficiency virus by unsterile injections and the emergence of epidemic human immunodeficiency virus in Africa." *Philos Trans R Soc B Biol Sci* 356(1410):911–920.

McMichael, A. J., and Hanke, T. 2003. "HIV vaccines 1983–2003." *Nat Med* 9(7):874–880.

Miedema, F. 2002. "Review of science fictions: A scientific mystery, a massive coverup, and the dark legacy of Robert Gallo." *Nat Med* 8(7):655.

Montagnier, L. 2003. "A History of HIV discovery." *Science* 298:1727–1728.

Niccolai, L. M., et al. 2010. "Estimates of HIV incidence among drug users in St. Petersburg, Russia: Continued growth of a rapidly expanding epidemic." *Eur J Public Health* doi: 10.1093/eurpub/ckq115.

Novembre, J., Galvani, A. P., and Slatkin, M. 2005. "The geographic spread of the CCR5Δ32 HIV-resistance allele." *PLoS Biol* 3(11):1954–1962.

Pace, M. J., et al. 2011. "HIV reservoirs and latency models." *Virology* 411:344–354.

Paul, S. M., et al. 1998. "Emerging issues in AIDS: Variant strains of human immunodeficiency virus and its subtypes." *Med Lab Obs* 30(2):32–36.

Piedade, J., et al. 2000. "Longstanding presence of HIV-2 infection in Guinea-Bissau (West Africa)." *Acta Tropica* 76:119–124.

Poulsen, A. G., Aaby, P., Jensen, H., and Dias, F. 2000. "Risk factors for HIV-2 seropositivity among older people in Guinea-Bissau, a search for the early history of HIV-2 infection." *Scand J Infect Dis* 32(2):169–175.

Pusch, O., et al. 2005. "Bioengineering lactic acid bacteria to secrete the HIV-1 virucide cyanovirin." *J Acquir Immune Defic Syndr Hum Retrovirol* 40(5):512–520.

Rambeau, A., et al. 2001. "Phylogeny and the origin of HIV-1." *Nature* 410:1047–1048.

Relf, M. V., et al. 2011. "Essential nursing competencies related to HIV and AIDS." *J Assoc Nurses AIDS Care* 22(1S):e5–e40.

Rotblatt, H., et al. 2005. "HIV transmission in the adult film industry—Los Angeles, California, 2004." *MMWR* 54(37):923–926.

Schim, M. F., et al. 2001. "HIV-2 does not protect against HIV-1 infection in a rural community in Guinea-Bissau." *AIDS* 15:2303–2310.

Simon, V., and Ho, D. D. 2003. "HIV-1 dynamics *in vivo:* implications for therapy." *Nat Rev Microbiol* 1(3): 181–190.

Spaulding, A., et al. 2002. "Human immunodeficiency virus in correctional facilities: a review." *Clin Infect Dis* 35:305–312.

Steigbigel, R. T., et al. 2008. "Raltegravir with optimized background therapy for resistant HIV-1 infection." *N Engl J Med* 359(4): 354.

Stephens, J. C., et al. 1998. "Dating the origin of the CCR5-Δ32 AIDS-resistance allele by coalescence of haplotypes." *Am J Hum Genet* 62:1507–1515.

Stevenson, M. 2003. "HIV-1 pathogenesis." *Nat Med* 9:853–860.

Stremlau, M., et al. 2004. "The cytoplasmic body component TRIMa restricts HIV-1 infection in old world monkeys." *Nature* 427:848–852.

Taussig, J., et al. 2006. "HIV transmission among male inmates in a state prison system—Georgia, 1992–2005." *MMWR* 55(15):421–426.

Teixeira, C., et al. 2011. "Viral surface glycoproteins, gp120 and gp41, as potential drug targets against HIV-1: Brief overview one quarter of a century past the approval of Zidovudine, the first anti-retroviral drug." *Eur J Med Chem* 46:979–992.

Thorne, C., et al. 2010. "Central Asia: Hotspot in the worldwide HIV epidemic." *Lancet Infect Dis* 10:479–488.

Trono, D., et al. 2010. "HIV persistence and the prospect of long-term drug-free remissions for HIV-infected individuals." *Science* 329:174–180.

Vandamme, A.-M. 2006. "Anti-HIV yoghurt?" *AIDS Rev* 8:44.

Wang, H., et al. 2011. "Knowledge of HIV seropositivity is a predictor for initiation of illicit drug use: Incidence of drug use initiation among female sex workers in a high-prevalence area in China." *Drug Alc Depend* doi:10.1016/j.drugalcdep.2011.02.006.

Wayengera, M. 2011. "Targeting persistent HIV infection: Where and how, if possible?" *HIV AIDS Rev* 10:1–8.

Weiss, R. A. 2001. "Polio vaccines exonerated." *Nature* 410:1035–1036.

Working Group on estimation of HIV prevalence in Europe 2011. "HIV in hiding: Methods and data requirements for the estimation of the number of people living with undiagnosed HIV." *AIDS* 25(8):1017–1023.

Worobey, M., et al. 2010. "Island biogeography reveals the deep history of SIV." *Science* 329:1487.

Zhu, T., et al. 1998. "An African HIV-1 sequence from 1959 and implications for the origin of the epidemic." *Nature* 391:594–597.

Popular Press

Barnett, T. 2006. *AIDS in the Twenty-First Century.* New York: Palgrave McMillion.

Burkett, E. 1996. *The Gravest Show on Earth: America in the Age of AIDS.* London: Picador.
Crewdson, J. 2003. *Science Fictions: A Scientific Mystery, a Massive Cover-up and the Dark Legacy of Robert Gallo.* Boston: Back Bay Books.
Culshaw, R. 2007. *Science Sold Out: Does HIV Really Cause AIDS?* Berkeley, CA: North Atlantic Books.
Curtis, T. 1992. "The Origin of AIDS." *Rolling Stone Magazine,* pp. 54–59, 61, 106, 108.
Duesberg, P. H. 1997. *Inventing the AIDS Virus.* Washington, DC: Regnery Publishing.
Engel, J. 2006. *The Epidemic: A Global History of AIDS.* Washington D.C.: Smithsonian Books.
Epstein, H. 2008. *The Invisible Cure: Why We Are Losing the Fight Against AIDS.* London: Picador.
Farber, C. 2006. *Serious Adverse Events: An Uncensured History of AIDS.* Hoboken, NJ: Melville House.
Gaur, A. H., et al. 2009. "Practice of feeding pre-masticated food to infants: A potential risk factor for HIV transmission." *Pediatrics* 124:2.
Hooper, E. 1999. *The River: A Journey to the Source of HIV and AIDS.* London: Penguin.
Iliffe, J. 2006. *The African AIDS Epidemic: A History.* Athens, Ohio: Ohio University Press.
Jefferson, D. 2006. "How AIDS changed America." *Newsweek,* May 15, 2006, pp. 36–53.
Johnson, E. "Magic," and Novak, W. 1992. *My Life.* New York: Random House.
Laffoon, B., et al. 2010. "HIV transmission through transfusion—Missouri and Colorado, 2008." *MMWR* 59(41):1335–1339.
Markel, H. 2004. *When Germs Travel: Six Major Epidemics That Have Invaded America Since 1900 and the Fears They Have Unleashed.* New York: Pantheon.
Nolen, S. 2008. *28: Stories of AIDS in Africa.* Virginia: Walker & Company.
Pisani, E. 2009. *The Wisdom of Whores: Bureaucrats, Brothels and the Business of AIDS.* New York: W. W. Norton & Company.
Rom, M. 1997. *Fatal Extraction: The Story Behind the Florida Dentist Accused of Infecting His Patients with HIV and Poisoning Public Health.* Hoboken, NJ: John Wiley & Sons.
Sherman, I. W. 2007. *Twelve Diseases That Changed Our World.* Washington DC: ASM Press.
Shilts, R.1987. *And the Band Played On: Politics, People and the AIDS Epidemic.* New York: St. Martin's Press.

Video Productions

AIDS Evolution of an Epidemic. 2008. Howard Hughes Medical Institute.
Umoja, the Village Where Men are Forbidden. 2008. Films for the Humanities.
NOVA: Epidemic—Ebola, AIDS, Bird Flu and Typhoid. 2007. WGBH, Boston.
HIV & Me: A Global Exploration with Stephen Fry. 2007. Films for the Humanities.
HIV & Me: Fear, Ignorance, and Education. 2007. Films for the Humanities.
HIV & Me: Medical Advances and Setbacks. 2007. Films for the Humanities.
Rampant: How a City Stopped a Plague. 2007. Films for the Humanities.
Fighting AIDS. 2006. CBS. 60 Minutes.
Helping the Youngest Victims of AIDS: Spotlight on South Africa. 2006. Films for the Humanities.
Out of Control: AIDS in Black America. 2006. Films for the Humanities.
Yesterday. 2006. HBO, Home Video.
Portraits in Human Sexuality: Sexual Orientation. 2006. Films for the Humanities.
Age of AIDS. 2006. PBS, Frontline.
RX For Survival: A Global Health Challenge. 2005. PBS.
AIDS and Other Epidemics. 2004. Films for the Humanities.
Three Girls I Know: Intimate Stories of Personal Responsibility. 2004. Films for the Humanities.
The United Nations: Working for All of Us. 2004. Films for the Humanities.
Absolutely Positive. 2004. New Video Group.
Breaking the Silence: Lifting the Stigma of HIV/AIDS in Ethiopia. 2003. Films for the Humanities.
Deadly Desires. 2003. Films for the Humanities.
Pandemic: Facing AIDS. 2003. HBO.
Breaking the Silence: Lifting the Stigma of HIV/AIDS in Ethiopia. 2003. Films for the Humanities.
One Week. 2002. First Look Pictures.
AIDS: A Global Crisis. 2002. Films for the Humanities.
The Bottom Line: Privatizing the World. 2002. Films for the Humanities.
New Blood. 2002. Films for the Humanities.
Pediatric AIDS. 2002. Films for the Humanities.
Exposed: The Continuing Story of AIDS. 2001. Films for the Humanities.
And the Band Played On. 2001. HBO Home Video.
Longtime Companion. 2001. MGM.
Changing Faces. 2000. Films for the Humanities.
Surviving AIDS. 1999. NOVA Production, PBS.
AIDS in America History. 1998. Films for the Humanities.
Philadelphia. 1997. Sony Pictures.

eLearning

go.jblearning.com/shors2

The site features eLearning, an online review area that provides quizzes and other tools to help you study for your class. You can also follow useful links for in-depth information, or just find out the latest virology and microbiology news.

CHAPTER 17

Hepatitis Viruses

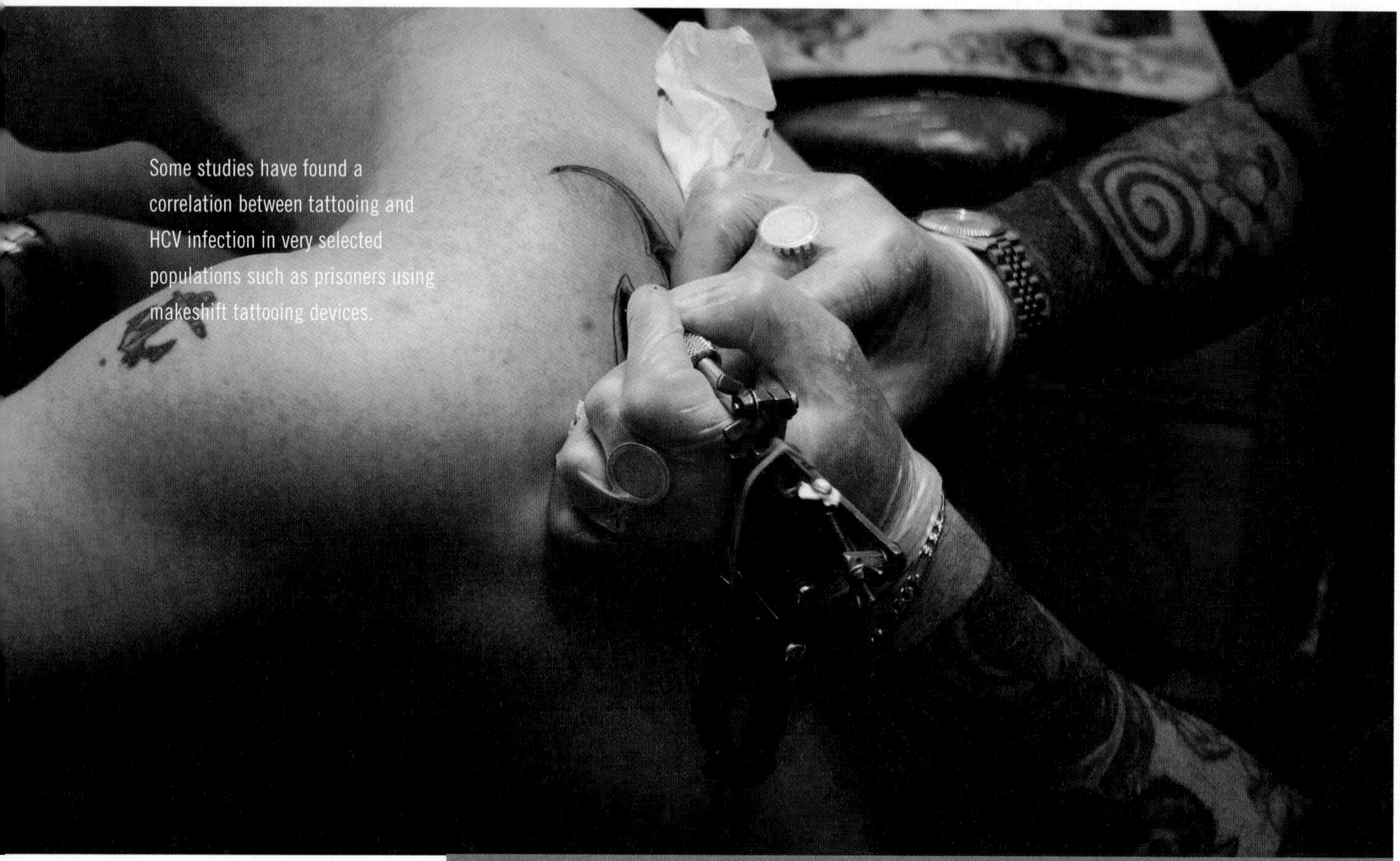

Some studies have found a correlation between tattooing and HCV infection in very selected populations such as prisoners using makeshift tattooing devices.

OUTLINE

"Love is like a virus. It can happen to anybody at any time."

Maya Angelou, American poet, educator, civil rights activist

CASE STUDY: CONTAMINATED SALSA

In 2003, over 600 individuals who dined at a Mexican restaurant in Monaca, Pennsylvania, fell ill with hepatitis A. At least 13 of the food service workers also became ill. Three individuals died and over 100 were hospitalized. Epidemiologists from the Centers for Disease Control and Prevention (CDC) were charged with determining the cause of the outbreak, which was determined to be the largest of its kind in the United States. Their ongoing investigation tested foods prepared at the restaurant and stool samples from staff and restaurant customers. The CDC determined that restaurant workers were not the source of the outbreak; it was the uncooked green onions that were used in making salsa.

17.1 The History of Viral Hepatitis

Hepatitis means "inflammation of the liver." The classic symptom of hepatitis is **jaundice**, a yellowing of the skin and eyes caused by too much **bilirubin** in the blood (refer to the clinical features of the virus for further explanation). Hepatitis is caused by a number of viruses and on occasion is caused by alcohol or prescription drugs. Viral hepatitis is one of the most important public health threats worldwide. One of the most notorious cases of viral hepatitis is that of the popular American stunt performer, Robert Craig "Evel" Knievel, Jr. Knievel was diagnosed with a hepatitis C infection in 1993. He contracted hepatitis C from contaminated blood received during one of his 14 reconstructive surgeries used to repair broken bones acquired during a stunt. Knievel received a liver transplant in 1999 and died in 2007 of pulmonary fibrosis at the age of 69. In 2002, actress from the television series *Baywatch* and former Playboy model Pamela Anderson claimed she contracted hepatitis C from sharing a tattoo needle with her ex-husband, rock musician Tommy Lee.

Epidemics of jaundice or hepatitis have been recorded throughout history, especially during times of war; for example, outbreaks occurred among Napoleon's troops during the Egyptian and Russian campaigns that took place in the early 1800s. The crowding and unsanitary conditions of military encampments created an ideal environment for hepatitis A viruses to be transmitted. In 1821, Napoleon Bonaparte died in exile on the Island of St. Helena. There are numerous theories about his cause of death, including stomach cancer, arsenic poisoning, inappropriate medical treatment, and hepatitis—an endemic disease at St. Helena.

Approximately 70,000 troops fell ill with hepatitis during the U.S. Civil War. Hepatitis was referred to as "camp jaundice" during the trench warfare battles in the Mediterranean theater in World War I (**FIGURE 17-1**). During World War II, there were at least two significant outbreaks of hepatitis. The first occurred in 1942, when approximately 330,000 soldiers were vaccinated against yellow fever virus. Of these vaccinees, 50,000 experienced jaundice and 62 died. It was believed that a virus was present in the serum used to manufacture the vaccine. After those lots of vaccines were destroyed and a human serum-free vaccine was manufactured, the cases of jaundice stopped. Years later it was confirmed that these soldiers had **serum hepatitis** caused by hepatitis B virus.

Later jaundice epidemics during World War II occurred during the invasion of North Africa and subsequent invasion of the Italian mainland in the Mediterranean theater of operations (MTO). Even though the antibiotic penicillin was available at this time to treat bacterial infections and medicine in general had improved, diseases still proliferated in combat situations, making treatment of the troops difficult. Malaria and hepatitis epidemics incapacitated 55,000 American assault troops, thus delaying D-Day. The viral hepatitis outbreaks were caused by **infectious hepatitis** (hepatitis A virus) during the occupation of recaptured land, amid battlefields that were littered with human excrement and corpses. Hepatitis A virus is found in the feces of people infected with hepatitis A virus and is spread by fecal–oral transmission. Unsanitary conditions were common at the frontlines of war. There were no bathroom facilities at the battle lines, and troops could not move behind the lines when they needed to go to the bathroom. Hence, feces got on their boots and clothing and into the bottom of the next shell hole or foxhole they dashed into. Soldiers detected the proximity of the frontlines because of the increasing intensity of odors. Infectious hepatitis was the greatest cause of disabling disease among U.S. and British forces in the MTO.

FIGURE 17-1 Many soldiers fighting in World War I contracted viral hepatitis during trench warfare, where infectious hepatitis infection was common. Field sanitation during heavy combat was inadequate, and hepatitis spread by a fecal–oral mode of transmission.

Today hepatitis C is a major problem in U.S. military veterans. Approximately 8% to 9% of Veteran Affairs (VA) medical center patients are positive for hepatitis C antibodies. Some VA hospitals have had as high as 10% to 20% of patients with hepatitis C antibodies.

Hepatitis C in Vietnam-era veterans, who are at the front lines of the hepatitis C epidemic, is an ongoing problem, and most studies have focused on this group. There are multiple risk factors for hepatitis C virus transmission among veterans in this time period, including:

- combat casualties who survived multiple blood transfusions in an era prior to hepatitis C virus screening of the blood supply
- intravenous drug experimentation (heroin use)
- blood exposure through sharing of razors or other nonsterile instruments
- history of tattooing
- history of prostitution (or more than 15 sexual partners)
- blood/body fluid exposure to healthcare and combat personnel
- receipt of immune globulin for hepatitis A prophylaxis in an era prior to hepatitis C virus screening of the blood supply

Challenges remain in the treatment of hepatitis C infection and prevention. Testing and improved treatments will benefit all individuals who suffer from hepatitis C infection.

To date, there are eight members of the group of viruses referred to as the hepatitis viruses. **FIGURE 17-2** is a flow diagram of how these viruses were originally distinguished from each other.

17.2 Epidemiology

Hepatitis can be caused by viral infections, trauma, alcohol abuse, or drug-induced toxicity. Viral hepatitis is the most important cause of hepatitis on a global scale. Yellow fever virus, herpes simplex viruses, cytomegaloviruses, and Epstein-Barr virus can cause ***secondary hepatitis***, but a range of unrelated human pathogens termed the "**hepatitis viruses**" cause the vast majority of virally induced ***primary hepatitis*** cases. Primary hepatitis occurs immediately after an initial infection by a hepatitis virus in contrast to secondary hepatitis. Secondary

FIGURE 17-2 Historical perspective illustrating the separation of viruses that cause primary hepatitis.

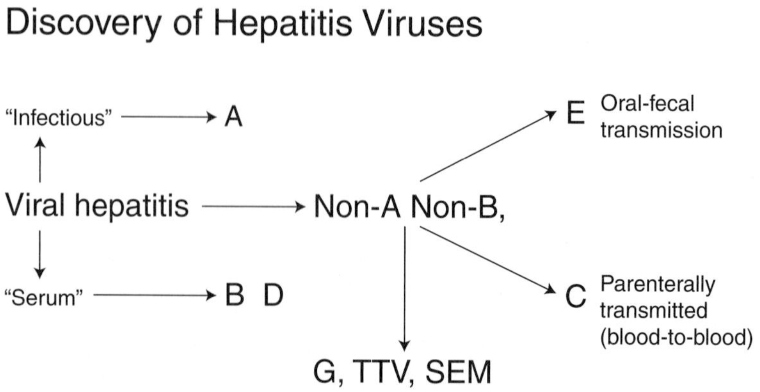

hepatitis symptoms follow after another type of infection, disease (e.g., autoimmune disease), treatment (e.g., drug toxicity), or other causes (e.g., alcohol abuse). This chapter focuses on the hepatitis viruses. *These viruses are quite diverse genetically and taxonomically.* They are grouped together only because they share a liver tissue tropism, causing inflammation and necrosis of the liver.

Each hepatitis virus is named with a letter of the alphabet with the exception of the newly identified transfusion transmission virus (TTV) and SEN viruses. Although the hepatitis viruses cause similar disease, they are widely varied in their origin and are evolutionarily distinct (**TABLE 17-1**).

For years there were three main hepatitis viruses: hepatitis A virus (HAV), the cause of infectious hepatitis; hepatitis B virus (HBV), the cause of serum hepatitis; and hepatitis "non-A, non-B" viruses. Diagnostic tests were developed to identify HAV and HBV in the 1970s. A new "antigen" found in liver specimens of some patients infected with HBV led to the identification of hepatitis D virus (HDV) in 1977. HDV was only seen in a fraction of patients with HBV and it was often associated with a more severe course of hepatitis. Hepatitis non-A, non-B viruses were considered the cause of all other not identified viral causes of hepatitis. In 1989, hepatitis C virus (HCV), the agent that caused non-A, non-B hepatitis, was discovered using modern molecular techniques. Since that time, additional non-A, non-B hepatitis viruses have been discovered: hepatitis E, G, and TTV. Hepatitis A, B, and C viruses cause over 90% of acute viral hepatitis in the United States. **TABLE 17-2** lists these hepatitis viruses and their modes of transmission, source, and how they are prevented. The geographical distribution of hepatitis viruses A–E is shown in **FIGURES 17-3a–e**.

Hepatitis A commonly occurs in developing parts of the world where sewage disposal and food hygiene are unsatisfactory. HBV is spread via a **parenteral** (blood-to-blood) mode of transmission, such as through tattooing, body piercing, and drug abuse. Transmission can also occur sexually and via fomites (e.g., sharing contaminated razors). High to intermediate prevalence of HBV also occurs in developing countries, Western Europe, newly independent states of the former Soviet Union, Central America, and the Caribbean. Variants of HBV have been described, and the current hepatitis B vaccine does not protect immunized individuals against these variant HBVs.

HCV is spread primarily by direct contact with human blood. There is a higher prevalence of hepatitis C infection among blood donors in the Eastern Mediterranean, Western Pacific, and Southeast Asia compared to some countries in North America and Europe. Reporting of data is lacking from African countries.

Hepatitis D virus requires the presence of hepatitis B virus in order to replicate. In general, then, the global pattern of hepatitis D infection corresponds to the prevalence of hepatitis B virus infection; for example, in countries where chronic hepatitis B cases are low, there is a low prevalence of hepatitis D virus. Hepatitis E is spread by the fecal–oral mode of transmission, often through contaminated water. It is endemic and sometimes epidemic in the developing world.

17.3 Clinical Features

Hepatitis A

Most hepatitis A outbreaks are sporadic and associated with contaminated food or water supplies. Shellfish may become contaminated with sewage

TABLE 17-1 Human Hepatitis Viruses Nomenclature and Characteristics

Virus Member	Abbreviation	Enveloped or Naked	Genome	Family
Hepatitis A virus	HAV	Naked	+ssRNA	*Picornaviridae*
Hepatitis B virus	HBV	Enveloped	DNA (partially double-stranded)	*Hepadnaviridae*
Hepatitis C virus	HCV	Enveloped	+ssRNA	*Flaviviridae*
Hepatitis D (δ) virus	HDV	Enveloped	–ssRNA	Not assigned (defective virus, requires HBV for viral assembly)
Hepatitis E virus	HEV	Naked	+ssRNA	*Hepeviridae*
Hepatitis G virus	HGV	Enveloped	+ssRNA	*Flaviviridae*
Transfusion transmission virus or torque teno virus	TTV	Naked	ssDNA	*Circoviridae*
SEN virus	SENV	Naked	ssDNA (circular)	*Circoviridae*

TABLE 17-2 Comparison of Human Hepatitis Virus Infections

Virus Member	Source of Virus	Mode of Transmission	Chronic Infection	Prevention
Hepatitis A virus	Feces	Fecal–oral	No	Pre-/postexposure immunization
Hepatitis B virus	Blood or blood-derived bodily fluids	Blood-to-blood contact Sexual Sweating?*	Yes	Pre-/postexposure immunization
Hepatitis C virus	Blood or blood-derived bodily fluids	Blood-to-blood contact Sexual	Yes	Blood donor screening; risk behavior modification
Hepatitis D (δ) virus	Blood or blood-derived bodily fluids	Blood-to-blood contact	Yes	Pre-/postexposure modification; risk behavior modification
Hepatitis E virus	Feces	Fecal–oral	No	
Hepatitis G virus	Blood or blood-derived bodily fluids	Blood-to-blood contact	?	Blood donor screening
Transfusion transmission virus	Blood or blood-derived bodily fluids	Blood-to-blood contact	?	? Blood donor screening
SEN virus	Blood or blood-derived bodily fluids	Posttransfusion hepatitis	?	? Blood donor screening

*New study by Selda Bereket-Yucel. 2007. "Risk of hepatitis B infections in Olympic wrestling." *Br J Sports Med.* 41:305–310.

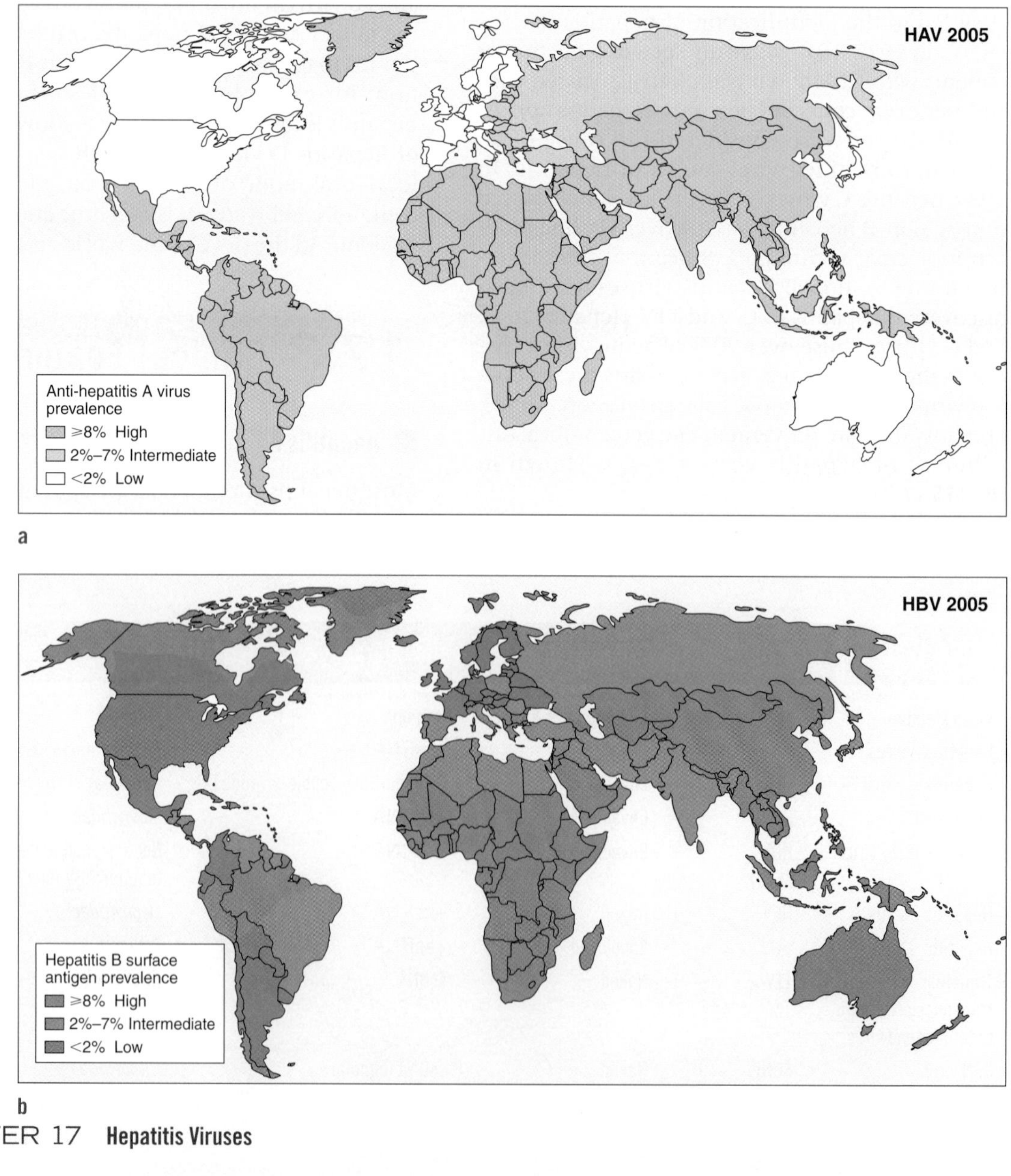

FIGURE 17-3 Geographical prevalence of human hepatitis viruses A–E. **(a)** Global prevalence of hepatitis A (based on 2005 data, WHO). Adapted from CDC. *Health Information for International Travel 2008.* Elsevier, 2007. **(b)** Global prevalence of hepatitis B (based on 2005 data, WHO). Adapted from CDC. *Health Information for International Travel 2008.* Elsevier, 2007.

(continued)

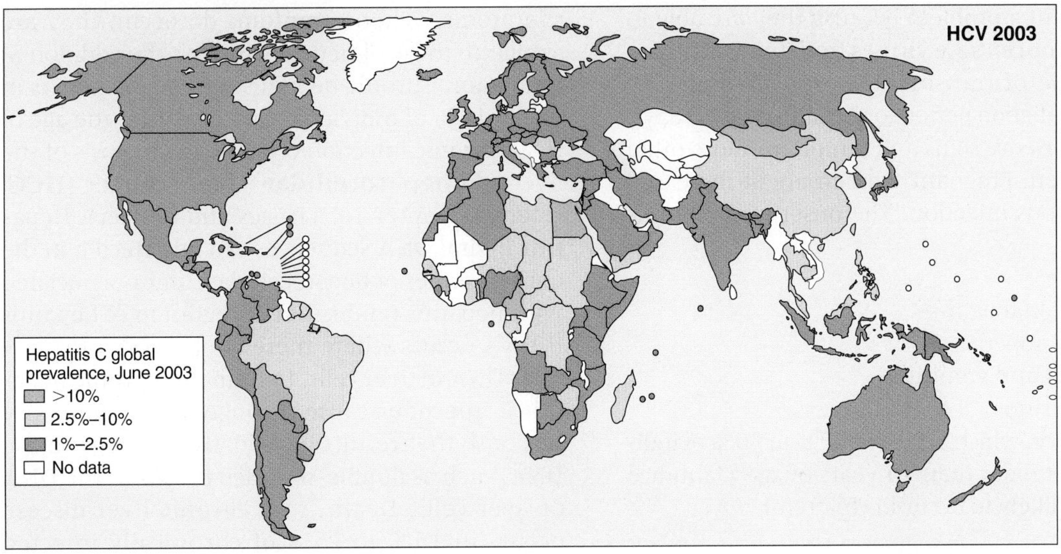

c

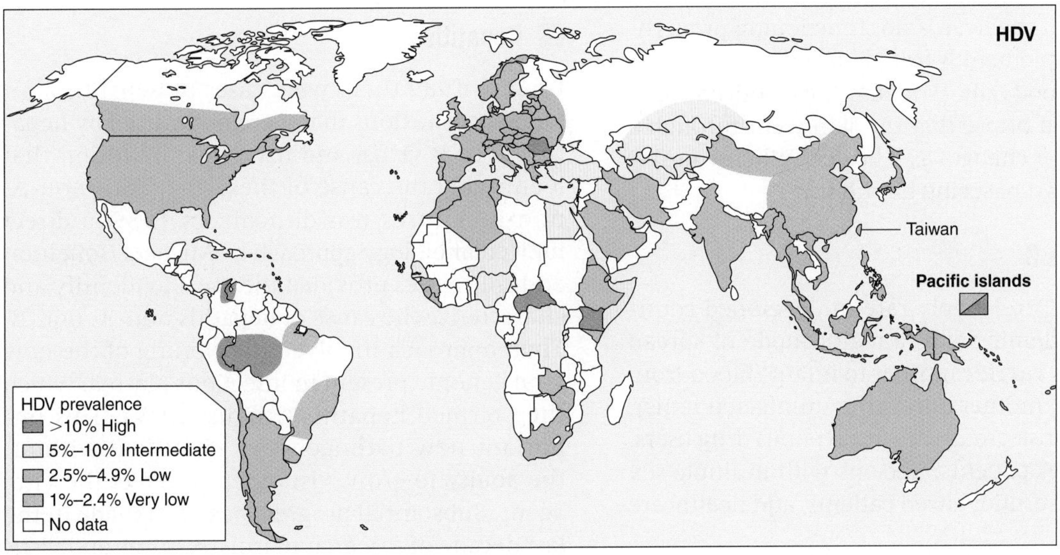

d

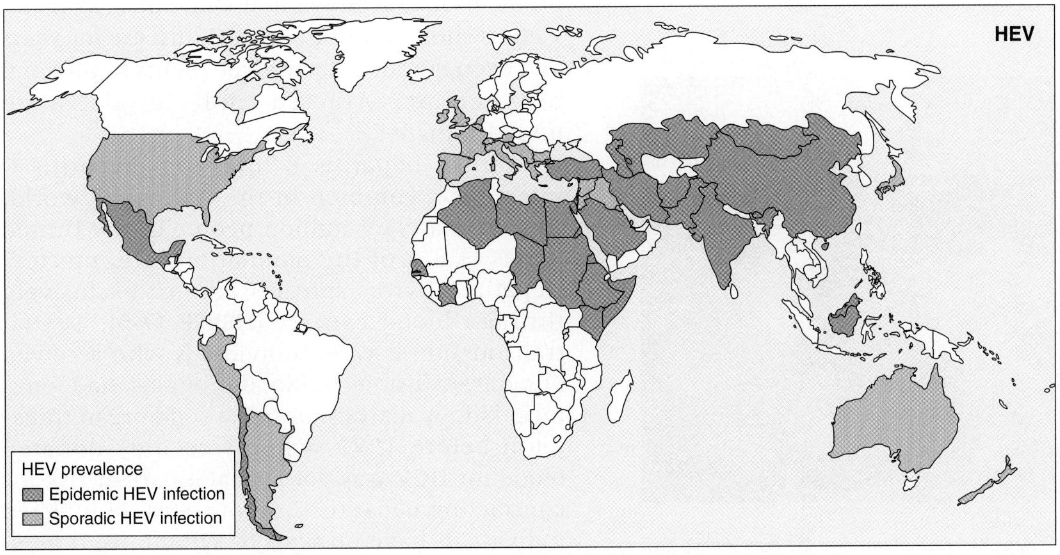

e

FIGURE 17-3 (Continued) Geographical prevalence of human hepatitis viruses A–E. **(c)** Global prevalence of hepatitis C (based on 2003 data, WHO). Adapted from World Health Organization. *International Travel and Health 2007*. WHO, 2007. **(d)** Global prevalence of hepatitis D (CDC data). Adapted from the Centers for Disease Control, *Epidemiology and Prevention of Viral Hepatitis A to E: Hepatitis D Virus*: Slide 6. **(e)** Global prevalence of hepatitis E (CDC data). Adapted from Panda, S. K., Thakral, D., and Rehman, S. *Rev Med Virol* 17 (2006): 151–180.

and represent a problem because they are able to retain and concentrate viruses from the water. The major mode of transmission is fecal–oral. The average incubation period for hepatitis A is 30 days. Adults experience signs and symptoms more often than children. Pregnant women are at increased risk for serious infection. The onset of symptoms include:

- fatigue
- abdominal pain
- loss of appetite
- nausea and vomiting
- dark urine
- jaundice, which occurs in 70% to 80% of individuals older than 14 years of age (Jaundice is less likely to occur in children.)

Signs of jaundice are a yellow color in the skin, mucous membranes, or eyes (**FIGURE 17-4**) and occurs when the liver is not functioning properly. The yellow pigment is from bilirubin, a byproduct of old red blood cells. (You may have noticed when you've had a bruise that the skin went through a series of color changes as it healed. When you saw yellow you were seeing bilirubin.)

Hepatitis B

Hepatitis B is relatively rare in developed countries. In endemic areas, a major mode of spread occurs from **carrier** mother to infant (blood from the infected mother enters the uninfected fetus). Other high-risk groups are intravenous drug users, hemodialysis patients, persons with multiple sex partners, institutionalized patients, and healthcare workers.

The average incubation period for hepatitis B is 80 days. About 30% of individuals have no signs or symptoms. When symptoms do occur, they are similar to those of hepatitis A, with the addition of joint pain. Chronic hepatitis B infection occurs in 5% to 10% of individuals infected after the age of five. Chronic infection can lead to cirrhosis of the liver and **hepatocellular (liver) cancer (HCC**; refer to Chapter 10, Viruses and Cancer). Hepatitis B viral DNA sequences can be detected in the chromosomes of hepatocellular tumors associated with hepatitis B infection. Integration of hepatitis B DNA occurs where there are breaks in the cellular DNA of liver cells. In humans, chronic hepatitis B infections cause an ongoing inflammatory response that results in oxidative damage to the DNA, such as double-stranded breaks in the DNA of liver cells. Death from chronic liver disease occurs in 15% to 25% of chronically infected individuals.

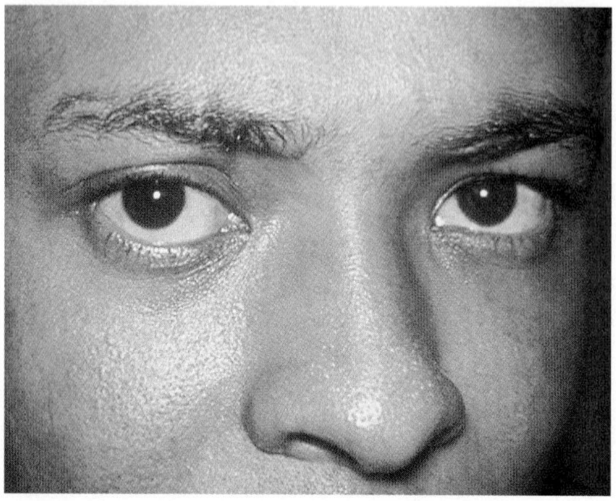

FIGURE 17-4 This person has jaundice of the conjunctiva and facial skin that manifested after a hepatitis A infection.

Hepatitis C

Prior to 1989 there were cases of hepatitis after blood transfusions that were not caused by hepatitis A or B virus, and investigators found that identifying the cause of these "hepatitis non-A, non-B" viruses was difficult. In 1989, a direct molecular biology approach by Michael Houghton and colleagues provided the tools to identify and characterize the cause of hepatitis non-A, non-B. Their approach involved the cloning of the new viral genome present in infectious plasma, which they termed hepatitis C virus (HCV). This important new pathogen was identified without the ability to grow, visualize, or detect the viral agent. Substantial progress has been made in the last decade of research to isolate, grow, and visualize HCV, which is now a major public health problem. It causes a "silent epidemic," as many victims show few or no signs of disease for years and even decades. Early assessments of the total number of infected individuals were underestimated.

Unlike hepatitis B infection, hepatitis C infection is common in the developed world. Approximately 4 million people in the United States (1.8% of the population) are infected. Hepatitis C virus spread is almost exclusively through blood contact (**FIGURE 17-5**). Sexual transmission is rare. Individuals who received blood transfusions or blood products, had long-term kidney dialysis, or had a solid organ transplant before 1992 (when screening donated blood for HCV was not available) are at risk for contracting hepatitis C. About 80% of infected individuals have no signs or symptoms. If present, signs and symptoms are similar to other hepatitis infections. Between 55% and 85% of infected persons experience a chronic infection, resulting in chronic liver disease. Chronic infection can lead to cirrhosis of the liver and HCC in 5% to 20% of persons (see Chapter 10, Viruses and Cancer). The average incubation period is six to seven weeks.

Hepatitis D

Hepatitis D virus (HDV) is a defective virus that requires coinfection with HBV for virion assembly but not for replication. HDV infections occurs with two types of HBV infections: coinfection and superinfection. A **coinfection** refers to a simultaneous onset of infection by separate pathogens, as by hepatitis B and D viruses. A **superinfection** occurs when a chronic HBV carrier is infected with HDV. Symptoms of hepatitis D infection are indistinguishable from those of hepatitis B infection. Less than 5% of cases of acute coinfection with HBD and HDV result in a chronic HDV infection. In some people, HDV superinfection leads to the clearance of hepatitis B surface antigen (HBsAg) and HDV replication ceases, allowing the acute liver disease to resolve. In other people, superinfection in which HDV and HBV persist, liver disease is severe and there is an increased risk of fulminant liver failure or progression to cirrhosis. The highest risk factor for hepatitis D infection in the Western world is intravenous drug use (17% to 90% of addicts test positive for hepatitis D virus).

Hepatitis E

Hepatitis E virus is endemic in the developing countries of Asia and Africa and rare in industrialized nations. Sporadic cases in the developed world almost always occur in travelers returning from endemic areas. The mode of transmission and symptoms are similar to hepatitis A virus (i.e., through fecal–oral transmission). Person-to-person transmission is rare. The average incubation period is 40 days. Outbreaks are often associated with fecally contaminated water supplies. Prevention strategies include the avoidance of drinking water and beverages with ice of unknown purity, uncooked shellfish, and uncooked vegetables not peeled or prepared by the traveler.

Infection with hepatitis E virus causes a more severe illness than hepatitis A virus. It causes mortality in 1% to 3% of persons and in 15% to 25% of infected pregnant women. Even though hepatitis E is rarely diagnosed in industrialized nations, antihepatitis E virus antibodies have been found in a significant proportion (up to 28% in some areas) of healthy individuals. Sporadic cases have been reported in the United States, northern England, Italy, and France. Considerable evidence suggests that hepatitis E may be a **zoonotic disease** (e.g., transmitted from animals to people). Human hepatitis E virus, for which isolates are genetically similar to pig hepatitis E strains, has been isolated from pigs in the United States. It has been shown that pig hepatitis E viruses can cross species barriers and infect nonhuman primates. Human antihepatitis E virus antibodies cross-react and bind to pig antihepatitis E virus capsid proteins.

Hepatitis Non-A–E Agents

As described earlier, hepatitis A–E viruses have not accounted for all cases of hepatitis. Hence, hepatitis "non-A–E" agents have been implicated.

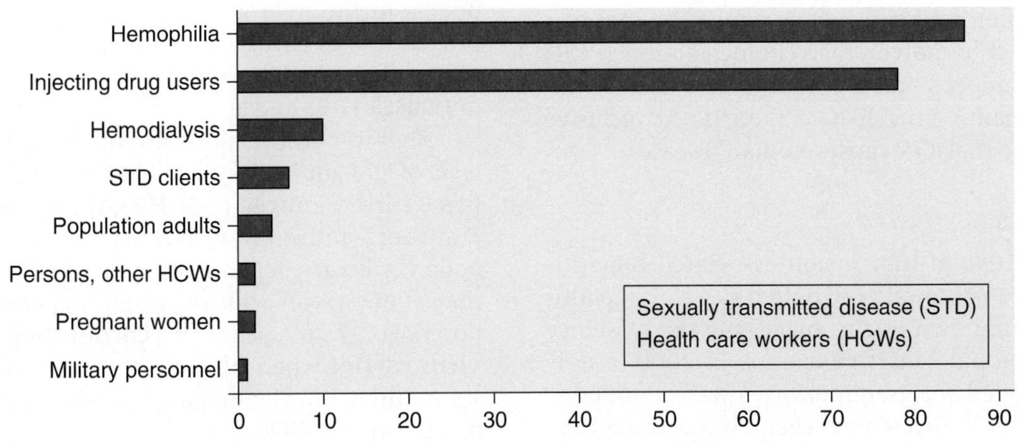

FIGURE 17-5 Hepatitis C virus prevalence by selected groups in the United States. Adapted from the Centers for Disease Control and Prevention, *Epidemiology and Prevention of Viral Hepatitis A to E: Hepatitis C Virus*: Slide 24.

Hepatitis F

Hepatitis F virus does not exist. During the 1990s, a handful of cases were reported from France. Experiments demonstrated that monkeys could be infected with the isolated hepatitis F virus from patient stools. This research was not reproducible and the scientific community no longer lists hepatitis F as a cause for infectious hepatitis.

Hepatitis G

Hepatitis G virus (HGV) was first cloned from the plasma of a surgeon in 1996. HGV, which is transmitted parenterally, is a flavivirus and is a distant relative of hepatitis C virus. HGV is distributed globally. In the United States, 1.7% of volunteer blood donors have antibodies against HGV. The prevalence of HGV is higher in patients infected with HCV or HIV or patients who have had multiple transfusions or organ transplantation, and patients undergoing hemodialysis. HGV RNA has been detected in patients with non-A–E acute hepatitis, in patients with cirrhosis of unknown origin, in patients with chronic hepatitis, and in patients with primary hepatocellular carcinoma (HCC; liver cancer). The clinical significance of hepatitis G virus is not well understood.

Transfusion Transmission Virus or Torque Teno Virus (TTV)

In 1997, TTV was found using **representational difference analysis** (**RDA**) in Japanese patients who had posttransfusion hepatitis of unknown etiology (hepatitis non-A–G). RDA was used to compare the DNA from sera of hepatitis patients before and after they experienced elevated levels of **alanine aminotransferase** (**ALT**). Elevated ALT levels are an indicator of liver inflammation that can be the result of viral hepatitis. The procedure detected extraneous viral DNA that was amplified by polymerase chain reaction (PCR) and sequenced and reported to be a new candidate hepatitis virus. To date, there is no conclusive evidence that TTV causes human disease.

SEN Virus

About 10% of transfusion-associated hepatitis and 20% community-acquired cases of hepatitis are of unknown origin, suggesting the existence of additional causative agents. In 2000, a new virus, called SEN (sentinel virus) was identified as a hepatitis non-A–E viral agent. It was thought to be associated with posttransfusion hepatitis. Little is known about the clinical significance of SEN virus.

17.4 Laboratory Diagnosis of Viral Hepatitis Infections

Diagnosis of viral hepatitis is based on:

- symptoms
- blood tests for liver enzymes
- viral antibodies (e.g., IgM)
- viral genetic material (e.g., reverse transcriptase PCR [RT-PCR] or PCR)

Patient symptoms of **acute hepatitis** are fatigue, abdominal pain, loss of appetite, nausea and vomiting, and darkening of urine, followed by jaundice. Serology (enzyme-linked immunosorbent assay [ELISA] tests detecting viral antibodies or antigens), liver enzyme tests, and molecular techniques to detect viral genomes (e.g., RT-PCR or PCR) are used to confirm diagnosis of acute viral hepatitis. Blood samples are tested for the presence of elevated levels of two liver enzymes: **aspartate aminotransferase** (**AST**) and ALT. These enzymes are normally found in the liver but spill into the blood if the liver is damaged, thus raising the enzyme levels in the blood. Nucleic acid tests to detect viral genomes are only available in specialized laboratories. Patients with **chronic hepatitis** are harder to diagnose because these patients do not have nausea or jaundice until liver damage is quite advanced.

Hepatitis A infections are reliably diagnosed by the presence of anti-HAV IgM. Hepatitis E infections are clinically indistinguishable from hepatitis A infections. Diagnosis of hepatitis E infections is made by serology tests that test for anti-HEV IgM or by RT-PCR. Tests for HEV are not widely available. Hepatitis A and E viruses do not persist in the liver and there is no evidence of progression to chronic liver damage. The courses of hepatitis A and hepatitis E infections are shown in **FIGURES 17-6a** and **b**.

Hepatitis B infections are diagnosed by the level of IgM antibodies produced toward the hepatitis B surface antigen (anti-**HBsAg**) and hepatitis B anticore-antigen (anti-**HBcAg**). Levels of antibodies will vary depending upon whether or not the patient has an acute or chronic hepatitis infection (**TABLE 17-3**). A patient is a **chronic hepatitis B virus carrier** when HBsAg is present longer than six months, but the diagnosis is often suspected much earlier (**FIGURE 17-7**).

Acute hepatitis C cannot be reliably diagnosed by serology tests because patients usually take up to three months to become positive. Acute hepatitis

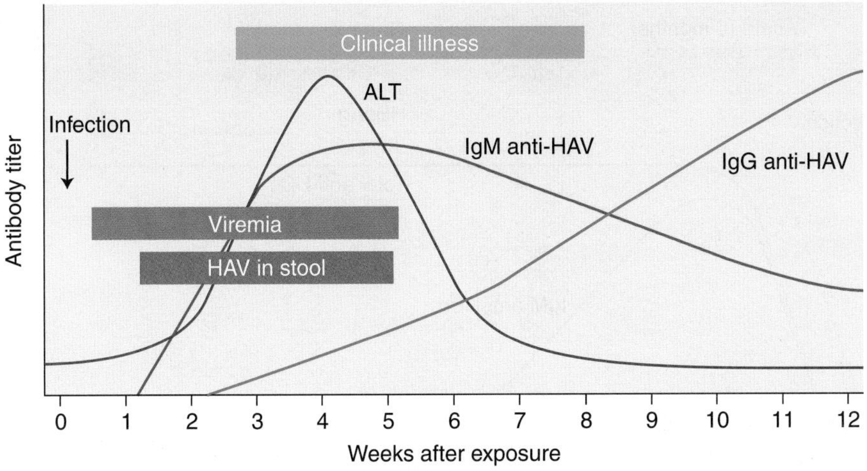

a

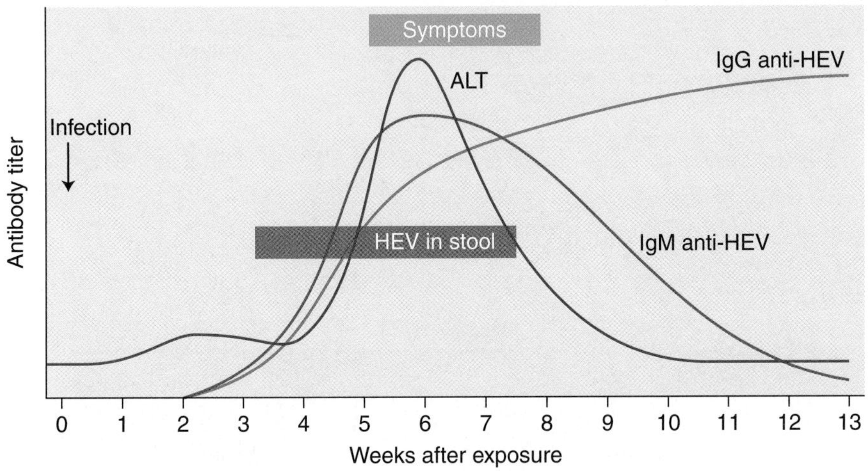

b

FIGURE 17-6 (a) Events in hepatitis A virus infection. Adapted from the Centers for Disease Control and Prevention, *Epidemiology and Prevention of Viral Hepatitis A to E: Hepatitis A Virus*: Slide 6. **(b)** Events in hepatitis E virus infection. "ALT" represents levels of alanine aminotransferase activity. Adapted from the Centers for Disease Control and Prevention, *Epidemiology and Prevention of Viral Hepatitis A to E: Hepatitis E Virus:* Slide 3.

C is diagnosed by blood tests detecting serum ALT levels that are seven times higher than the normal upper limit of liver enzyme levels. In addition, nucleic acid testing for hepatitis C virus RNA in serum is also an indicator of infection. Early diagnosis is beneficial because it is believed that early treatment can prevent chronic infection with hepatitis C virus. Chronic hepatitis C virus infection can occur in as high as 80% of untreated patients (Figure 17-7c). Diagnosis is based on antihepatitis C virus antibodies and hepatitis C virus RNA levels (Table 17-3 and Figure 17-7b).

TABLE 17-3 Diagnostic Testing for Hepatitis A–C Virus Infections

Type of Viral Infection	Diagnostic Tests
Acute hepatitis A	IgM anti-HAV positive result
Acute hepatitis B	IgM anti-HBcAg positive, anti-HBsAg positive
Chronic hepatitis B	Anti-HBsAg positive, IgM and IgG anti-HBcAg positive, IgM anti-HBeAg negative
Acute hepatitis C	High serum alanine aminotransferase levels (7$\times$ normal limit), nucleic acid testing for HCV-RNA
Chronic hepatitis C	Anti-HCV positive, HCV RNA $\geq$6 months apart

17.5 Screening the Blood Supply for Viral Hepatitis Agents

The American Red Cross has been screening for hepatitis B surface antigen since 1971. In 1987, a test for anti-HBcAg was added to the screening protocol. From 1986 to 2003, the U.S. blood supply was also screened for elevated levels of ALT. In 1990, testing began for antihepatitis C virus, and in 1999 a nucleic acid test for hepatitis C virus-RNA was added to blood donation screening in order to reduce the risk of disease transmission.

FIGURE 17-7 (a) Events in hepatitis B virus chronic infection. Adapted from the Centers for Disease Control and Prevention, *Epidemiology and Prevention of Viral Hepatitis A to E: Hepatitis B Virus*: Slide 10. **(b)** Events in hepatitis C virus chronic infection. Adapted from the Centers for Disease Control and Prevention, *Epidemiology and Prevention of Viral Hepatitis A to E: Hepatitis C Virus*: Slide 9. **(c)** Natural history of hepatitis B and C virus infections.

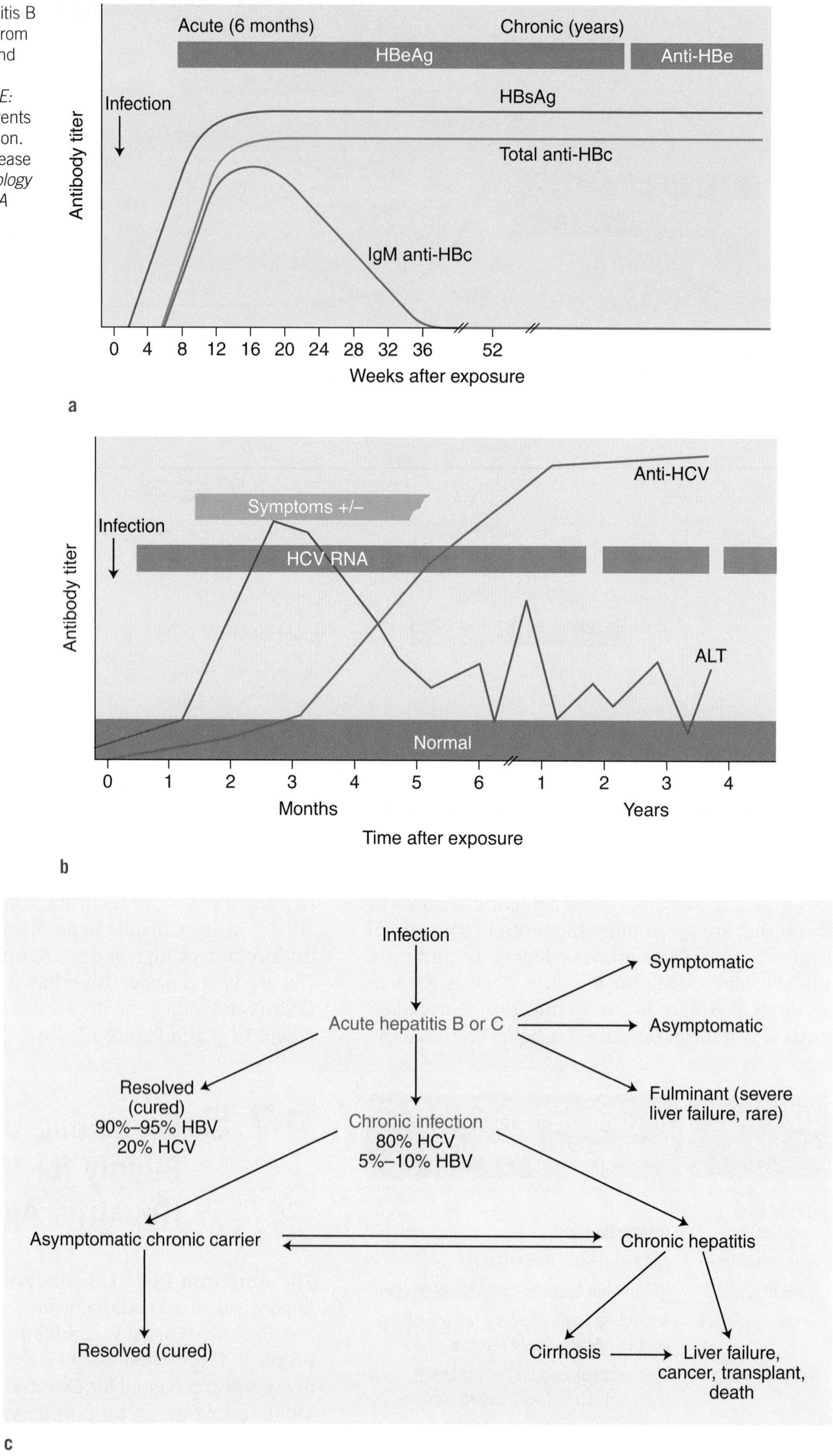

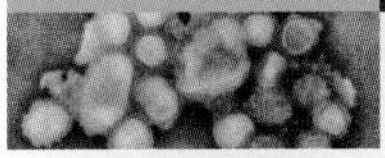

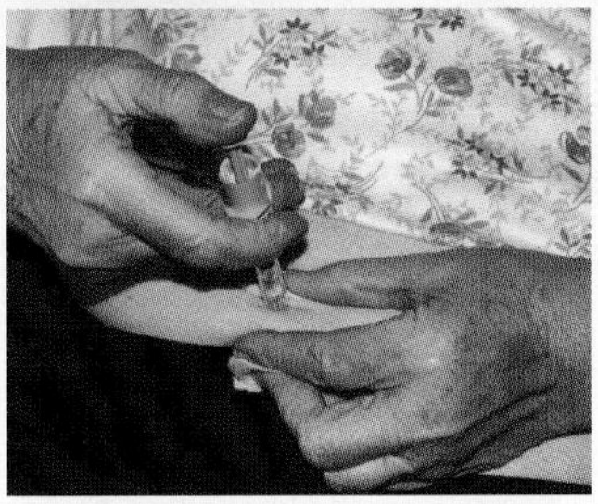

FIGURE VF 17-1 Interferon α is used to treat hepatitis C infections. Unfortunately, it has only limited success.

Approximately 170 million people worldwide are chronically infected with hepatitis C virus, which exclusively infects humans and chimpanzees. The only therapy available to treat hepatitis C virus infections is a combination of ribavirin and interferon α (**FIGURE VF 17-1**). Interferon therapy is effective in just a fraction of patients due to the adverse side effects associated with interferon treatment. There is no vaccine for hepatitis C virus. Hepatitis C virus-associated end-stage liver disease is now the leading cause for liver transplantation in the United States.

The hepatitis C virus was discovered in 1989. It was difficult to study hepatitis C viruses because cell culture and small animal models to propagate and research the virus in the laboratory were not available at that time; thus, understanding the virus life cycle and how the virus damages the liver was hampered. Testing potentially useful drugs or development of a vaccine against these infections was not possible.

In 2005, major breakthroughs by three different research groups made it possible to grow hepatitis C viruses in cell culture, paving the way for the development of new antiviral therapies and vaccines. A critical advance in this breakthrough was the isolation of the entire viral genome from an infected individual who was suffering from fulminant hepatitis C. The genome was cloned and sequenced and designated JFH1. RNA transcripts from this genome were transfected into a human liver hepatoma cell line (Huh-7.5 cells). With this method, infectious particles of hepatitis C virus were generated at relatively high titers. Chimpanzees were intravenously inoculated with the viruses produced in the Huh-7.5 cells and proved to be infectious (**FIGURE VF 17-2**). The development of this tissue-culture system has accelerated the pace of hepatitis C research.

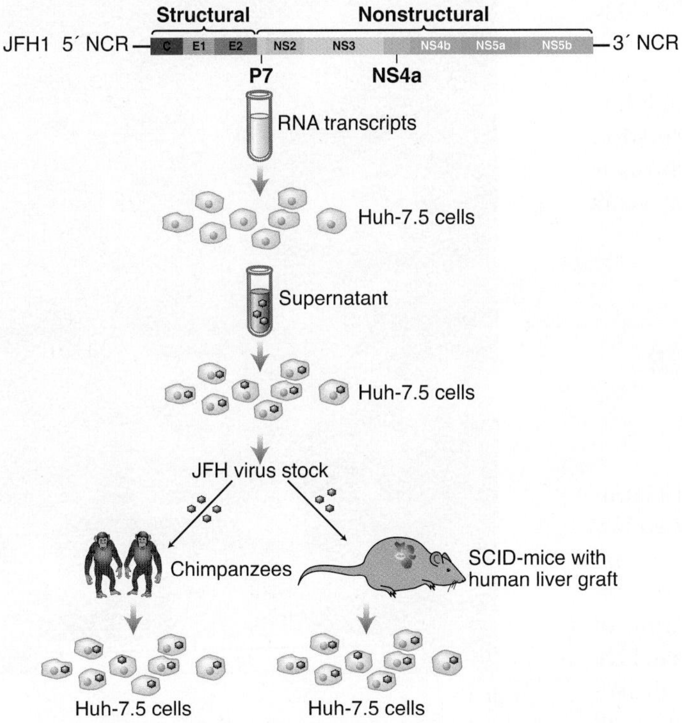

FIGURE VF 17-2 RNA transcripts of the JFH1 virus were transfected into Huh-7.5 cells. New viruses from the infected Huh-7.5 cells were present in the supernatant of the cell cultures. The supernatant was filtered and used to infect naïve (uninfected) Huh-7.5 cells in order to generate a stock of virus. The virus stock was then used to infect chimpanzees and SCID mice that contained human liver grafts. Viruses were recovered from the infected chimpanzees and mice and efficiently were able to infect naïve Huh-7.5 cells. NCR = noncoding region, * = HCV. Adapted from Bukh, J., and Purcell, R. H. *Proc Natl Acad Sci USA* 103 (2006): 3500–3501.

References

Bukh, J., and Purcell, R. H. 2006. "A milestone for hepatitis C research: A virus generated in cell culture is fully viable in vivo." *Proc Natl Acad Sci USA* 103(10):3500–3501.

Lindenbach, B. D., et al. 2005. "Complete replication of hepatitis C virus in cell culture." *Science* 309:623–626.

Wakita, T., et al. 2005. "Production of infectious hepatitis C virus in tissue culture from a cloned viral genome." *Nat Med* 11:791–796.

Zhong, J., et al. "Robust hepatitis C virus infection in vitro." *Proc Natl Acad Sci USA* 102:9294–9299.

Blood is not screened for other hepatitis viruses and is not entirely risk-free, but it is very safe. The U.S. blood supply in particular is among the safest in the world.

17.6 Pathogenesis: Chronic Hepatitis

Chronic hepatitis occurs when the inflammation of the liver is active, persists for more than six months, and is detectable by increased ALT levels in serum. Hepatitis B, C, D, and G viruses can cause chronic hepatitis infections. A patient who is **viremic** and has an abnormal ALT level should have a liver biopsy, which is the only way to assess the degree of inflammation of the liver and the stage of liver disease. Chronic liver damage may result in **cirrhosis**, which is characterized by the formation of fibrous tissues, nodules, and scarring that interfere with liver cell function and blood circulation (**FIGURE 17-8**). A late complication of cirrhosis is hepatocellular carcinoma, which takes many years to develop.

17.7 Hepatitis Virus Life Cycles

The hepatitis viruses belong to five different families, designated by terms with the suffix *viridae* (Table 17-1). Families are groups of viruses that share common characteristics. Although there are five families listed in the right-hand column of Table 17-1, one of the viruses are unassigned. Five of eight hepatotropic viruses (hepatitis viruses A–E) discussed in this chapter have been well characterized.

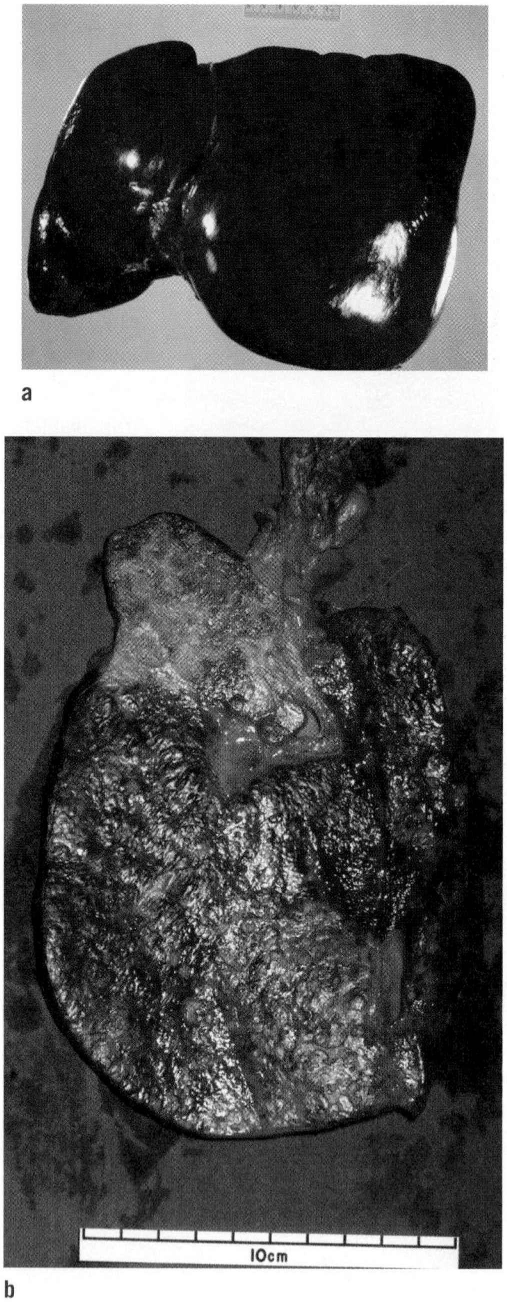

FIGURE 17-8 (a) A healthy human liver. **(b)** A human liver diseased by viral hepatitis and cirrhosis.

Virus Structure

Hepatitis viruses A–E have a spherical shape and icosahedral capsid symmetry ranging from 28 to 50 nm in diameter. Virus particles are either enveloped or nonenveloped. Electron micrographs of hepatitis A–D viruses are shown in **FIGURE 17-9**. Hepatitis B, C, and D are enveloped and relatively sensitive to many physical and chemical agents, whereas hepatitis A is not enveloped and can remain infectious on inanimate surfaces in the environment for a month. Hepatitis A viruses can also retain their infectivity in the hostile environment of the gastrointestinal tract, as they are acid and bile resistant.

Virus Replication

Hepatitis viruses enter the bloodstream and are carried to the liver, where they infect the **hepatocytes** of the liver. As the viruses spread and replicate in the liver, hepatocytes may become damaged. The damage may result in cirrhosis, impairing liver function (**FIGURE 17-10**). The liver has many functions, such as playing a role in metabolism and the

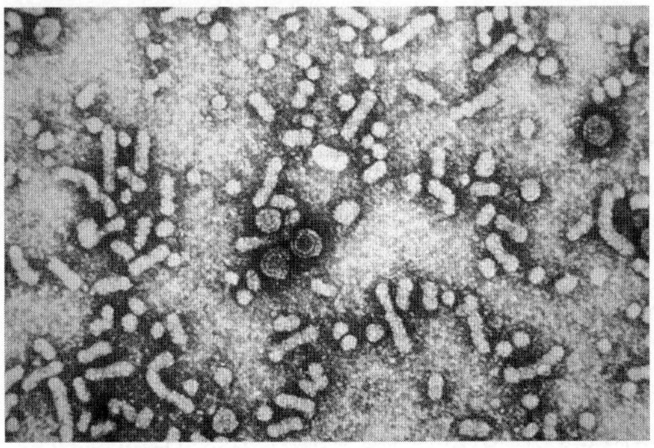

a

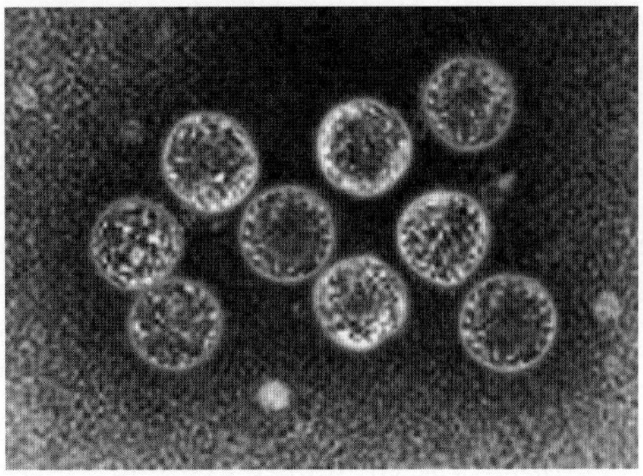

b

FIGURE 17-9 Electron micrographs of hepatitis viruses. **(a)** Hepatitis B virus. **(b)** Hepatitis C virus.

filtering of toxins and waste products in the blood. The liver produces **bile**, which aids in digestion.

The clinical importance of hepatitis F and G viruses and TTV is uncertain and poorly understood; thus, the remainder of this chapter will summarize the life cycles of hepatitis viruses A–E.

Hepatitis A Virus

Hepatitis A virus is a very stable, naked picornavirus that contains a +ssRNA genome that is 7.5 kb in length. The viral genome contains a small protein, VPg, that is covalently linked to the 5′ end of the

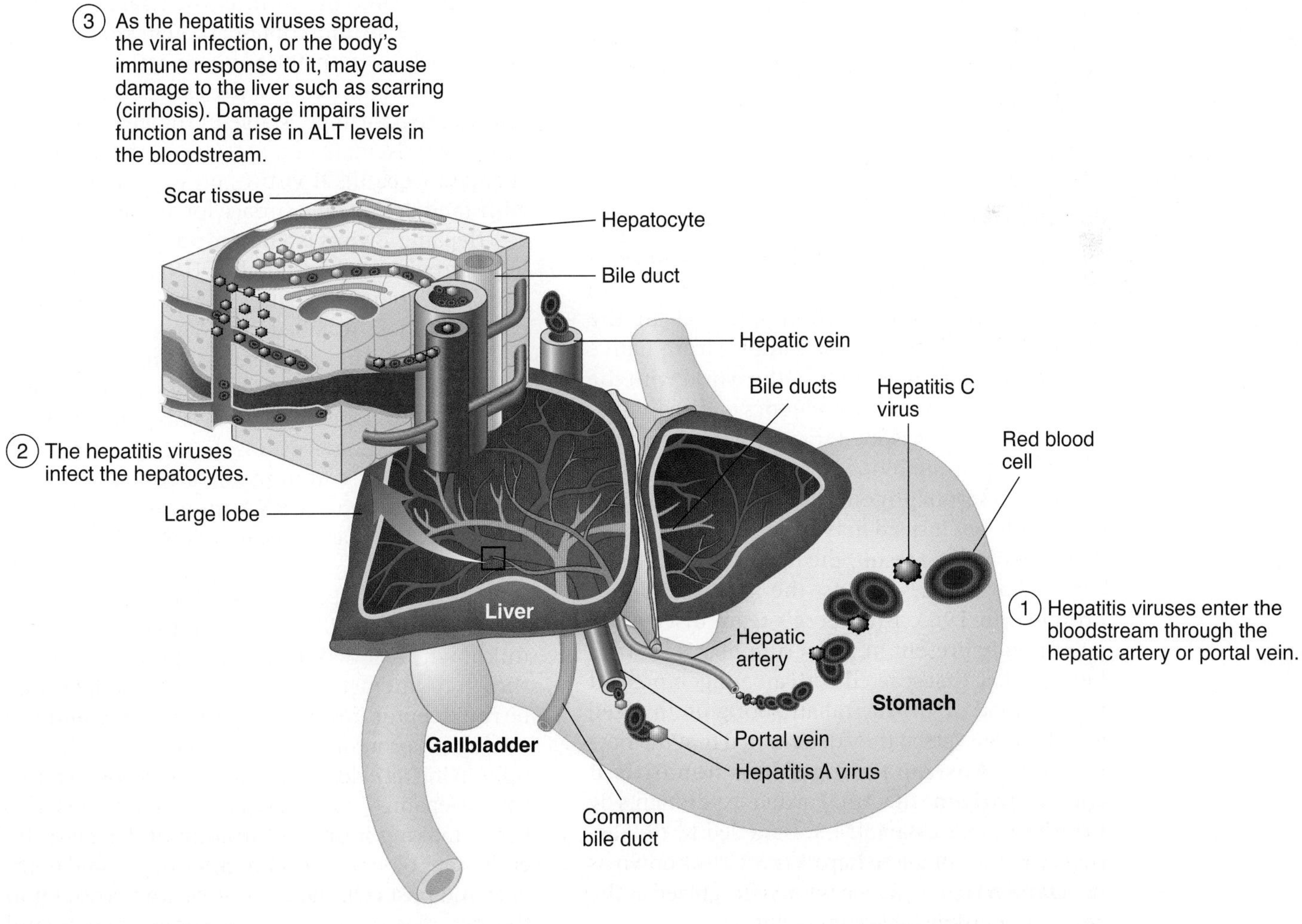

3 As the hepatitis viruses spread, the viral infection, or the body's immune response to it, may cause damage to the liver such as scarring (cirrhosis). Damage impairs liver function and a rise in ALT levels in the bloodstream.
Scar tissue
Hepatocyte
Bile duct
Hepatic vein
Bile ducts
Hepatitis C virus
Red blood cell
2 The hepatitis viruses infect the hepatocytes.
Large lobe
Liver
1 Hepatitis viruses enter the bloodstream through the hepatic artery or portal vein.
Hepatic artery
Stomach
Portal vein
Hepatitis A virus
Gallbladder
Common bile duct

FIGURE 17-10 Hepatitis A and E viruses are ingested and gain entry into liver cells by way of the bloodstream via the hepatic artery or the portal vein. The portal vein carries blood containing digested food from the small intestine to the liver, thus facilitating the entry of hepatitis A virus into the liver. Bloodborne hepatitis viruses, such as hepatitis C virus, enter the liver via the hepatic artery. Adapted from the American Medical Association. *Complete Medical Encyclopedia.* Random House Reference, 2003.

genome and a poly-(A) tail at the 3′ terminus of the genome. The 5′ terminus also contains a nontranslated region that contains **hairpins** and **pseudoknots**. The pseudoknots may function in the recognition of the genome by the viral RNA replicase. The hairpin secondary structures act as an **internal ribosome entry site** (IRES) that is necessary for cap-independent translation of the viral RNA (**FIGURE 17-11a**).

Hepatitis A viruses bind receptors located on the outside of hepatocytes. A candidate cellular receptor that interacts with the virus particle has been identified. It is a mucin-like glycoprotein referred to as the **hepatitis A virus cellular receptor 1** (huHAVcr-1). After entry, the virus particles undergo penetration and uncoating with the release of viral RNA into the cytoplasm of the cell. The +ssRNA viral genome acts directly as an mRNA for the synthesis of a large polyprotein. The polyprotein is processed by a viral 3C protease into structural and nonstructural proteins of the virus. A viral RNA-dependent RNA polymerase encoded by 3D synthesizes new –ssRNA intermediates that are used to create new progeny genomic +ssRNAs that will be packaged into new virus particles. Newly assembled particles are transported to the surface of hepatocytes and exported (Figure 17-11b).

Hepatitis B Virus

Dr. Baruch Blumberg and colleagues began characterizing and comparing serum proteins from **hemophiliacs** and nonhemophiliacs in the 1950s. They collected blood samples from native populations in remote parts of the world with the goal of learning about the genetics of disease susceptibility. They reasoned that hemophiliacs who received multiple blood transfusions would be exposed to serum proteins they had not inherited genetically but instead may have "inherited" from their donors. As a result, the hemophiliacs would produce antibodies against the inherited donor proteins. In 1963, Blumberg's team discovered an antibody present in the serum of a New York hemophiliac that reacted with an antigen present in the blood of an Australian aborigine infected with hepatitis. Later, they determined that the new unknown **Australia antigen** was **hepatitis B surface antigen** (HBsAg). Further experiments by British virologist David M.S. Dane led to the discovery of the complete hepatitis B virus known as the **Dane particle**, which is now recognized as the agent responsible for serum hepatitis.

Through research and clinical observations, it is known that three types of virus-associated particles are present in the blood of individuals suffering from hepatitis B. The most abundant hepatitis B-like particle present in highly viremic carriers of hepatitis B virus is a spherical particle that is approximately 17 to 25 nm in diameter. These particles appear as small vesicles that contain small HBsAg proteins (**FIGURE 17-12**). Less numerous are filamentous particles that are approximately 17 to 20 nm in diameter and up to 200 nm in length (Figure 17-12). The filamentous particles contain small, **medium**, and **large HBsAg proteins** and are not infectious. The large amounts of free HBsAg present in the serum of hepatitis patients correlates with viremia and in many cases indicate chronic infection.

Dane particles are infectious. These enveloped particles measure 42 nm in diameter and contain small, medium, and large HBsAg proteins. In addition to the HBsAg proteins, a DNA-based core that is 27 nm in diameter is located within the virus particle (Figure 17-12). The core that surrounds the DNA genome contains two proteins: HBcAg and HBeAg. Most of HBeAg is secreted and not actually incorporated into the particle. The viral DNA polymerase (reverse transcriptase) protein kinase C and heat shock protein 90 are associated with the genome within the core of the particle. Hepatitis B viruses are amazingly stable in the environment. They are resistant to organic solvents, heat, and pH changes. Hepatitis B viruses present in blood can withstand drying on a surface for at least a week.

Hepatitis B virus is a **hepadnavirus** (*hepa* from hepatotropic and *dna* from their DNA genome). The circular genome of hepatitis B virus is composed of partially double-stranded DNA. The completed full-length strand is approximately 3.2 kb in length and the shorter strand is approximately 1.7 kb in length (**FIGURE 17-13**). One of the molecular biology hallmarks of hepadnaviruses is that they use a replication strategy common to retroviruses: they replicate through an RNA intermediate, sometimes referred to as a pregenomic RNA by **reverse transcription**.

The host cell receptor used for the attachment and entry of hepatitis B viruses into hepatocytes is unknown. A few potential receptors have been shown to interact with HBsAg, such as the transferrin receptor, human liver endonexin, and the asialoglycoprotein receptor molecule, but there is no convincing evidence of their connection to infectivity. Hepatitis B viruses replicate poorly in cell cultures, making it difficult to fully understand the early steps of viral infection. After fusion and entry with the host cell, the viral cores are released into the cytoplasm where the genome uncoats and enters the nucleus through the nuclear pores.

In the nucleus, host enzymes ligate the ends of the genome. DNA synthesis is completed and gaps

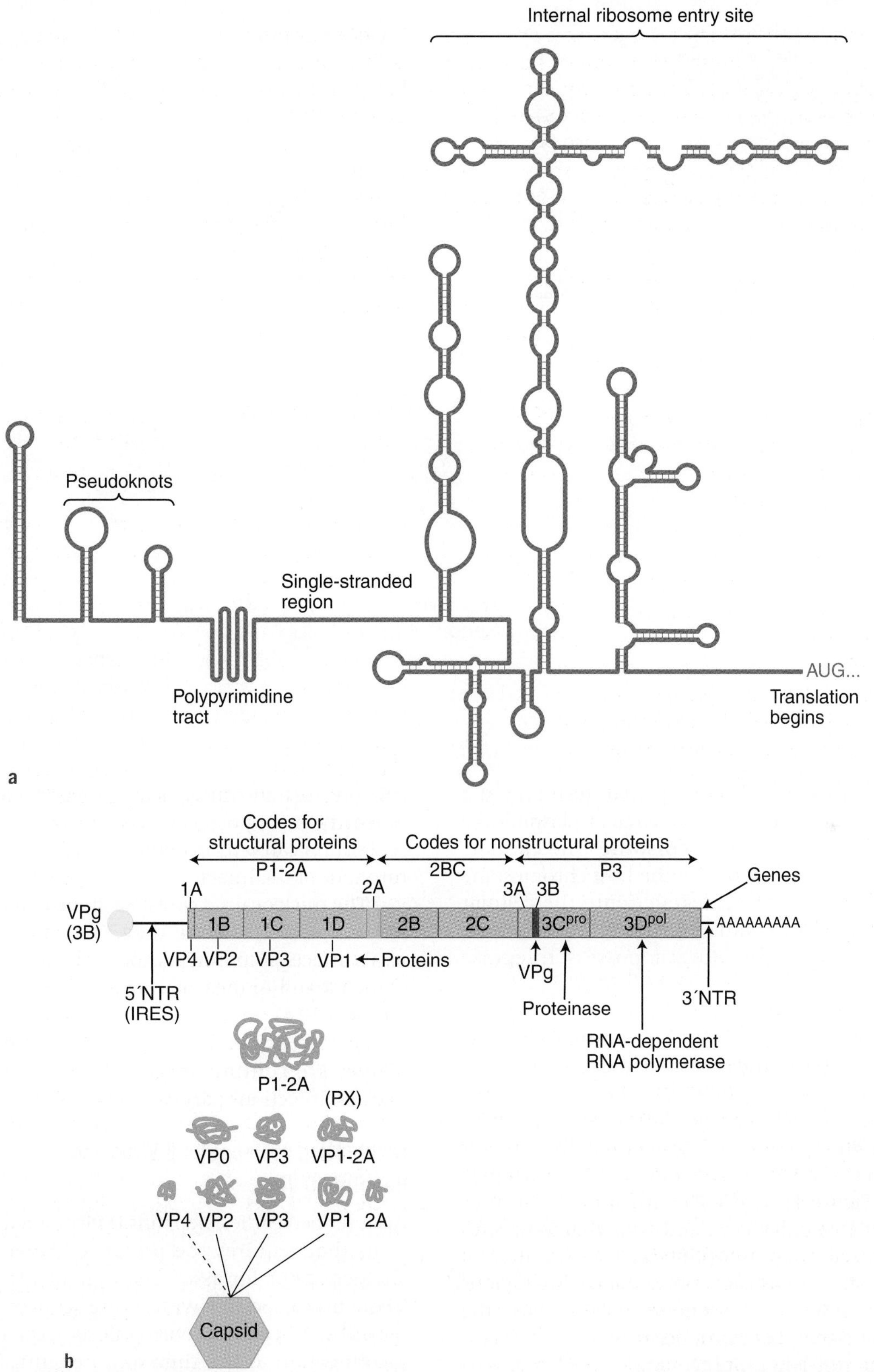

FIGURE 17-11 (a) Secondary structure of the 5′ noncoding region of hepatitis A virus genome. The pseudoknots are involved in the recognition of the viral RNA-dependent RNA polymerase or replicase, whereas the internal ribosome entry site allows for cap-independent translation of the viral mRNA transcripts into a polyprotein product. The function of the polypyrimidine tract is unknown. Adapted from Thomas, H. C., Lemon, S., and Zuckerman, A. J. *Viral Hepatitis,* Third Edition. Blackwell Publishing, 2005. **(b)** Organization of the hepatitis A virus genome. Genes are listed within the genome and the gene product names are listed above the genes. The genome is translated into a large polyprotein that is processed by the viral 3C protease into three distinct precursors that are proteolytically processed to yield smaller precursors and mature viral proteins, etc. Adapted from Thomas, H. C., Lemon, S., and Zuckerman, A. J. *Viral Hepatitis,* Third Edition. Blackwell Publishing, 2005.

FIGURE 17-12 Structure of hepatitis B virus particles. Only the Dane particle is infectious. Classic HBsAg, which contains the S domain only, is also called the S-protein. Two other proteins share the C-terminal S domain, but differ by length and structure of their N-terminal (pre-S) extensions. The large L protein contains the pre-S1, the pre-S2 region, and the S region, whereas the medium M protein contains the pre-S2 and the S region only. HBsAg is the most abundant of the S-related antigens. The L and M proteins are expressed at levels of about 5% to 15% and 1% to 2% compared with S protein. Adapted from Thomas, H. C., Lemon, S., and Zuckerman, A. J. *Viral Hepatitis,* Third Edition. Blackwell Publishing, 2005.

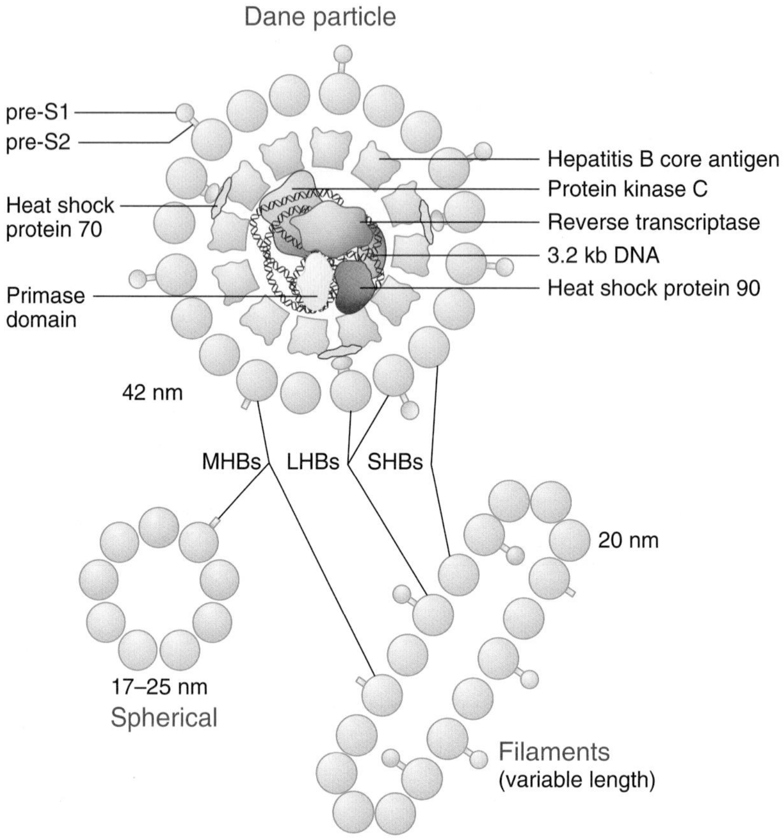

in both DNA strands are repaired, resulting in a completed covalently closed circular plasmid-like DNA molecule called an **episome**. The episome replicates independently of the host chromosome in contrast to being integrated into the cellular DNA, as is the case with retroviruses. This is because hepatitis B viruses do not possess integrase activity.

The episome acts as a template for two classes of viral transcripts: viral **pregenomic** RNA transcripts and **genomic** RNAs that are transcribed by the host's RNA polymerase II. Hepatitis B virus DNA has a very compact coding organization with four partially overlapping open reading frames (ORFs) that are translated into seven known proteins (Figure 17-13). There are at least four promoters, two enhancers, and several binding sites recognized by transcription factors identified in the hepatitis B virus genome. Noncoding regions are not present on the genome. **TABLE 17-4** lists the ORFs and their gene products.

The full-length, pregenomic +ssRNA is synthesized and complexed with the viral polymerase and protein kinase C protein into core particles. Host-cell heat shock proteins associate with the viral reverse transcriptase, allowing it to fold into an active conformation. The active viral reverse transcriptase converts this pregenomic RNA into DNA inside the particles. Unlike retroviruses, the HBV reverse transcriptase activity occurs by **protein-priming** as opposed to RNA-priming. Precise replication can only occur in the specialized environment of the intact nucleocapsids.

The nucleocapsid cores reach the endoplasmic reticulum (ER), where they associate with the viral surface glycoproteins and bud into the lumen of the ER and/or the Golgi apparatus out of the cell (**FIGURE 17-14**). Empty envelopes containing viral surface proteins embedded in the host cell's lipid bilayer are continually being shed along with mature, infectious particles.

Coinfection of Hepatitis B Virus and Hepatitis D Virus

As described earlier, hepatitis D virus is a defective virus that requires the presence of hepatitis B "helper" virus for assembly. Hepatitis D virus is frequently associated with severe acute or chronic hepatitis. A hepatitis D virus infection can occur as a coinfection at the time of a hepatitis B virus infection or as a **superinfection** in individuals who are carriers of hepatitis B virus. In the latter scenario, a superinfection is defined as a hepatitis D virus infection following a previous infection by hepatitis B virus. Hepatitis D virus probably makes use of the same cellular receptor as hepatitis B virus in order to attach to and enter hepatocytes.

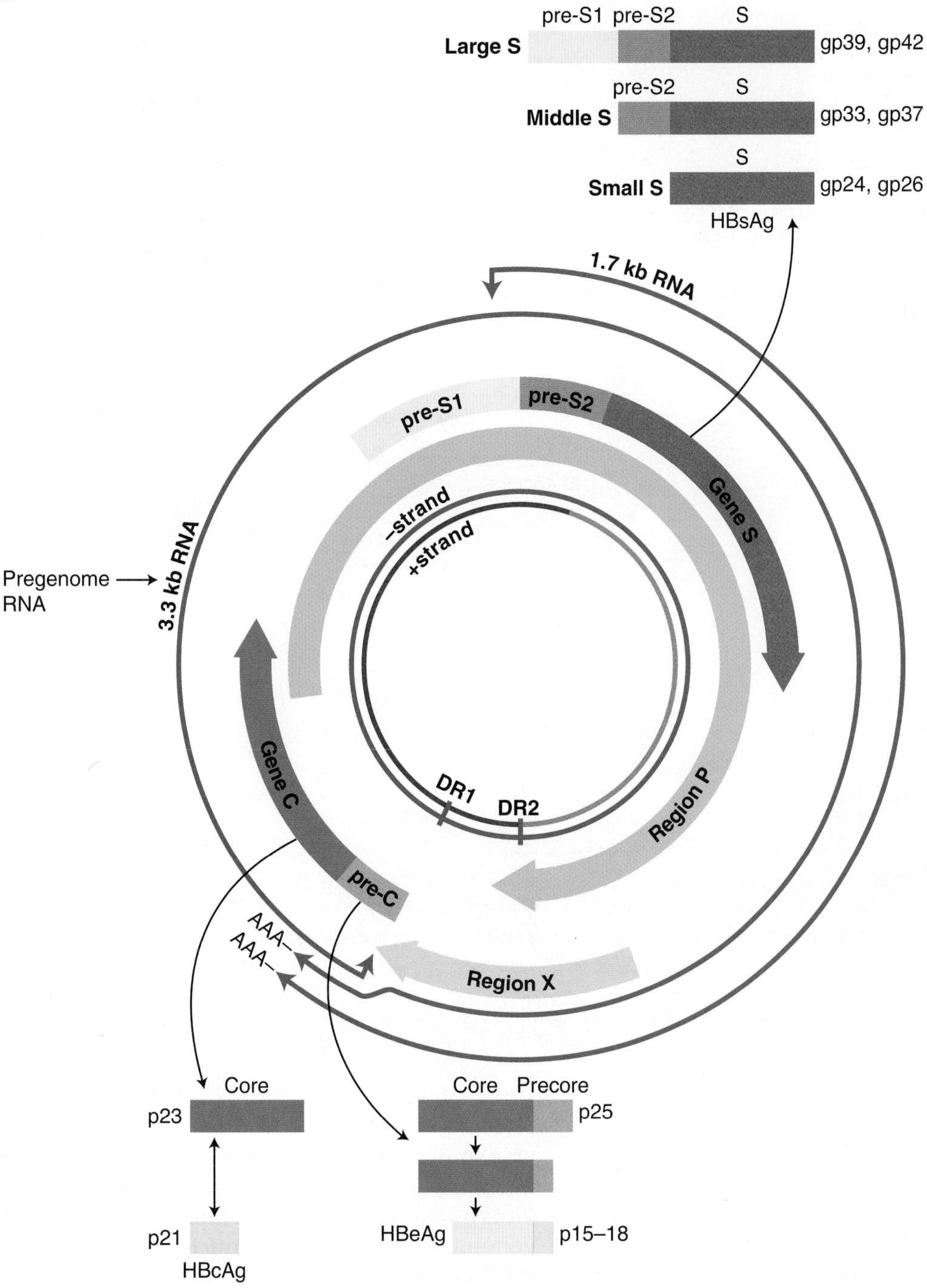

FIGURE 17-13 Genomic organization of the hepatitis B virus. Adapted from Cohen, J., and Powderly, W. G. *Infectious Diseases*, Second Edition. Mosby, 2003.

TABLE 17-4 Hepatitis B Virus ORFs and Their Gene Products

ORF	Gene Products
1 (S)	Large, middle, and small hepatitis B surface (envelope) antigens (HBsAg)
2 (C)	Hepatitis B secreted precore antigen (HBeAg) and hepatitis B core (capsid) antigen (HBcAg)
3 (P)	Viral polymerase (DNA-dependent DNA polymerase, reverse transcriptase, with RNase H activity)
4 (X)	Transcriptional *trans*-activating protein kinase

Hepatitis D virus is a spherical particle that is approximately 36 nm in diameter. The surface or outer coat of the virus consists of a lipid bilayer that contains small, medium, and large HBsAgs, and envelopes the –ssRNA circular genome and the delta antigen. Hepatitis D virus can only form infectious particles in a cell coinfected with hepatitis B virus because hepatitis B virus provides the HBsAg that is required for reinfection into another cell. Once inside the cell, hepatitis D virus can replicate in the absence of hepatitis B virus.

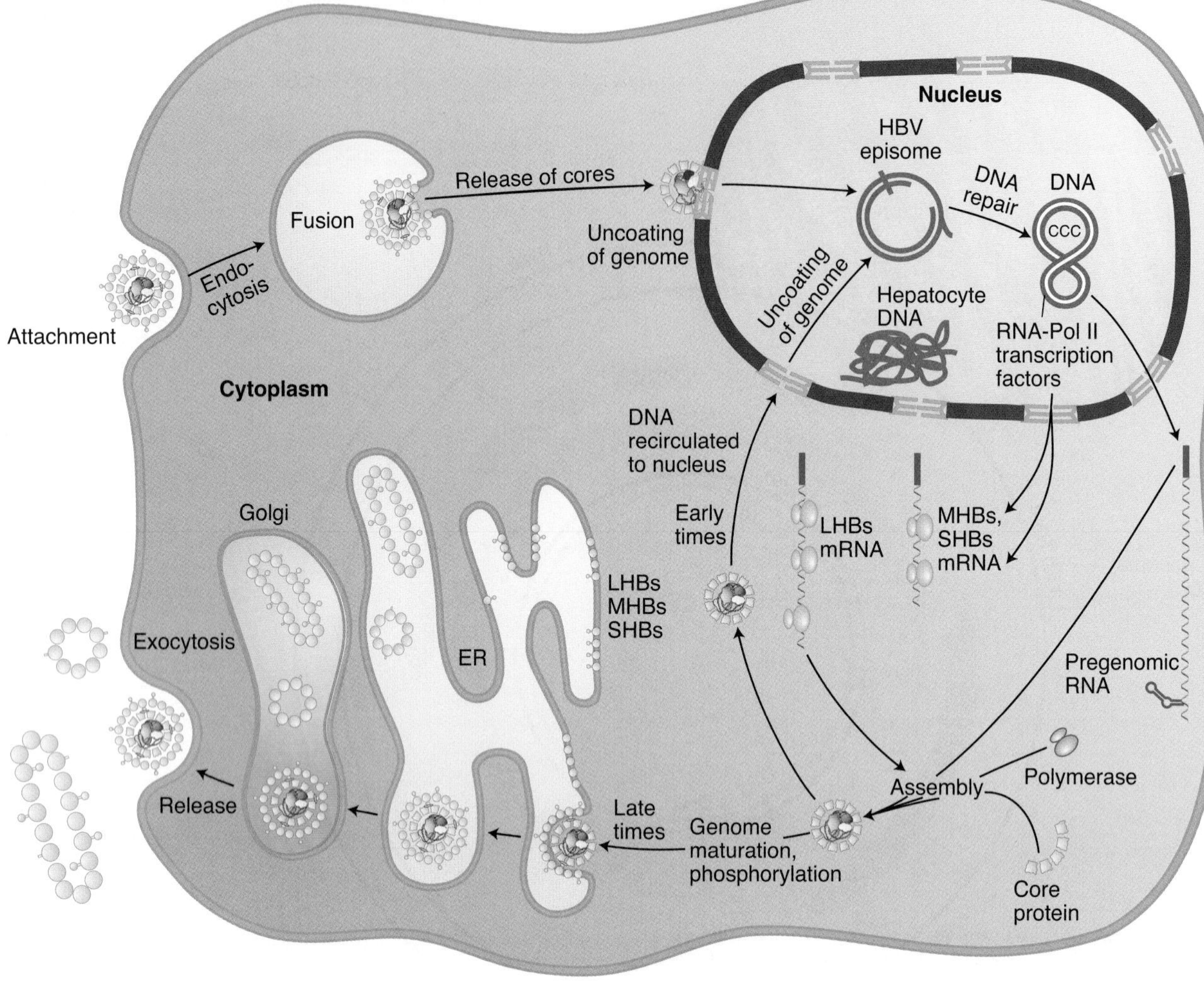

FIGURE 17-14 Life cycle of hepatitis B virus. The HBeAg and HBxAg synthesis are omitted for the sake of simplicity. HBsAg proteins are inserted into the host's endoplasmic reticulum membrane. Adapted from Thomas, H. C., Lemon, S., and Zuckerman, A. J. *Viral Hepatitis*, Third Edition. Blackwell Publishing, 2005.

The genome is 1.7 kb in length, which is similar to plant viral satellites and viroids (see Chapter 19). It codes for two different types of the same protein (the **delta antigen**) that are 195 or 214 amino acids in length. About 70 molecules of delta antigen surround and stabilize the viral RNA genome. The large delta antigen contains four different functional domains: an RNA binding domain, a nuclear localization signal, a virus assembly signal, and a coiled-coil sequence that is responsible for oligomerization of the protein. The delta antigen can form either a homo- or heterodimer. The small form of the delta antigen is required for hepatitis D virus RNA replication, whereas the large form is required for the suppression of hepatitis D virus RNA replication and the assembly and export of virion particles. The small delta antigen is missing the virus assembly signal (**FIGURE 17-15a**).

The genome has a high degree of intramolecular base-pairing, giving it a double-stranded property similar to that found in plant viroids under natural conditions. When the envelope of the virus is removed inside of an infected cell, the viral genome is targeted to the nucleus. Unlike most RNA viruses, though, it does not encode its own RNA-dependent RNA polymerase for genome replication. Instead, it is dependent upon the cellular RNA polymerase II and other host factors to replicate its genome. The hepatitis D virus genomic RNA is a **ribozyme** that cleaves itself, and ribozyme activity is necessary for hepatitis D virus replication. HDV genome replication is believed to occur via a **rolling circle mechanism** similar to that of viroids (see Chapter 19). Figure 17-15b shows the structure of hepatitis D virus RNAs, and a model for the hepatitis D virus replication cycle is shown in **FIGURE 17-16**. The hepatitis D RNA –sense (negative-sense) genome acts as a template for synthesis of antigenomic RNA (+sense, or positive-sense) and the viral mRNAs. The antigenomic RNA

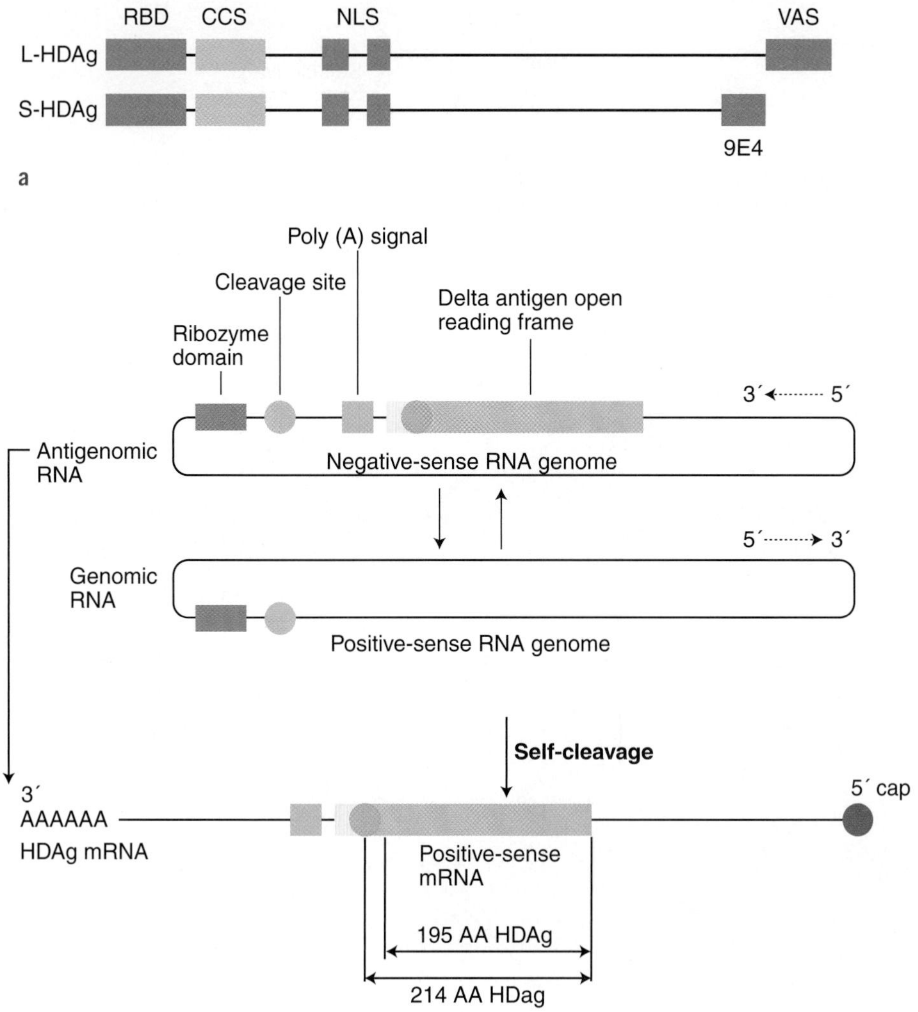

FIGURE 17-15 (a) Hepatitis D virus functional domains of the small and large forms of the delta antigen. RBD = RNA-binding domain, CCS = coiled-coil sequence, NLS = nuclear localization sequence, VAS = virus assembly signal. Adapted from Thomas, H. C., Lemon, S., and Zuckerman, A. J. *Viral Hepatitis*, Third Edition. Blackwell Publishing, 2005. **(b)** Structure of hepatitis D virus RNAs. Adapted from University of South Carolina, School of Medicine. "Virology: Hepatitis viruses." *Microbiology and Immunology.* University of South Carolina, 2008. http://pathmicro.med.sc.edu/virol/hepatitis-virus.htm.

multimers are self-cleaved (step 2 in Figure 17-16) and recircularize, serving as a template for multimers of genomic RNAs, which are self-cleaved and circularized (step 4 in the figure). The single mRNA transcribed from the viral genome exits the nucleus to produce the small and large forms of the delta antigen (step 7 in the figure). The ribonucleoprotein complexes (RNPs) assemble and are exported to the ER located in the cytoplasm of the cell, where they associate with the HBsAgs (steps 8 and 9 in the figure). The HBsAgs are required for the packaging of the hepatitis D virions. Mature hepatitis D viruses are produced and secreted outside of the cell.

Hepatitis C Virus

Hepatitis C virus is an icosahedral-shaped, enveloped +ssRNA flavivirus that causes hepatitis non-A, non-B. Little is known about the ultrastructure of hepatitis C virus because it has been extremely difficult to visualize the virus particles directly from infected serum and tissues. The hypothetical structure of hepatitis C virus is shown in **FIGURE 17-17**. It is the only flavivirus that is not transmitted by arthropod vectors (e.g., mosquitoes). Yellow fever virus is the prototype of viruses in the *Flaviviridae* family. Hepatitis C viruses are also similar to picornaviruses (e.g., poliovirus), with the main exception that they are enveloped. Since the cloning of the viral nucleic acid in 1989, considerable progress has been made in characterizing the virus genome and proteins.

The hepatitis C virus genome is about 9.2 kb in length and contains an IRES and one long ORF that encodes a polyprotein precursor. The viral RNA is translated by host ribosomes by cap-independent translation. The polyprotein precursor is cleaved into structural and nonstructural proteins by cellular and viral proteases. Hepatitis C virus *NS5b* gene encodes its own RNA-dependent RNA polymerase. There is substantial sequence variation in the hepatitis C virus genome because *NS5b* lacks proofreading capability.

A better understanding of the viral life cycle, especially the entry process into host cells is needed. Viral entry requires clathrin and more than one receptor. CD81, an integral membrane protein belonging to a family of **tetraspanins**, scavenger

FIGURE 17-16 Proposed model for hepatitis D virus life cycle. Adapted from Cunha, C., Freitas, N., Mota, S., "Developments in hepatitis delta research," *Internet J Tropical Med* 1 (2003).

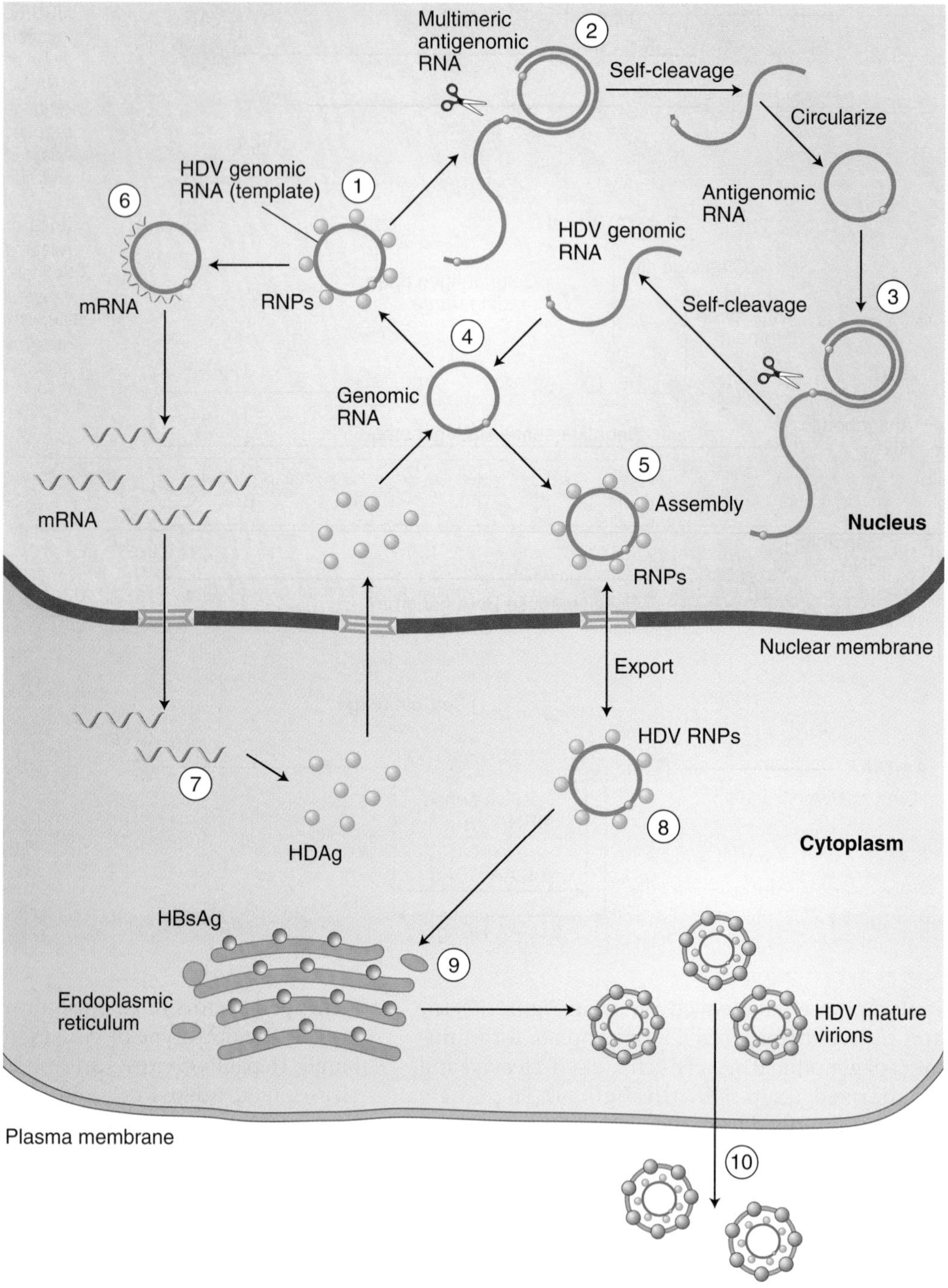

receptor class B type 1 (SR-B1), and claudin-1 have recently been identified as co-receptors based on their interaction with hepatitis C viral structural E2 or E1 proteins.

The lack of patient response to interferon therapy has been correlated with mutations in the *E2* gene. The E1 and E2 proteins presumably self-assemble to form the virion (Figure 17-17).

Hepatitis E Virus

Hepatitis E virus is transmitted by the fecal–oral route, often through contaminated water. Like hepatitis A virus, it has a +ssRNA genome. It is a member of the *Hepeviridae* family. Hepatitis E infection is clinically indistinguishable from hepatitis A infection, but its particles are much less stable than those of hepatitis A. Hepatitis E viruses are approximately 32 to 34 nm in diameter, nonenveloped, and icosahedral-shaped. Its genome is approximately 7.2 kb in length and contains short 5′- and 3′-end noncoding regions. The 3′ end contains a poly (A) tail.

The genome has three ORFs and all three coding frames are used to express different proteins.

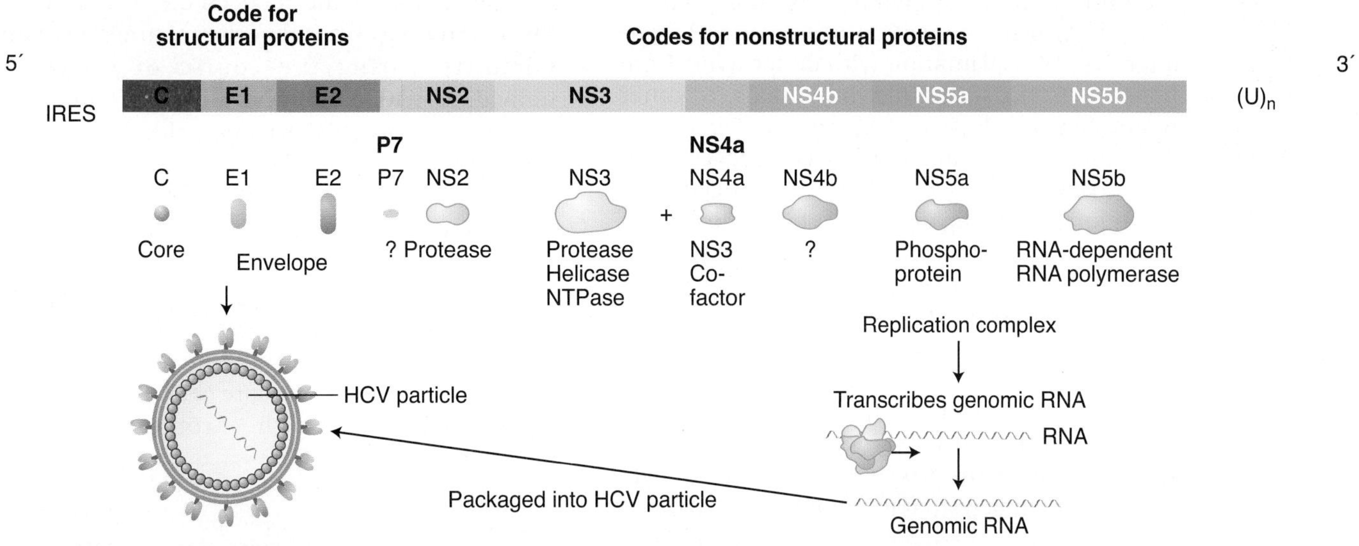

FIGURE 17-17 Hypothetical structure of the hepatitis C virus. Nonstructural proteins are labeled "NS" followed by a number. They have diverse functions and form the replication complex. There is no known function for the p7 protein, and it has not been classified as structural or nonstructural. Adapted from Roingeard, P. *Biol Cell* 96 (2004): 103–108.

ORF1 codes for nonstructural proteins that possess methyltransferase, protease, RNA helicase, and RNA-dependent RNA polymerase (replicase) motifs. ORF2 does not overlap with ORF1 and is located at the 3′ end of the genome and presumably codes for the only capsid (structural) protein. ORF3 begins at the end of ORF1. It overlaps with ORF2 and codes for a small phosphoprotein of undefined function (**FIGURE 17-18**). Presumably hepatitis E virus replicates solely in the cytoplasm of the cell. Overall, our knowledge of hepatitis E virus replication is poor. The lack of a suitable cell culture system to propagate the virus is an obstacle in deciphering its molecular biology.

17.8 A Quick Note About the Pathophysiology of Chronic Hepatitis Virus Infections

The biggest difference between enveloped and naked hepatitis viruses is that the enveloped viruses (hepatitis B, C, D, and G) cause persistent and chronic infections. These viruses have devised a strategy to escape detection and elimination following natural infection and thus can hide from

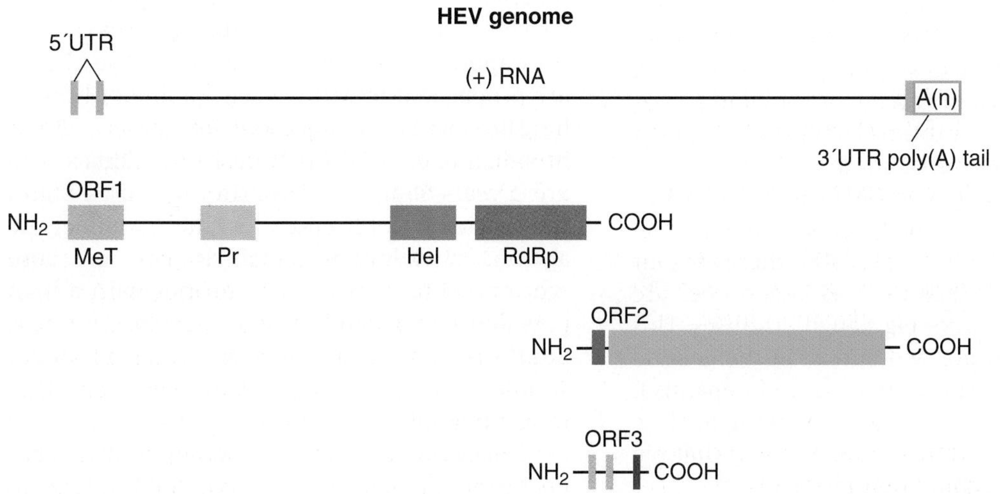

FIGURE 17-18 Genome organization of hepatitis E virus. Three ORFs encode the viral proteins: ORF1 (a nonstructural, replicative polyprotein), ORF2 (a capsid protein), and ORF3 (unknown function). Adapted from Jameel, S. *Exp Rev Mol Med* 6 (1999): 1–16.

the host's immune system. For example, reports have described reactivation of hepatitis B virus after liver transplantation with donor livers from hepatitis B virus-immune individuals. In other words, hepatitis B virus DNA was recovered from "immune" patients and donors. This suggests that hepatitis B virus may exist in the hepatocytes in a latent form for a long time. It has been shown that hepatotropic viral clearance is associated with a strong virus-specific cytotoxic T cell and T helper cell response. Antibody-mediated responses have been inadequate in clearing the infection.

Evidence is accumulating that the liver damage caused by hepatitis B and C viruses during the course of infection may result from the **autoimmune reaction** directed against hepatocyte membrane antigens that is initiated by viral infection. Damage to the liver is mediated by inflammatory cytokines, which are small proteins that communicate between cells and may play a role in the liver. They play an important role in the defense against viral infections. Both hepatitis B and C viruses may have the capacity to modulate cytokine gene expression and responses. The chronic infections caused by these viruses may result in an on-going inflammatory response that activates the process of liver fibrosis. More studies are needed to develop strategies useful in combating the effects of hepatitis B and C infections on the liver.

17.9 Genetic Diversity Among the Hepatitis Viruses

There is only one known **serotype** of hepatitis A viruses, but there are at least seven different **genotypes** worldwide. Individuals who are infected by hepatitis A virus in one part of the world are not susceptible to reinfection by a different hepatitis A virus genotype in another part of the world. Immune globulin and vaccine prepared in a variety of countries protect travelers from hepatitis A irrespective of their destination.

The majority of hepatitis E virus genomic isolates have only been partially sequenced. Based on what genomic sequence is available, there are four major genotypes. Hepatitis E viral genomes are present in sewage from pig slaughterhouses. This suggests that they are widespread in the general swine population. There are reports of hepatitis E virus antibodies among farmers working in close contact with pigs, raising the possibility that pig hepatitis E viruses could infect humans.

A common theme between the chronic hepatitis B and C viruses is that they synthesize many **variants** during the course of infection. Individuals with chronic hepatitis B or C infection produce 10^{10} to 10^{11} virus particles daily, which indicates that the viruses are actively replicating. This high replication activity by the viral RNA-dependent RNA polymerases that lack proofreading function explains, in part, the rapid emergence of viral variants. It also may explain the frequent immune escape phenomena. Mutations in the hepatitis B virus promoter of the core gene or in the precore region of the core gene are frequently found in patients with chronic hepatitis B infection.

17.10 Management and Prevention of Hepatitis A–E Viruses

Hepatitis A

No specific treatment for hepatitis A infections is available and management is supportive. Spread of hepatitis A infections can be reduced by handwashing and proper sanitary disposal of human feces. Prevention by passive immunization of individuals who likely had been exposed to or in close contact with a person with hepatitis A is achieved by intramuscular injection of human immunoglobulin that contains hepatitis A antibodies. In general, if administered before exposure or within two weeks after exposure, immunoglobulin is 85% effective in preventing hepatitis A.

Hepatitis A vaccines were developed in a manner similar for that of poliovirus vaccines. Formalin-inactivated, cell culture-produced whole-virus vaccines such as Havrix, Vaqta, and Avaxim are now approved in much of the world. The viruses are propagated in MRC-5 cells (a line of human lung fibroblast cells). Injections are typically administered at 0, 6, and 12 months in individuals who are 2 years of age or older. (In the United States the vaccine is not licensed for children under the age of 1.) In addition to high-risk persons, those who travel to or work in countries with a high prevalence of hepatitis A (e.g., Mexico, Central or South America, Asia, Africa, and Eastern Europe) should be vaccinated. Live attenuated vaccines (like the Sabin-type of oral vaccine) have been developed and tested in animals and humans but are not yet available, as they have not been able to induce a satisfactory immune response when administered orally.

Hepatitis B

The basic aim of treating hepatitis B patients is to stop viral replication and prevent end-stage liver disease. Six licensed drug therapies are used against hepatitis B infections: **interferon α-2b, lamivudine, adefovir dipivoxil, entecavir, telbivudine**, and **tenofovir** therapy. Interferon is an **immune modulator** and is administered as an injection (see Chapter 7 for more about interferon and innate immunity). About 46% of patients respond to interferon α-2b (Intron A) in the first year of treatment. (A response is the cessation of viral replication.) There are three drawbacks to interferon α-2b treatment: it has a large number of side effects (see Chapter 7), it is expensive, and it is not effective in a high percentage of patients.

Lamivudine (Epivir), adefovir dipivoxil (Hepsera), entecavir (Baraclude), telbivudine, and tenofovir (Viread) are nucleoside analogs that are potent inhibitors of DNA replication. Telbivudine (Tyzeka) interferes with plus-strand synthesis of HBV DNA, thereby inhibiting the HBV polymerase. Lamivudine was approved for treatment in 1999 and was first used to treat HIV patients, its mechanism of action being the inhibition of reverse transcriptase. Advantages of lamivudine are that it has few side effects compared to interferon α-2b and has been shown to dramatically inhibit hepatitis B virus DNA replication. The disadvantage is that replication rapidly returns when therapy is stopped. Lamivudine-resistant strains of hepatitis B virus develop at a rate of 15% to 20% per year of therapy.

Adefovir dipivoxil therapy has been used since 2002 as a first-line monotherapy, combination therapy, or a therapy in patients who have lamivudine-resistant infections. It can be administered orally and patients tolerate it well.

Entecavir was approved for use in 2005. It cannot be used as monotherapy in patients with HBV/HIV infection who need treatment for HBV infection. Telbivudine was approved in 2006 but has not been widely embraced for therapy against HBV. Tenovir was first approved as an HIV reverse transcriptase inhibitor and has similar activity against HBV. It was approved for use in hepatitis B patient treatment in 2008. To date, it has had superior results compared to adefovir and is expected to replace adefovir as a treatment for chronic hepatitis B.

Antibodies against HBsAg protect individuals from acute and chronic hepatitis B infection if used shortly after exposure to hepatitis B virus. Today immune globulin is used as an adjunct to hepatitis B vaccine in preventing hepatitis B virus transmission from an infected mother to fetus. If untreated, 70% to 90% of infants born to HBeAg-positive mothers become infected at birth and develop a chronic hepatitis B infection. Hepatitis B immunoglobulin is also used for postexposure prophylaxis after accidental needlestick or other medical-related injuries in persons who work with infectious body fluids that contain HBsAg.

Early **hepatitis B vaccines** were prepared by harvesting the 17- to 25-nm particles from the plasma of chronically infected individuals. The particles were purified and inactivated by heat, formaldehyde, urea, or pepsin. These **plasma-derived vaccines** have been available since 1982. Subsequently, vaccine manufacturers used genetic engineering to express HBsAg in yeast (*Saccharomyces cerevisiae*) and mammalian cells. This led to the development of **recombinant DNA vaccines**.

Recombinant DNA vaccines were produced by inserting the *S* gene that codes for the HBsAg into a DNA plasmid. The plasmids were then transformed into yeast or transfected into mammalian cells. The HBsAg was expressed and purified by chromatography and filtration (**FIGURE 17-19**). Under these conditions, purified HBsAg self-assembles into the spherical particles closely resembling the 17- to 25-nm particles found in the serum of people with chronic hepatitis. This vaccine is not an infectious particle, making this a very safe vaccine.

The CDC recommends that all newborns receive a birth dose of hepatitis B vaccine before leaving the hospital unless a physician provides a written order to defer the birth dose. The CDC also recommends that all children age 19 and younger receive the vaccine series. Hepatitis B can cause liver cancer, and, thus, an added benefit of the vaccine is that it prevents hepatitis B-related liver cancer. (This vaccine was the first to prevent a cancer.) From 1990 to 2002, the incidence of hepatitis B decreased 67% in the United States. Hepatitis B vaccination programs have been implemented with considerable success. Unfortunately, the recombinant hepatitis B vaccine is expensive to purchase, which restricts its availability in developing nations. For this reason, oral hepatitis B vaccine candidates are being produced and delivered in plants such as potatoes and soybeans. A number of candidates are suitable for early stage clinical trials.

Hepatitis C

Hepatitis C infections progress to liver disease in up to 85% of patients. There are no vaccines available to prevent hepatitis C infection. Like other RNA viruses, hepatitis C has a high mutation rate, resulting in genetic diversity. So far, six main HCV

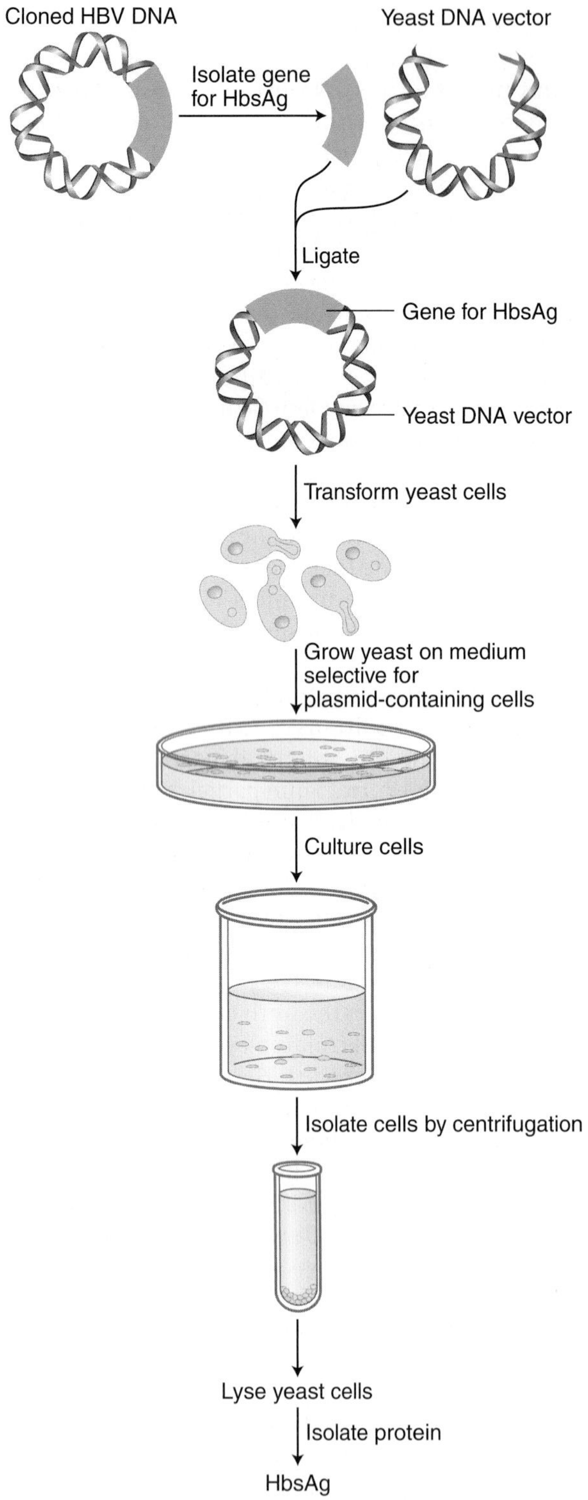

FIGURE 17-19 Flowchart showing the steps involved in preparing a recombinant DNA hepatitis B vaccine in yeast. Adapted from Marks, D. B., Marks, A. D., and Smith, C. M. *Basic Medical Biochemistry: A Clinical Approach.* Lippincott Williams & Wilkins, 1996.

genotypes have been described. **Genotype** refers to the genetic makeup of the virus. Recently new variants, assigned as types 7–11, have been recognized. Genotypes are further broken down into subtypes designated by a lowercase letter. For example, there are three subtypes of hepatitis C type 1 viruses, which are designated 1a, 1b, and 1c. There are general global patterns of genotypes. Genotype 1, 2, and 3 are the most common genotypes seen in North America and Western Europe (**FIGURE 17-20**). *HCV genotype determines the duration of treatment and expected efficacy of antiviral activity.*

The therapies available for treatment are **pegylated interferon α-2a** or **pegylated interferon α-2b** and/or **ribavirin**. Pegylation involves the attachment of a polyethylene glycol molecule to the active interferon molecule. Pegylation improves the half-life of interferon, limiting its degradation in the body thereby making it long-acting. Interferon is a glycoprotein produced by the body as part of the immune system response toward viral infection (see Chapter 7). Hence, interferon treatment may prevent hepatitis C virus replication and the production of new viral particles in infected individuals. Interferon is administered to the patient as weekly injections. Ribavirin is an oral nucleoside analog that inhibits replication of a number of viruses. Response to therapy is measured by determining the levels of serum transaminases and the detection of hepatitis C virus RNA in serum.

In order to prescribe a treatment plan with the highest probability of success, a person must have his or her particular HCV genotype and subtype identified. Once the genotype is identified, it must be tested again; genotypes do not change during the course of infection. A combination of ribavirin/pegylated interferon α-2b treatment over a 6- to 12-month period has increased overall cure rates from 20% to 40%. This therapy is challenging because of the side effects caused by interferon and ribavirin. Ribavirin is **teratogenic** (i.e., it interferes with normal embryonic development) and therefore cannot be taken by patients who are pregnant. The main side effect of ribavirin is anemia.

Patients with genotypes 2 or 3 are almost three times more likely than patients with genotype 1 to respond to therapy with α interferon or a combination of α interferon and ribavirin. A 24-week course of combination treatment is usually adequate for those with genotypes 2 and 3. A 48-week course of combination treatment is typically adequate for those with genotype 1. Data concerning genotype 4 treatment are mixed. Genotype 5 patients respond similarly to those with genotype 1 toward combination treatment. Patients with genotype 6 have an intermediate response level, between that seen

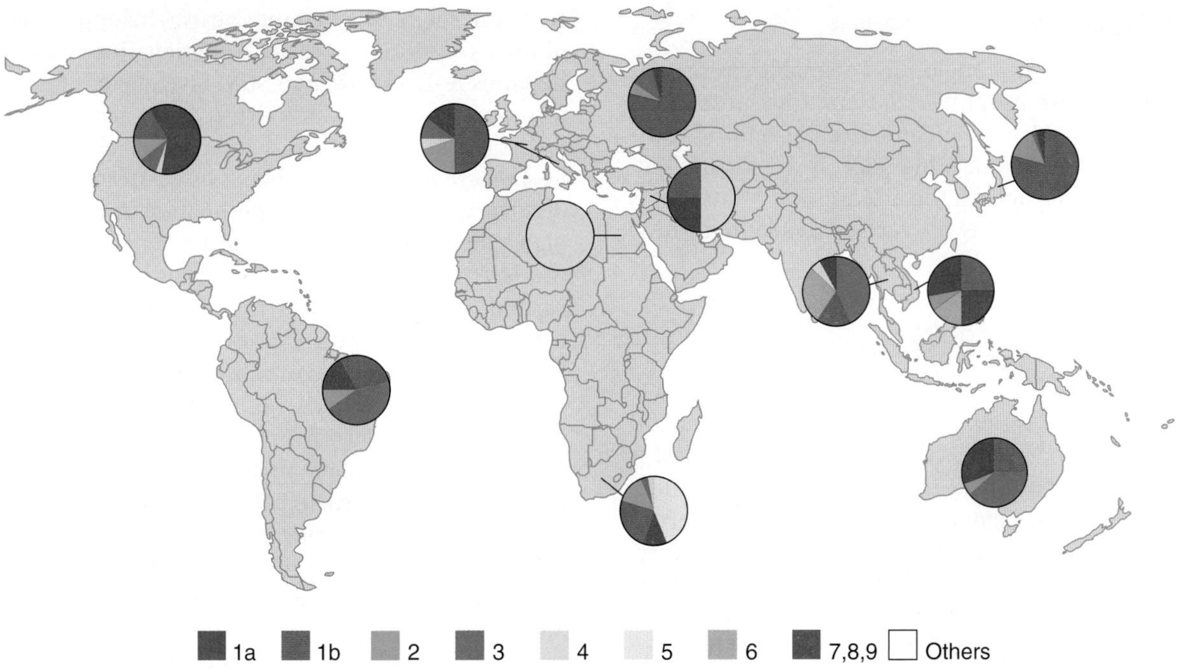

FIGURE 17-20 Geographic distribution of the hepatitis C virus genotype.

with genotype 1 and genotypes 2 or 3. The response to standard treatment is not well established for genotypes 7–11. Patients need to be monitored closely for complications or symptoms of adverse reactions of combination therapy.

HCV affects 170 million people worldwide. Standard HCV treatment using ribavirin/pegylated interferon α-2b treatment is effective in about 50% of patients with genotype 1. That being said, there is a need for more promising orally administered antiviral drugs that interrupt HCV replication. Currently in preclinical and clinical developments are inhibitors of HCV entry, NS3/4 protease inhibitors, NS5A inhibitors, NS5B polymerase inhibitors and agents targeting host factors, and inhibitors of HCV assembly and release.

In 2011, Victrelis (boceprevir) and Incivek (telaprevir) were enthusiastically approved by the FDA as two new treatment options for hepatitis C infection. Boceprevir binds to the active site of NS3 (an HCV protease). Telaprevir targets the NS3/4a protease of HCV. Targeting its protease activity cripples HCV replication. The viral proteases are responsible for processing the nonstructural polyprotein located in the N-terminus of the HCV polyprotein. Inhibition of the proteases results in unprocessed HCV polyprotein, which ultimately affects the viral proteins involved in genome replication. Both of these drugs increase the already difficult side effects of standard therapy, such as anemia. However, the drugs do not need to be taken for a full 48-week course of treatment. Clinical trials were promising for these drugs. Adding either drug to the standard regimen boosted sustained viral response rates 20% to 45%. Because these drugs target HCV specifically, they are a major improvement over current hepatitis C therapy.

It is recommended that telaprevir be given for the first 12 weeks of standard therapy in combination with interferon and ribavirin. Treatment with boceprevir is more complicated. Patients who have not had any treatment for HCV would receive four weeks of standard treatment (ribavirin plus pegylated interferon), then 24 weeks of standard treatment plus boceprevir (for a 28-week total treatment length). If patients still had detectable HCV levels after 8 weeks of boceprevir, treatment would be extended with an additional 20 weeks of standard treatment after finishing boceprevir (for a total treatment length of 48 weeks). Patients who have been treated previously for hepatitis C infection would receive 4 weeks of standard treatment, then 32 weeks of standard treatment plus boceprevir (for a 36-week total treatment length). If patients still had detectable HCV levels after 8 weeks of boceprevir, treatment would be extended with an additional 12 weeks of standard treatment after finishing boceprevir (for a total treatment length of 48 weeks).

Hepatitis D

Hepatitis D infections are maintained by reducing hepatitis B virus replication (e.g., interferon α-2b, lamivudine, and adefovir dipivoxil therapy). There is no immunoglobulin or vaccine available against hepatitis D virus.

Hepatitis E

There is no immunoglobulin available for pre- or postexposure prophylaxis. Hepatitis E infections are usually self-limited, and hospitalization is generally not required. Prevention is the most effective approach against the disease, and treatment is supportive. There is no vaccine available against hepatitis E virus.

Summary

Viral hepatitis is a major global health problem, and the clinical significance and economic costs of it have made it one of the most important of all infectious diseases. Epidemics of infectious and serum hepatitis have been recorded throughout history, especially during times of war. Infectious hepatitis is caused by the fecal–oral mode of transmission, whereas serum hepatitis is an infection of the liver caused by a bloodborne virus. A group of unrelated human pathogens termed the "hepatitis viruses" cause virally induced primary hepatitis. Primary hepatitis occurs immediately after an initial infection by a hepatitis virus in contrast to secondary hepatitis. This chapter focuses on the hepatitis viruses. *These viruses are quite diverse genetically and taxonomically.* They are grouped together only because they share a liver tissue tropism, causing inflammation and necrosis of the liver.

In general, the signs and symptoms of hepatitis are:

- fatigue
- abdominal pain
- loss of appetite
- nausea and vomiting
- dark urine
- jaundice

Viral hepatitis occurs endemically in all parts of the world. Epidemics of enterically transmitted hepatitis often occur in countries in Asia and Africa where sanitation is suboptimal. The nomenclature of hepatitis viruses is based on the order of discovery of new agents that cause hepatitis rather than modes of transmission or clinical problems associated with that viral agent. The hepatitis viruses belong to five different families. There are eight hepatitis viruses: hepatitis A–E, G, TTV, and SEN. Five of eight hepatotropic viruses (hepatitis viruses A–E) account for 90% of viral hepatitis cases and are well characterized. There are still other hepatitis viruses awaiting discovery, but it appears that other non-A–E viruses play a minor role in pathogenicity.

Hepatitis viruses enter the bloodstream and are carried to the liver where they infect the hepatocytes of the liver. As the viruses spread and replicate in the liver, hepatocytes may become damaged. The damage may result in cirrhosis, which is an impairing of liver function. The liver has many functions, such as playing a role in metabolism and the filtering of toxins and waste products in the blood. It produces bile, which aids in digestion. The biggest difference between enveloped and naked hepatitis viruses is that enveloped viruses (hepatitis B, C, D, and G) cause persistent and chronic infections. These viruses have devised a strategy to escape detection and elimination following natural infection and can hide from the host's immune system. In contrast, hepatitis A and E viruses are both self-limited, fecally spread diseases.

Proper sanitation practices are the best way to prevent outbreaks of hepatitis A and E viruses. Once an individual is infected, treatment is supportive. Passive immunization and a vaccine are available for hepatitis A virus.

Hepatitis B and C viruses are well-recognized causes of chronic hepatitis, cirrhosis, and hepatocellular carcinoma and are capable of triggering autoimmune reactions. Treatments for these two viruses consist of two classes of drugs: immune modulators (such as interferon) and nucleoside analogs that block viral RNA or DNA replication. The side effects of these drugs make it a challenging therapy for patients.

DNA technology has made the development and production of a safe hepatitis B vaccine possible. It is the first anticancer vaccine because it prevents hepatitis B and thus hepatocellular carcinoma. Infant vaccination against hepatitis B virus is having a huge impact on carrier rates of hepatitis B in adolescents and young adults.

CASE STUDY QUESTIONS

These questions relate to the Case Study presented at the beginning of the chapter.

1. What is the incubation period of hepatitis A?
2. What are the signs and symptoms of hepatitis A?
3. How did CDC epidemiologists eliminate restaurant workers as the cause of this outbreak?
4. How is hepatitis A virus transmitted? How is it prevented?
5. In this instance, would health officials recommend that consumers of green onions obtain postexposure prophylaxis? Why or why not?
6. Trace-back investigations determined that one or more farms in Mexico were the source of the green onions. Hypothesize how this fresh produce became contaminated. Be sure to address the cultivation and harvesting practices of fresh fruits and vegetables.
7. Compare and contrast hepatitis A and E infections.

References

Centers for Disease Control and Prevention (CDC). 1999. "Prevention of hepatitis A through active or passive immunization: Recommendations of the Advisory Committee on Immunization Practices (ACIP)." *MMWR* 48(RR12):1–37.

Dato, V., et al. 2003. "Hepatitis A outbreak associated with green onions at a restaurant—Monaca, Pennsylvania, 2003." *MMWR* 52(47):1155–1157.

Pugh, J. 2004. "Learning about food safety from deadly hepatitis A virus outbreak." *ASM News* 70(2): 51–52.

United States Food and Drug Administration (US FDA)/ Center for Food Safety and Applied Nutrition (CFSAN). 1998. *Guide to Minimize Microbial Food Safety Hazards for Fresh Fruits and Vegetables.* Washington, DC.

Wasley, A., Fiore, A., and Bell, B. P. 2006. "Hepatitis A in the era of vaccination." *Epidemiol Rev* 28: 101–111.

Wheeler, C., et al. 2005. "An outbreak of hepatitis A associated with green onions." *N Engl J Med* 353(9): 890–897.

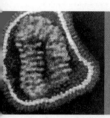

CASE STUDY 1: BOB MASSIE AND BLOODBORNE INFECTIONS

Bob Massie, a hemophiliac, was used to daily blood transfusions and frequent trips to the hospital (**FIGURE CS 17-1**). Despite his physical illness, he earned degrees at Princeton University and Yale University's Divinity School, and led an active and full life as an Episcopal priest, scholar, writer, and political activist.

At the beginning of his first semester at Yale in the fall of 1978, Massie fell ill with an undiagnosable illness that was so severe he withdrew from school and went home for the rest of the year. Years later, in 1984, he was diagnosed as being HIV-positive while participating in a hemophilia research study.

Massie had contracted HIV when receiving contaminated factor VIII blood clotting factors. At the time of his diagnosis doctors told him that his life expectancy was one to five years. As the years passed, though, it was clear that Massie's case was unusual because his viral loads remained nearly undetectable. Scientists studied his immune response as they tried to determine why he was resistant to the effects of HIV infection. In 1999, PBS broadcasted a NOVA program called "Surviving AIDS" that featured Bob Massie. Massie had never shown any symptoms of AIDS and for years lived a full and vigorous life.

Unfortunately, Massie contracted another bloodborne pathogen, hepatitis C virus, which caused a chronic hepatitis infection. He likely contracted the infection when he received large-pool clotting factor from plasma donors during a time when it was not virucidally treated (the 1980s through the mid-1990s). Bob Massie received a liver transplant in 2009 and is now cured of hemophilia.

1. What virus(es) cause chronic hepatitis?
2. What are the symptoms of chronic hepatitis?

(continued)

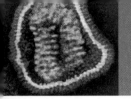

CASE STUDY 1: BOB MASSIE AND BLOODBORNE INFECTIONS (continued)

FIGURE CS 17-1 Bob Massie was infected with HIV and HCV after receiving contaminated clotting factor VIII.

3. What diagnostic tests are used to detect viral hepatitis?
4. How is chronic hepatitis C managed?
5. Explain why Bob Massie is no longer a hemophiliac. Research the mechanism.
6. Bob Massie is now taking immunosuppressive drugs to prevent organ rejection. Because Massie was also infected with HIV, has his health status changed regarding his HIV status? Why or why not? Perform an Internet search on Massie to determine his current health status to answer these questions.
7. After a liver transplant, is Massie cured of HCV infection? Explain your answer.

References

Bob Massie, http://www.bobmassie.org/

Surviving AIDS. 1999. NOVA, PBS, WGBH Science Unit: Boston. Accessed May 17, 2011, from http://www.pbs.org/wgbh/nova/aids/

CASE STUDY 2: HEPATITIS E

Outbreaks of viral hepatitis E have been plaguing the African country of Chad in recent years. Villages in Chad have few roads and during the rainy season floodwaters create huge seasonal muddy rivers called *wadis,* making travel even more difficult. There are no boats or horses to help people cross wadis. People have to swim to cross them. There are overcrowded refugee camps in Chad that struggle to accommodate the stressed civilians. Clean drinking water and good hygiene practices are lacking.

1. How is hepatitis E virus transmitted? What is the likely source of hepatitis E virus contamination in Chad?
2. How is hepatitis E diagnosed?
3. How is hepatitis E treated? How is it prevented?
4. Outbreaks of hepatitis E in Hong Kong have been traced to contaminated shellfish. Why are shellfish a common reservoir of hepatitis A and E viruses? Explain your answer.
5. There is evidence that hepatitis E virus is a zoonosis. Hepatitis E virus has been isolated from what animal(s), and how would the virus be transmitted to humans?

References

ProMED-mail. August 22, 2005. "Hepatitis E—Chad (Eastern)." Archive number 20050822.2477.

Chau, T. N., et al. 2006. "Epidemiology and clinical features of sporadic hepatitis E as compared with hepatitis A." *Am J Gastroenterol* 101(2):292–296.

Sadler, G. J., et al. 2006. "UK acquired hepatitis E—An emerging problem?" *J Med Virol* 78:473–475.

Resources

Primary Literature and Reviews

Abraham, P. 2005. "TT viruses: how much do we know?" *Indian J Med Res* 122:7–10.

Andenaes, S., Lie, A., and Degre, M. 2000. "Prevalence of hepatitis A, B, C, and E antibody in flying airline personnel." *Aviat Space Environ Med* 71(12):1178–1180.

Beck, J., and Nassal, M. 2007. "Hepatitis B replication." *World J Gasteroenterol* 13(1):48–64.

Bereket-Yucel, S. 2007. "Risk of hepatitis B infections in Olympic wrestling." *Br J Sports Med* 41:306–310.

Bill, C. A., and Summers, J. 2004. "Genomic DNA double-strand breaks are targets for hepadnaviral DNA integration." *Proc Natl Acad Sci USA* 101(30): 11135–11140.

Bourliere, M., et al. 2000. "Covert transmission of hepatitis C virus during bloody fisticuffs." *Gastroenterology* 119(2)507–511.

Briggs, M. E., et al. 2001. "Prevalence and risk factors for hepatitis C virus infection at an urban Veterans Administration medical center." *Hepatology* 34(6): 1200–1205.

Bukh, J., and Purcell, R. H. 2006. "A milestone for hepatitis C research: A virus generated in cell culture is fully viable *in vivo*." *Proc Natl Acad Sci USA* 103(10): 3500–3501.

Centers for disease control and prevention (CDC). 2005. "A comprehensive immunization strategy to eliminate transmission of hepatitis B virus infection in the United States." *MMWR* 54(RR16):1–23.

Chau, T. N., et al. 2006. "Epidemiology and clinical features of sporadic hepatitis E as compared with hepatitis A." *Am J Gastroenterol* 101(2):292–296.

Cheung, R. C. 2000. "Epidemiology of hepatitis C virus infection in American veterans." *Am J Gastroenterol* 95(3):740–747.

Cheung, R. C., et al. 1997. "Hepatitis G virus is it a hepatitis virus?" *West J Med* 167(1):23–33.

Choo, Q. L., et al. 1989. "Isolation of a cDNA clone derived from a blood-borne non-A, non-B viral hepatitis genome." *Science* 244(4902):359–362.

Cormier, E. G., et al. 2004. "CD81 is an entry coreceptor for hepatitis C virus." *Proc Natl Acad Sci USA* 101(19): 7270–7274.

Deka, N., Sharma, M. D., and Mukerjee, R. 1994. "Isolation of the novel agent from human stool samples that is associated with sporadic non-A, non-B hepatitis." *J Virol* 68(12):7810–7815.

Dooley, D. P. 2005. "History of U.S. military contributions to the study of viral hepatitis." *Mil Sci* 170(4):71–76.

Enomoto, M., et al. 2009. "Emerging antiviral drugs for hepatitis C virus." *Rev Rec Clin Trials* 4:179–184.

Evans, M. J., et al. 2007. "Claudin-1 is a hepatitis C virus co-receptor required for a late step in entry." *Nature* 446:801–805.

Fang, J., Chow, V., and Law, J. 1997. "HCV infection: worldwide genotype distribution." *Clin Liver Dis* 1:503.

Fernandez, A. F., Esteller, M. 2010. "Viral epigenomes in human tumorigenesis." *Oncogene* 29:1405–1420.

Flisiak, R., Parfieniuk, A. 2010. "Investigational drugs for hepatitis C." *Exp Opin Investig Drugs* 19(1):63–75.

Freeman, G. 1946. "Epidemiology and incubation period of jaundice following yellow fever vaccinations." *Am J Trop Med* 26:15–32.

Green, R. C. 1987. "The epidemics that delayed D-Day." *Nurs RSA* 2:24–25.

Guha, C., et al. 2005. "Cell culture models and animal models of viral hepatitis. Part II: hepatitis C." *Lab Anim* 34(2):39–47.

Hilleman, M. R., et al. 1983. "The preparation and safety of hepatitis B vaccine." *J Infect* 7(Suppl 1):3–8.

Hollinger, F. B. 1998. "Overview of viral hepatitis." *Circumpolar Health* 96:276–279.

Huang, Z., et al. 2005. "Virus-like particle expression and assembly in plants: Hepatitis B and Norwalk viruses." *Vaccine* 23(15):1851–1858.

Hyams, K. C., et al. 2001. "Prevalence and incidence of hepatitis C virus infection in the U.S. Military: A sero-epidemiologic survey of 21,000 troops." *Am J Epidemiol* 153(8):764–770.

Kobayashi, N., et al. 2003. "Clinical significance of SEN virus infection in patients on maintenance haemodialysis." *Nephrol Dial Transplant* 18:343–352.

Koutsoudakis, G., et al. 2006. "Characterization of the early steps of hepatitis C virus infection by using luciferase reporter viruses." *J Virol* 80(11):5308–5320.

Koziel, J. M. 1999. "Cytokines in viral hepatitis." *Semin Liver Dis* 19(2):157–169.

Lemon, S. M., et al. 1992. "Why do human hepatitis viruses replicate so poorly in cell cultures?" *FEMS Microbiol Lett* 79(1–3):455–459.

Lin, K. W., and Kirchner, J. T. 2004. "Hepatitis B." *Am Fam Physician* 69(1):76–82.

Lindenbach, B. D., et al. 2005. "Complete replication of hepatitis C virus in cell culture." *Science* 309:623–626.

Linnen, J., et al. 1996. "Molecular cloning and disease association of hepatitis G virus: A transfusion-transmissible agent." *Science* 271:505–508.

Matsuo, E., et al. 2006. "Characterization of HCV-like particles produced in a human hepatoma cell line by a recombinant baculovirus." *Biochem Biophys Res Comm* 340(1):200–208.

Meng, X.-J., et al. 1997. "A novel virus in swine is closely related to the human hepatitis E virus." *Proc Natl Acad Sci USA* 94:9860–9865.

Meng, X.-J., et al. 1998. "Genetic and experimental evidence for cross-species infection by swine hepatitis E virus." *J Virol* 72(12):9714–9721.

Nainan, O. V., et al. 2006. "Hepatitis C virus genotypes and viral concentrations in participants of a general population survey in the United States." *Gastroenterology* 131(2):478–484.

Nishizawa, T., et al. 1997. "A novel DNA virus (TTV) associated with elevated transaminase levels in post-transfusion hepatitis of unknown etiology." *Biochem Biophys Res Commun* 242:92–97.

Norman, J. E., et al. 1993. "Mortality follow-up of the 1942 epidemic of hepatitis B in the U.S. army." *Hepatology* 18(4):790–797.

Panda, S. K., Thakral, D., and Rehman, S. 2007. "Hepatitis E virus." *Rev Med Virol* 17(3):151–180.

ProMED-mail. 2007. "Hepatitis E, porcine carriers—United Kingdom (England)." *ProMED mail* 2007 18 Feb, 20070218.0615.

Resino, M. J., et al. 2010. "Hepatitis C virus (HCV) treatment uptake and changes in the prevalence of HCV genotypes in HIV/HCV-coinfected patients." *J Viral Hepat* 18(5):325–330.

Roingeard, P., et al. 2004. "Hepatitis C virus ultrastructure and morphogenesis." *Biol Cell* 96(2):103–108.

Roome, A. J., et al. 2000. "Hepatitis C virus infection among firefighters, emergency medical technicians, and paramedics—selected locations, United States, 1991–2000." *MMWR* 49(29):660–665.

Roque-Afonso, A. M., et al. 2002. "Antibodies to hepatitis B surface antigen prevent viral reactivation in recipients of liver grafts from anti-HBC positive donors." *Gut* 50(1):95–99.

Sadler, G. J., et al. 2006. "UK acquired hepatitis E—emerging problem?" *J Med Virol* 78:473–475.

Sagir, A., et al. 2004. "SEN virus infection." *Rev Med Virol* 14(3):141–148.

Springfeld, C., et al. 2000. "TT virus as a human pathogen: significance and problems." *Virus Genes* 20(1):35–45.

Streatfield, S. J. 2005. "Oral hepatitis B vaccine candidates produced and delivered in plant material." *Immunol Cell Biol* 83(3):257–262.

Suzuki, T., et al. 2007. "Hepatitis C life cycles." *Adv Drug Deliv Rev* 59(12):1200–1212.

Tavanez, J. P., et al. 2002. "Hepatitis delta virus ribonucleoproteins shuttle between the nucleus and the cytoplasm." *RNA* 8:637–634.

Taylor, J. M. 2003. "Replication of human hepatitis delta virus: Recent developments." *Trends Microbiol* 11(4): 185–190.

Wakita, T., et al. 2005. "Production of infectious hepatitis C virus in tissue culture from a cloned viral genome." *Nat Med* 11:791–796.

Wim, H. M., et al. 2001. "Hepatitis E virus sequences in swine related to sequences in humans, the Netherlands." *EID* 7(6):970–976.

Zhong, J., et al. "Robust hepatitis C virus infection *in vitro*." *Proc Natl Acad Sci USA* 102:9294–9299.

Popular Press

Blumberg, B. S. 2003. *Hepatitis B: The Hunt for a Killer Virus.* Princeton, NJ: Princeton University Press.

Pierce, J. R., and Writer, J. 2005. *Yellow Jack: How Yellow Fever Ravaged America and Walter Reed Discovered Its Deadly Secrets.* New York: John Wiley & Sons.

Sledge, E. B. 1981. "The stench of battle." In *Classics of Naval Literature: With the Old Breed at Peleliu and Okinawa.* Annapolis, MD: Naval Institute Press.

Tiollais, P., and Buendia, M.-A. 1991. "Hepatitis B virus." *Scientific American,* April, 1991, pp. 116–123.

Turkington, C. 1998. *Hepatitis C: The Silent Killer.* Columbus, OH: McGraw-Hill/Contemporary.

Wills, C. 1996. *Yellow Fever Black Goddess: The Coevolution of People and Plagues.* Reading, MA: Helix Books, Addison-Wesley.

Video Productions

Bad Blood produced by Marilyn Ness, 2010. Duart Film and Video. http://badblooddocumentary.com/

Understanding Hepatitis: 3-part Series Covering Hepatitis A–C. 2007. Films for the Humanities & Sciences.

Food-Borne Illness. 2006. Films for the Humanities & Sciences.

Understanding Hepatitis. 2006. Information Television Network.

Old Foes, A New Threat. 2004. Films for the Humanities & Sciences.

Blood and War. 2002. Films for the Humanities & Sciences.

New Blood. 2002. Films for the Humanities & Sciences.

Tainted Blood. 2002. Films for the Humanities & Sciences.

Fighting Hepatitis C. 2001. Films for the Humanities & Sciences.

eLearning

go.jblearning.com/shors2

The site features eLearning, an online review area that provides quizzes and other tools to help you study for your class. You can also follow useful links for in-depth information, or just find out the latest virology and microbiology news.

CHAPTER 18

New Viruses and Viruses That Are Reemerging

A flemish nun walks among the graves of hospital staff that had perished during the 1976 Ebola hemorrhagic fever outbreak in the Democratic Republic of Congo (formerly Zaire). A total of 318 people (88%) died of the infection. The outbreak was the first recognition of the disease. This chapter discusses factors that contributed to the emergence of new viruses like this.

"It's the perfect setup. Then you put air travel in and it could be around the world overnight."

Michael T. Osterholm, Director of the Center for Infectious Disease Research and Policy (CIDRAP), Associate Director of the Department of Homeland Security's National Center for Food Protection and Defense (NCFPD), and Professor in the School of Public Health at the University of Minnesota

OUTLINE

CASE STUDY: SICK HORSES

In 2008, a viral outbreak in which five horses came down with serious encephalitic infections occurred in Brisbane, Australia. The horses were treated at a veterinary clinic. Four horses died of the infection and the remaining horse was humanely killed after it recovered in accordance with the established national practice. A 33-year-old equine veterinarian and 21-year-old veterinary nurse at the clinic became ill. They suffered from a fever, chronic pain (myalgia) and headaches. Both of them were exposed to the infected horses.

What are the odds that a natural pandemic could kill a million people in the United States? What kind of viral outbreaks will we see in the future? Will there continue to be **emerging** and **reemerging** viruses? Emerging can be defined as new or recently identified viruses to humans. Reemerging viruses are viruses once thought to be under control from a public health perspective but that are making a "comeback" or reappearing and causing increased incidence or geographic range of infections in exposed human populations. In developed countries, public health measures such as sanitation, sewage treatments, vaccination programs, and access to good medical care have virtually eliminated "traditional" viral diseases such as smallpox, yellow fever, measles, and poliomyelitis. Since 2009, the following viral outbreaks were the focus of attention by the news media:

- A novel 2009 H1N1 pandemic influenza A virus emerged in Mexico.
- Imported "measles" cases occurred among returning U.S. travelers overseas.
- A new highly lethal hemorrhagic virus called "Lujo" infected people in Zambia and South Africa.
- **Chikungunya fever virus** caused extensive outbreaks in New Caledonia, Thailand, and South India (especially the Krishnagiri District).
- A polio outbreak in Tajikistan was imported from India.
- Various outbreaks of **norovirus** on cruise ships, in hospitals and hotels, and at public gatherings occurred.
- **Bluetongue virus**, an insect-borne infection of ruminants such as cows, sheep, goats, and deer, spread from Africa to Northern European farms.
- A nipah virus outbreak in Bangladesh did not involve pigs, but instead transmission was directly linked to humans via contaminated fruit or palm juice.
- Ranaviruses caused mass deaths of frog populations in the Netherlands.

Human behavior plays a key role in influencing disease emergence and patterns. Moreover, there are three different general types of factors that contribute to the emergence or reemergence of viral diseases:

- viral
- human
- ecological

18.1 Viral Factors: Evolution and Adaptation

TABLE 18-1 lists emerging and reemerging viruses, the majority of which contain RNA genomes. Viruses evolve by rapidly ongoing processes. Some genes in viruses mutate rapidly, whereas others may mutate but you do not see the product because the mutations are lethal; for example, a gene that codes for a viral polymerase (*pol*) has a very specific shape that is related to its function. A regular mutation in the *pol* gene is not tolerable, whereas in the case of a gene that encodes a capsid protein, it could be tolerated and not lethal to the virus.

The RNA viruses possess high mutation rates due to the lack of proofreading capability of the viral RNA-dependent RNA polymerases used to transcribe or replicate the viral genomes. In general, the **point mutation rate** of an RNA virus is one mutation per every 10^4 to 10^5 nucleotides. In contrast, the mutation rate of DNA viruses is one mutation per every 10^8 to 10^{11} nucleotides. The mutation rate of the human immunodeficiency virus (HIV) is two mutations per every five genomes. The mutation rate for severe acute respiratory syndrome coronavirus (SARS-CoV) was estimated to be three mutations per every RNA

TABLE 18-1 Emerging or Reemerging* Human Viruses Since 1970

Year	Virus
1972	Norwalk virus (now known as norovirus)
1973	Rotavirus
1975	Parvovirus B-19
1975	Enterovirus 71
1976	Ebola
1977	Hantaan virus
1980 (southern Texas)	Dengue fever virus*
1980	Human T cell leukemia virus I
1982	Human T cell leukemia virus II
1983	Human immunodeficiency virus
1988	Human herpesvirus type 6
1988	Hepatitis E virus
1989	Hepatitis C virus
1989	Guanarito virus (Venezuelan hemorrhagic fever virus)
1989 (United States)	Measles virus*
1990	Sabia virus
1993	*Sin nombre* virus
1994	Hendra virus
1995	Human herpesvirus type 8
1996	Hepatitis G virus
1996 (Uganda)	O'nyong'nyong fever virus*
1997	Transfusion transmission virus or torque teno virus (TTV)
1997 (Hong Kong)	Avian influenza A H5N1*
1999 (New York City)	West Nile virus*
1999	Nipah virus
2001	Human metapneumonia virus (hMPV)
2002	SARS-CoV
2003 (multistate outbreak in the United States)	Monkeypox*
2005	Human T cell lymphotropic virus III
2005	Bocavirus
2005 (Amish community in Minnesota)	Poliovirus*
2006 (midwestern United States)	Mumps virus*
2006 (La Reunion Island, France)	Chikungunya virus
2007 (United States)	Adenovirus 14 variant
2007 (Asia)	Zika virus
2008 (Africa)	Lujo virus

*Reemerging viruses are indicated with an asterisk. All other viruses listed are new or emerging.

genome in every round of replication. Point mutations in the genome may result in amino acid changes in the encoded protein products, affecting its function. Mutations in genes can result in drug resistance (e.g., changes in the viral reverse transcriptase or polymerases, proteases) or changes that prevent host antibodies from binding to the surface proteins of the virus, thereby evading the host's immune system (e.g., **antigenic drift** occurring in influenza viruses).

In addition to point mutations during replication, there are two other mechanisms of genetic variation that play a role in the evolution of viruses: **recombination** and **reassortment**. Recombination occurs when two viruses infect the same cell and a new chimeric or *hybrid* genome is formed via a process of intermolecular exchange of the viral genomes (**FIGURE 18-1a**). Recombination can occur in either segmented or nonsegmented viruses. A model regarding the recombination of RNA virus genomes is based on the premise that hybrid RNAs are formed when the viral RNA-dependent RNA polymerase complex switches, mid-replication, from one RNA molecule to another. Examples of RNA viruses with high recombination rates are SARS-CoV and other togaviruses, such as Eastern equine encephalitis virus and Western equine encephalitis virus. Recombination can significantly contribute to the establishment of new virus families. Viruses can steal genes from their hosts through a recombination process with the cellular chromosome. These cellular genes may be used or sometimes the virus does not utilize the genes. Retroviruses are notorious for pirating cellular genes (see Chapter 10, Viruses and Cancer).

Gene swapping or reassortment occurs when viruses that contain segmented genomes coinfect the same cell (Figure 18-1b). Influenza A viruses are notorious for this type of genetic adaptation. The genome of influenza A consists of eight single-stranded RNA (ssRNA) segments. When the virus surface glycoprotein gene coding for hemagglutinin (H) or neuraminidase (N) undergo reassortment, a new strain is formed. Progeny viruses contain a genome that has gene(s) from both or multiple parent viruses. When this gene swapping occurs, it is referred to as **antigenic shift** (see Chapter 12, Influenza Viruses).

18.2 Human Factors

The genetic adaptation of viruses does not appear to be the major driving force that causes viral emergence. Human demographics drive viral dis-

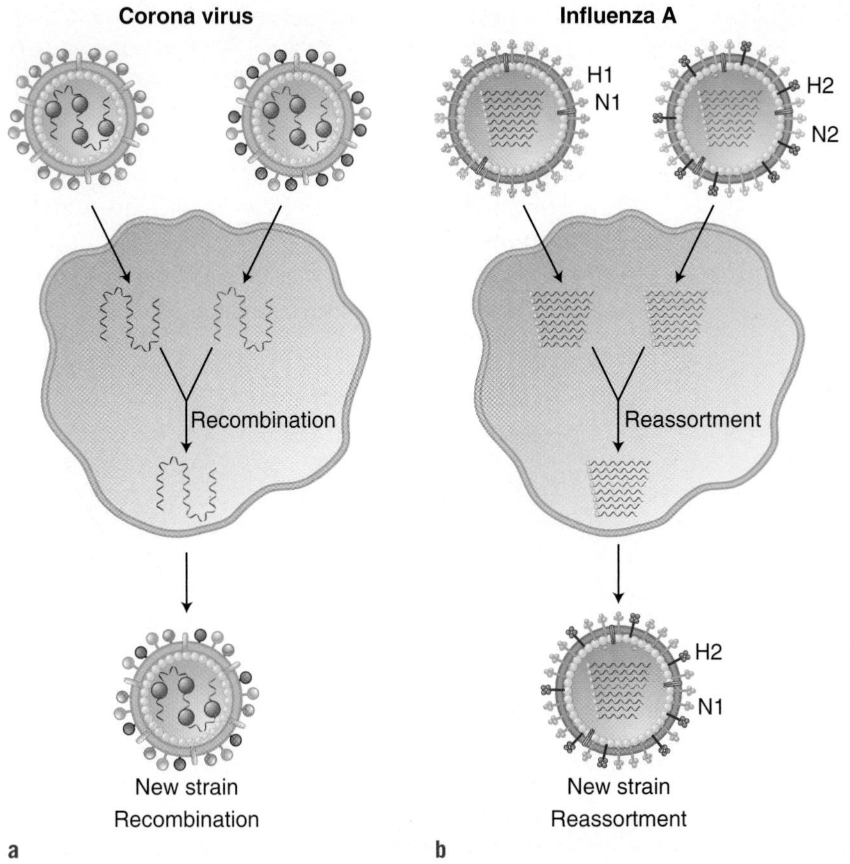

FIGURE 18-1 **(a)** Two coronaviruses coinfect the same cell in which the viral genomes undergo recombination. This results in the genome of a new strain. Adapted from from E. Ka-Wai Hui, E., *Microbes and Infection* 8 (2006): 905–916. **(b)** Two different strains of influenza A viruses (H1N1 and H2N2, in this case) coinfect the same host animal or bird. The H or N genes are swapped to form a new strain. Adapted from Ka-Wai Hui, E., *Microbes and Infection* 8 (2006): 905–916.

ease emergence more than virus evolution or ecological factors. **Human demographics** means physical population characteristics. Demographics includes such characteristics as:

- sex/gender
- race/ethnicity
- age
- mobility/travel
- occupation
- sexual behavior
- socioeconomic status (the economic, social, and physical environments in which individuals live and work)
- geographic location
- family size

Population Growth: Density and Crowding

A virus emerges when it reaches a new population and then disseminates to new places. The dramatic increase in the human population on the planet is correlated with higher human densities in a given area. The world human population growth curve is shown in **FIGURE 18-2a**. Human populations lead to **urbanization**: the migration of people from the countryside to cities. In 2010, there were at least 26 **megacities** in the world, which contained at least 10 million inhabitants. **Figure 18-2b** illustrates the geographic distribution of humans in the world in 2010. China, India, other parts of Asia, and Africa are the most populated. The largest megacity is the greater Tokyo area, which has a population of 34.2 million. The nine largest megacities after Tokyo are, in decreasing order of population as of 2010 are:

- Guangzhou, China (24,900,000)
- Delhi, India (23,900,000)
- Seoul, South Korea (24,500,000)
- Mumbai (Bombay), India (23,300,000)
- Mexico City, Mexico (22,800,000)
- New York City, United States (22,200,000)
- São Paulo, Brazil (20,800,000)
- Manila, Philippines (20,100,00)
- Shanghai, China (18,800,000)

Higher population densities favor the spread of viral diseases; for example, dense populations result in the combined problems of:

- crowding
- sanitation
- drinking water
- pollution
- healthcare facilities

When sanitary conditions are poor, people are more susceptible to respiratory and gastrointestinal infections.

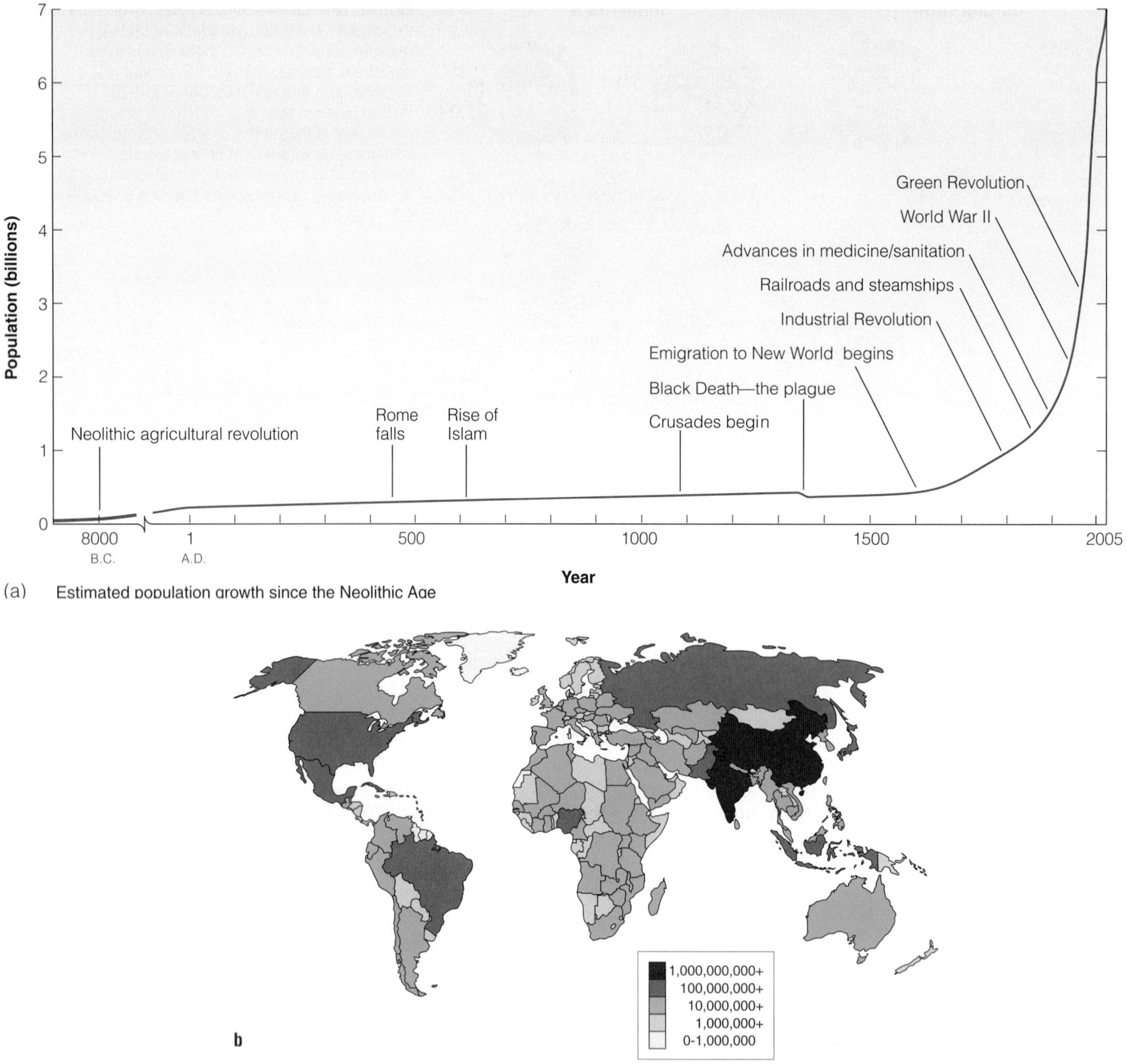

FIGURE 18-2 (a) Estimated population growth since the Neolithic Age. World population over time (8000 B.C. to 2005 A.D.). Global population is rising at an exponential rate. U.S. Census Bureau, International Programs Center, 2001. **(b)** Distribution of world population in 2010.

Human Movement (Travel and Migration): Viruses Live in a World Without Borders

Humans are constantly on the move either for planned migration (e.g., for leisure, work, and commerce) or unplanned migration (e.g., because of war or natural disasters). The speed of air travel allows infected travelers to reach any part of the world in less than 24 hours, possibly starting a global pandemic. The spread of viral diseases to new areas can happen at any time. Viruses present in rural populations of the world such as Africa or Asia may show up in a more developed part of the world such as Europe or the United States. The **SARS pandemic** of 2003 demonstrates the spreading of a viral disease by international air travel (**FIGURE 18-3**). Within a month, almost 20 countries had reported cases of SARS.

It is difficult to predict when the next influenza pandemic will occur. The **H5N1 avian influenza virus** has raised concerns about a potential human pandemic for several reasons:

- It is highly virulent.
- It is spread by migratory birds.

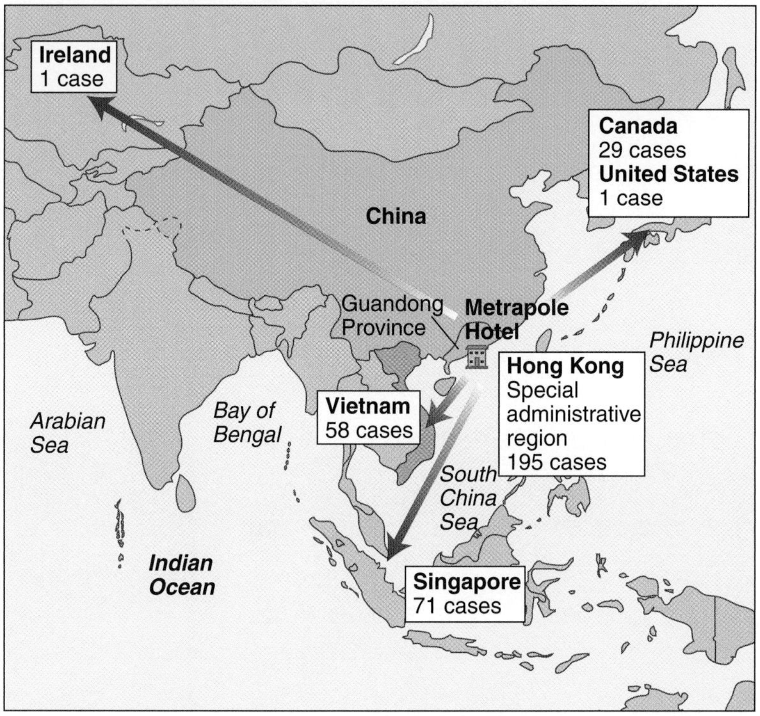

FIGURE 18-3 Spread of SARS-CoV from the Metropole Hotel in Hong Kong as of March 28, 2003. Adapted from National Intelligence Council, "SARS down but still a threat," *Intelligence Community Assessment.*

- It can be transmitted from birds to mammals (e.g., from birds to domestic cats) and in some limited circumstances to humans.
- It continues to evolve (i.e., due to antigenic shift and drift).

Since 2003, human cases of H5N1 avian influenza have been reported in 15 countries (**TABLE 18-2** and **FIGURE 18-4**). Avian influenza occurs naturally among wild waterfowl, which usually do not get sick from it. The virus is shed in bird feces, saliva, and nasal secretions. H5N1 avian influenza is highly contagious and nearly 100% fatal in domesticated birds such as chickens, turkeys, and ducks, often within 48 hours. Persons in direct contact with sick birds or surfaces contaminated with secretions and/or excretions from infected birds on farms or live animal markets are most susceptible to the virus. At the time of this writing, the virus did not efficiently spread from person to person. Half of the humans infected with the H5N1 influenza strain, however, have died. To date, the risk of avian influenza to travelers is very low.

The U.S. Geological Survey's National Wildlife Health Center, located in Madison, Wisconsin, monitors samples taken from migratory waterfowl for avian influenza viruses. They speculate that the H5N1 influenza strain will enter North America through migratory birds in the Pacific flyway, which follows the eastern Pacific coastline from the western Arctic, including Alaska and the Aleutian Islands, down the Rocky Mountain and Pacific coastal regions of Canada, the United States, and Mexico, to where it blends with other flyways in Central and South America. This monitoring can provide an early warning to agriculture, public health, and wildlife communities if the virus is detected among migratory birds. There are still other possible routes of the virus entering North America, such as through illegal importation of infected domestic birds or by an infected traveler.

Cruise vacations are an increasingly popular leisure activity. From 1993 to 1998, the number of cruise ship passengers in the United States increased by 50%. In North America alone, about 6.8 million passengers board cruise ships each year. In recent years, there has been an increase in gastrointestinal illness caused by **norovirus** on cruise ships, which in essence are floating minicities with a constantly changing population of hundreds of thousands of people in a confined space. Between 2001 and 2004, the number of norovirus outbreaks on cruise ships increased tenfold. In 2002 a new norovirus variant emerged in hospitals in the United Kingdom and Ireland and on U.S. cruise ships, and two new variants emerged in 2006 in the Netherlands. Noroviruses cannot be cultured in the laboratory and there is no animal model available to study these viruses, thus making it difficult to determine what genetic changes in a virus trigger outbreaks or cause more severe symptoms. The Centers for Disease Control and Prevention (CDC) started the Vessel Sanitation

TABLE 18-2 The World Health Organization's (WHO) Report of Human Cases and Deaths of Avian Influenza (2003 to April 6, 2011)

	2003		2004		2005		2006		2007		2008		2009		2010		2011		Total	
Country	Cases	Deaths	Cases	Deaths	Cases	Deaths	Cases	Deaths	Cases	Deaths	Cases	Deaths	Cases	Deaths	Cases	Deaths	Cases	Deaths	Cases	Deaths
Azerbaijan	0	0	0	0	0	0	8	5	0	0	0	0	0	0	0	0	0	0	8	5
Bangladesh	0	0	0	0	0	0	0	0	0	0	1	0	0	0	0	0	1	0	2	0
Cambodia	0	0	0	0	4	4	2	2	1	1	1	0	1	0	1	1	3	3	13	11
China	1	1	0	0	8	5	13	8	5	3	4	4	7	4	2	1	0	0	40	26
Djibouti	0	0	0	0	0	0	1	0	0	0	0	0	0	0	0	0	0	0	1	0
Egypt	0	0	0	0	0	0	18	10	25	9	8	4	39	4	29	13	18	5	137	45
Indonesia	0	0	0	0	20	13	55	45	42	37	24	20	21	19	9	7	5	4	176	145
Iraq	0	0	0	0	0	0	3	2	0	0	0	0	0	0	0	0	0	0	3	2
Lao People's Democratic Republic	0	0	0	0	0	0	0	0	2	2	0	0	0	0	0	0	0	0	2	2
Myanmar	0	0	0	0	0	0	0	0	1	0	0	0	0	0	0	0	0	0	1	0
Nigeria	0	0	0	0	0	0	0	0	1	1	1	0	0	0	0	0	0	0	1	1
Pakistan	0	0	0	0	0	0	0	0	3	1	0	0	0	0	0	0	0	0	3	1
Thailand	0	0	17	12	5	2	3	3	0	0	0	0	0	0	0	0	0	0	25	17
Turkey	0	0	0	0	0	0	12	4	0	0	0	0	0	0	0	0	0	0	12	4
Vietnam	3	3	29	20	61	19	0	0	8	5	6	5	5	5	7	2	0	0	119	59
Total	4	4	46	32	98	43	115	79	88	59	45	33	73	32	48	24	27	12	543	318

Total number of cases includes number of deaths. All dates refer to onset of illness.

Source: Data from the World Health Organization. WHO reports only laboratory-confirmed cases.

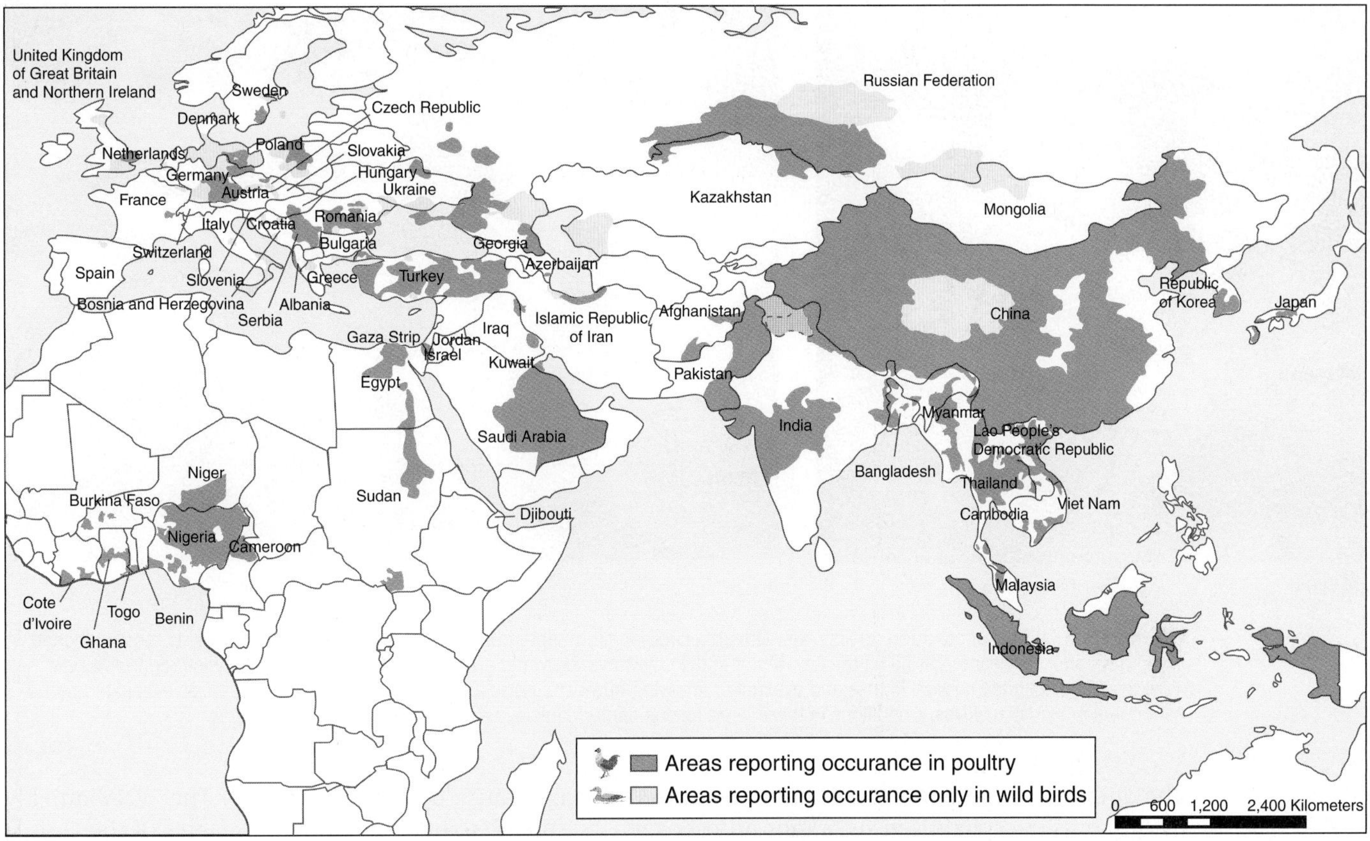

FIGURE 18-4 Nations with confirmed cases of H5N1 avian influenza in poultry and wild birds as of April 12, 2007. Courtesy of the World Health Organization/CDC.

Program (VSP) in the early 1970s as a result of several disease outbreaks on cruise ships. The mission of the VSP was to protect the health of passengers and crew by minimizing the risk of gastrointestinal illness on cruise ships. Routine inspections are now performed on the following:

- water supply—storage, distribution, protection, and disinfection
- spas and pools—filtration and disinfection
- food—protection during storage, preparation, and service
- employee hygiene practices
- general cleanliness and physical condition of the ship—cleanliness and absence of insects and rodents
- training programs—environmental and public health practices

Inspections scores are posted on the CDC's website.

War, famine, and natural disasters have played a large part in introducing infectious diseases into humans. Refugee camps and temporary shelters are often characterized by crowded living conditions, poor sanitation, a lack of access to clean water, and populations with a poor nutritional status and limited access to medical care. Under these conditions, people are highly susceptible to infectious agents; for example, hurricane Katrina evacuees in the New Orleans Superdome and the Houston Astrodome and Reliant Center experienced a norovirus outbreak. For some time, there were no working toilets and little clean water in these facilities (**FIGURE 18-5**).

Aircrafts are a potential means of unintentionally moving virus-infected animals or arthropod vectors such as mosquitoes from country to country and continent to continent. Researchers have demonstrated that mosquitoes, houseflies, and flour beetles can survive in the wheel bays of Boeing 747B airplanes. The insects were able to withstand the extreme temperatures in one-way and round-trip flights. The leading theory is that the 1999 **West Nile virus** (**WNV**) outbreak in New York was attributed to airplanes carrying West Nile virus–infected *Culex pipiens* mosquitoes when they flew across the Atlantic Ocean from Israel. At the time, aircraft cabin spaces were not routinely sprayed with insecticides on intercontinental flights entering the United States.

Prior to 1999, West Nile encephalitis had never been reported in the Western hemisphere and no one had considered that it could be introduced.

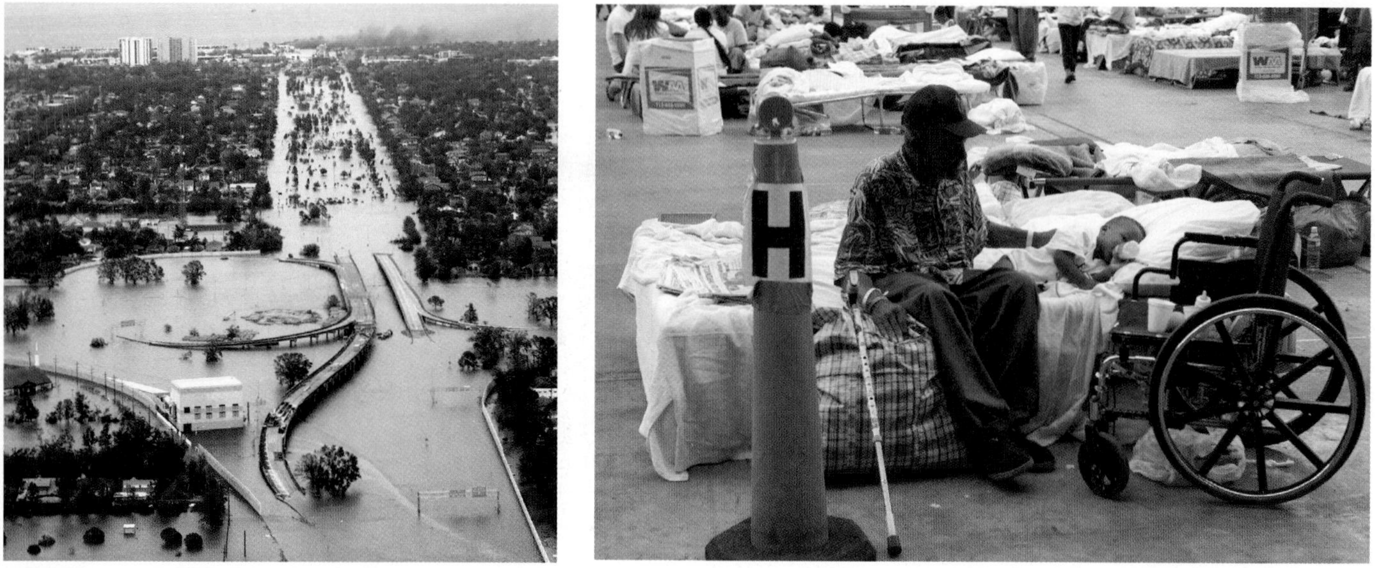

a b

FIGURE 18-5 (a) This photograph depicts New Orleans, Louisiana, on September 5, 2005. These homes and streets remain flooded seven days after hurricane Katrina hit the city. Much of the city was underwater and without electricity. **(b)** Many injured individuals were rescued from their flooded homes and evacuated following hurricane Katrina. This photograph depicts families evacuated to the Astrodome in Houston, Texas. Conditions in these large rescue centers quickly became overcrowded and filthy.

The outbreak was originally thought to be caused by St. Louis encephalitis virus, a mosquito-borne illness endemic to the United States. A correct diagnosis was made after a veterinarian at the Bronx Zoo correlated the death of native birds with human cases of encephalitis. (St. Louis encephalitis virus commonly infects birds but does not cause disease outbreaks.) The WNV quickly spread across the contiguous states of the United States and into Canada, the Caribbean, and Mexico (**FIGURES 18-6a** and 18-6b). The main reservoir of the West Nile virus is birds. Mosquitoes pick up the virus while taking a blood meal from infected birds. From there, the virus moves to

FIGURE 18-6 (a) Map of human West Nile encephalitis cases and avian, animal, or mosquito infections in 1999, the first year of the outbreak in the United States. Data courtesy of the Division of Vector-Borne Infectious Diseases/CDC. **(b)** Map of human West Nile encephalitis cases and avian, animal, or mosquito infections in 2006. The virus spread quickly across the United States. Data courtesy of the Division of Vector-Borne Infectious Diseases/CDC. **(c)** West Nile virus transmission cycle. Data courtesy of the Division of Vector-Borne Infectious Diseases/CDC.

(continued)

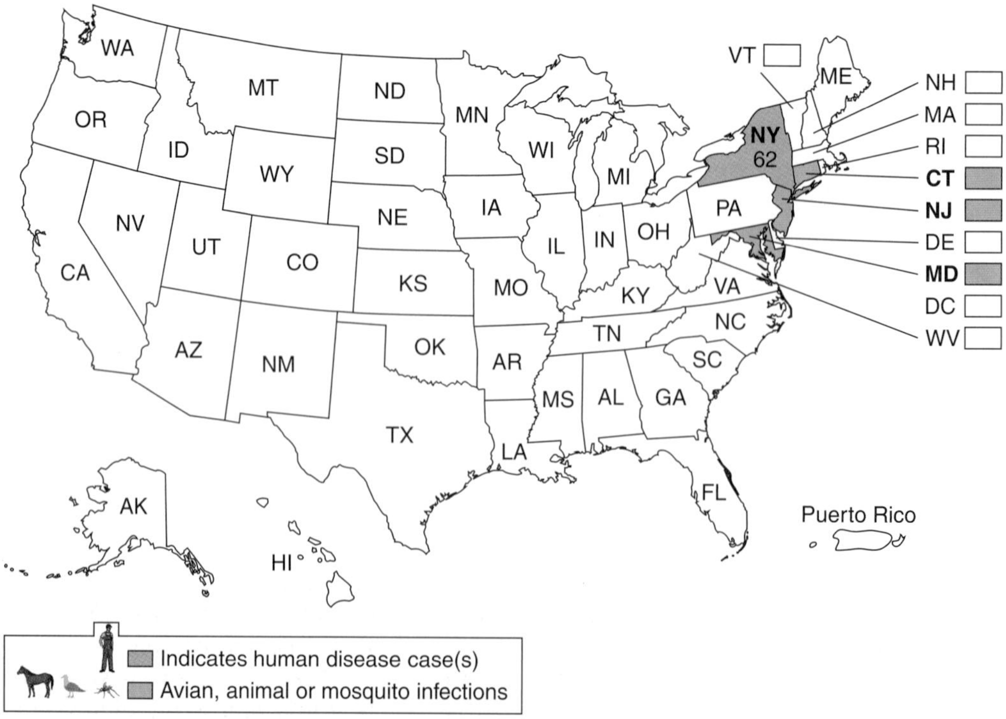

a

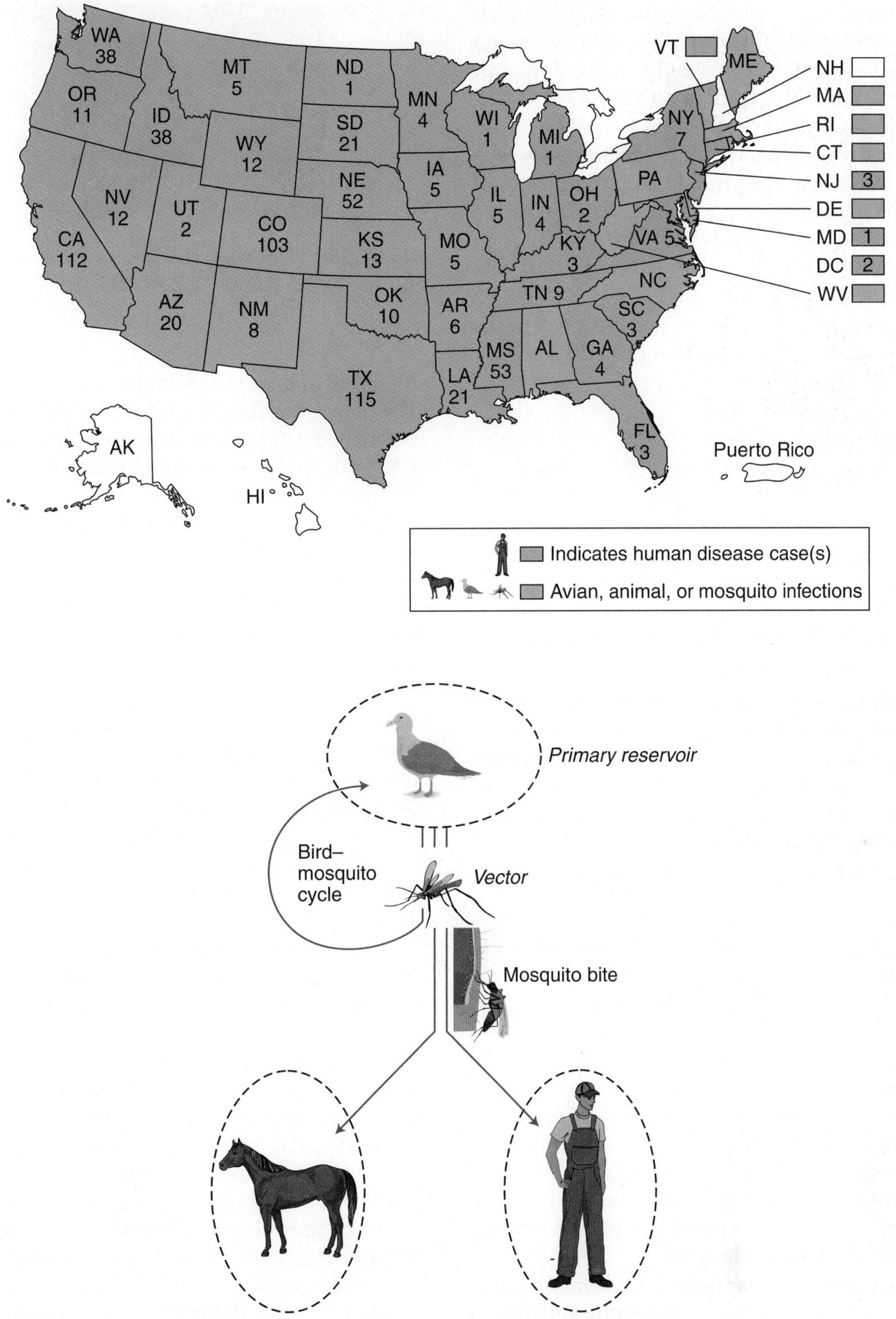

c

FIGURE 18-6 (Continued)

horses and humans through a mosquito bite. Humans and horses are "**dead-end hosts**" that cannot transmit the virus to further species (Figure 18-6c). Between 1999 and 2010, there were 30,702 human encephalitis cases and 1220 (4%) human deaths in the United States. Between 1999 and 2010, there were 25,889 cases of equine encephalitis. By 2004, 15,000 horses had died of West Nile encephalitis.

Investigation of a new reservoir of WNV began in 2007. Stable flies were suspected of being involved in the West Nile virus deaths of hundreds of pelican chicks at the Medicine Lake National Wildlife Refuge in northeast Montana. Montana State University entomologist Greg Johnson observed stable flies feeding on the blood of sick and dying pelicans. Stable flies look like house flies, but they have a painful bite. Johnson captured 1300 stable flies and divided them into 60 groups. Eighteen of those groups tested positive for West Nile virus. This was the first evidence to suggest that stable flies might transmit West Nile virus at the pelican colony.

In 2002, four novel routes of West Nile virus transmission to humans were reported:

- blood transfusion
- organ transplantation
- transplacental transfer
- breastfeeding

Twenty-three people in the United States were reported to have acquired WNV infection after receiving blood components from 16 West Nile virus–infected blood donors in 2002. Per the recommendation of the Food and Drug Administration (FDA), blood-collection agencies have since implemented West Nile virus nucleic acid amplification tests to screen all blood donations in order to prevent transmission of the virus through contaminated blood products.

Human Exposure to Zoonotic Diseases: Pasture, Hunting, and Fishing Practices

Zoonoses are infectious diseases that are transmissible from animals to humans or from humans to animals. The word is derived from the Greek words *zoon* (animal) and *nosos* (disease). There are approximately 208 human pathogenic viruses or prions (prions are discussed in Chapter 19). Of these human pathogens, 77 (37%) are considered emerging or reemerging, and about 73% of the emerging or reemerging viral or prion pathogens are known to be zoonotic. In many parts of the world, the close proximity of wildlife to humans is not uncommon. Under these conditions, viruses may cross the **species barrier**.

An example of a reemerging virus crossing the species barrier is the 2003 **monkeypox outbreak** in the United States. The first human cases of monkeypox were reported in 1970. It is a rare viral disease endemic in monkeys in the rainforests of Central and West Africa. Rodents are the reservoir of the monkeypox virus; humans and monkeys are incidental hosts. In 2003, 762 rodents, including Giant Gambian pouched rats, were legally imported from Ghana in West Africa to Texas. The Gambian rats were infected with monkeypox and were housed in close proximity to native pet prairie dogs

FIGURE 18-7 Prairie dogs spread monkeypox, a rare zoonotic disease, in 2003. It was the first occurrence of monkeypox in the United States.

(**FIGURE 18-7**) in a pet distributor's premises in Illinois. The prairie dogs became infected with monkeypox before being purchased at an exotic pet swap meet. The virus spread from the infected prairie dogs to 71 people through bites or through close contact with the rodents' blood, body fluids, or rash. To prevent further transmission of monkeypox, the U.S. government banned the importation of all rodents from West Africa and banned the transport, sale, or release of pet prairie dogs. The smallpox vaccine can prevent people from getting monkeypox as well as smallpox.

Since the beginning of human history, people have lived in close contact with animals—usually as hunters and as farmers. This practice is most common in developing countries where **free-range** farming—a method of farming in which the animals or birds are allowed to roam freely instead of contained in any manner (**FIGURE 18-8a**)—is used. There are 40 million small rural farms in Asia alone.

In southern China, especially the Guangdong province, wild animals are captured and sold in live markets or kept in cages in restaurants waiting for consumption. In the Chinese culture, fresh foods made from wild animals are considered a delicacy called "wild taste." The most common wild animals sold are beavers, badgers, civet cats, domestic cats, hares, raccoon dogs, snakes, and pangolin. The captured animals are kept in the same cage or in stacked cages where the animals—including those that are infected—often fight. These animals' blood, secretions, feces, and urine contaminate the cages, and the food handlers who subsequently butcher the animals have direct contact with the fluids.

The **SARS outbreak** of 2003 was an example of a virus that was transmitted from live-market animal to a human. An infected masked palm civet cat at a restaurant infected chefs, food handlers, and others who lived near the live-animal market. Further research suggested the natural reservoir of SARS-CoV, which has killed 774 people worldwide, is the Chinese horseshoe bat. These bats are commonly found at markets in southern

a

b

FIGURE 18-8 (a) These egg-laying chickens are free-range. The poultry house is located behind the chickens. **(b)** Ducks loaded onto a weighing scale at a wholesale, live poultry market in the Hatay province located 15.5 miles outside of Hanoi, Vietnam. Vietnam has experienced avian influenza outbreaks in poultry since 2004. Human cases of avian flu were first reported in Vietnam in 2004 as well.

China. Live poultry are also housed in crowded and unhygienic conditions at traditional markets (Figure 18-8b). Under these conditions **avian influenza** may spread from poultry to humans.

In Africa, bushmen live in temporary shelters and hunt animals for food. Sometimes they slaughter pathogen-infected nonhuman primates. During slaughter, they come in contact with raw bushmeat, blood, and bodily fluids of the butchered animal. Simian immunodeficiency virus (SIV) is found in nonhuman primates (such as SIV_{cpz} in chimpanzees). It crossed the species barrier into humans in the form of HIV. **HIV-1** is most closely related to SIV_{cpz}, whereas **HIV-2** is most similar to SIV_{sm}, which is found in the sooty mangabey (recall the theories of the origin of HIV, which were described in Chapter 16).

Media reports about a deadly Ebola-like virus that kills at least 15 different fish species made the headlines of U.S. newspapers from 2006 to 2007. **Viral hemorrhagic septicemia** (VHS) virus is a serious pathogen of fresh and saltwater fish and is causing an emerging disease in the Great Lakes region of the United States and Canada. Infected fish show signs that include bulging eyes, hemorrhaging in the eyes, gills, skin, and at the base of the fins, bloated abdomens, and inactive or overactive behavior. Genetic tests show that the strain of VHS found in the Great Lakes probably originated in Atlantic Ocean fish located near New Brunswick, Canada. This places the origin of the new strain near the start of the St. Lawrence Rivers shipping route that leads to the Great Lakes. Besides the Great Lakes, at the time of this writing, this highly contagious fish pathogen has been found in fish populations in the Niagara River and inland lakes in New York such as Lakes Skaneateles, and Winnebago and Little Lake Buttes des Morts near Oshkosh, Wisconsin.

On October 24, 2006, the United States Department of Agriculture's Animal and Health Inspection Service (APHIS) issued a federal emergency ordering the banning of importation of 37 species of live fish from two Canadian provinces into the country and the interstate movement of the same species from the eight states bordering the Great Lakes. This is a major fish health crisis. Sport fisherman and recreational boaters were being asked to adhere to good biosecurity practices of fishing or boating in waters where VHS has been found. Examples of some biosecurity and preventative measures encouraged include:

- Rinse boats and any other equipment that have been in VHS waters with hot water or drying it for at least five days.
- Do not dump unused bait into lakes, streams, rivers, or on shore.
- Never move fish eggs or live fish to other waters.
- Report fish die-offs to the local Department of Natural Resources (DNR) representative or conservation warden.
- Clean and disinfect all fishing gear.

Agricultural Practices, Deforestation, and Dam Construction

The extension of farmland into unused land exposes farmers to zoonotic diseases (especially rodents that carry viruses), new animal species (and their viruses), and disease-carrying arthropods. As one example, there was a fivefold increase in **Argentine**

VIRUS FILE 18-1

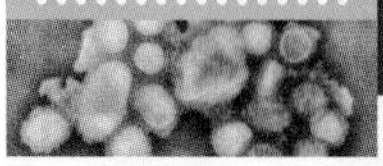

Retroviruses Crossing the Species Barrier in Nature: Hunters in Africa Infected with Retroviruses via Bushmeat

The origin of HIV has been debated. Genetic studies have traced the virus to its relative SIV. Scientists proposed that HIV arose from SIV in chimpanzees. Presumably, this crossover occurred in humans living in African communities where people hunted chimpanzees for food. The virus was transmitted to humans via contact with blood and body fluids of animals during hunting or butchering (**FIGURE VF 18-1a**). At first this theory was not proved and there was no documentation of transmission of SIV to laboratory and zoo workers who had close contact with nonhuman primates. Now, however, this theory has been backed up by a study carried out by U.S. and Cameroonian researchers who collected blood samples from 1800 volunteer participants of a community-based HIV-prevention campaign in Cameroon (Figure VF 18-1b). Sixty-one percent (1099) of the participants had exposure to nonhuman primates (monkeys, chimpanzees, or gorillas) through hunting, butchering, or as pets and reported exposure to fresh primate blood and body fluids. Hands, guns, bow and arrows, wire snares, and machetes were used to hunt the primates.

Blood samples were screened for antibodies against simian foamy virus (SFV), which is also known as spumaretrovirus. SFV is transmitted at a higher rate (2.5%) than normal primate retroviruses to zoo workers and researchers. SFV is also endemic in most Old World nonhuman

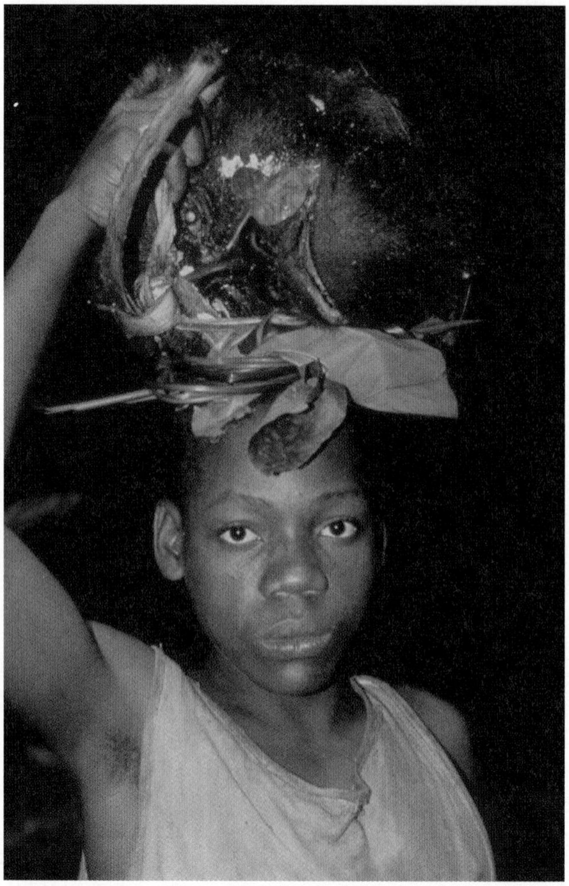

a

FIGURE VF 18-1 (a) Child carrying the head of a gorilla from a recent bushmeat hunt. **(b)** The gorilla head is removed during preparation of the meat. Bushmeat hunting is common in sub-Saharan African forests.

(continued)

VIRUS FILE 18-1 Retroviruses Crossing the Species Barrier in Nature: Hunters in Africa Infected with Retroviruses via Bushmeat (continued)

b

FIGURE VF 18-1 (Continued)

primates. Thus far, though, no humans have gotten ill from SFV, and the virus doesn't make nonhuman primates sick (as is the case with SIV).

SFV antibodies were detected in 10 of the 1099 blood samples (a 1% transmission rate). Via polymerase chain reaction (PCR) analysis, it was determined that the 10 SFV-infected individuals had been infected via three different primates: De Brazza's guenon, mandrill, and gorilla. The De Brazza's guenon and mandrill are also naturally infected with SIV.

These results show that retroviruses actively cross into human populations and that people in west Central Africa are currently infected with SFV. Contact with nonhuman primates during hunting and butchering can play a partial role in the emergence of human retroviruses. The urban demand for bushmeat and greater access to primate habitats by logging roads has increased the frequency of human exposure to primate retroviruses and other disease-causing pathogens. In addition to preserving endangered old world primates, reducing hunting activities will reduce the frequency of cross-species transmission of retroviruses and other biological agents.

References

Wolfe, N. D., et al. 2004. "Naturally acquired simian retrovirus infections in Central Africa." *Lancet* 363:932–937.

hemorrhagic fever cases occurring at the forest-farmland margins in the northwest province of Buenos Aires, South America, from 1958 to 1974. This region is the richest farmland in Argentina and farmers soon began to clear plains (called pampas) and plant maize. During the clearing practices, rodents that carry the **Junin virus**—the causative agent of Argentine hemorrhagic fever—began to flourish. The Junin virus was shed in the urine of the rodents and aerosolized by agricultural machinery. In similar agricultural situations, increases in Bolivian hemorrhagic fever caused by the Machupo virus and Lassa fever virus, via their natural rodent hosts, were recorded.

Deforestation is the removal of trees in forested areas without sufficient reforestation. There are many causes, ranging from natural wildfires or animal grazing to human activity. Forestry practices such as logging, mining, dam construction, and petroleum exploration all contribute to human-caused deforestation. In 2005, outbreaks of vampire bat-related human rabies in the rural parts of the states of Para and Maranhao were correlated with the deforestation of the Amazon rainforests in Brazil (**FIGURE 18-9**). The deforestation of the Amazon region caused thousands of the common vampire bats (*Desmodus* sp.) carrying rabies to move across northern Brazil.

Vampire bats, which are native to the Americas, and in particular Brazil, Chile, and Argentina, feed on blood. The common vampire bat, which is **nocturnal**, rarely bites humans, instead feeding on sleeping cattle and horses. Vampire bats live almost exclusively in dark places such as caves, old wells, and hollow trees, and are common carriers of rabies virus. The abandoned gold mines in Maranhao provided a favorable habitat for the vampire bat populations. During the 2005 outbreak, there was a failure of a generator that left people without electricity for six weeks. Many homes in the marshlands did not have screening in windows and had large gaps in the floor and ceiling, making it easy for vampire bats to gain access. The vampire bats carrying rabies may behave abnormally, which could explain in part the uncharacteristic behavior of biting humans during these Brazilian outbreaks. It was speculated that the vampire bats bit several hundred people and that at least 23 human rabies deaths were reported. Over 1350 people received postexposure rabies vaccine.

Another example of an outbreak resulting from deforestation activity is the **Nipah virus outbreak** that occurred in the north peninsula of Malaysia from 1998 to 1999. The virus caused severe fever and encephalitis in 265 patients, killing 104 (40%). The Nipah virus infects humans, pigs, dogs, and cats. Most patients in Malaysia were pig farmers and the infections were linked to exposure to sick pigs. The Nipah virus was transmitted to the pigs via bat droppings from infected Malayan flying fox and variable flying fox fruit bats, which feed on fruit and nectar. They play an important role in the rainforest as seed dispersers

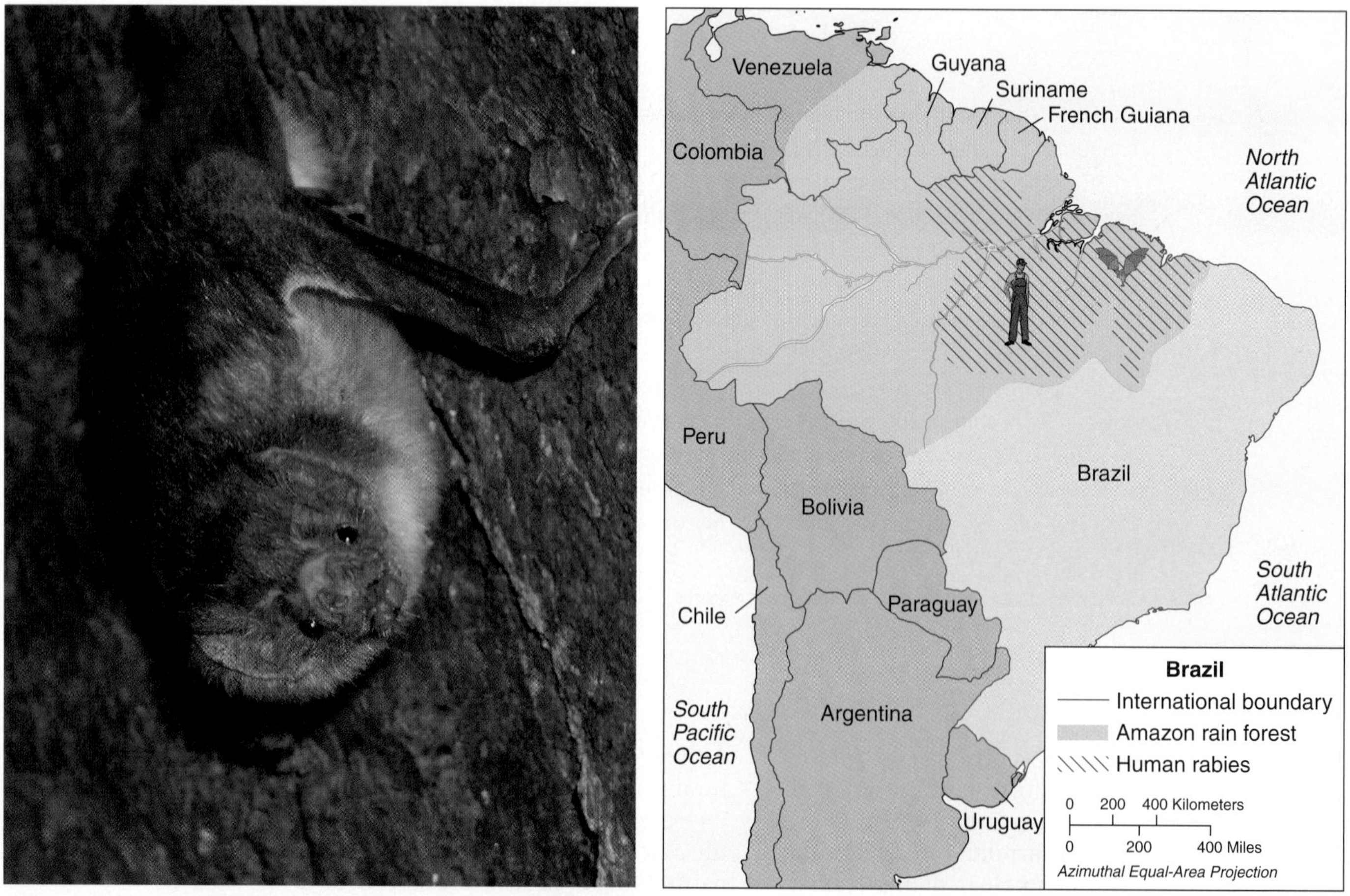

a b

FIGURE 18-9 (a) The common vampire bat (*Desmodus* sp.) was responsible for at least 23 human rabies deaths in Brazil in 2005. **(b)** Map of South America showing the Amazon region in relationship to human rabies transmitted by displaced vampire bats carrying the rabies virus.

and pollinators. Both types of bats carry the Nipah virus. Many pig farms were located next to orchards, where the fruit trees were a natural attraction for Malayan flying fox and variable flying fox bats. The pigpens were likely contaminated with bat droppings that contained the Nipah virus, providing a means for transmission to the pigs. The virus spread quickly because the farm was overcrowded with sick domestic pigs, and the virus ultimately aerosolized and spread to humans and other animals (**FIGURE 18-10a**). Nearly a million pigs were culled as a measure to stop the outbreak (Figure 18-10b).

A similar situation to the Nipah virus outbreak occurred in Australia from 1994 to 1995. A virus relative of Nipah virus, termed the **Hendra virus (formerly called equine morbillivirus)** was infecting racehorses, horse trainers, and stablehands. (The name Hendra comes from the location of the outbreak: Hendra, a suburb of Brisbane, Australia.) Hendra virus causes severe flulike symptoms and neurologic disease in horses and humans. The natural reservoir of the Hendra virus is the flying fox fruit bat (*Pteropus* sp.). Humans became infected when exposed to tissues and secretions from infected horses (**FIGURE 18-11**), which had been infected with Hendra virus from bat placental fluid. Two of three humans infected and at least 14 horses died during this outbreak. A subsequent outbreak occurred in 2004.

Severe outbreaks of **porcine reproductive and respiratory syndrome** (PRRS) in China occurred in domestic pigs during the summer of 2006. An arterovirus virus causes PRRS, which is sometimes referred to as "blue-ear pig disease." The virus attacks macrophages, crippling the immune system of pigs. The pigs are susceptible to secondary infections. Adult pigs (sows) usually recover and develop immunity. Symptoms of PRRS include reproductive failure (e.g., weak or stillborn piglets) in breeding stock and a respiratory illness (pneumonia) in piglets. The virus and secondary infections can kill piglets, whose ears turn blue due to **cyanosis**. Cyanosis is a bluish discoloration of the skin that is caused by a circulatory or lung problem that occurs when there is a

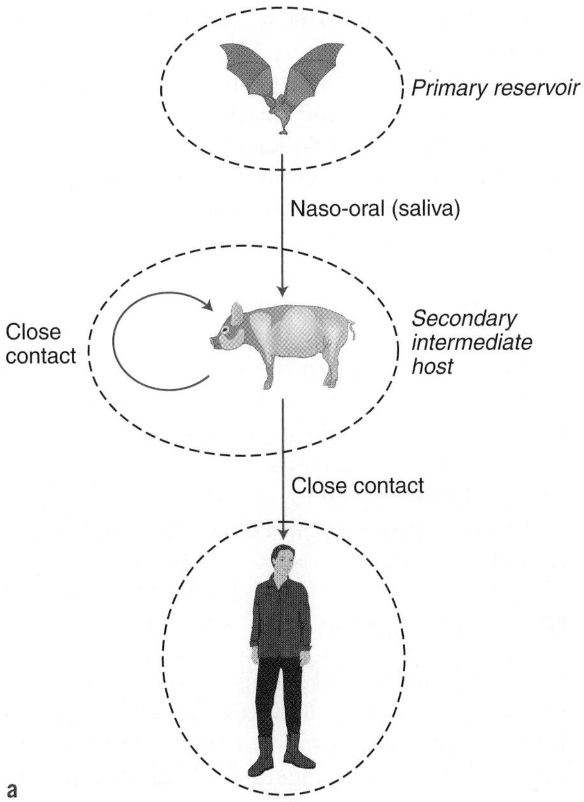

a

b

FIGURE 18-10 (a) Routes of transmission for Nipah virus. Adapted from Ka-Wai Hui, E., *Microbes and Infection* 8 (2006): 905–916. **(b)** Pigs were buried alive in mass graves to curb the Nipah virus outbreak.

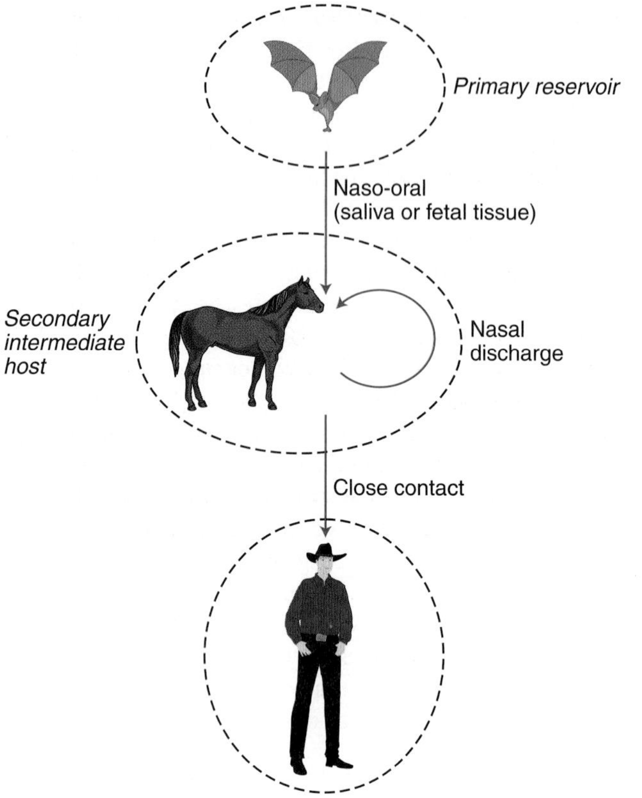

FIGURE 18-11 Routes of transmission for the Hendra virus. Adapted from Ka-Wai Hui, E., *Microbes and Infection* 8 (2006): 905–916.

lack of oxygen in the blood of the lungs or skin blood vessels (in this case, the vessels of the pigs' ears). Cyanosis occurs in PRRS-infected piglets suffering from secondary infections. The virus spreads rapidly from farm to farm. The PRRS virus is spread by pig-to-pig contact through contaminated mucus, manure, urine, and semen (artificial and natural) and possibly by airborne transmission. Infected live pigs are the major source of transmission to healthy pigs.

PRRS is becoming one of the most economically significant diseases, reducing the number of live pigs and prompting pork shortages and inflation of pork prices. The outbreak in China caused the pork economy to sizzle when prices of pork and its byproducts skyrocketed over 85%. The country's consumer price index increased to 6.5% during August 2007, after it rose to a ten-year high of 5.6% in July 2007. China is one of the largest exporters of pork and pork products in the world. In 2007, Chinese officials received international criticism for the safety of its food and other exports.

The PRRS virus began causing outbreaks in the United States during the late 1980s, when it was known as "mystery swine disease." PRRS first appeared in Europe in 1990. It is now present in almost all pig-breeding countries. At the time of this writing, the strain of PRRS virus circulating in China evolved into a highly pathogenic and lethal virus, killing sows and piglets. Millions of pigs were killed by the virus or culled in an effort to prevent spread of PRRS. These outbreaks were considered a threat to the global industry. Chinese officials were criticized for the lack of information made available to the public, comparing this to the 2003 SARS outbreak. Besides being a threat to the food supply, virologists raised concerns that this virus may cross the species barrier from pigs to humans. Central to containing these outbreaks is monitoring the evolution of PRRS virus for its ability to replicate in alternate hosts.

Dam Construction

Most of the world's dams for irrigation and flood control were constructed in the 1930s, and nearly every large river worldwide has at least one dam. Several mosquito-borne outbreaks of viral diseases have occurred following dam construction. In 1987, an outbreak of **Rift Valley fever** in Mauritania occurred during construction of a dam on the Senegal River (West Africa). Rift Valley fever causes serious complications: meningoencephalitis (an inflammation of the brain) and lesions of the retina, which leave victims with at least some permanent loss of vision. About 1% to 3% of infected individuals experience hemorrhagic symptoms and kidney failure.

In the case of the outbreak in Mauritania, which resulted in 200 deaths, the dam caused flooding in the lower Senegal River area. In 1993, another Rift Valley fever outbreak occurred during the construction of the Aswan dam in Egypt, when approximately 600 to 1500 human cases of Rift Valley Fever were recorded.

More than 40 different species of mosquitoes carry Rift Valley fever virus. Floods and heavy rainfall contributed to the spread of Rift Valley fever virus because mosquito populations increase dramatically during these periods. Dam construction results in a large amount of standing water that provides a breeding ground for mosquitoes. These arthropods serve as vectors that catalyze the spread of the virus.

Global Commerce

Today the buying and selling of foods, especially on a large scale, occur between nations. People no longer depend upon food produced locally. Imported fruits and vegetables from developing countries may contain viral contaminants through the use of raw human sewage as fertilizers. Studies have shown that viruses may persist for weeks or even months on vegetable crops and in soils that have been irrigated or fertilized with human sewage wastes. Hepatitis A viruses, noroviruses, enteroviruses, astroviruses, rotaviruses, and coronaviruses have caused foodborne disease outbreaks associated with the consumption of raw fruits and vegetables. The most commonly contaminated foods associated with foodborne viral outbreaks are iceberg lettuce, strawberries, raspberries, green onions, diced tomatoes, shredded cabbage, and raw oysters from contaminated bays (**FIGURE 18-12a**).

Honeybees in the United States are mysteriously dying of colony collapse disorder (CCD). Beekeepers are baffled by the cause of CCD. Approximately 23% of beekeeping operations in the United States suffered from CCD in 2006 to 2007. It has resulted in an average loss of 45% of each affected business. One hypothesis is that CCD is caused by a novel infectious agent.

To test the novel-infectious-agent hypothesis, a team of beekeepers, entomologists, and molecular biologists analyzed nucleic acids from pooled samples of bees from healthy colonies and samples of bees with CCD using a **metagenomic survey**. This survey analyzed genomes of microflora present in normal hives and CCD hives. Both DNA and RNA was extracted from the bees to determine if RNA viruses or other pathogens were present in the samples. All of the bee hives contained a rogues gallery of pathogens. Israeli acute paralysis virus (IAPV), however, was found in 25 of the 30 CCD hives and in only one of the healthy hives. These results suggest that IAPV is strongly associated with CCD. Where did IAPV come from?

Since 2004, beekeepers have struggled to keep up with the demand for almond pollination in California. To meet this need, millions of dollars worth of bees from Australia were imported into the United States. Hives with CCD contained imported bees from Australia or were located near operations with Australian bees. Bees imported from Australia also tested IAPV-positive. The sudden loss of honeybees, particularly among imported hives, was linked to IAPV that may have

a b

FIGURE 18-12 (a) This man is shopping for fresh produce in a supermarket. Some of the produce is likely imported from other countries and may serve as a vehicle for human disease. **(b)** This African green monkey is the same species that was involved in the 1967 Marburg hemorrhagic fever outbreak in Marburg, Germany, and Belgrade, Yugoslavia.

arrived with bees from Australia. Subsequently, a study demonstrated that IAPV was in the United States long before these recent CCD outbreaks.

In 2010, the murder mystery was solved using a mass-spectrophotometry proteomics approach to survey pathogens in honeybee colonies. Researchers detected a DNA virus and two RNA viruses that were not detected in honeybees in the United States prior to these CCD outbreaks. Through inoculation experiments, they concluded that CCD is caused by a co-infection with a DNA virus known as invertebrate iridescent virus-6 virus (IIV-6) and a fungal pathogen, *Nosema ceranae*.

Importing animals for biomedical research or for exotic pet trading has been a cause for concern. In 1967, African green monkeys were imported simultaneously to Marburg, Germany, and Belgrade, Yugoslavia, for research and preparation of poliovirus vaccine (Figure 18-12b). The monkeys were infected with **Marburg virus**, which causes a hemorrhagic fever in humans and has a high mortality rate. The first people infected were laboratory workers who had contact with monkey blood or organs or with prepared cell cultures.

With a few exceptions, the import of exotic species from around the world into the United States has been a largely unregulated industry. The **monkeypox outbreak** in the United States in 2003 discussed previously demonstrates how quickly an unusual virus can spread through the exotic-pet trade. The outbreak made many people consider the potential for human health risks related to importing exotic species.

Human Social Behavior: Sex, Intravenous Drugs, Cultural Traditions, and Research Laboratories

Having multiple sex partners increases the risk of acquiring sexually transmitted viral diseases such as HIV, hepatitis B and C, and human papillomaviruses. As discussed earlier, HIV originated during a cross-species transmission of SIV from nonhuman primates to humans. The HIV pandemic began when the virus began to spread from human to human, primarily via homosexual contact or by sharing contaminated needles used for **intravenous drug use**. Sub-Saharan Africa is home to about 64% of the world population living with HIV, and South Africa is the epicenter of the AIDS epidemic. Transmission is primarily through heterosexual contact, and more women than men are infected with HIV. The Caribbean is the second-most HIV-affected region of the world. Transmission there is also primarily heterosexual.

Cultural traditions can have a role in spreading viruses. Consider, for example, how many traditions involve physical contact or the sharing of a cup or other drinking and eating vessels. An example of this is a traditional belief that led to the 2005 Marburg hemorrhagic fever outbreak in the Angolan province of Uige, which is the deadliest one on record. The outbreak in Uige, which is located in northeastern Angola (**FIGURE 18-13**), lasted from March through August. According to information collected by health officials, the first cases appeared among Itoumbi villagers who had gone elephant hunting in the bush. "All the victims were great hunters. They touched a monkey they found dead in the forest and then ate it." Main symptoms of the infection were fever, hemorrhage (bleeding), vomiting, cough, diarrhea, and jaundice (yellow skin and eyes).

There were a total of 374 cases and 329 deaths (88% mortality). The majority of deaths occurred in the main hospital in Uige, and in fact the hospital was the source of the infection for many peoples. Hospital workers were administering medicine to patients by injection, using shared/contaminated syringes. Family members adhered to traditional funeral rites. They washed bodies of the deceased loved ones, and kissed the body to show respect and love for the deceased and to keep vengeful spirits away. This custom allowed the rapid spread of the virus from blood-to-blood contact. Safety protocols (e.g., barrier techniques for infection control) were breached. Relatives of the deceased asked doctors and nurses to respect these customs while doing what was necessary to prevent further infections.

The country had been ravaged by a 27-year civil war, and thus hospitals lacked the basic

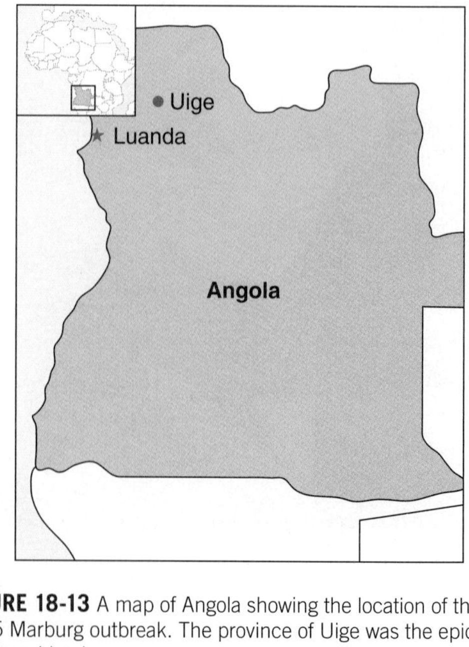

FIGURE 18-13 A map of Angola showing the location of the 2005 Marburg outbreak. The province of Uige was the epicenter of this epidemic.

equipment, supplies, transportation means, and infrastructure. At the beginning of the outbreak, at least 16 doctors, nurses, and traditional healers died from the disease after caring for sick patients. Officials from the CDC, World Health Organization (WHO), Health Canada, and Medecins sans Frontieres (MSF, or Doctors Without Borders) rushed in with medicines and protective gear to stem the epidemic, but convincing some communities to change their traditions in order to protect themselves from the communicable disease proved difficult. The locals feared the isolation wards because infected people in hospitals had died there. The families cared for the sick at home and then hid the bodies from the officials sent in to help. On at least three occasions, relief workers were stoned and their vehicles damaged by hostile residents who blamed them for the spread of the disease. Building trust was central to being able to collect and bury corpses appropriately. Religious leaders in Uige joined the public information campaign to stop the use of unsafe injections, an important component in the spread of the Marburg virus. Eventually trust was secured between the community and a collaborative effort brought the outbreak under control.

Applied biomedical and biodefense research on dangerous viral pathogens takes place in specialized Biosafety Level 3 and 4 (BSL-3 and -4) laboratories. These laboratories have the most stringent safety and security requirements (**FIGURE 18-14**). There are currently only four operational BSL-4 laboratories in the United States, which are located at the following institutions:

- CDC in Atlanta, Georgia
- U.S. Army Medical Research Institute for Infectious Diseases at Fort Detrick in Frederick, Maryland
- Southwest Foundation for Biomedical Research in San Antonio, Texas
- University of Texas at Galveston

Georgia State University in Atlanta has a small BSL-3/BSL-4 glove box facility. In addition, a small BSL-4 facility exists on the National Institutes of Health (NIH) campus in Bethesda, Maryland, but it is currently being operated only at a BSL-3 level for research on important emerging infectious disease pathogens.

The chances of a viral pathogen "escaping" from a BSL-3 or BSL-4 laboratory are remote. In 2005, though, a serious mistake occurred when a H2N2 influenza A virus present in diagnostic test kits was sent to thousands of laboratories in 19 different countries. This Asian strain killed millions of people in a pandemic of 1957–1958, and any person born after 1968 would have little or no immunity to this virus. Within days, all packages containing the deadly strain were recovered and destroyed and no one was infected by the virus during this mishap. This serious mistake created a heightened awareness of the management and operation of laboratories that contain viral pathogens that are potential bioweapons. Well-established guidelines aimed to protect laboratory workers and the surrounding community were updated and implemented.

FIGURE 18-14 Biomedical research laboratories contain some of the deadliest human pathogens on this earth. The viruses are stored in secured freezers or stored in liquid nitrogen tanks like the one shown in this photograph.

Human Medicine and Susceptibility to Infection

Health-related infections can occur in a number of ways, such as:

- unsafe medical practices (e.g., reusing syringes and the overuse of antimicrobials)
- blood transfusions
- poor hospital practices
- immunosuppression
- xenotransplantation (see Chapter 9)
- breakdown in public health measures including lack of vaccination programs at local, state, national, and global levels; problems with basic sanitation and infection-control procedures

Typically one thinks of poor hospital practices in developing countries regarding the spread of bloodborne pathogens. Health care in the United States, however, is not without problems. In 2006 CDC viral hepatitis expert Dr. Joseph Perz presented the challenges that exist today in outpatient and long-term care facilities in the United States at an international conference on emerging and infectious diseases. A disturbing trend in the last five years involving patient-to-patient spread of bloodborne diseases is being recognized. According to Perz, a common theme is that unsafe injection practices and failure to adhere to aseptic technique have resulted in the investigation of several outbreaks of bloodborne illnesses in outpatient settings. Bloodborne infections are not just a problem of the developing world. This represents an element of patient safety in the United States that must be remedied as soon as possible.

The good news is that blood supply safety in the United States is excellent, but the poor hospital practices and the blood supply of developing countries still create efficient ways of transmitting bloodborne viruses. Most notably, HIV and hepatitis C virus infections are now prevalent in Chinese communities. The origin of the HIV and hepatitis C epidemic in central China began in the late 1980s and early 1990s through a contaminated blood supply. The Chinese have a cultural belief that losing blood is unhealthy; for this reason, the blood supply is often short due to the lack of blood donors. As a means to supplement their income, Chinese villagers sell their blood for plasma. It was standard practice to pool the blood from dozens of sellers. The plasma was separated and red blood cells were reinfused into sellers. No HIV or hepatitis C virus screening was performed on the red blood cells. About one million paid blood donors contracted HIV and/or hepatitis C virus through this procedure as well as through the reuse of syringes and nonsterilzed equipment. China has since taken steps to halt illegal plasma collection and to improve blood-banking methods.

The population of individuals who are immunocompromised is on the rise. In the United States alone there are about 10 to 20 million immunocompromised people. This was the direct result of the use of immune-suppressive drugs during organ or bone marrow transplants, kidney dialysis, chemotherapy, and chronic corticosteroid treatment for such diseases as inflammatory bowel disease or rheumatoid arthritis. That being said, the number of opportunistic infections is on the rise. Immune suppressants encourage the following viral infections:

- human cytomegalovirus
- varicella zoster
- herpes simplex I
- Epstein-Barr virus
- human herpesvirus type 6
- human herpesvirus type 7
- adenoviruses
- hepatitis viruses

The large-scale use of drugs to treat these opportunistic infections led to the emergence of drug-resistant strains and other phenotypes of new strains that dominate the population. In turn, transmission of drug-resistant and new viral strains can occur.

Vaccine-preventable diseases such as measles and mumps are reemerging on college campuses. The safety of vaccines has been questioned and some parents are not allowing their children to be vaccinated. The first cases of outbreaks were in unvaccinated or immunocompromised individuals. In 2006 there was a multistate mumps outbreak in the Midwest. From January to May of that year there were 2597 reported cases of mumps in eight different states (**FIGURE 18-15a**). No deaths were reported. The majority of cases were in Iowa (Figure 18-15b). Eleven of the infected individuals traveled by aircraft on 33 commercial flights that were operated by 8 different airlines. At least 2 persons acquired mumps during air travel.

A 2005 measles outbreak occurred in Indiana after an unvaccinated 17-year-old Indiana resident working as a missionary in an orphanage and hospital in Romania contracted measles. The 17-year-old returned to the United States with symptoms of the measles. Subsequently, the largest measles outbreak occurred at the Romanian orphanage. Upon her return, the teenager (patient zero) attended a church function in which unvaccinated individuals were present. A total of 34 persons contracted measles through direct contact with the teenager. Vaccination and isolation mea-

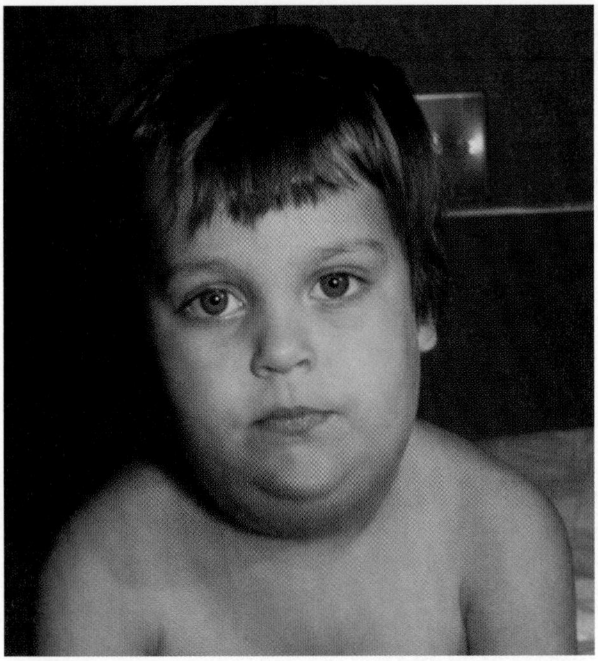

a

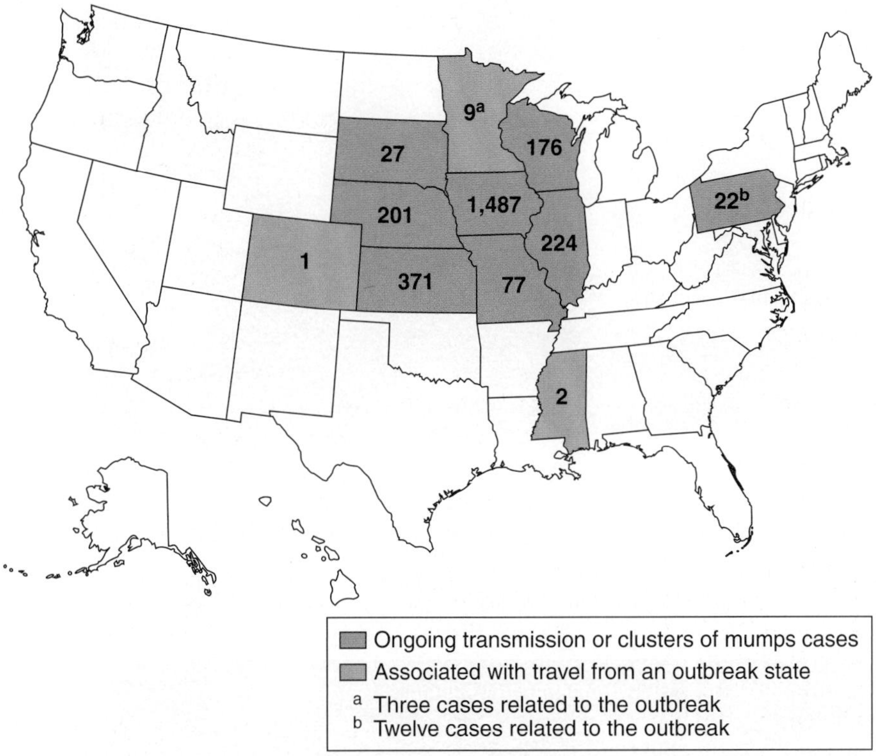

b

FIGURE 18-15 (a) A child with mumps. Note the swelling in his neck due to lymphatic buildup caused by mumps virus infection of the parotid salivary glands. **(b)** Mumps cases (2597) linked to multistate outbreak in 2006. Reproduced from Centers for Disease Control and Prevention, *MMWR* 55 (2006): 559–563.

sures were undertaken to prevent further measles transmission.

The first outbreak of poliomyelitis in the United States in 26 years occurred in a central Minnesota Amish community in 2005. For religious or social reasons, Amish avoid immunizations. The Amish live as isolated communities with little risk of passing on or being exposed to infectious diseases outside of the community. In this instance, an unvaccinated immunosuppressed Amish infant contracted an oral vaccine-derived poliovirus strain. Three additional children in the Amish community were infected. The original source of the poliovirus was likely a person who received the live oral poliovirus vaccine in another country. A WHO global initiative to eradicate polio

VIRUS FILE 18-2 Human Bocavirus (HBoV)—An Emerging Viral Pathogen?

Lower respiratory tract infections are the leading cause for hospitalization of infants and young children (**FIGURE VF 18-2**) and account for approximately 250,000 hospitalizations annually in the United States. Respiratory syncytial virus (RSV) is a leading viral agent in this group of patients. In about 12% to 39% of cases, though, no known agent has been identified as the cause of the lower respiratory tract infection.

In 2005, a group of researchers in Sweden identified a new virus present in respiratory secretions of children with lower respiratory tract infections. The new virus was named **human bocavirus** (**HBoV**). Bocaviruses belong to the family *Parvoviridae,* which includes the smallest known viruses at a mere 18 to 26 nm in diameter. The viruses are nonenveloped and icosahedral-shaped with ssDNA genomes and are very resistant to environmental forces.

The HBoV DNA was amplified from pooled samples of respiratory tract secretions collected from Swedish patients. This type of molecular screening is improving the speed at which new viruses can be identified because these viruses have not yet been cultivated in the laboratory. Follow-up studies screening for the presence of HBoV in children suffering from lower respiratory tract infections in other countries is ongoing (**TABLE VF 18-1**).

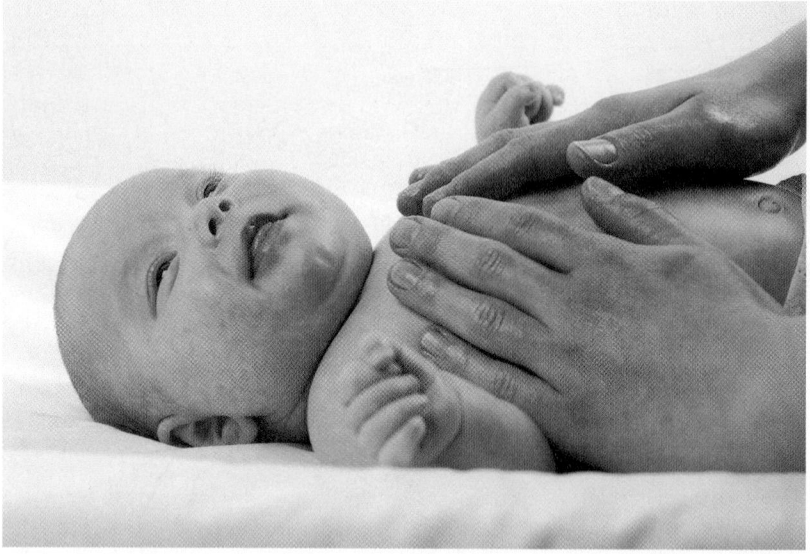

FIGURE VF 18-2 Child hospitalized with a lower respiratory tract infection. Respiratory tract infections are among the leading causes of death in children under 5 years of age.

TABLE VF 18-1 HBoV Detection in Children Suffering from Lower Respiratory Tract Infections

Country	Patient Sample Size	% HBoV Positive by PCR
Canada	1209	1.5%
France	589	4.4%
Germany	835	10.3%
Japan	318	5.7%
Korea	515	11.3%
United States (California)	1474	5.6%

VIRUS FILE 18-2 Human Bocavirus (HBoV)—An Emerging Viral Pathogen? (continued)

These studies affirm that viral sequences from other viral respiratory pathogens (coinfections) have also been found in up to 39% of HBoV-positive samples. As a result of the high rates of coinfections with HBoV and other respiratory tract pathogens, the clinical significance of HBoV is unknown.

References

Allander, T., et al. 2005. "Cloning of a human parvovirus by molecular screening of respiratory tract samples." *Proc Natl Acad Sci USA* 102(36):12891–12896.

Arnold, J. C., et al. 2006. "Human bocavirus: Prevalence and clinical spectrum at a children's hospital." *Clin Infect Dis* 43(3):283–288.

Choi, E. H., et al. 2006. "The association of newly identified respiratory viruses with lower respiratory tract infections in Korean children, 2000–2006." *Clin Infect Dis* 43(5):585–592.

Foulongne, V., et al. 2006. "Human bocavirus in French children." *EID* 12(8):1251–1253.

Song, J. 2010. "Novel bocavirus in children with acute respiratory tract infection." *EID* 16(2):324–327.

Vicente, D., et al. 2007. "Human bocavirus, a respiratory and enteric virus." *EID* 13(4):636–637.

Weissbrich, B., et al. 2006. "Frequent detection of bocavirus in German children with respiratory tract infections." *BMC Infect Dis* 6:109.

continues. Countries like Sudan that were once declared polio-free are experiencing outbreaks. The distrust of Western vaccines hampers the efforts by healthcare workers to immunize children most susceptible to the virus.

18.3 Ecological Factors: Climate Variability

Many emerging infectious disease outbreaks occur after environmental disruptions such as hurricanes, tsunamis, extreme floods, increased rainfall, and droughts. Global climate changes are expected to increase, which likely will broaden the range of many diseases, especially those that are vector-borne. Mosquitoes, mice, and other carriers of viruses are surviving warmer winters and expanding their range, bringing their viruses with them (**FIGURE 18-16**). When pathogens adapt to new environments, the diseases they cause pose a serious threat to international health. What is one nation's problem will become every nation's problem.

El Niño is an abnormal warming of the surface ocean waters that causes weather changes (e.g., heavy rains) around the globe. From 1991 to 1993, the southern United States experienced an

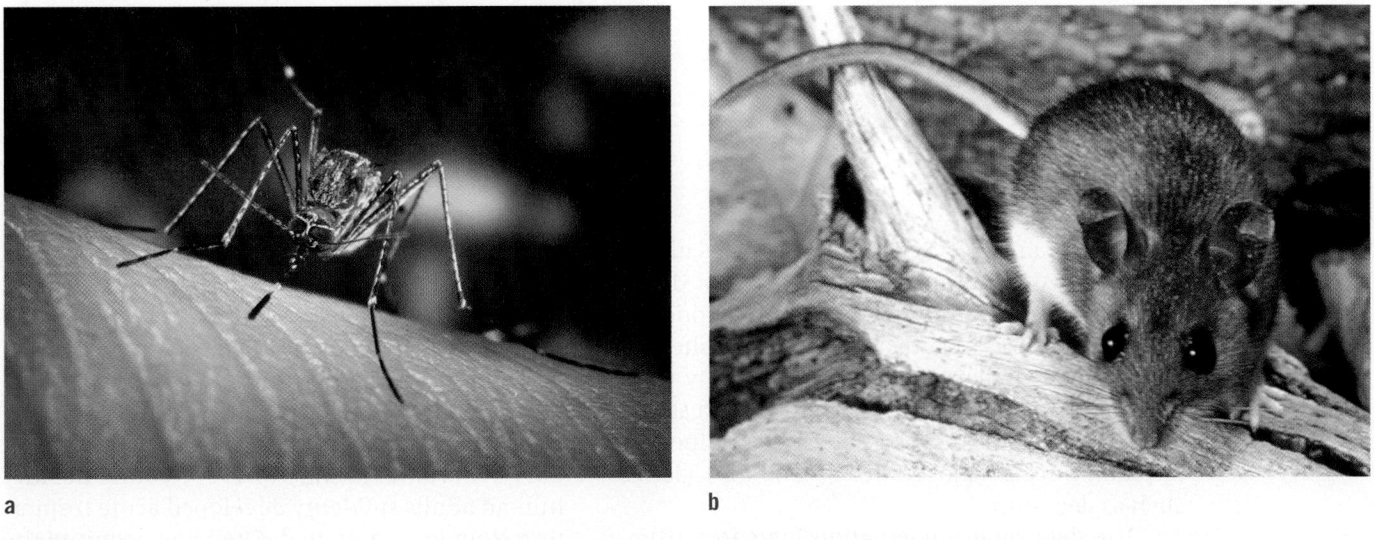

a b

FIGURE 18-16 Mosquito and rodent populations fluctuate during climate changes. **(a)** *Culex tarsalis* mosquito that is about to feed on its human host. This mosquito is currently the main vector of West Nile virus in the United States. **(b)** Deer mouse (*Peromyscus maniculatus*), which is the carrier of *sin nombre* hantavirus, the causative agent of hantavirus pulmonary syndrome.

TABLE 18-3 Reasons for the Emergence or Reemergence of Viral Diseases

Viral Pathogen Factor	HIV	Monkeypox	Chikungunya Virus	West Nile Virus	Norovirus	Nipah Virus	*Sin Nombre* Hantavirus	Avian Influenza	SARS-CoV	Mumps	Marburg Virus
Viral mutation	X		X				X	X			
Human travel and migration	X	X	X	X	X			X	X	X	
Human population growth and urbanization					X		X	X	X		
Humans and deforestation											
Humans and agriculture						X					
Human exposure to zoonotic diseases via hunting and pasturing	X					X	X	X	X		X
Humans and commerce		X						X			X
Human social behavior: Sex and drugs	X										
Human social behavior: Cultural traditions											X
Humans and modern medicine and unsafe practices	X										X
Humans and breakdown in public health measures	X		X						X	X	X
Ecological factors: Rain			X			X					
Ecological factors: Drought				X			X				

increase in rainfall caused by El Niño, particularly in the Four Corners area, which consists of Utah, Colorado, Arizona, and New Mexico. Rodents must forage large areas to obtain food and shelter. These states contain areas that are normally very dry, but when the El Niño rains hit this area at the beginning of 1991 there was increased production of piñon nuts, a major source of food for rodents such as deer mice.

The deer mouse population increased from less than one mouse per 2.5 acres during dry years to over five per 2.5 acres within eight months of the first rains. By the spring of 1993, there were 20 to 30 deer mice per 2.5 acres. Some of these rodents carried the ***sin nombre* hantavirus**. As the deer mouse population densities increased, the virus was transmitted from mouse to mouse. In 1993 about 30% of the deer mice carried the *sin nombre* hantavirus, the cause of hantavirus pulmonary syndrome. In May of 1993, young healthy human adults suddenly developed acute respiratory symptoms and died. They had acquired the *sin nombre* hantavirus infection by breathing aerosolized deer mice saliva, urine, or droppings that contained the excreted viruses. The deer mice were infesting their homes, contaminating floors and blankets.

Insect populations such as mosquitoes thrive during warming trends. Droughts reduce populations of predators of mosquitoes such as frogs, dragonflies, ladybugs, and lacewings. New York experienced a mild winter in 1998–1999, allowing mosquitoes to survive in hibernation. The dry spring and the summer heat waves of 1999 decreased mosquito predators and evaporated mosquito larvae breeding sites. The mosquitoes thrived in drainpipes and sewer puddles that contained concentrated organic compounds that attracted birds in search of water. This provided an opportunity for mosquitoes to thrive and become infected with WNV when feeding on sick birds. Infected mosquitoes can then transmit WNV to humans and animals while biting to take blood.

It is expected that extreme floods, droughts, and warming trends will be more common over the next 200 years. Viral pathogens will adapt to new environments in developing and industrialized countries. Global warming and air travel between continents will spread vectorborne diseases. It will take an international effort to control the impact resulting from these climate changes and disease epidemics.

Summary

In the past three decades, humans have faced many new and reemerging infectious diseases of viral origin. There are many reasons for the emergence of new viral diseases (**TABLE 18-3**). It is necessary to prepare to confront the unexpected. Viral, human, and ecological factors contribute to their emergence. Of these three factors, human behavior plays the most important role in influencing disease emergence and patterns. The three main factors have been dissected into further individual causes:

- Viral factors:
  - point mutations
  - certain genes mutate more rapidly
  - antigenic drift and shift
  - recombination
  - reassortment of viral genomes
- Human factors:
  - population growth and megacities (density/overcrowding)
  - human movement (travel and migration)
  - human exposure to zoonotic diseases (pasture and hunting practices)
  - human agricultural practices, deforestation, and dam construction
  - humans and global commerce
  - human social behavior (sex, intravenous drugs, cultural traditions, and research)
  - human medicine and susceptibility to infection
- Ecological factors:
  - Climate variability (El Niño rains, droughts)

A combination of factors contributes to emerging viral disease outbreaks. The future is hard to predict but we can be certain that we are going to continue to be challenged by new and reemerging viral diseases.

Viruses live in a world without borders. Organized community efforts will be needed to recognize, prevent, and control the future viruses that will challenge us.

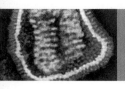

CASE STUDY QUESTIONS

These questions relate to the Case Study presented at the beginning of the chapter.

1. The horses and veterinary workers suffered from a Hendra virus infection. What is the natural reservoir of Hendra viruses?
2. What is the mode of transmission for the horses?
3. What is the mode of horse-to-human transmission?
4. Hendra virus 2 outbreaks involved horses and humans in 1994. What was different about the case study outbreak (2008) in contrast to the 1994 outbreak?
5. Research Hendra virus outbreaks occurring in Queensland and New South Wales. List the number of horses and humans involved. How many survivors were there? Describe their illnesses. How were the horses and humans treated? (*Hint:* the following reference may provide a useful start: Playford, E. G. 2010. *EID* 16(2):219–223.)

CASE STUDY 1: BORNA DISEASE

Borna disease was first described as a neurological illness of horses and sheep more than 200 years ago. It was named after the town of Borna in Saxony, Germany, where a large number of horses of a German cavalry regiment died in an 1885 epidemic. The diseased horses exhibited the following abnormal behaviors: running about excitedly, walking into walls, being unable to chew food, and standing with their heads and necks strained forward and down.

In later years, the Borna disease virus was isolated from the brains of infected horses. Brain homogenates from infected horses reproduced the disease in experimental naïve (uninfected) animals. Current research suggests a possible link between Borna disease virus and human neuropsychiatric disorders. Antibodies reactive with this virus were found in the sera of patients with psychiatric disorders such as schizophrenia. In 1996, researchers at the Robert Koch-Institut and Free University in Berlin, Germany, isolated Borna disease virus from patients with mood disorders.

1. Could Borna disease virus be an emerging zoonotic pathogen? Discuss your answer.
2. What population of people would you expect to be most vulnerable to this viral pathogen?
3. The field of Borna disease virus research is still fairly new and controversial. Perform a literature search and write an up-to-date summary of modern Borna disease virus research.

CASE STUDY 2: MYSTERIOUS PIG MORTALITIES

Access to proper sanitation in some parts of India does not exist. About 88% of the population of India has access to safe water and 31% has access to safe sanitation. Members of one Hindu caste known as untouchables are pig owners in charge of cleaning up the slums of some cities. They allow the pigs to roam and consume household garbage, including human waste. In February 2005, there were reports of 20 to 25 pigs dying each day in a slum of Agra as a result of an illness the untouchables called "pig flu." Officials stated that the pigs died of "cold and viral fever." There were over 100 rotting pig carcasses in the city slum at the time of the report.

1. If you were hired to investigate this outbreak, how would you rule whether or not an infectious agent caused this mysterious pig disease?
2. What types of specimens would you collect during the investigation and what would you test them for?
3. What factors in this instance may be contributing to the emergence of a viral pathogen?
4. What efforts would you use to improve sanitation?
5. How should carcass disposal be handled?

Reference

ProMED-mail. Unexplained deaths, pigs—India (Kerala). *ProMED-mail.* 20050215.0506.

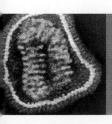

CASE STUDY 3: MYSTERIOUS ILLNESS AMONG COLORADO FIELD WORKERS

Two field workers in Boulder, Colorado, were admitted to a community hospital after complaining of fatigue, a headache, and fever. Hospital lab tests revealed the workers also had thrombocytopenia (an abnormal drop in the number of platelets involved in forming blood clots). One of the workers had 70,000 platelets/μl blood. Normal platelet counts are 450,000 platelets/μl blood. Thrombocytopenia is characterized by easy bruising and increased bleeding. The field workers had been trapping rodents for ecologic studies. They trapped mice, prairie dogs, squirrels and rabbits. One of the workers had been bitten by a vole during the field work.

1. Assuming these workers were infected during their fieldwork, generate a list of viruses that could potentially infect these two individuals.
2. What would be the primary mode of transmission for the viruses you listed above?
3. What kinds of methods would be used to determine if the mammals trapped were carriers of the virus that made the workers sick?
4. How could these viral infections have been prevented?

Reference

Torres-Perez, F., et al. 2010. *Emerg Infect Dis* 16(2): 308–310.

CASE STUDY 4: MYSTERIOUS PIG ILLNESS

During June 2009, about 30% of the nursing pigs in a barn in Argentina showed clinical signs of nasal discharge, coughing, fever, lack of appetite, and shortness of breath (dyspnea). Some of the pigs died of this unknown respiratory infection. Ten days before the outbreak, the farm manager and his wife had suffered from influenza.

1. Is direct human-to-pig transmission of influenza viruses possible? Explain your answer. (Need help? Refer to Chapter 12, Influenza Viruses).
2. What type of laboratory tests would you do to determine what virus(es) were making the nursing pigs ill?
3. Create a list of interview questions for the pig farmer and his wife that may help you to determine the cause of this outbreak.

Reference

Pereda, A. 2010. *Emerg Infect Dis* 16(2):304–306.

Resources

Primary Literature and Reviews

Arguin, P. M., et al. 2009. "Globally mobile populations and the spread of emerging pathogens." *Emerg Infect Dis* 15(11):1713–1714.

Arrigo, N. C., et al. 2010. "Cotton rats and house sparrows as hosts for North and South American strains of eastern equine encephalitis virus." *Emerg Infect Dis* 16(9): 1373–1380.

Arthur, R. R., et al. 1993. "Recurrence of rift valley fever in Egypt." *Lancet* 342 (8880):1149–1150.

Bahta, L., et al. 2005. "Poliovirus infections in four unvaccinated children—Minnesota, August–October, 2005." *MMWR* 54(Dispatch):1–3.

Basler, C., and Amarasinghe, G. K. 2009. "Evasion of interferon responses by ebola and Marburg viruses." *J Interferon Cytokine Res* 29:511–520.

Beasely, D. W. C. 2005. "Recent advances in the molecular biology of West Nile virus." *Curr Mol Med* 5:835–850.

Bromenshenk, J. J., et al. 2010. "Iridovirus and microsporidian linked to honey bee colony decline." *PLoS ONE* 5(10) e13181.

Busch, M., et al. 2003. "Update: Detection of West Nile virus in blood donations—United States, 2003." *MMWR* 52:911–919. [Erratum, *MMWR* 2003. 52:942.]

Centers for Disease Control and Prevention (CDC). 1999. "Update: Outbreak of nipah virus—Malaysia and Singapore, 1999." *MMWR* 48(16):335–337.

Chang, C. C., et al. 2002. "Evolution of porcine reproductive and respiratory syndrome virus during sequential passages in pigs." *J Virol* 76(10):4750–4763.

Chen, L. H., et al. 2009. "Illness in long-term travelers visiting geosentinel clinics." *Emerg Infect Dis* 15(11): 1773–1782.

Chironna, M., et al. 2003. "Prevalence rates of viral hepatitis infections in refugee Kurds from Iraq and Turkey." *Infection* 31:70–74.

Chua, K. B., et al. 2000. "Nipah virus: A recently emergent deadly paramyxovirus." *Science* 288:1432.

Cox-Foster, D. L., et al. 2007. "A metagenomic survey of microbes in honey bee collapse disorder." *Science* 318(5848):283–287.

Cramer, E. H., Gu, D. X., and Durbin, R. E. 2003. "Diarrheal disease on cruise ships, 1990–2000." *Am J Prev Med* 24:227–233.

Dato, V., et al. 2003. "Hepatitis a outbreak associated with green onions at a restaurant—Monaca, Pennsylvania, 2003." *MMWR* 52(47):1155–1157.

Diamond, M. S. 2009. "Virus and host determinants of West Nile virus pathogenesis." *PLoS Pathog* 5:6 e1000452.

Domenech-Sanchez, A., et al. 2009. "Gastroenteritis outbreaks in 2 tourist resorts, Dominican Republic." *Emerg Infect Dis* 15(11):1877–1878.

Fauci, A. S., Touchette, N. A., and Folkers, G. K. 2005. "Emerging infectious diseases: A 10-year perspective from the national institute of allergy and infectious diseases." *EID* 11(4):519–525.

Foy, B. D., et al. 2011. "Probable non-vector-borne transmission of zika virus, Colorado, USA." *Emerg Infect Dis* 17(5):[Epub ahead of print].

Gaidet, N., et al. 2007. "Avian influenza viruses in water birds, Africa." *EID* 13(4):626–629.

Gautret, P., et al. 2009. "Hajj pilgrims' knowledge about acute respiratory infections." *Emerg Infect Dis* 15(11): 1761–1762.

Gautret, P., et al. 2009. "Multicenter eurotravnet/geosentinel study of travel-related infectious diseases in Europe." *Emerg Infect Dis* 15(11):1783–1790.

Gershman, K. 2006. "Update: Multistate outbreak of mumps—United States, January 1–May 2, 2006." *MMWR* 55(Dispatch):1–5.

Gibbs, E. P. J. 2005. "Emerging zoonotic epidemics in the interconnected global community." *Vet Rec* 157: 673–679.

Giladi, M., et al. 2001. "West nile encephalitis virus in Israel, 1999: The New York connection." *EID* 7(4): 659–661.

Gupta, R. 2005. "Recent outbreak of rabies infections in Brazil transmitted by vampire bats." *Eurosurveillance* 10:11.

Hernandez, F., et al. 1997. "Rotavirus and hepatitis a virus in market lettuce (*latuca sativa*) in Costa Rica." *Int J Food Microbiol* 37(2–3):221–223.
Holland, R. A., Wikelski, M., and Wilcove, D. S. 2006. "How and why insects migrate?" *Science* 313:794–796.
Holmes, E. C., and Rambaut, A. 2004. "Viral evolution and the emergence of SARS coronavirus." *Philos Trans Soc Lond B Biol Sci* 359(1447):1059–1065.
Hughes, J. M. 2001 "Emerging infectious diseases: A CDC perspective." *EID* 7(3 Suppl):494–496.
Jones, K. E., et al. 2008. "Global trends in emerging infectious diseases." *Nature* 451:990–993.
Kamugisha, C., Cairns, K. L., and Akim, C. 2003. "An outbreak of measles in Tanzanian refugee camps." *J Infect Dis* 187:S58–S62.
Ka-Wai Hui, E. 2006. "Reasons for the increase in emerging and re-emerging viral infectious diseases." *Microbes Infect* 8:905–916.
Kumar, D., and Humar, A. 2005. "Emerging viral infections in transplant recipients." *Curr Opin Infect Dis* 18(4): 337–341.
Leroy, E. M., et al. 2005. "Fruit bats are reservoirs of ebola virus." *Nature* 438:575–576.
Lopman, B., et al. 2004. "Increase in viral gastroenteritis outbreaks in Europe and epidemic spread of new norovirus variant." *Lancet* 363:682–688.
Lu, Y., et al. 2005. "The needs of AIDS-infected individuals in rural China." *Qual Health Res* 15(9):1149–1163.
Maiztegui, J. T. 1975. "Clinical and epidemiological patterns of Argentine haemorrhagic fever." *Bull WHO* 52: 567–575.
MacPherson, D. W., et al. 2009. "Population mobility, globalization, and antimicrobial drug resistance." *Emerg Infect Dis* 15(11):1727–1732.
Morse, S. S. 1995. "Factors in the emergence of infectious diseases." *EID* 1(1):7–15.
Murray, K., et al. 1995. "A morbillivirus that caused fatal disease in horses and humans." *Science* 268(5207): 94–97.
Nabeth, P., et al. 2001. "Rift valley fever outbreak, Mauritania, 1998: Seroepidemiologic, virologic, entomologic, and zoologic investigations." *EID* 7(6): 1052–1054.
Oregon Department of Health Services, et al. 2007. "Acute respiratory disease associated with adenovirus serotype 14—Four states, 2006–2007." *MMWR* 56(45):1181–1184.
Palacios, G., et al. 2008. "A new arenavirus in a cluster of fatal transplant-associated diseases." *N Engl Med* 358(10):991–998.
Pealer L. N., Marfin, A. A., Petersen, L. R., et al. 2003. "Transmission of West Nile virus through blood transfusion in the United States in 2002." *N Engl J Med* 349: 1236–1245.
Petersen, L. R., and Roehrig, J. T. 2001. "West Nile virus: A reemerging global pathogen." *EID* 7:611–614.
Principi, N., et al. 2010. "Effects of coronavirus infections in children." *EID* 16(2):183–188.
ProMED-mail. 2004. "Poliomyelitis—Africa: Sudan." *ProMED-mail* 20041216.3324.
ProMED-mail. 2005. Marburg hemorrhagic fever—Angola. *ProMED-mail* 20050526.1458.
ProMED-mail. 2005. "Norovirus, post-hurricane—USA (Louisiana)." *ProMED-mail* 20050911.2693.
ProMED-mail. 2005. "Rabies, human, vampire bats—Brazil (Maranhao)." *ProMED-mail* 20051102.3203.
ProMED-mail. 2005. "Rabies, human, vampire bats—Brazil (Maranhao)." *ProMED-mail* 20051110.3287.
Qian, H.-Z., et al. 2006. "Co-infection with HIV and hepatitis C virus in former plasma/blood donors: Challenge for patient care in China." *AIDS* 20:1429–1435.
Ratnieks, F. L. W., and Carreck, N. 2010. "Clarity on honey bee collapse?" *Science* 327:152–153.
Reid, T. M., and Robinson, H. G. 1987. "Frozen raspberries and hepatitis A." *Epidemiol Infect* 98(1):109–112.
Ren, X., et al. 2008. "Viral paratransgenesis in the malaria vector *anopheles gambiae*." *PLoS pathog* 4:8:e1000135.
Rodas, J. D., and Salvato, M. S. 2006. "Tales of mice and men: Natural history of arenaviruses." *Rev Col Cienc Pec* 19(4):382–400.
Roehrig, J. T., Layton, M., Smith, P., et al. 2002. "The emergence of West Nile virus in North America: Ecology, epidemiology, and surveillance." *Curr Top Microbiol Immunol* 267:223–240.
Rosenblum, L. S., et al. 1990. "A multifocal outbreak of hepatitis a traced to commercially distributed lettuce." *Am J Pub Health* 80(9):1075–1079.
Rosenthal, M. 2006. "Transmission of bloodborne infections also occurs in the U.S." *Infect Dis News* August.
Russel, R. C. 1987. "Survival of insects in the wheel bays of a boeing 747B aircraft on flights between tropical and temperate airports." *Bull WHO* 65(5):659–662.
Sponseller, B., et al. 2010. "Influenza a pandemic (H1N1) 2009 virus infection in domestic cat." *Emerg Infect Dis* 16(3): 534–537.
Staggs, W., et al. 2005. "Import-associated measles outbreak—Indiana, May–June 2005." *MMWR* 54(42): 1073–1076.
Stokstad, E. 2007. "Genomics: Puzzling decline of U.S. bees linked to virus from Australia." *Science* 317(5843): 1304–1305.
Stone, R. 2010. "Rival teams identify a virus behind deaths in central China." *Science* 330:20–21.
Stone, R. 2011. "Breaking the chain in Bangladesh." *Science* 331:1128–1131.
Sutherst, R. W. 2004. "Global change and human vulnerability to vector-borne diseases." *Clin Microbiol Rev* 17(1):136–173.
Tian, K., et al. 2007. "Emergence of fatal PRRSV variants: Unparalleled outbreaks of atypical PRRS in China and molecular dissection of the unique hallmark." *PLoS ONE* 2(6):e526. DOI:10.1371/journal.pone.0000526.
Vivancos, R., et al. 2010. "Norovirus outbreak in a cruise ship sailing around the British Isles; investigation and multi-agency management of an international outbreak." *J Infect* 60:478–485.
Wendong, L. 2005. "Bats are natural reservoirs of SARS-like coronaviruses." *Science* 310:676–679.
Widdowson, M.-A., Monroe, S. S., and Glass, R. L. 2005. "Are noroviruses emerging?" *EID* 11(5):735–737.
Wolfe, N. D., Dunavan, C. P., Diamond, J. 2007. "Origins of major human infectious diseases." *Nature* 447: 279–283.

World Health Organization, et al. 2006. "The global HIV/AIDS pandemic, 2006." *MMWR* 55(31):841–844.

Yanni, E. A. 2009. "Health status of visitors and temporary residents, United States." *Emerg Infect Dis* 15(11): 1715–1720.

Popular Press

Abraham, T. 2007. *Twenty-First Century Plague: The Story of SARS.* Baltimore: The Johns Hopkins University Press.

Carborne, K. M., ed. 2002. *Borna Disease Virus and Its Role in Neurobehavioral Disease.* Washington, DC: ASM Press.

Cheney, A. 2006. *Body Brokers.* New York: Broadway Books.

Close, W. T. 2002. *Ebola Through the Eyes of the People.* Marbleton, WY: Meadowlark Springs Productions.

Crosby, M. C. 2006. *The American Plague: The Untold Story of Yellow Fever, the Epidemic That Shaped Our History.* New York: Berkley Publishing Group.

Dickerson, J. L. 2006. *Yellow Fever: A Deadly Disease Poised to Kill Again.* Amherst, New York: Prometheus Books.

Dworkin, M. S. 2009. *Outbreak Investigations Around the World: Case Studies in Infectious Disease Field Epidemiology.* Sudbury, MA: Jones and Bartlett Publishers, Inc.

Garrett, L. 2000. *Betrayal of Trust: The Collapse of Global Public Health.* New York: Hyperion.

Goodwin, M. 2006. *Black Markets: The Supply and Demand of Body Parts.* Cambridge, MA: Cambridge University Press.

Goudsmit, J. 2004. *Viral Fitness: The Next SARS and West Nile in the Making.* New York: Oxford Press.

Greenblatt, C., and Spigelman, M., eds. 2003. *Emerging Pathogens: Archaeology, Ecology & Evolution of Infectious Diseases.* Oxford, England: Oxford University Press.

Harper, D. R., and Meyer, A. S. 1999. *Of Mice, Men, and Microbes: Hantavirus.* San Diego: Academic Press.

Kimball, A. M. 2006. *Risky Trade.* Vermont: Ashgate Publishing Company.

Krauss, H., et al., eds. 2003. *Zoonoses: Infectious Diseases Transmissible from Animals to Humans,* 3rd ed. Washington, DC: ASM Press.

Krause, R. M., Gallin, J. I., and Fauci, A., eds. 2000. *Emerging Infections: Biomedical Research Reports.* San Diego: Academic Press.

Loh, C., and Civic Exchange, eds. 2004. *At the Epicentre: Hong Kong and the SARS Outbreak.* Hong Kong: Hong Kong University Press.

McKenna, M. 2004. *Beating Back the Devil: On the Front Lines with the Disease Detectives of the Epidemic Intelligence Service.* New York: Free Press.

McNeill, W. H. 1977. *Plagues and Peoples.* New York: Anchor Books Doubleday.

Morse, S. S. 1996. *Emerging Viruses.* Reprint Edition. Oxford, England: Oxford University Press.

Osterholm, M. T., and Swartz, John. 2000. *Living Terrors: What America Needs to Know to Survive the Coming Bioterrorist Catastrophe.* New York: Delacorte Press.

Peters, C. J., and Olshaker, M. 1997.*Virus Hunter: Thirty Years of Battling Hot Viruses Around the World.* New York: Anchor Books.

Preston, R. 1999. "West Nile Mystery." *The New Yorker* October: 90–108.

Preston, R. 1999. *Crisis in the Hot Zone: A Terrifying True Story.* New York: Anchor Books.

Preston, R. 2009. *Panic in Level 4: Cannibals, Killer Viruses and Other Journeys to the Edge of Science.* New York: Random House Trade Paperbacks.

Price-Smith, A. T. 2009. *Contagion and Chaos: Disease, Ecology, and National Security in the Era of Globalization.* Cambridge, MA: Massachusetts Institute of Technology Press.

Quammen, D. 2007. "Deadly contact: How animals and humans exchange disease." *National Geographic* October:79–105.

Scheld, W. M., Murray, B. E., and Hughes, J. H., eds. 2004. *Emerging Infections.* Washington, DC: ASM Press.

Service, M. 2008. *Medical Entomology for Students.* United Kingdom: Cambridge University Press.

Schlossberg, D., ed. 2004. *Infections of Leisure.* 3rd ed. Washington, DC: ASM Press.

Snodgrass, M. E. 2003. *World Epidemics: A Cultural Chronology of Disease from Prehistory to the Era of SARS.* Jefferson, NC: McFarland.

Villarreal, L. P. 2005. *Viruses and the Evolution of Life.* Washington, DC: ASM Press.

Walters, M. J. 2003. *Six Modern Plagues and How We Are Causing Them.* Washington, DC: Shearwater Books.

Wheelis, M., Rozsa, Lajos, and Dando, Malcolm, Eds. 2006. *Deadly Cultures: Biological Weapons Since 1945.* Cambridge, MA: Harvard University Press.

Zelicoff, A. P., and Bellomo, M. 2005. *Microbe: Are We Ready for the Next Plague?* New York: Amacom American Management Association.

Zimmer, C. 2011. *A Planet of Viruses.* Chicago: University of Chicago Press.

Video Productions

Viral Outbreak: The Science of Emerging Disease. 2011. Howard Hughes Medical Institute.

NOVA: Epidemic—Ebola, AIDS, Bird Flu and Typhoid. 2007. WGBH, Boston.

Mega Disasters: Alien Infection. 2007. A&E Productions.

Ready or Not: Avian Flu. Aired 4/10/2006. ABC News Primetime.

RX for Survival: A Global Health Challenge. 2006. PBS.

The Age of Viruses. 2005. Films for the Humantities.

H5N1: Killer Flu. 2005. Films for the Humanities.

Influenza. 2005. Films for the Humanities.

Emerging Diseases. 2004. Films for the Humanities

Infectious Diseases: More Mobility, Greater Danger. 2004. Films for the Humanities.

Outbreak: Stopping SARS. 2003. Films for the Humanities.

SARS: The True Story. 2003. Films for the Humanities.

The New Explorers: On the Trail of a Killer Virus. 2001. A&E Production.

The Next Plague: The Nipah Virus. 1998. Films for the Humanities.

eLearning

go.jblearning.com/shors2

The site features eLearning, an online review area that provides quizzes and other tools to help you study for your class. You can also follow useful links for in-depth information, or just find out the latest virology and microbiology news.

CHAPTER 19

What About Prions and Viroids?

These deer will be tested for chronic wasting disease, a prion disease of cervids.

> **"It has just stopped me cold from eating another burger!"**
>
> *Oprah Winfrey, during a segment on mad cow disease on her April 16, 1996, show*

OUTLINE

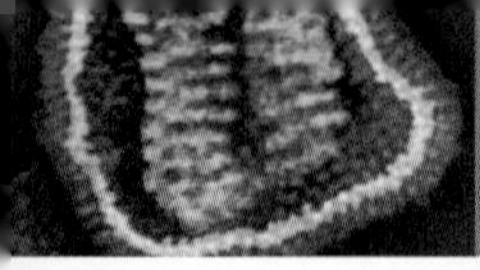

CASE STUDY: MYSTERIOUS ILLNESS IN A CAT OWNER

In 1990 the first cases of feline spongiform encephalopathy (FSE) were reported in the United Kingdom. In Italy in 1993, a cat owner and his cat suffered from simultaneous transmissible spongiform encephalopathies (TSEs). The seven-year-old short-haired, spayed female cat was usually fed canned food and slept under its owner's bed. In November of 1993, the cat became **ataxic** (lost its coordination) and suffered from bouts of frenzy and twitches. The cat was euthanized in mid-January, 1994. The clinical presentation of symptoms of the cat was different from previously reported feline spongiform encephalopathies.

The 60-year-old cat owner was admitted to a hospital in November 1993. He suffered from an unsteady ataxic gait, jerky movements, loss of speech, and was unable to swallow. He died in January 1994. There was no recollection of the cat biting her owner and the owner had no unusual dietary habits. The *PRNP* genes of the cat and owner were analyzed but no mutations were found. Both the cat and the owner were methionine homozygous at codon 129. Researchers have usually found clinical disease only occurs in patients homozygous for methionine at codon 129 of the *PRNP* gene. The brains of the cat and owner were analyzed. Both brains shared similar pathology: spongiosis.

Prions and **viroids** are *unconventional infectious agents*. They are not viruses, bacteria, or parasites, but instead are molecules that cause fatal diseases in humans and animals or cause a variety of effects in plants. Given their simple nature, these molecules find a home in virology textbooks. **Prions** *are proteins* that cause a group of diseases of the brain and nervous system called **transmissible spongiform encephalopathies** (**TSEs**). These diseases affect humans, sheep, goats, mink, deer, elk, antelope, and cats (**TABLES 19-1** and **19-2**). *None of the TSEs evoke a host immune response.* Instead, they cause a noninflammatory process that results in the vacuolation or **spongiosis** in the gray matter of the brain. The observation that there is no inflammation or immune defense associated with prion diseases is fundamentally important for understanding the prion model of disease. **Viroids** *are small pathogenic RNAs* that cause viruslike diseases in plants. They do not code for proteins and thus depend on plant host enzymes for their replication and other functions.

19.1 The "Mad" Diseases, Transmissible Spongiform Encephalopathies (TSEs): Kuru and Cannibalism

Deadly Feasts by Richard Rhodes is a compelling account of prion diseases. The author begins the book by describing **Kuru**, a fatal brain disorder that occurred among the South Fore people of the highlands of Papua New Guinea at epidemic levels during the 1950s and '60s. Kuru was known as "laughing sickness" because patients suffered from outbursts of laughter during the second stage of the disease. In the 1950s scholars Vin Zigas, Shirley Lindenbaum, and Daniel Carleton

TABLE 19-1 Human Prion Diseases

Disease	Cause	Lifespan After Symptoms Occur
Kuru	Cannibalism	3–6 months
Variant Creutzfeldt-Jakob disease (vCJD)	Consumption of contaminated beef (bovine prions)	13–14 months
Creutzfeldt-Jakob disease	Inherited germline mutation in *PrP* gene, sporadic (unknown cause), or iatrogenic	4–5 months
Gerstmann-Straussler-Scheinker disease (GSS)	Inherited germline mutation in *PrP* gene	2–6 years
Fatal familial insomnia (FFI)	Inherited germline mutation in *PrP* gene	12 months

TABLE 19-2	Animal Prion Diseases
Disease	**Animal Host**
Scrapie	Sheep and goats
Chronic wasting disease (CWD)	Deer and elk
Feline spongiform encephalopathy (FSE)	Cats
Bovine spongiform encephalopathy (BSE)	Cattle
Exotic ungulate encephalopathy (EUE)	Nyala and Greater Kudu (antelope in South Africa)
Transmissible mink encephalopathy (TME)	Mink

Gadjusek traveled to New Guinea to study this mysterious disease. Gadjusek's first impressions and observations of the Kuru epidemic were that the disease appeared to be "classical advancing Parkinsonism involving every age, overwhelming in females . . ." As field work progressed, Lindenbaum reported that in the early 1960s fewer than 10% of females survived past childbearing age, whereas males only had a 20% chance of dying of the disease. Gadjusek reported that Kuru victims suffered from three distinct stages of symptoms:

1. The **ambulant stage**: patients suffered from an unsteady gait, voice, hands, and eyes; tremors and shivering; slurred and deteriorating speech; loss of coordination in the lower extremities that slowly moved upward (**FIGURE 19-1**).
2. The **sedentary stage**: patients could no longer walk without support; experienced tremors and coordination problems with increased severity; suffered from jerky muscular movements, outbursts of laughter, depression, and mental slowing.
3. The **terminal stage**: patients could not sit without support; tremors and slurring of speech continued in severity. Patients suffered from incontinence, difficulty swallowing, and deep ulcerations.

The Fore people of Papua New Guinea began eating their dead relatives (**endocannibalism**) during the early 1900s. The Fore ate their relatives as a sign of love and respect, as part of their funeral rites, and everyone who died in the community was eaten. Human flesh was regarded as meat. The adult males preferred to eat domestic pigs or wild boars, and Kuru victims were considered excellent sources of food because the fat of the victims resembled pork.

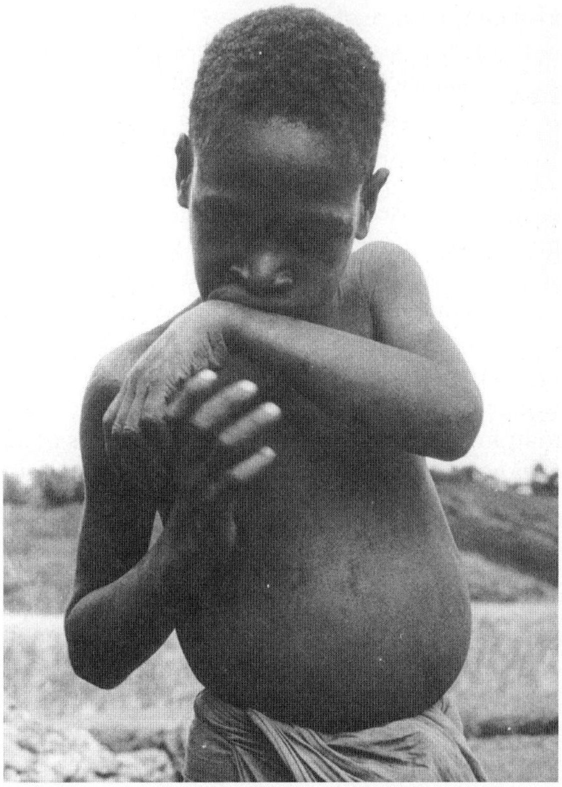

FIGURE 19-1 A Fore child with symptoms of the ambulant stage of Kuru.

Lindenbaum's description (in *Kuru Sorcery: Disease and Danger in the New Guinea Highlands*, 1978) of this practice is as follows:

> After a person's death, the maternal next of kin would dismember the corpse. She removed the arms and feet and stripped the muscle from the limbs. Subsequently, the chest was cut open and the internal organs were removed. The head was severed from the skull and broken open. The brain was scooped out of the skull. Bones were broken and marrow was sucked out. The wife was expected to eat the penis of her dead husband. The entire body was eaten, even the feces. Care was taken not to rupture the gallbladder because gallbladder contents seemed to spoil or ruin the taste of the body parts. A man's brain was eaten by his sister. A woman's brain was given to her son's wife or her brother's wife. Children under the age of ten and the elderly ate whatever the mothers ate and what was given to them: morsels of the brain or other internal organs. After the age of ten, the male children lived with the adult males. Males never ate the meat of females.

Scientists came up with two main hypotheses to explain the cause of Kuru:

1. Kuru was a hereditary disease.
2. A biological agent transmitted by endocannibalism caused Kuru.

Heredity was considered unlikely rather quickly because the disease appeared recently in the Fore population (between 1900 and 1920). It quickly spread through a population of 40,000 that could not have all descended from a single individual. During the early 1960s it reached epidemic numbers. About 1% of the South Fore people were dying of Kuru each year. Eight times more females than males contracted the disease. Females of all ages got the disease; old men and young boys got the disease as well. It was determined that the practice of endocannibalism began about ten years before the first appearance of Kuru.

James McConnell published a controversial paper regarding cannibalism in 1962. McConnell concluded that untrained flatworms (planaria) that were fed chopped-up trained flatworms retained some of the memories of the flatworms they ate. This research pointed Kuru researchers to the idea that cannibalism can affect the brain and raised the question of whether Kuru and cannibalism could be connected. Gajdusek's team inoculated the brains of healthy chimpanzees with brain suspensions made from Kuru patients. The chimpanzees were held in isolation and observed for up to five years. Three chimpanzees developed Kuru-like syndromes similar to that seen in human patients. The average time it took for the chimpanzees to become sick was between 18 and 21 months (this is considered a very long incubation period). These experiments led to the conclusion that a transmissible agent caused the Kuru-like disease, and that the agent was either a dormant or slow virus. In 1976, Gajdusek shared the Nobel Prize for Physiology or Medicine with Baruch S. Blumberg for their research on the origin and dissemination of infectious diseases. Blumberg gained notoriety for his work on the Australian antigen and the discovery of hepatitis B virus, whereas Gajdusek was awarded for work on the origin of a class of diseases now called TSEs.

In 1972, Stanley Prusiner initiated research to isolate the infectious agent that caused Kuru and similar diseases such as Creutzfeldt-Jakob disease (CJD) and scrapie, a disease affecting sheep. Ten years later, Prusiner and his colleagues isolated a single infectious protein from diseased hamster brains that they named a "prion," which stands for "proteinaceous infectious particle." The scientific community was very skeptical about this research because the infectious agent did not contain any nucleic acid or hereditary DNA or RNA. In 1984, Prusiner's team showed that the gene encoding the prion was found in the brains of all animals tested, including humans. It later became clear that there were two forms of the prion: the normal PrP^{C} proteins and the other protein that caused diseases such as scrapies, termed PrP^{SC}.

Another pioneer of prion research, Reed B. Wickner of the National Institutes of Health, reported in 1994 that a yeast prion called Ure2p could change shape and propagate that shape to change other yeast proteins. The Ure2p protein shared no sequence homology between the yeast and human protein. Wickner's group continues to contribute to the field of prion research using yeast as a model. His work was critical in the acceptance of Prusiner's work. In 1997 Prusiner was awarded the Nobel Prize in Physiology or Medicine for his "discovery of prions—a new principle of infection."

19.2 PrP and the "Protein Only" Hypothesis

Prions are infectious proteins that are highly resistant to routinely used methods of decontamination and steam sterilization. They are not inactivated by proteases, organic solvents, alkaline cleaners, ultraviolet light radiation, ethanol, formaldehyde, or extremely high temperatures (e.g., greater than 100°C; sterilization for one hour at 121°C in an autoclave does not kill prions). Researchers working with tissues, infectious waste, and instruments used in the processing of prion-contaminated samples decontaminate them in 1 N NaOH (sodium hydroxide) or undiluted fresh household bleach followed by autoclaving at 132°C for 4.5 hours. A study led by Paul Brown determined that incineration of prion-contaminated material for 15 minutes at 1000°C can destroy infectivity but it is not a practical solution during a mass culling of large outbreaks of bovine spongiform encephalophy (BSE), chronic wasting disease (CWD), or scrapie. Low infectivity remains after treatment at 600°C for 15 minutes.

There are two distinct conformations of the prion protein, PrP. PrP^{C} is found throughout tissues of the body in healthy people and animals (**FIGURE 19-2**). The normal cellular form of PrP, designated PrP^{C}, is sensitive to denaturing agents. The **"protein only" hypothesis** proposes that the

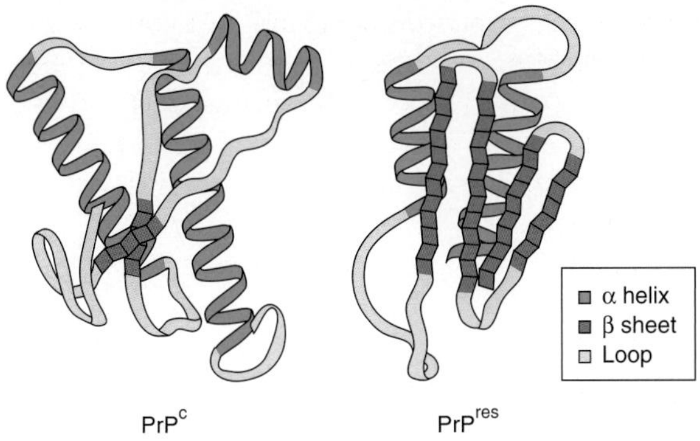

FIGURE 19-2 Tertiary structure of PrP^{C} and PrP^{res} proteins. The blue regions represent β sheets, the yellow regions are loops, and the green regions are α helices. The β sheets are thought to cause the amyloid clumping or aggregates in the brain.

abnormal, misfolded infectious form of PrP is able to initiate a reaction that causes the PrP^{C} to convert to the highly pathogenic and **resistant** or stable form termed **PrP^{res}**. Over time the PrP^{res} accumulates into clumps that damage or destroy nerve cells in the brain.

Purifying the infectious scrapie agent from hamster brains led to the amino acid sequence of the PrP^{SC} (or PrP^{res}) and, subsequently, to the identity of the encoding gene, *PRNP*. The *PRNP* gene is located on the short arm of chromosome 20 of humans. It codes for a protein of about 254 amino acids in length (**FIGURE 19-3**). It is unique, with no other proteins of similar homology to it in the database. Sequence analysis indicates that the prion protein is targeted via a secretory pathway to the cell surface of neurons and other cell types. A glycosylinositol phospholipid anchors it into the membrane. While in the membrane it may bind copper, and then is cycled back into the cell via endocytic vesicles where it may be degraded in lysozymes (**FIGURE 19-4**). Experiments suggest that the infectious prion finds its way to the endocytic vesicles, where it interacts with and changes the conformation of the wild-type noninfectious prion into the resistant form. These prions are resistant to degradation by the lysozomes. They accumulate, causing neurotoxicity.

Some researchers have hypothesized that other cellular factors stimulate the conversion of PrP^{C} to PrP^{res}. It has been shown that vertebrate host ssRNAs are required for the in vitro propagation of PrP^{res}. It is proposed that these specific ssRNAs bind avidly to and promote the conformational change in PrP^{C} to PrP^{res}.

Normal PrP^{C} is highly conserved in mammals and expressed predominantly in the brain. The exact function of PrP^{C} is unknown. Substantial scientific evidence suggests that it is a copper-binding protein. In the conversion to the aberrant form of the protein, copper-binding is lost. Other studies suggest that PrP^{C} can bind other metals such as manganese or zinc. Possible functions for PrP^{C} are roles in:

- signal transduction
- cellular differentiation
- cell adhesion
- copper transport
- resistance to the accumulation of destructive free radicals that can result in neuronal death

The Three Ways That TSEs Can Arise: Infection, Inherited, and Sporadic Forms

Sporadic CJD is the most frequent form of a human TSE. Its cause is unknown and patients usually die between the ages of 60 and 70 years old. Other forms of TSEs are inherited. Examples of genetic forms are genetic CJD, Gerstmann-Straussler-Scheinker disease, and fatal familial insomnia (Table 19-1). In 5% to 15% of TSE cases, mutations or insertions of nucleotides in the *PRNP* gene are found. A history of TSE or dementia is recorded in less than 50% of cases.

Creutzfeldt-Jakob disease can be inherited, sporadic, or infectious via diet (variant CJD) or acquired by iatrogenic means such as surgery, contaminated growth hormone injections, and corneal transplants. Kuru is an infectious form of TSE.

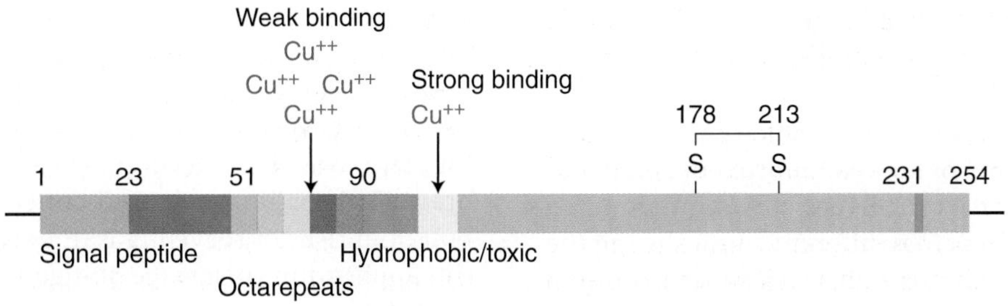

FIGURE 19-3 Biochemical analysis of the prion amino acid sequence. It contains a signal peptide that targets it to the surface of cells, a hydrophobic region that binds copper with a high affinity, and an octarepeat motif that contains amino acid residues that have low affinity for metals, especially copper. It is held together by a single disulfide bond. Adapted from Jones, I. "Prions show their metal." *Chembytes E-Zine*. The Royal Society of Chemistry, 2008.

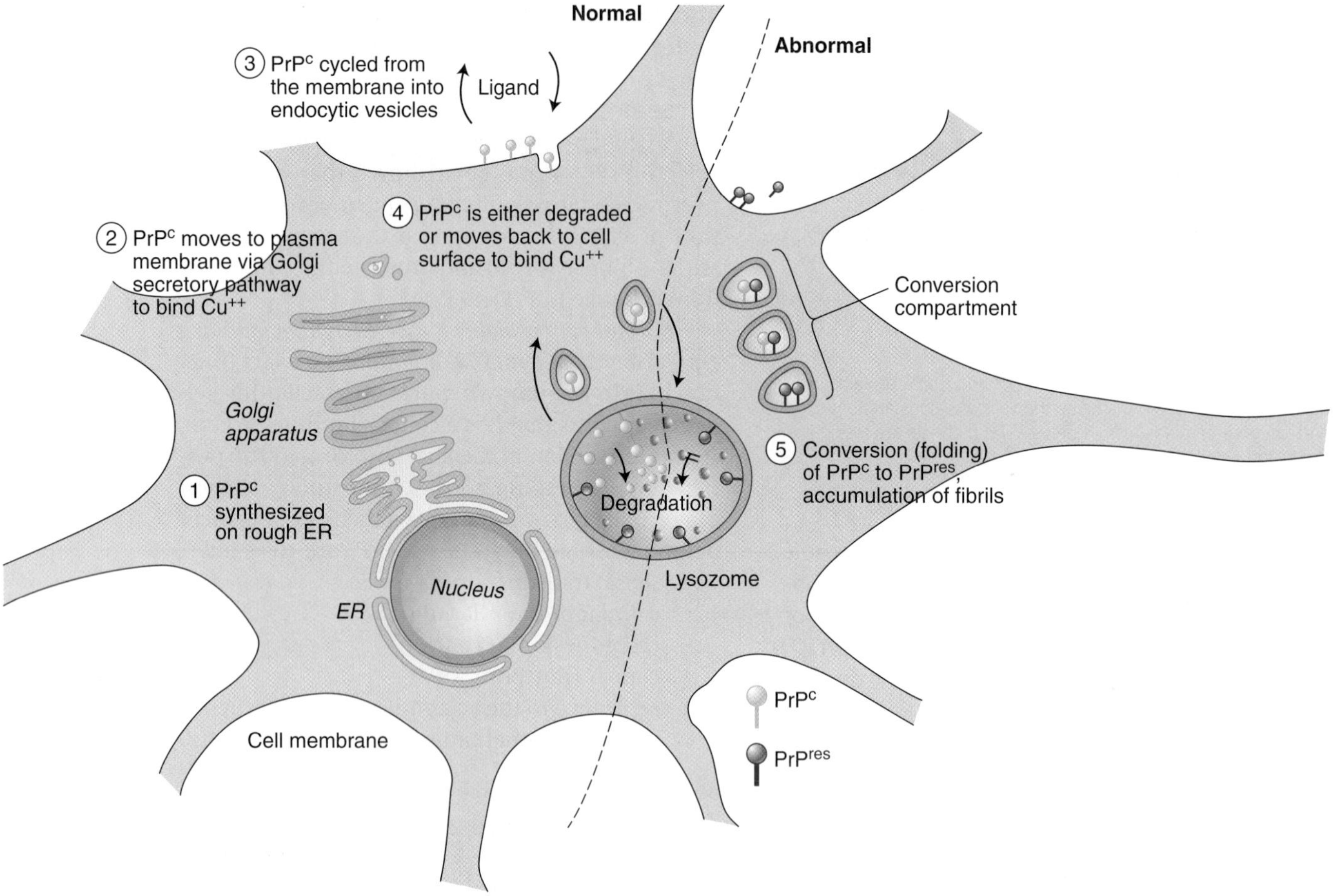
Normal
Abnormal
3 PrPᶜ cycled from the membrane into endocytic vesicles
Ligand
4 PrPᶜ is either degraded or moves back to cell surface to bind Cu⁺⁺
2 PrPᶜ moves to plasma membrane via Golgi secretory pathway to bind Cu⁺⁺
Conversion compartment
Golgi apparatus
5 Conversion (folding) of PrPᶜ to PrPʳᵉˢ, accumulation of fibrils
1 PrPᶜ synthesized on rough ER
Degradation
Lysozome
Nucleus
ER
PrPᶜ
Cell membrane
PrPʳᵉˢ

FIGURE 19-4 Hypothetical model showing PRP^{C} involved in the secretory pathway of cells. PrP^{res} causes a conformational change in PrP^{C}, resulting in the accumulation of prions resistant to lysozomal degradation. Adapted from Jones, I. "Prions show their metal." *Chembytes E-Zine.* The Royal Society of Chemistry, 2008.

19.3 Oral Transmission: How Do "Eaten" Prions Travel to the Brain to Cause Disease? Why Isn't Variant CJD More Common?

The majority of experiments designed to study transmission of bovine spongiform encephalopathy (BSE) and other TSEs are with animals of the same species or across different species using the intracranial route of transmission. It is known that oral transmission of TSE diseases is very inefficient compared with intracerebral inoculations. Neuroinvasion of the prions occurs via the splanchnic and vagus nerves (the sympathetic peripheral nervous system) present in the abdominal region of the body that extends to the brain. New research suggests that prions may also invade the brain via the **hypoglossal nerve of the tongue**. Food products that contain tongue may be a potential source of prion infection for humans.

The **infectious dose** through ingestion of prion-contaminated food is unknown. A 2001 study by Gunn estimated the human oral infectious dose (ID_{50}) to be 10^{13} BSE prion molecules. This is a very large infectious dose compared to known bacterial and viral pathogens. It was mistakenly believed that the minimum infectious dose for cattle was the ingestion of 1 gram of infected brain. This estimate caused concern about the potential risks of recycling BSE to cattle through the application of sewage sludge to agricultural land. A 4-mm mesh screen was placed over drains of slaughterhouses in the United Kingdom to prevent particles bigger than 0.1 gram from entering the sewage system. A 2005 study by Lasmezas, though, showed that cattle are infected by oral doses as small as 0.01 to 0.001 gram of brain.

Some studies suggest that very low doses through diet may be enough to cause a subclinical infection in humans. These individuals would be carriers without symptoms of the disease. The scale of the variant CJD epidemic was calculated to be a few thousand cases as an upper limit, but carriers might pose risk for contamination of blood products, surgical instruments, and tissue transplants. The variant CJD epidemic in the United Kingdom remains a public health concern and a subject of speculation.

19.4 Other Routes of Transmission: Iatrogenic Transmission, Including Prions in Blood

An **iatrogenic** disease is one that may be inadvertently caused by a physician or surgeon or by a contaminated medical or surgical instrument or diagnostic procedure. Iatrogenic transmission of CJD, which is rare, has been reported since the late 1970s. The following incidents have been associated with iatrogenic transmission of CJD:

- Recipients of a corneal graft acquired CJD from donors who developed CJD.
- Patients acquired CJD from contaminated depth electroencephalography (EEG) electrodes implanted into their brains.
- Patients acquired CJD from contaminated neurosurgical instruments.
- Patients acquired CJD after acquiring human growth hormone from donors who developed CJD.
- Patients acquired CJD after receiving dura mater grafts from donors who developed CJD.

The bloodborne transmission of variant CJD has long been suspected as possible for two reasons:

1. The variant CJD agent can be detected in lymphoid tissues, raising the possibility that it could also be found in circulating lymphocytes present in blood.
2. The prions may exist in the blood as it travels from the original site of infection to the brain.

Experimental studies have demonstrated the transmission of BSE to sheep by blood transfusion from asymptomatic infected sheep to healthy sheep, which means that the BSE agent is infectious in blood during the incubation period. Two highly probable bloodborne, person-to-person cases of transmission of variant CJD were reported in the United Kingdom. All cases of probable or definite variant CJD are reported to United Kingdom blood services by the national CJD surveillance unit. At least 48 individuals who received blood components from a total of 15 donors who later became variant CJD cases are being monitored as a precautionary step for their "at risk" status. Recipients of United Kingdom plasma products in the 1980s and early 1990s might also be at increased risk of variant CJD infection. For this reason, hemophiliacs in the United Kingdom have been notified that they are at an increased risk of variant CJD. **TABLE 19-3** lists the precautionary steps introduced in the United Kingdom to reduce the risk of variant CJD transmission by blood and blood products. Surveillance systems record the actual chain of exposure to blood.

TABLE 19-3 United Kingdom's Measures to Reduce the Spread of Variant CJD Associated with Blood or Blood Products

Year	Measure/Precautionary Step
1997	Blood components, plasma products, or tissues from an individual who contracted variant CJD were withdrawn or recalled
1998	Plasma imported from the United States for fractionation (extracting and purifying proteins in plasma for patients)
1998–1999	Leukodepletion of all blood used for transfusion
2002	Importation of fresh plasma from the United States for patients born after January 1, 1996
2004	Blood donation is not accepted from people who have received a blood transfusion in the United Kingdom since 1980
2005	Donors of blood to patients who have developed variant CJD following transfusion have been advised that they are "at risk" for variant CJD.

19.5 Clinical Signs and Symptoms of Variant Creutzfeldt-Jakob Disease (Variant CJD)

Over 50% of variant CJD patients die before the age of 30 (the average age of death is 28). Patients suffering from sporadic or classic CJD die at an average of 68 years of age. The symptoms of variant CJD begin with anxiety, memory loss, mood changes, depression, and withdrawal, followed by obvious neurologic signs that include muscle twitching or spasms (jerky movements) and posture and gait abnormalities (motor difficulties). Final symptoms are loss of speech, stupor, and a persistent vegetative state (coma) before death 14 months after symptoms appear.

19.6 Diagnosis of Variant CJD

Most patients with variant CJD are often referred to a psychiatrist because of the behavioral changes associated with the disease. A definite diagnosis of variant CJD can be made via prion positive immunostaining of biopsy material from tonsil, spleen, or lymph nodes to detect prion accumulation. Other methods to aid diagnosis in conjunction with clinical symptoms are EEG (looking for slow or negative brain wave activity), magnetic resonance imaging (MRI, looking for brain lesions), and testing cerebral spinal fluid (CSF) for elevated levels of neuronal, astrocytic, and glial proteins. Elevated levels of CSF proteins reflect a change in blood–brain barrier function. The CSF proteins are released into the CSF as a consequence of extensive brain tissue damage.

There is no detectable immune response to infection, and as a result no routine test is available to diagnose asymptomatic individuals or patients experiencing symptoms. This rare but significant disease is difficult to diagnose with accuracy before death.

The gold standard of diagnosis is the **postmortem examination of brain tissues** via the demonstration of abnormal protease resistant prion proteins by Western blot analysis and immunocytochemistry of PrP^{res} accumulation in brain tissues (**FIGURE 19-5**). Genetic studies may also be performed to examine whether the patient is homozygous for methionine at codon 129 of the prion protein.

19.7 Pathogenesis of TSEs

The incubation period of TSEs is long (20 to 56 years). Individuals who lived in the South Fore in Papua New Guinea who were exposed to mortuary feasts are still dying of Kuru today, some 39 to 56 years after the cessation of cannibalism.

The brain, spinal cord, and retina are heavily infected before symptoms or clinical signs of infection appear. The term **spongiform**, which

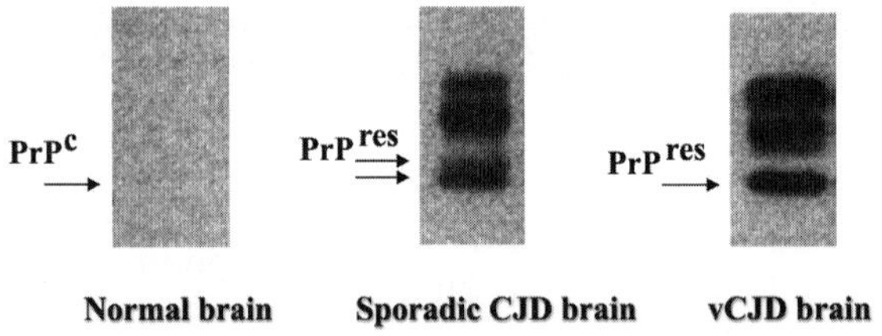

a

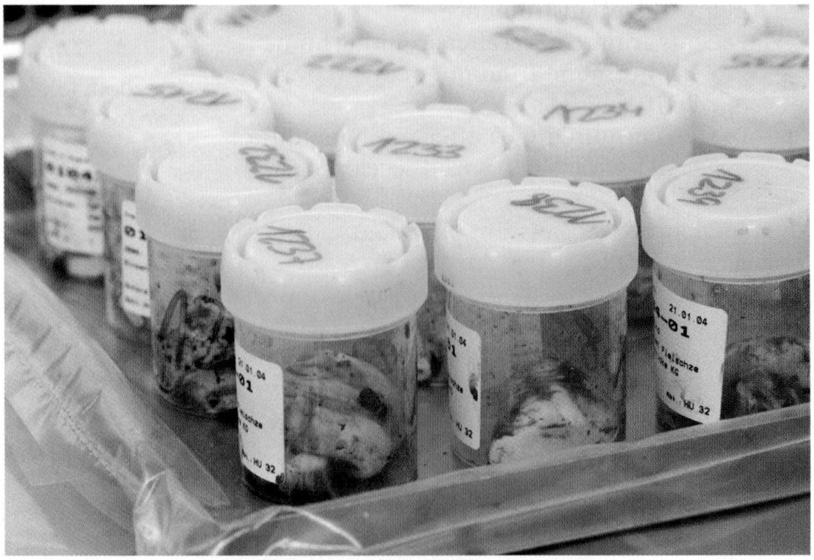

b

FIGURE 19-5 (a) Detection of PrP^{res} by Western blot analysis. Brain homogenates are treated with proteinase K. Proteinase K will digest PrP^C but not PrP^{res}. Digests are electrophoretically separated on a polyacrylamide gel and then blotted onto a membrane that is probed with tagged anti-PrP antibodies. The arrows indicate PrP^C or PrP^{res}. The sporadic CJD brain contains a nonglycosylated PrP^{res} protein (resulting in a double band). **(b)** Samples of human brain tissues from a diseased individual suspected of variant CJD.

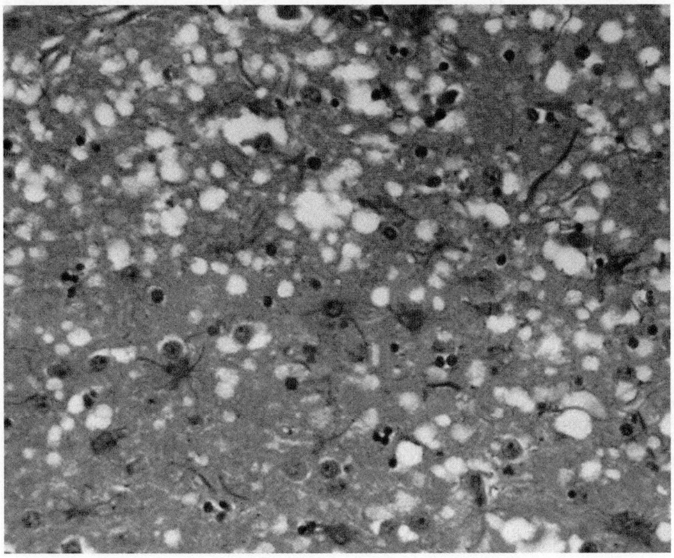

a

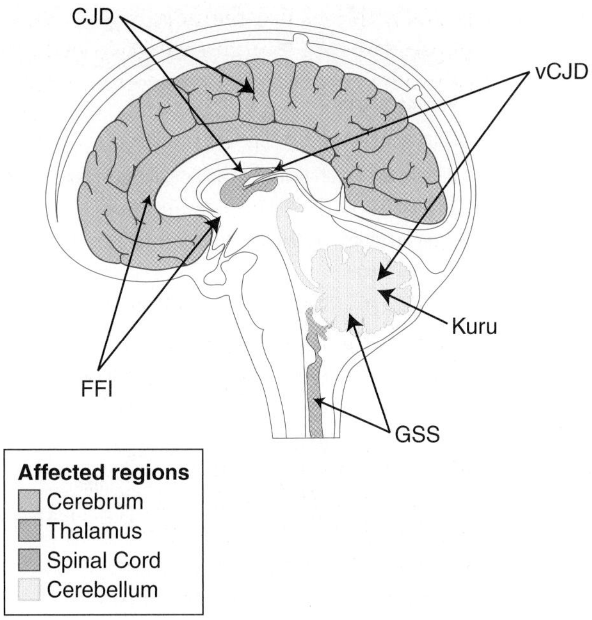

b

FIGURE 19-6 (a) Brain tissues showing histopathological changes found in bovine spongiform encephalopathy. Note the presence of vacuoles or spongiosis. **(b)** Diagram showing the major regions of the human brain affected by the different TSEs. Adapted from University of Illinois at Urbana-Champaign, *BSE Slide Show*, http://w3.aces.uiuc.edu/AnSci/BSE.

is sometimes referred to as **spongiosis**, comes from the microscopic observation of infected brains. Upon autopsy, slices of the brain contain vacuoles (clear zones) similar to a sponge (**FIGURE 19-6a**). Other classic histological changes of the brain are *neuronal loss*, **astrocytosis** (the spread of astrocytes to damaged tissues in the brain), and the formation of PrP threadlike structures that aggregate into **amyloid plaques**. The amyloid plaques accumulate in different regions of the brain (Figure 19-6b).

Depending upon what region of the brain is affected, different symptoms that are typical for the particular type of disease are evident; for example, memory is affected when the cerebral cortex is infected. The precise mechanisms that lead to brain damage are not known; however, evidence suggests that apoptosis causes neuronal death. No inflammation or immune defenses against prions exist, presumably because they are natural proteins (i.e., the body does not recognize prions as being foreign). The prion proteins become harmful only when they are converted to PrP^{res}.

19.8 Genetic Research and the Function of PrP^C

If the protein only hypothesis holds true, then animals that lack the *PRNP* gene would be resistant to scrapie infection. Charles Weissman's team of researchers created knock-out mice that do not express PrP^C. When they were injected with scrapie, the mice failed to develop symptoms of scrapie, an indication that the cellular form of PrP is required for propagation of the infectious from of PrP^{res}.

The idea that prions act as copper-binding proteins has been enticing to scientists. Some neurological disorders such as Menkes syndrome and Wilson disease are caused by altered copper metabolism. Could a change in copper binding be associated with the change to the pathogenic prion that is directly linked to TSEs? Weissman's group answered this question by introducing a *PRNP* gene that lacked the **octarepeat copper binding motif** into the prion knock-out mice (the gene that encodes the PrP protein has been disrupted in knock-out mice), and the mice became susceptible to scrapie infection. This suggests that copper-binding octarepeat motif is not necessary for the pathogenic effects of prions. Does this mean copper binding is not important in pathogenesis?

Researchers have found a different site on PrP^C that is outside of the octarepeat region. It lies within the hydrophobic region of PrP^C and binds copper strongly. This region is essential for disease propagation (see Figure 19-3). Rather than PrP^C copper binding being involved in transport, it may be involved in chelating free copper. Free copper radicals are toxic to cells and are usually quickly scavenged. Experiments in which radioactive copper was added to cells expressing *PRNP in vitro* showed that PrP^C bound the radioactive copper

TABLE 19-4	Polymorphisms of the Human *PRNP* Gene
Polymorphism	**Disease Phenotype**
Point mutation: 129 methionine or valine	Increases susceptibility to vCJD but does not cause disease directly
Double point mutations: 171 asparagine or serine coupled and 129 methionine or valine	Increases susceptibility to vCJD but does not cause disease directly
Point mutation: D178N (aspartic acid → asparagine) and 129 methionine	Fatal familial insomnia
Point mutation: V180I (valine → isoleucine)	Creutzfeldt-Jakob disease
Point mutation: T183A (threonine → alanine)	Creutzfeldt-Jakob disease
Point mutation: E200K (glutamic acid → lysine)	Creutzfeldt-Jakob disease
Point mutation: R208H (arginine to histidine)	Creutzfeldt-Jakob disease
Point mutation: V210I (valine → isoleucine)	Creutzfeldt-Jakob disease
Point mutation: M232R (methionine → arginine)	Creutzfeldt-Jakob disease
Point mutation: P102L (proline → leucine) and 129 methionine	Gerstmann-Straussler-Scheinker disease
Point mutation: P105L (proline → leucine)	Gerstmann-Straussler-Scheinker disease
Point mutation: A117V (alanine → valine) and 129 methionine	Gerstmann-Straussler-Scheinker disease
48 base pair insertion	Creutzfeldt-Jakob disease
96 base pair insertion	Creutzfeldt-Jakob disease
120 base pair insertion	Creutzfeldt-Jakob disease
192 base pair insertion	Gerstmann-Straussler-Scheinker disease
216 base pair insertion	Gerstmann-Straussler-Scheinker disease

at the cell surface and internalized it. These cells were more resistant to copper toxicity and oxidative stress compared to cells that did not express *PRNP*. Hence, PrP^C may play a role in cleaning up free radicals similar to other **cellular scavenging mechanisms**. The loss of PrP^C function may change in copper balance in the brain and other organs and cause oxidative damage throughout the brain. Further research is needed to determine if oxidative stress and copper activity are involved the pathogenesis of TSEs.

Human Genetics: Codon 129

At least 50 known mutations in the *PRNP* gene that result in spontaneous PRP^{res} formation have been identified in people who suffer from inherited TSEs such as classic CJD, fatal familial insomnia, or Gerstmann-Straussler-Scheinker syndrome. A few of these mutations or common variations, referred to as **polymorphisms** in the *PRNP* gene, are recognized as increasing the susceptibility to TSEs (**TABLE 19-4**).

VIRUS FILE 19-1 Point-Counterpoint: Is *Spiroplasma* Involved in TSEs? The Scientific Debate Continues

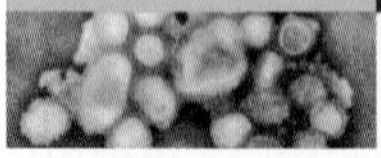

The beauty of science is that knowledge in the field is cumulative. The scientific method allows researchers to explain natural phenomena through the testing of hypothesis-driven experiments. As technology improves, the same questions asked 20 or 30 years ago can be revisited with new methods of experimentation in order to gain further knowledge.

In 1982 Prusiner's team showed that TSEs are caused by an infectious proteinaceous particle. These findings were unexpected: the idea that a protein can act as an infectious pathogen and cause the rapid deterioration of the central nervous system was only accepted after a long and grueling scientific debate.

A conventional pathogen, *Spiroplasma* sp., was postulated in the late 1970s to be associated with TSEs. Spiroplasma is a wall-less, fastidious bacterium that can pass through a

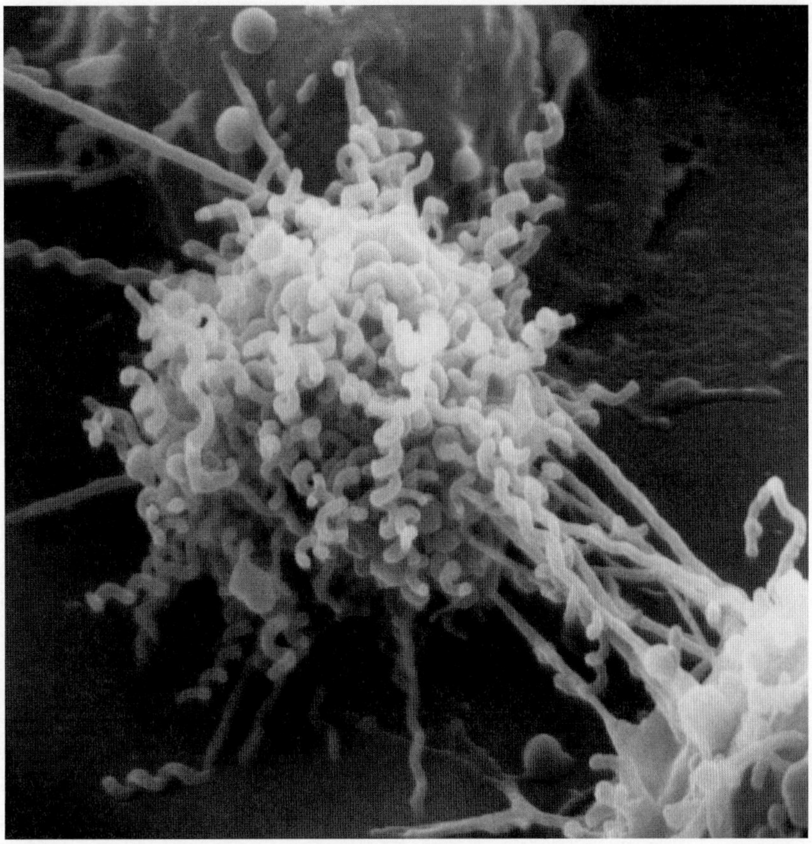

FIGURE VF 19-1 Scanning electron micrograph of a *Spiroplasma* (15,000×). They look similar to TSE fibrils.

0.05-μM filter and has an affinity for mammalian membranes. Spiroplasmas can evade the host immune system and "enter" cells, thus making their detection difficult. Pathologist Frank O. Bastian examined brain biopsies from CJD patients using electron microscopy in 1976. He observed spiral structures foreign to the tissue that had features of a newly reported *Spiroplasma*.

Subsequently researchers found a fibril protein in scrapie-infected brain tissues that was named scrapie-associated fibril (SAF). SAF proteins were identical in shape to the internal fibrils of spiroplasmas (**FIGURE VF 19-1**). Antibodies to SAF reacted with the fibrillar proteins from *Spiroplasma* in Bastian's experiments. Hence, the spiroplasmas contained fibrillar proteins that were morphologically and immunologically similar to the CJD and scrapie fibrillar proteins observed in diseased brain tissues. Bastian's team inoculated *Spiroplasma* proteins into rodents and was able to show a similar neuropathology to TSEs.

In 1983 a group of researchers were unable to confirm Bastian's findings. Dr. R. H. Leach's team was unable to culture spiroplasmas from the brain tissue of CJD patients, and the team also determined that antibodies that recognize *Spiroplasma* sp. were not detected in the sera of CJD patients. These negative results did not support Bastian's claims of CJD brains containing *Spiroplasma* structures.

In 2003 a group of Japanese researchers showed that the normal cellular form of PrP could serve as a receptor for the bacterium *Brucella*. The team implied that the abnormal, resistant

(continued)

VIRUS FILE 19-1 **Point-Counterpoint: Is *Spiroplasma* Involved in TSEs? The Scientific Debate Continues (continued)**

form of PrP was a byproduct of infection rather than the cause. A prior study by Tremblay in 1998 showed that rodents infected with the scrapie prions that were treated with tetracycline did not develop scrapie, which supports the idea that the prion may serve as a receptor for a bacterial infection. Presumably in these experiments, the antibiotic tetracycline inhibited the bacterial infection, thus preventing scrapie.

In 2001 Bastian's team was able to demonstrate via polymerase chain reaction (PCR) the presence of *Spiroplasma* 16S rDNA in TSE-infected brain tissues (13 of 13 CJD cases and 5 of 9 scrapie cases), thus suggesting that *Spiroplasma* is not necessarily the causative agent of TSEs but is a factor involved in the pathogenesis of TSEs. A team of researchers countered with a 2005 blind study in which the DNA samples prepared from the brains of scrapie-infected or -uninfected hamster brains were screened via PCR for *Spiroplasma* 16S rDNA sequences. No samples were positive for the bacterial *Spiroplasma* DNA. Alexeeva pointed out that PCR is an extremely sensitive method. Autopsy and field material are vulnerable to contamination with environmental DNAs that may interfere with the validation of the results obtained by researchers.

Is *Spiroplasma* involved in the pathogenesis of TSEs? Could bacteria cause TSEs by either triggering autoimmunity by molecular mimicry with bacterial antigens or as a source of chronic infection? How would you prove or disprove that *Spiroplasma* is responsible for the cytopathic effects such as spongiosis observed in the TSE-infected brains? This scientific investigation continues.

References

Alexeeva, I., et al. 2006. "Absence of spiroplasma or other bacterial 16S rRNA genes in the brain tissue of hamsters with scrapie." *J Clin Microbiol* 44(1):91–97.

Bastian, F. O. 1979. "Spiroplasma-like inclusions in Creutzfeldt-Jakob disease." *Arch Pathol Lab Med* 103(13):665–669.

Bastian, F. O., Dash, S., and Garry, R. F. 2004. "Linking wasting disease to scrapie by comparison of spiroplasma mirum ribosomal DNA sequences." *Exp Mol Pathol* 77(1):49–56.

Bastian, F. O., and Fermin, C. D. 2005. "Slow virus disease: Deciphering conflicting data on the transmissible spongiform encephalopathies (TSE) also called prion diseases." *Micro Res Tech* 66:239–246.

Bastian, F. O., and Foster, J. W. 2001. "Spiroplasma sp. 16S rDNA in Creutzfeldt-Jakob disease and scrapie as shown by PCR and DNA sequence analysis." *J Neuropath Exp Neuro* 60(6):613–620.

Bastian, F. O., Jennings, R. A., and Gardner, W. A. 1987a. "Antiserum to scrapie-associated fibril protein cross-reacts with spiroplasma mirum fibril protein." *J Clin Microbiol* 25:2430–2431.

Bastian, F. O., Jennings, R. A., and Hoff, C. J. 1987b. "Neurotropic response of spiroplasma mirum following peripheral inoculation in the rat." *Ann Inst Pasteur Microbiol* 138:651–655.

Leah, R. H., Mathews, W. B., and Will, R. 1983. "Creutzfeldt-Jakob disease. Failure to detect spiroplasmas by cultivation and serological tests." *J Neurol Sci* 59(3):349–353.

Tremblay, P., et al. 1998. "Doxycycline control of prion protein transgene expression modulates prion disease in mice." *Proc Natl Acad Sci USA* 95:12580–12585.

Watarai, M., et al. 2003. "Cellular prion protein promotes brucella infection into macrophages." *J Exp Med* 198(1):5–17.

A common polymorphism at codon 129 of *PRNP* appears to act as a genetic susceptibility factor. Either methionine or valine is encoded at this position. All people suffering from variant CJD acquired through the consumption of prion-contaminated beef products have been homozygous for methionine. Methionine homozygosity is generally associated with a rapidly progressive dementing disease, whereas a more prolonged course with an onset of ataxia (loss of motor coordination) is most often associated with the valine/valine genotype.

19.9 Steps Toward Treatment and Vaccination

The following compounds were tested in cell lines and animal models: amphotericin B derivatives, Congo red, the spice curcumin, anthracyclines, polyanions, porphyrins, polyamines, β-sheet breakers, and even small interfering RNAs. There have been promising prion therapy approaches *in vitro*, but none of these treatments has been successful in human clinical trials due to toxicity or moderate untoward effects.

Anti-PrP antibodies can inhibit prion replication in cultured cells. Passive immunity experiments in which infected mice were injected with anti-PrP antibodies produced a delay in the onset of symptoms. Any active vaccination experiments, however, have met without success because the mammalian system is tolerant to PrP^C. Studies in which antibodies against PrP^C were injected into the brains of mice caused neurotoxicity. Vaccination efforts therefore should proceed through exhaustive safety trials in animals prior to application in humans.

19.10 Species Barrier: Bovine Spongiform Encephalopathy (BSE) and Variant Creutzfeldt-Jakob Disease

Transmissibility of TSEs is easy among the same species. A high degree of homology in the amino acid sequence of the prion protein between two species increases the efficiency of transmission. It is also possible, though, for transmission to occur between different species.

The origin of BSE is unclear. It has been hypothesized that it likely came from scrapie-infected sheep. The disease was spread to cattle by the ingestion of scrapie-contaminated bone meal derived from sheep **offal** fed to young calves. (Offal consists of brain, spleen, thymus, tonsil, and guts—parts of the animal normally discarded.) When bone meal of rendered bovine protein from infected cattle was fed to cattle in order to increase their milk yield, a new disease was further amplified among cattle.

Scrapie was considered a rare disease of sheep that did not infect humans. It has been recognized as a disease of sheep and goats in Great Britain and other countries in Western Europe for more than 250 years. The first case of scrapie in the United States was diagnosed in 1947 in a Michigan flock of sheep. The sheep owner had imported sheep of British origin through Canada for several years. Today, only Australia and New Zealand are free of scrapie.

BSE was first observed in 1985. A dairy farmer in England noticed several cows with abnormal behavior: unsteady gait, kicking unexpectedly during milking, and aggressiveness—hence the term "raging" or "mad cow" disease. Other signs of the disease are:

- difficulty in rising from a lying position
- itching
- heightened sensory perception
- anorexia
- excessive licking
- decreased milk production

The symptoms last for two to six months before the animal dies. In 1986, an official diagnosis of BSE was made after brain tissues of slaughtered sick cows were similar to those of sheep suffering from scrapie. However, research suggests the first probable infections of BSE in cattle occurred during the 1970s, with two cases identified in 1986. The BSE epizootic in the United Kingdom peaked in January, 1993 at almost 1000 new cases per week. Over the next 17 years, the annual numbers of BSE dropped dramatically: 14,562 cases in 1995, 1443 in 2000, 225 in 2005, and 11 cases in 2010. Cumulatively, through the end of 2010, more than 184,000 cattle in the United Kingdom were confirmed cases of mad cow disease and 1 to 3 million in more than 35,000 herds were likely to have been infected with the BSE agent, most of which were slaughtered for human consumption before developing signs of the disease. Cattle and feed exported from the United Kingdom seeded smaller epidemics in other European countries. In 1989, the United Kingdom implemented control measures that included the culling of all sick animals and the banning of all bovine materials from entering food that was fed to cattle.

In humans, the onset of variant CJD cases was first described in the United Kingdom from 1994 to 1996. The British Minister of Health declared that ten people had died from a new form of CJD, a variant form. Transmission of variant CJD was (and is) believed to be caused by the ingestion of cattle products contaminated with the BSE agent. The interval between the most likely period for the initial exposure of the population to potentially BSE-contaminated food (1984–1986) and the onset of initial variant CJD cases (1994–1996) is consistent with known incubation periods for the human forms of prion disease. As of April 4th, 2011, 171 deaths of variant CJD patients were reported in the United Kingdom since the start of the epidemic. The epidemic in cattle is largely under control and any remaining risk to humans through the consumption of beef should be very small. Cases of BSE are diminishing in the United Kingdom, and only two cases of BSE originating in the United States have occurred.

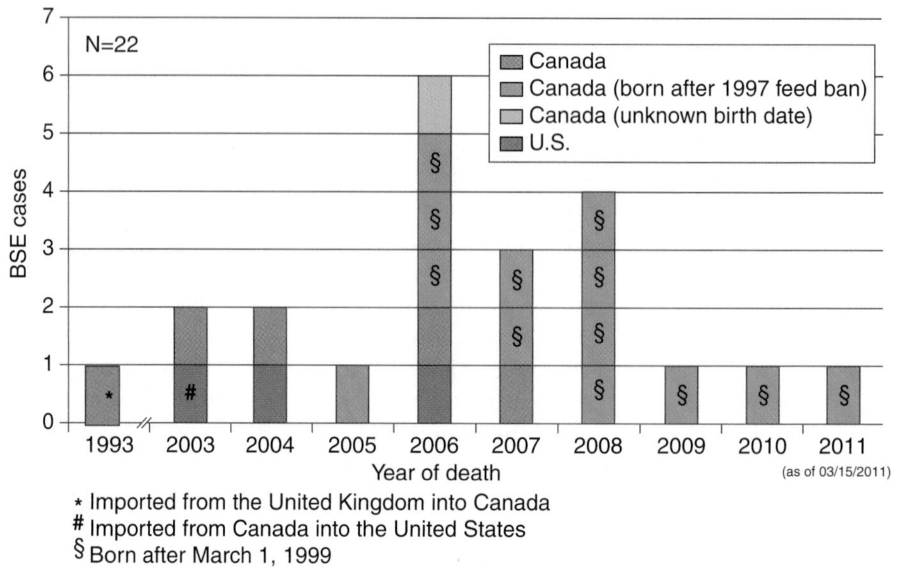

FIGURE 19-7 BSE cases in North America, 1993 to March 15, 2011. Adapted from Division of Viral and Rickettsial Diseases, Centers for Disease Control and Prevention, *BSE,* http://www.cdc.gov/ncidod/dvrd/bse.

Through March 15, 2011, BSE surveillance has identified 22 cases in North America: 3 cases in the U.S. and 19 in Canada. The cases in the United States were cows imported from the United Kingdom into Canada or Canada into the United States (**FIGURE 19-7**). Since March 2006, each of the 15 cattle reported with BSE in North America were born in Canada and identified through the Canadian BSE surveillance system. The Canadian Food Inspection Agency (CFIA) implemented an enhanced feed ban on July 12, 2007. On October 26, 2009, the U.S. Food and Drug Administration (FDA) followed by also issuing the enhanced feed ban. The **enhanced BSE-related feed ban** is a regulation or policy aimed to more effectively prevent and quickly eliminate BSE from Canada and the United States. Prior to the enhanced feed bans, the standard feed bans prohibited the feeding of most protein materials from mammals to ruminant animals. The enhanced feed ban prohibits most proteins, including the potentially BSE infectious tissues *not just from cattle feed but from all animal feeds, pet foods, and fertilizers.* This practice is intended to eliminate any possibility of cross-contamination that was likely responsible for BSE in cattle born after the introduction of the feed ban.

Strains of Bovine Spongiform Encephalopathy (BSE)

There is increasing evidence that there is more than one strain of BSE: the classic or **typical BSE strain** (C-type) and two atypical strains (H- and L-types). The typical BSE strain is responsible for the outbreak in the United Kingdom and most of the BSE cases in Canada. It is preventable through the elimination of BSE-contaminated feed and has been causally linked to variant CJD in humans. The typical strain has not yet been identified in the United States.

The atypical BSE strains were described in reports in 2004. Researchers assessed molecular and neuropathological (phenotypical) characteristics of seven cattle (3 from France and 4 from Italy). Western blot analysis of the PrP^{res} from brainstem tissues of sick cattle showed higher and lower molecular masses of unglycosylated PrP^{res} in these types. The types were named H-type and L-type. The L-type has now been identified in a number of different countries. In July 2007, the United Kingdom Spongiform Encephalopathy Advisory Committee announced that the atypical BSE strains may have risen spontaneously (although transmission through feed or the environment cannot be ruled out). Of the North American cases, three were linked to the H-type atypical strain of BSE and one was linked/identified as the L-type.

19.11 Chronic Wasting Disease (CWD)

History

The deer population is plagued by a new disease. Symptoms of the disease were first observed in mule deer grazing on northern Colorado wildlife research land in 1967. Symptoms were chronic weight loss (ribs showing), blank facial expression, excessive drooling and thirst, frequent urination, teeth grinding, holding head in a lowered position, nervousness, sluggish behavior, and isolation from the herd. The sick animals had poor hair coats and appeared emaciated, or starving and "wasting away" (**FIGURE 19-8**). Most animals

FIGURE 19-8 Deer suffering from CWD.

died within several months of illness onset, sometimes from **aspiration pneumonia**. Aspiration pneumonia is caused by breathing in foreign material (e.g., foods, vomit, liquids) from the mouth into the lungs. It can lead to a lung abscess (collection of pus in the lungs), an inflammatory reaction or a lung infection (pneumonia).

In 1978, researchers named it chronic wasting disease (CWD) and classified it as a TSE. In the mid-1980s, CWD was observed in free-range elk and deer in northern Colorado and adjacent parts of southern Wyoming. In 1997, it was observed in a farmed herd of elk in South Dakota. Since 1999, the Wisconsin Department of Natural Resources has been routinely testing the state's deer herd for CWD. In 2002, Wisconsin's first cases of CWD were reported. A map of CWD in North America is shown in **FIGURE 19-9**. Positive deer herds and farmed elk have been identified in 16 states at the time of this writing: Colorado, Illinois, Kansas, Minnesota, Missouri, Montana, Nebraska, New Mexico, New York, North Dakota, Oklahoma, South Dakota, Utah, West Virginia, Wisconsin, and Wyoming. CWD has been detected in wild

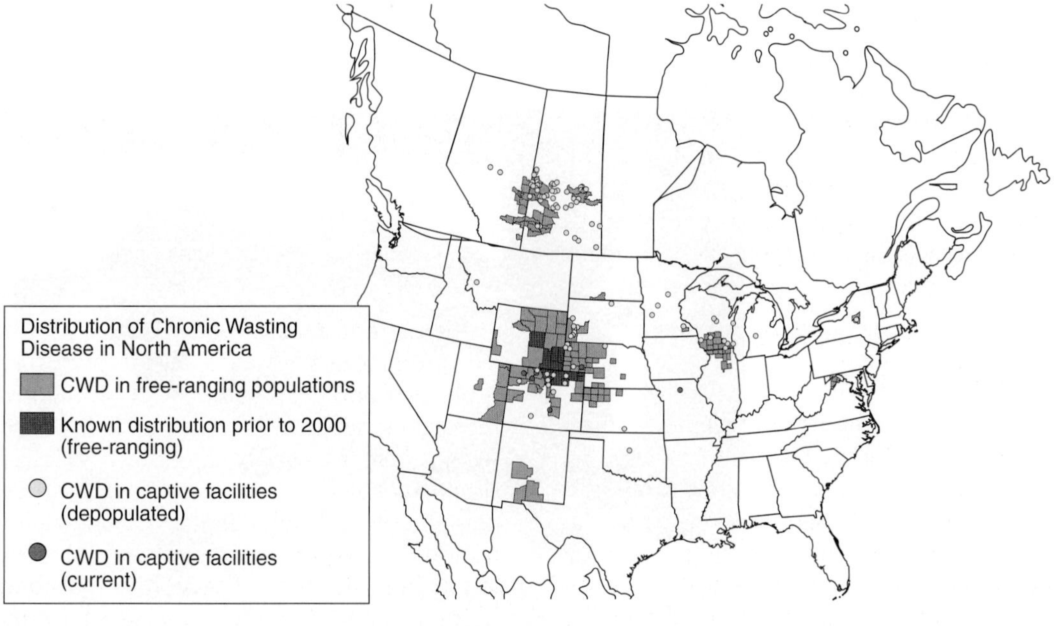

FIGURE 19-9 Map showing endemic locations of CWD in North America.

deer and farmed elk in the Canadian providences of Alberta and Saskatchewan. Outbreaks of CWD have occurred in South Korea as a result of the importation of subclinically infected animals from Canada. During 2005 testing in Colorado, 1 out of 288 moose tested positive for CWD. It was not surprising that CWD is rarely found in moose given their social habits. Moose are typically solitary and stay with other moose only in cow-calf pairs.

Animal-to-Animal Transmission

While human prion diseases are very rare, CWD incidence can be over 15% in free-ranging deer and over 90% in captive deer herds. The incubation period of CWD is one to five years. Typically, symptoms are not seen in deer younger than 16 months old. Transmission of CWD prions (PrP^{CWD}) between animals is much easier than that of BSE between cattle. Evidence suggests that infected deer or elk may transmit the CWD prions directly (animal-to-animal contact) or indirectly (through contaminated feed or water sources with saliva, urine, blood, feces, or decomposing tissues). Prions are shed through the saliva, feces, blood, and urine of the infected animal. CWD more likely can occur where deer or elk are crowded and congregate at artificial feeding, watering, and birthing sites. CWD is of particular concern because the PrP^{CWD} are horizontally transmissible and remain infectious in the soil. A controlled lab study showed that the PrP^{CWD} remained infectious for at least two years in a pasture. It is likely that the environment serves as a stable reservoir of infectious CWD prions. It is unclear if environmental transmission is sustained by mites, flies, predators, or scavengers (e.g., vultures, cougars). The stability of prions makes them difficult, if not impossible to eliminate from the environment.

Diagnosis and Management

Hunting season typically occurs in the fall. The deer hunt plays an important role in wildlife management. Deer cause agricultural damage, deer–vehicle collisions, forest damage by overgrazing, damage to ornamental plants and landscaping, airport runway safety issues, and decreased food supply and cover for other species. There are now CWD surveillance programs that identify CWD management zones. The gold standard for diagnosis of CWD is based on immunocytochemistry on tissues from the obex portion of the animals' brain in a laboratory. There is no certified testing for live animals. Managing CWD has become a major concern. The disease presented significant economic problems for game farmers after it was unintentionally introduced into farmed elk populations taken from CWD endemic areas.

Hunters' Precautions

State departments of Fish and Game are working hard to educate hunters, meat processors, taxidermists, and farmers. When field-dressing carcasses, hunters in CWD endemic areas are asked to wear latex/rubber gloves and carefully bone out the meat of the entire carcass. The catchphrase for returning home with deer and elk is "**no skull, no backbone**." This is based on the idea that the prions are concentrated in the nervous tissue spinal cord and antlers of the animal. Removing them would reduce exposure to prions. Hunters are cautioned not to consume brain, spinal cord, eyes, spleen, tonsils, or lymph nodes of deer (**FIGURE 19-10**). Proper field

FIGURE 19-10 Hunters field dressing a deer.

dressing will remove most, if not all, of these body parts. A 2009 study by Angers et al. showed evidence that **antler velvet** from CWD-infected elk contains prions. A highly innervated and vascularized skin layer, called velvet, covers the growing antlers of male cervids. The velvet is shed after the animal's testosterone increases and the antlers become ossified. Antler velvet has been used as a nutritional supplement for a variety of health remedies and health maintenance. Before this study, antler velvet was an unrecognized source of CWD prions in the environment.

Potential Transmission to Humans

CWD continues to spread among cervid populations in North America, creating concern that CWD may cross the species barrier to infect humans or domestic animals that may be eaten by humans. The most sensitive method to determine the susceptibility of TSE agents is by performing intracerebral injections of the PrP^{res} of interest into a healthy host. Unfortunately, this type of experiment does not mimic most natural situations. Testing the susceptibility of infection of TSE agents through oral infection is used in species barrier experiments to represent an oral route of transmission in laboratory investigations. In animal models excluding monkeys, oral transmission is generally 1000-fold less effective than direct intracerebral injection challenge, resulting in longer incubation periods and lower efficiency of disease transmission.

In the laboratory, susceptibility of PrP^{CWD} to nonagricultural and laboratory animals has been variable. For example, in experiments in which ferrets were intracerebrally injected with PrP^{CWD}, 100% were susceptible; however, the ferrets were not susceptible to oral infection. Mink were 25% susceptible to TSE^{CWD} infection but not susceptible to oral infection.

CWD was successfully transmitted and adapted to laboratory hamsters, transgenic mice expressing hamster PrP, and transgenic mice overexpressing mouse PrP. However, transgenic mice overexpressing human PrP were not susceptible to CWD by intracerebral injection of PrP^{CWD}. These experiments suggested a human species barrier against CWD infection.

Between January 1, 1997, and May 31, 2000, three young patients (a woman age 28, and two men ages 28 and 30) who regularly consumed venison were diagnosed with CJD. The cases were reported to the CDC out of concern about a possible zoonotic transmission of CWD. The cases were investigated but no strong evidence supported a causal link between CWD and their illness. None of the patients had eaten deer or elk meat harvested in CWD endemic areas of Colorado and Wyoming.

Research started by Richard F. Marsh in 1980 and published posthumously in 2005 showed that two squirrel monkeys injected intracerebrally with brain homogenate from a mule deer suffering in the late stages of CWD, developed a progressive TSE at 31 and 34 months post-infection. The CWD-infected squirrel monkeys were euthanized during the terminal stages of disease. Their brain tissues were analyzed and found to contain PrP^{res}. Histological examination of the brainstem and brain showed spongiform changes. This experiment provide some evidence that at least one species of nonhuman primate was susceptible to CWD, re-opening the possibility that humans may not be protected from CWD by a species barrier.

Race et al. followed Marsh's work by carrying out further studies at the Rocky Mountain Laboratories (Hamilton, Montana). Their experiments involved cynomolgus macaques and squirrel monkeys (**FIGURE 19-11**). Humans are evolutionarily more closely related to cynomolgus macaques than to squirrel monkeys, making them a more accurate model for a human species barrier. In their study, 13 squirrel monkeys and 6 cynomolgus macacques were injected intracerebrally with CWD-positive brain homogenates. They also tested the more natural oral route of infection by giving 15 squirrel monkeys and 9 cynomolgus macaques oral doses of CWD-positive brain homogenates. The monkeys were anesthetized and given the oral doses through a rubber gastric tube. The brain homogenates were prepared from CWD positive brains of free-ranging and captive mule deer, captive and wild white-tailed deer, and captive farmed elk located in Wyoming, Colorado, South Dakota, or Montana. The brain homogenates were pooled from eight forms of inoculum to make the CWD-positive brain homogenates.

A total of 11 out of 13 squirrel monkeys came down with a severe wasting syndrome similar to that of deer or elk within 33 to 53 months post-intracerebral injection. The monkeys were euthanized and their brains examined. The brains had spongiosis and brain regions contained PrP^{res} as determined by immunocytochemistry. To date, three of the squirrel monkeys given oral doses of CWD brain homogenates came down with a severe wasting syndrome by 69 months post-inoculation by oral dosage. Their brains had spongiosis and PrP^{res}.

At the time of this writing (8 years post-inoculation), none of the cynomolgus macaques have shown evidence of severe wasting disease

a   b

FIGURE 19-11 Nonhuman primates used in experimental CWD studies. **(a)** Squirrel monkey. **(b)** Cynomolgus macaque.

through the intracerebral or oral route of transmission. Because cynomolgus macaques are evolutionarily closer to humans, this evidence suggests that humans may also be resistant to CWD. However, in experiments in which cynomolgus macaques were intracerebrally injected with BSE brain homogenates, the human variant-CJD disease occurred three years post-inoculation and human sporadic CJD required five years. Oral inoculation of cynomolgus macaques with BSE brain homogenates took a minimum of five years before clinical disease was observed. Therefore, CWD transmission to cynomolgus macaques should not be ruled out.

Barrie et al. published an *in vitro* study in 2011 that used a protein-misfolding cyclic amplification technique (PMCA). PMCA mimics protein replication *in vitro* at an accelerated speed. The misfolded form of PrP^{C} made *in vitro* using PMCA is infectious in animals. New strains can be produced, adapted, and stabilized upon crossing species barriers *in vitro* by PMCA. Barrie's study showed that cervid PrP^{CWD} can trigger the conversion of human PrP^{C} into PrP^{res}, suggesting that CWD might be infectious to humans.

Human exposure to CWD-infected deer or elk meat in the past decades has likely occurred. The highest level of PrP^{res} in cervids occurs in the central nervous system and lymphatic tissues of CWD-infected animals. The contamination of knives, saws, and muscles with these tissues can easily occur when field dressing game. *Despite the high probability of hunters exposed, epidemiological studies of humans living in the CWD endemic states of Colorado and Wyoming during 1979–2006 have not shown any increases in human TSE cases.*

The early signs of CWD in deer and elk were subtle behavior changes, staring, depression, and weight loss without a loss of appetite. The original explanation was that the wasting was caused by a metabolic disorder. It is possible that a wasting syndrome in humans may be diagnosed as a metabolic disorder as well and the patient would never be tested for a TSE. On the contrary, the epidemiological data, along with the *in vivo* cynomolgus macaque studies, are consistent with the conclusion that a species barrier protects humans from CWD infection.

19.12 Plant Viroids: The Smallest Living Fossils of a Former RNA World?

Viroids are small RNA molecules that infect plants in the same manner as conventional plant viruses (see Chapter 20), but they are smaller in genome size (ranging from 246 to 401 nucleotides in length) and are not encapsidated in a protein coat. Viroid RNAs are positive sense, single-stranded, covalently closed circular pathogenic molecules. Short helical regions are interrupted by internal and bulging loops. There are several regions in which base-pairing occurs across adjacent portions of the RNA molecule, a process sometimes referred to as **internal-basing**. In nature, viroids assume a rodlike conformation (~37 nm) that is very stable against ribonucleases. Viroids do not code for any proteins, and some viriod RNAs even lack the AUG initiation codon. Viroid RNAs are classified as noncoding RNAs. As a result, viroids depend upon the plant host enzymes for their replication and other functions.

There are two families of viroids: *Avsunviroidae* and *Pospiviroidae*. Members of the *Avsunviroidae* family have a branched or "quasi" rodlike secondary structure, whereas members of the *Pospiviroidae* family possess a true rodlike secondary structure (**FIGURE 19-12a–c**). Viroids contain five structural or functional domains: central (C), pathogenicity (P), variable (V), and two terminal domains (T1 and T2; Figure 19-12d). Viroids of the *Avsunviroidae* family lack a C region but possess hammerhead **ribozyme** structures with self-cleaving properties. The C domain consists of conserved nucleotides in the upper and lower strands of the molecule. A secondary structure—the bulges—in the conserved C domain are critical for viroid replication and processing. The P domain is associated with symptoms of viroid-infected plants. The V domains contain the highest variability in nucleotide sequences between closely related viroids. The T domains are interchangeable between viroids and may play a role in viral movement.

Viroids do not replicate in the cytoplasm of cells like conventional plant RNA viruses. They instead traffic within the cell through the nuclear pores using a host nuclear localization protein that binds to viroid RNA. The viroids traffic from cell-to-cell through the plasmodesmata of the plant, traveling long distances through the phloem. All of these processes are associated with host proteins (see Chapter 20).

Viroids replicate in either the nucleus or the chloroplast of the plant. In the nucleus, the cellular DNA-dependent RNA polymerase III synthesizes the viroid RNA. The secondary structure of the viroid might provide binding signals to host factors directly or indirectly involved in the life cycle of the infectious agent. Members of the *Avunsviroidae* family accumulate and replicate in chloroplasts, whereas members of the *Pospiviroidae* family have a nuclear localization. It is proposed the viroid replication occurs by the **rolling circle**

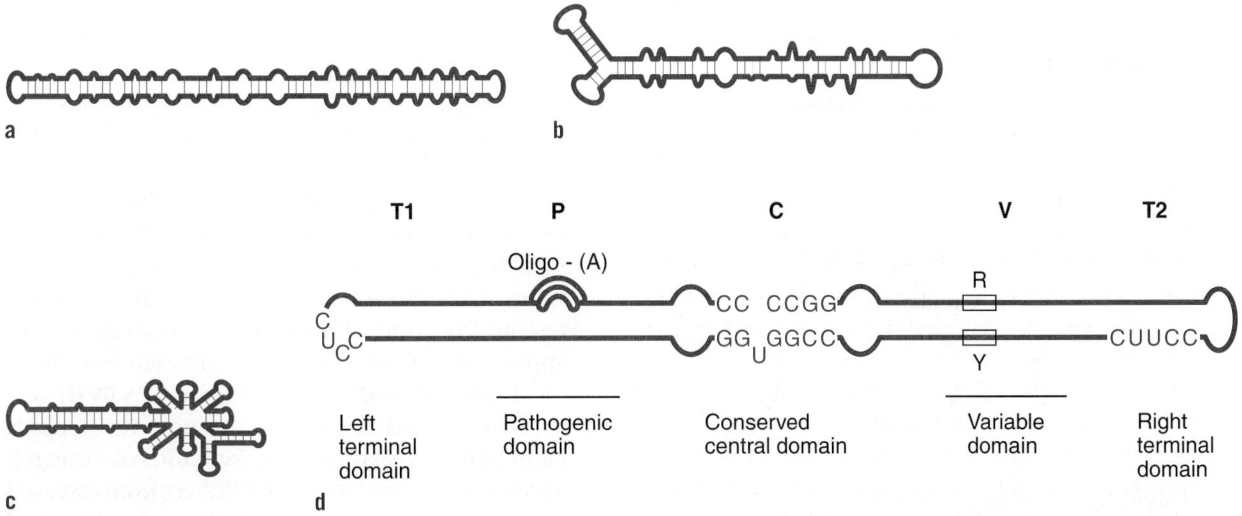

FIGURE 19-12 Models of viroid structures. **(a)** Rodlike secondary structure of a member of the Pospiviroidae family. **(b)** "Quasi" rodlike secondary structure of the Avsunviroidae family member that has two terminal hairpins in the left part of the RNA molecule. **(c)** A member of the Avsunviroidae family with a complex branched structures. **(a–c)** Adapted from Gora-Sochacka, A., *Acta Biochim. Pol.* 51 (2004): 587–607. **(d)** Typical regions of a viroid: Terminal ends (T1 and T2), pathogenicity (P), conserved central region (C), and a variable domain (V). Adapted from an illustration by W. Rohde and J. W. Randles, Max Planck Institute for Plant Breeding Research.

mechanism of replication. The host cell RNA polymerase synthesizes multiple copies of a linear complementary minus strand using the positive single-stranded RNA (ssRNA) viroid genome as a template. The new, longer plus-strand RNA is then synthesized by the host cell RNA polymerase. The multiunit +ssRNAs are cleaved via the **enzymatic self-cleavage** properties of the **hammerhead motifs** of the viroid RNAs. A host RNA ligase may cause circularization of the plus strands (**FIGURE 19-13**). Viroid strands of positive polarity localize to the nucleolus and nucleoplasm of host cells. Viroid strands of negative polarity are localized only to the nucleoplasm.

More than 40 viroid species and variants have been characterized (**TABLE 19-5**), and they cause at least 15 crop diseases. "Classical" viroids have been found only in plants. Human hepatitis delta virus RNA contains a viroid-like region with a secondary structure that has similar properties to viroid RNAs; it also appears to replicate by

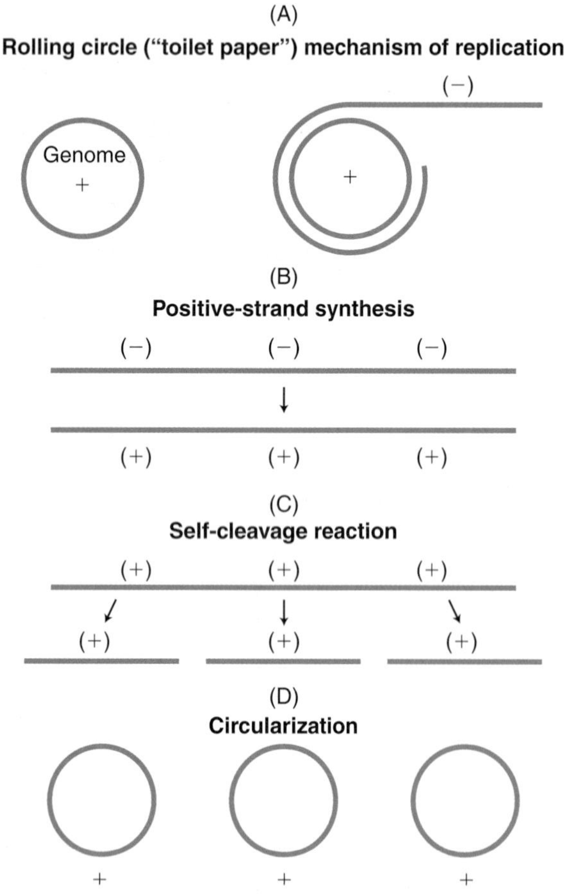

FIGURE 19-13 Rolling circle mechanism ("toilet paper" mechanism) of viroid RNA replication. The host cell RNA polymerase begins making multiple copies of the viroid genome much like toilet paper being unraveled from a dispenser. Adapted from Komoto, J., "The hammerhead ribozyme," *Introduction to the RNA World*, http://rnaworld.bio.ku.edu/class/RNA/RNA00/RNA_World_3.html

TABLE 19-5 Examples of Plant Viroid Diseases

Viroid	Abbreviation
Apple dimple fruit viroid	ADFVd
Apple scar skin viroid	ASSVd
Australian grapevine viroid	AGVd
Avocado sunblotch viroid	ASBVd
Chrysanthemum stunt viroid	CSVd
Citrus bent leaf viroid	CBLVd
Citrus viroid species III	CVd-III
Citrus viroid species IV	CVd-IV
Citrus exocortis viroid	CEVd
Coconut cadang-cadang viroid	CCCVd
Coconut tinangaja viroid	CTiVd
Coleus blumei viroids I, II, and III	CBVd I, II and III
Columnea latent viroid	CLVd
Grapevine yellow speckle viroids 1 and 2	GYSVd 1 and 2
Hop latent viroid	HLVd
Hop stunt viroid	HSVd
Iresine viroid	IRVd
Mexican papita viroid	MPVd
Peach latent mosaic viroid	PLMVd
Pear blister canker viroid	PBCVd
Potato spindle tuber viroid	PSTVd
Tomato apical stunt viroid	TASVd
Tomato planta macho viroid	TPMVd

some type of rolling circle mechanism. The RNA is larger (1.7 kb), though, is encapsidated, and encodes a virion-associated protein (hepatitis delta antigen; see Chapter 17). Viroids spread by mechanical transmission, vegetative propagation, and through pollen and seed (refer to Chapter 20). As a group, nothing distinguishes the disease symptoms produced by viroids from those caused by plant viruses. Infections from any type results in stunting, mottling, leaf distortion, and necrosis. Viroid diseases can cause a range of symptoms, from a slowly developing lethal disease to more mild and symptomless infections. Viroids cause a number of serious diseases in economically important plants including potato, tomato, coconut palm, grapevine, avocado, peach, apple, pear, citrus, coleus, and chrysanthemums.

Potato spindle tuber viroid (**PSTVd**) was the first viroid characterized. Theodor O. Diener, a plant pathologist at the U.S. Agricultural Research Service, discovered it in 1971. Tubers from diseased plants are long and narrow, with smooth skin and more eyes than are typically found in potatoes (**FIGURE 19-14**). The disease is spread easily by contact of healthy plants with diseased plants, by contaminated cultivating equipment, and through seed

FIGURE 19-14 Healthy tuber (top row) and diseased tubers caused by PSTVd (bottom row).

and pollen. All potatoes and tomatoes are susceptible to PSTVd.

Chrysanthemum stunt disease crippled and nearly destroyed the chrysanthemum industry prior to 1950. It was prevalent in U.S. and Canadian greenhouses with infections rates as high as 50% to 100%. **Chrysanthemum stunt viroid** (**CSVd**) is difficult to detect. It can be spread plant to plant by infected plant material that does not show symptoms, and it takes weeks for symptoms to appear after an infection. Symptoms of chrysanthemum stunt include shortened stems, uneven flowering (such as faded or irregular color), smaller than normal foliage, irregular flower sizes, and uneven maturation of parts within single flowers (**FIGURE 19-15**). Chrysanthemums are now grown in "clean" rooms using stringent precautions to prevent infection by CSVd.

19.13 Viroid Pathogenesis

The various effects of viroid infection are likely caused by more than one mechanism. Recent evidence suggests that pathogenic viroids activate a plant RNA-activated protein kinase, termed PKR. The plant PKR is analogous to the mammalian PKR that is activated by viral RNAs, which leads to the induction of the interferon pathway (see Chapter 7).

FIGURE 19-15 Chrysanthemums in the foreground of this photograph show symptoms of stunting and premature opening and color break caused by CSVd infection.

Summary

Prions and viroids are simple pathogenic molecules. Some scientists propose that they are the smallest living fossils. Could either one be related to the emergence of early life? Both types of molecules are very stable in the environment. Prions are proteinaceous proteins, whereas viroids are circular +ssRNAs that contain a lot of secondary structure. Neither one codes for proteins and both have self-replicating properties.

Prions cause transmissible spongiform encephalopathies (TSEs) in animal and human hosts. TSEs are rare, incurable neurodegenerative fatal diseases. They do not induce an immune response. Creutzfeldt-Jakob disease (CJD) and variant CJD (caused by eating prion-contaminated beef products) are the most common TSEs in humans. The incubation period of TSEs is long (20–56 years). The brain, spinal cord, and retina are heavily infected before symptoms or clinical signs of infection appear. The term spongiform (sometimes referred to as spongiosis) comes from the microscopic observation of infected brains. Upon autopsy, slices of the brain contain vacuoles (clear zones), similar to a sponge. Other classic histological changes of the brain are neuronal loss, astrocytosis, and the formation of PrP threadlike structures that aggregate into amyloid plaques.

Prions are highly resistant to routinely used methods of decontamination and steam sterilization. There are two distinct conformations of the prion protein, PrP. PrP^C is found throughout tissues of the body in healthy people and animals. The normal cellular form of PrP, designated PrP^C, is sensitive to denaturing agents. Its exact function is unknown. The "protein only" hypothesis proposes that the abnormal, misfolded infectious form of PrP is able to initiate a reaction that causes the PrP^C to convert to the highly pathogenic and resistant or stable form termed PrP^{res}. Over time the PrP^{res} accumulates into clumps that damage or destroy nerve cells in the brain. At least 50 known mutations in the *PRNP* gene that result in spontaneous PRP^{res} formation have been identified in people who suffer from inherited TSEs such as classic CJD, fatal familial insomnia, or Gerstmann-Straussler-Scheinker syndrome.

Bovine spongiform encephalopathy (BSE), a disease in cattle, evolved into a mass epidemic in Great Britain, causing the death of more than 200,000 cattle during the 1980s to early 1990s. The BSE agent has been shown to cause variant CJD in humans. Symptoms of TSEs in humans vary, but they commonly include personality changes, psychiatric problems such as depression, lack of coordination, and/or an unsteady gait. Patients may experience involuntary jerking movements called myoclonus, unusual sensations, insomnia, confusion, or memory problems. In the later stages of the disease, patients have severe mental impairment and lose the ability to move or speak. Death occurs within a few months to a few years. There is no treatment that can halt the progression of TSEs.

Chronic wasting disease (CWD) is a contagious, fatal wasting disease of deer, elk, and moose caused by prions that continues to emerge in new locations. At the time of the writing, outbreaks have occurred in 16 U.S. states, in 2 providences in Canada, and in imported cases in South Korea. The mode of transmission is not fully understood; however, evidence supports direct transmission through animal-to-animal contact and indirect transmission through exposure of the PrP^{CWD} in the environment, including contaminated feed and water sources. Surveillance of hunter-harvested deer and elk continues in order to identify endemic locations of the disease. Many humans consume deer and elk, which raises a cause for concern regarding transmission of CWD to humans. *In vitro* experiments using **protein misfolding cyclic amplification** (**PMCA**) implicate a possibility of CWD prions infecting humans. *In vivo* experiments challenging cynomolgus monkeys with CWD-positive brain homogenates and epidemiological studies to date suggest that a species barrier protects humans from CWD.

Viroids are small (about 300 nucleotides), infectious, nonencapsidated +ssRNAs that infect plants of economic importance. They reproduce by a process called the rolling circle model of replication. Viroids do not code for any proteins and are totally dependent upon and interact with host cell proteins for their replication and movement from cell to cell. They do have self-cleavage ribozyme activity. Nothing distinguishes plant infections caused by viroids from those caused by viruses. Despite their simplicity, viroids can cause varied symptoms such as stunting, mottling, necrosis, and leaf distortion, or no symptoms at all. These pathogens are difficult to control and can be catastrophic where large areas of planted food crops are destroyed by disease, thus threatening our food supplies.

CASE STUDY QUESTIONS

These questions relate to the Case Study presented at the beginning of the chapter.

1. Is it possible that the cat and its owner contracted the same TSE agent strain? How could you test for this in the laboratory?
2. List common sources of prion disease agents.

Reference

Zanusso, G., et al. 1998. "Simultaneous occurrence of spongiform encephalopathy in a man and his cat in Italy." *Lancet* 352:1116–1117.

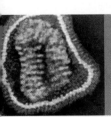

CASE STUDY 1: CJD-LIKE ILLNESS AMONG CONSUMERS OF SQUIRREL BRAIN

In 1997, five unrelated individuals from different towns who were suffering from a CJD-like disease attended a clinic in western Kentucky. The patients, two of whom were women, ranged in age from 56 to 78. All five of the patients had a history of eating squirrel brains. The neurologist who observed these patients suggested a possible association between eating squirrel brains and CJD in the British medical journal *The Lancet.*

Eating wild squirrels is common in rural Kentucky. The people usually eat squirrel brains or the meat, but not both. **Burgoo** is a vegetable stew that contains chopped squirrel meat. There are two different ways in which squirrel brains are eaten. One popular way is to scramble the squirrel brains in white gravy with eggs. Squirrel brains are also eaten during a gift-giving ritual in which someone brings the head of a squirrel to the home of a family. The fur is shaved off the head and the entire head is fried, the skull is cracked open at the dinner table, and the brains are scooped out and eaten.

1. Could these patients have contracted a CJD-like disease from eating beef and not squirrels? Explain your answer.
2. Typically, one CJD case is reported in Kentucky every ten years, so it was unusual to have a cluster of cases all in the same given year. In cases such as this, should doctors issue a warning that people should not eat squirrel brains? Why or why not?
3. Do a literature search to determine if there are reports of squirrels or other small mammals contracting transmissible spongiform encephalopathies.

References

Berger, J. R., Weisman, E., and Weisman, B. 1997. "Creutzfeldt-Jakob disease and eating squirrel brains." *Lancet* 350:642.

Bilger, B. 2000. "Squirrel and man: Is local custom worth dying for?" *The New Yorker,* July 17, 2000.

Kamin, M., and Patten, B. M. 1984. "Creutzfeldt-Jakob disease: Possible transmission to humans by consumption of wild animal brains." *Am J Med* 76:142–145.

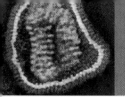

CASE STUDY 2: CJD-LIKE ILLNESS AMONG DEER HUNTERS

From 1976 to 2002, a group of hunters from the Midwest met at a cabin in northern Wisconsin to participate in wild game feasts. Elk, deer, antelope, and other game were eaten during these gatherings. Three of the hunters attending these feasts died of degenerative neurological diseases during 1993 to 1999. Their deaths were later investigated because of the increasing public health concerns about the possible transmission of TSEs to humans from deer and elk that suffer from chronic wasting syndrome. All three hunters ate venison frequently.

1. What did the investigators conclude in their report (refer to the *MMWR* reference)?
2. Another cluster of young male hunters (aged 28–30 years) from the United States who regularly consumed deer or elk meat died of fatal neurodegenerative diseases from 1997 to 1998. What did investigators conclude in this instance (refer to the *Archives of Neurology* reference below)?

References

Belay, E. D., et al. 2001. "Creutzfeldt-Jakob disease in unusually young patients who consumed venison." *Arch Neurol* 58:1673–1678.

Davis, J. P., et al. 2003. "Fatal degenerative neurologic illnesses in men who participated in wild game feasts—Wisconsin, 2002." *MMWR* 52(7):125–127.

CASE STUDY 3: HUMAN TO HUMAN CJD TRANSMISSION

Sporadic CJD occurs at a rate of about one case per million people. In 1985, a 52-year-old man began showing symptoms of a CJD, and he passed away in 1987. Five years later, his then-55-year-old widow experienced a sudden onset of confusion. Her condition deteriorated rapidly and she died one month after the onset of her symptoms. The brains of the husband and wife both contained the resistance prion protein, PrP^{res}. Neither one had a family history of neurological illnesses. The *PRNP* genes were normal.

1. Could this be a case of human-to-human transmission?
2. Were any surgical or dietary risks identified?
3. What is the probability that married couples would both die of CJD within years of one another?

Reference

Brown, P., et al. 1998. "Creutzfeldt-Jakob disease in a husband and wife." *Neurology* 50:684–688.

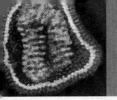

CASE STUDY 4: DECONTAMINATED SURGICAL INSTRUMENTS

On December 14, 2007, Froedtert Memorial Hospital, located in the Milwaukee, Wisconsin, area, suspended surgery after learning that an operation was performed on a patient who had Creutzfeldt-Jakob disease. About 100 patients had been operated on after this patient. Authorities contacted all of them. The hospital officials collected every potentially contaminated surgical instrument and cleaned them using special sterilization methods. Iatrogenic transmission of prion diseases by medical devices poses a serious infection control challenge to healthcare facilities.

1. Surface contamination of reusable medical devices is a concern. What are some classic decontamination procedures used in hospitals to ensure safety of medical and surgical instruments?
2. How should individuals treat skin contact with possible infectious prion material?

References

ProMed-mail. 1997. "Prion research safety guidelines—USA." *ProMED-mail* 1997; 3 Jun: 19970603.1151.

Fichet, G., et al. 2004. "Novel methods for disinfection of prion-contaminated medical devices." *Lancet* 364(9433):521–526.

Fichet, G., et al. 2007. "Investigation of a prion infectivity assay to evaluate methods of decontamination." *J Microbiol Methods* 70(3):511–518.

Lawson, V. A., et al. 2007. "Enzymatic detergent treatment protocol that reduces protease-resistant prion protein load and infectivity from surgical-steel monofilaments contaminated with a human-derived prion strain." *J Gen Virol* 88(Pt 10):2905–2914.

Lipscomb, I. P., et al. 2007. "Effect of drying time, ambient temperature and pre-soaks on prion-infected tissue contamination levels on surgical stainless steel: Concerns over prolonged transportation of instruments from theatre to central sterile service departments." *J Hosp Infect* 65(1):72–77.

Resources

Primary Literature and Reviews

Aguzzi, A., et al. 2004. "Progress and problems in the biology, diagnostics, and therapeutics of prion diseases." *J Clin Invest* 114:153–160.

Aguzzi, A., and Glatzel, M. 2006. "Prion infections, blood and transfusions." *Nat Clin Pract Neurol* 2(6):321–329.

Almond, J., and Pattison, J. 1997. "Human BSE." *Nature* 389:437–438.

Alper, T., et al. 1967. "Does the agent of scrapie replicate without nucleic acid?" *Nature* 214:764–765.

Angers, R. C., et al. 2009. "Chronic wasting disease prions in elk antler velvet." *EID* 15(5):696–703.

Animal and Plant Health Inspection Ser., et al. 2004. "Bovine spongiform encephalopathy in a dairy cow—Washington State, 2003." *MMWR* 52:1280–1285.

Baier, M., et al. 2003. "Prion diseases: Infectious and lethal doses following oral challenge." *J Gen Virol* 84: 1927–1929.

Balter, M. 2000. "Epidemiology: Tracking the human fallout from 'mad cow disease.'" *Science* 289(5484): 1452–1454.

Barnaby, C. H., et al. 2006. "Developing therapeutics for the diseases of protein misfolding." *Neurology* 66(Suppl 1): S118–S122.

Barria, M. A., et al. 2011. "Generation of a new form of human PrP^{SC} *in vitro* by interspecies transmission from cervid prions." *J Biol Chem* 286(9):7490–7495.

Bartz, J. C., et al. 2003. "Rapid prion neuroinvasion following tongue infection." *J Virol* 77(1):583–591.

Belay, E. D., et al. 2001. "Creutzfeldt-Jakob disease in unusually young patients who consumed venison." *Arch Neurol* 58:1673–1678.

Belay, E. D., et al. 2004. "Chronic wasting disease and potential transmission to humans." *EID* 10(6):977–984.

Belay, E. D., and Schonberger, L. B. 2005. "The public health impact of prion diseases." *Annu Rev Public Health* 26:191–212.

Berger, J. R., et al. 1997. "Creutzfeldt-Jakob disease and eating squirrel brains." *Lancet* 350:642.

Biacabe, A-G., et al. 2006. "Distinct molecular phenotypes in bovine prion diseases." *EMBO Rep* 5(1)110–114.

Biacabe, A-G., et al. 2008. "Atypical bovine spongiform encephalopathies, France, 2001–2007." *EID* 14(2): 298–300.

Bratosiewicz, J., et al. 2001. "Codon 129 polymorphism of the PRNP gene in normal polish population and in Creutzfeldt-Jakob disease, and the search for new mutations in PRNP gene." *Acta Neurobiol Exp* 61:151–156.

Brown, D. R. 2004. "Metallic prions." *Biochem Soc Symp* 71:193–202.

Brown, D. R., et al. 1999. "Normal prion protein has an activity like that of superoxide dismutase." *Biochem J* 344:1–5.

Brown, P. 2004. "Mad-cow disease in cattle and human beings." *Am Sci* 92:334–341.

Brown, P. 2004. "Infectivity studies of both ash and air emissions from simulated incineration of scrapie-contaminated tissues." *Environ Sci Technol* 38:6155–6160.

Casalone, C., et al. 2004. "Identification of a second bovine amyloidotic spongiform encephalopathy: Molecular similarities with sporadic Creutzfeldt-Jakob disease." *Proc Natl Acad Sci USA* 101(9):3065–3070.

Caughey, B., and Baron, G. S. 2006. "Prions and their partners in crime." *Nature* 443:803–810.

Caughey, B., and Kocisko, D. A. 2003. "A nucleic-acid accomplice?" *Nature* 425(6959):673–674.

Choi, C. J., et al. 2006. "Interaction of metals with prion protein: Possible role of divalent cations in the pathogenesis of prion diseases." *Neurotoxicology* 27(5): 777–787.

Collinge, J., et al. 2006. "Kuru in the 21st century—An acquired human prion disease with very long incubation periods." *Lancet* 367(9528):2068–2074.

Dabaghian, R. H., et al. 2004. "Prospects for the development of pre-mortem laboratory diagnostics for Creutzfeldt-Jakob disease." *Rev Med Virol* 14:345–361.

Daus, M. L., et al. 2011. "Presence and seeding activity of pathological prion protein (PrPTSE) in skeletal muscles of white-tailed deer infected with chronic wasting disease." *PLoS ONE* 6(4):e18345.

Davis, J. P., et al. 2003. "Fatal degenerative neurologic illnesses in men who participated in wild game feasts—Wisconsin, 2002." *MMWR* 52(7):125–127.

Deleault, N. R., et al. 2003. "RNA molecules stimulate prion protein conversion." *Nature* 425:717–720.

Dormant, D. 2002. "Prion diseases: Pathogenesis and public health concerns." *FEBS Lett* 529:17–21.

Farquhar, J., and Gajdusek, D. C., eds. 1981. *Kuru: Early Letters and Field-Notes from the Collection of D. Carleton Gajdusek.* New York: Raven Press.

Flores, R., et al. 2005. "Viroids and viroid-host interactions" *Ann Rev Phytopathol* 43:117–139.

Gajdusek, D. C. 1977. "Unconventional viruses and the origin and disappearance of kuru." *Science* 197(4307):943–960.

Gale, P. 2006. "BSE risk assessments in the UK: A risk tradeoff?" *J App Micro* 100:417–427.

Griffith, J. S. 1967. "Self-replication and scrapie." *Nature* 215:1043–1044.

Guillaume, E., et al. 2003. "A potential cerebrospinal fluid and plasmatic marker for the diagnosis of Creutzfeldt-Jakob disease." *Proteomics* 3:1495–1499.

Gwyther, C. L., et al. 2011. "The environmental and biosecurity characteristics of livestock carcass disposal." *Waste Management* 31:767–778.

Hartry, A. L., et al. 1964. "Planaria: Memory transfer through cannibalism re-examined." *Science* 146: 274–275.

Hill, A. F., et al. 1997. "Diagnosis of new variant Creutzfeldt-Jakob by tonsil biopsy." *Lancet* 349:99–100.

Hinckley, G. T., et al. 2008. "Persistence of pathogenic prion protein during simulated wastewater treatment processes." *Environ Sci Technol* 42:5254–5259.

Horonchik, L., et al. 2005. "Heparan sulfate is a cellular receptor for purified infectious prions." *J Biol Chem* 280(17):17062–17067.

Houston, F., et al. 2000. "Transmission of BSE by blood transfusion in sheep." *Lancet* 356(9234):999–1000.

Huillard d'Aignaux, J. N., et al. 2001. "Predictability of the UK variant Creutzfeldt-Jakob disease epidemic." *Science* 294:1729–1731.

Hunter, N., and Houston, F. 2002. "Can prion diseases be transmitted between individuals via blood transfusion: Evidence from sheep experiments." *Adv Transf Safety Dev Biol Basel, Karge*r 108:93–99.

Hu, W., et al. 2008. "Prion proteins: Physiological functions and role in neurological disorders." *J Neurol Sci* 264:1–8.

Kamin, M., and Patten, B. M. 1984. "Creutzfeldt-Jakob disease: Possible transmission to humans by consumption of wild animal brains." *Am J Med* 76:142–145.

Kim, T-Y., et al. 2005. "Additional cases of chronic wasting disease in imported deer in Korea." *J Vet Med Sci* 67(8):753–759.

Ladogana, A., et al. 2005. "High incidence of genetic human transmissible spongiform encephalopathies in Italy." *Neurology* 64:1592–1597.

Llewelyn, C. A., et al. 2004. "Possible transmission of variant Creutzfeldt-Jakob disease by blood transfusion." *Lancet* 363(9407):417–421.

Lupi, O., et al. 2006. "Are prions related to the emergence of early life?" *Med Hypotheses* 67:1027–1033.

Marsh, R. F., et al. 2005. "Interspecies transmission of chronic wasting disease prions to squirrel monkeys (*Saimiri sciureus*)." *J Virol* 79(21):13794–13796.

Masison, D. C., and Wickner, R. B. 1995. "Prion-inducing domain of yeast Ure2p and protease resistance of Ure2p in prion-containing cells." *Science* 270(5233):93–95.

Mathiason, C. K., et al. 2006. "Infectious prions in the saliva and blood of deer with chronic wasting disease." *Science* 314:133–136.

MaWhinney, S., et al. 2006. "Human prion disease and relative risk associated with chronic wasting disease." *EID* 12(10):1527–1535.

McConnell, J. V. 1962. "Memory transfer through cannibalism in planarians." *J Neuropsychiatry* 3:42–48.

McDonnell, G., and Burke, P. 2003. "The challenge of prion decontamination." *CID* 36:1152–1154.

Mead, S., et al. 2007. "Creutzfeldt-Jakob disease, prion protein gene codon 129VV, and a novel PrPSC type in a young British woman." *Arch Neurol* 64(12):1780–1784.

Mehrpour, M, Codogno, P. 2010. "Prion protein: From physiology to cancer biology." *Cancer Lett* 290:1–23.

Miller, M. W., et al. 2004. "Environmental sources of prion transmission in mule deer." *EID* 10(6):1003–1006.

Minor, P., et al. 2004. "Standards for the assay of Creutzfeldt-Jakob disease specimens." *J Gen Virol* 85:1777–1784.

Oomkes, C., and van Knapen, F. 2001. "Cows, cats, and FSE: Death penalty justified?" *Vet Q* 23(1):51–52.

Pattison, J. 1998. "The emergence of bovine spongiform encephalopathy and related diseases." *EID* 4(3):300–304.

Pearson, G. R., et al. 1993. "Feline spongiform encephalopathy: A review." *Vet Q* 33:1–10.

Peden, A. H., et al. 2004. "Preclinical vCJD after blood transfusion in a PRNP codon 129 heterozygous patient." *Lancet* 364(9433):527–529.

Peden, A. H., Ritchie, D. L., and Ironside, J. W. 2005. "Risks of transmission of variant Creutzfeldt-Jakob disease by blood transfusion." *Folia Neuropathol* 43(4):271–278.

Priola, S. A., Chesebro, B., and Caughey, B. 2003. "Biomedicine: A view from the top—prion diseases at 10,000 feet." *Science* 300:917–919.

Priola, S. A., Rianes, A., and Caughey, W. S. 2000. "Porphyrin and phthalocyanine antiscrapie compounds." *Science* 287:1503–1506.

ProMED-mail. 1997. "Prion research safety guidelines—USA." *ProMED-mail* 1997;03 Jun: 199770603.1151.

Prusiner, S. B. 1994. "Inherited prion diseases." *Proc Natl Acad Sci USA* 91:4611–4614.

Prusiner, S. B. 2001. "Shattuck lecture—Neurodegenerative diseases and prions." *N Engl J Med* 344(20):1516–1526.

Race, B., et al. 2009. "Susceptibilities of nonhuman primates to chronic wasting disease." *EID* 15(9):1366–1376.

Raeber, A. J., and Oesch, B. 2006. "Diagnostics for TSE agents." *Dev Biol (Basel)* 123:313–323.

Roma, A. A., and Prayson, R. A. 2005. "Bovine spongiform encephalopathy and variant Creutzfeldt-Jakob disease: How safe is eating beef?" *Cleveland Clin J Med* 72(3):185–194.

Saunders, S., et al. 2008. "Prions in the environment." *Prion* 2(4):162–169.

Sohn, H-J., et al. 2002. "A case of chronic wasting disease in elk imported to Korea from Canada." *J Vet Med Sci* 64(9):855–858.

Solforosi, L., et al. 2004. "Cross-linking cellular prion protein triggers neuronal apoptosis in vivo." *Science* 303: 1514–1516.

Tabler, M., and Tsagris, M. 2004. "Viroids: Petite RNA pathogens with distinguished talents." *Trends Plant Sci* 9(7):339–348.

Thackray, A. M., et al. 2003. "Subclinical prion disease induced by oral inoculation." *J Virol* 77(14):7991–7998.

Trifilo, M. J., et al. 2006. "Prion-induced amyloid heart disease with high blood infectivity in transgenic mice." *Science* 313:94–97.

Valleron, A.-J., et al. 2001. "Estimation of epidemic size and incubation time based on age characteristics of vCJD in the United Kingdom." *Science* 294:1726–1728.

Wadsworth, J., and Collinge, J. 2007. "Update on human prion disease." *Biochimica et Biophysica Acta* 1772: 598–609.

Weissmann, C., and Flechsig, E. 2003. "PrP knock-out and PrP transgenic mice in prion research." *Br Med Bull* 66:43–60.

Wickner, R. B. 1994. "As an altered URE2 protein: Evidence for a prion analog in *saccharomyces*." *Science* 264(5158): 566–569.

Wilham, J. M., et al. 2010. "Rapid end-point quantitation of prion seeding activity with sensitivity comparable to bioassays." *PLoS Pathog* 6(12):e1001217.

Will, R. G., et al. 1996. "A new variant of Creutzfeldt-Jakob disease UK." *Lancet* 347:921–925.

Wong, B.-S., et al. 2001. "Oxidative impairment in scrapie-infected mice is associated with brain metals perturbations and altered antioxidant activities." *J Neurochem* 79:689–698.

Zanusso, G., et al. 1998. "Simultaneous occurance of spongiform encephalopathy in a man and in his cat in Italy." *Lancet* 352:1116–1117.

Zeidler, M., et al. 1997. "New variant Creutzfeldt-Jakob disease: Neurological features and diagnostic tests." *Lancet* 350:903–910.

Zou, S., et al. 2008. "Transmission transfusion of human prion diseases." *Transfus Med Rev* 22(1):58–69.

Popular Press

Anderson, W. 2008. *The Collectors of Lost Souls: Turning Kuru Scientists into Whiteman.* Baltimore: Johns Hopkins University Press.

Cowley, G. 2001. "Cannibals to cows: The path of a deadly disease." *Discover Magazine,* March 12, 2001.

Kelleher, C. A. 2004. *Brain Trust: The Hidden Connection Between Mad Cow and Misdiagnosed Alzheimer's Disease.* New York: Paraview Pocket Books.

Klitzman, R. 2001. *The Trembling Mountain: A Personal Account of Kuru, Cannibals, and Mad Cow Disease.* New York: Perseus Publishing.

Lindenbaum, S. 1978. *Kuru Sorcery: Disease and Danger in the New Guinea Highlands.* New York: McGraw-Hill.

Max, D. T. 2007. *The Family That Couldn't Sleep: A Medical Mystery.* New York: Random House.

Prusiner, S. 1995. "The prion diseases." *Sci Am,* January, 1995.

Prusiner, S. B. 1991. "The molecular biology of prion diseases." *Sci Am,* June 1991.

Rhodes, R. 1997. *Deadly Feasts: Tracking the Secrets of a Terrifying New Plague.* New York: Simon and Schuster.

Sacks, O. 1997. "Eat, drink, and be wary." *The New Yorker,* April 14, 1997.

Schwartz, M. 2003. *How the Cows Turned Mad.* Berkeley, CA: University of California Press.

Shell, E. R. 1998. "Could Mad-Cow Disease Happen Here?" *Atlantic Monthly,* September, 1998.

Thornton, J. 2007. "Special Feature: The Quiet Spread of CWD." *Field & Stream,* May 10, 2007.

Yam, P. 2003. *The Pathological Protein: Mad Cow, Chronic Wasting, and Other Deadly Prion Diseases.* Germany: Springer.

Videos

New Killers. 2002. Films for the Humanities.

Foreign Body: Mad Cow Disease. 1998. Films for the Humanities.

Kuru: The Science and the Sorcery. 2010. Madman.com.au.

The Brain Eaters. 1998. NOVA PBS.

"Our Town," *The X-Files,* Season 2, Episode 24. Original airdate: May 12, 1995. Fox Television.

eLearning

go.jblearning.com/shors2

The site features eLearning, an online review area that provides quizzes and other tools to help you study for your class. You can also follow useful links for in-depth information, or just find out the latest virology and microbiology news.

CHAPTER 20

Plant Viruses

Photograph of the starchy tuberous roots of cassava. Cassava is mainly grown in South America, Africa, India, and parts of Southeast Asia. Cassava is the third largest carbohydrate source for human consumption in the world. Two plant viruses are threatening cassava crops. These viruses are currently labeled as Africa's biggest threat to food security.

"Nature's bioterrorism is far more fearsome than man-made bioterrorism."

Stanley Falkow, Professor of Molecular Microbial Pathogenesis, Stanford University

OUTLINE

CASE STUDY: PLUM POX

Pennsylvania is the leading producer of **stone fruits** in the United States. Stone fruits are fleshy fruits that contain a single seed or pit such as peaches, apricots, nectarines, plums, and cherries. In 1999, peach trees located on a commercial grove in Adams County, Pennsylvania, were infected with **plum pox virus**, which is also known as **Sharka virus**. The first symptoms appeared on the leaves of fruit of the fruit trees in the form of chlorotic (yellowing) and necrotic (browning) ring patterns. Infected fruits became fibrous and gummy with irregular brown and reddish spots. The fruit was low in sugar and not tasty. Fruits dropped prematurely (about three to four weeks before ripening), making them unsuitable for eating and processing.

Plum pox virus was first discovered in Bulgaria sometime between 1915 and 1918. Within 30 years, it spread throughout most eastern European countries and the former U.S.S.R. Following World War II, it spread to Western European countries, Egypt, Syria, and Cyprus; it was found in New Zealand and Chile in 1992 and in India in 1994. The virus was first detected in North America in 1998 and since 1999 has been detected in Michigan and New York and in orchards in Ontario, Canada. The strain of the virus found in Canada was identical to the strain found in Pennsylvania. Researchers and growers believe that plum pox virus may be the most damaging virus in fruit production.

20.1 History of Plant Viruses

Many still-life paintings and drawings created in Western Europe from 1600 to 1660 contained tulips that portrayed flower symptoms of a viral disease. These tulips had striped or variegated patterns that were considered prized as special varieties. It led to a craze called ***tulipomania*** in which bulbs of the infected tulips were traded for large amounts of money or goods (**TABLE 20-1**). These tulips were likely infected with tulip breaking virus, an infection that interferes with the pigmentation of the flower and results in a "breaking" or loss of pigment that causes the flower to appear marbled.

TABLE 20-1 Examples of the Goods Exchanged for One Bulb of Viceroy Tulip Variety During Tulipomania

Two lasts (four tons) of wheat
Four lasts (eight tons) of rye
Four fat oxen
Eight fat pigs
Twelve fat sheep
Two hogsheads of wine (126 gallons)
Four barrels of beer (252 gallons or 15 kegs)
1000 pounds of cheese
A bed with accessories
A full dress suit
A silver goblet (drinking cup)

By the end of the 1900s, the idea that infectious disease was caused by microorganisms was well accepted. Martinus Beijerinck's 1898 discovery of a plant virus that infected tobacco (refer to Chapter 1) was considered to be the birth of virology. Between 1900 and 1935, many plant diseases were thought to be caused by viral agents that could pass through porcelain filters that would normally trap bacterial pathogens.

Later it was discovered that some plants displayed virus-like symptoms that could not be transmitted to healthy plants by **mechanical transmission**, which is the most common technique for experimental infection of plants. It is usually performed by inoculating or rubbing virus-containing preparations into the leaves of susceptible, healthy plants. In 1922, Kunkel published the first paper to recognize that insects transmitted some plant viruses by showing that yellow stripe disease virus was transmitted to corn by a leafhopper. Within ten years, many insects were reported to be virus vectors of plant diseases. In 1939, Holmes published the *Handbook of Phytopathogenic Viruses*, a list of 129 known viral diseases of plants.

Many advances were made toward virology research during the 1930s through the 1950s and much of the attention focused on tobacco mosaic virus (TMV). Serological, electron microscopic, and X-ray crystallography were the major techniques used to distinguish different strains of virus and to explore virus structure. During the 1980s, research in plant virology focused on the diagnosis of virus diseases, especially using enzyme-linked immunosorbent assays (ELISA; see Chapter 7) and the application of molecular biology techniques to study the organization of viral genomes and viral replication strategies. Today there are approximately 900 plant virus species recognized by the **International Committee on Taxonomy of Viruses (ICTV)**, and plant virology research is focused on the manipulation of plant virus genomes in order to understand viral resistance for commercial gain in the agriculture business. Plant viruses are of considerable importance: At least 10% of global food production is lost to plant disease each year. In the United States alone, over $60 billion in crop losses occur each year.

20.2 Transmission of Plant Viruses

The infection of plant cells is not achieved by surface receptors; this is a major difference between viral infections of animals and those of plants. Infection of plants requires cell wounding or the transfer of extracellular fluid (e.g., sap). Viruses infect plants by horizontal or vertical transfer. **Horizontal transfer** occurs from plant to plant (within the same generation), such as when plants are touching each other. **Vertical transfer** is from parent to offspring such as infected seeds (between generations). Plant viruses are transmitted in a number of different ways, such as:

- mechanical means (human and environmental damage)
- soil transmission
- vegetative propagation (grafting)
- piercing and chewing insects and other vectors
- seed transmission
- pollen transmission

Plant viruses depend upon a break in the cell wall for entry. Mechanical transmission is a common way to experimentally infect plants by wounding the leaves with a virus-containing preparation (inoculum). It is a valuable method for plant virologists; however, not all viruses are mechanically transmissible and in general mechanical inoculation is very inefficient (at least 500 virus particles are needed to give rise to a visible infection).

In addition to human intervention, physical damage to plants can also occur through wind damage. Leaves of healthy and infected plants may rub together in the wind, allowing natural transmission to occur. Viruses present in the soil may be transmitted to leaves of new plants as windblown dust or rain-splashed mud strikes the leaves. Soil acts as an abrasive, causing damage that will give the virus access to the plant cells. **Soilborne** and **waterborne** viruses may also infect plants through slightly damaged rootlets.

Grafting, a process in which two young plants are joined together, is a form of vegetative propagation that also provides an opportunity for virus transmission. Typically a twig or bud (called a **scion**) is grafted onto a rootstock (**FIGURE 20-1**). The scion eventually develops into a new shoot. Grafting allows greenhouse workers to combine the best features of two different plants into one plant. For a long time, grafting was the only experimental method available to create virus infections in plants. It was a way for researchers to graft an infected scion onto a healthy plant's rootstock.

Due to the rigidity of the plant cell walls, most plant viruses have to be introduced into cells with the help of **vector organisms**. Many different organisms can act as vectors and spread plant viruses such as bacteria, fungi, nematodes, parasitic plants, and insects (e.g., aphids, leafhoppers, planthoppers, beetles, mites, mealybugs, whiteflies, and thrips). Several soilborne viruses are transmitted by fungi. Most plant viruses are transmitted by vectors, the majority of which are insects (**FIGURE 20-2**). The viruses are transmitted directly into cells when the mouthparts of an insect such as a beetle directly injure the plant through chewing or when injected into the plant through the piercing of mouthparts of a sap-sucking insect or a root-feeding nematode.

Viruses transmitted through seed are also transmitted through pollen grains. When a virus is transmitted by pollen, it may infect the seed and the seedling that will grow from that seed, or it may infect the plant through the fertilized flower. Sometimes, insects play an important role in pollen-mediated transmission. Plant-to-plant transmission of a virus by pollen is known to occur in fruit trees. Seeds may be contaminated by viruses adhering to the outside of the seed or due to infection of the living tissues of the plant embryo.

20.3 Symptoms of Plant Diseases Caused by Viruses

Some infected plants exhibit no symptoms. Plants infected by viruses can lead to the following symptoms, among others:

- dwarfing or stunting of plants
- leaf curling

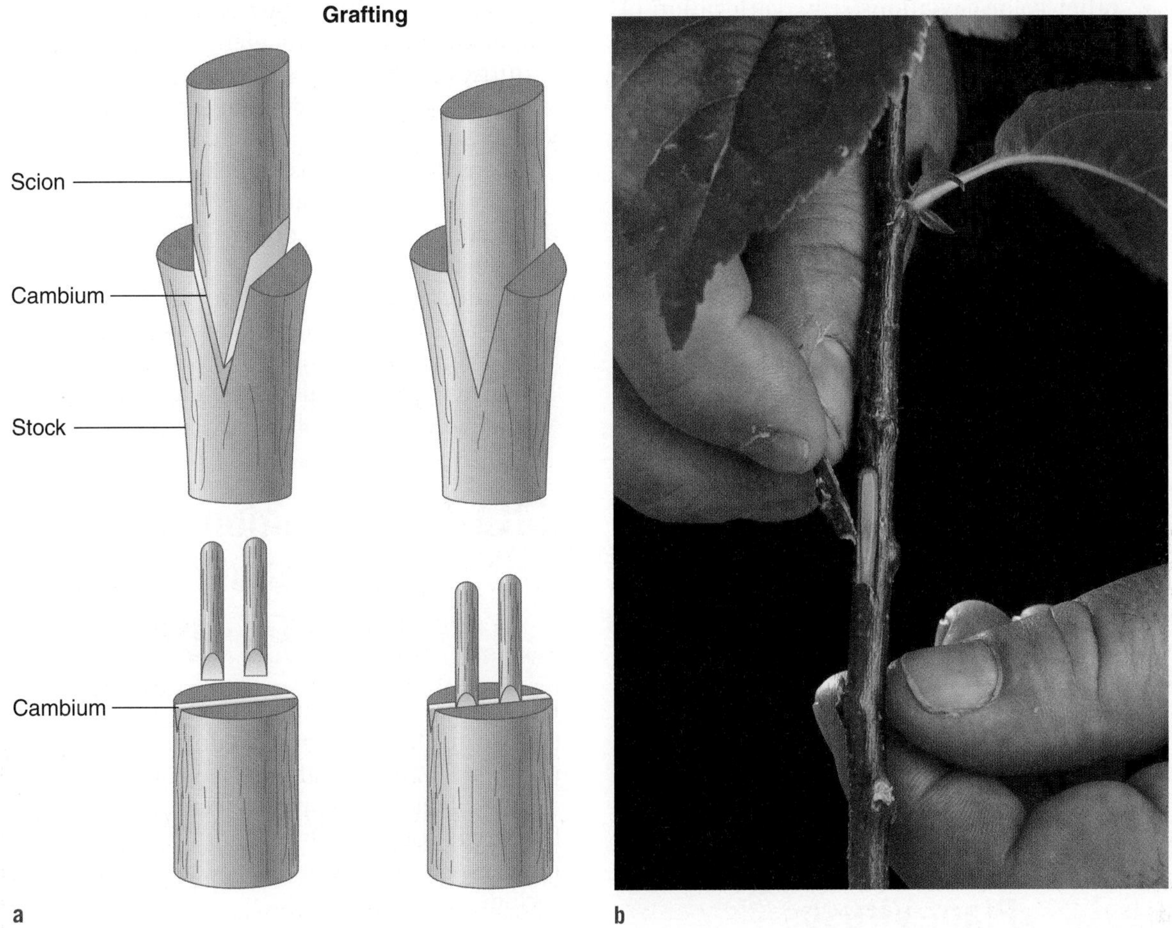

FIGURE 20-1 (a) Grafting. Adapted from James, V. G. "SparkNote on the life cycle of plants." *Biology Study Guides*. SparkNotes, LLC. http://www.sparknotes.com/biology/plants/lifecycle. **(b)** Grafting showing a chip bud being joined to a rootstock.

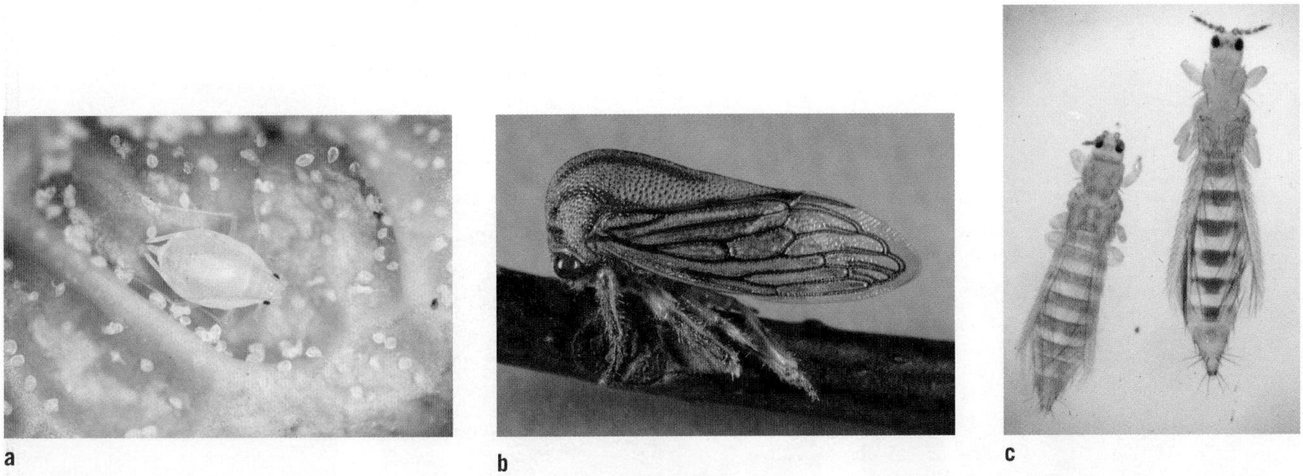

FIGURE 20-2 Insects form the largest and most significant vector group. Most notably this group includes: **(a)** aphid, **(b)** hopper, and **(c)** thrips.

- reduced yield
- fruit distortion
- chlorosis (yellowing)
- other color deviations (e.g., variegation of flower petals)
- "mosaic" patterns on leaves
- ring-shaped spots on leaves
- wilting and withering
- necrosis (black or grayish brown discoloration due to the death of cells and tissues)
- bark scaling

All of these symptoms mimic other problems of the plant, such as nutrient deficiencies or toxicities and infection by other pathogens (e.g., fungi, bacteria).

Viral infections rarely kill the plant. Most plant viruses multiply at the site of infection, giving rise to localized symptoms such as necrotic lesions on leaves. Viruses must pass various tissue barriers in order to cause systemic infections. Virus **movement proteins** must be compatible with the host in order for the virus to move systemically (refer to Section 20.8, Plant Virus Life Cycles). Once a virus infects a plant, the plant usually remains infected during its life. Examples showing viral disease symptoms are shown in **FIGURE 20-3**.

20.4 Diagnosis and Detection of Plant Viruses

In addition to observing and recording plant symptoms, there are several considerations or questions that need to be asked to determine what diagnostic approach should be applied to a particular plant disease. For example, is it important to know if the plant is infected with a virus or what virus is infecting the plant? Or what virus strain is infecting the plant? Costs and the sophistication of the technique must also be considered, particularly in the developing countries of the tropics. Most methods used are similar to those used in detecting animal and human viruses such as:

- detection of inclusion bodies in infected cells by light microscopy
- direct observation of virus particles by electron microscopy
- infectivity assays (e.g., inoculation of healthy plants with sap from a symptomatic plant; this is time consuming)
- use of serological probes (e.g., ELISA)
- DNA or RNA probes (e.g., PCR)

20.5 Prevention and Control of Plant Virus Diseases

Plant viruses cause important plant diseases that cause huge losses in crop production and quality in all parts of the world. Infected plants may show a range of disease symptoms. Plant viruses cannot be directly controlled by chemical application. The major means of control (depending on the disease) include:

- control of insect vectors (chemical or biological control)
- removal of alternate hosts (e.g., weeds)
- sanitation techniques (e.g., cleaning of farm implements, proper grafting techniques)
- use of certified virus-free seed or stock

a b c

FIGURE 20-3 (a) Viral necrosis to corn. **(b)** Yellowing of lettuce by closterovirus. **(c)** Ring and band mosaics due to plum pox potyvirus.

- growing resistant crop varieties
- plant quarantines to prevent disease establishment in areas where it does not occur (material is grown or kept isolated for longer periods for observation—absence of disease or pest) and testing. These procedures are time-consuming and costly, and they require skilled professionals.

20.6 Morphology of Plant Viruses

An envelope is a typical feature in animal viruses but is uncommon in plant viruses; for plant viruses, the typical feature is a naked helical rod. The rods may be:

- naked, long, helical flexuous rods (full of bends or curves) that are typically 10 nm wide by 480 to 2000 nm in length
- rigid helical rods of 15 nm by 300 nm in length
- short (bacillus-like) rods

Polyhedron-shaped plant viruses also exist that are 17 to 60 nm in diameter in addition to multipartite viruses that may contain the same component proteins but often differ in size depending on the length of the genome segment packaged. The only plant viruses that contain membranes are the plant rhabdoviruses and tospoviruses. Examples of these morphologies are shown in **FIGURE 20-4**.

20.7 Types of Plant Virus Genomes

The majority of all known plant viruses contain their genetic information encoded in positive-sense single-stranded RNA (ssRNA). Like animal viruses, there are also genomes of negative sense ssRNA, double-stranded RNA (dsRNA), ssDNA, and dsDNA (dsDNA is the rarest genome type in plant viruses). Plant viruses of segmented genomes (genomes that comprise more than one molecule of nucleic acid) also exist. They are referred to as **multipartite**

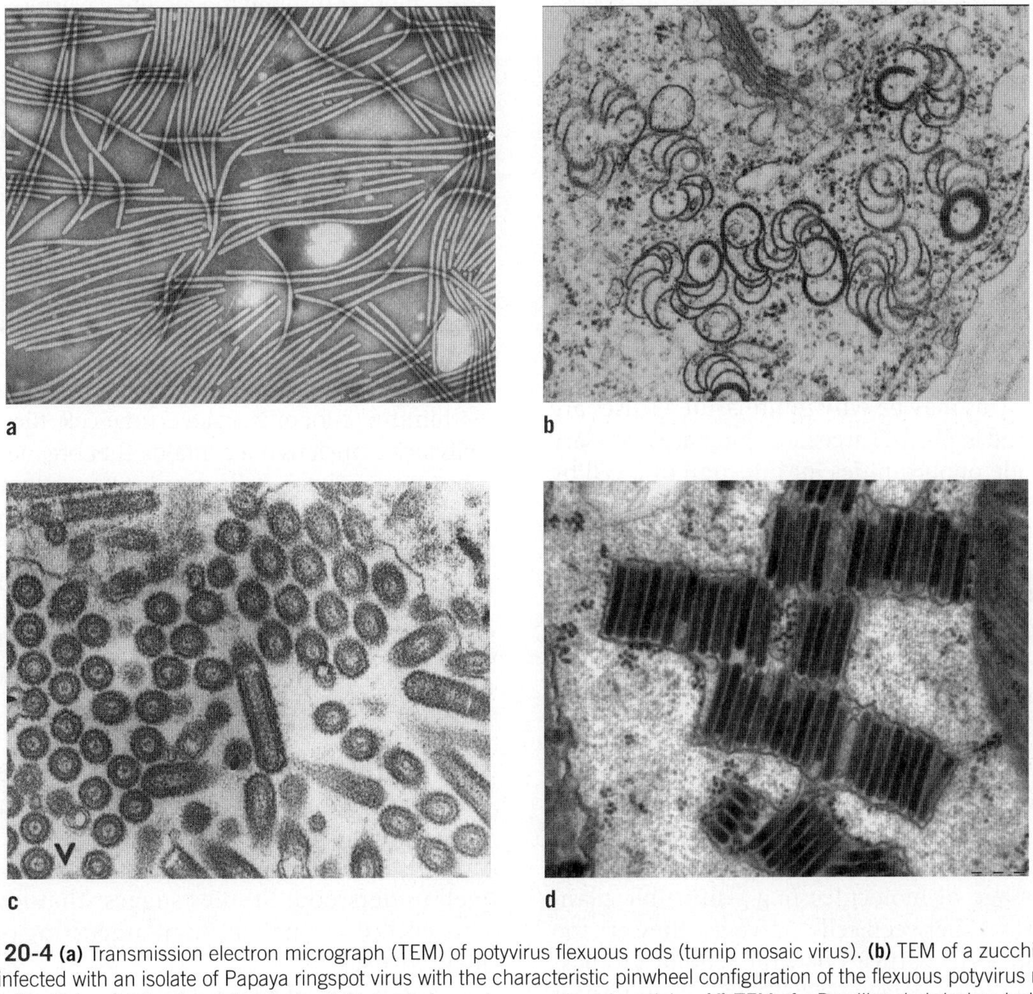

FIGURE 20-4 (a) Transmission electron micrograph (TEM) of potyvirus flexuous rods (turnip mosaic virus). **(b)** TEM of a zucchini squash infected with an isolate of Papaya ringspot virus with the characteristic pinwheel configuration of the flexuous potyvirus rods. **(c)** TEM of melilotus (white clover rhabdovirus). V stands for mature enveloped particles. **(d)** TEM of a Brazilian rhabdovirus isolate (maize mosaic virus) infecting parenchymal cells of maize.

FIGURE 20-5 (a) Segmented influenza virus particle. Each influenza particle contains eight molecules of negative sense ssRNA of different sizes and sequence. **(b)** Bipartite plant Gemini virus. Gemini (twin) viruses are usually found as two joined incomplete icosahedral particles. Their genome consists of two molecules of circular ssDNA of almost equal size but of different sequence.

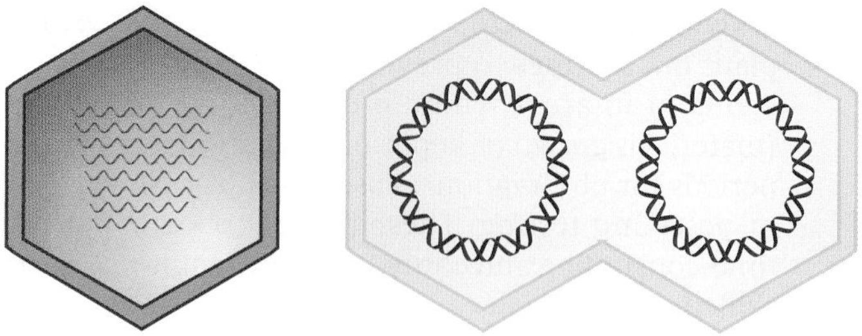

genomes that are either quadra-, tri-, or bipartite (i.e., four, three, or two molecules comprise the total genome of the virus).

Animal viruses with segmented genomes contain all of the nucleic acid molecules or segments within the same viral particle. *Unlike animal viruses, multipartite plant viruses package each nucleic acid molecule into a separate virus particle* (**FIGURE 20-5**). As a result, the sizes of each virus particle can also differ if the lengths of the segments differ.

What Are the Advantages and Disadvantages of Multipartite Genomes of Plant Viruses?

The advantage of this multipartite strategy is that it reduces the probability of losing the coding capacity of the genome if breakages of the genome occur. It also removes the requirement for accurate sorting or packaging of genomes into one virus particle.

The disadvantage of this multipartite strategy is that all the individual genome segments must be packaged into a separate virus particle. All of the discrete virus particles must be taken up by a single cell to establish a productive infection or the virus will be defective as a result of loss of genetic information. This may be why multipartite viruses are only found in plants. Large inoculums are necessary to provide opportunities for infection of a cell by more than one particle.

20.8 Plant Virus Life Cycles

All plant cells contain an impermeable cell wall composed of cellulose that blocks viral entry. Viruses can only enter through a break in the cell wall by a wound in the plant or by vector transmission, especially by insects. Communication between plant cell walls is achieved through microchannels called **plasmodesmata**. The plasmodesmata allows the passage of molecules (e.g., diffusible plant metabolites) between cells; however, they are too small for virus particles or viral genomes to pass through them unaided (**FIGURE 20-6**). Plant viruses have evolved to carry genes that code for specialized movement proteins that enable their movement from one infected cell to neighboring cells. The movement proteins function by

- associating with or coating the RNA or DNA genomic nucleic acid of the virus to form nucleoprotein complexes. These nucleoprotein complexes cause the plasmodesmata microchannels to dilate and a host translocation system allows the complex to be delivered in the cytoplasm of the neighboring cell.
- forming a tubular transport structure that is inserted into a plasmodesmata pore, allowing the passage of virus particles from one cell to another through the modified plasmodesmata (**FIGURE 20-7**).

Movement proteins appear to be derived from host plant genes that encode for **chaperonins** and plasmodesmata-associated proteins. Most plant viruses multiply at the site of infection, causing localized symptoms such as necrotic spots on leaves. A systematic (long distance) infection of the plant may occur if plant viruses can be transported longer distances through the vascular system via the **phloem**. The phloem consists of specialized living cells that transport sap. One type of specialized cell within the phloem is a sieve-tube cell. Sieve-tube cells lack a nucleus but contain other organelles and ribosomes. These cells conduct food materials throughout the plant. The end walls of the sieve tube cells have many smaller pores and enlarged plasmodesmata (**FIGURE 20-8**). Plant viruses "channel" through the plant cell wall via movement proteins without causing cellular lysis. As a result, all plant viruses spread directly from cell to cell within a leaf without an extracellular phase. Viral movement is localized.

Once viral particles have reached the cytoplasm, the genome is uncoated, exposing the viral genome to the replication processes. The uncoating step of the plant virus genome is not well understood. Studies suggest that low Ca^{2+} concentrations may facilitate uncoating of some plant viruses.

About 90% of all plant viruses contain positive sense ssRNA genomes (+ssRNAs); **FIGURE 20-9** lists

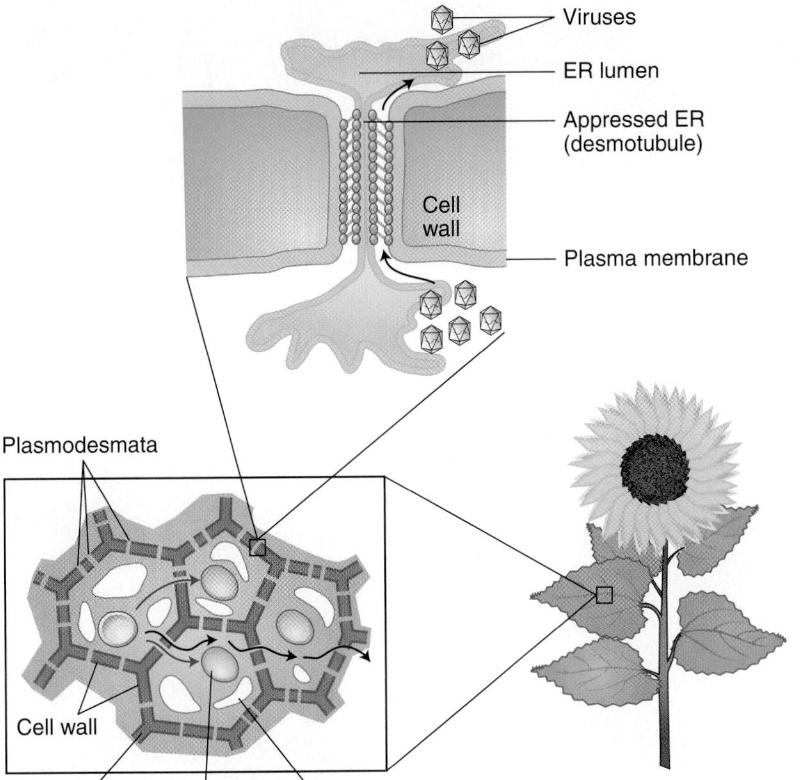

FIGURE 20-6 The sizes of some plant viruses compared to the plasmodesmata. Citrus tristeza virus (CTV; 2 μM × 10 nm), tobacco mosaic virus (TMV; 300 nm × 18 nm), cowpea mosaic virus (CPMV; 28 nm), potato virus Y (PVY; 750 nm × 11 nm), lettuce necrosis yellow virus (LNYV; 220 nm × 80 nm), tomato spotted wilt virus (TSWV; 80 nm), clover wound tumor virus (CWTV; 70 nm), cauliflower mosaic virus (CaMV; 50 nm). Adapted from Hull, R. *Matthews' Plant Virology,* Fourth Edition. Academic Press, 2001.

the plant virus genomes and taxonomy. The +ssRNA plant viruses take advantage of the host cell mRNA processing and translational machinery. In contrast to the 5′ capped host messenger RNAs (mRNAs), the 5′ end of plant viral +ssRNAs can take one of several forms:

- 5′ phosphate group
- a 5′ cap
- a virus-encoded polypeptide called VPg (viral protein genome-linked) that is covalently attached to the first nucleotide of the RNA

The 5′ untranslated regions (UTRs) of the viral RNAs differ significantly from cellular mRNAs by length and secondary structure. Some contain **internal ribosomal entry sites** (IRES) that allow for **cap-independent translation**. The 3′ end of viral RNAs may contain a poly(A) tail, tRNA-like structure, or simply a 3′ hydroxyl (OH) group.

The +ssRNA plant viruses encode their own RNA-dependent RNA polymerases for genome replication. They also need host factors for the formation of the replicase complex. During replication, the +ssRNA genome is copied into a –ssRNA intermediate that serves as a template for the production of genomic +ssRNAs. These genomic RNAs are used for translation, for replication, or as genomic RNA in new virions.

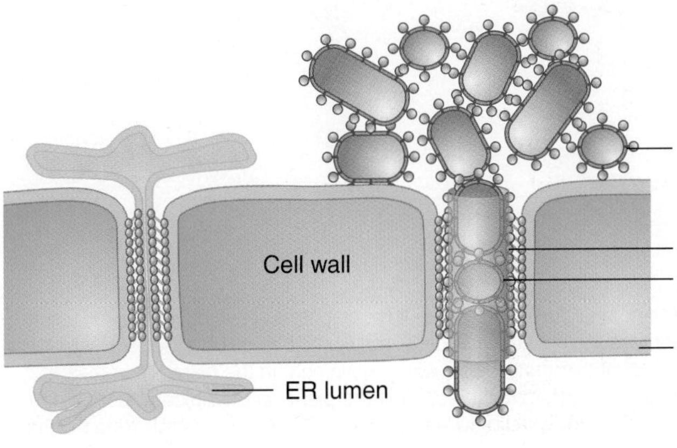

FIGURE 20-7 Tubule strategy for cell-to-cell movement of plant viruses. Adapted from Hull, R. *Matthews' Plant Virology,* Fourth Edition. Academic Press, 2001.

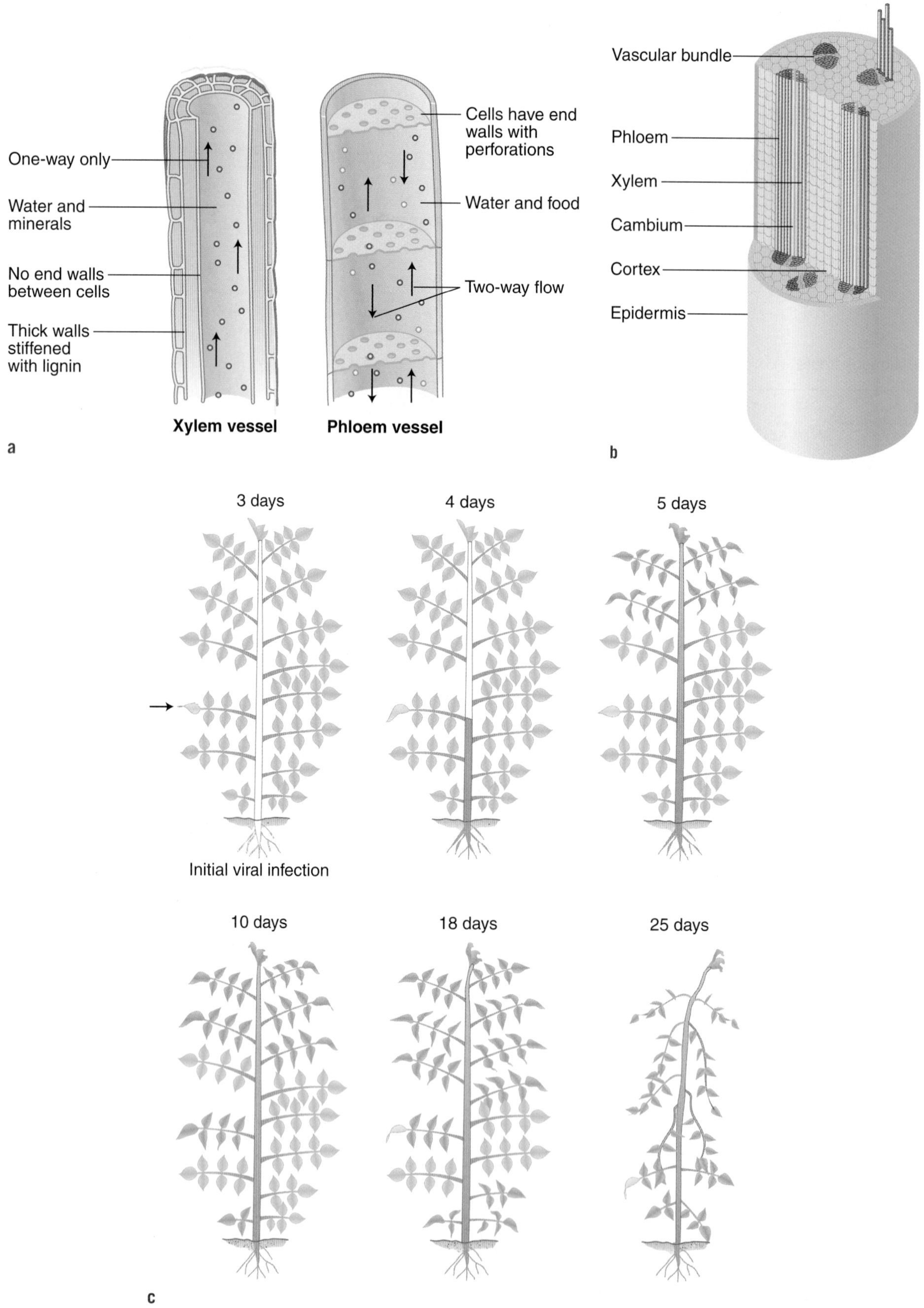

FIGURE 20-8 (a) The vascular plant anatomy, which delivers water and nutrients to the plant. It is composed of two main tissue types: the phloem and the xylem. Adapted from BBC, *Schools: Learning Resources for Home and School*, http://www.bbc.co.uk/schools/gcsebitesize/img/bixylemphloem.gif. **(b)** The xylem is made up of nonliving cells that are responsible for water movement in the plant. The phloem contains living cells that conduct nutrient movement. Viruses can cause a systematic infection of the plant via the phloem. Adapted from Hull, R. *Matthews' Plant Virology*, Fourth Edition. Academic Press, 2001. **(c)** Systematic spread of a virally infected plant. Note the viral spread (shaded areas) after the initial infection of the plant through the tissues of the phloem. Adapted from Fairchild Tropical Botanic Garden, *Teaching Resources*, http://www.fairchildgarden.org/index.cfm?section=education&page=teacherprofessionaldevelopment.

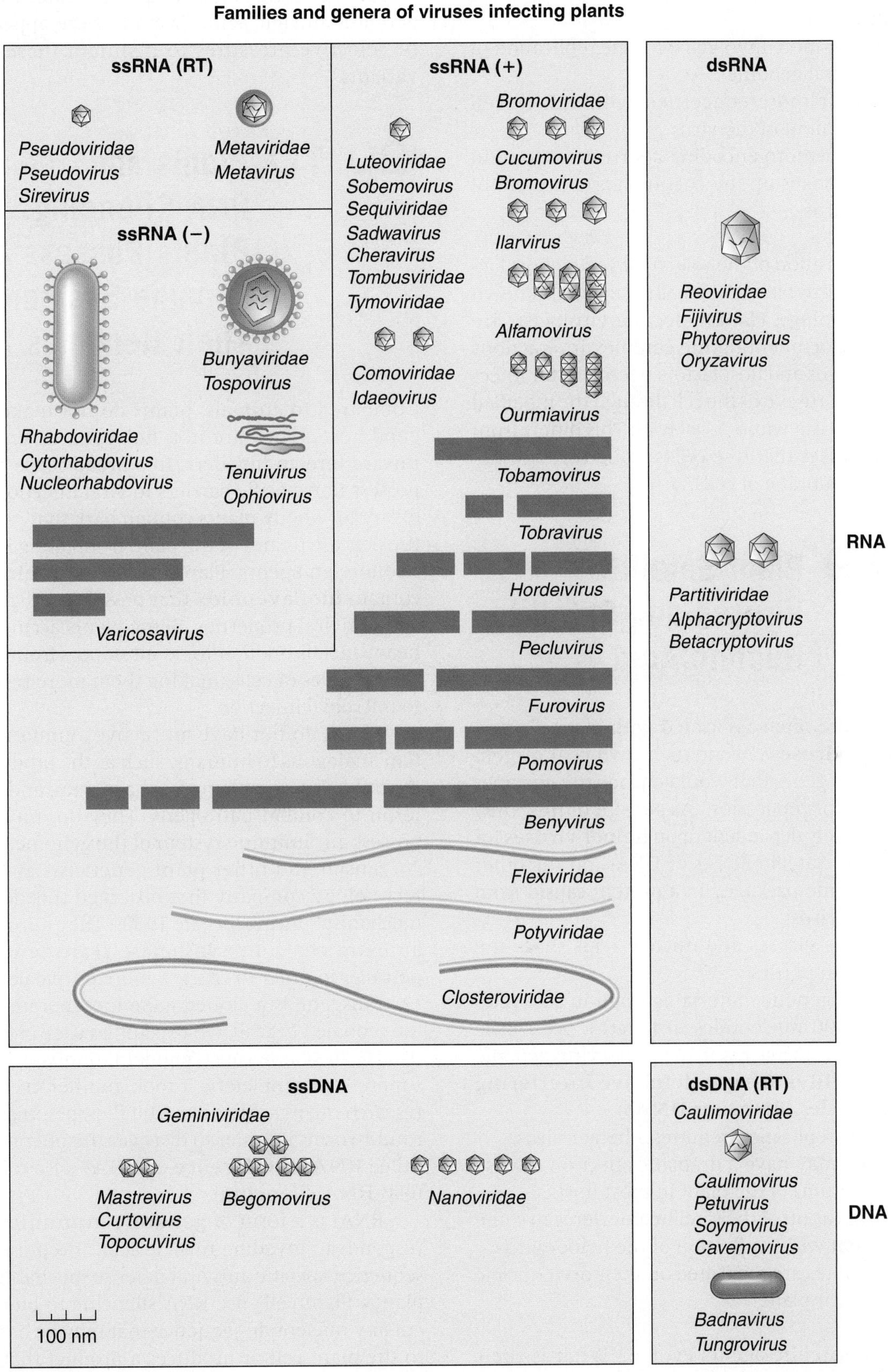

FIGURE 20-9 Morphology and genomes of plant viruses shown by genera and family. Adapted from Fauquet, C. M., et al., eds. *Virus Taxonomy: VIIIth Report on the International Committee on Taxonomy of Viruses.* Academic Press, 2005.

The number of genes that plant viruses have varies considerably. Most plant viruses have at least three genes:

1. one or more involved with the replication of the viral genome
2. one or more concerned with cell-to-cell movement of the virus
3. one or more encoding a structural protein that makes up the coat or capsid protein of the virus

The identification of the role(s) played by host factors in the virus life cycle remains a major unknown in plant virology. Plant molecular virologists are currently documenting the complex interactions between virus and host factors necessary for infection. Plant viruses exit the cell through the modified plasmodesmata without cell lysis. This differs from animal viruses that lyse cells or bud through the plasma membrane of cells.

20.9 Plant Satellite Viruses and Satellite Nucleic Acid

Satellites are often associated with plant viruses. **Satellite viruses** encode their own coat protein but lack the genes that would encode the enzymes necessary for replication. As a result of this, they are completely dependent upon **helper viruses** for replication. Satellite RNAs or DNAs, on the other hand, become packaged in a protein capsid from the helper virus.

Satellite viruses and nucleic acids share the following properties:

- Their genetic material is small in size (200 to 1500 nucleotides in length), with little sequence similarity to the helper virus genome (**satellites are not defective interfering particles, RNAs, or DNAs**).
- Their replication requires a helper virus.
- They may have a dramatic affect on disease symptoms of the plant in most hosts.
- Replication of the satellite interferes to some extent with replication of the helper virus.
- Satellites are replicated on their own nucleic acid template.

How did satellite agents evolve? What is their function? The most popular theory is that satellite viruses arose from an independent virus. During the course of infection satellite viruses lose genetic material, making them dependent upon a helper virus for their replication. Satellite nucleic acids may be a renegade nucleic acid that originated as plant genomic nucleic acid. Overall, there appears to be selective pressures to maintain these virus variants.

20.10 Plants and RNA Silencing: Plants Possess an Immune System of Their Genomes

Compared to animals, plants do not make very good hosts. They are not, however, defenseless toward foreign invaders. In a broad sense plants possess nonspecific barriers to viral infection. For example, woody plants contain bark that protects the delicate tissues of the plant from biting insects and foreign agents. Plants secrete **saponins** and contain **bioflavenoids** that possess antibacterial and antiviral properties. Some plants accumulate heavy metals or other toxic substances from soil in the vacuoles of cells, making them more resistant to microbial infection.

Plants do not have an "active" immune system analogous to humans, such as the production of antibodies, complement, phagocytes, and interferon to combat pathogens. They do, however, possess an "immune system of the genome." Rich Jorgensen and other plant geneticists at a biotechnology company first observed this defense mechanism during the late 1980s. They introduced an extra chalcone synthetase **transgene** into petunias in order to create a more purple petunia. (A transgene is a cloned gene incorporated into the genome.) Instead, white petunias were produced (**FIGURE 20-10a**) because pigmentation was lost or suppressed completely. It took another decade of research on organisms like fruit flies, petunias, and roundworms in order to discover the phenomena called **RNA interference** or **RNAi** (also referred to as **RNA silencing**).

RNAi is a form of **genomic immunity** that responds to invading nucleic acid. It represents a sequence-specific antiviral defense mechanism in plants. Plant cells use RNA silencing to bind and cut any nucleotide sequence that may be harmful to the plant cell or produces a product that may inhibit plant cell activities. The discovery was hailed by *Science* as the "breakthrough of the year" in 2002. Craig C. Mello and Andrew Z. Fire shared the Nobel Prize in Physiology or Medicine in 2006

VIRUS FILE 20-1 Silencing Genes

All genomes are targets for invading viruses and **transposons**. Transposons are sequences of DNA that can move around to different positions within the genome of a single cell, a process called transposition. About 45% of the human genome consists of remnant sequences of previous transposon and virus invasions. As a result, it is expected that organisms need to fight off invading nucleic acids so that they are not completely taken over by these molecular invaders. The immune system of vertebrates fights off invaders by antibody-mediated responses. In addition to the immune system, the genome defense mechanism protects the genome of the organism. This mechanism appears to be an ancient defense against viral invaders.

RNAi gene-silencing is conserved in plants, fungi worms, flies, and mammals. It has become an extremely powerful tool for silencing gene expression in vitro. It promises to revolutionize experimental biology and may have many important therapeutic applications.

RNAi technology can be applied to the following areas:

- as an applied experimental technique to "knock out" genes
- in functional genomic studies
- in studies to identify genomic regulators
- for determining the importance of a gene in a process (e.g., cell development and G. J. Pruss, and V. B. Vance, cell division)
- as therapeutic tools (especially in the areas of genetic disorders, cancer, and viral infection)
- in engineering disease-resistant crops
- in engineering improved nutritional and handling characteristics of fruits

RNAi technology has the potential to target any gene unique to a specific disease or viral infection, thus reducing host cell toxicity problems. The siRNAs can be synthesized chemically in the laboratory and then introduced into cells, opening up many opportunities for siRNAs as potential drug candidates. Studies to date have shown that siRNAs can inhibit infection by HIV-1, SARS-CoV, respiratory syncytial virus (RSV), parainfluenza virus (PIV), herpes simplex 2, poliovirus, and hepatitis C viruses (**FIGURE VF 20-1**).

In 2005, a Phase I clinical trial used an RNAi therapeutic to treat 26 patients for **age-related macular degeneration**. Macular degeneration is the leading cause of adult blindness. It is caused by the overexpression of vascular endothelial growth factor (VEGF). VEGF is a protein that induces the growth of leaky and inflamed blood vessels underneath the retina, resulting in the slow progression of vision loss (**FIGURE VF 20-2**). The new RNAi drug (Sirna-027) was directly injected into the eye of patients as a means to shut down VEGF genes. Patients received 100 to 1600 μg of Sirna-027. After three months of a single injection, 24 of 26 patients showed visual stabilization and four patients experienced significant improvement in vision. Only two patients experienced vision loss. Sirna-027 studies have proceeded to Phase II clinical trials.

In 2006, clinical trials to use RNAi to treat asthma and Huntington's disease began. Huntington's disease is a neurodegenerative disease caused by the accumulation of mutated proteins that contain polyglutamine repeats (up to about 37 to 40 glutamines in the mutated protein, whereas normal proteins contain about 35 glutamines). Patients with Huntington's disease gradually lose their ability to think, speak, or care for themselves. There is no cure or treatment. Other applications of RNAi technology include the treatment of HIV infection and AIDS lymphomas.

Is RNAi technology really that magical? Are there any technological hurdles? Initial experiments using RNAi technology to combat viral infections resulted in the activation of genes in the **interferon pathway**, causing cells to undergo **apoptosis** (cellular suicide). Interferons are cytokines that function as the host's first line of defense against viral infection in mammals. This problem was overcome by using shorter siRNA molecules that no longer induce the interferon response. Another obstacle is addressing viral sequence diversity. Viruses like HIV mutate so rapidly that they will escape RNA silencing. This has been addressed by targeting genes required for viral entry or replication. These sequences of these genes contain fewer mutations. Delivering the

(continued)

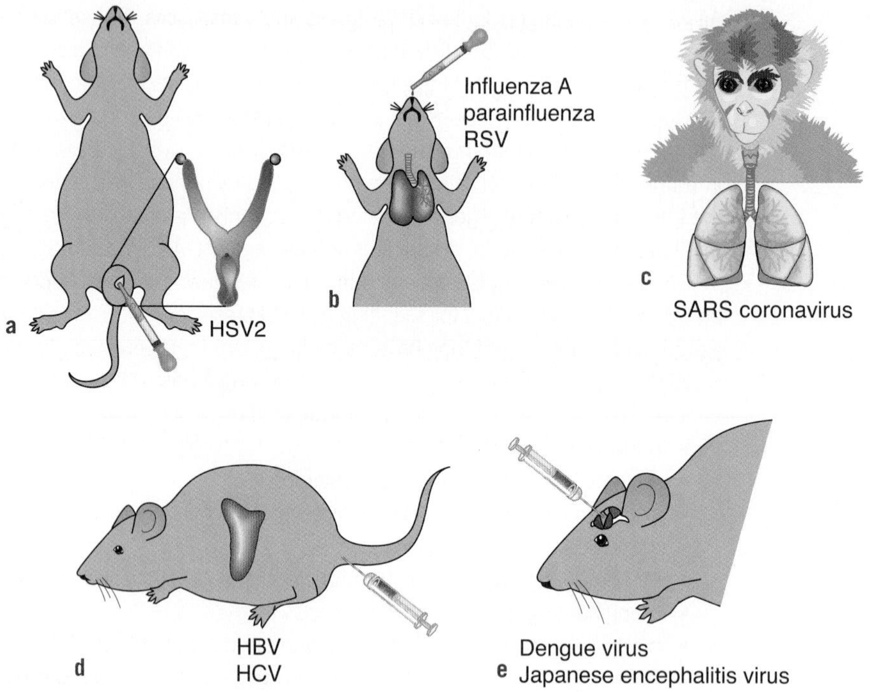

FIGURE VF 20-1 siRNA therapy to treat viral infections in different animal models. **(a)** Intravaginal application of siRNAs to inhibit sexual transmission of herpes simplex virus 2 in mice. **(b)** Intranasal administration of siRNAs to protect against RSV in mice. **(c)** Intranasal administration of siRNAs in rhesus monkeys to protect against SARS CoV infection. **(d)** Intravenous injection of siRNAs to inhibit hepatitis B and C viruses. **(e)** Intracerebral injection of siRNAs to protect against Dengue fever and Japanese encephalitis viruses. (a–e) Adapted from Dykxhoorn, D. M., and Lieberman, J., *PLoS Med.* 3 (2006): 1000–1004.

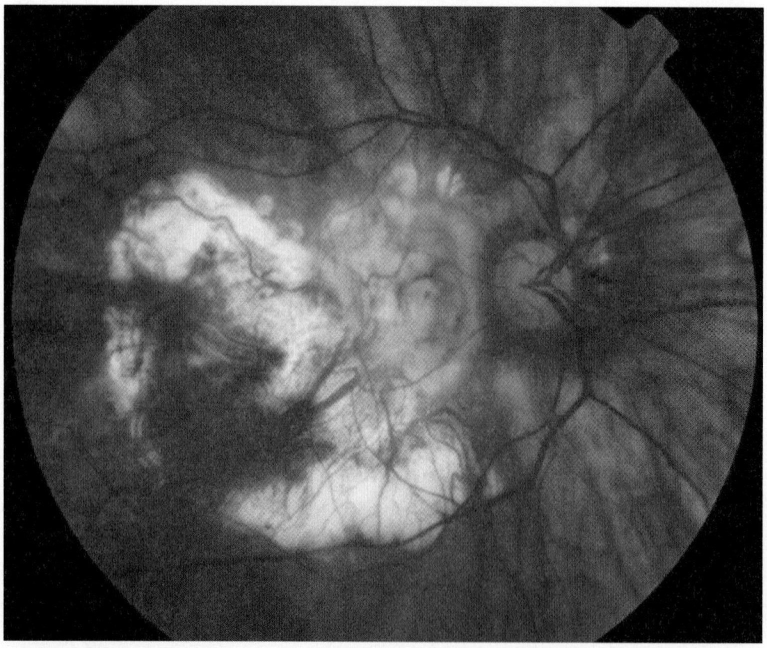

FIGURE VF 20-2 Eyeball showing the neovascularization that occurs in age-related macular degeneration.

siRNA molecules to where they are needed most remains a challenge. The siRNAs are rapidly degraded in the bloodstream and they do not home only on the diseased or infected cells. The solutions to these problems may make or break the field.

References

Adams, A. 2004. "RNAi inches toward the clinic." *Scientist* 18(6):32.

Bitko, V., et al. 2005. "Inhibition of respiratory viruses by nasally administered siRNA." *Nat Med* 11:50–55.

Check, E. 2005. "A crucial test." *Nat Med* 11(3):244–245.

Constans, A. 2006. "RNAi: Five tips to better silencing." *Scientist* 20(1):71–72.

Dykxhoorn, D. M., and Lieberman, J. 2006. "Silencing viral infection." *PLOS Medicine* 3(7):1000–1004.

Grimm, D., et al. 2006. "Fatality in mice due to oversaturation of cellular microRNA/short hairpin RNA pathways." *Nature* 441:537–541.

Howard, K. 2003. "Unlocking the money-making potential of RNAi." *Nat Biotechnol* 21:1441–1446.

Palliser, D., et al. 2006. "An siRNA-based microbicide protects mice from lethal herpes simplex virus 2 infection." *Nature* 439:89–94.

Plasterk, R. H. A. 2002. "RNA silencing: The genome's immune system." *Science* 296:1263–265.

Shen, J., et al. 2006. "Suppression of ocular neovascularization with siRNA targeting VEGF receptor 1." *Nat Gene Ther* 13:225–234.

Sledz, C. A., et al. 2003. "Activation of the interferon system by short-interfering RNAs." *Nat Cell Biol* 5(9):834–839.

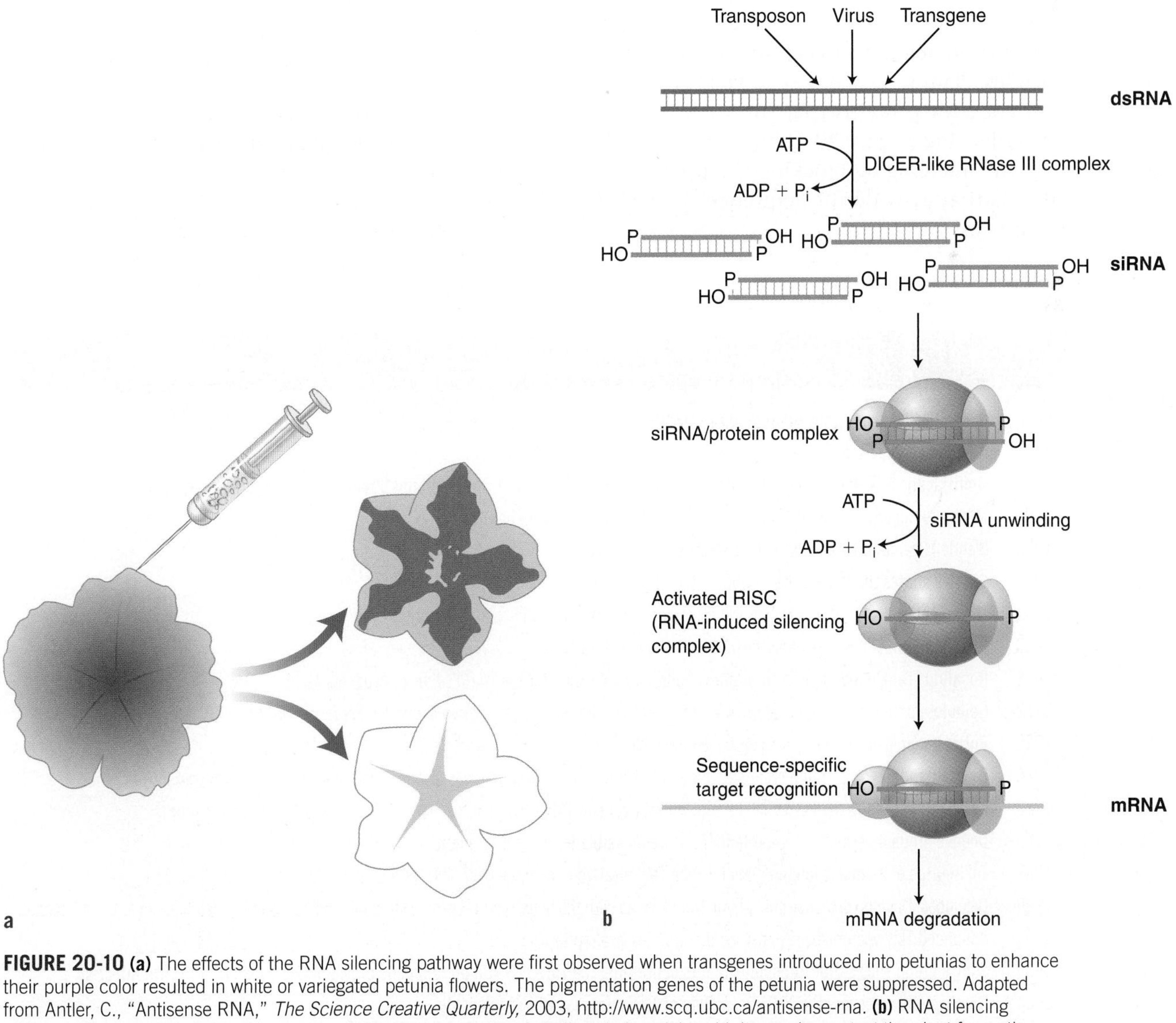

FIGURE 20-10 (a) The effects of the RNA silencing pathway were first observed when transgenes introduced into petunias to enhance their purple color resulted in white or variegated petunia flowers. The pigmentation genes of the petunia were suppressed. Adapted from Antler, C., "Antisense RNA," *The Science Creative Quarterly,* 2003, http://www.scq.ubc.ca/antisense-rna. **(b)** RNA silencing pathway of plants. This defense system protects the genome from invading viral nucleic acid. It can also protect the plant from other foreign nucleic acids such as transposons and transgenes introduced into plant cells by biotechnologists attempting to genetically modify plants. Adapted from Roth, B. M., Pruss, G. J., and Vance, V. B., *Virus Res.* 102 (2004): 97–108.

for their discovery of the mechanism of RNAi in *Clostridium elegans*.

The RNAi defense system is triggered by **dsRNA molecules**. Most plant viruses contain RNA genomes, and during the replication of RNA viruses by RNA-dependent RNA polymerase, dsRNA is a replicative intermediate during the viral infection. The viral dsRNA intermediate triggers a cascade of events. Specifically, it causes a host cell RNase III-like enzyme termed **dicer** (because it chops or "dices" the dsRNA) into 21- to 24-nucleotide dsRNA molecules called **short-interfering RNAs** (**siRNAs**). This is an adenosine triphosphate (ATP)-dependent step that likely requires additional proteins that have dsRNA binding and RNA helicase activity.

Subsequently, in another ATP-dependent step, the siRNAs are denatured and incorporated into an endonuclease silencing complex called **RNA-induced silencing complex** (**RISC**). The siRNAs act as a guide to bring the RISC into contact with complementary mRNAs, causing their degradation and silencing the expression of the protein it encodes. Plant viruses have evolved to carry genes that code for proteins that suppress or counter RNA silencing (Figure 20-10b). Many suppressors prevent silencing by blocking the production of siRNAs that provide the sequence specificity of the process.

20.11 Tobacco Mosaic Virus (TMV)

Plant viruses, as a rule, are named after the first plant on which they are found—hence, TMV infects tobacco. Most plant viruses have wide host ranges, which allows them to be used with well-characterized model plants such as tobacco (*Nicotinia spp.*) and *Arabidopsis.* A chapter on plant viruses would not be complete without discussing the first virus to be discovered, TMV. TMV is the most studied plant virus, having been examined for over 100 years. It has served as a model system for virology and molecular biology. A timeline of the history of TMV is shown in **TABLE 20-2**. Martinus Beijerinck (1851–1931), a young botanist trained in agricultural chemistry, is given credit for his TMV experiments and coining the term "virus." Like many early discoveries in the field of microbiology, Beijerinck's 1898 vision of viruses was not appreciated or confirmed during his lifetime. Another player in TMV's history is Rosalind Franklin (1920–1958), who hypothesized that the TMV particle was hollow and that its RNA genome was single-stranded. Franklin's scientific work on viruses and the structure of DNA contributed toward research resulting in two Nobel Prizes (**FIGURE 20-11**).

TABLE 20-2 Timeline of the History of Tobacco Mosaic Virus

Year	Spread of Transmissible Tobacco Virus (TMV)
1886	Adolf Mayer report describes a transmissible tobacco plant disease
1892	Dmitri Ivanoski demonstrates the infectious agent passes through a Chamberland filter candle (bacteria are trapped by such filters)
1898	Martinus Beijerinck was the first to call TMV a "contagium vivum fluidum" or a "virus"
1935	Wendell Stanley crystallizes TMV for electron microscopy
1946	Stanley shares the Nobel Prize in Chemistry for his work
1955	Heinz Fraenkel-Conrat and Robley Williams demonstrate that TMV "self-assembles" *in vitro*
1956	Rosalind Franklin *Nature* paper describes RNA in TMV particle
1958	Rosalind Franklin (working with Stanley) builds a TMV model for the World's Fair at Brussels
1972	Complete amino acid sequence of TMV's coat protein published in *American Chemical Society's Chemical Abstracts*
1982	Entire genome of TMV genome sequenced by Goelet et al.
1986	Dawson and Okada's group: create genomic cDNA library of TMV sequences and produce of infectious mRNA transcripts
1986	Powell-Abel's group demonstrate the coat protein gene of TMV confers virus-derived resistance in transgenic plants
1987	Doem's group determines 30K protein of TMV responsible for virus movement
1987	Takamatus et al. insert foreign gene into the TMV viral genome to develop TMV as a gene vector
1990	Golemboski's group demonstrates plants transformed with the nonstructural gene coding for a 54 kDa protein are resistant to virus infection
1995	Advanced TMV vectors express malarial or animal viral epitopes
1999	Song's group shows that transgenic plants containing the TMV 126 kDa protein gene are completely resistant to viral infection
2004	Ding's group discovers the 126 kDa protein of TMV suppresses RNA silencing

FIGURE 20-11 Rosalind Franklin died at the young age of 37 in 1958. She hypothesized that the TMV particle was hollow and that its RNA genome was single-stranded. Franklin's scientific work on viruses and the structure of DNA contributed toward research resulting in two Nobel Prizes.

TMV Structure

TMV particles are typically helical shaped with dimensions of 300 nm in length by 18 nm in diameter. TMV is one of the most stable viruses. It is resistant to chemical and physical agents that inactivate many other viruses. The coat protein (CP) of TMV is extraordinarily resistant to most proteases. Its stability is likely why the virus was the first one to be identified and purified to homogeneity. The exact structure of TMV has been worked out: the capsid contains 2130 identical CP subunits. Particles contain a single +ssRNA that is approximately 6400 to 6600 nucleotides in length. The protein:RNA ratio is 95%:5% (**FIGURE 20-12**). TMV virions can self-assemble. In other words, purified TMV RNA and coat protein will **self-assemble** into stable rodlike helical virus particles. This self-assembly process takes little free energy to occur.

Host Range and Transmission

TMV has a broad host range; over 550 species of flowering plants can be infected by TMV. Natural hosts that are important commercially include several tobacco species, peppers, tomatoes, and potatoes. There are high concentrations of TMV in cells of tobacco and solanaceous plants (potato family).

TMV is easily transmissible from crop to crop primarily by mechanical transmission or in the roots/soil from infected plants but is not known to spread by insect vector transmission. TMV is so stable that its infectivity can be maintained for about two years in soil that is not exposed to freezing and drying. The virions are stable in plant residues in the soil and in the composts of tomato plants. Tobacco mosaic is one of the most persistent diseases of tomatoes because the virus can remain infectious without a host for many years and is able to withstand high heat. The hands and clothing of smokers who come in contact with TMV-infected tobacco, particularly when rolling cigarettes, easily transmit TMV to susceptible plants while working in the fields. Cigarettes and other tobacco products such as chewing tobacco or snuff can be important sources of TMV and contribute to its spread, which is why smoking should not be permitted in greenhouses, especially propagation houses.

Symptoms and Diagnosis

Compared to other plant viruses, the symptoms of TMV-infected plants are not that distinct. They vary depending upon virus strains, plant age, and growing conditions. A common characteristic is yellow or light green coloration between the veins of young leaves followed by a "mosaic" or mottled pattern of light and dark green areas in the leaves (**FIGURE 20-13a**). Lower leaves are subjected to "mosaic burn," especially during hot and dry weather (Figure 20-13b). The virus almost never kills plants, but infection does lower the quality and quantity of the crop.

TMV-infected plants can easily be confused with plants affected by herbicides, mineral deficiencies, and other plant diseases. A plant pathologist and the use of electron microscopy are required for a positive identification of TMV in infected plants.

TMV Life Cycle

TMV enters the cell by mechanical transmission. Upon entry, after a few CP subunits are removed, the 5′ end of the genome appears and is exposed to the cytoplasm. This initiates uncoating. The Ca^{2+} concentrations in the cell may alter the CP subunits, causing a conformational change that results in the uncoating of the viral genome. TMV replicates in the cytoplasm of the host plant cell. The genome +ssRNA genome of TMV contains a 5′ terminal cap (m7GpppG) followed by an AU-rich

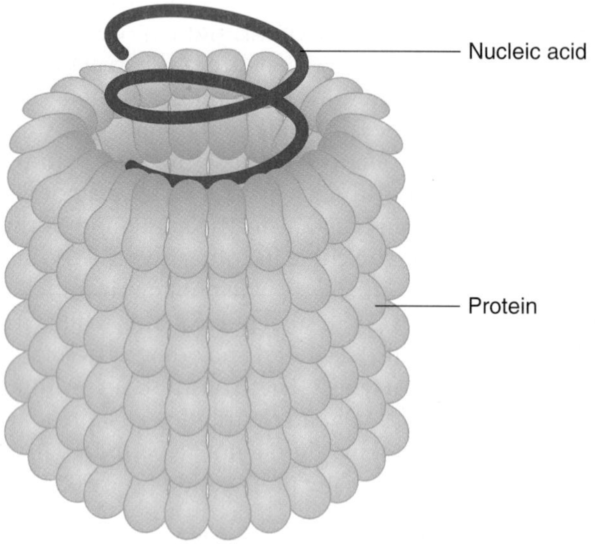

a

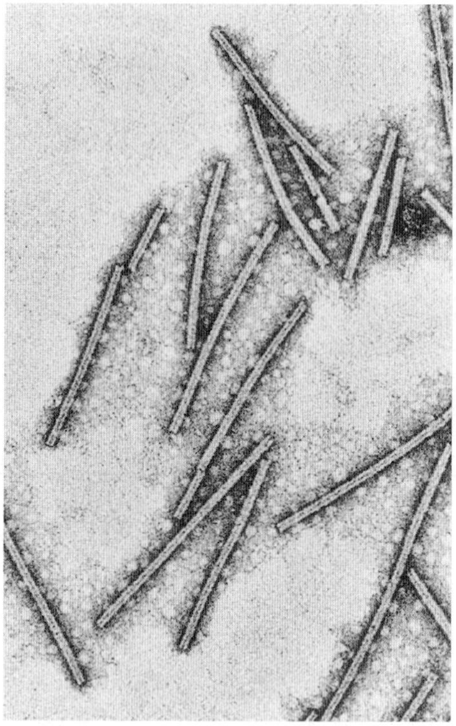

b

FIGURE 20-12 (a) Drawing of the structure of TMV. Adapted from University of South Carolina, School of Medicine. "Virology: Basic virology." Microbiology and Immunology. University of South Carolina, 2008. http://pathmicro.med.sc.edu/mhunt/intro-vir.htm. **(b)** Electron micrograph of TMV.

a

b

FIGURE 20-13 (a) Early symptoms of TMV-infected tobacco plants. **(b)** Later symptoms of TMV-infected tobacco plants.

UTR of about 70 nucleotides. The 3′ end of genome contains a UTR that can fold into a transfer-RNA–like (tRNA–like) structure that can accept histidine *in vitro*. The tRNA-like structure is important for –ssRNA synthesis during genomic replication. The TMV genome contains four open reading frames (ORFs; **FIGURE 20-14**). The first ORF codes for an **RNA-dependent RNA polymerase (RdRp)** that has helicase and methyltransferase activity. The RdRp is made up of two proteins of 126 kd or 183 kd, respectively. The 183 kd protein is produced as a readthrough of the UAG during translation. About 5% to 10% of the time, the 126 kd protein is produced because the ribosome stops translation at the UAG.

The remaining two ORFs code for a 30 kd movement protein (MP) and a 17.5 kd CP. Viral RNA is infectious and acts as a mRNA that is translated by the host machinery. After the viral RdRp is synthesized, it begins to replicate full-length –ssRNAs or three additional shorter –ssRNAs, termed **subgenomic RNAs**, which function as monocistronic RNAs. The smallest subgenomic RNA is the most abundant and translated into the viral CP. The middle-sized subgenomic RNA is translated at lower levels and expresses MP that is responsible for viral cell-to-cell spread. The largest subgenomic RNA codes for a 54 kd protein that has not been detected in infected cells or protoplasts. Plants that contain the transgene of the 54 kd protein, however, are resistant to TMV infection even though there were not any detectable levels of the 54 kd protein expressed in the transgenic plants. As a result of this, the 54 kd protein remains enigmatic.

It was recently proposed that TMV infection is spread quickly from cell to cell via intact **viral replication complexes (VRCs)**. The VRCs increase in size on the endoplasmic reticulum and in association with the protein cytoskeleton. The VRCs move along microtubules of filamentous actin and myosin and through the plasmodesmata as large bodies (**FIGURE 20-15**). Movement through the plasmodesmata is facilitated by the action of the viral MPs or VRCs or by the β 1,3-glucanase of the host cell. The VRCs spread to neighboring cells as VMCs (VRCs that move to adjacent cells are referred to as VMCs).

Control

It is nearly impossible to prevent TMV infection in nature because of the stability of the virus. TMV occurs wherever tobacco is grown (**FIGURE 20-16**). On average, TMV reduces crop yield by 30% to 35%, reducing the market value. As mentioned earlier in the chapter, the virus remains stable in dead plant matter in the soil and on contaminated seeds. In addition, TMV can exist for years in cigars and cigarettes made from TMV-infected leaves. The following control measures are recommended to prevent the spread of TMV when maintaining and transplanting seedlings in the greenhouse or field:

- Use only uncontaminated soils for seedling production in flowerbeds and greenhouses.
- Prohibit smoking during work.
- Ask workers to wash hands regularly.
- Remove and destroy all infected or suspected plants from nurseries.

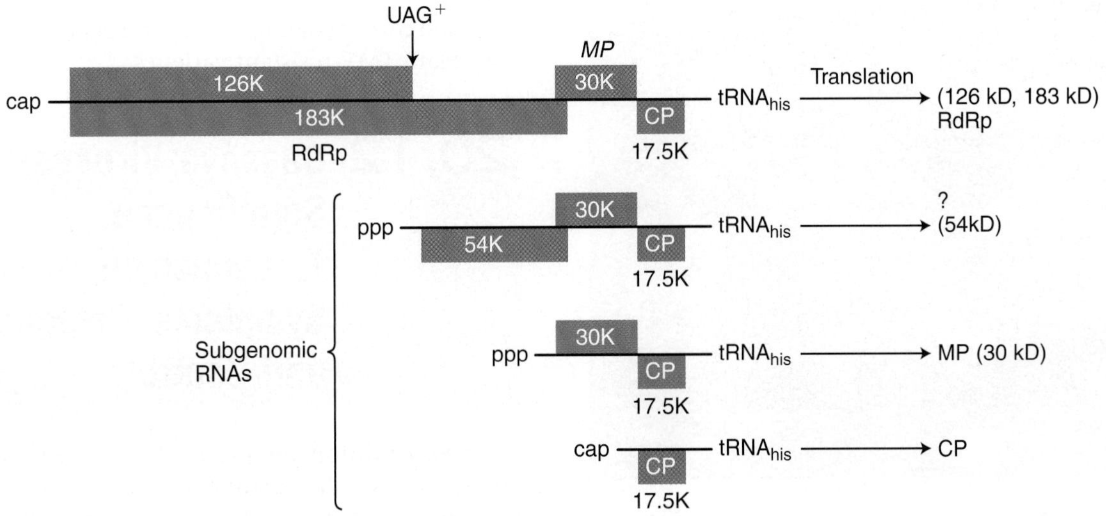

FIGURE 20-14 TMV viral genome and subgenomic RNAs and their translated products. Adapted from Knipe, D. M., and Howley, P. M., eds. *Fundamental Virology*, Fourth Edition. Lippincott Williams & Wilkins, 2001.

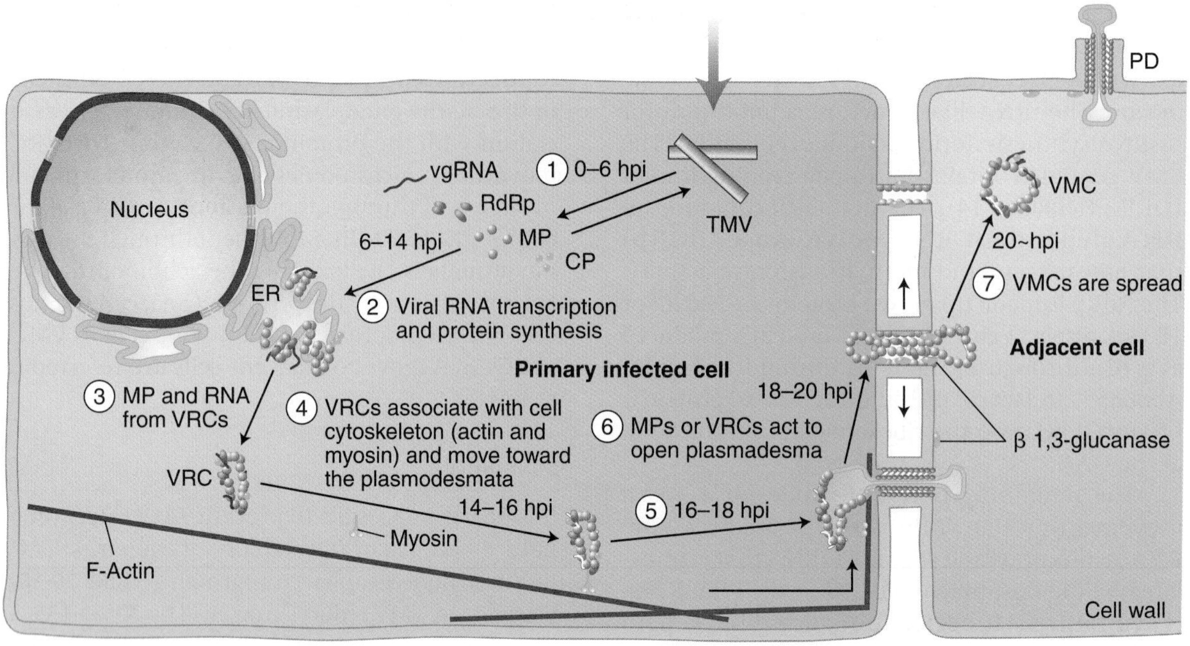

FIGURE 20-15 The life cycle and rapid cell-to-cell spread of TMV. Adapted from Kawakami, S., Watanabe, Y., and Beachy, R. N., *Proc. Natl. Acad. Sci. USA* 101 (2004): 6291–6296.

FIGURE 20-16 A farmer checks his tobacco plants on his farm in Virginia.

- When transplanting in fields, transplant seedlings into soils free of tobacco, tomato, and pepper plant residues.
- Do not plant tomato or pepper seedlings in the same flower beds.
- Spray plants with skim milk or buttermilk to reduce mechanical spread of TMV (milk changes the pH of the surface of the leaves so that they are less susceptible to mildew and viral infection).
- Destroy all weed hosts of TMV in tobacco and nearby fields.
- Rotate crops every three years with maize and corn (avoiding tomato and pepper crops).
- Plant TMV-resistant varieties.

20.12 Cassava Viruses: Significance, Transmission, Symptoms, Detection, and Control

Over 800 million people in Africa, Asia, and South America eat cassava. Cassava is the third most important source of carbohydrates for meals, after banana and sweet potato, in the tropics

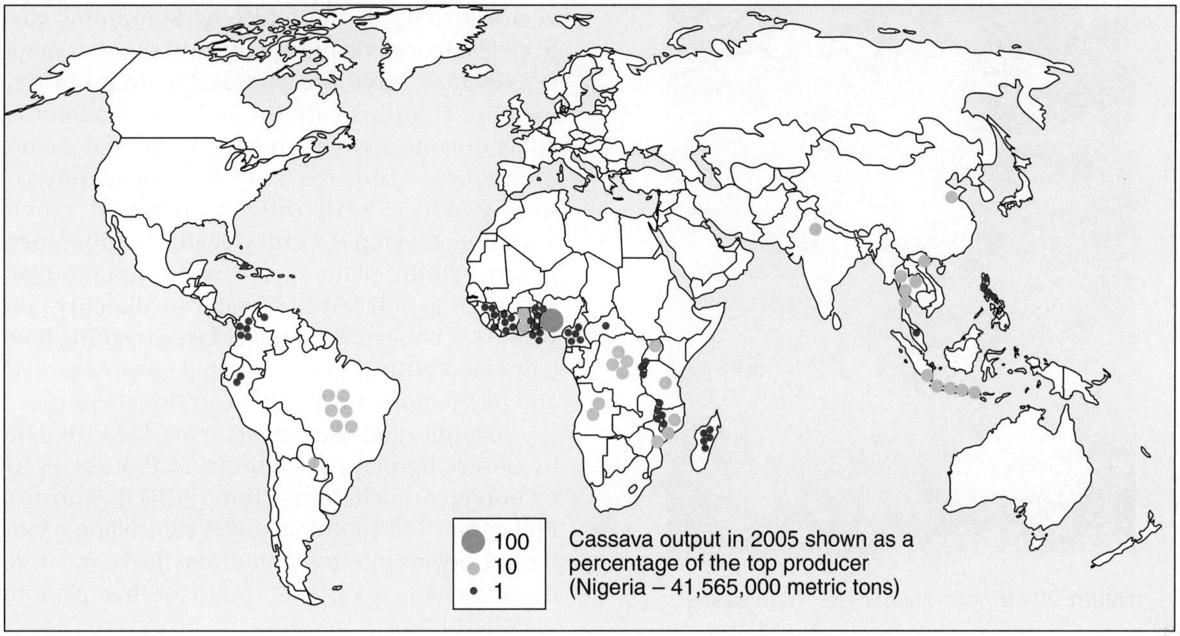

FIGURE 20-17 Locations where cassava is grown in the world in 2005.

(**FIGURE 20-17**). It is a root crop (tubers are harvested) that is valued as a starchy staple food and is sometimes used as a medicine. Archeologists have found evidence to suggest that ancient Mayans in South America grew and consumed cassava for its high-energy value. Cassava can grow in drought conditions and poor soils. The roots can stay in the ground for as long as three years before the tubers are harvested. It is believed that Portuguese traders introduced cassava into Africa in the 16th century. There are two major viral diseases of cassava: **cassava mosaic disease** and **cassava brown streak disease**. Cassava brown streak disease is emerging as a major threat to a crop that has been infected and damaged by cassava mosaic disease virus for about 100 years. *These two viruses have been labeled Africa's biggest threat to food security.* Loss of cassava can lead to hunger, which can lead to migration and conflict.

Cassava mosaic disease occurs in all cassava growing areas of sub-Saharan Africa (**FIGURE 20-18**) and the Indian sub-continent. It was first reported

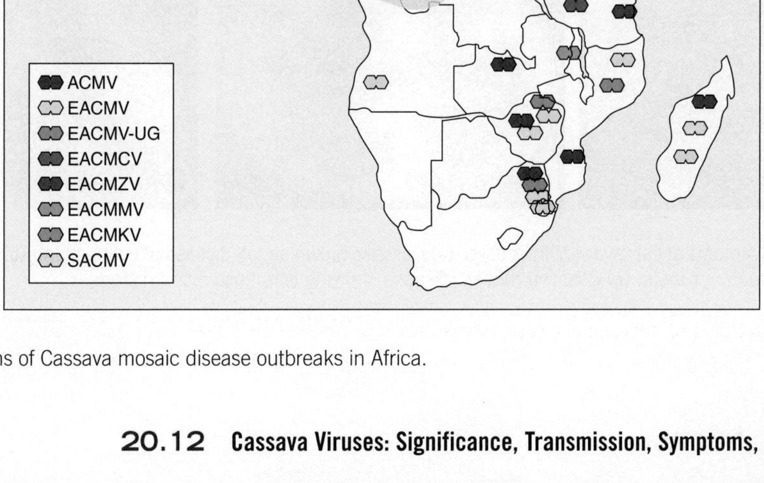

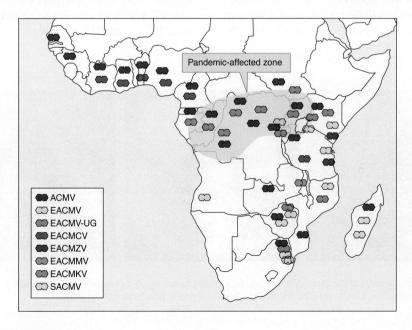

FIGURE 20-18 Locations of Cassava mosaic disease outbreaks in Africa.

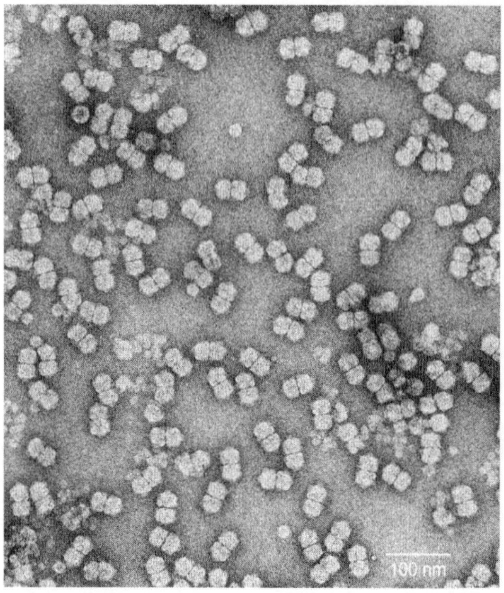

FIGURE 20-19 Cassava mosaic geminivirus particles.

in 1894 in present day Tanzania and reached all cassava growing areas in Africa by the 1930s. It affects a large proportion of all cassava and castor bean crops, and was responsible for annual financial losses of more than $1200 million U.S. dollars in damage estimates by the middle 1990s and 35 million tons of yield losses worth an estimated $1 billion USD in 2010. Yield losses from 20% to 95% can result.

Cassava mosaic disease is caused by a group of cassava mosaic geminiviruses in the *Geminiviridae* family, Begomovirus (**FIGURE 20-19**). The viruses are transmitted either through the use of cuttings from infected plants for planting material or by the vector activity of the whitefly. Symptoms vary from leaf to leaf, plant to plant, and shoot to shoot. Leaves may have the characteristic mosaic or feathery chlorosis patterns and deformation that affect discrete areas at an early stage of development. Leaf chlorosis may be pale yellow or nearly white with only a shade of green. Variation in symptoms may be due to differences in virus strain, plant age, or environmental factors such as soil fertility, water availability, and weather changes. Seasons of fast spreading have coincided with periods of rapid cassava growth and population of whiteflies carrying the virus.

Identification of cassava mosaic disease is done by observing plant symptoms, ELISA assays, or by polymerase chain reaction (PCR) diagnostics. Prevention and control involves controlling whiteflies, removing infected plants from fields, and using cassava-resistant varieties or disease-free planting materials.

Cassava brown streak disease was first reported and distinguished from Cassava mosaic disease in Tanzania during the 1930s (**FIGURE 20-20**). Cassava brown streak disease virus is a member of the genus Ipomovirus, belonging to the family *Potyviridae*. Cassava mosaic virus stunts the plant's growth but does not affect the tubers whereas the infection caused by *cassava brown streak disease virus is insidious, causing no or mild leaf distortion, making it hard to notice. Instead, the plants look healthy but the tubers of the plant become yellow/brown with a corky necrosis occurring in the starch-bearing tissues. It can affect the entire root, making it unfit for human consumption* (Figure 20-19). Cassava brown streak disease virus was shown to be transmitted by graft-inoculation from cassava to cassava, by whiteflies, and also mechanical transmission through plant sap.

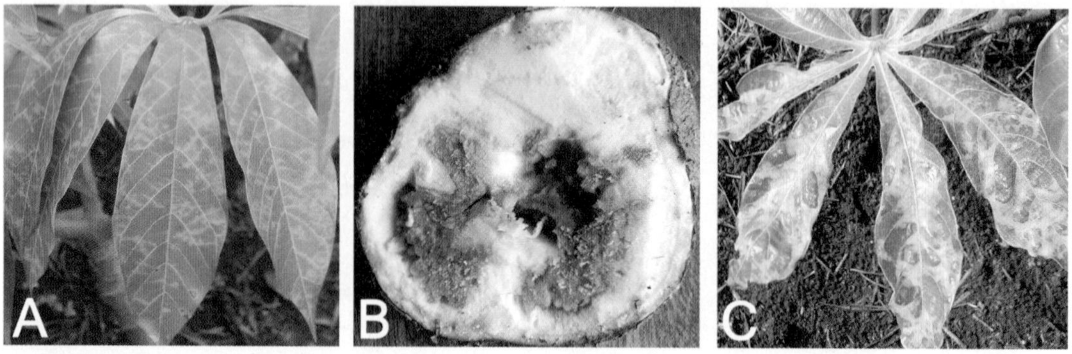

FIGURE 20-20 Symptoms of cassava viral infections. **(a)** Cassava brown streak disease affecting plant. **(b)** Cassava brown streak disease affecting tubers. **(c)** Cassava mosaic disease showing leaf mosaic and distortion.

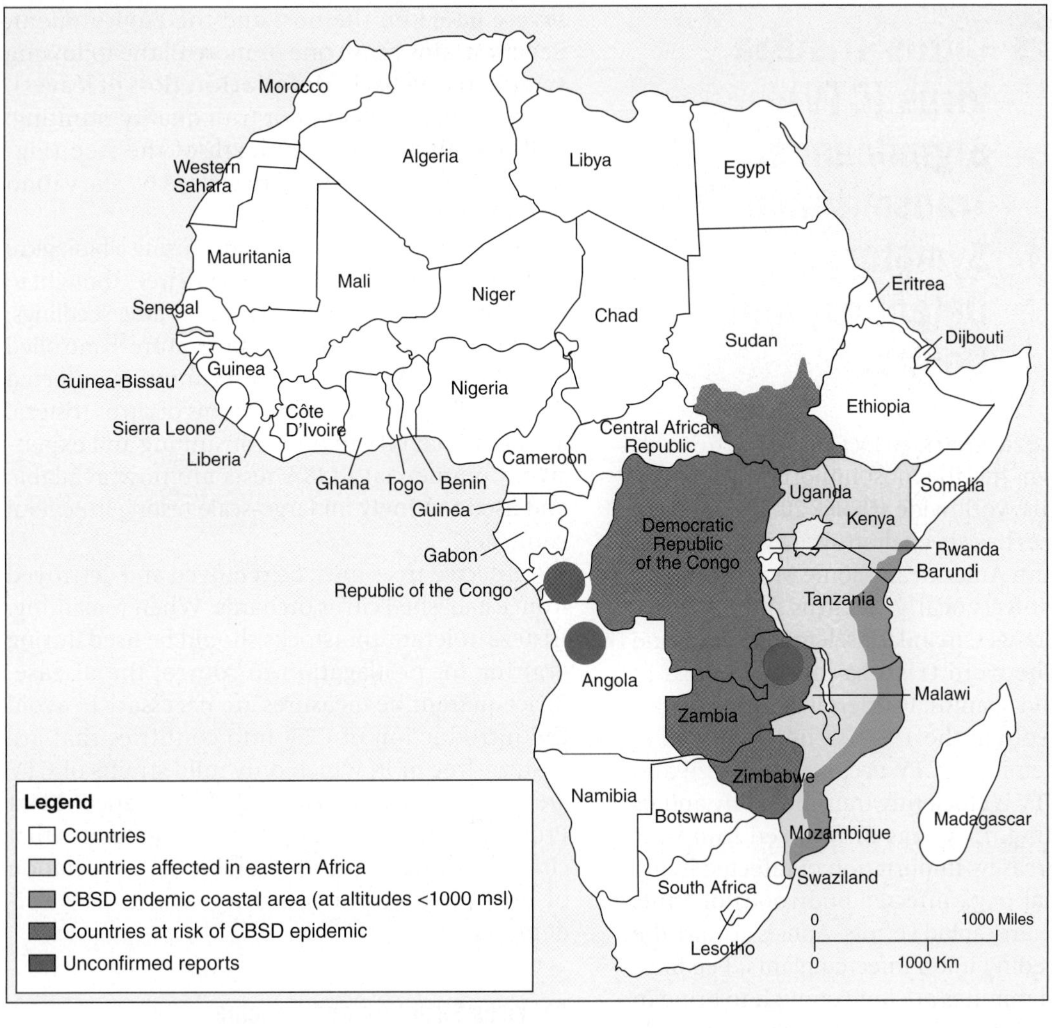

FIGURE 20-21 Cassava brown streak disease-endemic countries.

It was confined mainly on the coast lowlands of Tanzania and Mozambique, but in 2004 it appeared in the highland regions of Uganda, spreading to Lake Victoria Basin, Kenya, and Malawi of East Africa (**FIGURE 20-21**). Crop yields have dropped by up to 100%.

Reliance on symptoms is not enough to detect the cassava brown streak virus in the field because many of the young leaves remain symptomless and the older leaves may be masked by symptoms of other diseases, damage by mites, and senescence. Reverse transcriptase polymerase chain reaction (RT-PCR) tests are being developed to distinguish between the cassava brown streak virus and Ugandan cassava brown streak viruses. Early harvesting can help reduce the impact of the disease. The International Institute of Tropical Agriculture based in Nigeria is developing disease-tolerant varieties in which the leaves become diseased but the roots stay healthy. In 2008, the Bill and Melinda Gates Foundation began funding research to develop disease-resistant varieties of cassava. The Monsanto Fund and the United States Agency for International Development are also funding research to fight the disease.

Virus Structure and Genome

Cassava mosaic virus is a geminivirus that has a nonenveloped twinned (also called *geminate*) particle morphology of icosahedral symmetry. The particles are 20–30 nm in size, containing a bipartite, single-stranded, circular DNA genome termed DNA-1 and DNA-2. The total genome is approximately 5503 nucleotides in length. DNA-1 is 2779 nucleotides long. The DNA by itself is infectious.

Cassava brown streak virus is a monopartite virus. Particles are nonenveloped, filamentous, and usually flexuous with a length of approximately 750 nm (**FIGURE 20-22**). It is a positive sense single-stranded RNA virus with genome size of approximately 9069 nucleotides. It encodes a large polyprotein that autoproteolytically cleaves into 10 mature proteins. There is exceptional variability in the protein-encoding sequences of the 5′ end of the genome.

20.13 Citrus Tristeza Virus (CTV): Significance, Transmission, Symptoms, Detection, and Control

Citrus tristeza virus (CTV) occurs wherever citrus is grown. It is the most important pathogen affecting citrus worldwide (**FIGURE 20-23a**). Tens of millions of trees have been destroyed by the disease, mainly in South America and some Mediterranean countries. It infects nearly all citrus species, especially sweet oranges, mandarins, lemons, limes, and grapefruit. The word **tristeza** means sadness in Portuguese and Spanish. Farmers in South America coined this word in the 1930s, a time when devastating epidemics of CTV occurred in Brazil and Argentina. CTV is efficiently transmitted by aphids and by grafting. CTV was introduced into new geographic areas by importation of infected living citrus material (e.g., infected budwood) or fruits that may also carry aphid vectors. Aphids acquire the virus after feeding upon infected plants. Feedings as brief as ten minutes are long enough to bring in the virus from infected plants. Aphids are capable of transmitting the virus for 24 hours after leaving the plant.

Depending on the viral strain, citrus plants will exhibit a variety of symptoms from mild to severe based on the host and the environment. Severe strains cause one or more of the following symptoms: dieback, **defoliation** (loss of leaves), stem pitting, small or poor fruit quality, stunting, and overall decline and death of the tree (Figure 20-23b). Symptoms are caused by starvation of the roots.

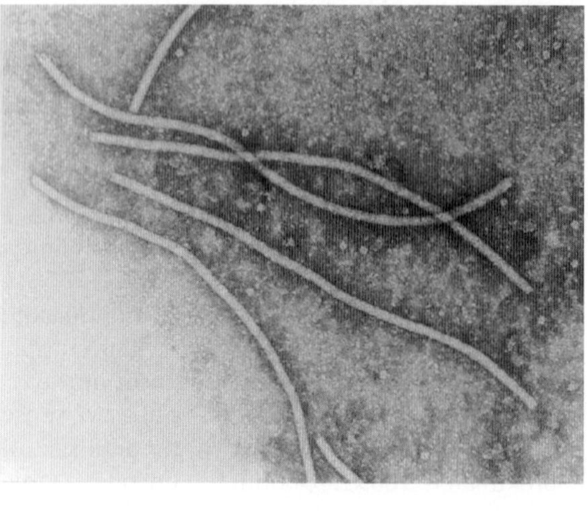

FIGURE 20-22 Cassava brown streak potyvirus particles.

CTV is detected and identified using a biological assay in which buds or patches from trees thought to be infected are grafted onto healthy lime seedlings, which then are grown in a temperature-controlled greenhouse. Within two to six months, infected lime seedlings will show symptoms of citrus tristeza, a process that is both time-consuming and expensive. Commercial ELISA tests are now available and used routinely for large-scale testing in several countries.

Infected trees must be removed and destroyed from established citrus orchards. When replanting, tristeza-tolerant rootstocks should be used during grafting for propagation to control the disease. Strict quarantine measures are necessary to avoid the introduction of CTV into countries that are tristeza-free or in which only mild strains of CTV are present. The European and Mediterranean Plant Protection Organization (EPPO) recommends that citrus-growing countries prohibit the importation of host plants from countries where CTV epidemics occur.

Virus Structure and Genome

CTV is especially difficult for researchers to work with because the work must be done with citrus trees. Growing citrus seedlings is time-consuming; therefore, the molecular biology of CTV has progressed slowly. CTV has been observed to be the largest known plant virus via electron microscopy (**FIGURE 20-24**). Particles are nonenveloped, long, flexuous or threadlike helical rods that are 2000 nm by 11 nm in diameter. Virions contain +ssRNA that is approximately 19,300 nucleotides in length. The entire genomes of several strains have been sequenced to date. The first CTV isolate sequence from Florida encodes 12 ORFs that potentially code for 17 protein products, including two papain-like proteases, a methyltransferase, and a helicase. Research to isolate, identify, and characterize CTV isolates with respect to pathogenicity, molecular structure, and genome organization is in progress. There is a need to adapt and develop new biological, serological, and molecular-based methods to detect CTV strains in the field.

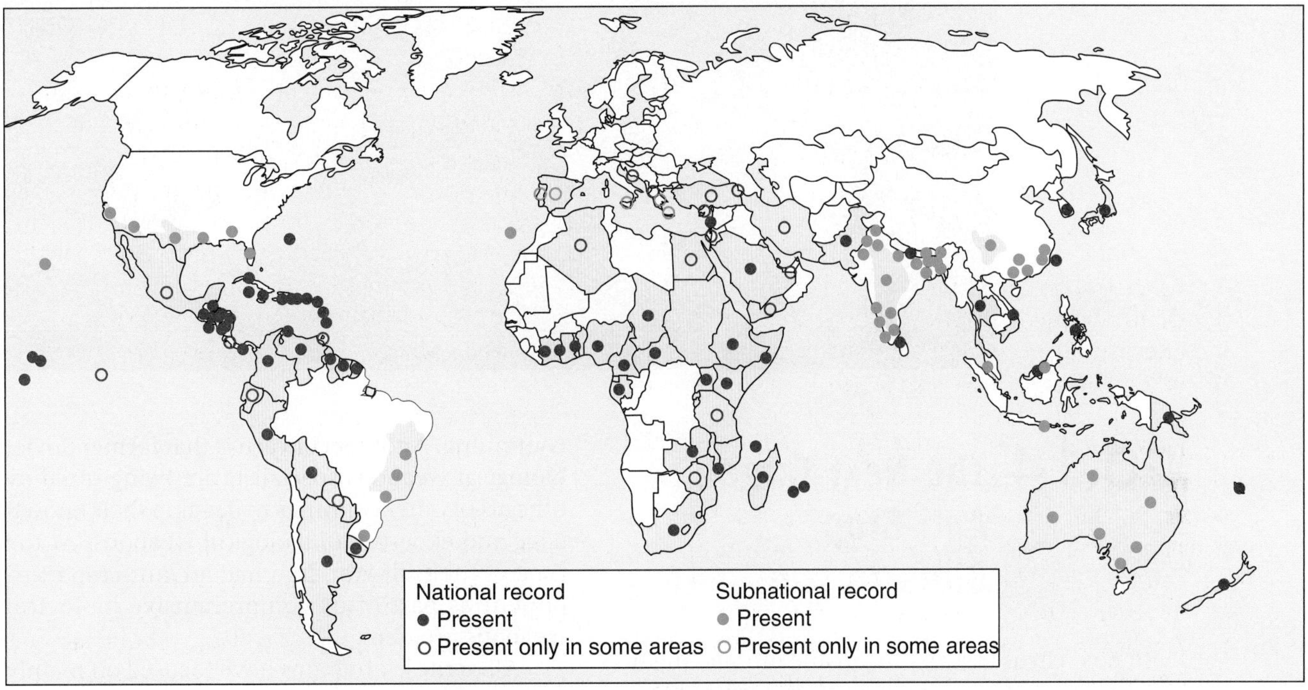

a

b

FIGURE 20-23 (a) Global distribution of CTV. Adapted from European and Mediterranean Plant Protection Organization, "Citrus tristeza closterovirus," *Distribution Maps of Quarantine Pests for Europe*. **(b)** The navel, sweet orange tree was grafted with a CTV-infected sour orange rootstock. Note the quick decline in which the limbs die back and the tree gradually dies (August 1999, near Lindsay, California).

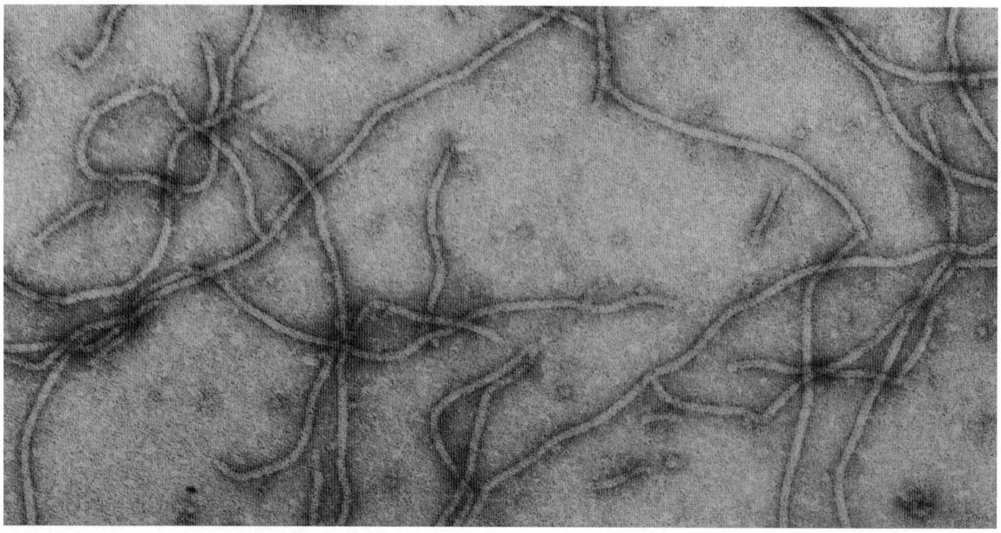

FIGURE 20-24 Electron micrograph of filamentous-shaped CTV isolated from an infected plant. The particles are approximately 2000 nm in length.

20.14 The Next Target: Anti-Crop Bioterrorism

Plum pox, citrus tristeza, sugarcane mosaic, sugar beet yellow, and cacao swollen shoot viruses have been responsible for heavy crop losses, famine, and massive economic losses. These disease epidemics resulted from human activity. It is not surprising that crops would be good targets for biological agents. Most crops are planted using varieties that are genetically identical or almost identical. If one plant is susceptible to disease, all plants in the same field or grove will be susceptible to disease. Developing countries with millions of people dependent upon such staples as rice or wheat could experience famine, starvation, and severe malnutrition if these crops were severely damaged by a deliberate anti-crop attack.

Anti-crop warfare research and development programs existed in the United Kingdom, France, Germany, and Japan before World War II. From 1946 to 1973, the United States maintained an active biological weapons program that included anti-crop agents. The former Soviet Union maintained an active anti-crop program until its dissolution in 1991. Most of the biological weapons facilities in Russia still function as civilian research facilities. However, foreigners are not allowed to visit some of the facilities occupied by the Russian Ministry of Defense (MOD) where some of the aforementioned activities may have taken place. There is a lack of transparency that causes some officials to question whether all of the stockpiles of biological weapons have been destroyed or whether the development of biological agents is continuing. Another concern is that former Soviet biological weapons specialists are being hired by officials of other countries to develop their knowledge and expertise of biological weapons. At the time of the Gulf War, Iraq had an anti-crop component as part of its comprehensive biological weapons program.

All of these programs have focused on mainly fungal and bacterial pathogens. Only a couple of plant viruses are listed as important anti-crop pathogens. Even though extensive research has been done regarding current anti-crop agents, no one seems to use them. Many scientists and governments view anti-crop bioterrorism as an inevitable event for which countries must be prepared.

Summary

Martinus Beijerinck, a botanist, is credited with coining the term *virus* while studying TMV-infected plants during the late 1800s. Even though the first virus to be discovered was tobacco mosaic virus (TMV), plant viruses are not nearly as well understood as their animal counterparts. Plant viruses continue to be a major threat in the production of vegetable and ornamental crops worldwide. Their control remains a major challenge in the 21st century.

The infection of plant cells is not achieved by surface receptors—a major difference between viral infections of plants as opposed to animals. Plant viruses depend upon a break in the cell wall for entry and are transmitted in a number of different ways, including:

- mechanical means (human and environmental damage)
- soil transmission
- vegetative propagation (grafting)

- piercing and chewing insects and other vectors
- seed transmission
- pollen transmission

The majority of plant viruses are transmitted by insect vectors (e.g., aphids, leafhoppers, planthoppers, beetles, mites, mealybugs, whiteflies, and thrips). Symptoms of viral plant diseases vary from mild to severe but rarely kill the plant. Examples of symptoms are:

- dwarfing or stunting of plants
- leaf curling
- reduced yield
- fruit distortion
- chlorosis (yellowing)
- other color deviations (e.g., variegation of flower petals)
- "mosaic" patterns on leaves
- ring-shaped spots on leaves
- wilting and withering
- necrosis (black or grayish brown discoloration due to the death of cells and tissues)
- bark scaling

In addition to symptomatic observations, methods used to detect and identify plant viruses are similar to those used in detecting animal viruses. Infectivity assays, ELISAs, electron microscopy, and PCR are used to diagnose diseases caused by plant viruses. The major means of control (depending on the disease) include:

- control of insect vectors (chemical or biological control)
- removal of alternate hosts (e.g., weeds)
- sanitation techniques (e.g., cleaning of farm implements and proper grafting techniques)
- use of certified virus-free seed or stock
- growing resistant crop varieties
- plant quarantines to prevent disease establishment in areas where it does not occur (material is grown or kept isolated for longer periods for observation to determine absence of disease or pest) and testing. These procedures are time-consuming and costly, and require skilled professionals.

Advances in control of viral plant diseases are focused on creating natural (breeding) and engineered resistance to plant viruses.

The majority of plant viruses are naked, helical-shaped +ssRNA viruses. Instead of having segmented genomes like influenza virus, some plant viruses have multipartite genomes in which the genome consists of more than one molecule of nucleic acid, but each genome molecule is packed into a separate virus particle. Thus, all discrete virus particles must enter the cell to establish a productive infection.

Viruses spread short distances (from cell to cell) with the aid of movement proteins that "channel" them through the plasmodesmata. This channeling is done without causing cellular lysis. A systematic (long distance) infection of the plant may occur if plant viruses can be transported longer distances through the vascular system via the phloem.

Plants do not have an "active" immune system analogous to humans, such as the production of antibodies to combat pathogens. Plants do, however, possess an "immune system of the genome" that is known as RNA interference (RNAi) or RNA silencing. This system is a sequence-specific antiviral defense mechanism. Plant cells use RNA silencing to bind and cut any nucleotide sequence that may be harmful to the plant cell or produces a product that may inhibit plant cell activities. The players involved in this pathway of defense are dsRNA, RNA-dependent RNA polymerase, a DICER protein, and siRNAs. The discovery of this new mechanism of gene regulation has revitalized interest in the development of nucleic acid-based technologies as therapeutic tools (especially in the areas of genetic disorders, cancer, and viral infection).

TMV, the cassava viruses, and citrus tristeza virus represent important viral plant pathogens. The cassava viruses have been labeled Africa's biggest threat to food security. Over 100 years of research has been conducted on TMV. TMV's stability in nature, the fact that it can easily be grown in large quantities in infected plants in a greenhouse, and the fact that it does not infect humans or animals has made it an attractive virus to investigate.

CASE STUDY QUESTIONS

These questions relate to the Case Study presented at the beginning of the chapter.

1. Speculate as to how plum pox virus is spread over short distances (within the same grove) and longer distances (e.g., to states outside of Pennsylvania and to Canada).
2. Infected trees must be destroyed to prevent the spread of disease. What other control measures do growers and agricultural officials put in place in order to prevent its spread?
3. Can plum pox virus be transmitted to humans?

References

ProMED-mail. 1999. "Plum pox virus, peaches—USA (Pennsylvania)." *ProMED-mail* 1999; 24, Oct; 19991024.1912.

ProMED-mail. 2000. "Plum pox virus, nectarines—Canada (Ontario)." *ProMED-mail* 2000; 26, June; 20000626.1057.

ProMED-mail. 2006. "Plum pox virus, plum—USA (Michigan): First report." *ProMED-mail* 2006; 14, Aug; 20060814.2277.

ProMED-mail. 2006. "Plum pox virus, plum—USA (New York): First report." *ProMED-mail* 2006; 06, Aug; 20060806.2181.

Resources

Primary Literature and Reviews

Aaziz, R., et al. 2001. "Plasmodesmata and plant cytoskeleton." *Trends Plant Sci* 6:326–330.

Ahlquist, P. 2002. "RNA-dependent RNA polymerases, viruses, and RNA silencing." *Science* 296:1270–1273.

Allicai, T., et al. 2006. "Re-emergence of cassava brown streak disease in uganda." Plant Disease 91(1):24–29.

Anonymous. 2010. "Sowing the seeds for the ideal crop." *Science* 327:802.

Anonymous. 2010. "What it takes to make that meal." *Science* 327:809.

Ash, C., et al. 2010. "Feeding the future." *Science* 327:797.

Bagasra, O., and Prilliman, K. R. 2004. "RNA interference: The molecular immune system." *J Mol Histol* 35: 545–553.

Barrett, C. B. 2010. "Measuring food insecurity." *Science* 327:825–828.

Baulcombe, D. 2004. "RNA silencing in plants." *Nature* 431:356–363.

Behlke, M. A. 2006. "Progress towards *in vivo* use of siRNAs." *Mol Ther* 13(4):644–670.

Beijerinck, M. W. 1898. "Over een contagium vivum fluidum als oorzaak van de vlekziekte der tabaksbladen." *Versl Gewone Vergad Wis Natuurkd Afd K Akad Wet Amsterdam* 7:229–235 (in Dutch).

Bos, L. 1999. "Beijerinck's work on tobacco mosaic virus: Historical context and legacy." *Phil Trans R Soc Lond B* 354:675–685.

Bustamante, P. I., and Hull, R. 1998. "Plant virus gene expression strategies." *EJB* 1(2):1–18.

Cousins, N. 2000. "The next target of bioterrorism: Your food." *Environ Health Perspect* 108(3):A126–A129.

Couzin, J. 2002. "Breakthrough of the year: Small RNAs make a big splash." *Science* 5602:2296–2297.

Deteris, A., et al. 2006. "Hierarchical action and inhibition of plant dicer-like proteins in antiviral defense." *Science* 313:68–71.

Dry, I. B., et al. 1997. "A novel subviral agent associated with a geminivirus: The first report of a DNA satellite." *Proc Natl Acad Sci USA* 94:7088–7093.

Ejeta, G. 2010. "African green revolution needn't be a mirage." *Science* 327:831–832.

Fedoroff, N. V., et al. 2010. "Radically rethinking agriculture for the 21st century." *Science* 327:833–834.

Fire, A., et al. 1998. "Potent and specific genetic interference by double-stranded RNA in *caenorhabditis elegans*." *Nature* 391:806–811.

Gebbers, R., and Adamchuck, V. I. 2010. "Precision agriculture and food security." *Science* 327:829–831.

Godfray, H. C. J., et al. 2010. "Food security: The challenge of feeding 9 billion people." *Science* 327:812–818.

Jones, J. D. G., and Dang, J. L. 2006. "The plant immune system." *Nature* 444:323–329.

Kamukondiwa, W. 1996. "Alternative food crops to adapt to potential climatic change in southern Africa." *Clim Res* 6: 153–155.

Karasev, A. V., et al. 1995. "Complete sequence of the citrus tristeza virus RNA genome." *Virology* 208:511–520.

Kawakami, S., et al. 2004. "Tobacco mosaic virus infection spreads cell to cell as intact replication complexes." *Proc Natl Acad Sci USA* 101:6291–6296.

Kunkel, L. O. 1922. "Insect transmission of yellow stripe disease virus." *Hawaii Plant Rec* 26:58–64.

Lazarowitz, S. G., and Beachy, R. N. 1999. "Viral movement proteins as probes for intracellular and intercellular trafficking in plants." *Plant Cell* 11:535–548.

Lee, S. K., et al. 2005. "Lentiviral delivery of short hairpin RNAs protects CD4 T cells from multiple clades and primary isolates of HIV." *Blood* 106:818–826.

Legg, J. P. 2003. "Cassava mosaic disease in Africa: Where are the epidemics?" *African Crop Science Conference Proceedings* 6:322–328.

Legg, J. P. 2008. "African cassava mosaic disease." *Encyclopedia of Virology*, 3rd ed., 30–36.

Lucas, W. J. 2006. "Plant viral movement proteins: Agents for cell-to-cell trafficking of viral genomes." *Virology* 344(1):169–184.

Mayer, A. 1886. "Ueber die mosaikkrankheit des tabaka." (English Translation: Concerning the Mosaic Disease of Tobacco.) *Phytopathol Classics* 7:11–24.

Mbanzibwa, D. R., et al. 2009. "Cassava brown streak virus (*Polyviridae*) encodes a putative Maf/HAM1 pyrophosphatase implicated in reduction of mutations and a P1 proteinase that suppresses RNA silencing but contains no HC-Pro." *J Virol* 83(13):6934–6940.

Mbanzibwa, D. R., et al. 2011. "Simultaneous virus-specific detection of the two cassava brown streak-associated viruses by RT-PCR reveals wide distribution in East Africa, mixed infections, and infections in *Manihot glaziovii.*" *J Virol Methods* 171:394–400.

Mello, C. C., and Conte, D., Jr. 2004. "Revealing the world of RNA interference." *Nature* 431:338–342.

Nelson, R. S., and Citovsky, V. 2005. "Plant viruses: Invaders of cells and pirates of cellular pathways." *Plant Physiol* 138:1809–1814.

Night, G., et al. 2011. "Occurance and distribution of cassava pests and diseases in Rwanda." *Agriculture, Ecosystems and Environment* 140:492–497.

Njock, T. E., and Ndip, R. N. 2007. "Limitation in detecting African cassava mosaic geminivirus in the lignified tissues of cassava stems." *Afr J Biotechnol* 6(20):2340–2347.

Ogwok, E., et al. 2010. "Transmission studies with cassava brown streak uganda virus (Potyviridae: Ipomovirus) and its interaction with abiotic and biotic factors in *Nicotiana benthamiana.*" *J Virol Methods* 169:296–304.

Okada, Y. 1999. "Historical overview of research on the tobacco virus genome: Genome organization, infectivity, and gene manipulations." *Phil Trans Soc Lond B* 354: 569–582.

Pennisi, E. 2010. "Armed and dangerous." *Science* 327: 804–805.

Plasterk, R. H. A. 2002. "RNA silencing: The genome's immune system." *Science* 296:1263–1265.

Power, A. G. 2000. "Insect transmission of plant viruses: A constraint on virus variability." *Curr Opin Plant Biotech* 3:336–340.

Ritzenthaler, C. 2005. "Resistance to plant viruses: Old issue, new answers?" *Curr Opin Biotechnol* 16:118–122.

Rocheleau, C. E., et al. 1997. "Want signaling and an APC-related gene specify endoderm in early *c. elegans* embryos." *Cell* 90:707–716.

Roossinck, M., et al. 1992. "Satellite RNAs of plant viruses: Structures and biological effects." *Microbiol Rev* 56(2): 265–279.

Roth, B. M., et al. 2004. "Plant viral suppressors of RNA silencing." *Virus Res* 102:97–108.

Ruiz-Ruiz, S., et al. 2006. "The complete nucleotide sequence of a severe stem pitting isolate of citrus tristeza virus from spain: Comparison with isolates from different origins." *Arch Virol* 151(2):387–398.

Silhavy, D., and Burgyan, J. 2002. "Effects and side-effects of viral RNA silencing suppressors on short RNAs." *Trends Plant Sci* 296:1263–1265.

Stone, R. 2010. "Dialing up knowledge—and harvests." *Science* 327:808.

Tester, M., and Langridge, P. 2010. "Breeding technologies to increase crop production in a changing world." *Science* 327:818–822.

Thivierge, K., et al. 2005. "Update on what makes virus RNAs different from host mRNAs: Plant virus RNAs. Coordinated recruitment of conserved host functions by (+) ssRNA viruses during early infection events." *Plant Physiol* 138:1822–1827.

Thresh, J. M., et al. 1994. "The viruses and virus diseases of cassava in Africa." *Afr Crop Sci J* 2(4):459–478.

Tompkins, S. M., et al. 2004. "Protection against lethal influenza virus challenge by RNA interference *in vivo.*" *Proc Natl Acad Sci USA* 101:8682–8686.

Uprichard, S. L. 2005. "The therapeutic potential of RNA interference." *FEBS Lett* 579:5996–6007.

van der Want, J. P., and Dijkstra, J. 2006. "A history of plant virology." *Arch Virol* 151(8):1467–1498.

Vince, G. 2010. "From one farmer, hope—and reason for worry." *Science* 327:798–799.

Waterhouse, P. M., et al. 2001. "Gene silencing as an adaptive defence against viruses." *Nature* 411:834–842.

Whitby, S., and Rogers, P. 1997. "Anti-crop biological warfare—Implications of the Iraqi and US programs." *Def Anal* 13(3):303–318.

Whitby, S. M. 2002. *Biological Warfare Against Crops.* Chippenham, England: Palgrave Publishers.

Winter, S., et al. 2010. "Analysis of cassava brown streak viruses reveals the presence of distinct virus species causing cassava brown streak disease in East Africa." *J Gen Virol* 91:1365–1372.

Xie, F. Y., et al. 2006. "Harnessing *in vivo* siRNA delivery for drug discovery and therapeutic development." *Drug Discov Today* 11:67–73.

Popular Press

Creager, A. N. 2002. *The Life of a Virus: Tobacco Mosaic Virus as an Experimental Model, 1930–1965.* Chicago: University of Chicago Press.

Khan, J. A., and Dijkstra, J. Eds. 2006. *Handbook of Plant Virology.* New York: Food Products Press.

Rogers, P., Whitby, S., and Dando, M. 1999. "Biological Warfare Against Crops." *Sci Am,* June 1999.

Scholthof, K.-B. G., Shaw, J. G., and Zaitlin, M. Eds. 1999. *Tobacco Mosaic Virus: One Hundred Years of Contributions to Virology.* St. Paul, MN: APS Press.

Sutic, D. D., Ford, R. E., and Tosic, M. T. Eds. 1999. *Handbook of Plant Virus Diseases.* New York: CRC Press.

Video Productions

Biotechnology on the Farm and in the Factory: Agricultural and Industrial Applications. 2009. Films for Humanities.

The Seeds of a New Era. 2003. Films for the Humanities.

China: Food for a Billion Plus. 2002. Films for the Humanities.

eLearning

go.jblearning.com/shors2

The site features eLearning, an online review area that provides quizzes and other tools to help you study for your class. You can also follow useful links for in-depth information, or just find out the latest virology and microbiology news.

CHAPTER 21

The Best for Last: Bacteriophages

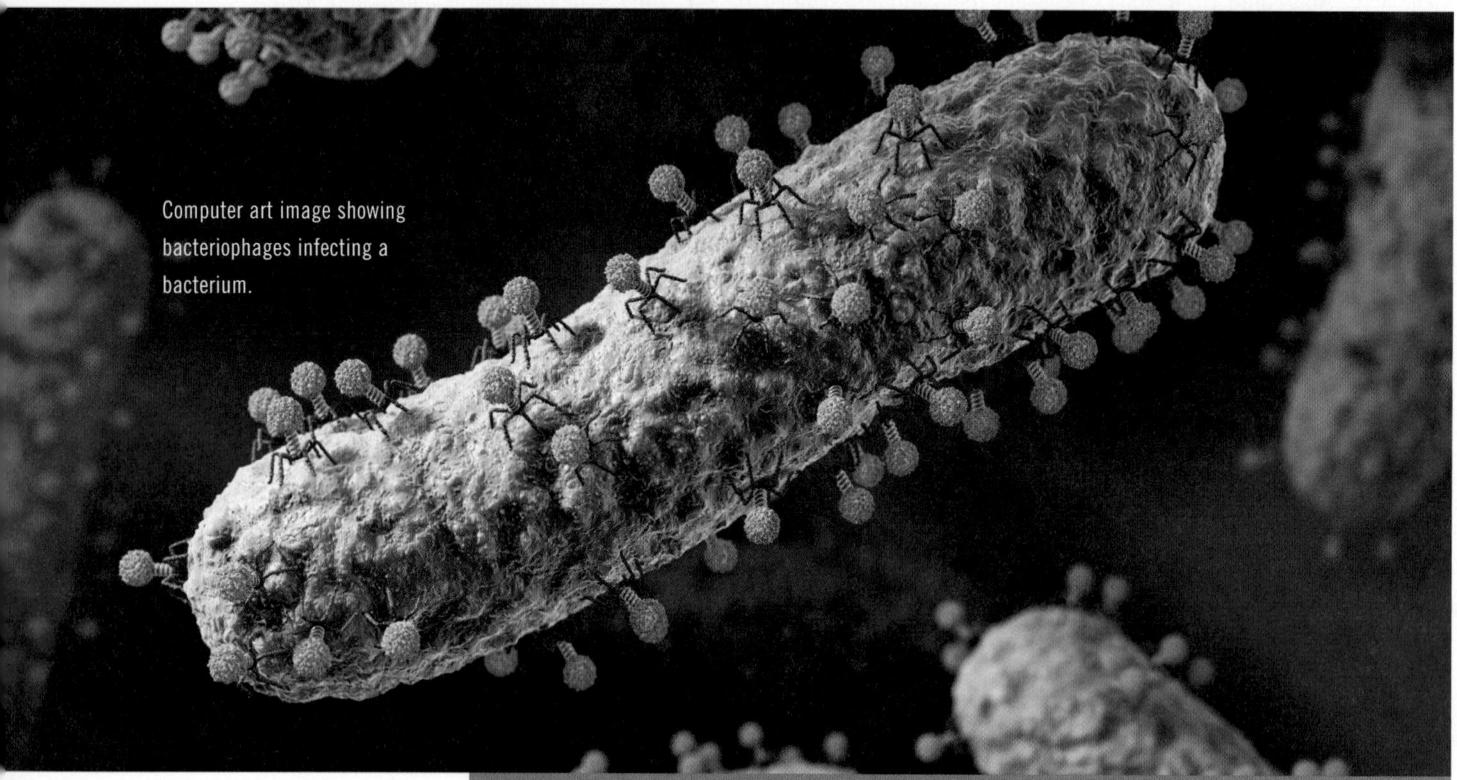

Computer art image showing bacteriophages infecting a bacterium.

OUTLINE

"Do you mean to say you think you've discovered an infectious disease of bacteria, and you haven't told me about it? My dear boy, I don't believe you quite realize that you may have hit on the supreme way to kill pathogenic bacteria . . . And you didn't tell me!"

Sinclair Lewis, *Arrowsmith*, 1924.

CASE STUDY: PHAGE THERAPY

Had it not been for the elegant experiments done by H. Williams Smith and his colleagues during the 1980s, phage therapy would not be in the headlines today. Smith reported successful veterinary applications of treating *Escherichia coli* infections in calves with bacteriophages. Use these publications to answer the questions that follow:

Smith, H. W., and Huggins, M. B. 1983. "Effectiveness of phages in treating experimental *escherichia coli* diarrhoea in calves, piglets and lambs." *J Gen Microbiol* 129:2659–2675.

Smith, H. W., Huggins, M. B., Shaw, K. M. 1987. "The control of experimental *escherichia coli* diarrhoea in calves by means of bacteriophages." *J Gen Microbiol* 133:1111–1126.

Smith, H. W., Huggins, M. B., and Shaw, K. M. 1987. "Factors influencing the survival and multiplication of bacteriophages in calves and in their environment." *J Gen Microbiol* 133:1127–1135.

1. Choose one of the preceding papers and outline the experiments performed by Smith and his colleagues.
2. How were the cattle treated for *E. coli* infections? What was the success rate?
3. Was phage therapy more successful than multiple doses of antibiotic therapy in Smith's experiments? Discuss your answer.
4. Were phage-resistant mutants of *E. coli* strains isolated from the animals used in these studies? Discuss your answer.

21.1 Bacteriophage Research History

Bacteriophages are viruses that infect bacteria. The term *bacteriophage* is a combination of the word bacteria and the Greek word *phagein,* meaning "to eat." They were discovered after the recognition of bacterial hosts in the 1880s, a time known as the golden age of microbiology. British bacteriologist **Frederick W. Twort** authored a paper published in 1915 that is considered to be the birth of modern bacteriophage research. He reported an observation termed *glassy transformation.* At the time, Twort was trying to grow vaccinia viruses on agar medium in the absence of host cells. He inoculated the solid medium with unfiltered pustule fluid used in smallpox vaccinations. Rather than vaccinia virus, though, micrococci colonies grew on the solid medium, where they appeared watery or glassy in some areas. Twort collected the watery colonies, which could not be grown on new medium, and filtered them with a **Chamberland filter**. The filtrate was added to fresh cultures of micrococci, causing them to undergo *glassy transformation.* He concluded that a filterable infectious agent killed the bacteria, and that in the process the agent multiplied itself, thus causing the glassy transformation. Twort did not state that it was a bacterial virus despite the fact that it was known that filterable agents infected plants and animals. The scientific community did not acknowledge Twort's paper for five years.

French-Canadian **Felix d'Herelle** is credited as the sole discoverer of bacteriophages (**FIGURE 21-1**). In a 1917 paper, he described the lysis of dysentery-causing bacteria grown in liquid medium. He hypothesized that "an invisible antagonist"—a bacteriophage—of the dysentery bacteria that could pass through a Chamberland filter caused

FIGURE 21-1 Felix d'Herelle (1873–1949) is credited with the discovery of bacteriophages and is considered the father of bacteriophage therapy.

the lysis. d'Herelle did not cite Twort's research in his publication. He continued research in two directions: determining the biological nature of bacteriophages and exploring the use of bacteriophages as a therapy to treat bacterial infections in a preantibiotic era.

d'Herelle believed that bacteriophages were an agent of immunity in fighting infectious diseases, a conclusion he reached by observing that the numbers of bacteriophages increased in patients recovering from bacterial dysentery and typhoid. He speculated that the lysis of pathogens by specific bacteriophages formed the mechanism of recovery. This idea supported his efforts to use bacteriophage therapy to treat a wide variety of bacterial diseases. The majority of the scientific world failed to accept the fact that bacteriophages were filterable viruses of bacteria. Instead, the leading theory regarding bacterial lysis was based on Jules Bordet's view that endogenous autolytic enzymes caused bacteria to be lysed during infection. Not until the early 1940s did d'Herelle's bacteriophage theory become universally accepted.

d'Herelle was quick to realize the therapeutic potential of bacteriophages. During the summer of 1919 in France, he attempted to use bacteriophages in field trials to control an epidemic of chicken typhoid caused by the bacterium *Salmonella gallinarum*. He inoculated chickens either orally or by injection with bacteriophages, and the overall results suggested that bacteriophage-treated flocks suffered fewer deaths and shorter epidemics that did not reoccur. He also conducted field trials in Indochina during an outbreak of highly fatal water buffalo pasteurellosis (sometimes referred to in French as *barbone* disease). Once again inoculations appeared successful. These positive results motivated d'Herelle to conduct human trials in the 1920s, a time when human trials were scientifically and ethically crude and not very safe by today's standards.

To prove that bacteriophage preparations (Shiga-bacteriophage **lysates**) were safe for human experiments, d'Herelle injected himself, his family members, and his coworkers. No harmful effects were observed. He applied this treatment to a 12-year-old boy suffering from a confirmed case of bacterial dysentery. The boy's symptoms disappeared after one dose and he fully recovered after a few days. Based on this success, three additional dysentery patients received similar treatment. All of them were cured.

The most notorious bacteriophage therapy experiments occurred when d'Herelle worked at the League of Nations quarantine station in Alexandria, Egypt. Four patients on a ship passing through the Suez Canal were suffering from laboratory-confirmed **bubonic plague**. d'Herelle injected all four individuals with antiplague bacteriophages into the bubos present in their lymph nodes and all four patients recovered rapidly. This extraordinary result was reported in 1925 in the French medical journal *Presse Medicale*. After this report, the British government invited d'Herelle to work at the Haffkine Institute in Bombay on bacteriophage therapies for plague and cholera in India. Cholera epidemics posed a major concern because they occurred regularly during religious festivals and pilgrimages in India. Reports from India during the 1920s and 1930s were positive regarding the oral administration of cholera-specific bacteriophages in patients. This treatment reduced the overall severity of symptoms and mortality from the disease.

d'Herelle traveled the world continuing bacteriophage therapy as a means of controlling cholera outbreaks. He accepted an appointment as a professor of protobiology at Yale University in 1928 and went on to play a vital role in the establishment of a bacteriophage institute in Tbilisi, Soviet Georgia (the current Republic of Georgia), in 1934. This institute exists today as the **George Eliava Institute of Bacteriophage, Microbiology and Virology**.

Bacteriophage therapy was popularized by Sinclair Lewis's Pulitzer Prize-winning 1924 novel *Arrowsmith*. The main character, a young physician named Martin Arrowsmith, travels to the West Indies to use bacteriophages as a therapy against bubonic plague.

For several reasons nearly all bacteriologists abandoned bacteriophage therapy in the 1930s. For one, published reports on bacteriophage therapy contained both negative and positive conclusions. It appeared to be a "hit or miss" therapy and the biological nature of bacteriophages was poorly understood. The main reason, though, for the abandonment of bacteriophage therapy was the development of antibiotics. Today, the Western world has a renewed interest in bacteriophage therapy as a treatment to combat antibiotic-resistant strains of pathogenic bacteria.

The History of the George Eliava Institute of Bacteriophage, Microbiology, and Virology

The Tbilisi Institute, founded by Professor George Eliava, opened its doors as a bacteriology laboratory in 1923. Professor Eliava met Felix d'Herelle in 1926 during a visit to the Pasteur Institute in Paris. After meeting at the Institute, d'Herrelle and Eliava exchanged information about bacteriophage therapy. As a result of these exchanges, d'Herelle visited the Institute during 1934–1935. Josef Stalin (the dictator of the Soviet Union from roughly 1928 to 1953) invited d'Herelle to the Institute to explore the possibility of bacteriophage therapy for military use. d'Herelle worked at the Institute off and on for the year, hoping to form a great collaboration with Eliava. Unfortunately, the KGB executed Eliava in 1937.[1] d'Herelle did not return to Tbilisi.

Even after Eliava's death, the Institute continued activities in bacteriological and bacteriophage research. In the 1940s, Institute researchers isolated bacteriophages to combat gas gangrene caused by *Clostridium perfringens.* The Institute merged with other research institutes and became the Institute of Vaccine and Sera in 1952. In 1988, the Institute was once again reorganized and was renamed the George Eliava Research Institute of Bacteriophage and included such notable researchers as Zemphira Alavidze and Amiran Meipariani. The Institute manufactured bacteriophage sprays, salves, ointments, pills, and patches and supplied these therapies to the entire Soviet Union—especially the Soviet military, which was, and continues to be, a major consumer of bacteriophage preparations. During the late 1980s, the Institute was at its peak in the production of bacteriophage products, employing 1200 staff and manufacturing 2 tons of phage products per day. Bacteriophage therapy was and continues to be a routine part of treatments in clinics and hospitals in Eastern Europe. These preparations are sold at pharmacies at incredibly low prices and are available without a doctor's prescription.

The Eliava Institute facilities were damaged during the Georgian civil war. When the Soviet Union collapsed in 1991, bacteriophage samples that were identified and catalogued were lost. The bacteriophage collection dwindled from approximately 5000 to 3500. Electrical outages ruined many refrigerated bacteriophage libraries, and the Institute nearly closed. Privatization was implemented in order to save the collection. Dr. Leila Kalandarishvili led a team of Georgian scientists in acquiring the production division of the Eliava Institute. Dr. Kalandarishvili took over the financial responsibilities (she was the sole owner of this new company) required to maintain the production equipment, purchased needed electrical generators, and drilled a well in order to obtain a reliable water source. The phage collections were named Biomedic.

In 1994, Biomedic merged with Biopharm. Dr. Kalandarishvili was able to secure more than 20 Georgian patents on phage products, obtain the commercial licenses, and meet registration requirements of the Georgian government. A 1997 BBC broadcast, *The Virus That Cures,* inspired energetic entrepreneurs from around the world to help save the Institute and its stocks.

In 2001, Dr. Kalandarishvili formed Advanced Biophage Technologies, LLC (ABT) in partnership with American businessman William B. McGall. Dr. Kalandarishvili and members of her scientific team continue to produce phage products using the trade names of "Pyophage," "Intestiphage," "Biosept," and other products for retail and over-the-counter sale.

At the time of this writing, many companies around the world are avidly pursuing bacteriophage therapy as an inexpensive, effective remedy to counter antibiotic resistance and hard-to-treat bacterial infections. (**FIGURE VF 21-1**)

The American Phage Group

During the 1940s, scientists from universities throughout the United States who were studying bacteria and bacteriophages formed the "Phage Group." Among the members of the group were Max Delbruck, Salvador Luria, and Alfred Hershey. These key members were educated as theoretical physicists or biologists, applying their approaches to the study of the physical nature of the gene. The Phage Group spent summers doing research experiments at Cold Spring Harbor Laboratory (in Long Island, New York) and designed an annual course to improve laboratory instruction in the techniques of bacteriophage research.

Their experiments centered upon studying the basic biology of bacteriophages, including the optimization of **one-step growth experiments** (see Chapter 5). The development of bacteriophage **plaque assays** contributed to the foundations of modern animal virology (**FIGURE 21-2**). By 1952, the Phage Group consisted of about 50 members. Important problems of bacteriophage research were tackled because researchers focused on a selected group of "authorized" phages, such as the Type (T) phages. During the 1950s Hershey and Chase showed by radiolabeling that the phosphorous (DNA) entered bacterial cells during T2 bacteriophage infection but that most of the sulfur (protein) did not (refer to Chapter 1). Delbruck, Luria, and Hershey shared the 1969 Nobel Prize in

[1]Eliava had become romantically involved with the same woman that the head of Stalin's police was in love with. Reports out of the Soviet Union suggested that the KGB assassinated Eliava, denouncing him as an enemy of the people.

VIRUS FILE 21-1 Bacteriophage Therapy Makes a Comeback

On a frigid day in December 2001, three men gathering wood in the Republic of Georgia found some canisters the size of paint pails that were so hot to the touch that they melted the surrounding snow. The men hauled them to their campsites where they used the canisters to warm themselves during a bitterly cold night. Unfortunately, the heat of the canisters was so intense that it caused severe deep burns on the backs of two of the men. The containers were old Soviet-era makeshift heaters packed with strontium 90, a powerful chemical that was used to power remote generators.

Within hours the two men began suffering from radiation poisoning. They were dizzy and started vomiting. And their skin started to peel. The bacterium *Staphylococcus aureus* infected their wounds. Antibiotics failed to prevent the pathogenic bacteria from invading deeply into the wounds and the two men were near septic shock. They were rushed to a hospital in Tbilisi where doctors placed patches containing bacteriophages on the wounds. The infections cleared up within a few weeks and the men were stable enough to undergo skin grafting abroad.

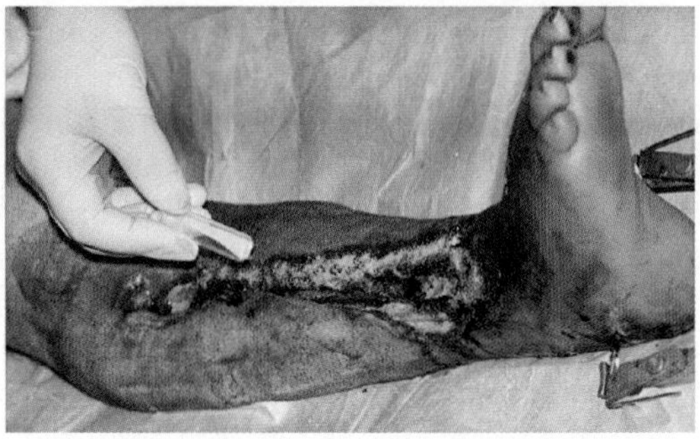

FIGURE VF 21-1 These wounds were treated with the bacteriophage powder PhagoBioDerm. There is rekindled interest in bacteriophage applications against antibiotic-resistant bacteria.

Bacteriophage therapy is making a comeback because of the accelerating crisis of antibiotic resistance. Several hurdles have confronted bacteriophage therapy, including:

- Contamination of bacteriophage stocks with bacterial toxins and bacterial debris causes symptoms ranging from mild fever to death in patients.
- Bacteriophages have a narrow host range.
- Bacteriophages are eliminated by the host's immune system (antibodies).
- Bacterial hosts are resistant to bacteriophage infection.
- **Lytic** bacteriophages are ideal for therapy, but half of all bacteriophages are **temperate** (their DNA is integrated into the host genome, leaving the host intact and alive).

In the 1990s, U.S. scientists began laboratory experiments to successfully address these hurdles. Pure preparations that remove toxins and bacterial debris from bacteriophage stocks can now be produced. Methods have been developed to isolate bacteriophage mutants that are hardier and have broader host ranges. The hardy strains possess the ability to avoid rapid elimination by the host's immune system. Since 1993, at least 13 companies located in the United States, Canada, Germany, and the United Kingdom were founded to develop bacteriophage therapeutics. Some of these companies are creating bacteriophage cocktails, whereas others are isolating specific bacteriophage components such as lytic enzymes aimed to target pathogenic bacteria. For therapeutic purposes, the Federal Drug Administration (FDA) would approve bacteriophages that are stably lytic. Experts believe that bacteriophage therapy that was once scorned may ultimately find a niche in Western medicine. It has the potential to save lives.

References

Carlton, R. M. 1999. "Phage therapy: Past history and future prospects." *Archivum Immunologiae et Therapiae Experimentalis* 47:267–274.

Koerner, B. I. 1998. "Return of a killer: Phages may once again fight tough bacterial infections." *U.S. News & World Report,* November 2, 1998.

Merril, C. R. et al. 1996. "Long-circulating bacteriophage as antibacterial agents." *Proc Natl Acad Sci USA* 93:3188–3192.

Stone, R. 2002. "Stalin's forgotten cure." *Science* 298:728–731.

Thiel, K. 2004. "Old dogma, new tricks—21st century phage therapy." *Nat Biotech* 22:31–36.

a

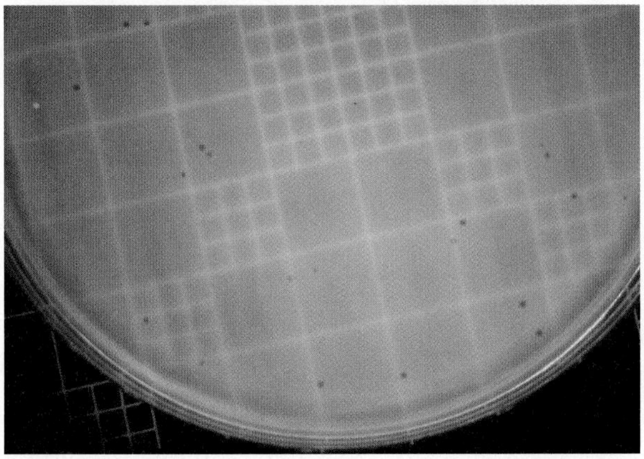

b

FIGURE 21-2 **(a)** Bacterial cells growing in log phase (left) and being lysed by bacteriophages (right). **(b)** Bacteriophage plaque assays from one-step growth experiments. The circular clearings present on the lawn of bacteria represent plaques or regions where the bacteriophage infected the bacterium and lysed the cells. Each plaque arises from a single infectious bacteriophage. The infectious particle that gives rise to a plaque is called a PFU (plaque-forming unit). **(c)** Illustration showing the formation of plaques within a lawn of bacteria.

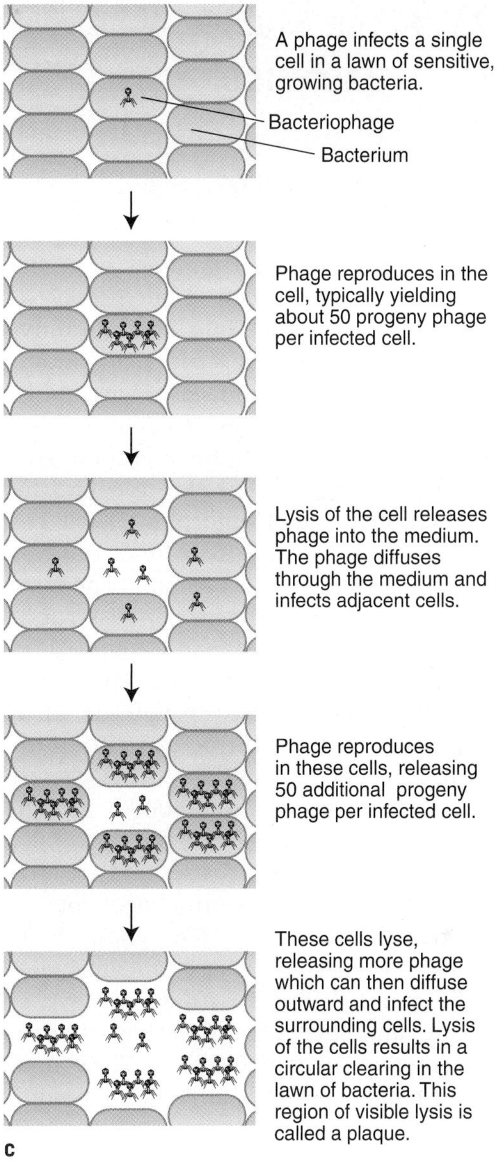
A phage infects a single cell in a lawn of sensitive, growing bacteria.
Bacteriophage
Bacterium
Phage reproduces in the cell, typically yielding about 50 progeny phage per infected cell.
Lysis of the cell releases phage into the medium. The phage diffuses through the medium and infects adjacent cells.
Phage reproduces in these cells, releasing 50 additional progeny phage per infected cell.
These cells lyse, releasing more phage which can then diffuse outward and infect the surrounding cells. Lysis of the cells results in a circular clearing in the lawn of bacteria. This region of visible lysis is called a plaque.

c

Physiology or Medicine for "their discoveries concerning the replication mechanism and the genetic structure of viruses."

21.2 Bacteriophage Ecology

Most bacteria live in the subsurfaces of soil and sediment and the open ocean, and are essential components of Earth's biota. Scientists at the University of Georgia (in Athens, Georgia) estimated the total number of bacteria on Earth to be 4–6 × 10^{30} cells. The scientific community agrees that there is a global population of about 1 × 10^{31} bacteriophages. Less than 1% of the observed bacteriophage by electron microscopy have been grown in culture. It is estimated there are about 10^8 genotypes. Bacteriophages may have the most gene diversity of all life. These values come from bacteriophage estimates in aquatic and soil environments.

The marine viruses are the best characterized of the phage ecology system. Marine bacteriophages infect and lyse bacteria in aquatic microbial communities. This is now recognized as the most important mechanism of recycling bacterial carbon in the marine environment. The dissolved organic carbon released by viral lysis of a bacterial population stimulates the growth of new bacterial populations that are part of the microbial food web in marine systems (refer to Chapter 1). Marine microbial ecology is a rapidly developing field.

21.3 The Biology of Bacteriophages: Composition and Structure

To date, over 5400 bacteriophages have been examined by means of electron microscope. The International Committee on the Taxonomy of Viruses (ICTV) recognizes 1 order, 13 families, and 31 genera of bacteriophages. There are four basic shapes or symmetries of bacteriophages: binary (phages that have a head and tail structure), icosahedral (also referred to as cubic), helical (filamentous), and pleomorphic (**FIGURE 21-3**). All bacteriophages contain a head or capsid structure that varies in size or shape. The majority of bacteriophages contain a tail structure (**FIGURE 21-4**) that varies in size and length. The tail structure is a helical, hollow tube. The genome of the bacteriophage passes through the tail and into the bacterial host during the initiation of infection. The genome length of binary bacteriophages varies between 17 and 500 kb, and their tail lengths range from 10 to 800 nm.

Three families contain viruses that are enveloped. About 96% of all bacteriophages contain a double-stranded DNA (dsDNA) genome. The remaining 4% of bacteriophages contain genomes of single-stranded DNA (–ssDNA), ssRNA, and dsRNA. Genomes may code for as few as three to five gene products or as many as 100. The genomes of bacteriophages contain unusual or modified bases that protect them from degradation by host nucleases during phage infection. The most famous group is the T-Type bacteriophages that are numbered T1 to T7. The best known individual bacteriophages are T4 and λ. With respect to bacteriophages, much research has focused on new viruses, and thus the 5000 others that are known have been neglected. It is inevitable that new bacteriophage families are likely to be discovered in unusual habitats, such as hypersaline lagoons, volcanic hot springs, and the mammalian rumen.

21.4 Overview of Bacteriophage Infection: Bacteriophages Possess Alternative Lifestyles: Lytic Versus Temperate Phages

Bacteriophages adsorb to the surface receptor molecules of bacterial cells during the first step of the infection. Receptors may be host pili, proteins, oligosaccharides, or lipopolysaccharides (LPS). **Adsorption** is mediated by specialized structures such as tail fibers. The tailed T4 bacteriophage anchors its tail fibers onto the LPS receptors of its host. This causes a conformational change resulting in a contraction of the tail sheath and penetration of the cell membrane. The DNA is subsequently

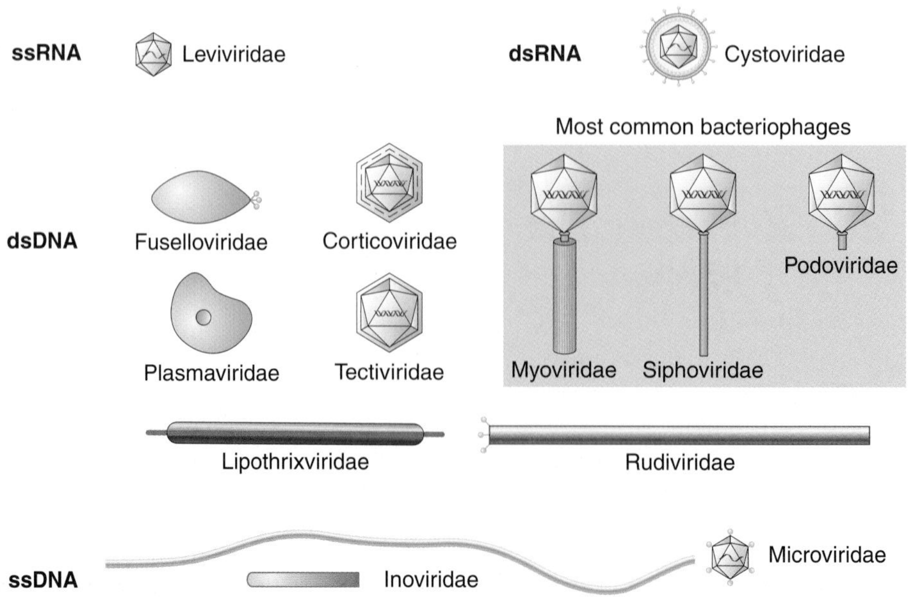

FIGURE 21-3 Diagram depicting the 13 different bacteriophage families grouped by their types of nucleic acid genomes. Tailed bacteriophages are the most prevalent worldwide and infect eubacteria and the archeabacteria. Adapted from Ackermann, H. W., *Arch. Virol.* 146 (2001): 843–857.

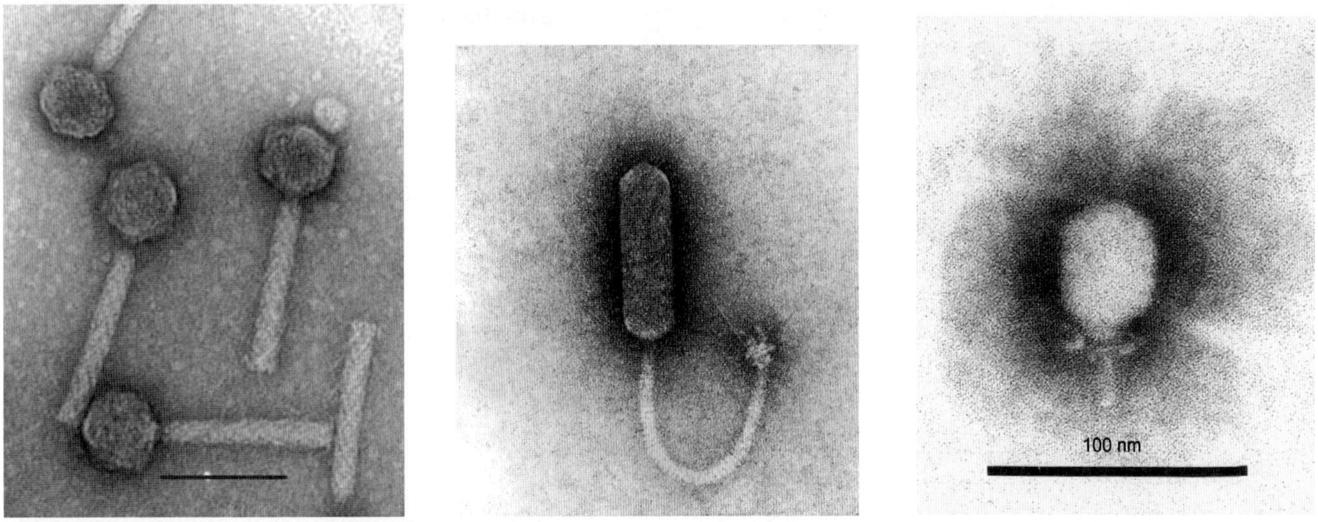

a b c

FIGURE 21-4 Images of tailed bacteriophages. **(a)** *Enterobacter* virus P2, *Myoviridae* family. **(b)** *Bacillus licheniformis* virus, *Siphoviridae* family. **(c)** Bacteriophage 3/K26 (infects *Kurthia zopfii*), Podoviridae family.

injected into the host cell through the tail tube (**FIGURE 21-5**).

Other bacteriophages may bind to a second receptor. Bacteriophages that do not have tails penetrate the host bacterium by producing polysaccharide-degrading enzymes that digest components of the bacterial envelope or cell wall. Bacteria develop resistance to bacteriophage infection when the host cell receptors are altered by a mutation. Losing some receptors offers no protection against infection by bacteriophages that bind to other host receptors. There is much interest in engineering new receptor-recognition elements into the tail fibers of well-characterized bacteriophages so that they can infect taxonomically distant hosts.

The bacteriophage genome **penetrates** through the tail tube into the host cell. This is not an injection process, as it has often been portrayed. Bacteriophage T4 packages lysozyme in the base of its tail from a prior infection and uses the

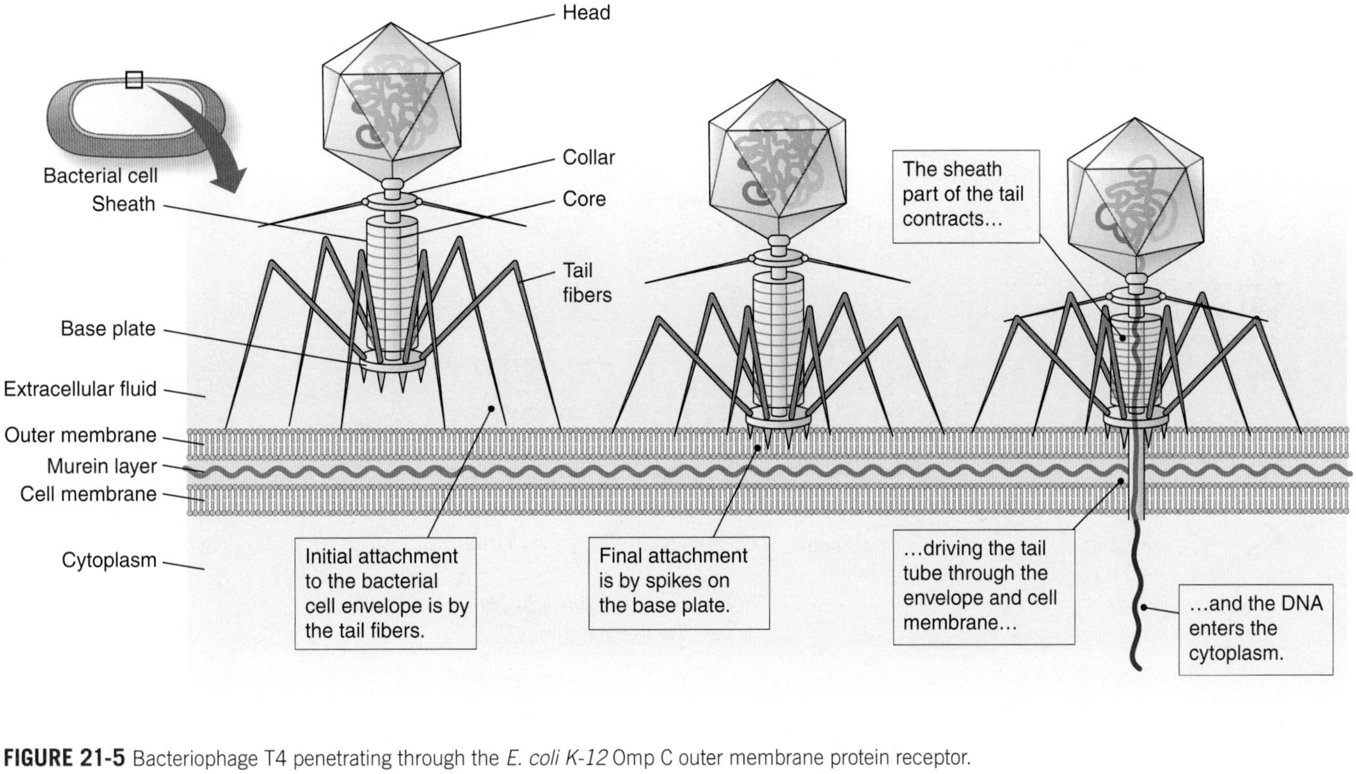

FIGURE 21-5 Bacteriophage T4 penetrating through the *E. coli K-12* Omp C outer membrane protein receptor.

enzyme to degrade a portion of the peptidoglycan of the bacterial cell wall. The DNA is drawn into the cell by a process that is not well understood for most bacteriophages. Once inside the host cell, the bacteriophage circularizes its DNA rapidly by means of sticky ends or termini, or the linear ends are modified and protected from the bacterial nucleases (Figure 21-5).

As soon a bacteriophage DNA genome enters the cell, the host RNA polymerase recognizes the viral DNA promoters and transcription of the early genes begins. Translation of these **early genes** is coupled with transcription. Examples of early proteins produced are repair enzymes (to repair the hole in the bacterial cell wall created during genomic penetration) and proteins that deactivate host proteases and block restriction enzymes (nucleases) that could degrade the bacteriophage DNA. During this time period, copies of the bacteriophage genome are replicated. Subsequently, a set of **late genes** are transcribed and translated. Examples of late gene products are structural proteins that make up the capsomers of the head and tail of the bacteriophage and lysozyme, which is used by the bacteriophage to escape the host during a **lytic infection**. **Lysozyme** is packaged into the tail of some bacteriophages in order to initiate a new infection.

How do the RNA bacteriophages regulate the amount of each protein synthesized during their replication? The secondary structure of the ssRNA phage genomes of phages such as **MS2** and **Q-beta** regulates both the quantities of different phage proteins that are translated, but in addition it also operates temporal control of a switch in the ratios between the different proteins produced in infected cells.

Assembly and Release Process: Lytic Infection

As soon as all bacteriophage "parts" have been produced, new bacteriophage particles are assembled. A copy of the phage DNA genome is packaged or "reeled into" a preassembled icosahedral head. The tail and accessory structures (such as the base plate and tail fibers) are assembled, and a few molecules of lysozyme or a comparable enzyme are packaged into the tail base plate.

The remaining lysozyme or **phage lysin**, **endolysins**, **muramidases**, or **virolysins** in the cell allow the viruses to escape or be released from the host. These enzymes hydrolyze specific bonds in the murein or peptidoglycan layer of the cell wall. This part of the cell wall is responsible for the rigidity and mechanical strength of bacterial cells. Another key enzyme, **holin**, is used by phages to create pores in the inner membrane of the host cell at the appropriate time to allow lysin or similar enzymes to reach the peptidoglycan, thus facilitating host lysis and bacteriophage release. Ultimately, the host cell is killed by cell lysis. This type of infection is referred to as a **lytic infection**. Lysis of the host cell is a tightly timed event (**FIGURE 21-6**).

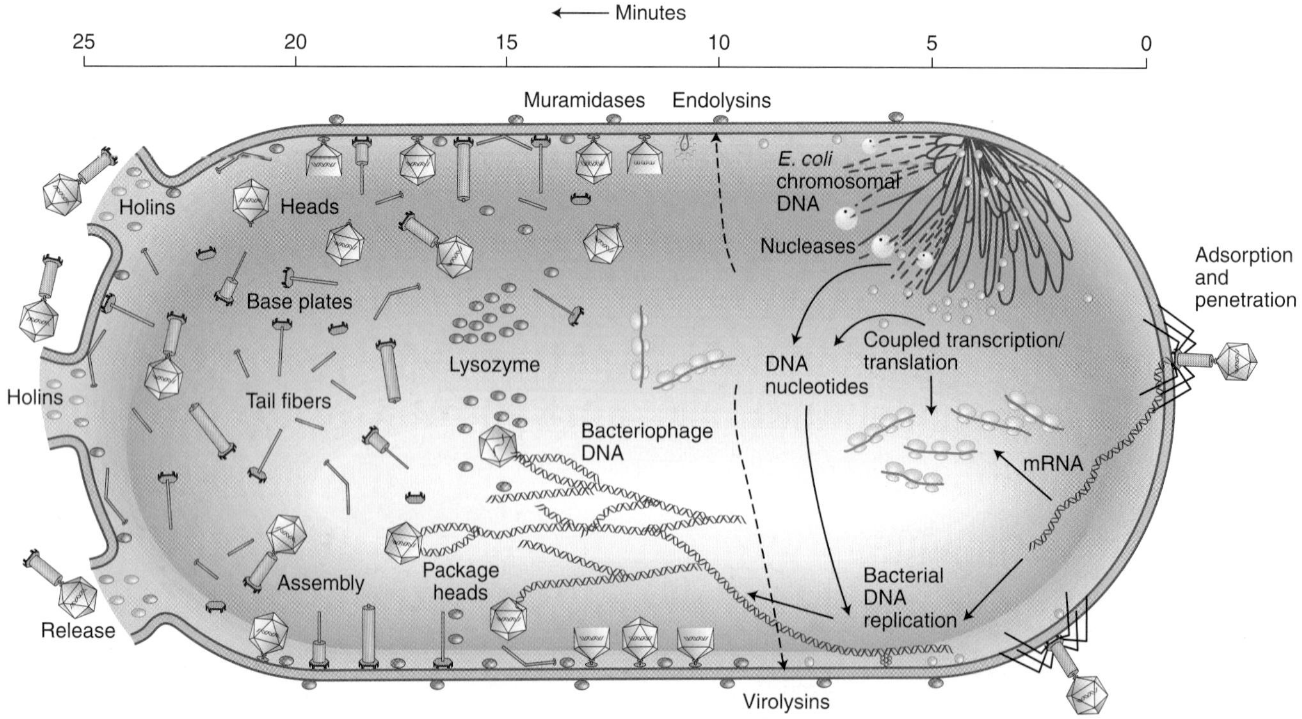

FIGURE 21-6 *E. coli* bacteriophage T4 lytic infection cycle. Adapted from Abedon, S. T., and Calendar, R. L., eds. *The Bacteriophages*, Second Edition. Oxford University Press, 2005.

Lysogenic (Temperate) Infections

Lysogenic or **temperate bacteriophages** infect their hosts, but instead of killing the host during a lytic infection their genome becomes integrated into a specific region of the host chromosome. After penetration, the viral DNA integrates into the host chromosome and replicates every time the cell copies its chromosomal DNA during cell division. Lysogenic cells divide, giving rise to lysogenic progeny. When the phage genome is integrated into a site in the bacterial chromosome, it is called a **prophage**. Some temperate bacteriophages encode **transposase**, which allows the phage to insert randomly into the chromosome. Other bacteriophages integrate into specific locations within the chromosome. During the prophage state, all of the bacteriophage genes are repressed except a gene encoding a **repressor** protein. The repressor protein prevents the synthesis of enzymes and proteins required for the lytic cycle.

If the repressor becomes inactivated (loses function), viral DNA is excised from the bacterial chromosome. The excised DNA (bacteriophage genome) acts like a lytic virus that is able to produce new viral particles that are released during cell lysis. This spontaneous **derepression** or **induction** happens about once in every 10,000 cell divisions (**FIGURE 21-7**). Temperate bacteriophages can carry host genes from one bacterial cell to another in a process called **transduction**.

21.5 Bacteriophages Create Pathogenic Bacteria in Nature

Bacteriophages have contributed to virulence of pathogenic bacteria. Sometimes a prophage may carry genes that alter the phenotype of a lysogenic bacterium in a process called **lysogenic conversion**. For example, the pathogen *Corynebacterium diphtheriae* produces a toxin responsible for diphtheria only if it carries the temperate bacteriophage called β. The *tox* gene is located near one end of the phage genome. Diphtheria is a very contagious and potentially life-threatening disease that usually attacks the mucous membranes of the throat and nose. The hallmark sign of this disease is a thick, gray covering in the back of the throat that can make breathing difficult. It is very rare today because of widespread vaccination. The diphtheria vaccine is produced by growing the lysogenic *C. diphtheria* in a liquid medium. The bacterial culture is filtered and the filtrate is treated with

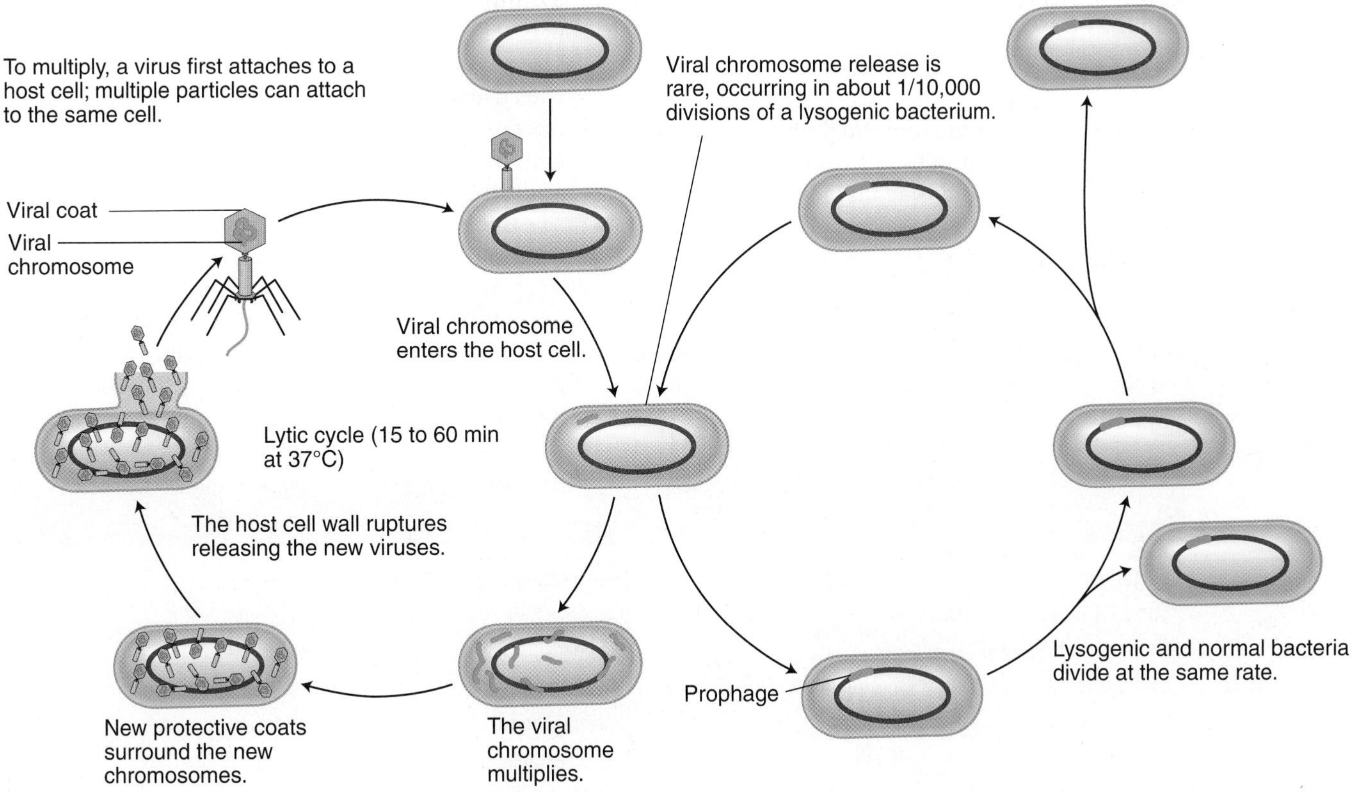
To multiply, a virus first attaches to a host cell; multiple particles can attach to the same cell.
Viral coat
Viral chromosome
Viral chromosome enters the host cell.
Lytic cycle (15 to 60 min at 37°C)
The host cell wall ruptures releasing the new viruses.
New protective coats surround the new chromosomes.
The viral chromosome multiplies.
Viral chromosome release is rare, occurring in about 1/10,000 divisions of a lysogenic bacterium.
Prophage
Lysogenic and normal bacteria divide at the same rate.

FIGURE 21-7 Infection cycle of a temperate bacteriophage. Adapted from Todar, K. "Introduction to viruses: Bacteriophage." *The Microbial World.* University of Wisconsin-Madison, Department of Bacteriology, 2006. http://textbookofbacteriology.net/themicrobialworld/Phage.html.

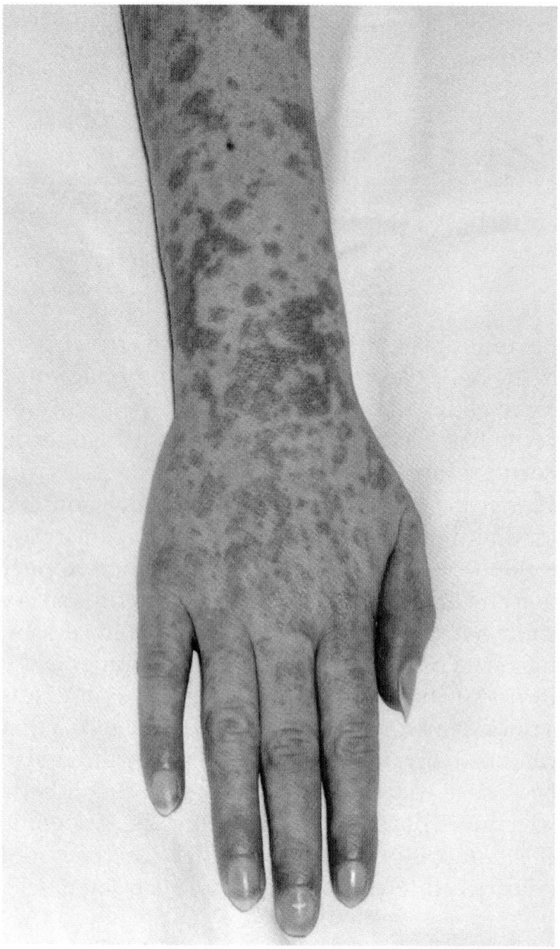

FIGURE 21-8 This child has the scarlet fever rash caused by a lysogenic *Streptococcus pyogenes* bacterium.

formaldehyde, converting the β phage toxin to a **toxoid**. (Single antigen diphtheria toxoid is not available.) The toxoid is available when it is combined with both tetanus toxoid and acellular pertussis vaccine as DTaP.

The infection of *Streptococcus pyogenes* with the temperate bacteriophage T12 results in the lysogenic conversion of a nontoxigenic strain to a pyrogenic- or **erythrogenic exotoxin**-producing bacterium. This type of toxin causes the rash of scarlet fever that sometimes occurs in people who have strep throat (**FIGURE 21-8**). The loss of the bacteriophage may render the bacterium nonpathogenic. The majority of bacteriophages possessing **virulence genes**, however, are temperate phages that form stable **lysogens**. A lysogen is a bacterial cell that has been infected with a temperate bacteriophage (one that does not destroy the cell).

The evolution of toxigenic pathogens has depended extensively on bacteriophage infections and exchanges of genetic information. Due to the lack of genetic and bacteriophage systems available to study the mechanisms of toxin gene transfer, scientists have barely scratched the surface in advancing knowledge of bacteriophage molecular biology and evolution. **TABLE 21-1** lists examples of virulence factors carried by bacteriophages.

21.6 Control of Bacteriophages in Industrial Fermentations

Bacteria are used in a variety of food fermentation processes because they are able to break down complex organic substances into simpler ones. Examples of commercial food fermentation products are yogurt, cheese, sauerkraut, and soy sauce. The pharmaceutical and biotechnology industries use bacteria on a large scale to produce products such as alcohols, vitamins, amino acids, enzymes, hormones, biopolymers, antibiotics, and other therapeutics (**FIGURE 21-9**).

Bacteriophages are ubiquitous in nature. Any bacterial strain can be infected by virulent bacteriophages or harbor one or more prophages. All industries that utilize bacteria to produce food or commodity chemicals are vulnerable to bacteriophage infections. Bacteriophage researcher Roger Hendrix provides us with the following analogy:

TABLE 21-1 Examples of Virulence Factors Carried on Bacteriophages

Bacterium	Bacteriophage	Toxin or Other Gene Product	Phenotype
Corynebacterium diphtheriae	β	Diphtheria toxin	Diphtheria
Streptococcus pyogenes	T12	Erythrogenic toxins	Scarlet fever
Clostridium botulinum	Clostridial phages (e.g., CE β)	Botulinum toxin	Botulism (food poisoning)
Vibrio cholerae	CTX	Chlorerae toxin	Cholera
Escherichia coli	H-19B	Shigalike toxins	Hemorrhagic diarrhea
Staphylococcus aureus	Φ315, Φ13, ΦMu50A	Enterotoxins P and A	Food poisoning
Mycoplasma arthritidis	MAV1	Outer membrane protein (*vir* gene)	Arthritis involving two or more joints?

FIGURE 21-9 Microbe fermentation vessel used to make enzymes for medical and industrial use. This vessel can hold up to 120,000 liters.

"If a bacteriophage were the size of a cockroach—these cockroaches would cover the surface of the earth in a layer that is 31,071 miles or 50,000 km deep!" This is a nightmare statistic to laboratory staff working in a commercial fermentation factory, where bacteriophages really are considered to threaten production.

Attacks by bacteriophages can cause serious problems in the production process and result in considerable economic losses. An entire fermentation batch may be lost due to bacteriophage contamination. Examples of common products and their bacterial hosts in fermentation-based industries are listed in **TABLE 21-2**. Approximately 1% to 10% of dairy product fermentation batches are lost to bacteriophage infection. This usually happens because the raw starting materials can be contaminated with undetectable numbers of bacteriophages.

Laboratory staff at fermentation facilities should be trained to expect and respond to bacteriophage outbreaks, as the problem is unavoidable. Effective cleaning and monitoring procedures, especially during the bacteriophage "seasons" of January to March and October to November is warranted. Measures used to control bacteriophages in the dairy industry are listed in **TABLE 21-3**. Dairy plants typically harbor 10^5 plaque-forming units per cubic meter of air. Effective control measures are needed to minimize bacteriophage infection.

21.7 Biofilms and Bacteriophages

Bacteria can adhere to solid surfaces to form a slimy, slippery coat called a **biofilm**. Biofilms are found on most wet surfaces in nature, such as on river rocks, on the walls of limestone caves, in pipes, on industrial equipment, in the hulls of ships, on teeth, in the middle ear in patients with persistent otitis media (ear infections), in the lining of the colon, on sutures, in the lungs of cystic fibrosis patients; on medical devices such as catheters, stents, heart valves and surrounding tissues of the heart,

TABLE 21-2 Common Products and Their Bacterial Hosts in Fermentation-Based Industries

Product	Bacterial Host
Yogurt	*Lactobacillus delbrueckii* subspecies *bulgaricus* *Streptococcus thermophilus*
Cheese	*Lactobacillus delbrueckii* subspecies *lactis* *Lactobacillus helveticus* *Lactobacillus lactis* *Propionibacterium freudenreichii* *Streptococcus thermophilus*
Sourdough bread	*Lactobacillus fermentum*
Sauerkraut	*Leuconostoc mesenteroides* *Leuconostoc fallax* *Lactobacillus plantarum*
Vinegar	*Acetobacter sp.*
Soy sauce	*Tetragenococcus halophila*
Chloramphenicol	*Streptomyces venezuelae*
Streptomycin	*Streptomyces griseus*
Tetracycline	*Streptomyces aureofaciens*
L-glutamic acid	*Brevibacterium lactofermentum*
Insecticide (BT)	*Bacillus thuringiensis*
Silage	*Lactobacillus plantarum*
Hormones, enzymes, and other biotechnology products	*Escherichia coli*
Butanol	*Clostridium beijerinckii*
Acetone	*Clostridium sp.*

TABLE 21-3 Bacteriophage Control Measures in the Dairy Food Industry

- Rotation of bacterial starter cultures
- Use of starter cultures that do not contain strains of similar bacteriophage sensitivity patterns
- Air filtration (HEPA)
- Preparation of starter cultures in a positively pressured room
- Cleaning and chlorinating vessels between fills
- Performing daily tests for bacteriophage detection
- Regular inspections of equipment surfaces to remove biofilms
- Spraying the environment with adequate disinfecting agents
- Rotation of manufacturing processes

FIGURE 21-10 Electron micrograph of *Staphylococcus aureus* bacteria found on the surface of a patient's catheter. The sticky-looking substance woven between the bacteria is composed of polysaccharides and is the biofilm that protects the bacteria from antimicrobial agents such as antibiotics. Magnified 2363×.

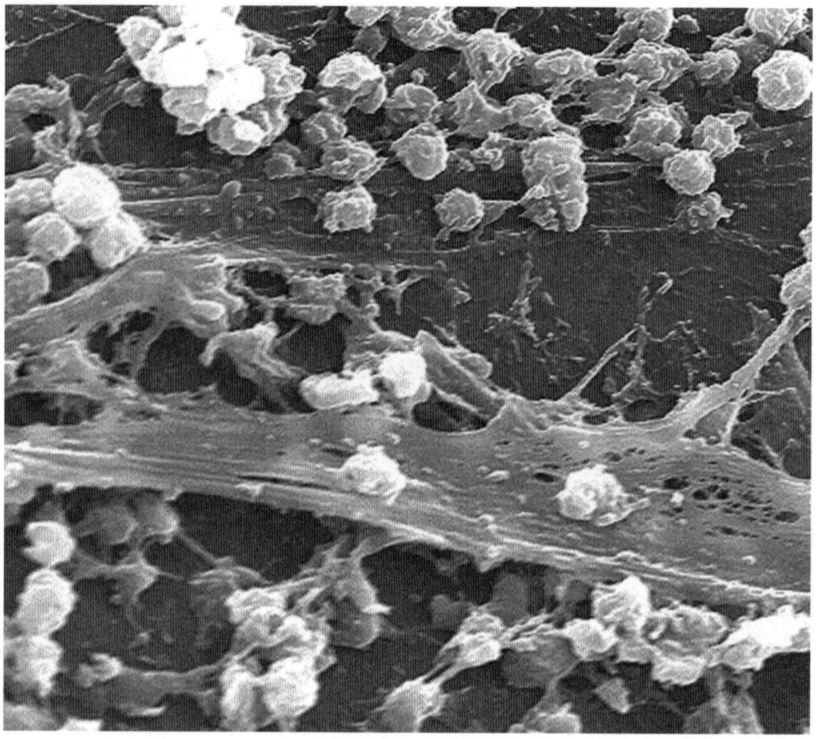

orthopedic devices, facial implants; and in chronic wounds. Biofilms are complex sessile (permanently attached) communities of bacteria that may consist of a single bacterial species to hundreds or even thousands of them. The biofilm bacteria produce extracellular polysaccharide polymers that surround and encase the microbes, facilitating their adhesion to surfaces (**FIGURE 21-10** and **21-11**). Biofilms can cause environmental problems or chronic bacterial infections in humans.

Development of bacterial biofilms has been intensely studied *in vitro* using confocal scanning laser microscopy to observe green fluorescent protein (GFP) tagged bacteria. For example, *Pseudomonas aeruginosa* produces a mature *in vitro* biofilm in five to seven days. In these experiments, free-moving (or motile by means of flagellar movement) GFP-tagged bacteria reversibly attach to a surface that may be covered by a layer of proteins such as a pellicle. *At this stage, the bacteria are still susceptible to*

FIGURE 21-11 Scanning electron micrograph of a biofilm of bacteria on the bristles of a used toothbrush.

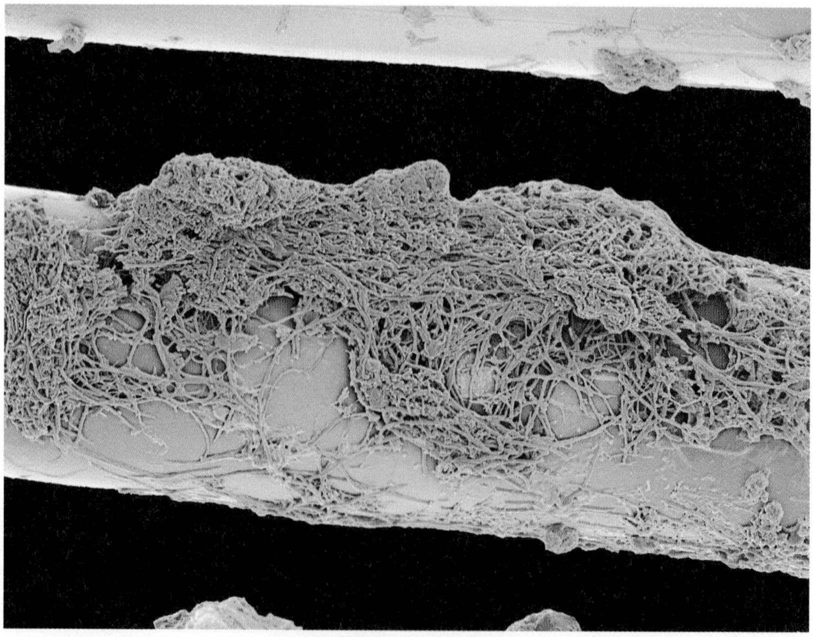

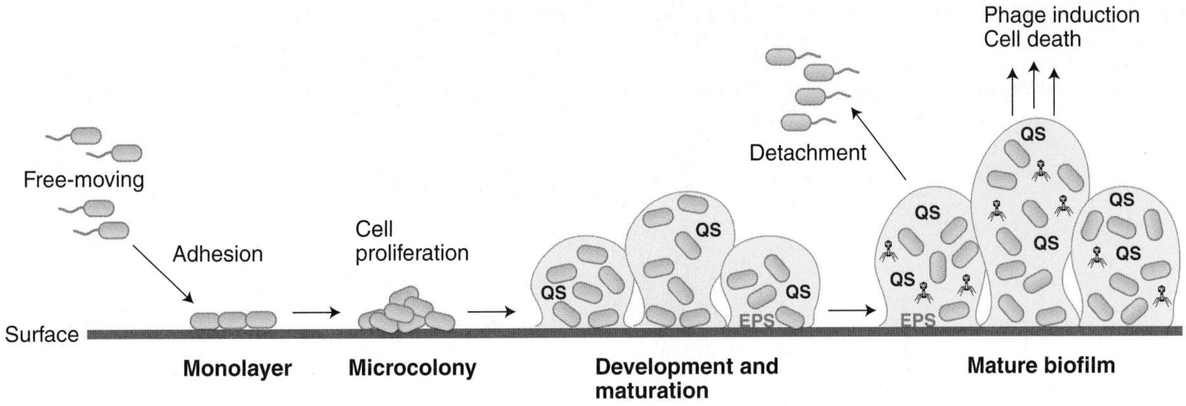

FIGURE 21-12 Illustration depicting the formation of a biofilm based on *in vitro* experiments. Adapted from Hoiby, N. 2010 "Antibiotic resistance of bacterial biofilms." *Int J Antimicrob Agents* 35:322–332.

antibiotics. Subsequently, the bacteria irreversibly bind to the surface within the next several hours and multiply, forming microcolonies on the surface. During microcolony formation, the bacteria produce the extracellular polysaccharide polymers around the microcolonies. *In vitro,* the biofilm forms a mushroom-like or tower-like structure in the mature biofilm. **Quorum sensing** occurs in the microcolonies and mature biofilm. Quorum sensing is a form of bacterial cell-to-cell communication. Bacteria use quorum sensing to regulate gene expression in response to fluctuations in population density. *At that stage, the biofilm has maximum tolerance (resistance) to antibiotic therapy.* Biofilm-growing bacteria may require up to 100 to 1000-fold higher minimal inhibitory concentrations of antibiotics compared to the free moving bacteria.

The mature biofilm can vary in thickness between 25 and 100 μm, depending upon the equilibrium bacterial growth and detachment of bacteria. Following this stage, distinct areas of the biofilm dissolve and bacterial cells are detached and free to move to another location, where new biofilms are formed. The liberation process may be caused by bacteriophage activity within the biofilm (**FIGURE 21-12**).

Biofilms present an important challenge to patient medical care. Bacteria commonly isolated from medical devices are *Enterococcus faecalis, S. aureus, S. epidermidis, Streptococcus viridans, E. coli, Klebsiella pneumoniae, Proteus mirabilis,* and *Pseudomonas aeruginosa.* These bacteria originate from either the skin of the patient or healthcare workers, tap water (e.g., when entry ports of catheters are exposed to tap water), or other sources in the environment. Biofilms are tenacious and highly resistant to antimicrobial treatment, making these infections difficult or impossible to treat.

Preventative strategies and treatment of biofilms on medical implants include using silver-impregnated catheters and cuffs, incorporating inline filters into catheters, coating devices with antibiotics, and disinfecting at the surface site. Interest is growing in the role of bacteriophages in disrupting or reducing the formation of biofilms on medical implants. Studies by Curtin and Donlan attempting to pretreat catheter surfaces with bacteriophages to control biofilms are promising (**FIGURE 20-13**). Further research in this area is needed, particularly as the problem of antibiotic resistance continues to grow and traditional antibiotic treatment becomes less effective.

21.8 FDA-Approved Listeria-Specific Bacteriophage Preparation on Ready-to-Eat Meats

Hot dogs, luncheon meats, and cheeses such as feta, Brie, Camembert, blue-veined cheeses, or Mexican-style cheeses such as queso blancho that are not made with pasteurized milk may contain *Listeria monocytogenes.* Luncheon meats are especially prone to contamination with *L. monocytogenes* during slicing and packaging at food processing plants. *L. monocytogenes* is a foodborne pathogen that is harmless in most healthy people but can cause **listeriosis** in pregnant women, newborns, and individuals with weakened immune systems.

Infections during pregnancy can lead to a miscarriage or stillbirth, premature delivery, or infection of the newborn. The symptoms of listeriosis include flulike symptoms such as nausea and diarrhea. If the bacterial infection spreads to the nervous system, individuals experience a headache, stiff neck,

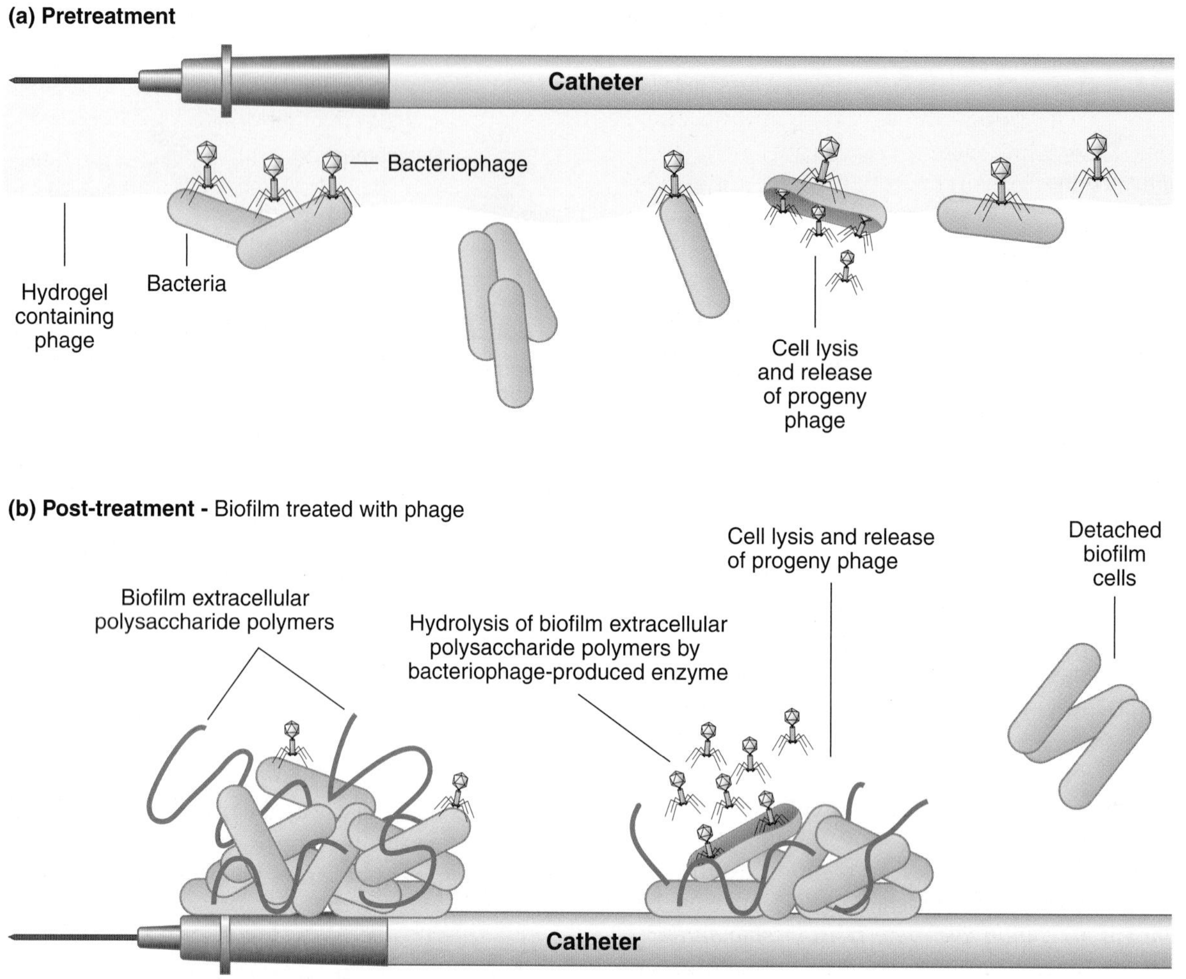

FIGURE 21-13 Illustrations showing applications of bacteriophages to treat biofilms. **(a)** Phages are embedded into a hydrogel present on a catheter as a pretreatment before the catheter is inserted into a patient. The bacteriophages should be able to infect and lyse bacterial cells, preventing the formation of biofilms on the catheter. **(b)** Phages are used to treat an existing biofilm present on a catheter, resulting in the hydrolysis of the extracellular polysaccharide polymers around the bacteria within the biofilm, liberating bacteria from the catheter, making them more susceptible to antibiotic and phage therapy. Adapted from Donlan, R. M. 2009. "Preventing biofilms of clinically relevant organisms using bacteriophage." *Trends Microbiol.* 17(2): 66–72.

confusion, loss of balance, and possibly convulsions. Listeriosis is 30% fatal in high-risk groups.

In 2006 the FDA recognized the use of a cocktail of six bacteriophages in the form of an application spray as GRAS (Generally Recognized as Safe). The product is marketed as LISTEX™ P100 and is produced by the Phage Technology Center in the Netherlands.

LISTEX will be used in the clean rooms of cheese, meat, and poultry processing plants to inhibit the growth of *Listeria* on products such as lunch meat and hot dogs. The phage preparation will not be declared as an ingredient on the label of a treated product.

Summary

No virology textbook should be without a chapter on bacteriophages. Bacteriophages—viruses that infect bacteria—have been an attractive "model" organism for scientists to learn about the basic biology of life. Bacteriophages are inexpensive and easy to work with in the laboratory. They have small genomes and reproduce rapidly, making them amenable to obtaining mutants identified by rapid screening techniques.

The field of bacteriophage biology began during what is sometimes called the golden age of microbiology (~1870–1920) and then languished for decades. Today it is once again thriving due to the use of phages to combat drug-resistant pathogens. Phage ecology is also a rapidly developing field. Over 5400 bacteriophages have been characterized by electron microscopy. The majority of bacteriophages are binary, in that they contain head and tail structures. About 96% of bacteriophages contain dsDNA genomes and the remaining 4% contain ssDNA, ssRNA, or dsRNA genomes. Bacteriophages may undergo a lytic or lysogenic infection. The host is killed (lysed) during a lytic infection and progeny phages are released. Lysogenic or temperate bacteriophages infect their hosts, but instead of killing the host their genome becomes integrated into a specific region of the host chromosome. The integrated viral genome is called a prophage. A prophage may carry genes that alter the phenotype of its host in a process called lysogenic conversion. Specifically, the evolution of several toxinogenic pathogens has depended extensively on bacteriophage infections and exchanges of genetic material. Examples include the following pathogenic bacteria: *C. diphtheriae* (causes diphtheria), *S. pyogenes* (causes strep throat and scarlet fever), and *C. botulinum* (causes food poisoning or botulism).

Felix d'Herrelle is credited with discovering bacteriophages and with the idea of bacteriophage therapy to treat bacterial diseases. He played a significant role in the establishment of the George Eliava Institute of Bacteriophage, Microbiology and Virology in Tbilisi in the Republic of Georgia. Bacteriophage therapy has been used by the Soviet Union since the 1920s. Bacteriophage sprays, salves, ointments, patches, and pills are still prepared and used in clinics and hospitals in the former Soviet Union and Eastern Europe. Today bacteriophage therapy is being investigated in the Western world as a potential therapy to defeat multidrug-resistant bacteria. Other applications of bacteriophages are to eliminate superbugs that form biofilms present on implanted medical devices and to treat ready-to-eat meats and soft cheeses with bacteriophages in order to eliminate the foodborne pathogen *L. monocytogenes.*

Bacteriophages are ubiquitous in nature. Bacteriophage outbreaks occur in the food fermentation, biotechnology, and pharmaceutical industries that use bacteria on a large scale to produce fermented foods, chemicals, enzymes, hormones, biopolymers, antibiotics, and therapeutics. Effective control measures are needed to minimize bacteriophage infections during the large-scale production of bacteria in these industries.

Resources

Primary Literature and Reviews

Ackermann, H. W. 2001. "Frequency of morphological phage descriptions in the year 2000." *Arch Virol* 146:84–857.

Beher, M. G., and Pugsley, A. P. 1981. "Coliphage which requires either the lamB protein or the ompC protein for adsorption to *escherichia coli* K–12." *J Virol* 38(1): 372–375.

Bernhardt, T., et al. 2002. "A protein antibiotic in the phage Qβ virion: Diversity in lysis targets." *Science* 292:2 326–2329.

Brussow, H. 2005. "Phage therapy: The *escherichia coli* experience." *Microbiology* 151:2133–2140.

Costerton, J. W., Stewart, P. S., and Greenberg, E. P. 1999. "Bacterial biofilms: A common cause of persistent infections." *Science* 284:1318–1322.

Curtin, J. J., and Donlan, R. M. 2006. "Using bacteriophages to reduce formation of catheter-associated biofilms by *Staphylococcus epidermidis.*" *Antimicrob Agents Chemother* 50(4):1268–1275.

Danovaro, R., and Serresi, M. 2000. "Viral density and virus-to-bacterium ratio in deep-sea sediments of the eastern Mediterranean." *Applied Environ Microbiol* 66(5): 1857–1861.

d'Herelle, F. 1917. "Sur un microbe invisible antagonistic des bacilles dysenterique. (On an invisible microbe antagonistic to dysentery bacteria.)" *C R Acad Sci Paris* 165:373–375.

d'Herelle, F. 1925. "Essai de traitement de la pest bubonique par le bacterio-phage. (Test of processing of the bubonic pest by the bacteriophage)." *Presse Med* 33: 1393–1394.

d'Herelle, F., Malone, R. H., and Lahiri, M. 1930. "Studies on Asiatic cholera." *Indian Med Res Memoirs* 14:1–161.

Donlan, R. M. 2001. "Biofilms and device-associated infections." *EID* 7(2):277–281.

Donlan, R. M. 2009. "Preventing biofilms of clinically relevant organisms using bacteriophage." *Trends Microbiol* 17(2):66–72.

Duckworth, D. H. 1976. "Who discovered bacteriophage?" *Bacteriol Rev* 40(4):793–802.

Furhrman, J. A. 1999. "Marine viruses and their biogeochemical and ecological effects." *Nature* 399:541–548.

Greer, G. G. 2005. "Bacteriophage control of foodborne bacteriat." *J Food Prot* 68(5):1102–1111.

Hoiby, N. 2010. "Antibiotic resistance of bacterial biofilms." *Int J Antimicrob Agents* 35:322–332.

Hudson, J. A., et al. 2005. "Bacteriophages as biocontrol agents in food." *J Food Prot* 68(2):426–437.

Hyman, P., and Abedon, S. T. 2010. "Bacteriophage host resistance and bacterial resistance." *Adv Appl Microbiol* 70:217–248.

Jones, D. T., et al. 2000. "Bacteriophage infections in the industrial acetone butanol (AB) fermentation process." *J Mol Microbiol Biotechnol* 2(1):21–26.

Kanamaru, S., et al. 2002. "Structure of the cell-puncturing device of bacteriophage T4." *Nature* 415:553–557.

Labrie, S., et al. 2010. "Bacteriophage resistance mechanisms." *Nat Rev Microbiol* 8:317–327.

Letellier, L., et al. 2004. "Main features on tailed phage, host recognition and DNA uptake." *Bioscience* 9:1228–1239.

Maki, D. G., and Tambyah, P. A. 2001. "Engineering out the risk of infection with urinary catheters." *EID* 7(2):1–6.
Miedzybrodzki, R., et al. 2007. "Phage therapy of staphylococcal infections (including MRSA) may be less expensive than antibiotic treatment." *Postepy Hig Med Dosw* (61):461–465.
Miller, E. S., et al. 2003. "Bacteriophage T4 genome." *Microbiol Mol Biol Rev* 67(1):86–156.
Pennazio, S. 2006. "The origin of phage virology." *Rivista di Biologia/Biology Forum* 99:103–130.
Serwer, P., et al. 2007. "Propagating the missing bacteriophages: A large bacteriophage in a new class." *Virol J* 4:21.
Shen, Y., et al. 2006. "Isolation and characterization of *listeria monocytogenes* isolates from ready-to-eat foods in florida." *App Environ Microbiol* 72(7):5073–5076.
Summers, W. C. 1993. "Plague and cholera in India: The bacteriophage inquiry of 1928–1936." *J Hist Med Allied Sci* 48:275–301.
Sutherland, I. W., et al. 2004. "The interaction of phage and biofilms." *FEMS Microbiol Lett* 232:1–6.
Twort, F. W. 1915. "An investigation on the nature of ultra-microscopic viruses." *Lancet* 2:1241–1243.
Weeks, C. R., and Ferretti, J. 1984. "The gene for type a streptococcal exotoxin (erythrogenic toxin) is located in bacteriophage T12." *Infect Immun* 46(2):531–536.
Weitz, J. S. 2010. "Evolutionary ecology of bacterial viruses." *Microbe* 3(4):171–178.
Whitman, W. B., Coleman, D. C., and Wiebe, W. J. 1998. "Prokaryotes: The unseen majority." *Proc Natl Acad Sci USA* 95:6578–6583.
Wunsche, L. 1989. "Importance of bacteriophages in fermentation processes." *Acta Biotechnol* 9(5):395–419.
Wurtz, M. 1992. "Bacteriophage structure." *Microscop Rev* 5:283–309.
Young, R. 1992. "Bacteriophage lysis: Mechanisms and regulation." *Microbiol Rev* 56(3):430–481.

Popular Press

Blakely, P. 1996. "The good virus." *Discover Magazine*, November issue, pp. 50–58.
Cairns, J., Stent, G. S., and Watson, J. D., eds. 1992. *Phage and the Origins of Molecular Biology*, Expanded Edition. Cold Spring Harbor, NY: Cold Spring Harbor Press.
Calendar, R., and Abedon, S. T., eds. 2006. *The Bacteriophages*, 2nd ed. Oxford and New York: University Press.
d' Herelle, F., and Smith, G. H. 2010. *The Bacteriophage, Its Role in Immunity.* Memphis, TN: General Books, LLC.
Hausler, T. 2006. *Viruses vs. Superbugs.* New York: Macmillan.
Koerner, B. 1998. "Return of a killer: Phages may once again fight tough bacterial infections." *U.S. News & World Report*, November issue, pp. 51–52.
Kutter, E., and Sulakvelidze, A. *Bacteriophages: Biology and Applications.* New York: CRC Press.
Lewis, S. 1998. *Arrowsmith*, with a new afterword by E. L. Doctarow, revised edition. New York: Signet Classics.
Shnayerson, M., and Plotkin, M. J. 2003. *The Killers Within: The Deadly Rise of Drug-Resistant Bacteria.* New York: Back Bay Books.
Stahl, F. W., and Hershey, A. D., Eds. 2000. *We Can Sleep Later: Alfred D. Hershey and the Origins of Molecular Biology.* Cold Spring Harbor, NY: Cold Spring Harbor Press.
Waldor, M. K., et al., eds. 2005. *Phages: Their Role in Bacterial Pathogenesis and Biotechnology.* Washington, DC: ASM Press.
Zimmer, C. 2011. *A Planet of Viruses.* Illinois, University of Chicago Press.

Video Productions

Arrowsmith. 1931. 2005. MGM.
The Virus That Cures. 1998. Films for the Humanities.

eLearning

go.jblearning.com/shors2

The site features eLearning, an online review area that provides quizzes and other tools to help you study for your class. You can also follow useful links for in-depth information, or just find out the latest virology and microbiology news.

Glossary

(–) sense Describes viral RNA that is complementary to the viral mRNA and, therefore, must be changed into (+) sense RNA before it can begin to translate itself into proteins within the host cell.

(+) sense Describes viral RNA that, upon entering the host cell, can begin to translate into proteins; it is of the same form as mRNA.

2′5′Oligo (A) synthetase An enzyme that activates the conversion of adenosine 5′-triphosphate (ATP) into oligoadenylates and pyrophosphate that, in the presence of double-stranded RNA, activate latent endoribonuclease (RNAse L), which cleaves single-stranded RNA.

5′m^{7}G mRNA cap structure A methylated guanine nucleotide that is a "cap" located on the 5′ end of messenger RNA (mRNA); during transcription, it is recognized by cellular ribosomes and thus plays an important role in the initiation of translation; the 5′m^{7}g cap helps protect the mRNA from degradation.

β galactosidase An enzyme that cleaves the chromogenic substrate X-gal, resulting in a blue color; useful in quantifying viruses in a plaque assay; *see* X-gal.

β propiolactone A chemical that inactivates viruses to use in vaccines; *see* Vaccine.

A

Aborted life cycle In viruses, a life cycle in which only the early viral genes are expressed.

Abortive replication A block in the life cycle of viruses that can infect cells such that early viral proteins are produced, but late proteins are not, and, therefore, full infectious particles are not produced in mammalian cells.

ACAM2000 Current U.S. smallpox vaccine. It was mas produced in Vero cells and has been stockpiled; currently, military personnel are the only population being vaccinated with ACAM2000.

Acquired immunity A response to a specific immune stimulus that involves immune defensive cells and frequently leads to the establishment of host immunity.

Acquired immunodeficiency syndrome (AIDS) A serious viral disease caused by the human immunodeficiency virus (HIV) in which the T (CD4) lymphocytes are destroyed and opportunistic illnesses occur in the patient.

Active immunity The immune system responds to antigen by producing antibodies and specific lymphocytes; *see* Antibody, Lymphocyte.

Acute disease A disease that develops rapidly, exhibits substantial symptoms, and lasts only a short time.

Acute hepatitis A case of hepatitis in which symptoms are severe and worsen quickly; *see* Hepatitis.

Acute infections Infections that have symptoms of sudden onset or last a short time.

Acute vascular rejection When the human immune system causes a rapid vascular injury in a transplanted organ within a few days to several weeks of the transplant; triggered by complement and antibody response toward the organ tissues.

ADA deficiency A deficiency in the enzyme adenosine deaminase; a condition that severely compromises the immune system by destroying immature lymphocytes so they fail to reach maturity.

Adaptive immunity An arm of the immune system responsible for antibody production and cell-mediated immunity.

Adenosine deaminase (ADA) genes Genes located on chromosome 20 that produce the enzyme adenosine deaminase, which is required for metabolic function of a variety of cells, especially T lymphocytes.

Adjuvant Chemical added to vaccine to enhance its effects (e.g., stimulates the immune response).

Affinity Strength at which an antibody molecule binds an epitope.

AIDS dementia complex A neurological complication of acquired immunodeficiency syndrome (AIDS) that may impair cognitive reasoning, behavior, or motor performance.

Alanine aminotransferase (ALT) An enzyme in the body that, when at elevated levels, indicates liver inflammation that can be the result of viral hepatitis.

Alveolar sacs Tiny subdivisions at the ends of the bronchioles in the lungs that are lined with immune cells called macrophages.

Amantadine An antiviral drug that blocks the M2 ion channel function, resulting in the incomplete release of the viral ribonuclear proteins, thus interfering with uncoating during influenza virus infections.

Ambisense Describes ssRNA that is both + and – sense on the same ssRNA segment.

Amyloid plaques Cluster of dead and dying protein fragments that accumulate in different regions in the brain; characteristic of Alzheimer's patients and those suffering from prion diseases.

Analytical epidemiology Determines the why and how of an epidemic in a given population.

Anesthesia Procedure in which the patient is administered a drug (anesthetic) to facilitate surgery.

Antibody A highly specific protein produced by the body in response to a foreign substance, such as a bacterium or virus, and capable of binding to the substance.

Antibody-dependent cell-mediated cytotoxicity (ADCC) A type of cytotoxicity in which natural killer cells, monocytes, macrophages, or neutrophils bind to target cells containing bound antibodies to kill them.

Antibody-dependent enhancement The rare condition in which certain antibodies can be beneficial to a virus.

Antigen A chemical substance that stimulates the production of antibodies by the body's immune system; also called immunogen.

Antigen-presenting cell (APC) A macrophage or dendritic cell that on its surface expresses major histocompatibility complex class II (MHC II) and exposes antigen peptide fragments to T helper (T_H) cells.

Antigenic drift A minor variation over time in the antigenic composition of influenza viruses.

Antigenic shift A major change in the antigenic composition of a virus (e.g., the reassortments between human and bird influenza viruses, the result of which potentially is pandemic strain of influenza A virus).

Antitoxin An antibody produced by the body that circulates in the bloodstream to provide protection against toxins by neutralizing them.

Antiviral state The cellular state occurring when host cells have secreted interferons in response to a viral invasion and interferons have bound to receptors on neighboring cells, signaling other cells to produce antiviral proteins.

Apoptosis A genetically programmed form of cell death. Apoptosis may be induced during viral infection, allowing for the removal of infected cells in the body.

Artificially acquired active immunity The production of antibodies by the body in response to antigens in a vaccination.

Artificially acquired passive immunity The transfer of antibodies formed in one individual or animal to another susceptible person.

Aseptic meningitis or encephalitis The term *aseptic* is used when medical laboratory technicians have been unable to culture an organism present in the cerebrospinal fluid (CFS) of patients suffering from a central nervous system (CNS) disease or syndrome. Encephalitis or meningitis is an inflammation of the tissues that cover the brain and/or spinal cord.

Assembly The process in which the immature virus particle is formed.

Astrocytosis The spread of astrocytes (star-shaped nerve cells) to damaged tissues in the brain.

Atherosclerosis A thickening of the artery walls that is mainly due to a fatty diet that is high in cholesterol.

Attenuate To reduce the ability of a virus to do damage to the exposed individual. Attenuated virus strains may be used to vaccinate individuals against wild-type viral pathogens.

Atypical pneumonia Any type of pneumonia of an unknown cause.

Australia antigen The original name for the hepatitis B surface antigen (hBsAg).

Autocleavage The process of a protein cleaving itself.

B

B lymphocyte (B cell) A white blood cell that matures into memory cells and plasma cells that secrete antibody.

Bacteriophage (phage) A virus that infects and replicates within bacterial cells.

Bacteriophage therapy Bacteriophages used in the curing of bacterial infections in humans.

Barrier technique A method of containing infection in which an impenetrable object or surface is placed between the infected and the uninfected substance or body (e.g., gloves, masks, and gowns are worn during surgery to prevent the spread of infection).

Bifurcated needle A two-pronged needle used in mass vaccination against the smallpox virus.

Bile A digestive fluid produced by the liver that aids in digestion of fats and gets rid of wastes in the body.

Bilirubin A yellow-pigmented byproduct of old red blood cells that causes jaundice.

Biofilm A complex community of microorganisms that form a protective and adhesive matrix that attaches to a surface, such as a catheter or industrial pipeline.

Bioflavenoids Brightly colored, water-soluble compounds with antibacterial and antiviral properties that are secreted by plants.

Biosafety level 3 (BSL-3) A standard set of procedures for handling, shipping, and disposing of moderately hazardous substances or microbes that may be aerosoled and cause infections in personnel and the environment (e.g., rooms with negative air pressure relative to adjacent facilities; working with hazardous materials in containment, and using a laminar flow hood).

Biosafety level 4 (BSL-4) A standard set of procedures for handling, shipping, and disposing of extremely dangerous substances or microbes to personnel and the environment (e.g., testing anthrax).

Bioterrorism The intentional or threatened use of biological agents to cause fear in or actually inflict death or disease upon a large population.

Bloodborne pathogen A disease-causing microorganism that is carried in the bloodstream and can be transmitted to others if proper precautions are not taken in handling the infected blood.

Blood–brain barrier A semipermeable interface between the bloodstream and central nervous system (CNS) tissues that protects the brain from potentially toxic substances or microbes.

Breakthrough virus A wild-type virus that causes a mild form of disease.

BUdR (5-bromodeoxyuridine) A toxic analog to thymidine in DNA; the vaccinia virus (smallpox vaccine) disrupts the activity of the thymidine kinase gene and renders host cells resistant to the effects of BudR; also an anticancer drug and a potential antiviral drug.

Bulbar polio Less common than spinal polio; results in weakness of muscles, thus impairing the ability to talk or swallow and requiring the need for an iron lung or respirator.

Bulbospinal polio A combination of bulbar and spinal paralysis.

Burgoo Thick spicy stew that contains whatever meat (e.g., squirrel brains) and vegetables available; often served in the southern United States.

Bushmeat The meat of wild animals, often used in reference to monkey or ape meat.

C

Cancer A group of over 100 diseases characterized by the uncontrolled growth of cells in the human body that can migrate to other sites in the body and grow; if the spread is not halted or controlled, cancer can cause death.

Cap-dependent translation The process by which the cellular ribosomes recognize the cap present on the 5′ end of mRNA transcripts (host and some viral) in order to direct their translation or synthesis of proteins.

Cap-independent translation The process by which the cellular ribosome bypasses the 5′ cap recognition requirement in order to direct the translation of viral mRNAs.

Capsid The protein shell that encloses and protects the nucleic acid genome of a virus.

Capsomere Many viral structural proteins that come together to form identical subunits, forming the capsid of a virus.

Carcinogen Any physical or chemical substance that causes cancer; *see* Cancer.

Carcinoma A type of cancer that begins in the skin or in tissue cells lining or covering the body's organs.

Carrier An individual who has recovered from a disease but retains the infectious agent in the body and continues to shed them.

Case-control method A process in experimentation in which two subject groups—one exposed to the tested factor and one not—are compared with each other scientifically.

Case definition A standard set of criteria used to identify which person has the disease being studied.

Caspases A family of proteases that are activated during apoptosis.

CCR5 A coreceptor present on macrophages that helps HIV to enter cells.

CCR5-Δ32 A genetic polymorphism of the *CCR5* gene that codes for a protein that is severely truncated and blocks HIV entry because the viral gp120/gp41 cannot attach to it to infect cells; it has been linked to the resistance to HIV infection.

CD4 A type of receptor expressed on the surface of T helper (T_H) cells in association with the T cell receptor (TCR) on their surface in order to recognize antigen bound to major histocompatibility complex class II (MHC II) molecules.

Cell cycle The life cycle of a cell from formation to division.

Cell cultures The cultivation of living cells outside of the host (in media that supports their growth) under controlled conditions (e.g., monkey kidney cells, HeLa cells, mouse L cells).

Cell-mediated immune response The body's ability to resist infection through the activity of T lymphocyte recognition of antigen peptides presented on macrophages and dendritic cells and on infected cells.

Cell tropism The affinity a virus has to target and infect a particular cell type.

Cellular oncogene A cancer-producing gene: a normal cell gene that can be activated by a corresponding oncogene in a virus so that the cell becomes cancerous.

Cellular scavenging mechanism Cellular molecules that remove harmful free radicals in order to detoxify their effects on the cell, which helps to protect the cell from DNA damage.

Cellular surface receptors Usually proteins, glycoproteins, carbohydrates, or lipids that allow the attachment proteins on a virus to infect cells.

Central nervous system (CNS) Refers to the brain and spinal cord.

Centrifugation culture (shell vial technique) Technique used in clinical labs for rapid diagnosis of viral infections; involves propagation of cells in shell vials that contain cover slips on the bottom of the vial. Clinical samples are added to the vials, the vials are centrifuged at a slow speed, incubated for 36 hours, and then analyzed for the presence of viral antigens.

Centrifugal rash A rash mainly covering the face and extremities as opposed to the entire body.

Chain of infection 1. A process from when an agent leaves its reservoir/host, is conveyed by a mode of transmission, and enters a portal to infect a susceptible host. 2. The practices used to control and prevent infectious diseases (e.g., proper sanitation, hand washing, and recognizing high-risk individuals).

Chain of transmission How infectious diseases can be spread from human to human (or animal/insect to human).

Chain termination The end effect of nucleoside and nonnucleoside analogs that prevent the elongation of viral DNA of an infected cell, halting their replication.

Chamberland filter A porcelain filter that removes microorganisms from a liquid.

Chemokine A small protein that prompts specific white blood cells to migrate to an infection site and carry out their immune system functions.

Chemotherapy The use of a chemical agent to selectively kill cancerous cells.

Chickenpox A mild infectious disease caused by the varicella zoster virus that usually afflicts children but may be severe in infants, adults, and persons with impaired immune systems.

Chorioallantoic membrane An extensively vascularized membrane covering the embryos of birds and fishes.

Chronic hepatitis Hepatitis lasting longer than six months that often does not show any symptoms until the late stages of the disease.

Chronic infection An infection lasting more than six months.

Chronic rejection When the human immune system rejects an organ within months to years following transplantation; probably the result of continued, prolonged multiple acute rejections.

Chronic wasting disease (CWD) Prion disease of cervids (e.g., deer, elk, moose); sick animals have poor coats and appear to be starving; *see* Prion.

Cidofovir An antiviral medication used for patients with AIDS and HIV, and for treating adverse reactions associated with the smallpox vaccine; also known as HPMPC or Vistide.

Ciliary action The sweeping motion of hairlike cilia of respiratory cells that helps to move particles.

Circulating recombinant forms (CRF) Two HIV viruses of different subtypes can infect the same cell in an individual, resulting in new HIV progeny hybrid viruses with recombined genomes. These HIV hybrid viruses often do not survive very long, but if they do infect more than one individual, they are referred to as CRFs.

Cirrhosis The formation of fibrous tissues, nodules, and scarring on the liver that interfere with its cell function and blood circulation.

Clade (*also,* subtypes) A group of organisms that have a common ancestor.

Clathrin A large, fibrous protein that forms a basketlike network around vesicles.

Clathrin-coated pits Specialized regions of the cell membrane that appear as invaginations that are coated on the cytoplasmic side of the membrane with dark material.

Clathrin-coated vesicles Transport vesicles coated with a latticelike network of the protein clathrin.

Cohort method A type of experimentation in which two similar populations are studied and participants are grouped by common characteristics.

Commerce Large-scale buying and selling of goods between cities and countries; food commerce today is a major cause of viral spread.

Common name A nickname by which a virus or disease may be known that is not as scientific as its taxonomic name.

Communicable disease A disease that is readily transmissible between hosts.

Communicable period The time period when an infected individual or animal is contagious and he/she can directly or indirectly infect another person, animal, or arthropod.

Complement system Over 30 different serum- and membrane-bound glycoproteins that act in sequence when activated by a viral infection to form a membrane attack complex that ultimately causes the destruction of the invading viruses and infected cells; the three complement system pathways are classical, lectin, and alternative.

Complex The shape used to describe viruses that do not fit the typical icosahedral or helical design.

Confidential name-based reporting A method of assessing and monitoring the AIDS epidemic. Patient names are forwarded to the local health department by healthcare providers when AIDS or HIV is diagnosed. Patient names are kept secure and confidential at the state health department.

Congenital effects (also, teratogenic effects) Severe congenital malformations in a fetus caused when some bloodborne viruses cross the placenta to reach the fetal circulation.

Conjunctiva The thin, transparent tissue that covers the outer surface of the eye.

Conjunctivitis (*also,* pinkeye) A general term for disease of the conjunctiva, the thin mucous membrane that covers the cornea and forms the inner eyelid.

Contact inhibition When normal cells stop replicating because they come in contact with another cell; in cancers, abnormal cells continue to divide and pile up on top of each other.

Contagious A disease whose agent passes with particular ease among hosts.

Containment books Record-keeping for smallpox outbreak investigations. These books provide basic data, including the cases of smallpox and corresponding dates of occurrence, which allow field workers to plot the spread of the disease.

Containment vaccination Vaccination of a population in a certain area around an outbreak of a disease.

Control group A group of subjects in an experiment who are not subject to the factor being tested.

Convalescence The recovery period after an illness.

Convalescent serum Antibody-rich serum obtained from a convalescing patient.

Coreceptor A second cellular receptor necessary for viral entry into a cell.

Cross-sectional study A type of research that looks at its subjects at a single point in time.

Crusades A time period of Christian religious expeditions and wars (1095–1270 AD).

CXCR4 (*also,* fusin) The T lymphocyte coreceptor that helps to allow HIV to enter cells.

Cyanosis Bluish discoloration of the skin and mucous membranes caused by a lack of oxygen in the blood. It can be associated with certain viral diseases (e.g., blue ear pig disease).

Cyanovirin An HIV cell fusion inhibitor that was isolated from cultures of blue green algae (cyanobacterium).

Cyclosporin A drug that suppresses the body's immune system; used to prevent organ rejection after transplantation.

Cytokine Small protein released by immune defensive cells that affects other cells and the immune response to an infectious agent.

Cytokine storm (*also,* systemic inflammatory response syndrome [SIRS]) An immune system that is overreacting toward the pathogen; cytokine-signaling immune cells such as macrophages and T cells travel to the site of infection causing damage to the body and resulting in multiple organ failure.

Cytopathic effects (CPEs) The visible changes in a host cell that take place because of a viral infection.

Cytoplasm The part of a eukaryotic cell in which processed mRNAs exit the nucleus and are translated.

Cytotoxic T cells T lymphocytic cells that express CD8 in association with T cell receptors (TCR) on their surfaces; they also recognize and destroy antigens that are associated with major histocompatibility complex class I (MHC I) molecules present on infected cells.

D

Dane particle A 42 to 47 nm mature spherical hepatitis B virus particle.

Dead-end hosts Virus hosts that cannot transmit the virus to further species.

Defective interfering particles Noninfectious virus particles coproduced with infectious viruses during infection. These particles contain defective genomes and may interfere with subsequent infections.

Defoliation Loss of leaves; a characteristic of viral infections in plants.

Dendritic cells Immune cells present in tissues that have contact with the environment and identify foreign invaders such as viruses and bacteria; once dendritic cells find foreign invaders, they migrate to the nearest lymph node via the lymphatics, where they will come in near proximity to other immune cells to eliminate the pathogen.

Deoxyribonucleic acid (DNA) The genetic material of all cells and many viruses.

Descriptive studies A type of research that is performed "after the fact"; used to generate a testable hypothesis or enough evidence to indicate sufficient cause for conducting lengthier and more costly analytical studies.

DICER A host cell RNAse III-like enzyme that chops or "dices" dsRNA.

Differential diagnosis A systematic method used by physicians to distinguish between two or more diseases with similar symptoms.

Dilated cardiomyopathy (DCM) Disease of the heart muscle that causes the heart to become enlarged and to pump less strongly; the heart becomes thin, weak, or floppy and is unable to pump blood efficiently around the body, causing fluid to build up in the lungs, which leads to congestion and the feeling of breathlessness.

Direct contact The form of disease transmission involving close association between hosts; *see also* Indirect contact.

Direct detection A method to diagnose viral infections that directly visualizes virus particles contained in a clinical specimen (e.g., electron microscopy).

Disease Any change from the general state of good health.

DNA-dependent DNA polymerases Enzymes that catalyze the synthesis of DNA using an RNA template.

DNA ligase An enzyme that binds together DNA fragments; it creates an ester bond between adjacent 3′OH and 5′PO_4 ends on cleaved DNA.

DNA microarray A tool consisting of a chip that contains an array of nucleic acids for use in probing a biological sample to determine gene expression. Large numbers of genes can be assessed and classified (e.g., the genes present in tumors of cancer patients). It also is used to identify pathogens causing a particular infection (i.e., the quick identifying or screening for types of influenza, severe acute respiratory syndrome [SARS]-CoV, respiratory syncytial virus [RSV], and other airborne bioterrorist pathogens).

DNA polymerase An enzyme that catalyzes DNA replication by combining complementary nucleotides to an existing strand.

DNA template The original strand of DNA to be replicated.

DNA vaccine A vaccine consisting of plasmid DNAs that have been manipulated to contain a gene encoding a viral antigen of interest; it raises an immune response to the expressed antigen carried on the plasmid DNA.

Donor One who supplies pieces of body tissue or organ(s) for use or transplantation into another person's body.

Downstream DNA element A region of DNA that is located further in the direction of transcription.

dsRNA RNA with two complementary strands that trigger the RNA interference (RNAi) defense system.

E

Eclipse phase The period after the infection in which the input, or inoculated virus, disappears. No virus particles are detected at this time.

Eczema vaccinatum A rare and severe side effect of smallpox vaccination that can occur in people with eczema; causes extensive lesions and may be fatal.

El niño An abnormal warming of the surface ocean waters that causes weather changes around the globe.

Emerging infectious disease A new disease or changing disease that is seen for the first time; *see also* Reemerging infectious disease.

Emerging virus A new or recently identified virus.

Emigration The movement of a population from the native land to settle in another location.

Enanthem A rash that affects mucous membranes; accompanies some viral infections.

Encephalitis Inflammation of the tissue of the brain or infection of the brain.

Encephalomyelitis Inflammation of the brain and spinal cord.

Encephalopathy General term for disease of the brain.

Endemic A disease that is constantly present in a specific area or region.

Endocannibalism A funeral ritual in which the bodies of dead members of a group or tribe are eaten by their relatives as a sign of love and respect.

Endogenous infection A disorder that starts with a microbe or virus that already was in or on the body as part of the microbiota; *see also* Exogenous infection.

Endogenous retroviruses (*also,* proviruses) A retrovirus formed when the retroviral DNA is inherited from parent to offspring of the infected host if germline cells (sperm and egg) contain the integrated viral genome.

Endolysin An enzyme that allows viruses to escape or be released from the host and hydrolyzes specific bonds in the murein or peptidoglycan layer of the cell wall.

Endosome A clathrin-coated vesicle shortly after formation, when its clathrin coat has been removed (*see* Clathrin).

Enhancers DNA elements outside of the promoter region that stimulate the frequency of transcription of genes by RNA polymerase II.

env A retrovirus gene that encodes a protein that is embedded within a lipid bilayer and surrounds the nucleoprotein core particle of the virus; these proteins are the "spikes" that bind to host cell receptors.

Envelope The flexible membrane of protein and lipid that surrounds many types of viruses.

Enveloped virus A human or animal virus with an additional lipid bilayer membrane that has been stolen from the host cell wrapped around the capsid of the virus particle.

Enzyme A reusable protein that while remaining unchanged itself brings about a chemical change.

Enzyme-linked immunosorbent assay (ELISA) A serological test in which an enzyme system is used to detect an individual's exposure to a pathogen.

Epidemic An unusually high number of cases in excess of normal expectation of a similar illness in a population, community, or region.

Epidemiological triad Shows the interactions among the host, agent, and environment that produce disease.

Epidemiology The study of how diseases affect whole communities; the scientific field linked to virology.

Episome A viral genome that is maintained in cells by autonomous replication.

Epitope (*also,* antigenic determinant) A section of an antigen molecule that stimulates a specific response and to which the response is directed; often consists of several amino acids or monosaccharides.

Eradication To destroy or remove a pathogen after a disease has been established (e.g., smallpox eradication).

Essential genes Genes essential for growth in cell cultures; they are located in the central part of the genome of poxviruses.

Etiological agent The disease-causing agent.

Etiology The assignment of the cause or origins of a disease.

Eukaryotic Refers to a cell or organism containing a cell nucleus with multiple chromosomes, a nuclear envelope, and membrane-bound compartments.

European Medicines Evaluation Agency (EMEA) A decentralized agency of the European Union that evaluates and supervises any medicines used for humans and animals.

Exanthemasubitum A maculopapular rash occurring on the skin surface.

Exanthematous Describes a disease accompanied by skin eruption.

Exhuming variola victims Disinterring the bodies of dead smallpox victims in order to study the variola virus.

Exogenous infection A disorder that starts with a microbe or viruses that enter the body from the environment; *see also* Endogenous infection.

Exotic animals/pets Rare or unusual animals kept as pets but are not commonly thought of as a pet (e.g., tiger, monkey, anaconda).

Exotic pet trade The import of exotic species from around the world to be sold as pets.

F

Family In biology, a taxonomic category of related organisms ranking below an order and above a genus; the second broadest classification in the taxonomy of viruses.

Fc fragment The portion of an antibody molecule that combines with phagocytes, mast cells, or complement. Fc stands for the "fragment crystallizable" portion of an antibody binding site.

Febrile response A rise in body temperature (fever); a defense mechanism against some invading viruses and other microbes.

Fixed strains Virus strains used in laboratory research as opposed to strains isolated directly from infected humans or animals; for example, rabies fixed strains.

Flaccid paralysis A loss of voluntary movement or paralysis because weakness or loss of muscle tone has caused damage to the nerves in the muscle (e.g., as a result of poliovirus infections).

Flavivirus A virus genus that includes the human pathogens West Nile virus, Dengue fever virus, and yellow fever virus.

Fluor A fluorescent compound, as a dye that is fluorescent in ultraviolet light and used in a method of viral study and isolation utilizing real-time polymerase chain reaction (PCR) and fluorescence resonance energy transfer (FRET).

Fluorescence The emission of one color of light after being exposed to light of another wavelength.

Fluorescence resonance energy transfer (FRET) When used in combination with real-time polymerase chain reaction (PCR), part of a virus detection method that emits a fluorescent signal proportional to the amount of PCR product in pathogens, and for viral-load monitoring.

Fluorescent antibodies Antibodies containing a fluorescent "tag." The tagged antibodies are used as a stain to detect specific viral antigens present in the clinical specimen or virus-infected cell cultures. Any unbound antibody is washed away and the specimen is observed with a fluorescent microscope.

Fomite An inanimate object, such as clothing or a utensil, that carries disease organisms.

Formalin A 37% to 40% solution of formaldehyde used as embalming fluid, in the inactivation of viruses, and as a disinfectant.

Free-ranging animals Animals or birds on farms that are allowed to roam freely instead of being contained in any manner.

FRET technology (*see,* Fluorescence resonance energy transfer).

Fusion protein A protein created by joining ("fusing") two genes together; these can be created in the laboratory for research but can also occur naturally.

G

gag An acronym for a group-specific antigen gene that encodes the internal structural proteins of retroviruses such as HIV.

Gastroenteritis An infection of the stomach and intestinal tract, often due to a virus.

Gay cancer Another name in the early 1980s for Kaposi's sarcoma, a rare skin cancer that had begun to appear in many young homosexual men.

Gene A segment of a DNA molecule that provides the biochemical information for a function product.

Gene therapy An experimental treatment that involves the introduction of genes into a person's cells to replace or compensate for defective genes in a person's body that are responsible for a disease or medical problem; genes can be carried and delivered by customized vectors such as viruses.

Genetic code The specific order of nucleotide sequences in DNA or RNA that encode specific amino acids for protein synthesis.

Genetic engineering The use of bacterial and microbial genetics to isolate, manipulate, recombine, and express genes.

Genetic recombination The process of bringing together different segments of DNA.

Genetically modified organism (GMO) An organism produced by genetic engineering.

Genome The complete set of genes in a virus or an organism.

Genus The second smallest classification in the taxonomy of viruses.

Germ theory The principle formulated by Pasteur and proven by Koch that microorganisms are responsible for infectious diseases.

Germline cells Sperm and egg cells.

Girus Exceptionally large viruses that are not filterable through filters with a 0.2 to 0.3-μm pore size.

Glassy transformation An important observation made by Frederick Twort in the discovery of bacteriophages; the change to a watery or glassy appearance undergone by micrococci when filtrate from other watery-appearing micrococci was applied to them.

Glioma Tumor that originates from the brain or spinal cord.

Glucocorticoids Class of steroid hormones that upregulate the expression of pro-inflammatory proteins.

Glycan microarray A new technology that focuses on detecting the virus–receptor specificity of new influenza strains, especially H5N1 viruses.

Glycosaminoglycans (GAGs) Long, chained molecules of disaccharides on the surfaces of cells to which the viral gb or gc envelope proteins of HSV attach or dock. The vaccinia virus A27L virion protein also interacts with GAGs.

Glycosylation The process of adding saccharide molecules to protein molecules.

Grafting A form of vegetative propagation in which two young plants are joined together; provides an opportunity for virus transmission.

Green Revolution A time period beginning in the early 1960s when agricultural productivity increased significantly due to the development of hybrid seeds, use of pesticides, and mechanization that allowed for scaled-up production of crops.

GRID (gay-related immune deficiency) The original name given to AIDS by the Centers for Disease Control in 1981.

H

H1N1 Influenza A strain that caused a pandemic in 2009. The epicenter of the pandemic was Mexico.

H5N1 Influenza A avian strain that is 60% lethal in humans, 99% lethal in domesticated birds (e.g., chickens and turkeys in poultry farms).

HAART (highly active antiretroviral therapy) The administration of at least three different compounds to jointly block viral replication by inhibiting the viral protease and/or reverse transcriptase of the HIV virus; was a possible approach to eliminate the virus from the body altogether.

Hairpin The structure on a viral genome that acts as an internal ribosome entry site (IRES), which is necessary for cap-independent translation of the viral RNA; hepatitis A and poliovirus contain hairpin structures in their genome.

Hammerhead motif A unit of plant viroid RNAs with enzymatic self-cleaving properties.

Hand, foot, and mouth disease A virulent disease often caused by enterovirus 71 and Coxsackie viruses. It is common in infants and children; symptoms are fever, sores in the mouth, and blisters on the hands and the bottom of toes.

HBsAg Hepatitis B surface antigen.

Helicase The enzyme that separates double-stranded DNA into single strands for replication.

Helper viruses Viruses that enable satellite viruses to replicate.

Hemadsorption test A test used to detect the presence of viruses that bind to red blood cells.

Hemagglutination The process in which there is a virus in solution and red blood cells are added to the solution; subsequently the virus can interact with the red blood cells to form a lattice-like structure, resulting in agglutination of red blood cells.

Hemagglutinin An agent (i.e., a virus or antibody) that causes red blood cells to clump together.

Hemophilia A congenital disorder in which blood cannot clot normally. The most common form of hemophilia is caused by a defective clotting factor known as factor VIII.

Hepadnavirus Virus capable of causing hepatitis B in humans and that replicates through an RNA intermediate, sometimes referred to as a pregenomic RNA by reverse transcription.

Heparin sulfate The most common kind of glycosaminoglycan; the preferred binding receptor for viral gb or gc envelope proteins of HSV.

Hepatitis Inflammation of the liver caused by a number of viruses or by alcohol or prescription drugs.

Hepatitis non-A, non-B The hepatitis C virus, which causes chronic hepatitis in 80% of infected individuals.

Hepatocellular cancer (HCC) Liver cancer; it may be associated with hepatitis B or C infections.

Hepatocyte A liver cell.

Herd immunity The premise that, if the majority of the population is protected from a disease through immunizations or genetic resistance, the chance of a major epidemic is highly unlikely; the proportion of a population that is immune to a disease.

Heterozygous Containing two different alleles for the same gene, one from each parent.

High efficiency particulate air (HEPA) filter A tightly meshed screen system that removes 99.97% of particles of 0.3 μm or higher from the air.

HIV home testing system A simple blood test to detect HIV; a "finger prick" blood sample is applied to a filter paper, packaged, and then mailed to a laboratory for testing.

Holin An enzyme used by bacteriophages to create pores in the inner membrane of the host cell to facilitate host lysis and bacteriophage release.

Homolog A gene sequence that is similar to a gene sequence in another species both structurally and in evolutionary origin.

Homozygous Having two identical alleles for a gene.

Hormone A chemical substance synthesized by one type of cell that travels through the bloodstream to affect and/or direct the function of another type of cell.

Host An organism on or in which a microorganism lives and grows or in which a virus replicates.

Host range The variety of species and types of cells that a disease-causing microorganism or virus can infect.

Human immunodeficiency virus (HIV) The virus that causes AIDS by attacking and compromising the body's immune system.

Hunter theory The theory that HIV was originally transmitted to human beings via infected chimpanzees that were killed and eaten by African hunters.

Hydrophobia An emotional condition ("fear of water") arising from the inability to swallow as a consequence of rabies.

Hyperacute rejection When the immune system quickly kills a transplanted organ and thus kills the recipient.

Hypocomplementemia A rare condition in which components of the complement system are lacking or reduced in concentration.

Hypoglossal nerve The nerve that innervates the muscles in the tongue.

Hypovolemic shock A type of shock in which severe loss of blood and fluid from the body makes the heart unable to pump enough blood to the body's organs.

I

Iatrogenic disease Physician-induced disease introduced during a medical treatment or procedure.

Icosahedron A symmetrical figure composed of 20 triangular faces and 12 points; one of the major shapes of some viral capsids.

IgA (*also,* immunoglobulin A) A class of antibody found in bodily secretions, especially the mucous membranes of the respiratory tract, the gastrointestinal tract, and other mucous surfaces, that helps to neutralize pathogens.

IgG (*also,* immunoglobulin G) The most abundant class of antibody present in serum; produced for very long periods of time; offers long-term protection against encounters with prior pathogens.

IgM (*also,* immunoglobulin M) A class of antibody that is the first immunoglobin to respond to a viral or other microbial antigen.

Immune modulator A substance that changes the immune system to boost immunity.

Immune–contraceptive vaccine A vaccine that causes the body of the female of a species to produce an immune reaction to sperm and egg proteins, thus causing infertility.

Immunocytochemistry Methods that use antibodies to detect or study host cell–virus interactions in cell cultures or tissues.

Immunofluorescence A method of detecting viral antigens within a cell or on its surface. The virus-specific antibodies containing a fluorescent tag bound to the Fc region of the antibody are allowed to react with the specimen that contains the virus, and any unbound antibody is washed away and the specimen is observed under a fluorescent microscope.

Immunogen Any substance capable of generating an immune response.

Immunoglobulin (Ig) The class of immunological proteins that react with an antigen; an alternate term for antibody.

Immunological memory The long-term ability of the immune system to remember past pathogen exposures.

Immunology The scientific study of how the immune system works and responds to nonself agents.

Immunomodulation The modification of some aspect of the immune system as part of a treatment, especially the suppression of an overactive inflammatory response.

Immunotherapy Treatment of a disease by enhancing or suppressing the actions of the body's immune system.

***In situ* hybridization** A method of locating nucleic acids in the place where they naturally occur.

Inactivation protocols Methods of killing but not destroying a virus so it is no longer able to infect cells; developed to determine what procedures are necessary to decontaminate or completely disinfect surfaces where viruses are present.

Incidence A measurement of morbidity; the number of new cases of a disease that occurs in a specified period of time in a susceptible population, expressed per 1000 persons.

Incidental host A host that becomes infected but is not required for the maintenance of a pathogen (e.g., humans are incidental hosts for West Nile virus).

Incineration To burn to ashes or cause something to burn to ashes.

Inclusion bodies Subtle intracellular abnormalities that only occur in virus-infected cells.

Incubation period The time that elapses between the entry of a pathogen into the host and the appearance of signs and symptoms of disease.

Indirect contact The mode of disease transmission involving nonliving objects; *see also* Direct contact.

Infectious disease A disorder arising from a pathogen invading a susceptible host and inducing medically significant symptoms.

Infectious dose The number of microorganisms or viruses needed to bring about infection.

Infectious hepatitis A hepatitis A virus infection; can cause symptoms such as fever, nausea, and jaundice.

Infectious mononucleosis A virus infection causing a sore throat, fever, and enlarged and tender glands. It frequently causes a prolonged period of tiredness.

Influenza hemagglutinin (HA) An enzyme composing one type of surface spike on influenza viruses that enables the viruses to bind to the host cell.

Innate immunity (also, nonspecific immunity) An inborn set of the preexisting defenses against infectious agents; includes the skin, mucous membranes, and secretions; protects humans against any pathogen, regardless of the species or type of microbe, but is not improved by repeated exposure to the pathogen.

Insertional activation The increase in expression of a *c-onc* due to the insertion of a viral promoter or enhancer element into the host's chromosome upon infection.

Integrase activity Enzyme that facilitates the integration of a virus into cellular DNA; HIV and hepatitis B viruses possess integrase activity.

Interference assays A method of virus detection in which the assays' ability to interfere with the growth of a second virus added to the same cell culture is examined.

Interferon An antiviral protein produced by body cells on exposure to viruses and that triggers the synthesis of antiviral proteins; proteins that are the first line of defense against all viral infections; induce nonspecific immune responses before specific ones begin.

Interferon (pegylated) An active interferon molecule that has an attached polyethylene glycol molecule fused to it; an antiviral used to treat hepatitis C infections.

Interleukin A chemical cytokine produced by white blood cells that causes other white blood cells to divide; *see also* Cytokine.

Internal base-pairing Base-pairing that occurs across adjacent portions of a nucleic acid molecule.

Internal ribosomal entry sites (IRES) Sites that allow the ribosome to enter the 59 end of the viral RNA independently of a cap structure at the 5′ end and cause translation.

Intervillous space Space between the vessels of the mother and growing embryo.

Invasive species A species that is not native to its environment, reproduces quickly, and often has adverse environmental and economic effects.

In vitro Taking place in an artificial environment outside of the body (e.g., in a laboratory vessel).

In vivo Biological process occurring in a living organism or natural environment.

Iron lung An artificial respirator in the shape of a cylindrical steel chamber or tank and containing pumped air, causing the lungs of a patient to expand; used from the 1930s to the 1950s to help victims of poliomyelitis in breathing.

Isolation A method to diagnose viral infections in which various methods and host systems are used to isolate viruses from clinical specimens; also the separation of ill/infected individuals from healthy individuals.

J

Jaundiced A condition in which bile seeps into the circulatory system, causing the complexion to have a dull yellow color.

K

Kaposi's sarcoma A type of cancer in immunocompromised individuals, such as AIDS patients, where cancer cells and an abnormal growth of blood vessels form solid lesions in connective tissue.

Killed vaccine A vaccine in which whole-virus particles have been inactivated.

Koch's postulates A set of procedures by which a specific organism (typically bacteria) can be related to a specific disease.

Kozak's rules Marilyn Kozak's hypothesis that the 40S ribosome will scan along the mRNA until it encounters an AUG in the best consensus sequence (usually, but not always, the first AUG); the consensus sequence is GCC A/G CCAUG(G).

Kuru A fatal brain disorder that occurred among the South Fore tribe of the highlands of Papua New Guinea at epidemic levels during the 1950s to 1960s; caused by tribal members eating infected human tissues as part of their funeral ceremonies.

L

Laminar vertical flow biosafety hood A piece of equipment used in virology laboratories that provides clean/sterile air to the working area and provides a constant flow of air out of the work area to prevent room air from entering it.

Langerhans cells Dendritic cells in the skin.

Latency A process in which the viral DNA (provirus) becomes inserted into the host's DNA and the proviral DNA replicates "silently" along with the cellular DNA without immediately causing disease.

Lateral bodies Part of the structure of poxviruses containing various enzymes essential for viral replication.

LATs Latency associated transcripts of herpesviruses; a limited group of viral genes expressed in the latent stage.

Leaky scanning A strategy that allows multiple viral proteins to be synthesized from a single messenger RNA (mRNA). The cell's ribosome may inefficiently initiate translation at the first AUG of the genetic code, but more often the 40S ribosome will bypass that first AUG and initiate translation farther downstream at an AUG in better context.

Ligand A molecule that binds to another to form a larger molecule.

Ligand-mediated fusion Fusion in which viral ligand rather than the host receptor mediates the event.

Ligase An enzyme that joins DNA molecules together by creating a phosphodiester bond between the two molecules.

Listeriosis An infection caused by a foodborne listeria bacteria that primarily affects pregnant women.

Live attenuated vaccine A traditional vaccine that stimulates the host to produce antibodies with highly weakened, but still live, viruses.

Localized infection An infection that, for one or more reasons, is restricted to a confined location.

Long-term nonprogressors A 1% subgroup of HIV-1 infected individuals who have not previously taken or received antiretroviral therapy and show no sign of disease.

LUCA (last universal common ancestor) A common single-celled ancestor from which all living organisms on Earth have evolved or descended from.

LTR (long terminal repeat) Long terminal repetitive sequences of DNA or RNA; located at the ends of the retrovirus genome.

Lymph A watery fluid that consists of water, salts, glucose, urea, proteins, and some immune cells (i.e., lymphocytes and macrophages).

Lymph node A bean-shaped organ located along lymph vessels that is involved in the immune response and contains phagocytes and lymphocytes.

Lymphadenopathy Swollen lymph nodes.

Lymphatic system The series of interconnected and interdependant spaces and channels between the organs and body tissues that circulate lymph (including fats and infection fighting cells [lymphocytes]), removing cell debris and foreign matter from the tissues and transferring them to the blood; part of the immune defense.

Lymphocytes White blood cells that are formed in the lymphoid organs and tissues and that make up 25% to 33% of all white blood cells in adult peripheral blood; lymphocytes exhibit exquisite recognition in distinguishing pathogens from host cells.

Lymphoma Cancer that arises from the lymphatic system.

Lysine One of the 20 essential amino acids present in many proteins.

Lysis or **Lyse** The rupture of a cell and the loss of its contents.

Lysogenic conversion When a prophage carries genes that alter the phenotype of a lysogenic bacterium.

Lysogenic cycle The events of a bacterial virus that integrate its DNA into the bacterial chromosome.

Lysogenic virus A bacteriophage that, when it infects its host, integrates its genome into a specific region of the host chromosome and replicates every time the cell copies its chromosomal DNA during cell division.

Lysozyme An enzyme found in tears and saliva that digests the peptidoglycan of gram-positive bacterial cell walls.

Lytic cycle A process by which a bacterial virus replicates within a host cell and ultimately destroys the host cell.

M

Macrophage A large cell derived from monocytes that is found within various tissues; macrophages are five to ten times larger than monocytes and contain more lysosomes; they can ingest viruses and other foreign invaders, dead cells, cell debris, and other cellular matter.

Macule A patch of skin that is pink-red but not usually elevated; associated with infectious disease.

Maculopapular rash In general, a large area that is red and has small, confluent bumps.

Major histocompatibility complex (MHC) A set of genes that controls the expression of MHC proteins; involved in transplant rejection.

Major histocompatibility complex (MHC) protein Any of a set of proteins at the surface of all body cells that identify the uniqueness of the individual.

Malaise A vague feeling of bodily discomfort, as at the beginning of an illness.

Malignancies Cancerous tumors that have the potential to spread to other parts of the body and invade and destroy tissue; if not stopped, they can lead to death of the organism.

MassTag PCR A sensitive polymerase chain reaction (PCR) method that allows for the nucleic acid detection of 20 to 30 different pathogens simultaneously in clinical samples.

Mass vaccination The vaccination of great numbers of people at one time.

Matrix protein A protein shell between the genome and capsid that is found in some viruses.

Maturation The stage of the virus life cycle in which the virus becomes infectious.

Mechanical immunity (*also,* "barrier immunity") The body's ability to prevent viruses from gaining entry into the body.

Mechanical vector A living organism, or an object, that transmits disease agents on its surface; *see also* Biological vector.

Megacity Cities that contain at least ten million inhabitants.

Membrane attack complex The end result of complement activation that inserts into lipid membranes of bacteria eukaryotic cells or viruses and causes osmotic lysis.

Memory cell A cell derived from B lymphocytes or T lymphocytes that reacts rapidly upon reexposure to antigen.

Meninges The covering layers of the brain and spinal cord.

Meningitis A general term for inflammation of the covering layers of the brain and spinal cord due to any of several bacteria, fungi, viruses, or protozoa.

Messenger RNA (mRNA) An RNA transcript containing the information for synthesizing a specific polypeptide.

Metagenomics A relatively new genetic strategy based on technical advances in sequencing protocols used to analyze the large genomic DNA fragments recovered from environmental samples.

Metastasis When a cell or clump of cells separates from a tumor and spreads to another location.

Miasmatic theory A popular theory of disease in the 19th century that holds that disease is caused by *miasma* or polluted gases that rise from swamps and decaying or putrid matter.

Microbiocide Any agent that kills microbe.

Microorganism (microbe) A microscopic form of life including bacterial, archaeal, fungal, and protozoal cells.

Mode of transmission Defines how an infectious disease is spread or passed on; can be direct or indirect.

Molecular epidemiology The study of the sources, causes, and mode of transmission of diseases by using molecular diagnostic techniques.

Molecular mimicry The immune response against the viral protein can break tolerance to the self-protein and result in an autoimmune disease.

Molluscum Pink, pearl-like lesions, often with a dimple in the center, that are a sign of *molluscum contagiosum,* a common poxvirus infection; a marker for severe immunodeficiency.

Monoclonal antibodies Clones of a single antibody produced by one parent immune cell that consists of antigen-stimulated B cells fused to myeloma cells; "pure" antibodies. A monoclonal antibody recognizes only one antigen.

Monocyte Precursors to macrophages; large white blood cells that circulate in the bloodstream, during which time they enlarge, migrate to tissues, and differentiate into macrophages.

Monolayers A confluent, single layer of cells in culture growing side by side on the bottom of a dish.

Monopartite (*also,* linear genome) A configuration in which the entire genome occupies one nucleic acid molecule.

Monotherapy Treating a disease using only one drug.

Morbidity Refers to the illness or disease state.

Morphogenesis The process of creating form and structure during growth and development; the formation of virus particles.

Mortality Refers to numbers of deaths correlated with a particular disease.

Mosaic patterns on plants A symptom of a viral infection in plants characterized by a pattern of parallel, discontinuous streaks.

Movement proteins Proteins in plant viruses that enable their movement from one infected cell to neighboring cells.

Multipartite genomes A situation in which the entire genome occupies several nucleic acid molecules.

Multiplicity of infection (MOI) The average number of viruses that infect a single cell in a specific experiment.

Muramidase An enzyme that allows viruses to escape or be released from the host and hydrolyzes specific bonds in the murein or peptidoglycan layer of the cell wall; produced by some bacteriophages.

Mutation A change in the characteristic of an organism or virus arising from a permanent alteration of a DNA or RNA sequence.

Myalgia Pain in muscles.

Myocarditis The inflammation or degeneration of the myocardium, often due to Coxsackie virus.

Myocarditis and dilated cardiomyopathy (DCM) A disease of the heart muscle that causes the heart to become enlarged and to pump less strongly; as a result the heart becomes thin, weak, or floppy, and unable to pump blood efficiently.

Myocardium The muscular tissue of the heart.

Myopericarditis Inflammation of the heart.

Myxomatosis A disease caused by a species-specific pox virus used to control the rabbit population in Australia.

N

Naïve T cell An immature T cell.

Naked virus A virus without a lipid bilayer membrane envelope around the capsid of the virus particle.

Natural killer (NK) cell A type of defensive body cell that attacks and destroys cancer cells and infected cells without the involvement of the immune system.

Natural selection The Darwinian theory of evolution, which holds that the organisms best adapted to their environment are the most likely to survive and produce the most offspring.

Naturally acquired active immunity A host response resulting in antibody production as a result of experiencing the disease agent.

Naturally acquired passive immunity A process resulting from the passage of antibodies to the fetus via the placenta or the milk of a nursing mother.

Necrosis The death of an area of live tissue or cells.

Negative strand (*also,* negative sense) RNA viruses whose genome cannot be directly transcribed into protein.

Negri bodies Round or oval bodies commonly found in the cytoplasm of brain neurons of a rabies-infected subject.

Neolithic agricultural revolution The change from nomadic hunting and gathering communities to settlement and the agricultural practice of using wild and domestic crops and animals (8000 BC).

Neuraminidase A protein that cleaves the sialic acid receptors on the outside of cells as the infectious influenza viruses are released, thus preventing the viruses from aggregating at the cell surface and releasing.

Neuraminidase inhibitor A compound that inhibits neuraminidase, thus preventing the release of influenza viruses from an infected cell.

Neuron The cell of the nervous system that transmits impulses in the body; main actors of the brain.

Neurotoxin A substance that is toxic to nerve tissue.

Neurotropic Describes viruses that infect the nervous system.

Neurovirulence The ability of viruses to cause disease in the nervous system.

Neutralization A type of antigen–antibody reaction in which the activity of a toxin is inactivated.

Neutralizing antibody An antibody that interferes with the infectivity of viruses or other pathogens.

Neutrophil The most common type of white blood cell; originates in the bone marrow and contains specialized granules that possess peroxidases, lysozyme, hydrolytic enzymes, collagenase, and lactoferrins that aid in their phagocytic activity in order to engulf and destroy foreign material, including bacteria and viruses that have entered the body.

Newer-generation vaccines Antiviral substances that incorporate advances based on recombinant DNA technology, thus providing advantages over traditional vaccines; *see* Vaccine.

Nocturnal Describes species that are active at night.

Nonessential genes Genes not essential for growth in cell cultures; they are located at the ends of the genome.

Nonneutralizing antibody An antibody that binds to viral antigens but does not interfere with infectivity; in some cases a nonneutralizing antibody can enhance infectivity.

Nonsegmented genome A type of viral genome that determines how the virus replicates; examples are rabies, herpes, and influenza viruses.

Nonspecific immunity Immunity that protects against any pathogen regardless of the species or type of microbe.

Nosocomial infection A disorder acquired during an individual's stay at a hospital.

Nuclease An enzyme that breaks the phosphodiester bonds of nucleic acids.

Nucleic acid A high-molecular-weight molecule consisting of nucleotide chains that convey genetic information and are found in all living cells and viruses; *see* DNA and RNA.

Nucleic acid-amplification tests (NAATS) Methods used to detect viral nucleic acids in order to diagnose and manage patients suffering from viral diseases (e.g., PCR or RT-PCR based tests).

Nucleocapsid A viral capsid and its associated genome.

Nucleoid The chromosomal region of a prokaryotic cell.

Nucleoside analogs Artificial molecules that resemble nucleosides and when incorporated into the RNA or DNA of viruses inhibit viral replication.

Nucleotide A component of a nucleic acid consisting of a carbohydrate molecule, a phosphate group, and a nitrogenous base.

Nucleus The part of a eukaryotic cell in which DNA replication and RNA transcription occurs.

O

Okazaki fragment A segment of DNA resulting from discontinuous DNA replication.

Oncogene A segment of DNA that can induce uncontrolled growth of a cell if permitted to function.

Oncogenesis The growth and development of a tumor.

Oncogenic Refers to any agent, such as a virus, that can cause tumors.

Oncology The scientific study of tumors and cancers.

Oncolytic virus Viruses that are able to infect and lyse cancer cells.

One-step growth experiments A type of scientific research to study a single replication cycle of viruses.

Open reading frame (ORF) A section of DNA that begins with an *atg* codon and ends with a stop codon (*taa, tag,* or *tga*) and that has the potential to be translated into an amino acid.

Opportunist A microorganism that invades the tissues when body defenses are suppressed.

Opportunistic infection A disorder caused by a microorganism or virus that does not cause disease but that can become pathogenic or life-threatening if the host has a low level of immunity.

Oral herpes labialis (*also,* cold sores or fever blisters) Infection of the mouth caused by a herpesvirus that leads to small painful blisters or lesions.

Oral rehydration solution A mixture of blood salts and glucose in water.

Oral–fecal route of transmission The means of viral infection in which viruses shed in feces are transferred into someone's mouth and are ingested with swallowed materials (e.g., food or drink) and then carried to the intestinal tract.

Orchitis A condition caused by the mumps virus in which the virus damages the testes.

Order In scientific classification, the broadest taxonomic category of viruses; contains one or more families.

ORFans Open reading frames that have no known homologs or known functions.

Ornithine transcarbamylase (OTC) deficiency A rare genetic disorder of the liver in which there is a lack of the ornithine transcarbamoylase enzyme, which is needed to help break down ammonia in the body. Gene therapy using adenoviruses was being studied as a possible treatment for the disease until the 1999 death of Jesse Gelsinger.

Orphan virus A virus that has not been etiologically linked to any human disease.

OTC deficiency *see* Ornithine transcarbamylase (OTC) deficiency

Outbreak A small, localized epidemic.

P

Pandemic A worldwide epidemic.

Panspermia hypothesis A hypothesis asserting that viruses or other microorganisms are raining down upon Earth and contaminating it.

Papilloma A wart.

Papule Small, solid, usually inflammatory (red) elevation of the skin that does not contain pus; a pink pimple on the skin.

Parenteral mode of transmission Blood-to-blood mode of transmission.

Parotid salivary glands The largest of the three sets of the glands that produce saliva that are located on either side of the head, above the jaw, and in front of the ear; the mumps virus commonly infects the parotid salivary glands.

Paroxysm A sudden intensification of symptoms (i.e., a severe bout of coughing).

Passive immunity The temporary immunity that comes from receiving antibodies from another source.

Pathogen A microorganism or virus that causes disease in a host organism.

Pathogen-associated molecular patterns (PAMPs) "Patterns" shared by viruses (e.g., viral genomes) that trigger immune responses.

Pathogenicity The ability of a disease-causing agent to gain entry to a host and bring about a physiological or anatomical change interpreted as disease.

Patient zero The first infected patient in an epidemiological investigation who is likely responsible for the spread of a particular infectious disease.

Pattern-recognition receptors (PRRs) Detect non-self molecules or "patterns" shared by viruses that are referred to as pathogen-associated molecular pathways (PAMPs).

PCR *See* Polymerase chain reaction.

Peptide vaccine A vaccine that uses chemically synthesized viral peptides to induce a protective immune response via using a specific domain of a protein that contains a single epitope or antigenic determinant.

Perinatal transmission Viral transmission that occurs during labor or vaginal delivery.

Period of convalescence The phase of a disease during which the body's systems return to normal.

Period of decline The phase of a disease during which symptoms subside.

Peripheral blood mononuclear cell Any blood cell having a nucleus (e.g., macrophages or lymphocytes).

Peripheral nervous system The nerves and neurons outside the central nervous system (i.e., the brain and spinal cord).

Persistent infection A continuing state of infection.

PERV *See* Porcine endogenous retrovirus.

PFU *See* Plaque forming unit.

pH-dependent fusion A mode of viral penetration that depends on an acidic pH for fusion between the viral envelope and the host cell membrane to occur.

pH-independent fusion A mode of viral penetration that does not depend on an acid pH for fusion between the viral envelope and the host cell membrane to occur; the remaining viral envelope remains as a "patch" on the cellular plasma membrane.

pH scale A scale from 1 to 14 that indicates the degree of acidity or alkalinity of a solution, with 1 representing the highest degree of acidity and 14 representing the highest degree of alkalinity.

Phage *See* Bacteriophage.

Phage typing A procedure of using specific bacterial viruses to identify a particular strain of a bacterium.

Phagocyte A white blood cell capable of engulfing and destroying foreign materials or cells, including bacteria and viruses.

Phagocytosis The engulfment and ingestion of foreign material (e.g., viruses, bacteria) by phagocytes.

Phenol (*also,* carbolic acid) A chemical derivative of benzene used in antiseptics and disinfectants to kill germs and to inactivate viruses for use in vaccines.

Phloem Specialized living cells that transport sap in plants.

Phosphodiester bond A chemical bond that forms the sugar-phosphate-sugar backbone of DNA and RNA.

Phosphorylation A mechanism of adding a phosphate molecule to a protein.

Phytoplankton Microscopic free-floating communities of cyanobacteria and unicellular algae.

PKR Protein kinase regulated; double strand RNA (dsRNA)-dependent serine/threonine kinase; functions as a kinase to phosphorylate serine or threonine residues on itself or on amino acid residues of other protein targets; involved in the interferon response.

Placard A notice from a state's board of health posted on entrances of the homes or dwellings of infected individuals.

Placenta An organ that serves as a protective interface between the mother and developing fetus. The fetus is connected by the umbilical cord to the placenta. It is involved in the exchange of nutrients and waste products and gas exchange between the blood supply (circulatory system) of the mother and fetus through the blood vessels of the umbilical cord.

Plaque A clear area on a lawn of bacteria or a monolayer of cells where viruses have destroyed the cells.

Plaque assay Used to quantitate the number of intracellular or extracellular virus particles present during a point of time in an infection by examining the plaques made by viruses *in vitro.*

Plaque forming unit (PFU) The measure of viruses that can form a plaque on a monolayer of the cells.

Plaque reduction assay A method for directly measuring the extent to which an antiviral drug inhibits the effects of viral infection in tissue culture in which cells infected with the virus are incubated in the presence of the antiviral compound.

Plasmid A small, closed-loop molecule of DNA apart from the chromosome that replicates independently and carries nonessential genetic information.

Plasmodesmata Microchannels in plant cell walls that allow the passage of molecules between cells, but that are too small for viruses to pass through unaided.

Pneumonia Infection of the alveoli and surrounding lung tissue.

Point mutation The replacement of one base in a DNA strand with another base.

pol A retrovirus gene that encodes reverse transcriptase and performs the conversion of RNA into DNA.

Poliomyelitis A disease caused by an enterovirus that, in severe cases, can weaken or paralyze its victims; poliomyelitis is rare today because of vaccination efforts.

Poliovirus receptor (PVR) A cell surface integral membrane protein and a member of the immunoglobin (Ig) superfamily of proteins; functions as an activating ligand that stimulates natural killer (NK) cells.

Poly(A) tail A series of adenine nucleotides that is added to the end of mRNA after transcription that may help protect the mRNA from degradation and affects the stability of some mRNAs.

Polyadenylated Describes an mRNA molecule with a poly(A) tail attached.

Polycistronic mRNA Messenger RNA that codes for more than one protein.

Polymerase chain reaction (PCR) A method of quickly replicating a piece of DNA a great number of times using a thermostable DNA polymerase in a test tube; can be used to detect the presence of viral nucleic acids.

Polymorphism In biology, the occurrence of different forms, stages, or types in individual organisms or organisms of the same species; common mutations in DNA.

Polyprotein A protein generated by a viral mRNA encompassing several genes, which is then cleaved into individual proteins via viral or cellular proteases.

Polythetic class A class whose members share several common properties but for which no individual members possess all properties.

Porcine endogenous retrovirus (PERV) Sequences for this virus are found in pig genomes, and it potentially may cause disease in humans; thus, it is a safety concern in xenotransplantation from pigs.

Portals of entry The points on the body where viruses can enter to cause infection; usually body openings or surfaces.

Portals of exit The points on the body from which viruses shed themselves to infect new hosts; usually body openings (e.g., oral or fecal exit) or surfaces.

Positive strand Refers to the RNA viruses whose genome consists of an mRNA molecule; + sense RNA.

Postexposure immunization The receiving of a vaccine after contracting the pathogen; used after a bite by a rabid animal.

Postexposure prophylaxis The provision of antiretroviral drugs to lower the risk of HIV infection after an individual has had a possible exposure to the virus.

Postherpetic neuralgia Pain that persists for months and sometimes years after a bout of shingles.

Postnatal transmission Occurs when the newborn is infected through breast milk, through transfused blood, by hands or instruments, or by respiratory route from infected persons.

Post-polio syndrome A condition that affects polio survivors years after recovery from an initial acute attack by the polio virus.

Posttranslational modifications Changes to eukaryotic proteins following translation (i.e., phosphorylation and glycosylation).

Pox A disease such as syphilis, chickenpox, or smallpox that is characterized by a rash that forms eruptive skin pustules that may leave pockmarks (a pitlike scar).

Prevalence Refers to the number of disease cases existing in a population at a specified time; does not take into account the duration of the disease.

Primary antibody response The first contact between an antigen and the immune system, characterized by the synthesis of IgM and then IgG antibodies; *see also* Secondary antibody response.

Primary viremia The stage of a viral infection in which the virus is still localized in the bloodstream.

Primase Cellular proteins that synthesize RNA primers.

Prion An infectious, self-replicating protein containing no genetic material that is responsible for a number of neurodegenerative brain diseases (*see* Transmissible spongiform encephalopathy in both humans and animals).

Prodromal period Refers to the first appearance of mild or nonspecific signs and symptoms of an illness.

Professional phagocytes Phagocytes that contain receptors on their surface to detect antigens normally not found in the body (e.g., pathogens, transplanted or foreign tissues).

Promoter Region on DNA to which RNA polymerase binds before initiating transcription of DNA into RNA.

Proofreading ability The ability of virus polymerases (or replicases) to correct errors or point mutations during the replication of their genomes.

Prophage The viral DNA of a bacterial virus that is inserted into the bacterial DNA and is passed from one generation to the next during binary fission.

Prophylaxis Measures taken to prevent the development of a disease.

Prospective study A study that looks at its subject forward in time.

Prostaglandins The substance produced by the brain's hypothalamus that raises the body's temperature, resulting in fever.

Protease An enzyme that cleaves the peptide bonds in proteins.

Protease inhibitor A compound that blocks the activity of the enzyme protease, thus inhibiting the replication of some viruses such as HIV.

Protein A chain or chains of linked amino acids used as a structural material or enzyme in living cells.

Protein array A spin-off of the DNA chip used in viral diagnostics in which proteins produced by an infectious microorganism are immobilized on DNA chips and subsequently scanned and interpreted by a microarray reader instrument.

Protein synthesis (*also,* translation) The process of forming a polypeptide or protein through a series of chemical reactions involving amino acids.

Proteoglycans Large molecules on cell surfaces consisting of protein and glycosaminoglycan.

Protooncogene Cellular genes that promote normal growth and division of cells.

Provirus (*also,* endogenous virus) A retrovirus formed when the retroviral DNA is inherited from parent to offspring of the infected host if germline cells (sperm and egg) contain the integrated viral genome.

PrP^{C} The normal cellular form of the prion protein.

PrP^{CWD} The infectious PrP (prion) that causes chronic wasting disease in cervids.

PrP^{res} The highly pathogenic and resistant or stable form of PrP^{C}.

PrP^{SC} The infectious PrP (prion) that causes scrapie.

Pseudoknots Structures on the genome of the hepatitis A virus that function in the recognition of the genome by the viral RNA replicase.

Pustule A small elevation on the skin containing pus.

Pyrogen Fever-inducing substance.

Q

Quantitative assay A measure of the amount of a substance found in a sample; a test to find the number of viruses in a given sample.

Quarantine The segregation from the general population of healthy persons who are not ill but have been exposed to an individual who suffers from a communicable disease.

R

Rabies vaccine A vaccine against the rabies virus that is used either pre- or postexposure to the virus.

Rapid progressors A very quick change in worsening disease status. Individuals infected with HIV-1 who develop AIDS within five years of infection.

Rash (exanthem) A change in the skin's appearance or texture, often produced after viruses have invaded the skin via blood vessels.

Reactivation Becoming infectious again (e.g., the reactivation of latent herpes simplex viruses).

Reactogenic Capable of causing a reaction, especially an immune reaction.

Real-time PCR A variation of polymerase chain reaction (PCR) that bypasses the need for gel electrophoresis, because fluorescence in the PCR reaction tube of amplified DNA can be measured instead of visualizing DNA fragments in a gel.

Reassortment A virus containing segmented genomes from two parent viruses, created after two viruses containing segmented genomes have infected the same cell.

Receptor-mediated endocytotic entry Occurs when an enveloped virus attaches to a receptor on the plasma membrane of the cell and the cell is stimulated to engulf the entire virus, forming an endocytotic vesicle.

Receptors On the surface of cells, they allow the receptor-binding protein or viral attachment protein particles on viruses to adhere to a cell.

Recombinant DNA technology Genetic engineering; the insertion of foreign genes into a virus in order to study the expression of these foreign genes within the cytoplasm of mammalian cells.

Recombinant subunit vaccine Vaccines consisting of purified antigenic components of viruses or other pathogens; safer than live vaccines.

Recombination A process that occurs when two viruses infect the same cell and a new chimeric or hybrid genome is formed via a process of intermolecular exchange of the viral genomes.

Reemerging viral diseases Diseases caused by viruses that are reappearing after a period of absence, and causing increased incidence or geographic range of infections in exposed human populations.

Replicase Describes RNA-dependent RNA polymerase whose second function is to synthesize the viral/progeny genome using the +ssRNA as a template.

Replication The process in which genetic material copies itself.

Replication fork The point where complementary strands of DNA separate and new complementary strands are synthesized.

Representational difference analysis A method to analyze differences between two given sets of DNA.

Reservoir Where the etiological agent lives, grows, and multiplies.

Respiratory hygiene/cough etiquette Measures used to minimize the transmission of respiratory pathogens: washing the hands often, sneezing into a paper tissue and disposing of it properly, and sneezing or coughing into the elbow.

Restriction fragment length polymorphisms (RFLP) Technique used to detect the differences between two very similar but not identical viral nucleic acid sequences in which the DNA is digested with restriction enzymes and the digested fragments are analyzed for their electrophoretic pattern; based on the concept that restriction enzymes recognize very specific sequences.

Reticulo-endothelial system Phagocytic cells constituting part of the immune system that accumulate in the lymph nodes and spleen.

Retinitis An infection in the eyes.

Retroid viruses (*also,* retroviruses) Any RNA virus that inserts a DNA copy of its genome into the chromosome of a host cell.

Retrospective study A type of research that looks at its subject backward in time.

Retrovirus A virus that contains an RNA genome and reverse transcriptase (e.g., HIV, HTLV).

Reverse genetics An experimental procedure that starts with cloned segments of DNA and uses this knowledge to introduce programmed mutations or gene sequences back into the genome in order to determine gene and protein function; used to systematically create candidate influenza vaccine against pandemic influenza strains.

Reverse transcriptase An enzyme that synthesizes a DNA molecule from the code supplied by an RNA molecule.

Reverse transcriptase inhibitor A compound that inhibits the action of reverse transcriptase, thus preventing the viral genome from being replicated.

Reverse-transcriptase polymerase chain reaction (RT-PCR) The most sensitive technique to measure and compare messenger RNA (mRNA) levels among samples; a process by which a cDNA is made from RNA by reverse transcription; the cDNA is then amplified by standard PCR protocols. Used in tests to measure HIV viral load and other diagnostic testing of viruses with RNA genomes.

Reye's syndrome A complication of influenza and chickenpox, characterized by vomiting and convulsions as well as liver and brain damage. It is associated with aspirin consumption by children for flu symptoms.

Rhinorrhea Nasal discharge.

Ribonucleic acid (RNA) The nucleic acid involved in protein synthesis and gene control; also the genetic information in some viruses.

Ribosomal frameshifting Occurs when the ribosome shifts into another reading frame and then continues translating the mRNA into protein in that new frame.

Ribosomal RNA (rRNA) An RNA transcript that forms part of the ribosome's structure.

Ribosome A cellular component of RNA and protein that participates in protein synthesis.

Ribozyme A segment of RNA that can catalyze the breakage and formation of RNA bonds at certain sites.

Rimantidine An antiviral drug that blocks the M2 ion channel function of influenza viruses, resulting in the incomplete release of the viral RNPs, thus interfering with uncoating.

RISC *See* RNA-induced silencing complex.

RNA-dependent RNA polymerases Enzymes that catalyze RNA using a RNA template.

Rivers' postulates A set of procedures, modifying Koch's postulates, by which a specific virus can be associated with a specific disease.

RNA-induced silencing complex (RISC) siRNAs that are denatured and incorporated into an endonuclease-silencing complex.

RNA interference (RNAi) (*also,* **RNA silencing**) A form of genomic immunity that responds to invading nucleic acid; double-stranded RNA inhibits gene expression in a sequence-specific manner.

RNA interference systems The application of RNA interference that uses double-stranded RNA to inhibit gene expression in a sequence-specific manner. It is useful in research and therapy development.

RNA polymerase The enzyme that synthesizes an RNA polynucleotide from a DNA template.

RNA polymerase II Eukaryotic enzyme synthesizes messenger RNA (mRNA) and some small nuclear RNAs (snRNA).

RNase Ribonuclease, an enzyme that degrades RNA.

RNase H A ribonuclease that breaks down the bonds in an RNA/DNA complex; an example is the pregenome in retroviruses that use reverse transcriptase to replicate.

RNase L An antiviral protein activated by 2′5′ oligo(A) synthetase to catalyze the degradation of both viral and host cellular RNAs, resulting in the shutdown of viral and host protein synthesis, a cellular altruistic event.

Rolling circle model of replication A method of DNA replication resulting in linear concatemers of DNA from a circular template.

Rous sarcoma virus A retrovirus capable of causing cancers in chickens.

RT-PCR *See* Reverse-transcriptase polymerase chain reaction.

Rumor registers In 1975, when smallpox cases decreased to very low levels globally, hospitals and health centers were asked to enter the names of suspected smallpox cases into a "rumor register" so that these cases could be checked by surveillance teams.

S

Sabin vaccine A type of polio vaccine prepared with attenuated viruses and taken orally.

Salk vaccine A type of polio vaccine prepared with inactivated viruses and injected into the body.

Saponins A substance secreted by plants that has antibacterial and antiviral properties.

Sarcoma A slow-growing cancer starting in a soft tissue (i.e., fat, muscle, fibrous tissue, blood vessels, or supporting tissue of the body); *see* Cancer.

Satellite viruses Viruses that encode their own coat protein but lack the genes that would encode the enzymes necessary for replication; as a result they are dependent upon helper viruses.

Scion A twig containing buds from a woody plant, used in grafting.

Secondary antibody response A second or ensuing response triggered by memory cells to an antigen and characterized by substantial production of IgG antibodies; *see also* Primary antibody response.

Seed lymph Cowpox lesions from infected calves used to make an early smallpox vaccine.

Segmented genome A viral genome that consists of two or more physically separate nucleic acid molecules that is usually packaged into a single virion.

Self-assembly Spontaneous formation of a well-defined, complex structure under equilibrium conditions from noncovalent interations between two or more molecular components; some viral particles (e.g., tobacco mosaic virus) can self-assemble in a test tube.

Sentinel chickens A surveillance program used to identify the presence of West Nile virus in susceptible birds before the disease is detected in humans, in which chickens are tested for the presence of antibodies against West Nile virus on a weekly basis.

Sentinel physician A doctor who conducts surveillances for influenza-like illness (ILI) in collaboration with a state's health department and the Centers for Disease Control and Prevention (CDC); data from large numbers of sentinel physicians are analyzed to help to provide a national picture of influenza virus and ILI activity in the United States.

Serological epidemiology A branch of science that investigates for the presence of virus antibodies, antigens, and other immunological substances in the sera of a population to determine the frequency and distribution of disease as well as to measure risks.

Serology A branch of science dealing with the measurement and characterization of antibodies, antigens, and other immunological substances in body fluids (serum) and plants that are virally infected.

Seroprevalence The number of persons in a population who test positive for a virus based on serology (blood serum) specimens; often presented as a percent of the total specimens tested or as a rate per 100,000 persons tested.

Serpins Serine protease inhibitors; these are produced by some poxviruses.

Serum (pl. sera) The fluid portion of the blood consisting of water, minerals, salts, proteins, and other organic substances, including antibodies; contains no clotting agents.

Serum hepatitis A type of hepatitis caused by the hepatitis B virus present in a human serum vaccine against the yellow fever virus.

Serum sickness A hypersensitivity reaction in which the body reacts toward antigens of nonhuman species present in antiserum.

Sexually transmitted disease (STD) (*also,* venerial disease) Describes over 20 different types of infections that can be transferred from person-to-person, or through sexual contact (exchange of body fluids).

Shell vial technique A clinical laboratory method for rapid diagnosis of viral infections using a combination of cell culture and centrifugation to detect viral antigens before cytopathic effects are present in a given clinical specimen.

Shingles A disease caused by the reactivation of the previously dormant varicella zoster virus during a period when the immune system is weakened.

Sialic acid A chemical found in receptors on cell surfaces that binds to viral H glycoprotein in influenza viruses, anchoring the virus to the cell; it is then removed by the viral N glycoprotein to prevent the viral particles from aggregating, preventing the clumping or aggregation of viruses to the cell surface.

Sign An indication of the presence of a disease, especially one observed by a doctor but not apparent to the patient; *see also* Symptom.

Single-dose vaccine The combination of several vaccines into one measured quantity.

siRNA Short-interfering RNAs; 21- to 24-nucleotide dsRNA molecules chopped by DICER molecules.

Slow infections Infections that have long incubation periods (measured in years); caused by slow viruses or prions, these types of infections lead to slow, progressive diseases.

Slow progressors HIV-1 infected individuals who remain free of AIDS for more than 15 years.

Smallpox recognition cards Cards featuring pictures of a case of smallpox that are distributed among populations in endemic countries to facilitate diagnosis and surveillance of the disease.

Species Taxonomic classification of a group of individual organisms that are capable of interbreeding to produce fertile offspring in nature; polythetic class of viruses that share the same replicating lineage and occupy a particular niche.

Species barrier The differences between species that make it impossible for an infectious disease to cross from one species to another.

Spike A protein projecting from the viral envelope or capsid that aids in attachment and penetration of a host cell.

Spinal polio Polio characterized by asymmetrical paralysis.

Spontaneous mutation A genetic error (mutation) that arises from natural phenomena in the environment.

Sporadic Describes disease outbreaks that have no pattern of occurrence in time or location.

Sputnik The first virophage ("virus eater") discovered in a mamavirus found in an amoeba in a water-cooling tower in Paris, France in 2008; *see* Virophage.

ssDNA Single strand DNA.

ssRNA Single strand RNA.

Stone fruits Fleshy fruits that contain a single seed or pit such as peaches, apricots, nectarines, plums, and cherries.

Strain Different isolates of the same virus that come from different geographical locations.

Street strains Rabies strains isolated from patients or animals that have experienced the disease via a bite from a rabid animal or other means.

Structural protein Proteins that make up the virus particle both inside and outside.

Subclinical disease A disease in which there are few or inapparent symptoms.

Subfamily The third broadest classification in the taxonomy of related organisms (here, viruses); between family and genus.

Subgenomic RNAs Short ssRNAs that function as monocistronic RNAs; a complex method of transcription used by viruses.

Subtypes (*also,* clades) A taxonomic group of viruses that evolve from a single common viral ancestor.

Super-spreading An individual who infects more than the average number of secondary cases.

Surfactant A synthetic chemical (i.e., a detergent) that emulsifies and solubilizes particles attached to surfaces by reducing the surface tension.

Surveillance Close observation of infection patterns of infectious diseases by experts in the field; systematic collection, analysis, and interpretation of health data to assist in the planning of prevention programs (e.g., vaccination).

Symptom An indication of some disease or other disorder that is experienced by the patient; *see also* Sign.

Syncytia Giant cells formed through the fusion of host cells; cytopathic effect of certain viral infections.

Syndrome A collection of signs or symptoms that together are characteristic of a disease.

Systemic infection An infection that disseminates to the deeper organs and systems of the body.

T

T lymphocyte (T cell) A type of white blood cell that matures in the thymus gland and is associated with cell-mediated immunity.

T lymphocyte helper (T_H) cells A type of immune cell that contains CD4 receptor molecules on its surface and is infected and destroyed by HIV.

TATA box A eukaryotic promoter sequence that defines the transcription start site.

Telomerase An enzyme that creates and relengthens telomeres on the ends of eukaryotic chromosomes.

Telomere DNA structures present at the ends of eukaryotic chromosomes that are not included in DNA replication; they protect the chromosome from fusion to other chromosomes and degradation by nucleases.

Temperate Refers to a bacterial virus that enters a bacterial cell and then the viral DNA integrates into the bacterial cell's chromosome.

Teratogenic effects Severe congenital malformations, including deafness, blindness, and congenital heart and brain defects.

Tetraspanin A family of membrane proteins found in all multicellular eukaryotes. CD81 is a tetraspanin that has recently been identified as a coreceptor for hepatitis C virus.

T_H cells (*also,* T helper cells) A subgroup of white blood cells (lymphocytes) that promote immune responses via a T_H1 response or a T_H2 response.

T_H1 response Occurs when T_H1 lymphocytes are made after microbes infect or activate macrophages, natural killer (NK) cells, and in response to viruses; T_H1 cells produce cytokines (e.g., interferon gamma [INF-γ]), lymphotoxin, and tumor necrosis factor-beta [TNF-β]) to fight against inflammation and boost cell-mediated immunity.

T_H2 response Produced in response to helminths, allergens, and extracellular microbes and toxins; T_H2 lymphocytes produce cytokines (interleukin[IL]-4, IL-5, IL-13) as part of the body's immune response to antibodies.

Thymidine kinase (cellular) Essential cellular enzyme that is expressed during cell division activity by allowing a cell to utilize an alternate metabolic pathway for incorporating thymidine into DNA. 5-bromodeoxyuridine (BudR) is a mutagenic nucleoside that substitutes for thymidine in DNA, acting as a chain terminator. It causes breaks in the chromosomes of cells.

Thymidine kinase gene (vaccinia) Nonessential gene present in the vaccinia virus genome; absence of thymidine kinase activity in host cells and disruption of the thymidine kinase gene in vaccinia virus renders host cells resistant to toxic effects of 5-bromodeoxyuridine (BudR). This selection enriches for cells that carry recombinant vaccinia virus.

Tissue culture infectious dose (TCID) A method used to study viruses in which cell monolayers are infected in replicates with serial dilutions of virus. After incubation, the infected cells are observed for cytopathic effects and are scored + (positive or infected) or – (negative or uninfected).

Tissue tropism Refers to the specific tissues within a host that a virus infects.

Titer The amount of antibody in a sample of serum that is determined by the most dilute concentration of antibody that will yield a positive reaction with a specific antigen.

Toxemia The presence of toxins in the blood.

Toxins Poisonous substances.

Toxoid A toxin whose harmful properties have been destroyed but still retains the capacity to produce an immune response; toxoids are used as vaccines.

Traditional vaccines Live or inactivated whole vaccines; vaccines not made via recombinant DNA technology; *see* Vaccine.

Transcriptase Describes RNA-dependent RNA polymerases that transcribe its –ssRNA genome into several different viral monocistronic +ssmRNAs that can be recognized by the host cell machinery.

Transcription The biochemical process in which RNA is synthesized according to a code supplied by the bases of a gene in the DNA molecule; the creation of complementary RNA from a DNA template.

Transcription factors Cellular proteins that facilitate the binding of RNA polymerase II to a gene promoter and determine the starting point of transcription.

Transfection A method that introduces the genome of a virus or viral gene into cultured cells.

Transfer RNA (tRNA) A molecule of RNA that unites with amino acids and transports them to the ribosome in protein synthesis.

Transformation The conversion of a normal cell into a malignant cell due to the action of a carcinogen or virus.

Transformation or focus assay An assay used to determine if a virus is capable of "transforming" or immortalizing cells rather than killing them.

Transgene A cloned gene incorporated into the genome.

Translation The biochemical process in which the code on the mRNA molecule is translated into a sequence of amino acids in a polypeptide.

Translational readthrough A strategy that viruses can use to control gene expression in which a stop translation signal may be ignored; a variation of frameshifting.

Transmissible spongiform encephalopathy A group of slow infection diseases caused by prions, including bovine spongiform encephalopathy and new variant Creutzfeldt-Jakob disease.

Transplacental transmission Occurs when viruses may cross the placenta, infecting the fetus.

Transposable genetic element Fragments of DNA called insertion sequences or transposons that can cause mutations.

Transposase An enzyme that is involved in the transfer of transposons randomly into the host chromosome; transposase genes are carried by some phages.

Transposon (*also*, jumping gene or transposable element) A segment of DNA that moves from one site on a DNA molecule to another, carrying information for protein synthesis.

Tulipomania A fad in Western Europe from 1600 to 1660 in which bulbs of tulips infected with a virus that made their petals striped were traded for large amounts of money or goods.

Tumor An abnormal uncontrolled growth of cells that has no physiological function; *see* Cancer.

Tumor-associated carbohydrate antigens A type of antigen that appears on the cell membranes of cancer cells.

Tumor necrosis factor A superfamily of proteins that induces necrosis of tumors cells and may cause inflammatory actions; blocking tumor necrosis factor in cells has shown to be beneficial in reducing inflammatory responses.

Tumor suppressor genes Genes that suppress or inhibit the conversion of a normal cell into a cancer cell.

Tumor virus A virus capable of inducing a tumor in its host.

Tumorigenicity The ability to form tumors if injected into an animal model.

Type (T) phages A group of dsDNA bacteriophages.

Type-species Different serotypes of the same species.

Typical progressors Persons infected with HIV-1 who display neither signs nor symptoms of AIDS for approximately ten years.

U

Ultrastructure The detailed structure of a cell, virus, or other object when viewed with an electron microscope.

Ultraviolet (UV) light A type of electromagnetic radiation of short wavelengths that damages DNA.

Uncoating Refers to the separation of viral nucleic acid from the capsid inside the infected eukaryotic host cell.

Universal precautions Safety measures taken to avoid contact with a patient's bodily fluids or toxic compounds (e.g., wearing gloves, goggles, and proper disposal of used hypodermic needles).

Universal vaccine Refers to a potential vaccine that would protect effectively against all strains of influenza viruses affecting humans.

Untranslated region (UTR) A section of a viral mRNA that is not translated by cellular ribosomes; in the poliovirus, it is located downstream of the cloverleaf structure.

Upstream DNA element In DNA sequences that regulate transcription; located in the direction from which the polymerase has come.

Urbanization The migration of people from rural locations to cities.

V

Vaccination Inoculation with weakened or dead microbes or viruses in order to generate immunity; *see also* Immunization.

Vaccine A preparation containing weakened or dead microorganisms or viruses, treated toxins, or parts of microorganisms or viruses to stimulate immune resistance.

Vaccine adverse events reporting system (VAERS) A reporting system designed to identify any serious adverse reactions to a vaccination.

Vaccine-associated paralytic polio (VAPP) The rare occurrence of poliomyelitis in a person who received the oral polio vaccine.

Vaccine lymph An early smallpox vaccine that was made by mixing vaccine pulp with glycerin and then emulsifying and clarifying the mixture via centrifugation.

Vaccine pulp Cowpox lesions scraped from the skin of calves and used to make an early smallpox vaccine.

Vaccinee One who has been vaccinated.

Vaccinia The prototype of poxviruses and the virus used in smallpox vaccines.

Vaccinia immunoglobulin Antiserum used to treat the adverse effects of the smallpox virus.

Vaccinia necrosum A severe reaction following a smallpox vaccination; progressive death of cells and living tissue (necrosis) at the vaccination site.

Vaccinifer Heifers used in the late 1800s for production of an early smallpox vaccine.

Vagus nerve The longest of the cranial nerves, stretching from the brain stem to the abdomen.

Variant A virus whose phenotype differs from the original wild-type strain but where the genetic basis is not known.

Variolation An 18th century method to inoculate someone with the smallpox virus to render that person resistant to infection; the intentional inoculation of dried smallpox scabs into the skin of an uninfected individual, causing a mild form of the disease and immunity against subsequent exposure to variola virus.

Vector 1. An arthropod that transmits the agents of disease from an infected host to a susceptible host. 2. A plasmid used in genetic engineering to carry a DNA segment into a bacterium or other cell.

Vesicle A fluid-filled skin lesion ranging up to one centimeter in size (i.e., in chickenpox) found in many eukaryotic cells.

Villous trophoblasts Cells that line the placenta that lack typical major histocompatibility complex (MHC) class I and II molecules.

Viral dsRNA A replication intermediate produced within cells during genome replication of RNA viruses.

Viral load Quantitation of viruses in the bloodstream of an organism.

Viral oncogene A viral gene responsible for causing cancer or the transformation of cells.

Viral replication complexes (VRCs) Viral complexes move along microtubules of filamentous actin and myosin and through the plasmodesmata as large bodies; virus-induced cytoplasmic inclusion bodies are comprised of the host and virus in plant cells to help spread a virus quickly through a plant (e.g., tobacco mosaic virus).

Viral zoonosis A viral disease transmittable between animals and human beings.

Viral-load monitoring The measure of the number of HIV-1 RNA genome copies per mL of plasma; a predictor of the disease progression to AIDS.

Viremia The presence and spread of viruses through the blood.

Virion A completely assembled virus outside its host cell.

Viroceptor Altered cellular receptors that have lost their transmembrane anchor sequences; consequently, these unanchored viral proteins are secreted from infected cells where they sequester ligands into the receptor portion of the protein.

Viroid An infectious RNA segment associated with certain plant diseases.

Virokine Secreted viral proteins that resemble host cytokines.

Virology The scientific study of viruses.

Virophage Biological entities smaller than viruses that infect larger viruses.

Virucide A drug or chemical that inactivates viruses.

Virulence The degree to which a pathogen is capable of causing a disease.

Virulence factor A molecule possessed by a pathogen that increases its ability to invade or cause disease to a host.

Virulent A virus or microorganism that can be extremely damaging when in the host.

Virus An infectious agent consisting of DNA or RNA and surrounded by a protein sheath; in some cases, a membranous envelope surrounds the coat.

VLP Virus-like particles.

W

Wart A small, usually benign, skin growth commonly due to a virus.

Wasting syndrome A loss of more than 10% of body weight due to fever or diarrhea for more than 30 days; a common problem in many AIDS patients.

Weapons of mass destruction Instruments or methods of attack designed specifically to cause bodily harm or death (nuclear, bacteriological, or chemical), with the potential to kill an extremely high number of people.

Western blot Method used to detect proteins in a given sample of tissue homogenate or extract. It uses electrophoresis to separate the proteins; proteins are then transferred to nitrocellose and probed with antibodies. This allows for the detection of a specific protein out of many thousands of different proteins. This confirmatory test is essential to exclude false-positive HIV test results.

Wild-type The form of an organism, virus, or gene isolated from nature; the typical infectious form of virus in nature that causes disease as distinguished from a mutant form that may cause a milder form or no disease in its host.

Window period The duration of time after infection until antibodies are detectable by serologic methods.

X

Xenotransplantation Any procedure that involves the use of live cells, tissues, or organs from a nonhuman animal source, transplanted or implanted into humans or used for clinical *ex vivo* perfusion.

Xenozoonosis A disease transmitted from an animal to a human recipient after transplantation of an animal organ.

X-gal A chemical (5-bromo-4-chloro-3-indolyl-β-D-galactopyranoside [BCIG]) used in biological assays (e.g., plaque assays) that when cleaved by the enzyme β-galactosidase turns blue; useful in quantifying viruses.

Xylem Water-conducting tissue that consists of mainly nonliving cells in higher plants.

Z

Zoonosis Refers to a disease being transmitted from an animal to a human.

Zoonotic pathogens Biological agents that cause disease and are naturally transmitted from animals to humans; pathways include aerosol inhalation, ingesting contaminated water or food, and hand-to-mouth contact.

Index

Note: Page numbers containing *f* or *t* refer to pages on which a figure or table appear.

B

C

D

E

F

I

N

O

Q

R

T

U

V

Photo Credits

Note: Unless otherwise indicated, photographs are under copyright of Jones and Bartlett, LLC, or have been kindly provided by the author.

Front Matter

Image of Dr. Teri Shors: Photographed by Dylan Stolley; **Timeline 1:** © National Library of Medicine; **Timeline 2:** © National Library of Medicine; **Timeline 3:** Courtesy of Dr. Lyle Conrad/CDC; **Timeline 4:** Courtesy of the CDC; **Timeline 5:** Courtesy of Frank Collins, Ph.D./CDC; **Timeline 6:** © AbleStock; **Timeline 7:** © AbleStock; **Timeline 8:** © Dreamshot/Dreamstime.com.

Chapter 1

Opener: © Dreamshot/Dreamstine.com; **1.1:** © Oliver Meckes/E.O.S./MPI Tubingen/Photo Researchers, Inc.; **1.3:** Courtesy of Ken Kloppenborg. **1.6a:** Courtesy of James Hicks/CDC.; **1.6b:** © nazira_g/ShutterStock, Inc.; **1.10a:** Courtesy of Dr. Frederick Murphy/CDC; **1.10B:** Courtesy of James Gathany/CDC; **1.11:** © SPL/Photo Researchers, Inc.; **1.13a:** Reproduced from J.D. Robertson. A Report on an Epidemic of Influenza in the City of Chicago in the Fall of 1918. (Chicago, 1918), p. 103.; **1.13b:** © National Archives and Records Administration; **1.14:** Courtesy of Los Angeles County DHS-Rancho Los Amigos national Rehabilitation Center, Donney CA.; **1.15a:** Courtesy of Franklin D. Roosevelt Presidential Library and Museum; **1.15b:** Courtesy of Franklin D. Roosevelt Presidential Library and Museum; **1.16:** © Wesley Bocxe/Photo Researchers, Inc.; **1.18a:** © Simon Fraser/Photo Researchers, Inc.; **1.18b:** © Barry Batchelor/PA Photos/Landov; **1.19:** Courtesy of the CDC.

Chapter 2

Opener: Courtesy of Cynthia Goldsmith, Jacqueline Katz, and Sherif R. Zaki/CDC; **Virus File 1:** Courtesy of Gary P. Williams, M.D., University of Wisconsin Medical School, Department of Pediatrics.

Chapter 3

Opener: © Science Photo Library/Alamy; **Case Study 1:** Courtesy of Didier Raoult, Rickettsia Laboratory, La Timone, Marseille, France; **3.3:** Courtesy of Dr. Fred Murphy, Sylvia Whitfield/CDC; **3.4a** Courtesy of David Bhella, University of Glasgow Centre for Virus Research; **3.4b:** Courtesy of Dr. Eric Blair, Biochemistry and Molecular Biology, University of Leeds; **3.6:** A reprint of Hurst, C. J., Benton, W. H., and Enneking, J. M. 1987. "Three-dimensional model of human rhinovirus type 14." TIBS 12:460.; **Virus File 1:** © Guido Picchio/AP Photos; **3.10a:** Courtesy of the CDC; **3.10b:** Micrograph from F.A. Murphy, School of Veterinary Medicine, University of California, Davis; **3.10c:** Micrograph from F.A. Murphy, School of Veterinary Medicine, University of California, Davis; **3.14:** Courtesy Dr. Ng Mah Lee, Mary, Department of Microbiology, National University of Singapore.

Chapter 4

Opener: © Sven Hoppe/ShutterStock, Inc.; **4.3c:** Micrograph from F. A. Murphy, School of Veterinary Medicine, University of California, Davis; **4.8a:** Courtesy of the CDC; **4.18:** Courtesy of Shmuel Rozenblatt, Tel Aviv University, Israel; **4.19:** © Richard Vogel/AP Photos.

Chapter 5

5.1a: Reproduced from The Lancet, Vol. 362, T. W. Geisbert et al, Treatment of Ebola virus infection with a recombinant inhibitor of factor VIIa/tissue factor: a study in rhesus monkeys, pp. 1953-1958, Copyright 2003, with permission from Elsevier. Photograph courtesy of Thomas W. Geisbert; **5.1b:** Reproduced from The Lancet, Vol. 362, T. W. Geisbert et al., Treatment of Ebola virus infection with a recombinant inhibitor of factor VIIa/tissue factor: a study in rhesus monkeys, pp. 953-1958, Copyright 2003, with permission from Elsevier. Photograph courtesy of Thomas W. Geisbert; **5.3:** Courtesy of John Strous, Medical Technology Program, UW-Oshkosh; **5.4:** Reproduced from P. R. Hazelton and H. R. Gelderblom, Emerg. Infect. Dis. 9 (2003): 294–303.; **5.5:** Reproduced from P. R. Hazelton and H. R. Gelderblom, Emerg. Infect. Dis. 9 (2003): 294–303.; **5.7:** © Hank Morgan/ Science Photo Library/Photo Researchers, Inc.; **5.8b:** Modified from J. H. Shelhamer, et. al., Ann. Intern. Med. 124 (1996): 585–599.; **Virus File 1:** © Stockbyte/Thinkstock; **5.18a:** Courtesy of Shmuel Rozenblatt, Tel Aviv University, Israel; **5.18b:** Courtesy of Shmuel Rozenblatt, Tel Aviv University, Israel; **5.20:** Courtesy of Dharam Ramnani, Virginia Urology Center Pathology Laboratory; **5.21:** Courtesy of Shmuel Rozenblatt, Tel Aviv University, Israel; **5.24:** Courtesy of Dr. Linda Cameron, University of Manitoba; **5.27a:** Courtesy of James Gathany/CDC; **5.27b:** Courtesy James Gathany/CDC; **5.27c:** Courtesy James Gathany/CDC; **Virus File 2:** Courtesy of Dr. Fred Murphy/CDC.

Chapter 6

Opener: © National Museum of Health and Medicine, Armed Forces Institute of Pathology, (Reeve 32486); **Case Study 1:** © Reuters/Romeo Ranoco/Landov; **6.3insert:** ©

Phototake, Inc./Alamy Images; **6.5a:** © J. Helgason/ShutterStock, Inc.; **6.5b:** Courtesy of Dr. Joseph Sowka, Nova Southeastern University; **Virus File 1a:** © Lyn Alweis, The Denver Post/AP Photos; **6.7:** © PhotoSpin/age footstock; **6.10:** Courtesy of the CDC; **6.11:** Courtesy of Dr. Herrmann/CDC; **Virus File 1b:** © Gilbert S. Grant/Photo Researchers, Inc.

Chapter 7

Opener: Reproduced from F. D. A. Fenner, et al, eds. Smallpox and its eradication. World Health Organization, 1988.; **Case Study 1:** © Stephen Mahar/ShutterStock, Inc.; **7.2:** © Stephen VanHorn/ShutterStock, Inc.; **7.3b:** © Tihis/ShutterStock, Inc.; **7.4:** © David Scharf/Photo Researchers, Inc.; **7.8:** Reproduced from The Story of Interferon: The Ups and Downs in the Life of a Scientist, K. Cantell, 1998 © World Scientific Publishing Company.; **7.9:** © Phototake, Inc./Alamy Images; **Virus File 1a:** Courtesy Dr. Erskine Palmer, Russell Regnery, Ph.D./CDC; **Virus File 1c:** Courtesy of the CDC; **7.22:** Courtesy of the CDC; **7.27:** Courtesy of APHIS/USDA; **7.33:** Courtesy of the CDC.

Chapter 8

Opener: © Mark Thomas/Photo Researchers, Inc.; **8.5b:** © National Library of Medicine; **8.6a:** © National Library of Medicine; **Virus File 1b:** Courtesy of Columbia University, Mailman School of Public Health; **Virus File 2:** © Mehmet Biber/Photo Researchers, Inc.; **Virus File 3:** Reproduced from Centers for Disease Control and Prevention, MMWR 5 (1981): 250-252.; **Virus File 4:** Courtesy of the CDC; 8.4: Courtesy of the University of Michigan Health System, Gift of Pfizer, Inc. UMHS.23; **8.12:** © National Library of Medicine; **8.14:** © David P. Lewis/ShutterStock, Inc.; **8.15:** Courtesy of INTREPID, NOAA, AVHRR/NASA.

Chapter 9

Opener: Courtesy of Department of Historical Collections, Health Sciences Library, SUNY Upstate Medical University; **9.1:** © National Library of Medicine **9.2:** © National Library of Medicine; **9.3:** © National Library of Medicine; **9.4a:** © National Library of Medicine; **9.4b:** Courtesy of the Archives of Ontario [ACC13993-8]; **9.5:** Courtesy of Upjohn Company; used with permission from Pfizer, Inc.; **9.8:** © M. English, M.D./Custom Medical Stock Photo.

Chapter 10

Opener: Courtesy of Dr. Cecil Fox/National Cancer Institute; **10.1:** Photo by Joe Wesley. Used with permission from Stars and Stripes. © 1965, 2011 Stars and Stripes; **Refresher Box 1b:** Courtesy of Robert Moyzis, University of California, Irvine, CA/U.S. Department of Energy Human Genome Program (http://genomics.energy.gov); **10.10:** Courtesy of Rober S. Craig/CDC; **10.11:** Courtesy of Dr. Steve Kraus/CDC; **10.12:** Courtesy of CDC. Used with permission of Patricia F. Walker, M.D., D.T.M. & H., Center for International Health & International Travel Clinic, HealthPartners; **Virus File 2:** © Wendy Nero/ShutterStock, Inc.; **10.13a:** Courtesy Dr. Wiesner/CDC; **10.13b:** Courtesy of Susan Lindsley/CDC; **Virus File 3:** Courtesy of Charles N. Farmer/CDC; **10.14:** The Protein Data Bank, ID: 3IYJ; Wolf, M., Garcea, R.L., Grigorieff, N., Harrison, S.C.: subunit interactions in bovine papillomavirus. *Proc. Natl. Acad. Sci.* USA pp. 107 (2010); **10.19:** Courtesy of David Bhella, Medical Research Council Virology Unit, University of Glasgow; **10.21a:** Courtesy of Dr. Erskine Palmer/CDC.

Chapter 11

Opener: Courtesy of the CDC; **Virus File 1:** Courtesy of Dr. Joseph J. Esposito/CDC; **11.2:** Courtesy of Library of Congress, Prints and Photographs Division [LC-USZ62-41892]; **11.4a:** Courtesy of Post-Polio Health International; **11.9:** © Minnesota Historical Society; **11.10:** © AP Photos; **11.13:** Courtesy of Mr. Stafford Smith/CDC; **11.15:** © Dr. P. Marazzi/Photo Researchers, Inc.

Chapter 12

Case Study 1: Courtesy and © Becton, Dickinson and Company; **12.5a:** Courtesy of Dr. Erskine Palmer/CDC; **12.5b:** Courtesy of Dr. Erskine Palmer/CDC; **12.16a:** © Bjorn Svensson/Alamy Images; **12.16b:** © Bjorn Svensson/Alamy Images; **12.17:** Courtesy of the National Museum of Health and Medicine, Armed Forces Institute of Pathology, Washington, D.C., NCP 1603; **12.18:** Courtesy of Dr. Johan Hultin; **Virus File 3b:** Courtesy of Seth Dixon, St. Jude Children's Research Hospital; **12.19:** Courtesy of James Gathany/CDC; **12.23:** © Vincent Yu/AP Photos; **12.26:** © Huang Benqiang, Xinhua/AP Photos.

Chapter 13

Opener: Courtesy of Library of Congress, Prints & Photographs Division, [reproduction number LC-USZ62-10610]; **13.1:** © National Library of Medicine; **13.5:** Courtesy of APHIS/USDA; **13.6:** Courtesy of APHIS/USDA; **13.8:** Courtesy of the CDC; **Refresher Box 1a:** © Du Cane Medical Imaging Ltd. / Photo Researchers, Inc.; **Refresher Box 1b:** © Airelle-Joubert/Photo Researchers, Inc.; **13.9:** Micrograph from F. A. Murphy, School of Veterinary Medicine, University of California, Davis; **13.10:** Courtesy of the CDC; **Virus File 1a:** © Morry Gash/AP Photos.

Chapter 14

Opener: Reproduced from Fenner, F. et al. *Smallpox and Its Eradication.* World Health Organization, 1988.; **Case Study 1:** © TessarTheTegu/ShutterStock, Inc.; **Virus File 1:** Courtesy of the National Archives of Australia: A1200, L44186; **Virus File 2:** Courtesy of Charles Drake; **14.1a:** Reproduced from Fenner, F. et al. Smallpox and Its Eradication. World Health Organization, 1988.; **14.1b:** Reproduced from Fenner, F. et al. *Smallpox and Its Eradication.* World Health Organization, 1988.; **14.2a:** Courtesy of Barbara Rice/NIP/CDC; **14.2b:** Courtesy of James Hicks/CDC; **14.2c:** Courtesy of the World Health Organization; **14.2d:** Courtesy of Stanley O. Foster, M.D., M.P.H./World Health Organization/CDC; **14.3:** Courtesy of Marshfield Clinic; **14.4:** © Jarrod Erbe/ShutterStock, Inc.; **14.6b:** Courtesy of the CDC.; **14.10a:** Courtesy of the WHO, Stanley O. Foster M.D., M.P.H./CDC; **14.10b:** Courtesy of Dr. J.D. Millar/CDC; **14.11:** Courtesy of James Gathany/CDC; **Virus File 1:** Courtesy of the National Archives of Australia: A1200, L44186; **14.13:** Courtesy of the WHO; **Virus File 2:** Courtesy of Charles Drake; **14.14:** Courtesy of John D. Millar/CDC; **14.15a:** Courtesy of Allen W. Mathies, MD/ California Emergency Preparedness Office (Calif/EPO), Immunization Branch/CDC; **14.15b:** Courtesy of Arthur E. Kaye/CDC.

Chapter 15

Opener: © Zephyr/Photo Researchers, Inc.; **Virus File 1:** Courtesy of Eileen McVey, NOAA Central Library/NOAA; **15.1:** Courtesy of Dr. N. J. Flumara and Dr. Gavin HArt/ CDC; **15.2:** Courtesy of CDC; **15.3:** © Mikhail Nekrasov/ ShutterStock, Inc.; **15.4:** Courtesy of the CDC; **15.5b:** Courtesy of E. L. Palmer/CDC; **15.13a:** © Dagfrida/ ShutterStock, Inc.; **15.13b:** © Stephen VanHorn/ ShutterStock, Inc.; **15.15:** © Scott Camazine/Alamy Images.

Chapter 16

Opener: © David Young-Wolff / PhotoEdit Inc.; **16.2a:** Courtesy of National Cancer Institute; **16.2b:** Courtesy of the CDC; **Virus File 1:** © PixAchi/ShutterStock, Inc.; **16.11:** Courtesy of the CDC; **16.13a:** Courtesy of Louisa Howard, Dartmouth College, Electron Microscope Facility; **16.17:** Courtesy of Dr. Edwin P. Ewing, Jr./CDC.

Chapter 17

Opener: © Connie Puntoriero/ShutterStock, Inc.; **Case Study 1:** © Josh Reynolds/AP Images; **17.1:** © National Library of Medicine; **17.4:** Courtesy of Dr. Thomas F. Sellers/ Emory University/CDC; **Virus File 1:** © Michelle Marsan/ ShutterStock, Inc.; **17.8a:** © Southern Illinois University/ Photo Researchers, Inc.; **17.8b:** © Medical-on-Line/Alamy Images; **17.9a:** Courtesy of CDC; **17.9b:** © James Cavallini/ Photo Researchers, Inc.

Chapter 18

Opener: Courtesy of the CDC; **18.5a:** Courtesy of Petty Officer 2nd Class Kyle Niemi/U.S. Coast Guard. Photo courtesy of U.S. Army; **18.5b:** Courtesy of Andrea Booher/ FEMA; **18.7:** Courtesy of Susan Mercado/CDC; **18.8a:** © CoolR/ShutterStock, Inc.; **18.8b:** © Reuters/Kham/Landov; **Virus File 1a:** © Karl Ammann; **Virus File 1b:** © Karl Ammann; **18.9a:** Courtesy of Alessandro Catenazzi; **18.10b:** © Andy Wong/AP Photos; **18.12a:** © Photos.com; **18.12b:** © Francois Etienne du Plessis/ShutterStock, Inc.; **18.14:** © Olivier Le Queinec/ShutterStock, Inc.; **18.15a:** Courtesy of the CDC; **Virus File 2a:** © Adam Borkowski/ ShutterStock, Inc.; **18.16a:** Courtesy of James Gathany/ CDC; **18.16b:** Courtesy of James Gathany/CDC.

Chapter 19

Opener: © Photoshot Holdings Ltd/Alamy; **19.1:** © National Library of Medicine/WHO; **19.5a:** Reproduced from P. Minor., J. Gen. Virol. 85 (2004): pp. 1777–1784. Image courtesy of P. Minor; **19.5b:** © vario images GmbH & Co.KG/ Alamy Images; **19.6a:** Courtesy of Dr. Al Jenny, APHIS, USDA/CDC; **Virus File 1:** © David M. Phillips/Photo Researchers, Inc.; **19.8:** Courtesy of Dr. Terry Kreeger, Wyoming Game and Fish Department; **19.10:** © Farlap/ Alamy; **19.11a:** © Boris Katsman/ShutterStock, Inc.; **19.11b:** © iStockphoto/Thinkstock; **19.14:** Courtesy of ARS Archive, USDA; **19.15:** Courtesy The Food and Environment Research Agency (Fera), Crown Copyright.

Chapter 20

Opener: © Jakub Pavlinec/Dreamstime.com; **20.1b:** Courtesy of Peggy Greb/USDA ARS; **20.2a:** Courtesy of Whitney Cranshaw, Colorado State University, Bugwood.org; **20.2b:** Courtesy of Larry R. Barber, USDA Forest Service, Bugwood.org; **20.2c:** Courtesy of Alton N. Sparks, Jr., University of Georgia, Bugwood.org; **20.3a:** Courtesy of Clemson University - USDA Cooperative Extension Slide Series, Bugwood.org; **20.3b:** Courtesy of J.K. Brown, University of Arizona, Bugwood.org; **20.3c:** Courtesy of OMAFRA/Used with permission from the Canadian Food Inspection Agency, 2011; **20.4a:** Courtesy of Elliot W. Kitajima, Ph.D., University of São Paulo, Brazil; **20.4b:** Courtesy of Elliot W. Kitajima, Ph.D., University of São Paulo, Brazil; **20.4c:** Reproduced from J. Ultstruct. Res. Vol. 29, Kitajima, E. W., Lauritis, J. A. and H. Swift, Morphology and intracellular localization of a bacilliform latent virus in sweet clover, p. 141, Copyright 1969, with permission from Elsevier. Image courtesy of Elliot W. Kitajima, Ph.D., University of São Paulo, Brazil; **20.4d:** Reproduced from E. W. Kitajima, Fitopatol. Bras. 1 (1976): 34; **Virus File 2:** © Phototake/ Alamy Images; **20.11:** © Science Source/Photo Researchers, Inc.; **20.12b:** Photo courtesy of ATCC; **20.13a:** Courtesy of R.J. Reynolds Tobacco Company Slide Set, R.J. Reynolds Tobacco Company, Bugwood.org; **20.13b:** Courtesy of Clemson University - USDA Cooperative Extension Slide Series, Bugwood.org; **20.16:** Courtesy of Ken Hammond/ USDA ARS; **20.19:** Reproduced from Dr. R. G. Milne, milne@ bifa.to.cnr.it.; **20.20:** From Alicai, T. et al. 2007. "Re-emergence of cassava brown streak disease in Uganda." Plant Disease 91(1): 24–29; **20.22:** Copyright Rothamsted Research Ltd.; **20.23b:** Courtesy of Raymond K. Yokomi, Ph.D., USDA, Agricultural Research Service, Crop Diseases, Pests & Genetics Research Unit; **20.24:** Courtesy of Dr. Ronald H. Brlansky, University of Florida, CREC.

Chapter 21

Opener: © David Mack/Photo Researchers, Inc.; **21.1:** © SPL/Photo Researchers, Inc.; **Virus File 1:** Courtesy Dr. Ramaz Katsarava, Center for Medical Polymers & Biomaterials and Georgian Technical University; **21.2a:** © David B. Fankhauser, PhD, University of Cincinnati Clermont College; **21.2b:** © David B. Fankhauser, PhD, University of Cincinnati Clermont College; **21.4a:** Courtesy of Maria Schnos and used with permission of Ross Inman, Department of Molecular Virology, Bock Laboratories, University of Wisconsin-Madisonx; **21.4b:** Photo courtesy of ICTVdB Picture Gallery. Used with permission from Hans-W. Ackermann, Department of Medical Biology, Laval University.; **21.4c:** Photo courtesy of ICTVdB Picture Gallery. Used with permission from Hans-W. Ackermann, Department of Medical Biology, Laval University.; **21.8:** © Medical-on-Line/Alamy Images; **21.9:** Courtesy of Shawn Walshaw-Wertz, Cherokee Pharmaceuticals LLC; **21.10:** Courtesy of Rodney M. Donlan, Ph.D. and Janice Carr/ CDC; **21.11:** © Steve Gschmeissner/Photo Researchers, Inc.